A Survey of Agricultural Economics Literature
VOLUME 4

The four volumes in *A Survey of Agricultural Economics Literature* have been prepared by and published for the American Agricultural Economics Association. The general editor of the survey volumes is Lee R. Martin.

Volume 1, *Traditional Fields of Agricultural Economics, 1940s to 1970s.* Lee R. Martin, editor.

Volume 2, *Quantitative Methods in Agricultural Economics, 1940s to 1970s.* George G. Judge, Richard H. Day, S. R. Johnson, Gordon C. Rausser, and Lee R. Martin, editors.

Volume 3, *Economics of Welfare, Rural Development, and Natural Resources in Agriculture, 1940s to 1970s.* Lee R. Martin, editor.

Volume 4, *Agriculture in Economic Development, 1940s to 1990s.* Lee R. Martin, editor.

A
SURVEY OF
AGRICULTURAL
ECONOMICS
LITERATURE
VOLUME 4

Agriculture
in Economic Development,
1940s to 1990s

Lee R. Martin, editor

Published by the University of Minnesota Press
for the American Agricultural Economics Association

Published by the University of Minnesota Press
2037 University Avenue Southeast, Minneapolis, MN 55414
Printed in the United States of America on acid-free paper

Library of Congress Cataloging-in-Publication Data

A Survey of agricultural economics literature.

    Includes bibliographies.
    Contents: v. 1. Traditional fields of agricultural economics, 1940s to 1970s. —
v. 2. Quantitative methods in agricultural economics, 1940s to 1970s. —[etc.]—
v. 4. Agriculture in economic development, 1940s to 1990s
    1. Agriculture–Economic aspects–Research. I. Martin, Lee R. II. American Agricultural Economics Association.
HD1410.5.S88   1977   338.1   76-27968
ISBN 0-8166-0801-6 (v. 1)
ISBN 0-8166-1942-5 (v. 4)

Contents

Foreword

In March 1968, C. E. Bishop, president of the American Agricultural Economics Association, appointed a committee to investigate the need for a major survey of the agricultural economics literature published from the 1940s to the 1970s. The committee found that an extensive assessment of this body of literature would indeed be of value to research workers, teachers, extension workers, and graduate students in economics and economic statistics, sociology, geography, political science, and anthropology, as well as teachers, research workers, extension workers, and graduate students in the different fields of technical agriculture. In the end the committee was assigned the responsibility for planning the project and commissioning authors to prepare the papers.

The original members of the committee were Glenn L. Johnson (Michigan State University), M. M. Kelso (University of Arizona), James E. Martin (Virginia Polytechnic Institute), M. L. Upchurch (Economic Research Service of the United States Department of Agriculture), and Lee R. Martin, chairman (University of Minnesota). Early in 1969, James E. Martin resigned from the committee, and several new members—John P. Doll (University of Missouri), Peter G. Helmberger (University of Wisconsin), J. Patrick Madden (Pennsylvania State University), and Edward W. Tyrchniewicz (University of Manitoba)—were appointed. As its first step, the committee tentatively identified the fields to be covered and commissioned highly regarded members of the profession to draw up outlines of the coverage to be undertaken in the different fields. These outlines were used in the selection of economists and sociologists to prepare the surveys and in negotiating agreements with prospective authors. Once the surveys were prepared, the committee again obtained assistance from highly competent members of the profession to make critical, constructive evaluations of each survey draft. In the stages of outline preparations and paper reviews, the committee sought to strike a representative balance among differing viewpoints in each field. For the preparation of the papers themselves, the committee obtained the services of outstanding rural social scientists with special competence in the respective fields.

In connection with the surveys published in this volume, substantial assistance was provided by the following individuals:

PART ONE. Agricultural Development in Sub-Saharan Africa: A Critical Survey. Preparation of outline: Carl K. Eicher. Review of paper: Dunstan Spencer, Derek Byerlee.

PART TWO. Agriculture in Economic Development: Theories, Findings, and Challenges in an Asian Context. Preparation of outline: John W. Mellor. Review of paper: Yujiro Hayami, Vernon Ruttan, Randy Barker, Dave Feeny, and Delane Welsch.

PART THREE. The Theory, Empirical Evidence, and Debates on Agricultural Development Issues in Latin America: A Selective Survey. Preparation of outline: G. Edward Schuh. Review of paper: Alain de Janvry, Richard A. King, William C. Thiesenhusen, Alberto Valdés, and Robert Stevens.

PART FOUR. Philosophic Foundations of Agricultural Economic Thought from World War II to the Mid-1970s. Preparation of outline: Glenn L. Johnson. Review of paper: W. Burl Back, M. M. Kelso, the late A.N. Halter.

This list includes only the official reviewers who acted on behalf of the association and the committee. Many other individuals who assisted the authors of the surveys in different ways are cited in the notes preceding most of the surveys. Authors were urged to incorporate into their surveys the comments and suggestions provided by the respective reviewers, but final decisions about the content of the surveys were left to the discretion of the authors.

The Committee on Publication of Postwar Literature Reviews arranged for publication of the four-volume set of literature surveys. Members of that committee were Emerson M. Babb (chairman), J. Patrick Madden, Lee R. Martin, and the late John C. Redman. Neil Harl provided valuable assistance to both committees in the publication phase.

On behalf of the members of the association, the Postwar Literature Review Committee, and the Publication Committee, I wish to express sincere gratitude to the authors of the surveys in the volume and to the advisers, reviewers, and others who participated in the planning and implementation of this project.

Finally, I would like to direct readers' attention to literature surveys of some closely related fields of agricultural economics—surveys that both complement and supplement the surveys in the four volumes.

The following surveys have been published in an Australian journal, the *Review of Marketing and Agricultural Economics*.

Weinschenck, G., W. Henricksmeyer, and F. Aldinger [1969]. "The Theory of Spatial Equilibrium and Optimum Location in Agriculture: A Survey." 37(1):3-70.

Renborg, U. [1970]. "Growth of the Agricultural Firm: Problems and Theories." 38(2):51-101.

Dillon, J. L. [1971]. "An Expository Review of Bernoullian Decision Theory in Agriculture: Is Utility Futility?" 39(1):3-80.

Gray, R. W., and D. J. S. Rutledge [1971]. "The Economics of Commodity Futures Markets: A Survey." 39(4):57-108.

Breimyer, H. F. [1973]. "The Economics of Agricultural Marketing: A Survey." 41(4):115-165.

Anderson, J. R. [1974]. "Simulation: Methodology and Application in Agricultural Economics." 42(1):3-55.

Richardson, R. A. [1976]. "Structural Estimates of Domestic Demand for Agricultural Products in Australia: A Review." 44(3):71-100.

Dillon, J. L., and C. Perry [1977]. "Multiattribute Utility Theory, Multiple Objectives and Uncertainty in *Ex Ante* Project Evaluation." 45(1):3-27.

Anderson, J. R. [1979]. "Impacts of Climatic Variability in Australian Agriculture." 47(3):147-177.

Greig, I. D. [1981]. "Agricultural Research Management and the *Ex Ante* Evaluation of Research Proposals: A Review." 49(2):73-94.

Kennedy, J. O. S. [1981]. "Applications of Dynamic Programming to Agriculture, Forestry and Fisheries: Review and Prognosis." 49(3):141-173.

Godden, D. [1982]. "Plant Variety Rights in Australia: Some Economic Issues." 50(1):51-95.

Randall, A. [1982]. "Economic Surplus Concepts and Their Use in Benefit Cost Analysis." 50(2):135-163.

Mules, T. J. [1983]. "Input-Output Analyses in Australia: An Agricultural Perspective." 51(1):9-30.

Sgro, P. M. [1983]. "A Selective Review of Developments in International Trade Theory: Commercial Policy and Free Trade." 51(1):31-50.

Colman, D. [1983]. "A Review of the Arts of Supply Response Analysis." 51:201-230.

Davidson, B. R. [1984]. "A Preliminary Benefit Cost Analysis of the Inland Diversion of the Coastal Rivers of New South Wales." 52(1):23-47.

McCarl, B. [1984]. "Model Validation: An Overview with Some Emphasis on Risk Models." 52(3):153-173.

Fisher, B. [1985]. "Frontiers in Agricultural Policy Research." 53(2):74-81.

Throsby, C. D. [1986]. "Agriculture in the Economy: The Evolution of Economists' Perceptions over Three Centuries." 54(3):5-48.

Gerritsen, R., and A. Murray [1987]. "Rural Policy Survey, 1986: The Battle for the Agenda." 55(1):7-23

Gerritsen, R., and J. Abbott [1988]. "Shifting to Certainty? Australian Rural Policy in 1987: A Review." 56(1):9-26.

Manegold, D. [1988]. "Recent Changes in EC Agricultural Policy: A 1987-88 Policy Review." 56(2):153-178.

Muller, J. D., *et al.* [1988]. "The Consumption Behaviour of Farmers: A Review of the Evidence." 56(2):179-193.

Another important set of literature surveys in agricultural economics is being published in the British *Journal of Agricultural Economics.* To date the following survey articles have been published:

Peters, G. H. [1970]. "Land Use Studies in Britain: A Review of the Literature with Special Reference to Applications of Cost–Benefit Analysis." 21:171-214.

Thornton, D. S. [1973]. "Agriculture in Economic Development." 24:225-287.

Josling, T. E. [1974]. "Agricultural Policies in Developed Countries: A Review." 25:229-264.

Barnard, C. S. [1975]. "Data in Agriculture: A Review with Special Reference to Farm Management Research, Policy, and Advice in Britain." 26:289-333.

Bateman, D. I. [1977]. "Agricultural Marketing: A Review of the Literature of Marketing Theory and Selected Applications." 27:171-225.

Hardaker, J. B. [1979]. "A Review of Some Farm Management Research Methods for Small-Farm Development in LDCs." 30:315-327.

Newby, H. [1982]. "Rural Sociology and Its Relevance to the Agricultural Economist: A Review." 33:125-165.

Le Vay, C. [1983]. "Agricultural Cooperation Theory: A Review." 34:1-44.

Fox, G. [1987]. "Models of Resource Allocation in Public Agriculture Research: A Survey." 38:449-462.

Gasson, R., et al. [1988]. "The Farm as a Family Business: A Review." 39:1-41.

Turner, R. H. [1988]. "Pluralism in Environmental Economics: A Survey of the Sustainable Economic Development Debate." 39:352-359.

Two literature surveys in natural resource economics have been published:

Fisher, A. C., and F. M. Peterson [1976]. "The Environment in Economics: A Survey." *J. Econ. Literature* 14:1-33.

Peterson, F. M., and A. C. Fisher [1977]. "The Exploitation of Extractive Resources: A Survey." *Econ. J.* 87:681-721.

Several surveys have appeared in the *Journal of Economic Literature*:

Johnston, B. F. [1970]. "Agricultural and Structural Transformation in Developing Countries: A Survey of Research." 8:369-404.

Perlman, M. [1981]. "Population and Economic Change in Developing Countries: A Review Article." 19:74-82.

Mellor, J. W., and B. F. Johnston [1984]. "The World Food Equation: Interrelations among Development, Employment, and Food Consumption." 22:531-574.

Shoven, J. B., and J. Whalley [1984]. "Applied General-Equilibrium Models of Taxation and International Trade: An Introduction and Survey." 22:1007-1051.

Sen, A. [1985]. "Social Class and Justice: A Review Article." 23:1764-1776.

Winston, C. [1985]. "Conceptual Development in the Economics of Transportation: An Interpretive Survey." 23:57-94.

Griffin, K., and J. Gurley [1985]. "Radical Analysis of Imperialism, the Third World, and the Transition to Socialism: A Survey Article." 23:1089-1143. R. Schenk [1985]. "A Comment." 24:676.

Other 1982-89 literature reviews related to agricultural development are listed below.

McPherson, M. F. [1982]. *Land Fragmentation: A Selected Literature Review*. Cambridge, MA, HIID, Discussion Paper No. 141.

Clayton, E. [1983]. *Agriculture, Poverty and Freedom in Developing Countries*. London, Macmillan.

Bartlett, J., and D. Gibbon [1984]. *Animal Draught Technology: An Annotated Bibliography*. London, Overseas Dev. Group of the University of East Anglia.

Cohen, A. [1984]. "Review Article: The Macroeconomic vs. the Microeconomic Approach to Development: The Fertility of Development Efforts as a Function of Institutional Inertia." *Econ. Dev. and Cultural Change* 32(2):423-430.

Kumar, T. M. V. [1984]. "Integrated Rural Development Planning and Energy Priorities: Participating Surveys in India Micro–Regions." In *Rural Energy to Meet Development Needs: Asian Village Approaches*, ed. N. Islam, R. Morse, and M. H. Soesastro. Boulder, CO, Westview Press, pp. 241-278.

Philippine Ministry of Energy [1984]. "Philippine Rural Energy Resource and Consumption Survey." In *Rural Energy to Meet Development Needs: Asian Village Approaches*, ed. N. Islam, R. Morse, and M. H. Soesastro. Boulder, CO, Westview Press, pp. 135-168.

Von Witzke, H. [1984]. "Poverty, Agriculture, and Economic Development: A Survey." *European Rev. Agr. Econ.* 11(4):439-453.

Clay, E. J., and H. W. Singer [1985]. *Food Aid and Development: Issues and Evidence (A Survey of the Literature since 1977 on the Role and Impact of Food in Developing Countries).* World Food Programme, Occasional Paper No. 3.

Feder, G., R. E. Just, and D. Zilberman [1985]. "Adoption of Agricultural Innovations in Developing Countries: A Survey." *Econ. Dev. and Cultural Change* 33(2):255-298.

Jones, S. F. [1985]. *Marketing Research for Agriculture and Agribusiness in Developing Countries: Courses, Training and Literature.* UK Tropical Dev. and Research Institute.

Kim, S. [1985]. "Models of Energy-Economy Interactions for Developing Countries: A Survey." *J. Energy and Dev.* 11(1):141-164.

Pevetz, N. [1985]. "Standortsgemasse Landwirtschaft und 'Okoentwicklung' in der Dritten Welt" (Agriculture Adapted to the Environment and 'Ecodevelopment' in the Third World: An Analysis of the Literature). *Monatsberichte über die Osterreichische Landwirtschaft* 32(12):741-751.

Hodge, I. D., and M. Whitby [1986]. "The UK: Rural Development, Issues and Analysis." *European Rev. Agr. Econ.* 13:391-413.

Rao, J. M. [1986]. "Agriculture in Recent Development Theory." *J. Dev. Economics* 22(1):41-86.

Rohrer, W. C. [1986]. "Developing Third World Farming: Conflict between Modern Imperatives and Traditional Ways." *Econ. Dev. and Cultural Change* 34(2):299-314.

Sands, D. M. [1986]. *The Technology Applications Gap: Overcoming Constraints to Small Farm Agriculture.* FAO Research and Technology Paper No. 1.

Timmer, C. P. [1986]. "Redesigning Rural Development from a Food Policy Perspective." *Econ. Dev. and Cultural Change* 34(4):855-860.

Whitby, M. [1986]. "Rural Development in Europe: Some Surveys of Literature: An Editorial Postscript." *European Rev. Agr. Econ.* 13(3):433-438.

Arnon, I. [1987]. *Modernization of Agriculture in Developing Countries: Resources, Potentials and Problems.* Chichester, West Sussex, UK, 2nd ed.

Clay, E. J., and J. Shaw, eds. [1987]. *Poverty, Development and Food.* Festschrift for H. W. Singer. Basingstoke, Hants, UK, Macmillan Press.

Kirk, C. [1987]. *People in Plantations. A Literature Review and Annotated Bibliography.* University of Sussex, IDS, Research Report No. 18.

Sarris, A. H. [1987]. *Agricultural Stabilization and Structural Adjustment Policies in Developing Countries.* FAO Econ. and Social Dev. Paper No. 65.

Shaw, A. B. [1987]. "Approaches to Agricultural Technology Adoption and Consequences of Adoption in the Third World: A Critical Review." *Geoforum* 18(1):1-19.

Brookfield, H. [1988]. " 'Sustainable Development' and the Environment: Review Article." *J. Dev. Studies* 25(1):126-135.

Groosman, A. J. A. [1988]. *Technology, Development and the Seed Industry in North-South Perspective: A Literature Survey.* Tilburg Institute of Dev. Research, Working Paper No. 38.

Hulme, D. [1988]. "Land Settlement Schemes and Rural Development: A Review." *Sociologia Ruralis* 28:42-61.

Lele, U. [1988]. "Empowering Africa's Rural Poor: Problems and Prospects in Agricultural Development." In *Strengthening the Poor: What Have We Learned?* ed. J. P. Lewis. Washington, ODC.

Rao, P. K. [1988]. "Planning and Financing Water Resource Development in the United States: A Review and Policy Perspective." *Amer. J. Econ. and Sociology.* 47(1):81-96.

Srinivasan, T. N. [1988]. "International Trade and Factor Movements in Development Theory, Policy, and Experiences." In *Trade and Development, Proceedings of the Winter 1986 Meeting of the International Agricultural Trade Research Consortium,* ed. M. D. Shane, Washington, USDA, ERS Staff Report.

Vandergeest, P. [1988]. "Commercialization and Commoditization: A Dialogue between Perspectives." *Sociologia Ruralis* 28(1):7-29.

Long, N., and van der Ploeg [1988]. "New Challenges in the Sociology of Rural Development. A Rejoinder to Vandergeest." *Sociologia Ruralis* 28(1):30-41.

Amanor, K. [1989]. *340 Abstracts on Farmer Participating Research.* London, ODI, Network Paper No. 5.

Herrmann, R. [1989]. *Wirkungen Nationaler Agrarpolitiken auf den Agrarhandel der Entwicklungsländer und Möglichkeiten der Handelsliberalisierung* (The Effects of National Agricultural Policies on the Agricultural Trade of Developing Countries and Possibilities of Liberating Trade). University of Kiel, Institut für Weltwirtschaft, Arbeits Papiere No. 374.

Norem, R. H., R. A. Yoder, and Y. Martin [1989]. "Indigenous Agricultural Knowledge and Gender Issues in Third World Agricultural Development." In *Indigenous Knowledge Systems: Implications for Agriculture and International Development,* ed. D. M. Warren *et al.* Ames, ISU Press.

Piccini, A. [1989]. "Dalle Carestie al Surplus: Teoria e Pratica dello Sviluppo Agricolo" (From Shortage to Surplus: The Theory and Practice of Agricultural Development). *Rivista di Politica Agraria* 7(2):3-11.

Rao, J. M. [1989]. "Agricultural Supply Response: A Survey." *Agr. Economics* 3(1):1-22.

Also worthy of note is Marguerite C. Burk's [1967] "Survey of Interpretations of Consumer Behavior by Social Scientists in the Postwar Period," *J. Farm Econ.* 49:1-31.

December, 1991

Lee R. Martin
Survey Editor

Abbreviations Used in Texts, Notes, and References

AAEA	American Agricultural Economics Association. Ames, Iowa
ACAR	*Associaçao de Crêdito e Assistência Rural (Association of Credit and Rural Assistance)*
ACE	American Council on Education. Washington
ACIAR	Australian Centre for International Agricultural Research. Canberra
ACM	Andean Common Market
ACMP	Andean Common Market Pact
ACP	Africa, Caribbean, Pacific. EEC, Brussels
ADB	Asian Development Bank. Manila
ADC	Agricultural Development Council. New York
AES	Agricultural Experiment Station
AIC	Agriculture Inputs Corporation
AID	United States Agency for International Development. Washington
AL	Alabama, United States
AMIRA	Groupe de Recherche pour l'Amélioration des Méthodes d'Investigation en Milieu Africain (Research Group on the Improvement of Survey Methods in Rural Africa). Paris
ANPES	Associação Nacionalde Programação Econômica e Social. São Paulo, Brazil
ANU	Australian National University. Canberra
APO	Asian Productivity Organization. Tokyo

APROSC	Agricultural Projects Services Center. Katmandu, Nepal
AR	Animation Rurale
ARE	African Rural Economy
ARTEP	Asian Regional Team for Employment Promotion. ILO, Bangkok
ARTI	Agrarian Research and Training Institute. Sri Lanka
ARU	Agricultural Research Unit. World Bank, Washington
ASEAN	Association of Southeast Asia Nations. Jakarta
ATIP	Agriculture Technology Improvement Project
BAE	Bureau of Agricultural Economics, USDA
BARC	Bangladesh Agriculture Research Council. Dhaka
BCST	Bananas, Cotton, Sugar, and Tobacco
BIDS	Bangladesh Institute of Development Studies. Dhaka
BRALUP	Bureau of Research Assessment and Land Use Planning. University of Dar es Salaam, Tanzania
BULOG	Badan Urusan Logistic (National Logistics Agency). Indonesia
CA	California, United States
CAA	Center for Agricultural Adjustments. ISU, Ames
CAAE	Chinese Association of Agricultural Economics
CAAS	Chinese Academy of Agricultural Sciences
CACM	Central American Common Market
CADU	Chílalo Agricultural Development Unit. Ethiopia
CAEA	Central American Economic Association. San José, Costa Rica
CAEC	Central American Economic Council
CAED	Center for Agricultural and Economic Development. ISU, Ames
CAP	Common Agricultural Policy. EC
CARDEN	Centre d'Analyse et de Recherche Documentaires pour l'Afrique Noire (Documentary Analysis and Research Center for Africa). Paris
CCMEP	Marketing and Price Stabilization Commission under CAEC
CD	Community Development
CEA	Commission Economique de l'Afrique des Nations-Unies (UN Economic Commission for Africa). Addis Ababa
CEDEAL	Centro Documentación Económica para América Latina (Economic Documentation Center for Latin America)
CEEMAT	Centre d'Études et d'Experimentation de Machinisme Agricole Tropicále (Center for Studies and Experimentation on Tropical Agricultural Machinery). Antony, France
CEPAL	Comisión Económica para a América Latina (Economic Commission for Latin America). Santiago, Chile (Same as ECLA)

CEPE   Centro de Estudios en la Pesquisa Económica (Center for
            Studies in Economic Research). Bogotá, Colombia
CEPE   Centro de Pesquisa em Educação (Center for Research in Ed-
            ucation). Ministério de Educação, Brasilia, Brazil
CEPI   Centre d'Étude et de Promotion Industrielle (Center for the
            Study of Industrial Promotion)
CERES   FAO magazine
CES   Constant Elasticity of Substitution
CFDT   Compagnie Française pour le Développement de Fibres Textiles
            (French Society for the Development of Textile Fibers). Paris
CFP   Companhia Para o Financiamento de Produção (Company for
            the Financing of Production). Brasilia, Brazil
CGIAR   Consultative Group on International Agricultural Research.
            World Bank, Washington
CGPRT   Centre for Research and Development of Coarse Grains,
            Pulses, Roots and Tuber Crops in the Humid Tropics of Asia
            and the Pacific. Bogor, Indonesia
CGT   Confederación General del Trabajo (General Confederation of
            Labor). Argentina
CHAC   Model: a study of the Mexican agricultural sector using LP
            and a partial equilibrium framework
CIA   United States Central Intelligence Agency
CIAT   Centro Internacional do Agricultura Tropical (International
            Center for Tropical Agriculture). Cali, Colombia
CIDA   Comité Interamericano de Desarollo Agricola (Interamerican
            Committee for Agricultural Development). Pan American
            Union, Washington
CIF   Cost, Insurance and Freight
CILSS   Comité Permanent Inter-États de Lutte contre la Sécheresse
            au Sahel (Permanent Interstate Committee of Drought Con-
            trol in the Sahel). Ouagadougou, Burkina Faso
CIMMYT   Centro Internacional de Mejoramiento de Maiz y Trigo (In-
            ternational Center for the Improvement of Maize and Wheat).
            Mexico, DF
CIP   Centro Internacional de la Papa (International Potato Center).
            Lima, Peru
CISEPA   Centro de Investigaciones Sociales, Economicas, Politicas y
            Antropologicas (Center for Social, Economic, Political and
            Anthropological Investigations). Lima, Peru
CMA   Center for Management in Agriculture. IIM, Ahmedabad, India
CMT   Corbo, de Melo, Tybout

CNRA	Centre National de Recherches Agronomiques (National Center for Agronomic Research). Dakar, Senegal
CO	Colorado, United States
CODESRIA	Council for the Development of Economic and Social Research in Africa. Dakar, Senegal
CODEVASP	Compania de Desenvolvimento do Valedo São Francisco (Valley of São Francisco Development Co.). Brasília, Brazil
CONASUPO	Compañia Nacional del Subsistencia Popular (National Council for Popular Subsistence). Mexico, DF
CORA	Agrarian Reform Corporation. Chile
CRED	Center for Research on Economic Development. Univ. of Michigan, Ann Arbor
CS	Consumer Surplus
CT	Connecticut, United States
CTA	Technical Centre for Agriculture and Rural Cooperation. Wageningen, Netherlands
DAE	Department of Agricultural Economics
DC	Developed Country
DEA	Departmento de Economía Agraria (Department of Agrarian Economics). Univ. Católica de Chile, Santiago
DMA	Division de Machinism Agricole (Agricultural Mechanization Division). Bamako, Mali
DRC	Domestic Resource Costs
DSE	Deutsch Stiftung für Internationale Entwicklung (German Foundation for International Development). Berlin
EAC	East African Community
EAP	Economically Active Population
EC	European Community
ECA	United Nations, Economic Commission for Africa. Addis Ababa, Ethiopia
ECIEL	Estudios Conjunctos sobre Integración Económica Latino-Americana (Joint Studies on Latin American Integration). Washington, D.C.
ECLA	United Nations Economic Commission for Latin America. Santiago, Chile (Same as CEPAL)
ECOWAS	Economic Community of West Africa States. Lagos, Nigeria
EDF	European Development Fund. EEC, Brussels
EDI	Economic Development Institute. World Bank, Washington
EEC	European Economic Community
EFSAIP	Evaluation of Farming Systems and Implements Project

EFTA	European Free Trade Association. Geneva
EMBRAPA	Emprésa Brasileira de Pesquisa Agropecuária (Brazilian Corporation for Agricultural Research). Brasilia, Brazil
EMENA	Europe, Middle East and North Africa Region. World Bank, Washington
ENDEF	Estudo Nacional de Despesa Familiar (National Study of Family Expenditures). Rio de Janeiro, Brazil
EPC	Effective Protection Coefficient
EPGE	Escola de Pós-Graduação em Econômia (Graduate School of Economics). FGV, Rio de Janeiro, Brazil
EPR	Effective Protection Rate
ERP	Effective Rate of Protection
ERS	Economic Research Service. USDA, Washington
ERS	Effective Rate of Subsidy
ESCAP	UN Economic and Social Commission for Asia and the Pacific. Bangkok
ESS	Economics and Statistics Service. USDA, Washington
ETENE	Escritório Técnico de Estudos Econômica do Nordeste (Technical Report on Studies of the Economy of the Northeast). Fortaleza, Brazil
ETTESS	École des Hautes Études in Sciences Sociales (School of Advanced Studies in Social Sciences). Paris
FAER	Foreign Agricultural Economic Report, issued by USDA
FAI	Fertilizer Association of India
FAO	Food and Agriculture Organization of the United Nations. Rome
FOA	Fund de Développement Agricole (Agricultural Development Foundation)
FGV	Fundação Getulio Vargas (Getulio Vargas Foundation). Rio de Janeiro, Brazil
FIPE	Fundação Institutio de Pesquisas Economicas (Foundation for the Institute of Economic Research). São Paulo, Brazil
FL	Florida, United States
FOB	Free on board
FSR	Farming Systems Research
FTC	Farmer Training Center
GATT	General Agreement on Tariffs and Trade. Geneva
GDP	Gross Domestic Product
GNFP	Gross National Farm Product
GNP	Gross National Product
GPO	United States Government Printing Office. Washington

HIID	Harvard Institute of International Development. Harvard Univ. Cambridge, MA
HMSO	Her Majesty's Stationery Office. London
HYV	High-Yielding Variety
IAAE	International Association of Agricultural Economists
IADB	Inter-American Development Bank. Washington
IADS	International Agricultural Development Service. Washington (Now part of Winrock International)
IAR	Institute for Agricultural Research. Ahmadu Bello Univ., Zaria, Nigeria
IARC	International Agricultural Research Center
IBGE	Instituto Brasileiro de Geografia Estatísticas (Brazilian Institute of Geography and Statistics). Rio de Janeiro, Brazil
IBRD	International Bank for Reconstruction and Development. Washington (Same as World Bank)
ICAE	International Council of Agricultural Economists
ICAR	Indian Council on Agricultural Research. New Delhi
ICIRA	Instituto de Capacitacióne Investigación en Reforma Agraria (Institute Authorized to Investigate Agrarian Reform). Santiago, Chile
ICRISAT	International Crops Research Institute for the Semi-Arid Tropics. Hyderabad, India
ICS	Institute for Contemporary Studies. San Francisco, CA
IDE	Institute of Developing Economies. Tokyo
IDEMA	Marketing Agency. Colombia
IDRC	International Development Research Center. Ottawa, Canada
IDS	Institute for Development Studies. Univ. of Nairobi. Kenya
IEA	Instituto de Economia Agricola (Institute of Agricultural Economics). São Paulo, Brazil
IED	International Economics Division. USDA. Washington, D.C.
IEMVT	Institut d'Élevage et de Médécin Vétérinaire de Pays Tropicaux (Institute of Livestock and Veterinary Medicine for Tropical Countries). Maisons-Alforêt, France
IER	Institut d'Économie Rurale (Institute of Rural Economy). Bamako, Mali
IER	Instituto de Economia Rural (Institute of Rural Economics)
IFDC	International Fertilizer Development Center. Muscle Shoals, AL
IFO	Institut für Land Wirtschaftsforschung (Institute for Agricultural Research). Munich, West Germany

IFPRI	International Food Policy Research Institute. Washington
IICA	Instituto Interamericano de Cooperaciôn para la Agricultura (Interamerican Institute for Agricultural Cooperation). San José, Costa Rica
IIM	Indian Institute of Management. Ahmedabad, India
IIMI	International Irrigation Management Institute. Kandy, Sri Lanka
IITA	International Institute of Tropical Agriculture. Ibadan, Nigeria
IL	Illinois, United States
ILO	United Nations International Labor Office. Geneva
ILCA	International Livestock Centre for Africa. Addis Ababa
ILPES	Instituto Latino Americano de Planification Economica y Social (Latin American Institute of Economic and Social Planning). Santiago, Chile
ILRAD	International Laboratory for Research on Animal Diseases. Nairobi, Kenya
IMF	UN International Monetary Fund. Washington
IN	Indiana, United States
INCAP	Instituto de Nutrition de Centro American y Panarama (Institute of Nutrition of Central America and Panama). Guatemala City
INCORA	Instituto Colombian de Reforma Agraria (Colombian Institute of Agrarian Reform). Bogota, Columbia
INEAC	Institut National pour l'Étude Agronomique de Congo (National Institute for Agronomic Study in the Congo). Yangambi, Zaire (formerly Belgian Congo)
INESPRE	Instituto de Estabilizaciôn de Precios (Institute of Price Stabilization). Santo Domingo, Dominican Republic
INPES	Instituto de Pesquisas Econômicas e Socials do IPEA (Institute of Economic and Social Research of IPEA). Rio de Janeiro, Brazil
INSEE	Institut National de la Statistique et des Études Économiqués (National Institute of Statistics and Economic Studies). Paris
IPC	Innovation Possibility Curve
IPE	Instituto de Pesquisas Econômicas (Institute of Economic Research). Univ. of São Paulo, Brazil
IPEA	Instituto de Planejamento Econômico e Social (Institute of Economic and Social Planning). Rio de Janeiro, Brazil
IPES	Instituto de Pesquisas Econômicas e Socials (Institute of Economic and Social Research)
IRAT	Institut de Recherches Agronomiques Tropicales et des Cul-

	tives Vivriéres (Institute of Tropical Agronomic Research and Food Crops). Montpellier, France
IRCT	Institut de Recherche de Coton et des Textiles Exotiques (Institute for Research in Cotton and Synthetic Textiles). Paris
IRD	Integrated Rural Development
IRDP	Integrated Rural Development Program
IRRI	International Rice Research Institute. Los Baños, Philippines
ISAE	Indian Society of Agricultural Economics
ISNAR	International Service for National Agricultural Research. The Hague, Netherlands
ISRA	Institut Sénégalais de Recherche Agricoles (Senegalese Institute of Agricultural Research). Dakar, Senegal
ISSER	Institute of Statistical, Social and Economic Research. Univ. of Ghana. Legon
ISU	Iowa State University. Ames
ITO	International Trade Organization
JASPA	Jobs and Skills Program for Africa
JCRR	Joint Commission on Rural Reconstruction. Taipei, Taiwan
K	Potassium (K_2O)
KDTA	Kenyan Tea Development Authority. Nairobi, Kenya
LACM	Latin American Common Market
LAFTA	Latin American Free Trade Association
LASC	Latin American Studies Center. MSU, East Lansing
LDC	Less Developed Country
LNL	Landless and Nearlandless
LP	Linear Programming
LSMS	Living Standards Measurement Study. Development Research Department, World Bank. Washington
LTC	Land Tenure Center. Univ. of Wisconsin. Madison
MA	Massachusetts, United States
MD	Maryland, United States
MFC	Marginal Factor Cost
MI	Michigan, United States
MIT	Massachusetts Institute of Technology. Cambridge, MA
MOTAD	LP technique to account for risk
MPP	Marginal Physical Product
MSU	Michigan State Univ. East Lansing
MTN	Multilateral Trade Negotiations
MVP	Marginal Value Product
N	Nitrogen
NARI	National Agricultural Research Institute

NBER	National Bureau of Economic Research. New York
NCAER	National Council of Applied Economic Research. New Delhi
NEDA	National Economic Development Authority. Government of the Philippines. Manila
NERP	Net Effective Rate of Protection
NERP3	NERP, Technology 3
NERP4	NERP, Technology 4
NERS	Net Effective Rate of Subsidy
NERS3	NERS, Technology 3
NERS4	NERS, Technology 4
NERS5	NERS, Technology 5
NESDB	National Economic and Social Development Board. Thailand
NIC	Newly Industrialized Country
NISER	Nigerian Institute of Social and Economic Research. Ibadan
NJ	New Jersey, United States
NNRP	Net Nominal Rate of Protection
NORAD	Norwegian Agency for Development. Oslo
NPC	Nominal Protection Coefficient
NRC	National Research Council. Washington
NRP	Nominal Rate of Protection
NS	New Series
NSF	National Science Foundation. Washington
NSS	National Sample Survey. Government of India. New Delhi
NY	New York, United States
OACV	Operation Arachide et Culturas Vivriéres (Groundnuts and Food Crops Operation)
OAS	Organization of American States. Washington
OAU	Organization of African Unity. Addis Ababa, Ethiopia
ODA	Official Development Assistance
ODC	Overseas Development Council. Washington
ODI	Overseas Development Institute. London
ODU	Yoruba (Nigeria): name of a defunct journal
OEA	Organización de los Estados Americanos (Organization of American States). San José, Costa Rica (Same as OAS)
OECD	Organization for Economic Cooperation and Development. Paris
OLC	Overseas Liaison Committee
OMVS	Organisation pour la Mise en Valeur de Fleuve Senegal (Senegal River Development Authority). Dakar, Senegal
ONAREST	Office National de la Recherche Scientifique et Technique (National Office for Scientific and Technical Research)

ONCAD	Office National de la Coopération et d'Assistance pour le Développement (National Office of Cooperation and Development Assistance). Dakar, Senegal
OPEC	Organization of Petroleum Exporting Countries. Vienna
ORD	Organisme Regional de Développement (Regional Development Organization). Burkina Faso
ORSTOM	Office de la Recherche Scientifique e Technique Outre-Mer (Overseas Scientific and Technical Research Office). Paris
OTA	Office of Technology Assessment. United States Congress. Washington
P	Phosphorus (P_2O_5)
PA	Pennsylvania, United States
PAN	Programa de Alimentaciôn Nacional (National Food Program). Argentina
PCARR	Philippine Council for Agriculture and Resources Research. Los Baños
PCARRD	Philippine Council for Agriculture and Resource Research and Development. Los Baños
PHN	Population, Health and Nutrition. World Bank. Washington
PIDE	Pakistan Institute of Development Economics
PIDS	Philippine Institute for Development Studies
PIMA	Personnel of the Integrated Program of Agricultural Marketing. MSU, East Lansing
PLANALC	Plan de Acción Conjunta para la Reactivación Agropecuaria en América Latina y el Caribe (Joint Action Plan for Reactivation of Crops and Livestock in Latin America and the Caribbean). San José, Costa Rica
PL-480	United States Public Law 480
PRC	People's Republic of China
RFF	Resources for the Future. Washington
RISD	UN Research Institute for Social Development. Geneva
SA	Secretariat da Agricultura (Secretary of Agriculture)
SADCC	Southern Africa Development Coordination Conference. Gaborone, Botswana
SAFGRAD	Semi-Arid Food Grain Research and Development. Ouagadougou, Burkina Faso
SAM	Mexican Food Supply System
SAREC	Swedish Agency for Research Cooperation with Developing Countries. Stockholm
SC	Supervised Credit

SEARCA	Southeast Asian Regional Center for Graduate Study and Research in Agriculture. Laguna, Philippines
SECAL	Société d'Études pour le Développement Économique et Social (Economic and Social Development Studies Corporation). Paris
SIAS	Scandinavian Institute of African Studies. Uppsala, Sweden
SIDA	Swedish International Development Authority. Stockholm
SOBER	Sociedade Brasileia de Economica Rural (Brazilian Society of Rural Economics). Brasilia, Brazil
SSI	Small-Scale Industry
SSRC	Social Science Research Council. New York
STABEX	Revenue Stabilization Scheme for Exports from Africa, Caribbean and Pacific Countries to EEC Countries. Funded by EEC
STRC	Scientific, Technical and Research Commission. OAU. Lagos, Nigeria
SUDENE	Superintêndencia para o Desenvolvimento do Nordeste (Superintendent for the Development of the Northeast). Recife, Brazil
SUNAB	Superintêndencia Nacional de Abastecimiento (National Superintendent of Supply). Brasilia, Brazil
TAC	Technical Advisory Committee. CGIAR
TCA	Technical Cooperation Administration. Washington (A predecessor of AID)
UN	United Nations. New York
UNCTAD	United Nations Conference on Trade and Development. Geneva
UNDP	United Nations Development Plan. New York
UNESCO	United Nations Educational, Scientific, and Cultural Organization. Paris
UNICEF	United Nations Children's Fund. New York
UNZALPI	Universities of Nottingham and Zambia, Agricultural Labor Productivity Investigation
UP	Uttar Pradesh, India
U.S.	United States of America
USDA	United States Department of Agriculture. Washington
VEW	Village Extension Workers
VT	Vermont, United States
WARDA	West African Rice Development Association. Boace, Côte d'Ivoire. (formerly in Liberia)
WHO	United Nations World Health Organization. Geneva
WIDER	World Institute for Development Economics Research. Helsinki, Finland

PART ONE. Agricultural Development in
Sub-Saharan Africa: A Critical Survey

The original version of this survey on the literature on the rural economies of sub-Saharan Africa was completed in 1982 and published by Michigan State University in English [Eicher and Baker, 1982]. The survey was subsequently translated by the IDRC of Canada and published in French [Eicher and Baker, 1985]. Because of time constraints, the original 1982 manuscript which covers the 1960–82 period has only been slightly updated.

Our initial mandate from the AAEA was to review the literature on agricultural development by agricultural economists in the forty-five countries of sub-Saharan Africa but we broadened this mandate. First, we review research by economic historians on the precolonial and colonial development experiences. Second, we present a brief overview of the literature by technical scientists on farming and livestock systems. Third, we go beyond the agricultural sector to appraise the literature on the rural nonfarm economy, including small-scale industry, fishing, processing, storage, migration, income distribution, and several other topics. Fourth, we present some of the key research findings of political scientists, anthropologists, sociologists, geographers, and technical scientists.

We have two major goals in preparing this survey. The first is to present a critical review of the major theoretical and policy debates and empirical findings on the development of rural economies in sub-Saharan Africa. The second is to identify the major research gaps and research directions for the 1990s. Because of lagging food production and widespread poverty, Africa is likely to receive even more research attention in the coming decades than it has in the past relative to Asia and Latin America, and it is important that research resources be used efficiently.

We would like to acknowledge invaluable help from the following reviewers: Vincent Barrett, Sara Berry, Derek Byerlee, Enyinna Chuta, Eric Clayton, John Cohen, Mike Collinson, Eric Crawford, Bob Deans, Christopher Delgado, W. Doppler, John Eriksen, B. Falusi, Don Ferguson, Russell Freed, Donald Heisel, Lane Holdcroft, Francish Idachaba, R. W. Palmer-Jones, M. C. Lathan, Uma Lele, Carl Liedholm, A. R. C. Low, R. E. McDowell, K. Meyn, Isaac Minde, M. Miracle, Wilford W. Morris, W. Mwangi, David Norman, O. Ogunfowora, Kenneth Robinson, Stephen Sanford, Tjaart Schillhorn, Kenneth Shapiro, John Staatz, Martin Upton, William Whelan, David Wilcock, and the late Pascal Fotzo. We are grateful to Lucy Wells, Jeannette Barbour, and Sherry Rich for typing assistance.

The research supporting this paper was financed by the Food Security in Africa Cooperative Agreement DAN-7790-A-4092-00 between the U.S. Agency for International Development (USAID) and Michigan State University's Department of Agricultural Economics.

Agricultural Development in Sub-Saharan Africa: A Critical Survey

Carl K. Eicher and Doyle C. Baker

Chapter I. Introduction

Sub-Saharan Africa is a vast subcontinent of forty-five countries,[1] heterogeneous endowments of resources, seven colonial histories, and uneven levels and opportunities for development (Figure 1).[2] The population of Africa in 1990 is about 500 million. Nigeria has one-fourth of the population and produces about one-half of the gross national product of the subcontinent. Population densities in Africa are extremely low relative to Asia. The Sudan, for example, is two-thirds the size of India but it has only twenty-three million people as compared with 800 million in India. The Republic of Zaire (formerly the Belgian Congo) is five times the size of France, but only 5 to 10 percent of its arable land is under cultivation.

Although the density of Africa's population is low relative to Asia, the distribution is uneven and there are parts of Africa that are near maximum population capacity given present agricultural technology and knowledge of how to deal with soil erosion and environmental problems. Moreover, sub-Saharan Africa is the only region in the world where the rate of natural growth of population increased over the 1960-90 period. The population growth rate increased from 2.5 percent in the 1960s to 2.7 percent in the 1970s[3] and is currently around 3.2 percent per year.

In Africa, 60 to 90 percent of the people are in the agricultural sector, in contrast to 30 to 50 percent of the population in most Latin American nations. Agricultural and mineral exports dominate most economies as they do in many

3

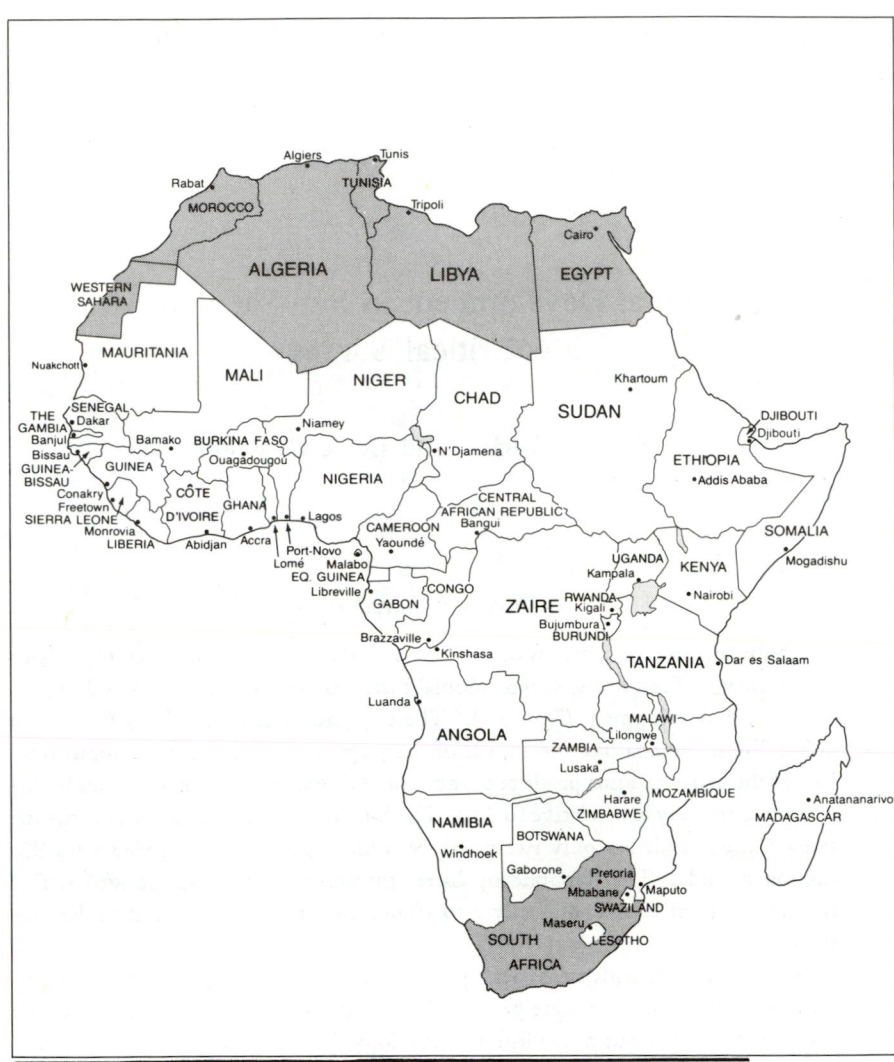

Figure 1: Countries and Capital Cities in Sub-Saharan Africa.

Latin American countries. But unlike much of Latin America, the size of the industrial sector is extremely modest in most African countries. Although the rate of growth of African cities is currently the highest in the world, Africa continues to be the least urbanized region. Overall, the population is about 70 percent rural in 1990 compared with about 80 percent in 1960. The percentage varies from 56 percent rural in Ghana to 93 percent in Burundi. For the next twenty-five to fifty years the majority of the population in most countries will continue to live in rural areas.[4]

Africa is the poorest part of the world's economy. Table 1 presents economic indicators for forty of the forty-five countries in Africa. Africa's poverty is illustrated by the point that the total GNP of the forty-five countries in 1985 was less than the total GNP of Spain, a nation with forty million people.

One of the most disturbing trends over the last twenty years has been declining per capita production of food crops. The USDA [1981] pointed out that Africa is the *only* region of the world where per capita food production declined from 1960 to 1980. As a result, the average per capita calorie intake was below minimum nutritional levels in *most* countries during this period. Many countries formerly self-sufficient in food have increased the ratio of food imports to total food consumption. The declining per capita food production has led to what many observers are now calling an "agrarian crisis" or "Africa's food crisis" [Eicher, 1982a].

The prospects for economic development in sub-Saharan Africa are not encouraging. Nearly all African nations are overwhelmingly dependent upon agriculture for the bulk of their national income, employment, and foreign exchange. Only a few countries such as Zimbabwe, Cameroon, and the Côte d'Ivoire have harnessed their agriculture as an "engine of growth" of the overall economy. But this bleak assessment should not overlook some major achievements during the first three decades of independence. The average life span is about 30 percent longer in Africa today than it was in 1960. A few countries—Botswana, Cameroon, Kenya, and the Côte d'Ivoire—have achieved respectable rates of economic growth over the past two decades. The achievements in education have been dramatic in several countries. But adult literacy is below 25 percent in many countries in sub-Saharan Africa.[5]

1. Scope of Review

This survey focuses on the rural economy which is broadly defined to include the agricultural sector plus rural nonfarm activities such as small-scale industry, trade, processing, and fishing. In addition, research on migration, employment, and income distribution is reviewed to understand the linkages between the agricultural and nonagricultural sectors. We have adopted a broad rural economy

Table 1. Economic indicators for forty countries in sub-Saharan Africa

	Population (millions) mid-1989	GNP per Capita Dollars 1989	GNP per Capita Annual average growth rate (percent) 1965-89	Agriculture annual average growth rate (percent) 1965-80	Agriculture annual average growth rate (percent) 1980-89	Average index of food production per capita (1979-81=100) 1987-89	Percentage of labor force in agriculture 1965	Percentage of labor force in agriculture 1989	Life expectancy at birth (years) 1989
Low Income									
1. Mozambique	15.3	80	−0.1	—	0.7	83	95	74	49
2. Ethiopia	49.5	120	−0.1	1.2	−0.4	89	92	87	48
3. Tanzania	23.8	130	−0.1	1.6	4.2	90	95	69	49
4. Somalia	6.1	170	0.3	—	3.8	97	80	64	48
5. Guinea–Bissau	0.9	180	—	—	—	—	—	—	40
6. Malawi	8.2	180	1.0	4.1	2.2	85	95	88	48
7. Chad	5.5	190	−1.2	−0.3	2.5	101	91	71	47
8. Burundi	5.3	220	3.6	6.7	3.1	98	98	95	49
9. Sierra Leone	4.0	220	0.2	3.9	2.4	89	85	68	42
10. Madagascar	11.3	230	−1.9	—	2.4	93	88	76	51
11. Gambia, The	0.8	240	0.7	—	—	—	—	—	44
12. Nigeria	113.8	250	0.2	1.7	1.3	96	83	65	51
13. Uganda	16.8	250	−2.8	1.2	2.2	87	93	90	49
14. Zaire	34.5	260	−2.0	—	2.6	94	74	61	53
15. Mali	8.2	270	1.7	−2.8	1.5	97	87	81	48
16. Niger	7.4	290	−2.4	−3.4	1.8	86	93	81	45
17. Burkina Faso	8.8	320	1.4	—	5.8	115	95	91	48
18. Rwanda	6.9	320	1.2	—	−1.4	77	97	93	49
19. São Tomé/Principe	0.1	340	—	—	—	—	—	—	66
20. Kenya	23.5	360	2.0	4.9	3.2	101	91	77	59
21. Benin	4.6	380	−0.1	—	4.2	114	89	63	51
22. Central Afr. Rep.	3.0	390	−0.5	2.1	2.9	90	73	54	51
23. Ghana	14.4	390	−1.5	1.6	0.9	109	74	67	55
24. Togo	3.5	390	0.0	1.9	5.7	89	89	75	54
25. Zambia	7.8	390	−2.0	2.2	4.1	97	77	51	54
26. Guinea	5.6	430	—	—	—	90	88	75	43
27. Lesotho	1.7	470	5.0	—	−0.8	80	94	80	56
28. Mauritania	1.9	500	−0.5	−2.0	1.5	88	90	55	46
29. Liberia	2.5	—	—	5.5	—	95	78	55	54
30. Sudan	24.5	—	—	2.9	—	87	87	78	50
Lower Middle-Income									
31. Senegal	7.2	650	−0.7	1.3	2.9	106	67	62	48
32. Zimbabwe	9.5	650	1.2	—	2.9	90	86	73	64
33. Cape Verde	0.4	780	—	—	—	—	—	—	66
34. Cote d'Ivoire	11.7	790	0.8	3.3	2.3	96	77	60	53
35. Swaziland	0.8	900	2.1	—	—	—	—	—	56
36. Congo, People's Rep.	2.2	940	3.3	3.1	3.2	98	66	60	54
37. Cameroon	11.6	1000	3.2	4.2	1.9	96	84	60	57
38. Botswana	1.2	1600	8.5	9.7	−4.0	68	96	74	67
39. Mauritius	1.1	1990	3.0	—	3.0	100	63	59	70
Upper Midle-Income									
40. Gabon	1.1	2960	0.9	—	—	81	79	55	53

Source: *World Development Report 1991*, Tables 1, 2, 4, and 31 and Box A.1, p. 271.
Key: . . . Not available.

perspective rather than the agricultural sector because research has shown that rural households in sub-Saharan Africa typically allocate from 25 to 40 percent of their labor to nonfarm activities.

We begin by identifying standard references on African agricultural development and presenting a descriptive overview of agricultural systems in sub-Saharan Africa. Our survey examines research on food and agricultural policy in historical perspective starting with the precolonial period (1800-1880s) and moving to the colonial period from the 1880s to 1960 including discussion of key theoretical perspectives on agricultural development which have had an important influence on policy makers and scholars working in Africa. We also cover policy debates during the postindependence period of the sixties and seventies. In the 1960s, most African governments stressed industrialization and large-scale farming, including what a World Bank mission to Tanzania called the "transformation approach" to modernizing African agriculture. Despite numerous experiments with large-scale farming and ranching schemes during the colonial and postindependence periods, the bulk of the land and labor in Africa is still devoted to small-scale farming and livestock herding by individual families.

In the late 1960s and early 1970s, many policy makers, planners, and foreign donors shifted their attention to small-scale farming and small-scale livestock projects. In light of this shift in policy, substantial attention is devoted to micro research on small holder crop and livestock production. Later, we shift to research on food and agricultural distribution systems, with emphasis on international trade, marketing, processing, storage, credit, cooperatives, consumption, and nutrition. Chapter VIII reviews research on equity and employment issues such as income distribution and inequality, population, migration, women in development, small-scale industry, and fishing.

Several caveats are in order. First, throughout this review, we shall emphasize research completed since 1970. The literature of the 1960s has already been covered in a number of surveys, including McLoughlin [1967] for East Africa and C. K. Eicher [1970] for West Africa. Second, we have concentrated on journal articles and books and have included only a few unpublished reports and theses.

Third, we have found it impossible to provide uniform coverage of the literature by topic and for each of the forty-five countries. We are struck by the unevenness of the research coverage and the number of publications by country. This unevenness is dramatized by comparing the number of publications in the references for Niger, Nigeria, Botswana, and Chad. For example, Sims and Kagan [1976] reported that there were four American and Canadian doctoral dissertations and master's theses on Niger and 714 for Nigeria over the 1886-1974 period. Shirley Eicher [1981] cited 1,280 publications in her bibliography on rural development in Botswana from independence in 1966 to 1981. If some countries such as Chad, Niger, and Rwanda seem to be undercited in our survey, it is partially because of language barriers and political instability, and partially because the research enterprise like the fashion industry is trendy. For example, almost every Western researcher tried to touch down in Tanzania sometime during the late 1970s to gain some first-hand impressions of the Ujama experiment and the villagization program. Kocher and Fleisher [1979] cite 761 publications on rural development in Tanzania from independence in 1960 through 1979. For this reason, one can infer that Tanzania has been vastly overstudied relative to countries such as Somalia, The Gambia, and Rwanda.

A final caveat is in order because the authors are from the same subdiscipline — agricultural economics. Since agricultural development is a technical, social, and political process, the literature of any subdiscipline such as agricultural economics falls short of capturing the complexity of the agricultural development process. Albert Hirshman sums up this dilemma in his analysis of the rise and decline of development economics over the past twenty-five years:

> Development economics started out as the spearhead of an
> effort that was to bring all-around emancipation from
> backwardness. . . . By now it has become quite clear that this
> cannot be done by economics alone. It is for this reason that
> the decline of development economics cannot be fully reversed:
> our subdiscipline had achieved its considerable luster and
> excitement through the implicit idea that it could slay the
> dragon of backwardness virtually by itself or, at least that its
> contribution to this task was central. We now know that this is
> not so; a consoling thought is that we may have gained in
> maturity what we have lost in excitement. [1981b, p. 23]

2. Standard References

Until the 1960s, social science research on Africa was dominated by anthropologists, historians, and geographers.[6] Anthropologists (mostly European) were noted for their ethnographic studies which were largely financed by colonial offices in the 1930s, 1940s, and 1950s. A handful of economists and started to pursue research in sub-Saharan African beginning in the 1950s and the number greatly expanded in the 1960s. But the technical and social science knowledge base for agriculture continues to be sparse and uneven. Except for a few countries such as Nigeria and Kenya, agricultural research is fragmentary and the scientific knowledge base, especially for food crops, is behind that of Latin America and Asia.

Despite the generally weak data base, many studies over the last twenty years have been extremely useful to policy makers, scholars, and donor agencies. Foremost is the two-volume study coordinated by John de Wilde for the World Bank.

De Wilde *et al.* [1967] drew on information from thirteen agricultural projects: five in Kenya, two each in Uganda and Mali, and one each in Tanzania, Burkina Faso, Chad, and the Côte d'Ivoire. The first volume, the synthesis, contains information about the distinguishing features of agriculture, response of farmers to incentives, labor allocation, mechanization, land tenure, extension, credit, and marketing institutions. Volume two includes case studies. A classic of the 1960s that unfortunately has received little attention is Jurion and Henry's *Can Primitive Farming Be Modernized?* [1969]. This book summarizes the extensive research of the Belgian scientists at the INEAC research station in northern Zaire (formerly Belgian Congo) in the 1930s.

Two important books based on many years of microeconomic research in Africa are Michael Collinson's *Farm Management in Peasant Agriculture* [1972] and Martin Upton's *African Farm Management* [1987]. Collinson drew on many years of farm level research experience in Tanzania to show how practical farm management studies can contribute to the needs of extension workers and local planners. Upton's text stresses the application of production economics to the study of farming systems.

Standard references of the mid-1970s include John Cleave's *African Farmers* [1974], the authoritative volume on labor use in African agriculture, and Uma Lele's *The Design of Rural Development* [1975] which summarizes problems encountered in seventeen major rural development projects in eastern and western Africa. Valuable sources on the diverse cropping systems and technical problems are Benneh [1972], Leakey and Wills, eds. [1977], and Morgan [1978]. Particular attention should be given to the classic by the late Hans Ruthenberg, *Farming Systems in the Tropics* [1980].

Agricultural development strategies and policy issues are covered in edited volumes by Bunting [1970], McLoughlin [1970], Amann [1973], and Ofori [1973]. Books with a policy orientation include *Agricultural Change in Tropical Africa*, a comparative study by Anthony *et al.* [1979]; a collection of papers edited by R. H. Bates and Lofchie entitled *Agricultural Development in Africa: Issues of Public Policy* [1980]; *Rice in West Africa: Policy and Economics*, edited by S. R. Pearson, Stryker, Humphreys, *et al.* [1981]; and a collection of thirteen case studies edited by Heyer, Roberts, and Williams, in *Rural Development in Tropical Africa* [1981].

Six reports and books are indispensable for examining the crisis in food and agriculture in Africa. The first is FAO's *Regional Food Plan* [1978] which outlines the background to Africa's food crisis and steps to meet it. The Food Plan was endorsed by the Organization for African Unity (OAU) in Arusha in 1978 and in Monrovia in 1979. The second is the *Lagos Plan of Action* [1980] which was endorsed by the African heads of state in Lagos in April 1980. The Lagos Plan calls for massive increases in foreign assistance and measures to increase food production. The third is the USDA's *Food Problems and Prospects in Sub-Saharan Africa*

[1981]. The fourth is the World Bank's *Accelerated Development in Sub-Saharan Africa: An Agenda for Action* [1981b]. The fifth is a collection of papers edited by Mellor, Delgado, and Blackie, *Accelerating Food Production Growth in Sub-Saharan Africa* [1987]. The sixth is a collection of papers edited by Mandivamba Rukuni and Carl K. Eicher, *Food Security for Southern Africa* [1987].

3. Overview of Agricultural Systems

There are numerous methods of classifying agricultural systems in sub-Saharan Africa. Ruthenberg [1980] identifies six approaches including: type of rotation, intensity of cultivation, water supply, cropping pattern and animal activities, implements used for cultivation, and the degree of commercialization. Benneh [1972] argues that since the "pivot of every system of agriculture is the technique used to restore soil fertility," the method of fertility restoration should be the basis for classifying agricultural systems. Morgan [1978] classifies systems on the basis of purpose (commercial *versus* subsistence), type of management, and enterprise class (single staple dominant, two or more staple dominant, mixed crops and livestock, and livestock). In this review, we shall distinguish farming systems and livestock systems since there is little mixed farming in Africa. While many farmers raise poultry and small ruminants, the large-ruminant livestock economy is based on herding as a distinct activity from farming.

FARMING SYSTEMS

We identify farming systems on the basis of three characteristics: ecological zone, intensity of rotation, and major crops.[7]

Ecological Zones. Knowledge of soil and water resources in various ecological regions is crucial for understanding farming and livestock systems. Ecological zones in West Africa can be divided into several classifications. We have chosen to subdivide the continent into three subzones: includes the Sahelian, Sudanian, and Guinean zones.[8]

The Sahelian zone, also referred to as the subdesert wooded savanna, has less than 500 mm of rainfall annually and the rainfall tends to be irregular in amount and distribution. In the northern part of the Sahelian zone, the rainy period is frequently less than fifty-five days, the minimum period needed for settled rainfed farming. While the northern Sahel is unsuitable for farming, herders migrate throughout the zone searching for pockets of grass which grow in riverbeds after the sporadic rains. Throughout the Sahelian zone, the open water evaporation rate generally exceeds 2,000 mm per year, in part attributable to desiccating winds such as the harmattan which originates in the Sahara in the dry season and moves southward.

The Sudanian zone (or arid wooded savanna) is characterized by open grassland and an annual average of 500 to 800 mm of rain. This zone has favorable climate and soils for farming, particularly the southern part of the zone which is intensively farmed. Areas with rainfall exceeding 605 mm are considered a transition zone from the arid to humid wooded savannas and are often referred to as the northern Guinean zone.

The Guinean zone proper has rainfall exceeding 1,000 mm per annum with the rainfall distributed over a 120- to 190-day period. This zone is the most extensive in West Africa and it has high agricultural potential. The Guinean zone includes the subarid wooded savanna, subhumid savanna, and the humid forest zone. The southern part of the Guinean zone, commonly referred to as the subhumid tropics or forest zone, is characterized by heavy rainfall during most of the year and reliance on swamp rice and root and tuber crops [Leakey and Wills, eds., 1977]. Parts of the Guinean zone are further classified as derived savanna. The derived savanna refers to former forest land which has been irreversibly transformed into a semiwooded landscape as a result of farming activities.

The above classification of ecological zones applies to a wide band across West Africa south of the Sahara. There is not a comparable pattern of ecological zones in Eastern and Central Africa where the zones are greatly influenced by variation in altitude. The semi-arid belt of the Sahelian and Sudanian zones in West Africa does continue into southern Sudan and south through Kenya and Tanzania, but the distribution of rain, and its seasonal character, changes [Ahn, 1977]. In many of the highlands of East Africa, Rwanda, and Burundi, soils are of volcanic origin and rainfall is generally sufficient to grow a wide range of crops. In contrast to West Africa where there is one rainy season from May to September, in the East African savanna, rainfall has a bimodal distribution which is variable in amount and length.

Intensity of Cultivation. The intensity of cultivation can be used to distinguish among shifting cultivation, fallow, and permanent cultivation systems. A simple criterion for classifying the intensity of a system is the relationship between crop cultivation and fallowing. Following Ruthenberg [1980], the index of land use intensity (R) is defined as the years of cultivation multiplied by 100 and divided by the sum of the number of years in cultivation plus the number of years in fallow. A value of zero is given for land in permanent fallow and a value of 100 for continuous cultivation.

Shifting cultivation[9] is a system of growing crops for a few years on selected fields and then allowing the land to rest and regenerate for a lengthy period—often exceeding twenty years [Ruthenberg, 1980]. Shifting cultivation is predominantly practiced in humid and semihumid climates such as in the forests of Central Africa and in the lowlands of West Africa. Migration is an integral feature of systems of shifting cultivation and nomadic pastoralism as households move periodically to more fertile land.

Fallow systems are generally defined as those in which one-third to two-thirds of the land is cultivated each year—the land use intensity (R) ranges between 33 and 66. Landholdings are usually clearly defined in fallow systems and the rural households have permanent housing or they move occasionally over short distances. The economic importance of fallow farming systems is far greater than that of shifting cultivation systems. A modified form of fallow cultivation systems found throughout sub-Saharan Africa is the ley system. Ruthenberg defines a ley system as one in which several years of arable cropping are followed by several years of grass and legumes for livestock. Regulated ley systems, which may include both pasture management and planting of grass leys, are found on large farms in Zambia, Zimbabwe, the Kenyan highlands, and in some settlement schemes in other countries.

In systems of permanent cultivation, arable crops are only rarely interspersed with fallow or leys. A land use system is considered permanent cultivation when the land use intensity (R) exceeds 70. The major problem of permanent cultivation systems is maintaining soil fertility. Efforts to maintain yields in the face of higher land use intensity have forced farmers to modify their land use practices. One of the most widely noted responses to a reduction in soil fertility is to farm fields closer to the compound more intensively, using household wastes and animal manure to enrich the soil.

Major Crops. The heterogeneity of ecological conditions in sub-Saharan Africa is reflected in the diversity of cropping patterns. In general, one or two staple food crops dominate cropping systems in each ecological zone and can account for as much as 70 to 80 percent of the area cultivated. After giving priority to planting their dominant staples, African farmers rely on a strategy of diversified production, incorporating food crops, cash crops, and nonfarm activities. While ecological conditions, principally the amount and distribution of rainfall, determine the dominant crop or crops in any region, there is variability in the combinations of secondary crops grown in any village by given farmers.

Cereal grains are the most important crops grown in the semi-arid regions of sub-Saharan Africa, both in terms of acreage and calories. The major cereal grains are sorghum, millet, maize, rice, wheat, and teff (in Ethiopia). In most parts of Africa, grains are intercropped with secondary crops such as cowpeas, beans, and vegetables. Intercropping (mixed cropping) entails growing two or more crops at the same time on the same field.

Sorghum and the various millets are among the most widely grown crops in Africa. [ICRISAT, 1980a, b, 1987; Matlon, 1987; Spencer and Sirakumar, 1987] Sorghum is concentrated in a belt south of the Sahara where rainfall ranges from 600 to 1,000 mm a year. In northern Nigeria, for example, sorghum and millets are grown on over 50 percent of the area cultivated [Beeden et al., 1976]. In the savanna region of West Africa,[10] sorghum is generally grown in a bush fallow land use system and is frequently intercropped with millet. Sorghum is generally dominant in areas susceptible to both drought and flooding while millets, which are extremely drought resistant and tend to store better than other cereal grains, play a major role in zones receiving low and uncertain rainfall. Even in cases where sorghum and millet are not intercropped, farmers often grow several varieties of each crop on different fields in the same year.[11]

Maize is the staple grain in many parts of Southern and Eastern Africa. Maize is the most important food crop grown in Zambia, Tanzania, Malawi, and Kenya and it is increasing in West Africa as a replacement for sorghum and millet [Taylor and Bailey, 1979]. Maize is also grown as an important secondary crop in many areas, particularly forest zones, and is eaten fresh. Under favorable conditions, maize tends to yield more than other cereal grains and is less susceptible to attacks by birds than sorghum and millet. Maize is performing well in West Africa in Cameroon, Togo, Benin, Côte d'Ivoire, Nigeria, and Sierra Leone.

While sorghum, millet, and maize are the primary food grains, rice and wheat are increasing in importance throughout Africa. Madagascar is Africa's largest rice producing nation. Rice is also pervasive in West Africa where it is the staple food in Liberia, Sierra Leone, Ghana, and Mali [Bingen, 1985; Kamuanga, 1981]. For example, a 1974–75 survey in Sierre Leone revealed that rice was grown on over 97 percent of the small farms [D. S. C. Spencer and Byerlee, 1976]. Rice is grown in four major cropping systems: upland or rainfed, in paddies, in mangrove swamps, and on bottomlands (bas fonds). Upland rice, which accounts for 75 percent of the rice cultivated in West Africa, is generally grown in a bush fallow system and is frequently intercropped with maize, millet, and cassava. Most rice is grown as a subsistence crop. Rice imports have increased dramatically over the last decade in response to growing urban demand [Pearson, Stryker, Humphreys, et al., 1981]. For example, West Africa is importing about 40 percent of its domestic rice consumption [D. C. S. Spencer and Nyateng, 1987; WARDA, 1988].

The major wheat producing areas are in Ethiopia, Kenya, Tanzania, Lesotho, Zimbabwe, and the Sudan, countries where it is often the major staple in highlands 2,100 meters above sea level. The majority of the wheat is grown on large farms and is sown and harvested mechanically. At lower elevations, there has been little expansion of wheat acreage because of low and unreliable yields. Wheat production is being promoted on government irrigation schemes in the Sudan, northern Nigeria, and in a number of other countries [Byerlee and Longmire, 1986].

In humid forest zones, farmers rely on plantains, roots, and tubers rather than grain for most of their calories. Cropping systems in forest zones are usually dominated by one or two crops. Plantains are the major staple food grown in parts of Uganda, Tanzania, Ghana, Côte d'Ivoire (formerly Ivory Coast), and

Zaire. Plantains may be planted in pure stands but often are intercropped as shade for coffee or with food crops such as maize or beans.

Cassava (manioc) is growing in importance for several reasons: it requires less labor than yams and cereals; it can be grown in a wide range of climatic conditions; has low soil nutrient requirements; is resistant to pests; and is regarded as a "famine crop" because some varieties can be left in the ground for twelve to eighteen months without a significant deterioration in quality [Fresco, 1986].

Many high-yielding varieties of yams are being grown in West Africa, particularly in Nigeria where 90 percent of the West African output and two-thirds of the world output of yams is grown. Yams are labor-intensive and are declining in importance in some regions because of rising wage rates.

The major grain legumes are groundnuts (peanuts), cowpeas, and pigeon peas. Cowpeas, which are the most commonly eaten legume, are generally intercropped with cereal grains. Groundnuts are an important cash crop throughout the semi-arid regions but they are also eaten boiled or roasted. Soybeans are a minor crop but are of increasing importance. Two problems with soybeans remain: consumer resistance, and storage of seeds during the dry season in semi-arid regions.

Since the colonial period, African farmers increasingly have produced certain crops primarily for sale (cash crops). Groundnuts and cotton are important cash crops in the semi-arid regions. Cotton plays a major role in Uganda, Tanzania, Malawi, the Sudan, Mali, Burkina Faso, Senegal, and northern Nigeria. The most important perennial cash crops are oil palm, cocoa, coffee, tea, pyrethrum flowers,[12] and rubber. Oil palm is important in Zaire and in the rain forest of West Africa, particularly in the Côte d'Ivoire and Nigeria. Cocoa is an important crop in Ghana, Nigeria, Cameroon, and the Côte d'Ivoire. Coffee is grown by small farmers in Uganda, the Côte d'Ivoire, Cameroon, Kenya, and Ethiopia. Coffee in West Africa is mostly of the Robusta type which is of lower drinking quality and is primarily used for making instant coffee. Most of the coffee in Kenya, Uganda, and Ethiopia is Arabica and is considered to be some of the best coffee produced anywhere in the world. Kenya is the center of a flourishing smallholder tea industry.

Farmers grow vegetables such as okra, onions, tomatoes, peppers, and leafy greens on small garden plots to supplement the staples in their diets. Vegetables may also be intercropped with staples. Vegetables are frequently grown on small plots along riverbeds during the dry season. While labor inputs are high for vegetable production, the returns to labor are about the same as most staple crops because of the high value of the vegetables.

Livestock Systems. Livestock is a major industry in Mauritania, Mali, Niger, and Chad in West Africa; Ethiopia, Somalia, and Sudan in East Africa; and Botswana in South Africa.[13] Cattle are the most important type of livestock in terms of milk and meat consumption but small ruminants (goats and sheep) are dominant in a few countries. For example, small ruminants account for about one-half of red meat consumption in rural Mali [Delgado, 1980] and in the humid zone of Nigeria [ILCA, 1979a].

Sheep and goat production is attractive to small farmers because of low initial investment (relative to cattle) and the corresponding small financial loss with the death of individual animals. Small ruminants can be fed roughage and crop by-products; managed by women and children; and sold to pay school fees or purchase grain and cloth. Until recently, sheep and goats were neglected by both technical researchers and social scientists [R. E. McDowell and Bove, 1977; ILCA, 1979a].

Donkeys and camels are primarily used for transportation and traction in the semi-arid and arid regions. Camels are an important source of milk in some countries. There is relatively little swine production in Africa, particularly in Moslem areas. Poultry products are common in rural households and are increasingly produced in specialized units around the large cities.

There are many arbitrary methods of classifying livestock production systems. We shall use the following: nomadism or total pastoralism, seminomadism or semipastoralism, sedentary pastoralism, and ranching—private, group, association, cooperative, and state.

Nomadic households are defined as those which do not have a permanent place of residence or do not practice regular cultivation. Nomadic and seminomadic households have historically migrated through arid and semiarid regions in search of water and grazing land and have played a major role in regional trade—especially in West and East Africa [Salzman, ed., 1980]. Pastures are traditionally considered to be available to all members of an ethnic group or are rented under land-use agreements. The priority of nomads is to feed their families a continuous supply of milk and meat from their herds. Nomadic pastoralism is presumed to be declining[14] in most countries due to the loss of prime grazing areas to seminomadic herders and farmers, overpopulated arid livestock zones forcing a reduction in distance covered by transhumance,[15] and recurring drought.

Seminomadic pastoralism is a system in which households establish a permanent place of residence which is kept for several years. Crops are cultivated as a supplementary food source but herds are moved on transhumance to assure sufficient forage and water. But in some semiarid areas, nomads have acquired use rights to dry season grazing areas such as in the Niger Delta in Mali [Gallais, 1975].

Sedentary pastoralism and mixed farming are systems of integrated crop and livestock production. Sedentary pastoralism refers to permanent settlers who grow a few crops but rely on livestock production as their dominant enterprise. Mixed farming refers to a system whereby crop farmers add a livestock enterprise either for fattening or for the use of oxen in cultivation and the subsequent fattening of the cull work oxen. The integration of crops and livestock has a long history dating from schemes introduced by colonial governments in the 1920s. Mixed farming is expanding but its success is dependent upon the presence of a cash crop such as cotton and groundnuts.

Systems of stratified production and ranching are alternatives to the dominant system of pastoral production [FAO, 1977]. Stratification involves removing young male stock from the arid and semi-arid pastoral areas for fattening in the intermediate rainfall zones and eventually for sale in urban centers. Stratification and industrialization feedlots are not widespread in West Africa because of the risk involved in securing a steady supply of animals for fattening and the fluctuation of the price of finished steers.

Ranching has been practiced for decades in Zaire, Angola, and eastern and southern Africa, and to a lesser degree in some countries in West Africa. During the 1960s and early 1970s, ranching was endorsed by many experts and donors because ranching was thought to: prevent overgrazing through the control of the number of animal units and the practice of on-ranch rotational grazing, prevent spreading of contagious disease, guarantee a supply of cattle to markets, and provide a demonstration function for small herders. But in recent years, the push to establish ranches has slowed in light of evidence that the number of jobs created and the returns on investment are low [von Kaufmann, 1976]. Moreover, concerns for equity are encouraging governments to help small herders form ranching associations and group ranches [Ayuko, 1980, 1981].

Dairy production is concentrated in Kenya, Tanzania, Ethiopia, and the Sudan.[16] Although colonial policy in Kenya reserved dairy production for large-scale farmers (mostly white), smallholder milk output has grown rapidly over the past twenty years and it now accounts for 40 percent of Kenya's milk output. Some nomadic groups such as the Fulani in western Africa supply milk to local villages.

Chapter II. Historical and Theoretical Perspectives

Since all the countries in this review, with the exception of Ethiopia and Liberia, are former colonies, we shall examine precolonial economic activity and agricultural policy during the colonial and postindependence periods. We then present an overview of the main theoretical perspectives and note the lag in developing theories of development based on African values, institutions, resource endowments, and data. Two streams of thought—neoclassical economics and dependency/political economy models—dominate political debates, research, and policy analysis in Africa [Staatz and Eicher, 1984].

1. Historical Perspective

PRECOLONIAL ECONOMIC ACTIVITY: 1800 TO 1880s

Until recently, African history was essentially colonial history and, more specifically, the political history of the colonial powers starting around the 1880s. Economic historians devoted little attention to the precolonial period because they assumed that the development of Africa began with the imposition of colonial rule. Many scholars such as Peter Bauer [1975] of the London School of Economics have argued that modern social and economic life in Africa began with colonial rule and the creation of trade linkages with Europe. But over the past twenty years, there has been an explosion of research on the history of the precolonial period from 1800 to 1880. There is now convincing evidence of extensive economic activity prior to the beginning of colonial rule in the 1880s. The standard reference to the economic history of the precolonial and colonial periods is McPhee's *The Economic Revolution of British West Africa* [1971]. Although McPhee called the rapid expansion in export crops a "revolution" which was facilitated by the imposition of British rule starting in the 1880s, he carefully traced the origins of the revolution to the 1820s.

A Swedish economic historian, Sundstrom, has made a major contribution to the economic history of the precolonial period. Sundstrom's book, published in Swedish in 1964, was virtually unnoticed until it was translated and published in English in 1974. He drew on English, French, German, Dutch, Portuguese, and Swedish sources in his penetrating analysis of the internal trade of salt, textiles, iron, copper, and brass in West, Central, and East Africa from the eighteenth to the twentieth centuries. Evidence that internal trade was widespread, and of great antiquity, is also supported by Dike's [1956] study on oil palm trade in the Niger delta of southern Nigeria in the nineteenth century and Lovejoy's [1980] research on long-distance kola nut[17] trade in the nineteenth century. A. Cohen [1971] traced the origins of long-distance livestock trade in West Africa to the eighth century and highlighted the central role of Islam as the binding force of a network of traders scattered over several countries. Baier [1977] pointed out that for many centuries long-distance traders have played a major role in the economies of West Africa by moving salt, dates, and livestock from the desert and savannah zones to the coast; and grain, cloth, gold, and manufactured articles from the coast to the savannah and desert zones. The net impact of the above studies is that they have essentially demolished the view that economic development in sub-Saharan Africa began with the arrival of colonial governments in the late 1880s.

A. G. Hopkins, an economic historian at the University of Birmingham, established himself as a controversial and influential scholar with the publication of *An Economic History of West Africa* [1973]. Hopkins used neoclassical economic theory and a modified formalist paradigm of economic anthropology in his analysis of the interaction between internal trade and external market forces over the 1800-1950 period. Hopkins's grand theme was that market forces played a key role in integrating West Africa into the world economy. Although colonial rule did not begin until the 1880s, Hopkins's book has been acclaimed by scholars in numerous disciplines but attacked by George Dalton, a Northwestern University economic anthropologist and leader of the substantive paradigm of economic anthropology. In Dalton's [1976] fifty-page review of Hopkins's book, he criticized the emphasis on the market as the major force in development in the nineteenth century. Dalton contends that reciprocity and redistribution were also important forces in the precolonial and colonial periods.

The slave trade is at the heart of West African history. The late Walter Rodney argued in *How Europe Underdeveloped Africa* [1974] that a European imposed pattern of exchange of slaves for manufactured goods from the fifteenth through the nineteenth centuries was the cause of the underdevelopment of Africa. Rodney argued that the slave trade reduced Africa's population growth rate to zero and caused a severe technological arrest (although Rodney neglected to explain how population growth leads to technical advance).

Now that economic historians have documented the presence of extensive international trade and the growth of cash crops before the advent of colonial rule, there is need for micro research on the agricultural history of particular crops, the origins and dynamics of internal trade, and the linkages between internal and international trade. Carolyn Barnes's study [1979] of the agricultural history of coffee production among the Gusii in Kenya over the 1933-48 period and Baier's analysis [1980] of the economic history of the central region of Niger from 1850 to 1960 are examples of the type of in-depth research which is needed.

THE COLONIAL PERIOD: 1880-1960

An analysis of agricultural development in the postindependence period since 1960 should be rooted in an understanding of colonial strategies. Colonial strategies varied widely throughout Africa and it is difficult to generalize about the impact of these strategies. British colonial policy in Kenya, for example, promoted extensive European settlements. M. P. K. Sorenson's [1967] assessment of Kenyan agriculture during the colonial period reveals that much of the best land was reserved for Europeans starting in the late 1890s. Head taxes were introduced to encourage small farmers to produce cash crops and to sell their labor to European plantations and mines. On the other hand, British colonial policy in Nigeria, Ghana, The Gambia, and Sierra Leone sharply restricted plantation development and settlement by white farmers. In fact, British colonial policy in Nigeria prevented foreign private firms from gaining long-term control over land. For example, Unilever, a large European trading firm, gave up trying to establish plantations in Nigeria and eventually opened plantations in the Belgian Congo (now Zaire). For a reference on British colonial policy in Africa, see Hancock [1942].

In contrast to British policy which restricted settlers and plantations in Nigeria, French policy encouraged Europeans to establish plantations to grow coffee and cocoa in the Côte d'Ivoire [B. Campbell, 1974]. In addition, the decrees of 1925 regulated forced labor and insured a steady flow of Ivorian workers for the European plantations. Gradually, Ivorians started to grow coffee and cocoa on small plots scattered throughout the forest but since they used European techniques they were called "planters." The Ivorian planters led the drive to independence after World War II and the planters remain a powerful political force in the Côte d'Ivoire today. For a discussion of French colonial policy in West Africa, see Newbury and Kanya-Forstner [1969].

In Tanganyika (now Tanzania), the 1905-12 thrust of German colonial and commercial policy was to develop plantation agriculture. Hut and poll taxes were imposed to force Africans to provide wage labor for the plantations. Iliffe [1979] has shown that by the end of German rule in 1912 an export orientation was firmly established. For an excellent discussion of poverty in historical perspective see Iliffe [1987].

An important assessment of colonialism is Duignan and Gann's collection of edited papers, *The Economics of Colonialism* [1975a]. Papers include an analysis of trade policy by Meier [1975], the emergence of cash crop exports by Hogendorn [1975], industrialization by Kilby [1975], French colonial policy by Thompson and Adloff [1975], agricultural research by Yudelman [1975], and British colonial policy by P. T. Bauer [1975]. The editors and most of the authors generally presented a procolonial assessment of the contribution of colonialism to African development. Studies of the colonial period include Hill [1963], Suret-Canale [1971], Wrigley [1965], de Wet [1977], Brett [1973], Dorward [1975], Yansane [1976], Baier [1977], Dumett [1971], Green and Hymer [1966], G. E. Brooks [1975], Howard [1978], and Kitching [1980]. For a discussion of the economy and apartheid in South Africa see S. Lewis [1988].

Since agricultural development programs in the postindependence period have been directly or indirectly influenced by colonial policies, approaches, and attitudes about agriculture's role in development, it is important to examine the following:[18]

(1) the degree to which Africans were excluded from, or "forced" to participate in, colonial development programs;

(2) the colonial record on the training of Africans;

(3) the colonial position on promoting research on export *versus* food crops; and

(4) who benefited from colonial land grants and export agriculture.

A growing number of scholars have revealed that Africans were systematically excluded from participation in many colonial development schemes and in producing certain export crops and improved cattle. R. E. Baldwin's [1966] study of 1920-60 export growth in Zambia (then Northern Rhodesia) reported that "the agricultural policy for most of the period covered was designed to benefit European settlers. African farmers were either ignored or discriminated against when their interests conflicted with those of the European population . . . and those few agricultural measures specifically directed toward helping African farmers were often poorly conceived and ineffective." Reviewing colonial policy in Kenya, Heyer, and Waweru [1976] stated that "there was an effective prohibition on African coffee growing until 1933 when it was allowed on a very limited and experimental scale in three districts relatively far removed from European coffee growing." For further evidence on how colonialism excluded Africans from growing a number of crops and from producing improved cattle, see Uchendu and Anthony [1975a, b].

A number of scholars have documented how the colonial governments compelled village chiefs to force farmers to grow selected crops (e.g. cotton), or to "contribute" labor to maintain roads. Magasa [1978] presents a damning account of the repressive policies adopted by the French colonial service to recruit labor for government projects such as the Office du Niger irrigation scheme in Mali. The colonial legacy of top-down approaches to agricultural change is still present in agricultural ministries throughout Africa.

Colonial governments gave little attention to the training of Africans. McKelvey's survey of "Agricultural Research in Africa" [1965] reported the almost total failure of colonial governments to develop institutions for training African agricultural scientists and managers. For example, by the time of independence in the early 1960s, there was only one Faculty (College) of Agriculture in French-speaking tropical Africa. Between 1952 and 1963, only four university graduates in agriculture were trained in Francophone Africa and 150 in English-speaking Africa. Johnston [1964] observed that there were only three African scientists working in all the experimental stations in the East African countries of Kenya, Uganda, and Tanzania in 1964.

One of the most important legacies of the colonial era is the bias against research on food production and small farmers and herders. In numerous countries, colonial regimes focused their research and development programs on export crops, plantations, land settlement schemes, and commercial farmers. For example, Hunt [1975a] reported that millet yields in 1972-73 in Kenya were at best only 50 percent of the maize yields and that "no [agronomic] work has been done on developing improved bulrush millet or fingermillet strains," in contrast to a large amount of research on maize during the colonial period to benefit the commercial farmers in the highlands areas with high rainfall. Heyer [1981] reported that colonial policy in Kenya, which established export crops in areas with the most favorable natural resource endowments, helps to explain some of the regional inequalities and the fragmented road and railroad systems. Although the number of European farms was never very large in Kenya, reaching a maximum of 3,600 holdings at the end of the 1950s, Heyer reported that "the development of Kenya's agriculture was profoundly affected by the presence of white farmers" because at the end of the 1950s European farms were responsible for an estimated 80 percent of the agricultural output that reached urban and export markets [1981, pp. 93-94].

Economic historians are grappling with the roots of rural poverty in Central and Southern Africa. Palmer and N. Parsons's collection of essays, *The Roots of Rural Poverty in Central and Southern Africa* [1977], has been described by Ranger [1978] as a "landmark in African agricultural and peasant studies." The theme of the book is the alleged "strangulation" of peasant (smallholder) farming by the capitalist market forces which first stimulated peasants to produce for the market and sell their labor to the mines but later resulted in the exploitation of farmers through the taxation of export crops. Although Ranger lauds *Roots*, he urges historians to shift from archival research to more field studies and oral history to document how small farmers in Central and Southern Africa were affected by specific colonial policies. Another important study is Bundy's *The Rise and Fall of the South African Peasantry* [1979].

Any definitive assessment of the colonial period is faced with the unanswerable question of what would have happened in the absence of colonialism. Until recently, numerous scholars accepted at face value the development programs initiated in the colonial period, while ignoring the broader question of possible exploitation. For example, two scholars recently lamented the narrow approach to their doctoral dissertation research. Brett, a British political scientist, reflected on his doctoral dissertation research on colonialism in Kenya as follows:

> In London I worked in the general ethos of the Institute of
> Commonwealth Studies where the tendency in Colonial
> History was towards a thorough empiricism in a framework
> which did not question the overall validity of Britain's colonial
> contribution to African development except where extensive
> resettlement (e.g., encouragement of white farmers to settle in
> Kenya) had been allowed. This environment certainly
> encouraged me to take at face value a great deal in the material
> which I was given and did not lead me to make any serious
> attempt to relate it to the broader questions of colonialism and
> underdevelopment which have now forced themselves into the
> center of all serious work in this general field. [1973, p. ix]

An American anthropologist, Sudarkasa, wrote in the preface to a book based on her dissertation on market women in Nigeria:

> When I wrote this study a decade ago, I tacitly gave support to the social science fiction that what happened in Nigeria (and in all of Africa) in the 20th century could be sanguinely described in terms of "modernization", and that the processes of European entrenchment and exploitation in Africa could be subsumed under the benign if not indeed beneficient concept of "development". The fact that I did not delve into the factors underlying the conditions described in this study is perhaps an indication of the success of my training in Western social science. In any case, suffice it to say that I would bring a decidedly different perspective to such an undertaking if I were doing it today. [1973]

In summary, although the full impact of the colonial policies on contemporary Africa will require further research, the debate has been joined. Research on the colonial period by neoclassical economists is now being supplemented by a growing number of political economy studies of trade, capital formation, and the uses of surplus. There is a rich agenda for research by economic historians of diverse ideological positions. For example, why has export growth in some countries, the Côte d'Ivoire for instance, been cumulative; and why do other countries such as Ghana, Zambia, and Senegal remain dependent on one major export crop after sixty years of export trade? Future research on the colonial period is likely to show that the net effect of the colonial policies and programs on the peoples of sub-Saharan Africa will lie somewhere between Peter Bauer's assertion that "virtually all components of modern social and economic life in Africa date from the colonial period" [1975, p. 653] and Kwame Nkrumah's assessment that "without exception they [the colonial powers] left us nothing but our resentment" [1963, p. xiii].

TRANSITION TO INDEPENDENCE

As African nations became independent in the late 1950s and early 1960s, most of them pursued mixed economies with a heavy emphasis on foreign aid,[19] nation building, industrial development, education, and diversification of their economies. A small number of countries such as Mali, Ghana, and Guinea shifted abruptly to revolutionary socialism in the early 1960s. But whether political leaders were espousing capitalism or socialism they generally all gave low priority to agriculture. African leaders generally viewed agriculture as a "backward" sector which could provide agricultural surpluses—taxes and labor—to finance structural change and industrial/urban development. Agricultural policies in many capitalist and socialist countries supported plantations, state farms, land settlement schemes, and the replacement of private traders and money lenders with government trading corporations and credit agencies. Moreover, the empirical record shows that all countries under all types of governments—civilian, military, capitalist, and socialist— have exploited and controlled the agricultural sector via harsh taxation policies and the underpricing of agricultural commodities; concentration of secondary schools, hospitals, and other social services in the capital city, or several large cities; top-down patterns of government-imposed agricultural development schemes; and unwillingness to transfer the administration of marketing, storage, and credit programs to groups of farmers. Among the reasons for the low priority given to agriculture in the 1960s were the following: industrialization was perceived to be the most expedient way to bring about structural change, a high rate of economic growth, and economic independence; investment in food production was assumed to be unnecessary because of low man to land ratios and surplus land; and export crops were perceived to be "colonial" crops which contributed to dependency and price and income instability.

During the last decade, almost all African governments shifted from industrialization, export promotion, and agricultural transformation to espousing the multiple goals of food self-sufficiency, improved nutrition, diversification of their economy, increased trade within Africa, economic growth, and increased access of the poor to employment, income, and social services. While nearly all African governments espouse the primary goal of self-sufficiency in food production, it takes resources to increase food production and it will involve some difficult production/equity conflicts (e.g., conflicts between regions with good agricultural potential and regions with poor potential). Thus, as African nations entered the fourth decade of independence, there was generally little debate over the goals of development. The contentious issues were priorities, financing, time frame, the role of agriculture, and the state in national development.

2. Theoretical Perspectives

Over the past thirty years scholars working on African development have been caught in a crossfire of imported models and theories of development. The lag in developing models of development by Africans and based on African data and institutions is linked to the token role which colonial governments gave to the training of Africans for positions in universities and research organizations.[20] As a result, the production of new knowledge about African development still remains to a substantial degree in the hands of expatriates except in a few countries such as Nigeria, Kenya, and Zimbabwe. Moreover, expatriates dominated planning agencies and social science research institutes in Africa throughout the 1960s and most of the 1970s. For example, the Economic Commission for Africa (ECA) in Addis Ababa relied heavily on expatriate advisors such as A. F. Ewing and Rene Dumont to provide vision for the 1960s. But these ECA advisors failed to articulate feasible development strategies for an agrarian-dominated continent. For example, A. F. Ewing reported that "industry is the sole means of raising the productivity of an economy" [1968, p. 11]. In the mid 1960s the Executive Secretary of the Economic Commission for Africa requested the noted French agronomist, Rene Dumont, to undertake a survey on how to transform African agriculture. In his report entitled *African Agricultural Development* [UN/ECA/FAO, 1966], Dumont recommended cooperative farming and lifelong claims to land rather than private ownership of land and he urged African nations to learn from the Socialists experiences in Eastern Europe, USSR, China, and Cuba. Dumont also moralized about the need for Africans to work harder and the need for austerity, integrity, education, and exemplary moral qualities.[21] An influential UN/ECA/FAO report on agriculture called for "an accelerated movement into market agriculture through government measures aimed at both individual producers and large-scale projects" [1964, p. 39]. But as we point out later, large-scale projects have generally been ineffective in achieving both efficiency and equity objectives.

The contrast between Africa and Latin America in producing new knowledge about development problems and strategies is striking. In Latin America in the 1950s and 1960s, Latin American scholars such as Prebisch, Dos Santos, Furtado, Sunkel, and Pinto produced a number of seminal papers and reports for the United Nations Economic Commission on Latin America (ECLA).[22] These scholars, under the leadership of Prebisch, came to the conclusion that models on development based on the experience of high-income countries in Western Europe and North America, including "classical" Marxist models, were not relevant to Latin American conditions and that attention should be directed to developing theories and models based on Latin American conditions. Moreover, Prebisch's [1959] influential thesis that the world economy is rigged against the Third World—declining terms of trade—was one of the central issues in the North-South meeting of twenty-two heads of state in Cancun, Mexico in October 1981. The Prebisch thesis remains an article of faith in the Third World.

As African nations became independent in the early 1960s and Western economists assumed important roles in preparation of development plans and service as policy advisors, naturally questions were raised about the usefulness of Western economic theory. One of the recurring debates in the 1960s centered on the economic motivation of farmers and traders in subsistence economies. Another major issue was the relevance of Western development models—dual-sector, labor surplus, land surplus, and stages of growth models.

ECONOMIC BEHAVIOR OF FARMERS AND TRADERS

Many Western economists have long contended that Africans were not "economic men" in the Western sense. For example, 100 years ago, Alfred Marshall, the founder of neoclassical economics, wrote about the savages "living under the dominion of impulse; scarcely ever striking out new lines for themselves; never forecasting the distant future; fitful in spite of their servitude to custom, governed by the fancy of the moment; ready at times for the most arduous exertions, but incapable of keeping themselves long to steady work" [Marshall, 1956]. A major implication of this view is that Western economic theory may have little to say about the behavior of farmers and traders in Africa, a theme that has been hotly debated in the literature of several disciplines, including sociology, anthropology, and political science.

One of the first challenges to the notion that Western economics can be applied in low-income countries came from the Dutch sociologist Boeke. On the basis of many years of research in Indonesia, Boeke [1953] advanced the concept of sociological dualism to describe the modern and traditional sectors in which the peasants in the traditional sector had limited needs, a value system based on prestige, exhibited fatalistic behavior, and created backward-bending supply curves of labor. As a result, Boeke asserted that Western-financed development interventions in the traditional sector would be constrained by these social and cultural barriers to change.[23]

In the 1950s and 1960s, the debate over the relevance of Western economics in developing countries preoccupied many economic anthropologists. During this period, economic anthropologists were divided into two schools of thought—the substantivists and the formalists. The substantivists contended that exchange in many low-income countries is carried out according to principles of reciprocity and redistribution. As a result, the substantivists contended that Western economic theory which stresses profit maximization has limited application in many parts of the developing world, including Africa. The substantive paradigm is conveyed through the writings of Polanyi et al. [1957] and Dalton [1962, 1978a].[24] The formalist paradigm was articulated by economic anthropologists such as Raymond Firth [1975] and Harold Schneider [1974]. The formalists contended that Western economic theory could be selectively applied to the Third World including Africa, because African farmers, traders, and migrants were believed to respond in a general way to economic incentives just as producers and consumers respond in high-income market economies.[25] Fortunately, the often sterile debate between the substantivists and formalists was relegated to the "dust-bin of history" during the 1970s. Anthropologists generally reached a consensus that many of the views of both schools were not mutually exclusive and that each school could provide hypotheses which shed insights on economic behavior in agrarian societies in the Third World.

The debate over the economic behavior of farmers has re-emerged among political scientists working in Asia. J. C. Scott [1976], for example, has advanced the view that peasants are concerned with village cohesion as a means for assuring economic security and survival. In Scott's view, the transition to a market-oriented economy is likely to increase the inequality and insecurity of the poorest families in a village. Joel Migdal [1974] also stressed the importance of a village orientation in decision making and the possible destruction of village safety nets as commercialization of agriculture proceeds. Scott's and Migdal's view of a "moral economy of subsistence farming" was challenged by Popkin in *The Rational Peasant* [1979]. Drawing extensively on historical data from Vietnam from precolonial times to the emergence of the Communist movement, Popkin contended that peasants make individual investment decisions which may be at the short-run expense of the village. Popkin was of the opinion that the commercialization of agriculture, rather than increasing inequality and insecurity, presents opportunities for peasants to increase their welfare.

The proposition that farmers and traders in Africa act in a manner which is inconsistent with the postulates of Western economic theory is an empirical question. During the colonial period, the view that Africa's farmers were not "economic men" was reflected in the target income concept which relates to the hypothesis that there was a backward-bending supply curve for labor, and the hypothesis that social and cultural factors are barriers to the adoption of innovations.

Target Income/Backward-Bending Supply Curve of Labor. The origins of the target income or backward-bending supply curve of labor hypothesis can be traced to the colonial period when plantation and mine owners reported that they could not fill their vacancies at going wage rates because Africans were "lazy" and mine workers would often return to their village after they had earned their target income. This view was implicitly and, in some cases, explicitly supported by colonial administrators. For example, head taxes were introduced in numerous colonies to force Africans to increase their supply of labor to plantations and mines. At independence, numerous scholars repeated the theme that Africans had limited wants and that they would not respond to market forces — higher wages, for example — after they had earned a specific amount of money or target income to pay for taxes, bride price, or consumer goods. It was assumed that a 5 percent increase in wages, for example, might lead to a reduction rather than an increase in the supply of labor. The policy implication was that higher wages and higher prices to farmers could not be relied upon to draw workers into the labor force and farmers into the market economy. But during the 1960s, economists marshalled an array of empirical evidence that Africans responded "normally" to economic incentives.

Two British economists, Bauer and Yamey [1959], were among the first researchers to present quantitative evidence on the response of small farmers to price incentives. Bauer and Yamey examined two of the statutory marketing boards which were established in Nigeria during and after World War II—the Nigerian Cocoa Marketing Board and the Nigerian Oil Palm Produce Marketing Board. Starting in 1947-48, these Boards offered large differentials in prices paid to farmers in order to encourage the production of higher grades of cocoa and oil palm. Bauer and Yamey's study revealed that the proportion of Grade I cocoa purchased by the Nigerian Cocoa Marketing Board increased from 47 percent in 1947-48 to 98 percent in 1953-54 following the introduction of premium prices for high quality cocoa. Bauer and Yamey demonstrated that the small Nigerian cocoa and oil palm producers were indeed "economic men" and that they responded to market incentives just as farmers in high-income countries.

William O. Jones's [1960] masterful survey of studies of migration, trade, production, and marketing in the precolonial and colonial periods concluded that "the economic drive is present in a great many Africans." Peter Kilby [1961] studied Nigerian factory workers and found that they surpassed their European counterparts in sheer physical exertion by as much as 50 percent when the proper financial rewards were held out to them. Elliott Berg [1961] added evidence to suggest that the target income/backward-bending supply curve of labor hypothesis should be discarded. Moreover, Miracle and Fetter [1970] argued that even if a backward-sloping, labor-supply schedule had been validated by empirical studies this behavior pattern could be consistent with orthodox economic behavior if all of the costs associated with migration and working were taken into account—costs such as uncertainty, disease, and long period of separation from their families. Miracle studied the emergence of the wage labor force in Kenya in the early 1900s and found that migrants who left the highlands of Kenya for work 300 miles away on the coast encountered substantial risks of contracting disease, not receiving wages when they became ill, being subject to brutality, and even death. Miracle concluded [1976] that the costs associated with migrants leaving their families could have produced a backward-bending supply curve of labor in this particular case. Helleiner [1975] and Miracle [1976] summarized the debate by noting that it is doubtful that the target income hypothesis was ever valid for a large number of Africans, but, even if it was, it is of limited validity today.

The Cultural Barrier Hypothesis. In the early 1960s, it was common to identify social and cultural factors as overriding barriers to the adoption of innovations and the achievement of development objectives. But the failure of the Community Development movement in Asia in the 1950s and 1960s[26] (which stressed social factors such as felt needs, participation, and self-help) along with numerous examples of change in spite of perceived social barriers led to a consensus that the cultural barrier hypothesis, like any single barrier theory, should be rejected.[27]

In a major study of agricultural change in six countries in sub-Saharan Africa, Uchendu, an anthropologist, and Anthony, an agronomist, concluded that there is no empirical support for the widely held belief that traditional values constitute a general barrier to change [Uchendu, 1968; Anthony and Uchendu, 1974; Uchendu and Anthony, 1975a, b]. This same point is reiterated by Hutton and R. Cohen in their reassessment of sociological approaches to the study of change among peasants in Latin America and Africa:

> Local suspicions, jealousies, ignorance, fatalism, passivity, and fears can play their part, just as they can in any human situation, but we no longer use them as general explanations; they can be relegated to their proper place, enabling us to understand why sometimes they impede change and sometimes they do not. [1975, p. 28]

Although social and cultural factors may not be a general barrier to change in rural areas in the Third World, it is obvious that these factors still play significant roles in shaping the overall pattern of development and who gains or loses in the process. For example, Parkin [1972] points out the enduring role of custom in village life of the Giryama society on the coast of Kenya over a twenty-five-year period. Parkin, a social anthropologist, showed how the Giryama shifted from primarily subsistence farming and herding to dependence on the cash economy through the production of copra which was prepared from coconuts. Parkin found that successful farmers deliberately kept themselves from appearing to rise above their neighbors by participating in such customs as bride-wealth transactions and reciprocal funeral obligations while at the same time they were purchasing land from small farmers who were forced to sell land to pay for the escalating costs of funerals and bride wealth.[28] Philip Mbithi [1977] similarly has shown that in East Africa the social environment, including observance of rituals and taboos, is a major influence on the timing of agricultural practices such as planting and harvesting.

Synthesis. In summary, the target income/backward-bending supply curve of labor hypothesis was discredited by empirical studies in the 1960s and 1970s.[29] As one looks back upon the debate, it is surprising to find the proponents of the hypothesis did not present empirical evidence to support their position. The debate was carried on for years through an exchange of hearsay and assertions. Nor is there any support for the view that social and cultural values are an immutable barrier to change. But we must be cautious about generalizing about the relative importance of social and cultural factors in some 1,000 ethnic groups in sub-Saharan Africa. There are many isolated villages where poor transportation and communication are still constraints on the operation of market forces. Moreover, there are still many nonmarket institutions such as the extended family, clans, and age groups which still play a significant role in shaping economic decisions. But there is unambiguous evidence that African farmers, traders, and migrants will respond to economic incentives when they are offered appropriate incentive structures.

WESTERN DEVELOPMENT MODELS

Although Western development economics is now thirty years old, one thing is clear—it is still unencumbered by evidence from Africa [Staatz and Eicher, 1984]. For example, although aggregate data from 101 countries are included in *Patterns of Development: 1950-70* [Chenery and Syrquin, 1975], the poor quality of data enabled them to use only eight African countries in their detailed analysis of the patterns of development. Moreover, dual-sector and stage-of-growth models which dominated Western development economics in the 1960s and 1970s were primarily based on patterns of development, resource endowments, institutional structures, and empirical findings from Asia and Latin America. For example, the well-known dual economy models of W. A. Lewis [1954], and Fei and Ranis [1964] depend on surplus labor, institutionally determined agricultural wage rates, and the assumption of a closed economy. While these models may have provided insight into the interaction of agricultural and industrial development in Asia and Latin America, they shed little light on the rural economies of Africa.

Vent-for-Surplus. Myint's vent-for-surplus model is used to explain the rapid growth of agricultural exports during the colonial and postindependence periods in Africa.[30] Myint, an economist at the London School of Economics, abandoned the traditional classical assumptions such as specialization and comparative advantage in certain crops in trying to explain the sudden surge in exports in some countries such as Burma, Nigeria, and Ghana in the late 1800s and early 1900s. Myint developed a vent-for-surplus model [1958] which directly attributes the export boom to improved local transport, access to overseas markets, and incentive goods from overseas. These factors provided a "vent" to tap the surplus productive capacity inherent in surplus land and family labor after subsistence food needs of farm families have been met.

The vent-for-surplus model is appealing because it is not a global theory of development like stages of growth and dual sector models. The model stresses the key role of effective demand from European markets in mobilizing surplus labor and land in underpopulated areas with a peasant (smallholder) type of production system.[31] It is presumed to be a costless type of growth which could be largely self-financed by small farmers and local traders by reducing their leisure time. Governments or private international firms only have to provide improved transport, communication, and access to overseas markets.

Szereszewski's [1961] and Polly Hill's [1963] research on the rapid growth of smallholder cocoa production in Ghana in the 1880s and 1890s are empirical tests of the vent-for-surplus model. Hill's meticulous field work in southern Ghana revealed that the emergence of cocoa production involved more than a response to growing European demand for cocoa and improved transport in Ghana. Hill found that migrant farmers were the source of innovation in cocoa farming through their leadership in organizing, financing, producing, and marketing cocoa. The spread of cocoa farming by migrants can be viewed as a process of indigenous capital formation in a land surplus economy. Migrants were also instrumental in the diffusion of cocoa production in the Côte d'Ivoire [Dupire, 1960] and in Western Nigeria [Berry, 1975]. Although Hogendorn [1975, 1978] and Berry agree with Myint that increased effective demand provided incentives for smallholders, they contend that Myint's vent-for-surplus model needs to be refined to devote more attention to the role of local institutions in facilitating capital formation and the spread of innovations. Also, the vent-for-surplus model of development does not explain why short spurts of export-led growth have resulted in cumulative growth and diversification of the economy in countries such as Australia, New Zealand, and Canada while the cocoa export boom in Ghana and copper exports in Zambia have stalled.

The vent-for-surplus model has been used directly or indirectly to study the following crops: cocoa in Ghana [Szereszewski, 1961; Hill, 1963]; cocoa in Nigeria [Berry, 1975]; groundnuts (peanuts) in Nigeria [Hogendorn, 1978]; rubber in Ghana [Dumett, 1971]; peanuts in Senegal and The Gambia [G. E. Brooks, 1975]; and kola nuts in West Africa [Lovejoy, 1980; Agiri, 1977]. For a summary of studies before 1914, see Hogendorn [1975, 1978]. For a radical critique of Hogendorn's use of vent-for-surplus model, see Freund and Shenton [1977] and Hogendorn's reply [1977].

What is the value of Myint's model [1958] in Africa in the 1990s? Whereas Myint's model highlights the potential of trade as an engine or handmaiden of growth, there are two major shortcomings in using the model for policy guidance in Africa. First, the model fails to stress the investment in research, rehabilitation, and replanting which is needed in order to maintain a country's share of world trade in a particular commodity. Ghana and Nigeria are examples of countries which have lost world market share of cocoa and oil palm to the Côte d'Ivoire, Malaysia, and Brazil over the past twenty years. Second, the model ignores the need to invest in increasing the productivity of the food subsector. This problem is now at the crisis stage in several African countries as the frontier is exhausted and investments in irrigation, land reclamation, and tsetse fly control are needed to intensify agriculture and increase food production.[32]

Labor Surplus. The concept of surplus labor as outlined by W. Arthur Lewis's [1954] model of development with unlimited supplies of labor has never been seriously applied by scholars working in sub-Saharan Africa because, as Lewis acknowledged, sub-Saharan Africa is known to be a land-abundant region relative to Egypt, India, Java, and Bangladesh.[33] This abundance of land historically has been reflected in land-extensive farming systems[34] in which long periods of fallow (often ten to twenty years) are interspersed with short periods of cultivation. Moreover, it was empirically shown in the 1960s that most African countries were faced with both seasonal labor surpluses during the dry season and labor shortages or bottlenecks[35] during some periods of the farming season (e.g., weeding). As a result, Byerlee and C. K. Eicher [1974] urged model builders to concentrate on developing models of African development which were able to deal with a seasonal labor shortage or surplus, rural nonfarm labor interactions, determinants of rural-urban migration, and capital and labor interactions between the agricultural, industrial sectors and international trade.

Land Surplus. While Myint's vent-for-surplus model hypothesizes that increased effective demand from international trade can stimulate the use of *both* surplus labor and land, there have been few attempts in sub-Saharan Africa to analyze the effects of surplus land on labor allocation and development outside the vent-for-surplus framework. Helleiner [1966b] and Hanson [1979] are among the few who have advanced models of development with unlimited supplies of land. Helleiner [1966b] concluded that because of the diversity of ecological zones and population densities in Nigeria it was impossible to classify Nigeria as either a land or labor surplus economy. Once it is recognized that both situations exist in a country such as Nigeria, the issue of labor mobility re-emerges as an important factor in development. Hanson [1979] developed a model of development with unlimited supplies of land, but did not present empirical information to test his model.

Surplus land models have been criticized by several scholars. Sunday Essang [1973] contended that even though a country has a high man to land ratio: land in bush fallow cannot be considered surplus since under existing technology long fallow periods are needed to regenerate soil fertility; aggregate statistics conceal the fact that much of the uncultivated land not in fallow is of low quality and often is not suitable for agricultural production because of unreliable rainfall; and tsetse flies, river blindness, other health problems, and poor transport (as in southern Sudan, for example) preclude the use of large blocks of land which appear as idle land in aggregate statistics. Essang further argued that development interventions based on a land surplus model can lead to bias in favor of large-scale farming,[36] and inadequate attention to land improvement measures.

On the basis of our review, we conclude that the concept of a land surplus economy has little heuristic value. Most economies are too large, complex, and diversified to be described as either land or labor surplus. Furthermore, the notion of surplus, whether of labor or land, overlooks complex institutional and administrative questions about who controls the access to land and whether supporting services and adequate incentives are in place to mobilize the surplus.

Assessment. As African nations entered the second decade of independence in the 1970s, the proponents of abstract theories and models of neoclassical economists were on the defensive. Many assumptions of their models were recognized to be irrelevant to Africa and the resulting policy prescriptions were not taken seriously. In retrospect, the major shortcomings of the dual sector models were their excessive macro orientation and the inability of these models to provide a convincing specification of the agricultural sector—the sector which employs 50 to 95 percent of the total labor force in African economies. Most models ignored structural, institutional, and managerial problems and the structure of demand and its relationship to income distribution and employment. Because of the failure of neoclassical models to deal with the key problems of employment, equity, and constraints on food production, it was necessary to go back to the basics, to build an understanding of African rural economies based on meticulous microeconomic research.[37]

During the late 1960s and early 1970s, several scholars developed models based on African resource endowments and institutions. Valuable theoretical frameworks were proposed to study migration [Todaro, 1969], rural small-scale industry [Liedholm and Chuta, 1976], and consumption [Robert P. King and Byerlee, 1978]. Byerlee and C. K. Eicher [1974] proposed a multisector rural economy model to examine the linkages between rural and urban sectors and small- and large-scale agricultural producers.

POLITICAL ECONOMY AND RADICAL PERSPECTIVE

Western development economics was challenged in Africa in the 1960s by the emergence and rapid growth of political economy, dependency, and radical models of development and underdevelopment.[38] The political economy models of development have their roots in the writings of Lenin on imperialism and in the post-World War II writings of the late Paul Baran. Baran, a Marxist economist at Stanford University, wrote a seminal article, "On the Political Economy of Backwardness" [1952]. In Baran's political economy model, he did not rule out the possibility of broad-based capitalistic development in the LDCs but he argued that in most underdeveloped countries it would be impossible to bring about broad-based development without violent change in social and political institutions and without a dynamic industrial sector. Although Baran was clearly ahead of his time in putting his finger on institutional and structural barriers to development and the need to put the effective demand of the masses at the center of development programs, his views on agriculture were naive and misleading. For example, Baran wrote that since the marginal product of labor tends to be zero in agriculture, "There is no way of employing it usefully in agriculture." Farmers "could only be provided with opportunities for productive work only by transfer to industry." Moreover, Baran advanced an anti small farm view when he wrote "very few improvements that would be necessary in order to increase productivity can be carried out within the narrow confines of small-peasant holdings."

Dependency Theory. The dependency interpretation of underdevelopment was first proposed in the 1950s by ECLA, under the leadership of Raul Prebisch. The basic hypothesis of this perspective is that underdevelopment is not a stage of development but is the result of the development of the world capitalist system. Although a number of different views of dependency have been put forward by scholars such as Sunkel [1973], Furtado [1973], A. G. Frank [1966], Galtung [1971], and others, the following definition of dependency by Dos Santos has been widely cited:

> By dependency we mean a situation in which the economy of certain countries is conditioned by the development and expansion of another economy to which the former is subjected.[39] [1970, p. 231]

In the 1960s, dependency theory was imported into Africa from Latin America. Over the past twenty years, Samir Amin has provided leadership in developing a Marxist version of dependency theory. Amin, an Egyptian by birth, and a national accounts specialist in economics, turned his early attention to an analysis of development in Mali, Guinea, and Ghana in 1965 and the Côte d'Ivoire in 1967. Amin subsequently elaborated on his dependency views through analyses of the precolonial, colonial, and postindependence periods in Africa. In *Accumulation on a World Scale* [1974c] and *Unequal Development* [1976], Amin presents an analytical framework of underdevelopment in Africa based on surplus extraction and the domination of the world capitalist system. Amin contends that social structures in the periphery are "truncated" and can only be understood in relation to the "world social structure" [1976, p. 294]. One of the cornerstones of Amin's analysis is the concept of the "social formation of peripheral capitalism."[40] Amin argues that peripheral formations are fundamentally different from those of the center because of their "extraversion."[41] It is notable that Amin contends that "despite their different origins the peripheral formations tend to converge toward a pattern that is essentially the same" [1974c, p. 378].

Amin has provided valuable insights into the development process [1970, 1972, 1974a, c, 1976], but his prescriptions for agriculture have been naive and have changed over time. For example, he attributes Africa's agrarian crisis to the predominance of agrarian capitalism which he argues takes two major forms: kulaks (farmers who employ wage labor), and organization of export production subject to a theocratic-political authority (such as the Mourides in Senegal). While both "formations" clearly are prevalent, they by no means dominate the structure of agricultural production in sub-Saharan Africa. During the 1960s, Amin favored animal traction, promoted industrial crops, and argued that traditional social values were a serious constraint on development at the village level. He also argued that the transition to privately owned small farms was a precondition for socialism. By the mid-1970s, Amin reversed himself and recommended the collectivization of agricultural production and he abandoned his support for animal traction and industrial crops. These shifts reflect, in our view, the weakness of deriving prescriptions for agricultural policy on the basis of global and abstract analyses of the world economy. For critical surveys of Samir Amin's work, see Sheila Smith [1980] and Schiffer [1981].

During the past decade, there have been several attempts to evaluate the contribution of dependency models in understanding the causes of poverty and underdevelopment in Africa. McGowan [1976] and Vengroff [1977] attempted to test dependency theory in Africa but came up with inconclusive results. Kleemeier [1978] criticized McGowan and Vengroff for attempting to use correlation analysis to infer causality. Palma [1978] raised several important questions about whether the dependency hypothesis can be empirically tested. Recent political economy research in Africa which focuses on agriculture include Wilcock [1978]; Ntangsi [1979]; Henn [1978]; and some of the selections in the book edited by Heyer, Roberts, and Williams [1981]. Collections of essays presenting political economy and dependency perspectives on African development include Arrighi and J. S. Saul [1968]; Gutkind and Wallerstein, eds. [1976]; Shaw and Heard, eds. [1979]; J. S. Saul [1979]; and M. A. Klein, ed. [1980]. For other influential writings, see Leys [1974]. For an assessment of the dependency literature in the Third World, see Tony Smith [1979]. During the 1980s there has been a sharp reduction in the influence of the dependency school in Africa.

Micro-Marxists. [42] A small group of French Marxist anthropologists — including J. Suret-Canale [1971], M. Godelier [1972], C. Meillassoux [1964, 1972], G. Dupre and P. Rey [1978], and C. Coquery-Vidrovitch [1978] — and a few political scientists such as Goran Hyden have rejected the notion that a global Marxist-Leninist ideology based on a generalization of the historical experience of Europe and America is sufficient to explain patterns of development in Africa. The French Marxist anthropologists also reject the views of anthropologists such as Firth [1975], Bohannan [1963], Dalton [1962, 1978b], and Polanyi, Arensberg, and H. W. Pearson [1957] who stress forms of exchange rather than modes of production and the theory of reproduction [Clammer, 1975]. Regarding Western economics, the Marxist anthropologists question the applicability of economic concepts derived from capitalism to economic systems with different characteristics. For example, Meillassoux [1981] argues that Western economists looking for simple economic explanations are often confused by the fact that the system of circulation of goods in precapitalist societies is conditioned by nonmaterial phenomena. [43]

The micro-Marxists attribute the roots of underdevelopment to the failure of capitalism to produce a dynamic transformation of precapitalist economies while mainstream Marxist scholars and dependency theorists stress the extraction of the surplus as the root cause of underdevelopment. Micro-Marxists argue that there is a need to understand the interaction between the precapitalist modes of production and the capitalist mode in each particular setting.[44] As Hyden has said:

> That modes of production differ in their articulation in the
> Third World countries has only recently become a subject of
> research. How these forms of articulation affect the
> development potential has not yet been fully explored.
> [1980, p. 4]

Thus, the focus of research by the micro-Marxists has been on identifying the characteristics of precapitalist modes of production in Africa. J. Suret-Canale [1971] made a seminal contribution when he applied the famous Asian mode of production to precolonial Africa.

Meillassoux's [1964] study of the village of Gouro in the Cote d'Ivoire is regarded by some scholars as the definitive benchmark exposition of a micro-Marxist analysis of precapitalist development and the transition from subsistence to commercial farming.[45] Meillassoux's basic theme is that agriculture is based on communities (roughly equivalent to households) which have the goal of self-sufficiency so they are not dependent on other social classes. Circulation of foodstuffs is controlled by a social hierarchy based on seniority. He argues that these communities, "agricultural self-sustaining formations," contain within themselves all the means necessary for providing the basic social and material needs of their members, but as production for external markets is grafted onto these self-sustaining formations, it is inevitable that a class society results. A collection of papers edited by Meillassoux which is based on this theme, *The Development of Indigenous Trade and Markets in West Africa* [1971], provides an important counterinterpretation to A. G. Hopkins's [1973] stress on the positive role of market forces in West Africa's economic history. Catherine Coquery-Vidrovitch [1978] has also made an influential contribution in her attempt to specify an African mode of production. Unlike many other micro-Marxists, Coquery-Vidrovitch stresses the historical importance of long-distance trade in African development. A key element of her model is the exclusive ascendancy of one group over external trade.

Hyden [1980, 1983, 1987] stresses the strength of precapitalist societies in the face of the expansion of the market. Based on his extensive experience in Tanzania, Hyden argues that smallholders can opt out of the market in response to a shift in price relatives or unfavorable domestic terms of trade [Hyden, 1983, 1987; Kasfir, 1986; Iliffe, 1987].

Assessment. The political economy literature attempts to link rural poverty and underdevelopment to historical forces, world capitalism, and surplus extraction. Political economy scholars also emphasize the linkages between colonial policies and contemporary underdevelopment and encourage agricultural scientists to move beyond the simple view that African agriculture is unproductive primarily because of the lack of new technology. But Amin and his followers who stress global theories of underdevelopment in Africa have underplayed the large number of internal policies and factors which also contribute to poverty and agricultural stagnation in Africa. The Achilles heel of dependency and political economy theorists in Africa is likely to be the same one which discredited Western dual sector models in the 1960s — abstract theorizing and the neglect of empirical research at the micro level. The question remains: Can political economy and dependency scholars move beyond their abstract models to develop models based on studies of the behavior of African farmers and herders, on African institutions, and on micro/macro linkages in order to provide policy guidance in a region in which the majority of the people are farmers?

A small group of micro-Marxists, primarily French anthropologists, have rejected the view that Marxist ideology can be applied in Africa without modification. These scholars are carrying out village studies with emphasis on Francophone countries in West Africa. The micro-Marxists have made important contributions to the study of the role of precapitalist modes of production in shaping the development process, the analysis of the transition to commercial farming, and the study of inequality. The ability of French Marxist anthropologists to ask what some scholars call the "key questions" about development undoubtedly explains the growing number of translations of Marxist works into English (from French and German) over the past decade.[46] But the micro-Marxists also must face the challenge of translating their insights into recommendations which can provide guidance to policy makers and donor agencies.

Chapter III. Food and Agricultural Policy

As African governments tried to establish their legitimacy in the postindependence period, they experimented with a wide range of development strategies, programs, and policies.[47] In this section, we review research on the major policy issues of the postindependence period, including agrarian capitalism *versus* socialism, agricultural planning and agricultural sector modelling, large- *versus* small-scale farming, marketing boards and food grain boards, agricultural prices, rural development programs, and accelerated food production campaigns.

1. Agrarian Capitalism and Socialism

One of the most important policy issues during the postindependence period has been the ideology of economic policy — capitalism or socialism. In the postindependence period beginning in the late 1950s, numerous countries shifted from capitalism to socialism after a few years of independence. For example, soon after

Ghana (formerly the Gold Coast) became independent in 1957, its dynamic leader, Kwame Nkrumah, shifted from capitalism to an ideology of "African socialism." Although difficult to define, African socialism in Ghana and many other countries included the establishment of state farms, government tractor hire stations, promotion of cooperative farming and farmer associations, and a number of policies and restrictions to reduce the influence of private traders.

The 1970s could be called the golden decade of socialism in Africa even though in many countries it was socialism in name far more than practice.[48] About one-fourth of the countries are still partially or firmly committed to socialist economic ideology.[49] The reasons for moving from capitalism to socialist economic ideology and back to capitalism (as in Ghana and Mali) and from capitalism to socialism (e.g., Tanzania, Guinea-Bissau, Mozambique, Ethiopia, and Zimbabwe) should be carefully analyzed. We shall focus on three countries— Ghana, Mali, and Tanzania—because their experience with agrarian socialism is well documented. One of the first skeptical views is Elliot Berg's [1964] analysis of socialism in Guinea. The case for socialism is put forward by Dumont [1966, 1969], Arrighi and J. S. Saul [1968], and Seidman [1972, 1977].

When Ghana became independent in 1957, it is reported that President Nkrumah had at his disposal £150 million sterling in reserves (from its cocoa marketing board) in London banks. In 1961, Nkrumah abruptly shifted from capitalism to a radical socialist strategy which equated modernization with industrialization and the mechanization of agriculture. Ghana established mechanized state farms because it was thought that "small-scale private farming is an obstacle to the spread of socialist ideas" [Killick, 1978, and Nweke, 1978b, 1979]. Also, see Nkrumah's analysis [1970] of class struggle in Africa.

Mali was the second country to move to socialism soon after its dramatic break in diplomacy with France. But socialism was shortlived; the rise and sudden demise of socialism in Mali is recorded in Zolberg [1967], W. I. Jones [1972, 1976], and G. Martin [1976]. For a prosocialist view, see Ernst [1977].

The vision of agrarian socialism in Tanzania is set forth in Nyerere's essay "Socialism and Rural Development" [Nyerere, 1967] and in Nyerere [1968]. For a remarkably candid assessment of some of the problems in achieving rural socialism in Tanzania, see Nyerere [1977]. Although many observers dismiss Tanzania's experiment, there were important gains in literacy and social services in the 1970s but some reverses during the 1980 to 1985 period. Valuable insights on socialism in Tanzania are found in Hyden [1980], a collection of essays edited by Mwansasu and Pratt [1979]; J. S. Barker [1979]; and Samoff [1981]. For a bibliography on Ujamaa villages, see McHenry [1981]. For a comparative study of four small socialist states, including Tanzania, see Morawetz [1980].

There are many important questions about Tanzania's experiment with agrarian socialism such as why did President Nyerere use coercion to round up farmers and move them into villages? Many pro-Nyerere scholars avoid this topic. Second, how serious were exogenous factors such as the drought and war with Uganda in undermining socialist programs at critical junctures? Third, were faulty economic policies the Achilles heel of socialism in Tanzania? Nyerere [1967] pointed out long ago that the worst enemy of socialism is a faulty economic policy. The ruling party took a number of steps in the early 1980s to increase incentives to farmers and remove restrictions on private trading. Even early admirers of President Nyerere, such as Rene Dumont, wrote, "Nyerere, through all his writings has made all Europe dream but the stark reality dispels all illusion" [Dumont and Mottin, 1980].

Agrarian socialism is in decline in Africa. Presently there are no countries in Africa where agrarian socialism is performing well. But there is growing awareness that neither agrarian socialism nor agrarian capitalism can assure successful agricultural development [Eicher, 1982a]. As Gerry Helleiner pointed out in his perceptive article on "Socialism and Economic Development" [1972], all countries — capitalist or socialist — must break common economic constraints, including capital formation, foreign exchange, human resources, and institutional or technical bottlenecks. All governments following either economic ideology must develop agricultural institutions and incentive structures to solve the most basic prerequisite of development — achieving a reliable food surplus. For an uneven but valuable assessment of socialism in the Third World, see the volume edited by Desfosses and Levesque [1975]. For a recent reassessment of socialism in sub-Saharan Africa, see the volumes edited by Rosberg and Callaghy [1979] and Sklar [1988].

References on agrarian socialism are as follows:

General: Desfosses and Levesque, eds. [1975]; W. A. Lewis [1978a]; Morawetz [1980].

Africa: E. Berg [1964]; Friedland and Rosberg, eds. [1964]; Arrighi and J. S. Saul [1968, 1973]; Rosberg and Callaghy, eds. [1979].

Ghana: Amin [1965]; Miracle and Seidman [1968a,b]; Killick [1978].

Guinea: Amin [1965]; Derman and Derman [1973].

Guinea-Bissau: Goulet [1978]; Urdang [1980].

Mali: Amin [1965]; Zolberg [1967]; Ernst [1977]; W. I. Jones [1972, 1976]; G. Martin [1976]; Bingen [1985].

Mozambique: Isaacman [1979]; J. S. Saul [1979].

Tanzania: Nyerere [1967,1968]; Dumont [1969]; Helleiner [1972]; van Hekken and van Velzen [1972]; Mwansasu and Pratt, eds. [1973]; Lofchie [1978]; J. S. Saul [1977]; J. S. Barker [1979]; Coulson, ed. [1979, 1982]; McHenry [1979, 1981]; von Freyhold [1979]; J. M. Due [1980]; Dumont and Mottin [1980]; Hyden [1980, 1983, 1987]; F. Ellis [1980]; Samoff [1981]; Zalla [1981]; Lele [1984a].

2. Planning and Agricultural Sector Modeling

Planning was launched by colonial officers following World War II. Nigeria prepared a ten-year plan in 1946. As countries became independent, in the late 1950s and early 1960s, almost all countries launched medium-term plans which focused on high rates of growth of GNP as the target and indicator of development and appealed for foreign aid.[50] For example, the government of Mali was proud that the 11 percent growth target in its first Plan was the highest in any African plan in the sixties [Zolberg, 1967]. On the eve of Nigeria's civil war, Gusten [1967] expressed the obsession of economists over growth rates and macro planning in the 1960s in his "Can the Nigerian Economy Grow at 6 Percent per Annum in the Near Future—A Pre-Planning Exercise?"

Most assessments of planning over the 1960-80 period concluded that failures have far outweighed achievements. Professor Aboyade, a distinguished Nigerian economist and architect of Nigeria's Second Five-Year Plan, concluded that "for most of tropical Africa, planning over the past two decades has been little more than false hope" [1973]. Rimmer [1969], an economist at the University of Birmingham with considerable experience in Ghana, noted, "any resemblance between development plans and the actual course of economic change in African and other poor nations is purely coincidental." Shen [1977] reviewed development plans in twenty-two tropical African countries and identified weak institutions for plan implementation. In a survey of planning in developing countries, Killick [1976b] concluded that planning has not lived up to its expectations. But Helleiner [1972] pointed out that one should not equate the preparation of plans with economic planning. Although many elaborate national plans were prepared by foreign experts in the 1960s, most were dropped or ignored soon after they were published. Helleiner believes that the critical issue is not whether countries achieve the growth rates spelled out in plans but how effective they are in the slow process of improving the data base, training people, and strengthening institutions.

Turning to agriculture, major agricultural sector assessments have been carried out in Nigeria, the Côte d'Ivoire, Sierra Leone, Ghana, Zambia, and the Sahel over the past twenty years. For a model of Senegal's agricultural sector, see Labonne and Legagneux [1977]. World Bank-sponsored modelling efforts have been completed by Goreux [1977] in the Côte d'Ivoire and by Blitzer [1979] in Zambia. Nigeria has been heavily studied.[51] A four-year Nigerian agricultural sector analysis was carried out in the late 1960s by the Consortium for the Study of Nigerian Rural Development under the leadership of Glenn L. Johnson. Thirty-two working papers and a final report by Glenn L. Johnson *et al.* [1969] were published by the Consortium. Building on these findings, Johnson, Manetsch, and colleagues developed a generalized simulation approach to agricultural sector analysis in Nigeria [Manetsch *et al.*, 1971]. Byerlee [1973] later developed a simulation approach to trace the indirect employment and income distribution effects of alternative agricultural development strategies in Nigeria. Byerlee's ten-sector dynamic macro model was linked to an employment-income model and to the agricultural sector model developed by Manetsch *et al.* [1971].

Most of these modelling efforts have been ineffective because of the static nature of the models, the lack of micro data, and the lack of African participation in the conduct of the studies [See Seifert and Kamrany, 1974]. In most countries until more Africans are trained, agricultural sector models which cost millions of dollars and depend on foreign researchers are likely to be perceived by Africans as "academic toys," rather than as productive tools which can help Africans improve planning and decision making.

3. Large *versus* Small Farms

A continuing debate in agricultural policy has been the economics of smallholders *versus* large farms, including plantations, state farms, land settlements, and river basin settlements. In the 1960s, the debate over large- *versus* small-scale agriculture became known (especially in Eastern Africa) as the transformation *versus* the improvement approaches. The transformation approach featured a wide variety of large-scale farming (plantations, settlements, and state farms) and processing plants; it was designed to bypass the lengthy process of improving small farms within the existing village structure. The major ingredients of the transformation strategy were capital-intensive technologies, such as tractor mechanization, central management (often expatriates), and removing people from villages and training an unskilled labour force.

SMALLHOLDER FARMING: A DESCRIPTIVE OVERVIEW

Smallholder crop cultivation is the predominant farming system in sub-Saharan Africa. Smallholder farming is primarily characterized by reliance on family labor, a small stock of physical capital, and abundant land relative to Asian countries.[52] Family labor is the most important factor of production, with family labor inputs ranging from 80 to 90 percent of total labor inputs [Byerlee, 1980]. Farming households generally have six to ten family members and it is common for households to include more than one nuclear family. Adult male farmers work an average of five hours per day or 1,000 to 1,500 hours per year in farming activities, but the number of hours of labor devoted to off-farm activities, such as rural small-scale industries, is substantial. This is in stark contrast to Egypt and many Asian countries where the total adult hours in farming ranges from 2,500 to 3,000 hours per year [Cleave, 1974]. Women play an important role in farming, processing, and marketing, but the extent of their participation varies greatly by activity, ethnic group, and religion. Children are an important source of labor for tasks such as weeding, the collection of firewood, bird scaring, carrying water, and taking care of sheep, goats, and cattle.

Most small farmers till their land with human labor and hand tools, including metal hoes, cutlass or machete, digging sticks, and knives. Although the shift from hand cultivation to animal traction cultivation (oxen and donkeys) has been promoted for more than fifty years, animal traction is still a minor source of farm power in almost all countries in sub-Saharan Africa [Pingale, Bigot, and Binswanger, 1987]. Capital investments in housing, storage, and perennial crops are mainly created by family labor using local materials. Although the separation of farming and livestock production is common throughout sub-Saharan Africa, there is a slow but discernible adoption of livestock enterprises by sedentary farmers.

Cash expenses generally represent a small proportion of the value of production. Purchased inputs—seed, fertilizer, and chemical pesticides—are not widely used by farmers. Most fertilizer is applied in the form of organic manure.

The land area controlled[53] by a typical smallholder varies considerably but it is generally far larger than in Asia. In a study of smallholder cotton production in northwestern Tanzania, K. Shapiro [1978] reported that the typical farm controlled about twenty-five acres of land of which four were in cotton, five and one-half in food crops, and the remainder in fallow and grazing land. Likewise, the typical smallholder in Sierra Leone controlled about forty acres of land but actually farmed only a small portion of the forty acres because the idle land was regaining its fertility in the bush fallow system [D. S. C. Spencer and Byerlee, 1976].

The area cultivated per family ranges from two to ten acres throughout sub-Saharan Africa [De Wilde et al., 1967; Upton, 1973; Cleave, 1974; Winch, 1976; D. S. C. Spencer and Byerlee, 1976; Heyer and Waweru, 1976]; D. S. C. Spencer, 1986; and Johnston 1986]. Because of the widespread reliance on hand tools and the lack of a landless labor class, the area cultivated per farm family depends on the size and composition of the family labor force.

While crop production is the major activity of smallholders, off-farm activities such as trading, small-scale industry, livestock, and fishing are important activities throughout Africa. Microeconomic research has shown that farmers devote a significant amount of their time to off-farm activities [Luning, 1967; Liedholm and Mead, 1987] and that there is an inverse seasonality with a large percentage of the total hours worked off-farm during the dry season(s). Although small farmers strive to meet their own food needs, 20-30 percent of the staple food production is marketed in most countries.[54]

LARGE-SCALE FARMING: AN OVERVIEW

Large-scale farming in sub-Saharan Africa dates to the colonial period with the introduction of plantations and large European farms which produced for export markets. Today, large-scale farming accounts for a substantial portion of export crop production in only a few countries in sub-Saharan Africa.

At independence, Western advisors generally endorsed large farms and plantations because it was assumed that they would achieve economies of scale, that they would be convenient vehicles for newly independent governments to "bring rapid development" to selected rural areas, and that they would provide rural employment for the growing number of school leavers.[55] The rationale of the transformation approach is reflected in the recommendation by the World Bank Mission to Tanganyika (later Tanzania) that the Government of Tanzania should support land settlement schemes because "quicker progress towards these ends is likely to be made, within the limitations of the resources available for government action, by planned settlement of empty areas than through exclusive concentration on improvement of methods (small farms) in settled areas" [IBRD, 1961, p. 131].[56] The world famous Gezira scheme in the Sudan [Gaitskell, 1959],[57] tea plantations in East Africa, Firestone rubber estates in Liberia, and Unilever estates in the Belgian Congo (now Zaire) were often cited as examples of the superiority of large-scale agriculture. But proponents of large schemes often overlooked or glossed over the horrendous failures of large-scale schemes such as the East African groundnut scheme introduced by the British colonial service in Tanganyika after World War II, the failure of the Mokwa settlement scheme in northern Nigeria in the 1950s [K. D. S. Baldwin, 1957], and the mixed results with land settlement schemes and state farms in Africa [Chambers, 1969] and throughout the world [W. A. Lewis, 1964; FAO, 1976b; Higgs, 1978].[58]

The state farm was a type of farm organization included in the transformation strategy of the 1960s. State farms were adopted by the Governments of Sierra Leone and Ghana in the late 1950s and early 1960s. Sierra Leone established state farms to produce export crops in twelve regional provinces, but the farms were abandoned within a few years. In Ghana, state farms and tractor mechanization formed the centerpiece of Nkrumah's socialist strategy of development [Miracle and Seidman, 1968a]. Ghana's dismal record of tractor mechanization and state farms was documented by Kline et al. [1969], Nweke [1978a,b], and Killick [1978].

Plantations were another type of large-scale farming pursued by African leaders in the 1960s. Since it is almost impossible to gain access to data on private and government plantations, it is difficult to pass judgment on the economic, social, and political costs and returns on plantations. Since British colonial land policy prevented plantation development in Nigeria, plantations were insignificant until the constitutional change of 1951 permitted the establishment of plantations by both private and foreign capital [Essang and Ogunfowora, 1975]. Saylor and C. K. Eicher [1970] found that government plantations in Nigeria were generally unprofitable because of the lack of technical data, poor management, and high turnover in unskilled labor (frequently 100 percent per year), etc. The number of private plantations increased during the 1951-65 period in Nigeria, but marketing board taxes reduced the rate of return on oil palm and rubber plantations to almost zero by the early 1960s [Glenn L. Johnson, 1968].

SMALL-SCALE AND LARGE-SCALE FARMING IN NIGERIA

Since both the transformation (large-scale) and improvement (small-scale) strategies were pursued in different regions of Nigeria in the 1960s, Nigeria provides a unique case study of the results of these two alternative strategies. During Nigeria's First Development Plan (1962-68), the three regions in the southern part of the country (Western, mid-Western, and Eastern) devoted some 70 percent of their capital and recurrent budgets in agriculture to the transformation approach (farm settlements, school leaver farms, and plantations). On the other hand, the Northern region pursued an improvement strategy during the 1962-68 Plan to help small farms through subsidized fertilizer, credit, and farmer training centers. As Nigeria approached independence in the late 1950s, political leaders in the three regions in the southern part of the country were not in a mood to wait for research results on whether to pursue large-scale or small-scale agricultural development strategies. For example, in 1959, one year before independence, a policy paper issued by the government of the Western Region noted that "while scholars conduct unbiased research . . . agricultural development must go on" [Nigeria, 1959, p. 9]. The policy paper noted that political leaders from western Nigeria had visited the Gezira scheme in the Sudan and Moshav settlements in Israel and concluded that experience in other countries has shown that " . . . a system of co-operative farm settlements would be a major step in the agricultural development of the Region" [ibid.].

In a detailed analysis of the western Nigeria settlement scheme, Roider [1971] found that after six years of operation, the government had spent $11,200 per settler, or double the amount originally projected, while yields ranged from 25 percent (cotton) to 65 percent (rice) of the yields estimated in the feasibility study. Similar settlement schemes were tried in the other two regions (Eastern and mid-Western), but by the end of the 1960s it was obvious that the settlement schemes had failed in all three regions in southern Nigeria [Andreou, 1981]. The reasons for the failure of schemes in southern Nigeria were almost identical to the findings of W. Arthur Lewis [1964] in his review of settlement schemes and Nelson's [1973] study of twenty-three schemes in Latin America: lack of technical and microeconomic data, superficial planning, overinvestment in housing and social services, inappropriate mechanical technology, and lack of participation by settlers.

C. K. Eicher and Glenn L. Johnson [1970] evaluated the consequences of pursuing the transformation *versus* the improvement strategies and concluded that smallholder improvement programs, rather than land settlements or plantations, should form the backbone of Nigeria's agricultural strategy over the 1969-85 period. These findings were reinforced by Wells's authoritative book [1974] on agricultural policy and plan implementation during Nigeria's First Plan (1962-68).

ASSESSMENT

Small-scale farming has many advantages relative to large-scale farming in Africa. Although there is still substantial support among African politicians and policy makers for large-scale agriculture, few donors support large-scale farms and ranches today. Large-scale farms are still of central importance in the Sudan, Zimbabwe, Zambia, and Swaziland. Although foreign private investors launched large-scale food production schemes in several countries, many have experienced unforeseen technical and economic problems. For example, Uniroyals' large scale food production complex in Liberia was terminated in the late 1960s. A $2.1 million foreign-financed maize farm and grain storage complex in Central Ghana also experienced numerous difficulties. In the Cameroon in 1979, a 4,000-bectare government wheat mechanized scheme with thirty-five tractors did not produce enough—twenty-two kilos of wheat per hectare—to recover its seeding rate of 100 kilos per hectare. However, commercial wheat production is profitable in Zimbabwe and the yields are among the highest in the world.[59]

Although it is understandable why governments do not publicize the failure of large-scale schemes, we have gleaned enough information to conclude that in most countries, large-scale, capital-intensive food production complexes cannot compete with African smallholders in producing staple foods.[60] To be sure, there is scope for a few multinational firms to produce fresh fruit and cut flowers for European markets [FAO, 1976e]. But there is very little empirical research on multinational firms [Widstrand, 1975; Sklar, 1976].

Most large-scale farming and land settlement schemes in Africa have been failures over the past fifty years. But settlement is still being debated because two-thirds of the remaining arable land in the world is in Africa. The challenge is to learn from past settlement schemes. Research has shown that the role of the government should shift from planner, financier, and manager to providing agronomic research, disease control, and a minimum of infrastructure. Families should build their houses and clear their own land in order to drive down capital cost per settler.

4. Marketing Boards and Food Grain Boards

The extraction of the agricultural surplus for urban/industrial development has been a common objective of both colonial and postindependence rule. Colonial authorities imposed various taxes (head taxes, hut taxes) and compulsory planting of selected export crops to stimulate the production of export crops and to capture the agricultural surplus. Shortly after World War II, the British colonial government introduced marketing boards in their East and West African colonies following the relatively successful record of marketing boards in Australia and New Zealand since the 1930s. The objective of the marketing boards was to stabilize producer prices and foreign exchange earnings and to reduce interseasonal price movements.

In the 1960s and 1970s, numerous African governments introduced grain boards to control producer prices and food grains and to channel food to the urban centers.[61] Boards usually accumulate and carry stocks to mitigate both intra- and inter-annual fluctuations in price and supply, and develop distribution systems to facilitate the transfer of grain from surplus to deficit regions.[62]

MARKETING BOARDS

The introduction of marketing boards was followed by a wave of studies by economists. Leading the charge against the marketing boards was Peter Bauer who contended in his *West African Trade* [1954] that marketing boards failed to stabilize producer prices and reduce seasonal price variations, and that the boards dampened producer incentives by paying producers one-half to two-thirds of world prices of exports such as cocoa, oil palm, coffee, and rubber. Later, A. G. Hopkins [1973] showed that export producers in Ghana lost 41 percent and Nigerian producers lost 27 percent of their potential gross income through marketing board taxes over the 1947–61 period.

Helleiner's meticulous study, "The Fiscal Role of the Marketing Boards in Nigeria" [1964], introduced new criteria for examining the performance of the boards. Instead of being preoccupied with the issues of whether marketing boards stabilized producer prices, producer incomes, and foreign exchange earnings, Helleiner contended that the more important issue is whether the boards are effective in performing the fiscal role of capturing the agricultural surplus for the development of *both* urban and rural areas. Helleiner was of the opinion that no single taxing scheme could simultaneously achieve multiple goals of stabilizing producer prices, foreign exchange earnings, and interseasonal price variations. Helleiner concluded that, on balance, marketing boards were the best mechanism for mobilizing the agricultural surplus in subsistence economies like Nigeria in the 1950s and 1960s because of the lack of administrative capacity to impose other taxes such as land and income taxes and the lack of other sources of revenue such as petroleum, gold, phosphate, and timber. But Helleiner's qualified endorsement of marketing boards was based on a shaky microdata base. He was unable to show that the rates of return on government investments (financed by marketing board surpluses) in plantations, hotels, airlines, and industrial estates were unambiguously higher than if Nigerian farmers had received higher prices for their export crops and had reinvested their expanded earnings in farming (such as the new hybrid oil palms) or if the boards had indirectly returned the surplus to farmers through fertilizer subsidies and agricultural research. Idachaba [1973] concluded that marketing boards in Nigeria substantially dampened producer incentives and restricted output and employment generation in agriculture. Olayide, Ogunfowora, and Essang [1974] found widespread inefficiencies in Nigerian marketing boards and recommended alternative structures such as producer and marketing cooperatives. For other research on marketing boards, see Storm [1976] for Senegal; for Tanzania, see Kriesel *et al.* [1970]; in Ghana, see Kotey, Okali, and Rourke, eds. [1974]; and in West Africa, see Blandford [1979]. For an overview on food marketing boards, see William O. Jones [1987].

During the past thirty years, taxes on export crops via marketing boards have provided a convenient way in many countries to capture and transfer the agricultural surplus to finance government airlines, hotels, factories, plantations, and in a few cases, subsidized inputs such as fertilizer for farmers. Whether export taxes and marketing boards should have been scrapped or continued over the past thirty years needs to be analyzed on a country-by-country basis. The central questions have been the level of taxes imposed by the boards, the use of agricultural surpluses, and whether there were fiscal alternatives to the boards. Clearly, countries such as Nigeria, Ghana, and Tanzania had few fiscal alternatives to marketing boards in the 1950s and 1960s because they lacked the administrative capacity to register land and collect land taxes, and they lacked mineral and petroleum exports. But Nigeria scrapped its marketing boards in 1987. Marketing boards, like import substitution in Latin America in the 1950s, served a useful role at a particular stage of economic history of *some* African countries. A few countries, such as Angola, Nigeria, and Gabon have been able to find alternative sources of revenue from petroleum. But the problem of raising government revenues in countries without mineral or petroleum exports remains a central issue in African development.

FOOD GRAIN BOARDS

Numerous African governments have established government grain boards and given them monopoly power over the domestic distribution of food grains, making private trade in grains illegal [Buccola and Sukume, 1988]. An important argument in favor of government monopolies in grain trade is that it allows governments to subsidize trade in remote, less productive areas. It is often stated that boards need to control the distribution of 20 to 25 percent of total grain production to affect prices [Becker, 1974; L. Sorenson, *et al.*, 1975; Grolleaud and Kohler, 1979], but it is common for private, and often illegal, trade in food grains to handle 85 percent or more of the marketed surplus in countries with grain boards. As a result, even where grain boards have been made legal monopolists, monopoly power is often more legal fiction than fact [Lele and Candler, 1981]. The inability of grain boards to dominate the flow of grain makes it almost impossible for them to stabilize prices and to transfer grain from surplus to deficit regions.

Many researchers have argued that the same issues plague food grain boards as export marketing boards faced in the 1950s and 1960s: inefficiency, dampened producer incentives, corruption, and a cadre of thousands of employees who are idling away their time [W. O. Jones, 1987].[63] There is substantial evidence that administrative and operating costs of most grain boards are quite high, reflecting inefficient operations [L. Sorenson, *et al.*, 1975; Temu, 1975; Wilcock, 1978; CILSS/Club du Sahel, 1977; Grolleaud and Kohler, 1979; Blandford, 1979; Lele and Candler, 1981]. While some boards have managed to reduce per unit costs over the years [e.g., The Gambia Produce Marketing Board in The Gambia, Blandford, 1979], the operating costs of most have escalated mainly because of high overhead, thousands of employees, and fleets of trucks. The marketing margins of grain boards have tended to be high, clearly above those of private traders for comparable services [E. Berg, 1975, 1986a]. Large margins effectively mean that a large percentage of the total sales value accrues to government boards rather than producers.[64] This factor is particularly important since high margins necessitate payment of low farm gate prices if retail prices are to remain at a politically acceptable level [Buccola and Sukume, 1988].

Peter Temu [1975] showed that not only have mark-ups by grain boards in Tanzania been high relative to private traders, they have been extremely variable, largely because of the inability of boards to control their costs. Many governments have adopted a dual policy of setting guaranteed producer prices in order to reduce the uncertainty faced by farmers and controlling consumer prices of key staple foods. The margins available to boards are a residual of the two controlled prices and frequently do not cover costs in any given year [CILSS/Club du Sahel, 1977, 1979].

Several researchers have found that improvements in operational efficiency should enable most grain boards to carry out their duties with smaller margins. L. Sorenson *et al.* and Temu show, for example, that administrative costs are a large component of the margins received by grain boards. Several researchers have argued that extensive duplication of functions among grain boards underlies some of these costs [Kriesel *et al.*, 1970]. There are widely acknowledged problems of grain loss during storage and transportation. Supplementary urban grain storage with a network of rural storage centers may reduce transport costs by eliminating the transfer of grain from rural areas to capital cities at harvest and back to rural areas in times of crisis. Some grain boards are not taking advantage of opportunities to offset their costs by generating revenues through spatial price arbitrage. Thodey [1969] shows, for example, that price differences in Ethiopia exceeded transportation costs over half of the time on routes between markets connected with Addis Ababa.

There is now evidence that grain boards are often a constraint on the production and marketing of food [W. O. Jones, 1987]. Wilcock [1978], for example, argued that the activities of grain boards in the Sahel have contributed to the misallocation of resources because of overcentralization, high operating costs, duplication of effort, and dysfunctional incentive systems. Heyer and Waweru [1976] reported that the high degree of regulation and control of food marketing has resulted in low producer prices and high consumer costs in Kenya. The Senegal government's decision to abolish its grain board (ONCAD) in late 1980 is a rare example of a country admitting that its board was ineffective. E. Berg [1975, 1980, 1986a] has argued that the absence of a suitable substitute for the private trader in primary markets is a major constraint on food self-sufficiency in the Sahel. While most researchers have argued for minimizing the role of governments in food grain trade, Berg contends that even mixed marketing arrangements (government and private) are unworkable. Berg rejects what he calls the "imprudent peasant/monopolized market" model which he feels dominates the views of African planners, offering as an alternative a "prudent peasant/competitive market" model. Of three potential alternatives governments might consider—including continuation of the status quo, increased public control, and competitive liberalization—Berg firmly believes only competitive liberalization will work to the benefit of both producers and consumers [Berg, 1986a].

One of the primary results of creating legal monopolies for food grain boards has been the emergence of illegal parallel markets [Collins, 1976]. Temu [1975] argued that the failure of grain boards to control domestic food grain distribution systems even in countries where they are given legal monopoly power can largely be traced to the disincentive effects of low and uncertain producer prices by grain boards.

5. Agricultural Prices

The manipulation of agricultural prices is a standard government technique to influence the level and composition of agricultural production and the transfer of economic surplus to urban areas. The primary mechanism is to set the prices of major agricultural commodities and inputs administratively, and to enforce the administered prices through buying and selling operations of parastatals and licensed buying agents. African governments also influence relative agricultural prices through indirect pricing policies, including export taxes, subsidized credit and fertilizer,[65] and overvalued exchange rates. The major reasons for intervening in agricultural pricing have been to stabilize prices and production, to foster self-sufficiency, to generate tax revenue, to curb the profits of middlemen, and to control the cost of living for urban consumers.

There have been few studies of the aggregate impact of pricing policies. Helleiner's [1964, 1966a] studies of the fiscal role of marketing boards in Nigeria and his [1968a] review of pricing strategies in Tanzania are among the exceptions. Helleiner argues that agricultural pricing policies should insure that the structure of price and tax incentives is working in the right direction. Helleiner concluded that "clearly, the income distribution policy implicit in Tanzania's wage and agricultural pricing policies has worked to the increasing disadvantage of the smallholder agricultural sector" [1968a]. Olayide, Ogunfowora, and Essang [1974] used systems simulation to test the effects of marketing board pricing on the Nigerian economy, the authors concluded that depressed producer prices had reduced the growth of the economy and they recommended the elimination of licensed buying agents and centralized fixing of prices. On the question of input subsidies in Nigeria, see Idachaba's pioneering articles [1973, 1977].

The Byerlee et al. [1983] analysis of employment-output conflicts and factor price distortions in Sierra Leone reveals four facts: There is a wide choice of technology available not only between large-scale and small-scale sectors, but within each sector; the choice of technology is sensitive to relative factor prices; administratively established prices, wages, taxes, and subsidies have favored the adoption of larger-scale, capital-intensive techniques in agricultural production, processing, fishing, and industry; and a continuation of present government policies will have a serious adverse effect on rural employment and national income.

The World Bank [1981b] presented the view that pricing policies are a root cause of the food and agricultural crisis in Africa. For example, in 1980, Tanzania's total exports of major agricultural commodities (which account for nearly two-thirds of the total value of all exports) were 28 percent lower than in 1966 and export earnings fell from 25 percent of the GDP in 1966 to 11 percent in 1979. Ghana has long been the world's leading cocoa exporter but Ghana's cocoa production declined from a peak of 566,000 metric tons in 1965 to 210,000 tons in 1987 and Ghana's position in world trade of cocoa fell to third in 1979, behind the Côte d'Ivoire and Brazil.[66] For the political economy scholar who criticizes neoclassical economists for their fetish over getting prices right (removing subsidies and taxes), empirical evidence from Ghana illustrates what happens when prices are distorted. Ghana's drive to promote large-scale rice farming in northern Ghana in the mid-1970s through taxes, subsidies, and an overvalued foreign exchange rate brought forth imported rice combines in a low wage economy with substantial unemployment and rural underemployment [Winch, 1976]. This same mistake was repeated in 1981 in Ghana's drive to promote state farms and mechanized cotton harvesting. In mid-1981, the official exchange rate was 2.6 cedis per dollar but in neighboring Togo, it took thirty-three cedis to purchase one dollar. State farms used the official exchange rate of 2.6 and imported $100,000 cotton pickers in a low wage economy. This illustrates the impact of overvalued exchange rates on agriculture. Getting prices right is not the answer to Ghana's economic problems but a move in this direction is a prerequisite for economic recovery. For example, the exchange rate has been devalued from 2.75 cedis per dollar in 1983 to 163 in August, 1987. Real GDP increased 10.7 percent in 1984, and 6.1 percent in 1985 [Eicher, 1988a].

In summary, there is now substantial evidence that agricultural pricing policies have an adverse effect on: the gap between rural and urban income, the incentive to produce food and export crops, the ability of governments to establish and maintain food reserves, and, employment opportunities in farming, processing, and rural industries. Surveys of agricultural pricing policy include de Wilde's [1980] case studies of Kenya, Tanzania, and Ghana; R. H. Bates [1981]; R. H. Bates and Lofchie, eds., [1983]; Mukui, ed. [1979]; Lele and Candler [1981]; Bovet and Unnevehr [1981]; and World Bank [1981b]. For a devastating critique of pricing policies in Tanzania, see F. Ellis [1982]. For an excellent analysis of agricultural price policy, see Ghai and L. D. Smith [1987].

6. Rural Development Programs

Rural development programs have received a great deal of attention from policy makers, scholars, and donor agencies over the past two decades.[67] The historical experience with designing, implementing, and evaluating rural development programs in Africa is vast and would require a separate review. In this section, we shall mention only a few of the important references, starting with the community development literature of the 1950s.

COMMUNITY DEVELOPMENT AND ANIMATION RURALE

During the 1950s and early 1960s, rural development was primarily promoted through community development (CD) in English-speaking countries and animation rurale (AR) programmes in French-speaking nations. CD emerged from experiences in the United States in the 1930s and 1940s and from England in the form of Fabian socialism after World War II. CD programs were introduced in the British colonies in the late 1940s with emphasis on building bridges, schools, and health clinics. CD was viewed as a peaceful way to mobilize people to help themselves in meeting their felt needs. India, Pakistan, the Philippines, and Korea were the primary CD laboratories of the 1950s. Holdcroft [1984] summarizes the reasons for the rapid rise and fall of CD in Asia in the 1950s as follows:

> (1) CD architects overlooked class conflict in rural areas and assumed that agents working at the village level could mobilize people to help themselves. But, in practice, a CD worker was an all purpose worker

with few technical skills in agriculture and did not have the means to help farmers gain access to credit and land.

(2) Since increasing agricultural production was not the main objective of CD programs, when food crises emerged, governments shifted resources to ministries of agriculture to step up food production.

(3) CD was zealously promoted as a separate strategy from agricultural development and, as a result, there was rivalry between the old line ministries—agriculture, health, and education—and newly established ministries and departments of community development.

CD was transferred to the Caribbean (e.g., Jamaica) and to several English-speaking countries in Africa by the British Colonial service in the 1950s. For an overview and assessment of CD in Nigeria, see I. C. Jackson [1956]. But CD never received the prominence in Africa in the 1950s that it did in Asia.

Since 1960, a number of Francophone countries in Africa have adopted animation rurale (AR) as a means of mobilizing and educating their rural populations [Charlick, 1980]. AR essentially has played the same role in Francophone Africa as did CD in the former British colonies. For a review of animation rurale programs, see H. J. C. Elliott [1974], Lele [1975], and Charlick [1980]. Gellar, Charlick, and Y. Jones [1980] point out that, except for a brief period in Senegal, AR has been used as a technique for fostering local organization and for non-formal education rather than as a comprehensive strategy of rural change. Two of the key elements of AR are dialogues with villagers and the development of a network of local animators to increase local organizational capability. In his review of AR in Cameroon, Burkina Faso, Senegal, and Niger, Charlick [1980] argues that in most AR programs animation agents promote the programs of other technical services instead of helping facilitate development from the bottom up and that nearly all important decisions concerning development programs were made by government agents and then imposed on villagers. By the end of the 1960s, the broadly defined populist mission of AR programs had, in most countries, given way to a more limited role in which AR agencies focused on a few tasks such as informal education [Gellar, Charlick, and Y. Jones, 1980].

Although CD or AR programmes are still present in most African countries, they are a minor force in African development today. In retrospect, both AR and CD movements had two basic flaws: they underplayed technical constraints, and they were too optimistic about the degree of national political support for decentralized development [Gellar, Charlick, and Y. Jones, 1980; Birgegard, 1987].

INTEGRATED RURAL DEVELOPMENT

The 1970s was the golden era of integrated rural development (IRD). IRD projects were introduced in the late sixties and early seventies throughout the Third World in the wake of the Green Revolution in Asia. IRD in Africa received a boost at an IRD conference in Kericho, Kenya in 1966, when political leaders emphasized the need to give attention to employment generation and rural development projects [See Sheffield, ed., 1967]. Well-known IRD projects in Africa include the Chialo Agricultural Development Unit (CADU), Wolamo Agricultural Development Unit (WADU) and minimum package programs (MPP) in Ethiopia; the action priority zones program (ZAPI) in Cameroon; the Lilongwe land development program (LLDP) in Malawi; and the Special Rural Development Program in Kenya. These projects were based on the assumption that a critical minimum effort was necessary for a noticeable impact on target populations in a short time. The projects were concentrated in a limited area and administered through semiautonomous agencies.

Uma Lele [1975] reviewed seventeen IRD projects in Eastern, Southern, and West Africa and found that most of them were based on inadequate knowledge of technical possibilities of small farm conditions and exhibited little understanding of the local institutional environment. She argued that the key to success of IRD projects is the systematic acquisition of local knowledge and flexibility in the course of implementation. H. Dupriez [1979] reviewed several IRD projects funded by the EEC. J. M. Cohen [1980a] surveyed IRD and found that the common denominators of IRD projects are a focus on small farmers and an attempt to promote improvements in the quality of rural life and off-farm opportunities.

Interest in IRD projects began to decline in the late 1970s as the pendulum of donors shifted to strategies to increase food production by small holders. The decline of IRD does not reflect a retreat on equity goals as much as growing recognition that pilot IRD programs rarely if ever could be replicated on a broad scale, most governments cannot afford to finance these projects after donor assistances is phased out, and IRD (like CD in the 1950s) was not achieving a reliable food surplus [J. M. Cohen, 1987].[68]

For an assessment of rural development projects in Kenya, see Heyer, Ireri, and Moris [1971]; University of Nairobi [1975]; and Heyer, Maitha, and Senga, eds. [1976]. For IRD in Nigeria, see the volume edited by Olayide, Eweka, and Bello-Osagie [1980] and an overview by F. S. Idachaba [1981]. For a discussion of rural development projects in Ethiopia, see J. M. Cohen [1975]; Tecle [1975]; B. Aklilu [1980], and J. M. Cohen and Isaksson [1988]. Hyden [1980] presents an insightful analysis of the Ujamaa program in Tanzania, which stresses rural mobilization and collectivization as an approach to rural development. Tanzania's rural development experience is covered in bibliographies by Kocher and Fleisher [1979] and McHenry [1981]. Atayi and Knipscheer [1980] evaluate the ZAPI program in Cameroon. For an overview of rural development policies in Botswana, see Picard [1979]. For a skeptical view of IRD, see Ruttan [1975] and Birgegard [1987]. For an upbeat view see J. M. Cohen [1987].

In a recent assessment of the World Bank's experience in IRD from 1965-86, the Bank concluded that the failure rate of IRD projects was twice as high in Africa as in other regions [World Bank, 1988c]. The Bank has appointed a task force to review what can be done in the area of rural development.

7. Accelerated Food Production Campaigns

During the 1970s, many countries launched accelerated food production programs to reverse the decline in per capita food production and to reduce the dependence on food imports. For example, Ghana launched an "Operation Feed Yourself Program" in 1972, and invested heavily in large-scale farms which turned out to be expensive and ineffective [Nweke, 1978b]. Moreover, the program was placed under the Ministry of Agriculture while research on small farmers was in another ministry so coordination was difficult [Girdner et al., 1980].

In the early 1960s, Nigeria was a net exporter of food (mainly oil palm and groundnuts) but by the early 1970s, Nigeria was importing food and in 1981, the import bill was $1.3 billion. In evaluating Nigeria's National Accelerated Food Production Program, Abalu and D'Silva [1980b] link the food crisis to repressive marketing board policies. Nigeria established a Green Revolution Committee to explore how to speed up food production [Nigeria, 1980; Idachaba, 1980b, 1981; Idachaba et al., 1981].

Sudan is one of the most glaring examples of the failure of a country to mobilize its agricultural sector to feed its people. In the mid-1970s, it was frequently asserted that the Sudan could become the "bread basket of the Middle East" by drawing on OPEC loans and investments to develop its vast reserve of idle land [Kiss, 1977]. But the issue today is not one of exporting food but feeding its twenty-three million people and repaying its foreign debt. Sudan's role as the bread basket of the Middle East remains a dream.

Economists from the Food Research Institute at Stanford University and WARDA studied the following aspects of the rice industry in five countries in West Africa:[69] the private and social profitability of producing rice; the effect of governmental policies on the production, consumption, and trade of rice; and the potential for increased interregional trade. The major conclusions of the Stanford/WARDA study are: Mali and Sierra Leone can achieve rice self-sufficiency while promoting efficient use of resources, and could profitably export rice with some production and processing techniques;[70] the Côte d'Ivoire and Liberia should emphasize other more profitable crops and import rice during periods of deficit; and Senegal needs to concentrate rice production in particular regions (such as the Casamance region in southern Senegal) if it wants to increase food security without a high cost in efficiency [S. R. Pearson, Stryker, Humphreys, et al., 1981].[71]

West Africa is now importing about 40 percent of its rice consumption each year, up from 25 percent in 1970 [D. S. C. Spencer and Nyateng, 1987]. Research on rice is a high priority in the 1990s.

Maize is the bright spot on the food production front in Eastern and Southern Africa. Malawi, a small country with a poor natural resource base, exported maize for seven years between 1979 and 1984, even though its rate of rural malnutrition was extremely high. When Zimbabwe became independent in 1980, it inherited a dual agrarian structure of roughly 5,000 commercial farms producing 90 percent of the staple food (white maize) and 700,000 communal (smallholder) farmers. However, Zimbabwe's communal farmers tripled maize production from 1980 to 1986 and they now produce roughly 50 percent of the marketed surplus of maize. Rohrbach [1988] reports that no single factor is responsible for Zimbabwe's maize revolution. About one-third of the increased communal maize production came from area expansion following the end of the war in 1979 and two-thirds from increased yields that were a function of higher producer prices, expansion of hybrid maize seed production, increased access to credit, improved extension services, and lower marketing costs.[72]

In summary, over the past three decades in Africa, crash food production and integrated rural development projects have generally displayed the following characteristics: lack of clear objectives, excessive self-promotion by international funding agencies, inappropriate pressure on field teams for early success, a truncated pilot phase, a need to redesign projects after a few years of experience, and difficulty in replicating "successful" projects on a regional or national scale [Eicher, 1988a].

8. Synthesis

A consensus emerged in the late 1960s that macro planning and the transformation approach to agricultural development were not working in many countries. Therefore, most governments turned to small farmer strategies of agricultural development and many initiated integrated rural development projects. Because of lagging food production in most countries during the 1970s, support for IRD began to wane in the late 1970s and numerous countries launched crash food production campaigns. The empirical record of such campaigns is not encouraging [C. K. Eicher, 1988a,b]. The prospects of dramatically increasing food production in the short run are poor in most countries in the face of the long-term neglect of agriculture. Long-term food and agricultural development strategies must be built on a solid foundation of technology generation geared to micro environments. We now turn to a survey of microeconomic research on smallholder farming.

Chapter IV. Smallholder Farming

The purpose of this chapter is to identify some of the major institutions carrying out microeconomic studies on smallholder farming, discuss methodological issues in conducting rural surveys, evaluate analytical techniques used by agricultural economists in analyzing survey data, and present empirical findings for the postindependence period. A word of caution is in order. We obviously cannot go into depth in all of the forty-five countries included in this review. The publications cited are illustrative of research completed over the past twenty-five to thirty years and they will help identify research gaps and define research agendas for the 1990s.

1. Institutions Carrying Out Farm Level Studies

Prior to the 1960s, research by agricultural scientists in sub-Saharan Africa focused on export crops and commercial farming. Little was known about the socioeconomic aspects of subsistence farms except for studies by anthropologists, geographers, and a handful of studies by agricultural economists.[73] In the 1960s, a number of research institutes were established and there was a large increase in the research on the economics of smallholder farming. This overview identifies some of the major institutions carrying out farm level studies.

Several African governments and universities initiated farm management and socioeconomic surveys in the 1960s. The Farm Economic Survey Unit (FESU) in Kenya began conducting research on large estates in 1958 and undertook its first survey of smallholders in the 1961-62 cropping season. For a review of the FESU farm survey program, see MacArthur [1968]. In Kenya, the Institute of Development Studies and the Department of Agricultural Economics of the University of Nairobi have made major contributions to knowledge about Kenya's agricultural economy [Heyer, Maitha, and Senga, eds., 1976], small farmers [Heyer, 1971], extension and diffusion of innovations [Ascroft *et al.*, 1972], rural development [Heyer, Ireri and Moris, 1971], dairy industry [Hopcroft and Ruigu, 1976; Ruigu, 1978], and fertilizer [Mwangi, 1978].

In Tanzania, M. P. Collinson [1962-64] carried out surveys of small farms in Sukumuland district in the 1960s. The Economic Research Bureau, the Department of Agricultural Economics, and the Bureau of Resource Assessment and Land Use Planning (subsequently renamed the Institute of Resource Assessment) of the University of Dar es Salaam carried out a wide range of farm level studies. For bibliographies on Tanzania, see Kocher and Fleisher [1979] and McHenry [1981].

In Malawi, three sets of surveys were conducted from 1962 to 1965. One focused on a few progressive farmers in each of twelve districts. Later, a random survey of cotton growers was carried out in three villages. The third approach relied on weekly visits to farmers. Each enumerator was responsible for only eight farms [Catt, 1966]. In the late 1960s, the Agricultural Economic Survey (AES) program was given primary responsibility for farm level research in Malawi. See Farrington [1975b] for a review of the survey approach used in the AES.

In Uganda, D. Belshaw, D. Pudsey, M. Hall, and J. Cleave carried out a number of surveys in the 1960s. For reviews and assessments of farm level research in East Africa through the early 1970s, see M. Hall [1970], Collinson [1972], and Cleave [1974].

There has been a long history of research in Zambia (formerly Northern Rhodesia) dating from surveys initiated by the Department of Agriculture in the 1930s in collaboration with the Rhodes-Livingston Institute of Social Studies [W. Allan, 1965; A. Richards, 1932, 1939]. The University of Zambia sponsors research through its Institute of African Studies [Colson, 1971; Quick, 1978] and the Rural Development Studies Bureau [Honeybone and Marter, 1975; Marter, 1978]. In Zimbabwe (formerly Southern Rhodesia), farm level studies were carried out by R. W. M. Johnson [Massell and R. W. M. Johnson, 1968]. The University of Zimbabwe's Department of Agricultural Economics and Extension has been carrying out a wide range of studies of irrigation, tobacco, and marketing [Rukuni and Eicher, 1988].

Since 1970, the Ministry of Agriculture in Botswana has been conducting an annual farm management survey. In 1976, sampling procedures were revised and sample size increased, so there is now a valuable time series of enterprise data. Farm level studies have also been carried out by several on-farm technical research programs [D. C. Baker, 1987].

Turning to West Africa, the Department of Agricultural Economics at the University of Ibadan in Nigeria carried out a large number of farm management and marketing studies in the 1960s under the leadership of the late H. A. Oluwasanmi, Martin Upton, and Q. B. O. Anthonio. Rufus Adegboye [1969, 1977] was one of the first agricultural economists to point out the strategic importance of research on land tenure issues. An innovative socioeconomic survey was carried out in Uboma village in eastern Nigeria in 1963-64 and reported in Oluwasanmi et al. [1966]. Ibadan researchers also carried out numerous studies of marketing boards and supply response during the 1960s.

A village studies research program was initiated by David Norman and colleagues at Ahmadu Bello University (ABU) in northern Nigeria in 1964. The studies produced by ABU researchers have made a major contribution to our understanding of intercropping, constraints on small farm production and marketing, and consumption patterns [Norman, 1972; Goddard, 1972; Abalu, 1974; Simmons, 1976a, c]. For a description of the survey approach used, see Norman [1973], Abalu and D'Silva [1980a], and Norman, Simmons and Hays [1982].

The Institute of Statistical, Social, and Economic Research (ISSER) of the University of Ghana, Legon, has carried out important studies on food production [C. K. Brown, 1972], consumption [Dutta-Roy, 1969], the cocoa industry [Kotey, Okali, and Rourke, eds., 1974], and demography [Caldwell, 1969; Caldwell *et al.*, eds., 1975].

In Francophone West Africa, the Senegalese Institute of Agricultural Research (ISRA) in Senegal and the Institute of Rural Economy (IER) in Mali have been active in village studies and in forging a link between farmers and research stations. In the 1970s, ISRA established "experimental units" or research zones in the groundnut basin in an effort to demonstrate the potential of introducing new technology that had been developed on the main research station at Bambey. For an overview of the experimental unit program in Senegal, see ISRA [1977], Faye and Niang [1977], Benoit-Cattin [1980], Fall [1980], and Bingen and Faye [1987]. IER publications in Mali include IER [1977] and Traore [1980].

The Food Research Institute of Stanford University carried out an interdisciplinary study of agricultural change in six English-speaking countries in sub-Saharan Africa in 1965-66 (Kenya, Uganda, Tanzania, Zambia, Ghana, and Nigeria). Survey results were reported in Anthony and Uchendu [1970, 1974] and Uchendu and Anthony [1975a,b]. A summary is reported in Anthony *et al.* [1979]. In southern Africa, the University of Nottingham collaborated with the University of Zambia in conducting farm surveys in two areas in Zambia in the late 1960s. The focus of the UNZALPI[74] project was to identify approaches for increasing the labor productivity of small farmers. Survey results were reported in C. M. Elliott *et al.* [1970]. Tench [1975] also used UNZALPI survey data. The University of Reading sponsored farm level research on Botswana in the mid-1960s, in Malawi in collaboration with the AES, and in Ghana with the University of Legon. Thornton [1973] summarized the findings of the village development project in southeast Ghana. Palmer-Jones [1974] reported the results of a study of tea production and marketing in Malawi. In Botswana, Kansas State University has assisted the Department of Agricultural Research with a wide range of farm management and institutional studies since 1982. Findings are summarized in ATIP [1986a,b] and D. C. Baker [1987].

In Sierra Leone, a national farm survey was carried out by Njala University College and Michigan State University in 1973-75. The results were presented in D. S. C. Spencer and Byerlee [1976, 1977] and Byerlee, C. K. Eicher, *et al.* [1977, 1983]. A survey of 480 rural households in the Eastern Region of Burkina Faso was carried out by a MSU research team over the 1978-80 period and the results are reported in Barrett *et al.* [1982], Lassiter [1981], and Wilcock [1981].

Purdue University has conducted numerous surveys in West Africa to evaluate the economics of sorghum and millet production in the Sahelian countries, and the costs and benefits of small- and medium-sized irrigation perimeters. Major findings of research in the 1970s are presented in Sanders [1989]. The University of Michigan has carried out studies in West Africa, focusing on livestock [K. H. Shapiro, ed., 1979; Ariza-Nino and Steedman, 1979, 1980]; grain marketing [E. Berg, 1980; Sherman, 1981]; and mixed farming in Burkina Faso [Delgado, 1979a, 1980] and in Niger [Eddy, 1979].

The University of Bordeaux began multidisciplinary survey work in the Maradi Region of Niger in 1977. De Miranda and Billaz [1980] discuss their survey methodology and preliminary findings.

Farm level research in sub-Saharan Africa has also been assisted by a network of international and regional research institutes. WARDA was established in 1960 to promote cooperation in the development of rice improvement programs in fifteen French- and English-speaking countries. In 1988 WARDA was rearranged and its headquarters moved from Liberia to Bouake, Côte d'Ivoire. [WARDA, 1988]

The Institute for Economic Research (IFO) has published more than 100 monographs on farm level studies in their "Afrika-Studien" series covering many of the English-speaking African countries. Most of the studies deal with agriculture and nearly one-half are in English. See, for example, Ruthenberg [1968], Gusten [1968], Kraut and Cremer, eds. [1969], Roider [1971], Chambers and Moris, eds. [1973], and Lagemann [1977].

The Scandinavian Institute of African Studies (SIAS) at Uppsala has published a wide range of studies on a variety of topics including: co-operatives [Widstrand, 1972], tobacco production [Boesen and Mohele, 1979], women [Bukh, 1979], multinational firms [Widstrand, 1975], and rural development [Chambers, 1974].

The Office de la Recherche Scientifique et Technique Outre-Mer (ORSTOM) in Paris has a network of around twenty research centers covering most of the Francophone countries in sub-Saharan Africa. Publications by ORSTOM researchers include Kohler [1971, 1972] in Burkina Faso; Ancey [1974] and Ancey, Michotte, and Chevassu [1974] in the Côte d'Ivoire. Also, Delpechi and Gastellu [1974]; Copans et al. [1972] and Rocheteau [1975] in Senegal. See Couty and Hallaire [1980] for an overview of ORSTOM studies.

The International Institute of Tropical Agriculture (IITA), established in 1968 at Ibadan, Nigeria, has carried out technical and socioeconomic research on the major food crops in the humid tropics. Studies by IITA researchers include K. L. Robinson [1974]; Flinn, Jellema, and K. L. Robinson [1975]; Lagemann, Flinn, and Ruthenberg [1976]; Lagemann [1977]; Okigbo and Greenland [1977]; Bachmann and Winch [1979]; Fotzo and Winch [1978]; Diehl and Winch [1979]; Flinn and Lagemann [1980]; Menz [1980]; and Zuckerman [1977, 1979a,b,c].

In 1976, CIMMYT launched a program in Eastern and Southern Africa under the leadership of Michael Collinson to help improve national research systems through training and farming systems research on the problems of small farms. CIMMYT's cooperative studies in Kenya, Tanzania, and Zambia are reported in CIMMYT [1977a, b, 1978] and Collinson [1982, 1988].

ICRISAT started a cooperative research program in Burkina Faso in 1975 with a sorghum breeder followed by a plant pathologist, millet breeder, two agronomists, a striga specialist, and an entomologist. A production economist was added in 1979 and farm level research was started in 1980 [Matlon, 1987].[75]

In addition to the above research programs and projects, farm level studies include the following:

Africa: Upton [1987], ICRISAT [1987].

East Africa: Heyer [1966, 1971, 1972], Catt [1970], Hutton [1973], Richards, Sturrock, and Fortt [1973], K. H. Shapiro [1973], Saylor [1974], Vail [1973], Humphrey [1975], Hunt [1975a], Gerhart [1975], and Clayton [1983].

Central and Southern Africa: Massell and R. W. M. Johnson [1968], Atayi and Knipscheer [1980], Tollens [1975], Weinrich [1975], Perrault [1978], Marter [1978], Kinsey [1978], Rukuni and Eicher, eds., [1987], Rohrbach, [1988], Low [1986], and Francis [1988].

West Africa: Meillassoux [1964], Welsch [1965], Pelissier [1966], Luning [1967], Upton [1967], Hill [1968], Smock and Smock [1972], Capron [1973], Monnier, *et al.* [1974], Maymard [1974], Berry [1975], Kleene [1976], Winch [1976], Marchal [1977], Matlon, *et al.* [1979a,b], Matlon, 1987], Faulkingham [1977], Reyna [1977], Sawadogo [1977], D. S. C. Spencer and Sivakumar [1987], Zuckerman [1977], Venema [1978], Clayton [1983], Matlon and D. S. C. Spencer [1984].

2. Methodological Issues in Rural Surveys

ISSUES IN SURVEY DESIGN

Since most small farmers are illiterate and do not keep farm account books,[76] three methods have been used to generate farm level information: case studies, infrequent surveys, and cost route or multiple visit surveys [D. S. C. Spencer, 1972]. The case study or model farm approach provides descriptive information on a single farm or a number of farms purposively selected to be representative or to reflect the practices of progressive farmers. Infrequent visit surveys[77] entail visiting a farm once or a few times to collect a range of stock (inventory) data and information about current practices. In the cost route (or multiple visit) approach,[78] farmers are visited regularly by an enumerator over an entire cropping season or full year, generally one to three times weekly and from 50 to 150 times a year. The rationale for using the cost route approach is that it is an effective way to capture flow (input/output) data on the magnitude and variability of labor— the most important input on small farms.

During the 1960s, researchers in East Africa used all three approaches—case studies, infrequent surveys, and cost route surveys—to collect farm-level information [M. Hall, 1970]. Clayton [1963], for example, used a model farm approach, drawing on data from government farms. Researchers in the Farm Economic Survey Unit in Kenya used a model farm approach in their whole-farm studies, based on data drawn from interviews with progressive farms [MacArthur, 1968]. Heyer [1966] used a case study approach, relying on intensive observations on a small sample of farmers in Kenya. Researchers in Tanzania used farm business type surveys on a large number of randomly selected farmers [Collinson, 1962-64]. In Uganda, researchers from Makerere University and the Ministry of Agriculture used a cost route approach, interviewing randomly selected farmers three times weekly for the entire crop year [Pudsey, 1967].

During the late 1960s and early 1970s, the case study approach was largely abandoned by agricultural economists in English speaking countries and researchers shifted to surveys and random sampling to insure input/output data more representative of typical farm-level conditions. The prevailing opinion was that the cost route approach should be used if there were time, money, and adminstrative capacity [Kearl, ed., 1976]. The rationale for cost route surveys can be traced to the 1960s when labor was identified as a major constraint on smallholder production in Africa. Advocates of the cost route approach such as D. S. C. Spencer [1972] were aware of difficulties that single visit surveys had in coming to grips with the seasonality of labor use and argued that frequent visits by enumerators were necessary to capture the seasonality of labor use and to control measurement errors on the labor variable. The cost route method was widely used in farm management surveys during the 1970s [Norman, 1972; Matlon et al., 1979a, b; Winch, 1976; Fotzo and Winch, 1978; K. H. Shapiro, 1973; Elliott et al., 1970; D. S. C. Spencer and Byerlee, 1976; Zuckerman, 1979a]. In order to reduce the cost of multiple visit surveys, multistage sampling procedures were employed. In this approach, stock data were collected in a reconnaissance survey of a large population, often several hundred households, and in the second stage, a smaller sample was randomly selected for repeated surveying in order to collect flow data over a period which often covers twelve months [Norman, Newman, and Ouedraogo, 1981].

Cost route or multiple visit surveys have thus far provided the most reliable data on input flows, particularly labor inputs, but this type of survey is substantially more costly per farm interview than single visit surveys. As a result, there is a trade-off between sample size and visiting frequency. While it has been recognized that single visit surveys will likely have high measurement errors for variables such as labor, the cost route or frequent interview methodology has a number of inherent problems. For example, the cost of interviewing the same farmer 50 to 150 times a year is extremely high, and there is a problem of sustaining the interest of the farmer during repeated interviews. Moreover, it often requires six to twelve months to plan a cost route study, a year to carry it out, and sometimes two to three years to analyze and publish the results. Concern with the cost of cost route surveys and the need to generate rapid results has led to a search for survey methodologies which can produce results in a few months rather than two to three years.

Starting in the late seventies, there was a discernible shift from cost route to infrequent visit surveys. These infrequent visit surveys, including both informal and formal single visit surveys, are popular among some advocates of farming systems research such as Byerlee, Collinson, et al. [1980], and Collinson [1981, 1982]. Although there is clearly a need to expand the use of infrequent visit or informal surveys for more rapid identification of major constraints on production, it is not possible to collect reliable data on labor use by crop, and by activity (e.g., weeding, ridging), through single visit surveys during a few critical periods in the year. During the 1980s, the validity of using the household as the unit of analysis became an important topic among Anglophone researchers as well. Anthropologists in particular argued that studies focused on agricultural households misrepresent the importance of both intra-household dynamics and interhousehold exchanges [J. L. Moock, ed. 1986].

Two approaches have emerged to deal with the conflicting goals of quick turnaround time and reliable labor data. One approach is to incorporate the best features of infrequent visits and cost route surveys into an activity approach which was tried by the late Pascal Fotzo in his 1980-81 study of rice production in eastern Burkina Faso. Fotzo recorded data once on each activity (e.g., land preparation, planting, and weeding) involved in the production of bas fond rice over a six-month period.[79] A second approach is to prepare case studies for a small sample of farmers (an average of thirty) from only one or two villages [Matlon, Eponou, et al., 1980; D. C. Baker, 1987]. The recent emphasis, by some farming systems research teams, on combining survey research with anthropological type case studies has been stimulated by increasing acceptance of the view that an understanding of farming systems requires more than the input of economists and more than the computation of averages from a farm survey. But whatever method is used to collect data most researchers collect far too much information [Farrington, 1975b; Upton, 1987].

The approach to farm-level studies has been significantly different in Francophone than in English-speaking countries. In Francophone Africa, farm-level researchers have generally carried out case studies on a small number of farms rather than large-scale rural surveys. Although we classify the approach often used by French researchers and French-trained researchers as a case study approach, we do not mean that statistical surveys were never used. For example, Boutillier et al. [1962] used a single visit questionnaire in which farmers were asked to recall information for the preceding year; they also collected daily information for a seven-day period twice during the year on such variables as expenditures and food consumption. But where formal attempts have been made to quantify key variables, sample sizes are generally extremely small and households are often selected on a nonrandom basis with the limited goal of deriving numbers for descriptive purposes, as opposed to sampling for statistical inference.[80] Copans et al. [1972] and Delpechi and Gastellu [1974], for example, report surveys in which the samples were ten to twelve adults representing two to four compounds.

Methodological debates in Francophone countries have focused on alternative conceptual frameworks for understanding social structure and social relationships and how they affect the organization of production. One of the major issues debated during the seventies was whether studying the production unit (exploitation agricole) is sufficient to explain the dynamic forces affecting the decisions of small farmers or whether studies at different levels — clans, villages, age groups, etc. — can provide more reliable information. The term exploitation agricole refers to a production unit where people work together on the main food grain field and eat as a group from the output of the main field [Ancey, 1975]. While English-speaking researchers undertaking survey research have generally relied on the household as the unit of analysis (often defined as "people who eat from the same pot"), researchers in Francophone countries consider the question of the proper decision unit to investigate to be an important unresolved issue [Monnier et al., 1974; Maymard, 1974; Ancey, 1975; Kleene, 1976; Couty, 1979; Gastellu, 1980]. For example, Ancey [1975] argued that a single sampling unit such as an exploitation agricole was not broad enough to capture the multidimensional relationships affecting decision making on African farms. He contended that different groups, main field production, residence, lineage, village, and supravillage — and the decisions made by these various units — often overlap and compete with others. Based on survey experience in Senegal, Gastellu [1980] argued that the starting place in survey research is to identify the units of production, consumption, and accumulation, rather than relying on a single concept such as exploitation agricole. He proposed that, in order to identify these groups, researchers should evaluate who makes the main decisions in each activity, evaluate the syntax of the local language in order to discern obligations and expectations between people, and focus on identifying patterns of privileged exchanges — food, gift, and labor, as well as money.

The value of using the individual rather than the entire production unit as a unit of analysis has been stressed by Winter [1975], Rocheteau [1975], Kleene [1976], Couty [1979], and Raynaut [1980]. Raynaut argued, for example, that, by taking into account differences between individuals, researchers can better distinguish the roles that physical and technical factors play in socioeconomic strategies of different household members.

During the 1980s, the validity of using the household as the unit of analysis became an important topic among Anglophone researchers as well. Anthropologists in particular argued that studies focused on agricultural households misrepresent the importance of both intrahousehold dynamics and interhousehold exchanges [J. L. Moock, ed., 1986].

REFERENCES ON SURVEY DESIGN

In presenting results, researchers generally have devoted little space to justifying the approaches they followed in collecting and analyzing survey data. But the choice of data collection and analysis procedures may importantly influence survey results. For example, the decision to use open-ended as opposed to closed-ended instruments can exert a major influence on the results obtained. Open-ended questions allow farmers to identify problems in their own words but tend to introduce intractable problems in the analysis and interpretation of data. On the other hand, more structured schedules may reduce ambiguity in interpretation of data but the choice of wording will frequently bias the replies.[81] A large body of wisdom about designing rural surveys is now available. A review of these reports should be the starting place for researchers contemplating farm-level research.

Collinson [1972] reviewed methodological problems in collecting and analyzing farm management data for planning purposes, drawing primarily on his early experience in Tanzania. D. S. C. Spencer [1972] drew on his experience in conducting farm management and marketing studies in Sierra Leone to discuss methodological problems in collecting flow data. Norman [1973] reviewed his experience in directing farm-level surveys in northern Nigeria. A volume edited by Kearl [1976] contains a valuable discussion of field data collection methods used by the Agro-Economic Survey Unit in Malawi with emphasis on labor flow information. Zuckerman [1979a, b] discusses the approach used by IITA researchers in western Nigeria in the early 1970s. For a discussion of infrequent (informal) surveys, see Byerlee, Collinson, et al. [1980], Collinson [1982], and Byerlee and Tripp [1988].

A discussion of methodological issues in rural surveys in several Francophone countries can be found in AMIRA's "notes de travail."[82] Ancey [1975] presented a valuable picture of the complex organization of small farms. Winter [1975] and Thenevin [1978] identified different survey techniques in gaining information for economic planning. Couty [1979] presented a general overview of methodological issues encountered using socioanthropological approaches to farm-level research. De Miranda and Billaz [1980] discussed multidisciplinary surveys of farms in Maradi, Niger.

DATA PROCESSING

Processing of survey data has posed a major problem for researchers throughout Africa. There has been a tendency to collect a wide range of data, paying little attention to how the data are to be analyzed until after data collection is finished. As a result, portions of the collected data are frequently never even keypunched, let alone analyzed. Moreover, the effort required to validate and aggregate data into files for analysis has often led to delays of one to three years before preliminary results are published. Researchers are slowly starting to realize that data processing must be considered an integral part of the entire survey design, data collection, and data analysis chain of events. Data processing can take as many resources as data collection but most manuals on survey methodology pay little attention to processing problems.

Several major decisions have to be made at or before the beginning of data cleaning and validation. Often, little attention is given to the following two critical issues: how to stratify sample households into appropriate groups for subsequent analysis, and how to convert labor into a homogenous unit in order to make labor files more manageable [Norman, 1972]. Several approaches have been used for stratification. In many cases, particular regions or villages have been purposively selected at the data collection phase, reducing the need for *ex post* stratification. Clayton [1964] and Collinson [1972] advocated stratification by land per resident because capital use by traditional farmers is limited. Researchers at Ahmadu Bello University stratified their surveys in northern Nigeria: by village or section, and by land per resident. The IITA 1970-71 survey in western Nigeria used age of the farmer to stratify their surveys [Zuckerman, 1979a]. Researchers have often relied on land tenure patterns, ethnic groups, or agroecological zones to stratify samples in farms. For example, agroclimatic zones were the main means of stratifying survey data by Barrett *et al.* [1982] in Burkina Faso and in Sierra Leone by D. S. C. Spencer and Byerlee [1976]. Ancey [1977] classified twenty-five farm models in the Sahel on the basis of farm size and ecology. Poulain *et al.* [1979] used ecologically homogenous zones, and within each zone distinguished among households according to land availability and livestock. D. C. Baker [1987], and others working in countries with relatively homogenous agroclimatic conditions such as Botswana, have advocated use of household circumstances such as gender of household head, cattle assets, and traction access to stratify households within zones or villages.

There has been a dichotomy between Francophone and Anglophone countries in the treatment of labor [Norman, Newman, and Ouedraogo, 1981]. Studies in

Francophone countries generally have relied on a stock measure of labor called the *actif*, primarily because of their heavy emphasis on case studies. An *actif* is often defined as a person between ages fifteen and fifty-four or fifty-nine The *actif* concept tends to understate the stock of labor available for agricultural activities, particularly in peak seasons when most persons over ten years work in fields. If the proportion of labor coming from individuals under fifteen and over fifty-four is significantly different among subsamples of households, the net return per *actif* will be overestimated in the households where non-*actifs* make a major contribution to farming activities. But the *actif* concept does not make any allowance for differences in productivity and labor use by seasons [Ancey, 1974]. Examples of studies which have used the *actif* approach include Kohler [1971] and Monnier *et al.* [1974] in Senegal; and Barrett *et al.* [1982] in Burkina Faso.

In Anglophone Africa, researchers have generally aggregated labor units by using the concept of "man-equivalents." The concept was applied in Tanzania in order to discount the inputs of women and children since it was assumed their labor was less productive than that of men [Collinson, 1962-64; Ruthenberg, 1968]. More commonly, indices reflecting productivity differences have been used to convert flow inputs by women and children into man-hours or man-days. Most studies have used fairly arbitrary approaches for deciding on the weights to be used in aggregating labor inputs and, as Collinson [1987] noted, there has been controversy over how to derive appropriate weights. Norman [1972] contended that a work study approach is an objective way to assign weights. D. S. C. Spencer and Byerlee [1976] used wages in rural labor markets to derive relative weights. In a 1978-79 survey in eastern Burkina Faso, weights were obtained by asking a sample of farmers to estimate the relative productivity of different classes of labor defined by age and sex for each major cropping activity [Barrett *et al.*, 1982].

During the 1980s, the practice of weighting labor to create "man-equivalents" was challenged with the growing sensitivity to the gender division of labor in African farming systems. Many researchers started to report labor findings which were disaggregated by gender and age, and to calculate labor returns to actual labor hours rather than to arbitrarily defined person-units of labor.

The need to speed up the processing of survey data is one of the dominant factors affecting survey design and data processing strategies. Quicker turnaround times necessitate better coordinated data collection and processing. Norman and Palmer-Jones [1977], for example, called for a more standardized economic methodology for research on cropping systems. Microcomputers are invaluable but they cannot solve the turnaround problem unless researchers stop collecting excess data [Candler and Slade, 1981].[83]

3. Applications of Analytical Techniques

BUDGETING

Farm budgeting was introduced in a large number of countries in English-speaking Africa in the 1960s.[84] Complete farm budgets were used to derive performance standards for extension programs [MacArthur, 1968]. During the 1970s, researchers shifted to enterprise budgeting except in areas where there was extensive intercropping. Enterprise budgeting was used to compare the costs and returns of different crop and nonfarm enterprises [Lagemann, 1977; Ruthenberg, 1980; D. S. C. Spencer, Byerlee, and Franzel, 1979; Lassiter, 1981], and to compare costs and returns in producing the same crop with different techniques [Winch, 1976; D. S. C. Spencer and Byerlee, 1976; Lang, 1979]. During the late 1970s and in the 1980s numerous researchers have used partial budgeting to evaluate changes in enterprise mix or production techniques [Fotzo and Winch, 1978; ATIP, 1986a,b]. We shall concentrate on whole farm and enterprise budgeting.

Numerous methodological issues need to be addressed in constructing budgets from cross-sectional data [Dillon and Hardaker, 1980]. Four major problems have repeatedly surfaced in deriving budgets, making it difficult to interpret and compare the results of different studies. First, no standard approach has been used in deciding what to include in farm budgets. For example, many researchers have only included yields of the main crops, arbitrarily excluding secondary crops, livestock, and nonfarm activities. Few researchers have taken into account changes in stocks or have attempted to estimate the opportunity costs of farm-supplied inputs such as manure. Furthermore, there has been little consistency in the treatment of land. Some researchers included all land under the control of the farm, other studies used only land owned or cultivated in the year of the survey.

Another problem in budgeting studies has been the valuation of inputs and outputs. Because small farmers generally purchase few inputs and retain a major proportion of farm output for family consumption, returns are highly dependent on the value researchers assign to inputs and outputs. Both measuring and valuing family labor pose major problems in constructing farm and enterprise budgets. A few researchers have presented results using a range of assumptions such as not including the cost of any family labor during different times of the year, including the cost of only hired labor and inputing a cost for all labor [Norman, Pryor, and Gibbs, 1979]. More commonly, researchers have treated hired labor as a variable cost, included a charge for land, and then computed a net return to labor and management as a residual [Ruthenberg, 1980; D. S. C. Spencer, Byerlee, and Franzel, 1979]. Even this approach is fraught with difficulties. How does one value labor in situations where farmers rely on reciprocal labor exchanges, providing meals for guest workers as partial payment? Another issue is whether to include time walking to and from fields. Most studies have ignored walking time even though in some cases it can be substantial [Cleave, 1974].

The problem of valuing capital services is equally difficult. The most common practice has been to exclude costs of capital for hand implements. Straight line depreciation is often used for major capital items. D. S. C. Spencer and Byerlee [1976] and D. S. C. Spencer, Byerlee, and Franzel [1979] used a capital recovery formula to convert capital stocks to annual service flows taking into account the average life of the equipment and a discount rate assumed to approximate the social opportunity costs of capital.

The value of outputs has usually been derived by multiplying average yields over the sample of farms by the average annual price in local markets. Use of a simple average of observed prices throughout the year can substantially misrepresent the actual average prices received by farmers because prices can fluctuate by more than 100 percent during the year. Moreover, sales are often concentrated in the postharvest period when prices are lowest. Another common problem in valuing outputs is whether to attribute the value of intercropped fields to individual crops or to treat the crop combination as a single enterprise.

A third problem is interpreting budgets that have been constructed from average input/output relationships on surveyed farms. Averages, as Upton and Casey [1974] and Dillon and Hardaker [1980] point out, can obscure wide variations in soil characteristics, managerial ability, class differences, access to inputs, and other variables which influence the returns to any given farmer. Moreover, as Ruthenberg [1980] has warned, budgets built from averages of survey data generally show a more diversified production pattern than one is likely to observe on any individual farm. Ruthenberg contended that data from a modal or typical farm are often more useful than survey data from farm budgets. G. E. Dalton [1973] pointed out that while net income figures derived from farm budgets provide useful descriptive information, they do not provide insights into the factors responsible for poor performance of a single enterprise or the reasons for low productivity on an individual farm.

Finally, budgets based on cross-sectional data do not account for changes over time and space. Because of inflation, overvalued exchange rates, and changes in relative prices of crops over time, figures such as the cost of plowing with tractors or the returns to labor in rice farming are generally not comparable in different countries and in different years. Also, net income figures do not provide adequate information on the welfare of rural households unless data are also available on the cost of living. Because of periodic droughts and wide variations in the timing of rains, weather can greatly influence the returns to land and labor in any given year. Finally, land clearing activities and time spent planting perennial crops add value to production in future years, but trying to identify the returns to that labor poses an accounting nightmare.

Despite our stress on data limitations and problems of valuation, farm budgeting is a valuable tool of analysis. Budgeting and comparative farm indices were found to be invaluable in farm survey research in the 1960s in large part because they could be constructed relatively quickly and the training required for building and interpreting budgets was limited relative to programming and regression methods. Ruthenberg's [1980] invaluable farming systems book attests to the descriptive value of comparative farm budgets. Enterprise budgets have contributed to our understanding of relative costs and returns in fallow and permanent cultivation systems, from improved *versus* traditional production practices, from perennial and annual crops, and for farming and nonfarm activities.

REGRESSION ANALYSIS

Econometric studies of smallholder production relationships can broadly be divided into three categories: farm production functions; analyses of the determinants of levels of resource use, and supply response studies.

Farm Production Functions. Production function studies in the 1960s were few and far between and of uneven quality. Interest in this topic was stimulated by the desire to evaluate smallholder productivity and efficiency in order to test Schultz's "poor but efficient" hypothesis [T. W. Schultz, 1964]. Production functions estimated from controlled survey data and data from experiment stations have also been used to analyze factors influencing yields of recommended technical packages and to identify optimal levels of resources use for new practices.

Among the earliest reported attempts to estimate production functions for smallholders from survey data are the case studies of Tanzanian agriculture reported in Ruthenberg [1968], Massell and R. W. M. Johnson's [1968] analysis of African farmers in Zimbabwe, Upton's [1967] research in southwestern Nigeria, Welsch's [1965] study of rice farmers in eastern Nigeria, and Lunings's [1967] study of sorghum-millet-groundnut farmers in northern Nigeria. During the 1970s, efforts to evaluate determinants of farm household income using regression techniques have been carried out in numerous countries, including *Zambia* [C. M. Elliott *et al.*, 1970; Tench, 1975], *Nigeria* [Upton, 1970; Norman, 1972; Norman, Fine, *et al.*, 1976; Matlon, Eponou, *et al.*, 1979a,b; Osuntogun, 1978; Norman, Pryor, Gibbs, 1979; Mijindadi, 1980], *Kenya* [Saylor, 1974; Wolgin, 1975; Rukandema, 1978; P. R. Moock, 1981], *Tanzania* [K. H. Shapiro, 1973; K. H. Shapiro and Muller, 1977], *Sierra Leone* [D. S. C. Spencer and Byerlee, 1976], the *Côte d'Ivoire* [Lang, 1979], *Malawi* [Farrington, 1975b, 1977a], and *Ghana* [Prakah-Asante, 1976].

After almost two decades of experience in estimating smallholder farm production functions using survey data, there are still many unresolved problems. Since production function studies have relied on survey data of uneven quality, overcoming data limitations has been one of the major determinants of the approaches followed in specifying production functions. Major specification problems have included: choice of a functional form, and measurement and aggregation of inputs and outputs. The choice of functional form has received surprisingly little attention in discussions about the specification of production functions. Most researchers have used Cobb-Douglas functions, generally with little or no discussion of why alternative functional forms were not considered. A few researchers have presented comparative results using linear, quadratic, and square root functional forms [Ruthenberg, 1968; Luning, 1967; Tench, 1975; K. H. Shapiro, 1973]. D. S. C. Spencer and Byerlee [1976] used a constant elasticity of substitution (CES) production function to evaluate returns to scale and the elasticity of substitution between labor and capital services in rice production in Sierra Leone. In a series of studies conducted by researchers at the Institute for Agricultural Research, Zaria, Nigeria, several functional forms were used: linear for yield functions, quadratic or linear for estimating the relationship between profits and yields, and Cobb-Douglas to evaluate the marginal value products of inputs [e.g., Norman, Beeden, et al., 1976a, b].

Almost without exception, farm production functions have been estimated using ordinary least squares. Three exceptions include Massell's use of analysis of covariance [Massell and R. W. M. Johnson, 1968; Massell, 1967b], D. S. C. Spencer and Byerlee's [1976] use of nonlinear maximum likelihood estimation of their CES function, and Strauss's [1981] household firm model using survey data from Sierra Leone. The production component of Strauss's simultaneous equation model is Cobb-Douglas in inputs and constant elasticity of transformation in outputs. The model was estimated using a tobit model in order to account for some products not being produced by all households.

In general, the dependent variable in farm production functions has been total farm income from all cropping activities.[85] Limitations of survey data have prevented estimation of crop-specific production functions. Yields from separate fields or from individual crops in crop mixtures have typically been converted to value terms using an average of annual prices in local markets. In a few cases, return per acre, instead of total farm income, has served as the dependent variable. In the few cases where production functions have been estimated for specific crops, one or a few crops either accounted for most of the area cultivated, or were important sources of cash income [Massell and R. W. M. Johnson, 1968; D. S. C. Spencer and Byerlee, 1977; Saylor, 1974]. Yields of specific crops have also served as the dependent variable in studies using experiment station data or data from controlled surveys [Norman, Beeden, et al., 1976a,b; Norman, Hayward, Hallam, 1974, 1975; Flinn and Lagemann, 1980].

Data problems have forced researchers to rely on inadequate representation of aggregate input categories—land, labor, capital—in estimating interfarm production functions.[86] In some early studies, stock variables were used to represent flow inputs in cases where flow data were not available. Ruthenberg [1968], for example, used family man-equivalents as a proxy for labor. Massell and R. W. M. Johnson [1968] used the value of farm implements at undepreciated replacement cost to represent capital. Upton [1967] represented capital investments by acres of tree crops and number of livestock. Even in cases where flow variables have been included, it has been common to use undiscriminating units such as man-equivalent days [Upton, 1973; Lang, 1979].[87]

Because of specification and measurement problems, the coefficient of multiple determination of most farm income studies based on aggregate physical farm inputs has been low and the standard deviations of individual coefficients have been large. Researchers have followed two major approaches in trying to increase the amount of income which is explained by their models. First, several researchers have abandoned efforts to rely exclusively on variables representing underlying physical production functions and have included indices of social and personal characteristics as well as physical input variables [e.g., Upton, 1967, 1970; C. M. Elliott et al., 1970; K. H. Shapiro, 1973; Saylor, 1974; Tench, 1975; Rukandema, 1978]. Upton [1967] was one of the first researchers to analyze the relationships between production, consumption, and social characteristics. Upton used Guttman scaling to rank farmers by factors such as progressiveness and correlation analysis was then used to evaluate the strength of relationships between social characteristics and production.

One of the first studies to incorporate social variables formally into functions of the value of farm output was based on farm survey data collected in Zambia under the UNZALPI Project. C. M. Elliott et al. [1970] incorporated indices to represent factors such as education, general awareness, and knowledge of farming and attempted to attribute proportions of the explained production per acre to social variables instead of the usual farm inputs. Tench [1975] presented a valuable discussion of conceptual and statistical issues involved in evaluating smallholder productivity in Zambia. In addition to estimating a basic Cobb-Douglas model relying on aggregate input categories for land, labor, and capital, Tench explored alternative formulations, including inclusion of several indices of social and personal factors, and regressing the value of output on a set of variables representing only social and personal factors. Surprisingly, Tench found that the multiple coefficient of determination based on social and personal indices was nearly as large as the coefficient based on the usual physical inputs. Tench [1975] also used the technique of principal components analysis to identify a set of social, personal, and technical variables associated with variation in the value of farm output. Using survey data from Tanzania, Saylor [1974] used stepwise regression to analyze the relationship between output per hectare and per unit of labor. Saylor also used several technical variables such as the time of planting, plant densities, and ridging as well as social variables such as modernity and contact with extension workers.

Researchers have also attempted to improve the specification of their models by disaggregating land, labor, and capital into more specific components. Upton [1967] and Norman, Pryor, and Gibbs [1979] decomposed labor into family labor and hired labor. More commonly, labor has been decomposed by either age-sex category [Shapiro, 1973, 1978; Tench, 1975] or by cropping operation [K. H. Shapiro, 1973; Saylor, 1974; Rukandema, 1978]. Specification of land has been improved by incorporating dummy variables for soil quality and fertilizer use. Capital has been divided into cash expenditures and depreciation on capital equipment [Norman, 1972].

One of the major methodological issues in specifying farm production functions has been whether to include a variable for management. Attempts have been made by Massell and R. W. M. Johnson [1968], Upton [1970], and K. H. Shapiro [1973] to derive proxies for management in order to reduce the specification bias that arises when management is correlated with included variables. Massell and R. W. M. Johnson used survey data from two areas in Zimbabwe, the Cheweshe Reserve and Mt. Darwin District, to fit Cobb-Douglas production functions for each of the three major crops—groundnuts, millet, and corn. Farmers were interviewed weekly for the entire crop year. Massell and Johnson attempted to reduce management bias through the analysis of covariance and by incorporating dummy variables to distinguish three groups of small farms by management level. A discussion of the procedures used in the analysis of farms in the Cheweshe Reserve and major findings was also presented in Massell [1967a]. Massell represented management differences using dummy variables, and later drew on data from the Mt. Darwin District to illustrate his method for reducing management bias through the analysis of covariance [Massell, 1967b]. Massell treated each firm-product combination as a separate observation, making the assumption that the relative efficiency of farmers is constant across all the crops they grow.

Upton [1970] used the technique of principal components analysis to derive a single index as a proxy for management. K. H. Shapiro [1973] derived several indices for social and personal characteristics using Guttman scaling and then combined the indices into a single index whereby farmers were ranked according to their degree of "modernization." Because there was a high correlation between the modernization ranking and a farmer's ranking in terms of technical efficiency, Shapiro argued that the modernization index could be used as a proxy for management efficiency. In estimating production functions for his sample of farmers in Geita District, Tanzania, Shapiro experimented with several specifications, by including and then excluding the management variable. Shapiro concluded that the effect of including management was to raise the productivity of labor and to decrease the productivity of land. With J. Muller, Shapiro [1977] expanded on his analysis by incorporating indices of information in their analysis of farmer efficiency.

Although the above research by Massell, Upton, and Shapiro represents important methodological contributions, researchers have largely abandoned the use of a management variable and have tried to incorporate management indirectly through variables such as the timing of key activities (e.g., date of planting, date of weeding) and by specifying variables reflecting the quality of inputs to account for differences in "technical efficiency" among farmers. In production function studies for maize, cotton, and sorghum: sowing date, plant density, amount of fertilizer, and time from sowing to weeding have been found to have significant relationships with yields [Norman, Beeden, et al., 1976a, b; Norman, Hayward, and Hallam, 1974, 1975]. In their research on maize production in Nigeria, Flinn and Lagemann [1980] found that coefficients of variables representing the timing of field activites—especially the date of planting— proved to be more significant and more stable than variables representing the level of labor inputs.

In summary, due to data limitations, and consequent estimation procedures and specification problems, the results of production function studies should be considered rough approximations to be treated with caution, as several researchers have acknowledged [Ruthenberg, 1968; Massell and R. W. M. Johnson, 1968; K. H. Shapiro, 1973; Norman, Beeden, et al., 1976b; Upton, 1979, 1987].

Determinants of Resource Use. The second major use of regression analysis has been to identify factors affecting the level of resource use by farmers. Applications have focused on determinants of the amount of land cultivated or labor used in farming activities; or the rate of diffusion, probabilities of adoption of a recommended input such as fertilizer, or an improved variety.

In general, regression analyses of factors affecting labor or land use have depended on somewhat *ad hoc* selection of variables to be included as regressors. A limited amount of the variation in resource use is generally explained in these applications. The major purpose of these regressions has been to test hypotheses in order to improve an understanding of farming system relationships. Using simple linear regression, Upton [1967] and Luning [1967] demonstrated a positive and significant relationship between the number of people available for farming activities and the area cultivated. Norman, Pryor, and Gibbs [1979] estimated several regressions relating to levels of resource use, including hours worked per hectare, total hours worked on the family farm, and hours worked per hectare in northern Nigeria. Man-hours per acre were found to be inversely related to the number of acres cultivated but positively related to the proportion of higher quality land. Also, as might be expected, the authors were able to show that total family hours devoted to work on the farm was directly related to the size of family and area cultivated and inversely related to the use of hired labor. On the other hand, days worked per male adult on the farm was inversely related to the number of male adults available to work.

Farrington [1975a] used farm survey data in Malawi collected in the 1970-72 cropping seasons to test the hypotheses that the length of working day varies according to the type of worker and the type of crop operation. The main regressors were proxies for the energy requirements of a cropping operation and the urgency with which the task had to be performed. Farrington found that energy requirements had a stronger influence on the length of the working day than did urgency, but the model explained only a small amount of the observed variation in the length of the working day.

Researchers have included a wide range of independent variables in studies of adoption patterns. Gerhart's [1975] study of maize diffusion in Kenya took into account such factors as population density, proximity to a research station, average annual rainfall, education, knowledge of credit, number of extension visits, and farm size; and found that agroclimatic zone was the most important variable in explaining adoption. Binary dependent variable models have been used to evaluate factors affecting the decision to adopt improved technologies. Falusi [1974/75], for example, used a multivariate probit model to analyze factors affecting the decision to use fertilizer in Nigeria. Aklilu [1980] used a logit model in his study of fertilizer adoption in Ethiopia.[88]

Supply Functions. Since the mid-1960s, numerous studies have been carried out on the magnitude and direction of smallholders' supply response for a wide variety of cash crops. Evidence on supply response for food crops continues to be scarce and largely impressionistic. Supply response studies of smallholder cash crops have focused on two questions. First, is there evidence of positive short- and long-run price elasticities? Second, are price elasticities high enough that statutory marketing boards could use pricing policies to influence the composition of smallholder production? Positive price elasticities have been reported in every case, although the results have not always been statistically significant. Elasticities have been found to be low with long-run price elasticities being somewhat higher than short-run elasticities. For summaries of empirical findings, see Askari and Cummings [1976] and Helleiner [1975].

Numerous methodological problems have plagued supply response studies and debates about appropriate estimating procedures have been heated.[89] Many methodological weaknesses can be traced to data limitations. For example, almost all time series data for estimating supply functions have come from marketing board records, but these figures do not distinguish between aggregate supply response and board purchases. Moreover, reliance on marketing board data has made it difficult to take planting dates of perennial crops into consideration. As a result, most studies have focused on marketed output as opposed to more refined analyses of acreage and yield response. Although many studies have incorporated a weather index (usually rainfall) or a trend variable to reflect technical change [Olayide, 1972; Oni, 1969a], few have been able to take into account labor, land, or liquidity constraints even though each could have a major impact on yields and marketed output [Berry, 1976; Adegeye, 1976].

G. K. Helleiner [1975] noted the following common methodological problems in his review of supply response studies: prices received by producers may differ significantly from reported marketing board prices, few studies have convincingly defended the underlying models, in several cases only one of the possible forms of supply relationships has produced significant results and the form of the lead equation for the same crop has usually been different in different studies, and only a few studies have taken income and liquidity constraints into consideration. Berry [1976] points out that because researchers have rarely bothered to explore the possibility that their statistical findings may be consistent with more than one model, empirical findings have shown little more than the fact that farmers do respond to market forces.

Through the early 1970s, the major methodological issues in supply response studies were the validity of using marketed output as opposed to planting decisions in constructing supply response models for perennial crops, and the question of how to handle producers' price expectations for both perennial and annual crops. Several researchers acknowledged the desirability of using acreage as the dependent variable but used output as the dependent variable since acreage figures were not available [e.g., Bateman, 1965]. R. Stern [1965] used data on cocoa plantings in Nigeria from before 1945 to estimate a planting response model but, as Berry [1975] points out, Stern's model was crude, relying on five-year moving averages of plantings and the real price of cocoa. Berry [1976] used data from Nigeria, from 1911 to 1944, to evaluate an alternative model of cocoa supply. By doing this, she showed that the rate of planting was more closely related to farmers' income from cocoa sales than it was to current price. Berry concluded that liquidity constraints were more important to Nigerian cocoa producers than future price expectations as long as the opportunity costs of planting more trees are not prohibitive.

Kenya is one of the few countries where time series data on plantings of cash crops and smallholder yields are available. Maitha [1969] derived a tree stock demand function from an aggregate CES production function to explain annual variations of mature acreage of coffee. Farmers were assumed to determine their acreage so as to maximize the present value of net receipts. Maitha subsequently presented an alternative model [1970] by using output per acre as the dependent variable.[90] Etherington [1973] formulated a polyperiod production function for tea, incorporating such variables as past plantings, maintenance, and harvesting inputs, and the botanical characteristics of the crop. Output in any one year was then treated as a function of the distribution of past plantings and the estimated yield from each cohort group. Etherington contended that actual output could be closely predicted by introducing climatic and/or economic factors specific to a given year.

Several approaches have been used to represent producers' price expectations. Bateman's [1965] research in Ghana was one of the first supply studies of a perennial crop to use a Nerlovian adaptive expectations model. Alibaruho [1974] also used a Nerlovian model in his study of cotton supply in Uganda. More commonly, either proxy variables or lagged producer prices have been used to represent expected prices [e.g., Maitha, 1974]. Ady [1968] used world prices in a West African comparative study on the grounds that, even though producer prices were determined by marketing boards, the world price affected producers' price expectations. Olayide [1972] included current world prices in his study of Nigerian cash crops for the same reason but this was criticized by Blandford [1973] as being a poor substitute for expected producer price.

R. Stern [1965], as mentioned above, used a five-year moving average of real cocoa prices in his analysis of cocoa plantings in Nigeria. Fredrick [1969] and Oni [1969a] used simple lagged prices to represent price expectations in studies on cotton in Uganda and Nigeria, respectively. Both included prices of major substitute crops as well. Fredrick also estimated cotton output as a function of the price ratio of cotton to coffee. Saylor [1967] and Alibaruho [1974] also used price ratios to represent price expectations.

During the 1970s and 1980s, the attention of researchers shifted from grinding out elasticities using basic lagged price models to developing more refined models of farmers' supply response. Several refinements have been suggested, including incorporating asymmetric price response [Olayemi and Oni, 1972; Olayemi, 1976]; taking into account production and prices of food crops in estimating cash crop supply functions [Alibaruho, 1974]; simultaneous equations models [Ford, 1971; Blandford, 1973]; distinguishing between marketable surplus and total output response [Medani, 1975; Livingstone, 1977c; Gemmill, 1979]; taking into account nonprice factors in estimating the determinants of output response [Adegeye, 1976; Helleiner, 1975; Abalu, 1975]; making price expectations endogenous [Phillips and Abalu, 1987], and disaggregating acreage *versus* output responses to account for output response constraints [Ngambeki and Idachaba, 1985].

In summary, supply response studies have provided irrefutable evidence that smallholders are economic decision makers and they incorporate prices into their production decisions. But the discovery of a positive correlation between the price of a commodity and sales to a marketing board tells us very little about small farmer decision making [Helleiner, 1975; Berry, 1976]. Although some refinements have been made in supply response modeling in recent years, we agree with Helleiner [1975] that research on the price responsiveness of smallholders has reached a point of diminishing returns.

Supply response studies by commodity are as follows: cotton [Oni, 1969a; Fredrick, 1969; Aldington, 1971; Olayide, 1972; Alibaruho, 1974; Chembeze and Womack, 1987]; cocoa [Bateman, 1965; R. Stern, 1965; Ady, 1968; Behrman, 1968; Olayide, 1972; Olayemi and Oni, 1972; Berry, 1976]; coffee [Ady, 1968; Fredrick, 1969; Maitha, 1969, 1970]; palm oil and kernels [Helleiner, 1966a; Saylor, 1967; Oni, 1969b; Olayide, 1972]; tobacco [Dean, 1966]; rubber [Olayide, 1972; Olayemi, 1976]; groundnuts [Olayide, 1972; Abalu, 1974; Owosekun, 1976]; Phillips and Abalu [1987]; rice [Ngambeki and Idachaba, 1985]; and maize [Muir-Leresche, 1984].

PROGRAMMING MODELS

The first application of linear programming (LP) in African agriculture was Clayton's [1961] study of the effect of resource constraints on the profitability of typical farms in the Central Province of Kenya. The study was based on secondary data and was limited in the number of activities and constraints considered since Clayton had to derive optimal plans by hand. Despite methodological problems, Clayton made a valuable contribution in identifying family labor, rather than land, as the major constraint on increasing farm output. In a follow-up study using a computer in England, Clayton [1963] used parametric programming to indicate the effect of differing resource endowments on farm profitability and to derive a normative supply curve.

Heyer's [1966] programming analysis of Kenyan agriculture represented a major improvement over Clayton's work by using input/output data collected from farmers and introducing a range of activities for crops to reflect differing production intensities. Heyer's research also identified labor as the major constraint on smallholder farming — clearly demonstrating the need to look beyond returns to land in evaluating potential technical packages. Later, Heyer [1971] refined her LP model and showed that the introduction of cash crops — cotton and a quick maturing variety of maize — would have a modest impact on the incomes of subsistence farmers in semiarid zones of Kenya.

During the 1970s, LP models emerged as one of the most important tools to study smallholder farming. Major applications have included:

(1) identification of constraints on smallholder farming [Clayton, 1961, 1963; Heyer, 1966, 1971; Atta-Konadu, 1974; Abalu, 1975; Richard, Fall, and Attonaty, 1976; Perrault, 1978; Delgado, 1979a,b; Traore, 1980; Crawford, 1982];
(2) derivation of normative supply and input demand functions [Ogunfowora, 1972; Ogunfowora and Norman, 1973; Mwangi, 1978; Metson, 1978];
(3) estimation of frontier production functions on the basis of cross-sectional farm survey data [K. H. Shapiro, 1973; Mijindadi, 1980];

(4) evaluation of the profitability of new technologies [Ogunfowora and Norman, 1974; Vail, 1973; Adesina, Abbott and Sanders [1988]; E. Hopkins, 1974; I. J. Singh, 1976; Kinsey, 1978; Etuk, 1979]; and

(5) identification and evaluation of management strategies [Heyer, 1972; Low, 1974; Farrington, 1976].

In general, the goal of researchers conducting LP studies of the first type—identification of farm level constraints—has been to provide direction to policy makers rather than attempt to recommend improved plans for individual farmers. Models generally have been based on an abstract "representative" or typical farm. Researchers have typically used average values obtained through farm surveys to determine both resource constraints and input/output coefficients for the representative farm. Coefficients for new technologies usually have been based on experiment station trials, sometimes discounted to approximate responses under farm-level conditions. Most models have been designed to maximize expected profits from cropping activities and have included subjective constraints, particularly a minimum subsistence food requirement, as well as resource constraints.

In addition to the well-known limitations of LP [Dillon and Hardaker, 1980], three major problems have been encountered in the studies reviewed. First, the assumption of profit maximization has led to model results which bear little resemblance to observed patterns of resource allocation—even when constrained by minimum food requirements and by setting resource constraints to the average of observed values. Second, few researchers have considered the problems involved in inferring policy implications from the results of individual farm models.[91] Third, most LP models have only taken into account optimal plans for annual crops over a one-year horizon, ignoring interannual resource flows, household activities, interaction between farming and animal husbandry, and the cash flow problems associated with the introduction of large capital purchases such as animal traction, which includes oxen and equipment. Studies without these limitations include Abalu [1975], Delgado [1979a, b], and Crawford [1982]. Delgado evaluated the possibility of mixed farming in southern Burkina Faso as an alternative to the current "entrusting system" between the Mossi who specialize as farmers and the Fulani who specialize as herders. Delgado found that the opportunity cost for either group engaging in mixed farming could not justify abandoning the "entrusting system." Abalu [1975] used a dynamic programming model to identify the optimal sequence of investments in perennial crops on public, cooperative, and traditional farms in Cameroon over a twenty-year horizon. The objective function was to maximize the present value of the twenty-year stream of benefits from investments in perennial crops, subtracting a constant proportion of the income generated each year for consumption expenditures. Abalu found land to be the only limiting constraint.

Crawford [1982] creatively combined multiperiod programming with simulation of stochastic variables in an attempt to assess the effect of variable crop yields, physical resource limitations, and family size on growth of incomes and consumption over a multiyear horizon. An optimal solution was found for a one-year horizon at the beginning of the period using a multiperiod model.

The second category of applications, derivation of normative supply and input demand functions, has been employed to predict farmer responses in cases where time series data are not available [e.g., Clayton, 1963; Ogunfowora, 1972] or in which existing use of an input is limited [Ogunfowora and Norman, 1973; Mwangi, 1978]. These studies have generally relied on a representative or "benchmark" farm. Resource demand and output supply functions are derived through parametric programming and, in some cases, continous functions have been obtained through regression analyses of derived output-price observations [e.g., Ogunfowora, 1972]. In general, the aggregation bias inherent in using representative farms has not been adequately addressed in these studies. Odero-Ogwel and Clayton [1973], in an attempt to construct a regional planning model for Nyeri District, Kenya, tried to reduce aggregation bias by identifying relatively homogenous groups of farmers defined by similar factor endowments.

The third major application, estimation of a frontier production function, has been used by K. H. Shapiro [1973] and Mijindadi [1980] in order to derive an index of technical efficiency. In both studies, the coefficients of a frontier Cobb-Douglas production function were obtained following the method presented by Timmer [1970]. An index of technical efficiency was then derived for each farm by dividing actual output by the potential output for that farm given the level of resource use, where the potential output for each farm was calculated by multiplying the resource levels of that farm by the coefficients of the frontier production function.

Attempts to use programming models to evaluate new technologies have been one of the most useful applications of individual farm models, particularly during the 1980s.[92] Vail [1973], for example, attempted to identify what package of innovations, consistent with farmers' values and financial resources, had the greatest promise for increasing net farm income in Uganda, taking into account different seasonal rainfall patterns and cash crop prices. Similarly, in northern Nigeria, Ogunfowora, and Norman [1974] used a programming model to evaluate the relative stability of sole and mixed cropping systems under different resource endowments and husbandry practices in an effort to determine whether technical researchers should give more attention to mixed cropping in designing appropriate technical packages. Adesina, Abbott, and Sanders [1988] used a MOTAD model to evaluate the implications of risk aversion, liquidity and seasonal labor contraints on the adoption of fertilizer in Niger.

Finally, increasing dissatisfaction with the unrealistic policy prescriptions obtained from models designed to maximize profits has stimulated the development of models which more nearly reflect decision processes of small farmers. Thus far, models in this category primarily have been used for methodological research, although some researchers [Low, 1974] have presented policy recommendations. Heyer [1972], using a game theoretic framework, evaluated differing strategies for dealing with risk and uncertainty, including maximin, minimax, and attempting to maximize expected returns during best, worst, and average years. She then compared model results for different states of nature with farm plans observed in her farm survey.

Farrington [1976] attempted to provide a better test of the approach used by Heyer [1972] by drawing on survey data collected from two sites in Malawi over consecutive cropping seasons.[93] Farrington identified optimal farm plans for good years and bad years. Comparing the results with actual farmer practices, Farrington found that farmers adhere to a cropping pattern which, although producing returns somewhat lower than potential in both good and bad years, gave acceptable returns in both types of years. He concluded that such a strategy was consistent with long-run profit maximization.

Farrington's research touched off a debate between Palmer-Jones and Low over the appropriate use of LP to analyze optimal resource use in smallholder agriculture [Palmer-Jones, 1977a, 1979; Low, 1978]. On the basis of survey work in northern Nigeria, Palmer-Jones [1977a] contended that the value of LP in analyzing small holder decision making was limited because farmers adjust the quality and timing of operations as the cropping season progresses. Palmer-Jones also criticized the use of individual farm models for appraising new technologies since the effects of a new technology on the social system might result in an increase in total output but a reduction in the welfare of the target groups.

LP is moving toward more sophisticated farm level models.[94] The unanswered question is whether models can be developed and adapted for specific environments at a justifiable cost.[95]

4. Research on Selected Topics

SOCIAL AND CULTURAL ISSUES

Agricultural development is a product of the interaction of technical, market, and social variables. While agricultural scientists concentrate on the technical side and economists on market variables, anthropologists and sociologists examine how variables such as social structure, extended family, and custom inhibit or facilitate the process of change.[96] Since the literature on social and cultural factors and agricultural development is vast, we shall examine a few studies on three issues: class analysis and rural change; social impact analysis, and the role of indigenous knowledge in promoting development.

Class Analysis. Class analysis has been a theme stressed by a small but growing number of researchers in their examination of how classes affect the pattern of development and who benefits from technical change, development projects, and foreign aid. A basic question is whether there are significant, almost invisible, class differences within villages. Polly Hill [1968] challenged the economists' view of uniform poverty or what she called the assumption of the amorphous peasantry in her study of a Moslem village in northern Nigeria. She concluded that rural inequality may be the norm in the village she studied and (by implication) throughout northern Nigeria. She contended that inequality was caused by historical and political forces and she speculated that inequality could be a source of change if the large farmers and traders would help pull up the disadvantaged in the village. To be sure, determining the presence or absence of rural inequality requires far more empirical data than provided by Polly Hill's study of one village, but she has encouraged economists to reassess their typical assumption that poverty is uniform and that there is a classless rural society. Other studies on rural inequality include: Post [1972]; Berry [1980]; and Derman and Derman [1973].

French researchers have been in the vanguard in probing class formation and inequality in the transition from subsistence to semisubsistence and commercial production. Foremost is Meillassoux's [1964] classic study of the economic anthropology of the village of Gouro in the Côte d'Ivoire. Raynaut [1973, 1977] stressed the social disruption of villages and family relationships during the transition from subsistence to commercial farming in central Niger. Kohler [1971] is typical of the French researchers who argue that farm and village problems can only be understood as part of the overall rural social structure. He contended that quantitative economic analysis is justified only when farmers are substantially engaged in commercial production.

Studies of class analysis have shed new insights on the process of change in rural Africa. The relationship between class and development projects is explored in Steeves's [1978] study of the Kenya Tea Development Authority and some of the case studies in the volume *Rural Development in Tropical Africa*, edited by Heyer, Roberts, and Williams [1981]. Kitching's *Class and Economic Change in Kenya* [1980] provides a reappraisal of Kenya's economic history from 1905 to 1970. Class domination in Africa is analyzed in relation to class formation and consolidation in Sklar [1976].

Social Impact Analysis. The analysis of the social consequences of technical change and resettlement projects has been explored by many anthropologists, sociologists, and a few economists over the past two decades. Standard references on the social impact of resettlement projects are Chambers [1969], Roider [1971], Colson [1971], and Scudder and Colson [1979]. Since 1956, Scudder and Colson have carried out longitudinal research in Zambia on the resettlement of the 57,000 Gwembe people following the construction of the Kariba Dam in Zambia and Zimbabwe (then Rhodesia).

Pioneering social impact studies in Ethiopia are G. Ellis's [1972] analysis of tractor mechanization in the Ada district and J. M. Cohen's [1975] analysis of the impact of the CADU Rural Development Project on tenant farmers in the early 1970s. Both Ellis and Cohen showed that tenant farmers and landless laborers were displaced by tractor mechanization, a fact that was suppressed by the Ministry of Agriculture. Another pioneering study is the D. S. C. Spencer, May-Parker, and Rose [1976] analysis of the social and economic consequences of moving from hand-pounding to capital-intensive rice processing plants in Sierra Leone. The study revealed that a possible outcome of shifting from hand pounding to totally mechanized rice processing would be the elimination of the equivalent of 40,000 full-time rural jobs held mostly by women.

Unfortunately, many of the social impact studies are impressionistic. For example, Franke and Chasin's [1980] *Seeds of Famine* is the product of a five-month tour of Sahelian countries. Swift [1981] notes that Franke and Chasin give an unduly rosy picture of conditions in the Sahel in precolonial times and fail to provide solid data on important segments (e.g., livestock economy) of the contemporary economic conditions in the region. The authors do not provide hard facts to support their grand hypothesis that the spread of the market economy underdeveloped the Sahel and that recovery programs are laying the seeds of future famines. Boesen and Mohele [1979] analyzed the expansion of smallholder tobacco production in Tanzania for the international market and concluded that smallholders were being exploited by international market forces. The failure of the authors to discuss how changes in relative prices and tax policy affected tobacco and alternative crops such as maize raises serious questions about the validity of their findings. For the social impact analysis to gain credibility, it should be incorporated into the predesign phase of projects (*ex ante* phase) and it must be adequately funded. If these two guidelines are not followed, then "quick and dirty" *ex post* social impact studies will be ignored by planners and policy makers [Collinson, 1982].

Indigenous Knowledge. In the late 1970s, there was an explosion of literature on indigenous knowledge systems, including "traditional" farming practices. Examples of research on indigenous knowledge include A. W. Johnson's [1972] study of experimentation by traditional farmers; Vermeer's [1970, 1979] pioneering studies of experimentation among Tiv farmers in the middle belt of Nigeria; Coward's [1977] analysis of sorghum and millet farmers in Burkina Faso; research on intercropping by Norman [1974], Abalu [1976], and Belshaw [1979]; the IDRC seminar on intercropping [Monyo, Ker, and M. Campbell, eds., 1976];

and the valuable collection of papers, *Indigenous Knowledge Systems* [1980], edited by Brokensha, Warren, and Werner. Indigenous knowledge can be a valuable input into research on farming systems, integrated pest management, soil fertility, and livestock systems [Roling, 1988]. A better understanding of the goals and decision making of African farmers will also help bury pejorative terms such as traditional farming and traditional farmers. Unfortunately, these terms imply passive farmers who would suddenly become modern farmers if they had access to new technology [Cleave, 1977]. But, as we have pointed out, there are many structural barriers and macro policies which are responsible for poverty, stagnation, and the nonadoption of technical packages.

LAND TENURE

Land tenure in sub-Saharan Africa can be characterized as a communal tenure system of public ownership and private use-rights of land. In this system, communities control access to land and individuals appropriate the use of land, control its produce, and have descent rights to the land.[97] The combination of private use-rights and communal control over access to land allows families to have continued use of the same land over time, and the right to buy and sell land and rights to trees (e.g., oil palm and cocoa) through a system of pledging [Adegboye, 1977.]

There is an emerging rural land market in most countries. Individual freehold tenure is well established in the Buganda region of Uganda. A. I. Richards, Sturrock, and Fortt [1973] argued that the individualized tenure system has greatly facilitated the establishment of commercial agriculture in Buganda. Individual tenure is often criticized because it is thought to contribute to increasing farm size, inequality, and the creation of a landless class [Van Hekken and van Velzen, 1972].

Unlike Asia and Latin America, landlord and landless classes are not present in most of Africa because of low population density, abundant land, and colonial policies which prevented white settlers and foreign firms from acquiring titles to land, especially in West Africa. A notable exception was the landlord/tenant system in Ethiopia prior to the 1974 revolution and subsequent redistribution of land. The control over land is a crucial political and economic issue in countries such as Zimbabwe where an estimated 4,200 commercial farmers occupy one-half of the land and produce an estimated 50 percent of the marketed surplus of maize and cotton and 99 percent of the tobacco [Zimbabwe, 1982; Rukuni and Eicher, eds., 1987]. In Zambia an estimated 300 commercial farmers produced about 40 percent of the marketable surplus of maize in 1980. The economic question in countries such as Zimbabwe and Zambia involves the trade-off in transferring land from commercial farmers to small farmers and the potential reduction in the marketable surplus and the increase in employment and political participation among smallholders.

Until recently, most studies of land tenure in sub-Saharan Africa concluded that communal land tenure institutions were flexible and not an immediate constraint on increasing agricultural production. The token priority given to research on land tenure over the past twenty years was justified in our judgment by the large amount of idle land available in almost all countries. But the view that land tenure problems are not a constraint on production is outdated. Land tenure and land use policy issues will be of strategic importance in the 1990s as the frontier phase is exhausted and the intensification of agriculture proceeds.

We recommend that research on land use and land tenure receive greatly increased attention because of growing population; the emergence of land markets; questions of access to land for the landless and small farmers in newly independent countries such as Zimbabwe; the development of river basins such as the OMVS in Senegal, Mauritania, and Mali; tenure issues involved in irrigation schemes [A. Adams, 1977b]; and the movement from nomadic to sedentary livestock production systems. Instead of pursuing separate studies of land use and land tenure, both topics should be studied as integral parts of multidisciplinary research on river basins, irrigation, and livestock schemes. For surveys of land tenure in Africa, see J. M. Cohen [1980b] and Noronha [1987]. Additional references include: Biebuyck, ed. [1963]; Bohannan [1963]; Chambers [1969, 1970]; K. H. Parsons [1971]; Harbeson [1973]; Hoben [1973]; Uchendu [1967]; T. J. Anderson, ed. [1976]; Faye and Niang [1977]; CILSS/Club du Sahel [1978b]; Higgs [1978]; and Riddell, K. H. Parsons, and Kanel [1978].

LABOR USE

Microeconomic research on labor utilization has focused on the organization of work on family farms, including the number of hours worked by members of rural households and the seasonal pattern of labor use. When evaluating the number of hours worked, it has been common to classify labor inputs by sex, age, and season. Many researchers have also differentiated between labor from family members and hired labor. Comparisons of the number of hours worked by members of rural households by sex, age, and seasons have provided a relative measure of utilization or underutilization of labor by region, ecological zone, and farming system.[98] Few studies have provided insights into the adjustments in labor use resulting from the introduction of cash crops and new technologies.[99]

Farm level surveys throughout sub-Saharan Africa have consistently shown that farmers have low annual labor inputs in agricultural production by international standards. Cleave [1974] reviewed fifty micro studies in countries of both high and low man-land ratios and found that male adults were working an average of 1,000 hours per year in agricultural production as compared with 2,500 to

3,000 hours per year in Egypt and many Asian countries. Some researchers have reported even fewer hours spent on farming activities—as low as 500 to 600 hours per year [Haswell, 1953; Norman, 1972]. Farmers in humid areas often work more hours in farming—1,400 to 1,500 hours per year—than those in semiarid areas [Byerlee, 1980]. Also, labor inputs tend to be lower in permanent cultivation farming systems than in systems relying on bush fallow rotation due to reduced requirements for land clearing and preparation. Ruthenberg [1980] points out that perennial crops tend to require higher labor inputs per hectare than rainfed annual crops.

The number of hours worked per day, month, and year differ among African farming systems. Many researchers have reported that farmers work from around 4.5 to 7 hours per day, or about half of the daylight hours [e.g., Massell and R. W. M. Johnson, 1968; Haswell, 1953]. In general, the hours worked per year in farming tend to be lower in arid regions because of the short growing season. The length of the working day has been shown to be affected by the time of the year, the sex of the laborer, the crop, and the cropping operation [Cleave, 1974; Farrington, 1975b]. The number of hours worked per month can reach as high as 200 and 250 hours during peak seasons but in many areas farming activities stop completely during the dry season. The number of days worked per year in farming by adults has generally been found to vary between 150 and 250 days [Cleave, 1974].

The gender division of labor has received much attention, both in terms of total hours contributed to farming activities and the extent of specialization by gender. In nearly every study reviewed by Cleave [1974], there was some gender division of labor by crops and tillage activities performed. In many areas, women grow crops for household consumption while men grow crops for sale. Even where women are primarily responsible for a field, men commonly carry out the heavy tasks such as clearing and preparing a field and plowing. Women, on the other hand, often weed and harvest fields controlled by men. The individual control of fields and implications for household labor use has been examined by Kleene [1976], and Rocheteau [1975]. Several case studies on the gender division of labor are included in Poats, Schmink, and Spring, eds. [1988].

A major debate during the 1980s has been whether males or females work longer hours on farming activities. During the 1970s, the majority of studies showed that males spend more hours on fieldwork, even though females work more hours overall because of additional household activities. In Sierra Leone, for example, women worked about 900 hours per year in farming compared to men who worked an average of 1,450 hours [D. S. C. Spencer, 1976; Byerlee, C. K. Eicher, et al. 1977]. In the Moslem area in northern Nigeria, women contributed less than five percent of the total hours worked per year in farming [Luning, 1967; Matlon, Eponou, et al., 1979a,b]. At the same time, there were studies from other areas which showed that women devote more time to farming than men [Haswell, 1953; Massell and R. W. M. Johnson, 1968].

During the 1980s, as more evidence has accumulated from eastern and southern Africa, it appears that men tend to work longer hours than women in farming systems dominated by hand-hoe field preparation, and of course, in Moslem areas. In eastern and southern Africa, where land preparation has been mechanized in many areas, women generally work substantially more hours than men on both farming and household activities.

At first glance, the figures on the annual hours worked imply that there is a substantial pool of surplus labor in rural areas. But microeconomic research has identified three important factors which largely account for the relatively low annual hours worked in farming during the year: seasonality of labor demand, nonfarm competition for farm labor, and failure to account for invisible labor inputs such as walking to and from fields.

After more than twenty years of farm level research, there is now overwhelming evidence that seasonal labor is a constraint on the expansion of production with hand-hoe technologies [De Wilde et al., 1967; Luning, 1967; Cleave, 1974; Collinson, 1972; D. S. C. Spencer and Byerlee, 1976; Singh, 1976; K. H. Shapiro, 1978; and Upton, 1987]. Cleave [1974] presented several examples in which the availability of family labor during peak agricultural season was identified as a crucial factor determining the level of farm output and income. D. S. C. Spencer and Byerlee [1976] evaluated the magnitude of the seasonal variation in labor inputs in Sierra Leone by computing coefficients of variation for male labor inputs in farming by microclimates. The coefficient of variation for male labor inputs was above 0.50 in the northern region and less than 0.25 in the southern regions receiving more rainfall. Norman, Pryor, and Gibbs reported [1979] that nearly 60 percent of farm labor inputs in their survey in northern Nigeria were concentrated in the four-month peak season. In a recent study of millet and sorghum farmers in Burkina Faso, Richard Swanson, an anthropologist, reports [1981, p. 35] that "the days immediately following the first rains are characterized by an average of over eight to ten hours planting per day per man-unit. The reason planting drops off three to five days after the first big rains is because the soils have already become too dry for proper germination and establishment of seedlings."

Heyer [1971] and Norman [1972] show that the shadow price of labor during peak seasons is at least four times higher than the prevailing wage rate. Several production function analyses have shown that the MVP of labor is large and significant during peak season activities. Moreover, most regressions of the amount of land cultivated show that the number of hours worked in peak season activities can explain most of the variation in output. In general, seasonal labor peaks tend to be more acute in systems of permanent cultivation than in fallow systems, except in cases where irrigation enables crops to be grown in the dry season [Ruthenberg, 1980].

Nonfarm activity is now recognized as a major use of rural labor. Pudsey [1967] and his colleagues in Uganda were among the first to document the large amount of time spent on nonfarm activities. There is some evidence that as much as 50 percent of working time in some seasons of the year may be spent in rural nonfarm activities [Cleave, 1974]. Hoben [1973] reported that in Ethiopia, court hearings, funerals, and other social and religious ceremonies are very time-consuming for rural people. C. M. Elliott et al. [1970] provided evidence that males other than the head of household spend as much as four to five times as many hours on nonfarm activities as they do in farming.

There is evidence that even at times of maximum use of family labor in farming up to three to four hours a day per adult may be spent on nonfarm activities [Liedholm and Chuta, 1976]. Norman [1972], for example, found that farmers spent 31 percent of their working time in off-farm employment during the peak month of farming activities.[100] Norman speculated that because some activities such as trading need to be maintained on a year-round basis the farmers might forego higher short-term earnings in agriculture during peak seasons because they can earn higher annual returns by allocating a portion of their labor throughout the year to trading and other nonfarm activities. Hunter [1967], Zalla, et al. [1981], and others have shown that some farmers are forced to work off the farm when they encounter a cash and food shortage during the hungry season and do not have access to credit to buy enough food to last until harvest time.

In addition to nonfarm employment activities, household maintenance activities are a major constraint on the time available for farming activities. D. C. Baker [1987] and many others have shown that the time required for collecting firewood, fetching water, cooking, and washing far outstrip the time needed for farming activities. Child rearing and compound repair also are major, but often ignored, time-using activities.

Finally, the failure of researchers to include time spent going to and from fields and markets has made it appear that farmers spend less time on farming-related activities than they actually do. Time spent walking to and from fields can account for as much as 20 to 25 percent of the hours worked in farming during the day. The underrepresentation introduced by not including walking time to fields is not as severe in East and Central Africa where in most places there is a dispersed settlement pattern. But even in East Africa, the time spent going to markets may be substantial.

SUPPLY OF LABOR

Substantially less research has been conducted on factors determining the supply and productivity of labor (health, nutrition, education, family size) than on the demand for labor. In general, the agricultural labor supply is determined by the potential stock of labor (family and nonfamily), expected returns to labor and labor opportunity costs, and the physical health and nutritional level of a laborer.

The size of the rural household is the key determinant of the potential stock of labor for agricultural activities. The supply of labor for particular activities also is affected by many agricultural tasks that are traditionally considered to be age- and gender-specific. In East Africa, women are heavily involved in food production, while in West Africa women play an important role in food production and processing, trading, weaving, and other nonfarm activities. While children from ten to fifteen years of age are an important source of farm labor in many parts of Africa, they generally work fewer hours than adults and tend to specialize in tasks such as tending livestock, wood gathering, and bird scaring. Although gender roles affect the potential labor supply, there is growing evidence that gender may not be a major constraint on agricultural production. Women and children, for example, increase their labor inputs severalfold during peak seasons to help overcome labor bottlenecks. Also, gender roles change over time [S. Young, 1977]. But cultural factors influence the supply of agricultural labor in other ways than through gender- and age-specific roles [Mbithi, 1977]. For example, Reyna's study [1977] of the Barma of Chad illustrated how marriage payments can delay the age of marriage of males, having an important effect on the supply of male labor. Copans *et al.* [1972] have described the relationships between religious obligations and work patterns among the Mouride in Senegal.

Hired labor has historically played a minor role in increasing the supply of labor on African farms and in alleviating seasonal labor bottlenecks. Since there are relatively few landless laborers in most countries in Africa, hired labor must be provided by other farmers or migrants from other areas. Migration within ecological areas is limited because small farms often reach their peak demand for labor at about the same time. As a result, farmers in Africa generally purchase less hired labor than in other regions of the third world. Norman [1972], D. S. C. Spencer and Byerlee [1976], and Byerlee [1980] have shown that the percentage of labor inputs supplied by hired labor is usually below 20 percent on small farms. The supply of labor, however, can change dramatically in a short period of time when there are opportunities and incentives. For example, the dramatic increase in wage rates in Nigeria in the late seventies induced a large volume of migration from northern Cameroon to northern Nigeria.

One of the major constraints on the supply of labor to agricultural activities is the possibility of earning higher returns in nonfarm activities. Because the returns to agricultural labor are around ten cents to fifty cents an hour, the value of leisure or other nonfarm activities does not have to be very high before labor is diverted to nonfarm jobs. In one of the few formal tests of the hypothesis that relative returns to alternative employment opportunities determine the allocation of labor between farm and nonfarm activities, Minford and Ohs [1976] used multiple regression analysis to evaluate determinants of the supply of agricultural labor in Malawi. Based on time series data from 1948 to 1968, they found significant (0.05) relationships between the returns to nonfarm employment opportunities and the amount of labor used in farming.

Physical health and the nutritional level of rural people logically must impact on the supply of labor but, with the exception of C. M. Elliott's [1970] study in Zambia and R. M. Brooks, Latham, and Cromptons's [1979] pilot study in Kenya, there has been little research on the impact of health, nutrition, temperature, and humidity on labor productivity. In many countries, there is a "hunger" (*soudure*) period before the harvest and an accompanying significant reduction in caloric intake. Although improved transportation and emergency food programs have reduced the frequency of famine, labor productivity will be low during the hungry season because of the inability of some families to acquire enough calories either through their own food stocks or through off-farm employment.

H. Brandt's [1979] study of peasant work capacity in Africa was designed to test the hypothesis that there is an upper limit to labor inputs in agriculture which is determined by climate, health, and nutritional status. Although the hypothesis could not be directly tested because data were not available, Brandt argues that an annual labor input of 1,100 to 1,200 hours per year in farming represents an upper limit for many farmers unless new production techniques are introduced and/or improvements are forthcoming in the health and nutrition of the workers. For an assessment of the relationships between disease and development, see Hughes and Hunter [1972] and Hunter [1981]; and for an overview of human nutrition in Africa, see Latham [1980] and Alan Berg [1987].

The relationship between education and agricultural change has received surprisingly little attention by researchers in Africa relative to Asia and Latin America. A survey by John Hanson [1980], entitled "Is the School the Enemy of the Farm?", notes that the effect of schooling on the diffusion of agricultural innovations and agricultural change in Africa is hard to measure because of the lack of bench-mark data and the interdependent nature of education, health, age, and access to credit. But Hanson concludes that, on balance, the primary school is a positive force in improving the lives of rural people. One of the few rigorous studies of the effect of schooling on male and female farm managers is P. R. Moock's [1976] study in western Kenya.[101] Moock found that the impact of the number of years of schooling on maize output was greater for female than male-headed farm households, and that women do not seem to benefit as much as men do from extension contact, perhaps due to the almost all male extension system in 1976. P. R. Moock [1981] evaluated the effect of education on the efficiency of input use by a sample of 101 male farmers in western Kenya. Moock found that a farmer who completed four or more years of school can be expected to obtain higher maize yields than farmers without formal education but that extension contact appears to substitute for formal education in the acquisition of knowledge relevant to maize production.

ALLOCATIVE EFFICIENCY

In his seminal book, *Transforming Traditional Agriculture*, T. W. Schultz [1964] contended that in "traditional agriculture,"[102] farmers producing with "age-old techniques" are generally efficient in the use of their resources even if they are poor. He drew on village studies in Asia and Latin America to test the allocative efficiency hypothesis by using a simple test of whether the marginal value product (MVP) of an input derived from a value productivity function for the entire farm is different from its marginal factor cost (MFC) as represented by the market price for that input. Only a few studies have evaluated farmer efficiency by comparing the MVP of each input in different uses. In general, farmers have been found to be efficient in the use of their present resources. Specifically, empirical studies have proven the null hypothesis that farmers allocate resources so that the MVP of inputs equals their MFC [Welsch, 1965; Luning, 1967; Upton, 1967; Massell and R. W. M. Johnson, 1968; Norman, 1972; K. H. Shapiro, 1973; Norman, Fine, *et al.*, 1976; Norman, Beeden, *et al.*, 1976a, b]. The finding that small farmers are efficient but poor has important policy implications because it indicates that additional agricultural output must come through technical change and not through a reallocation of resources.

Several researchers have expressed reservations about the above method to evaluate farmer efficiency and the validity of the conclusions drawn on the basis of those methods. Luning [1967], for example, contended that the MVP of an input should not be expected to equal its MFC in any empirical test using survey data since the estimated average production function actually reflects a hybrid of several production functions, and the regression coefficients usually have large standard errors, making it unlikely that a t-test will indicate that the MVP and the MFC of an input will be shown to be significantly different. In their research on the efficiency of Zimbabwean farmers, Massell and R. W. M. Johnson [1968] argued that efficient resource allocation on average is necessary but not a sufficient condition for efficiency on individual farms. As a result, there may be considerable scope to improve resource allocation on individual farms regardless of findings relative to an entire sample of farmers.

K. H. Shapiro [1973] has challenged some of that empirical evidence supporting the efficient but poor hypothesis. He argued that failure to control Type II errors in tests of allocative efficiency has led to unjustified acceptance of the conclusion that there are not significant differences between the MVP and MFC of inputs.

Ruthenberg [1968] and Wolgin [1975] pointed out, among others, that first-order conditions for profit maximization do not necessarily constitute an appropriate test of whether farmers are efficient resource allocators. Ruthenberg argued that even though farmers in Tanzania appeared to be using too much labor and too little land, their resource allocation pattern could be explained by constraints on access to land for high return cash crops, and by a "heritage of traditional behavior" favoring subsistence production. Wolgin [1975] showed analytically that if farmers are risk averse it should not be expected that they would equate the MVP to MFC, nor would the MVPs of an input be equated in all uses. He argues that two appropriate tests for maximizing expected utility are the ratio of the MVPs of each pair of inputs should be equal in all uses, and rank of the MVPs of an input in different uses should be the same as the rank of the marginal contribution to risk of each use.[103] In an empirical test of his risk averse hypothesis using a sample of Kenyan farmers, Wolgin found that in 86 percent of the cases evaluated, the ratios of the MVPs of two inputs in different uses were equal. Regarding his second test, Wolgin found that in 87 percent of the cases, the ranking of crops on the basis of the MVPs of inputs was the same as their ranking in terms of increment to risk.[104]

Several studies reveal that efficiency issues must be evaluated on a more disaggregated basis than a simple comparison of an average MVP of an input with its MFC throughout the year. For example, Heyer's LP model [1971] showed that the shadow price for early planting labor in Kenya was nearly four times the prevailing wage rate for hired labor and that shadow prices for medium planting and early weeding labor were around twice the wage rate. Norman [1972] found that the shadow price of peak season labor was around four times the local wage rate in northern Nigeria.

The use of an average production function over a sample of farms is questionable in light of the evidence that when crop-specific production functions can be estimated, the MVPs of inputs used on different crops are generally different [Massell and R. W. M Johnson, 1968; Tench, 1975; Wolgin, 1975]; production functions for different villages are generally different and therefore tests of efficiency should not be based on a pooled sample of farmers from many villages [Massell and R. W. M. Johnson, 1968; Ruthenberg, 1968; Tench, 1975]; and assessments of relative technical efficiency among farmers indicate that farmers are often not producing on the same production surface [K. H. Shapiro, 1973; Mijindadi, 1980].

The few attempts in Africa to estimate the potential gains from reallocation of resources have shown that the potential increase in farm income from equating the MVP of all inputs in all uses would generally be under 10 percent. [Massell and R. W. M. Johnson, 1968; K. H. Shapiro, 1973] On the other hand, there is some evidence that the potential gain could be substantial if all farmers could operate on the production function of the most "technically efficient" farmers [Shapiro, 1973; Mijindadi, 1980]. But structural problems such as limited access to high-quality land, inability to carry out cropping operations when desired because cash requirements force low-income farmers to sell their labor to other farmers during peak farming seasons, and market imperfections are largely responsible for the lower productivity achieved by most farmers. It is therefore unlikely that deviations from a frontier production function will ever be eliminated.

The main implications of studies of allocative efficiency for researchers are that future production function studies should be based on disaggregated data, taking into account at a minimum the differences among villages and among crops; give increased attention to the analysis of efficiency under uncertainty; and consider the divergence between purchase and sale prices of outputs and the acquisition and salvage prices of inputs.

The conclusion for policy makers is that although there is scope for individual farmers to increase their output through the reallocation of their resources, technical change is vital for increasing aggregate farm output.

RETURNS TO LAND AND LABOR

During the 1960s and 1970s, many researchers generated data on the returns to land because of the difficulty of collecting labor data throughout the entire agricultural year. More recently, researchers have emphasized comparing returns to labor from different crops or across types of farming systems. In surveys where detailed labor data have not been collected, some researchers have estimated returns per man-day or returns per hour by assuming a standard work day of six to eight hours [Ruthenberg, 1980].

One of the first comparative studies of returns to land and labor for different types of farming systems in Africa was Ruthenberg's [1968] collection of ten case studies in Tanzania.[105] The case studies showed that: the highest average returns to land were realized by tobacco-maize farmers followed by rice-sugar cane farms; farms under permanent cultivation, usually based on bananas, consistently had a higher return to land than did farming systems based on semipermanent cultivation of cotton, millet, or maize; and returns per worker were highest on tobacco-maize farms, followed by banana-coffee farms and then by farmers relying on cotton-millet-maize.

The most comprehensive summary of returns to land and labor in different enterprises and farming systems in sub-Saharan Africa is found in Ruthenberg's collection of representative farm and enterprise budgets in *Farming Systems in the Tropics* [1980]. Ruthenberg selected data from more than forty surveys covering nineteen African countries. Ruthenberg's findings about returns to land and labor in farming systems defined according to the intensity of rotation are as follows:

(1) Output per man-equivalent is likely to decrease but returns per man-hour are usually higher with increasing permanency of croppings.

(2) Returns per hour and per acre are more variable in bush fallow cultivation systems than in shifting cultivation systems.

(3) Returns to land are often higher in permanent cultivation systems than in rotation systems but returns to labor are generally lower.

(4) Marginal returns to labor tend to be lower in permanent cultivation systems than in fallow systems since no unused land is available to absorb additional labor productively.

In terms of different cropping enterprises, Ruthenberg found:

(1) The returns to land and labor were generally higher and more stable from perennial crops than for annual crops.

(2) Several crops with high labor requirements such as sugarcane have higher returns per acre but lower returns per hour than staple food crops such as maize or sorghum.

(3) Crops such as manioc and sweet potatoes tend to have high yields of low quality calories while crops such as millet, sorghum, and maize have lower yields (in terms of calories) but better quality calories.

(4) In forest zones, bananas and plantains return more calories per unit of land with less labor than do root crops.

The most comprehensive data on returns to labor in farming and nonfarming activities for any country in Africa are available from a 1973-74 national survey in Sierra Leone which was carried out under the direction of Dunstan Spencer. The major findings reported in D. S. C. Spencer and Byerlee [1976] and D. S. C. Spencer, Byerlee, and Franzel [1979] were:

(1) In general, the lowest returns were received for annual food crops but the returns to rice varied widely among the five dominant rice production systems.

(2) The returns per man-hour for tree crops were generally from two to four times higher than those of annual crops.

(3) For some cropping enterprises such as onions, peppers, and tomatoes, the returns to land were high but the returns to labor were relatively low because of high labor inputs.

(4) Manioc had relatively low returns to land but because of low labor requirements the returns to labor were higher than annual crops such as groundnuts.

(5) Cocoa had around the same return to land as coffee but much higher returns to labor because of its lower labor requirements.

(6) Nonfarm enterprises generally had higher returns to labor than did either annual or perennial crops.

In the semihumid and forest zones in Nigeria, K. L. Robinson [1974] found that returns to labor from tree crops such as cocoa and oil palm were around two to three times higher than those for annual crops, including rice, maize, and cowpeas; and the returns to labor from yams and manioc (cassava) were slightly higher than those for most cereal crops. In contrast, Boateng, Ratchford, and Blase [1987] showed that food crop production was more profitable than cocoa, and this is why farmers devote resources to food. Olayemi [1974] showed that the net returns per acre were generally similar for rice, cocoyams, yams, and cocoa in western Nigeria.

Several estimates of returns to land and labor are available for the semiarid region of West Africa. Matlon, Eponou, *et al.* [1979a,b] and Norman, Pryor, and Gibbs [1979] demonstrated that returns per acre and per hour were higher for crops grown in mixtures than sole stands. On the basis of a survey of 480 farms in the Eastern Region of Burkina Faso, Lassiter [1981] found that the returns to rice per acre were around 150 percent higher than those for maize, the crop with the second highest returns. Returns per acre for sorghum and millet were around one-third of those for rice even though sorghum and millet accounted for 80 percent of the area cultivated. Both cotton and groundnuts, the traditional cash crops in the area, had very low returns per acre. Based on Botswana data from a season with less than 350 mm of rain, D. C. Baker [1987] showed that returns to cowpeas were higher than those for sorghum, particularly when the value of leaves was considered. Maize failed completely, as it often does in Botswana. The highest returns were for groundnuts and jugo beans (bambara nuts), despite high labor requirements, because the local market prices increased so much due to the drought.

Information on returns to labor and land for rice can be found in Winch [1976], Fotzo and Winch [1978], and Lang [1979]. Fotzo and Winch examined rice production systems in northwest Cameroon and found that the returns to labor were higher in the traditional system after government subsidies had been removed from mechanized systems. Lang compared the returns to land and labor in several systems of rice production in West Africa and found a wide variation in the returns to upland rice; the return per man-day was nearly nine times higher in parts of western Nigeria than in northern Sierra Leone.

LAND INTENSIFICATION

Land use patterns in Africa have changed dramatically over the past thirty years [Noronha, 1987; Feder and Noronha, 1987]. Shifting cultivation systems have largely given way to semipermanent and permanent cultivation systems and an intensification of farming. Numerous factors have stimulated intensification, including increasing population pressure; growth of rural and urban purchasing power; government policies and programs such as settlement schemes; changing price relationships among crops, particularly the relative prices of export crops and food crops; and improved production technologies [Netting, 1974; Ruthenberg, 1980].

Population pressure has been widely acknowledged to be the most important explanation of agricultural intensification.[106] While some countries still have substantial amounts of idle land, there are large variations between and within countries. Several countries have reached and exceeded the point where the traditional land-extensive systems are able to meet the food requirements of the local population and maintain environmental stability [Hunter and Ntiri, 1978; Binswanger and Pingali, 1988].

Historically, land intensification has taken place in isolated pockets of high population density. Studies by Netting [1968] of the Kofyar in Nigeria and by Boulet [1975] in the Mandara Mountains, Cameroon, reveal that hill and mountain people can survive on a poor natural resource base. The two most important adjustments noted were to increase the use of animal manure, and develop concentric cropping zones in which nearby fields are worked intensively and are heavily manured while more distant fields receive little attention. Concentric land use patterns have also been noted in heavily populated plains such as the peanut basin in Senegal and in parts of Nigeria [Goddard, 1972; Lagemann, 1977; Ruthenberg, 1980]. One of the most dramatic examples of response to land shortage is the case of the Wakara who have lived for more than a century on a small island in Lake Victoria [Ruthenberg, 1980].

Due to rapid rates of population growth throughout Africa, land intensification is no longer being viewed as an isolated and exceptional event. Public policy is focused on land intensification in countries such as Rwanda, Senegal, and in Kenya where a 4.0 percent rate of growth of population will likely double the size of the population in eighteen years [Shah and Willekens, 1978; Toksoz, 1981].

The link between population growth, land use, and the structure of agricultural production has been studied, particularly by researchers in Francophone Africa [e.g., Tourte, 1974; Delpechi and Gastellu, 1974; Marchal, 1977; ORSTOM, 1979]. Research shows that intensification is accompanied by a reduction in the length of fallow and farm size and an increase in labor inputs per unit of land. For example, Faye and Niang [1977] and Hunter and Ntiri [1978] found that the major difference between sparsely settled and high density settlement areas is that farms in any given village area tend to become smaller and fields controlled by households become more widely scattered as density increases. M. J. Mortimore [1967] observed the following changes with increasing population density in the Kano close-settled zone in northern Nigeria: disappearance of fallow and more individualized tenure, rising value of land, increasing fragmentation, more use of fertilizer, and an increase in off-farm employment.

Various ethnic groups have shown a remarkable ability to adapt to growing population pressure. Faulkingham and Thorbahn [1975] conducted field research in Tudu, a large village in south central Niger in 1974-75 and concluded that the local population has adapted their system of consumption and production to maximize the chance of survival in the face of periodic environmental degradation. Nukunya's [1975] case study in southeastern Ghana shows that local farmers have become relatively wealthy despite high population density and poor soil because they have modified their agricultural production technologies to the requirements of the area. Villages have often been able to adjust to increasing pressure on land by seasonal migration and the formation of subvillages [Faulkingham, 1977].

Although the evidence is overwhelming that African farmers have adjusted their cultivation techniques in the face of reduced fallow and falling yields, there is considerable controversy over the ability of food production to keep up with population growth. Boserup [1965] hypothesizes that population growth is a major factor stimulating the adoption of improved agricultural practices. Although Boserup's model has been generally rejected as a model of agricultural development,[107] some support for the Boserup hypothesis is reported by Datoo [1973] in his study of 216 households in a mountain region of Tanzania. Datoo found that farmers in high density areas were using significantly more improved production techniques than those in areas of medium and low density. Moreover, Ruthenberg [1980] found that in areas of high population density there was a tendency toward increased use of high yielding varieties which produce a greater proportion of edible dry-matter content; reliance on crops which result in higher output with higher inputs—such as shifting from millet to maize and from grains to root crops; and increasing the length of the growing season through multiple cropping.

Lagemann's [1977] study in eastern Nigeria presents some of the rare quantitative data on the impact of increasing population density on land use and soil fertility. Lagemann studied three villages in high population density areas of eastern Nigeria and found that physical measures of soil fertility (organic carbon and nitrogen) declined as population density increased and the length of fallow was reduced. More dramatic was the reduction in crop yields as the years in fallow declined. Cassava yields fell dramatically from 10.8 tons to 2.0 tons per hectare as the length of fallow was reduced from 5.3 to 1.4 years. Moreover, the length of fallow was found to explain 60 percent of the variation in cassava yields. In addition to a reduction in soil fertility, there was an increase in soil erosion, acidity, and weeds as the length of fallow was reduced. Although Lagemann found many examples of farmers introducing changes in farming practices as population pressure increased—intensification of production on compound fields, using mulch, and increasing dependence on off-farm employment for income generation—the evidence is clear that population pressure was running far ahead of technical change in the three villages studied by Lagemann in eastern Nigeria. In Zaire, Fresco [1986] documents the reductions in fallow period as population density increases.

Stimulated by Boserup's hypothesis of population-led agricultural development, Pingali, Bigot, and Binswanger [1987] have recently completed a major comparative study of *Agricultural Mechanization and the Evolution of Farming Systems in Sub-Saharan Africa*. The implications of their study for research priorities are reported in "Technological Priorities for Farming in Sub-Saharan Africa" [Binswanger and Pingali, 1988].

5. Synthesis

There is a rich, but largely unacknowledged, heritage of research on smallholders. Prior to 1960 most research focused on communities but paid little attention to technical practices or the economics of production. During the mid-1960s, there was a shift in orientation toward the application of production economics concepts to African circumstances and on corresponding interest in proper data collection procedures for support of economic analyses.

The decade between the mid-1960s and the mid-1970s was the heyday of multiple-visit farm level surveys and the economics of smallholder farming. Regression analysis was used to investigate output and labor supply response, and allocative efficiency. Mathematical programming focused on relationships between observed and optimum enterprise combinations, primarily as a means of assessing decision criteria and resource constraints.

During the late 1970s, researchers working on African smallholder farming systems faced an identity crisis. Progress had been made in field research methods, but the standard multiple-visit survey methods were increasingly being challenged as costly and inefficient [Collinson, 1981]. Data analysis procedures lagged behind those being used in the North America and Europe, largely due to data limitations, and African applications of production function and mathematical programming analyses were being questioned [K. H. Shapiro and Muller, 1977; Upton, 1979; Collinson, 1981].

By 1980, the multiple-visit survey was replaced by informal, single-visit survey methods and the farming systems approach. Smallholder research in the 1980s has been dominated by descriptive studies of household resources and farming practices. The primary source of information has been single-visit surveys, generally carried out in the context of farming systems research programs. This has been reflected in a paucity of citations on African smallholder research in either the *American Journal of Agricultural Economics* or the *Journal of Agricultural Economics*. Less than ten articles on African smallholders appeared in both journals combined between 1982 and 1988. The few articles that did appear generally dealt with issues such as price supply response which were well outside the mainstream of smallholder research [e.g., D. O. A. Phillips and Abalu, 1987; Ngambeki and Idachaba, 1985]. In contrast, there has been a burst of studies on African smallholder farming systems in journals such as *Agricultural Systems* [e.g., Arua and Obidiegwu, 1988].

The dominant form of economic analysis during the 1980s has been budgeting, primarily to assess enterprise profitability [Boateng, Ratchford, and Blase, 1987; Upton, 1985] or determine the benefits from specific changes in production practices [Abalu and Etuk, 1986; Joseph, 1987]. Few results have been published in the leading economics journals but there have been innumerable monographs published by research departments in nearly all African countries. These data form a rich, but as yet unexploited, basis for comparative analysis.

Regression analysis and mathematical programming have been used selectively. Regression models have been used to assess determinants of production outcomes, with an emphasis on technical practices rather than levels of resource use. There have been few production function studies, either enterprise-specific or farm-level functions. The most notable use of mathematical programming has been a few attempts to project the whole farm consequences of technology adoption [e.g., Adesina, Abbott, and Sanders, 1988]. Anthropologists have reasserted their role in smallholder research, accompanied by calls for increased attention to the social and cultural factors affecting household resources and farm decision making. Research on gender is increasing in importance. Professor Jean Due of the University of Illinois estimates that, on an average, 30 percent of the rural households in Africa are headed by women. This is a preliminary figure and it varies widely by country and subregion.

In the 1990s, it can be expected that smallholder research will no longer be equatable with farm management research because of the growing awareness that the welfare of smallholders is as much determined by markets, credit, input supply and international price movements as it is by the quality of production technologies. Also, future studies are likely to have a policy orientation and an emphasis on family food security rather than an emphasis on farmer decision-making about crop and livestock production.

Chapter V. Technical Change

Technical scientists have focused on helping farmers through research on improved varieties and agronomic practices, including spacing, timing of planting, weeding, and the application of fertilizer, herbicides, and pesticides; and research on mechanical technology, including hand tools, animal traction, and tractor mechanization. Most agronomic research has been carried out on experiment stations and has focused on increasing yields, yield stability, and insect and disease resistance. Research on animal traction, tractor mechanization, and selective mechanization of particular tasks has been dominated by two groups of researchers over the past twenty-five years: engineers and economists. The engineers have concentrated on how mechanization influences variables such as yields, acreage, timeliness, and cropping intensity [Kline, Green, et al., 1969; Giles, 1975]. Economists have focused on the financial and economic profitability of alternative types of mechanization and the employment and income distribution consequences. [Gemmill and C. K. Eicher, 1973; Binswanger, 1978; Binswanger and Pingali, 1988].

Although research on plant breeding, agronomic practices, and mechanization has been extensive, African agriculture is less mechanized and has been less affected by new technologies than other areas of the world. There has been a long history of research recommendations being rejected by farmers and endless debates about the need to reorganize national research systems. Interest in why technologies were or were not being adopted stimulated social science research on the diffusion of innovations beginning in the 1960s. Since the mid-1970s, there has been growing interest in irrigation and in farming systems research to complement commodity research programs. The following topics will be reviewed: evolution of approaches to technical research, agronomic research, irrigation, mechanization, agricultural extension, and farming systems research (FSR).

1. Historical Perspective

As a broad generalization, from the beginning of the colonial period in the 1880s until the 1920s and 1930s, agricultural growth was based on exploiting Africa's natural resources and unskilled labor [C. K.Eicher, 1967]. Since the 1920s, there have been seven major turning points or shifts in research strategies in Africa. These turning points represent a progressive movement from a natural resource base to a science-based strategy of agricultural research development.[108]

The first turning point occurred in the 1920s when national research stations were established in many of the then colonial territories of Africa. During the colonial period, the focus of agricultural research was on expanding the production of export crops such as cotton, groundnuts, oil palm, and cocoa.[109] Colonial governments established global research networks in order to increase the productivity of research on export crops. For example, the Empire Cotton Growing Corporation was launched in British colonies in 1921. Anthony et al. [1979, p. 252] point out that "this organization was able to recruit agricultural scientists of high caliber by offering career opportunities that were not dependent on the research programs within a single colony. Cotton research teams were assigned by the Corporation to experiment stations in Kenya, Malawi, Nigeria, Sudan, Swaziland, Tanzania, Uganda, and Zambia." The country research teams were linked to the Imperial Research Institute in Trinidad. The French also established a number of global networks beginning in 1921, followed by a Belgian network for their colonies—Belgian Congo (now Zaire), Rwanda, and Burundi. These networks were a forerunner to the IARCs beginning with IRRI in the Philippines in 1960.

The second turning point came in the 1950s with the introduction of regional research stations serving several countries in a common ecological zone. For example, the British established a system of specialized research institutes for its four English-speaking colonies in West Africa. These included the West African Institute for Social and Economic Research (WAISER) in Ibadan, the West African Rice Research Station in Sierra Leone, the West African Cocoa Research Institute in Ghana, and the West African Oil Palm Research Institute in Nigeria [C. K. Eicher, 1970].

The performance of the global and regional research institutes in Africa over the 1920-60 period was mixed, partly because many did not have a critical mass of scientific talent, a few were placed in poor locations, and harsh taxation policies in some countries dampened economic incentives to adopt new technology. But some research institutes were highly productive such as the oil palm research institutes which were set up in the Belgian Congo in 1926, in Nigeria in 1939, and in three French-speaking colonies in West Africa beginning in 1947. These institutes were linked with oil palm institutes in Malaysia. Priority in the oil palm research network was directed to crossing African and Asian oil palms to produce highly productive hybrid oil palm varieties. Hybrid oil palms were introduced on plantations and small farms in Nigeria in the early 1960s. Although the hybrid varieties outyielded "wild" oil palms by several hundred percent, Glenn L. Johnson [1968] pointed out that the adoption of this genetic breakthrough was held back by the harsh taxes of the Oil Palm Marketing Board in the 1960s which extracted one-third to one-half of potential smallholder revenue from oil palm produce.

The third turning point in agricultural research came in the independence era of the early 1960s when some of the regional institutes were allowed to atrophy (e.g., the West African Rice Research Institute in Sierra Leone), while others were incorporated into national research systems. For example, the West African Cocoa Research Institute in Ghana became the Cocoa Research Institute of Ghana, while Nigeria converted the West African Oil Palm Research into the Nigerian Institute for Oil Palm Research. In the mid-1970s, the well established East African Agricultural Research Organization with substations in Kenya, Uganda, and Tanzania was allowed to languish and finally dissolve with the breakup of the East African Community in 1978.

The fourth turning point came in the mid-1960s with decisions to reactivate the colonial concept of a regional institute to serve a region such as West Africa. For example, the former West African Rice Research Institute with its headquarters in Sierra Leone was reactivated in 1970 as WARDA, and its headquarters was moved to Monrovia. In 1988 WARDA was reorganized and its headquarters was moved to Bouake in the Côte d'Ivoire. The mission of WARDA is to assist its sixteen member countries in carrying out research on rice improvement geared to the specific agroecologies of West Africa [WARDA, 1988]. Several IARCs were established in Africa in the 1960s starting with the IITA near Ibadan, Nigeria, in 1969, followed by ILCA in Ethiopia in 1973, and the ILRAD in Nairobi.

The fifth turning point came in the mid-1970s in response to the Sahelian drought of the late 1960s and early 1970s and rising food imports. Crash research programs were launched to discover how to expand food crop production. Some of the French regional research systems (e.g., IRAT stations) were converted into national research systems.[110] IRAT is doing research on sorghum, millet, maize, rice, tubers, and vegetables. ICRISAT started a cooperative program in Burkina Faso in 1975, followed by a decision in 1981 to set up a major ICRISAT Sahelian Center outside Niamey, Niger.

The sixth turning point came in the late 1970s and early 1980s when farming systems research projects or programs were launched in most African countries. The views of farmers and extension agents were given greater weight in research, and the research shifted from development of new farming systems to incremental changes in existing varieties and practices. The former domination of agricultural research by on-station technical experimentation gave way to multidisciplinary on-farm research. However, farming systems research started to decline in Africa in the mid-1980s.

The seventh turning point will take place in the coming decades with the closing of the frontier and the intensification of agricultural production through irrigation and a movement to double and triple cropping. The intensification phase will require large investments in agricultural research and training because research on a number of problem areas, such as irrigation, was almost nonexistent until the 1970s. Also, intensification increases insect and disease problems and can lead to micronutrient problems. This discussion of turning points of agricultural research points out the gradual increase in the role of science and technical change in African agriculture.

A major lesson to be learned from the history of agricultural research in sub-Saharan Africa is that long-term (twenty-five to fifty-year) investments will be necessary to develop effective *and* sustainable national agricultural research services [Eicher, 1988c]. Francophone countries are generally one to two decades behind Anglophone countries in the Africanization of their national agricultural research systems.

2. Agronomic Research

The number of published technical reports on agronomic research is staggering relative to published economic studies. But it is difficult to evaluate technical research because much of it is highly location-specific and much of it has never been synthesized and put into farmer recommendations.[111] As a result, when one asks a basic question such as should governments subsidize fertilizer for food crops, one is forced to go from experiment station to experiment station throughout Africa in order to review annual reports and assemble unpublished input/output data on fertilizer response in fertilizer trials. Attempts to synthesize agronomic research results can be found in Leakey, ed. [1970] and Acland [1971] for East Africa, and Irvine [1969] and Kassam [1976] for West Africa. Leakey and Wills, eds. [1977] is a valuable reference for food crops. Important journals reporting results of agronomic research are *Experimental Agriculture*, *Tropical Agriculture*, *Tropical Science*, and *L'Agronomie Tropicale*.

This review of agronomic research covers: crop improvement programs; soil fertility and fertilizer; and research on selected managerial practices.

CROP IMPROVEMENT

Research on crop improvement forms the cornerstone of agronomic research programs. The main approaches are to identify, screen, test, and multiply better local varieties and promising new varieties from throughout the world, coupled with selective breeding for desired characteristics. Progress has been uneven by ecological zone and by crops—sorghum, millet, maize, rice, wheat and triticale, cotton, groundnuts, cowpeas, cassava, and yams.

Sorghum and Millet. Research on sorghum and millet is of central importance in the semiarid regions of Africa because these are the dominant staple foods.[112] Millet and sorghum are often grown in mixtures. Millet and cowpea mixtures are common in low rainfall areas and millet and sorghum are often grown together in higher rainfall areas.

Local sorghum varieties throughout sub-Saharan Africa are generally photosensitive,[113] long season (120 to 140 days), reasonably tolerant to striga,[114] and up to five meters in height. The photosensitive quality is desired because it enables local sorghums to escape head mold and provide grain of good quality even though planting may be spread over several weeks. Varieties with long stalks are favored by farmers since sorghum generally has multiple uses: the grain for family consumption, while the stalk is used for housing and animal fodder during the dry season. But most long-stalk, local varieties have low yield potential [Andrews, 1975; Arrivets, 1976] and day-length sensitivity has restricted the adaptation of local varieties to different latitudes. Also, sorghum diseases are a major problem in Africa and throughout the world. For an excellent survey of sorghum diseases, see ICRISAT [1980b].

Sorghum selection and breeding programs have been active in Africa for over forty years. Major breeding objectives have focused on: shorter stalks (around two meters); increased resistance to disease, pests, and drought; hard grains with good storing and eating qualities; and higher and more stable yields. Yields of unimproved sorghum varieties are estimated to average only 600 to 700 kg/ha under traditional management practices [Etasse, 1977]. Yields several times higher have been achieved on experiment stations using improved husbandry practices. In Nigeria, for example, long-season improved local varieties have yielded from 2,000 to 4,000 kg [Norman, Beeden, et al., 1976b]. However, in countries such as Botswana, where rainfall is extremely limited, on-station yields of the best varieties often are only one ton, while on-farm yields average less than 300 kg.

Two categories of millets are grown in Africa, a short-season (75-100 days) nonphotoperiodic millet and a long season (120-180 days) photoperiodic millet. Short-season millets are generally grown as a "hungry season" crop since they can be harvested a month or more before sorghum is ready. Long-season millets have desirable taste and good storage qualities. Average yields of millets have been estimated to be 580 kg/ha in West Africa, ranging from a low of 290 in Mauritania to 690 kg in Mali [Kassam, 1976]. In East Africa, yields of finger millet range from 450 to 900 kg/ha [Acland, 1971]. Yields of local millet can reach 1,200 to 1,650 kg/ha under improved practices. Less progress has been made on improved millet varieties than on sorghum [ICRISAT, 1987; Matlon, 1987; Spencer and Sivakumar, 1987].

Major problems have been encountered in importing high-yielding sorghum and millet varieties from other continents. While some hybrid varieties have achieved high yields on experiment stations, many have not performed well under farm conditions [Herdt, 1987; C. K. Eicher, 1984, 1986; D. S. C. Spencer, 1986]. For example, if farmers plant short-season hybrid sorghum varieties early in the rainy season, as they do with local varieties, the hybrid varieties are often afflicted with head mold because the grain matures before the end of the rainy season. But if farmers delay planting short-season varieties, less time is available for planting cash crops and there is the possibility of sorghum crop failure if the rains end early. Also, many hybrid varieties have had low germination rates and are susceptible to striga, the root parasitic, and to pests such as army worms. Moreover, since farmers often plant several varieties geared to different soils on their scattered plots, a single hybrid variety will not meet all the needs of small farmers.

ICRISAT's cooperative research program in West Africa illustrates the problems in introducing sorghum varieties from other continents. When ICRISAT initiated its program in 1975, it was hoped that rapid improvements could be made by the direct import of hybrid sorghum varieties from India followed by a few years of local adaptive trials. But the hybrid sorghum varieties from India did not perform well in trials in Burkina Faso, Niger, and Mali [Eicher, 1984]. In

1981 ICRISAT decided to de-emphasize the *direct* transfer of sorghum varieties from India to the Sahel and to establish a major Sahelian research station in Niger and to develop a ten to twenty year research program for millet and sorghum.

The Director-General of ICRISAT recently summarized ICRISAT's experience in importing improved sorghum and millet varieties from India to the Sahel in the mid-1970s: "The results were generally disappointing. Very little of the introduced material was adapted to West African conditions. It was obvious that the project had to go back to basics to plan a longer term program" [Swindale, 1984, p. 77].

Maize. Maize is the most promising crop in many temperate regions in Africa. Research has developed improved open pollinated varieties, synthetics (composites), and hybrids [Gelaw, ed., 1986; Blackie, 1989]. Maize breeding was launched in Kenya and Zimbabwe[115] in the early 1930s, primarily for the benefit of European farmers. Heyer and Waweru [1976, p. 203] report that in Kenya, "The development of hybrids for the high rainfall areas, and synthetics (composites) for the medium and low rainfall areas, has been extremely successful and widely applied." Since maize seed is divisible (can be sold in small amounts), improved seed has been made available to both large and small farms in Kenya. Gerhart [1975] reported that in district variety trials in Kenya, hybrid seed increased yields 30 to 80 percent depending mainly on altitude. In Zimbabwe, research on hybrid maize was started in 1932 and in 1960, twenty-eight years later, a high yielding hybrid (SR-52) was released that revolutionized maize growing in Southern Africa. Subsequently, shorter season hybrids were developed for lower rainfall areas. Eicher [1984], Collinson [1982], CIMMYT [1977a, 1978], and Rohrbach [1988] have shown that improved agronomic practices are central to improving farm-level maize yields.[116]

In West Africa, farm-level maize yields have averaged around 800 kg/ha with a low of 500 kg in Benin and a high of 1,100 in Ghana [Kassam, 1976]. In an economic analysis of improved maize in West Africa using a controlled experiment on farmers' fields, Norman, Beeden, *et al.* [1976a] showed that improved maize varieties yielded several times more grain per hectare than improved sorghum and millet. Maize yields of more than 5,000 kg/ha were obtained using oxen cultivation and yields were relatively stable in low rainfall conditions. Norman, Beeden, *et al.* concluded that maize has great potential in the Sudanian zone of West Africa. Flinn and Lagemann [1980] analyzed farm use of a recommended maize package in Imo State, southern Nigeria, and found that the proposed package was "too risky and not sufficiently superior to present practices" to be attractive to farmers.[117]

Rice. Although research on improved rice varieties has been under way for over fifty years, average farm-level yields of rice are only around 1,000 to 1,200 kg in swamps (bas fonds) and flood plains. Yields of upland rice are around 500 to 800 kg [Chabrolin, 1977] and yields of paddy average only 500 to 600 kg in East Africa [Acland, 1971].

Improved cultivars (cultivated varieties) have demonstrated high yield potential in the range of three to six tons for upland rice and ten tons for paddy. A major problem in the development of improved rice varieties is that the life of an individual variety is often limited because of pests and diseases. For example, new strains of blast have emerged as rapidly as breeders have developed varieties resistant to the old strains. While cultivars with a high degree of disease resistance have been developed, these have had relatively low yield ceilings and poor grain quality. For overviews of rice in West Africa, summaries of rice cropping systems, and genetic improvement programs, see Buddenhagen and Persley, eds. [1978], D. S. C. Spencer and Nyateng [1987], and WARDA [1988].

Wheat and Triticale. Wheat is a minor crop in Africa but it is of intense interest to policy makers as urbanization proceeds and growing imports of wheat and wheat flour are needed to meet the rapid shift in consumer preferences to bread. For example, about 75 percent of the wheat consumed in Africa in 1988 is imported. In West Africa 90 to 95 percent of the wheat consumed is imported because heat tolerant wheat varieties are not available for West African conditions and the profitability of producing wheat is low. Satisfactory wheat yields are being achieved in the highlands of Ethiopia and Kenya but CIMMYT reports that wheat yields in the highlands of East Africa have lagged behind the yield breakthroughs achieved in more favorable areas such as the irrigated plains of India and Mexico [CIMMYT, 1981, p. 50].[118] The highland environment is conducive to disease; and stem and stripe rust are common yield constraints. Although wheat has been promoted for many years in West Africa, the profitability of producing wheat is low because of many technical problems. Most of the government wheat schemes in West Africa require heavy subsidies [Byerlee and Longmire, 1986]. For a study of wheat production potential in Africa, see Choudhri [1987].

Although research on wheat is increasing in Africa, there have been few farm studies of the economics of wheat production. At present, wheat cultivation is profitable in the cool highlands of Ethiopia, Kenya, Tanzania, and in Zimbabwe under irrigation. Large-scale irrigated wheat production in northern Nigeria has required large subsidies. But there are no solid studies on the economics of farm level wheat production in West Africa. We are of the opinion that maize is more promising than wheat in Africa and that there should be a large increase in technical research expenditures on maize relative to wheat.

Triticale is a relatively new crop which was developed by crossing wheat and rye. Triticale has a higher protein content than wheat and it is used mainly for livestock feed in high-income countries.[119] Triticale is being used for human consumption in East Africa as a blend with wheat flour to make bread. Triticale grows well on acid soils in Kenya and Tanzania and it has had few disease problems in East Africa to date. Triticale outperforms wheat in marginal environments such as the drier tropical highlands and acid soils.

Cotton. Cotton was introduced in East Africa from the U.S. around the turn of the century and much work has been done on it both genetically and through agronomic research. Breeding and selection initially focused on jassid resistance and then turned to bacterial blight. More recently research has focused on yields of seed cotton, increased length and strength of fiber, and higher ginning percentages (the ratio of lint to seed cotton).

But there is a gap between yields on farmers' fields and yields in experiment station trials [Arnold, ed., 1976]. In East Africa, yields average from 220 to 450 kilos of seed cotton on small farms. A similar range of yields with indigenous practices has been reported in West Africa [Kassam, 1976] and in Nigeria by Norman, Hayward, and Hallam [1974, 1975]. Indigenous practices usually entail little or no insecticide or fertilizer and late (July) planting. Improved cotton packages are yielding 800 to 900 kilos of seed cotton per hectare in Nigeria while yields on experiment stations are in the range of 1,200 to 1,300 kilos [Acland, 1971]. Because cotton yields are often extremely low without complementary inputs of spraying and fertilization, many countries have established parastatal organizations to promote cotton through vertically integrated programs (input supply, controlled management, marketing).

The problems in introducing improved cotton varieties under farm conditions have been illustrated in research in Nigeria and Malawi. Based on multiyear trials in northern Nigeria, Norman, Hayward, and Hallam [1975] found that sole-cropped cotton was marginally profitable in the first year of trials but was clearly unprofitable in subsequent trials when rainfall was less than average. Also, yields were found to be highly sensitive to the selection of a spraying regime. An important finding by Norman, Hayward, and Hallam was that farmers planted cotton later than recommended dates, despite reduction in cotton yields, in order to insure that sufficient millet and sorghum could be grown to meet family food needs. Farrington [1977b] addressed the issue of how many times to spray cotton on small farms in Malawi. Although results of spraying trials conducted at experiment stations on improved varieties often show that as many as twelve sprays of insecticide should be used to maximize yields [e.g., Davies, 1976], small farmers applied insecticides at a much lower rate.

Legumes. Research on grain legumes has concentrated on groundnuts and secondarily on cowpeas. Groundnuts are the major cash crop in the West African savanna, reaching their greatest concentration in the northern Guinea and Sudanian zones. In West Africa, average yields are around 710 kg/ha or about 500 kg/ha of kernel [Kassam, 1976]. With improved management, including pest control, ex-

perimental yields with improved cultivars range from three to 3.5 tons per hectare and have exceeded five tons in the Guinea zone. Two of the major breeding
goals for groundnuts are to increase the oil content and the protein content
[Rachie and Silvestre, 1977]. Additional breeding criteria have included seed dor-
mancy to increase the flexibility for leaving groundnuts in the ground, shorter
season, and resistance to rosette virus.

Yields of cowpeas in sub-Saharan Africa are extremely low. Average yields are
estimated to be around 250 kilos per hectare in West Africa and slightly higher,
350–450 kg/ha, in East Africa [Acland, 1971; Kassam, 1976]. But on-farm yields
are often as low as 100 kilos of dry seed [Rachie and Silvestre, 1977]. Some major
genetic factors limiting cowpea yields are lodging and limited resistance to pests
and diseases. In West Africa, improved cultivars have yielded 1.5 to 2.5 tons in
the Sudanian zone under experimental conditions and over three tons in the
Guinea zone. For these high yields to be realized, improved varieties have been
sole-cropped at high population densities with chemical pest control. In East Africa, with improved management but without insecticides, yields have reached
2,200 kg/ha.

The major objectives of cowpea improvement programs have been to develop
lines that combine stable yields with resistance to pests, disease, and drought
stress. Cowpeas are being tested both as an intercrop and under sole cropping.
IITA at Ibadan has assumed worldwide leadership for research on cowpeas. Insect
damage is a major constraint on sole-cropped cowpea production and pest control accounts for most of the increases reported in experiments relative to local
varieties grown with traditional practices [Hays and Raheja, 1977; Ejiga, 1977;
Bingen, Hall and Ndoye, 1988].

Root and Tuber Crops. Tropical root crops—especially cassava—have been neglected relative to research on cereal crops [Terry, Oduro, and Caveness, eds.,
1981]. As a result, root crops are still largely unimproved and should have a large
potential for future improvement [Coursey and Booth, 1977]. Root crops already
are at substantial advantage relative to cereal grains in forest zones where, for example, yams can produce more protein per unit of area and cassava more energy
per unit of labor than rice.

Cassava (manioc) was initially grown in the forest areas of West Africa but
since the turn of this century it has been slowly moving into the northern savanna
zones [W. O. Jones, 1959]. In East Africa, cassava is widely grown in areas below
1,500 meters. Local varieties of cassava have low yield potential even under improved practices [Coursey and Booth, 1977]. Estimates of cassava yields in Africa
range from 6.7 to ten tons of fresh tubers. The most common estimate of seven
tons is one-half the average of fourteen tons/ha achieved in Latin America [On-

wueme, 1978]. For an in-depth study of cassava production in Zaire, see Fresco [1986].

The major breeding objectives have been resistance to pests and disease (particularly mosaic disease and bacterial blight), increased yield potential, higher starch and protein content, lower fiber, and trying to combine early maturing with good storing properties. Major strides have been made in increasing the yield potential of cassava in Africa. Selected varieties have yielded twenty to forty tons per hectare of fresh tubers under experimental conditions [Coursey and Booth, 1977].

IITA is the leading center for yam breeding and selection. The main breeding objectives are to increase yields, tolerance to disease, and protein content; to develop semierect plants so they will not need staking; and to shorten the growing season. Research at the IITA has indicated yield potential of selected varieties of thirty to fifty tons/ha in favorable climate and yields have exceeded sixty tons/ha [Kassam, 1976]. A major constraint on yam improvement is that many species of yams have been propagated vegetatively for such a long time that they flower irregularly and have reduced ability for sexual reproduction [Onwueme, 1978]. See Bachmann and Winch [1979] and Diehl and Winch [1979] for descriptions of yam-based farming systems in Nigeria; Lawani and Odubanjo [1976] for a bibliography on yams; and Onwueme [1978] and Terry, Oduro, and Caveness, eds. [1981] for research strategies for root crops.

SOIL FERTILITY AND FERTILIZER

Soil Resources. Soil deficiencies are a major constraint on African agriculture [Ahn, 1977]. Soils in the savanna are extremely weathered, the chemical status is poor due to deficiencies in phosphate and organic nitrogen, and the amounts of phosphorus and sulphur mineralized annually are often below requirements of high crop yields. Soil deficiencies often are attributed to poor parent material. The nutrient content of soils also has been greatly reduced over the years through leaching during heavy rains and because nitrogen, organic material, and other elements such as sulphur have been lost through the traditional system of clearing land by fire. The porosity of many African soils is low, leading to problems with high water run-off during intense rains and a tendency for soils to compact under wet conditions. The degree of hardness of soils and resistance to penetration is often five to ten times higher in the dry season than in rainy seasons. Descriptions of soils in Africa can be found in Ahn [1969, 1977], M. J. Jones and Wild [1975], Charreau [1977], Kowal and Kassam [1978], and Obeng [1978].

Concern with soil erosion has increased dramatically in recent years. For surveys of research on soil erosion and conservation, see Fournier [1967], FAO [1973], Greenland and Lal, eds. [1977], Fauck [1977], and Roose [1977].

Fertilizer Use. Fertilizer use is extremely low in Africa. For example, in 1986 only 3.7 million metric tons of fertilizer (NPK) was used in all of Africa, or less than 3 percent of the fertilizer used worldwide. Fertilizer is almost totally restricted to cash crops such as tobacco, cotton, tea, and groundnuts. Less than one kilo of fertilizer per hectare of arable land in crops was applied in Nigeria in 1970 as compared with eighty-three in the United States and over 200 kilos per hectare in Europe [Falusi, 1976]. In a valuable assessment of technical and economic research on fertilizer in West Africa, Zalla, Diamond, and Mudahar [1977] reported that fertilizer use averages less than two kilos of nutrients per hectare in West Africa. Mudahar [1980] reported that the highest use of fertilizer is in Zimbabwe followed by Kenya, Liberia, Senegal, Sudan, and Zambia with rates ranging from ten to twenty-five kilos per hectare. For a summary of soil fertility and fertilization research, see Padwick [1983].

Yield Response. Fertilizer research on experiment stations has produced varied results. In West Africa, research has shown that rice, cotton, and maize varieties cultivated as sole crops have responded well to fertilizers while responses for sorghum and millet have been somewhat lower [Zalla, Diamond, and Mudahar, 1977]. Amon and Adetunji [1973] found in research on maize, yams, and cassava over the 1964–69 period in the Savanna zone of Nigeria that maize significantly responded significantly to small dressings of N, P, and K but that higher treatment levels did not significantly increase yields. Bigot [1977], in a review of tests in the Côte d'Ivoire over an eight-year period, found that when rainfall was average or better a composite NPK fertilizer had a positive effect on cotton and yam yields.

Acland [1971] reported that rice, cassava, and plantains have shown little response to fertilization in East Africa while sorghum has responded well to manure, nitrogen, and phosphorus. Acland also found that cotton has generally responded well to nitrogen but that phosphorus and potassium have little effect on yields. Nitrogen has been found to have a large impact on maize yields in East Africa (ten to fifteen kilos of maize per kilo of nitrogen) but the effect appears to be highly dependent on the level of crop management.

Rock and soluble superphosphates have received particular attention in fertilizer research since nearly all African countries must import sources of nitrogen while phosphate deposits are available in several countries.[120] Long-term phosphate trials in Samaru, Nigeria revealed that rock phosphates were more promising than superphosphates because of the high cost of superphosphates [M. J. Jones, 1973]. Zalla, Diamond, and Mudahar [1977] reported that the application of 40 to 160 kilos of rock phosphate per hectare gave significant increases in yields in several trials in West Africa and that phosphate rock was generally 50 to 90 percent as effective in increasing yields as calcium phosphates or triple superphosphates. For a comparison of results with phosphate deposits from several West African countries, see Truong Binh, Pichot, and Beunard [1978].

An alternative means of maintaining soil fertility without turning to imported fertilizer is the use of manure. The modest research to date on this topic has focused on how manure influences soil productivity as farmers move from shifting to continuous cultivation. Based on a review of research on cotton, sorghum, and groundnuts carried out over two decades in Nigeria, Lombin and Abdullahi [1977] concluded that soil fertility and productivity could be maintained under continuous cultivation and application of manure but they add that few farmers have sufficient manure to keep their land under continuous cultivation. This may explain why farmers throughout Africa often use manure primarily on relatively high-valued crops near the household compound [See, for example, Lassiter, 1981; Lagemann, 1977].

The generally positive findings about the effectiveness of fertilizers in increasing yields must be treated with caution because of the lack of research conducted under management levels and resource constraints on small farms. Moreover, many fertilizer trials have been carried out on new plant varieties which generally have the capacity to make effective use of relatively larger quantities of plant nutrients. Robinson and Falusi [1974] point out, for example, that recommended fertilizer application rates for improved varieties are often three to four times the recommended application rates for traditional varieties and as much as ten times the average amounts actually applied by small farmers. As a result, more attention is needed in quantifying the farm level response to fertilizers over a number of years. The limited data on farm response show that responses of sorghum and millet are generally about one-half those realized at experiment stations. Farm level yields of rice have been much closer to experiment station trials but on-farm results have been extremely variable [Zalla, Diamond, and Mudahar, 1977].

Profitability of Fertilizer. There have been few attempts to evaluate the profitability of using fertilizers or to identify optimal use rates. Moreover, only a small proportion of the available studies refer to responses achieved under farm conditions. Falusi [1976] reported that demonstration trials in Nigeria have shown that returns from fertilizing crops like groundnuts, rice, and yams are relatively favorable while returns to fertilizer on sorghum have been only marginally profitable. Falusi also noted that expected returns were highly variable, especially for upland food crops. Falusi and Williams [1981] reported that at the subsidized price of fertilizer in Nigeria, the value/cost ratio for all major food crops exceeded 5, with the highest returns to fertilizers being realized on root crops—cassava and yams. The evidence indicated that fertilizer use would be profitable even if fertilizer was not subsidized.

In Uganda, H. L. Foster [1978] found that in 3,000 trials of N and P on large farms, yield responses of cotton were generally large enough to make the use of

N and P profitable. Foster showed that it was profitable to use fertilizer on groundnuts and that groundnut yields without fertilizer had a coefficient of variation of 20 to 30 percent. The profitability of fertilizer on both cotton and groundnuts was strongly influenced by: soil organic content, amount of previous cultivation, and average hours of sunshine. The response of cotton was also influenced by the soil pH. Vadlamundi and Thimm [1974] analyzed the economics of fertilizer trials on maize in Kenya using quadratic production functions with N and P as the variable inputs. Their economic analysis included deriving the Marginal Physical Products (MPPs) of each nutrient for given levels of the other nutrient, identifying yield isoquants, and presenting nutrient isoclines for a range of price ratios. Montgomery [1977] synthesized experiment station results in Mali and Burkina Faso and found that for both sorghum and millet a light dose of fertilizer was profitable while a heavy dose was profitable only for millet. But farmers use almost no fertilizer on millet in these countries today.

Several methodological issues have been raised in attempts to evaluate the profitability of using fertilizer and other purchased inputs. Lang and Bartsch [1977] discussed a method for evaluating the economics of improved practices, including use of fertilizer, in the face of uncertainty about rainfall. By statistically estimating the relationship between yield responses and rainfall, they were able to project the probability of yield responses using historical data on rainfall patterns. In an illustration of their method, they showed that fertilizer use was highly profitable in the Côte d'Ivoire in most years, particularly for yams, rice, and cotton. Pieri, Ganry, and Siband [1978] argued that three major factors must be considered in evaluating the profitability of fertilizer use: economic profitability, the incentive to use fertilizers, and the need to restore minerals removed to maintain soil quality over time.[121] Flinn [1975] argued that research on the economics of new inputs and practices should use a nonzero discount rate in establishing the MFCs of resources in order to represent the time dimension of investing in variable inputs, use effective prices received and paid by farmers (including transport costs), and concentrate on input levels which yield stable returns rather than on identifying maximum profits, since slightly increased returns may greatly increase risk.

Demand for Fertilizer. Ogunfowora and Norman [1973] found that the demand for fertilizer was influenced by the availability of working capital, followed by fertilizer costs and output prices. Mwangi's study in Kenya [1978] also showed that capital availability and fertilizer prices were more important determinants of fertilizer demand than product prices. But in northern Nigeria, Etuk [1979] found that the optimum level of fertilizer use was relatively insensitive to a change in the price of fertilizer. The lack of working capital was found by K. L. Robinson and Falusi [1974] to be a major constraint on fertilizer use in Nigeria. Falusi [1974-75] showed that the variables representing the wealth of farmers had the biggest effect on the decision to use fertilizer in Nigeria. Falusi [1976] reported that in a survey of Nigerian farmers carried out in 1971 over half of the farmers cited a lack of money or credit as the major reason they did not use fertilizer; 40 percent said they did not apply more fertilizer because it was not available. Falusi and Williams [1981] identified the following constraints on fertilizer use in Nigeria: low returns to investment in land in bush fallow farming systems, absence of fertilizer-responsive varieties, moisture stress in drier areas, and inadequate extension support. Kelly [1988] found that the low demand for fertilizer in Senegal arises from farmers' failure to apply the correct amount of fertilizer at the right time on the right soils.

RESEARCH ON MANAGERIAL PRACTICES

Historically, agronomic research programs have focused on identifying optimal practices to maximize yields of improved varieties under different soil and climatic conditions. In the 1960s, the large gap between yields achieved with improved practices on experiment stations and those on farmers' fields gave rise to the view that indigenous practices had to be abandoned before major breakthroughs could be made in farm yields. But in the 1970s, both social and technical scientists developed increased respect for and interest in the value of indigenous practices of small farmers. The focus of most agronomic research programs in sub-Saharan Africa now encompasses alternative management practices, including indigenous practices. We shall review research on: intercropping, plowing, crop rotations, sowing dates, and planting densities.

Intercropping. Intercropping refers to the indigenous practice of producing two or more crops on a field at the same time. According to Belshaw [1979], extensive research was undertaken on intercropping in Africa as early as the 1930s. But over the 1930-60 period, colonial administrators and researchers considered intercropping to be irrational and urged farmers to substitute sole cropping and planting in rows for intercropping. Intercropping research was reactivated in some countries in the 1960s and now has become a major component of several national programs. The potential technical advantages of intercropping include: legume intercrops fix nitrogen; intercrops which spread and cover the entire ground surface reduce weeds and the time which must be spent on weeding; mixtures can be grown at higher densities than equivalent areas of sole-cropped fields; fewer insect and disease problems relative to sole cropping; and differences in the heights and age of crops in mixtures at maturity enable a farmer to exploit a limited area of high quality soil with minimal competition between crops.

Research on the economics of intercropping was initiated at Ahmadu Bello University in northern Nigeria in the mid-1960s. The results have shown that, although the returns to individual crops often are lower when grown in mixtures, farmers achieve higher gross and net returns per hectare for crops grown in mixtures. Norman [1974] estimated that the average return per hectare of intercropping was 35 percent higher than sole-cropped fields. Ogunfowora and Norman [1974] and Andrews [1972,1974] showed that there is an advantage to intercropping even when using improved varieties that have been bred to be sole-cropped. E. F. I. Baker [1978-80] reported that in a series of trials, mixed systems of cereals and groundnuts; cereal, groundnuts and maize; cereals and cotton; and mixtures of cereals all increased returns per hectare relative to sole-cropped fields.

Intercropping has received more qualified support in other sub-regions in Africa. Mercer-Quarshie [1979] for example, found in a test of fourteen mixtures of five varieties of sorghum grown in eleven different agroclimatic zones in northern Ghana that each mixture yielded less than the best individual crop in that mixture but that the average yield of each mixture was higher than the mean yields of its individual crops. Moreover, the mean yields of mixtures were more stable. Mercer-Quarshie concluded that the stability of yields is a more important reason for intercropping than increases in yield or gross income. In a series of intercropping trials in Kenya, Fisher [1977] reported mixed results for maize and beans and maize and potatoes. Fisher found that in short rainfall seasons, maize and beans and maize and potatoes competed for water, and yields from mixtures were less than yields from an equivalent area of each crop grown as a sole crop. But in a follow-up study, Fisher [1979] reported that in the long rain season, there was a clear advantage of maize and bean intercropping. Fisher concluded that mixtures are more efficient where yields of pure stand are low but that there is little difference between sole and intercropping where the yields of pure stands are high.

In summary, research in Africa and Asia has shown that intercropping is a desirable practice for many smallholders because it serves as a hedge against crop failure, increases the variety of food for a family with a land constraint [Flinn and Lagemann, 1980], and can increase the returns to labor during the peak seasons. [Okigbo and Greenland, 1977; Monyo, Ker, and Campbell, eds., 1976]. Research stations are starting to incorporate intercropping as a standard component of their experimental design.

But intercropping may not be appropriate in extremely arid environments where crops such as cowpeas outcompete dominant food crops such as sorghum for limited amounts of soil moisture. Also, intercropping is not necessarily desirable in areas where large fields are cultivated, particularly when indeterminant varieties are grown, because of the difficulty of harvesting crops in a timely manner. Nevertheless, intercropping often does have advantages relative to sole cropping and should be included as a standard component in experimental design.

Plowing. Research on plowing has focused on the impact of deep plowing *versus* minimum or no tillage on yields and soil quality in the long run. Charreau [1977] contended that plowing has a major beneficial impact on yields of crops grown on soils with a clay content of under 20 percent; this applies to much of the arable land in West Africa [M. J. Jones and Wild, 1975]. Charreau further argued that the benefits of plowing have often not been realized due to low quality plowing. Strong support for the benefits of plowing was presented in a series of reports on trials in Senegal by Charreau and Nicou [1971]. IITA/SAFGRAD [1980] reported that in trials on maize, zero tillage and hand hoeing gave significantly lower yields (2,000 kg/ha) than did oxen plowing. Barrett *et al.* [1982] found that plowing increased yields of sorghum and groundnuts in eastern Burkina Faso and that there was even a greater increase if phosphate rock was incorporated. Chopart and Nicou [1976] provided evidence that plowing increases yields and tends to increase drought resistance because it increases the porosity of soils and enables plants to establish stronger and deeper root systems. Ahn [1977] pointed out that there is a general consensus among French soil scientists, who have played a leading role in soil research in West Africa, that plowing can have a major positive effect on yields. On-farm research in Botswana has even shown that the addition of a second plowing can increase yields relative to single-plowed plots by more than 75 percent [ATIP, 1986a,b].

It has been common to evaluate plowing in conjunction with fertilization and incorporation of straw to increase the organic content of soils. On the basis of trials over a five-year period on rice in Senegal, Beye [1977] reported that the combined effect of nitrogen with plowing under of straw had a significantly positive effect on yields. The effect of nitrogen alone was also positive but less than the combined treatment. On the other hand, Bigot [1977] found in the Côte d'Ivoire that plowing had little effect on yields of cotton, maize, rice, and yams relative to hand hoes and that the beneficial effects of plowing did not overcome the effect of bad rainfall.

In general, there appears to be evidence that plowing, particularly when complemented by incorporation of organic material or fertilizers, can have a positive effect on yields, depending on soil properties and quality of plowing. But the question of the impact of plowing on soil quality in the long run remains unanswered.

Crop Rotation. Crop rotation has long been practiced to take advantage of different levels of soil fertility, to counteract weeds and to take advantage of residual soil fertility either from fertilizer on a previous crop or from a legume in the rotation. For example, crops such as yams are often planted soon after a plot is out of fallow since they have high nutrient requirements while cassava is generally the last crop planted in a rotation. Rotations of sorghum and millet have also been used to counter striga. An important topic is the impact of continuous cultivation of a single crop or a rotation of crops on soil quality over a ten to fifteen year period [Fauck, Moreaux, and Thomann, 1969; Charreau, 1972; Nicou, 1978; IRAT, 1980].

Sowing Dates and Planting Density. Since yields of most crops decline with late sowing, an important research issue is the relative sensitivity of different crops to late sowing. The sowing date of cotton has received a large amount of attention because small farmers consistently plant cotton later than recommended dates [Norman, Hayward, and Hallam, 1974]. Results have shown that the time of sowing does have a significant effect on yields of both cotton and maize. In general, late sowing of photosensitive varieties can reduce yields which cannot be offset by changes in soil preparation and fertilization [ICRISAT, 1980c; Kassam and Andrews, 1975]. The timing of planting tends to be less significant in regions with extremely erratic rainfall, since early-planted plots can die due to post-planting drought while late-planted plots survive due to late season sporadic rains.

Plant density has also received attention in agronomic trials because African farmers traditionally plant at intervals exceeding those used on experiment stations and recommended by extension services. This indigenous practice represents an adjustment to the low fertility of many Africa soils and the need to conserve soil moisture to fill grain after the rains have stopped. There appears to be an important interaction betwen sowing date and plant density, at least for local varieties of certain crops such as sorghum and millet which tiller. Early planting of tillering varieties of sorghum at wide spacings stimulates vigorous tillering, resulting in increased yields relative to later planting at higher densities [ICRISAT, 1980c]. For research findings on sowing dates and plant densities, see ICRISAT [1980c] and IRAT [1980].

SYNTHESIS

Every country in sub-Saharan Africa has national research programs which conduct a wide range of agronomic experiments. Historically, investment in research on export crops such as cocoa, tea, and coffee has greatly exceeded that on food crops. Not unexpectedly, progress in increasing yields, yield stability, and disease resistance has been greater on export crops than food crops with the exception of a few crops such as maize in Kenya, Zimbabwe, and Malawi [D. S. C. Spencer, 1986; Eicher, 1986b].

Many factors are responsible for the lack of progress in generating food crop technology for small farms in Africa. First, there is a gap between resource endowments of experiment stations and small farms. For example, soils on research stations often have a history of better management, including previous applications of fertilizer and dry season conservation practices. Experiment station plots are usually plowed and seeded at optimal times, weeding often exceeds levels practiced by small farmers, and complementary inputs such as insecticides and fertilizers which are routinely used on experiment stations are often not available to farmers in village markets. As a result, many of the technical recommendations presented to farmers have proven to be overly optimistic.

Second, many of the technical packages which increase yields and yield stability call for practices which are not consistent with the goals of farmers or their "prevailing wisdom" about optimal cultivation practices under low population densities or environmental uncertainty. For example, researchers frequently have recommended early planting of cash crops in rows even though most farmers have traditionally intercropped and planted food crops before cash crops, believing that these practices increase the probability that household food requirements can be met even in low rainfall years. As a result, farmers have selectively adopted some of the components of technical packages such as an improved variety, applying a small amount of fertilizer, or changing planting dates rather than adopting the entire package. Even where entire packages have been adopted, farmers generally have done so sequentially over a period of several years. Thus, there is a continuing need for on-farm research to understand the goals, resource endowments, and constraints faced by farmers in designing on-station research [Byerlee and Tripp, 1988]. There is also a need to target research in land abundant areas in Africa and realize that research priorities in land abundant areas of Africa will be dramatically different from labor surplus areas in Asia [Binswanger, 1986].

3. Irrigation

Irrigation plays a minor role in Africa except for large-scale projects in the Sudan and in Madagascar where there is a history of irrigation by small farmers. With these exceptions, the percent of cultivated land under irrigation is probably less than 5 percent in most of the other countries.[122] This compares with around 35 percent in India. But irrigation is important in the river valleys of Zimbabwe, Somalia, Ethiopia, and Mozambique and in parts of Mali, Senegal, and northern Nigeria. Also, there are numerous indigenous irrigation techniques which have been finely honed to local ecological conditions. For example, one system is planting crops (usually rice) in small swamps [Welsch, 1965] or what are called *bas fonds* in Francophone Africa. Another is the use of small streams on mountains. The Chagga people on the hills of Mt. Kilimanjaro in northern Tanzania have developed an intricate network of small streams and ditches which crisscross the entire mountain and support a banana, coffee, and dairy cattle farming system. In many parts of Africa, the *shadoof* is still used to lift water to irrigate small patches of vegetables. Flood recession farming is of major importance in Mali where river bottom land is planted to crops as the flood water recedes. But indigenous water control systems are footnotes in African agriculture.

The token role of irrigation in sub-Saharan Africa is understandable because the marginal cost of bringing idle rainfed land under cultivation has been substantially less than the cost of leveling and preparing land for irrigation. For example, the World Bank [1981b, p. 79] reports that irrigation projects in Niger,

Mauritania, and northern Nigeria all had costs of more than $10,000 per hectare in 1980 prices. A recent CILSS/Club du Sahel report [1980b] tells us that "the cost of irrigation development in the Sahel is running between $5,000 to $20,000 per hectare". On the other hand, small perimeters are being built in Senegal with only several hundred hours of family labor per hectare.

During the past fifteen years, two issues brought about a dramatic increase in the need for greater priority to research on irrigation. The 1968-74 drought in the Sahelian region of West Africa, Sudan, Somalia, and the Great African famine in 1984-85 stimulated interest in irrigation as a means of reducing the dependency on rainfed agriculture [AID, 1976; Club du Sahel, 1977; Eicher, 1986a]. The frontier stage is almost exhausted in countries such as Kenya and Senegal and, as a result, increases in agricultural output will require investments to intensify production, including multiple cropping on irrigated land [Carruthers, ed., 1983; Fell, 1983; Hotes, 1983; FAO, 1986a; Moris and Thom, 1987; Levine and Bailey, 1987; P. Bloch, 1986, 1987].

POTENTIAL FOR IRRIGATION

The potential for increasing the proportion of arable land is unknown because of the lack of soil mapping, hydrological surveys, and soil and agronomic research for irrigated farming. For example, Carruthers and Weir [1976] estimated that the potential land available for irrigation in Kenya was 230,000 hectares in 1976, but this figure was adjusted upward in 1977 and 1978 as more technical information became available. In 1979, the Government of Kenya estimated that 540,000 hectares of land were available for irrigation [Toksoz, 1981]. Regardless of which figure one accepts, the potential for expanding irrigation in Kenya is large since only 16,000 hectares, or between 5 and 10 percent of the potential, is under irrigation. The FAO has estimated that the Sahelian region of West Africa has about twelve million acres of land which is potentially available for irrigation if water is available [FAO, 1976b]. The Club du Sahel, however, estimated that only 80,000 hectares were under irrigation in the mid-1970s [1977, p. 27]. The gap between 80,000 and 12,000,000 hectares is staggering and even if the FAO estimate is cut in half to six million hectares the potential for irrigation is enormous. In summary, even though estimates of the potential land for irrigation are some of the weakest data for any subsector, the technical potential for irrigation is large but highly variable by country and estimates likely will be increased as more information becomes available [CILSS/Club du Sahel, 1978b].

COUNTRY AND REGIONAL EXPERIENCES

Irrigated farming dominates the agricultural sector in the Sudan with its Gezira and Rahad schemes. The Gezira scheme of 740,000 hectares was established in 1925 in an area south of Khartoum. Cotton is the main export crop supplemented with millet, groundnuts, wheat, and rice. Cotton exports are handled through the government's Gezira Board. Gaitskell [1959] is a standard reference on the Gezira scheme. Although the Gezira scheme is often cited as the most successful large-scale irrigation project in Africa, large irrigation projects in the Sudan have been plagued by the lack of participation of farmers in decision making, lack of flexibility in choosing crops, and difficulties in adjusting the size of farms in response to changes in the life cycle of tenant families [Levine and Bailey, 1987]. For a radical critique of the Gezira scheme, see Barnett [1977, 1979, 1981]. Because of a limit on the amount of water which the Sudan and Egypt can each draw from the Nile River, rainfed farming is now receiving increased priority in the Sudan [ILO, 1976]. By contrast, in other countries policy makers are concerned with speeding the transition from rainfed to irrigated farming.

Irrigation in northern Nigeria has its roots in colonial policy. Soon after independence in 1960 the government pushed ahead with irrigation in northern Nigeria [Wells, 1974] even though the FAO estimated that the cost of irrigated wheat production was $168 per ton as compared with an imported cost of eighty-four dollars per ton. [FAO, 1966, pp. 180-181] For a historical perspective, see Palmer-Jones's "How Not to Learn from Past Irrigation Mistakes" [1981]. The World Bank [1981b, p. 80] reports that Nigeria is shifting emphasis away from large-scale irrigation to small-scale irrigation based on groundwater development by hand operated and small motor-driven pumps. For a pessimistic view of the Kano River project, see Wallace [1981]. For an assessment of small-scale irrigation in western Nigeria, see Ansell and Upton [1979].

In Francophone West Africa, many irrigation projects were developed by the French colonial service and the remains of these projects are visible in almost every country. The Office du Niger project on the Niger River in Mali was the centerpiece of French colonial irrigation policy in West Africa. It was started some seventy years ago to grow cotton and rice when it was observed that rainfed cotton often failed in the Sudanian zone [De Wilde et al., 1967; W. I. Jones, 1976]. A French parastatal— Office du Niger—was created in 1921 to carry out the development of the one million hectare scheme. The colonial service relied heavily on forced labor (until 1945) to develop the infrastructure [Magasa, 1978]. During the peak settlement period, only a small proportion of the projected one million hectares was settled but the Office du Niger was given a reprieve by the 1968-74 drought. Several donors are now financing the rehabilitation of the infrastructure. The first farm-level study of the economics of rice production in the Office reveals that rice yields at the farm level are lower, 1.7 metric tons per hectare, than previously assumed and that government taxes are a major constraint on the profitability of rice production by small farms [Kamuanga, 1981; Bingen, 1985].

Four river basin complexes are being developed in the Sahelian region of West Africa; the Office du Niger in Mali, the Volta River Valley Project in Burkina

Faso, the Lake Chad River Basin Commission (LCBC), and the Senegal River Valley Development Authority (OMVS). The $900 million OMVS project on the Senegal River is the largest river basin project under way in West Africa at this time; Mali, Mauritania, and Senegal recently completed dams at the estuary (Diama) and upstream (Manatali) and are financing related investments to regulate the flow of the Senegal River in order to arrest flooding and the incursion of saltwater, provide hydroelectric power, expand river navigation and fishing, and regulate and increase the supply of water for irrigated crop and livestock production. The OMVS optimistically projects that it will be possible to expand the present 30,000 hectares of land under irrigation to 375,000 over the next twenty to thirty years. Clearly, there is a need for a large amount of technical, social, and economic research to guide the development of irrigated farming in the OMVS project area. And there is a need to keep in mind that many big schemes have failed in the past [See K. D. S. Baldwin, 1957].

After the 1968-74 drought in the Sahel, there was a great deal of optimism about the role of irrigated farming in "drought-proofing" the Sahel. But the CILSS/Club du Sahel reported [1980b] in a remarkably candid assessment that because of numerous technical and administrative problems, the projected expansion of irrigation in the Sahel is falling behind schedule. In summary, irrigation is not a panacea for the recovery of the Sahel during this century [Fell, 1983].

ECONOMICS OF IRRIGATION

Studies of the economics of irrigation in Africa are summarized in Blackie, ed. [1984a] and Moris and Thom [1987]. Chambers and Moris, eds. [1973] traced the history of the Mwea irrigation project in Kenya for two decades and report that it was financially successful for small farmers and that it has produced some of the highest rice yields in the world. Carruthers and Weir [1976] examined five government schemes for small farmers, including the Mwea scheme, and recommended the expansion of small-scale irrigation and horticulture research in Kenya. But Toksoz [1981, pp. 32-33] contends that the per hectare cost of land reclamation (through drainage and flood control) appears to be less than one-fourth to one-half the cost of irrigation projects. Toksoz recommends that priority be given to land reclamation projects and the expansion of small-scale irrigation in Kenya.

Sparling [1981] surveyed the sparse literature on the economics of irrigation in the Sahel[123] and tentatively concluded that small-scale perimeters will be more profitable than large-scale perimeters and large irrigation perimeters are unprofitable for rice production. Sparling contends that the social profitability of small perimeters almost surely will exceed their private profitability because of the multiplier effects (indirect effects) for those who provide increased goods and services to farmers whose incomes have risen. See Bell and Hazell's [1980] study of the key role of indirect benefits in a rice irrigation project in Malaysia.

Although there is growing evidence that small-scale irrigation is more successful than large projects, there are cases where large-scale irrigation systems or combinations of small- and large-scale irrigation can meet both efficiency and equity goals. For example, Hazlewood and Livingstone's [1978] linear programming analysis of large- and small-scale irrigated farming in Tanzania showed that the area under rice can be maximized by continuous production both on large state farms and through small-scale village production. The complementarity of the two production systems depends on different monthly water requirements stemming from assumed cropping patterns on large and small farms. J. A. Smith [1978] emphasized another complementarity in a study of large-scale irrigated sugar schemes in Kenya. Smith argued that estates can produce seed cane for smallholders enabling the smallholders to undertake cash cropping. Moreover, once nearby smallholders are in full production, estates can be used to regulate the flow of cane to processing mills, assuring the financial feasibility of processing mills.

SYNTHESIS AND RESEARCH DIRECTION

Although research on the economics of irrigation is fragmentary, the results to date support a small-scale irrigation strategy with priority given to increasing flood recession farming, groundwater development with small pumps, land reclamation through drainage and water control, and small perimeters which are developed and maintained by family labor. Second priority should be given to improving the performance of existing large perimeters and river basin complexes by investments to rehabilitate and upgrade the infrastructure on schemes such as the Office du Niger in Mali. Once again these are preliminary guidelines for Africa. There can be important exceptions, of course, when one examines the economics of alternative irrigation systems in a given area.

The big issues to be addressed by researchers are the economics of rainfed *versus* irrigated farming in site-specific locations, the choice between large (500 to 1,000 hectares and up) *versus* small irrigation perimeters [Moris and Thom, 1987] and whether irrigated land should be managed by government agencies and farmed by tenant farmers or managed by associations of farmers and farmed by small farmers.[124] The scope of irrigation research in Africa should include historical studies of indigenous irrigation systems, lessons from past mistakes [Palmer-Jones, 1981], the social impact of resettlement components of River Basin projects [Scudder, 1973; Scudder and Colson, 1972], and the potential of land reclamation [Toksoz, 1981] as an alternative to investment in irrigation. Policy makers and researchers in Africa can learn a great deal from the Asian experience with indigenous systems of irrigation management [Coward, 1977], distribution of water in canal projects [Bromley, Taylor, and Parker, 1980], problems of implementing and managing irrigation schemes [Bottrall, 1981], and the role of social science research in helping to design, implement, and evaluate irrigation and land reclamation projects [Coward, ed., 1980; Moris, 1987].

IRRIGATION REFERENCES

Literature Reviews: Des Bouvrie and Rydzewski [1977]; Bromley, Taylor, and Parker [1980]; Sparling [1981]; Moris [1987]; Moris and Thom [1987].

Asian Experience: Coward [1977, 1980]; Bottrall [1981]; Barker, Herdt and Rose [1985].

Africa: Gaitskell [1959]; Welsch [1965]; FAO [1966]; de Wilde *et al.* [1967]; Thornton and Wynn [1968]; Chambers [1969, 1970]; Chambers and Moris, eds. [1973]; Wells [1974]; Steedman *et al.* [1976]; FAO [1976b]; W. I. Jones [1976]; Carruthers and Weir [1976]; Club du Sahel [1977]; A. Adams [1977b]; Palmer-Jones [1977b, 1981]; T. Barnett [1977, 1981]; Hazlewood and Livingstone [1978]; Weiler [1979]; Ansell and Upton [1979]; Bell and Hazell [1980]; Coward, ed. [1980]; Diallo [1980]; Toksoz [1981]; CILSS/Club du Sahel [1980a,b]; Blackie [1982]; Kamuangua [1981]; Fortmann and Roe [1981]; Wallace [1981], Blackie, ed. [1984a]; Moris, Thom and Norman [1984]; Carruthers, ed. [1983]; Fell [1983]; P. Bloch [1986, 1987]; Moris and Thom [1987], FAO [1986a]; Levine and Bailey [1987]; Simpson [1988].

4. Animal and Tractor Mechanization

For centuries, farming has been carried out in Africa with human labor and the machete, which was used to clear the bush in shifting cultivation systems. Various types of short handle hoes are used in land preparation for weeding and for making heaps and ridges. The importance of the hand hoe is illustrated in Kenya where it is estimated that 84 percent of the arable land in 1980 was cultivated by hand hoe, 12 percent by oxen, and 4 percent by tractor. These ratios are fairly common in Africa.

Since the early 1900s, colonial and later independent African governments tried to help farmers replace the machete and hoe cultivation with oxen, donkey, and tractor mechanization.[125] The rationale for mechanization ranges from increasing agricultural output and profit to relieving the drudgery in rural life. But Africa's history is littered with discontinued animal traction schemes sponsored by missionaries, colonial governments, and more recently by foreign aid programs. Even though Africa's experience with oxen and tractor mechanization has generally been unsuccessful, the mechanization of agriculture is almost inevitable. The crucial policy question today is what types of mechanical power are appropriate in low-wage economies with abundant land, seasonal labor bottlenecks, and present and projected energy prices. The task for researchers is to aid in developing improved tools and implements to raise labor productivity in agriculture, to break seasonal labor bottlenecks, and to promote the transition from hand to animal and tractor cultivation in as socially desirable a manner as possible.

METHODOLOGICAL ISSUES

There are major methodological problems in carrying out research on mechanization [Gemmill and C. K. Eicher, 1973; Pingali, Bigot, and Binswanger, 1987]. The first is the need to shift more research from *ex post* to *ex ante* while insuring that previous *ex post* results are incorporated into the design of *ex ante* research. *Ex ante* research can project some of the likely economic, social, and technical consequences of alternative farming systems, including alternative approaches to performing specific tasks in agricultural production and harvesting. But *ex ante* research needs to be supplemented with on-farm research on both mechanical and biochemical solutions to particular tasks such as weeding. For example, the use of herbicides is emerging as a cost-effective way to control weeds in maize fields.

The second methodological problem involves studying mechanization as a dynamic process to capture the selective and sequential process of mechanization. Unfortunately, most mechanization research in Africa has relied on cross-sectional data to determine whether it is economically and socially desirable to replace one complete technology set (hand hoe) with another package (such as oxen-powered or tractor mechanization) rather than a particular task, weeding, for example, which could be performed by hand, oxen, tractor, or herbicides. I. J. Singh's [1976] research on mechanization in Tanzania provided insights into the mechanization of particular tasks in the farming system. Singh allowed his programming model to specify the lowest cost technique—hoe, oxen, or tractor—to undertake each task in the farming system.

The third problem is to capture the nonagricultural use of equipment—especially the use of oxen carting and tractors for off-farm transportation. Rarely are these important benefits quantified by researchers in Africa. The fourth problem is to capture the consumption of such benefits of mechanization as reduction of drudgery. These benefits may be substantial—in maize and rice processing, for example—and they should also be included in the benefit stream. The fifth problem is examining alternative energy sources.[126] The sixth problem is designing research to gain an understanding of the organizational, logistical, and managerial problems involved in setting up and maintaining a support system for animal traction programs and tractor hire schemes. These problems are rarely addressed by academic researchers because they need to be studied over a substantial period of time (five to ten years), they are politically sensitive, and there are few academic rewards for this type of research.

ANIMAL TRACTION

When one refers to animal-powered mechanization in Africa, the term usually refers to oxen cultivation because in practice it has been difficult to leapfrog from hoes to tractors. For example, Pingali, Bigot, and Binswanger [1987] report that only three of seventeen projects in Africa were successful in moving directly from hand hoe to tractors. The remaining fourteen failed. Various terms are used to describe cultivation with oxen such as animal traction, bullock cultivation, and mixed farming. Oxen are primarily used for plowing except in Senegal, Mali, and a few other countries where seeders and other implements are being used. Normally, two oxen are used in plowing but six to eight oxen are common in Botswana. Although oxen are a dominant factor in subsistence farming in Asia, oxen cultivation in Africa is primarily practiced by farmers who produce export crops such as cotton and groundnuts. Horses are not common in most of Africa because of the prevalence of tsetse fly (sleep sickness). Although heavy soils restrict donkey cultivation to sandy soils, donkey cultivation is increasing in importance following the sharp rise in the price of oxen in recent years.

The rationale for animal traction is the potential increase in yields through improved seed bed preparation, deeper plowing, more timely planting and weeding, and moisture conservation; the potential increase in the acreage cultivated; income generation through off-farm transportation; the reduction in drudgery; and the longer-term benefit of improving soil fertility through application of manure from the animals, deeper plowing, and plowing under crop residues. But animal traction is justified in most feasibility studies on the basis of the presumed increases in acreage and yields—both short-term considerations relative to the improvement of soil fertility.

Animal traction was introduced by the French[127] and British colonial services in the early 1900s with a big push occurring in the 1920s and 1930s as part of the drive to expand cotton and groundnut exports. In 1922, a mixed farming campaign was introduced by the British Colonial Service in northern Nigeria whereby farmers were given credit to purchase two oxen and equipment. The planners assumed that oxen could supplement human energy, expand the area under cultivation, and generate a cash surplus with the two cash crops—cotton and groundnuts. In 1928, research on animal traction implements was started by French researchers at the IRAT station in Bambey, Senegal.

The spread of animal traction has been closely linked to the introduction and expansion of cash crops. For example, oxen cultivation increased in the Côte d'Ivoire from 700 to 8,000 pairs over the five-year period in the mid-1970s as part of a smallholder cotton program. Although these figures are impressive, similar "waves" of animal traction have appeared in other African countries over the past fifty years only to disappear or recede during periods of drought, changes in government policies, and the failure to provide veterinary support services. In 1981, the major concentration of animal traction was in Senegal, Mali, Botswana, and to a lesser extent in Tanzania, Uganda, and northern Nigeria.

As one reviews the historical experience, a major question is whether farmers selectively adopt individual implements to perform a particular task or replace one system with a totally new system of mechanization. Over the past twenty years, numerous African governments and foreign donors have been promoting a total oxen cultivation package with a tool bar and attachments such as a plow, seeder, and ridger, or tractors to replace hoe farming. A few governments include carts as part of the package. We shall examine the individual implement *versus* the package approach below because the widespread failure of foreign aid financed animal traction programs over the past twenty years may be tied to an approach—total animal traction package—which runs counter to the historical diffusion of mechanization not only in Africa but in high-income countries. For historical reviews of animal traction, see Kline *et al.* [1969]; Weil [1970]; Uchendu and Anthony [1975b]; de Wilde *et al.* [1967]; Migot-Adholla [1972]; Okai [1975]; Gaury [1977]; Monnier [1975]; Oluwasanmi [1975]; The Gambia [1976]; Zerbo and Le Moigne [1977]; Sargent *et al.* [1981]; Barrett *et al.* [1982]; Le Moigne [1980]; T. R. Whitney [1981]; Jaeger [1986]; Pingali, Bigot, and Binswanger [1987].

Surprisingly, although animal traction has been promoted for more than sixty years in Africa, research results on the impact of animal traction at the farm level are often impressionistic. Basic data on the yield and acreage effects of animal traction at the farm level are still inadequate in many countries because research on animal traction usually takes place on experiment stations rather than on farmers' fields: most research does not isolate the effects of oxen from other inputs such as the planting date, spacing, timing, quality of weeding, and fertilizers; most of the research has been done by foreign researchers who are unable to provide the continuity which is needed to develop, test, and adapt implements over time; and there is a lack of longitudinal data on the effect of oxen cultivation on yields, acreage, and soil fertility under differing weather conditions over time. Although there are many *ex post* studies of animal traction schemes, we are aware of only Andrew Ker's [1973] comparison of tractor and oxen-powered cultivation under farm level conditions in Uganda over a twelve-year period.

Gemmill's [1971] study of 132 farmers in one region of Malawi revealed that oxen power was no more timely than hand labor because neither system could be used until the arrival of the first rains and that oxen were adopted by farmers to reduce drudgery and gain prestige. In Uganda, Vail [1973] used linear programming to evaluate the effect of introducing Indian-type oxen cultivation and found that with appropriate training, oxen could be used to draw locally developed mechanized seeders and weeders. Vail argued that in East Africa the failure to develop mechanical technology appropriate for small farmers is partially a function of the overemphasis on testing individual pieces of equipment rather than research on the total farming system, including farmer preferences, managerial ability, risk, and assets.

In a review of the literature on twenty-seven animal traction projects or rural development projects with an animal traction component in Francophone West Africa, Sargent et al. [1981] found that many studies have been undertaken by agronomists who typically estimate what they call hypothetical "maximum potential benefits" of animal traction based on technical coefficients from experiment station trials or demonstration farms. For example, Ramond [1971] derived estimates of the maximum potential benefits from animal traction based on research on forty-one demonstration farms in the groundnut basin in Senegal. Tourte et al. [1971] and Monnier [1972] showed that single-row oxen traction generated maximum potential benefits equal to about double the net farm income obtainable from donkey traction in Senegal. But Tourte et al. and Monnier did not consider the potential cash flow problems in their one-year budgets. Cash flow problems typically emerge during the first (two to four) years when farmers are learning how to handle oxen and when the acreage and yield effects are likely to be low. The studies by Monnier, Tourte, et al. and Ramond which estimated maximum potential benefits typically inflate the projected returns and the long-run economic profitability from animal traction that would likely be achieved under farm-level conditions. Le Moigne [1980, p. 219] presented this optimistic assessment of animal traction in Francophone Africa: "Timely sowing can generate yield gains of 50 percent on groundnuts and sorghum; several mechanized weedings can increase yields of 50 percent for groundnuts and up to 175 percent for millet; and that gross margins of animal traction cultivation 'present a good picture' because they range from $286 to $476 per hectare."

The crucial question is not what are the maximum potential benefits of animal traction on experiment stations and on demonstration farms but under farm conditions. Sargent et al. [1981] found in their review of twenty-seven projects that most of them had not lived up to expectations because of high cost of animals and equipment, low acreage and yield effects, and the lack of reliable institutional support. The Gambia [Mettrick, 1978] found that the impact of oxenization has been largely confined to the groundnut crop with a small increase in area cultivated but no increase in yields. In a study of forty farmers in southeastern Mali, T. R. Whitney [1981] found that animal traction farmers increased their acreage by 39 percent over hoe farming but there was no change in yields.

Over the 1975-80 period, the government of Burkina Faso provided donkey and animal traction packages to 1,200 farmers through subsidized loans at 5.5 percent interest rates with a one-year grace period and a four-year repayment period. Based on a year-long (1978/79) farm management survey of 355 hand hoe farming households and 125 animal traction farmers (both donkey and oxen), Barrett *et al.* [1982] found:

1. There is a slow learning curve for farmers using donkeys or oxen for the first time. It takes about three to four years before a farmer knows how to use a complete package of donkey and/or oxen equipment.[128]

2. The acreage and yield effects (through deeper plowing) of animal traction were modest but labor inputs per acre were reduced by as much as 20 to 25 percent. Moreover, the observed yield effects of animal traction were small compared with spectacular increases from the addition of local rock phosphate.

3. There was substantial appreciation in the value of the oxen (through weight gains) because the animals were usually purchased at three years of age and sold when they were seven or eight years old.[129] This suggests that studies of animal traction should include income from the sale of oxen for beef because it can be an important part of the returns from oxen traction.

4. Animal traction is risky in a semiarid environment in the Sahel. Although a government loan scheme was introduced to pay farmers about 90 percent of the value of animals who died because of illness, snake bites, and other physical problems, the value of the animal not covered by insurance was $150 or more than ten times the annual cash costs incurred by hoe farmers. Moreover, the government does not have a crop loan program to protect farmers from weather-induced crop failure.

5. Farmers adopting animal traction experienced major cash flow problems. Although internal rates of return on the animal traction package were positive over a ten-year period, the net returns in the first four years for oxen farmers were *below* the net returns before the adoption of animal traction.

6. The economics of animal traction are problematic for subsistence farmers producing primarily food crops—millet and sorghum. These findings are almost identical to the findings on oxen-drawn wheeled tool carriers on millet and sorghum farms in southern India [Binswanger, Ghodake, and Thierstein, 1980]. Unless a cash crop such as cotton can be introduced into the project in eastern Burkina Faso, it is questionable whether the farmers can repay their subsidized loans.

The research in Burkina Faso and numerous other studies reveal that there is no single factor responsible for the repeated failure of animal traction programs [Jaeger, 1986]. A major problem is that breaking one constraint usually induces another constraint. For example, although animal power can reduce labor inputs in land preparation, animals are not widely used for weeding. As a result, weeding often becomes a major bottleneck. Further, animals are generally underutilized since there is a shortage of well-adapted equipment available, particularly for weeding, harvesting, and threshing. Other problems plaguing animal traction programs include poor nutrition, lack of dry season fodder, disease, high mortality rates, uncertain supply of tools, problems in destumping and consolidating fields,[130] lack of reliable markets, and failure to establish training programs for potential animal owners.[131] Finally, government policies on oxen and tractor support systems vacillate over time. For example, although the government of Tanzania officially adopted animal mechanization as a major thrust in its Ujamaa program starting in the late 1960s, it quietly shifted its emphasis from ox power to "power mechanization as an instrument of change and modernization of agriculture" [FAO, 1975b].

The costs of oxen, donkeys, and equipment packages are rising and the profitability of total packages of animal traction equipment is in question in much of Africa. In 1977 in West Africa, a donkey traction package (donkey, western hoe, supereco seeder, and donkey cart) cost about $500 while an oxen traction package was about $1,000 (f.o.b. Dakar) for two oxen, arara tool bar, ox cart, and supereco seeder [Zerbo and Le Moigne, 1977, p. 281]. In 1980 in eastern Burkina Faso, the cost of a good three year-old pair of oxen was about $350, while a donkey cost about $75 [Barrett, et al., 1982].

The presence of a cash crop is a central determinant of successful animal traction schemes. This finding is illustrated by research in northern Nigeria [Tiffen, 1976], the groundnut basin in Senegal, the cotton zones in southern Mali, the Côte d'Ivoire, and in northern Cameroun. For example, in a World Bank financed cotton project in the Côte d'Ivoire, oxen cultivation generated a return of 500 CFAF ($2.50) per day of family labor in 1980 which was 40 percent higher than the return to hand cultivation of cotton.

In summary, there are technical, economic, and logistical constraints on the spread of total packages of animal traction in Africa. Under present factor and product price ratios and the token research under way on improving implements, it appears that animal traction will be restricted to "pockets of cash crop production" in Africa just as tractors in India have been primarily restricted to large farmers in the Punjab [Binswanger, 1978]. African governments and donors should consider examining the economic history of high-income countries where mechanization historically has proceeded by replacing one tool or implement with another.[132] Complete animal traction packages cannot serve as an engine of growth of agriculture in Africa. Selective mechanization—farmers replacing one implement at a time—should be facilitated and major attention should be directed to improved agronomic practices which can complement animal traction.

TRACTOR MECHANIZATION

Most African governments started to import tractors in the 1950s and 1960s as part of large-scale agricultural schemes. Still, the number of tractors in Africa in 1981 was small and the number in operation at any one time was insignificant. For example, Otieno, Muchiri, and Johnston [1975] report that only about 40 percent of the tractors of private contractors in Kenya are operational at any one time. For overviews of tractor mechanization, see M. Hall [1968]; Kline, Green, et al. [1969]; Gemmill and C. K. Eicher [1973]; Westley and Johnston, eds., [1975]; Clayton [1975]; ILO [1976]; Winch [1976]; Muchiri [1979]; Hunt [1975b]; Monnier [1975]; Purvis [1968b]; Kolawole [1972,1974]; Kinsey [1978]; Nweke [1979]; and Wuyts [1981].

Tractor hire services—both government and private—are theoretically attractive because of their perceived ability to spread the fixed costs of the tractor and equipment over a large number of small farmers. Although the demand for tractor hire services has been artificially increased by government subsidies, Gemmill and C. K. Eicher [1973] reported that most government tractor hire schemes in Africa have not been successful because of high operating costs on fields which are small, scattered, and irregular in shape. Purvis [1968b] and Kolawole's [1972] studies of government tractor hire schemes in western Nigeria revealed that there was little financial advantage to participating farmers because of frequent breakdowns and a shortage of operators resulting in delayed planting. A survey of 907 tractors in a government tractor hire scheme in northern Ghana in 1968 revealed that 78 percent of the tractors were broken and waiting repairs [Kline, Green, et al., 1969]. In Uganda, tractors in the government's tractor hire scheme were being used only 450 hours per year after ten years of operation of the scheme [I. J. Singh, 1976].

Numerous studies of the economics of private tractor ownership and government tractor hire schemes show that while government subsidies help make mechanization financially rewarding to individual farmers, the schemes generally have high social costs in terms of required government subsidies and, in some cases, the displacement of tenants. For example, D. S. C. Spencer and Byerlee [1976] reported that 85 percent of the cost of the government tractor hire scheme in Sierra Leone was subsidized in the mid-1970s. In Ethiopia, J. M. Cohen [1980a] reported that farmers in the CADU rural development project received duty-free importation of machinery and spare parts, fuel-tax waivers, and subsidized credit at 7 percent interest rates. It follows that mechanization programs should be evaluated on the basis of financial returns to farmers (the main concern of engineers), economic returns to society (taking taxes and subsidies into account), and the social impact in terms of employment and income distribution.

The employment and income distribution effects of tractor mechanization schemes started to appear in the literature in the 1970s. Winch's [1976] study of rice production in northern Ghana provided evidence on the financial and economic returns to alternative systems of mechanization. Shepard [1981], a sociologist, analyzed agricultural development projects in northern Ghana, including the rice farming scheme in the Tamale area studied by Winch, and contended that "the benefits of state investment in agriculture, at least in the north, have accrued largely to capitalist [meaning large farmers] and part-time or absentee farmers" [pp. 187-188; see also Goody, 1980].

Mechanization is a major policy issue in the Sudan—a country with only twenty-three million people and two-thirds the land area of India.[133] Since the clay soils throughout the country can be tilled by hand only with great difficulty, there is a technical case for tractor land preparation. Mechanized farming was started in the Sudan in 1945 and it spread rapidly in the 1960s with the establishment of the Government's Mechanized Farming Corporation. Mechanized farming is highly subsidized and the financial returns to large-scale commercial farmers are high. For example, the ILO [1976] estimated that subsidized mechanized farms of 1,500 acres earned $5,000 to $6,000 gross income before taxes. This is ten to twenty times larger than that of farmers who cultivate small farms with hoes. The ILO report alleged that there are serious problems with wind erosion and loss of soil fertility on the large government and private mechanized sorghum farms. Research is needed on the economic, technical (e.g., erosion and environmental problems), and social impact of subsidized rainfed mechanization.

I. J. Singh's [1976] programming model of 290 farming households in the main grain, cotton, and groundnut growing regions bordering Lake Victoria (Sukumaland) in Tanzania merits careful scrutiny. Singh generated data on hoe, oxen, and tractor cultivation under traditional and improved packages. Singh's model revealed that hoe cultivation is the most economically efficient method of cultivation for 78 percent of the farmers (farms with less than six hectares); and that hoe cultivation supplemented by oxen cultivation is the most efficient method for the other 22 percent of the holdings (farms with more than six hectares).

Singh observed that many countries in Africa are in a transition period from hand to oxen and tractor mechanization. As a result we should expect some farming tasks can be most efficiently performed by hand while others could be performed by oxen or tractors.

In a study in the main maize growing area in southern Zambia, Kinsey [1978] also found that there was no one optimal choice of technology which could satisfy the large commercial (mostly European) farmers and the small African farmers. Using data collected during the 1970-71 growing season, Kinsey's LP analysis revealed that there were numerous conflicts between the criterion of private profitability of commercial farmers using tractors and tractor hire services and the small farmers using hoes and oxen. Kinsey concluded that subsidies promoting tractor mechanization should be discontinued and that research attention should shift to developing technologies which were intermediate between tractor mechanization and hoe cultivation.

Binswanger's [1978] careful review of tractor mechanization studies in India, Pakistan, and Nepal is a valuable reference for researchers in Africa. Binswanger found that (except in special cases) there were no significant increases in yield, cropping intensity, or timeliness arising from the replacement of bullocks with a tractor. He further added that "at best such benefits may exist but they are so small that they cannot be detected and statistically supported, even with massive survey research efforts." Binswanger recommended that subsidies on tractors in the three countries studied be removed and speculated that with the quantum jump in fuel prices since 1973, tractor mechanization would be feasible in the future only under conditions of rapidly rising wage rates and increased prices of bullock implements.

SYNTHESIS

Mechanization—especially total tractor mechanization of all farming tasks—has wide support throughout Africa; it is a dream of many politicians, donors, and engineers who want to "modernize" African agriculture. For example, President Nyerere is quoted as saying, "We are using hoes. If two million farmers in Tanzania could jump from hoe to the oxen plough, it would be a revolution. It would double our living standard, triple our product. This is the kind of thing China is doing" [W. E. S. Smith, 1971].

Our review has shown that both animal and tractor mechanization are plagued with technical, economic, and institutional problems. We believe that tractors will remain as a minor power source in most countries in sub-Saharan Africa in the 1990s because of the post-1973 quantum jump in fuel prices and the myriad of technical problems (low use rates and lack of maintenance). But if rural wage rates increase it will be financially profitable for farmers and economically profitable for governments to carry out some farming tasks such as land preparation with tractors. For a few countries such as the Sudan, the mechanization of some tasks such as land preparation and planting will probably be desirable from a national policy perspective. As a broad generalization, however, hand and oxen power will undoubtedly remain the major power sources in the 1990s and should receive the bulk of research support.

Research on agricultural mechanization has been modest, *ad hoc*, and usually not conducted under actual farming conditions. For the 1950-80 period, the focus of much of the engineering research was on testing imported tractors and equipment. There is a need to shift the emphasis to on-farm research as an integral part of research by multidisciplinary teams on farming systems in specific locations. There are numerous experiments with small tractors under way in Africa.[134] For example, small Japanese walking tractors have been tested and rejected in northern Nigeria, the Swazi tractor has been tested in cotton projects in the Côte d'Ivoire and in southern Mali and French tractors are being used on pilot projects in Zimbabwe. For a landmark comparative study of mechanization and the evolution of farming systems in Africa, see Pingali, Bigot, and Binswanger [1987].

5. Agricultural Extension

Over the past thirty years, most countries have expanded the size of their extension staffs but the ratio of agents to farmers varies widely within and between countries.[135] Boyce and Evenson [1975] showed that expenditures on agricultural extension in Africa in 1974 were 2.2 percent of the value of agricultural production, which was more than twice the percentage of any other region. Nevertheless, extension programs in Africa have been plagued by the same problems affecting extension throughout the developing world, including too few agents (in some countries), low pay, poor training, insufficient logistical support, dilution of efforts, low status, lack of effective linkages with research units, and inappropriate technical packages. For overviews of the problems of extension services in developing countries, see Benor and Harrison [1977], Stavis [1979], de Wilde *et al.* [1967], E. Hopkins [1974], Lele [1975], Chambers [1974], de Vries [1976, 1978], Leonard [1977], Roling [1988],[136] and Howell [1988].

Agricultural extension services were established throughout Africa during the colonial period to expand the production of export crops such as cotton, groundnuts, coffee and tea. In many cases, extension contact consisted of little more than issuing improved seeds [Moris, 1973]. Extension agents focused almost exclusively on progressive farmers. Colonial extension agents in most countries had conflicting roles as they often collected taxes and were responsible for enforcing prohibitions on growing certain crops, and facilitating the recruitment of "forced" labor for roads, mines, and plantations. As a result, the colonial period left a legacy of distrust for government extension agents. This distrust continues to plague comtemporary extension programs.

Following independence, the focus of extension services shifted from coercion to persuasion but the focus was still on export commodities and progressive farmers. Over the past twenty years, most extension services in Africa have been ill-equipped, and undertrained relative to their counterparts in Asia or Latin America. One of the frequent recommendations of international agencies during the 1960s was to step up training programs to increase the number of agents and improve the extension/farmer ratio [e.g., FAO, 1966]. But the effectiveness of extension services, even in countries such as Kenya which has one of the lowest farmer/agent ratios, has continued to be limited. Starting in the 1970s, many donors jumped on the bandwagon to support farming systems research in an attempt to improve research-extension linkages and to improve the relevance of on-station research.

Following a pattern established under colonial governments, most extension services are oriented toward technical problems and pay little attention to farm management issues or to the social constraints faced by rural households. Belshaw [1968], for example, argued that a failure to understand the socioeconomic structure of smallholder farming is a major reason why resources invested in extension services generally have a low return. Watts [1969] similarly argued that extension advice is too oriented toward maximizing agricultural production and gives little consideration to the overall income and welfare of farming households.

Lele [1975] contended that farmers can only be convinced of the value of extension advice through personal contact.[137] Philip Mbithi [1973] has labeled this approach to extension a "message model" which is based on the assumption that knowledge and the desire to change always originate in the research and extension system, on a tendency to promote selected packages, or messages, at different times with no logical continuity between messages, and on a tendency to stereotype farmers and to group them in homogenous masses. Another rural sociologist with extensive experience in East Africa, J. R. Moris [1973], has called the dominant extension model a "hub-and-wheel system" because a "center-post" person generally takes responsibility for all phases of extension activities in a given area. This person supervises several field agents who have little education or training. Moris claims that many of the success stories of the colonial period can be traced to the skill and perseverance of the individuals who were able to fill the center-post role.

During the 1970s, several researchers carried out assessments of national extension systems in order to identify ways of making general extension services more effective. The extension system in Kenya is one of the most studied programs in Africa. Political scientist David Leonard [1977] presented a valuable overview of the problems faced in establishing and supervising field agents in Kenya. Additional evaluations of extension in Kenya were presented in Leonard, ed. [1973]; Heyer, Maitha, and Senga, eds. [1976]; University of Nairobi [1975]; and Chambers [1974].

Atsu [1974] reviewed approaches to extension in Ghana with an emphasis on the Focus and Concentrate program. The Focus and Concentrate program emerged in 1968 in response to the shortage of personnel and funds. The program was based on focusing resources on progressive farmers in a limited area. Atsu shows that the proposed package had little effect on yields and net returns and the program had little spread-effect to other farmers. The Focus and Concentrate approach proved to be bankrupt and has few supporters in Africa. D. C. Baker [1987] showed that despite substantial changes in the organization of extension in

Botswana during the mid-1970s, most of the same problems plague the extension service in the mid-1980s. As a result, extension agents implemented few demonstrations, and most demonstrations and extension messages dealt with one or two themes which were not closely related to farmers' problems.

M. Schulz [1976] reviewed the organization of extension services in Ethiopia immediately preceding and following the socialist revolution in 1974. The extension service in Ethiopia historically had been understaffed relative to most other African countries. For example, in 1968 there were around 500 field level staff in Ethiopia compared with over 5,000 in Kenya, with one-half the population. In 1971, the Extension Program and Implementation Division (EPID) was established. EPID was based on two principles: a well-defined technical package based on fertilizer, improved seeds and credit; and a program for phasing in extension activities in conjunction with farm surveys over a four-year period in each new area where extension services were to be established. For additional assessments of extension in Ethiopia under the Chilalo Agricultural Development Units (CADU), and Wolamo Agricultural Development Units (WADU) and the various minimum package programs, see Tecle [1975], J. M. Cohen [1975, 1987] and J.M. Cohen and Isaksson [1988].

One of the most effective alternatives to national extension programs has been the commodity-specific parastatal and private companies which integrate extension with input supply and marketing. While these agencies are invariably organized on the basis of a lead commercial crop, many advocates of this approach contend there are usually spread-effects to other crops. The commodity-specific parastatals are common in Francophone Africa.

The rapid expansion of tea production by small farmers in Kenya in the 1960s and 1970s illustrates the effectiveness of extension programs administered by specialized commodity agencies. Etherington [1971] showed how institutional innovations by the Kenya Tea Development Authority (KTDA) were crucial to the spread of tea production and the growth of the vertically integrated tea industry with 150,000 smallholders in 1988. First, the propagation of tea stumps in government tea nurseries enabled new varieties to be rapidly multiplied for small farmers. Second, the KTDA built "tea roads" throughout tea-producing areas to insure the delivery of inputs and the rapid movement of output to tea processing plants. Finally, the Tea Authority built processing plants because tea leaves must be processed within a few hours after they are plucked and a single tea plant could serve several thousand smallholders.[138]

The activities of the CFDT in Mali, Burkina Faso, and other Francophone countries is often cited as another example of a successful commodity-based extension and input supply program. The CFDT has concentrated on cotton and the efficiency of its operations has resulted in a rapid expansion of cotton acreage in the Sudanian zone of West Africa [De Wilde et al., 1967]. The major criticism of specialized commodity programs is that they have often had little impact on other crops. For a assessment of the CFDT program in Mali, see de Wilde *et al.* [1967, Vol. 2]. In the late 1980s, the cotton program in the Mali Sud (Southern) area has 80,000 smallholder families.

TRAINING AND VISIT

The Training and Visit (T-V) extension approach developed by Daniel Benor in India has received strong endorsement by the World Bank and other donor agencies as a system for improving extension services in low-income countries [Benor and Harrison, 1977]. In the T-V approach, village extension workers (VEW) carry out an intensive series of weekly or biweekly visits with farmers on a fixed schedule.[139] While initial results of the T-V approach in India and Turkey appear promising, it is unlikely the approach can be applied in Africa without major modifications. The required level of highly trained manpower, not to mention the transport and communication infrastructure and financial resources required for the proposed visit and supervision schedule, are beyond the reach of most African countries. Moreover, Benor [1987] proposes to unify the extension services under the T-V approach but this is inconsistent with the historical record in Africa which shows that commodity parastatals have consistently outperformed generalized extension services. For a critique of T-V in Africa, see Howell [1988].

Studies of the diffusion of agricultural innovations were popular in many developing countries in the "Green Revolution Decade" of the 1960s. Innovation was thought to be the best single indicator of the multifaceted dimension called modernization, the individual-level equivalent of development [Rogers, 1976b]. Research on the diffusion of new technologies was justified because it was assumed that technology was the prime mover in development. In most diffusion studies, farmers were interviewed in one-shot interviews in order to trace the acceptance of a particular innovation. Correlation analysis was usually used to assess the correlation between attributes of individuals such as age and education and the spread of the innovations. Research on the correlation between extension and diffusion of technology also was carried out because it was thought that information on the pattern of diffusion could be of direct help to extension workers in speeding up the adoption of new technology.

As part of the Special Rural Development Program in Kenya, Ascroft, Roling, *et al.* [1973] conducted a survey of the impact of the extension service in Tetu District. They divided farmers into several groups on the basis of their progressiveness. The progressiveness index was based on the number of recommended practices adopted and the time at which they were adopted. Their survey showed that more progressive farmers tended to have larger farms, better access to extension, and were more likely to have a fully titled farm. For example, they found that agricultural extension officers visited 100 percent of the most progressive farmers but only 41 percent of the laggards. A surprising 37 percent of the laggards received no visits from any extension officers compared to only 8 percent of the population as a whole. But Ascroft, Roling, *et al.* pointed out that 81 percent of the progressive farmers initiated contact with agricultural extension officers while only 17 percent of the laggards did and concluded that the imbalance in extension contact could at least in part be attributed to farmer demand.[140]

An important exception to the generalization that research and extension programs disproportionately help larger and richer farmers is found in the diffusion of hybrid maize in Kenya. Gerhart [1975] and Heyer and Waweru [1976] reported that whereas the bulk of agricultural research expenditures in Kenya has been geared to export crops and to large farmers, research on maize has developed varieties for different ecological zones which have been quickly adopted by large and small farmers throughout the country. For example, during the 1960s, the area in hybrid maize grew from 400 acres to over 800,000 acres and it "proceeded at a rate somewhat more rapid than hybrid corn was adopted by American farmers thirty years earlier" [Gerhart, 1975]. Gerhart observed that the new hybrid varieties were first adopted by large farmers followed by the bulk of small farmers after a few years.[141]

In summary, diffusion studies have provided valuable information on the influence of institutions, particularly extension services, on the adoption of innovations and farmer assessment of new technology. But many researchers became disillusioned with diffusion studies. Roling [1970], for example, argued that variables such as age, education of the farmers, and the ratio of extension workers to farmers were unable to explain the behavior of noninnovators. Rogers [1976b] surveyed 1,800 diffusion studies in developing countries and concluded that many of the studies were too narrowly conceived, they ignored important structural barriers to change, and they did not study noninnovators. We agree with Roling [1988] that there is a need for more research on communication as a process and more research on knowledge systems, field experiments and political variables.

6. Farming Systems Research (FSR)

Several experienced researchers [Belshaw and Hall, 1972; Palmer-Jones, 1977a; and Collinson, 1981] contend that much of the microeconomic information collected in the sixties and seventies was of limited relevance to small farmers in Africa for the following reasons:

    (1) Most studies failed to address the information needs of small farmers in the context of their goals and management strategies.

    (2) There was a large gap between the values, interests, and education

of researchers and extension agents on the one hand, and small farmers on the other.

(3) Many researchers studied only one or at most a few farm enterprises such as maize or sorghum.

(4) Most studies failed to take into account the impact of social and political institutions on household decision making.

(5) Research findings rarely were disseminated in a form usable by farmers.

In light of these difficulties, numerous researchers recommended that more research should be pursued within a cropping and farming systems framework (FSR) [CGIAR, 1978; Norman, 1980; Gilbert, Norman, and Winch, 1980; Byerlee, Collinson, et al., 1980; Collinson, 1981, 1982; and CIMMYT Economics Staff, 1984].

The primary goal of FSR is to design research programs which are holistic, interdisciplinary, and cost-effective in generating technology which is appropriate to the production and consumption goals of rural households in specific microenvironments. The focus on a systems approach to the study of farming systems is a key feature which distinguishes FSR from old style farm management research.[142] However, there are more similarities than differences between farm management research and FSR. The ability of systems scientists to model the complex interactions within cropping systems and between cropping and livestock subsystems is one of the potential advantages of FSR.

Research on farming systems accepts the following as a point of departure: FSR is a complement to and not a substitute for strong national commodity research programs; FSR should focus on the rural household as a production and consumption unit; farming practices that have evolved over generations are assumed to be well-honed and adapted to the goals of rural households as joint consumption/production units; both farm and nonfarm activities should be analyzed in FSR; the farmer should be an active partner in carrying out trials of new varieties and agronomic practices; and in most national research systems there is a need to increase farm level experimentation by both FSR teams and commodity researchers.

Four or five basic steps are generally recommended in carrying out FSR: conduct informal (reconnaissance) surveys of farmers to identify their problems and constraints on achieving their multiple goals, conduct on-farm trials of promising techniques and varieties, analyze results of farm trials in several locations and evaluate the economics of promising interventions, assess the experiences of farmers in using the recommended practices and varieties, and extend promising interventions to a broader group of farmers [Norman, 1980; Byerlee, Collinson, et al., 1980].

In the 1970s, farming systems research was initiated in several IARCs, including IRRI in the Philippines, CIAT in Colombia, IITA in Nigeria, ICRISAT in India, and CIMMYT[143] in Mexico. FSR programs to date have stressed cropping systems research and have relied on two key disciplines—agronomy and economics. For example, CIMMYT[144] stresses the role of agronomists and economists in informal surveys [Byerlee, Collinson, et al., 1980]. Much of the FSR by the IARCs has focused on cropping subsystems and the production decisions of farmers. Little attention has been given to livestock, marketing, and off-farm employment and linkages between macroeconomic policies (e.g., exchange rate on farmers).[145]

Farming systems research expanded rapidly in Africa from 1976 to 1985. Several conferences were turning points in the evolution of FSR. First, a conference on intercropping in Tanzania emphasized the need to study indigenous farm practices such as intercropping rather than promoting western practices such as sole cropping and row cropping [Monyo et al., 1976]. Second, under the leadership of David Norman, a conceptual framework for FSR was spelled out at a conference at the Institut d'Economie Rurale in Mali [IER, 1977]. Third, a symposium on the Experimental Unit approach in Senegal highlighted the importance of farm level experimentation [ISRA, 1977]. Finally, a conference on land use in Burkina Faso represented a major synthesis of social, economic, geographical, and technical perspectives on farm level studies in Africa [ORSTOM, 1979]. Numerous papers in the ORSTOM conference proceedings directly or indirectly endorsed a systems approach to agricultural research.

While FSR received much attention in the early 1980s, FSR is not a new approach to agricultural research in Africa. There is an invisible literature on FSR in Africa which can provide a perspective on current FSR programs. A major research effort on farming systems was reported in *Agricultural Research for Development*, an indispensable book edited by Arnold [1976]. Arnold analyzed the experience of the Cotton Research Corporation (then the Empire Cotton Growing Corporation) in setting up a new research station at Namulonge in Uganda which was charged with developing a cotton research program for Uganda by drawing on the cotton breeding work in Trinidad and the practical research on cotton at the Barberton Research Station in South Africa. Three important insights can be gleaned from the Namulonge experience for FSR today. First, a decision was taken at the Namulonge research station to evaluate complete farming systems in a multiyear context (five years), including crop rotations, instead of relying on conventional test plot research. This decision was later to dominate the research program of the station and make a major contribution to research on tropical agriculture. Second, Arnold reported that "devising a highly productive farming system on the Ancient Ugandan soils took, not five years, but fifteen" [p. vii]. Third, Arnold reported, "Soil science had only a small place in the original plans" but that "when it became clear that soil fertility was at the heart of the problem of raising productivity . . . the Ministry of Overseas Development was enlisted to increase the soil fertility research program." In summary, the CGIAR [1978] report which urged the international research centers to move more research off the stations was simply reinforcing what had been learned in Uganda in the 1950s—on-farm research is not a luxury but a valuable input in shaping the research programs of experiment stations and national research systems.

Other examples of FSR in Africa which predate contemporary programs include studies conducted by multidisciplinary teams in the 1960s. Foremost of these is the Uboma study in eastern Nigeria in which a socioeconomic and nutritional survey was carried out in Uboma Village in 1964, followed by experimentation in agronomy, health, nutrition, and livestock [Oluwasanmi et al., 1966].

Another FSR pilot program which predates contemporary FSR is the Experimental Unit program in Senegal [See ISRA, 1977; Faye and Niang, 1977; Benoit-Cattin, 1977a]. ISRA established an Experimental Unit in a small number of pilot villages in an experimental zone in the groundnut basin. The purpose of the program was to develop technological packages with an emphasis on increasing yields without degrading soil resources. The first two phases—analytical and socioeconomic studies, and agronomic experimentation—were pursued by researchers of different disciplines and commodity specialists. The unique aspect of the approach came in the third phase when potential modifications in existing farming systems were tested by farmers in the villages in the Experimental Unit Zone. Testing was first done under strict research supervision with the best farmers.[146] Successful innovations were then presented as a total package for demonstration and pre-extension in each agroecological zone. During later phases of testing, farmers were given managerial responsibility so that the end result of the testing sequence was a recommended modification in the farming system which could be widely diffused by the extension system in the zone and beyond. The program was phased out in Senegal because it was costly and it had not been able to come up with recommended changes in farming systems which had been used by the Extension Service and widely accepted by farmers. Also, see Boutillier et al. [1962] for multidisciplinary research effort in Senegal.

As initially conceived, the primary justification for FSR was to improve farming systems productivity. As FSR spread in Africa, four additional contributions emerged: development of better agricultural research methods, improvements in institutional performance, documentation of factors affecting existing farming systems, and promotion of farmers' perspectives on research and extension programs. While FSR has represented a step toward more relevant research, it has not yet lived up to the expectations of African policy makers and donors.[147] Several factors account for the limited success of FSR in Africa:

1. The concept of a multidisciplinary team in practice generally turned out to be a two-discipline team—an economist and an agronomist. The contribution of other disciplines such as soil science should be heeded [Arnold, ed., 1976]. Also, anthropologists can help identify the "indigenous knowledge" underlying current practices and constraints on recommended practices [Brokensha, Warren, and Werner, eds., 1980].
2. Much FSR has, in fact, been cropping systems research, reflecting the impact of narrow commodity mandates of the IARCs such as CIMMYT. By the mid-1980s, there were several calls to broaden the scope of research to include livestock [Butler, ed., 1984], food consumption, and nutrition [M. Smith, 1984].
3. FSR has overemphasized short-run, marginal improvements in farming systems based on farmers' perceptions of immediate constraints. Correspondingly, FSR has underemphasized the importance of strong commodity research programs and has given too little attention to policy issues [Eicher, 1987].
4. Most FSR has been carried out on a project basis, independent of on-station research programs and isolated from policy debates.

Eicher [1988d] reported that FSR was in decline in Africa. FSR has been too narrow to have a significant impact. It is now clear that FSR is not a panacea. It must be supplemented with vigorous commodity research and must have strong links with planners and policy analysis. FSR also must be perceived by African researchers to be in their professional self-interest in terms of status, salaries, and promotion.

FSR studies include the following: East Africa, Collinson [1982, 1988]; Tanzania, CIMMYT [1977b]; Kenya, CIMMYT [1977a]; Zambia, CIMMYT [1978]; Nigeria, Norman [1980] and Menz [1980]; Niger, de Miranda and Billaz [1980]; and Burkina Faso, Swanson [1980]. Collinson [1982, 1987, 1988] discusses FSR in Eastern and Southern Africa. For a discussion on the role of social scientists in FSR, see Byerlee and Tripp [1988]. Other studies include Merrill-Sands [1986], Simmonds [1986], and Maxwell [1986].

Chapter VI. Livestock

Until recently, social science research on livestock was dominated by anthropologists.[148] Only in the past few years have economists taken a serious interest in livestock. Moreover, technical scientists have historically concentrated on cattle but in the past few years they have broadened their agenda to include small ruminants—sheep and goats. Anthropologists have been preoccupied with pastoralism (nomadic and seminomadic herding) and defending the pastoralists (herders). But the structure of livestock production is changing in Africa.[149] The majority of the Fulani in West Africa is sedentarized or in the process of becoming sedentarized and cultivating subsistence crops at permanent homesteads in Senegal, Mali, Burkina Faso, Cameroon, and Central African Republic, as are the Arabs in Chad.[150] The implication of this structural change is the need to focus research on seminomadic herding, mixed farming, and small ruminants.

1. Behavior of Herders: Empirical Evidence

One of the major arguments in the literature over the past sixty years is whether herders in Africa are economically rational in the western sense of being profit maximizers. The debate on herders' motives has been narrowly conceived in terms of which objective herders strive to meet—wealth and prestige or profit. The hypothesis that a primary motive for accumulating large herds is to gain prestige and serve as a store of wealth is important to policy makers because this motive can work against livestock improvement programs which are striving to increase off-take rates and thereby reduce herd size and overgrazing.

The thesis that herders may not be profit maximizers has its origins in research by the American anthropologist, M. Herskovitz's [1926] well-known diagnosis of the "cattle complex," the term used to draw attention to the highly significant social and religious roles of cattle among the East African pastoralists and the role that large cattle herds may play in serving as a store of wealth and prestige in these societies. But there have been few rigorous tests of the wealth hypothesis. Numerous anthropologists in the 1960s and 1970s reported that herders were insensitive to opportunities to sell cattle when prices were favorable and that herders' negative supply response to price seemed to imply nonprofit maximizing behavior. For example, in a study of the Kel Adrar Tuareg pastoral group in northern Mali, Jeremy Swift [1975] observed that the nomadic Kel Adrar did not seem to respond to normal market forces, they had strictly limited cash needs (to pay taxes and buy salt, tea, sugar, and tobacco); and they met their target income when cattle prices increased by selling fewer animals. But Swift later [1977] shifted his position and stressed the shrewd ability of the Kel Adrar to cope with the Sahelian drought by herd diversification, herd movement, storing feed, and sharing animals.

The dynamics of herder behavior has recently been enriched by research on the relationship between the demography of herd size and the survival of pastoral families under harsh ecological conditions. L. H. Brown [1971], former agricultural officer in Kenya, showed that pastoral families who lived off milk, meat, and grain (obtained by trading milk for grain) required much larger herds to survive during periods of drought than anthropologists had previously believed necessary.[151] Later, Brown [1977] developed a simulation model of the factors influencing herd size for an average pastoral family of eight people in Kenya.[152] The results of Brown's simulation model show that a family of eight must maintain a herd size of thirty to thirty-five adult cattle to insure family survival in a semiarid area. Brown's findings show that objectives such as prestige, wealth, or income are likely to be secondary to the survival objective.

An excellent book by two Swedish anthropologists, Dahl and Hjort [1976], entitled *Having Herds: Pastoral Growth and Household Economy*, adds further support for Brown's stress on the importance of large herds for survival purposes. Dahl and Hjort contended that low off-take rates observed in many countries are not a function of cultural factors but technical and economic reasons such as disease, death, and low productivity of the rangeland. From their research in northern Kenya, the authors simulated alternative herd sizes required for the survival of pastoral families under conditions of adversity. They concluded that it is rational for a herder to accumulate a large herd to produce milk for his family as well as to provide an insurance fund and portable bank.

A major weakness of the Dahl/Hjort model is that it does not permit exchange to take place except to allow enough grain to be purchased to meet the minimum physiological requirements of the herding family. Although Dahl and Hjort admit that revisions in their model would be necessary if it were possible to convert some of the produce to the herd into agricultural goods by trade or barter [Dahl and Hjort, 1976, p. 178], they do not stress the fact that almost all of the pastoral groups in Africa are actively engaged in trading milk and cattle for grain except in sparsely populated areas where there are limited market opportunities. For example, Harold K. Schneider [1979] cites numerous studies [e.g., D. G. Bates and Lees, 1977] which show that pastoralists throughout Africa purchase grain (primarily millet) and other products. The Brown and Dahl/Hjort models are abstract and insightful but their predictive value is likely to be very low.

There has been little rigorous research in Africa by economists on the motives for keeping cattle. This is understandable because of the paucity of data on herd numbers, prices, etc. Also, the few economists who have studied livestock have been slow to draw on research findings from Europe, Latin America, and North America which point up the pitfalls in generating short- and long-run supply response coefficients for a subsector which is fraught with cyclical patterns of building up inventories — a process that can take years and one that can affect the validity of short-run supply response coefficients. This topic has been explored for many years by economists outside of Africa.

Reutlinger's [1966] study of the beef industry in the United States pointed out that the cow and calf components of beef supply enabled the producer to slaughter for consumption or retain calves to build up inventories. Thus it was perfectly feasible to generate negative short-run supply elasticities among profit-maximizing beef producers in periods when inventories were being built up. Tryfos [1974] reported similar results in his study in Canada. Jarvis [1974] showed that cattle can be used as capital goods or consumption goods in his study of the Argentine cattle industry. Jarvis developed a number of microeconomic models in which producers acted as portfolio managers and held cattle as long as their capital value in production exceeded their slaughter value. His model showed that a negative short-run supply response was consistent with a positive long-run supply response when producers were withholding cattle to build up herds.

Turning to studies of supply response in Africa, Khalifa and Simpson [1972] reported a positive supply response to changes in prices for one market in the Sudan but Low [1980] questioned the reliability of the data and the validity of the findings of Khalifa and Simpson. In a recent study in Swaziland, Doran, Low, and Kemp [1979] argued that herders keep cattle as a store of wealth (status and prestige) rather than profit maximization and that they are little motivated by market incentives. The authors pointed out that the quality of data in Swaziland is exceptionally good[153] for testing the income *versus* the wealth hypothesis because data on herd size and slaughter are recorded for twenty-seven years (1950-76) as part of the closely supervised cattle trade with the Republic of South Africa. The authors tested the wealth *versus* income hypothesis by regressing the off-take rate against rainfall and real cattle prices. The regression results revealed an inverse relationship between cattle sales and cattle prices and rainfall and led the authors to accept the wealth/status hypothesis as a motive for keeping cattle. The authors concluded that the government's strategy of reducing overgrazing by improving breeding and pastures will not work even if these interventions would improve the quality of cattle and bring higher prices in South African markets.

In a recent critique, Jarvis [1980] raised serious questions about the Doran, Low, and Kemp findings by pointing out that on theoretical grounds a store of wealth hypothesis could be explained either in terms of a negative short-term price response or the communal grazing system. Jarvis noted that the communal grazing system by itself is capable of explaining the overgrazing, advanced-age slaughter, and the other herd characteristics cited by Doran, Low, and Kemp as evidence for the store of wealth motive. Moreover, Jarvis showed that Swaziland government programs which offer incentives to increase beef production will have the desired effect whether or not the store of wealth effects hold. In their reply, Low, Kemp, and Doran [1980] agreed that communal grazing could also explain the wealth hypothesis and that the advanced age of cattle and constant herd composition do not provide convincing evidence for the store of wealth hypothesis.

Turning to West Africa, Delgado [1979a] investigated the behavior of Fulani herders and Mossi sedentary farmers in Burkina Faso. Delgado rejected the wealth (status) hypothesis and showed that the Fulani are more integrated into the market economy than the crop farmers. He found that the Fulani herders have entered the market economy over time and now sell three-fourths of the value of their annual production of livestock and livestock by-products as compared with the sedentary Mossi farmers who sell one-sixth of the value of their crops (mainly millet and sorghum).

In summary, research on the behavior of livestock herders in Africa is about at the same point where research was on the economics of crop production some twenty years ago—many assertions and a sparse supply of facts.[154] L. H. Brown's [1971] and Dahl and Hjort's [1976] findings that pastoralists need to maintain large herds for the survival of the pastoral family under harsh ecological conditions are important contributions of the 1970s. Communal grazing is now recognized as an important contribution to overgrazing. There is now agreement that cattle perform a number of social, ritual, and economic functions and that the relative ranking of these functions will vary widely according to ethnic group, country, ecological conditions, etc. What was earlier alleged to be ultra-conservative behavior of herders is now viewed as prudence. Ayuko summarizes Kenya's experience with livestock projects over the past forty years as follows:

> One of the most significant lessons arising from the experience
> of these projects is that the alleged ultraconservatism of
> pastoralists towards proposed technological interventions is
> more appropriately viewed as prudence. Pastoralists welcome
> modern technology when they perceive its beneficial
> relationship to the basis of their economy and culture and they
> will accept change at their own pace, if it is introduced under
> their control, that is initiated and directed by senior elders who
> are aware of the advantages the proposed change will bring,
> and provided it does not undermine their established culture.
> [1980, pp. 22-23]

2. Livestock Production: Major Issues for Researchers

African governments and donors are channeling millions of dollars each year into livestock projects throughout Africa in the absence of a sound knowledge base.[155] For example, the CILSS/Club du Sahel's livestock strategy [IEMVT, 1980] for the development of livestock in the eight Sahelian countries is little more than a shopping list of projects for donors.[156] The project-by-project approach to livestock development is doomed to failure in the absence of a coherent strategy of the livestock subsector which, by necessity, must be rooted in a strategy for agricultural development, including a basic knowledge of the interaction between the cropping and livestock subsectors. The following discussion reviews research on key problem areas and points to needed research to expand the knowledge base. A major point which emerges from this section is the need for a substantial increase in research on technical problems and livestock policy for the subsector as a whole.

IMPROVING THE DATA BASE

The starting point for a discussion of livestock interventions is to realize that government censuses of livestock are among the most unreliable of official data in Africa. Planners need data on national herd size, off-take rates, and mortality rates. The difficulties of generating reliable production data are understandable in light of the large percentage of cattle in seminomadic and to a lesser degree in nomadic systems. An example from Tanzania points up the need to question published data about national herd size and off-take rates. MacKenzie [1976] reports that the Government of Tanzania decided to emphasize marketing rather than production interventions in preparing its 1973 application for a $27 million loan from the World Bank because it assumed that the national herd size was adequate. But subsequently, the unofficial results of the 1971-72 National Livestock Census revealed a national herd size of only 9.4 million. The difficulty of generating reliable cattle counts in Kenya through survey research led by Aldington and Wilson [1968] points out some problems in using hide figures to estimate herd size in his beef subsector study in Nigeria. Other techniques for generating information on herd size include data obtained from rinderpest vaccination campaigns, slaughter figures, and aerial counts. Some authorities argue that data from cattle vaccination campaigns provide estimates on the size of the national cattle herd which may be as reliable as official data on the acreage and yields of subsistence crops. The head tax on cattle is another reason why it is difficult to carry out survey research to determine herd sizes.[157]

Another data gap is returns to labor in various livestock activities. While there is now universal agreement on labor constraints in performing certain cropping activities in Africa, there is a large void on returns to labor in livestock. For example, because of overgrazing associated with deep bore holes with motor-driven pumps, a number of animal specialists are recommending a shift to shallow, lined wells from which water can be drawn with a bucket, thereby reducing the number of cattle which can be maintained. But in the discussions of shallow wells, labor is usually assumed to be free—a questionable assumption as we pointed out earlier.

HERDERS OR RANCHERS?

A perennial debate since the colonial period is whether governments and donors should concentrate on assisting herders who are subsistence-oriented or on promoting ranches which produce exclusively for the market.[158] Those who support the herder strategy stress the advantages of a migratory pastoral system in response to seasonal changes in rainfall and drought, the accumulated expertise of herders, and the general failure of ranching schemes.

But there is a long history of antiherder sentiments. For example, Allan [1965] contended that "nomadic pastoralism is inherently self-destructive." Two authorities on range management in East Africa, Pratt and Gwynne, eds. [1977], assert: "In most cases . . . the people [pastoralists] are tied to a way of life that limits their own development and that leads to overstocking of the land." A leading Nigerian animal scientist, Professor V. A. Oyenuga, threw his prestige behind ranches by concluding that "the required production level can no longer be met by adhering solely to the traditional practice. . . . It calls for well-managed, heavily capitalized ranch systems supplied with feedlots on an intensive management basis" [Oyenuga, 1973, p. 395].

Large-scale goverment ranches were tried in numerous countries over the 1960-75 period with the support of the World Bank and several bilateral donors. But commercial and government ranches have performed poorly in most countries. Ranching has generally failed because of high investment costs and low returns. Von Kaufmann [1976] reports that ranches in Kenya are usually large; they receive massive loan support and government subsidies and they benefit the few. Odell and Odell [1980] document the difficulties the Government of Botswana experienced in setting up commercial ranches in the early 1970s.

The failure of numerous commercial and government ranches in the past two decades, supplemented with increasing population pressure and concerns for equity, have forced many governments to de-emphasize commercial ranches and move toward group ranches,[159] grazing blocks (associations),[160] and assistance to small herders. But group ranches and grazing blocks are extraordinarily complex; they require systematic predevelopment analysis ranging from surveys of land and water use and sociocultural aspects of traditional grazing systems to the enactment of legislation to allow adjudication of group landholding rights [Ayuko, 1980]. For example, Odell and Odell noted "that the failure of commercial ranches in Botswana was followed by a failure of the government to establish forty group ranches in the mid-1970s because of enormous logistical, managerial, technical, ecological, social, economic, and political problems" [1980]. The Odells noted, however, that these failures were invaluable in stimulating land use planning and in forcing central planners to realize that communities themselves must play a more active role in land use planning, including sanctions to prevent overgrazing. Doherty's [1979] preliminary analysis of group ranching in Narok District, Kenya, suggests that efforts to promote group ranching in Kenya appear to have increased factionalism among the Masai. In summary, the evidence (economic, political, and social) is overwhelmingly against ranches in Africa. The challenge is for African governments to help herders improve the productivity of their herds and assist them in experimenting with grazing associations and other systems by which group action can maintain agreed-upon stocking levels.

ANIMAL BREEDING

The technical coefficients for livestock are unfavorable in most African countries [ILCA, 1978, 1979a, b, c]. Depending on breed and level of nutrition, the first birth ranges from two to five years. Calving rates, defined as the percentage of cows giving birth during one year out of the total number of cows and grown heifers, are low with rates ranging from 40 to 60 percent among pastoralists. Calf mortality rates commonly are 20 to 30 percent and often exceed 50 percent during disease outbreaks or periods of low rainfall. The annual off-take rate (yield) of herds is generally 10 percent or less per year.[161] Most herds are composed of indigenous Zebu and Sanga cattle, which do not have the genetic potential of cattle in temperate zones either in terms of weight gains or milk production. But the Zebu and Sanga are well adapted to environmental conditions and the low levels of management found in many parts of Africa. Comparative studies of breeding and feeding problems are starting to clarify the nature of these technical constraints. A study by ILCA and the Institute of Rural Economy [ILCA, 1978] in Mali reported that the body weight of Maure and Peul cattle increased by only 7.5 percent per year over the 1966 to 1975 period, the mortality rate of all calves from birth to age three was 26 percent, and that 56 percent of all females born were required as herd replacements.[162]

CATTLE FEEDING AND ANIMAL NUTRITION

The lack of a reliable feed supply is a major technical constraint on animal production. Numerous studies in West Africa have shown that regardless of the stocking rate and the type of forage it is impossible to prevent cattle, sheep, or goats from losing weight during the dry season without supplemental feeding. A study of supplemental feeding of small herds in Senegal by Calvet, Friot, and Gueye [1976] revealed that there was a "substantial" reduction of dry season weight loss achieved by the feeding of 300 g of groundnut cake per animal unit per day but the authors did not investigate the economics of feeding groundnut cake.[163]

A major issue in increasing feed supplies is pasture improvement. The tender shoots or twigs of shrubs and trees (browse) are increasingly recognized as a source of feed for domestic and wild animals, especially during drought or seasonal periods of nutrition stress [De Leeuw, 1965; Brinckman and De Leeuw, 1979; ILCA, 1980b]. The deliberate burning of grazing land has been the subject of a great deal of speculation but research has shown that burning of the mature grass can improve both protein content and nutritive value. Although there are benefits of burning—it prepares land for cultivation, destroys mature grass, improves protein and nutritive value of new grass flush, and destroys parasites — there are obvious negative features such as destruction of forests and removal of soil cover which might increase erosion. But the consensus is that when the range or savannah is used primarily for grazing, there is probably no alternative to periodic burning [See De Leeuw, 1965; Van Raay, 1975; and Breman and Cisse, 1977].

H. Ruthenberg [1974] and Doppler [1980] provide some of the first estimates of the economic potential of investments in permanent pastures in semihumid areas in West Africa. Doppler concludes that the economic incentive for introducing or expanding beef production in a semihumid zone in Togo is small and risky. He draws on data from two ranches in Togo for his simulation model and concludes that the costs of pasture development are too high for investing in pasture improvement on ranches.

A number of animal scientists such as R. E. McDowell [1978] stress the strategic role of improving animal nutrition as a prerequisite for increased weight gains and reduced mortality from disease. A simulation model of the livestock economy in northern Nigeria [Manetsch et al., 1971] demonstrated that improvement in animal nutrition was the most significant intervention to increase livestock production. A study of herders with White Fulani cattle on the Jos plateau in Nigeria found that disease was not a serious problem and that the major factor affecting productivity was poor nutrition, especially in the dry season [Pullan and Grindle, 1980]. All herds studied were relatively unproductive as regards fertility, growth rates, off-take, and milk production with the exception of one herd which received a significant amount of dry season supplementation. But again, the economics of supplementation was not investigated. In summary, research on supplemental feeding, animal nutrition, and pasture improvement under herder conditions is in its infancy. One authoritative reviewer, Stephen Sanford, summed it up [1977] as follows: "Range science applied to Africa is fraudulent because it has never done adequate local homework before pontificating (northern Nigeria is a slight exception to this rule)."[164]

CATTLE DISEASES AND PARASITES

The seven major cattle diseases and parasites in Africa in order of importance are as follows: rinderpest, contagious pleuropneumonia, clostridial diseases, internal parasites, trypanosomiasis, East Coast fever, and foot and mouth disease.[165] Trypanosomiasis at present is receiving the most attention by researchers because rinderpest, contagious pleuropneumonia, and clostridial diseases are controlled by vaccination[166] and because tsetse flies virtually preclude the use of approximately one third of the African continent, including some of the best watered and fertile land. Trypanosomiasis is one of the major constraints on intensified livestock production and rainfed farming in sub-Saharan Africa. The flies are the vectors which transmit several species of trypanosomes, causing human sleeping sickness and trypanosomiasis in livestock [J. Ford, 1971; ILCA, 1979c; FAO, 1981a].

Present tsetse control measures include clearing of vegetation which harbors flies, spraying of insecticides by helicopter, ground spraying, breeding and release of sterile male flies, and the use of traps. Specialists believe that tsetse flies can be economically controlled only in combination with sedentary agriculture/livestock systems because it does not pay to clear land only for livestock grazing.[167] Moreover, the cleared land will be overtaken by tsetse unless there is a minimum human population density which farms it and keeps it cleared. For a valuable analysis of Uganda's attempt to control tsetse flies in the 1960s, see Talbot [1972] and Jahnke [1974] and a spraying program in northern Nigeria in the late 1950s [Putt and A. R. Ellis, 1980]. An ILCA [1979c] study reports that trypanotolerant N'Dama and West African shorthorn cattle are economically attractive in tsetse-infested environments where other breeds can only be maintained under high levels of management based on chemoprophylaxis and therapy. For example, the N'Dama and shorthorns have been successfully introduced into Central African Republic, Gabon, and Congo where tsetse had prevented the introduction of the Zebu or Sanga cattle from surrounding areas.

RANGE MANAGEMENT AND LAND USE PLANNING

Many governments, donor agencies, and scholars argue implicitly or explicitly that overstocking is the basic cause of soil erosion, desertification, and degradation of the range in many African nations.[168] Interest in the overgrazing and desertification was stimulated by the livestock losses during the 1968-74 Sahelian drought. For twenty-five years preceding the drought, the livestock population in the Sahel increased significantly because of above average rainfall, expansion of public water facilities, and improved veterinary services [Bernus and Savonnet, 1973; Gallais, ed., 1977]. Because of livestock migration across national boundaries during and following the drought, the losses for the Sahelian region are rough estimates. The FAO [1975a] estimates that the total cattle population in the ECOWAS countries declined 23 percent (35.5 million in 1971 to 27.3 million in 1974) over the 1971-74 period. Farmer/herder conflicts were numerous as a result of the abrupt shifts in migration streams during and following the drought.[169] For example, many herdsmen from Burkina Faso and Mali drove their herds south in search of water and better grazing land.

Research is showing that desertification is a complex and long-run process [Glantz, ed., 1987]. In the early 1970s, a few alarmists claimed that desert encroachment in the Sahel was advancing at the rate of up to 100 miles per year. But Bernus and Savonnet [1973] have pointed out that the boundary between the Sahel and the desert to the north was elastic and although three major droughts over the past 100 years had induced desert encroachment, there was a "retreat" of the desert after normal rainfall resumed and the range recovered. Lovejoy and Baier's [1975] analysis of this drought/recovery cycle reveals that migration from the drought-stricken areas was carried out in an organized fashion. It is now accepted that "only over periods longer than a decade can desertification be clear, distinguished from the less lasting effect of drought" [Warren and Maizels, 1977].[170] But even though desertification may not be occurring in the technical sense, the issue of environmental deterioration is a serious problem in many parts of Africa.

The argument that controlled grazing is a prerequisite for improving pasture productivity is echoed throughout Africa by policy makers and donors. A popular view is that overgrazing in the Sahel is a classic case of "A Tragedy of the Commons." Controlled grazing schemes have been more aggressively pursued in East Africa than in West Africa. The best-known schemes have involved the Kamba in Kenya and the Masai in Tanzania and Kenya [Talbot, 1972]. For a historical perspective on Masai pastoralism, see Jacobs [1975, 1978]. Over the 1946-61 period, the British colonial government introduced large-scale grazing schemes and range demonstrations among the Masai in Kenya and Tanzania. In fact, much of the cost of the grazing schemes was covered by the Masai through self-tax levies. Nevertheless, Jacobs [1975] noted that the schemes were considered a failure for the following reasons:

> In their failure the grazing schemes clearly demonstrate the
> futility of water development and management measures
> without control of the livestock population. The danger of
> water development and range improvement projects in general
> without adequate management provisions cannot be too
> strongly stressed for almost invariably the result is the
> deterioration or destruction of the range resource involved.

In West Africa, there were colonial systems of controlled grazing supervised by local chiefs (e.g., in the Niger Delta in Mali) which broke down with national independence when national governments took over the rights to control land use.

The three-way competition for land—livestock, game and forest reserves, and crops—has led to recent decisions to establish national land use planning capabilities and legislation in Tanzania and Botswana (Tanzania's 1964 Land Use Planning Act and Botswana's 1975 Tribal Lands Grazing Policy) [See Botswana, 1975; and Hinchey, ed., 1978.]. The purpose of the legislation is to encourage socially optimum use of grazing land and natural resources. Research is urgently needed on the multifaceted dimensions of land use. Hitchcock [1978], an anthropologist, has provided a wealth of information on indigenous systems of keeping cattle in Botswana. This type of ethnographic information will be invaluable in developing socially optimum methods of livestock management in the 1990s.

SMALL-SCALE FATTENING SCHEMES

Small-scale fattening schemes may be a desirable alternative to continued open grazing or ranching. There are two basic types of fattening schemes. The first is to remove the range steers from the range for a sixty to ninety day fattening period. The second and more common fattening scheme is part of a mixed farming system whereby a crop farmer becomes a mixed farmer by adding livestock for oxen cultivation and then fattening one or two of the cull oxen with supplementary feeding during the dry season. Available research on mixed farming is sparse but preliminary findings suggest that a mixed farming strategy is profitable only if it can be linked to cash crops such as groundnuts and cotton. In the Tenkodogo region of Burkina Faso, Delgado [1979a,b] found that small farmers who added a livestock enterprise and became mixed farmers found it difficult to graze the animals in their unfenced fields, faced a seasonal labor conflict because major labor requirements for managing livestock coincided with the harvesting of millet and sorghum in November, and lacked dry season fodder for their animals. Delgado concluded that government policies should be directed toward improving the present "entrusting" system whereby nomadic (Fulani) herders care for animals owned by sedentary (Mossi) farmers rather than encouraging farmers to add a livestock enterprise and become mixed farmers.[171]

Mali has experimented with two types of small-scale fattening programs—one unsuccessful and one successful. The unsuccessful scheme was a cull oxen model which was started in the mid-1970s under the assumption that a massive expansion in animal traction would provide cull oxen for fattening by small farmers. For a variety of reasons, this plan was dropped. A successful model of fattening of thin range steers has been developed in Mali since 1975 as part of an AID Livestock I project. This program is a dry season fattening scheme (*embouche paysanne*) in which small farmers receive supplementary feed, extension, veterinary care, and credit for ninety-day feeding programs for approximately two head of cattle per farm [Delgado, 1980].

Eddy's [1979] LP study of mixed farming among the bush Tuareg in Niger illustrates how mixed farmers in harsh terrain have developed an intricate, mixed farming system of cattle, goats, and grain production to cope with drought and fluctuations in grain prices. In spite of government appeals for the Tuareg to produce more cattle and less grain in the pastoral zone, they still produce enough grain for their family consumption because of their experience in being forced to purchase grain for family survival at greatly escalated prices during the 1968-74 drought. In short, mixed farming, like intercropping, is an effective way to spread risk. A fattening scheme in Niger is discussed by Wardle [1979].

3. Small Ruminants

Whereas small ruminants (sheep and goats) in Australia, New Zealand, and the Middle East have a threefold objective—meat, milk, and wool—small ruminants are used almost exclusively for meat and milk production in Africa. In Muslim areas, sheep production is mainly aimed at producing rams which are consumed or sold for slaughter during religious festivals, and other ceremonies. Since the research base on small ruminants is modest and because small ruminants are produced almost exclusively by small herders and farmers in Africa, more attention should be given to research on small ruminants [see McDowell and Bove, 1977; Mathewman, 1979; McDowell and Hildebrand, 1980]. ILCA has a comparative research program underway on small ruminants [1979a].

4. Research Direction

Both technical scientists and economists need to step up research on livestock in order to catch up with the impressive knowledge base on livestock that has been put into place by anthropologists over the past fifty years. But research, in our judgment, should move from such narrow disciplinary interest as supply response studies to problem-solving research under field conditions. For example, there are few rigorous *field* studies of the technical, economic, and social issues involved in mixed farming, including the economics of supplementary feeding during the dry season. Another example is the need for research on the land tenure issues involved in the transition from nomadic to seminomadic and sedentary production systems. It goes without saying that multidisciplinary research is essential in addressing these problems.

5. Livestock References

Bibliographies and Literature Reviews: Ferguson and Sleeper [1976]; Dahl and Hjort [1979]; Ergas [1979].

Worldwide: R. E. McDowell [1972]; McCown, Haaland, and De Haan [1979].

Africa, General: Allan [1965]; de Wilde *et al.* [1967]; D. G. Bates and Lees [1977]; L. H. Brown [1971, 1977]; Dahl and Hjort [1976, 1979]; J. Ford [1971]; Goe and McDowell [1980]; Darling and Farvar [1972]; AID [1980]; Konczacki [1978]; R. E. McDowell [1972]; R. E. McDowell and Bove [1977]; R. E. McDowell and Hildebrand [1980]; Monod, ed. [1975]; Galaty *et al.*, eds. [1981]; Salzman, ed. [1980]; Warren and Maizels [1977]; FAO [1981a]; ILCA [1979d,1980a,b]; Felton and Ellis [1978].

West Africa: Bernus and Savonnet [1973]; Ariza-Nino and C. Steedman, eds. [1979, 1980]; Ferguson, ed. [1976]; Sargent *et al.* [1981]; Ruthenberg [1974]; ILCA [1979a,b,c]; Kafando [1972]; Shapiro, ed. [1979]; Doppler [1980].

Sahelian West Africa: SEDES [1974]; Breman and Cisse [1977]; FAO [1977]; Gallais, ed. [1977]; Horowitz [1977]; IEMVT [1980]; Riesman [1978)]; Swift [1977].

East Africa: Jacobs [1975]; Pratt and Gwynne, eds. [1977]; Schneider [1979]; Talbot [1972]; Helland [1980].

Angola: de Carvalho [1974].

Botswana: Botswana [1975]; Hitchcock [1978]; Hjort and Ostberg [1978]; Odell and Odell [1980]; Sanford [1977]; Fortmann and Roe [1981]; S. F. Eicher [1981].

Burkina Faso: Barrett *et al.* [1982]; Delgado [1979a,b]; Gooch [1979]; Herman [1979]; Riesman [1977]; Tyc [1975].

Cameron: J. Holtzman, J. Staatz, and M. Weber [1980]; Zalla *et al.* [1981].

Côte d'Ivoire: Staatz [1979].

Kenya: Aldington and Wilson [1968]; D. Campbell [1979]; R. K. Davis [1971]; Doherty [1979]; Hopcroft and Ruigu [1976]; Ruigu [1978]; P. Spencer [1973]; von Kaufmann [1976]; Dahl [1979].

Mali: Delgado [1980]; Gallais [1975]; ILCA [1978].

Mauritania: Grayzel [1977]; Swift [1975]; Toupet [1977]; Vermeer [1981].

Niger: Horowitz [1972]; Bernus [1974a]; M. S. Diarra [1975]; Dupire [1962]; Eddy [1979].

Nigeria: Stenning [1959]; A. Cohen [1965]; de Leeuw [1965]; Ferguson, ed. [1976]; Fricke [1979]; Putt and P. R. Ellis [1980]; Manetsch *et al.* [1971]; Awogbade [1979]; van Raay [1975]; Dunbar [1970]; Brinckman and de Leeuw [1979].

Swaziland: Low and Kemp [1977]; Doran, Low, and Kemp [1979]; Low, Kemp, and Doran [1980]; Fowler [1981].

Tanzania: MacKenzie [1976]; Jacobs [1978]; Grindle [1980]; Zalla [1981].

Togo: Doppler [1980].

Uganda: Jahnke [1974]; Eilam [1973].

Pastoral Migration: Stenning [1959]; I. M. Lewis [1975]; Dahl and Hjort [1979]; Hjort [1979].

Economic Theory and Behavior of Pastoralists: Herskovitz [1926]; Reutlinger [1966]; Khalifa and Simpson [1972]; Jarvis [1974]; Tryfos [1974]; Doran, Low, and Kemp [1979]; Low, Kemp, and Doran [1980]; Schneider [1979]; Jarvis [1980]; Livingstone [1977b]; Low [1980]; Crotty [1980].

Chapter VII. Trade, Marketing, Credit, and Consumption

In this chapter, we turn from agricultural and livestock production to empirical research on food and agricultural distribution systems, including international trade, local trade and marketing, transportation, processing, and storage; credit and rural financial markets; cooperatives; and consumption and nutrition.

1. International Agricultural Trade

Trade policies are particularly important in Africa since most African economies are open with foreign trade often accounting for around a quarter of GNP.[172] Research on international agricultural trade policies has focused on: export linkages, the role of intra-African trade, regional integration, special trade agreements, and food aid.

EXPORT LINKAGES

The central issue in research on policies to promote export commodities such as coffee, cocoa, cotton, and tea has been the long-run impact of these policies on the domestic economy. Drawing on arguments first presented by Prebisch and ECLA researchers, many researchers and policy makers have criticized policies continuing the reliance on trading patterns established during colonial rule and on primary commodities which face low-income elasticities of demand [I. G. Stewart and Ord, eds., 1965]. For example, Clower et al. [1966] argued that Liberia experienced "growth without development" in the 1960s because of the weak linkages between production for export and the domestic economy. Several case studies in S. R. Pearson and Cownie [1974] project the net social gains, domestic resource costs, and linkage effects of export-oriented policies. Case studies include cotton in Uganda [Jamal, 1974], cocoa and coffee in Côte d'Ivoire [J. D. Stryker, 1974b], cocoa in Ghana [S. L. Gordon, 1974], and coffee in Ethiopia [T. Haile-Mariam, 1974], the results show that weak forward and backward linkages have limited the potential impact of export-led growth. But the studies also show that exports can contribute to national income via final demand linkages when most of the revenue goes to smallholder producers such as tea production in Kenya. Recent studies of the effects of export promotion in agricultural trade include Jabara and R. L. Thompson's [1980] analysis of Senegal's dependence on groundnuts, Franco's [1981] appraisal of cocoa prices in Ghana, Priovolos's [1981] study of coffee trade in the Côte d'Ivoire, and Stein's [1979] analysis of trade in East Africa. See also the studies using a dependency framework cited in the discussion of historical and theoretical perspectives.

INTRA-AFRICA TRADE

Trade within Africa has played a central role in studies of African history [A. G. Hopkins, 1973]. Although many African leaders have called for greater African unity and a concomitant increase in intra-Africa trade [Mboya, 1967] the amount of intra-Africa trade is still modest today. Studies supporting the case for increased intra-African trade include Vinay [1968], Chileshe [1977], and Akinwumi and Adegeye [1977]. Vinay acknowledges the large amount of clandestine trade among African countries [see also Collins, 1976] but goes on to develop a case for substantially increasing the proportion of international trade which is carried out within Africa. But the question of barriers to increased intra-African trade involves more than the level of import and export duties. There are major institutional barriers to increased trade, including the lack of agreement on common grades for agricultural products such as maize and sorghum, language barriers, and financial and communication networks which link African countries to Europe rather than to each other. Chileshe's work [1977] is a useful source of data on the level of intra-Africa trade and the potential role of trade among African countries. Acknowledging several problems in trying to quantify intra-Africa trade, Chileshe nevertheless argues that intra-Africa trade is substantially higher than it has generally been thought to be. Chileshe points out that although it is increasingly being recognized by African policy makers that more intra-African trade would be mutually beneficial, there is agreement that existing international marketing agreements (e.g., the GATT agreement) are major constraints on the expansion of intra-Africa trade.

REGIONAL INTEGRATION

Organizations to promote intra-Africa trade have had a checkered history dating from the Federation of Nyasaland (Malawi and Rhodesia) to the East African Community and more recently to the West African Community (ECOWAS). The most recent failure is the East African Community (EAC, comprising Kenya, Tanzania, and Uganda) which was established in 1967 and "unofficially" dissolved in 1978. The EAC faced many of the common problems associated with integration schemes in other continents, including disagreements over how to distribute gains and wavering political support. Ndegwa [1968] evaluated the long-term importance of import substitution in the economic development of East Africa. Hazlewood [1979] analyzed the factors leading to breakup of the East African Community; he concluded that Kenya, Uganda, and Tanzania were more closely integrated on an informal basis before independence than in later years and that the formal treaty proved to be divisive. Key problems leading to the demise of the EAC in 1978 were: dominance of Kenya's position in the industrial field and the inability to agree on the location of new industrial plants to allow Uganda and Tanzania an opportunity to "catch up";[173] disagreements over the distribution of gains, particularly through transfer taxes among member countries; continuing balance of payment problems; weak coordination of regional transportation; inadequate investment program by the East African Development Bank; and different development strategies pursued by Kenya and Tanzania. A. Weber and Hartmann [1977] criticized the EAC for failing to establish common agricultural policies to the detriment of trade in agricultural commodities within the EAC.

SPECIAL TRADE AGREEMENTS

An important avenue open to countries to increase their gain from exports is to enter into special trade agreements such as commodity agreements. But commodity agreements have met with little success, largely because no single African country is a dominant supplier of a particular commodity. For an assessment of commodity agreements, see Gwyer [1973]. Etherington [1972] used an oligopolistic framework to assess the potential benefit of Kenya's participation in an International Tea Agreement. He argued that because East African countries are small, they could expand production with little impact on price and they would be in a strong bargaining position with India and Ceylon who need a tea agreement to stabilize export revenues.

An alternative to commodity agreements is broadly defined concessionary trade agreements, such as the 1975 Lomé agreement between the EEC and 48 African, Caribbean, and Latin American countries. The Lomé convention represented an attempt by the EEC to compensate for the loss of bilateral privileges by developing countries, particularly former colonies, when the European Economic Community was formed. The Lomé convention was designed to promote trade cooperation, to stabilize export earnings, and to accelerate financial and technical aid. Under the agreement, up to 94 percent of the agricultural exports of the developing country members were to enter the EEC duty free. While it is too early to assess the long-term impact of the Lomé agreement, it has been criticized on the grounds that liberalization is far from complete and that nontariff barriers remain largely intact [S. Harris *et al.*, 1978; Akinwumi and Adegeye, 1977]. For a radical critique of the Lomé convention, see Galtung [1976].

FOOD AID

Until recently, almost all food aid in Africa was for emergency purposes but food aid for development is now becoming firmly entrenched in a number of countries, such as the Sudan, Zambia, Ghana, Senegal, Tanzania, and Zaire. The debate about food aid basically turns on the possible disincentive effects of food aid on domestic production and saving. If the food aid is sold on the open market, it may depress prices and, therefore, dampen producer incentives. The availability of food aid is also thought to allow governments to avoid making the necessary investments in research, extension, and credit to expand food production. Although they primarily rely on evidence from sub-Saharan Africa, Maxwell and Singer's [1979] valuable survey of food aid suggests the following criteria for the effective use of food aid: there should be a clear need for food assistance, food aid should substitute for commercial imports, food aid programs should be incorporated into poverty-oriented programs, there should be a guarantee that food aid will be available when promised, and food aid and financial aid should be coordinated.

One of the few attempts to assess the impact of food aid in sub-Saharan Africa is C. Stevens's [1979] analysis of Botswana, Lesotho, Burkina Faso, and Tunisia. The focus is on the effect of food aid as a form of development assistance rather than emergency relief. A study of food aid in rural Kenya concluded that the food for work program increased agricultural production, income, capital investment, employment, and marketable surplus [Bezuneh, Deaton and Norton, 1988]. For a recent study of food aid in the third world see Singer, Wood, and Jennings [1987].

2. Local Trade and Food Marketing

Marketing research during the 1950s and 1960s concentrated on marketing boards and on the movement of export crops. Few researchers were concerned with the economics of local trade or food marketing. W. O. Jones [1972] contended that the neglect of food marketing during this period may have been due to the fact that local markets were adequately performing their major task (from the perspective of policy makers) of assuring food supplies to urban markets. But during the 1960s and 1970s, urban centers grew rapidly and the view spread among planners and policy makers that unreliable local markets encourage farmer self-sufficiency which inhibits the provision of stable food supplies to urban centers;[174] farmers are, for the most part, thought to be price takers subject to the whims of traders who collude to lower producer prices;[175] and government intervention in food marketing was necessary to ensure regular supply of food to cities.

Over the past twenty years, research on local trade and food marketing has been a medium-priority research topic. Riley and Weber [1979] reported that over fifty U. S. Ph.D. dissertations were completed on marketing in sub-Saharan Africa during the 1970s. But research coverage has been uneven both geographically and in the range of topics covered. Moreover, there has been considerable variation in research findings in different studies and among different countries. For summaries and assessments of the literature, see W. O. Jones [1972, 1987], Whetham [1972], Wilcock [1978], CILSS/Club du Sahel [1977], Lele [1977], Couty [1977], and B. Harriss [1979a, b]. For annotated bibliographies, see Arditi [1975], CILSS/Club du Sahel [1977], and Riley and Weber [1979].

STUDIES BY GEOGRAPHERS AND ANTHROPOLOGISTS

Much of the research on local traders and markets has been carried out by geographers and anthropologists who have described marketing channels, including the origin and timing of markets, the role of ethnic groups in determining market patterns, and the impact of local markets on social relations. For the most part, anthropological case studies which dominated marketing research in the 1950s and 1960s paid little attention to economic issues such as the volume of food trade, price determination, operating margins, or the capacity of markets to handle expanded production [W. O. Jones, 1972]. A classic collection of twenty-eight case studies on local marketing is *Markets in Africa* [1962] edited by Bohannan and Dalton. A shortcoming of the case studies was a failure to examine the linkages between rural periodic markets and regional and national marketing systems.

Research by anthropologists and geographers on the origins of local markets has revealed that the reasons for the development of local markets and the length of time that formal markets have existed vary greatly throughout Africa. For example, Hodder found that traditional markets in Yorubaland, Nigeria have a long history and that their growth was closely associated with long-distance trade [Hodder and Ukwu, 1969]. Wood [1974] provided contrary evidence in East Africa where only a small proportion of the local markets in Kenya have been in existence for more than forty years. Handwerker [1974] found, on the basis of interviews of 783 market sellers in Liberia, that the origins and changes in market places in Liberia were linked to changes in social organization. For an overview of the origins of local markets in Uganda, see Good [1970].

Much of the literature by economic geographers on the spatial organization and evolution of market networks has focused on the utility of central place theory for explaining patterns of rural development [C. A. Smith, 1976a, b] and on the functioning of periodic markets. Most rural markets are periodic with the periodicity occurring on a fixed interval of days for each market. Some of the advantages of periodic markets are: they are spatially distributed so as to reduce the distance traveled; enough people attend markets to make a trip worthwhile for traders and assemblers; and periodic markets often serve as an initial collection point for the flow of food to towns and urban centers. E. P. Scott's [1972, 1978] studies in northern Nigeria illustrated the role of periodic markets in linking exchange systems and in stimulating regional economic development. Additional studies of periodic markets are McKim [1972], J. B. Riddell [1974], and R. T. Smith, ed. [1978].

An important concern in ethnographical studies has been the relationship between marketing, social relations, and who gains and who loses as a larger percentage of agricultural production is sold. A widely cited collection of twenty essays analyzing the impact of marketing on production and social relationships is found in Meillassoux, ed. [1971]. Raynaut's [1973, 1977] studies of millet exchange at the village level in Niger stressed the relationship between markets, class formation, and wealth. In general, anthropologists have emphasized how commercialization of agricultural production and marketing undermines village social structure and promotes class conflict.[176]

COMPETITIVENESS AND EFFICIENCY OF LOCAL TRADE

Research on the economics of private trade and local markets began to receive substantial attention in the mid-1960s. Most of the research on private trade has used case studies to evaluate marketing costs of the flow of food from rural to urban centers and the extent to which markets are constrained by imperfections and inefficiencies such as too many intermediaries, inadequate infrastructure, barriers to entry, exploitive pricing practices by middlemen, and excessive variability in quality and quantity of supply.

The efficiency of distribution systems which rely on a long chain of small traders received early attention by researchers because many policy makers contended that a long chain of middlemen increases marketing cost. But research has shown that the length of the chain is a false issue because the number of intermediaries involved in retail food trade in rural areas is generally limited to farm families themselves or at most two to three intermediaries [Thodey, 1969; W. O. Jones, 1972; Whetham, 1972]. Trade in staple foods destined for urban centers may involve several intermediaries but is often conducted separately from local trade, occasionally even in a separate market place [Bauer, 1954; Whetham, 1972]. Several researchers also have argued that the long chain of traders leading to urban centers in most countries is an efficient use of available resources [CILSS/Club du Sahel, 1977; Wilcock, 1978].

The process of price formation through "haggling" has been of particular interest to researchers, especially anthropologists. While low incomes of both farmers and traders make the potential gain from haggling attractive, the importance of haggling appears to have been greatly overemphasized by researchers. In most cases, both buyer and seller have a reasonably good idea of the range of prices prevailing in the market on any given day. Thodey [1969] and Gladwin and Gladwin [1971] have shown that this range tends to be considerably smaller for traders dealing in larger units, reducing the impact of haggling on major transactions. Whetham [1972] noted also that buyers and sellers are often well known to each other and that final bargains reflect degrees of kinship and social obligations as well as estimates of the equilibrium price.

The most important issue in the market imperfection debate has been the extent to which traders influence prices through collusion, manipulation of weights and measures, and misreporting of market information. Collusive practices are often attributed to the presence of groups which are organized along product lines such as the Hausa who control the Kola trade [Lovejoy, 1980] and livestock trade [J. M. Cohen, 1971] throughout West Africa. Support for the belief that markets are subject to collusive practices is found in Bauer and Yamey [1959] who showed a positive correlation between producer price and the number of buyers in the northern Nigerian peanut trade. In his landmark study, Bauer [1954] contended that some barriers to entry can stem from tribal affiliations and trader groups which regulate prices and establish conditions and fees for entry. Anthonio [1968] argued that a shortage of stalls in Ibadan markets allowed yam wholesalers to set prices above competitive levels. In an insightful article, M. P. Miracle [1968] argued that price formation among farmers was competitive in West Africa but that there were large departures from competition in the distribution system for all major products beyond local assembly markets. Miracle argued that even markets with large numbers of buyers and sellers may be influenced by cartelization and that cartels wield monopsony and monopoly power through their control over capital resources. In Miracle's view, there was considerable collusion among traders of seasonal commodities which can be stored and of nonseasonal commodities (e.g., Kola) not locally produced.

W. O. Jones [1972] challenged the validity of many of the studies reporting market imperfections in an important analysis of staple food marketing in Nigeria, Sierra Leone, and Kenya. The study was designed to provide a definitive account of price formation and the role of foodsheds in serving urban markets. In addition to impressions gained through personal interviews, three major tests of marketing efficiency were used to evaluate the extent of market imperfections: bivariate coefficients of correlation of prices were calculated to test whether prices in nearby markets move together, reflecting the extent of market integration; price differentials between markets were assessed in comparison to transportation costs, and seasonal price fluctuations were compared with storage costs. Jones found in his study[177] that inadequate physical infrastructure did not appear to restrict marketing; there was no evidence that producers and consumers are exploited by middlemen; there were few signs that family ties impair functioning of markets; market entry appeared to be relatively unrestricted; marketing chains tended to be short; but markets were weakly integrated. Jones concluded that governments should restrict their role to improving the integration of markets and the ability of markets to respond to changes in demand and supply. The major findings presented in W. O. Jones [1972] are reviewed and assessed in W. O. Jones [1980a].

W. O. Jones's finding that rural markets in Africa are weakly integrated has been noted in other studies throughout Africa. Ongla [1978], for example, in a case study of markets in Yaounde, Cameroon found significant price variations between markets and that problems in transportation were the primary cause of high marketing margins. Despite widespread evidence of weak market integration, most of the research carried out in the 1970s supported both Jones's conclusion that African marketing systems are reasonably efficient and competitive in the face of numerous obstacles *and* his policy recommendation that governments should restrict their role to improving marketing intelligence and roads [Hays, 1975; CILSS/Club du Sahel, 1977; Ejiga, 1977; Ongla, 1978; Hays and McCoy,

1978; Southworth, W. O. Jones and S. R. Pearson, 1979; E. Berg, 1980]. This point of view is forcefully presented in the World Bank report [1981b, p.58] which states: "The central problem in marketing and input supply is the very general tendency to give too large a set of responsibilities to public sector institutions, and too few to other agents — individual traders, private companies, and farmers' cooperatives."

The findings of neoclassical economists such as W. O. Jones, Ejiga, Hays, and others were generally unchallenged until Barbara Harriss's [1979b] devastating critique questioned the relevance of the neoclassical paradigm and raised several important questions about research methodology. Specifically, she contended that correlation coefficients are an inadequate tool for demonstrating either market integration or competition;[178] the simplifying assumptions made in analyzing marketing margins over time and space were very crude, leading to problems in interpreting results; that the majority of economic analyses of agricultural marketing has displayed a "serious lack of logical relationship between the data presented and the conclusions derived;" attempts to synthesize the conclusions of various studies, such as by W. O. Jones [1972] and E. Berg's synthesis of a report prepared for the Club du Sahel [CILSS/Club du Sahel, 1977] seem to be guilty of oversimplification;[179] and the "fetishism of competition" which has dominated African marketing research is ideologically motivated: it is anti-interventionist and pro-infrastructural. Harriss concluded that preoccupation with the debate over the competitiveness of rural markets has diverted attention from the structural interrelationships between production, exchange, and consumption. She urged future research to devote more attention to structural factors such as the extractive role of markets.

In summary, most efforts to evaluate market performance through an analysis of intermarket and interseasonal price movements indicate that private trading systems do suffer from the following imperfections: high transaction costs due to lack of information; high physical handling costs due to inadequate transportation infrastructure; instability in supply channels is pervasive since there tends to be little control over weather, disease, and storage losses; and lack of product homogeneity. But collusive practices do not appear to be widespread and there is little or no evidence that either cartels or monopsonistic buyers at assembly markets exert a significant downward pressure on producer prices.

In our judgment, Barbara Harriss [1979b] is correct in her observation that the debate over governmental control *versus* private traders has diverted attention from research on the range of political economy factors that affect the role of marketing in the development process. Research by neoclassical economists on market imperfections and inefficiencies has frequently generated few policy relevant conclusions because it has relied too much on idealized models. More attention needs to be given to the interrelationship between agricultural production and marketing in a dynamic context, examining the linkages between production, assembly, processing, distribution, and consumption such as Sherman's [1981] research. Marketing research within the political economy framework outlined by Wilcock [1978] and the social-institutional analyses by French researchers [e.g., Raynaut, 1977; Kohler, 1971] have provided insights which complement research carried out within the structure-conduct-performance paradigm.

LIVESTOCK MARKETING

Trade between herders and farmers is one of the oldest forms of exchange in sub-Saharan Africa. Herders specializing in cattle historically have relied on sedentary farmers for their staple grain which is usually millet. This exchange has taken several forms, including informal trade around watering and camping places in the Sahelian zone, permanent market places in the Sudanian zone, and reciprocal arrangements between farmers and herders involving rights to grazing, passage, and manure as well as grain and livestock products. While formal market exchanges have increased in recent years, trading arrangements have broken down as population pressure has forced farmers to move into areas traditionally used by herders and as nomadism has given way to sedentary herding.

Over the last twenty-five years, West African governments have struggled with the basic livestock marketing policy question of whether they should stress improvements in traditional processing and marketing systems or should they invest in slaughter plants in producing zones and in the transportation and distribution of fresh and chilled meat.[180] Unfortunately, research on livestock marketing has been a low priority for economists and answers to this question are just starting to emerge. Some of the early studies by economists include: Aldington and Wilson's [1968] study of beef marketing in Kenya and Ferguson's [1967] study in Nigeria.

Complementarities between cattle breeding and livestock production in the savannah zone and meat markets on the West African coast have given rise to long-distance trade in livestock [A. Cohen, 1965; Couty, 1977]. Population and income growth have led to increasing demand for cattle in the southern forest zones where livestock production is limited by the tsetse fly. Since urban markets for meat on the coast of West Africa are long distances from arid and semiarid production zones in the interior countries of Chad, Niger, Burkina Faso, and Mali, it is necessary to transport cattle, or alternatively meat, to the coastal markets. In West Africa, historically cattle have been moved to urban markets by trek (on foot) but recently cattle have been shipped in trucks and by railroad because it was assumed that trekking resulted in large weight losses in cattle and often death as cattle moved through tsetse zones on their way to coastal markets. Abner Cohen [1965], an anthropologist, examined the role of a network of Hausa dealers who moved around 75,000 cattle a year some 600 miles from northern

Nigeria to Ibadan. As part of a major CRED study of livestock marketing in West Africa,[181] Staatz [1979] studied long-distance cattle trade between producing zones in Burkina Faso and Mali and consuming centers in the Côte d'Ivoire. Surprisingly, Staatz found that, depending on the time of the year, quality of the grass, the state of health of animals, and the speed at which cattle are trekked, weight losses of animals which had been trekked 400 to 800 miles to market could be quite low. Staatz also showed that margins for cattle merchants and butchers were relatively small, and that a lack of market infrastructure did not hinder the market forces in regulating supply and demand. Staatz also studied the location of slaughter plants and found that it was more profitable for producers to slaughter cattle in the consumption zones on the West African coast rather than in the interior producing zones in Mali and Burkina Faso because of the lack of local demand for by-products.

Trekking has also been the major means of marketing cattle in Eastern and Southern Africa. However, Sanford [1977] noted though that trekking is in danger of dying out in Botswana. Cattle corridors have also come under pressure in East African countries such as Kenya due to increased population and expansion of farming. Researchers such as Staatz and Sanford generally agree that governments should attempt to facilitate trekking by marking corridors and establishing water points.

Some researchers are starting to take a subsector approach in their analysis of production, marketing, and processing rather than studying marketing in isolation. Holtzman, Staatz, and M. Weber [1980], for example, recently evaluated the livestock production and marketing subsystem in the Northwest province of Cameroon and Ruigu [1978] analyzed the milk system in Kenya.

FISH MARKETING

While it is not uncommon for farmers to fish on a part-time basis, most of the fish consumed is supplied by individuals who specialize in fishing. Frequently, an entire village or groups of villages near major lakes and marine fisheries are "fishing villages," relying on market exchange to obtain their staple grain. In most cases, fishermen sell their fish to traders as soon as they are landed and traders are responsible for drying the fish and getting them to market before they spoil. In many places, such as the Lake Chad basin on the borders of Niger, Chad, Cameroon, and Nigeria, traditional fish distribution systems have been active for decades [Couty, 1964; Couty and Duran, 1968]. The prosperity of this trade appears to have fallen in recent years due to such factors as falling yields, increased supplies from marine fisheries, and an expansion of frozen fish marketing in urban centers [Couty, 1977]. Spoilage rates are often quite high in traditional fish marketing channels, necessitating large marketing margins. Hoffman et al. [1974]

showed, for example, that most species of East African freshwater fish spoil within twenty-four hours if they are not dried or frozen.

One of the best approaches to improving the productivity and profitability of small-scale fishing is to focus on improved vertical coordination of fishing, fish processing, and fish marketing. Unfortunately, there has been little research on the efficiency of traditional drying methods or on the economics of alternative marketing strategies. Krone's [1970] analysis of frozen fish marketing in several West African countries showed it is unlikely that local fisheries would benefit from the adoption of frozen preservation techniques. Krone reported that the prices of smoke-dried fish are cheaper than frozen fish and that fish marketing margins vary considerably, often accounting for 50 percent of consumer prices.

Valuable insights into the marketing strategies used by fish traders were provided in a series of studies in Ghana. Gladwin and Gladwin [1971] evaluated fish sellers' decisions on the Cape Coast in terms of the expected profits of sales in alternative markets. Later C. Gladwin [1975] used a hierarchical decision model to predict decisions about when and where fish sellers market their fish. Gladwin's approach is based on the view that sellers make decisions sequentially rather than simultaneously, weighing advantages and disadvantages (profitability) of alternative markets. Traders know by experience which markets usually entail the greatest risk but they do not know what the prices will be on any given day in distant markets. Therefore, traders consider markets in terms of increasing levels of risk and go to the least risky market in which it is expected that all their expenses will be covered. Gladwin also pointed out that since there is no storage at markets and the quantity brought to markets is necessarily predetermined, the price received on any given day will be determined solely by demand. Gladwin's hierarchical model was able to predict the decision about when and where to market in 90 percent of the cases for which it was tested in 1973. Quinn [1978] studied the behavior of marketing women in a village only twelve miles from the one studied by the Gladwins in Ghana. Quinn acknowledged the predictive capability of C. Gladwin's model but disagreed with the thesis that marketing women form conditional probabilities in making their decisions. Rather, Quinn argued, most fish retailers subjectively introduce heuristics that eliminate the need for recall, summarization, and computation. The main implication of Quinn's amendment to the C. Gladwin model is that strategies of market women are more sensitive to risk than indicated by Gladwin.

3. Transportation

Africa's rudimentary transportation network poses one of the major constraints on the distribution of agricultural commodities and the integration of African economies. Commodities are generally transported to local assembly markets using headloads, baskets, bicycles, and donkeys. Transport by animals is gradually giving way to trucks. Transportation problems are particularly severe in countries such as Mali, Niger, Chad, and Mauritania with vast land areas and roads that are often impassable during the rainy season [Grolleaud and Kohler, 1979]. Because of the poor infrastructure, much of the value added in retail prices of agricultural commodities, particularly food grain, is due to transportation costs. Whetham [1972] contends that transportation may be the largest single item in trading costs. L. Sorenson *et al.* [1975] estimate that by the time maize has travelled sixty kilometers from a village in Zaire, its price has increased by over 25 percent because of transportation costs. Larger margins are not uncommon with farm gate prices frequently being as little as half the price found in nearby assembly markets and towns.

Studies of transportation costs in sub-Saharan Africa must be location-specific. For example, Thodey [1969] compared rates per ton/kilometer over eighteen different routes in Ethiopia and found that charges on the highest cost route were over six times those of the lowest cost route. Inability to secure backhauls is considered to be a major problem throughout Africa. Probably the greatest influence on cost other than distance, however, is the condition of the road. Thodey [1969] estimated that even under the assumption that an all-weather road serves ten kilometers on either side of the road, only 10 percent of Ethiopia was served by year-round roads.

There is little hope that transportation problems will be alleviated in the foreseeable future. Rail and barge rates tend to be significantly lower than truck rates but few locations are served by these modes. Clearly, transportation infrastructure should be a major item on the agenda of African governments and donor agencies over the next twenty to thirty years. [Ahmed and Rustagi, 1987]. One of the critical problems is that African governments often underfund recurrent costs to maintain the roads which are often built with foreign aid. Peter Heller's [1979] "Underfinancing of Recurrent Development Costs" and the CILSS/Club du Sahel's [1980b] report on recurrent costs should be closely examined.

4. Processing

A wide range of processing technologies with varying factor intensities and technical efficiencies is available in most countries. Most research on processing has been undertaken by consulting firms in the form of feasibility studies for governments and donor agencies. Unfortunately, many of these hastily conducted feasibility studies have been carried out by engineers and management consultants who assess the private profitability (financial returns) of the new technology to the individual firm while ignoring transportation and handling costs, as well as the equity (employment and income distribution) consequences of alternative processing techniques. Another shortcoming of many studies is the failure to compare alternative processing techniques. For example, studies of rice processing in West Africa by Tempelman [1972], Oni and Olayemi [1975], Rosenboom and Parker [1975], and Goodwin [1975] analyzed costs and returns of only one processing technique.

The two major issues in processing research are the efficiency and equity effects of alternative techniques. One of the first studies to deal with these two questions was W. Miller's [1965] analysis of five palm oil processing technologies in Nigeria—hand processing, the small screw press, the small hydraulic hand press, the intermediate mill, and large capital-intensive mill. The study revealed that there was a large difference in labor and capital intensity of the technologies. Capital investment per firm in 1964 ranged from eleven dollars in hand processing to $156,000 for the capital-intensive mill. Hand pounding was less technically efficient but more economically profitable than alternative technologies because of lower transportation costs, lower priced family labor, and more consistent use of fruit with a high oil content. For an exchange of views on Miller's findings, see Purvis [1968a], and Kilby [1967, 1968, 1969].

Miracle [1966] reported that hand pounding maize with mortar and pestle is the primary processing technique used for maize throughout sub-Saharan Africa. Miracle cited studies of maize pounding in Angola, Malawi, and Sudan which showed that an average of twelve to thirteen hours was needed to pound thirty pounds of maize. Since this yields an estimated four to five days' worth of food, women must spend three or more hours a day in pounding their family's daily grain intake. In light of this time commitment, Miracle stated that it is not surprising that "Africans are quick to adopt grinding mills wherever they can be obtained."

An important issue in the choice of processing techniques— consumer taste— was addressed by Francis Stewart [1979] in a study of maize processing in Kenya. Stewart found that small-scale hammer mills were far superior to large roller mills in terms of employment and surplus generated but they were not competitive because the flour produced by the small mills was considered inferior by consumers to the flour produced by the large mills. Stewart attributed the demand for high quality flour to advertising by large flour mills and because of a skewed income distribution which induces a demand for higher priced flour.

Eastman [1980] has compared hand pounding *versus* mechanized hulling and grinding in a report on a new flour milling system in Botswana, Ghana, Nigeria, Senegal, and Sudan. Eastman argues that the advantages of mechanized processing include the demand for sorghum and millet flour will increase if a flour with acceptable taste, texture, and color can be developed; mechanical processing is needed to meet urban demand for more highly processed foods; time spent on hand pounding—two to three hours a day per family—could be more profitably spent on other activities; and flour produced with dry milling techniques has a longer shelf life than those produced using traditional wet milling processes. But

Eastman acknowledged that mill-ground flour has a lower nutrient content—25 percent less fat, 10 percent less crude fiber, 15 percent less ash, and slightly lower protein—than flour produced with hand processing methods. A weakness of Eastman's assessment is that he underplays the potential loss of income-earning opportunities of those currently engaged in hand pounding.

The argument that mechanized processing displaces labor engaged in hand pounding was challenged by Uhlig and Bhat [1979] in their study of maize milling which drew primarily on Kenyan production patterns and price relationships. They contended that because capital-intensive techniques are associated with an increase in scale and capital expenditure, primarily to serve urban markets, large mills do not actually displace labor. While Uhlig and Bhat provide a framework for evaluating alternative techniques for processors oriented toward urban foodsheds, based on flour quality and production rates as well as technical efficiency, they assessed only the private profitability of alternative techniques, and failed to consider transportation and handling costs, and equity issues.

Emmy Simmons [1975] studied food processing and preparation among Moslem women in villages in northern Nigeria. She noted that home food processing and preparation provides employment for thousands of rural people in northern Nigeria—primarily women. Since the women are in purdah, the prepared food is sold by their children in the villages. While women only earned three to nine cents an hour in processing, the return on invested capital generally ranged from 16 to 40 percent because of the low initial investment. With 90 percent of the women in the villages studied engaged in food processing and preparation, Simmons was concerned over the possible loss of employment opportunities as modern processing replaced home processing.

A definitive study of processing in Africa is an analysis of the choice of rice processing technology in Sierra Leone which was carried out under the direction of Dunstan Spencer and reported in D. S. C. Spencer, May-Parker, and Rose [1976]. The Spencer study broadened Timmer's [1972] framework for studying rice processing in Indonesia by incorporating the costs of assembling raw materials from farmers prior to processing and the costs of distribution of the final product to consumers throughout Sierra Leone. Spencer's research team collected engineering and socioeconomic data throughout Sierra Leone on all rice processing technologies in use in 1973-74. A surprising finding was the relatively high technical efficiency of hand pounding of rice, a finding which has been underplayed by salesmen of modern processing plants, and by some donors. The technical efficiency (pounds of clean rice per hundred pounds of husked rice) was found to be 68.4 percent for hand pounding; 67.5 percent for small steel cylinder mills; 70.0 percent for small rubber roller mills; 64.0 percent for large disc sheller mills; and 72.0 percent for large rubber roller mills.

Spencer fed the technical data into an LP model to generate information on the output, employment, imports, exports, and foreign exchange effects of each of the five technology options. The surprising finding was that if the government's 1974 policy of subsidized interest rates (10 percent) was allowed to continue, then machine processing would rapidly replace hand pounding which would lead to a large displacement of female labor, estimated to be 40,000 person years of employment. This is a staggering figure in a small country with a total population of about three million people. Spencer concluded that no one processing technique but rather a combination of hand pounding and small scale mills would be appropriate depending on the location (i.e., the wage rates and transportation costs) in the country. Spencer's work will be a standard reference for years to come; it demonstrates how to generate information on which technology is appropriate for processing rather than advancing facile statements that Africa needs appropriate technology.

5. Storage

Three issues have stimulated research on storage over the past ten years: urbanization, drought, and growing food imports. Research on storage is sparse and largely impressionistic. The common impression is that storage losses are around 25 percent even though, as we shall see, the actual figures can range from three to forty. For an overview on storage, see National Research Council [1978].

METHODOLOGICAL ISSUES

Estimation of on-farm storage losses is as much an art as a science. The NRC defines "loss" as a measure of a reduction in weight in the amount of food available for consumption. Losses are distinguished from damage, with the latter being a measure of the proportion of grain having been infested, broken, or molded. One of the main problems in deriving reliable estimates is that losses in traditional storage vary considerably by rainfall, type of grain, and type of storage. For example, losses tend to be higher in humid areas and when grain is stored in houses rather than in separate granaries where insecticides can be used. Another problem is the lack of a standard approach for determining when damage is so extensive that grain cannot be consumed. A final problem is the sampling procedure by which grain is selected for testing. Weighing the grain in an entire granary is clearly not feasible. Exposure of grain to open air during the process of weighing will bias subsequent weighings since exposure affects the rate of insect infestation. Guggenheim [1978] found that the degree of insect infestation in household granaries in Mali was much greater near the exposed surface than in grain near the center of the granary. There is clearly a need for some consensus on procedures to follow in estimating losses [J. M. Adams and G. W.

Harman, 1977]. See FAO [1980] for an overview of methods for assessing posthtarvest food grain losses.

ON-FARM STORAGE

The greatest proportion of national grain stocks in most countries is held by rural households. Granaries are often little more than a section of the house in which a family lives. Most grain is stored in containers ranging from mud-brick silos to woven straw granaries. While grain for consumption is generally kept in large containers, often raised off the ground, seed grain is frequently sealed in gourds or clay containers and kept in the house. The volume of cereal stored is largely determined by decisions made by family members. Little is known about the volume of holdings, physical weight losses by crop and by ecological zone, time release patterns, investments in storage facilities, and techniques for minimizing storage losses [CILSS/Club du Sahel, 1977].

Guggenheim, an anthropologist, describes an on-farm storage system among the Dogon people in Mali in "Of Men, Millet and Mice: Traditional and Invisible Technology Solutions to Post-Harvest Losses in Mali" [1978]. Guggenheim found that the Dogon stored millet left on the head for up to four years. Storage losses by weight were found to be extremely low—less than 5 percent—because of the low humidity, design of granaries, use of fumigants such as ashes and smoke, constant inspection, and rotation. Hamilton [1975] described two major types of traditional storage in Botswana, including mud/wattle cribs and reed and grass baskets but these methods have largely given way to storage in jute bags.

Grain storage in Tanzanian villages is influenced by the socioeconomic position of the family. In high-income farming households nearly all grain is stored in jute bags in unoccupied rooms, often with concrete floors. Woven reed containers (one-half to three-quarter ton capacity) are used to store rice. Poor farmers generally store their maize piled on a ceiling platform in the house, usually above the kitchen fire. While losses are high in this system, the expense required to construct an elevated, free standing granary called a *dungu* could not be justified in light of the small harvests of poor households.

Studies by Hays [1975], Hays and McCoy [1978], and Ejiga [1977] in northern Nigeria were among the first studies of on-farm storage by economists. Hays found that 80 percent of millet and sorghum produced in the area studied was stored on farms for later consumption. Eighty-five percent of the households owned at least one dried earth granary (*rumbu*, pl. *rumbuna*) and the remaining households stored grain in their houses. Hays found that the capacity of *rumbuna* ranged from 2.5 to 3.2 metric tons and slightly exceeded the minimum annual requirements of a rural household. Hays [1975] also found that while timing of sales was dictated by the need for money two-thirds of the time, there was little evidence of postharvest distress sales. A major finding by Hays is that on-farm storage losses of millet and sorghum were extremely low—3 to 4 percent— because of the low humidity in the savannah environment in northern Nigeria. But in the same environment, Ejiga [1977] found that the damage due to insect infestation of cowpeas was 35 percent of stored grain. Although there is some promising research underway to reduce on-farm storage losses— storage in black plastic bags—high insect damage is a major constraint on the cowpea production. Kamuanga and D. S. C. Spencer [1981] found in a study of thirty farmers in the Office du Niger irrigation scheme in Mali that the postproduction losses of rice were as follows: grain not harvested and left in fields, 6 percent; machine thresh- ing, 3 percent; household storage, 9 percent; and hand pounding and processing, 2 percent; giving a total of 20 percent.

OFF-FARM STORAGE

Cereals are stored in off-farm facilities for four general purposes: pipeline stocks for milling and distribution, interseasonal storage to stabilize intra-annual supplies and prices, buffer stocks to stabilize interannual supplies, and emergency reserves. In most countries, these functions are being shared by private traders and public agencies. Off-farm storage facilities include bag storage and bulk silos. Grain is usually stored in jute bags if it enters the market. Little grain is handled in bulk due to a lack of infrastructure.

Public storage of food grains has historically been of little importance because of the small percentage of people living in cities and the token volume of grain imports. But the need for public sector storage changed overnight during and following the 1968-74 drought in the Sahel and Eastern Africa. All Sahelian coun- tries have now assigned the procurement and management of cereal storage to either autonomous public agencies or to existing government departments. The need for public storage is obviously influenced by the proximity of a country to surplus supplies. Reports by the CILSS/Club du Sahel [1977, 1979] and Grol- leaud and Kohler [1979] argue, for example, that interior Sahelian countries must take a different approach toward public storage than coastal countries. For coun- tries such as Niger, six months can pass between donor agreement to ship food and the delivery of assistance. Coastal countries, on the other hand, can afford to hold smaller reserves. Hamilton [1975] points out that a country such as Bots- wana has less need for public storage because it has access to food from surplus- producing countries in the region.

Public investment in off-farm storage of food grain represents a major cost. The cost of accumulating and renewing stocks is influenced by the quality of off- farm storage facilities. In order to minimize the cost of buffer stock and emer- gency reserve programs, storage facilities are needed which enable grain to be held for two or more years. Unfortunately, the loss record on long-term govern- ment storage has been poor [E. Berg, 1980]. The FAO [1981b] has recently completed a study for the CILSS/Club du Sahel of the cost of building and maintaining a comprehensive grain storage system for the Sahel.[182] The report has touched off a debate among donors and there are many critics who question the need for the system and further question the wisdom of donors paying for most of the annual $14–15 million cost of maintaining the scheme. The World Bank questions buffer stock schemes by noting "buffer stocks are an expensive and risky road to food security" [1981b, p. 69].[183]

STORAGE COSTS

While storage cost estimates must be treated cautiously, some general observations can be made about the costs of storage in on-farm granaries, bag warehouses, and bulk silos. Where significant bulking and transfer is not needed, traditional storage will most likely be the least costly. Bag warehousing is often the least expensive when the grain must pass through a number of intermediaries and be transported via a number of different modes. But for long-term storage, net costs per ton are often lower in modern silos than in warehouses using bagged storage [Grolleaud and Kohler, 1979].

Wilcock [1978] and L. Sorenson et al. [1975] proposed that the least cost approach to storage is often one that relies on a combination of different storage techniques. If village level stocks are desired, either by private traders to facilitate intertemporal price arbitrage or by public agencies as a local buffer stock, improved granaries similar to those used at the farm level will generally be the lowest cost option. The bulk of stocks for long-distance trade, intra-annual stabilization schemes, and for interregional transfer to deficit areas could be held in bag warehouses. Finally, stocks held for interannual price and supply stabilization and for emergency shortfalls could be held in bulk silos.

STORAGE LOSSES

The main empirical findings are: on-farm losses of millet and sorghum are quite small—often under 5 percent annually—in the low humidity savannah regions, losses of maize are low in Eastern Africa, but high (30 to 40 percent) in the rain forest area in West Africa, and losses in publicly operated warehouses generally are much higher than traditional on-farm granaries. In summary, losses by weight are often lower than the 25 percent figure frequently cited by the FAO but damage from insect infestation—especially cowpeas—is extremely high with 30–50 percent figures being common.

Storage losses of perishable staples are high: cassava, yams, sweet potatoes, white potatoes, taro, bananas, plantains, and breadfruit. Booth [1974] reports that postharvest losses of tropical root crops are enormous. The main problem with storing roots and tubers is that they have a high moisture content relative to grains and they continue to respire and metabolize at a faster rate than cereals.

Research on storage of perishables should receive increased attention. The inability to store and transport perishables constitutes a major constraint on increased specialization of production.

IMPROVEMENTS IN TRADITIONAL GRANARIES

Even though storage losses in traditional granaries are lower than had been previously assumed, losses of even 5 to 10 percent may be significant from the perspective of an individual household. Nevertheless, it is not clear how much room there is for improvement of traditional storage. Hays [1975] argues that there is limited scope for reducing losses in northern Nigeria because of the cost effectiveness of the traditional *rumbu*. Many researchers have argued that selective use of insecticides in on-farm granaries can reduce insect losses. Increased crib drying has been successful in reducing losses in the humid regions of West Africa and Zambia [NRC, 1978].

6. Credit

Dale Adams and Douglas Graham [1981] recently pointed out that foreign donors have spent in excess of $5 billion on agricultural credit projects in the Third World in the past several decades. Rural financial markets in Africa are dominated by informal lenders, including merchants, traders, friends, relatives, and money lenders. Rural people rely heavily on informal rather than formal sources of credit because many loans are used for consumption purposes such as ceremonial obligations and school fees [Vasthoff, 1968; L. Miller, 1977]. In fact, numerous studies have revealed that nonfarm use of credit generally accounts for over half of the funds borrowed. Osuntogun's [1980] study of 220 small farmers in the cocoa zone in western Nigeria revealed that 60 percent of the credit received from the government financed cooperative credit scheme was used for nonfarm uses in 1977. Only a small portion (7.8 percent) of this total was used for ceremonial purposes while 42 percent was used to pay school fees of their children. Adegboye [1969] studied the "pledging" of assets such as cocoa trees as collateral for loans from informal borrowers and found that children's education ranked first among reasons why Yoruba farmers in western Nigeria pledged cocoa trees for loans. The relative importance of loans for consumption purposes is often highly location-specific and seasonal in nature. For example, Roger King's [1975, 1981] studies of credit in northern Nigeria revealed that most small farm loans were used to pay for farming inputs, including hired labor.

Formal credit for agriculture has been generally channeled to relatively high-income areas, to export rather than food crop producers, and to classes in rural society who have land, power, and privilege [L. Miller, 1977]. Lele [1975] reported that in 1971 a total of 88 percent of the loans outstanding of the Agricultural Finance Corporation in Kenya had gone to large farmers. Moreover, in the early 1970s, smallholders in Kenya received 25 percent of short- and medium-term agricultural credit but they produced 50 percent of the marketed output. [Donaldson and von Pischke, 1973] Also see Winch's [1976] study of rice production in northern Ghana and Kinsey's [1978] study of maize production in Zambia. The modest flow of formal credit for food production has been documented by Lele [1975], L. Miller [1977], and others. The issues and problems in rural credit and financial markets in Africa are similar to those identified by Donald [1976], D. W. Adams [1978] and Howell, ed. [1980] in other regions of the Third World.

THEORETICAL FRAMEWORK

The theoretical framework for the analysis of rural financial markets in Africa has been almost exclusively neoclassical with emphasis on supply and demand factors following the expositions of Bottomley [1963, 1975] and D. W. Adams [1978]. Bottomley and Adams attempted to explain abnormally high interest rates for private loans in rural areas in terms of a premium for lender's risk (to cover the alleged high default rate) and other factors such as the greater administration costs to lend to small borrowers. A challenge to the neoclassical approach has come from Bhaduri [1977] in his Marxist analysis of interest rates in precapitalist agriculture in India. Bhaduri notes that official government credit surveys in India have substantially underestimated interest rates paid by farmers because some of the interest paid is concealed in various modes of repayment in kind. In Bhaduri's model, he examines how the lender influences the default rate through his selection of the interest rate to be charged and the valuation on the collateral to be offered but he presents no data to test his model. Von Pischke [1980] urged economists to examine credit within a political economy framework.

INFORMAL LENDERS

Money lenders, traders, farmers, and other informal lenders play a major role in supplying credit to small farmers because they provide loans on short notice, with little or no collateral, and few restrictions on how funds are used. But the average amount of each loan is small. Longer-term loans and large loans usually require collateral or a guarantor. There has been a dearth of comparative studies of informal and formal lending institutions. Tapsoba's [1981] comparative study of formal and informal credit systems serving small farmers in eastern Burkina Faso revealed that there were two types of informal lenders at the village level—noncommercial lenders who lent money without interest charges to members of families and friends, and commercial lenders who lent primarily within the village and to a few people from neighboring villages.

Despite evidence that informal lenders fill a valuable role in supplying short-term credit in rural areas, money lenders have been widely condemned for charging exorbitant interest rrates. The literature is full of assertions about money lenders charging 100 to 150 percent annual interest rates in Africa.[184] Linsen-meyer [1976] found in Sierra Leone that the effective annual interest rate was 168 percent but the actual yield received by money lenders was 43 percent after deducting for late payments and defaults. This finding suggests that money lenders in Sierra Leone face high risk, notwithstanding their local knowledge of villages and their clients, and that risk is an important element in the high costs of capital [Byerlee, C. K. Eicher, et al., 1983]. But this issue needs further research. The challenge for policy makers is to find out what can be done to incorporate some of the desirable features of money lenders (e.g., lending for seasonal consumption needs) into formal lending programs.

GOVERNMENT CREDIT INSTITUTIONS

A common response of African governments to the limited flow of credit to farmers from commercial intermediaries and to alleged usurious interest rates of money lenders has been to establish government credit institutions, generally in the form of national agricultural banks. Although these institutions enable funds to be directed toward target groups, they have generally been unable to recover enough of their loaned funds to continue operation without government subsidies. Tapsoba [1981] and other researchers argue that low repayment rates reflect a widely held view among borrowers that government credit is a gift.[185] Agricultural credit banks have experienced several additional problems, including high administrative costs, poor coordination, inadequate supply of loanable funds, and poorly trained personnel [Von Pischke, 1980; J. M. Due, 1980, 1981].

SHOULD INTEREST RATES BE SUBSIDIZED?

In order to reduce the power and influence of money lenders and channel more formal credit into agriculture, many African countries subsidize interest rates by charging 6-12 percent annual interest rates on loans even though inflation is running at 10-15 percent per year.[186] The policy of encouraging an expansion of rural credit through artificially low interest rates has been strongly criticized by D. W. Adams [1978] and D. W. Adams and D. H. Graham [1981] as being counterproductive because subsidized credit reduces the incentive to mobilize rural savings and prolongs the dependency on foreign aid for financing credit projects. Moreover, financial intermediaries tend to extend the bulk of their loans to large farmers in order to minimize their administrative costs. The net effect of artificially low interest rates appears to be one of helping larger farmers who often have poorer repayment rates than small farmers.[187]

Although it is increasingly being recognized that low interest rates are counterproductive, few African countries have abandoned the policy of subsidizing interest rates. Moreover, rural credit is more complex than the rate of interest. D. W. Adams [1978] and others in the collection by Howell [1980] have shown that other factors than interest rates—ease of securing credit, the timing of credit, transaction costs, uncertainty about when funds will be delivered, and collateral requirements—often discourage small farmers from seeking loans from formal institutions despite low interest rates. For example, L. Miller [1977] found that 65 percent of 249 farmers surveyed in Nigeria said that they would be "willing" to pay 15 percent interest and that 20 percent would be willing to pay 30 percent interest rates on their loans. But these data must be treated with caution because they are based on hypothetical conditions.

The relative impact of interest rates, credit limits, and form of credit disbursement on small farmers' net cash flows and farming patterns has been tested in an evaluation of a government credit program in the Cameroon. To test the on-farm effects of varying credit limits, interest rates, and other loan terms, Kamajou and C. B. Baker [1980] used LP to model small farm borrowers and found that disbursements in kind had little impact on choice of production techniques but cash loans had a positive impact on farm output and income due to reduced requirements for cash reserve; increased interest rates had little impact on the optimal solution, including pattern of land use, intensity of input use, and borrowing activities until interest rates exceeded 24 percent; and increased credit limits are more important than interest rates in improving farm income because higher limits allow a reduction in required cash reserves, releasing cash for productive investments. The Kamajou and Baker study provides strong support for D. W. Adams's [1978] conviction that interest rates should be raised, taking into account the opportunity cost of capital.

CREDIT COMPONENTS OF TECHNICAL PACKAGES

In light of the problems that governments have had in operating subsidized credit programs with their own field agents, many researchers have recommended that credit should be extended as part of a package administered by area and regional rural development programs.[188] The crucial aspects of extending credit through these programs include the following: the soundness of the technical package, the farmer's ability to implement the package, supporting services, the ability of the borrower to repay, marketing opportunities, and the system of reporting and control. Several studies have shown that Integrated Rural Development (IRD) programs addressing these issues have resulted in improved credit services to farmers and improved loan repayment rates [Belloncle, 1974; Roger King, 1975, 1981]. For example, Dennis Anderson [1975] found that credit default in the Lilongwe rural development project in Malawi was reduced by increased participation of farmers in the credit allocation process, tying credit and marketing together, and making dividend payments a function of village repayment rates.

In a study of a government credit program in eastern Burkina Faso, Tapsoba [1981] found that the program was floundering after four years of experience. A total of about 1,200 four-year loans had been extended at 5.5 percent interest for small farmers to purchase oxen and donkeys and animal traction equipment. The collection ratio had varied between 22 and 54 percent and the percentage of portfolio in arrears has risen steadily from 2 percent in 1977 to 28 percent in 1980. The study shows that in the absence of a proven biological package for the staple foods (sorghum and millet) it is unwise to shower subsidized credit on farmers and encourage them to buy animals (donkeys and oxen) and animal-powered equipment.

RURAL SAVINGS

Although mobilizing rural savings as a source of loan funds has been stressed repeatedly in recent years, most countries have not pursued this because it has been generally assumed that rural people are too poor to save. D. W. Adams, however, has argued [1978] that substantial voluntary rural savings capacities exist throughout the Third World and that subsidized loans destroy the incentives to mobilize rural savings. Haggblade [1978] found in the Cameroon that the informal savings association led to the establishment of a formal savings institution. For a survey of informal savings, see Alberici and Baravelli's [1973] survey of savings banks and institutions.

POLICY DIRECTION

Research on credit in Africa has shown that credit programs can help small farmers, especially if credit is tied to profitable technical packages (usually export crops) and to marketing organizations which can deduct credit repayments from the sales of loan recipients. For example, there are 80,000 families producing cotton in Southern Mali and 150,000 smallholders producing tea under the Kenya Tea Development Authority.

Credit programs for food production have a high failure rate because of weather instability (one crop failure in five years is common), unprofitable technological packages, corruption, mismanagement, and failure to communicate the terms of loans to farmers. The overriding problem in many countries is the lack of a profitable technical package for food crops and not the absence of credit. Unfortunately, in many countries, government credit programs will continue to fail not because of the usual problems of mismanagement and corruption but because of the tendency for donor agencies to "move money" through credit projects ahead of a solid and profitable technical package—especially for food grains such as millet and sorghum—as illustrated by Barrett *et al.* [1982] in Burkina Faso and other researchers. D. W. Adams and Graham [1981, p. 362] sum up this dilemma by observing that "if these other problems are not properly dealt with, credit (subsidized or not) will not make any difference. Credit by itself cannot raise the rate of return on farm investments."

RESEARCH DIRECTION

Research on credit has been narrowly defined and has usually relied on single-visit surveys without probing the social and political environment in which farmers are operating. Research should use a neoclassical political economy framework which examines the economic, political, and institutional forces influencing the performance of rural financial markets and credit programs [Von Pischke, 1980]. At the micro level, research on credit should be incorporated into research on the total farming system and the agenda broadened to include formal and informal credit, consumption and production credit, and rural savings.

7. Cooperatives

Over the past thirty years, many Western development specialists have felt that cooperatives were "good" for developing countries. This theme is conveyed in papers in the volume edited by Anschel, Brannon, and E. D. Smith, eds. [1969]. Moreover, in the postindependence period of the 1960s, it was widely thought by some African leaders such as President Nyerere [1967] that cooperatives could be established by building on traditional extended families and communal support systems. Finally, Scandinavian aid for cooperatives in East Africa is tied to the belief that cooperatives have performed well in Scandinavia and that this type of institution is good for Africa.

Cooperatives—especially for African smallholders—are in their infancy. For example, although the first cooperative society in northern Rhodesia (now Zambia) was formed in 1914, the cooperative movement was an informal affair for Europeans until 1948 when African smallholders were invited to participate [Quick, 1978]. Likewise, cooperatives for smallholders were first established in the 1930s in Tanzania and in Kenya as recently as the 1950s. Despite early optimism, the record on cooperatives has been one of almost uniform failure under civilian, military, capitalist, and socialist governments except in a few cases involving an export crop such as tea, cotton, or tobacco.

There are dozens of explanations why cooperatives fail. Quick points out that "with a few exceptions most authors who study cooperatives advance a 'blame the peasants' view and assert that the failure of cooperatives lies in the culture, values and attitudes and habits of rural villagers who are asked (by the government) to form these institutions" [1978, p. xi]. But as Quick points out, this is a naive view. For example, in 1965, President Kaunda enthusiastically launched a major campaign to develop cooperatives throughout Zambia as the cornerstone of the government's rural development program but seven years later "almost everyone in Zambia felt that the cooperative movement had been a costly and disappointing failure" [Quick, p. xi]. The failure was not caused by cultural factors and attitudes of cooperative members but by a complex set of bureaucratic mistakes by the government's Department of Cooperatives.

Two collections of essays on cooperatives and rural development in East Africa edited by Widstrand [1971, 1972] identified the following reasons for the failure of cooperatives: failure to appreciate the difference between African collective values and requirements for a formal cooperative, insensitivity of government-initiated cooperatives to local conditions and social structure, a widespread tendency for cooperatives to be dominated by wealthier members of rural communities, excessive government control, shifts in emphasis of government programs, poor management, and corruption. In an important study of cooperatives in northern Nigeria, Roger King [1975, 1981] attributed the failure of cooperatives to reliance on a top-down approach which is insensitive to local problems.

One of the most consistent criticisms of cooperatives has been that government initiated cooperatives have tended to reinforce existing social structures or even to aggravate class divisions rather than act as a vehicle of equitable change. Miracle [1969] argued that equity among members and the protection of the interests of members can contribute to the success of cooperatives. But Roger King [1981] argues that, in general, cooperatives are an institutional form that lend themselves to being used by wealthier members in countries following a market-oriented policy of agricultural development. In Senegal, D. B. Cruise-O'Brien [1975] reached a similar conclusion in his study of small farmers and cooperatives.

Goran Hyden observed that problems of cooperatives stem in large part from what Ekeh [1975] describes as a conflict between the moral imperative of the modern organization—the cooperative—and those of the primary social organization—the village, the clan, or the lineage. For example, in Kenya, Hyden [1978/1979] found that when there was a conflict over whether a cooperative leader should discharge his public responsibility to look after the cooperative or to "use" the cooperative for personal gain or the gain of his clan or village, unfortunately the latter prevailed in all too many cases. He also found that "most leaders and followers consider it quite normal to 'use' the cooperative in order to strengthen the position of their clan or village." Hyden concluded that until African society becomes more differentiated and a new form of social stratification emerges "wholesale introduction of cooperatives based on principles practiced in more developed countries is still far from being the solution to Africa's rural development problem" [1978/79, p. 57].

Cooperatives make heavy demands on the scarcest resource in rural Africa—skilled managerial and administrative talent. The years required to develop skilled manpower have been consistently underestimated. For example, when Tanzania's Second Five-Year Plan was released in 1969, the government announced that Tanzania would be self-sufficient in high level manpower by 1980. But because of the large outflow of trained people over the past twenty years, this target may be achieved by the year 2000.

In summary, the evidence on cooperatives leaves little room for optimism. While governments will undoubtedly continue to support cooperatives on an ideological basis, there is no evidence to suggest that cooperatives are less exploitive or more efficient than the existing system of private trading and marketing. While cooperative authorities recognize that it may take thirty to forty years to develop successful programs, this time frame falls on deaf ears of politicians who promote crash programs. Assessments of marketing and producer cooperatives can be found in Dumont [1966]; Ellman [1977]; Lele [1975, 1981]; Lele and Candler [1981]; Mensah [1977]; Apthorpe [1972]; and Feldman [1969]. For references in Francophone countries, see N. S. Hopkins [1976]; Storm [1976]; G. Martin [1976]; Charlick [1980]; W. I. Jones [1972]; Wilcock [1978]; E. Berg [1980]; and Belloncle [1974]. Tanzania's attempt to develop cooperatives over the past fifty years has been closely scrutinized in Hyden [1973a,b, 1978/1979], J. S. Barker [1979], and Due [1980]. For an analysis of Ghana's thirty-year attempt to build cooperatives, see Miracle and Seidman [1968b], and Killick [1978].

8. Consumption

In Africa, consumption expenditures (cash and consumed home production) are dominated by food, and food consumption is dominated by cereal grains and tubers. Traditionally, millet, sorghum, and maize have accounted for well over 80 percent of the grain consumed. Sorghum and millet have been the primary staples in the arid zones of West Africa while maize has dominated diets in Kenya, Tanzania, and most of Southern Africa. Tubers, particularly cassava, are the basic staple in the humid forest zone of West Africa and in most of Central Africa. For an overview of consumption patterns by region, see USDA, ERS [1981, pp. 40-55].

THEORETICAL ISSUES

Consumption analysis is perhaps the most underdeveloped component of development economics, partly because most Western development economists over the past twenty to twenty-five years have assumed that supply factors occupy the central role in explaining patterns of economic growth. To the extent consumption was considered, it was treated as a drain on savings and therefore a constraint on capital formation. But consumption linkages are now recognized to have an important impact on sectoral growth and employment generation as Hirschman [1977], Robert P. King and Byerlee [1978], and Bell and Hazell [1980] have shown. During the past decade, researchers have turned their attention to the relationships between income and the factor intensity of consumption patterns and the relative demand for locally produced *versus* imported products. Soligo [1973] hypothesized that low-income households consume goods and services requiring more labor and less capital and foreign exchange than do higher-income households. Robert P. King and Byerlee [1978] hypothesized that low-income rural households have higher relative demand for labor-intensive, rural, and home-produced products than do urban or higher income rural households. These hypotheses imply that more equitable income distribution may stimulate increased employment opportunities in rural areas via consumption linkages.

Planners concerned with food policy analysis need estimates of the sensitivity of demand to incomes and prices. It has generally been assumed that Engel's Law (that the income elasticity of demand for food is below one and falls as incomes increase) applies in low-income countries. Moreover, the income elasticity of demand for staple foods which provide the bulk of calories is usually hypothesized to be low and to fall as incomes increase while those for meats, milk, imported foodstuffs, and nonfood items are hypothesized to exceed one at all income levels. The influence of price has received little attention in consumption studies in Africa, largely because there have been few cases in which it was possible to estimate price elasticities. Consistent with consumer theory, it has commonly been hypothesized that the demand for staples such as millet and sorghum is price inelastic while that for foods such as rice, wheat, meats, beverages, services, and nonfood items is price elastic. Researchers in Africa have hypothesized that several factors other than income and price have an important effect on consumption patterns, including the size and composition of households, location with respect to a large town or city, ethnic group, education, and the extent to which production is oriented toward the market.

METHODOLOGICAL ISSUES

There is a paucity of data on consumption patterns and linkages in Africa, partially because of difficulties in generating reliable data. For example, Farnsworth [1961] argued that consumption estimates based on highly aggregated food balance sheets were significantly distorted because: reliable data came primarily from exports and government-controlled commercial transactions, agricultural production was invariably underestimated, and regional differences are so great that estimates of national consumption patterns can seriously misrepresent the consumption patterns of most of the population.

Most estimates of consumption patterns in Africa have been based on cross-sectional expenditure surveys which have often taken into account only cash transactions. As a result, few studies have been able to identify seasonal variations in consumption patterns or the change in consumption patterns as production for the market increases. For evaluations of the advantages and disadvantages of cross-sectional surveys for consumption analysis in Africa, see Howe [1966], Massell [1969], and Simmons [1976c]. Methodological issues in collecting consumption data are discussed by Winter [1975] and Sarah Lynch [1980]. Issues to consider in the selection of functional forms of demand equations and estimating techniques are discussed in Massell [1969]; Robert P. King and Byerlee [1978]; V. E. Smith, Strauss, and Whelan [1980]; V. E. Smith, Strauss, and Schmidt [1981]; and Strauss [1981].

EMPIRICAL FINDINGS

Consumer expenditure studies were conducted in a number of African cities in the 1960s primarily to generate data to construct consumer price indices for government/trade union wage negotiations. These studies provided some insights into the determinants of expenditure patterns and they generated expenditure elasticities[189] for groups of goods and services. In a review of several urban consumption studies, Poleman [1961] found that expenditures on food were around 60 percent of all expenditures. Several studies found that the percentage of income spent on food in urban areas may often be less than 60 percent [Howe, 1968; Ostby and Gulilat, 1969]

In an expenditure study in Nairobi, Massell and Heyer [1969] found that the income elasticity of demand for all food was low (0.48) but that meals away from home and rice had much higher income elasticities of demand and that the income elasticity of demand for most nonfood items exceeded unity. Massell and Heyer further showed that the household-size elasticity for food was low (0.36) because, they hypothesized, households tended to substitute cheaper food for more expensive food as the number of consumers increased. Several urban budget studies indicated that consumers adjust their pattern of consumption in response to occupation, employment status, age of head of household, education, and tribal affiliation but the results were not consistent across surveys [Howe, 1966, 1968; Poleman, 1961].

During the 1960s, pioneering rural consumption surveys were carried out in a few countries such as Rwanda and Burundi [Leurquin, 1960]; Ghana [Dutta-Roy and Mabey, 1968; Dutta-Roy, 1969]; and Uganda [Massell and Parnes, 1969]. Information on rural consumption patterns was also presented in two USDA-sponsored studies in Ghana [Ord et al., 1964] and Nigeria [I. G. Stewart, Ogley, and Wright, 1962]. Most of the 1960s rural consumption surveys found that the proportion of expenditures (cash and imputed) for food in rural areas was higher than in urban areas. But, not unexpectedly, cash expenditures on food were often quite low since most households produced the majority of their food needs. Of cash expenditures, clothing was usually the largest item of expenditure, followed by taxes, fuel and lights, beer, and schooling. Agricultural equipment and supplies were rarely found to be a major expenditure item. In general, combined expenditures on food, clothing, and shelter accounted for around 80 percent of expenditures for most rural households. Gifts and ceremonial expenditures were a surprisingly important expenditure, accounting for as much as 5 to 10 percent of annual expenditures.

In terms of the relationship between expenditure patterns and income, it has been consistently found that the share of expenditures for staple foods declines with increasing incomes and that expenditure elasticities for meat and other livestock products are the highest of all food items. Higher-income rural households also tend to spend a greater proportion of their incomes on housing, household supplies, hired labor, and consumption of beer and tobacco than do lower-income households. However, a notable finding of several household budget surveys in the 1960s was that most increases in expenditures stem from quality changes in bundles consumed rather than from a different pattern of expenditure. Because rural consumers tend to substitute foods such as meat, milk, rice, and purchased meals for coarse grains as their incomes increase, the proportion of expenditure going to food does not decrease as fast as otherwise might be the case. This has led some researchers [e.g., Ord et al., 1964] to question whether Engel's Law holds over the range of income found in rural areas.

Consumption analysis received sporadic attention in the 1970s. Humphrey and Oxley [1976] collected data on nearly 7,500 urban and rural households in Malawi and found little similarity between expenditures of rural and urban households, whether in terms of absolute levels of expenditures or their rankings. Expenditure elasticities for rural households were estimated to be 0.706 for food, 0.968 for total durable goods, and 1.546 for household construction. This expenditure elasticity for food is lower than those estimated by Leurquin [1960] for Rwanda and Burundi (0.9) and Robert P. King and Byerlee [1978] for Sierra Leone (0.93). On the basis of interviews with 120 families in three villages in northern Nigeria, Emmy Simmons [1976c] found that the overall elasticity for food was only around 0.4 but this was largely attributable to the extremely low elasticity for sorghum (0.14), the staple food in the area.

Robert P. King and Byerlee's national consumption survey in Sierra Leone was the first study in Africa, to our knowledge, to have estimated the factor intensity of rural consumption patterns by income group.

King and Byerlee found that at all income levels the greatest proportion of expenditures was on rice, the staple food which was produced by 97 percent of the farmers in their national sample in Sierra Leone. They also found that the marginal propensity to consume increased by income level with higher income households consuming more livestock products, beverages, tobacco, transport, services, and ceremonial activities. The marginal propensity to consume at higher-income levels was the highest for expenditures on services and ceremonial activities. [See also Byerlee, C. K. Eicher et al., 1983.] King and Byerlee concluded that there was weak support for Soligo's [1973] hypothesis that low-income households consume goods and services requiring less capital and foreign exchange and more labor than higher-income households.[190] King and Byerlee also compared consumption patterns between rural and urban consumers and found that rural households consumed more labor-intensive goods, that rural-urban consumption linkages were poorly developed, and that the urban centers produced consumer goods purchased almost exclusively by urban consumers.

In follow-up analyses of the Sierra Leone survey data, V. E. Smith, Lynch, et al. [1979] disaggregated estimates of food consumption and related consumption per consumer equivalent to income, the number of consumer equivalents in a household, the dependency ratio, ecological zone, market orientation, and the proportion of labor devoted to rice. V. E. Smith, Strauss, and Whelan [1980] used tabular analysis to evaluate the effect of nonprice factors on consumption. They found that region, ethnic group, household composition, and orientation toward market production all affect consumption patterns but that the effects differ among foods and among income classes of households. V. E. Smith, Strauss, and Schmidt [1981] presented quantity and share equations and derived expenditure elasticities for fourteen major foods plus six groups of foodstuffs. Strauss [1981] made a major contribution in estimating elasticities by using a household-firm model with a quadratic expenditure system. The series of studies by V. E. Smith and his colleagues has made it clear that it is necessary to move toward more disaggregated analysis of consumption patterns since general categories such as food or even cereal grains may hide more than they reveal about how consumption patterns are likely to change over time; and nonprice factors, particularly the pattern of household production, have an important influence on consumption patterns and it is long overdue for researchers working in Africa to begin analyzing the relationships between the production and consumption decisions of rural households.

Starting in the early 1980s, there was a resurgence of interest in food consumption research. Studies completed in the 1980s include a number of chapters in *Food Policy*, edited by Gittinger, Leslie, and Hoisington [1987]; as well as Delgado and Miller [1985]; Delgado [1986]; Shapouri, Dommen, and Rosen [1986]; FAO [1987c]; Delgado and Reardon [1987]; Kennedy and Cogill [1987]; and V. Quinn, Cohen, et al. [1988].

SYNTHESIS

Consumption analysis has received too little attention in Africa. Consumption linkages are important in the development process and have taken on additional importance in light of rapidly rising food imports which reached ten million tons during the Great African Famine of 1985 and declined to eight million tons in 1988. Empirical research has shown that most of the standard hypotheses of consumer theory can generally be expected to hold for rural households in Africa, provided it is recognized that nearly all rural households are low-income households and therefore food expenditures will dominate the consumption bundles of most rural households. The small body of consumption studies conducted over the last decade has made it clear that there is a wide range of nonprice factors which affect consumption patterns in Africa. Moreover, there is substantial evidence that expenditure analysis should be carried out on disaggregated categories of food and nonfood items and should take into consideration the relationships between the production and consumption decisions of households. Because consumption and production decisions are strongly related, we feel that expenditure analysis should be an integral component of future farm surveys in Africa. For a bibliography of food consumption surveys, see FAO [1981c].

9. Nutrition

Until the 1970s, nutrition research in Africa was dominated by medical workers, nutritionists, geographers, and anthropologists.[191] Most nutrition researchers have not collected adequate data on household incomes, relative prices, and home consumption to analyze the impact of economic policies on nutrition. During the 1960s, a few economists such as Joy [1967] started raising questions about nutrition in Africa but it was not until the 1970s that research on the economics of nutrition got underway. For one of the few economic studies of nutrition at the village level, see Emmy Simmons's [1976a] analysis of caloric and protein intake in three villages in northern Nigeria. Simmons found that the average diet provided 2,264 calories and fifty-five to sixty-five grams of protein. Sorghum and millet accounted for 70 percent of the caloric intake and most of the supply of protein. She showed that there was a significant positive relationship between nutrient intake, estimates of amount of grain in storage, and the family labor hours spent on farm work. One of the first comprehensive studies of the economics of nutrition on a national scale was V. E. Smith's [1975] study in Nigeria. More recently, V. E. Smith and his MSU colleagues have examined the effect of price and income changes on nutritional levels in Sierra Leone drawing on survey data collected by Dunstan Spencer in 1974–75 [V. E. Smith, Lynch, et al., 1979; V. E. Smith, Strauss, et al., 1980].

Three major issues dominated research on nutrition in Africa in the 1960s and 1970s: seasonal hunger, impact of cash crop expansion on nutritional status, and strategies for alleviating malnutrition.

SEASONAL HUNGER

Nutritionists have long pointed out that nutrient availability varies by season. The concept of seasonal hunger (*soudure*) has been examined by colonial administrators, anthropologists, and geographers. Seasonal hunger refers to a decline in food intake for one to three months before harvest because food stocks have been exhausted and rural households lack the financial resources to purchase adequate food in the market. Colonial administrators, for example, directly and indirectly forced farmers to plant crops such as cassava in order to create a potential food reserve because cassava could remain in the ground for one to three years. Unfortunately, there are few hard data on food consumption, body weights, and labor productivity during the hungry period. The literature on seasonality and food availability has been drawn together by Annegers [1973] for West Africa. In their survey of seasonal hunger in the Third World, Longhurst and Payne [1979] report that there is no consensus on how to deal with seasonal hunger because of the scattered nature of the empirical studies and the complex methodological problems.

Haswell's [1953] study in The Gambia in West Africa revealed that body weights of rural people declined during the hungry season. In her restudy [1975], she presented data to suggest that rural people in The Gambia were more vulnerable during the hungry season than twenty years earlier because a larger percentage of total family calories is purchased today. Hunter [1967] studied seasonal hunger among the Nangodi people in northeast Ghana and found that an average member of the adult community lost 6.4 pounds of body weight during the hungry season preceding harvest. Nearly a quarter of the adults lost more than 10 percent of their body weight and even in the period following harvest 6 percent of the active adult population remained underweight. The impact of drought on nutrition has been explored by a number of researchers and research teams in numerous African countries [Kloth *et al.*, 1976].

IMPACT OF CASH CROP EXPANSION ON NUTRITIONAL STATUS

A major theme of research on nutrition has been the impact of the expansion of nonfood cash crops on the nutritional status of rural families and plantation workers. There are two opposing views on this issue. On the one hand, it is argued that if nonfood cash crop expansion results in higher net returns to farmers, then the rural households can maintain their level of food consumption and nutritional status by purchasing more food or other items with their incremental income. On the other hand, it is argued that an expansion of cash crops might reduce the land used for producing food crops which will lead to a reduction in home food production and in the nutritional status of farm families. Unfortunately, the available research rarely quantifies the impact of expanded cash crop production on acreage planted, output, and food prices in the area. Moreover, the definition of cash and food crops is fuzzy. Although crops such as coffee and cotton are grown primarily for the market as cash crops, some food crops such as maize and groundnuts can be consumed or sold as a cash crop. In this section, we shall examine the impact of nonfood cash crops such as coffee, cocoa, and cotton on the nutritional status of the households producing these commodities.

Perisse [1962] examined the cash crop hypothesis in three ecological zones in Togo and found some evidence to suggest that the expansion of cocoa and coffee reduced food production. But Perisse offered no quantitative data on shifts in acreage output or changes in the price of food crops. Collis, Dema, and Omololu [1962] assessed the impact of expanded cocoa production on the nutrition of rural households in a survey in western Nigeria and found that people living in cocoa-producing villages were in worse condition from a nutrition point of view than villages not producing cocoa. Collis, Dema, and Omolulu concluded that nutrition education programs were needed. Idusogie [1969] examined cash cropping in Nigeria and came up with inconclusive results.

The cash crop hypothesis was recently tested by the Ministry of Health in Kenya as part of a broader study of child nutrition in rural areas. A major indicator used to evaluate the nutritional status of children was "height for age" which is less subject to short-term fluctuations than a "weight for height" standard. The study revealed that in farming systems emphasizing one or more of five cash crops—coffee, tea, cotton, pyrethrum, and sugar cane—there was little evidence with the exception of sugar cane to support fears that cash crop cultivation is detrimental to nutritional well-being [Kenya, 1979a]. In the case of sugar, apparently children consumed so many empty sugarcane calories that they were unable to meet their other nutritional needs. Also in Kenya, Keller, Muskat, and Valder [1969] used regression analysis to test the relationship between a series of economic variables and two dependent variables: anthropometric measurements and the adequacy of diet. Keller found that calorie adequacy was positively correlated with total cash income, income from the sale of agricultural products, land size, and expenditure for food and clothing.

Lev [1981] drew on farm management and food consumption data collected by Zalla [1981] and examined the impact of coffee production on food consumption among the Meru people in northern Tanzania. Lev found that the simple food/cash crop dichotomy did not hold because the Meru have developed a farming system where coffee is always intercropped with bananas. Lev found in his regression analysis that the coffee/banana intercropping had a positive influence on the nutrition adequacy ratios. In Kenya, Kennedy, and Cogill [1987] studied the impact of commercialization on nutrition by examing the outcome of smallholders adding a sugar enterprise to their farm plan. The results showed that both the smallholders producing sugar and nonsugar producers were meeting 94 percent of their energy requirements. The authors concluded that "clearly, household food security has not been jeopardized by the shift from subsistence to cash (sugar) cropping" [Kennedy and Cogill, 1987].

The results of the above studies provide little support for the hypothesized inverse relationship between export crop expansion and malnutrition. But in several cases, the hypothesis was not rejected; it merely could not be shown to be statistically significant. Also, there have been few attempts to evaluate changes in nutrition over time in a given community following the introduction of cash crops. Although the cash crop hypothesis seems to be of doubtful validity, it remains to be rigorously tested in different agroecological zones in Africa.

STRATEGIES FOR ALLEVIATING MALNUTRITION

Historically, policies designed to alleviate malnutrition have focused on increasing the protein content of local diets, assuming that protein was the most important cause of malnutrition; and establishing nutrition clinics. Many child-feeding programs focused on children under age five since it was common for children to suffer from both kwashiorkor, a lack of protein, and marasmus, a lack of protein and calories. Limited success and high costs of nutrition clinics have made it apparent that a national malnutrition strategy based on widespread use of clinics is not feasible in Africa [Pinstrup-Andersen, 1981].

Research in the 1970s has shown the major cause of malnutrition is generally a lack of calories (energy) rather than a lack of protein except for people in ecological zones where the main staple (cassava, yams, bananas) is low in protein [Sukhatme, 1970; Reutlinger and Selowsky, 1976]. A consensus has emerged that in the short run malnutrition is unlikely to be solved in the normal course of social and economic development in market-oriented economies [Pinstrup-Andersen, 1981]. A recent study done in Rwanda by Braun *et al.* [1991] concludes that "decreasing household calorie consumption is important but alone does not solve the nutritional status problem." Other factors such as sanitation and the education of women are also important determinants of family food security.

Chapter VIII. Migration, Employment, and Equity Issues

Starting in the mid-1960s, policy makers, donors, and scholars began to acknowledge that equity objectives were frequently not being achieved in development programs even in countries achieving relatively high growth rates [Clower *et al.*, 1966; C. K. Eicher, Zalla, *et al.*, 1970]. The concern over equity issues has stimulated a large body of research on the following topics: income distribution, population growth, national and international migration, women in development, and rural employment problems. This chapter reviews the empirical evidence on these topics.

1. Income Distribution and Inequality

When Simon Kuznets [1955] published his seminal paper on income distribution some thirty-five years ago, he apportioned the subject into "perhaps 5 percent empirical information and 95 percent speculation." On the basis of our review, a figure of 99 percent speculation is probably closer to the mark in Africa. For example, Jain [1975] reported data on the size distribution of income in ten countries in Africa with the earliest data reported for 1958. A. O. Phillips [1975] reviewed data sources for Ghana, Kenya, Tanzania, and Nigeria and reported that data are rarely available for more than fifteen years. Currently, data on income distribution are available for only about one-third of the countries in Africa. Moreover, most of these data are based on cross-sectional surveys in urban areas. For a recent collection of studies on income distribution see Haggblade, Hazell, and Brown [1987] and Haggblade and Hazell [1988].

HISTORICAL PERSPECTIVE

Little is known about the distribution of income in a historical perspective. Colonial policies exacerbated inequality in some countries through numerous decisions to assist plantation owners and European settlers in producing cash crops and grade (improved) cattle. Although the sources of many forms of inequality can be traced to colonial policies and settlement patterns, inequality was obviously present in the precolonial era.[192] For an analysis of the origins of inequality in Zambia, see R. E. Baldwin [1966]; in northern Nigeria, Hill [1968, 1972]; in Kenya, Heyer [1981]; and in Central and Southern Africa, the collection by Palmer and N. Parsons, eds. [1977].

Two books by economists document how contemporary problems of inequality are often linked to colonial policies. The first is R. E. Baldwin's [1966] analysis of Zambian development, a dual economy dominated by copper for export, and by Europeans who provided the skills for the copper industry and commercial farms which produced maize for African mine workers. The second is a Northwestern University study by Clower et al. [1966], *Growth without Development: An Economic Survey of Liberia*, which showed that Liberia's impressive growth rates of 7-9 percent in the 1960s were not benefiting the masses because the underlying social and economic structure was channeling the benefits to urban people, bureaucrats, and foreign firms which controlled mining, timber, and rubber concessions. The rice riots in 1979 and the *coup d'état* of 1980 dramatize the underlying inequality in Liberia and point to the political risk of growth without development.

METHODOLOGICAL ISSUES

Kuznets [1976] and Knight [1976] have identified some of the problems in carrying out research on income distribution. The first issue is the need to clarify the objective of the research. Is the major concern one of measuring absolute or relative poverty? Although most research has relied on cross-sectional data to measure relative inequality, cross-sectional data are not very useful in understanding the causes of inequality. The ILO report [1972] on Kenya and Hazlewood [1978] urge researchers to shift their attention from relative inequality to absolute poverty, including attention to changes in absolute income, malnutrition, and employment.

The second issue is one of definition of the recipient unit to be studied. Although there is widespread agreement that the household should be the recipient unit, there is great diversity in size and complexity of African households. Kuznets [1976] recommends taking account of different sizes of households by using income per person or per consumer unit as the basis of comparisons. The next and more difficult step is to take the life cycle of households into consideration in comparisons of income distribution. This requires demographic data to be generated, allowing the lifetime income of a cohort of households to be studied over time. The third problem is measurement of wages and income in subsistence and semisubsistence households. Should firewood be included in national accounts and estimates of rural income? How should unpaid family labor and subsistence food production be valued? This problem has plagued national account surveys. For example, until 1978, it was impossible to use Kenya's system of national accounts to draw inferences about changes in functional distribution of income because wages of the self-employed and the value of unpaid family labor were not included in the computation of national accounts. For a debate on national accounts and income distribution in Kenya, see Hodd [1976]; a critical note by House and Killick [1978]; Hodd's reply [1978]; and for the resolution of the debate, see Hodd, House, and Killick [1978]. The fourth problem is dealing with seasonality of activities such as fishing, farming, and livestock grazing. How can rural incomes of nomadic and seminomadic herders be estimated in Botswana and Mauritania when herders are constantly on the move? In The Gambia, a large percentage of the rural labor force is composed of seasonal migrants from neighboring countries. Data on remittances are essential for measuring the social welfare effects of international migration. These problems help explain why many of the research findings on rural income distribution should be treated with caution.

EMPIRICAL EVIDENCE

Jain's [1975] summary of research on the size distribution of income covers ten countries in sub-Saharan Africa with the earliest data from Chad in 1958, but data on rural income are available in only three of the ten countries. The data summarized by Jain and studies of income distribution in Botswana [Botswana, 1976]; Tanzania [van Ginneken, 1976]; Malawi and Rhodesia [R. A. Jones and R. J. Robinson, 1976]; Sierra Leone [Byerlee, C. K. Eicher, et al., 1977]; and Nigeria

[Matlon, Eponou, *et al.*, 1979a, b] show that incomes are generally more equally distributed in sub-Saharan Africa—especially West Africa—than in Latin America, and that rural incomes in Africa are more equally distributed than urban incomes. But these data must be treated with caution because the Gini coefficients for almost all of the countries were computed by piecing together numerous household budget surveys (mainly urban), farm management studies, and data from national accounts. For example, in Zambia, R. E. Baldwin [1966] computed a Gini coefficient of 0.48 for Zambia based on his 1959 survey. Subsequent studies by van der Hoeven [1977] suggest that by pooling various *ad hoc* surveys one can conclude that it is "almost certain that incomes in Zambia over the 1960–70 period have been more unevenly distributed." C. M. Elliott [1980] examines the unresolved conflicts in growth and equity policies for rural development in Zambia.

At independence in 1966, Botswana was one of the poorest countries in Africa with a per capita income of fifty dollars. But with the discovery and exploitation of vast mineral resources, notably diamonds, the annual rate of growth of the GNP has been the highest in the world (11.0 percent in real terms) from 1973 to 1985) and its per capita income climbed to $840 in 1985. Livestock dominates the rural economy and it is difficult to generate reliable national accounts and income distribution data from nomadic and seminomadic herders. In 1973, Robert McNamara, then President of the World Bank, gave a speech in Nairobi and stressed the need to give more attention to research on rural development and income distribution. As a direct follow-up to McNamara's speech, the government of Botswana carried out a Rural Income Distribution Survey (RIDS) in 1974–75 [Botswana, 1976]. Botswana now has the most comprehensive data set on income distribution in Africa. The RIDS covered 1,765 households, including 1,115 rural households who were interviewed monthly, supplemented by estimates from the Central Statistics Office (CSO) for 593 rural households and sixty-two nomadic households. The results of the RIDS revealed that the lowest 40 percent of the rural households received only 12 percent of the income; the Gini coefficient of 0.52 indicated a high degree of inequality which is believed to be a function of the unequal holdings of livestock. Szal [1979] has advanced a number of proposals to reduce inequality within rural areas, notably the reduction of local and income taxes, school fees, etc.

A comparative study of rural income distribution in Sierra Leone and Nigeria was carried out in the mid-1970s. Rural incomes were defined as the return to household land, labor, and management in all farm and nonfarm occupations. In both countries, rural households were interviewed twice a week over fifty-two weeks. In Sierra Leone, the Gini coefficients were found to be 0.34 at the village level, 0.38 at the resource region level, and 0.39 for the nation [Matlon, Eponou, *et al.*, 1979b]. Norman, Pryor, and Gibbs [1979] reported Gini coefficients for rural arreas of northern Nigeria in the range of 0.30 to 0.40. The Matlon, Eponou, *et al.* [1979a] study of income distribution in three villages in northern Nigeria generated a Gini coefficient of 0.28 at the village level, indicating a high degree of equality. When income from Moslem women engaged in food processing and trading activities was incorporated into the estimates of income of the sampled rural households, the Gini coefficient was 0.24. The high degree of rural equality in Sierra Leone and Nigeria appears to be a function of a relatively egalitarian land tenure system, and the absence of technical change. Nevertheless, the research reveals that there is a serious degree of absolute poverty among the poorest 30 percent of the rural population in both Sierra Leone and in Nigeria [Matlon, Eponou, *et al.*, 1979b]. Large-scale rural surveys which produce Gini coefficients have been appropriately criticized by scholars such as Palmer-Jones and Polly Hill. For example, in Polly Hill's [1968] case study in northern Nigeria, "The Amorphous Peasantry," she takes economists to task for assuming that "poverty is roughly uniform as between farmers" and she questions whether survey research is an appropriate tool for understanding rural inequality. We agree that although cross-sectional data from rural surveys can provide snapshots of income distribution, they cannot get at the roots of inequality.

Income distribution has been an important topic for researchers and policy makers in Kenya. The historical origins of rural inequality were examined by Heyer [1981] and C. Barnes [1979]. The ILO Employment Mission [1972] to Kenya devoted major attention to income distribution problems. The Government of Kenya's response [Kenya, 1973] to the ILO report has been criticized by Ng'ethe [1980] because it accepts the *status quo* and does not come to grips with the cause of rural inequality—inability of the poor to gain access to land, credit, and government services.

RESEARCH AGENDA

Research on rural income distribution and inequality is in its infancy; much remains to be done, and undoubtedly this is a high priority research topic. Computation of Gini coefficients is only a small portion of the research needed to understand the causes of inequality. The starting point for improved research is to face up to the methodological issues discussed above. There is need for research on asset ownership and how ownership and control of assets influence income distribution. Since the control of resources (land and capital) is often a function of historical forces, it behooves economists to pursue research on asset ownership in historical perspective. Unfortunately, Western-trained agricultural economists are noted for their lack of interest in economic history. There is also a need to examine the impact of government policy on asset ownership and the linkages between political power and accumulation of wealth and the influence of wealth accumulation on political power. Finally, there is a need to analyze how various classes and groups in society (women, landless, tenants, etc.) are affected by technical change, by migration, and by various government policies such as subsidized credit.

Berry [1976], Post [1972], van Hekken and van Velzen [1972], and Sklar [1976] offer perceptive comments on how to use class analysis as an organizing theme for research on inequalilty. Hazlewood [1978] stresses the need to keep research focused on both growth and income distribution rather than jumping on the income distribution "band wagon" and formulating policy prescriptions on the basis of results from narrowly conceived studies.

Studies on income distribution and inequality include:

> *Africa General*: Jain [1975]; W. A. Lewis [1978a]; Rweyemamu, ed. [1980].
> *West Africa*: Berry [1980]; Post [1972]; Matlon, Eponou, *et al.* [1979a, b].
> *Botswana*: Botswana [1976]; Szal [1979].
> *Ghana*: Hill [1963, 1970]; A. O. Phillips [1975]; Winch [1976]; Ewusi [1977].
> *Côte d'Ivoire*: Lee [1980].
> *Kenya*: ILO [1972]; A. O. Phillips [1975]; Hunt [1975a]; Holtham and Hazlewood [1976]; Heyer [1981]; Heyer and Waweru [1976]; Hodd [1976]; House and Killick [1978]; Hodd [1978]; Hodd, House, and Killick [1978]; Hazlewood [1978]; C. Barnes [1979]; House and Killick [1980]; Ng'ethe [1980].
> *Liberia*: Clower *et al.* [1966].
> *Malawi*: Ghai and Radwan [1980].
> *Niger*: Raynaut [1977].
> *Nigeria*: Hill [1968]; Essang [1972]; Aboyade [1973]; A. O. Phillips [1975]; Matlon, Eponou, *et al.* [1979a, b]; Bienen and Diejomaoh, eds. [1981].
> *Sierra Leone*: Byerlee, C. K. Eicher, *et al.* [1977].
> *Tanzania*: van Hekken and van Velzen [1972]; A. O. Phillips [1975]; van Ginneken [1976].
> *Zambia*: R. E. Baldwin [1966]; Maimbo and Fry [1971]; Van der Hoeven [1977]; Blitzer [1979]; Kinsey [1978]; C. M. Elliott [1980]; Turok, ed. [1979].
> *Zimbabwe*: R. A. Jones and R. J. Robinson [1976].

2. Population

Demographic research was started in most countries after independence in 1960. Since few countries maintain an accurate register of births and deaths, the quality of demographic data is extremely uneven. The most important demographic characteristics are the following:

1. The population is young with nearly half under fifteen compared to only about one-quarter in Europe and North America.
2. Fertility levels are high and almost unchanged since 1960. The 1980 crude birth rate was forty-eight per thousand as compared with forty-nine in 1960.
3. Fertility levels are extremely heterogeneous within and between countries. Lesthaeghe, Ohadike, Kocher, and Page [1981] report that the fertility of Africa represents a "mosaic of strongly contrasting levels of fertility." This again points out the fallacy of discussing averages—such as crude birth rates— in a subcontinent of such complexity and diversity.
4. Although mortality has been declining in recent decades, it is still high by international standards. The overall crude death rate is around eighteen per thousand.
5. The average expectation of life at birth is below fifty years.

Africa is the region with the highest rate of population growth in the world. The pattern and pace of population growth are important issues facing African policy makers.[193] First, as Byerlee and C. K. Eicher [1974] point out, any discussion of the rural employment problem must be viewed in light of the high population growth rate and the inability of urban areas to generate jobs for the expanding rural labor force. Second, Kocher [1979] has shown that governments have a harder time providing social services when the population is growing rapidly. Third, many demographic experts agree that there are areas in Africa which are now being affected by population pressure [Fresco, 1986; Cantrelle, ed., 1974; Caldwell et al., eds., 1975]. Fourth, village studies have shown that greater spacing of children may increase the chance of survival during times of stress. Faulkingham and Thorbahn [1975], for example, found that malnutrition was rarely a problem among the men of Tudu, a village in Niger, but it was common among women and children. When Faulkingham [1977] conducted follow-up research in the same village, he found that 25 percent of all children age one to five died during the drought of the early 1970s when food supplies were short.[194]

Although many agree that food production is a critical issue for African governments, few African policy makers feel that there is a population problem. On the contrary, most African governments continued to restrict access to modern birth control devices until the early 1970s. Some countries such as Mauritania even had a pronatal policy because they feel that their country is underpopulated. Moreover, numerous African scholars, including the late Okediji [1972] and Amin and Okediji [1974], believe that concern with population growth and promotion of family planning is Western-inspired and another "false start" for Africa.

During the 1970s, there was a shift in public policy on population in many African countries. In 1973, only nine countries supported family planning (primarily for reasons of health and as a human right) but by 1978, twenty countries had adopted this policy [World Bank, 1981b, p. 112]. Nevertheless, assessments of family planning programs have consistently indicated that Africans desire large families and that their interest in contraception primarily exists for the purpose of child spacing and premarital contraception. It is fairly clear that family planning programs were introduced and promoted in sub-Saharan Africa in the 1960s and 1970s far ahead of the knowledge base about the determinants of fertility among different ethnic and religious groups. For example, family planning is a delicate political issue in a country like Senegal where 95 percent of the people are Moslem.

James Kocher [1979] has stressed the importance of understanding the relationship between socioeconomic development and fertility before introducing family planning programs. On the basis of interviews with the adult members of 1,500 rural households in Tanzania, Kocher concluded that the prospects for reducing fertility through family planning programs were crucially linked to Tanzania's stage of socioeconomic development. Kocher contends that in countries where 80-90 percent of the population is in the rural sector it would be unwise to promote family planning until the educational level of women is increased and problems of health and disease have been tackled.

Two demographers recently compared the Asian and African family planning experience and concluded that the slowness to adopt family planning in Africa is "not explained by the African countries being at an earlier stage of socioeconomic development" [Caldwell and Caldwell, 1988, p. 19]. The Caldwells contend that the essential differences in family planning are caused by differences in African family structures, as well as economic and religious attitudes towards fertility that severely limit the ability of African states to implement forceful family planning programs that have been used in China, India, and Indonesia.

3. Migration

The migration of rural people for work in plantations, mines, and factories has been a major catalyst for social change in Africa. Common themes in African history include long distance migration to mines in Southern Africa, to plantations in West Africa, and the role of "strangers" in African societies [Shack and Skinner, eds., 1979].

Historically, migration has been viewed favorably in the development literature because it was perceived to contribute to reducing intra- and interregional wage differentials, and in transferring new crops and ideas over wide regions. For example, Mabogunje [1972], observed that two or three million West Africans leave their homes and businesses every year in search of profitable economic opportunities across ethnic or national boundaries. Migrants have been character-ized as innovators, risk takers, and entrepreneurs. Polly Hill's [1963] pioneering research revealed that migrants were the risk takers in settling land and mobiliz-ing capital in Ghana's cocoa boom in the late 19th century. Parkin, ed. [1975] re-ported that rural and urban areas in Eastern and Southern Africa are influenced by a "vast criss-crossing of people, ideas, and resources." Vermeer [1979] found that farmers on the Jos plateau in Nigeria secured new crops and varieties of plants from traders as far away as Liberia. These new crops (e.g., pepper and spices) were tested on intensive "garden plots" near the compound (main housing unit) of rural families and the more promising crops and varieties were then introduced into the farming system.

In the early 1960s, rural-to-urban migration of young school leavers was per-ceived by many policy makers to be excessive, contributing to the explosive rate of growth of urbanization (8-10 percent per year) and the alleged urban unem-ployment of 10-30 percent. A central policy question which followed was whether rural to urban migration should and could be controlled by national governments.

THEORETICAL PERSPECTIVES

There are three broad theoretical perspectives on migration: structural-func-tionalist, neoclassical economics, and political economy. The structural-func-tional approach by anthropologists, sociologists, and geographers has a long his-tory starting with Schapera [1947] and followed by Mitchell [1959], Gugler [1969], Hutton [1973], and Parkin, ed. [1975]. The structural-functionalist ap-proach examines the individual decision to migrate within a broad pattern of so-cial relationships and social-structural conditions, including some economic vari-ables [Parkin, ed., 1975].

Neoclassical economists treat migration as an economic phenomenon in which the migrant weighs the costs and returns from present and future employ-ment opportunities. A turning point in migration research by neoclassical econ-omists came with Todaro's [1969] "expected incomes" model of migration based on his research in Kenya in the mid-1960s. Todaro's seminal contribution has provided a framework for much of the econometric work on migration in the past decade. Todaro's model was extended by J. R. Harris and Todaro [1970] by explicitly specifying the "elasticity" of migration (the induced migration) re-sponse to changes in urban-rural wage differentials and urban employment prob-abilities. For a discussion of extensions of the Todaro model, see Todaro [1980].

Neoclassical models of migration play up the role of economic variables in explaining migration but these models do not shed much light on the net social loss or gain from internal or international migration. Moreover, as Fields points out, even if "economic factors are primarily responsible for migration behavior, which economic variables are included and how they are specified makes a great deal of difference in the explanatory power of the economic model" [1980, p. 392].

In the political economy approach, the historical expansion of capitalism is viewed as the main explanation of migration and it is assumed that while migration may improve the private economic return of the individual migrant, the net short- and long-term social and economic effects of migration may be negative in the source area and positive in the receiving area. Amin [1974b], for example, asserted that "migration impoverishes the home area and proletarianizes the migrants." Plange [1979] examined migration from northern Ghana to plantations and gold mines during the 1900-40 period and concluded that migration contributed to poverty and underdevelopment of the northern region. In the ten-country region of Southern Africa, the theme of exploitation has dominated the long history of research on migration from Malawi, Botswana, Lesotho, and Swaziland to the gold mines in the Republic of South Africa starting with Schapera's [1947] classic study and continuing with Wilson's [1972] historical analysis of the wages of black workers in the gold mines. These scholars contend that migration facilitates capital accumulation in the Republic of South Africa by drawing on "labor reserves" in countries such as Lesotho, Botswana, and Malawi. These latter countries still supply up to one-fourth of their adult male labor to the mines in the Republic of South Africa on short-term (usually nine months) contracts. Census data show most migrants working in the mines enter the mines at an early age and return to their homes in rural areas once a year and generally return to live permanently with their families by late middle age.

METHODOLOGICAL ISSUES

Byerlee [1972] reviewed several hundred migration studies in Africa and reported that the bulk of research on migration over the 1950-70 period was carried out by sociologists, geographers, and demographers relying on census data and cross-sectional surveys of migrants in urban areas. Unfortunately, most of these studies did not quantify the determinants of migration and were inconclusive about the role of economic variables in the migration process. Byerlee also found that information rarely has been generated on both male and female migrants for an entire region, or country, including remittances and return migration. In addition, few migration studies have ever generated accurate data on rural incomes. As a result, many of the comparisons of rural-urban income differentials (and some of the policy conclusions by economists who advocate reducing the rural-urban income gap) must be treated with skepticism. The conceptual problems in defining and measuring income are spelled out by Knight [1972].

The starting point for research on rural to rural and rural to urban migration is to focus on the conceptualization of the migration decision-making process in rural families, viewing migration as a holistic process. Factors affecting the decision to migrate can be analyzed in terms of monetary costs and returns related to incomes and employment in the rural and urban labor markets; and nonmonetary costs and returns relating to risk, attitudinal characteristics, social ties, and expectations. But to generate these data requires more resources than are usually available to a single researcher studying migration.

A study making a major contribution to methodology was conducted in 1974/75 in Sierra Leone by Byerlee, Tommy, and Fatoo [1976]. The major policy question in their national migration survey was: Why was rural to urban migration proceeding at a high rate when unemployment was alleged to be high (30 percent) in urban areas and the probability of obtaining an urban job was perceived to be low? Instead of using census data to compare migration streams over a ten-year period or interviewing migrants in urban areas and asking them why they left rural areas, Byerlee and his colleagues drew on information on rural incomes and labor use from a companion farm management survey of 500 rural households who were being interviewed twice a week over a twelve-month period [D. S. C. Spencer and Byerlee, 1976]. In addition, Byerlee and his colleagues collected demographic data and work histories from 2,000 persons in rural areas; a total of 800 of the 20,000 who had migrated were "traced"[195] and interviewed in urban areas in order to determine how they phased into the urban labor force. Detailed data were also collected on migrants returning to rural areas. The Sierra Leone study is one of the few national studies in Africa which analyzed both male and female migrants and quantified both gross and net rates of migration.

SEASONAL AND RURAL TO URBAN MIGRATION

Who Migrates? Seasonal migration is of central importance in providing labor for farming throughout sub-Saharan Africa. Dupire's [1960] study is a classic on the role of "strange farmers" (seasonal migrants) in the Côte d'Ivoire.

Invariably every migration study concludes that the typical rural-urban migrant is younger and better educated than the average rural resident. For example, Rempel's [1971] survey in Kenya and Byerlee, Tommy, and Fatoo [1976] in Sierra Leone showed about one-half of all rural-urban migrants to be between the ages of fifteen and twenty-four.

Rates of Migration. Studies in Sierra Leone and Ghana are among the few which shed light on gross and net migration rates. Byerlee, Tommy, and Fatoo [1976] computed gross and net migration rates from their nationwide survey and found that gross migration rates greatly exaggerated the magnitude of rural to urban migration in Sierra Leone because roughly two of three migrants returned to rural areas after five years. Caldwell [1969] reported a similar percentage of return migrants in his study of migration in Ghana.

Determinants of Migration. Unfortunately, most econometric studies of migration in Africa provide policy makers with limited advice on the key question—what determines migration? Most migration studies are so poorly designed and limited in scope that they cannot determine whether urban housing, public services, or the "bright lights" of the cities are more important in attracting migrants than differential wage rates or employment opportunities. Since most econometric studies of urban amenities do not measure the migrant's utilization of these services, the outcome of research on this determinant is fuzzy. We do know that migrants respond to economic incentives and that friends, relatives, and distance from the sending area can be important determinants.

Studies of the Todaro hypothesis of the importance of expected income in migration decisions provide preliminary support that the job probability variable has "independent" statistical significance and adds to the overall explanatory power of the regressions [Todaro, 1980, pp. 380-381]. Although Rempel's [1971] study in Kenya found no consistent evidence that migrants responded to rural-urban income differentials, Knowles and Anker's [1981] sample of 1974 households in seven of Kenya's eight provinces provided some support for the expected income hypothesis. Likewise, studies by House and Rempel [1980] for Kenya, and Barnum and Sabot [1976] for Tanzania provide some evidence that an autonomous expansion of urban jobs might induce rural-to-urban migration and add to urban unemployment.[196]

Remittances. While micro research is providing some evidence on the magnitude of urban-to-rural income transfers, the empirical results are still soft on this topic. In a comprehensive survey of the literature, Rempel and Lobdell [1978, p. 205] reported that remittances account for "between ten and twenty percent of migrants' urban incomes in Africa and a somewhat higher percentage in the Asian subcontinent." Caldwell [1969] provided crude estimates of urban to rural remittances in Ghana and concluded that they were approximately 10 percent of urban earnings in Accra. In Sierra Leone, Byerlee, Tommy, and Fatoo [1976] reported that urban-rural remittances account for about 5 percent of urban earnings. In urban areas of Sierra Leone, about 17 percent of the income of working migrants was used to support friends and relatives (mostly unemployed migrants trying to phase into the urban labor force).

Level of Urban Unemployment. In the early 1960s, it was widely asserted that the level of open unemployment was 20-30 percent in many African cities and that the level of unemployment was increasing. But empirical studies later revealed that the level of urban unemployment was in the range of 10-15 percent and that the important policy issue was not unemployment but massive underemployment in the urban informal sector and in rural areas.

WHO BENEFITS FROM MIGRATION?

In general, migration studies show that migrants improve their income by moving but there are sharp differences by educational level. For example, in the Sierra Leone study, Byerlee, Tommy, and Fatoo [1976] found that the wage rate for educated migrants was three times the average rural wage rate but the unskilled urban migrant (working in the informal sector) was earning a wage only slightly higher than the rural wage after differences in the cost of living were taken into account. We have already pointed out that remittances are a major benefit of migration, ranging from 10 to 20 percent of the migrant's urban income in Africa.[197] What is the social welfare impact of migration in both the sending and receiving areas?[198] The neoclassical cost/benefit framework which concentrates on the private returns of migration is inadequate for evaluating the net welfare impact of migration. But can the political economy paradigm provide better answers to policy makers? The absence of quantification in political economy research on migration is noteworthy. It takes more than a string of assertions to prove that migration is the cause of underdevelopment in the source region. For example, Samir Amin [1974a] asserts that cost-benefit analysis is "an ideological defense, which takes the place of science, attempts to justify migratory phenomena by pretending that they are in the interest of both regions . . . " In an extremely balanced and constructive article, Knight and Lenta [1980] assess research by both neoclassical and political economy scholars on whether migration has underdeveloped the labor reserves in Southern Africa and conclude that there is no clear answer to the question on the basis of present research. Research on quantifying the social costs and returns to migration is urgently needed.

SYNTHESIS

Although seasonal and rural-rural migration has occurred on a substantial scale in Africa, interregional wage differentials still exist within most African nations. For example, in Sierra Leone in 1974, the unskilled wage differential was almost 2 to 1 between the highest and lowest wage regions in the country. Rempel and House [1978] report there are still substantial differences among regions in the wage paid for unskilled labor in Kenya. The persistence of wage differentials will stimulate migration in the future. Rural areas will have to absorb the bulk of the growth in the labor force over the next ten to twenty years. Although the relative percentage of population in agriculture will decline in most countries, the absolute number of people in agriculture will likely increase in most countries over the next ten to twenty years.

This review adds support to Yap's [1977] finding about the limitations of most migration studies for policy purposes in the Third World. Although hundreds of migration studies have been carried out in Africa, few are well designed, comprehensive, and quantitative. The results of most subnational, point-to-point, and cross-sectional studies yield information which is generally of limited value to policy makers. *We believe that research on migration is of low priority for the 1990s.* Migration studies include the following:

Bibliographies and Literature Reviews: Byerlee [1972]; Yap [1977]; Todaro [1976, 1980].

Africa General: Mitchell [1959]; E. Berg [1965]; J. A. Jackson [1969]; Meillassoux [1974]; Amselle, ed. [1976]; Van Binsbergen and Meilink, eds. [1978]; Rempel and Lobdell [1978]; Shack and Skinner, eds. [1979]; Peek and Standing [1979]; Swamy [1981]; Fortmann [1981]; Seidman [1981].

Southern Africa: Wolpe [1972]; Wilson [1972]; Elkan [1980]; Bromberger [1979]; Knight and Lenta [1980].

West Africa: Kuper, ed. [1965]; Mabogunje [1972]; Amin [1974a, b]; Caldwell [1975]; LeBris, Rey, and Samuel [1976]; J. B. Riddell [1978]; Guyer [1980a]; Zachariah and Conde [1980]; Byerlee [1980].

Central and Eastern Africa: Parkin, ed. [1975].

Botswana: Schapera [1947]; B. Brown [1980]; J. Harris [1981].

Burkina Faso: Skinner [1965]; Kohler [1972]; ORSTOM [1975]; Remy [1977]; Songre [1973]; Gregory [1979]; Coulibaly, Gregory, and Piche [1980].

Gambia: De Jonge *et al.* [1978].

Ghana: Beals and Manezes [1970]; Hill [1963]; Caldwell [1969]; Knight [1972]; Hill [1978]; Plange [1979]; Schwimmer [1980].

Kenya: Rempel [1971]; George Johnson [1971]; Fields [1975, 1980]; Knowles and Anker [1981]; Rempel and House [1978]; House and Rempel [1980].

Lesotho: Van der Wiel [1977]; Murray [1977]; Eckert and Wykstra [1979].

Mauritania: Dussauze-Ingrand [1974].

Niger: Faulkingham and Thorbahn [1975].

Nigeria: Mabawonku [1978]; Essang and Mabawonku [1974].

Senegal: Rocheteau [1975]; Colvin *et al.* [1981]; A. Adams [1977a,b]; P. David [1980]; De Jonge *et al.* [1978].

Sierra Leone: Byerlee, Tommy, and Fatoo [1976].

Tanzania: Barnum and Sabot [1976]; Collier [1979].

Togo: LeBris, Rey, and Samuel [1976].

Uganda: Hutton [1973].

Zambia: R. H. Bates [1976]; Cliffe [1978]; Mwanza [1979].

Sahel: Caldwell [1975].

Migration Theory: Todaro [1969, 1976, 1980]; J. Harris and Todaro [1970]; Knight [1972]; Byerlee [1974]; Byerlee and C. K. Eicher [1974];

Griffin [1976]; Blomquist [1978]; Rempel and Lobdell [1978]; Gerold-Scheepers and Van Binsbergen [1978]; Lipton [1980]; Sabot, ed. [1981]. *School Leavers*: Hutton [1973]; Callaway [1964].

4. Rural Employment

Growing unemployment in many African cities, explosive rates of urbanization, and rising urban wages created an awareness in the early 1960s that migration and employment generation had to be addressed by policy makers. For example, President Nyerere of Tanzania observed that economic policies and projects seemed to be geared to improving the lives of the urban people who were employed by the government and trade unions. By the late 1960s, the growth of industrial employment was found to be lagging behind the rate of growth of industrial output in many African nations indicating that migration was in excess of the absorptive capacity in urban areas [C. R. Frank, 1971]. The implications of these findings were that the industrial/urban sectors could not generate adequate jobs for rural to urban migrants and that attention should be directed to slowing down migration and generating more productive employment in rural areas [C. K. Eicher, Zalla, *et al.*, 1970]. These problems formed the agenda for two major conferences in East Africa. The 1966 Kericho, Kenya conference on employment and rural development [Sheffield, ed., 1967] and Tanzania's 1967 Arusha Declaration [Nyerere, 1967, 1968, 1977] were manifestations of a search for a development strategy which addressed rural mobilization, equity, employment generation, and redressing the balance of rural and urban power.

Research on rural employment was a high priority research topic in the 1970s in countries such as Kenya, Nigeria, Ghana, Sierra Leone, and Botswana.[199] Research on employment and income distribution in English-speaking countries has tended to be micro and quantitative (e.g., computing Gini coefficients and factor price distortions). Research on employment in French-speaking nations generally has been historical, macro, and non-quantitative.

ILO country studies of employment in Kenya [ILO, 1972] and the Sudan [ILO, 1976] were pursued within a modified neoclassical (redistribution with growth) paradigm. The ILO team wisely rejected the concept of unemployment and the calculation of unemployment trends because they contended that few people have the luxury to be unemployed in Kenya's low wage economy. The ILO mission identified the urban "informal sector" to be its major conceptual advance. The informal sector includes petty traders and artisans who generally earn low wages and returns. Although the ILO Kenya report contains an immense body of useful information, it made little contribution to a deeper understanding of the causes of rural poverty and underdevelopment and of how to generate more employment in rural areas. Leys [1973] appraised the ILO report on Kenya from a political economy perspective and described it as a bland, innocuous report emphasizing redistribution with growth while glossing over the underlying structural causes of poverty and underdevelopment. For additional studies on employment in Kenya, see Clayton [1975]; F. Stewart [1979]; Child [1977]; Rempel and House [1978]; Ghai and Godfrey, eds. [1979]; and Knowles and Anker [1981].

In Sierra Leone, major studies of migration, small-scale industry, fishing, and processing were carried out in 1974–75 in conjunction with a nationwide farm survey by a team of researchers from Njala University College, University of Sierra Leone, and Michigan State University under the direction of D. S. C. Spencer. A summary of findings is reported in Byerlee, C. K. Eicher, et al. [1983]. Specific studies include D. S. C. Spencer [1976]; D. S. C. Spencer and Byerlee [1976]; D. S. C. Spencer, May-Parker, and Rose [1976]; Linsenmeyer [1976]; Robert P. King and Byerlee [1978]; Byerlee, Tommy, and Fatoo [1976]; Byerlee, Eicher, et al. [1977]; and Liedholm and Chuta [1976].

5. Women in Development

Research on women in African development has a long history. The importance of women in farming was recognized over fifty years ago by Baumann [1928] in his classic article, "The Division of Work According to Sex in African Hoe Culture," which appeared in Volume I of *Africa*. Kaberry's [1952] study of women in the Cameroon is a standard reference. One of the first studies to present empirical data on the differentiation of adult male and female activities was *Nigerian Cocoa Farmers* [1956] by Galletti, K. D. S. Baldwin, and Dina.

Research on women in Africa mushroomed following the publication of Ester Boserup's *Woman's Role in Economic Development* [1970][200] which revealed that women play significant roles in farming in the Third World and that Africa could be described as the "region of female farming *par excellence.*" She drew on several case studies and surveys to show that women often "do more than half of the agricultural work; in some cases they were found to do around 70 percent and in one case nearly 80 percent of the total" [p. 22]. She also showed that women play a major role in local trade in Africa, particularly in West Africa.[201]

Boserup's timely analysis has led to a number of polemics on the adverse effects of development on women [Tinker, 1976],[202] but has also sparked a large amount of serious research on women as illustrated in the bibliographies by Buvinic et al. [1976] and Mascarenhas and Mbilinyi [1980].

EMPIRICAL RESULTS

Sudarkasa's [1973] study of Yoruba market women in Nigeria revealed that women have played a pervasive role in trading in West Africa but in East Africa men of Asian background dominated trading for many years. Ann Seidman argues that technological change has contributed to the deterioration of the status of women in Africa but she does not present hard data to support her assertion [1981, p. 122].

Dunstan Spencer [1976] evaluated the impact of development interventions on women's workload by interviewing twenty-three rural households in Sierra Leone twice a week for one year; fourteen of the twenty-three were rural households participating in a World Bank financed rice project in the eastern province while the other nine households were nonparticipants selected at random in the same province. Spencer's results revealed that the workload of the women in households participating in the rice project increased slightly while the workload of the men and male children was substantially increased during the first three years of the project. Spencer rejected the hypothesis that women's workload increases relative to men as commercialization of agriculture proceeds but noted that his sample was small and more research was needed on the impact of technical change on men and women in different ethnic groups and farming systems. Spencer's research is noteworthy because he moved beyond the typical single-visit survey and studied labor allocation of men, women, and children in a micro-environment through repeated interviews over a twelve-month period.

Barrett et al. [1982] reported that in eastern Burkina Faso men worked more total hours in farming during the 1978-79 survey year than women but women worked more hours per year when household tasks were taken into consideration.

The impact of male migration on families left behind in the village is central to the analysis of migration. Staudt [1975] interviewed 212 small-scale farm households in densely populated western Kenya in 1975 and found that 40 percent of the rural households were headed by females because the male had temporarily or permanently migrated to urban areas in search of work. P. R. Moock [1976] interviewed 152 maize farmers in western Kenya and found that one-third of the male heads of rural households were away from home engaged in or searching for work. Moock compared the technical efficiency of male *versus* female managed small-scale maize farms and found that education of women had a more significant impact on maize output than the education of males, that male educational achievement was correlated with success in off-farm employment, and that female farm managers did not seem to benefit as much as the males do from extension contact. The latter finding may be explained by the fact that most of the extension agents were predominantly male at the time of Moock's survey.

Emmy Simmons's studies [1975, 1976b] in northern Nigeria illustrate the conceptual problems involved in estimating the income of Moslem households where women are secluded in their compounds (homes) during the daylight hours through a form of *purdah*. Simmons [1975] studied women who prepared processed food in their compounds which was sold by their children in the village markets. She stressed the need to include the income of females engaged in such activities as trading and food processing in farm management surveys and she observed that female enumerators were needed to gain access to women in *purdah*.

A recent compilation of ten studies of gender issues in farming systems research is a major contribution to the literature [Poats, Schmink, and Spring, eds., 1988].

NEEDED RESEARCH

Over the past ten years, numerous studies have shown that a significant portion of the labor inputs in agriculture in Africa comes from women, except for the physically demanding tasks such as the brushing and felling of trees. Women dominate some activities such as weeding, food processing, and trading, and participate in almost all other farming activities, depending on the farming systems and social and climatic factors. *But it is a vast overgeneralization to argue that women produce 70 or 80 percent of the food in Africa!* There is no empirical evidence to support such sweeping generalizations for all of Africa.

Research, in our judgment, should move beyond descriptive and anecdotal studies to more quantitative research on the rural household, including the roles of women and men in different farming and livestock systems and off-farm employment and the roles of women and men in household decision making. For studies on women, see Mbilinyi [1972]; Eilam [1973]; Robertson [1974]; B. A. Clark [1975]; Chuta [1978]; Hafkin and Bay, eds. [1976]; Lancaster [1976]; Sudarkasa [1973]; D. S. C. Spencer [1976]; S. Young [1977]; R. J. Gordon [1978]; Achola Pala [1976]; Bukh [1979]; Staudt [1978-79]; Issard [1979]; Tripp [1978]; Urdang [1980]; and Guyer [1980a,b]; Savane [1981]; Fortmann [1981]; Seidman [1981]; Goody and Buckley [1973]; and Poats, Schmink, and Spring, eds. [1988].

6. Rural Small-Scale Industry

Small-scale industry[203] accounts for most industrial employment and output throughout Africa. For example, Liedholm and Chuta [1976] estimated that 95 percent of the people engaged in industrial production in Sierra Leone were working in firms with less than fifty employees. Moreover, farm management surveys have shown that as much as 25 to 50 percent of the annual labor supply of rural households in sub-Saharan Africa is spent on off-farm activities such as small-scale industry, rural public works, and trading. Since rural industries and agricultural production are clearly linked through both factor and product markets, it is important to review research on rural industries.

THEORETICAL ISSUES

Most studies of industrial production in the 1960s in Africa focused on urban large-scale firms, accepting the analytical framework of dual sector models. Later dual sector models were modified by Hymer and Resnick [1969], Byerlee and C. K. Eicher [1974], and House and Killick [1980] to include the urban informal sector. Hymer and Resnick [1969] incorporated off-farm activities as a separate sector; they hypothesized that the products of rural small-scale industries were inferior goods—have negative income elasticities of demand. But Hymer and Resnick provided no empirical support to test their model.

The role of rural small-scale industries in economic development over time will be largely determined by the linkages between the rural area and regional, national, and international markets; by the composition of demand for products of the sector; by the efficiency of the sector relative to larger-scale modes of production; and by the factors affecting the supply responsiveness of small firms. One of the most important theoretical relationships between rural industry and agriculture is the allocation of rural household labor to farming and off-farm employment. Data from throughout Africa show that the amount of time devoted to nonagricultural activities is significantly related to seasonal labor requirements in agriculture. In northern Nigeria, for example, Luning [1967] found that the percentage of people primarily engaged in nonagricultural activities dropped from 65 percent in the slack season to only 6 percent during the peak farming season. Similarly, Norman [1969] found that 27 percent of family labor in villages surveyed in northern Nigeria was devoted to nonfarm activities even during peak farming months, down from nearly 80 percent during slack months.

EMPIRICAL RESULTS

Research on rural small-scale activities was undertaken in Nigeria by Kilby [1962], Callaway [1964], and Luning [1967]; Tanzania by Schadler [1968]; Botswana by Lewycky [1977]; Ghana by Steel [1977]; Kenya by Child [1977], K. King [1977], Forsyth [1977], and F. Stewart [1979]; Cameroon by Steel [1979]; Sierra Leone by Liedholm and Chuta [1976]; Nigeria by Mabawonku [1978]; Burkina Faso by Wilcock [1981]; and Kenya, Tanzania, and Zambia by Gulhati [1981].

Research has shown that the primary orientation of rural small-scale industries is toward production and provision of goods and services for local markets. The composition of small-scale industries seems to be quite similar throughout Africa. The most prominent activity in terms of employment is tailoring, followed by carpentry, blacksmithing, baking, and vehicle repair activities. Traditional activities such as blacksmithing and weaving are more important in villages, while tailoring and vehicle repair assume greater importance in small towns [Liedholm and Mead, 1987].

The three sources of demand for the products of rural small-scale industries are: local demand from rural and urban consumers, export markets, and demand arising from backward and forward linkages. The growth of small-scale industries will depend upon the sign and size of the income elasticities of demand for goods produced for each of these three markets. Liedholm [1973] reviewed the few consumer budget surveys available in sub-Saharan Africa and found that the income elasticity of demand for these goods and services was positive [Leurquin, 1960; Massell, 1969]; he concluded that more research was needed on the relative income elasticities of locally produced *versus* imported goods. The Hymer-Resnick "inferior goods" hypothesis [1969] of low expenditure elasticities was tested in a nationwide consumption survey in Sierra Leone by Robert P. King and Byerlee [1978]. The results showed that the expenditure elasticity coefficient for rural small-scale industry products was positive and rather high (0.9). Although the King-Byerlee results are cross-sectional and for only Sierra Leone, they suggest that the Hymer-Resnick prognosis of poor market prospects for small-scale industries should be questioned until more research is completed.[204]

The belief that rural small-scale industries have a major role to play in African economic development has been considerably enhanced by recent evidence that rural small-scale industries are efficient and profitable and because small-scale firms generally have low capital/labor ratios, jobs can be generated with small capital outlays. In Sierra Leone, for example, the capital per worker for small-scale industries was about $400 as compared with $7,300 for large-scale firms with fifty or more workers [Liedholm and Chuta, 1976]. Steel [1977] found that the original capital cost per worker in Ghana averaged $435 in firms with family labor and no wage workers, while it was approximately $9,000 per worker in firms with over 100 workers. Wilcock [1981] reported that average initial capital per firm in Eastern Burkina Faso was $435. Child [1977] found that the cost of capital per job in the modern sector of Kenya was three times higher than in small-scale firms. The output/capital ratio for small-scale industries was found to be higher than that of large-scale industries in Sierra Leone [Chuta and Liedholm, 1979].

Available evidence from both Kenya and Sierra Leone shows that small-scale firms can generate high rates of financial and economic returns. From a dynamic perspective, there appears to be no reason why small-scale industries should have difficulty in responding to increases in demand in most African countries during the foreseeable future. In general, small firms are easy to establish because the costs of entry are low. For example, one-half of the firms Child surveyed in Kenya started with an initial investment of $140 or less. In Sierra Leone the mean initial investment was less than $90 [Liedholm and Chuta, 1976]. Most studies, in fact, show that established small-scale firms tend to be overcapitalized and, therefore, to have excess capacity. Given sufficient adjustment time, trained labor should not be a constraint because small-scale firms in most countries tend to train their own labor through various apprentice systems [Callaway, 1964; Kilby, 1962; Child, 1977; K. King, 1977; Van Rensburg, 1978; Mabawonku, 1978; Steel, 1979; Wilcock, 1981].

POLICY DIRECTION

Research on industrialization in Africa has pointed up the strategic importance of rural small scale industry [Skarstein and Wangwe, 1986; Liedholm and Mead, 1987; Meier and Steel, eds., 1989]. The growing evidence that small-scale industries generate more employment and output per unit of capital than their large-scale counterparts should be seriously heeded by policy makers. Moreover, studies of the failure of capital-intensive techniques are widespread.[205]

On the demand side, it is obvious that the growth of agricultural production and the income of farmers is of strategic importance in providing the demand for rural small-scale industries. As a result, efforts to promote rural small-scale industry should be an integral part of a strategy to raise rural incomes. Piecemeal attempts (training programs and credit) to promote rural small-scale industry in Africa in the 1980s are likely to be ineffective unless they are part of a broader rural mobilization strategy. The centerpiece of a rural mobilization strategy should be raising agricultural production and incomes which in turn will provide the effective demand for the products of rural small-scale industries.

RESEARCH DIRECTION

More research is needed on the following questions [Liedholm and Mead, 1987]. What are the savings and reinvestment rates and patterns of both small- and large-scale industries in both the short and long run? What can be done to promote backward and forward linkages between rural small-scale industries and agricultural production, processing, and large-scale industries? What are the determinants of entrepreneurship? What types of educational and managerial assistance are needed to help rural small-scale industries?

7. Fisheries

Fishing is both economically and nutritionally important because it is often the least expensive source of animal protein in Africa and its overall nutritional value compares favorably to beef and eggs [Deelstra, White, and Wiggins, 1974]. Africans consume somewhat over nine kilograms of fish per person annually compared with an estimated six kg in Latin America and nine kg for Asians, but there is wide variation in per capita fish consumption. FAO data for thirty-four sub-Saharan countries indicate, for example, that fish contribute more than 40 percent of all animal protein in twelve countries, between 20 and 40 percent in thirteen countries, and under 20 percent in nine countries [FAO, 1976d].

The annual catch in Africa has been estimated to be around 3.5 million tons. About 60 percent of the catch comes from marine (offshore) fisheries and about 40 percent from inland fisheries. The primary marine fisheries are located on the west coast of Africa between the Tropic of Cancer and Capricorn, and off the east coast of Somalia and Kenya. Of the approximately 1.4 million metric tons of fish taken annually from freshwater fisheries, around half is from lakes and reservoirs with the rest being caught in rivers and floodplains [FAO, 1976d].

In West Africa, fish are usually salted, dried, or smoked for preservation during transport and distribution. In eastern African countries such as Kenya, 60 to 70 percent of fish are sold fresh for local consumption. Frozen fish, both domestic and imported, are becoming increasingly popular in urban areas, but are not likely to supplant dried fish among rural consumers [Krone, 1970]. There is some evidence that fish has been substituted for meat in recent years because of rising meat prices [Staatz, 1979]. There is also evidence, however, that demand varies widely for different types of fish [R. H. Bates, 1976]. The income elasticity of demand for fish is thought to be quite high, around 0.9 to 1, but evidence is scanty. V. E. Smith, Strauss, and Schmidt [1981] showed that the expenditure elasticity for fresh fish in Sierra Leone in 1974–75 ranged from 0.88 to 1.36 depending on the mean expenditure level while expenditure elasticities for dried fish ranged from 0.51 for low expenditure households up to 1.92 for high expenditure households.

Despite the importance of fish in African diets and the role of fishing as a source of employment and income, Africans appear to be underexploiting their fishery resources relative to other regions. While some of the major lakes such as Lake Victoria and Lake Tanganyika are heavily exploited [Oduro-Otieno et al., 1978], the FAO [1976d] estimates that the catch from the major lakes of Africa can be doubled. The supply of fish is constrained by an extremely low level of productivity among artisan (small-scale) fishermen and by insufficient large-scale fleets owned by Africans. For overviews on inland and marine fisheries, see P. B. N. Jackson [1971]; Msangi and Griffin, eds. [1974]; Crutchfield and Lawson [1974]; FAO [1976d]; and Kollberg [1979].

SMALL-SCALE (ARTISAN) FISHING

Small-scale or artisan fishing accounts for as much as 95 percent of the annual catch in Africa. Since most fishing is carried out on a seasonal basis, estimates of the number of people engaged in fishing are unreliable. Even full-time fishermen, such as the Addi canoe fishermen off the Ghanian coast, generally go to sea an average of only 150 to 160 days a year [Mansvelt-Beck and Sterkenburg, 1976]. Half of the annual catch is generally taken during the two or three most active fishing months. Part-time fishermen often fish during the flood season to supplement their diets, especially before harvesting staple food crops. While most small-scale firms primarily rely on family labor, it is common to hire additional labor with the proportion of man-hours supplied by hired labor increasing with firm size. In Sierra Leone, 90 percent of wages were paid in kind [Linsenmeyer, 1976], while in Kenya cash wages are the standard practice.

A wide range of technologies is employed by fishermen in exploiting marine and freshwater fisheries. Small one-man canoes are often used in shallow water in conjunction with cast nets. One of the most profitable types of small-scale fishing relies on large, traditional boats equipped with outboard motors. Linsenmeyer's [1976] economic study of alternative technologies in Sierra Leone revealed that the returns per person are approximately the same for seven combinations of boats and nets used in small-scale fishing. He also found that the returns were similar for paddled and motorized canoes despite the greater capital costs of larger canoes because they could go further to sea, getting a larger, more consistent catch.

Christensen [1977] analyzed rapid technological change in the Fanti fishing economy on the coast of Ghana. Christensen, an anthropologist, returned to an area he had studied twenty-five years earlier and found that the Fanti combated declining yields by replacing traditional canoes with larger motorized canoes, leading to concentration of ownership in the hands of a few individuals who are able to secure capital for initial investments and for subsequent maintenance. Because of the larger investment for equipment, owners of the motorized boats now receive a much larger share of the catch. Christensen noted that, increasingly, market women provide capital for boats and motors. As a result, female control of the industry is enhanced and the industry is being divided into classes of owners and crews, an ominous sign for the future of small-scale fishing.

RESEARCH DIRECTION

Policy makers and researchers need to examine the economics of fish *versus* meat in meeting protein needs, small- *versus* large-scale fishing, inland *versus* off-shore fishing, and fish ponds. Numerous governments and some donor agencies are explicitly and implicitly promoting large-scale fishing via subsidies, tariff structures, research, and technical assistance. Since small-scale fishermen (like small farmers) are unorganized and are mainly seasonal producers, they have had little voice in bringing about a change in government investment, research, and extension programs in support of small-scale production. Research on small-scale fishing should be carried out in conjunction with research on fish processing and marketing since increased productivity will likely have a minimal impact on the profitability of small-scale fishing unless marketing is also improved.

8. Recovery of the Sahel

The six-year drought from 1968 to 1974 had a devastating impact on the lives of millions of the people in the Sahelian region of West Africa.[206] Sahel is an Arabic word meaning the edge of the desert. In ecological terms, the Sahel is the belt of land along the southern edge of the Sahara desert from the Atlantic Ocean to Lake Chad, with annual rainfall varying from 150 mm to 500 mm [Swift, 1977].

In political terms, the Sahel refers to eight countries: six French-speaking countries—Mauritania, Senegal, Mali, Burkina Faso, Niger, and Chad; one English-speaking country—The Gambia; and one Portuguese-speaking country—Cape Verde Islands with a population of 300,000 off the shore of Senegal.

Sahelian countries are poor as shown in Table 1 (p. 6). The staple foods in the region are millet and sorghum except in Senegal and The Gambia where rice is important in local diets. Cotton and groundnuts are major export crops, while livestock dominates the economies of Mauritania and Mali. The Sahel is noted for its large variation in the amount, timing, and geographical spacing of rainfall. The extreme variation in rainfall is illustrated in northern Senegal where the theoretical carrying capacity for cattle ranged from 187 per 1,000 ha in a "normal" good year, to 87 in a "normal" bad year, and zero in 1972—a disastrous year at the peak of the drought [Swift, 1977, p. 458].

The Sahelian countries established a permanent secretariat called CILSS in Ouagadougou, Burkina Faso during the drought in order to coordinate their requests for emergency relief. The drought brought forth a massive relief effort—mainly food grain—from Western Europe and North America. Major donor nations organized a Secretariat (Club du Sahel) in the OECD Headquarters in Paris to coordinate donor assistance for relief and recovery of the Sahel. CILSS and the Club du Sahel have a close working relationship; they have jointly published a number of excellent studies. The history of the drought and relief operations is covered in Dalby and Church, eds. [1973]; Bernus and Savonnet [1973]; Sheets and Morris [1974], *Disaster in the Desert*; Caldwell [1975]; indispensable collections by Copans, ed. [1975], Glantz, ed. [1976], Dalby, Church, and Bezzaz, eds. [1977], and Glantz, ed. [1987]. For a skeptical view of foreign aid, see Meillassoux [1974], "Development or Exploitation: Is the Sahel Famine Good Business?"

As the relief effort phased down in 1974 and 1975, the Sahelian nations through CILSS and the major donors through the Club du Sahel jointly adopted food self-sufficiency[207] and self-sustaining economic development as the goals of a long-term recovery program for the 1975-2000 period. The FAO's strategy paper, *Perspective Study on Agricultural Development in the Sahelian Countries: 1975-1980* [1976c], was a basic document used by CILSS and the Club du Sahel in developing their long-term recovery plans. The Club du Sahel's strategy is found in Club du Sahel [1977] and CILSS/Club du Sahel proposals for the recovery of the Sahel include CILSS/Club du Sahel [1977; 1978a, b; 1979; 1980a, b]. The U.S. strategy for the recovery of the Sahel was published as AID [1976] and Shear and Clark [1976]. The U.S. strategy for river basin development is spelled out in AID [1978].

RESEARCH FINDINGS

Climate. During the drought, two critical questions were debated. First, was the 1968-74 drought a manifestation of a permanent shift in the climate in the Sahel resulting from clearing of land, burning, and overgrazing? This question still cannot be answered because benchmark data are not available. The second question was the probability of the reoccurrence of drought in the 1980s and 1990s — a crucial question for recovery plans. Economic historians, Lovejoy and Baier [1975] and Baier [1976, 1977], studied the occurrence of drought in the Sahel over the past 200 years and found that the period was marked by wet and dry rhythms of irregular length and by two major droughts over the 1900-1968 period. Baier, Lovejoy, and other researchers concluded that there was no evidence to suggest that a major drought will recur at any greater frequency than it has over the past several hundred years.

Desert Encroachment. Throughout the 1968-74 drought, there was considerable debate on the question of desert encroachment. The international press played up the number of hectares of farmland that were being permanently "lost" each year as the Sahara desert moved southward and "converted" grazing and farm land into desert. Glantz, ed. [1976, 1987], Lovejoy and Baier's [1975], and M. S. Diarra's [1975] seminal articles on the "drought/recovery cycle" revealed that the pastoralists and farmers have historically moved back and forth between the Sahelian and Savannah ecological zones like a finely tuned accordion; the authors believed that the return of normal rainfall in the Sahel would convert the encroached land back to grazing and farm land in the same pattern that appeared over the past several hundred years. The research of M. S. Diarra [1975]; Baier [1976, 1977]; Lovejoy and Baier [1975], and Horowitz [1972, 1977] stressed the positive role of the mobility of people and livestock in response to drought; they cautioned against establishing fenced ranching schemes and they encouraged governments to facilitate livestock transhumance through the development of water points and marked corridors.

Demography. Despite widespread reports in the international press about the death of hundreds of thousands of people during the drought, Caldwell [1975] reported that the loss of human life was modest and that the remarkable feature of the drought was the ability of the population to survive a six-year drought. The impact of the drought on the demography of a village in Niger is reported by Faulkingham and Thorbahn [1975] and Faulkingham [1977]. The impact of the drought on farming and livestock is recorded in an indispensable collection by Gallais, ed. [1977]; Swift [1977]; and an AID workshop on livestock [AID, 1980].

Overcoming Food Dependency. The USDA/ERS [1981] has shown that lagging food production in the Sahel cannot be reversed overnight and that rice and wheat import dependency cannot be reduced in the short run. In our judgment, the recovery of the Sahel will be a long and painful process because there is little that can be done in the short run (five to ten years) to overcome the Sahel's dependency on rainfall when only 1-5 percent of the arable land in the region is presently under irrigation; proven biologically stable and economically profitable millet and sorghum packages are generally not available for the big four: wheat, rice, sorghum, and millet; action programs—seed multiplication, credit, animal traction, and extension—are generally running ahead of the research base for cereal crops; the shortage of technical, managerial and administrative manpower and the entrenched position of convenience foods such as wheat and rice in urban diets. In short, there is no "Green Revolution" on the horizon for food crops in the Sahel.

ASSESSMENT

Both CILSS and the Club du Sahel are to be applauded in laying out a long-term development strategy for twenty to twenty-five years. To date, the emphasis has been on improving smallholder rainfed farming in the 1980s and a gradual increase in land under irrigation [Horowitz and Painter, eds. 1986]. In several countries, a few projects are producing encouraging results and should be replicated. But there are a number of questions about the "food first strategy," the extractive pricing policies of national governments, grain boards [CILSS/Club du Sahel, 1979], and the proposed grain storage program for the Sahel [FAO, 1981b]. As a political slogan, a "food first" strategy has been effective in mobilizing support from donors for recovery projects. But there is a danger that CILSS and member governments are promoting crash programs before extension agents have a biologically sound and financially profitable food package ready to extend in rainfed farming areas [Tapsoba, 1981; Sanders, 1989]. Since returns per hour of labor are generally substantially higher in cash crops and off-farm employment than in food crops throughout West Africa, there is a need for Sahelian countries to shift to a more balanced food/cash crop strategy, taking into account variation in resource endowments by country, opportunities for intraregional trade, etc. [E. Berg, 1986a]. The 1974-87 recovery phase to date should be most appropriately called a "pilot phase." It is too soon to evaluate whether the strategies for the long-term recovery and development of the Sahelian region will be effective [Glantz, ed. 1987].

Chapter IX. Synthesis and Research Priorities for the 1990s

1. Synthesis[208]

At independence in 1960, sub-Saharan Africa (Africa) was a net exporter of food. Three decades later, despite vast physical potential to produce food, Africa is importing around eight million tons of food each year and around 100 million people do not get enough to eat. To compound these problems, Africa's population of around 500 million is expected to double and reach a billion in twenty to twenty-five years.

Africa's poverty is captured in a single statistic: the total GNP of the forty-five countries in sub-Saharan Africa in 1985 was slightly less than the total GNP of Spain, a nation of forty million. [World Bank, 1987a].[209] Sixteen of the twenty poorest countries in the world are African. Since 70 percent of the people in Africa live in rural areas, it follows that raising the income of rural people is a prerequisite for improving the African standard of living. Because poverty is a central cause of hunger and malnutrition, it also follows that per capita income growth is a primary way of increasing access to food and reducing malnutrition.

During the first three decades of independence, agrarian stagnation has been the Achilles heel of African development. The record of foreign-financed food and agricultural projects is poor. The World Bank reported that 40 percent of all Bank-financed projects in Africa that were evaluated in 1985 were judged to have unsatisfactory or uncertain results compared with 10 to 15 percent in Asia and Latin America [World Bank, 1987a]. The average economic rate of return (ERR) for agriculture projects in Africa was less than 6 percent, far below the usual cut-off rate of 10 to 12 percent. The failure rate in agriculture was twice as great as in other sectors such as health and education.

But Africa's agrarian crisis is neither unique nor unexpected. In fact, in Latin America, industrialization ground to a halt during the Peron regime in Argentina in the 1950s because of a food crisis. The same pattern of events was repeated during India's food crisis of the mid-1960s, and again in China in the late 1960s. It is now generally acknowledge that Africa's food crisis was quietly building up for two decades before it exploded onto television screens around the world during the Great African Famine of 1984–85 [Eicher, 1982a]. In 1984–1985, a million people died in Ethiopia and half the nations of Africa requested emergency food aid.[210] The famine belatedly forced Africa to face up to the same question that Argentina, India, and China had been forced to confront: "What is to be done about agriculture?"[211]

After a third of a century of independence, there is a growing awareness that many African countries may be generations, and a few may be centuries, behind Asia and Latin America in terms of their stage of scientific and human development and political and institutional maturity. This is a sensitive topic that was shunned in the 1960s and 1970s and is only slowly starting to be brought into the open. For example, the respected Africanist, Colin Legum, recently observed that as colonial powers withdrew from the continent in 1960, they "left behind them a series of national states, but very few nation-states. The level of development of the continent's nation-state was still roughly equivalent to that of Europe or China in the fourteenth and fifteenth centuries — and certainly no later than the seventeenth century" [Legum, 1985, p. 24]. Africa's institutional and scientific gap relative to Asia and Latin America is illustrated by the following comparisons:

> African states started independence with an extremely small pool of trained scientists and managers relative to Asia and Latin America. For example, at independence in 1960, Zaire had only sixteen university graduates [J. Coleman, 1984]. Mauritania had two civil servants with a degree in agriculture at independence.

> Africa's stock of scientific and managerial personnel is about one-fifth the number per million as that of Asian people [K. H. Shapiro, 1985 and UNESCO, 1988]. About one fourth of the researchers in national agricultural research services (NARS) and teachers in faculties of agriculture in universities in Africa are expatriates.

> The first university in Nigeria, the University of Ibadan, was established in 1948. By contrast, the first three Indian universities (Madras, Bombay, and Calcutta) were established in 1857. The University of Bombay conferred its first Ph.D. in Economics in the 1930s. The capacity of African universities to offer quality post-graduate training in agriculture is decades behind that of their counterparts in Asia and Latin America.

> Many African universities and national research services are weaker today than at independence. The tree crop research institutes in Nigeria and Ghana are overstaffed, ineffective and lagging scientifically behind their counterparts in Malaysia and Indonesia. At independence in 1960, there were 420 European (mostly Belgian) scientists and technicians (more than half were university graduates) and a Congolese labor force of 12,000 to support seventeen research stations, fourteen experimental plantations and a veterinary laboratory in Zaire [Drachoussoff, 1965, p. 188]. Today there are forty-three national scientists in Zaire's National Agricultural Research Service supplemented by fifty-six national and eleven expatriate scientists in a separate research and extension project in the Ministry of Agriculture. Because of the lack of government support for research operating costs, Zaire's research stations are mainly used to produce food for the stations' labor force.

> Livestock projects have been plagued with technical, managerial and financial problems. "As is often the case, where the World Bank led, other donors followed . . . an estimated $625 million of international funds have been invested in sub-Saharan African livestock development

since the late 1960s . . . more often than not the effect has been disappointing" [Dyson-Hudson, 1985, p. 158].

Input delivery systems (e.g. seed multiplication, fertilizer) are rudimentary and monopolized by the state.[212]

STATUS OF AFRICAN AGRICULTURE AFTER THREE DECADES OF INDEPENDENCE

African agriculture possesses the following characteristics:

Range of Farming Systems. African agriculture is noted for its diversity, complexity and an immense range of agroclimates and land, water, soil and labor endowments, leading to an immense variety of farming systems [DeWilde *et al.*, 1967; Ruthenberg, 1980; FAO, 1986b; Binswanger and Pingali, 1988; David Shapiro, 1988; R. Singh, 1988].

Smallholder Production. Smallholder farming is the dominant type of farming system. About 25 to 50 percent of the working time of farm household members is spent in off-farm jobs during slack periods of the farming season [C. K. Eicher and D. C. Baker, 1982].

Female-Headed Farm Households. The proportion of smallholder households headed by women has been underestimated by many researchers. Professor Jean Due reports that about 30 percent of the farm households in Africa are headed by women but there is wide variation within countries and between regions such as the Sahel and SADCC [J. M. Due, 1988]. In some countries such as Lesotho where many men are working in the mines of South Africa for six to nine months per year, the percentage of rural households headed by females is about 50 to 60.

Labor Force. Over the next forty years the percentage of population employed in agriculture will probably decline slowly (68 to 58 percent), but the size of the agricultural labor force will likely *triple* because of rapid population growth and the failure of the industrial sector to open up enough new jobs [Pingali, Bigot, and Binswanger, 1987].

Landless Labor Force. A landless labor class is emerging in a number of countries with high rural population densities.

Low Level of Purchased Inputs. Farmers are using low levels of purchased inputs. The average fertilizer consumption rate is 6.4 kg nutrients/ha, the lowest of any region in the world. Africa accounts for less than 3 percent of the fertilizer used worldwide [IFDC, 1986].

Food/Population Trends. From 1970 to 1984, food production in Africa grew at about half the population growth rate. Fertility rates will likely remain high and the total fertility rate will likely increase in the short run in some countries. There is no policy lever that can be pulled to slow population growth over the next ten to twenty years [Caldwell and Caldwell, 1988].

Sources of Growth of Food Production. From 1960 to 1980, about 80 percent of the increase in food production in Africa came from area expansion and the balance from an increase in yields [Paulino, 1987, p. 30]. During this same period, 100 percent of the increase in food output in West Africa came from area expansion. The cheapest source of increased food production in most land abundant countries is area expansion in the foreseeable future. But in countries where most of the arable land is under cultivation such as Rwanda, Kenya, Niger, and Senegal, the increase in required food supplies will have to come from yield-increasing innovations or food imports.[213]

Availability of Improved Technology. The scientific community is not in agreement on the stock of improved food crop and livestock technology "on the shelf," awaiting diffusion to farmers. For example, Dunstan Spencer, a Sierre Leonean authority on African agriculture, reports that probably less than 2 percent of total sorghum, millet, and upland rice area in West Africa is sown with cultivars (varieties) produced through modern genetic research [D. S. C. Spencer, 1986, p. 224]. On the other hand, the FAO recently asserted that "except in arid and semiarid areas without irrigation, food production can be roughly doubled with existing technology. Thus, the immediate need is to provide adequate supplies of fertilizer, improved seeds, tools" [FAO, 1986b, p. 61].[214]

Three basic forces are exerting pressure on most African countries to increase food supplies from expanded domestic production or imports. First, there is a need to meet the food needs accompanying Africa's rapid population growth rate of 3.2 percent—the highest of any region in the world. Second, there is a need to stabilize and/or reduce Africa's food grain imports that are currently running at around eight million metric tons per year. Third, increased production is needed to accommodate some of the growth in per capita income that will be spend on food.

2. Research Priorities

> *National priorities in research are a noble objective.*
> *International priorities would be still nobler.*
> —T. W. Schultz [1983]

Africa is a vast subcontinent of forty-five countries, 1,000 ethnic groups and seven colonial histories. The immensity of Africa is captured in a single statistic: the land area of the entire continent is larger than the combined land area of Western Europe, North America and China. Because of Africa's immensity, diversity of natural resources, and uneven prospects for development; *there is little to be gained from a search for Africawide research priorities.* For example, during Zambia's food riots in late 1986, neighboring Zimbabwe was struggling to finance its "maize mountain" that was equivalent to two years of domestic consumption. The disparity between Zimbabwe's maize mountain and Zambia's empty harvest cannot be blamed on colonialism, the weather or other acts of God because both are former British colonies producing the same staple food, maize, in the same agroecology, and under the same general rainfall pattern. The food crisis in Zambia is part of the overall economic crisis brought about by a string of policy failures *par excellence!* For several decades the government of Zambia has pursued a copper-led development strategy while paying lip service to agriculture. The striking contrast between the food situations in Zimbabwe and Zambia illustrates the need for scholars to identify country-specific problems and carry out research to generate a local knowledge base to replace the grand slogans such as "getting prices right" [Rukuni and Eicher, 1987, 1988].

The economic crisis in Africa is first and foremost an agrarian crisis. And since the agrarian crisis in almost all African states is a failure of the food and agricultural sectors, rather than a food crisis *per se,* it follows that the research agenda of the 1990s should be expanded to discover how to raise rural productivity and rural incomes across the board. To illustrate this point, in 1986, the Government of Kenya identified seven "essential" commodities that formed the core of its food and agricultural policy: coffee and tea for improving farm incomes and as a source of export earnings; maize, wheat, milk, and meat (mainly beef and poultry) for food security; and horticultural crops for both export and home consumption."[215] This example points out the need to move beyond sterile debates on food *versus* cash crops and identify a mix of commodities to achieve multiple policy goals, including raising rural incomes in order to generate the effective demand to enable rural people to secure their food needs.

Without question, research on food production should continue to receive high priority in the 1990s. But there is also a need for rural social scientists to *discover* livestock and *rediscover* export crops as important and challenging research topics. Many of the classics in the social science literature of the 1950s and 1960s should be reread as background to the research agenda for the 1990s. Classic studies of export crops include: Galletti, Baldwin, and Dina's *Nigerian Cocoa Farmers* [1956]; Polly Hill, *The Migrant Cocoa Farmers of Southern Ghana* [1963]; and John de Wilde *et al., Experiences with Agricultural Development in Tropical Africa* [1967] which included case studies of pyrethrum, cotton, tea, coffee. Other classics include Peter Bauer's study of market liberalization in his *West African Trade* [1954]. Turning to the environment, Pierra de Schlippe's *Shifting Cultivation in Africa* [1956] discussed alternating strips of forests and food crops in Zaire, a forerunner to IITA's current research on alley farming. Likewise, Rene Dumont's *False Start in Africa* [1966] was twenty-five years ahead of the Brundtland report [World Commission on Environment and Development, 1987] in identifying Africa's fragile environment as a critical research and policy issue. The effect of cash cropping on family nutrition was studied by R. E. Baldwin *et al.* [1956]. Food consumption researchers will benefit from a careful review of several classics by Food Research Institute scholars: Bruce Johnston, *The Staple Food Economies of Western Tropical Africa* [1958]; William O. Jones, *Manioc in Africa* [1959], and Farnsworth, "Defects, Uses and Abuses of National Food Supply and Consumption Data" [1961]. Finally, the reader is directed to the World Bank's long-term perspective study that looks back thirty years and looks ahead thirty years to outline a policy agenda for doubling Africa's food production growth rate from 2 to 4 percent per annum [World Bank, 1989].

Eight research topics should be given high priority in the 1990s. The first topic is agricultural and industrial sector interactions.

AGRICULTURAL AND INDUSTRIAL SECTOR INTERACTIONS

Since Africa has so many countries at different stages of political, economic and institutional maturity, there is a need for African scholars to develop an array of models that address the central problems of agricultural development in different countries and subregions such as Central Africa, Eastern Africa, etc. In developing these models, agricultural economists should move beyond their current fixation on technical change and FSR and deal with a broader set of issues such as the interactions between the agricultural and industrial sectors over time, population growth and the development of human capability and sustainable African institutions. Although Professor T. W. Schultz omitted population growth from his classic study *Transforming Traditional Agriculture* [1964], social scientists in Africa cannot assume away population growth because of the increasing population pressure on the national resource base and the projected tripling of the size of agricultural labor force over the next forty years.

Researchers can gain valuable insights from Asia's experience where impressive rates of industrialization and overall economic growth have not been accompanied by increases in real wage rates of labors except for the urban organized sector [Hayami, 1988]. In fact, Hayami reports that one of the basic forces underlying the declining rural wage rate in many Asian countries is increasing population pressure on the land. The Asian experience reinforces the need for scholars working on African agricultural development to devote substantial attention to demographic change and the design of agrarian institutions to cope with rapid population growth.

The social science community addressing Africa's agrarian crisis need to address three interrelated transformations: agricultural, demographic, and the industrial. Substantially more attention should be devoted to historical studies of the role of political and institutional factors in influencing the speed and nature of these transformations [Bairoch, 1973; Berry, 1984; Hayami, 1988; Kanon, 1985; Johnston, 1986; Bonnen, 1987; and de Janvry, 1987]. The public debate on population growth is slowly emerging in a few countries such as Niger, Kenya, Zimbabwe, and Nigeria. The policy debate on population will likely follow public debate over the next ten to twenty years. The debate on the role of the state in the demographic transformation in Africa is one to two decades behind the debate in Asia.

The debate on Africa's industrial transformation has come full circle over the past thirty years [Meier and Steel, 1989]. The basic industry (iron and steel, petro-chemicals, metallurgy, etc.) development strategy promoted by ECA and OAU in the 1970s and early 1980s has failed in Nigeria, Tanzania and many other countries. The ECA and OAU have recently urged African countries to reorder priorities and pursue industrial policies in support of agriculture [OAU and ECA, 1986] and rural small scale industry [Liedholm and Mead, 1987 and Chuta and Liedholm, 1990].

In summary, there is a need for social and technical scientists to start addressing the three interrelated transformations: agricultural, demographic, and industrial. In Latin America, ECLA fostered debate and research on the critical development issues of the 1950s and 1960s under the leadership of Professor Prebisch. By contrast, ECA has provided inadequate intellectual leadership on the critical development issues during Africa's first three decades of independence. The African Development Bank in Abidjan has concentrated on financing investments in infrastructure, industries, public services, and agriculture.

To deal with these tough institutional, social organizational, and human capability issues, there is a need for African scholars to carry out studies of the economic history of agrarian change in Africa. Drawing from both neoclassical and political economy literature, scholars should focus on developing neoclassical political economy models[216] based on empirical evidence from Africa. To aid in the development of African models of agrarian change, economists in African universities will have to be weaned away from their preoccupation with short-term structural adjustment problems and industrialization and become deeply involved in the debate on food and agricultural policy options at this stage of Africa's institutional and political fragility. The recent publications of Professor Aboyade [1988], Philip Ndegwa [1985], and Professor Benno Ndulu [Lundahl and Ndulu, 1987] are by macroeconomists now addressing agrarian problems.

MACROECONOMICS OF FOOD AND AGRICULTURAL POLICY

Presently one-half to two-thirds of the rural social scientists in Africa are working on micro issues such as village studies, diffusion of technology, impact assessment and FSR. Although these studies can contribute valuable information to planners, policy makers and donors, we are living in an increasingly interdependent world economy. Professor Ed Schuh reminds us that "a decade ago, the value of a nation's currency was largely ignored as an issue of domestic food and agricultural policy. Today, it is probably the most important price in the economy." [1990, p. 140].

When the pendulum among donors shifted from project to policy-based lending in the early 1980s, agricultural economists in Africa were focused on micro studies. Typically, "quick and dirty" agricultural sector assessments or updates were carried out by teams of short-term consultants in order to provide the policy context for agricultural loans. But short-term missions contribute little to building local capacity for policy analysis [Weber, Staatz, et al., 1988]. Visiting missions invariably do not have enough time or resources for primary data collection or to dig into the tough institutional issues. For example, a recent ILO/JASPA mission to Somalia under the leadership of Professor Victor Diejomaoh concluded that:

> the major problem with the Somali economy is not "getting
> prices right" but "getting development going." The Somali
> economy operates under relatively de facto free markets and
> tinkering with prices for the most part can only have a marginal
> impact on performance. . . . Unfortunately fundamental
> questions of increasing rural productivity are not getting adequate
> attention in Somalia at the moment with devaluation and
> liberalization having been given centre stage. [ILO, 1987, p. x]

Twenty-five years ago Vernon Ruttan and Abe Weisblat [1965] carried out a study of Asian agricultural economists who were trained in the United States. They concluded that "too little emphasis has been placed on preparing agricultural economists from abroad for work on problems of national significance, to giving them training in macroeconomic theory."

Is it not peculiar that after thirty years of independence, that there are few, if any, African agricultural economists who make their living by teaching and carrying out research on agricultural trade and exchange rate problems? There is a need to train a first generation of African agricultural economists to work on the macroeconomics of food and agricultural policy, including structural adjustment issues [Wolgin, 1990].

Insights for the research agenda on the macroeconomics of food and agriculture policy are found in Aboyade [1988]; Elliot Berg [1986a, 1987]; Bryant, ed.

[1988]; Lundahl and Ndulu [1987]; Koester [1987]; Idachaba [1988]; Lipumba, Msambichaka, and Wangwe, eds. [1984]; Rado [1986]; Dione and Staatz [1988]; Bevan, Collier, and Gunning [1987]; Akoto [1987]; Staatz [1988]; Skarsten and Wangwe [1986]; Waelti [1988]; Eicher [1988c]; Paulino and Sarma [1988]; Weber, Staatz, *et al.* [1988]; Dione [1989]; Staatz *et al.* [1989]; and Mosley and Smith [1989].

INSTITUTIONAL INNOVATIONS FOR AGRARIAN CHANGE

Three decades of independence have produced a large knowledge base on why agricultural policies and projects are *not* performing well at this stage of Africa's economic history and institutional fragility. There is consistent evidence that human capability and institutional barriers to development have been skirted in the drive to increase foreign aid flows to African agriculture—especially during the rapid build-up of aid for agricultural projects over the 1973–83 period. Starting in 1983, the pendulum shifted from project to policy-based lending. But regardless of whether foreign aid was focusing on projects or policies, the end result was the same: the prime movers of agricultural development, scientific, institutional, social organizational issues, and long-run human capability are being neglected by most African policy makers and donors. There is a need for a fundamental re-examination of the assumptions about Africa's stage of economic history, and the basic development strategies that have been pursued by African states and donors over the past three decades [Eicher, 1988b]. Research is also needed on and the development of sustainable institutions for African agricultural development [Eicher, 1988c].

Donors are confused on how to assist Africa in developing its human capability and agricultural institutions at this early stage of development. A recent joint UNDP/IBRD technical mission dug deeply into the mode of delivering aid to Somalia, a country riven with clan wars and generations behind most Asian countries in terms of its level of scientific, institutional and administrative maturity. The joint team reported that donors were collectively pumping US$ 100 million into Somalia each year to support 1200 expatriates on technical assistance contracts and overseas training for Somali nationals in a revolving door type of operation [UNDP and IBRD, 1985]. Nevertheless, this model of foreign advisors and overseas training is not achieving the ultimate objective, "the development of national capacity through the permanent transfer of skills and know-how to Somali nationals and national institutions" [UNDP and IBRD, 1985, p. 2]. In short, the basic foreign assistance model is not addressing the long-term problem of developing *sustainable* Somali institutions.

The capacity of many African countries to deal with food and agricultural stagnation is constrained by their political instability and their early stage of scientific, administrative, and institutional development.[217] Moreover, in some countries, there is a fundamental lack of political commitment to come to grips with poverty, malnutrition, and access to food. Because of these political and institutional issues, "the boundaries of traditional economics make it a rather limited tool to understand food battles and their outcomes" [Sen, 1984, p. 89]. Without question, hunger, malnutrition, poverty and famine in Africa are just as much a function of political, macroeconomic, and institutional barriers, as the lack of technology and effective input supply systems. Illustrations from the current agrarian chaos in Zambia, Sudan, Ethiopia, Somalia, Chad, and Nigeria are too numerous to conclude otherwise.

But the tough institutional issues are currently receiving relatively little attention by social scientists [Eicher, 1989, 1990a]. This pattern of inaction does not come as a surprise. For example, the late Gunnar Myrdal of Sweden reports that when he was carrying out research for his classic *Asian Drama* over a ten year span in the 1960s, the most difficult issue was learning how "to deal with the political issues of changing institutions, which were then, as now, avoided by most ordinary economists in their writings on development" [Myrdal, 1984, p. 154].

Today the study of institutions has moved to center stage in both industrial countries and the Third World, including Africa. In fact Professor Glenn L. Johnson of Michigan State University contends that "institutional limitations are presently the most serious constraining factor" for the agriculture of developed and newly industrializing countries and that the less developed countries "are now constrained more by existing institutions and human capital stocks than by technologies and stocks of biological and physical capital" [Glenn L. Johnson, 1988, p. 1].

What is the agenda for research on institutions? The sensitive nature of research on institutions in Africa requires researchers to walk on two legs in carrying out this type of research. First, we recommend that institutional change be included as an integral part of studies of marketing, irrigation, food production, and other topics in order to increase the probability that the research is relevant to Africa. Second, we recommend that researchers examine both institutional successes[218] as well as institutional failures.[219] Research on institutional innovations is needed to examine why a large number of institutions have proven to be strikingly effective. Examples include the Kenya Tea Development Authority that serves 150,000 Kenyan smallholders [Lamb and Muller, 1982], the Zimbabwe Smallholder Cotton Marketing Board [Abbott, 1987a], the Botswana Meat Commission, the West Cameroon Coffee Cooperative Union, and the celebrated CFDT/IRCT's institutional network that oversees smallholder cotton production in nine countries in Francophone West Africa. Cotton research is carried out by IRTC researchers in France, Côte d'Ivoire and eight satellite countries in Francophone Africa. The CFDT is the cotton management/extension organization with four decades of experience in West Africa. The World Bank recently reviewed the Francophone cotton production model in Burkina Faso, Côte d'Ivoire and Togo—and declared it a "striking success" [World Bank, 1988c, p. 29]. The Bank report also noted that cotton yields in Ghana, an Anglophone country not participating in the IRTC/CFDT network, were one-half of those of their common neighbors, Burkina Faso, Côte d'Ivoire, and Togo.[220]

Recent studies of institutions include the following: [Chambers, 1983; Bonnen, 1987; Lipumba, 1984; Lipumba, Msambichaka, and Wangwe, eds., 1984; Cernea, 1985; Williamson, 1985; North, 1984, 1987; B. Child, Muir, and Blackie, 1985; Hazlewood, 1985; Land Tenure Center, 1985; Elliot Berg, 1986a; Bloch, 1986, 1987; Nellis, 1986; Abbott, 1987a,b; Elliot Berg, 1987; Feder and Noronha, 1987; Benor, 1987; Bingen and Fay, 1987; Birgegard, 1987; Idachaba, 1987, 1988; Eicher, 1988c, 1990a; Glenn L. Johnson, 1988; Ruttan, 1988; McDermott, 1988; Howell, 1988; Aboyade, 1988; Weber, Staatz, et al., 1988; Roling, 1988; Byerlee and Tripp, 1988; Pickering, ed., 1988; World Bank, 1988d, 1988e; Leonard, 1991].

RURAL POVERTY AND INEQUALITY

Rural poverty and inequality are increasing in Africa and demand high priority research attention in the 1990s. During Africa's first three decades of independence, rural poverty *per se* was a low-priority research topic in Africa relative to Asia [Lele, 1983]. Because of the land-abundant nature of most African economies, it was assumed that population growth would automatically bring more land under cultivation. But over the past three decades, rapid population growth has intensified the competition for land and the number of landless laborers is growing. In Asia, the leadership in research on poverty and rural inequality has come from economists such as the late Raj Krishna, M. L. Dantwala, V. Dandekar, K. N. Raj, R. M. Sundrum [1987], S. Chakravarty [1988], and many others. Because the agricultural economics departments in India were so micro oriented and so weak in the 1950s and 1960s, they were unable to get the national debate focused on food, hunger, poverty, and employment until economists started to work on these problems.

In Africa, agricultural economists and economists have generally ignored the study of poverty and inequality *per se*. One of the important exceptions to this generalization is the landmark study *Growth without Development* by a group of Northwestern University economists [Clower, Dalton, et al., 1966]. The authors concluded that Liberia's high economic growth rate produced by rubber, iron ore, and timber exports over the 1950-60 period had not led to an improvement in the welfare of Liberians because of structural and social barriers. The book foreshadowed Liberia's current political and economic difficulties. For an overview of poverty in historical perspective, see *The African Poor* by Professor John Iliffe [1987].

The problem of inequality in Africa was first identified by two anthropologists, Polly Hill [1968] and the late Lloyd Fallers [1973] of the University of Chicago. Anthropologist David Brokensha [1987] recently reviewed the literature on rural inequality since 1973 and reached the following conclusions:

(1) In the precolonial and even in the early colonial period, there was inequality, but it was a different kind: people were more mobile, poverty was not necessarily permanent, there a common culture was shared.

(2) Social and economic inequality are increasing.

(3) Today we see the beginnings of a class formation, with an ominous increase in landless rural people who have few or no opportunities for employment. There are signs of self-perpetuating rural elites. Many writers now divide rural people into distinct categories. . . . Such categories have been common in Asia for many years but are relatively new in Africa. The poorest are beginning to include a growing number of landless, also a new phenomenon for Africa.

(4) Off-farm employment is usually a major factor in determining position on the socioeconomic scale.

(5) Women suffer disproportionately more than do men, especially in *de jure* female-headed households. Louise Fortmann [1984] compared women farmers to male farmers in Botswana, showing that the former had less draft power (oxen), and access to extension workers, and lacked resources generally.

(6) Fallers [1973] was clearly overoptimistic in assuming that kinship and other traditional structures would effectively mitigate the worst consequences of inequality, and that society would remain relatively "open" [Brokensha, 1987].

Research on technological change and equity issues has been modest in Africa relative to research on equity effects of the Green Revolution in Asia over the past two decades. Three reasons explain the low priority given to equity issues in Africa. First, the Green Revolution has had little impact to date. For example, the two main Green Revolution commodities (wheat and rice) were grown on about 750,000 hectares of land in sub-Saharan Africa in 1983 [Dalrymple, 1986a,b]. This represents only about one-fourth of the cropped land in one medium-sized country such as Zimbabwe.[221] Second, rather than digging into issues such as land tenure, social structure, and class formation, many agricultural economists have approached inequality rather mechanically by generating Gini coefficients even though they are known for their unreliability [Jain, 1975]. Third, some rural inequality topics such as land distribution are ultrasensitive for researchers — especially in one party states. For example, research on land tenure has dried up in

Kenya over the past five years because of the political risk involved. And as Kenya's population doubles from twenty to forty million over the next twenty years, land problems will become even more sensitive for researchers. For recent research on inequality see: [Clayton, 1983; J. M. Cohen and Isakkson, 1988; Fortmann, 1984; Ghai and Smith, 1987; Haggblade and Hazell, 1988; Heisey, 1985; Hill, 1986; Johnston, 1986; Putterman, 1986; Mumbengegwi, 1988; Poats, Schmink, and Spring, eds., 1988; Vyas and Casley, 1988; Nafziger, 1988; and Mehretu, 1989].

THE ECONOMICS OF PRODUCING FOOD, LIVESTOCK, AND EXPORT COMMODITIES

Since a growing percentage of food production is sold as a cash crop within the country of production, the simple distinction between food and cash (export) crops is no longer valid in Africa. Today, a cash crop can be a food or nonfood commodity and it may be sold within the country or in international markets. For example, both maize and cotton in Zimbabwe are cash crops and they are sold in Zimbabwe and overseas. Therefore, we propose dividing crop production into food and export commodities with the latter referring to a range of commodities such as cotton, tobacco, coffee, tea, palm oil, cocoa, fresh fruits, vegetables, and spices. The research agenda on food production is outlined in D. S. C. Spencer [1986]; Johnston [1986]; Eicher [1986a]; and in the collection of papers edited by Mellor, Delgado, and Blackie [1987]. We have already discussed research on food production in sections on smallholder farming, technical change, and livestock.

The current research base on the economics of export crop production is surprisingly inadequate considering the importance of these commodities as agents of rural modernization. For example, to our knowledge, there is no up-to-date study of the economics of tea production in Kenya even though 150,000 smallholder families belong to the Kenya Tea Development Authority.[222] Since poverty is a central cause of malnutrition and family food insecurity, it follows that smallholder export crop production has the potential of generating jobs and income to help families secure their required food supplies. A critical question to address is why are tea yields on smallholder farms about one-half of those on private tea estates? Likewise, with the exception of a recent World Bank [1988d] survey of cotton production in three countries in Francophone West Africa, there has been surprisingly little research on cotton production since Shapiro's [1973] study in Tanzania. However, 70,000 smallholders are producing cotton in southern Mali under the Mali-Sud program [Dione, 1989]. In Zimbabwe, 45,000 people are employed by the cotton industry, including smallholders, seasonal laborers for picking, and employees in the factories. In Zimbabwe the average smallholder can make US$ 300 profit per hectare in cotton production and he can cultivate one to two hectares of cotton on a typical farm of five hectares [Abbott, 1987a, p. 55]. But the reasons for the decline of export crop production should also be studied. For example, the West African tree crop economies, with the exception of Côte d' Ivoire, have lost world market shares of oil palm and cocoa to Asian producers such as Malaysia, and Indonesia over the past fifteen years. In summary, there is a need for a large increase in research in the 1990s on the economics of producing export crops and livestock.

Rural production studies should address both the efficiency and equity issues as part of research on export crops and livestock. Why is this necessary? First, there is a need to generate a knowledge base for understanding both the efficiency and equity impacts of proposed policies and interventions. It is more cost effective to carry out one dual-purpose survey rather than two independent studies of the economics of production and benefit incidence. Rural production studies should also devote attention to the projected impact of increased production on net food sellers and net food buyers. For example, several recent studies have shown that about one-half the farmers in some African countries are net food buyers at some time during the year and that the percentage of smallholders who are net food sellers is surprisingly concentrated [Weber, Staatz, et al., 1988]. For example, an estimated 10 percent of smallholders selling maize in Zimbabwe accounted for 70 percent of all smallholder maize sales [Rohrbach, 1988]. The implication of this finding is that a policy of raising official maize prices might benefit a few smallholders with adequate land and resources but it may result in higher domestic food prices for farmers and rural and urban people who are food buyers.

The experience of India and other Asian countries shows that once large farmers reap the benefits of new technology and gain political power, they can counter the efforts of the government to help the poor and the landless. Professor Dantwala of the University of Bombay has pointed out that India's achievement of national food self-sufficiency has been offset by its inability to come to grips with 200 million people who lack adequate land or income to produce or purchase enough food to meet their family food needs [Dantwala, 1985].[223] Finally, because of the gross inability of Africa's industrial sector to generate sufficient jobs, the relative share of Africa's rural population will likely decline at a snail's pace or from 68 to 58 percent over the forty-year period, 1985-2025. The number of people in Africa's rural labor force may triple over this time span. This means that the rural sector must, by necessity, provide the bulk of the new jobs in most African economies until fertility rates slow down and/or industrial expansion accelerates to expand employment opportunities.

Because of the demographic explosion and the inability of the industrial sector to generate enough jobs for the newcomers to the rural labor force, most of the additions to the national labor force will have to be "parked" in rural areas for the next ten to thirty years. Therefore rural employment and income generation emerge as compelling research topics in the 1990s. For background information see *Employment Generation in African Agriculture* by Eicher, Zalla, *et al.* [1970], and Liedholm and Mead [1987].

The equity effects of alternative crop and livestock production systems are treated in more depth by researchers in Asia and Latin America than in Africa to date. For a comparative perspective on Asia see Dantwala's [1985] commentary on India and Hayami's [1988] perspective on Asia. Recent studies on food, livestock and export production in Africa include:

Abalu and Etuk [1986]; Arua and Obidiegwu [1988]; ATIP [1986a, b]; D. C. Baker [1987]; Bingen [1985]; Bingen, Hall, and Ndoye [1988]; Binswanger and Pingali [1988]; Borlaug [1988]; Byerlee and Longmire [1986]; Byerlee and Tripp [1988]; Collinson [1987, 1988]; Delgado and Mellor [1984]; Eicher [1988a, b]; FAO [1987c]; Fortmann [1984]; Fresco [1986]; Ghai and Smith [1987]; Heisey [1985]; Hubbard [1986]; ICRISAT [1987]; Jaeger [1986]; Johnston, Hoben, *et al.* [1987]; Joseph [1987]; Low [1986]; Matlon [1987]; Matlon, Cantrell, *et al.*, eds. [1984]; Norman, Simmons, and Hays [1982]; Norman, Baker, *et al.* [1988]; Pingali, Bigot, and Binswanger [1987]; Richards [1985, 1986]; Rohrbach [1988]; Rukuni and Eicher [1987, 1988]; Sanders [1989]; D. Shapiro [1988]; R. Singh [1988]; D. S. C. Spencer and Nyateng [1987]; and Upton [1987].

IRRIGATION: TECHNICAL, ECONOMIC, INSTITUTIONAL AND EQUITY ISSUES

About 5 percent of the arable land in Africa is under irrigation compared with about 35 percent in India. Africa is generations behind Asia in designing and implementing irrigation schemes.[224] For example, Senegal recently borrowed almost a billion dollars to build two large dams on the Senegal River in the late 1980s. The government only recently commissioned studies on the size of farms and type of farmer irrigation organizations. By contrast, there are written records of farmer irrigation associations in Northern Thailand for about 700 years [Surarerks, 1986].[225]

Presently, Africa has an extremely short supply of trained irrigation specialists relative to countries like Thailand where there are literally thousands of irrigation specialists throughout the government. For example, in 1988 the Government of Botswana has only one local irrigation specialist with graduate training even though the government is committed to a major expansion of irrigation.

For studies of irrigation in Africa see Hotes [1983]; Carruthers, ed. [1983]; Fell [1983]; Blackie, ed. [1984a]; FAO [1986a]; Bloch [1986, 1987]; Levine and Bailey [1987]; Rukuni [1988]. For a state-of-the-art report on irrigation in sub-Saharan Africa see the 635–page study by Moris and Thom [1987]. See Dhawan [1985] for a state-of-the-art paper on India's experience with irrigation. See Olivares (1987) for a study of irrigation in the Sudan, Botswana, Zambia, Zimbabwe, and Kenya.

FOOD MARKETING AND CONSUMPTION

Research on food marketing systems and food consumption is in its infancy in Africa because of the perceived urgency to give priority to research on increasing food production. However, when one considers that Zaire's capital city of Kinshasa has grown from 200,000 at independence in 1960 to 3,000,000 in 1985, the challenge of feeding Africa's cities is formidable. Guyer's [1987] collection of essays on *Feeding African Cities* is a good overview of the challenge. The bulk of marketing research over the past decade has been in the form of consultancy reports on the shortcoming of government grain marketing boards. Elliot Berg [1986a] summarizes this literature for the Sahel. Few studies are available on the marketing problems of smallholders. FSR teams usually dealt with farm level production problems and assumed that marketing specialists will deal with the marketing problems of smallholders. On the other hand, 80 to 90 percent of the energy of marketing specialists has been consumed by studies of government grain boards.

Vast investments will have to be made in food marketing systems as Africa's population doubles over the next twenty to twenty five years. Shaffer, Weber, Riley, and Staatz [1985] have spelled out the basic issues to take into account in research on food marketing systems. Marketing boards are surveyed in a collection of papers edited by Arhin, Hesp, and Van der Laan, eds. [1985] while W. O. Jones [1987] provides a valuable overview of food marketing boards. Studies of marketing systems in the 1980s include B. Child, Muir, and Blackie [1985]; Ahmed and Rustagi [1987]; Abbott [1987a]; Elliot Berg [1987]; Guyer [1987]; Von Braun and Puetz [1987]; Dione and Staatz [1988]; Stanning [1988]; Buccola and Sukume [1988]; Morris [1988a] and Staatz *et al.* [1989].

Turning to research on food consumption, many Africans are hooked on an unsustainable western food consumption profile aided and abetted by colonial policies and reinforced by the ready availability of food aid. Professor Ojetunji Aboyade of Nigeria is one of the first influential Africans to speak out on unsustainable Western tastes. He observes that the greatest danger is that these "consumption patterns come to be associated in the minds of people as being increasingly coterminous with the essence of economic development" [Aboyade, 1988, p. 16].[226]

Africa imported three out of every four tons of wheat consumed in the late 1980s. Zambia is importing 85 percent of its wheat requirements. About 40 percent of the rice consumed in West Africa is imported. West Africa's rice imports totalled 25 percent of domestic consumption requirements in the early 1970s and around 40 percent in the mid-1980s [D. S. C. Spencer and Nyateng, 1987].

Senegal became hooked on importing rice some fifty years ago when French colonial policy encouraged Senegal to specialize in groundut (peanut) production for France and import rice from the French colony of Indo-China. From 1935 to 1939 Senegal imported an average of 75,000 tons of rice per year [Founou-Tchui-goua, 1981]. When Senegal became independent in 1960, it continued to import rice—mainly broken rice from Thailand. Over the five-year period from 1982 to 1986, Senegal imported an average of *1,000 tons of rice per day* from Thailand, Pakistan, the U.S., and Japan [FAO, 1987b].[227]

African policy makers should be encouraged to face up to the inescapable conclusion that the food consumption profile in many countries is *unsustainable* given the domestic resource base, available technology, government revenue, and the foreign exchange earning potential. Theoretically, this poses a fundamental economic dilemma of paying for the present consumption pattern. But most countries can invoke the soft option and rely on what Susan George aptly calls food aid "subscriptions" [George, 1987]. Lesotho and Ethiopia illustrate Africa's food aid dependency. Lesotho imported food aid for fourteen consecutive years, rain or shine, from 1972 to 1986. Ethiopia has relied on food aid for twenty-two consecutive years, 1966 to 1991 [Shapouri, Dommen, and Rosen, 1986; FAO, 1988].

West Africa has the most unsustainable food consumption profile on the subcontinent. West Africa's food economy has been the poorest subregional performer in terms of food production over the past two decades, and it is the subregion with the most pessimistic medium-term food outlook. Four reasons fuel this pessimism. The first is the rapid shift in food consumption patterns to rice, wheat, or what are becoming known as the "fast foods" (convenience foods) of West Africa.[228] In the past twenty years, for example, per capita rice consumption in West Africa has doubled from twelve to twenty-four kg [WARDA, 1988], p. 1]. Rice consumption in the Côte d'Ivoire shot up from thirty-two kg to fifty-two kg over the 1960-85 period. Second, agronomic research on the rural staple foods (sorghum and millet) has been disappointing.[229] Although the French started research on millet in Senegal in 1931, there has been little progress on either millet or sorghum in the Sahel. In 1988, for example, about 1 percent of the millet and sorghum area in Mali was planted according to recommended modern practices (sole cropping) while 99 percent of the area under these crops was intercropped—a traditional practice of planting several different crops without fertilizer on the same field at the same time.

The third reason for pessimism about sustaining West Africa's food consumption profile is the lack of profitable wheat packages[230] and improved rice varieties for rainfed areas which account for about 60 percent of the area planted to rice in West Africa. Fourth, the fall in the production of export crops in many countries has resulted in a fall in farm income and foreign exchange earnings that could have been used to pay for food imports. For example, the groundnut pyramids in

Nigeria, Senegal, and Niger of the 1960s have been relegated to the history books. France has replaced groundnut imports with cheaper sources of edible oil such as sunflower oil from Eastern Europe and rapeseed oil from Canada. Nigeria and Ghana's oil palm industry is no longer competitive with Malaysia and Indonesia. The Côte d' Ivoire and the Cameroon are among the few countries in West Africa with a productive rain forest that can finance wheat and rice imports with agricultural exports (cocoa and coffee) earnings.

What is to be done about Africa's unsustainable food consumption profile? The first step is to study the problem and develop a medium-term (ten to fifteen years) plan of attack rather than introducing knee-jerk policies which cannot be enforced, such as banning all wheat imports overnight. In some countries exchange rate adjustments will be needed to reduce commercial food imports. More research is needed on traditional food plants as outlined by Professor Bede Okigbo [Okigbo, 1986]. A quantum increase in research on food science is needed to develop substitutes such as "cassava bread"[231] for wheat and rice, new crops such as sunflower oil to replace palm oil for home cooking, and to find cheaper sources of calories such as sweet potatoes and Irish potatoes as population density increases.[232] For example, Nigeria took the first step in restructuring its food demand by banning the imports of wheat, wheat flour, rice, and maize on January 1, 1988.[233] Senegal has imposed a 25 percent tax on imported rice. The long and painful restructuring West Africa's unsustainable food consumption profile is under way.

The most optimistic food outlook over the next decade is in Eastern and Southern Africa where there is a backlog of improved varieties of white maize [Eicher, 1984, 1986a]. Consumers have a continuing preference for white maize, the traditional staple food. For example, white maize contributes about 50 percent of the total calories in the average diet in Zimbabwe and Zambia. Scientists at the ICRISAT/SADCC regional research center in Zimbabwe are also reasonably optimistic about developing improved sorghum varieties in the short to medium term because plant, insect, and disease pressures are much lower than in the Sahel.

The message that emerges from this discussion is that the food situation in Africa is influenced by both colonial policies and contemporary food demand and supply factors. Food policy analysts must, by necessity, include both food demand and supply issues in their analyses instead of assuming that Africa's food gap can be closed by action on the supply side, stepping up food production, for example. More research is urgently required on food consumption, marketing and food systems [Shaffer, Weber et al. 1985].[234]

FOOD SECURITY POLICY OPTIONS

Currently 100 million people or one-fifth of Africa's population do not have access to enough resources to produce, or income to acquire, adequate food to meet their food security needs. The purpose of this section is to lay out a research agenda for three food security research topics: drought and famine; family food security; and national food security policy options.

Food security can be defined in various ways. We prefer a simple definition: the ability of all individuals in a nation to secure adequate food throughout the year for a healthy and nutritious diet.[235] Food security has two interlocking components: food availability through domestic production, storage, and food imports; and access to food through home production, purchase in the market, or food transfers [World Bank, 1988a]. The appropriate balance of research on the food availability or food access sides of the food security equation must be decided on a country-specific basis within a subregional context such as the Sahel or Southern Africa in order to include intraregional trade and storage possibilities.

DROUGHT AND FAMINE

The first research topic is drought and famine. Since the 1968–74 Sahelian drought and the Great African famine of 1984–1985 and the publication of Professor Sen's seminal book on *Poverty and Famines* [1981], there has been an explosion of research on drought[236] and famine.[237] Sen's treatise has revolutionized the study of famine and hunger and his ideas are slowly making an impact on famine relief programs. Without question, Sen's work has helped contribute to a growing international awareness of the multiple causes of hunger and famine and the need for policy makers to devote more attention to the food access (demand) side of the hunger equation. Sen's analysis challenges the conventional wisdom of many agriculturalists who argue that famine is mainly caused by a precipitate decline in food production and that it can be cured by investments in food production and grain storage to increase food availability. But as Sen hammers home the fundamental relationship between poverty and famine, the reader may get the impression that food *availability* and food *access* are mutually exclusive, rather than complementary approaches to combating hunger and famine. Sen's book has stimulated research on food access issues such as: food entitlements, access to food, poverty reduction and food transfers through food for work, cash for work, or outright food gifts (e.g., soup kitchens).

But it is possible that Sen and others with limited field experience in Africa are drawing unqualified lessons from Asia's famine prevention experience for Africa [Sen, 1987a, b; McAlpin, 1987; Dreze, 1988]. For example, Sen reports that "in the Sahel there is a need for a mechanism for directly tackling the problem of vulnerability through public institutions guaranteeing food entitlement. The last category includes not merely distribution of food when the problem becomes acute, but *more permanent arrangements* for entitlement through social security and employment protection" (underlining supplied) [Sen, 1981, p. 129]. But one can legitimately pose the question: Which government in the Sahel has the capacity to finance social security and employment protection programs? Half the countries in Africa are carrying out structural adjustment programs. Many governments such as Malawi are swamped with refugees and they have little time or resources to develop permanent social security and employment protection programs. Most Asian countries are generations ahead of Africa in terms of institutional capacity to combat drought and famine. For example, the Famine Codes in India were introduced over 100 years ago. Sri Lanka provided a subsidized rice ration in the sixties and seventies. The employment guarantee schemes have been in operation in the Maharashtra State in India for decades. In short, Asia has a long history of government experience in developing *appropriate* institutions—local, state, and national—to deal with famine.

Botswana is the first African country to have "broken the famine cycle" which is defined as a process of preventing starvation during a prolonged drought. With heavy rains in 1988, it is an appropriate time to analyze why no one died during the 1982-87 drought, the second drought in sixty-five years, which lasted for five consecutive years [Holm and Morgan, 1985; Quinn, Cohen, *et al.*, 1988]. During 1987, the most intense period of the famine, around 60 percent of Botswana's one million inhabitants received some type of government antifamine assistance. Because Botswana's semiarid land base is ideally suited to livestock production, it normally produces about 40 percent of its staple food requirements (sorghum) and imports the balance in the form of food aid and commercial imports with foreign exchange earned from diamond and livestock exports. But at the peak of the drought in 1987, Botswana imported 95 percent of its staple food—sorghum.

Care should be exercised in drawing lessons from Botswana's famine prevention experience for other African states because it has two characteristics that are in scarce supply in Africa. The first is rapid and sustained economic growth. Botswana had the fastest growing economy in the world from 1973-86 when its inflation-adjusted rate of GNP growth grew at a brisk rate of 10.8 percent per year [World Bank, 1988b]. Botswana's growth is propelled by diamond, mineral, and livestock exports and its dynamic economy provides government revenue and foreign exchange to import food on commercial terms, as well as resources to finance cash for work and school feeding programs during periods of drought. Second, Botswana has a democratic government and strong opposition parties committed to combating drought and hunger and helping rural people. Botswana also has a responsive government newspaper and private papers like the *Botswana Guardian*. In short, Botswana has a pluralistic government, unusual financial capacity, and a cheap and reliable famine early warning system—opposition parties, a free press, and open markets—to deal with drought.

Botswana's experience in combating five consecutive years of drought and breaking the famine cycle has a few general lessons (not policy designs) for other African countries. First, countries with a high degree of crop and livestock production instability need to develop a permanent but cost effective institutional capacity to deal with drought. Second, famine prevention is too complex and multifaceted to be left to a single ministry such as a Ministry of Agriculture or the Ministry of Health. Botswana's Inter-Ministerial Committee, for example, coordinates inputs from six ministries. Third, there is a need to tackle both food availability and food access programs. But, as mentioned earlier, Botswana's dynamic economy has generated government revenue and foreign exchange earnings from diamonds and livestock that few other African countries can match—except perhaps oil-rich Gabon—for famine prevention. Hence, Botswana's experience is a "special case" that must be interpreted with care [Eicher, 1990b].

Because many countries such as Botswana, Ethiopia, the Sudan, and those in the Sahel states are plagued with drought and weather instability, it is imperative to devote substantial famine research and policy attention to both food availability and family food security issues such as irrigation, water management, sorghum, millet, cassava, and cowpea improvement.

Professor Sen has rendered a valuable service in calling for more research on the access to food. But attention is also needed on food availability—production, storage, and food imports [Eicher, 1985]. We reinforce this need for research on food production and food availability by citing the case of Senegal. There is no way that Senegal, a nation with a cereals self-sufficiency index of around 40 percent (1983–85), is going to solve its food problem through food aid and tinkering with the food access side of the hunger equation. *The rationale for increasing food production in food deficit countries such as Senegal is compellingly unavoidable.* In summary, there is a need to conceptualize research on drought and famine as an integrated research agenda. The famine literature of the 1980s included: Devereux [1988]; Eicher [1985, 1990b]; Glantz, ed. [1987]; Holm and Morgan [1985]; Jansson, Harris, and Penrose [1987]; Mariam [1986]; McAlpin [1987]; Mellor and Gavian [1987]; Okojie [1987]; P. Richards [1986]; Sen [1981, 1987a, b]; and Drèze and Sen [1989].

FAMILY FOOD SECURITY

Africa has a rural dominated society and economy. Seven out of every ten people are engaged in rural production. Moreover, since subsistence farmers and nomadic herders are few and far between in Africa today, the research emphasis should focus on semisubsistence farmers and seminomadic herders. Moreover, since the majority of the poor in Africa are engaged in semisubsistence food and/or livestock production, it follows that there is justification for a major research effort by technical and social scientists on the causes of family food insecurity in semisubsistence households. Because of the prevalence of drought in many countries such as Botswana, Somalia, Ethiopia, the Sudan, and in the Sahel, it follows that research on family food security should give special emphasis to families in low rainfall areas where sorghum, millet, and cowpeas are the major staple foods.

Since poverty is a central cause of family food insecurity, it follows that the research agenda should examine new sources of nonfarm income and employment generation through the sale of export commodities, including crops and small ruminants. C. K. Eicher and D. C. Baker [1982] reported that 25 to 50 percent of the labor force of adult males in farm households is spent in off-farm jobs during nonpeak farming seasons. Moreover, Reardon, Matlon, and Delgado [1988] reported that one-half to three-quarters of the average rural household income comes from noncropping sources in Burkina Faso.

For research on the consequences of the commercialization of agriculture on family food security, see the section on nutrition in VII. For research on the role of rural small-scale industry in generating income to help families purchase food see Liedholm and Mead [1987], and Allal and Chuta [1988]. For research on sorghum, millet, and cowpeas, see ICRISAT [1987]; and Matlon [1987]. For a summary of recent research on family food security by members of the MSU Food Security Research Network, see Weber, Staatz, et al. [1988]; Rukuni and Eicher [1987, 1988]; and Dione and Staatz [1988].

NATIONAL FOOD SECURITY POLICY OPTIONS

When policy-based lending was initiated by the World Bank around 1983, the social science research community in Africa was heavily committed to micro studies of agricultural projects scattered across Africa's landscape. In focusing on projects, researchers were simply following Price Gittinger's sage advice of the 1970s that "projects were the cutting edge of development" [Gittinger, 1972]. In the 1980s, many mainstream economists reported that policy is the cutting edge of development. This point of view was articulated in the influential book *Food Policy Analysis* by Timmer, Falcon, and Pearson [1983] and its publication coincided with the World Bank's abrupt swing from projects to policy-based lending in Africa around 1983. Although one can quibble with the Asian orientation of *Food Policy Analysis*,[238] it represents the mainstream textbook on food policy analysis in the Third World.

Studies of national food security policy options should address the key policy question: What is the most cost-effective mix of domestic food production and storage, trade and/or food aid to meet national food security needs in both the short and long run?

Blanket endorsement of concepts such as food first, food self-reliance, and food self-sufficiency do not answer this crucial question. However, food self-sufficiency can be a useful operational concept if it is supported with underlying economic analysis. For example, if Botswana wants to increase its self-sufficiency index of sorghum from 30 to 80 percent through subsidized credit, mechanization, and irrigation projects, researchers should find out what these programs will cost in real terms. How much additional employment will be generated? What is the political value of reducing the ratio of food dependence? These are hard political economy questions that can only be answered by in-depth research [Rukuni and Eicher, 1988, pp. 134-135].

Studies of national food security policy options in the 1980s are reported in the following: [World Bank, 1986; Gittinger *et al.*, eds., 1987; Rukuni and Eicher, 1987, 1988; Staatz, 1988; Staatz *et al.* 1989; Weber, Staatz, *et al.*, 1988; Delgado and Mellor, 1984; Delgado and Miller, 1985; Delgado and Reardon, 1987; Dione and Staatz, 1987; J. M. Due, 1986; F. Martin, 1988; Eicher, 1990b; Sen, 1984, 1987a]. For a proposed research agenda on food security see Sen [1987b].

Notes

1. The definition of sub-Saharan Africa normally includes forty to forty-eight countries depending on the number of offshore islands (e.g., Seychelles, Mauritius) one wishes to include. We have excluded the Republic of South Africa and some of the offshore islands and arrived at an arbitrary list of forty-five countries to include in this survey.

2. Henceforth Africa shall be used to mean sub-Saharan Africa.

3. The population picture has a bright side, however. The life expectancy at birth has increased from an estimated thirty-eight years in 1950 to almost fifty years in 1980—and the crude death rate has fallen from an estimated twenty-seven per 1,000 in 1950 to eighteen per 1,000 per year in 1980.

4. In Zimbabwe only about 15 to 20 percent of the newcomers to the national labor force over the next decade are expected to find employment in the industrial and urban sectors.

5. Compared with other regions of the world, the cost of education per pupil as a percentage of GNP per person is the highest of any region of the world. The high cost of education per student is largely a function of teachers' salaries which typically account for 75 percent of educational cost [Hanson, 1980].

6. William O. Jones's "Economic Man in Africa" [1960], is a pioneering statement by an agricultural economist.

7. The division into small and large farms is used in this survey.

8. The three-tier classification scheme has been refined by several researchers, most notably by J. Phillips [1959] who uses a seven-zone classification: Desert-Southern Saharan Fringe, Subdesert Wooded Savanna, Arid Wooded Savanna, Subarid Wooded Savanna, Mild Subarid Wooded Savanna, Subhumid Wooded Savanna, and the Derived Savanna. Northern Guinea is the most intensively farmed zone of West Africa because rainfall is more reliable than more northern zones and the zone free of tsetse.

9. "Shifting cultivation" is a particular type of swidden agriculture in which homesteads are frequently moved to remain close to fields only cultivated for a few years. Standard references are de Schlippe [1956], Nye and Greenland [1960], Allan [1965], and Fresco [1986].

10. The term "savanna" zone or region encompasses the semiarid regions of West Africa, including the Sahelian zone, the Sudanian zone, and the northern parts of the Guinean zone where cereals are the main staple.

11. Some advantages of growing multiple varieties are security against crop failure, diet diversity, and spreading seasonal labor requirements.

12. Grown mainly by small farmers in Kenya, Tanzania, Rwanda, and Zaire, pyrethrum is used to make an environmentally safe pesticide.

13. References on livestock are R. E. McDowell [1972], Dahl and Hjort [1979], Ergas [1979], and ILCA [1978, 1979 a, b, c, d, 1980a, b], Dyson-Hudson [1985], and World Bank [1985a].

14. Stephan A. Sanford points out that it is unwise to generalize because to test this proposition one needs data on the area used by nomads, number of nomads, number of livestock, and length of nomadic moves. Moreover, the length of moves and the area of land used by nomads vary enormously from year to year depending on rainfall [Personal Communication, March 17, 1981].

15. Transhumance refers to the pattern of *regular* movement of cattle and herders in search of grazing land. For a classic study of transhumance in northern Nigeria and Niger, see Stenning [1957].

16. The small number of economic studies of dairying includes Zalla's study in Tanzania [1981], Hopcroft and Ruigu [1976], Ruigu [1978], and Stoltz's [1979] studies in Kenya.

17. A nut with a caffeine base, kola is chewed as a stimulant and used to welcome guests.

18. Several other issues also could be examined including the colonial infrastructure strategy which developed railroads and roads to link favorable natural resource zones to coastal trading centers. This explains why there are still no major rail and road links along the coasts of East and West Africa.

19. For example, one of Africa's most respected economists, the late Tom Mboya [1967] of Kenya, laid out a development strategy for Africa which called for "a massive inflow of capital over perhaps 30 years and an equally massive inflow of technical assistance personnel over 10 to 15 years." Kenya's President Kenyatta encouraged investors "to bring prosperity" to Kenya.

20. The lack of African staff is still a critical problem in many African universities.

21. The shortcomings of African leaders and the lack of hard work in African society are common themes in Dumont [1966, 1969] and Dumont and Mottin [1980].

22. For a synthesis of Latin American experience see Alain de Janvry [1981].

23. But as Benjamin Higgins [1959] and other scholars observed, Boeke did not present solid evidence to defend his case.

24. Pryor [1977] used econometric techniques to test sixty hypotheses about "primitive and peasant societies" in a number of case studies and found that most of the hypotheses of the substantive school were not supported. But one has to question the use of econometric techniques to test the validity of the substantive paradigm.

25. The debate between the formalists and the substantivists has been summarized by Posner [1980, pp. 608-609] as follows: "the formalists spend their time looking for explicit markets in primitive societies and the substantivists spend their time showing how resources in primitive societies are mostly allocated by nonmarket means."

26. See Holdcroft [1984].

27. See Streeten [1972b] for an insightful note on the vacuity of theories of single barriers to change.

28. We are indebted to Sara Berry for calling Parkin's study to our attention.

29. For further evidence supporting the concept of economic man in Africa, see the overview of "Supply [Response] Functions" studies in the section on smallholder farming, and the discussion of research in the section on livestock.

30. Starting in the late 1800s and early 1900s, there was a rapid growth in the production and international trade of crops such as cocoa, coffee, oil palm, and rubber throughout Africa. Many scholars contend that cash crop expansion through international trade was the "engine of growth" of African economies during the colonial period.

31. The economic historian's emphasis on the role of international demand for export crops stands in sharp contrast to the modest attention devoted to demand parameters in Western growth models.

32. W. Arthur Lewis [1978a, b] contended that, while international trade did serve as an "engine of growth" in the 19th century, this is not its proper role. He contended that technological change (especially in food production) is the engine of growth in the Third World today but acknowledges that trade can serve as a handmaiden of growth.

33. In an article on the proletarianization of the peasantry in Rhodesia, Arrighi [1970] criticized W. A. Lewis for viewing unlimited supplies of labor as a given rather than being produced by the colonizers or capitalists. But Lewis specifically addressed these issues by noting that in Africa the colonial governments impoverished the peasantry "by taking away the people's land or by demanding forced labor in the capitalist sector, or by imposing taxes to drive people to work for capitalist employers" [W. A. Lewis, 1954, p. 410]. Still, as Hirschman points out, these practices were not central to Lewis's model of unlimited supplies of labor because "a decline in infant mortality could have the same effect in augmenting labor supply as a head tax" [1981b, p. 16].

34. We prefer using the term "land extensive" farming system rather than land surplus models of development because even though farmers in Sierra Leone [D. S. C. Spencer and Byerlee, 1976, 1977] or Tanzania [K. H. Shapiro, 1978] have twenty to forty hectares of land under their control they actually cultivate only a small portion of that land (three to five hectares) in any one year. The small area under cultivation is a function of the shortage of family labor at critical periods in the production process and the need to keep the bulk of the land in fallow in order for the bush to regenerate and restore soil fertility.

35. For a discussion of seasonal labor bottlenecks, see de Wilde et al. [1967]; C. K. Eicher, Zalla, et al. [1970]; Cleave [1974]; and R. A. Swanson [1981].

36. In a subsequent article, Essang [1977] points out that the oil boom in Nigeria has provided the foreign exchange earnings and government revenues for a resurgence of government-directed large-scale farming schemes—especially large irrigated schemes and River Basin Development. Likewise, surplus land is cited as a justification for mechanized farming in the Sudan.

37. For example, Hayami and Ruttan [1971] noted that there was a need to step up micro research in the 1970s in order to provide the data necessary for a convincing specification of the agricultural sector. This has left Western economists open to the challenge from radical scholars that their micro studies are ahistorical, overstress technical and infrastructural constraints, and give too little attention to the influence of the world economy. For a critique of "conventional development research" and the role of Western social scientists in Africa, see Amin et al. [1978].

38. The emergence of political economy models in the mid-1960s occurred at the same time that a number of countries (Ghana, Mali, Guinea, Tanzania) shifted from an ideology of capitalism to socialism.

39. For critiques of the dependency school of thought in Latin America, see Cardoso and Faletto [1979] and de Janvry [1981].

40. Social formations are defined as "concrete, organized structures that are marked by a dominant mode of production and the articulation of a complex group of modes of production that are subordinated to it" [1976, p. 16]. The characteristics of peripheral capitalism are described in Amin [1976, pp. 333-364].

41. Extraversion is defined as the dominance of the exporting sector over the economic structure as a whole, which is subjected to, and shaped by, the requirements of the external market [Amin, 1976, p. 203].

42. Hirschman [1977] coined the term micro-Marxist to describe radical scholars who concern themselves with "specific events and country 'experiences.' "

43. See also M. Sahlins's *Stone Age Economics* [1974].

44. Some of the dependency scholars share this view even though their stress is on the manner in which dependent development is conditioned by the world economy rather than on the specific articulation of precapitalist modes of production.

45. Terray [1972] wrote a 100-page critique of Meillassoux's Gouro village study.

46. A collection of papers by nine French Marxist anthropologists (including Copans, Godelier, Roy, Coquery-Vidrovitch, and Meillassoux) is available in English in a paperback volume edited by David Seddon [1978]. Also see Meillassoux, *Maidens, Meal, and Money: Capitalism and Domestic Community* [1981]. See Raymond Firth [1975] for an analysis of "Social Anthropology and Marxist Views on Society."

47. For overviews of agricultural policy issues for Nigeria, see Byerlee [1973], Wells [1974], Essang [1977], Idachaba [1980 a,b, 1981, and 1988], Idachaba, Akinwumi, *et al.* [1981], and Nigeria [1980, a,b]; for Zambia, see Dodge [1977] and Turok, ed. [1979]; for Kenya, see Heyer, Maitha, and Senga, eds. [1976]; for Tanzania, see Coulson, ed. [1979]; and for Sierra Leone, see Byerlee, C. K. Eicher, *et al.* [1983]. Studies of agricultural policy in West Africa are Club du Sahel [1977], CILSS/Club du Sahel [1978 a,b; 1979; 1980 a,b], and S. R. Pearson, Stryker, Humphreys, *et al.* [1981].

48. For example, in 1965, the Government of Kenya stated its commitment to socialism [Kenya, 1965], but it has been decidedly capitalist since the 1965 pronouncement, even though it has fragments of state control. For example, in 1981, the government set the prices of cereals and the National Cereals Produce Board handled the marketing of maize (thereby excluding private traders). Moreover, parastatals such as the KTDA are common throughout the agricultural sector. But these fragments of state control do not add up to a socialist-controlled agriculture.

49. Benin, Guinea, Guinea-Bissau, Congo (Brazzaville), Ethiopia, Tanzania, Zimbabwe, Mozambique, Angola, Mauritius, Equatorial Guinea.

50. The volume edited by Helleiner [1968b] is a standard reference on agricultural planning in East Africa during the sixties. For an analysis of Kenya's agricultural planning since independence, see Heyer, Maitha, and Senga, eds. [1976]; Leys [1974]; and Holtham and Hazlewood [1976].

51. Nigeria's planning experience has been well documented. Stolper's influential book *Planning without Facts* [1969] stressed the lack of data in preparing Nigeria's first Five-Year Plan. Other analyses of Nigeria's planning processes during the 1960s can be found in Aboyade [1973], Dean 1972, Kilby [1969], Gusten [1967], Wells [1974], C. K. Eicher and Liedholm, eds. [1970], and C. K. Eicher and Glenn L. Johnson [1970].

52. For example, small farms in the Semiarid Tropics (SAT) Zone of West Africa have several times as much land at their disposal as farmers in the SAT Zone in Southern India.

53. Even though a farmer does not "own land" in the sense of having freehold title to land, farmers in most countries have "control" or use rights to land for their lifetime; these use rights can be passed on to heirs.

54. For example, from 30 to 40 percent of small farm production was sold in Kenya [Heyer and Waweru, 1976], 48 percent in Sierra Leone [Byerlee, C. K. Eicher, et al., 1977], and 24 percent of the total value of farm production in northern Nigeria [Norman, Pryor, and Gibbs, 1979].

55. A term used to describe recent graduates of primary and secondary schools who are trying to enter the labor force.

56. The Government of Tanganyika (now Tanzania) followed World Bank advice and started twenty-three settlement schemes over the 1963-67 period which featured heavy capital investment, government management, and little participation by the settlers [Ingle, 1972]. But the schemes met with little success and in 1967 Tanzania dropped the transformation approach and shifted to a rural socialist strategy which focused on helping small farmers through the Ujamaa program and later the village development scheme [Nyerere, 1967].

57. For a radical critique of the Gezira scheme see Barnett [1977].

58. We do not want to give the impression that all settlement schemes were large-scale and all were failures. Numerous examples of settlement schemes for smallholders are found in Eastern Africa, including the famous one million acre settlement scheme in Kenya which was conceived by R. Swynnerton and launched in 1953 to transfer land from large white farms to small holders [Kenya, 1954]. For an appraisal of Kenya's settlement schemes, see MacArthur [1975] and Clayton [1978]. For a discussion of spontaneous settlements in Kenya, see Mbithi and Barnes [1975]; in Senegal, Rocheteau [1975]; and in the Sahelian countries of West Africa, see CILSS/Club du Sahel [1978b].

59. See Byerlee and M. Morris [1987].

60. Research and empirical findings on the economics of smallholder farming are reviewed in the section entitled smallholder farming.

61. While it is common to refer to these boards as government boards, most food grain boards have a quasi-commercial character and some degree of autonomy from the government.

62. See Arhin et al., eds. [1985] and William O. Jones [1987] for recent surveys of the literature on marketing boards.

63. When Senegal abolished its grain board—ONCAD—in late 1980, it eliminated about 4,500 employees. Ghana's cocoa marketing board had 107,000 employees on its payroll during its peak period.

64. The World Bank [1981b, p. 59] reports, for example, that charges for marketing, storage, and transportation in Kenya accounted for 34 percent of the f.o.b. border price for maize, 23 percent for wheat, and 48 percent for rice during 1972-79. These figures are typical of government grain board margins in other countries.

65. For example, the government subsidized 80 percent of the fertilizer prices paid by farmers in northern Nigeria in 1981 [World Bank, 1981b].

66. Although these figures display an alarming loss of export markets for Tanzania and Ghana, the decline is overstated because of smuggling. For example, cocoa from Ghana has been smuggled into the Côte d'Ivoire because the price is several times higher than in Ghana. In Tanzania, in 1973-74, it was widely known that farmers smuggled maize across the border in response to lower government producer prices of maize. But even if official export figures are understated by 10-15 percent for some commodities such as cocoa in Ghana, the above figures do illustrate how Tanzania's and Ghana's export positions have eroded.

67. Rural development programs are broadly defined to include increased rural welfare as well as increased agricultural productivity. See J. M. Cohen [1980a].

68. See the lack of attention given to regions of poor resource endowments and to equity objectives in the World Bank's [1981b] study of African development.

69. The study focuses on the Ivory Coast, Liberia, Senegal, Mali, and Sierra Leone.

70. The Stanford WARDA study showed that the social profitability of rice production in Sierra Leone was positive because of extremely low rural wage rates. But over the 1974–81 period, rural wage rates doubled from 1 Leone to 2 Leones per day and rice imports increased to 41,000 tons in 1980.

71. A word of caution is in order. The five countries studied have one-third of the total population of Nigeria. It is risky to discuss West African rice trade without including Nigeria. For an up-to-date picture of rice in West Africa, see D. S. C. Spencer and Nyanteng [1987] and WARDA [1988].

72. For a discussion of the SADCC strategy to achieve food security in southern African see Rukuni and C. K. Eicher [1987, 1988].

73. A few studies provided information on the economics of agricultural practices. See, for example: Haswell [1953]; Galletti, Baldwin, and Dina [1956]; Leurquin [1960]; and Boutillier et al. [1962].

74. Universities of Nottingham and Zambia Agricultural Labor Productivity Investigations.

75. ICRISAT has published literature reviews on production systems [Norman, Newman, and Ouedraogo, 1981], marketing in the semiarid tropics of West Africa [Harriss, 1979 a,b], socioeconomic constraints on the development of semiarid tropical agriculture [ICRISAT, 1980a], and millet [ICRISAT, 1987].

76. There have been occasional attempts to use literate children to keep rudimentary records [MacArthur, 1968] but this approach has largely been abandoned in Africa.

77. There are numerous terms such as reconnaissance, exploratory, rapid appraisal, single visit, informal, and farm business surveys for what are essentially infrequent visit type of surveys. The farm business survey terminology is a western concept which was used in some countries in the 1960s but it was subsequently dropped.

78. Cost route derives its name from the repeated nature of the survey over the course of a year in order to derive data to compute costs and returns of production.

79. While the activity approach is promising, it may not reduce survey costs unless there is detailed information available on the cropping calendar since farmers have to be interviewed to see if an activity is completed.

80. French researchers have often expressed the view that sample surveys are a tool of statisticians which may be selectively used to supplement the qualitative understanding of farmers provided by social science researchers using techniques such as participant observation, recording life histories, and constructing genealogies [Couty, 1979; Benoit-Cattin, 1980].

81. Additional survey design issues which may influence survey results include: selection of the sampling frame, procedures used for gaining knowledge of local farming practices in order to design questionnaires, approaches for securing support and cooperation of interviewees, choice of direct measurement techniques — primarily for field size, yields, and intensity of labor use — to supplement recall information, alternative methods for gathering information about sensitive issues such as the size of land holdings or livestock, buildings, and credit, and methods for making field data checks to reduce inconsistency and to verify recorded responses.

82. AMIRA is an informal working group of researchers from ORSTOM, INSEE, and the French Ministry of Cooperation who have had extensive farm-level research experience in Francophone Africa. The group was formed in 1975 with the specific purpose of debat-

ing the refining data collection methods. Four main issues are addressed in the AMIRA papers: what information to collect, for what uses, and for what objectives; how to collect the relevant information; how to process and analyze the data; and how to use the information to improve decision making [Winter, 1978].

83. Additional information on microcomputers is available from the Farm Management Division, FAO, Rome, and from Dr. Michael Weber, DAE, MSU.

84. *Kenya* [MacArthur, 1968]; *Tanzania* [Collinson, 1962-64; Ruthenberg, 1968]; *Uganda* [Pudsey, 1967]; *Malawi* [Catt, 1966]; *Nigeria* [Upton and Petu, 1964; Upton, 1967]; and *Senegal* [Boutillier *et al.*, 1962].

85. We refer to the research cited in this section as production function studies even though many studies are not based on physical production functions of particular crops. This terminology has been consistently used in the literature in Africa.

86. In several cases, capital has been left out because there was little variation in the use of capital across farmers, little capital was used by farmers except a hand hoe, or farmers who used fertilizers or insecticides used them improperly [Ruthenberg, 1968; K. H. Shapiro, 1973; Lang, 1979].

87. Both the length and intensity of the working day vary significantly during the year [Norman, 1972; Cleave, 1974; Farrington, 1975a; K. H. Shapiro, 1978].

88. See also the discussion of agricultural extension in the section on technical change.

89. See, for example, Ford's [1971] critique of Maitha [1969, 1970]; Gemmill's [1979] critique of Medani [1975] and Medani's [1979] reply; and Blandford's [1973] comment on Olayide [1972] and Olayide's [1974] rejoinder.

90. Both models were estimated independently using ordinary least squares and combined to derive an elasticity of price for total output. Ford [1971] contended that Maitha's failure to estimate acreage and yield functions simultaneously led to an overestimate of price elasticities of yields for estates (plantations) and for the industry as a whole.

91. For example, Palmer-Jones [1979] has criticized LP studies because of the aggregation problems.

92. Although, it again must be noted that the models used to date have varied greatly in their sophistication. Care must be used in interpreting policy recommendations.

93. Heyer [1972] was forced to supplement her survey data with data from experiment stations in order to estimate input/output coefficients under different states of nature.

94. There have been few attempts to build simulation models at the farm level. The trade-off between the value of more refined farm models and their costs requires analysis.

95. See Hardaker [1979] for further discussion of alternative analytical techniques used in farm management research in developing countries.

96. For an appraisal of the relevance of anthropology to the study of economic development see Vernon Ruttan [1988].

97. For a discussion of inheritance and women's labor in Africa, see Jack Goody and Joan Buckley [1973].

98. Conventional measures of labor utilization, such as labor force participation, have not been widely used in analyzing rural labor markets because most rural people are self-employed in producing largely for home consumption, almost all the adult rural population participates in the labor force at some time of the year, and at any given time, a negligible proportion of the rural labor force is unemployed and seeking work [Byerlee and C. K. Eicher, 1974].

99. The dynamics of the organization of work on family farms has been a major topic of interest to French researchers [Kohler, 1971; Copans, Couty, *et al.*, 1972; Rocheteau, 1975].

100. Moreover, Norman did not include activities such as gathering, grass cutting, and firewood collection in calculating total time worked.

101. In this area, about one-third of the farm households are headed by women because the men are living in cities.

102. Traditional agriculture is defined as a farming system where no new factors have been introduced in a long time.

103. In other words, if the marginal value product of labor is higher in crop A than in crop B, then crop A should be more risky than crop B. This finding is consistent with portfolio theory.

104. D. L. Young [1979] pointed out that Wolgin made a mathematical error in the derivation of his model. Wolgin [1979] acknowledged the error but contended his findings are unchanged.

105. The late Professor Hans Ruthenberg designed and supervised the studies so the same format could be used in deriving gross returns per farm and per acre, production expenditures and net returns per acre, and per man equivalent in each study. Unfortunately, costs of production were not specified, labor was measured as a stock, and yield estimates were sometimes taken from experiment stations rather than from the sampled farmers.

106. Dependency theorists such as Amin argue that rapid population growth is merely a symptom of the transition to the social formation of peripheral capitalism.

107. Population growth by itself cannot be taken seriously as a model of agricultural change in the short run because of the recorded cases in history where people have starved (the Bengal famine, 1943) and where the number of landless has increased before innovation was forthcoming. Boserup [1981] analyzed long-term trends in population growth and technological change.

108. The history of agricultural research is documented by McKelvey [1965] and Yudelman [1975]. The status of agricultural research in the late 1960s is reviewed in the proceedings of the Abidjan Conference on Agricultural Research Priorities [NRC, 1968]. For the 1970 period, see NRC [1974, 1978]. For the 1980s, see Eicher [1986b, c].

109. Colonial governments invested few resources in food crop research because it was assumed that surplus land would automatically be brought under cultivation by subsistence farmers in line with population growth.

110. For example, the IRAT stations in Senegal became known as ISRA—Senegal Institute of Agricultural Research. In Niger, the IRAT station became known as INRAN—Nigerian Institute of Agricultural Research.

111. For an excellent overview of issues in translating agronomic research into farmer recommendations, see Perrin et al. [1976].

112. For references on sorghum see Doggett [1970]; ICRISAT [1980a, b].

113. Photosensitivity, day length sensitivity, and photoperiodicity are synonymous terms to indicate that the biological development of a plant, and in particular flowering and seeding, are governed by the length of the day.

114. Stiga is a weed which attaches itself to the roots of millet and sorghum and reduces plant growth and yields.

115. Zimbabwe, then Southern Rhodesia, was the first country after the United States to develop and release hybrid maize for commerical production [Eicher, 1984; Rohrbach, 1988].

116. For a discussion of long-term maize trends in West Africa, see Longmire and M. Morris [1987].

117. The use of herbicides for weed control on maize has promise of becoming one of the few "self-spreading" innovations in rainfed farming.

118. Two of the most technically unfavorable regions for producing wheat—West Africa and Southeast Asia—are the regions where per capita bread consumption is increasing the most rapidly. For a historical perspective on wheat breeding in East Africa, see Guthrie and Pinto [1970]. For a discussion of long term trends in wheat production and consumption in Africa see Byerlee and Longmire [1986] and Byerlee and M. Morris [1987].

119. Most of the estimated one million acres in world production are in the USSR and North America.

120. Since potassium levels are high in most African soils, the addition of potassium has generally had a relatively minor effect on the yields of most crops.

121. They derive a feasible region where all constraints are met and show that the optimal level of fertilizer use will vary depending on which factors are taken into account.

122. Because of the lack of an agreed-upon definition of "irrigation cultivation" there is no consensus on the percent of land in sub-Saharan Africa under irrigation except that it ranges from 3 to 5 percent.

123. See Diallo [1980] and Weiler [1979].

124. Maas and R. L. Anderson point out that the single most important finding of their comparative study of six major irrigation projects in the United States and Spain was the importance of "allowing water users to control their destinies as farmers, the extent to which the farmers of each community, acting collectively, have determined both the procedures for distributing a limited water supply and the resolution of conflicts with other groups over the development of additional supplies" [1978, p. 366]. This experience provides an alternative to the top-down centralized style of operating public irrigation projects in Africa, whereby the farmers are tenants or "quasitenants" of the state.

125. Mechanization is defined as any form of power used to assist or replace hand labor in agriculture, including donkey power, oxen power, tractors, combines, and mechanical threshers.

126. For an important study of energy in the Sahelian region of West Africa, see CILSS/Club du Sahel [1978a].

127. For a history of animal traction in French-speaking countries, see Hasif [1978] and Le Moigne [1980].

128. The slow learning curve has important implications for the evaluation of animal traction schemes. Researchers should be cautioned against "writing off" animal traction schemes after the first two to three years of results.

129. The appreciation for a pair of oxen was estimated at about $100 per year which more than covered all animal traction related costs (e.g., supplementary feeding during the dry season, veterinary supplies, etc.) in the 1978-79 survey year.

130. The Experimental Units in Senegal have addressed these problems by giving incentives to farmers to destump and consolidate their fields [ISRA, 1977; Faye and Niang, 1977].

131. These problems are covered in the following: Venema [1978]; Schulman [1979]; Weil [1970, 1980]; and Goe and R. E. McDowell [1980].

132. Paul A. David [1975] has shown that there was a fifteen-year time lag between the availability of the reaper for wheat harvesting in the western part of the U.S. in 1840 and the widespread diffusion of the reaper starting in the mid-1850s. The diffusion of the reaper was "held back" until there was a large increase in rural wage rates over the 1840-1855 period.

133. In 1978, the government raised minimum wages for both men and women in rural areas from sixteen to twenty-eight Sudanese pounds per day, an example of how a government policy promotes the substitution of capital for labor.

134. See Westley and Johnston, eds. [1975] for the proceedings of a major workshop on farm equipment innovations in Kenya.

135. The various field agents which together form the extension service often come from parastatals and governmental departments including agriculture, livestock, education, fisheries, forestry, health, and community development. Our discussion will focus on agricultural extension agents.

136. Roling [1988] argues that extension agents in developing countries are a marginal source of information for farmers.

137. Lele also contends that since extension agents generally are young and receive little training, farmers often know more than the agents.

138. For a definitive account of the KTDA, see Lamb and Muller [1982].

139. The VEWs are closely supervised by agricultural extension officers who in turn are supervised by subdivision extension officers. At the subdivision level, there is a team of subject matter specialists who assist in formulating the extension message for that period. Each VEW receives one day of in-service training each week and is expected to serve around 500 to 800 farm families. The hierarchical system of supervision in the Training and Visit extension approach can extend through as many as six to seven levels before reaching the headquarters of a zone or region.

140. The study by Ascroft, Roling, et al. [1973] also provided support for the effectiveness of FTC in Kenya. Nearly one-half (48 percent) of the progressive farmers and only 5 percent of the laggards had attended a FTC. Moreover, in an experiment to see whether extension services can effectively concentrate on average farmers rather than progressive farmers, they found that of 798 nonadopters of hybrid maize who attended a three-day training program, 97 percent began planting hybrid maize.

141. Gerhart found that agroclimatic zone was the single most important factor explaining the adoption of maize, followed by the risk associated with different cropping patterns and access to credit.

142. The problem of identifying groups of farms which are sufficiently homogeneous to serve as recommendation domains continues to be one of the main challenges facing FSR researchers. The extent to which small farmers are homogeneous and can therefore be treated as a group has been long debated [e.g., Hill, 1968; Collinson, 1972; Heyer, 1981]. Another major problem is the issue of sufficient conditions for aggregation [e.g., Odero-Ogwel and Clayton, 1973]. For discussion of these issues in the context of FSR, see Crawford [1982] and CIMMYT [1984].

143. CIMMYT researchers (emphasizing maize and wheat research) are using an FSR approach in collaboration with national agricultural services in Latin America and Africa [Collinson, 1982, 1988].

144. Although the words FSR do not appear in the title of the manual "Planning Technologies Appropriate to Farmers," it is an important reference for many FSR teams.

145. See Eicher [1987; 1988a, b].

146. The top-down approach—research station to the best farmers in the Experimental Unit—has been questioned by many FSR researchers [Byerlee, Collinson, et al., 1980 and CIMMYT, 1984].

147. Many of the FSR approaches have been used by researchers in the U.S. for decades. But FSR is a new approach in many Third World countries where research has been primarily pursued on a commodity-by-commodity basis on research stations.

148. The rich body of ethnographic literature on herders and pastoral systems includes: Dupire [1962]; Stenning [1959]; Gallais [1975]; Horowitz [1972]; Bernus [1974]; Jacobs [1975]; Dyson-Hudson [1972]; Monod, ed. [1975]; and Toupet [1977].

149. See the introduction for an overview of livestock systems.

150. Personal communication, June 25, 1980, from Klaus Meyn, a German livestock specialist with a great deal of experience in West Africa.

151. Brown reports that pastoralists in Kenya such as the Masai, Samburu, Boran, and Somali live largely on milk with meat eaten mainly in the dry season when milk yields fall and livestock mortality increases—sometimes blood is also consumed.

152. Brown made the following assumptions: herding families consume 75 percent milk and 25 percent meat; calving rate of 70 percent; fourteen cows are needed in milk during the year; the number of young stock reared is limited to that required for replacement; and that one-half of the family members are children below fourteen years of age. The assumed family of eight (6.5 adult equivalents) requires about 15,000 calories per family. R. E. McDowell notes that Brown's model has a calving rate of 70 percent which is too high, especially for a pastoral herding situation [Personal Communication, June 5, 1980]. Stephan Sanford adds, "What is unrealistic is not the figure of 70 percent but any single figure. Both on ranches and in pastoral herding situations there are very wide fluctuations between years according to weather, disease, etc." [Personal communication, March 17, 1981]. The World Bank uses a figure of 40-60 percent calving rate for pastoral herders in West Africa.

153. But Swaziland is an atypical country where many rural males earn substantial incomes from seasonal employment in South Africa. Thus the behavior of rural households with livestock may be different from full-time pastoralists in West Africa.

154. The need to learn from past mistakes was a common theme at AID's workshop on pastoralism [AID, 1980].

155. A few economists are pursuing systematic and comparative research on livestock. A major study of livestock production and marketing in West Africa by CRED at the University of Michigan is available in a synthesis by K. H. Shapiro, ed. [1979] and reports by individual team members: Delgado [1979a]; Eddy [1979]; Herman [1979]; and Staatz [1979].

156. The CILSS/Club du Sahel livestock strategy for the Sahel was prepared by IEMVT in France, and the CILSS/Club du Sahel Livestock Team [IEMVT, 1980].

157. Most countries impose a head tax on cattle and sometimes on sheep and goats. Livestock owners, of course, try to evade these taxes and are suspicious of government officials and researchers trying to gain knowledge about their herds. For information on cattle taxes, see van Raay [1975] and Stenning [1959].

158. For a study of an attempt by a Liverpool subsidiary (African Ranches, Ltd.) to establish a 16,000 acre ranch in northern Nigeria in 1914, see Dunbar [1970]. The ranch was operated from 1914 until it failed in 1923 and was turned over to the colonial government.

159. A group ranch is operated by a group of people who jointly have freehold title or the use rights of land and agree to continue to own cattle individually and herd animals collectively at agreed upon stocking levels.

160. A grazing block or association is a delimited pastoral area where some infrastructure improvements are made—especially improved water sources. A system of rotational grazing is overseen by a government grazing manager.

161. It is difficult to compare off-take rates in different countries because different assumptions are often used in computing the coefficients. The rate can vary widely depending upon at what point you aggregate data. For example, in the computation of the off-take rate, do you include the number of animals consumed by herding families and the number sold or only the animals sold? Do you include animal deaths? Finally, the choice of the base year is difficult because of the lack of reliable livestock censuses. When one considers these problems, one understands why off-take rates should be taken with a "grain of salt." The

off-take rate for livestock in the U.S. is around 20 to 25 percent per year as compared with 12 to 16 percent per year on commercial ranches in Kenya and 8 to 12 percent for pastoral herds throughout Africa. But an off-take rate by itself does not tell us very much.

162. One should keep in mind that these data include the 1968-74 Sahelian drought period and, as a result, they might be biased downward.

163. Unfortunately, most research on supplementary feeding and animal nutrition has been carried out on experiment stations or on enclosed ranches in countries such as Zimbabwe and South Africa.

164. Personal communication, March 17, 1981.

165. There is a wide variation in the ranking of the types of diseases in Asia, Latin America, and Africa. In Latin America, the number one cattle disease is foot and mouth disease, while in Africa it is number seven on the list.

166. For a cost-benefit analysis of vaccination programs in Africa, see ILCA [1979d] and Felton and Ellis [1978]. For a study of the economics of tick control in Tanzania, see Grindle [1980].

167. Some 1981 estimates of tsetse eradication by insecticides are twenty dollars per ha for eradication and five dollars per ha per year for maintenance. These costs cannot be recovered by livestock profits alone but they could be recovered by mixed farming or agricultural production.

168. This is a worldwide problem. For a discussion of mistakes in land use planning for pastoral zones in Australia, see M. D. Young [1979].

169. Farmer/herder conflicts are particularly serious in West Africa and they will intensify with increasing population pressure. See van Raay's [1975] analysis in northern Nigeria and M. S. Diarra's [1975] and Horowitz's [1972, 1977] studies of Fulani herders and Hausa farmers in northern Niger. Population pressure in the sorghum and millet belt in southern Niger is forcing Hausa farmers to move their cultivation northward into lower rainfall areas traditionally used by herders, resulting in a severe threat to herders.

170. See Paylore, ed. [1976], Paylore and Mabbutt [1980], Hinchey, ed. [1978], and Glantz, ed. [1987] for bibliographies on drought and desertification.

171. There are two major problems in Delgado's linear programming study. First, because there were only a few farmers in his sample using animal traction, Delgado had to rely on questionable animal traction coefficients from IRAT experiment station results rather than from actual farm conditions. Second, he assumes that adult rather than child labor of the sedentary farmers is needed to care for animals when millet and sorghum are being harvested. However, Barrett, et al. [1982] reported that children of a fairly young age manage herds of goats and cattle in Burkina Faso. See also Stenning [1959] and Dyson-Hudson [1972] for discussions of child labor.

172. The World Bank report [1981b] contends that the deteriorating trade position of most African countries in the 1970s reflects a failure of trade and exchange rate policies to provide incentives for agricultural production and exports.

173. For example, Coulson [1977a] points out the duplication of fertilizer plants in East Africa and the power struggle between Kenya and Tanzania over the location of factories.

174. The instability of local markets is one of the key problems in the transition from subsistence to commerical agriculture [Abercrombie, 1961].

175. The view of traders as exploitive has uncertain roots but bias against traders is an old theme, as shown in the case studies in Bohannan and Dalton, eds. [1962]. Bias against private traders has also led the majority of African governments to establish monopoly control over the procurement and distribution of inputs. See the World Bank [1981b] for a discussion of the problems associated with government control of input supplies.

176. This view has also been frequently expressed by some French researchers such as Kohler [1971] and Ancey, Michotte, and Chevassu [1974].

177. Jones's study is primarily based on fieldwork carried out by V. Alvis, P. E. Temu, E. H. Gilbert, R. J. Mutti, D. N. Atere-Roberts, and A. Whitney.

178. B. Harriss [1979b] showed that in several cases the correlation coefficients reported by Anthonio, Gilbert, Ejiga, and Hays were low and in some cases negative, and that they were in general based on weak data.

179. B. Harriss [1979b] supported her position by citing several instances in which specific findings reported by country researchers presented a much less optimistic picture of the functioning of rural markets than reported in later volumes summarizing a number of country studies.

180. The basic policy issue is more complex because it involves determining the appropriate mix of state and private market involvement in trade. It is not a case of pure government control *versus* free trade although many researchers have talked in such terms [John Staatz, personal communication].

181. Follow-up studies on livestock marketing in West Africa are presented in Ariza-Nino and Steedman, eds. [1979, 1980].

182. The FAO study outlines a strategy for the establishment of a coordinated system of national and regional reserve stocks.

183. The World Bank [1981b] and Buccola and Sukume [1988] note that the annual cost of buffer stock schemes may amount to 15 to 20 percent of the value of stocks because stocks need to be turned over every two to three years in order to avoid deterioration, and that administration of these schemes is demanding.

184. For an important debate on money lenders in India, see Bhaduri [1977].

185. There is much reality to this perception. For example, in May 1981, the government of Senegal announced that it was "writing off" all outstanding loans for seed and fertilizer and suspending the need for farmers to repay equipment loans until the financial records of the recently dissolved grain board (ONCAD) were put in order.

186. Research in the Third World generally shows that annual real interest rates of around 25 percent are needed to cover interest, default, and the administrative costs of small farmer credit programs.

187. There are virtually no hard data on repayment rates by size of farm in Africa.

188. Von Pischke [1980] proposed a general model of "supply-leading finance" whereby funds are made available to target groups in advance of demand, in an effort to stimulate the adoption of proposed innovations.

189. Expenditure elasticities were generally computed because data were not available for the computation of income elasticities of consumption items.

190. Robert P. King and Byerlee hypothesized that the relatively uniform income distribution in rural Sierra Leone may explain the differences in their findings and the results of Soligo's test of the factor intensity hypothesis in Pakistan.

191. Research on nutrition can be traced to Audrey Richards's [1932, 1939] studies of the Bemba in Northern Rhodesia (now Zambia). For an overview of nutrition and health in East Africa, see the Kraut and Cremer, eds. [1969] study of Kenya and Tanzania which drew on anthropometric measurements, food consumption survey data, demographic information, and clinical assessments of nutrition. An overview of nutrition problems and guidelines for nutrition workers is Latham's [1980] *Human Nutrition in Tropical Africa*. For a policy overview on nutrition in the Third World, see Alan Berg [1987]. A basic reference on nutritional surveillance techniques is WHO [1976].

192. Derman and Derman [1973] present a ethnographic study of the evolution of an inegalitarian serf village in Guinea under colonial, postindependence, and socialist conditions.

193. For example, Kenya's rate of population growth of around 4 percent means that population will double in eighteen years. The three-way race for land—to be used for food production, tourism, and livestock—is an explosive political issue.

194. Contrary to reports in the Western press, there is little evidence that there were comparable death rates across all age strata of the rural population. Caldwell [1975] found that the actual number of people who died during the Sahelian drought was surprisingly low.

195. Byerlee and his colleagues found it was relatively easy to "trace" and locate migrants living in rural areas.

196. Bannerjee and Kanbur report that much of the effect of rural development programs "will depend on the position of 'pre-development' income distribution curve vis-à-vis the propensity to migrate curve and the intensity of the development effort" [1981, p. 23].

197. Studies of the massive outmigration from the Mossi plateau in Burkina Faso to coastal countries such as Ghana and Ivory Coast include: Skinner [1965]; Kohler [1972]; ORSTOM [1975]; Remy [1977]; Songre [1973]; and Coulibaly, Gregory, and Piche [1980].

198. See Bromberger [1979].

199. For a further discussion see the sections on smallholder farming, migration, and mechanization.

200. Boserup's 1970 book on women touched off a stream of papers and workshops and special projects to help women. Boserup's book stimulated research on women just as Todaro's [1969] model spurred research on migration in the 1970s.

201. Boserup argued that men usually monopolize the use of new equipment and this tendency is frequently reinforced by a bias in extension programs in favor of men. As a result, there may be a relative decline in the productivity of women and "the corollary of the relative decline in women's labor productivity is a decline in their relative status" [p. 53].

202. For example, Tinker claims that in virtually all countries and in all classes women have lost ground relative to men. She attributes this "deplorable phenomenon" to development planners, who use "mythical stereotypes as the base for their development plans." Even regarding subsistence societies, Tinker claims that women's roles "often add up to near serfdom" [1976, pp. 22-24].

203. The definition of rural small-scale industry varies from country to country, within countries, and among government agencies. Chuta and Liedholm [1979] found fifty different definitions used in seventy-five countries. We have defined small-scale as establishments employing fewer than fifty people.

204. The finding that the expenditure elasticities of demand for rural small-scale industry products are high but under one, does not conflict with Resnick's observation that, in a long-run historical perspective, there has been a tendency for artisans and rural small-scale firms to account for a declining proportion of the national product. See Resnick's [1970] description of the decline of SSI in Burma, Philippines, and Thailand over the period 1870-1938.

205. But it is necessary to go beyond comparisons of small- and large-scale industry and analyze the returns to farming and rural small-scale industry [SSI]. For example, although SSI may be labor-intensive relative to large-scale firms, the SSI firms may be capital-intensive relative to farming activities. Moreover, if the capital requirements for establishing SSI are large relative to the incomes of poorer rural households, it may open the

door for higher-income rural households to engage in rural SSI. Therefore, policies to stimulate demand for SSI products or to partially subsidize the costs of SSI firms may tend to worsen rural income inequality even though they may narrow the gap between rural and urban areas.

206. For bibliographies, see Joyce and Beudot [1976-77]; the *Sahel Bibliographic Bulletin* [1977-1981] and Glantz, ed. [1987].

207. Food self-sufficiency implied regional and not national food self-sufficiency in all basic foods [Club du Sahel, 1977].

208. This section is adapted from Eicher [1986a, 1988 a, b].

209. Another comparison can be made with California. In 1985, the total GNP of forty-five African countries was about one-third of the total GNP of the State of California.

210. The official government report is 300,000 deaths but knowledgeable observers conclude that the number was around a million.

211. For a discussion of the food and industrialization crises in Argentina, India, and China, see T. W. Schultz [1965].

212. The two oldest, most reliable, and profitable seed companies in Africa are the Kenya Seed Company and the Seed Co-op Company of Zimbabwe Ltd. The latter is the largest seed company in Africa, including the Republic of South Africa. Both have been nurtured for decades by commercial farmers. In Zimbabwe a number of commercial farmers organized a seed maize association in 1940 which led to the foundation of the Seed Maize Co-op in 1969 followed by a Crop Seed Co-op in 1979. In 1983 the Seed Co-op Company of Zimbabwe was formed through a merger of the two co-ops. Since hybrid maize seed has to be replaced every year, the shift from open pollinated to hybrid seed is being delayed by ineffective government seed companies throughout Africa.

213. Rwanda has the highest population density in Africa.

214. The stock of on-shelf, improved, farmer-tested, and profitable food crop technology is more limited than FAO's claim, but there is wide variation by crop agroecology. Examples of on-shelf food crop technology include white maize in eastern and southern Africa, hybrid sorghum in Sudan, potatoes in Rwanda, cassava in Nigeria, maize in Ghana, Sierra Leone and Mali, and wheat in the cool highlands of Ethiopia, Kenya, northern Tanzania, and in Zimbabwe where it can be grown in the winter (May-September) under irrigation.

215. Quoted in Waelti [1988].

216. For a discussion see Colander, ed. [1984] and Srinivasan [1985].

217. The stage of institutional maturity (socioeconomic development) of African countries, relative to Asia and Latin America, requires further study and debate. Two demographers recently compared the Asian and African family planning experience and concluded that the slowness to adopt family planning in Africa is "not explained by the African countries being at an earlier stage of socioeconomic development" [Caldwell and Caldwell, 1988, p. 19]. The authors contended that essential differences in family structures, economies and religious attitudes towards fertility severely limit the ability of African states to implement forceful family planning programs. These differences have at times influenced the programs of China, India, and Indonesia. For a discussion of cultural endowments see Ruttan [1988].

218. Success stories include the privately-financed agricultural research such as the Rattray-Arnold Research Station in Zimbabwe and the Tea Research Foundation of Central Africa (headquarters in Malawi). For successful institutions providing marketing services see the collection of twenty-six case studies of private indigenous, transnational, co-oper-

ative, and parastatal marketing enterprises in the book edited by John Abbott [1987a] and Leonards's [1991] study of four Kenyan leaders in institution building.

219. For failures of institutions, see the studies of export marketing boards in the collection edited by Arhin, Hesp, and Van der Laan, eds. [1985] and the study of food marketing boards by Professor William O. Jones [1987].

220. But this comparison should be qualified because Ghana's policy climate reduced incentives and the real incomes of rural people. For example, Ghana's world market share of cocoa fell from 36 to 17 percent from the early 1970s to the mid 1980s.

221. Zimbabwe has about three million hectares planted to crops every year.

222. Etherington's [1973] study of tea in Kenya is a classic. For a recent World Bank desk study of the KTDA see Lamb and Muller [1982].

223. Dantwala [1985] and others have pointed out that 200 of the 800 million in India lack land or income to meet their basic food needs.

224. As early as 1880, fifty percent of the rice area cultivated in Java and Madura in Indonesia was irrigated [Barker, Herdt, Rose, 1985, p. 98].

225. For a synthesis of experience with irrigation in Africa, see Eicher and Baker [1982], Hotes [1983]; Blackie, ed. [1984a]; Moris [1987]; and Olivares [1987].

226. The western consumption patterns include: "Imported long-grain rice, wheat, wine, cheese, dairy products, automobiles, electronic gadgets, high quality clothing, and holidays abroad" [Aboyade, 1988, p. 16].

227. Total rice imports (commercial and food aid) averaged 369,000 tons per year for the five year period [FAO, 1987b].

228. Fast foods have lower net energy costs and faster preparation time. John Staatz of MSU recently pointed out that it takes about one-half less energy expense to prepare rice than the traditional millet or sorghum in Mali.

229. For information on the status of millet research in West Africa see Matlon [1987], D. S. C. Spencer and Sivakumar [1987], Stoop [1987], and Sanders [1989].

230. Wheat is a temperate crop that needs cool evenings when it is flowering and tillering. CIMMYT has two wheat specialists posted in Ethiopia to assist national researchers carry out research in the cool highlands of Ethiopia, Kenya, Tanzania, and in Zimbabwe under irrigation during the winter months (May-September). Because of the lack of heat-tolerant wheat varieties and the high cost of wheat production, CIMMYT does not have a wheat specialist posted in West Africa. For an assessment of the economics of wheat production in the tropics, see Byerlee and Longmire [1986].

231. It is being described as "Bread without Wheat" [Satin, 1988].

232. As population density increases in Rwanda, farmers are growing more sweet potatoes, a cheaper source of calories per unit of land than beans, the historical staple.

233. For a comparative study of food production and consumption trends in Brazil and Nigeria, see Paulino and Sarma [1988]. Aboyade reports that Nigeria's oil boom and inaction on exchange rate adjustment "literally wiped out domestic food production in Nigeria" [Aboyade, 1988, p. 246]. Idachaba [1988] reports that Nigeria's cheap food policy has hastened the shift to imported convenience foods and undermined local food production.

234. The ICRISAT/SADCC research center based at Matopos outside Bulawayo, Zimbabwe recently hired a food scientist to identify changing consumer preferences for sorghum and millet and develop new sorghum and millet food products.

235. The definition used by the World Bank [1985a] and Eicher and Staatz [1986].

236. For a survey of research on drought, see the excellent collection of papers in the book edited by Glantz, *Drought and Hunger in Africa* [1987].

237. For an overview of the literature on famine, see Sen [1981, 1984, 1987a, b]; Mariam [1986]; Mellor and Gavian [1987]; Jansson, Harris, and Penrose [1987]; Dreze [1988]; and Eicher [1985, 1986b].

238. The authors draw heavily on Indonesia, China, and Sri Lanka for their empirical examples. The authors also assume a data base for food policy analysis that is not available in Africa. For example, the authors report that the starting point for food policy analysis is a "food balance sheet, which most countries now publish on an annual basis" [Timmer, Falcon and Pearson, 1983, p. 22]. But we are not aware of a single African country that publishes a food balance sheet on a regular basis. Nevertheless, *Food Policy Analysis* contains a valuable analytical framework for food policy analysis in Africa.

References

Abalu, G. O. I. [1974]: "Supply Response to Producer Prices: A Case Study of Groundnut Supply in Northern Nigeria." *Nigerian J. Econ. Soc. Studies* 16:419-427.

———— [1975]: "Optimal Investment Decisions in Perennial Crop Production: A Dynamic Linear Approach." *J. Agr. Econ.* 26:383-393.

———— [1976]: "A Note on Crop Mixtures under Indigenous Conditions in Northern Nigeria." *J. Dev. Studies* 12:213-220.

Abalu, G. O. I., and B. D'Silva [1980a]: "Socioeconomic Analysis of Existing Farming Systems and Practices in Northern Nigeria." In ICRISAT, 1980a, pp. 3-10.

———— [1980b]: "Nigeria's Food Situation: Problems and Prospects." *Food Policy* 5(1):49-60.

Abalu, G. O. I., and E. G. Etuk [1986]: "Traditional *versus* Improved Groundnut Production Practices: Some Further Evidence from Northern Nigeria." *Experimental Agriculture* 22:33-88.

Abatena, H. [1987]: "Hunger and Starvation in Africa." *Rural Sociologist* 7(3):175-186.

Abbott, J. [1987a]: *Agricultural Marketing Enterprises for the Developing World*. Cambridge, Cambridge Univ. Press.

———— [1987b]: "Institutional Reform of Marketing and Related Services to Agriculture, with Particular Reference to Africa." *Agr. Econ.* 1(2):143-157.

Abercrombie, K. C. [1961]: "The Transition from Subsistence to Market Agriculture in Africa South of the Sahara." *Monthly Bull. Agr. Econ. and Stat.* 10(2):1-7. Reprinted in Whetham and Currie, eds., 1967, 1-11.

Aboyade, O. [1973]: *Incomes Profile. An Inaugural Lecture Delivered at the University of Ibadan, Nigeria*. Ibadan, Nigeria, Univ. of Ibadan Bookshop.

———— [1985]: *Administering Food Producer Prices in Africa: Lessons from International Experiences*. Washington, IFPRI.

———— [1988]: "Structural Adjustments and the African Food Economy." Keynote address presented at a seminar of the Kellogg International Fellowship Program in Food Systems. Harare, Zimbabwe. February 8.

Acland, J. D. [1971]: *East African Crops: An Introduction to the Production of Field and Plantation Crops in Kenya, Tanzania, and Uganda*. London, Longmans.

Adams, A. [1977a]: *Le long voyage des gens du fleuve*. Paris, Maspero.

———— [1977b]: "The Senegal River Valley: What Kind of Change?" *Rev. African Polit. Econ.* 10:33-60. Reprinted in Heyer, Roberts, and Williams, eds., 1981, pp. 325-353.

Adams, D. W. [1978]: "Mobilizing Household Savings through Rural Financial Markets." *Econ. Dev. and Cultural Change* 26:547-560.

Adams D. W., and D. H. Graham [1981]: "A Critique of Traditional Agricultural Credit Projects and Policies." *J. Dev. Econ.* 8:347–366.

Adams, J. M., and G. W. Harman [1977]: *The Evaluation of Losses in Maize Stored on a Selection of Small Farms in Zambia with Particular References to the Development of Methodology.* London, Tropical Products Institute.

Adams, M. E., and J. Howell [1979]: "Developing the Traditional Sector in the Sudan." *Econ. Dev. and Cultural Change* 27:505–518.

Adegboye, R. O. [1969]: "Procuring Loans through Pledging of Cocoa Trees." *J. Geographical Assoc. of Nigeria* 12(1/2).

——— [1977]: "Land Tenure." In Leakey and Wills, eds., 1977, pp. 313–327.

Adegeye, A. J. [1976]: "Producer Prices, Farmers' Responses and Productivity Interactions: Some Relevant Issues." *J. Rural Econ. and Dev.* (Ibadan) 10(2):20–27.

Adesina, A., P. Abbott, and J. Sanders [1988]: "*Ex-ante* Risk Programming Appraisal of New Agricultural Technology: Experiment Station Fertilizer Recommendations in Southern Niger." *Agr. Systems* 27:23–34.

Ady, P. [1968]: "Supply Functions in Tropical Agriculture." *Oxford Bull. Econ. and Stat.* 30:157–188.

Agboola, S. A. [1968]: "The Introduction and Spread of Cassava in Western Nigeria." *Nigerian J. Econ. and Soc. Studies* 10(3):369–385.

Agiri, B. [1977]: "The Introduction of Nitida Kola into Nigerian Agriculture, 1880–1920." *African Econ. Hist.* 3:1–14.

Agricultural Technology Improvement Project [ATIP, 1986a]: *Farming System Research Activities at Mahalapye: Summary of Activities, 1982-85.* Gaborone, Botswana, Ministry of Agriculture, Research Report No. 1.

——— [ATIP, 1986b]: *Farming System Research Activities at Francistown: Summary of Activities, 1983-85.* Gaborone, Botswana, Ministry of Agriculture, Research Report No. 2.

Ahmed, A. G. M. [1979]: "The Relevance of Contemporary Anthropology." In Huizer and Mannheim, eds., 1979, pp. 171–185.

Ahmed, R., and N. Rustagi [1985]: *Agricultural Marketing and Price Incentives: A Comparative Study of African and Asian Countries.* Washington, IFPRI.

——— [1987]: "Marketing and Price Incentives in African and Asian Countries: A Comparison." In Elz, 1987, pp. 104–118.

Ahn, P. M. [1969]: *West African Agriculture. Vol. I: West African Soils.* London, Oxford Univ. Press.

——— [1977]: "Soil Factors Affecting Rainfed Agriculture in Semi-Arid Regions with Particular References to the Sahel Zone of Africa." In Cannell, ed., 1977, pp. 128–165.

AID [1976]: *Proposal for a Long-Term Comprehensive Development Program for the Sahel: Report to the U.S. Congress. Part I. Major Findings and Programs: Part II. Technical Background Papers.* Washington.

——— [1978]: *Toward a Rational U.S. Policy on River Basin Development in the Sahel: Proceedings of a Colloquium.* Washington.

——— [1980]: *Workshop on Pastoralism and African Livestock Development.* Washington, Bureau for Program and Policy Coordination, Program Evaluation Report No. 4.

——— [1984a]: "Implementation Problems Delayed Impact of Western Sudan Agricultural Research Project While Financial Uncertainties Threaten Long-Term Benefits." Nairobi, Regional Inspector General for Audit. February 24.

——— [1984b]: "Inadequate Design and Monitoring Impede Results in Sahel Food Production Projects." Washington, Regional Inspector General for Audit. January 31.

_____ [1985]: *Plan for Supporting Agricultural Research and Faculties of Agriculture in Africa.* Washington, African Bureau.

Ajaya, S. S., and L. B. Halstead, eds. [1979]: *Wildlife Management in Savannah Woodland: Recent Progress in African Studies.* London, Taylor and Frances.

Ake, C. [1981]: *A Political Economy of Africa.* London, Longmans.

Akinola, A. A. [1986]: "An Application of Bass's Model in the Analysis of Diffusion of Cocoa-Spraying Chemicals among Nigerian Cocoa Farmers." *J. Agr. Econ.* 37(3):395-404.

Akinwumi, J. A., and A. J. Adegeye [1977]: "Prospects and Problems of a Common Agricultural Policy among the West African States." In Dams and Hunt, eds., 1977, pp. 396-418.

Aklilu, B. [1980]: "The Diffusion of Fertilizer in Ethiopia: Pattern, Determinants, and Implications." *J. Developing Areas* 14:387-399.

Akoto, Owusu [1987]: "Agricultural Development Policy in Ghana." *Food Policy* 12(3):243-254.

Aladejana, A. [1970]: *The Marketing Board System: A Bibliography.* Ibadan, Nigeria, NISER.

Alberici, A., and M. Baravelli [1973]: *Savings Banks and Savings Facilities in African Countries.* Milan, Italy, Cassa di Risparmio delle Provincie Lombarde.

Aldington, T. [1971]: *Producer Incentives as a Means of Promoting Agricultural Development: A Case Study of Cotton in Kenya.* Nairobi, Univ. of Nairobi, IDS, Discussion Paper No. 105.

Aldington, T. J., and F. A. Wilson [1968]: *The Marketing of Beef in Kenya.* Nairobi, Univ. of Nairobi, IDS, Occasional Paper No. 3.

Alibaruho, G. [1974]: "Regional Supply Elasticities in Uganda's Cotton Industry and the Declining Level of Cotton Output." *Eastern Africa Econ. Rev.* 6(2):35-56.

Allal, M., and E. Chuta [1988]: *Cottage Industries and Handicrafts: Some Guidelines for Employment Promotion.* Geneva, ILO.

Allan, W. [1965]: *The African Husbandman.* Edinburgh, Oliver and Boyd.

Alverson, H. [1979]: "Arable Agriculture in Botswana: Some Contributions of the Traditional Social Formation." *Rural Africana* NS 4-5:33-48.

Amann, V. F., ed. [1973]: *Agricultural Policy Issues in East Africa. Proceedings of the East African Agricultural Economics Society Conference, Nairobi, June 1971.* Kampala, Uganda, Makerere Univ. Printery.

Amie, L., and J. G. Disney [1973]: "Quality Changes in West African Marine Fish during Iced Storage." *Tropical Science* 15(2):125-38.

Amin, S. [1965]: *Trois expériences Africaines de développement: le Mali, la Guinée et le Ghana.* Paris, Presses Universitaires de France.

_____ [1967]: *Le développement du capitalisme en Côte d'Ivoire.* Paris, Editions de Minuit.

_____ [1970]: "Development and Structural Change: The African Experience, 1950-70." *J. International Affairs* 24:203-223.

_____ [1972]: "Underdevelopment and Dependence in Black Africa—Origins and Contemporary Forms." *J. Modern African Studies* 10:503-524.

_____ [1973]: "Transitional Phases in Sub-Saharan Africa." *Monthly Review* 25(5):52-57.

_____ , ed. [1974a]: *Modern Migrations in Western Africa.* London, Oxford Univ. Press for International African Institute.

_____ [1974b]: "Introduction." In Amin, ed., 1974a, pp. 3-124.

_____ [1974c]: *Accumulation on a World Scale: A Critique of the Theory of Underdevelopment.* New York, Monthly Review Press.

_____ [1976]: *Unequal Development: An Essay on the Social Formations of Peripheral Capitalism.* New York, Monthly Review Press. Translation of *Le développement inégal*, Paris, Les Editions de Minuit, 1973.

Amin, S., C. Atta-Mills, A. Bujra, G. Hamid, and T. Mkandawire [1978]: "Social Sciences and the Development Crisis in Africa, Problems and Prospects." *Africa Dev.* 3(4):23-45.

Amin, S., and F. Okediji [1974]: "Land Use, Agriculture and Food Supply, and Industrialization—Introduction." In Cantrelle, ed., 1974, pp. 409-423.

Amon, B. O. E., and S. A. Adetunji [1973]: "The Response of Maize, Yams, and Cassava to Fertilizers in a Rotation Experiment in the Savannah Zone of Western Nigeria." *Nigerian Agr. J.* 10(1):91-105.

Amselle, J.-L., ed. [1976]: *Les Migrations Africaines: Réseaux et Processus Migratoires.* Paris, Maspero.

Ancey, G. [1974]: *Relations de voisinage ville-campagne. Une analyse appliquée à Bouake: sa couronne et sa région (Côte d'Ivoire).* Paris, ORSTOM, Memoires No. 70.

_____ [1975]: *Niveaux de décision et functions objectif en milieu Africain.* Paris, INSEE, AMIRA Note de Travail No. 3.

_____ [1977]: "Recensement et description des principaux systèmes ruraux Sahéliens." *Cahiers ORSTOM, Ser. Sci. Hum.* 14(1):3-18.

Ancey, G., J. Michotte, and J. Chevassu [1974]: *L'économie de l'espace rural de la région de Bouake.* Paris, Travaux et Documents de l'ORSTOM No. 38.

Anderson, D. [1975]: *Fluctuations of Maize and Groundnut Yields in the Lilongwe Land Development Program.* Washington, World Bank, Studies in Employment and Rural Development No. 28.

Anderson, D., and M. W. Leiserson [1980]: "Rural Nonfarm Employment in Developing Countries." *Econ. Dev. and Cultural Change* 28:227-48.

Anderson, David, and R. Grove, eds. [1987]: *Conservation in Africa: People, Policies, and Practice.* Cambridge, England, Cambridge University Press.

Anderson, T. J., ed. [1976]: *Land Tenure and Agrarian Reform in Africa and the Near East: An Annotated Bibliography.* Boston, G. K. Hall.

Andrae, G., and B. Beckman [1985]: *The Wheat Trap: Bread and Underdevelopment in Nigeria.* London, Zed Books.

Andreou, P. [1981]: "Agricultural Development Effort in Nigeria—An Economic Appraisal of the Western State Settlement Scheme." *Agr. Systems* 7(1):11-20.

Andrews, D. J. [1972]: "Intercropping with Sorghum in Nigeria." *Experimental Agriculture* 8:139-150.

_____ [1974]: "Responses of Sorghum Varieties to Intercropping." *Experimental Agriculture* 10:57-63.

_____ [1975]: "Sorghum Varieties for the Late Season in Nigeria." *Tropical Agriculture* 52:21-30.

Annegers, J. F. [1973]: "Seasonal Food Shortages in West Africa." *Ecology of Food and Nutrition* 2:251-257.

Anschel, K. R., R. H. Brannon, and E. D. Smith, eds. [1969]: *Agricultural Cooperatives and Markets in Developing Countries.* New York, Praeger.

Ansell, A., and M. Upton [1979]: *Small Scale Water Storage and Irrigation: An Economic Assessment for South West Nigeria.* Reading, Univ. of Reading, DAE and Management.

Anthonio, Q. B. O. [1968]: "The Marketing of Staple Foodstuffs in Nigeria: A Study in Pricing Efficiency." Unpublished Ph.D. dissertation, Univ. of London.

Anthony, K. R. M., B. F. Johnston, W. O. Jones, and V. C. Uchendu [1979]: *Agricultural Change in Tropical Africa.* Ithaca, Cornell Univ. Press.

Anthony, K. R. M., and V. C. Uchendu [1970]: "Agricultural Change in Mazabuka District, Zambia." *Food Research Institute Studies* 9:215-266.

_____ [1974]: *Agricultural Change in Geita District, Tanzania*. Nairobi, East African Literature Bureau.

Apthorpe, R. [1972]: *Rural Cooperatives and Planned Change in Africa: An Analytical Overview*. Geneva, RISD.

Arditi, C. [1975]: *Les Circuits de Commercialisation des Produits du Secteur Primaire en Afrique de l'Ouest, Analyse Bibliographique*. Paris, Ministère de la Coopération.

Are, L. A., ed. [1975]: *Socio-Economic Aspects of Rice Cultivation*. Monrovia, Liberia, WARDA, Seminar Proceedings 3.

Arhin, K., P. Hesp, and L. Van der Laan, eds. [1985]: *Marketing Boards in Tropical Africa*. London, Kegan Paul International.

Ariza-Nino, E., and C. Steedman, eds. [1979, 1980]: *Livestock and Meat Marketing in West Africa*, 5 vols. Ann Arbor, Univ. of Michigan, CRED.

Arnold, M., ed. [1976]: *Agricultural Research for Development: The Namulonge Contribution (Uganda)*. Cambridge, England, Cambridge Univ. Press.

Arrighi, G. [1970]: "Labour Supplies in Historical Perspective: A Study of the Proletarianization of the African Peasantry in Rhodesia." *J. Dev. Studies* 6:197-234. Reprinted in Arrighi and Saul, 1973, pp. 180-234.

Arrighi, G., and J. S. Saul [1968]: "Socialism and Economic Development in Tropical Africa." *J. Modern African Studies* 6(2):141-169. Reprinted in Arrighi and Saul, 1973, pp. 11-43.

_____ [1973]: *Essays on the Political Economy of Africa*. New York Monthly Review Press.

Arrivets, J. [1976]: "Exigences minerales du sorgho — étude d'une variété voltaique a grand tige." *L'Agronomie Tropicale* 31(1):29-46.

Arua, E., and M. Obidiegwu [1988]: "Socio-economic Assessment of the World Bank Rice Project in Eastern Nigeria." *Agr. Systems* 27(2):99-115.

Ascroft, J., F. Chege, J. Kariuki, N. Roling, and G. Ruigu [1972]: "Does Extension Create Poverty in Kenya?" *East Africa J.* 9(3):28-33.

Ascroft, J., N. Roling, J, Kariuki, and F. Chege [1973]: *Extension and the Forgotten Farmer: First Report of a Field Experiment*. Wageningen, Netherlands, Afdelingen voor sociale wetenschappen aan de landbouwhogeschool.

Asefa, S., ed. [1988]: *World Food and Agriculture: Economic Problems and Issues*. Kalamazoo, MI, Upjohn Institute.

Askari, H., and J. Cummings [1976]: *Agricultural Supply Response: A Survey of the Econometric Evidence*. New York, Praeger.

Atayi, E. A., and H. C. Knipscheer [1980]: *Survey of Food Crop Farming Systems in the "ZAPI-EST," East Cameroon*. Ibadan, Nigeria, IITA/ONAREST.

Atsu, S. Y. [1974]: *The Focus and Concentrate Programme in the Kpandu and Ho Districts: Evaluation of an Agricultural Extension Programme*. Legon, Univ. of Ghana, ISSER, Tech. Pub. Series No. 34.

Atta-Konadu, Y. K. [1974]: "Economic Optima in Resource Allocation for Smallholder Subsistence Farming in Ghana." Unpublished Ph.D. dissertation, MSU, East Lansing.

Awogbade, M. O. [1979]: "Fulani Pastoralism and the Problems of the Nigerian Veterinary Service." *African Affairs* 78:493-506.

Ayuko, L. [1980]: *Ranch Organizational Structures. Summary of Papers presented at an ILCA Conference in Addis Ababa, 25-29 February 1980*. Addis Ababa.

_____ [1981]: *Organization, Structures and Ranches in Kenya*. London, ODI, Pastoral Network Paper No. 11b.

Bachmann, E., and F. E. Winch [1970]: *Yam Based Farming Systems in the Humid Tropics of Southern Nigeria.* Ibadan, Nigeria, IITA.

Baier, S. B. [1976]: "Economic History and Development: Drought and the Sahelian Economies of Niger." *African Econ. Hist.* 1:1-16.

_____ [1977]: "Trans-Saharan Trade and the Sahel: Damergu, 1870-1930." *J. African Hist.* 18:37-60.

_____ [1980]: *An Economic History of Central Niger.* Oxford, England, Clarendon Press.

Baier, S. B., and D. J. King [1974]: "Drought and the Development of Sahelian Economies: A Case Study of the Hausa and Tuareg." *Land Tenure Center Newsletter* 54, Univ. of Wisconsin, Madison.

Bairoch, P. [1981]: "Agriculture and the Industrial Revolution, 1700-1914." In Cipolla, ed., 1973, pp. 453-506.

Baker, D. C. [1981]: "Impact on Rural Inequality of Promoting Rural Non-farm Activities: Evidence from Sierra Leone." Paper presented at the 1981 AAEA Meetings, Clemson Univ., July 26-29.

_____ [1987]: "Arable Farming Development Priorities in the Central Agricultural Region, Botswana: A Farming Systems Analysis." Unpublished Ph.D. dissertation, MSU, East Lansing.

Baker, E. F. I. [1978-1980]: "Mixed Cropping in Northern Nigeria. I-IV." *Experimental Agriculture* 14:293-298; 15:33-40, 41-48; 16:316-370.

Baldwin, K. D. S. [1957]: *The Niger Agricultural Project: An Experiment in African Development.* Cambridge, MA, Harvard Univ. Press.

Baldwin, R. E. [1956]: "Patterns of Development in Newly Settled Regions." *Manchester School Econ. and Soc. Studies* 24:161-179. Reprinted in Eicher and Witt, eds., 1964, pp. 238-251.

_____ [1966]: *Economic Development and Export Growth: A Study of Northern Rhodesia, 1920-1960.* Berkeley, Univ. of California Press.

Bannerjee, B., and S. M. Kanbur [1981]: "On the Specification and Estimation of Macro Rural-Urban Migration Functions with an Application to Indian Data." *Oxford Bull. Econ. and Stat.* 43(1):7-30.

Baran, P. A. [1952]: "On the Political Economy of Backwardness." *Manchester School Econ. and Soc. Studies* 20:66-84.

Barker, J. S. [1979]: "The Debate on Rural Socialism in Tanzania." In Mwansasu and Pratt, eds., 1979, pp. 95-124.

Barker, R., R. W. Herdt with B. Rose [1985]: *The Rice Economy of Asia.* Washington, RFF.

Barnes, C. [1979]: "An Experiment with Coffee Production by Kenyans, 1933-1948." *African Econ. Hist.* 8:198-209.

Barnes, C. T. [1976]: "Problems of Transformation — Marine Fisheries in Tanzania." *Eastern Africa J. Rural Dev.* 9(1-2):168-199.

Barnett, T. [1977]: *The Gezira Scheme: An Illusion of Development.* London, Cass.

_____ [1979]: "Why Are Bureaucrats Slow Adopters? The Case of Water Management in the Gezira Scheme." *Sociologia Ruralis* 19(1):60-70.

_____ [1981]: "Evaluating the Gezira Scheme: Black Box or Pandora's Box." In Heyer, Roberts, and Williams, eds. 1981, pp. 306-324.

_____ [1984]: "Small-Scale Irrigation in Sub-Saharan Africa: Sparse Lessons, Big Problems, Any Solutions?" *Public Administration and Dev.* 4(2):21-47.

Barnum, H. D., and R. H. Sabot [1976]: *Migration, Education, and Urban Surplus Labour: The Case of Tanzania.* Paris, OECD.

Barrett, V., G. Lassiter, D. Wilcock, D. Baker, and E. Crawford [1982]: *Animal Traction in Eastern Upper Volta: A Technical, Economic and Institutional Analysis.* East Lansing, DAE, MSU International Development Papers, No. 4.

Barwell, C. [1975]: *Farmer Training in East-Central and Southern Africa.* Rome, FAO.

Bateman, M. J. [1965]: "Aggregate and Regional Supply Functions for Ghanaian Cocoa, 1946-1962." *J. Farm Econ.* 47:384-401.

Bates, D. G., and S. H. Lees [1977]: "The Role of Exchange in Productive Specialization." *Amer. Anthropologist* 79(4):824-841.

Bates, R. H. [1976]: *Rural Responses to Industrialization: A Study of Village Zambia.* New Haven, CT, Yale Univ. Press.

―――― [1981]: *Markets and States in Tropical Africa: The Political Basis of Agricultural Policies.* Berkeley, Univ. of California Press.

Bates, R. H., and M. F. Lofchie, eds. [1980]: *Agricultural Development in Africa: Issues of Public Policy.* New York, Praeger.

―――― [1983]: *Essays on the Political Economy of Rural Africa.* Berkeley, Univ. of California Press.

Bauer, P. T. [1954]: *West African Trade: A Study of Competition, Oligopoly and Monopoly in a Changing Economy.* London, Cambridge Univ. Press. Reprinted, with a new preface, by Routledge and Kegan Paul, London, 1963.

―――― [1975]: "British Colonial Africa: Economic Retrospect and Aftermath." In Duignan and Gann, eds., 1975a, pp. 632-654.

Bauer, P. T., and B. S. Yamey [1959]: "A Case Study of Response to Price in an Underdeveloped Country." *Econ. J.* 69:800-805.

Baumann, H. [1928]: "The Division of Work According to Sex in African Hoe Culture." *Africa* 1(3):289-319.

Beals, R. E., and C. F. Manezes [1970]: "Migrant Labour and Agricultural Output in Ghana." *Oxford Econ. Papers* 21(1):109-127.

Becker, J. A. [1974]: *An Analysis and Forecast of Cereals Availability in the Sahelian Entente States of West Africa.* Washington, AID.

Beckford, G. [1972]: "Strategies for Agricultural Development: Comment." *Food Research Institute Studies* 11:149-154.

Beckman, B. [1976]: *Organizing the Farmers: Cocoa Politics and National Development in Ghana.* New York, Holmes and Meier.

Beeden, P., D. W. Norman, W. J. Kroeker, D. H. Pryor, H. M. Hays, and B. Huizinga [1976]: *The Feasibility of Improved Sole Crop Cotton Production Technology for the Small-Scale Farmer in the Northern Guinea Savanna Zone of Nigeria.* Zaria, Nigeria, Ahmadu Bello Univ., IAR, Samaru Miscellaneous Paper 60.

Behnke, R. [1985]: "Measuring the Benefits of Subsistence *versus* Commercial Livestock Production in Africa." *Agr. Systems* 16(2):109-135.

―――― [1987]: "Cattle Accumulation and the Commercialization of the Traditional Livestock Industry in Botswana." *Agr. Systems* 24(1):1-29.

Behrman, J. R. [1968]: "Monopolistic Cocoa Prices." *Amer. J. Agr. Econ.* 50:702-719.

Bell, C. L. G., and P. B. R. Hazell [1980]: "Measuring the Indirect Effects of an Agricultural Investment Project on Its Surrounding Region." *Amer. J. Agr. Econ.* 62:75-86.

Belloncle, G. [1974]: *Étude sur le crédit agricole dans trois villages de la région de Maradi (Niger).* Rome, FAO, Agricultural Credit Case Studies, Working Paper No. 5.

Belshaw, D. G. R. [1968]: "Agricultural Extension, Education and Research." In Helleiner, ed., 1968b, pp. 57-78.

———— [1979]: "Taking Indigenous Technology Seriously: The Case of Inter-Cropping in East Africa." *IDS, Sussex, Bulletin* 10:24-27. Reprinted in Brokensha, Warren, and Werner, eds., 1980, pp. 197-203.

Belshaw, D. G. R., and M. Hall [1972]: "The Analysis and Use of Agricultural Experimental Data in Tropical Africa." *Eastern Africa J. Rural Dev.* 5:39-72.

Benneh, G. [1972]: "Systems of Agriculture in Tropical Africa." *Econ. Geography* 48(3):244-257.

Benoit-Cattin, M. [1977a]: "Analyse économique pluriannuelle d'un groupe de carres suivis. Unités expérimentales du Sénégal — 1969-1975 — Methodes et principaux résultats." *L'Agronomie Tropicale* 32-4:413-426.

———— [1977b]: *Project terres neuves II. Rapport sur la suivi agro-socioéconomique de la campagne, 1976-77.* Bambey, Sénégal, CNRA.

———— [1980]: "Approche socio-économique des exploitations agricoles." In ICRISAT, 1980a, pp. 410-414.

Benor, D. [1987]: "Training and Visit: Back to Basics." In W. Rivera and S. Schram, eds., 1987, pp. 137-148.

Benor, D., and J. Q. Harrison [1977]: *Agricultural Extension: The Training and Visit System.* Washington, World Bank.

Berg, A. [1981]: *Malnourished People: A Policy Review.* Washington, World Bank.

———— [1987]: *Malnutrition: What Can be Done?* Baltimore, Johns Hopkins Univ. Press.

Berg, E. [1961]: "Backward-Sloping Labor Supply Functions in Dual Economies: The Africa Case." *Quart. J. Econ.* 75:468-492.

———— [1964]: "Socialism and Economic Development in Tropical Africa." *Quart. J. Econ.* 78:549-573.

———— [1965]: "The Economics of the Migrant Labor System." In Kuper, ed., 1965, pp. 160-181.

———— [1975]: *The Recent Economic Evolution of the Sahel.* Ann Arbor, Univ. of Michigan, CRED.

———— [1980]: "Reforming Grain Marketing Systems in West Africa: A Case Study of Mali." In ICRISAT, 1980a, pp. 147-172.

———— [1986a]: *Cereals Policy Reform in the Sahel.* Paris, OECD, CILSS.

———— [1986b]: "The World Bank's Strategy." *Africa in Economic Crisis.* In Ravenhill, ed., 1986, pp. 44-59.

———— [1987]: "Obstacles to Liberalizing Agricultural Markets in Developing Countries." In Elz, ed., 1987, pp. 22-27.

Berg, R. J., and J. S. Whitaker, eds. [1986]: *Strategies for African Development.* Berkeley, Univ. of California Press.

Bernard, H. R., and P. J. Pelto, eds. [1972]: *Technology and Social Change.* New York, Macmillan.

Bernstein, H. [1979]: "African Peasantries: A Theoretical Framework." *J. Peasant Studies* 6(4):421-443.

Bernus, E. [1974]: "L'évolution récente des relations entre éleveurs et agriculteurs en Afrique tropical: L'exemple du Sahel Nigerien." *Cahiers ORSTOM, Ser. Sci. Hum.* 11(2):137-145.

Bernus, E., and G. Savonnet [1973]: "Les problèmes de la sécheresse dans l'Afrique de l'Ouest." *Presence Africaine* 88:112-138.

Berry, S. S. [1975]: *Cocoa, Custom and Socio-Economic Change in Rural Western Nigeria.* Oxford, England, Clarendon Press.

_____ [1976]: "Supply Response Reconsidered: Cocoa in Western Nigeria, 1909-44." *J. Dev. Studies* 13(1):4-17.

_____ [1980]: "Rural Class Formation in West Africa." In R. H. Bates and Lofchie, eds., 1980, pp. 401-424.

_____ [1984]: "The Food Crisis and Agrarian Change in Africa." *African Studies Rev.* 27:98-112.

Bevan, D., P. Collier, and J. Gunning [1987]: "Consequences of a Commodity Boom in a Controlled Economy: Accumulation and Redistribution in Kenya, 1975-1983." *World Bank Econ. Rev.* 1(3)8-16.

Beye, G. [1977]: "Étude de l'action de doses croissantes d'azote en presence ou en absence de paille de riz enfouie sur le développement et les rendements de riz en Basse-Casamance." *L'Agronomie Tropicale* 32(1):41-50.

Bezuneh, M., B. J. Deaton, and G. W. Norton [1988]: "Food AID Impacts in Rural Kenya." *Amer. J. Agr. Econ.* 70:181-191.

Bhaduri, A. [1977]: "On the Formation of Usurious Interest Rates in Backward Agriculture." *Cambridge J. Econ.* 1(4):341-352.

Bhatt, V. V. [1988]: "Growth and Income Distribution in India: A Review." *World Dev.* 16:641-47.

Bibliographie des travaux en langue Francaise sur l'Afrique au sud du Sahara sciences humaines et sociales [1979]. Paris, CEA-CARDAN.

Biebuyck, D., ed. [1963]: *African Agrarian Systems*. London, Oxford Univ. Press.

Bienen, H., and V. P. Diejomaoh, ed. [1981]: *The Political Economy of Income Distribution in Nigeria*. New York, Holmes and Meier.

Bigot, Y. [1977]: "Fertilisation, labour et espèce cultivée en situation de pluviosite incertaine de centre de Côte-d'Ivoire—Synthèse des principaux résultats d'un test de différentes systèmes culturaux 1967 à 1974." *L'Agronomie Tropicale* 32(3):242-247.

Bingen, R. J. [1985]: *Food Production and Rural Development in the Sahel: Lessons from Mali's Operation Riz-Segou*. Boulder, CO, Westview Press.

Bingen, R. J., and J. Faye [1987]: *Agricultural Research and Extension in Francophone West Africa: The Senegal Experience*. East Lansing, MSU, DAE. International Development Papers, Reprint No. 13.

Bingen, R. J., A. E. Hall, and M. Ndoye [1988]: "California Cowpeas and Food Policy in Senegal." *World Dev.* 16:857-865.

Binswanger, H. [1978]: *The Economics of Tractors in South Asia, An Analytical Review*. New York, ADC; Hyderabad, India, ICRISAT.

_____ [1986]: "Evaluating Research Systems Performance and Targeting Research in Land Abundant Areas of Sub-Saharan Africa." *World Dev.* 14:469-475.

_____ [1987]: *Agricultural Mechanization: Issues and Options*. Washington, World Bank.

Binswanger, H., R. D. Ghodake, and G. E. Thierstein [1980]: "Observations on the Economics of Tractors, Bullocks and Wheeled Tool Carriers in the Semi-Arid Tropics of India." In ICRISAT, 1980a, pp. 199-211.

Binswanger, H., and P. Pingali [1988]: "Technological Priorities for Farming in Sub-Saharan Africa." *World Bank Research Observer* 3(1):81-98.

Birgegard, L-E [1987]: "A Review of Experiences with Integrated Rural Development (IRD)." Uppsala, International Rural Development Centre, Swedish Univ. of Agricultural Sciences.

Blackie, M. J. [1982]: "A Time to Listen: A Perspective on Agricultural Policy in Zimbabwe." *Zimbabwe Agr. J.* 79(5):151-156.

———, ed. [1984a]: *African Regional Symposium on Smallholder Irrigation.* Harare, Univ. of Zimbabwe, Department of Land Management.

——— [1984b]: "A Regional Support Program for Faculties of Agriculture in the SADCC Region." Report prepared for AID's Southern Africa Regional Program, Harare, Zimbabwe.

——— [1989]: "Maize in East and Southern Africa." Lilongwe, Malawi: The Rockefeller Foundation.

Blandford, D. [1973]: "Some Estimates of Supply Elasticities for Nigeria's Cash Crops. A Comment." *J. Agr. Econ.* 24:601-604.

——— [1979]: "West African Export Marketing Boards." In Hoos, ed., 1979, Cambridge, MA, Ballinger, pp. 121-150.

Blitzer, C. R. [1979]: "Development and Income Distribution in a Dual Economy: A Dynamic Simulation Model for Zambia." *J. Dev. Econ.* 6:407-429.

Bloch, P. [1986]: *Land Tenure Issues in River Basin Development in Sub-Saharan Africa.* Madison, Univ. of Wisconsin, LTC.

——— [1987]: *The Dynamics of Land Tenure: The Case of the Bakel Small Irrigated Perimeters.* Madison, Univ. of Wisconsin, LTC.

Block, M., ed. [1975]: *Marxist Analyses and Social Anthropology.* London, Malaby Press.

Blomquist, A. G. [1978]: "Urban Job Creation and Unemployment in LDCs: Todaro *vs.* Harris and Todaro." *J. Dev. Econ.* 5:3-18.

Boateng, M., C. B. Ratchford, and M. Blase [1987]: "Profitability Analysis of a Farming System in Africa." *Agr. Systems* 24:81-93.

Boeke, J. H. [1953]: *Economics and Public Policy as Exemplified by Indonesia.* New York, Institute of Pacific Relations.

Boesen, J., and A. T. Mohele [1979]: *The "Success" Story of Peasant Tobacco Production in Tanzania: The Political Economy of a Commodity Producing Peasantry.* Uppsala, Sweden, SIAS.

Bohannan, P. [1963]: "Land, 'Tenure,' and Land Tenure." In Biebuyck, ed., 1963, pp. 101-115.

Bohannan, P., and G. Dalton, eds. [1962]: *Markets in Africa.* Evanston, IL, Northwestern Univ. Press.

Bond, M E. [1983]: "Agricultural Responses to Prices in Sub-Saharan African Countries." *IMF Staff Papers* 30:703-726.

Bonnen, J. T. [1987]: "U.S. Agricultural Development: Transforming Human Capital, Technology and Institutions." In Johnston, *et al.*, 1987, pp. 267-300.

Booth, R. H. [1974]: "Post-Harvest Deterioration of Tropical Root Crops: Losses and their Control." *Tropical Science* 16(2):49-63.

Borlaug, N. E. [1988]: "History of the Sasakawa-Global 2000 Initiatives for Increasing Agricultural Production in Sub-Saharan Africa." Paper presented at the Workshop, *Reviewing the African Agricultural Projects, Nairobi.* Atlanta, Georgia, The Carter Center.

Boserup, E. [1965]: *The Conditions of Agricultural Growth: The Economics of Agrarian Change under Population Pressure.* London, Allen and Unwin.

——— [1970]: *Woman's Role in Economic Development.* New York, St. Martin's Press.

——— [1980a]: "The Position of Women in Economic Production and in the Household, with Special Reference to Africa." In Presvelan and Spijkers-Swart, eds., 1980, pp. 11-16.

——— [1980b]: "Food Production and the Household as Related to Rural Development." In Presvelan and Spijkers-Zwart, eds., 1980, pp. 35-40.

———— [1981]: *Population and Technological Change: A Study of Long-Term Trends.* Univ. of Chicago Press.

Botswana [1975]: *National Policy on Tribal Grazing Land.* Gaborone, Government Printer, Paper No. 2 of 1975.

———— [1976]: *The Rural Income Distribution Survey in Botswana, 1974/75.* Gaborone, Government Printer.

Bottomley, A. [1963]: "The Premium for Risk as a Determinant of Interest Rates in Underdeveloped Rural Areas." *Quart. J. Econ.* 77:637-647.

———— [1975]: "Interest Rate Determination in Underdeveloped Rural Areas." *Amer. J. Agr. Econ.* 57:279-291.

Bottrall, A. [1981]: *Comparative Study of the Management and Organization of Irrigation Projects.* Washington, World Bank, Staff Working Paper No. 458.

Boulet, J. [1975]: *Magoumaz: pays Mafa (Nord Cameroun) étude d'un terroir de montagne.* Paris, Mouton and ORSTOM, Atlas des structures agraires au sud du Sahara 11.

Boutillier, J.-L., P. Cantrelle, J. Causse, C. Laurent, and Th. N'Doye [1962]: *La moyenne vallée du Sénégal: Étude socio-économique.* Paris, République Française, Ministére de la Coopération et INSEE.

Bovet, D., and L. Unnevehr [1981]: *Agricultural Pricing in Togo.* Washington, World Bank, Staff Working Paper No. 467.

Boyce, J. K., and R. E. Evenson [1975]: *National and International Research and Extension Programs.* New York ADC.

Brandt, H. [1979]: *Peasant Work Capacity and Agricultural Development.* Berlin, German Development Institute, Occasional Paper No. 55.

Bratton, M. [1977]: "Structural Transformation in Zimbabwe: Comparative Notes from the Neo-Colonisation of Kenya." *J. Modern African Studies* 15:591-611.

———— [1981]: "Development in Zimbabwe: Strategy and Tactics." *J. Modern African Studies* 19:447-475.

———— [1986]: "Farmer Organization and Food Production in Zimbabwe." *World Dev.* 14:376-384.

Braun, Joachim von, H. de Haen, and J. Blanken [1991]: *Commercialization of Agriculture under Population Pressure: Effects on Production, Consumption, and Nutrition in Rwanda.* Washington, IFPRI.

Breman, H., and A. M. Cisse [1977]: "Dynamics of Sahelian Pastures in Relation to Drought and Grazing." *Oecologia* 28(4):301-315.

Brett, E. A. [1973]: *Colonialism and Underdevelopment in East Africa: The Politics of Economic Change, 1919-1939.* London, Heinemann.

Brinckman, W. L., and P. N. De Leeuw [1979]: "The Nutritive Value of Browse and its Importance in Traditional Pastoralism." In Ajaya and Halstead, eds., 1979, pp. 101-122.

Brokensha, D. W. [1988]: "Inequality in Rural Africa: Fallers Reconsidered." *Manchester Papers on Dev.* 3(2):1-21.

Brokensha, D. W., D. M. Warren, and O. Werner, eds. [1980]: *Indigenous Knowledge Systems and Development.* Lanham, MD, Univ. Press of America.

Bromberger, N. [1979]: *Mining Unemployment in South Africa— 1946-2000.* Geneva, ILO, World Employment Program.

Bromley, D. W., D. C. Taylor, and D. E. Parker [1980]: "Water Reform and Economic Development: Institutional Aspects of Water Management in Developing Countries." *Econ. Dev. and Cultural Change* 28:365-387.

Brooks, G. E. [1975]: "Peanuts and Colonialism: Consequences of the Commercialisation of Peanuts in West Africa, 1830-70." *J. African Hist.* 16:29-54.

Brooks, R. M., M. C. Latham, and W. T. Crompton [1979]: "The Relationship of Nutrition and Health to Worker Productivity in Kenya." *East African Medical J.* 56(9):413-421.

Brown, B. [1980]: *Women, Migrant Labor and Social Change in Botswana*. Boston, Boston Univ., African Studies Center, Working Paper No. 41.

—— [1983]: "The Impact of Male Labour Migration on Women in Botswana." *African Affairs* 82:367-388.

Brown, C. K. [1972]: *Some Problems of Investment and Innovation Confronting the Ghanian Food Crop Farmer*. Legon, Univ. of Ghana, ISSER, Technical Publication No. 24.

Brown, C. P. [1970]: "The Malawi Farmers Marketing Board." *Eastern Africa Econ. Rev.* 2:37-52.

Brown, L. H. [1971]: "The Biology of Pastoral Man as a Factor in Conservation." *Biological Conservation* 3(2):93-100.

—— [1977]: "The Ecology of Man and Domestic Livestock." In Pratt and Gwynne, eds., 1977, pp. 34-40.

Bryant, C., ed. [1988]: *Poverty Policy and Food Security in Southern Africa*. Boulder, CO, Lynne Rienner.

Buccola, S., and C. Sukume [1988]: "Optimal Grain Pricing and Storage Policy in Controlled Agricultural Economies: Application to Zimbabwe." *World Dev.* 16:361-371.

Buddenhagen, I. W., and G. J. Persley, eds. [1978]: *Rice in Africa: Proceedings of a Conference held at the IITA, Ibadan, Nigeria, 7-11 March 1977*. London and New York, Academic Press.

Bukh, J. [1979]: *The Village Women in Ghana*. Uppsala, SIAS.

Bundy, C. [1979]: *The Rise and Fall of the South African Peasantry*. Berkeley, Univ. of California Press.

Bunker, S. [1987]: *Peasants against the State: The Politics of Market Control in Bugisu, Uganda, 1900-1983*. Urbana, Univ. of Illinois Press.

Bunting, A. H., ed. [1970]: *Change in Agriculture*. New York, Praeger.

Bush, R. [1988]: "Hunger in Sudan: The Case of Darfur." *African Affairs* 87(346):5-23.

Butler, F. C., ed. [1984]: *Proceedings of KSU's 1983 Farming Systems Research Symposium: Animals in the Farming Systems*. Manhattan, KS, Kansas State Univ., FSR Paper No. 6.

Buvinic, M., C. S. Adams, G. S. Edgcomb, and M. Koch-Weser [1976]: *Women and Rural Development: An Annotated Bibliography*. Washington, ODC for American Association for the Advancement of Science.

Byerlee, D. [1972]: *Research on Migration in Africa: Past, Present and Future*. East Lansing, MSU, DAE, African Rural Employment Paper No. 2.

—— [1973]: *Indirect Employment and Income Distribution Effects of Agricultural Development Strategies: A Simulation Approach Applied to Nigeria*. East Lansing, MSU, DAE, African Rural Employment Paper No. 9.

—— [1974]: "Rural-Urban Migration in Africa: Theory, Policy and Research Implications." *International Migration Rev.* 8:543-566.

—— [1980]: "Rural Labor Markets in West Africa with Emphasis on the Semi-Arid Tropics." In ICRISAT, 1980a, pp. 348-356.

Byerlee, D., M. P. Collinson, D. Winkelmann, R. Perrin, S. Biggs, E. Moscardi, J. C. Martinez, L. Harrington, and A. Benjamin [1980]: *Planning Technologies Appropriate to Farmers: Concepts and Procedures*. Mexico, CIMMYT.

Byerlee, D., and C. K. Eicher [1974]: "Rural Employment, Migration, and Economic Development: Theoretical Issues and Empirical Evidence from Africa." In Islam, ed., 1974, pp. 273-313.

Byerlee, D., C. K. Eicher, C. Liedholm, and D. S. C. Spencer [1977]: *Rural Employment in Tropical Africa: Summary of Findings.* East Lansing, MSU, DAE, ARE Working Paper No. 20.

——— [1983]: "Employment-Output Conflicts, Factor-Price Distortions, and Choice of Technique: Empirical Results from Sierra Leone." *Econ. Dev. and Cultural Change* 31:315-336.

Byerlee, D., and J. Longmire [1986]: "Wheat in the Tropics: Whether and When?" *CERES* 19(3):34-39.

Byerlee, D., and M. Morris [1987]: "The Political Economy of Wheat Consumption and Production with Special Reference to Sub-Saharan Africa." Paper presented at the Univ. of Zimbabwe Third Annual Conference on Food Security Research in Southern Africa, Harare.

Byerlee, D., and E. H. Polanco [1986]: "Farmers' Stepwise Adoption of Technological Packages: Evidence from the Mexican Altiplano." *Amer. J. Agr. Econ.* 68:519-527.

Byerlee, D., J. L. Tommy, and H. Fatoo [1976]: *Rural-Urban Migration in Sierra Leone: Determinants and Policy Implications.* East Lansing, MSU, DAE, ARE Paper No. 13.

Byerlee, D., and R. Tripp [1988]: "Strengthening Linkages in Agricultural Research Through a Farming Systems Perspective: The Role of Social Scientists." *Experimental Agriculture* 24:137-151.

Caldwell, J. C. [1969]: *African Rural-Urban Migration: The Movement to Ghana's Towns.* New York, Columbia Univ. Press.

——— [1975]: *The Sahelian Drought and its Demographic Implications.* Washington, ACE, OLC.

——— [1976]: "Toward a Restatement of Demographic Transition Theory." *Population and Dev. Rev.* 2:321-366.

Caldwell, J. C., N. O. Addo, S. K. Gaisie, A. Igun, and P. O. Olusanya, eds. [1975]: *Population Growth and Socio-Economic Change in West Africa.* New York: Columbia Univ. Press.

Caldwell, J. C., and P. Caldwell [1985]: "Cultural Forces Tending to Sustain High Fertility in Tropical Africa." *World Bank PHN TropicalNote 85-16.*

——— [1988]: "Is the Asian Family Planning Program Model Suited to Africa?" *Studies in Family Planning* 19(1):19-28.

Callaway, A. [1964]: "Nigeria's Indigenous Education: The Apprenticeship System." *ODU, Journal Univ. Ife* 1(1):62-79.

Calvet, H., D. Friot, and I. S. Gueye [1976]: *Revue d'élevage et de médicin vétérinaire des tropicaux* 29:59-66.

Campbell, B. [1974]: "Social Change and Class Formation in a French West African State." *Canadian J. African Studies* 8(2)285-306.

Campbell, D. [1979]: "Development or Decline: Resources, Land Use and Population Growth in Kajiado District." Nairobi, Univ. of Nairobi, IDS Working Paper No. 352.

Candler, W., and R. Slade [1981]: "Collection of Reliable Farm Level Data in LDCs." *J. Agr. Econ.* 30(1):65-70.

Cannell, G. H., ed. [1977]: *Proceedings of an International Symposium on Rainfed Agriculture in Semi-Arid Regions, April 17-22, 1977, Riverside, California.* Riverside, Univ. of California and Corvallis, Oregon State Univ., Consortium of Arid Lands Institute.

Cantrelle, P., ed. [1974]: *Population in African Development, Vol. 1.* Dolhain, Belgium, Ordina Editions for the International Union for the Scientific Study of Population.

Capron, J. [1973]: *Communaute villageoise Bwa, Mali—Haute-Volta.* Paris, Musée de l'Homme.

Cardoso, F. H., and E. Faletto [1979]: *Dependency and Development in Latin America*. Berkeley, Univ. of California Press.

Carruthers, I., ed. [1983]: *Aid for the Development of Irrigation*. Paris, OECD.

Carruthers, I., and A. Weir [1976]: "Rural Water Supplies and Irrigation Development." In Heyer, Maitha, and Senga, eds., 1976, pp. 288-312.

Catt, C. [1966]: "Surveying Peasant Farmers—Some Experiences." *J. Agr. Econ.* 17:99-100.

Catt, D. C. [1970]: *Progress in African Agriculture: An Economic Study in Malawi*. Aberdeen, Scotland, School of Agriculture, Miscellaneous Publication No. 11.

Cernea, M., ed. [1985]: *Putting People First: Sociological Variables in Rural Development*. New York, Oxford University Press.

CGIAR, Technical Advisory Committee [1978]: *Farming Systems Research at the IARCs*. Rome, TAC Secretariat.

Chabrolin, R. [1977]: "Rice in West Africa." In Leakey and Wills, eds., 1977, pp. 7-25.

Chakravarty, S. [1988]: "Development Experience in South Asia." *Asian Dev. Rev.* 6(1):22-49.

Chambers, R. [1969]: *Settlement Schemes in Tropical Africa: A Study of Organisations and Development*. London, Routledge and Paul.

———, ed. [1970: *The Volta Resettlement Experience*. London, Pall Mall.

——— [1974]: *Managing Rural Development: Ideas and Experience from East Africa*. Uppsala, Sweden, SIAS.

——— [1983]: *Rural Development: Putting the Last First*. London, Longman.

Chambers, R., and J. R. Moris, eds. [1973]: *Mwea: An Irrigated Rice Settlement in Kenya*. Munich, Weltforum Verlag.

Charlick, R. [1972]: "Participatory Development and Rural Modernization in Hausa Niger." *African Rev.* 2(4):499-524.

——— [1980]: "Animation Rurale: Experience with 'Participatory' Development in Four West African Nations." *Rural Dev. Participation Rev. (Cornell Univ.)* 1(2):1-6.

Charreau, C. [1972]: "Problèmes posés par l'utilisation agricole des sols tropicaux par des cultures annuelles." *L'Agronomie Tropicale* 27:905-929.

——— [1977]: "Some Controversial Technical Aspects of Farming Systems in Semi-Arid West Africa." In Cannell, ed. 1977, pp. 313-360.

Charreau, C. and R. Nicou [1971]: "L'amélioration du profil cultural dans les sols sableaux et sablo-argileux de la zone tropicale séche ouest Africaine et ses incidences agronomiques." *L'Agronomie Tropicale* 26:209-255, 565-631, 903-978, 1184-1247.

Chayanov, A. V. [1925]: *Peasant Farm Organization*. Moscow, Cooperative Publishing House. Reprinted in Thorner, Kerblay, and R. E. F. Smith, eds., 1966, pp. 29-317.

Chembezi, D., and A. Womack [1987]: "An Analysis of Supply Response among Cotton Growers in Malawi." *Agr. Systems* 23:79-94.

Chenery, H. B., M. S. Ahluwalia, C. L. G. Bell, J. H. Duloy, and R. Jolly, eds. [1974]: *Redistribution with Growth*. London, Oxford Univ. Press.

Chenery, H. B., and M. Syrquin [1975]: *Patterns of Development, 1950-70*. London, Oxford Univ. Press.

Chigaru, P. R. N. [1984]: "Future Directions of the Department of Research and Specialist Services, Ministry of Agriculture." Harare, Zimbabwe.

Child, B., K. Muir, and M. Blackie [1985]: "An Improved Maize Marketing System for African Countries: The Case of Zimbabwe." *Food Policy* 10:365-373.

Child, F. C. [1977]: *Small-Scale Rural Industry in Kenya*. Los Angeles, Univ. of California, African Studies Center, Occasional Paper No. 17.

Chileshe, J. H. [1977]: *The Challenge of Developing Intra-African Trade.* Kampala, Uganda, East African Literature Bureau.

Chopart, J.-L., and R. Nicou [1976]: "Influence du labour sur le développement radiculaire de différentes plantes cultivées au Sénégal. Conséquences sur leur alimentation hydrique." *L'Agronomie Tropical* 31:7-28.

Choudhri, M. B. [1987]: *Wheat Production Potential in Africa.* Rome, FAO.

Christensen, J. B. [1977]: "Motor Power and Women Power: Technological and Economic Change among the Fanti Fishermen of Ghana." In M. E. Smith, ed., 1977, pp. 71-95.

Christy, F. T. [1976]: "Effective Fisheries Management with Regional Diversities." *Dev. Digest* 14(3):26-37.

Chuta, E. [1978]: *The Economics of the Gara (Tie-Dye) Cloth Industry in Sierra Leone.* East Lansing, MSU, DAE, ARE Working Paper No. 25.

Chuta, E., and C. Liedholm [1979]: *Rural Non-Farm Employment: A Review of the State of the Art.* East Lansing, MSU, DAE, Rural Development Paper No. 4.

Chuta, E., and C. Liedholm [1990]: "Rural Small Scale Industry: Empirical Evidence and Policy Issues." In Eicher and Staatz, 1990, pp. 327-341.

Chuta, E., C. Liedholm, O. Roberts, and J. L. Tommy [1981]: *Employment Growth and Change in Sierra Leone Small Scale Industry: 1974-1980.* East Lansing, MSU, DAE, ARE Working Paper No. 37.

CILSS/Club du Sahel, Working Group on Marketing, Price Policy, and Storage [1977]: *Marketing, Price Policy, and Storage of Food Grains in the Sahel: A Survey.* 2 vols. Ann Arbor, Univ. of Michigan, CRED.

CILSS/Club du Sahel [1978a]: *Energy in the Development Strategy of the Sahel: Situation — Perspectives — Recommendations.* Paris, OECD.

———— [1978b]: *La mise en valeur des 'terres neuves' au Sahel. Synthèse du séminaire de Ouagadougou, 10-13 Octobre 1978.* Paris, OECD.

———— [1979]: *Cereals Policy in Sahel Countries. Papers presented at the Nouakchott Colloquium, Mauritania, 2-6 July 1979.* Paris, OECD.

CILSS/Club de Sahel, Working Group on Recurrent Costs [1980a]: *Recurrent Costs of Development Programs in the Countries of the Sahel: Analysis and Recommendations.* Paris, OECD, Club du Sahel, and Ouagadougou, CILSS.

CILSS/Club du Sahel, Working Group on Irrigation [1980b]: *The Development of Irrigation Agriculture in the Sahel: Review and Perspectives.* Paris, OECD, Club du Sahel, and Ouagadougou, CILSS.

CIMMYT, Eastern African Economics Programme [1977a]: *Demonstrations of an Interdisciplinary Approach to Planning Adaptive Agricultural Research Programmes: Part of Siaya District, Nyanga Province, Kenya.* Nairobi, Kenya, CIMMYT with Egerton College and the Kenya Ministry of Agriculture, Report No. 1.

———— [1977b]: *Demonstrations of an Interdisciplinary Approach to Planning Adaptive Agricultural Research Programmes: The Drier Areas of Morogoro and Kilosa Districts, Tanzania.* Nairobi, Kenya, CIMMYT with Tanzania Ministry of Agriculture and Univ. of Dar es Salaam, Faculty of Agriculture, Report No. 2.

———— [1978]: *Demonstrations of an Interdisciplinary Approach to Planning Adaptive Agricultural Research Programmes: Part of Serenji District, Central Province, Zambia.* Nairobi, Kenya, CIMMYT with Zambia Ministry of Agriculture and Univ. of Zambia, Rural Development Studies Bureau, Report No. 3.

CIMMYT [1981]: *CIMMYT Review, 1981.* Mexico, DF.

———— [1984]: "The Farming Systems Perspective and Farmer Participation in the Development of Appropriate Technology." In Eicher and Staatz, eds., 1984, pp. 378-388.

Cipolla, C. M., ed. [1973]: *The Fontana Economic History of Europe: The Industrial Revolution*. Glasgow, Collins.

Clammer, J. [1975]: "Economic Anthropology and the Sociology of Development: 'Liberal' Anthropology and its French Critics." In Oxaal, Barnett, and Booth, 1975, pp. 208-228.

Clark, B. A. [1975]: "The Work Done by Rural Women in Malawi." *Eastern Africa J. Rural Dev.* 8:80-91.

Clayton, E. S. [1961]: "Technical and Economic Optima in Peasant Agriculture." *J. Agr. Econ.* 14:337-377.

_____ [1963]: *Economic Planning in Peasant Agriculture: A Study of the Optimal Use of Agricultural Resources by Peasant Farmers in Kenya*. Ashford, Kent, Wye College, Department of Economics.

_____ [1964]: *Agrarian Development in Peasant Economies: Some Lessons from Kenya*. Oxford, England, Pergamon Press.

_____ [1975]: "Programming Rural Employment Opportunities in Kenya." *Internal Labour Rev.* 112:149-161.

_____ [1978]: *A Comparative Study of Settlement Schemes in Kenya*. London, Univ. of London, Wye College, Occasional paper No. 3.

_____ [1981]: "Monitoring, Management and Control of Irrigation Projects: The Example of Mwea, Kenya." *Water Supply and Management (Oxford)* 5(1):107-115.

_____ [1983]: "Agricultural Development and Farm Income Distribution in LDCs." *J. Agr. Econ.* 34(3):349-359.

Cleave, J. H. [1968]: "Food Consumption in Uganda." *Eastern Africa J. Rural Dev.* 1(1):70-87.

_____ [1974]: *African Farmers: Labour Use in the Development of Smallholder Agriculture*. New York, Praeger.

_____ [1977]: "Decision-Making on the African Farm." *Contributed Papers Read at the 16th ICAE*. Nairobi, Oxford Univ. Press, pp. 157-177.

Cliffe, L. [1976]: "Rural Political Economy of Africa." In Gutkind and Wallerstein, eds., 1976, pp. 112-130.

_____ [1978]: "Labour Migration and Peasant Differentiation: Zambian Experiences." *J. Peasant Studies* 5(3):326-346. Reprinted in Turok, ed., 1979, pp. 149-169.

_____ [1987]: "Discussion: The Debate on African Peasantries." *Dev. and Change* 18:625-635.

Clower, R. W., G. Dalton, M. Harwitz, and A. A. Walters [1966]: *Growth Without Development: An Economic Survey of Liberia*. Evanston, IL, Northwestern Univ. Press.

Club du Sahel [1977]: *Strategy and Programme for Drought Control and Development in the Sahel*. Paris, OECD.

_____ [1983]: *Chairman's Report: Special Meeting on the Role of the Club du Sahel*. Paris, OECD.

Cohen, A. [1965]: "The Social Organization of Credit in a West African Cattle Market." *Africa* 35(1):8-20.

_____ [1971]: "Cultural Strategies in the Organization of Trading Diasporas." In Meillassoux, ed., 1971, pp. 266-281.

Cohen, J. M. [1975]: "Effects of Green Revolution Strategies on Tenants and Small-Scale Landowners in the Chilalo Region of Ethiopia." *J. Developing Areas* 9:335-358.

_____ [1980a]: "Integrated Rural Development: Clearing Out the Underbrush." *Sociologia Ruralis* 20(3):196-211.

_____ [1980b]: "Land Tenure and Rural Development in Africa." In R. H. Bates and Lofchie, eds., 1980, pp. 349–400.

_____ [1987]: *Integrated Rural Development: The Ethiopian Experience and the Debate*. Uppsala, SIAS.

Cohen, J. M., and N.-I. Isaksson [1988]: "Food Production Strategy Debates in Revolutionary Ethiopia." *World Dev*. 16:323-348.

Colander, D. ed. [1984]: *Neo-Classical Political Economy: The Analysis of Rent-Seeking and DVP Activities*. Cambridge, MA, Ballinger Publishing Co.

Coleman, G. [1983]: "The Analysis of Memory Bias in Agricultural Labour Data Collection: A Case Study of Small Farmers in Nigeria." *J. Agr. Econ*. 34:79-86.

Coleman, J. [1984]: "Professorial Training and Institution Building in The Third World: Two Rockefeller Foundation Experiences." *Comparative Education Rev*. 28(2):180-202.

Collier, P. [1979]: "Migration and Unemployment: A Dynamic General Equilibrium Analysis Applied to Tanzania." *Oxford Econ. Papers* 31(2):205-236.

_____ [1983]: "Malfunctioning of African Rural Factor Markets: Theory and a Kenyan Example." *Oxford Bull. of Econ. and Stat*. 48:141-172.

Collins, J. D. [1976]: "The Clandestine Movement of Groundnuts across the Niger-Nigeria Boundary." *Canadian J. African Studies* 10(2):259-278.

Collinson, M. P. [1962–64]: *Farm Management Survey Report*. Nos. 1, 2, 3, 4. Ukiriguru Research Station, Tanganyika Ministry of Agriculture.

_____ [1968]: "The Evaluation of Innovations for Peasant Farming." *Eastern Africa J. Rural Dev*. 1(2):50-59.

_____ [1972]: *Farm Management in Peasant Agriculture: A Handbook for Rural Development Planning in Africa*. New York, Praeger.

_____ [1981]: "Micro-Level Accomplishment and Challenges for the Less Developed World." In Glenn L. Johnson and A. Maunder, eds., 1981, pp. 43-53.

_____ [1982]: *Farming Systems Research in Eastern Africa: The Experience of CIMMYT and Some National Agricultural Research Services, 1976-81*. East Lansing, MSU, DAE, International Development Papers, No. 3.

_____ [1987]: "Farming Systems Research: Procedures for Technology Development." *Experimental Agriculture* 33:365-386.

_____ [1988]: "The Development of African Farming Systems: Some Personal Views." *Agr. Administration and Extension* 29(1):7-22.

Collis, W. R. E., I. Dema, and A. Omololu [1962]: "On the Ecology of Child Health and Nutrition in Nigerian Villages: Part I. Environment, Population and Resources." *Tropical and Geographical Medicine* 14:140-163.

Colson, E. [1971]: *The Social Consequences of Resettlement: The Impact of the Kariba Resettlement upon the Gwembe Tonga*. Manchester, Manchester Univ. Press for the Univ. of Zambia.

Colvin, L. G., C. Ba, B. Barry, J. Faye, A. Hamer, M. Soumah, and F. Sow [1981]: *The Uprooted of the Western Sahel: Migrants' Quest for Cash in the Senegambia*. New York, Praeger.

Commins, S., M. Lofchie, and R. Payne, eds. [1986]: *Africa's Agrarian Crisis: The Roots of Famine*. Boulder, CO, Lynne Rienner.

Commission of the European Communities [CEC, 1984]: "Food Strategies: Review and Prospects." *Policy Paper, SEC 84(1692)*, Brussels, mimeo.

Copans, J., ed. [1975]: *Sécheresses et famines du Sahel: I. Écologie, dénutrition, assistance. II. Paysans et nomads*. Paris, Maspero.

Copans, J., Ph. Couty, J. Roch, and G. Rocheteau [1972]: *Maintenance sociale et changement économique au Sénégal. I. Doctrine économique et pratique du travail chez les Mourides.* Paris, Travaux et documents de l'ORSTOM No. 15.

Copans, J., J. Monod, K. Gough, and J. Pouillon [1970-71]: "Les responsabilities sociales et politiques de l'anthropologie." *Les Temps Modernes* 27(293-294):1121-1201.

Coquery-Vidrovitch, C. [1978]: "Research on an African Mode of Production." In Seddon, ed., 1978, pp. 261-288.

Cornia, G., R. Jolly, and F. Steward [1988]: *Adjustment with a Human Face:* Vol. II. *Ten Country Case Studies.* Oxford, England, Clarendon Press.

Coulibaly, S., J. Gregory, and V. Piche [1980]: *Les migrations Voltaiques: importance et ambivalence de la migration Voltaique, Tome 1.* Ottawa, IDRC.

Coulson, A. [1977a]: "Tanzania's Fertiliser Factory." *J. Modern African Studies* 15:119-125.

———— [1977b]: "Agricultural Policies in Mainland Tanzania." *Rev. African Polit. Economy* 10:74-100.

————, ed. [1979]: *African Socialism in Practice: The Tanzanian Experience.* Nottingham, UK, Spokesman.

———— [1982]: *Tanzania: A Political Economy.* London, Oxford Univ. Press.

Coursey, D. G. [1967]: *Yams: An Account of the Nature, Origins, Cultivation and Utilisation of the Useful Members of the Disoscoreaceae.* London, Longmans.

Coursey, D. G., and R. H. Booth [1977]: "Root and Tuber Crops." In Leakey and Wills, eds., 1977, pp. 75-96.

Couty, P. [1964]: *Le commerce du poisson dans le Nord-Cameroun.* Paris, ORSTOM, Memoires No. 5.

———— [1977]: "Recent Studies on Traditional Agricultural Marketing in the Sudan and Sahel Zones of Africa." In Cannell, ed., 1977, pp. 628-653.

———— [1979]: *Des éléments aux systèmes. Reflexions sur les procédés de généralisation dans les enquêtes de niveau de vie en Afrique.* Paris, INSEE, AMIRA note de travail no. 28.

Couty, P., and P. Duran [1968]: *Le commerce du poisson au Tchad.* Paris, ORSTOM, Memoires No. 23.

Couty, P., and A. Hallaire [1980]: *De la carte aux systèmes—les études agraires de l'ORSTOM au Sud du Sahara (1960-1980).* Paris, INSEE, AMIRA note de travail No. 29.

Coward, E. W. [1977]: "Irrigation Management Alternatives: Themes from Indigenous Irrigation Systems." *Agricultural Administration* 4(3):223-237.

————, ed. [1980]: *Irrigation and Agricultural Development in Asia: Perspectives from the Social Sciences.* Ithaca, NY, Cornell Univ. Press.

Cowen, M. [1981]: "Commodity Production in Kenya's Central Province." In Heyer, Roberts, and Williams, eds., 1981, pp. 121-142.

Crawford, E. W. [1982]: *A Simulation Study of Constraints on Traditional Farming Systems in Northern Nigeria.* East Lansing, MSU, DAE International Development Papers, No. 2.

Crotty, R. [1980]: *Cattle, Economics and Development.* Farnborough, England, Commonwealth Agricultural Bureau.

Cruise O'Brien, D. B. [1975]: *Saints and Politicians: Essays in the Organisation of a Senegalese Peasant Society.* London, Cambridge Univ. Press.

Cruise O'Brien, R., ed. [1979]: *The Political Economy of Underdevelopment: Dependence in Senegal.* Beverly Hills, CA, Sage.

Crutchfield, J. A., and R. Lawson [1974]: *West African Marine Fisheries; Alternatives for Management.* Washington, RFF.

D'Agostino, V. C. [1988]: "Coarse Grain Production and Transactions in Mali: Farm Household Strategies and Government Policy." Unpublished M.S. thesis, MSU, DAE, East Lansing.

Dahl, G. [1979]: *Suffering Grass: Subsistence and Society of Waso Borana*. Stockholm, Univ. of Stockholm.

Dahl, G., and A. Hjort [1976]: *Having Herds: Pastoral Growth and Household Economy*. Stockholm, Univ. of Stockholm, Studies in Social Anthropology No. 2.

————— [1979]: *Pastoral Change and the Role of the Drought*. Stockholm, Sweden, SAREC, Report No. 2.

Dalby, D., and R. J. Harrison Church, eds. [1973]: *Drought in Africa*. London, Univ. of London, School of Oriental and African Studies.

Dalby, D., R. J. Harrison Church, and F. Bezzaz, eds. [1977]: *Drought in Africa*. Vol. 2. London, International African Institute.

Dalrymple, Dana G. [1986a]: *Development and Spread of High-Yielding Wheat Varieties in Developing Countries*. AID, Washington.

————— [1986b]: *Development and Spread of High-Yielding Rice Varieties in Developing Countries*. AID, Washington.

Dalton, G. E. [1962]: "Traditional Production in Primitive African Economies." *Quart. J. Econ.* 76:360-378.

————— , ed. [1971]: *Studies in Economic Anthropology*. Washington, Amer. Anthropological Association.

————— [1973]: "Adoption of Farm Management Theory to the Problems of the Small-Scale Farmer in West Africa." In Ofori, ed., 1973, pp. 114-129.

————— [1976]: "A Review of *An Economic History of West Africa* by A. G. Hopkins." *African Econ. Hist.* 1:51-101.

————— [1978a]: "Is Economic Anthropology of Interest to Economists?" *Amer. Econ. Rev., Papers and Proceedings* 68:23-27.

————— , ed. [1978b]: *Research on Economic Anthropology*, Vol. 1. Greenwich, CT, JAI Press.

Dams, T., and K. Hunt, eds. [1977]: *Decision-Making and Agriculture*. Lincoln, Univ. of Nebraska Press.

Dantwala, M. L. [1985]: "Growth and Equity in Agriculture." *Indian J. Agr. Econ.* 42(2):149-159.

Darling, F. F., and M. A. Farvar [1972]: "Ecological Consequences of Sedentarization of Nomads." In Farvar and Milton, eds., 1972, pp. 671-682.

Datoo, B. A. [1973]: *Population Density and Agricultural Systems in the Uluguru Mountains, Morogoro District*. Dar es Salaam, Univ. of Dar es Salaam, BRALUP, Research Paper No. 26.

Dauber, R., and M. Cain, eds. [1981]: *Women and Technological Change in Developing Countries*. Boulder, CO, Westview.

David, P. [1980]: *Les Navetanes: histoire des migrants saisonniers de l'arachide en Sénégambie des origines à nos jours*. Dakar-Abidjan, Les Nouvelles Éditions Africaines.

David, P. A. [1975]: "The Mechanization of Reaping in the *Ante Bellum* Midwest." In *Technical Choice, Innovation and Economic Growth: Essays on American and British Experience in the Nineteenth Century*. Cambridge, England, Cambridge Univ. Press, pp. 195-232.

Davies, J. C. [1976]: "Trials of Spraying and Cultural Practices on Cotton in Uganda. II. Use of Extended Protection." *Experimental Agriculture* 12(2):163-176.

Davis, R. K. [1971]: "Some Issues in the Evolution, Organization and Operation of Group Ranches in Kenya." *Eastern Africa J. Rural Dev.* 4(1):22-33.

Dean, E. R. [1966]: *The Supply Response of African Farmers: Theory and Measurement in Malawi.* Amsterdam, North Holland.

_____ [1972]: *Plan Implementation in Nigeria: 1962-66.* Ibadan, Nigeria, Oxford Univ. Press.

De Carvalho, E. C. [1974]: " 'Traditional' and 'Modern' Patterns of Cattle Raising in Southwestern Angola: A Critical Evaluation of Change from Pastoralism to Ranching." *J. Developing Areas* 8(2):199-226.

Deelstra, H. A., D. White, and D. S. Wiggins [1974]: "Nutritive Value of Fish of Lake Tanganyika. I. Amino Acid Composition." *African J. Tropical Hydrobiology and Fisheries* 3(2):161-166.

De Janvry, A. [1975]: "The Political Economy of Rural Development in Latin America: An Interpretation." *Amer. J. Agr. Econ.* 57:490-499.

_____ [1981]: *The Agrarian Question and Reformism in Latin America.* Baltimore, Johns Hopkins Univ. Press.

_____ [1986]: "Food Security and the Integration of Agriculture: Options and Dilemmas." *CERES* 19(1)33-37.

_____ [1987]: "Dilemmas and Options in the Formulation of Agricultural Policies in Africa." In Gittinger *et al.*, eds., 1987, pp. 485-496.

De Janvry, A., and E. Sadoulet [1989]: "Investment Strategies to Combat Rural Poverty: A Proposal for Latin America." *World Dev.*, 17.

Dejene, T., and S. E. Smith [1973]: *Experiences in Rural Development: A Selected, Annotated Bibliography of Planning, Implementation, and Evaluating Rural Development in Africa.* Washington, ACE, OLC.

De Jonge, K., J. van der Klei, H. Meilink, and R. Storm [1978]: *Les migrations en basse Casamance (Sénégal).* Leiden, Netherlands, Afrika-Studiecentrum.

De Leeuw, P. N. [1965]: "The Role of Savanna in Nomadic Pastoralism." *Netherlands J. Agr. Science* 13(2):178-189.

Delgado, C. [1979a]: *Livestock versus Foodgrain Production in Southeast Upper Volta: A Resource Allocation Analysis.* Ann Arbor, Univ. of Michigan, CRED.

_____ [1979b]: *The Southern Fulani Farming System in Upper Volta: A Model for the Integration of Crop and Livestock Production in the West African Savannah.* East Lansing, MSU, DAE, ARE Paper No. 20.

_____ [1980]: "Livestock and Meat Production, Marketing, and Exports in Mali: A Review of the Evidence." In Ariza-Nino and Steedman, eds., 1980, pp. 211-439.

_____ [1986]: "A Variance Components Approach to Food Grain Market Integration in Northern Nigeria." *Amer. J. Agr. Econ.* 68:970-79.

Delgado, C., and J. W. Mellor [1984]: "A Structural View of Policy Issues in African Agricultural Development." *Amer. J. Agr. Econ.* 66:665-70.

Delgado, C., and C. P. J. Miller [1985]: "Changing Food Patterns in West Africa: Implications for Policy Research." *Food Policy.* 10(1):55-62.

Delgado, C., and T. Reardon [1987]: "Policy Issues Raised by Changing Food Patterns in the Sahel." A paper presented at an IFPRI/ISRA Conference, Dakar.

Delpechi, B., and J.-M. Gastellu [1974]: *Maintenance sociale et changement économique au Sénégal. II. Pratique du travail et rééquilibres sociaux en milieu Serer.* Paris, ORSTOM, Travaux et Documents No. 34.

De Miranda, E., and R. Billaz [1980]: "Methodes de recherche en milieu Sahelien: les approches écologiques et agronomiques d'une pluridisciplinaire: l'exemple de Maradi au Niger." *L'Agronomie Tropicale* 35:357-373.

Dequekes, J. [1983]: "Cotton: The Organization of Research in West Africa." *Courier* 82:76-79.

Derman, W., and L. Derman [1973]: *Serfs, Peasants, and Socialists—A Former Serf Village in the Republic of Guinea.* Berkeley, Univ. of California Press.

Des Bouvrie, C., and J. R. Rydzewski [1977]: "Irrigation." In Leakey and Wills, eds., 1977, pp. 161-193.

De Schlippe, P. [1956]: *Shifting Cultivation in Africa: The Zande System of Agriculture.* London, Routledge and Paul.

Desfosses, H., and J. Levesque, eds. [1975]: *Socialism in the Third World.* New York: Praeger.

Development Associates [1984]: *Evaluation of the Sahel Regional AID Coordination and Planning Project. Report Prepared for AID.* Washington.

Devereux, S. [1988]: "Entitlements, Availability and Famine: A Revisionist View of Wollo, 1972-74." *Food Policy* 13(3):270-282.

Devres, J. [1984]: *Assessment of the Agricultural Research Resources in the Sahel,* 2 vols. Washington, Devres.

――――― [1985]: *Agricultural Research Resource Assessment in the SADCC Countries: Regional Analysis and Strategy,* Vol. 1. Washington, Devres.

De Vries, J. [1976]: "On the Effectiveness of Agricultural Extension: A Case Study of Maize Growing Practices in Iringa, Tanzania." *Eastern Africa J. Rural Dev.* 9(1-2):37-56.

――――― [1978]: "Agricultural Extension and Development—Ujamaa Villages and the Problems of Institutional Change." *Community Dev. J.* 13:11-19.

De Wet, J. M. J. [1977]: "Domestification of African Cereals." *African Econ. Hist.* 3:15-32.

De Wilde, J. C. [1980]: "Price Incentives and African Agricultural Development." In R. H. Bates and Lofchie, eds., 1980, pp. 46-66.

De Wilde, J. C., P. F. M. McLoughlin, A. Guinard, T. Scudder, and R. Maubouche [1967]: *Experiences with Agricultural Development in Tropical Africa.* Vol. 1. *The Synthesis.* Vol. 2. *The Case Studies.* Baltimore, Johns Hopkins Univ. Press.

Dhawan, B. D. [1985]: "Questionable Conceptions and Simplistic Views about Irrigated Agriculture of India." *Indian J. Agr. Econ.* 40(1):1-13.

Diallo, M. [1980]: "Comparative Analysis of Capital Intensive and Labor Intensive Rice Irrigation Perimeters in the Senegal River Valley." Unpublished M.S. thesis, MSU, East Lansing.

Diarra, F. A. [1971]: *Femmes Africaines en devenir: les femmes Zarma du Niger.* Paris, Éditions Anthropos.

Diarra, M. S. [1975]: "Les problèmes de contact entre les pasteurs peul et les agriculteurs dans le Niger central." In Monod, ed., 1975, pp. 284-297.

Diehl, L., and F. E. Winch [1979[]]: *Yam Based Farming Systems in the Southern Guinea Savannah of Nigeria.* Ibadan, Nigeria, IITA.

Dike, K. O. [1956]: *Trade and Politics in the Niger Delta, 1830-1885.* Oxford, England, Oxford Univ. Press.

Dillon, J. L., and J. B. Hardaker [1980]: *Farm Management Research for Small Farmer Development.* Rome, FAO.

Dione, Josue [1989]: "Food Security Policy Reform in Mali and The Sahel." Paper presented at the International Economic Association, IXth World Congress, Athens, Greece, August 28-September 1.

Dione, J., and J. Staatz [1988]: "Market Liberalization and Food Security in Mali." In Rukuni and Bernsten, eds., 1988, pp. 143-170.

Dodge, D. [1977]: *Agricultural Policy and Performance in Zambia: History, Prospects, and Proposals for Change.* Berkeley, Univ. of California.

Doggett, H. [1970]: *Sorghum.* London, Longmans.

Doherty, D. A. [1979]: *A Preliminary Report on Group Ranching in Narok District.* Nairobi, Kenya, Univ. of Nairobi, IDS Working Paper No. 350.

Donald, G. [1976]: *Credit for Small Farmers in Developing Countries.* Boulder, CO, Westview Press.

Donaldson, G. F., and J. D. von Pischke [1973]: "A Survey of Small Farmer Credit in Kenya." *AID Spring Review of Small Farmer Credit.* Vol. VII. *Kenya.* Washington, Agency for International Development.

Doppler, W. [1980]: *The Economics of Pasture Improvement and Beef Production in Semi-Humid West Africa.* Eschborn, German Agency for Technical Cooperation.

Doran, M. H., A. R. C. Low, and R. L. Kemp [1979]: "Cattle as a Store of Wealth in Swaziland: Implications for Livestock Development and Overgrazing in Eastern and Southern Africa." *Amer. J. Agr. Econ.* 61:41-47.

Dorjahn, V. R., and B. R. Isaac, eds. [1979]: *Essays on the Economic Anthropology of Liberia and Sierra Leone.* Philadelphia, Institute for Liberian Studies, Monograph Series No. 6.

Dorner, P., ed. [1977]: *Cooperative and Commune: Group Farming in the Economic Development of Agriculture.* Madison, Univ. of Wisconsin Press.

Dorward, D. C. [1975]: "An Unknown Nigerian Export: TIV Benniseed Production, 1900-1960." *J. African Hist.* 16:431-459.

Dos Santos, T. [1970]: "The Structure of Dependence." *Amer. Econ. Rev.* 40:231-236.

Doyle, C. J. [1974]: "Productivity, Technical Change, and the Peasant Producer: A Profile of the African Cultivator." *Food Research Institute Studies* 13:61-76.

Drachoussoff, V. [1965]: "Agricultural Change in the Belgian Congo: 1945-1960." *Food Research Institute Studies* 5:137-201.

Drèze, J. [1988]: *Famine Prevention in India.* London, London School of Economics, DEP No. 3.

Drèze, J. and A. Sen [1989]: *Hunger and Public Action.* Oxford, Clarendon Press.

Due, J. F. [1979]: "The Problems of Rail Transport in Tropical Africa." *J. Developing Areas* 13(4):375-393.

Due, J. M. [1980]: *Costs, Returns, and Repayment Experiences of Ujamaa Villages in Tanzania, 1973-1976.* Lanham, MD, Univ. Press of America.

_____ [1981]: "Allocation of Credit to Ujamaa Villages in Tanzania and Small Farms in Zambia." *African Studies Rev.* 22(3):33-48.

_____ [1986]: "Agricultural Policy in Tropical Africa: Is a Turnaround Possible?" *Agr. Econ.* 1:19-34.

_____ [1988]: "Personal Correspondence," April 20.

Duignan, P., and L. H. Gann, eds. [1975a]: *Colonialism in Africa, 1870-1960. Vol. 4. The Economics of Colonialism.* Cambridge, England, Cambridge Univ. Press.

_____ [1975b]: "The Pre-Colonial Economies of Sub-Saharan Africa." In Duignan and Gann, eds., 1975a, pp. 33-67.

_____ [1975c]: "Economic Achievements of the Colonizers: An Assessment." In Duignan and Gann, eds., 1975a, pp. 673-696.

Dumett, R. (1971): "The Rubber Trade of the Gold Coast and Asante in the Nineteenth Century: African Innovation and Market Responsiveness." *J. African Hist.* 12(1):79-101.

Dumont, R. [1966]: *False Start in Africa.* New York, Praeger.

_____ [1969]: *Tanzanian Agriculture after the Arusha Declaration.* Dar es Salaam, Government Printer.

_____ [1977]: "Preface." *African Environment* 21(3):3-5.

Dumont, R. and M.-F. Mottin [1980]: *L'Afrique Étranglée: Zambie, Tanzanie, Séngal, Côte d'Ivoire, Guinée-Bissau, Cap Vert.* Paris, Éditions du Seuil.

Dunbar, G. S. [1970]: "African Ranches, Ltd., 1914-1931: An Ill-Fated Stockraising Enterprise in Northern Nigeria." *Annals, Assoc. Amer. Geographers* 60(1):102-123.

Dupire, M. [1960]: "Planteurs autochtones et étrangers en basse-Côte d'Ivoire Orientale." *Études Eburnéennes* 8:7-237. Abidjan, Côte d'Ivoire.

_____ [1962]: *Peuls nomades: étude descriptive des WoDaaBe du Sahel Nigerien.* Paris, Institut d'Ethnologie, Musée de l'Homme, Trav. and Mem., No. 64.

Dupre, G. and P. P. Rey [1978]: "Reflections on the Relevance of a Theory of the History of Exchange." In Seddon, ed., 1978, pp. 171-208.

Dupriez, H. [1979]: *Integrated Rural Development Projects Carried Out in Black Africa with EDF Aid: Evaluation and Outlook for the Future.* Luxembourg, Office for Official Publications of the European Communities.

Dussauze-Ingrand, E. [1974]: "L'émigration sarakollaise du guidimaka vers la France." In Amin, ed., 1974a, pp. 239-257.

Dutta-Roy, D. K. [1969]: *The Eastern Region Household Budget Survey.* Legon, Univ. of Ghana, ISSER.

Dutta-Roy, D. K. and S. J. Mabey [1968]: *Household Budget Survey in Ghana.* Legon, Univ. of Ghana, ISSER.

Dyson-Hudson, N. [1972]: "The Study of Nomads." *J. Asian and African Studies* 7(1,2):30-47.

_____ [1985]: "Pastoral Production Systems and Livestock Development Projects: An East African Perspective." In Cernea, ed., 1985, pp. 155-186.

Easterlin, R. A., ed. [1980]: *Population and Economic Change in Developing Counties.* Univ. of Chicago Press.

Eastman, P. [1980]: *An End to Pounding: A New Mechanical Flour Milling System in Use in Africa.* Ottawa, Canada, IDRC

Eberhart, S. A., L. H. Penny, and M. N. Harrison [1973]: "Genotype by Environment Interactions in Maize in Eastern Africa." *East African Agr. and Forestry J.* 39(1):61-71.

Eckert, J. and R. Wykstra [1979]: *The Future of Basotho Migration to the Republic of South Africa.* Maseru, Lesotho, Ministry of Agriculture; and Colorado State Univ., Department of Economics, LASA Research Report No. 4.

Eddy, E. [1979]: *Labor and Land Use on Mixed Farms in the Pastoral Zone of Niger.* Ann Arbor, Univ. of Michigan, CRED.

EEC [1988]: *Food Security Policy: Examination of Recent Experience in Sub-Saharan Africa.* Brussels, Commission Staff Paper.

EFSAID [   ]:

Egbert, A. C. [1978]: *Agricultural Sector Planning Models: A Selected Summary and Critique.* Washington, World Bank, Staff Working Paper No. 297.

Egger, K. and B. Martens [1987]: "Theory and Methods of Ecofarming and their Realization in Rwanda, East Africa." In Glaeser, ed., 1987, pp. 150-175.

Eicher, C. K. [1967]: "The Dynamics of Long-Term Agricultural Development in Nigeria." *J. Farm Econ.* 49:1158-1170.

_____ [1969]: "Reflections on Capital Intensive Farm Settlements in Southern Nigeria." In Anschel, Brannon, and E. D. Smith, eds., 1969, pp. 327-346.

_____ [1970]: *Research on Agricultural Development in Five English-Speaking Countries in West Africa.* New York, ADC.

_____ [1982a]: "Facing Up to Africa's Food Crisis." *Foreign Affairs* 61(1):151-174.

_____ [1982b]: *Reflections on the Design and Implementation of the Senegal Agricultural Research Project*. East Lansing, MSU, DAE.

_____ [1983]: "West Africa's Agrarian Crisis." *West African J. Agr. Econ.* 3(1):13-38.

_____ [1984]: *International Technology Transfer and the African Farmer: Theory and Practice*. Harare, University of Zimbabwe, Department of Land Management, Working Paper 3/84.

_____ [1985]: "Famine Prevention in Africa: The Long View." In *Food for the Future: The Philadelphia Society for Promoting Agriculture. Bicentennial Forum Proceedings: 1785-1985.* Philadelphia, PA, pp. 82-101.

_____ [1986a]: "Transforming African Agriculture." *The Hunger Project Papers*, No. 4, San Francisco, CA, pp. 1-32.

_____ [1986b]: "Strategic Issues in Combating Hunger and Poverty in Africa." In R. J. Berg and Whitaker, ed., 1986, pp. 242-275.

_____ [1986c]: "Western Science and African Hunger." Foreign Francqui Lecture, Catholic Univ. of Leuven, Belgium.

_____ [1987]: "Food Security Research Priorities in Sub-Saharan Africa." In *Food Grain Production in Semi-Arid Africa: Proceedings of an International Drought Symposium held at the Kenyatta International Center, Nairobi, Kenya, 19-23 May, 1986.* Ouagadougou, Burkina Faso, OAU/STRC-SAFGRAD, pp. 3-24.

_____ [1988a]: "An Economic Perspective on the Sasakawa-Global 2000 Initiative to Increase Food Production in Sub-Saharan Africa." Paper presented at the workshop, *Reviewing the African Agricultural Projects*. Nairobi, March 18.

_____ [1988b]: "Food Security Battles in Sub-Saharan Africa." Plenary address presented at The VII World Congress for Rural Sociology, June 26-July 2, Bologna, Italy.

_____ [1988c]: "Sustainable Institutions for African Agricultural Development." Paper prepared for an ISNAR/DSE/CTA Seminar at Feldafing, Federal Republic of Germany, 21-28 September 1988.

_____ [1988d]: "Ending African Hunger: Six Challenges for Scientists, Policymakers, and Politicians." In Asefa, ed., 1988, pp. 123-144.

_____ [1989]: *Sustainable Institutions for African Agricultural Development*. Working Paper No. 19. The Hague, ISNAR.

_____ [1990a]: "Building African Scientific Capacity for Agricultural Development." *Agr. Econ.* 4:(2):117-143.

_____ [1990b]: "Africa's Food Battles." In Eicher and Staatz, eds., 1990, pp. 503-530.

Eicher, C. K., and D. C. Baker [1982]: *Research on Agricultural Development in Sub-Saharan Africa: A Critical Survey*. East Lansing, MSU, DAE, International Development Papers, No. 1.

_____ [1985]: *Étude critique de la recherche sur le développement agricole en Afrique subsaharienne*. East Lansing, DAE, MSU International Development Papers No. 1.

Eicher, C. K., and Glenn L. Johnson [1970]: "Policy for Nigerian Agricultural Development in the 1970's." In Eicher and Liedholm, eds., 1970, pp. 376-392.

Eicher, C. K., and C. Liedholm, eds. [1970]: *Growth and Development of the Nigerian Economy*. East Lansing, MSU Press.

Eicher, C. K., and F. Mangwiro [1987]: "A Critical Assessment of the FAO Report on SADCC Agriculture." In Rukuni and Eicher, eds., 1987, pp. 47-61.

Eicher, C. K., and W. Miller [1963]: *Observations on Smallholder Oil Palm Production in Nigeria*. Enugu, Univ. of Nigeria, Economic Development Institute.

Eicher, C. K., M. Sargent, E. K. Tapsoba, and D. C. Wilcock [1976]: *An Analysis of the Eastern ORD Rural Development Project in Upper Volta.* East Lansing, MSU, DAE, African Rural Employment Working Paper No. 9.

Eicher, C. K., and J. M. Staatz, eds. [1984]: *Agricultural Development in the Third World.* Baltimore, Johns Hopkins Univ. Press.

———— [1990]: *Agricultural Development in the Third World.* Second Edition. Baltimore, Johns Hopkins University Press.

———— [1986]: "Food Security Policy in Sub-Saharan Africa." In Maunder and Renborg, eds., 1986, pp. 215-229.

Eicher, C. K., and L. W. Witt, eds. [1964]: *Agriculture in Economic Development.* New York, McGraw-Hill.

Eicher, C. K., T. Zalla, J. Kocher, and F. Winch [1970]: *Employment Generation in African Agriculture.* East Lansing, MSU, International Agriculture Research Report No. 9.

Eicher, S. F. [1981]: *Rural Development in Botswana: A Select Bibliography, 1966-80.* Washington, African Bibliographic Center.

Eilam, Y. [1973]: *The Social and Sexual Roles of Hima Women: A Study of Nomadic Cattle Breeders in Nyabushozi Country, Ankola, Uganda.* Manchester, Manchester Univ. Press.

Ejiga, N. O. O. [1977]: "Economic Analysis of Storage, Distribution and Consumption of Cowpeas in Northern Nigeria." Unpublished Ph.D. dissertation, Cornell Univ., Ithaca, NY.

Ekeh, P. P. [1975]: "Colonialism and the Two Publics in Africa: A Theoretical Statement." *Comparative Studies in Society and History* 17(1):91-112.

Elkan, W. [1959]: "Migrant Labour in Africa: An Economist's Approach." *Amer. Econ. Rev., Papers and Proceedings* 49(2):188-202.

———— [1980]: "Labor Migration from Botswana, Lesotho, and Swaziland." *Econ. Dev. and Cultural Change* 28:583-596.

Elliott, C. M. [1970]: "Effects of Ill Health on Agricultural Productivity in Zambia." In Bunting, ed., 1970, pp. 647-655.

———— [1980]: *Equity and Growth — Unresolved Conflict in Zambian Rural Development Policy.* Geneva, ILO, World Employment Research Program, Working Paper No. 30.

Elliott, C. M., J. E. Bessell, R. A. J. Roberts, and N. Vanzetti [1970]: "Some Determinants of Agricultural Labour Productivity in Zambia." UNZALPI Report No. 3, mimeo.

Elliott, H. J. C. [1974]: *Animation Rurale and Encadrement Technique in the Ivory Coast.* Ann Arbor, Univ. of Michigan, CRED.

Ellis, F. [1980]: "Agricultural Pricing Policy in Tanzania, 1970-1979: Implications for Agricultural Output, Rural Incomes and Crop Marketing Costs." Univ. of Dar es Salaam, Economic Research Bureau.

———— [1982]: "Agricultural Price Policy in Tanzania." *World Dev.* 10:263-284.

Ellis, G. [1972]: "Man or Machine: Beast or Burden — A Case Study of the Economics of Agricultural Mechanization in Ada District, Ethiopia." Unpublished Ph.D. dissertation, Univ. of Tennessee, Knoxville.

Ellman, A. [1977]: "Group Farming Experiences in Tanzania." In Dorner, ed., 1977, pp. 239-275.

Elz, D., ed. [1987]: *Agricultural Marketing Strategy and Pricing Policy.* Washington, World Bank.

Ergas, A. [1979]: "Livestock Production and Marketing in the Entente States of West Africa: Annotated Bibliography." In K. H. Shapiro, ed., 1979.

Ernst, K. [1977]: *Tradition and Progress in the African Village: The Non-Capitalist Transformation of Rural Communities in Mali.* London, C. Hurst.

Essang, S. [1972]: "Impact of the Marketing Board on the Distribution of Cocoa Earnings in Western Nigeria." *Nigerian Geographical J.* 15:103-115.

————— [1973]: "The 'Land Surplus' Notion and Nigerian Agricultural Development Policy." *West African J. Agr. Econ.* 2(1):58-78.

————— [1977]: "Impact of Oil Production on Nigerian Agricultural Policy." *Indian J. Agr. Econ.* 32(2):24-32.

Essang, S. M., and A. F. Mabawonku [1974]: *Determinants and Impact of Rural-Urban Migration: A Case Study of Selected Communities in Western Nigeria.* East Lansing, MSU, DAE, African Rural Employment Paper No. 10.

Essang, S. M., and O. Ogunfowora [1975]: *Plantation Agriculture and Labour Use in Southern Nigeria.* Ibadan, Nigeria, Univ. of Ibadan, Rural Development Paper No. 15.

Etasse, C. [1977]: "Sorghum and Pearl Millet." In Leakey and Wills, eds., 1977, pp. 27-39.

Etherington, D. M. [1971]: "Economies of Scale and Technical Efficiency: A Case Study of Tea Production." *Eastern Africa J. Rural Dev.* 4(1):72-87.

————— [1972]: "An International Tea Trade Policy for East Africa: An Exercise in Oligopolistic Reasoning." *Food Research Institute Studies* 11:89-108.

————— [1973]: *An Econometric Analysis of Smallholder Tea Production in Kenya.* Nairobi, East African Literature Bureau.

Etuk, E. [1979]: "Microeconomic Effects of Technological Change on Smallholder Agriculture in Northern Nigeria: A Linear Programming Analysis." Unpublished Ph.D. dissertation, MSU, East Lansing.

Evangelou, P. [1984]: *Livestock Development in Kenya's Maasailand: Pastoralists' Transition to a Market Economy.* Boulder, CO, Westview Press.

Ewing, A. F. [1964]: "Industrialization and the U.N. Economic Commission for Africa." *J. Modern African Studies* 2:351-363.

————— [1968]: *Industry in Africa.* London, Oxford Univ. Press.

Ewusi, K. [1977]: *Economic Inequality in Ghana.* Legon, Univ. of Ghana, ISSER.

Fall, M. [1980]: "Socioeconomic Aspects Involved in Introducing New Technology Into the Senegalese Rural Milieu." In ICRISAT, 1980a, pp. 45-51.

Fallers, Lloyd [1972]: *Inequality: Social Stratification Reconsidered.* Univ. of Chicago Press.

Falusi, A. O. [1974-75]: "Application of Multi-variate Probit to Fertilizer Use Decisions: Sample Survey of Farmers in Three States in Nigeria." *J. Rural and Econ. Dev.* (Ibadan) 9(1):49-66.

————— [1976]: "Economics of Fertilizer Use in Nigeria with Particular Reference to Food Crops." Report on the FAO/NORAD/FDA *National Seminar on Fertilizer Use Development in Nigeria.* AGL/MISC/76/3. Rome, FAO, pp. 100-107.

Falusi, A. O., and L. B. Williams [1981]: *Nigeria Fertilizer Sector: Present Situation and Future Prospects.* Muscle Shoals, AL, IFDC Technical Bulletin T-18.

FAO [1966]: *Agricultural Development in Nigeria, 1965-80.* Rome.

————— [1973]: *Shifting Cultivation and Soil Conservation in Africa.* Rome, Soils Bulletin No. 24.

————— [1974]: *A Selected Bibliography on Food Habits (Socio-Economic Aspects of Food and Nutrition)—Part I, Tropical Africa.* Rome, Occasional Paper, No. 10.

————— [1975a]: *Production Yearbook: Africa South of the Sahara (1974).* Rome.

————— [1975b]: "Assistance in Agricultural Mechanization in Tanzania." FAO Mission Report, September-October 1974, Rome.

————— [1976a]: *Bibliography on Land Settlement.* Rome.

_____ [1976b]: *Perspective Study on Agricultural Development in the Sahelian Countries, 1975-1990.* Vol. I. *Main Report.* Vol. II. *Statistical Annex.* Vol. III. *Summary and Conclusions.* Rome.

_____ [1976c]: "Prospects of Freshwater Fisheries Development for Africa." In *Report of the Ninth FAO Regional Conference for Africa, Freetown, Sierra Leone, 2-12 November 1976.* Rome, pp. 84-94.

_____ [1976d]: *Scope for Expanding Fruit and Vegetable Exports by Air from African Countries.* Rome.

_____ [1977]: *The Pastoral Systems in the Sahel: Basic Socio-Demographic Data Connected with the Conservation and Development of Arid and Semi-Arid Rangelands.* Rome.

_____ [1978]: *Regional Food Plan for Africa. Report of the Tenth FAO Regional Conference for Africa, 18-29 September, 1978, Tanzania.* Rome.

_____ [1980]: *Assessment and Collection of Data on Post Harvest Food Grain Loss.* Rome, Economic and Social Development Document No. 13.

_____ [1981a]: *Report of FAO Panel of Experts on Development Aspects of the Program for the Control of Animal Trypanosomiasis and Related Development.* Rome.

_____ [1981b]: *A Grain Reserve System for the Sahel.* Rome.

_____ [1981c]: *Bibliography of Food Consumption Surveys.* Rome, Food and Nutrition Paper No. 18.

_____ [1986a]: *Irrigation in Africa South of the Sahara.* Rome, Investment Center, Paper No. 5.

_____ [1986b]: *African Agriculture: The Next 25 years.* 5 Vols. Rome.

_____ [1987a]: *The Fifth World Survey.* Rome.

_____ [1987b]: *Rice Import Statistics, 1982-86.* Rome.

_____ [1987c]: *Improving Food Crop Production on Small Farms in Africa.* Proceedings of an FAO/SIDA Seminar held in Harare, Zimbabwe, 2-17 March. Rome.

_____ [1988]: *Food Aid in Figures, 1987.* Rome, Report No. 5, Rome.

FAO, Department of Fisheries [1973]: "Brief Review of the Current Status of the Inland Fisheries of Africa." *African J. Tropical Hydrobiology and Fisheries,* Special Issue 1:3-19.

FAO and WHO [1976]: *Food and Nutrition Stategies in National Development.* Geneva, WHO Technical Report Series No. 584.

Farnsworth, H. C. [1961]: "Defects, Uses and Abuses of National Food Supply and Consumption Data." *Food Research Institute Studies* 2:179-201.

Farrington, J. [1975a]: "Factors Influencing the Length of the Working Day in Malawi Agriculture." *Eastern Africa J. Rural Dev.* 8:61-79.

_____ [1975b]: *Farm Surveys in Malawi: The Collection and Analysis of Labour Data.* Reading, England, Univ. of Reading, DAE.

_____ [1976]: "A Note on Planned *versus* Actual Farmer Performance under Uncertainty in Underdeveloped Agriculture." *J. Agr. Econ.* 27:257-260.

_____ [1977a]: "Efficiency in Resource Allocation—A Study of Malawi Smallholders' Performance." *Tropical Agriculture* 54(2):97-106.

_____ [1977b]: "Research-Based Recommendations *versus* Farmers' Practices: Some Lessons from Cotton-Spraying in Malawi." *Experimental Agriculture* 13(1):9-15.

Farvar, M. T., and J. P. Milton, eds. [1972]: *The Careless Technology: Ecology and International Development.* Garden City, NY, Natural History Press.

Fauck, R. [1977]: "Soil Erosion in the Sahelian Zone of Africa: Its Control and its Effect on Agricultural Production." In Cannell, ed., 1977, pp. 371-396.

Fauck, R., C. Moureaux, and C. Thomann [1969]: "Bilans de l'évolution des sols de Sefa-Casamance (Sénégal) après quinze années de culture continué." *L'Agronomie Tropicale* 24(3):263-301.

Faulkingham, R. H. [1977]: "Ecological Constraints and Subsistence Strategies: The Impact of Drought in a Hausa Village: A Case Study from Niger." In Dalby, Church, and Bezzaz, eds., 1977, pp. 148-158.

Faulkingham, R. H., and P. F. Thorbahn [1975]: "Population Dynamics and Drought: A Village in Niger." *Population Studies* 29(3):463-477.

Faure, H., and Jean Yves Gac [1981]: "Will the Sahelian Drought End in 1985?" *Nature* 291:475-478.

Faye, J., T. Gallali, and R. Billaz [1977]: "Peasant Agronomy: A Challenge to Planners' Models?" *African Environment* 2(4) and 3(1):37-46.

Faye, J., and M. Niang [1977]: "An Experiment in Agrarian Restructuration and Senegalese Rural Space Planning." *African Environment* 2(4) and 3(1):143-153.

Feder, G., R. Just, and D. Zilberman [1985]: "Adoption of Agricultural Innovations in Developing Countries: A Survey." *Econ. Dev. and Cultural Change* 33:255-298.

Feder, G., and R. Noronha [1987]: "Land Rights Systems and Agricultural Development in Sub-Saharan Africa." *World Bank Research Observer* 2(2):143-169.

Fei, J. C. H., and G. Ranis [1964]: *Development of the Labor Surplus Economy: Theory and Policy.* Homewood, IL, Irwin.

Feldman, D. [1969]: "The Economics of Ideology: Some Problems of Achieving Rural Socialism in Tanzania." In Leys, ed., 1969, pp. 85-111.

Fell, A. [1983]: "An Overview of Irrigation Strategy and Results in the Sahel." In Carruthers, ed., 1983, pp. 108-123.

Felton, M. R., and P. P. Ellis [1978]: *Studies on the Control of Rinderpest in Nigeria.* Reading, England, Univ. of Reading, Department of Agriculture and Horticulture, Study No. 23.

Ferguson, D. C. [1967]: *The Nigerian Beef Industry.* Ithaca, NY, Cornell Univ., International Agricultural Development Bulletin No. 9.

———, ed. [1976]: *A Conceptual Framework for the Evaluation of Livestock Production Development Projects and Programs in Sub-Saharan West Africa.* Ann Arbor, Univ. of Michigan, CRED.

Ferguson, D. S., and J. Sleeper [1976]: "A Selected Bibliography of West African Livestock Development." In Ferguson, ed., 1976.

Fieldhouse, D. K. [1986]: *Black Africa, 1945-1980: Economic Decolonization and Arrested Development.* London, Allen and Unwin.

Fields, G. [1975]: "Rural-Urban Migration, Urban Unemployment and Underemployment, and Job Search Activity in LDCs." *J. Dev. Econ.* 2:165-188.

——— [1980]: "Internal Migration in Developing Countries: Comment." In Easterlin, ed., 1980, pp. 390-394.

Firth, R. [1975]: "The Skeptical Anthropologist: Social Anthropology and Marxist Views on Society." In Block, ed., 1975, pp. 29-60.

Fisher, N. M. [1977]: "Studies in Mixed Cropping. I. Seasonal Differences in Relative Productivity of Crop Mixtures and Pure Stands in the Kenyan Highlands." *Experimental Agriculture* 13:177-184.

——— [1979]: "Studies in Mixed Cropping. III. Further Results with Maize-Bean Mixtures." *Experimental Agriculture* 15:49-58.

Flinn, J. C. [1975]: "Economic Considerations in the Conduct of Cooperative Agricultural Research." *Eastern Africa J. Rural Dev.* 8:105-117.

Flinn, J. C., B. M. Jellema, and K. L. Robinson [1975]: "Problems of Increasing Food Production in the Lowland Humid Tropics of Nigeria." *Zeitschrift fur Auslandische Landwirtschaft* 14(1):37–48.

Flinn, J. C., and J. Lagemann [1980]: "Evaluating Technical Innovations under Low Resource Farmer Conditions." *Experimental Agriculture* 16:91–101.

Ford, D. J. [1971]: "Long-Run Price Elasticities in the Supply of Kenyan Coffee: A Methodological Note." *Eastern Africa Econ. Rev.* 3(1):65–67.

Ford, J. [1971]: *The Role of Trypanosomiasis in African Ecology: A Study of the Tsetse Fly Problem*. Oxford, England, Clarendon Press.

Forsyth, D. J. C. [1977]: "Appropriate Technology in Sugar Manufacturing." *World Dev.* 5:189–202.

Fortmann, L. [1981]: "The Plight of the Invisible Farmer: The Effect of National Agricultural Policy on Women in Africa." In Dauber and Cain, eds., 1981, pp. 205–214.

——— [1984]: "Economic Status and Women's Participation in Agriculture: A Botswana Case Study." *Rural Sociology* 49:452–464.

Fortmann, L., and E. Roe [1981]: *Dam Groups in Botswana*. London, ODI, Pastoral Network Paper 12b.

Foster, G. M., T. Scudder, E. Colson, and R. V. Kemper, eds. [1979]: *Long-Term Field Research in Social Anthropology*. New York, Academic Press.

Foster, H. L. [1978, 1980]: "The Influence of Soil Fertility on Crop Performance in Uganda. I. Cotton. II. Groundnuts." *Tropical Agriculture (Trinidad)* 55:255–268 and 57:29–42.

Fotzo, P. [1983]: "The Economics of Bas-Fond Rice Production in the Eastern Region of Upper Volta: A Whole Farm Approach." Unpublished Ph.D. dissertation, MSU, East Lansing.

Fotzo, P. T., and F. E. Winch [1978]: *The Economics of Rice Production in the North-West Province of Cameroon: Some Policy Considerations*. Ibadan, Nigeria, IITA.

Founou-Tchuigoua, B. [1981]: *Fondements de l'économie de traité au Sénégal*. Paris, Silex.

Fournier, F. [1967]: "Research on Soil Erosion and Soil Conservation in Africa." *African Soils* 12:53–96.

Fowler, M. [1981]: *Overgrazing in Swaziland: A Review of the Technical Efficiency of the Swaziland Herd*. London, ODI Pastoral Network Paper 12d.

Francis, P. [1988]: "Ox Draught Power and Agricultural Transformation in Northern Zambia." *Agr. Systems* 27:35–49.

Franco, G. R. [1981]: "The Optimal Producer Price of Cocoa in Ghana." *J. Dev. Econ.* 8(1):77–92.

Frank, A. G. [1966]: "The Development of Underdevelopment." *Monthly Rev.* 18(4):17–31.31.

Frank, C. R. [1971]: "The Problem of Urban Unemployment in Africa." In Ridker and Lubell, eds., 1971.

Franke, R. W., and B. H. Chasin [1980]: *Seeds of Famine: Ecological Destruction and the Development Dilemma in the West African Sahel*. Montclair, NJ, Allanheld and Osmun.

Fredrick, K. D. [1969]: "The Role of Market Forces and Planning in Uganda's Economic Development, 1900-1938." *Eastern Africa Rev.* 1:47–62.

Fresco, L. [1986]: *Cassava in Shifting Cultivation. A Systems Approach to Agricultural Technology Development in Africa*. Amsterdam, Royal Tropical Institute.

Freund, W. M., and R. W. Shenton [1977]: "Vent-for-Surplus Theory and the Economic History of West Africa." *Savanna* 6(2):191–196.

Fricke, W. [1979]: *Cattle Husbandry in Nigeria: A Study of Its Ecological Conditions and Social-Geographical Differentiations.* Heidelberg, West Germany, Univ. of Heidelberg, Geographischen Arbeiten No. 52.

Friedland, W. H., and C. G. Rosberg, eds. [1964]: *African Socialism.* Stanford, CA, Stanford Univ. Press.

Friedrich, K. H. [1977]: *Farm Management Data Collection and Analysis: An Electronic Data Processing Storage and Retrieval System.* Rome, FAO.

Fumagalli, C. T. [1978]: "An Evaluation of Development Projects among East African Pastoralists." In P. Stevens, Jr., ed., 1978, pp. 49-63.

Furtado, C. [1973]: "The Concept of External Dependence in the Study of Underdevelopment." In Wilber, ed., 1973, pp. 118-127.

Gaitskell, A. [1959]: *Gezira: A Story of Development in the Sudan.* London, Faber and Faber.

Galaty, J. G., D. Aranson, P. C. Salzman, and A. Chovinard, eds. [1981]: *The Future of Pastoral Peoples.* Ottawa, IDRC.

Galbraith, J. K. [1985]: "Ideology and Agriculture." *Harpers Magazine.* 270(2):15-16.

Gallais, J. [1975]: "Traditions pastorales et développement: problèmes actuels dans la région de Mopti (Mali)." In Monod, ed., 1975, pp. 354-368.

————, ed., [1977]: *Stratégies pastorales et agricoles des Sahéliens durant la sécheresse, 1969-74.* Bordeaux, Centre d'Études de Géographie Tropicale.

Galletti, R., K. D. S. Baldwin, and I. O. Dina [1956]: *Nigerian Cocoa Farmers: An Economic Survey of Yoruba Cocoa Farming Families.* London, Oxford Univ. Press.

Galtung, J. [1971]: "A Structural Theory of Imperialism." *J. Peace Research* 2:81-116.

———— [1976]: "The Lome Convention and Neo-Capitalism." *African Rev.* 6(1):33-42.

Gambia, The [1976]: *Cultivation and Ox-Drawn Implements.* Banjul, The Gambia, Department of Agriculture, Technical Bulletin No. 1.

Gamble, D. P. [1979]: *A General Bibliography of The Gambia up to 31 December, 1977.* Boston, G. K. Hall.

Gastellu, J. M. [1980]: "Mais où sont donc ces unités économiques que nos amis cherchent tant en Afrique?" *Cahiers ORSTOM, Ser. Sci. Hum.* 7(1-2):3-12.

Gaury, C. E. [1977]: "Agricultural Mechanization." In Leakey and Wills, eds., 1977, pp. 273-293.

Gbetibouo, M., and C. L. Delgado [1984]: "Lessons and Constraints of Export Crop-Led Growth: Cocoa in Ivory Coast." In Zartman and Delgado, eds., 1984, pp. 115-147.

Geertz, C. [1978]: "The Bazaar Economy: Information and Search in Peasant Marketing. *Amer. Econ. Rev., Papers and Proceedings* 68(2):28-32.

Gellar, S., R. B. Charlick, and Y. Jones [1980]: *Animation Rurale and Rural Development: The Experience of Senegal.* Ithaca, NY, Cornell Univ., Rural Development Committee.

Gelaw, B., ed. [1986]: *To Feed Ourselves: A Proceeding of the First Eastern, Central and Southern Africa Regional Maize Workshop.* CIMMYT, Mexico, D.F.

Gemmill, G. T. [1971]: "The Economics of Farm Mechanization in Malawi." Lilongwe, Malawi, Bunda College of Agriculture, mimeo.

———— [1979]: "Elasticity of the Marketable Surplus of a Subsistence Crop at Various Stages of Development: Comment." *Econ. Dev. and Cultural Change* 28:175-178.

Gemmill, G., and C. K. Eicher [1973]: *A Framework for Research on the Economics of Farm Mechanization in Developing Countries.* East Lansing, MSU, DAE, ARE Paper No. 6.

George, S. [1987]: *Food Strategies for Tomorrow.* San Francisco, The Hunger Project.

Gerhart, J. [1975]: *The Diffusion of Hybrid Maize in Western Kenya*, abridged by CIMMYT. Mexico City, CIMMYT.

Gerold-Scheepers, T., and W. M. J. van Binsbergen [1978]: "Marxist and Neo-Marxist Approaches to Migration in Tropical Africa." In van Binsbergen and Meilink, eds., 1978, pp. 21-35.

Ghai, D., and M. Godfrey, eds. [1979]: *Essays on Employment in Kenya*. Nairobi, Kenya Literature Bureau.

Ghai, D., and S. Radwan [1980]: *Growth and Inequality: Rural Development in Malawi, 1964-1978*. Geneva, ILO, World Employment Research Program, Working Paper No. 35.

Ghai, D., and L. D. Smith [1987]: *Agricultural Prices, Policy, and Equity in Sub-Saharan Africa*. Boulder, Lynne Rienner.

Gilbert, E. H., D. W. Norman, and F. E. Winch [1980]: *Farming Systems Research: A Critical Appraisal*. East Lansing, MSU, DAE, Rural Development Paper No. 6.

Giles, G. W. [1975]: "The Reorientation of Agricultural Mechanization for Developing Countries. Part I: Policies and Attitudes for Action Programs." *AMA: Agricultural Mechanization in Asia* 6(2):15-25.

Girdner, J., V. Olorunsola, M. Froming, and E. Hanson [1980]: "Ghana's Agricultural Food Policy—Operation Feed Yourself." *Food Policy* 5(1):14-25.

Gittinger, J. P. [1972]: *Economic Analysis of Agricultural Projects*. Baltimore, Johns Hopkins Univ. Press.

Gittinger, J. P., J. Leslie, and C. Hoisington, eds. [1987]: *Food Policy*. Baltimore, Johns Hopkins Univ. Press.

Gladwin, C. H. [1975]: "A Model of the Supply of Smoked Fish from Cape Coast to Kumasi." In Plattner, ed., 1975, pp. 77-127.

Gladwin, H., and C. Gladwin [1971]: "Estimating Market Conditions and Profit Expectations of Fish Sellers at Cape Coast, Ghana." In Dalton, ed., 1971, pp. 122-142.

Glaeser, B., ed. [1987]: *The Green Revolution Revisited: Critiques and Alternatives*. London, Allen and Unwin.

———, ed. [1976]: *The Politics of Natural Disaster: The Case of the Sahel Drought*. New York, Praeger.

Glantz, M. H., ed. [1987]: *Drought and Hunger in Africa: Denying Famine a Future*. Cambridge, England, Cambridge Univ. Press.

Goddard, A. A. [1972]: "Land Tenure, Land Holding and Agricultural Development in the Central Sokoto Close-Settled Zone, Nigeria." *Savanna* 1(1):29-41.

Godelier, M. [1972]: *Rationality and Irrationality in Economics*, translation by B. Pearce. New York, Monthly Review Press.

Goe, M. R., and R. E. McDowell [1980]: *Animal Traction: Guidelines for Utilization*. Ithaca, NY, Cornell Univ., International Agricultural Mimeograph 81.

Gooch, T. [1979]: *An Experiment with Group Ranches in Upper Volta*. London, ODI, Pastoral Network Paper No. 9b.

Good, G. M. [1970]: *Rural Markets and Trade in East Africa: A Study of the Functions and Development of Exchange Institutions in Ankole, Uganda*. Univ. of Chicago, Department of Geography, Research Paper No. 128.

Goodell, G. [1984]: "Bugs, Bunds, Banks, and Bottlenecks: Organizational Contradictions in the New Rice Technology." *Econ. Dev. and Cultural Change* 31:23-41.

Goodwin, J. B. [1975]: "An Analysis of the Effect of Price Distortion on the Development of the Rice Milling Industry in Ghana." Unpublished Ph.D. dissertation, Univ. of Maryland, College Park.

Goody, J. R., ed. [1958]: *The Development Cycle in Domestic Groups*. Cambridge, England, Cambridge Univ. Press.

———, ed. [1975]: *Changing Social Structure in Ghana: Essays in the Comparative Sociology of a New State and an Old Tradition*. London, International African Institute.

——— [1980]: "Rice-burning and the Green Revolution in Northern Ghana." *J. Dev. Studies* 16(2):136–155.

Goody, J., and J. Buckley [1973]: "Inheritance and Woman's Labour in Africa." *Africa* 43(2):108–121.

Gordon, R. J. [1977]: *Mines, Masters and Migrants: Life in a Namibian Compound*. Johannesburg, Raven Press.

——— [1978]: *The Women Left Behind: A Study of the Wives of the Migrant Workers of Lesotho*. Geneva, ILO.

Gordon, S. L. [1974]: "The Role of Cocoa in Ghanaian Development." In Pearson and Cownie, 1974, pp. 67–91.

Goreux, L. M. [1977]: *Interdependence in Planning: Multi-Level Programming Studies of the Ivory Coast*. Baltimore, Johns Hopkins Univ. Press.

Goulet, D. [1978]: *Looking at Guinea-Bissau: A New Nation's Development Strategy*. Washington, ODC.

Gray, C., and A. Martens [1983]: "The Political Economy of the 'Recurrent Cost Problem' in the West African Sahel." *World Dev.* 11:101–117.

Grayzel, J. A. [1977]: "The Ecology of Ethnic-Class Identity among an African Pastoral People: The Doukoloma Fulbe." Unpublished Ph.D. dissertation, Univ. of Oregon, Eugene.

Green, R. H., and S. H. Hymer [1966]: "Cocoa in the Gold Coast: A Study in the Relations between African Farmers and Agricultural Experts." *J. Econ. Hist.* 26:299–319.

Greenland, D. J., and R. Lal, eds. [1977]: *Soil Conservation and Management in the Humid Tropics*. Chichester, New York, Wiley.

Gregory, J. W. [1979]: "Underdevelopment, Dependence and Migration in Upper Volta." In Shaw and Heard, eds., 1979, pp. 73–94.

Griffin, K. [1976]: "On the Emigration of the Peasantry." *World Dev.* 4(5):353–361.

Grindle, W. [1980]: *Economic Losses from East Coast Fever in Sukamaland, Tanzania*. Edinburgh, Univ. of Edinburgh, Centre for Tropical Veterinary Medicine.

Grolleaud, M., and D. F. Kohler [1979]: *Cereals Storage: Survey, Reflections, and Suggestions*. Paris, OECD.

Grove, A. T. [1974]: "Desertification in the African Environment." *African Affairs* 73:137–151.

Guggenheim, H. [1978]: "Of Millet, Mice, and Men: Traditional and Invisible Technology Solutions to Post-Harvest Losses in Mali." In Pimentel, ed., 1978, pp. 109–162.

Gugler, J. [1969]: "On the Theory of Rural-Urban Migration: The Case of Sub-Saharan Africa." In J. A. Jackson, 1969, pp. 134–155.

Gulhati, R. [1981]: *Industrial Strategy for Late Starters: The Experience of Kenya, Tanzania and Zambia*. Washington, World Bank, Working Paper No. 457.

Gulland, A. [1973]: "Resource Studies in Relation to the Development of African Inland Fisheries." *African J. Tropical Hydrobiology and Fisheries* 1:21–25.

Gusten, R. [1967]: "Can the Nigerian Economy Grow at 6 Percent per Annum in the Near Future? A Pre-Planning Exercise." *Nigerian J. Econ. Soc. Studies* 9:11–32.

——— [1968]: *Studies in the Staple Food Economy of Western Nigeria*. New York, Humanities Press.

——— [1984]: "African Agriculture: Which Way Out of the Crisis?" *Rural Africana*: 55–62.

Guthrie, E. J., and F. F. Pinto [1970]: "Wheat Improvement in East Africa." In Leakey, ed., 1970, pp. 88–98.

Gutkind, P. C. W., and I. Wallerstein, eds. [1976]: *The Political Economy of Contemporary Africa*. Beverly Hills, CA, Sage.

Guyer, J. I. [1980a]: "Food, Cocoa, and the Division of Labor by Sex in Two West African Societies." *Comparative Studies in Society and History* 22:355-373.

_____ [1980b]: *Household Budgets and Women's Incomes*. Boston Univ., African Studies Center, Working Paper No. 28.

_____ [1987]: *Feeding African Cities: Studies in Regional Social History*. London, Manchester University Press.

Gwyer, G. D. [1973]: "Three International Commodity Agreements: The Experience of East Africa." *Econ. Dev. and Cultural Change* 21:465-476.

Hafkin, N. J., and E. G. Bay, eds. [1976]: *Women in Africa: Studies in Social and Economic Change*. Stanford, CA, Stanford Univ. Press.

Haggblade, S. [1978]: "Africanization from Below: The Evolution of Camerounian Savings Societies into Western Style Banks." *Rural Africana* 2:35-55.

Haggblade, S., and P. B. R. Hazell [1988]: *Prospects for Equitable Growth in Rural Sub-Saharan Africa*. Washington, World Bank, AGRAP Economic Discussion Papers, No. 3.

Haggblade, S., P. B. R. Hazell, and J. Brown [1987]: "Farm/Non-Farm Linkages in Rural Sub-Saharan Africa: Empirical Evidence and Policy Implications." Washington, World Bank, Report No. ARU 67.

Haile-Mariam, T. [1974]: "The Impact of Coffee on the Economy of Ethiopia." In Pearson and Cownie, 1974, pp. 117-134.

Hall, A. E., G. H. Cannell, and H. W. Lawton, eds. [1979]: *Agriculture in Semi-Arid Environments*. New York: Springer-Verlag.

Hall, M. [1968]: "Agricultural Mechanization in East Africa." In Helleiner, ed., 1968b, pp. 81-116.

_____ [1970]: "A Review of Farm Management Research in East Africa." *Agr. Econ. Bulletin for Africa* 12:11-24.

Hamilton, A. G. [1975]: *A Review of Post-Harvest Technology: Botswana*. Ottawa, Canadian Univ. Service Overseas.

Hancock, W. K., Sr. [1942]: *Survey of British Commonwealth Affairs. Vol. II. Problems of Economic Policy, 1918-1939, Part 2*. London, Oxford Univ. Press.

Handwerker, W. P. [1974]: "Changing Household Organization in the Origins of Market Places in Liberia." *Econ. Dev. and Cultural Change* 22:229-248.

_____ [1981]: "Productivity, Marketing Efficiency and Price Support Programs: Alternative Paths to Rural Development in Liberia." *Human Organization* 40(1):27-39.

Hansen, A., and D. McMillan, eds. [1986]: *Food in Sub-Saharan Africa*. Boulder, CO, Lynne Reinner.

Hansen, B. [1979]: "Colonial Economic Development with Unlimited Supply of Land: A Ricardian Case." *Econ. Dev. and Cultural Change* 27:611-628.

Hanson, J. [1980]: *Is the School the Enemy of the Farm?* East Lansing, MSU, DAE, ARE Paper No. 22.

Harberger, A. C., ed. [1984]: *World Economic Growth*. San Francisco, CA, Institute for Contemporary Studies.

Harbeson, J. W. [1973]: *Nation Building in Kenya: The Role of Land Reform*. Evanston, IL, Northwestern Univ. Press.

Harbison, F., and C. Myers [1964]: *Education, Manpower, and Economic Growth*. New York, McGraw-Hill.

Hardaker, J. B. [1979]: "A Review of Some Farm Management Research Methods for Small Farm Development in LDCs." *J. Agr. Econ.* 30(3):315-331.

Harlow, V., and E. M. Chilver, eds. [1965]: *History of East Africa. II.* London, Oxford, Univ. Press.

Harper, M. [1975]: "Sugar and Maize Meal: Cases of Inappropriate Technology from Kenya." *J. Modern African Studies* 13:501-509.

Harris, J. [1981]: *A Conceptual Framework for the Study of Migration in Botswana.* Boston, Boston Univ., African Studies Center, Working Paper No. 42.

Harris, J., and M. P. Todaro [1970]: "Migration, Unemployment and Development: A Two-Sector Analysis." *Amer. Econ. Rev.* 60:126-142.

Harris, S., K. Porris, G. Ritson, and E. Tollens [1978]: *The Re-negotiation of the ACP-EEC Convention of Lomé, with Special Reference to Agricultural Products.* London, Commonwealth Secretariat.

Harrison, M. N. [1970]: "Maize Improvement in East Africa." In Leakey, ed., 1970, pp. 21-59.

Harriss, B. [1979a]: "Going against the Grain." *Dev. and Change* 10:363-384. Also in ICRISAT, 1980a, pp. 265-288.

———— [1979b]: "There Is Method In My Madness: Or Is It Vice Versa? Measuring Agricultural Market Peformance." *Food Research Institute Studies* 17:197-218.

Hart, K. [1973]: "Informal Income Opportunities and Urban Employment in Ghana." *J. Modern African Studies* 11:61-89.

———— [1982]: *The Political Economy of West African Agriculture.* Cambridge, England, Cambridge Univ. Press.

Harwitz, M. [1978]: "Improving the Lot of the Poorest: Economic Plans in Kenya." In P. Stevens, Jr., ed., 1978, pp. 65-74.

Hasif, E. [1978]: "L'emploi de la traction animale dans les exploitations agricoles." Paper given at CILSS/IER meeting, February. Bamako, Mali, IER.

Haswell, M. R. [1953]: *Economics of Agriculture in a Savannah Village: Report on Three Years Study in Genieri Village and its Lands: The Gambia.* London, HMSO, Colonial Research Study No. 8.

———— [1975]: *The Nature of Poverty.* London, Macmillan.

Hayami, Y. [1988]: "Asian Development: A View from the Paddy Fields." *Asian Dev. Rev.* 6(1):50-63.

Hayami, Y., and V. W. Ruttan [1971, 1985]: *Agricultural Development: An International Perspective.* Baltimore, Johns Hopkins Univ. Press.

Hays, H. M. [1975]: *The Marketing and Storage of Food Grains in Northern Nigeria.* Zaria, Nigeria, Ahmadu Bello Univ., IAR, Samaru Misc. Paper No. 50.

Hays, H. M., and J. H. McCoy [1978]: "Food Grain Marketing in Northern Nigeria: Spatial and Temporal Performance." *J. Dev. Studies* 14:182-192.

Hays, H. M., and A. K. Raheja [1977]: "Economics of Sole Crop Cowpea Production in Nigeria at the Farmers' Level Using Improved Practices." *Experimental Agriculture* 13(2):149-154.

Hazlewood, A. [1975]: *Economic Integration: The East African Experience.* London, Heinemann.

———— [1978]: "Kenya: Income Distribution and Poverty—An Unfashionable View." *J. Modern African Studies* 16:81-96.

———— [1979]: "The End of the East African Community: What Are the Lessons for Regional Integration Schemes?" *J. Common Market Studies* 18(1):40-58.

———— [1985]: "Kenyan Land-Transfer Programmes and their Relevance for Zimbabwe." *J. Modern African Studies* 23:445-461.

Hazlewood, A., and I. Livingstone [1978]: "Complementarity and Competitiveness of Large- and Small-Scale Irrigated Farming: A Tanzanian Example." *Oxford Bull. Econ. and Stat.* 40(3):195-208.

Hedlund, S., ed. [1987]: *Incentives and Economic Systems.* London, Croom Helm.

Heinemann, E., and S. Biggs [1985]: "Farming Systems Research: An Evolutionary Approach to Implementation." *J. Agr. Econ.* 36(1):59-65.

Heisey, P. [1985]: "Employment and Income in Botswana's Arable Agriculture." Unpublished Ph.D. dissertation, Univ. of Wisconsin, Madison.

Helland, J. [1980]: *Five Essays on the Study of Pastoralists and the Development of Pastoralism.* Bergen, Norway, Univ. of Bergen, Social Anthropological Institute, Occasional Paper No. 20.

Helleiner, G. [1979]: "AID and Dependence in Africa: Issues for Recipients." In Shaw and Heard, eds., 1979, pp. 221-245.

Helleiner, G. K. [1964]: "The Fiscal Role of the Marketing Boards in Nigerian Economic Development, 1947-61." *Econ. J.* 74:582-610.

—————— [1966a]: *Peasant Agriculture, Government and Economic Growth in Nigeria.* Homewood, IL, Irwin.

—————— [1966b]: "Typology in Development Theory: The Land Surplus Economy (Nigeria)." *Food Research Institute Studies* 6:181-194.

—————— [1968a]: "Agricultural Export Pricing Strategy in Tanzania." *Eastern Africa J. Rural Dev.* 1(1):1-17.

——————, ed. [1968b]: *Agricultural Planning in East Africa.* Nairobi, East African Publishing House.

—————— [1972]: "Socialism and Economic Development in Tanzania." *J. Dev. Studies* 8:183-204.

—————— [1975]: "Smallholder Decision Making: Tropical African Evidence." In L. G. Reynolds, ed., 1975, pp. 27-52.

Heller, P. [1979]: "The Underfinancing of Recurrent Development Costs." *Finance and Dev.* 16(1):38-42.

Henn, J. [1978]: "Peasants, Workers and Capital: The Political Economy of Labour and Incomes in Cameroun." Unpublished Ph.D. dissertation, Harvard Univ., Cambridge, MA.

Herdt, R. W. [1987]: "Technology Transfer as Development Aid: Discussion." *Amer. J. Agr. Econ.,* 69:938-39.

Herman, L. [1979]: *The Livestock and Meat Marketing System in Upper Volta: An Evaluation of Economic Efficiency.* Ann Arbor, Univ. of Michigan, CRED.

Herskovitz, M. J. [1926]: "The Cattle Complex in East Africa." *Amer. Anthropologist* 28:230-272, 361-380, 494-528, and 633-664.

Heyer, J. [1966]: "Agricultural Development and Peasant Farming in Kenya." Unpublished Ph.D. dissertation, Univ. of London.

—————— [1971]: "A Linear Programming Analysis of Constraints on Peasant Farms in Kenya." *Food Research Institute Studies* 10:55-67.

—————— [1972]: "An Analysis of Peasant Farm Production under Conditions of Uncertainty." *J. Agr. Econ.* 23:135-146.

—————— [1981]: "Agricultural Development Policy in Kenya from the Colonial Period to 1975." In Heyer, Roberts, and Williams, eds., 1981, pp. 90-120.

Heyer, J., D. Ireri, and J. Moris [1971]: *Rural Development in Kenya.* Nairobi, East African Publishing House.

Heyer, J., J. K. Maitha, and W. M. Senga, eds. [1976]: *Agricultural Development in Kenya.* Nairobi, Oxford Univ. Press.

Heyer, J., P. Roberts, and G. Williams, eds. [1981]: *Rural Development in Tropical Africa.* New York, St. Martin's Press.

Heyer, J., and J. K. Waweru [1976]: "The Development of the Small Farm Areas." In Heyer, Maitha, and Senga, eds., 1976, pp. 187-221.

Higgins, B. [1959]: *Economic Development: Principles, Problems and Policies.* New York, Norton.

Higgs, J. [1978]: *Land Settlement in Africa and the Near East—Some Recent Changes.* Rome, FAO, ARRD/CS/19.

Hill, P. [1963]: *The Migrant Cocoa Farmers of Southern Ghana: A Study in Rural Capitalism.* Cambridge, England, Cambridge Univ. Press.

―――― [1966]: "A Plea for Indigenous Economics: The West African Example." *Econ. Dev. and Cultural Change* 15:10-20.

―――― [1968]: "The Myth of the Amorphous Peasantry: A Northern Nigerian Case Study." *Nigerian J. Econ. Soc. Studies* 10:239-261.

―――― [1970]: *Studies in Rural Capitalism in West Africa.* Cambridge, England, Cambridge Univ. Press.

―――― [1972]: *Rural Hausa: A Village and Setting.* Cambridge, England, Cambridge Univ. Press.

―――― [1975]: "The West African Farming Household." In Goody, ed., 1975, pp. 119-136.

―――― [1978]: "Food-Farming and Migration from Fante Villages." *Africa* 48(3):220-229.

―――― [1986]: *Development Economics on Trial: The Anthropological Case for a Prosecution.* Cambridge, England, Cambridge Univ. Press.

Hinchey, M T., ed. [1978]: *Proceedings of the Symposium on Drought in Botswana, June 5-8, 1978.* Gaborone, Botswana Society in collaboration with Clark Univ. Press.

Hirschman, A. O. [1977]: "A Generalized Linkage Approach to Development, with Special Reference to Staples." *Econ. Dev. and Cultural Change* 25(supplement):67-98.

―――― [1981a]: *Essays in Trespassing: Economics to Politics and Beyond.* Cambridge, England, Cambridge Univ. Press.

―――― [1981b]: "The Rise and Decline of Development Economics." In Hirschman, 1981a, pp. 1-24.

Hitchcock, R. K. [1978]: *Kalahari Cattle Posts: A Regional Study of Hunter-Gatherers, Pastoralists, and Agriculturalists in the Western Sandveld Region, Central District, Botswana,* 2 Vols. Gaborone, Botswana, Ministry of Local Government and Lands.

Hjort, A. [1979]: *Savanna Town: Rural Ties and Urban Opportunities in Northern Kenya.* Stockholm, Univ. of Stockholm, Studies in Social Anthropology.

Hjort, A., and W. Ostberg [1978]: *Farming and Herding in Botswana.* Stockholm, SAREC.

Hoben, A. [1973]: *Land Tenure among the Amhara of Ethiopia: The Dynamics of Cognatic Descent.* Univ. of Chicago Press.

Hodd, M. [1976]: "Income Distribution in Kenya (1963-72)." *J. Dev. Studies* 12:221-228.

―――― [1978]: "Income Distribution in Kenya: A Reply." *J. Dev. Studies* 14:375-377.

Hodd, M., W. J. House, and T. Killick [1978]: "Income Distribution in Kenya: A Controversy Resolved." *J. Dev. Studies* 15(1):117.

Hodder, B. W., and U. I. Ukwu [1969]: *Markets in West Africa: Studies of Markets and Trade among the Yoruba and Ibo.* Ibadan, Nigeria, Ibadan Univ. Press.

Hoffman, A., J. G. Disney, A. Pinegar, and J. D. Cameron [1974]: "The Preservation of some East African Freshwater Fish." *African J. Tropical Hydrobiology and Fisheries* 3(1):1-14.

Hogendorn, J. [1975]: "Economic Initiative and African Cash Farming: Pre-Colonial Origins and Early Colonial Development." In Duignan and Gann, eds., 1975a, pp. 283-328.

―――― [1977]: "Vent-for-Surplus Theory and the Economic History of West Africa: A Reply." *Savanna* 6(2):196-199.

―――― [1978]: *Nigerian Groundnut Exports: Origins and Early Development.* Zaria, Nigeria, Ahmadu Bello Univ. Press.

Holdcroft, L. [1984]: "The Rise and Fall of Community Development, A Critical Assessment." In Eicher and Staatz, eds., 1984, pp. 46-58.

Holm, J., and R. Morgan [1985]: "Coping with Drought in Botswana: An African Success." *J. Modern African Studies* 23(3):463-482.

Holtham, G., and A. Hazlewood [1976]: *Aid and Inequality in Kenya: British Development Assistance to Kenya.* London, Croom Helm.

Holtzman, J. J. Staatz, and M. Weber [1980]: *An Analysis of Livestock Production and Marketing Subsystems in the Northeast Province of Cameroon.* East Lansing, MSU, DAE, Rural Development Working Paper No. 11.

Honeybone, D., and A. Marter [1975]: *An Evaluation Study of Zambia's Farm Institutes and Farmer Training Centers.* Lusaka, Univ. of Zambia, Rural Development Studies Bureau.

Hoos, S., ed. [1979]: *Agricultural Marketing Boards — An International Perspective.* Cambridge, MA, Ballinger.

Hopcroft, P. N., and G. M. Ruigu [1976]: *Dairy Marketing and Pricing in Kenya: Are Milk Shortages the Consequences of Droughts or Pricing Policies?* Nairobi, Univ. of Nairobi, IDS Discussion Paper No. 237.

Hopkins, A. G. [1973]: *An Economic History of West Africa.* New York, Columbia Univ. Press.

Hopkins, E. [1974]: "Operation Groundnuts: Lessons from an Agricultural Extension Scheme." *IDS Bulletin* 5(4):59-66.

Hopkins, N. S. [1976]: "Participatory Decision Making and Modern Cooperatives in Mali: Notes Toward a Prospective Anthropology." In Nash, Dandler, and Hopkins, eds., 1976, pp. 99-111.

Horowitz, M. [1972]: "Ethnic Boundary Maintenance among Pastoralists and Farmers in the Western Sudan (Niger)." *J. Asian and African Studies* 7(1):104-114.

―――― [1977]: "Les stratégies adaptatives au Sahel avant et après la sécheresse." In Gallais, ed., 1977, pp. 219-233.

Horowitz, M., and T. Painter, eds. [1986]: *Anthropology and Rural Development in West Africa.* Boulder, CO, Westview Press.

Hotes, F. [1983]: "The Experience of the World Bank." In Carruthers, ed., 1983, pp. 126-140.

House, W. J., and T. Killick [1978]: "Hodd on Income Distribution in Kenya: A Critical Note." *J. Dev. Studies* 14:370-374.

―――― [1980]: *Social Justice and Development Policy in Kenya's Rural Economy.* ILO, World Employment Research Program, Working Paper No. 31.

House, W. J., and H. Rempel [1980]: "The Determinants of Interregional Migration in Kenya." *World Dev.* 8:25-36.

Howard, R. [1978]: *Colonialism and Underdevelopment in Ghana.* London, Croom Helm.

Howe, C. [1966]: "The Use of Sample Household Expenditure Surveys in Economic Planning in East Africa." *Oxford Bull. Econ. and Stat.* 28:199-209.

———— [1968]: "An Analysis of African Household Consumption and Behavior in Kenya and Uganda." *East African Econ. Rev.* 4:51-62.

Howell, J., ed. [1980]: *Borrowers and Lenders: Rural Financial Markets and Institutions in Developing Countries.* London, ODI.

———— [1985]: *Recurrent Costs and Agricultural Development.* London, ODI.

———— [1988]: *Training and Visit Extension in Practice.* London, ODI.

Hubbard, M. [1986]: *Agricultural Exports and Economic Growth: A Study of Botswana's Beef Industry.* London, Routledge.

Hughes, C., and J. M. Hunter [1970]: "Disease and 'Development' in Africa." *Social Science and Medicine* 3(4):443-493.

———— [1972]: "The Role of Technological Development in Promoting Disease in Africa." In Farvar and Milton, eds., 1972, pp. 69-101.

Huizer, G., and B. Mannheim, eds. [1979]: *The Politics of Anthropology: From Colonialism and Sexism toward a View from Below.* The Hague, Mouton.

Humphrey, D. H. [1975]: "Socio-Economic Aspects of Rural Development in Malawi: A Report of Some Survey Findings." *Eastern Africa J. Rural Dev.* 8(1-2):46-60.

Humphrey, D. H., and H. S. Oxley [1976]: "Expenditure and Household Size Elasticities in Malawi: Urban-Rural Comparisons, Demand Projections and a Summary of East African Findings." *J. Dev. Studies* 12(2):252-269.

Humphreys, C. P., and S. R. Pearson [1979-80]: "Choice of Technique in Sahelian Rice Production." *Food Research Institute Studies* 17:235-277.

Hunt, D. M. [1975a]: *Growth versus Equity: An Examination of the Distribution of Economic Status and Opportunity in Mbere, Eastern Kenya.* Nairobi, Univ. of Nairobi, IDS Occasional Paper No. 11.

———— [1975b]: "The Introduction of Single Axle Tractors on Peasant Coffee Farms in Masaka Division, Southern Uganda." *Eastern Africa J. Rural Dev.* 8:246-264.

———— [1976]: *Chayanov's Model of Peasant Household Resource Allocation and its Relevance to Mbere Division, Eastern Kenya.* Nairobi, Univ. of Nairobi, IDS Working Paper No. 276.

Hunter, J. M. [1967]: "Seasonal Hunger in a Part of the West African Savanna: A Survey of Bodyweights in Nangodi, North East Ghana." *Institute of British Geographers, Transactions* 41:167-183.

———— [1981]: "Progress and Concerns in the World Health Organization Onchocerciasis Control Program in West Africa." *Social Science and Medicine* 15:261-275.

Hunter, J. M., and G. K. Ntiri [1978]: "Speculations on the Future of Shifting Agriculture in Africa." *J. Developing Areas* 12(2):183-208.

Hutton, C. [1973]: *Reluctant Farmers? A Study of Unemployment and Planned Rural Change in Uganda.* Nairobi, East African Publishing House.

Hutton, C., and R. Cohen [1975]: "African Peasants and Resistance to Change: A Reconsideration of Sociological Approaches." In Oxaal, Barnett, and Booth, 1975, pp. 105-130.

Hyden, G. [1973a]: *Agricultural Credit in Three Village Areas in North-Eastern Tanzania: A Case Study.* Rome, FAO, Agricultural Credit Case Study, Working Paper No. 2.

———— [1973b]: *Efficiency versus Distribution in East African Cooperatives: A Study in Organizational Conflicts.* Nairobi, East African Literature Bureau.

———— [1978/79]: "Cooperatives and Local Leadership Patterns." *Rural Africana* 3:43-59.

———— [1980]: *Beyond Ujamaa in Tanzania: Underdevelopment and an Uncaptured Peasantry.* Berkeley, Univ. of California Press.

———— [1983]: *No Shortcuts to Progress: African Development Management in Perspective.* London, Heinemann.

———— [1987]: "Discussion: Final Rejoinder." *Dev. and Change* 18:661–667.

Hymer, S., and S. Resnick [1969]: "A Model of an Agrarian Economy with Nonagricultural Activities." *Amer. Econ. Rev.* 59:493–506.

ICRISAT [1980a]: *Proceedings of the International Workshop on Socioeconomic Constraints to Development of Semi-Arid Tropical Agriculture, 19-23 February 1979, Hyderabad, India.* Pantacheru, India.

———— [1980b]: *Proceedings of the International Workshop on Sorghum Diseases, 11-15 December 1978, Hyderabad, India.* Pantacheru, India.

———— [1980c]: *Annual Report, 1980.* Ouagadougou, Burkina Faso, Ministry of Rural Development.

———— [1987]: *Proceedings of the International Pearl Millet Workshop, 7-11 April 1986.* Pantacheru, India.

Idachaba, F. S. [1973]: "Marketing Board Crop Taxation and Input Subsidies: A Second-Best Approach." *Nigerian J. Econ. Soc. Studies* 15:317–324.

———— [1977]: "Pesticide Input Subsidies in African Agriculture: The Nigerian Experience." *Canadian J. Agr. Econ.* 25:88–103.

———— [1980a]: *Agricultural Research Policy in Nigeria.* Washington, IFPRI.

———— [1980b]: "Food Policy in Nigeria: Towards a Framework of Analysis." *Agricultural Research Bulletin* 1(1):1–47.

———— [1981]: *Farm Input Subsidies for the Green Revolution in Nigeria: Lessons from Experience.* Ibadan, Nigeria, Univ. of Ibadan, DAE, Food Policy Technical Paper No. 2.

———— [1987]: "Agricultural Research in Nigeria: Organization and Policy." In Ruttan and Pray, eds., 1987, pp. 333–362.

———— [1988]: "Marketing and Pricing Policy Interventions in Nigeria." Rome, FAO. (Forthcoming).

Idachaba, F. S., J. A. Akinwumi, C. E. Olumese, L. O. Ologide, S. A. Adetunju, and T. A. Taylor, eds. [1981]: *The Crop Subsector in the Fourth National Development Plan, 1981-85. Proceedings of a Workshop organized by the Federal Department of Agriculture, August 29-30, 1979.* Lagos, Nigeria.

Idusogie, E. O. [1969]: "A Critical Review of the Role of Cash Cropping on the Nutrition of Nigerian Peoples." Unpublished Ph.D. dissertation, Univ. of London.

IEMVT [1980]: *Elements for a Livestock Development Strategy in Sahel Countries.* Paris, OECD, Club du Sahel.

IITA/SAFGRAD [1980]: *SAFGRAD and IDRC-Upper Volta National Cowpea Improvement Program: Report 1979.* Ouagadougou, Burkina Faso.

ILCA [1978]: *Evaluation of the Productivities of Manure and Peul Cattle Breeds at the Sahelian Station, Niono, Mali.* Addis Ababa.

———— [1979a]: *Small Ruminant Production in the Humid Tropics.* Addis Ababa.

———— [1979b]: *Livestock Production in the Sub-humid Zone of West Africa: A Regional Review.* Addis Ababa.

———— [1979c]: *Trypanotolerant Livestock in West and Central Africa,* 2 Vols. Addis Ababa.

———— [1979d]: *Towards an Economic Assessment of Veterinary Inputs in Tropical Africa.* Addis Ababa.

———— [1980a]: *Pastoral Development Projects.* Addis Ababa.

———— [1980b]: *Economic Aspects of Browse Development.* Addis Ababa.

Iliffe, J. [1979]: *Modern History of Tanganyika.* Cambridge, England, Cambridge Univ. Press.

_____ [1987]: *The African Poor: A History*. Cambridge, England, Cambridge Univ. Press.

ILO [1970]: *Socio-Economic Conditions in the Ifo, Otta and Ilaro Districts of the Western State of Nigeria*. Geneva.

_____ [1972]: *Employment, Incomes and Equality: A Strategy for Increasing Productive Employment in Kenya*. Geneva.

_____ [1973]: *Employment in Africa: Some Critical Issues*. Geneva.

_____ [1976]: *Growth, Employment and Equity: A Comprehensive Strategy for the Sudan*. Geneva.

_____ [1981]: *First Things First: Meeting the Basic Needs of the People of Nigeria*. Addis Ababa, Ethiopia, Job and Skills Program for Africa, Report to the Government of Nigeria by a JASPA Basic Needs Mission.

_____ [1987]: *Generating Employment and Incomes in Somalia*: Provisional Report of an ILO/JASPA Inter-Disciplinary Employment and Project Identification Mission to Somalia. Addis Ababa.

Ingle, C. R. [1972]: *From Village to State in Tanzania: The Politics of Rural Development*. Ithaca, NY, Cornell Univ. Press.

Institut d'Économie Rurale [IER, 1977]: *Pour une programme de recherche sur les systémes de production agricole*. Bamako, Mali.

International Agricultural Development Service [IADS, 1977]: *Senegal Agricultural Research Project: Report of a Senegalese-IADS Team to the General Delegation for Scientific and Technical Research*. Dakar, Senegal.

International Bank for Reconstruction and Development [IBRD, 1961]: *The Economic Development of Tanganyika*. Baltimore, Johns Hopkins Univ. Press.

International Fertilizer Development Center [IFDC, 1986]: *Annual Report, 1986*. Muscle Shoals, AL.

IRAT [1980]: *Campagne, 1979. Résumé des travaux, Conclusions*. Ouagadougou, Burkina Faso.

Irvine, F. R. [1969]: *West African Agriculture, Vol. 1: West African Crops*. London, Oxford Univ. Press.

Isaacman, A. [1979]: "Transforming Mozambique's Rural Economy." *Rural Africana* NS 4-5:97-113.

Islam, N., ed. [1974]: *Agricultural Policy in Developing Countries*. New York, Macmillan.

ISRA [1977]: *Recherche et developpment agricole: les unités experimentales du Senegal: compte rendu du seminaire tenu au CNRA de Bambey du 16 au 21 mai 1977*. Dakar, Senegal.

Issard, W. [1979]: *Rural-Urban Migration of Women in Botswana*. Gaborone, Botswana, Ministry of Finance, Central Statistical Office.

Ita, E. O. [1975]: "A Conceptualized Fishing System for the African Environment with a Comparative Approach." *African J. Tropical Hydrobiology and Fisheries* 4(1):141-147.

Jabara, C. L., and R. L. Thompson [1980]: "Agricultural Comparative Advantage under International Price Uncertainty: The Case of Senegal." *Amer. J. Agr. Econ.* 62:188-198.

Jackson, I. C. [1956]: *Advance in Africa: A Study of Community Development in Eastern Nigeria*. London, Oxford Univ. Press.

Jackson, J. A. [1969]: *Migration*. Cambridge, England. Cambridge Univ. Press.

Jackson, P. B. N. [1971]: "The African Great Lakes Fisheries: Past, Present, and Future." *African J. Tropical Hydrobiology and Fisheries* 1(1):35-49.

Jacobs, A. H. [1975]: "Maasai Pastoralism in Historical Perspective." In Monod, ed., 1975, pp. 406-425.

_____ [1978]: *Development in Tanzania Maasailand: The Perspective Over 20 Years, 1957-1977*. Dar es Salaam, AID.

Jaeger, W. K. [1986]: *Agricultural Mechanization: The Economics of Animal Draft Power in West Africa*. Boulder, CO, Westview Press.

Jahnke, H. E. [1974]: *The Economics of Controlling Tsetse Flies and Cattle Trypanosomiasis in Africa Examined for the Case of Uganda*. Munich, West Germany, Weltforum Verlag, No. 48.

——— [1982]: *Livestock Production Systems and Livestock Development in Tropical Africa*. Kiel, Kieler Wissenschaftsverlag Vauk.

Jain, S. [1975]: *Size Distribution of Income: A Compilation of Data*. Washington, World Bank.

Jamal, V. [1974]: "The Role of Cotton in Uganda: Economic Development." In Pearson and Cownie, 1974, pp. 135-154.

Jansson, K., M. Harris, and A. Penrose [1987]: *The Ethiopian Famine*. London, Zed Books.

Jarvis, L. S. [1974]: "Cattle as Capital Goods and Ranchers as Portfolio Managers: An Application to the Argentine Cattle Sector." *J. Polit. Economy* 82:489-520.

——— [1980]: "Cattle as a Store of Wealth in Swaziland: Comment." *Amer. J. Agr. Econ.* 62:606-613.

Jequier, N., ed. [1976]: *Appropriate Technology: Problems and Promises*. Paris, OECD.

Jha, D. [1987]: "Strengthening Agricultural Research in Africa: Some Neglected Issues." *Quart. J. International Agriculture* 26(3):265-275.

Johnson, A. W. [1972]: "Individuality and Experimentation in Traditional Agriculture." *Human Ecology* 1:149-159.

Johnson, George [1971]: "The Structure of Rural-Urban Migration Models." *Eastern Africa Econ. Rev.* 3:21-28.

Johnson, George, and W. E. Whitlaw [1974]: "Urban-Rural Income Transfers in Kenya: An Estimated Remittance Function." *Econ. Dev. and Cultural Change* 22:473-479.

Johnson, Glenn L. [1968]: "Removing Obstacles to the Use of Genetic Breakthrough in Oil Palm Production: The Nigerian Case." In NRC, 1968, Vol. II, pp. 365-375.

——— [1988]: "The Urgency of Institutional Changes for LDC, NIC and DC Agriculture." Paper presented at a Symposium on Future U.S. Development Assistance. Winrock, AR.

Johnson, Glenn L., and A. Maunder, eds. [1981]: *Rural Change: The Challenge for Agricultural Economists. Proceedings of the Seventeenth ICAE, 3-12 September, Banff, Canada*. Westmead, England.

Johnson, Glenn L., O. J. Scoville, G. K. Dike, and C. K. Eicher [1969]: *Strategies and Recommendations for Nigerian Rural Development, 1969-85*. East Lansing, MSU, DAE, Consortium for the Study of Nigerian Rural Development, CSNRD 33.

Johnson, Harry G. [1967a]: *Economic Nationalism in Old and New States*. Univ. of Chicago Press.

Johnson, Harry G. [1967b]: *Economic Policies toward Less Developed Countries*. Washington, Brookings Institution.

Johnston, B. F. [1958]: *The Staple Food Economies of Western Tropical Africa*. Stanford, CA, Stanford Univ. Press.

——— [1964]: "The Choice of Measures for Increasing Agricultural Productivity: A Survey of Possibilities in East Africa." *Tropical Agriculture* 41(2):91-113.

——— [1986]: "Agricultural Development in Tropical Africa: The Search for Viable Strategies." In R. J. Berg and Whitaker, eds., 1986, pp. 155-183.

Johnston, B. F., and W. Clark [1982]: *Redesigning Rural Development: A Strategic Perspective*. Baltimore, Johns Hopkins Univ. Press.

Johnston, B. F., A. Hoben, D. Dijkerman, and W. Jaeger [1987]: *An Assessment of AID Activities to Promote Agricultural and Rural Development in Sub-Saharan Africa*. Washington, AID Evaluation Special Study No. 54.

Jones, M. J. [1973]: *A Review of the Use of Rock Phosphate Fertilizers in Francophone West Africa*. Zaria, Nigeria, Ahmadu Bello Univ., IAR, Samaru Misc. Paper No. 43.

Jones, M. J., and A. Wild [1975]: *Soils of the West African Savanna*. Farnham Royal, England, Commonwealth Agricultural Bureau.

Jones, R. A., and R. J. Robinson [1976]: "Income Distribution and Development: Rhodesia and Malawi Compared." *Rhodesian J. Econ.* 10:91-102.

Jones, W. I. [1972]: "The Mise and Demise of Socialist Institutions in Rural Mali." *Geneva-Africa* 11(2):19-44.

——— [1976]: *Planning and Economic Policy: Socialist Mali and her Neighbors*. Washington, Three Continents Press.

Jones, W. O. [1959]: *Manioc in Africa*. Stanford, CA, Stanford Univ. Press.

——— [1960]: "Economic Man in Africa." *Food Research Institute Studies* 1:107-134.

——— [1972]: *Marketing Staple Foods in Tropical Africa*. Ithaca, NY, Cornell Univ. Press.

——— [1974]: "Regional Analysis and Agricultural Marketing Research in Tropical Africa: Concepts and Experience." *Food Research Institute Studies* 13:3-28.

——— [1980a]: "Agricultural Trade within Tropical Africa: Historical Background." In R. H. Bates and Lofchie, eds., 1980, pp. 10-45.

——— [1980b]: "Agricultural Trade within Tropical Africa: Achievements and Difficulties." In R. H. Bates and Lofchie, eds., 1980, pp. 311-348.

——— [1987]: "Food-Crop Marketing Boards in Tropical Africa." *J. Modern African Studies* 25:375-402.

Joseph, N. S. [1987]: "An *Ex Ante* Economic Appraisal of Mono-Cropping, Mixed Cropping and Inter-Cropping of Annual and Perennial Crops." *Agri. Systems* 24:67-80.

Joy, L. [1967]: "The Economics of Food Production." *African Affairs* 65:317-327.

Joyce, S., and F. Beudot [1976-77]: *Elements for a Bibliography of the Sahel Drought*. Vol. 1 [1976] and Vol. 2 [1977]. Paris, OECD Development Center.

Jurion, F., and J. Henry [1969]: *Can Primitive Farming Be Modernized?* Translated from the French by AGRA Europe. Brussels, SERDAT.

Kaberry, P. M. [1952]: *Women of the Grassfields: A Study of the Economic Position of Women in Bamenda, British Cameroons*. London, HMSO.

Kafando, T. W. [1972]: *Contribution à l'étude du développement intégré du Liptako-Gourma: Introduction à l'étude de systèmes agro-pastoraux*. Dakar, Sénégal, African Institute for Economic Development and Planning.

Kamajou, F., and C. Baker [1980]: "Reforming Cameroon's Government Credit Programs: Effects on Liquidity Management by Small Farm Borrowers." *Amer. J. Agr. Econ.* 62:709-718.

Kamarck, A. M. [1976]: *The Tropics and Economic Development: A Provocative Inquiry into the Poverty of Nations*. Baltimore, Johns Hopkins Univ. Press.

Kamuanga, M. [1981]: "Farm Level Study of the Rice Production System at the Office du Niger in Mali: An Economic Analysis." Unpublished Ph.D. dissertation, MSU, East Lansing.

Kamuanga, M., and D. S. C. Spencer [1981]: *Losses of Rice in West Africa: The Case of the Office du Niger in Mali*. Monrovia, Liberia, WARDA, Occasional Paper No. 5.

Kanon, D. [1985]: *Développement ou appauvrissement*. Paris, Economica.

Kasfir, N. [1986]: "Review Article: Are African Peasants Self-Sufficient?" *Dev. and Change* 17:337-357.

Kassam, A. H. [1976]: *Crops of the West African Semi-Arid Tropics*. Hyderabad, India, ICRISAT.

Kassam, A. H., and D. J. Andrews [1975]: "Effects of Sowing Date on Growth, Development and Yield of Photosensitive Sorghum at Samaru, Northern Nigeria." *Experimental Agriculture* 11:227-240.

Kay, G. B. [1972]: *The Political Economy of Colonialism in Ghana: A Collection of Documents and Statistics, 1900-1960*. Cambridge, England, Cambridge Univ. Press.

Kearl, B., ed. [1976]: *Field Data Collection in the Social Sciences: Experiences in Africa and the Middle East*. New York, ADC.

Keller, W., E. Muskat, and E. Valder [1969]: "Some Observations Regarding Economy, Diet and Nutritional Status of Kikuyu Farmers in Kenya." In Kraut and Cremer, eds., 1969, pp. 241-266.

Kelly, V. [1988]: *Farmers' Demand for Fertilizer in the Context of Senegal's New Agricultural Policy: A Study of Factors Influencing Farmers' Fertilizer Purchasing Decisions*. East Lansing, DAE, MSU International Development Papers, Reprint No. 19.

Kennedy, E., and B. Cogill [1987]: *Income and Nutritional Effects of the Commercialization of Agriculture in Southwestern Kenya*. Washington, IFPRI, Research Report 63.

Kenya [1954]: *A Plan to Intensify the Development of African Agriculture in Kenya (Swynnerton Plan)*. Nairobi, Government Printer.

_____ [1965]: *African Socialism and its Application to Planning in Kenya*. Nairobi, Government Printer.

_____ [1973]: *Sessional Paper No. 10 of 1973 on Employment*. Nairobi, Government Printer.

_____ [1979a]: *Child Nutrition in Rural Kenya*. Nairobi, Ministry of Economic and Community Affairs, Central Bureau of Statistics.

_____ [1979b]: *Development Plan, 1979-83*. Nairobi, Government Printer.

Ker, A. D. R. [1973]: "The Development of Improved Farming Systems Based on Ox Cultivation." In Amann, ed., 1973.

Keregero, K. J. B., J. de Vries, and C. D. S. Bartlett [1976]: *Farmer Resistance to Extension Advice: Who Is To Blame? A Case Study of Cotton Production in Mara Region, Tanzania*. Morogoro, Tanzania, Univ. of Dar es Salaam, DAE.

Khalifa, A. H., and M. C. Simpson [1972]: "Perverse Supply in Nomadic Societies." *Oxford Agrarian Studies* 1:46-56.

Kilby, P. [1961]: "African Labor Productivity Reconsidered." *Econ. J.* 71:273-291.

_____ [1962]: *The Development of Small Industry in Eastern Nigeria*. Lagos, Nigeria, AID.

_____ [1965]: "Patterns of Bread Consumption in Nigeria." *Food Research Institute Studies* 5:3-18.

_____ [1967]: "The Nigerian Oil Palm Industry." *Food Research Institute Studies* 7:177-204.

_____ [1968]: "The Nigerian Oil Palm Industry: A Reply." *Food Research Institute Studies* 8:199-203.

_____ [1969]: *Industrialization in an Open Economy: Nigeria, 1945-66*. Cambridge, England, Cambridge Univ. Press.

_____ [1975]: "Manufacturing in Colonial Africa." In Duignan and Gann, eds., 1975a, pp. 470-520.

Killick, T. [1976a]: *The Economies of East Africa: A Bibliography*. Boston, G. K. Hall.

_____ [1976b]: "The Possibilities of Economic Planning." *Oxford Econ. Papers* 28:161-184.

_____ [1978]: *Development Economics in Action: A Study of Economic Policies in Ghana*. New York, St. Martin's Press.

_____ [1980]: "Trends in Development Economics and their Relevance to Africa." *J. Modern African Studies* 18(3):367-386.

King, K. [1977]: *The African Artisan: Education and the Informal Sector in Kenya.* London, Heinemann.

_____ [1986]: "Manpower, Technology and Employment in Africa." In R. J. Berg and Whitaker, eds., 1986, pp. 422-450.

King, Robert P., and D. Byerlee [1978]: "Factor Intensities and Locational Linkages of Rural Consumption Patterns in Sierra Leone." *Amer. J. Agr. Econ.* 60:197-206.

King, Roger [1975]: "Experiences in the Administration of Co-operative Credit and Marketing Societies in Northern Nigeria." *Agricultural Administration* 2(3):195-208.

_____ [1981]: "Cooperative Policy and Village Development in Northern Nigeria." In Heyer, Roberts, and Williams, eds., 1981, pp. 259-280.

Kinsey, B. H. [1976]: *Economic Research and Farm Machinery Design in Eastern Africa.* Norwich, England, Univ. of East Anglia.

_____ [1978]: "Agricultural Technology and Rural Development in the Rainfed Maize Area of Southeastern Zambia." Unpublished Ph.D. dissertation, Stanford Univ., Stanford, CA.

Kiss, J. [1977]: *Will Sudan Be An Agricultural Power?* Budapest, Hungarian Academy of Sciences, Institute for World Economics.

Kitching, G. [1980]: *Class and Economic Change in Kenya: The Making of an African Petite Bourgeoise, 1905-1970.* New Haven, CT, Yale Univ. Press.

Kleemeier, L. L. [1978]: "Empirical Tests of Dependency Theory: A Second Critique of Methodology." *J. Modern African Studies* 16(3):701-704.

Kleene, P. [1976]: "Notion d'exploitation agricole et modernization en milieu Wolof—Saloum (Senegal)." *L'Agronomie Tropicale* 31:63-82.

Klein, M. A., ed. [1980]: *Peasants in Africa: Historical and Contemporary Perspectives.* Beverly Hills, CA, Sage.

Kline, C. K., D. A. G. Green, R. L. Donahue, and B. A. Stout [1969]: *Agricultural Mechanization in Equatorial Africa.* East Lansing, MSU, Institute of International Agriculture, Research Report No. 9.

Kloth, T. I., W. A. Burr, J. P. Davis, G. Epler, C. A. Kolff, R. I. Rosenberg, N. W. Staehling, J. M. Lane, and M. H. Nichaman [1976]: "Sahel Nutrition Survey, 1974." *Amer. J. Epidemiology* 103(4):383-390.

Knight, J. B. [1972]: "Rural-Urban Income Comparisons and Migration in Ghana." *Oxford Bull. Econ. and Stat.* 34:199-228.

_____ [1976]: "Explaining Income Distribution in Less Developed Countries: A Framework and an Agenda." *Oxford Bull. Econ. and Stat.* 38:161-177.

Knight, J. B., and G. Lenta [1980]: "Has Capitalism Underdeveloped the Labor Reserves of South Africa?" *Oxford Bull. Econ. and Stat.* 42:157-201.

Knowles, J. C., and R. Anker [1981]: "An Analysis of Income Transfers in a Developing Country: The Case of Kenya." *J. Dev. Econ.* 8:205-226.

Kocher, J. E. [1979]: *Rural Development and Fertility Change in Tropical Africa: Evidence from Tanzania.* East Lansing, MSU, DAE, ARE Paper No. 19.

Kocher, J. E., and B. Fleisher [1979]: *A Bibliography on Rural Development in Tanzania.* East Lansing, MSU, DAE, Rural Development Paper No. 3.

Koester, U. [1987]: "Trade in Agricultural Products among African Countries." *Quart. J. International Agriculture* 26(4):190-206.

Kofi, T. A. [1972]: "International Commodity Agreements and Export Earnings: Simulation of the 1968 'Draft International Cocoa Agreement.' " *Food Research Institute Studies* 11:177-201.

Kohler, J. M. [1971]: *Activités agricoles et changements sociaux dans l'Ouest-Mossi (Haute Volta).* Paris, ORSTOM, Memoires No. 46.

———— [1972]: *Les migrations des Mossi de l'Ouest (Haute-Volta).* Paris, ORSTOM.

Kolasa, K. M. [1980]: "The Nutritional Situation in Sierra Leone." *Rural Africana* 7:55-67.

Kolawole, M. I. [1972]: "An Application of Queuing Theory to Tractor Contracting Operations in Western Nigeria." *Bull. Rural Econ. and Sociology* 7:155-185.

———— [1974]: "Economic Aspects of Private Tractor Operations in the Savanna Zone of Western Nigeria." *Savanna* 3(2):175-183.

Kollberg, S. [1979]: *East African Marine Research and Marine Resources.* Stockholm, Sweden, SAREC.

Konczacki, Z. A. [1978]: *The Economics of Pastoralism: A Case Study of Sub-Saharan Africa.* London, Cass.

Korte, R. [1969]: "The Nutritional and Health Status of the People Living on the Mwea-Tebere Irrigation Settlement." In Kraut and Cremer, eds., 1969, pp. 267-334.

Kostinko, G., and J. Dione [1980]: *An Annotated Bibliography of Rural Development in Senegal: 1975-1980.* East Lansing, MSU, DAE, ARE Paper No. 23.

Kotey, R. A., C. Okali, and B. E. Rourke, eds. [1974]: *The Economics of Cocoa Production and Marketing: Proceedings of Cocoa Economics Research Conference, Legon, April 1973.* Legon, Univ. of Ghana, ISSER.

Kowal, J. M., and A. H. Kassam [1978]: *Agricultural Ecology of Savanna: A Study of West Africa.* Oxford, England, Clarendon Press.

Kraut, H., and H. D. Cremer, eds. [1969]: *Investigations into Health and Nutrition in East Africa.* New York: Humanities Press.

Kriesel, H. C., C. K. Laurent, C. Halpern, and H. E. Larzelere [1970]: *Agricultural Marketing in Tanzania: Background Research and Policy Proposals.* East Lansing, MSU, DAE.

Krone, W. [1970]: *Frozen Fish Marketing in West African Countries — A Case Study in Fish Marketing Development.* Rome, FAO Fisheries Report No. 96.

Krueger, A. O., and V. W. Ruttan [1983]: *The Development Impact of Economic Assistance to LDCs,* 2 vols. Minneapolis, Univ. of Minnesota, Economic Development Center.

Kudhongania, A. W., and A. J. Cordone [19974]: "Past Trends, Present Stocks and Possible Future State of the Fisheries of the Tanzania Part of Lake Victoria." *African J. Tropical Hydrobiology and Fisheries* 3(2):167-181.

Kuper, H., ed. [1965]: *Urbanization and Migration in West Africa.* Berkeley, Univ. of California Press.

Kuznets, S. [1955]: "Economic Growth and Income Inequality." *Amer. Econ. Rev.* 45:1-28.

———— [1976]: "Demographic Aspects of the Size Distribution of Income: An Exploratory Essay." *Econ. Dev. and Cultural Change* 25:1-94.

Labonne, M., and B. Legagneux [1977]: *Réflexions sur l'agriculture Sénégalais: étude préalable à une modelisation.* Montpellier, France.

La Clau, E. [1971]: "Feudalism and Capitalism in Latin America." *New Left Rev.* 67:19-38.

Lagemann, J. [1977]: *Traditional African Farming Systems in Eastern Nigeria: An Analysis of Reaction to Increasing Population Density.* Munich, West Germany, Weltforum Verlag.

Lagemann, J. J., C. Flinn, and H. Ruthenberg [1976]: "Land Use, Soil Fertility and Agricultural Productivity as Influenced by Population Density in Eastern Nigeria." *Zeitschrift fur Auslandische Landwirtschaft* 5(2):206-219.

Lamb, G., and L. Muller [1982]: *Central Accountability and Incentives in a Successful Development Institution: The Kenya Tea Development Authority*. Washington, World Bank.

Lancaster, C. W. [1976]: "Women, Horticulture and Society in Sub-Saharan Africa." *Amer. Anthropologist* 78:539-564.

Land Tenure Center [1985]: *A Colloquium on Issues in African Land Tenure*. Madison, University of Wisconsin.

Lang. H. [1979]: *The Economics of Rainfed Rice Cultivation in West Africa: The Case of the Ivory Coast*. Saarbrucken, West Germany, Verlag Breitenbach.

Lang, H., and R. Bartsch [1977]: "Évaluation de l'intérêt économique de méthodes culturales améliorées en conditions d'incertitude climatique présentée à l'exemple de la région Centre en Côte-d'Ivoire." *L'Agronomie Tropicale* 32(3):248-256.

Lassiter, G. C. [1981]: *Cropping Enterprises in Eastern Upper Volta*. East Lansing, MSU, DAE, ARE Working Paper No. 35.

Lateef, N. V. [1980]: *Crisis in the Sahel: A Case Study in Development Cooperation*. Boulder, CO, Westview Press.

Latham, M. C. [1980]: *Human Nutrition in Tropical Africa*. Rome, FAO.

Laurent, C. K. [1968]: "The Use of Bullocks for Power on Farms in Northern Nigeria." *Bull. Rural Econ. and Sociology* 3(2):235-262.

Lawani, S. M., F. M. Alluri, and E. N. Adimorah [1979]: *Farming Systems in Africa: A Working Bibliography, 1930-1978*. Boston, G. K. Hall.

Lawani, S. M., and M. O. Odubanjo [1976]: *A Bibliography of Yams and the Genus Dioscorea*. Ibadan, Nigeria, IITA.

Leakey, C. L. A., ed. [1970]: *Crop Improvement in East Africa*. Farnham Royal, England, Commonwealth Agricultural Bureau.

Leakey, C. L. A., and J. B. Wills, eds. [1977]: *Food Crops of the Lowland Tropics*. Oxford, England, Oxford Univ. Press.

Le Bris, E., P.-P. Rey, and M. Samuel [1976]: *Capitalisme négrier: la marché des paysans vers le prolétariat*. Paris, Maspero.

Lee, E. [1980]: *Export-Led Rural Development: The Ivory Coast*. Geneva, ILO, World Employment Research Program, Working Paper 32.

Legum, C. [1985]: "Africa's Search for Nationhood and Stability." *J. Contemporary African Studies* October, pp. 21-45.

Lele, U. [1975]: *The Design of Rural Development: Lessons from Africa*. Baltimore, Johns Hopkins Univ. Press.

———— [1977]: "Considerations Related to Optimum Pricing and Marketing Strategies in Rural Development." In Dams and Hunt, eds., 1977, pp. 488-516.

———— [1981]: "Rural Africa: Modernization, Equity and Long-Term Development." *Science* 211:547-553.

———— [1983]: "Problems of Rural Development: Some Contrasts between Asia and Africa." In Nobe and Sampath, eds., 1983, pp. 237-252.

———— [1984a]: "Tanzania: Phoenix or Icarus?" In Harberger, ed., 1984, pp. 159-195.

———— [1984b]: "Food Security in Developing Countries: National Issues." In C. K. Eicher and Staatz, eds., 1984, pp. 207-221.

Lele, U., and W. Candler [1981]: "Food Security: Some East African Considerations." In Valdés, ed., 1981, pp. 102-122.

Le Moigne, M. [1980]: "Animal Draft Cultivation in French Speaking Africa." In ICRISAT, 1980a, pp. 213-220.

Leonard, D. K., ed. [1973]: *Rural Administration in Kenya: A Critical Appraisal*. Nairobi, East African Literature Bureau.

———— [1977]: *Reaching the Peasant Farmers: Organization Theory and Practice in Kenya*. Univ. of Chicago Press.

———— [1991]: *African Successes: Four Public Managers of Kenyan Rural Development*. Berkeley, University of California Press.

Lesthaeghe, R., P. O. Ohadike, J. Kocher, and H. J. Page [1981]: "Child-Spacing and Fertility in Sub-Saharan Africa: An Overview of Issues." In H. J. Page and Lesthaeghe, eds., 1981, pp. 3–23.

Leurquin, P. P. [1960]: *Le niveau de vie des populations rurales du Ruanda-Urundi*. Paris, Editions Nauwelaerts.

Lev, L. [1981]: "The Effect of Cash Cropping on Food Consumption Adequacy among the Meru of Northern Tanzania." Unpublished M.S. thesis, MSU, East Lansing.

Lever, B. G. [1970]: *Agricultural Extension in Botswana*. Reading, England, Univ. of Reading, DAE, Development Study No. 7.

Levine, G., and C. Bailey [1987]: "Water Management in the Gezira Scheme." *Water Resources Dev.* 3(2):115–126.

Lewis, I. M. [1975]: "The Dynamics of Nomadism: Prospects for Sedentarization and Social Change." In Monod, ed., 1975, pp. 426–442.

Lewis, J. V. D. [1978]: "Small Farmer Credit and the Village Production Unit in Rural Mali." In P. Stevens, Jr., ed., 1978, pp. 29–48.

Lewis, S. [1988]: *Economics and Apartheid: The Impact of South Africa's Economic Policies*. New York: Council on Foreign Relations.

Lewis, W. A. [1954]: "Economic Development with Unlimited Supplies of Labor." *Manchester School Econ. and Soc. Studies* 22:139–191.

———— [1955]: "The Economic Development of Africa." In Stillman, ed., 1955, pp. 97–112.

———— [1964]: "Thoughts on Land Settlement." In C. K. Eicher and Witt, eds., 1964, pp. 299–310.

———— [1978a]: "Socialism and Economic Growth." In Dalton, 1978b, ed., pp. 325–338.

———— [1978b]: *The Evolution of the International Economic Order*. Princeton, NJ, Princeton Univ. Press.

Lewycky, D. [1977]: *Tapestry Report from Oodi Weavers*. Gaborone, Botswana.

————, ed. [1969]: *Politics and Change in Developing Societies: Studies in the Theory and Practice of Development*. Cambridge, England, Cambridge Univ. Press.

Leys, C. [1973]: "Interpreting African Underdevelopment: Reflections on the ILO Report on Employment, Incomes and Equity in Kenya." *African Affairs* 72:419–429.

———— [1974]: *Underdevelopment in Kenya: The Political Economy of Neo-Colonialism, 1964-71*. Berkeley, Univ. of California Press.

Liedholm, C. [1973]: *Research on Employment in the Rural Non-Farm Sector in Africa*. East Lansing, MSU, DAE, ARE Paper No. 5.

Liedholm, C., and E. Chuta [1976]: *The Economics of Rural and Urban Small-Scale Industries in Sierra Leone*. East Lansing, MSU, DAE, ARE Paper No. 14.

Liedholm, C., and D. Mead [1987]: *Small-Scale Industries in Developing Countries: Empirical Evidence and Policy Implications*. East Lansing, MSU, DAE, International Development Papers No. 9.

Lijoodi, J. L., and H. Ruthenberg [1978]: "Income Distribution in Kenya's Agriculture." *Zeitschrift fur Auslandische Landwirtschaft* 17(2):115–128.

Linsenmeyer, D. A. [1976]: *Economic Analysis of Alternative Strategies for the Development of Sierra Leone Marine Fisheries*. East Lansing, MSU, DAE, ARE Working Paper No. 18.

Lipton, M. [1978]: *Botswana: Employment and Labour Use in Botswana*, 2 Vols. Gaborone, Botswana, Government Printer.

_____ [1980]: "Migration from Rural Areas of Poor Countries: The Impact on Rural Productivity and Income Distribution." *World Dev.* 8:1-24.

Lipumba, N. [1984]: "The Economic Crisis in Tanzania." In Lipumba, Msambichaka and Wangwe, eds., 1984, pp. 19-46.

Lipumba, N., L. A. Msambichaka, and S. Wangwe, eds. [1984]: *Economic Stabilization Policies in Tanzania*. Dar es Salaam, Univ. of Dar es Salaam.

Livingstone, I. [1977a]: "An Evaluation of Kenya's Rural Industrial Development Programme." *J. Modern African Studies* 15:495-504.

_____ [1977b]: *Economic Irrationality among Pastoral Peoples in East Africa: Myth or Reality.* Nairobi, Univ. of Nairobi, IDS Discussion Paper No. 245.

_____ [1977c]: "Supply Responses of Peasant Producers: The Effect of Own-Account Consumption on the Supply of Marketed Output." *J. Agr. Econ.* 28:153-159.

Lofchie, M. F. [1975]: "Political and Economic Origins of African Hunger." *J. Modern African Studies* 13:551-568.

_____ [1978]: "Agrarian Crisis and Economic Liberalisation in Tanzania." *J. Modern African Studies* 16:451-475.

Lombin, G., and A. Abdullahi [1977]: *Long-Term Fertility Studies at Samaru, Nigeria. I. Effect of Farm-Yard Manure on Monocropped Cotton, Sorghum and Groundnuts and a Rotation of the Three Crops under Continuous Cultivation.* Zaria, Nigeria, Ahmadu Bello Univ., IAR, Samaru Misc. Paper 72.

Long, N. [1974]: *An Introduction to the Sociology of Rural Development.* Boulder, CO, Westview Press.

Longhurst, R., and P. R. Payne [1979]: *Seasonal Aspects of Nutrition: Review of Evidence and Policy Implications.* Sussex, England, Univ. of Sussex, IDS Discussion Paper 145.

Longmire, J., and M. Morris [1987]: "Prospects for Production and Imports of Wheat and Maize in West Africa." Paper presented at an IFPRI/ISRA Conference, Dakar, July 15-17.

Lovejoy, P. [1980]: *Caravans of Kola: The Hausa Kola Trade, 1700-1900.* Zaria, Nigeria, Ahmadu Bello Univ. Press.

Lovejoy, P. E., and S. Baier [1975]: "The Desert-Side Economy of the Central Sudan." *International J. African Hist. Studies* 8(4):551-581.

Low, A. R. C. [1974]: "Decision Taking under Uncertainty: A Linear Programming Model of Peasant Farmer Behaviour." *J. Agr. Econ.* 25(3):311-321.

_____ [1978]: "Linear Programming and the Study of Peasant Farming Situations — A Reply." *J. Agr. Econ.* 29:189-190.

_____ [1980]: *The Estimation and Interpretation of Pastoralists' Price Responsiveness.* London, ODI, Pastoral Network Paper No. 10c.

_____ [1986]: *Agricultural Development in Southern Africa: Farm-Household Economics and the Food Crisis.* London: James Curry.

Low, A. R. C., and R. L. Kemp [1977]: "Destocking Dynamics: The Implications of a Swaziland Example." *Oxford Agrarian Studies* 6:26-43.

Low, A. R. C., R. L. Kemp, and M. H. Doran [1980]: "Cattle as a Store of Wealth: Reply." *Amer. J. Agr. Econ.* 62:613-617.

Lundahl, M., and B. J. Ndulu [1987]: "Market-Related Incentives and Food Production in Tanzania: Theory and Experience." In Hedlund, ed., 1987, pp. 191-228.

Luning, H. A. [1967]: *Economic Aspects of Low Labour-Income Farming*. Wageningen, Netherlands, Centre for Agricultural Publications and Documentation, Agricultural Research Report No. 699.

Lynch, S. G. [1980]: *An Analysis of Interview Frequency and Reference Period in Rural Consumption Expenditure Surveys: A Case Study from Sierra Leone*. East Lansing, MSU, DAE, Rural Development Working Paper No. 10.

Lystad, R. A., ed. [1965]: *The African World: A Survey of Social Research*. New York, Praeger.

MacArthur, J. D. [1968]: "Experience with Farm Management Analysis Techniques in Kenya." *Eastern Africa J. Rural Dev.* 1(2):22-32.

————— [1975]: "Benefits of Hindsight: Aspects of Experience in the High and Low Density Settlement Programme in Kenya." *Eastern Africa J. Rural Dev.* 8(1-2):1-45.

————— [1978]: "Appraising the Distributional Aspects of Rural Development Projects: A Kenya Case Study." *World Dev.* 6:167-194.

MacKenzie, W. [1976]: "The Livestock Development Program in Tanzania: Phase II." *Eastern Africa J. Rural Dev.* 9(1-2):92-115.

McAlpin, M. [1987]: "Famine Relief Policy in India: Six Lessons for Africa." In Glantz, ed., 1987, pp. 393-413.

McCown, R. L., G. Haaland, and C. de Haan [1979]: "The Interaction between Cultivation and Livestock Production in Semi-Arid Africa." In A. E. Hall, Cannell, and Lawton, eds., 1979, pp. 297-332.

McDermott, J. K. [1988]: "Agricultural Technology Innovation Institutions in Cameroon." Paper presented at the Cameroon Conference, Gainesville, Univ. of Florida, Center for African Studies.

McDowell, M. [1976]: *Lesotho Labour in South African Mines*. Johannesburg, Agency for Industrial Mission Conference Report.

McDowell, R. E. [1972]: *The Improvement of Livestock Production in Warm Climates*. San Francisco, W. H. Freeman and Co.

————— [1978]: "Feed Resources of Small Farms." Paper presented at a Seminar on Improvement of Farming Systems, February 20-March 1, 1978. Bamako, Mali, IER.

McDowell, R. E., and L. Bove [1977]: *The Goat as a Producer of Meat*. Ithaca, NY, Cornell Univ., International Agriculture Mimeograph No. 56.

McDowell, R. E., and P. E. Hildebrand [1980]: *Integrated Crop and Animal Production: Making the Most of Resources Available to Small Farms in Developing Countries*. New York, Rockefeller Foundation.

McGowan, P. J. [1976]: "Economic Dependency and Economic Performance in Black Africa." *J. Modern African Studies* 14(1):25-40.

McHenry, D. E., Jr. [1979]: *Tanzania's Ujamaa Villages: The Implementation of a Rural Development Strategy*. Berkeley, Univ. of California, Institute of International Studies.

————— [1981]: *Ujamaa Villages in Tanzania: A Bibliography*. Uppsala, Sweden, SIAS.

McIlwaine, J. H. [1979]: *Theses on Africa, 1963-1975, Accepted by Universities in the United Kingdom and Ireland*. London, Mansell Information.

McIntyre, J. [1981]: *Food Security in the Sahel: Variable Import Levy, Grain Reserves and Foreign Exchange Assistance*. Washington, IFPRI, Research Report 26.

McKelvey, J. J., Jr. [1965]: "Agricultural Research." In Lystad, ed., 1965, pp. 317-351.

McKim, W. [1972]: "The Periodic Market System in Northeastern Ghana." *Econ. Geography* 48(3):333-344.

McLoughlin, P. F. M. [1967]: *Research on Agricultural Development in East Africa*. New York, ADC.

————, ed. [1970]: *African Food Production Systems: Cases and Theory*. Baltimore, Johns Hopkins Univ. Press.

McNamara, R. S. [1973]: *Address to the Board of Governors: Nairobi, Kenya, September 24, 1972*. Washington, IBRD.

McPhee, A. [1971]: *The Economic Revolution of British West Africa*, 2nd ed. London, Frank Cass. First edition, 1926.

Maas, A., and R. L. Anderson [1978]: . . . *And the Desert Shall Rejoice: Conflict, Growth, and Justice in Arid Environments*. Cambridge, MA, MIT Press.

Mabawonku, W. [1978]: *Economic Evaluation of Apprenticeship Training in Western Nigerian Small-Scale Industries*. East Lansing, MSU, DAE, ARE Paper No. 17.

Mabogunje, A. L. [1972]: *Regional Mobility and Resource Development in West Africa*. Montreal, McGill-Queen's Univ. Press.

———— [1981]: *The Development Process: A Spatial Perspective*. New York, Holmes and Meier.

Magasa, A. [1978]: *Papa-commandant a jeté un grand filet devant nous. Les exploites des rives du Niger, 1900-1962*. Paris, Maspero.

Maimbo, F. J. M., and J. Fry [1971]: "An Investigation into the Change in the Terms of Trade between the Rural and Urban Sectors of Zambia." *African Social Research* 12:95-110.

Maitha, J. K. [1969]: "A Supply Function for Kenyan Coffee." *Eastern Africa Econ. Rev.* 1:63-72.

———— [1970]: "Productivity Response to Price: A Case Study of Kenyan Coffee." *Eastern Africa Econ. Rev.* 2:31-37.

———— [1974]: "A Note on Distributed Lag Models of Maize and Wheat Production Response: The Kenyan Case." *J. Agr. Econ.* 25:183-188.

Mandaza, I., ed. [1986]: *Zimbabwe: The Political Economy of the Transition, 1980-86*. Dakar, Senegal, CODESRIA.

Manetsch, T. J., M. L. Hayenga, A. N. Halter, T. W. Carroll, M. H. Abkin, D. R. Byerlee, K.-Y. Chang, G. Page, E. Kellogg, and Glenn L. Johnson [1971]: *A Generalized Simulation Approach to Agricultural Sector Analysis with Special Reference to Nigeria*. East Lansing, MSU, DAE.

Mansvelt-Beck, J., and J. J. Sterkenburg [1976]: *The Fishing Industry in Cape Coast, Ghana: A Social-Economic Analysis of Fishing in a Medium-Sized African Town*. Utrecht, Netherlands, Geografisch Instituut Rijksuniversiteit Utrecht.

Marchal, J.-Y. [1977]: "Évolution des systèmes agraires: l'exemple du Yatenga (Haute Volta)." *Environment Africaine* 2(4) and 3(1):75-88.

Mariam, M. W. [1986]: *Rural Vulnerability to Famine in Ethiopia, 1958-77*. London, Intermediate Technology Publications.

Marshall, A. [1956]: *Principles of Economics, An Introductory Volume*, 8th ed. London, Macmillan.

Marter, A. [1978]: *Cassava or Maize: A Comparative Study of the Economics of Production and Market Potential of Cassava and Maize in Zambia*. Lusaka, Univ. of Zambia, Rural Development Studies Bureau.

Martin, F. [1988]: "Food Security and Comparative Advantage in Senegal: A Micro-Macro Approach." Unpublished Ph.D. dissertation, MSU, East Lansing.

Martin, G. [1976]: "Socialism, Economic Development, and Planning in Mali, 1960-1968." *Canadian J. African Studies* 10(1):23-46.

Mascarenhas, O., and M. Mbilinyi [1980]: *Women and Development in Tanzania: An Annotated Bibliography*. Addis Ababa, ECA.

Massell B. F. [1967a]: "Farm Management in Peasant Agriculture: An Empirical Study." *Food Research Institute Studies* 7:205-214.

———— [1967b]: "Elimination of Management Bias from Production Functions Fitted to Cross-Section Data: A Model and an Application to African Agriculture." *Econometrica* 35:495-508.

———— [1969]: "Consistent Estimation of Expenditure Elasticities from Cross-Section Data on Households Producing Partly for Subsistence." *Rev. Econ. and Stat.* 51:136-142.

Massell, B. F., and J. Heyer [1969]: "Household Expenditure in Nairobi: A Statistical Analysis of Consumer Behavior." *Econ. Dev. and Cultural Change* 17:212-234.

Massell, B. F., and R. W. M. Johnson [1968]: "Economics of Smallholder Farming in Rhodesia: A Cross-Section Analysis of Two Areas." *Food Research Institute Studies 8* (Supplement):1-74.

Massell, B. F., and A. Parnes [1969]: "Estimation of Expenditure Elasticities from a Sample of Rural Households in Uganda." *Oxford Bull. Econ. and Stat.* 31:313-329.

Mathewman, R. W. [1979]: *A Survey of Small Livestock Production at the Village Level in the Derived Savanna and Lowland Forest Zones of S. W. Nigeria.* Reading, England, Univ. of Reading, Department of Agriculture and Horticulture.

Matlon, P. [1987]: "Making Millet Improvement Objectives Fit Client Needs: Improved Genotypes and Traditional Management Systems in Burkina Faso." In ICRISAT, 1987, pp. 233-245.

Matlon, P., R. Cantrell, D. King, and M. Benoit-Cattin, eds. [1984]: *Coming Full Circle: Farmers' Participation in the Development of Technology.* Ottawa, IDRC.

Matlon, P., T. Eponou, S. Franzel, D. Byerlee, and D. C. Baker [1979a]: *Income Distribution among Farmers in Northern Nigeria: Empirical Results and Policy Implications.* East Lansing, MSU, DAE, ARE Paper No. 18.

———— [1979b]: *Poor Rural Households, Technical Change, and Income Distribution in Developing Countries: Two Case Studies from West Africa.* East Lansing, MSU, DAE, ARE Working Paper No. 29.

———— [1980]: *Local Varieties, Planting Strategies and Early Season Activities in Two Villages of Central Upper Volta.* Ouagadougou, Burkina Faso, ICRISAT.

Matlon, P., and D. S. C. Spencer [1984]: "Increased Food Production in Sub-Saharan Africa: Environmental Problems and Inadequate Technological Solutions." *Amer. J. Agr. Econ.* 66:671-676.

Maunder, A., and U. Renborg, eds. [1986]: *Agriculture in a Turbulent World Economy. Proceedings of the Nineteenth ICAE, Malaga, Spain, 26 August-4 September, 1985.* London, Gower.

Maxwell, S. J. [1986]: "The Social Scientist in Farming Systems Research." *J. Agr. Econ.* 37:25-36.

Maxwell, S. J., and H. W. Singer [1979]: "Food Aid to Developing Countries: A Survey." *World Dev.* 7:225-247.

Maymard, J. [1974]: "Structures Africaines de production et concept d'exploitation agricole. Première partie: un exemple de terroir Africain; les confins Diola-Manding aux bords du Sonngrongron (Sénégal)." *Cahiers ORSTOM, Ser. Biologie* 24:27-64.

Mbilinyi, M. J. [1972]: "The 'New Woman' and Traditional Norms in Tanzania." *J. Modern African Studies* 10:57-72.

Mbithi, P. M. [1973]: "Agricultural Extension as an Intervention Strategy: An Analysis of Extension Approaches." In Leonard, ed., 1973, pp. 76-96.

———— [1977]: "Human Factors in Agricultural Management in East Africa." *Food Policy* 2(1):27-33.

Mbithi, P. M., and C. Barnes [1975]: *The Spontaneous Settlement Problem in the Context of Rural Development in Kenya*. Nairobi, East African Literature Bureau.

Mboya, T. [1967]: *A Development Strategy for Africa: Problems and Proposals*. Nairobi, Kenya, Ministry of Economic Planning and Development.

Medani, A. I. [1975]: "Elasticity of the Marketable Surplus of a Subsistence Crop at Various Stages of Development." *Econ. Dev. and Cultural Change* 23:421-429.

_____ [1979]: "Elasticity of the Marketable Surplus of a Subsistence Crop at Various Stages of Development: Reply." *Econ. Dev. and Cultural Change*. 28:179-181.

Mehretu, A. [1989]: *Regional Disparity in Sub-Saharan Africa: Structural Readjustment of Uneven Development*. Boulder, CO, Westview Press.

Meier, G. M. [1975]: "External Trade and Internal Development." In Duignan and Gann, eds., 1975a, pp. 427-469.

Meier, G. M., and D. Seers, eds. [1984]: *Pioneers in Development*. New York. Oxford Univ. Press.

Meier, G. M., and W. F. Steel, eds. [1989]: *Industrial Adjustment in Sub-Saharan Africa*. New York, Oxford Univ. Press.

Meillassoux, C. [1964]: *Anthropologie économique des Gouro de Côte d'Ivoire*. Paris, Mouton.

_____ , ed. [1971]: *The Development of Indigenous Trade and Markets in West Africa*. London, Oxford Univ. Press for International African Institute.

_____ [1972]: "From Reproduction to Production." *Economy and Society* 1(1):93-105.

_____ [1974]: "Development or Exploitation: Is the Sahel Famine Good Business?" *Rev. African Polit. Economy* 1:27-33.

_____ [1981]: *Maidens, Meals and Money: Capitalism and the Domestic Community*. Cambridge, England, Cambridge Univ. Press.

Mellor, J. W., ed. [1979a]: *India: A Rising Middle Power*. Boulder, CO, Westview Press.

_____ [1979b]: "The Indian Economy: Objectives, Performance and Prospects." In Mellor, ed., 1979a, pp. 85-110.

Mellor, J. W., C. L. Delgado, and M. J. Blackie, eds. [1987]: *Accelerating Food Production in Sub-Saharan Africa*. Baltimore, Johns Hopkins Univ. Press.

Mellor, J. W., and S. Gavian [1987]: "Famine: Causes, Prevention and Relief." *Science* 235:539-545.

Mensah, M. C. [1977]: "An Experience of Group Farming in Dahomey: The Rural Development Cooperatives." In Dorner, ed., 1977, pp. 277-286.

Menz, K. M. [1980]: "Unit Farms and Farming Systems Research: The IITA Experience." *Agr. Systems* 6:45-51.

Mercer-Quarshie, H. [1979]: "Yields of Local Sorghum (Sorghum Vulgare) Cultivars and their Mixtures in Northern Ghana." *Tropical Agriculture* 56(2):125-133.

Merrill-Sands, D. [1986]: "Farming Systems Research: Clarification of Terms and Concepts." *Experimental Agriculture* 22:87-104.

Metson, J. [1978]: "Normative Supply Response in a Mixed Farming System: A Study of Dairying and Maize Production in Nandi District, Kenya." Unpublished Ph.D. dissertation, Univ. of East Anglia, Norwich, England.

Mettrick, H. [1978]: *Oxenisation in The Gambia: An Evaluation*. London, Ministry of Overseas Development.

Migdal, J. S. [1974]: *Peasants, Politics and Revolution: Pressures toward Political and Social Change in the Third World*. Princeton, NJ, Princeton Univ. Press.

Migot-Adholla, S. E. [1972]: "The Politics of Mechanization in Sukumaland (Tanzania)." In Widstrand, ed., 1971, pp. 81-104.

Mijindadi, N. B. [1980]: "Production Efficiency on Farms in Northern Nigeria." Unpublished Ph.D. dissertation, Cornell Univ., Ithaca, NY.

Miller, L. [1977]: *Agricultural Credit and Finance in Africa.* New York, Rockefeller Foundation.

Miller, W. [1965]: "An Economic Analysis of Oil Palm Fruit Processing in Eastern Nigeria." Unpublished Ph.D. dissertation, MSU, East Lansing.

Minford, P. and P. Ohs [1976]: "Supply Response of Malawi Labour." *Eastern Africa Econ. Rev.* 8:15-34.

Miracle, M. P. [1966]: *Maize in Tropical Africa.* Madison, Univ. of Wisconsin Press.

——— [1968]: "Market Structure in Commodity Trade and Capital Accumulation in West Africa." In Moyer and Hollander, eds., 1968, pp. 209-227.

——— [1969]: "An Evaluation of Attempts to Introduce Cooperatives and Quasi-Cooperatives in Tropical Africa." In Anschel, Brannon, and E. D. Smith, eds., 1969, pp. 120-139.

——— [1976]: "Interpretation of Backward-Sloping Labor Supply Curves in Africa." *Econ. Dev. and Cultural Change* 24:399-406.

Miracle, M. P., and B. Fetter [1970]: "Backward-Sloping Labor-Supply Functions and African Economic Behavior." *Econ. Dev. and Cultural Change* 18:246-251.

Miracle, M. P., D. S. Miracle, and L. Cohen [1980]: "Informal Savings Mobilization in Africa." *Econ. Dev. and Cultural Change* 28:701-724.

Miracle, M. P., and A. W. Seidman [1968a]: *State Farms in Ghana.* Madison, Univ. of Wisconsin, LTC Paper No. 43.

——— [1968b]: *Agricultural Cooperatives and Quasi-Cooperatives in Ghana, 1951-65.* Madison, Univ. of Wisconsin, LTC Paper No. 51.

Mitchell, J. C. [1959]: "The Causes of Labour Migration." *Bull. International African Labour Institute* 6(1):12-47.

Mittendorf, H. J. [1985]: "Mobilization of Personal Savings for Agricultural and Rural Development in Africa." *Mondes en développement* 13(5-151):275-291.

Monnier, J. [1972]: "Relations entre mécanisation, dimensions et systèmes d'exploitation." *Machinisme Agricole Tropicale* 38:33-44.

——— [1975]: "Farm Mechanization in Senegal and its Effects on Production and Employment." *Report of the FAO/OECD Expert Panel on the Effects of Farm Mechanization on Production and Employment.* Rome, FAO, pp. 215-250.

Monnier, J., A. Diagne, D. Sow, and T. Sow [1974]: *Le travail dans l'exploitation agricole Sénégalaise.* Bambey, CNRA.

Monod, T., ed. [1975]: *Pastoralism in Tropical Africa.* London, Oxford Univ. Press.

Montgomery, R. [1977]: *The Economics of Fertilizer Use on Sahelian Cereals: The Experience in Mali and Upper Volta.* Abidjan, Côte d'Ivoire, AID, Regional Economic Development Support Office.

Monyo, T., A. D. R. Ker, and M. Campbell, eds. [1976]: *Intercropping in Semi-Arid Areas.* Ottawa, Canada, IDRC.

Moock, J. L. [1984]: "Overseas Training and National Development Objectives in Sub-Saharan Africa." *Comparative Education Rev.* 28(2):221-240.

———, ed. [1986]: *Understanding Africa's Rural Households and Farming Systems.* Boulder, CO, Westview Press.

Moock, P. R. [1976]: "The Efficiency of Women as Farm Managers: Kenya." *Amer. J. Agr. Econ.* 58:831-835.

——— [1981]: "Education and Technical Efficiency in Small-Farm Production." *Econ. Dev. and Cultural Change* 29:723-740.

Morawetz, D. [1974]: "Employment Implications of Industrialization in Developing Countries: A Survey." *Econ. J.* 84:491-542.

_____ [1980]: "Economic Lessons from Some Small Socialist Developing Countries." *World Dev.* 8(5/6):337-369.

Morgan, W. B. [1978]: *Agriculture in the Third World: A Spatial Analysis.* Boulder, CO, Westview Press.

Moris, J. R. [1973]: "Managerial Structures and Plan Implementation in Colonial and Modern Agricultural Extension: A Comparison of Cotton and Tea Programmes in Central Kenya." In Leonard, ed., 1973, pp. 97-131.

_____ [1987]: "Irrigation as a Privileged Solution in African Development." *Dev. Policy Rev.* 5(2):99-123.

Moris, J. R., and D. Thom [1987]: *African Irrigation Overview: Main Report.* Logan, Utah State Univ., Water Management Synthesis II Project.

Moris, J. R., D. Thom, and R. Norman [1984]: *Prospects for Small-Scale Irrigation Development in the Sahel.* Logan, Utah State Univ., Water Management Synthesis Report No. 26.

Morris, M. [1988a]: "Parallel Rice Markets: Policy Lessons from Northern Senegal." *Food Policy* 13:257-269.

_____ [1988b]: *Comparative Advantage and Policy Incentives for Wheat Production in Zimbabwe.* Mexico, CIMMYT Working Paper 88/02.

Mortimore, M. J. [1967]: "Land and Population Pressure in the Kano Close-Settled Zone, Northern Nigeria." *Advancement of Science* 23(118):677-686.

Mosley, Paul, and Lawrence Smith [1989]: "Structural Adjustment and Agricultural Performance in Sub-Saharan Africa, 1980-87." *J. Int. Dev.* 1(3):321-355.

Moyer, R., and S. Hollander, eds. [1968]: *Markets and Marketing in Developing Economies.* Homewood, IL.

Msangi, A. S., and J. J. Griffin, eds. [1974]: *International Conference on Marine Resources Development in Eastern Africa, April 4-9, 1974.* Tanzania, Univ. of Dar es Salaam; Narragansett, Univ. of Rhode Island.

Muchiri, G. [1979]: "Development of Tillage and Equipment Systems in Kenya; With Special Reference to the Work at the University of Nairobi." *Appropriate Technology for Tillage Operations Workshop.* Zaria, Nigeria, Ahmadu Bello Univ., IAR.

Mudahar, M. S. [1980]: "Principal Policy Issues Facing the Fertilizer Sector in Africa: A Perspective." Muscle Shoals, AL, IFDC, draft.

Mueller, E. [1984]: "The Value and Allocation of Time in Rural Botswana." *J. Dev. Econ.* 15:329-360.

Muir-Leresche, K. [1984]: "Crop Price and Wage Policy in the Light of Zimbabwe's Development Goals." Unpublished D. Phil. thesis, University of Zimbabwe, Harare.

Mukui, J. T., ed. [1979]: *Price and Marketing Controls in Kenya.* Nairobi, Univ. of Nairobi, IDS, Occasional Paper No. 32.

Mumbengegwi, C. [1988]: "The Political Economy of a Small-Farmer Agricultural Strategy in SADCC." In Bryant, ed., 1988, pp. 158-177.

Murray, C. [1977]: "High Bridewealth, Migrant Labour and the Position of Women in Lesotho." *J. African Law* 21(1):79-96.

Mwandemere, H. [1984]: "Reorganization of Agricultural Research in Malawi." In *SADCC,* 1984, pp. 61-72.

Mwangi, W. M. [1978]: "Farm Level Derived Demand Responses for Fertilizer in Kenya." Unpublished Ph.D. dissertation, MSU, East Lansing.

Mwansasu, B. U., and C. Pratt, eds. [1979]: *Towards Socialism in Tanzania*. Toronto, Univ. of Toronto Press.

Mwanza, J. M. [1979]: "Rural-Urban Migration and Urban Employment in Zambia." In Turok, ed., 1979, pp. 26-36.

Myint, H. [1958]: "The 'Classical Theory' of International Trade and the Underdeveloped Countries." *Econ. J.* 68:317-337.

Myrdal, G. [1984]: "International Inequality and Foreign Aid in Retrospect." In Meier and Seers, eds. 1984, pp. 149-172.

Nafziger, E. W. [1988]: *Inequality in Africa: Political Elites, Proletariats, Peasants, and the Poor.* Cambridge, Cambridge Univ. Press.

Nash, J., J. Dandler, and N. S. Hopkins, eds. [1976]: *Population Participation in Social Change: Cooperatives, Collectives, and Nationalized Industry.* The Hague, Netherlands, Mouton.

National Research Council [NRC, 1968]: *Conference on Agricultural Research Priorities for Economic Development in Africa*, 3 Vols. Washington, National Academy of Sciences.

National Research Council, Committee on African Agricultural Capabilities [NRC, 1974]: *African Agricultural Research Capabilities*. Washington, National Academy of Sciences.

―――― [NRC, 1978]: *Post-Harvest Food Losses in Developing Countries*. Washington, National Academy of Sciences.

Ndegwa, P. [1968]: *The Common Market and Development in East Africa*. Nairobi, East African Publishing House.

―――― [1985]: *Africa's Development Crisis and the Related International Issues*. Nairobi, Heineman.

―――― [1986]: *The African Challenge: In Search of Appropriate Development Strategies*. Nairobi, Heinemann.

Ndzinge, L., L. Marsh, and R. Greer [1984]: "Herd Inventory and Slaughter Supply Response in Botswana Beef Cattle Producers." *J. Agr. Econ.* 35:97-107.

Nellis, J. [1986]: *Public Enterprises in Sub-Saharan Africa*. Washington, World Bank, Discussion Paper No. 1.

Nelson, M. [1973]: *The Development of Tropical Lands: Policy Issues in Latin America*. Baltimore, Johns Hopkins Univ. Press.

Netting, R. [1968]: *Hill Farmers of Nigeria: Cultural Ecology of the Kofyar of the Jos Plateau*. Seattle, Univ. of Washington Press.

―――― [1974]: "Agrarian Ecology." *Annual Review of Anthropology* 3:21-57.

Newbury, C. W., and A. S. Kanya-Forstner [1969]: "French Policy and the Origins of the Scramble for West Africa." *J. African Hist.* 10(2):253-276.

Newman, M., I. Oeudraogo, and D. Norman [1980]: "Farm Level Studies in the Semi-Arid Tropics of West Africa." In ICRISAT, 1980a, pp. 241-263.

Ngambeki, D. S., and F. S. Idachaba [1985]: "Supply Response of Upland Rice in Ogun State of Nigeria: A Producer Panel Approach." *J. Agr. Econ.* 36:239-49.

Ng'ethe, N. [1980]: "Income Distribution in Kenya: The Politics of Mystification." In Rweyemamu, ed., 1980, pp. 191-213.

Ngoddy, P. O. [1977]: "Gari Mechanization in Nigeria: The Competition between Intermediate and Modern Technology." In Jequier, ed., 1976, pp. 260-275.

Nicholls, W. H. [1964]: "The Place of Agriculture in Economic Development." In Eicher and Witt, eds., 1964, pp. 11-44.

Nicou, R. [1978]: "Étude de succession culturales au Sénégal. Résultats et méthodes." *L'Agronomie Tropicale* 33(1):51-61.

Nigeria, Federal Ministry of Agriculture and Rural Development, Western Region [1959]: *Future Policy of the Ministry of Agriculture and Natural Resources.* Sessional Paper No. 9.

Nigeria, Federal Ministry of Agriculture [1980]: *The Green Revolution: A Food Production Plan for Nigeria. Final Report.* Lagos.

Nkrumah, K. [1963]: *Africa Must Unite.* New York, Praeger.

———— [1970]: *Class Struggle in Africa.* New York, International Publishers.

Nobe, K., and R. Sampath [1983]: *Issues in Third World Development.* Boulder, CO, Westview Press.

Nordic Delegation [1984]: *Policies in Agriculture and Rural Development: A Nordic View.* Position paper presented at the SADCC conference at Lusaka.

Norman, D. W. [1969]: "Labour Inputs of Farmers: A Case Study of the Zaria Province of the North Central State of Nigeria." *Nigerian J. Econ. Soc. Studies* 11:3-14.

———— [1972]: *An Economic Study of Three Villages in Zaria Province. 2. An Input-Output Study. Vol. 1. Text.* Zaria, Nigeria, Ahmadu Bello Univ., IAR, Samaru Miscellaneous Paper No. 37.

———— [1973]: *Methodology and Problems of Farm Management Investigations: Experiences from Northern Nigeria.* East Lansing, MSU, DAE, African Rural Employment Paper No. 8.

———— [1974]: "Rationalizing Mixed Cropping under Indigenous Conditions: The Example of Northern Nigeria." *J. Dev. Studies* 11:3-21.

———— [1980]: *The Farming Systems Approach: Relevancy for the Small Farmer.* East Lansing, MSU, DAE, Rural Development Paper No. 5. Also available in French and Spanish.

Norman, D. W., D. C. Baker, G. Heinrich, and F. Worman [1988]: "Technology Development and Farmer Groups: Experience from Botswana." *Experimental Agriculture* 24:321-332.

Norman, D. W., P. Beeden, W. J. Kroeker, D. H. Pryor, H. M. Hays, and B. Huizinga [1976a]: *The Feasibility of Improved Sole Crop Maize Production Technology for the Small-Scale Farmer in the Northern Guinea Savanna Zone of Nigeria.* Zaria, Nigeria, Ahmadu Bello Univ., IAR, Samaru Miscellaneous Paper No. 59.

Norman, D. W., P. Beeden, W. J. Kroeker, D. H. Pryor, B. Huizinga, and H. M. Hays [1976b]: *The Feasibility of Improved Sole Crop Sorghum Production Technology for the Small-Scale Farmer in the Northern Guinea Savanna Zone of Nigeria.* Zaria, Nigeria, Ahmadu Bello Univ., IAR, Samaru Miscellaneous Paper No. 60.

Norman, D. W., J. C. Fine, A. D. Goddard, W. J. Kroeker, and D. H. Pryor [1976]: *A Socio-Economic Survey of Three Villages in the Sokoto Close-Settled Zone. 3. Input-Output Study. Vol. 1. Text.* Zaria, Nigeria, Ahmadu Bello Univ., IAR, Samaru Miscellaneous Paper No. 64.

Norman, D. W., J. A. Hayward, and H. R. Hallam [1974]: "An Assessment of the Cotton Growing Recommendations as Applied by Nigerian Farmers." *Cotton Growing Rev.* 51:266-288.

———— [1975]: "Factors Affecting Cotton Yields Obtained by Nigerian Farmers." *Cotton Growing Rev.* 52:30-37.

Norman, D. W., M. D. Newman, and I. Oeudraogo [1981]: *Farm and Village Production Systems in the Semi-Arid Tropics of West Africa.* Hyderabad, India, ICRISAT, Research Bulletin No. 4, Vol. 1.

Norman, D. W., and R. W. Palmer-Jones [1977]: "Economic Methodology for Assessing Cropping Systems." In *Symposium on Cropping Systems Research and Development for the Asian Rice Farmer.* Los Banos, Philippines, IRRI.

Norman, D. W., D. H. Pryor, and C. J. N. Gibbs [1979]: *Technical Change and the Small Farmer in Hausaland, Northern Nigeria.* East Lansing, MSU, DAE, ARE Paper No. 21.

Norman, D. W., E. Simmons, and H. Hays [1982]: *Farming Systems in the Nigerian Savannah: Research Strategies for Development.* Boulder, CO, Westview Press.

Noronha, R. [1987]: *A Review of the Literature on Land Tenure in Sub-Saharan Africa.* Baltimore, Johns Hopkins Univ. Press.

North, D. C. [1984]: "Government and the Cost of Exchange in History." *J. Econ. Hist.* 44(2):255-64.

———— [1987]: "Institutions, Transactions Costs and Economic Growth." *Econ. Inquiry* 25(5):419-428.

Ntangsi, J. V. [1979]: "The Political Economy of Rural Development in Cameroon." Unpublished Ph.D. dissertation, Univ. of California, Berkeley.

Nukunya, G. K. [1975]: "The Effects of Cash Crops on an Ewe Community." In Goody, ed., 1975, pp. 59-71.

Nweke, F. I. [1978a]: "Agricultural Credit in Ghana: Priorities and Needs for Domestic Food Production." *Canadian J. Agr. Econ.* 26(3):38-46.

———— [1978b]: "Direct Governmental Production in Agriculture in Ghana, Consequences for Food Production and Consumption, 1960-66 and 1967-75." *Food Policy* 3(3):202-208.

———— [1979]: "Farm Mechanization and Farm Labour in Ghana: An Analysis of Efficiency and Impact on Domestic Food Production and Policy Issues." *Zeitschrift fur Auslandische Landwirtschaft* 18(2):171-187.

Nyantenz, V. K. [1972]: *The Storage of Foodstuffs in Ghana.* Legon, Univ. of Ghana, ISSER, Technical Publication Series No. 18.

Nye, P. H., and D. H. Greenland [1960]: *The Soil under Shifting Cultivation.* Farnham Royal, England, Commonwealth Agricultural Bureau.

Nyerere, J. K. [1967]: *Socialism and Rural Development.* Dar es Salaam, Tanzania, Government Printer. Also in Nyerere, 1968, pp. 106-144.

———— [1968]: *Ujamaa: Essays on Socialism.* Oxford, England, Oxford Univ. Press.

———— [1977]: *The Arusha Declaration. Ten Years After.* Dar es Salaam, Tanzania, Government Printer. Reprinted 1977 in abridged form in International Dev. Rev. 19(1):2-7.

———— [1984]: "Interview." *Third World Quarterly* 6(4):815-838.

Obeng, H. B. [1978]: "Major Soils of West Africa and their General Suitability for Crop and Livestock Production." *African J. Agr. Sciences* 5(1):71-83.

Odell, M. L., and M. J. Odell, Jr. [1980]: *The Evolution of a Strategy for Livestock Development in the Communal Areas of Botswana.* London, ODI, Pastoral Network Paper No. 10b.

Odero-Ogwel, L., and E. Clayton [1973]: *A Regional Programming Approach to Agricultural Sector Analysis: An Application to Kenya Agriculture.* Ashford, Kent, Wye College.

Oduor-Otieno, M. L., R. S. Karisa, J. O. O. Othiambo, and T. C. I. Ryan [1978]: *A Study of the Supply Function for Fish in the Kenya Waters of Lake Victoria and on the Kenya Coast.* Nairobi, Univ. of Nairobi, IDS, Working Paper No. 346.

OECD [1984]: *Development Cooperation, 1984 Review.* Paris.

Office of African Unity [OAU, 1980]: *The Lagos Plan of Action for the Implementation of the Monrovia Strategy for the Economic Development of Africa.* Lagos, Nigeria.

Office of African Unity and Economic Commission for Africa [OAU and ECA, 1986]: *Africa's Submission to the Special Session of the UN General Assembly on Africa's Economic and Social Crisis.* Addis Ababa.

Ofori, I. M., ed. [1973]: *Factors of Agricultural Growth in West Africa. Proceedings of an International Conference held at Legon, April 1971.* Legon, Univ. of Ghana, ISSER.

Ogunfowora, O. [1972]: "Conceptualizing Increased Resource Demand and Product Supply Inducing Policies in Peasant Agriculture." *Nigerian J. Econ. Soc. Studies* 14:191-201.

Ogunfowora, O., and D. W. Norman [1973]: "Farm-Firm Normative Fertilizer Demand Response in the North Central State of Nigeria." *J. Agr. Econ.* 24:301-309.

_____ [1974]: *An Optimization Model for Evaluating the Stability of Sole Cropping and Mixed Cropping Systems under Changing Resource and Technology Levels.* Zaria, Nigeria, Ahmadu Bello Univ., IAR.

Okai, M. [1975]: "The Development of Ox Cultivation Practices in Uganda." *Eastern Africa J. Rural Dev.* 8(1-2):191-214.

Okali, C., and R. A. Kotey [1971]: *Akokoaso: A Resurvey.* Legon, Univ. of Ghana, ISSER.

Okedi, J. [1974]: *Fishery Resources: Their Exploitation, Management and Conservation in Africa.* Jinja, Uganda, Eastern Africa Freshwater Fisheries Research Organization.

Okediji, F. O. [1972]: "Family Planning in Africa: Overcoming Social and Cultural Resistance." *Revue Internationale de l'Education de la Santé (Génève)* 15:199-206.

Okidi, C. O., ed. [1978]: *Management of Coastal and Offshore Resources in Eastern Africa: Papers Presented at the Workshop held at the Institute for Development Studies, University of Nairobi, April 26-29, 1977.* Nairobi, Univ. of Nairobi, IDS, Occasional Paper No. 28.

_____ [1979]: *Kenya's Marine Fisheries: An Outline of Policy and Activities.* Nairobi, Univ. of Nairobi, IDS, Occasional Paper No. 30.

Okigbo, B. N. [1986]: "Broadening the Food Base in Africa: The Potential of Traditional Food Plants." *Food and Nutrition* 12(1):4-17.

Okigbo, B. N., and D.J. Greenland [1977]: "Intercropping Systems in Tropical Africa." In Papendick, Sanchez, and Triplett, eds., 1977, pp. 63-101.

Okojie, P. [1987]: "Africa and the Famine Movement." *J. African Marxists* 16(June):80-89.

Okurume, G. [1973]: *Foreign Trade and the Subsistence Sector in Nigeria: The Impact of Agricultural Exports on Domestic Food Supplies in a Peasant Economy.* New York, Praeger.

Olayemi, J. K. [1974]: "Costs and Returns to Cocoa and Alternative Crops in Western Nigeria." In Kotey, Okali, and Rourke, eds., 1974, pp. 48-58.

_____ [1976]: "Irreversible Price Response Patterns: The Case of Nigerian Rubber Producers." *J. Rural Econ. and Dev.* (Ibadan) 10(2):28-35.

Olayemi, J. K., and S. A. Oni [1972]: "Asymmetry in Price Response: A Case Study of Western Nigerian Cocoa Farms." *Nigerian J. Econ. Soc. Studies* 14:347-356.

Olayide, S. O. [1972]: "Some Estimates of Supply Elasticities for Nigeria's Cash Crops." *J. Agr. Econ.* 23:263-276.

_____ [1974]: "Some Estimates of Supply Elasticities for Nigeria's Cash Crops: A Rejoinder." *J. Agr. Econ.* 25(2):181-190.

Olayide, S. O., J. A. Eweka, and V. E. Bello-Osagie, eds. [1980]: *Nigerian Small Farmers: Problems and Prospects in Integrated Rural Development.* Ibadan, Nigeria, Univ. of Ibadan.

Olayide, S. O., and T. J. O. Ogunfiditimi [1980]: "Agricultural Extension and Nigerian Small Farmers." In Olayide, Eweka, and Bello-Osagie, eds., 1980, pp. 257-271.

Olayide, S. O., O. Ogunfowora, and S. M. Essang [1974]: "Effects of Marketing Board Pricing Policies on the Nigerian Economy: A Systems Simulation Experiment." *J. Agr. Econ.* 25:289-309.

Olayide, S. O., and J. K. Olayemi [1978]: "Economic Aspects of Agriculture and Nutrition: A Nigerian Case Study." *Food and Nutrition Bull.* 1(1):32-39.

Olivares, Jose [1987]: *Options and Investment Priorities in Irrigation Development: Final Report.* Washington, Agriculture and Rural Development, Dep't., World Bank, August.

Oloya, J. J. [1969]: *Coffee, Cotton, Sisal, and Tea in the East African Economies.* Nairobi, East African Literature Bureau.

Oluwasanmi, H. A. [1975]: "Effects of Farm Mechanisation on Production and Employment in Nigeria." *Report of the FAO/OECD Expert Panel on the Effects of Farm Mechanisation on Production and Employment.* Rome, FAO, pp. 51-70.

Oluwasanmi, H. A., D. M. Lang, W. M. Corbett, J. S. Oguntoyinbo, C. Okali, I. S.
Dema, N. O. Osamo, R. O. Adegboye, M. Upton, and W. H. O. Ezeilo [1966]:
Uboma: A Socio-Economic and Nutritional Survey of a Rural Community in Eastern Nigeria.
Bude, Cornwall, England, Geographical Publications, World Land Use Survey.

Omari, C. K. ed. [1976]: *The Strategy for Rural Development: Tanzanian Experience.* Nairobi,
East African Literature Bureau.

Ongla, J. [1978]: "Structure, Conduct and Performance of the Food Crop Marketing System in Cameroon: A Case Study of Yaounde and Adjacent Areas." Unpublished Ph.D.
dissertation, Univ. of Florida, Gainesville.

Oni, S. A. [1969a]: "Econometric Analysis of Supply Response among Nigerian Cotton
Growers." *Bull. Rural Econ. and Sociology* 4:203-225.

———— [1969b]: "Production Response in Nigerian Agriculture: A Case Study of Palm
Produce, 1949-1966." *Nigerian J. Econ. Soc. Studies* 11:81-92.

Oni, S. A., and J. K. Olayemi [1975]: "The Economics of Rice Milling in Kwara and
North-Western States of Nigeria: A Comparative Analysis." In Are, ed., 1975.

Onitiri, H. M. A., and D. Olatunbosun, eds. [1974]: *The Marketing Board System. Proceedings of an International Conference.* Ibadan, Nigeria, Ibadan Univ. Press.

Onwueme, I. C. [1978]: *The Tropical Tuber Crops: Yams, Cassava, Sweet Potato, and Cocoyams.* New York, Wiley.

Oram, P. A. [1985]: "Agricultural Research and Extension: Issues of Public Expenditure."
In Howell, ed., 1985, pp. 59-82.

Ord, H. W., F. M. Andic, R. M. Bostock, and Y. Yannoulis [1964]: *Ghana: Projected Level
of Demand, Supply and Imports of Agricultural Products in 1965, 1970, and 1975.* Washington, USDA, ERS.

ORSTOM [1975]: *Les migrations de travail Mossi,* 11 Vols. Ouagadougou, Burkina Faso.

———— [1979]: *Maîtrise de l'espace agraire et développement en Afrique tropicale: logique paysanne
et rationalité technique. Actes du colloque de Ouagadougou, 4-8 Decembre 1978.* Paris, ORS-
TOM Memoires No. 89.

Ostby, I., and T. Gulilat [1969]: "A Statistical Study of Household Expenditure in Addis
Ababa." *Eastern African Econ. Rev.* 1:63-74.

Osuntogun, A. [1978]: "The Impact of Co-operative Credit on Farm Income and the Efficiency of Resource Use in Peasant Agriculture: A Case Study from Three States of
Nigeria." *African J. Agr. Sciences* 5(2):1-6.

———— [1980]: "Farm Level Credit among Co-operative Farmers in Nigeria." In Howell,
ed., 1980, pp. 259-272.

Otieno, J., G. Muchiri, and B. F. Johnston [1975]: "Expanded Use and Local Manufacture
of Appropriate Farm Equipment in Kenya's Strategy for Rural Development." In Westley and Johnston, eds., 1975.

Ouedraogo, I., M. D. Newman, and D. W. Norman [1982]: *The Farmer in the Semi-Arid
Tropics of West Africa: Partially Annotated Bibliography.* Pantacheru, India, ICRISAT.

Owosekun, A. [1976]: "Agricultural Supply Response. II. A Case Study of the Supply of
Groundnuts to the Marketing Board of the Northern States of Nigeria." *Eastern Africa J.
Rural Dev.* 9(1/2):130-139.

Oxaal, I., T. Barnett, and D. Booth [1975]: *Beyond the Sociology of Development: Economy
and Society in Latin America and Africa.* London and Boston, Routledge and Kegan Paul.

Oyenuga, V. A. [1973]: "Intensive Animal Production on a Subsistence Scale." In Reid, ed., 1973, pp. 393–400.

Padwick, G. W. [1983]: "Fifty Years of Experimental Agriculture II: The Maintenance of Soil Fertility of Tropical Agriculture: A Review." *Experimental Agriculture* 19:293-310.

Page, H. J., and R. Lesthaeghe, eds. [1981]: *Child Spacing in Tropical Africa: Tradition and Change.* New York: Academic Press.

Page, J. M., Jr. [1979]: *Small Enterprises in African Development: A Survey.* Washington, World Bank, Staff Working Paper No. 363.

Pala, A. [1976]: *African Women in Rural Development: Research Trends and Priorities.* Washington, ACE, OLC.

Palma, G. [1978]: "Dependency: A Formal Theory of Underdevelopment or a Methodology for the Analysis of Concrete Situations of Underdevelopment?" *World Dev.* 6(7/8):881-924.

Palmer, R., and N. Parsons, eds. [1977]: *The Roots of Rural Poverty in Central and Southern Africa.* London, Heinemann.

Palmer-Jones, R. W. [1974]: *Production and Marketing of Tea in Malawi.* Reading, England, Univ. of Reading, DAE and Management.

_____ [1977a]: "A Comment on Planned *versus* Actual Farmer Performance under Uncertainty in Underdeveloped Agriculture." *J. Agr. Econ.* 28(2):177-179.

_____ [1977b]: "Irrigation System Operating Policies for Mature Tea in Malawi." *Water Resources Research* 13(1):1-7.

_____ [1979]: "'Linear Programming and the Study of Peasant Farming: A Rejoinder." *J. Agr. Econ.* 30(2):199-204.

_____ [1981]: "How Not to Learn from Pilot Irrigation Projects: The Nigerian Experience." *Water Supply and Management (Oxford)* 5(1):81-105.

_____ [1985]: "Harvesting Policies for Tea in Malawi." *Experimental Agriculture* 21:357-368.

Papendick, R. I., P. A. Sanchez, and G. B. Triplett, eds. [1977]: *Multiple Cropping.* Madison, WI, American Society of Agronomy.

Parkin, D. J. [1972]: *Palms, Wine and Witnesses: Public Spirit and Private Gain in an African Farming Community.* San Francisco, Chandler Publishing

_____ , ed. [1975]: *Town and Country in Central and Eastern Africa.* London, Oxford Univ. Press.

Parsons, K. H. [1971]: *Customary Land Tenure and Development of African Agriculture.* Madison, Univ. of Wisconsin, LTC.

Paulino, L. A. [1987]: "The Evolving Food Situation." In Mellor, Delgado and Blackie, eds., 1987, pp. 23-35.

Paulino, L. A., and J. S. Sarma [1988]: *Analysis of Trends and Projections of Food Production and Consumption in Brazil and Nigeria.* Washington, IFPRI.

Paylore, P., ed. [1976]: *Desertification: A World Bibliography.* Tucson, Univ. of Arizona, Office of Arid Land Studies.

Paylore, P., and J. A. Mabbutt [1980]: *Desertification: World Bibliography Update, 1976-80.* Tucson, Univ. of Arizona, Office of Arid Land Studies.

Pearse, A. [1980]: *Seeds of Plenty, Seeds of Want: Social and Economic Implications of the Green Revolution.* New York, Oxford Univ. Press.

Pearson, S. R., and J. Cownie [1974]: *Commodity Exports and African Economic Development.* Lexington, MA, Heath.

Pearson, S. R., and W. D. Ingram [1980]: "Economies of Scale, Domestic Divergences, and Potential Gains from Economic Integration in Ghana and the Ivory Coast." *J. Polit. Economy* 88:994–1008.

Pearson, S. R., and R. K. Meyer [1974]: "Comparative Advantage among African Coffee Producers." *Amer. J. Agr. Econ.* 56:310–313.

Pearson, S. R., J. D. Stryker, C. P. Humphreys, P. Rader, E. A. Monke, D. S. C. Spencer, K. Craven, A.H. Tulrey, J. McIntire, and J. M. Page, Jr. [1981]: *Rice in West Africa: Policy and Economics.* Stanford, CA, Stanford Univ. Press.

Peek, P., and G. Standing [1979]: "Rural–Urban Migration and Government Policies in Low Income Countries." *International Labor Rev.* 118(6):747–762.

Pelissier, P. [1966]: *Les paysans du Sénégal. Les civilisations agraires du Cayor à la Casamance.* St.-Yrieix, Imprimerie Fabregue.

Perisse, J. [1962]: "L'alimentation des populations rurales du Togo." *Annales de la Nutrition et d'Alimentation* 16(6):1–58.

Perrault, P. T. [1978]: "Banana-Manioc Farming Systems of the Tropical Forest: A Case Study of Zaire." Unpublished Ph.D. dissertation, Stanford, Univ., Stanford.

Perrin, R. K., D. L. Winkelmann, E. R. Moscardi, and J. R. Anderson [1976]: *From Agronomic Data to Farmer Recommendations: An Economics Training Manual.* Mexico, CIMMYT.

Peters, P. [1983]: "Gender, Developmental Cycles and Historical Process: A Critique of Recent Research on Women in Botswana." *J. Southern African Studies* 10:100–122.

Phillips, A. O. [1975]: *Review of Income Distribution Data: Ghana, Kenya, Tanzania and Nigeria.* Princeton, NJ, Princeton Univ., Woodrow Wilson School.

Phillips, D. O. A., and G. O. I. Abalu [1987]: "Price Expectations Formation and Revision in the Nerlovian Framework with Application to Nigerian Farmers." *J. Agr. Econ.* 38(3):491–95.

Phillips, J. [1959]: *Agriculture and Ecology in Africa. A Study of Actual and Potential Development South of the Sahara.* London, Faber and Faber.

Picard, L. A. [1979]: "Rural Development in Botswana: Administrative Structures and Public Policy." *J. Developing Areas* 13:283–300.

Pickering, D., ed. [1988]: *African Agricultural Research and Technological Development: Proceedings of a High Level Meeting in Feldafing, Federal Republic of Germany,* September 24–27, 1987. Washington, D.C., World Bank.

Pieri, C., F. Ganry, and P. Siband [1978]: "Proposition pour une interpretation agro-économique des essais d'engrais. Exemples des fumures azotée et potassique du mil au Sénégal." *L'Agronomie Tropicale* 33(1):32–39.

Pimentel, D., ed. [1978]: *World Food, Pest Losses and the Environment.* Boulder, CO, Westview Press.

Pingali, P., Y. Bigot, and H. P. Binswanger [1987]: *Agricultural Mechanization and the Evolution of Farming Systems in Sub-Saharan Africa.* Baltimore, Johns Hopkins Univ. Press.

Pinstrup-Andersen, P. [1981]: *Nutritional Consequences of Agricultural Projects: Conceptual Relationships and Assessment Approaches.* Washington, World Bank, Staff Working Paper No. 456.

Pisani, E. [1988]: *Pour L'Afrique.* Paris, Éditions Odile Jacob.

Plange, Nii-K. [1979]: " 'Opportunity Cost' and Labour Migration: A Misinterpretation of Proletarianisation in Northern Ghana." *J. Modern African Studies* 17:655–676.

Plattner, S., ed. [1975]: *Formal Methods in Economic Anthropology.* Washington, American Anthropological Association.

Poats, S., M. Schmink, and A. Spring, eds. [1988]: *Gender Issues in Farming Systems Research and Extension.* Boulder, CO, Westview Press.

Polanyi, K., C. M. Arensberg, and H. W. Pearson [1957]: *Trade and Market in Early Empires.* New York, Free Press.

Poleman, T. T. [1961]: "The Food Economies of Urban Middle Africa: The Case of Ghana." *Food Research Institute Studies* 2:121-174.

Popkin, S. L. [1979]: *The Rational Peasant: The Political Economy of Rural Society in Vietnam.* Berkeley, Univ. of California Press.

Posner, R. A. [1980]: "Anthropology and Economics." *J. Polit. Economy* 88:608-616.

Post, K. [1972]: "Peasantization and Rural Political Movements in Western Africa." *Archives Europeennes de Sociologie* 13:223-254.

Poulain, J.-F., M. Sedogo, F. Ouali, and P. Morant [1979]: "La Démarche système en agronomie: essais de définition des zones homogenes en Haute-Volta et propositions de systèmes de cultures vulgarisalles." In ORSTOM, 1979.

Prakah-Asante, K. [1976]: "The Economics of Size of Food Crop Farms in the Mampong Enura Agricultural District of Ghana." Unpublished Ph.D. dissertation, Washington State Univ., Pullman.

Pratt, D. J., and M. D. Gwynne, eds. [1977]: *Rangeland Management and Ecology in East Africa.* London, Hodder and Stoughton.

Prebisch, R. [1959]: "Commerical Policy in Underdeveloped Countries." *Amer. Econ. Rev.* 49:251-273.

Presvelan, C., and S. Spijkers-Zwart, eds. [1980]: *The Household, Women and Agricultural Development.* Wageningen, H. Veenman and Zanen.

Priovolos, T. [1981]: *Coffee and the Ivory Coast: An Econometric Study.* Lexington, MA, Lexington Books.

Pryor, R. A. [1977]: *The Origins of the Economy: A Comparative Study of Distribution in Primitive and Peasant Economies.* New York, Academic Press.

Pudsey, D. [1967]: *An Economic Survey of Farming in a Wet, Long Grass Area of Toro.* Entebbe, Uganda, Department of Agriculture.

Pullan, N. B., and R. J. Grindle [1980]: "Productivity of White Fulani Cattle on the Jos Plateau, Nigeria. IV. Economic Factors." *Tropical Animal Health Proceedings* 12:161-170.

Purvis, M. J. [1968a]: "The Nigerian Oil Palm Industry: A Comment." *Food Research Institute Studies* 8:191-197.

———— [1968b]: *A Study of the Economics of Tractor Use in Oyo Division of the Western State.* East Lansing, MSU Consortium for the Study of Nigerian Rural Development, Report No. 17.

———— [1970]: "New Sources of Growth in a Stagnant Smallholder Economy in Nigeria: The Oil Palm Rehabilitation Scheme." In Eicher and Liedholm, eds., 1970, pp. 267-281.

Putt, S. N. H., and P. R. Ellis [1980]: *The Social and Economic Implications of Trypanosomiasis Control: A Study of its Impact on Livestock Production and Rural Development in Northern Nigeria.* Reading, England, Univ. of Reading.

Putterman, L. [1986]: *Peasants, Collectives and Choice: Economic Theory and Tanzania's Villages.* Greenwich, CT, JAI Press.

Quick, S. A. [1977]: "Bureaucracy and Rural Socialism in Zambia." *J. Modern African Studies* 15:379-400.

———— [1978]: *Humanism or Technocracy: Zambia's Farming Cooperatives, 1965-72.* Lusaka, Univ. of Zambia, Institute for African Studies.

Quinn, N. [1978]: "Do Mfantse (Ghana) Fish Sellers Estimate Probabilities in their Heads." *Amer. Ethnologist* 5:206-226.

Quinn, V., M. Cohen, J. Mason, and B. N. Kgosidintsi [1988]: "Crisis-Proofing the Economy: The Response of Botswana to Economic Recession and Drought." In Cornia, Jolly and Steward, 1988, pp. 3-27.

Rachie, K. O., and P. Silvestre [1977]: "Grain Legumes." In Leakey and Wills, eds., 1977, pp. 41-74.

Rado, E. [1986]: "Notes Toward a Political Economy of Ghana Today." *African Affairs* 85(341):563-572.

Raikes, P. [1978]: "Rural Differentiation and Class Formation in Tanzania." *J. Peasant Studies* 5:285-325.

Ramond, C. [1971]: *L'introduction des themes intensifs dans les exploitation traditionneles: conséquences économiques.* Bambey, Sénégal, CNRA.

Ranger, T. [1978]: "Growing from the Roots: Reflections on Peasant Research in Central and Southern Africa." *J. Southern African Studies* 5(1):99-133.

Ravenhill, J., ed. [1986]: *Africa in Economic Crisis.* New York, Columbia Univ. Press.

Raynaut, C. [1973]: *Structures normatives et relations électives: étude d'une communaute villageoise haoussa.* Paris, Mouton.

———— [1977]: "Circulation monetaire et évolution des structures socio-économiques chez les Haoussas du Niger." *Africa* 47:160-171.

Reardon, T., P. Matlon, and C. Delgado [1988]: "Coping with Household-level Food Insecurity in Drought-Affected Areas of Burkina Faso." *World Dev.* 16:1065-1074.

Reid, R. L., ed. [1973]: *Proceedings of the Third World Conference on Animal Production.* Melbourne, Australia.

Rempel, H. [1971]: "Labour Migration into Urban Areas and Urban Employment in Kenya." Unpublished Ph.D. dissertation, Univ. of Wisconsin, Madison.

Rempel, H., and W. J. House [1978]: *The Kenya Employment Problem: An Analysis of the Modern Sector Labour Market.* Nairobi, Oxford Univ. Press.

Rempel, H. and R. Lobdell [1978]: "The Role of Urban-to-Rural Remittances in Rural Development." *J. Dev. Studies* 14:324-341.

Remy, G. [1977]: *Enquête sur les mouvements de population à partir du pays mossi.* Paris, ORSTOM.

Resnick, S. A. [1970]: "The Decline of Rural Industry under Export Expansion: A Comparison among Burma, Philippines, and Thailand, 1870-1938." *J. Econ. Hist.* 30:51-73.

Reusse, E. [1976]: "Economic and Marketing Aspects of Post-Harvest Systems in Small Farmer Economies." *Monthly Bull. Agr. Econ. and Stat.* 25(10):1-10.

Reutlinger, S. [1966]: "Short Run Beef Supply Response." *Amer. J. Agr Econ.* 48:909-919.

———— [1977]: "Malnutrition: A Poverty or Food Problem?" *World Dev.* 5:715-724.

———— [1985]: "Food Security and Poverty in LDCs." *Finance and Dev.* 22(4):7-11.

Reutlinger, S., and M. Selowsky [1976]: *Malnutrition and Poverty: Magnitude and Policy Options.* Baltimore, Johns Hopkins Univ. Press.

Reyna, S. [1977]: "Marriage Payments, Household Structure, and Domestic Labour Supply among the Barma of Chad." *Africa* 47:81-88.

Reynolds, L. G., ed. [1975]: *Agriculture in Development Theory.* New Haven, CT, Yale Univ. Press.

Richard, J. F., M. Fall, and J. M. Attonaty [1976]: *Le model "4S" programme linéaire pour les exploitations agricoles du sine saloum-sud au Sénégal et calculs de budgets automatises.* Bambey, Senegal, IRAT/ISRA.

Richards, A. [1932]: *Hunger and Work in a Savage Tribe.* London, Routledge.

_____ [1939]: *Land, Labour and Diet in Northern Rhodesia.* London, Oxford Univ. Press.

Richards, A. I., F. Sturrock, and J. M. Fortt, eds. [1973]: *Subsistence to Commercial Farming in Present Day Buganda: An Economic and Anthropological Survey.* Cambridge, England, Cambridge Univ. Press.

Richards, P. [1985]: *Indigenous Agricultural Revolution: Ecology and Food Production in West Africa.* London, Hutchinson.

_____ [1986]: *Coping with Hunger: Hazard and Experiment in an African Rice-Farming System.* London, Allen and Unwin.

Riddell, J. B. [1974]: "Periodic Markets in Sierra Leone." *Annals Assoc. of Amer. Geographers* 64(4):541-548.

_____ [1978]: "The Migration to the Cities of West Africa: Some Policy Considerations." *J. Modern African Studies* 16:241-260.

Riddell, J. C., K. H. Parsons, and D. Kanel [1978]: "Land Tenure Issues in African Development: A Position Paper." Madison, Univ. of Wisconsin.

Ridker, R., and H. Lubell, eds. [1971]: *Employment and Unemployment Problems of the Near East and South Asia.* Delhi, Vikas Publishing.

Riesman, P. [1977]: *Freedom in Fulani Social Life: An Introspective Ethnology,* translation by M. Fuller. Univ. of Chicago Press.

_____ [1978]: *The Fulani in a Development Context: The Relevance of Cultural Tradition for Coping with Change and Crisis.* Abidjan, Côte d'Ivoire, AID.

Riley, P., and M. T. Weber [1979]: *Food and Agricultural Marketing in Developing Countries: An Annotated Bibliography of Doctoral Research in the Social Sciences, 1969-79.* East Lansing, MSU, DAE, Rural Development Working Paper No. 5.

Rimmer, D. C. [1969]: "The Abstraction from Politics: A Critique of Economic Theory and Design with Reference to West Africa." *J. Dev. Studies* 5:190-204.

Rivera, W., and S. Schram, eds. [1987]: *Agricultural Extension Worldwide: Issues, Practices and Emerging Priorities.* London, Croom Helm.

Roberts, R. A. J. [1972]: "The Role of Money in the Development of Farming in the Mumbawa and Katete Areas of Zambia." Unpublished Ph.D. dissertation, Univ. of Nottingham, England.

Robertson, C. [1974]: "Economic Women in Africa: Profit-Making Techniques of Accra Market Women." *J. Modern African Studies* 12:657-664.

Robinson, K. L. [1974]: "Th Economics of Increasing Staple Food Production in West Africa." Ibadan, Nigeria, IITA, unpublished paper.

Robinson, K. L., and A. O. Falusi [1974]: *The Present and Potential Role of Fertilizer in Meeting Nigeria's Food Requirements.* Ithaca, NY, Cornell Univ., International Agricultural Monograph No. 46. Also in *Nigerian Agr. J.*, 1974, 2(3):100-107.

Robson, P. [1968]: *Economic Integration in Africa.* London, Allen and Unwin.

Rocheteau, G. [1975]: "Pionniers mourides au Sénégal: colonisation des terres neuves et transformations d'une économie paysanne." *Cahiers ORSTOM, Ser. Sci. Hum.* 12(1):19-53.

Rodney, W. [1974]: *How Europe Underdeveloped Africa.* Washington, Howard Univ. Press.

Rogers, E. M. [1976a]: "Communication and Development: The Passing of the Dominant Paradigm." *Communication Research* 3:213-240.

_____ [1976b]: "Where We Are in Understanding the Diffusion of Innovations?" In Schramm and Lerner, eds., 1976, pp. 204-222.

Rohrbach, D. D. [1988]: "The Growth of Smallholder Maize Production in Zimbabwe: Causes and Implications for Food Security." Unpublished Ph.D. dissertation, MSU, East Lansing.

Roider, W. [1971]: *Farm Settlements for Socio-Economic Development—The Western Nigeria Case*. Munich, West Germany, Weltforum Verlag.

Roling, N. G. [1970]: "Adaptations in Development: A Conceptual Guide to the Study of Non-Innovative Response of Peasant Farmers." *Econ. Dev. and Cultural Change* 19:71-85.

―――― [1988]: *Extension Science: Information Systems in Agricultural Development*. Cambridge, England, Cambridge University Press.

Roling, N. G., J. Ascroft, and F. Chege [1976]: "The Diffusion of Innovations and the Issue of Equity in Rural Development." *Communication Research* 3:155-170.

Roose, E. J. [1977]: "Adaptation des méthodes de conservation des sols aux conditions oecologiques et socio-économique de l'Afrique de l'Ouest." *L'Agronomie Tropicale* 32(2):132-140.

Rosberg, C. G., and T. G. Callaghy, eds. [1979]: *Socialism in Sub-Saharan Africa: A New Assessment*. Berkeley, Univ. of California Press, Institute of International Studies.

Rosenboom, H. P., and R. E. Parker [1975]: "Economics of Small-Scale Rice Mills." In Are, ed., 1975.

Ross, C. G. [1980]: "Grain Demand and Consumer Preferences in Senegal." *Food Policy* 5(4):273-281.

Rothschild, B. J., ed. [1972]: *World Fisheries Policy: Multidisciplinary Views*. Seattle, Univ. of Washington Press.

Roy, D. K., and S. J. Mabey [1968]: *Household Budget Survey in Ghana*. Legon, Univ. of Ghana, Institute of Statistics.

Ruigu, G. M. [1978]: "An Economic Analysis of the Kenya Milk Subsystem." Unpublished Ph.D. dissertation, MSU, East Lansing.

Rukandema, F. M. [1978]: "Resource Availability, Utilization and Productivity on Small-Scale Farms in Kakamega District, Western Kenya." Unpublished, Ph.D. dissertation, Cornell Univ., Ithaca, NY.

Rukuni, M. [1988]: "The Evolution of Smallholder Irrigation Policy in Zimbabwe: 1928-1986." *Irrigation and Drainage Systems* 2:199-210.

Rukuni, M., and C. K. Eicher, eds. [1987]: *Food Security for Southern Africa*. University of Zimbabwe, DAE and Extension, UZ/MSU Food Security Project.

―――― [1988]: "The Food Security Equation in Southern Africa." In Bryant, ed., 1988, pp. 133-157.

Rukuni, M., and R. H. Bernsten, eds. [1988]: *Southern Africa: Food Security Policy Options*. University of Zimbabwe, DAE and Extension, UZ/MSU Food Security Project.

Ruthenberg, H. [1968]: *Smallholder Farming and Smallholder Development in Tanzania: Ten Case Studies*. Munich, West Germany, Weltforum Verlag.

―――― [1974]: "Artificial Pastures and their Utilization in the Southern Guinea Savanna and the Derived Savanna of West Africa. Tour d'Horizon of an Agricultural Economist." *Zeitschrift fur Auslandische Landwirtschaft* 13(3):216-231 and 312-330.

―――― [1980]: *Farming Systems in the Tropics*, 3rd edition. London, Oxford Univ. Press.

Ruttan, V. [1975]: "Integrated Rural Development Programs: A Skeptical Perspective." *International Dev. Rev.* 17(4):9-16.

―――― [1982]: *Agricultural Research Policy*. Minneapolis, Univ. of Minnesota Press.

―――― [1988]: "Cultural Endowments and Economic Development: What Can We Learn from Anthropology?" *Econ. Dev. and Cultural Change* 36(3, Supplement):S247-S257.

Ruttan, V. W., and Carl Pray, eds. [1987]: *Policy for Agricultural Research*. Boulder, CO, Westview Press.

Ruttan, V. W., and A. M. Weisblat [1965]: "Some Issues in the Training of Asian Agricultural Economics Graduate Students in the U.S." *J. Farm Econ.* 47:1024-1026.

Rweyemamu, J. F., ed. [1980]: *Industrialization and Income Distribution in Africa.* Dakar, Senegal, CODESRIA.

Sabot, R. H. [1979]: *Economic Development and Urban Migration, Tanzania, 1900-1971.* Oxford, England, Clarendon Press.

———, ed. [1981]: *Migration and the Labor Market in Developing Countries.* Boulder, CO, Westview Press.

Sahel Bibliographic Bulletin, Vols. 1-5 [1977-1981]. East Lansing, MSU Library.

Sahlins, M. [1974]: *Stone Age Economics.* Chicago, Aldine.

Saila, S. B., and P. M. Roedel, eds. [1979]: *Stock Assessment for Tropical Small-Scale Fisheries: Proceedings of an International Workshop Held September 19-21, 1979.* Kingston, Univ. of Rhode Island.

Salzman, P. C., ed. [1980]: *When Nomads Settle: Processes of Sedentarization as Adaptation and Response.* New York, Praeger.

Samoff, J. [1981]: "Crises and Socialism in Tanzania." *J. Modern African Studies* 19:279-306.

Sanders, John H. [1989]: "Agricultural Research and Cereal Technology Introduction in Burkina Faso and Niger." *Agricultural Systems* 30:139-154.

Sands, M., and R. E. McDowell [1979]: *A World Bibliography on Goats.* Ithaca, NY, Cornell Univ., International Agriculture Mimeograph No. 70.

Sanford, S. [1977]: *Dealing with Drought and Livestock in Botswana.* London, ODI.

Sargent, M., J. Lichte, P. Matlon, and R. Bloom [1981]: *An Assessment of Animal Traction in Francophone West Africa.* East Lansing, MSU, DAE, ARE Working Paper No. 34.

Satin, M. [1988]: "Bread without Wheat." *New Scientist,* April, pp. 56-69.

Saul, J. S. [1977]: "Tanzania's Transition to Socialism?" *Canadian J. African Studies* 11:313-339.

——— [1979]: *The State and Revolution in Eastern Africa.* New York, Monthly Review Press.

Savane, M A. [1981]: *Implications for Women and their Work of Introducing Nutritional Considerations into Agricultural and Rural Development Projects.* Rome, 7th Session of the ACC Sub-Committee on Nutrition, United Nations.

Savonnet, G. [1962]: "La colonisation du pays Koulango (Haute-Côte d'Ivoire) par les Lobi de Haute-Volta." *Cahiers d'Outre-Mer* 57:25-46.

——— [1976]: "Inégalités de développement et organisation sociale (exemple empruntes au sud-ouest de la Haute-Volta)." *Cahiers ORSTOM,* Ser. Sci. Hum. 13(1):23-40.

Sawadogo, A. [1977]: *L'Agriculture en Côte d'Ivoire.* Paris, Presses Universitaires de France.

Saylor, R. G. [1967]: *The Economic System of Sierra Leone.* Durham, NC, Duke Univ. Press.

——— [1974]: "Farm Level Cotton Yields and the Research and Extension Services in Tanzania." *Eastern Africa J. Rural Dev.* 7:46-60.

Saylor, R. G., and C. K. Eicher [1970]: "Plantations in Nigeria: Lessons for West African Developments." In Bunting, ed., 1970, pp. 497-530.

Scandizzo, P., and C. Bruce [1980]: "Methodologies for Measuring Agricultural Price Intervention Effects." Washington, World Bank, Staff Working Paper No. 394.

Schadler, K. [1968]: *Crafts, Small-Scale Industries, and Industrial Education in Tanzania.* Munich, West Germany, Weltforum Verlag.

Schapera, I. [1947]: *Migrant Labour and Tribal Life: A Study of Conditions in the Bechuanaland Protectorate.* London, Oxford Univ. Press.

Schiffer, J. [1981]: "The Changing Post-War Pattern of Development: The Accumulated Wisdom of Samir Amin." *World Dev.* 9:515-537.

Schneider, H. K. [1974]: *Economic Man: The Anthropology of Economics*. New York, Free Press.

―――― [1975]: "Economic Development and Anthropology." In Siegel, Beals, and Tyler, eds., vol. 4, 1975, pp. 271-292.

―――― [1979]: *Livestock and Equality in East Africa: The Economic Basis for Social Structure*. Bloomington, Indiana Univ. Press.

Schramm, W., and D. Lerner, eds. [1976]: *Communication and Change in the Developing Countries*. Honolulu, Univ. of Hawaii, East-West Center Press.

Schuh, G. E. [1990]: "The New Macroeconomics of Food and Agricultural Policy." In Eicher and Staatz, eds., pp. 140-153.

Schulman, R. [1979]: "Strategy for the Advancement of Animal Traction in Mali." Bamako, Mali, AID, mimeograph.

Schultz, T. W. [1964]: *Transforming Traditional Agriculture*. New Haven, CT, Yale Univ. Press.

―――― [1965]: *Economic Crises in World Agriculture*. Ann Arbor, University of Michigan Press.

―――― , ed. [1978]: *Distortions of Agricultural Incentives*. Bloomington, Indiana Univ. Press.

―――― [1980]: "Nobel Lecture: The Economics of Being Poor." *J. Polit. Economy* 88(4):639-651.

―――― [1983]: "An Unpersuasive Plea for Centralized Control of Agricultural Research: On a Report of the Rockefeller Foundation." *MINERVA* 21(1):141-43.

Schulz, M. [1976]: *Organizing Extension Services in Ethiopia before and after the Revolution*. Saarbrucken, West Germany, Verlag de SSIP-Schriften Breitenbach.

Schwimmer, B. E. [1980]: "The Organization of Migrant Farmer Communities in Southern Ghana." *Canadian J. African Studies* 14(2):221-238.

Scott, E. P. [1972]: "The Spatial Structure of Rural Northern Nigeria: Farmers, Periodic Markets and Villages." *Econ. Geography* 48:316-332.

―――― [1978]: "Subsistence, Markets and Rural Development in Rural Hausaland." *J. Developing Areas* 12(4):449-469.

Scott, J. C. [1976]: *The Moral Economy of the Peasant: Rebellion and Subsistence in Southeast Asia*. New Haven, CT, Yale Univ. Press.

Scudder, T. [1973]: "The Human Ecology of Big Projects: River Basin Development and Resettlement." In Siegel, Beals, and Tyler, eds., vol. 2, 1973, pp. 45-61.

Scudder, T., and E. Colson [1972]: "The Kariba Dam Project: Resettlement and Local Initiative." In Bernard and Pelto, eds., 1972, pp. 39-69.

―――― [1979]: "Long-Term Research in Gwembe Valley, Zambia." In Foster, Scudder, Colson, and Kemper, eds., 1979, pp. 227-254.

Seddon, D., ed. [1978]: *Introduction aux comptes économique de la production animal, application aux pays Sahélians. Tome 1: étude méthodologique*, 2nd ed. Paris.

SEDES [1974]: *Introduction aux comptes économique de la production animal, application aux pays Sahélians. Tome 1: étude méthodologique*, 2nd ed. Paris.

Seidman, A. [1972]: *Comparative Development Strategies in East Africa*. Nairobi, East African Publishing House.

―――― [1977]: "The Economics of Eliminating Rural Poverty." In Palmer and Parsons, eds., 1977, pp. 410-421. Reprinted in Turok, ed., 1979, pp. 37-48.

―――― [1981]: "Women and the Development of "Underdevelopment": The African Experience." In Dauber and Cain, eds., 1981, pp. 109-126.

Seifert, W. W., and N. W. Kamrany [1974]: *A Framework for Evaluating Long-Term Strategies for the Development of the Sahel-Sudan Region.* Vol. 1. *Summary Report: Project Objectives, Methodologies and Major Findings.* Cambridge, MIT, Center for Policy Alternatives.

Sen, A. [1981]: *Poverty and Famines.* Oxford, Clarendon Press.

_____ [1984]: "Food Battles: Conflicts in the Access to Food." *Food and Nutrition (FAO)* 10(1):81-90.

_____ [1987a]: *Africa and India: What Do We Have To Learn From Each Other?* Helsinki, WIDER, WP 19.

_____ [1987b]: *Research for Action: Hunger and Entitlements.* Helsinki, WIDER.

Sender, J. and S. Smith [1986]: *The Development of Capitalism in Africa.* London, Methuen.

Senegal, Government of ___ [1977]: *Food Investment Strategy, 1977-1985.* Dakar, Ministry of Rural Development.

Shack, W. A., and E. P. Skinner, eds. [1979]: *Strangers in African Societies.* Berkeley, Univ. of California Press.

Shaffer, J. D., M. T. Weber, H. M. Riley, and J. Staatz [1985]: "Designing Marketing Systems to Promote Development in the Third World Countries." *Agricultural Markets in the Semi-Arid Tropics: Proceedings of the International Workshop, 24-28 October, 1983.* Pantancheru, India, ICRISAT.

Shah, M., and F. Willekens [1978]: *Rural-Urban Population Projections for Kenya and Implications for Development.* Laxenburg, Austria, International Institute for Applied Systems Analysis.

Shapiro, D. [1988]: "Farm Size, Household Size and Composition, and Women's Contribution to Agricultural Production: Evidence from Zaire." State College, Pennsylvania State Univ., Department of Economics, Working Paper.

Shapiro, K. H. [1973]: "Efficiency and Modernization in African Agriculture: A Case Study in Geita District, Tanzania." Unpublished Ph.D. dissertation, Stanford Univ., Stanford.

_____ [1978]: "Water, Women and Development in Tanzania." Paper presented at the *Third Annual Conference of the International Water Resources Association.* Sao Paulo, Brazil.

_____ , ed. [1979]: *Livestock Production and Marketing in the Entente States of West Africa: Summary Report.* Ann Arbor, Univ. of Michigan, CRED.

_____ [1985]: "Strengthening Agricultural Research and Educational Institutions in Africa." Hearings, the Subcommittee on Foreign Operations, Senate Committee on Appropriations, U.S. Congress, Washington.

Shapiro, K. H., and J. Muller [1977]: "Sources of Technical Efficiency: The Roles of Modernization and Information." *Econ. Dev. and Cultural Change* 25:293-310.

Shapouri, S., A. Dommen, and S. Rosen [1986]: *Food Aid and the African Food Crisis.* Washington, USDA, FAER No. 221.

Shaw, T. W., and K. A. Heard, eds. [1979]: *The Politics of Africa: Dependence and Development.* New York, Africana Publishing Co.

Shear, D., and B. Clark [1976]: "International Long-Term Planning for the Sahel." *International Dev. Rev.* 4:15-20.

Sheets, H., and R. Morris [1974]: *Disaster in the Desert: Failures of International Relief in the West African Drought.* Washington, Carnegie Endowment for International Peace.

Sheffield, J. R., ed. [1967]: *Education, Employment and Rural Development: Proceedings of a Conference Held at Kericho, Kenya.* Nairobi, East African Publishing House.

Shen, T. Y. [1977]: "Macro Development Planning in Tropical Africa: Technocratic and Non-Technocratic Causes of Failure." *J. Dev. Studies* 13:413-427.

Shepard, A. W. [1981]: "Agrarian Change in Northern Ghana: Public Investment, Capital-ist Farming and Famine." In Heyer, Roberts, and Williams, eds., 1981, pp. 168-192.

Sherman, J. [1981]: *Crop Disposal and Grain Marketing in the Manga Region of Upper Volta: A Case Study.* Ann Arbor, Univ. of Michigan, CRED.

Siegel, B. J., A. R. Beals, and S. A. Tyler, eds. [1973, 1974, 1975]: *Annual Review of An-thropology.* Vols. 2, 3, 4. Palo Alto, CA, Annual Review, Inc.

Simmonds, N. [1986]: "A Short Review of Farming Systems Research in the Tropics." *Ex-perimental Agriculture* 22:1-13.

Simmons, E. B. [1975]: "The Small-Scale Rural Food Processing Industry in Northern Ni-geria." *Food Research Institute Studies* 14:147-161.

———— [1976a]: *Calorie and Protein Intakes in Three Villages of Zaria Province, May 1970-July 1971.* Zaria, Nigeria, Ahmadu Bello Univ., IAR, Samaru Miscellaneous Paper No. 55.

———— [1976b]: *Economic Research on Women in Rural Development in Northern Nigeria.* Washington, ACE, OLC.

———— [1976c]: *Rural Household Expenditures in Three Villages of Zaria Province.* Zaria, Ni-geria, Ahmadu Bello Univ., IAR, Samaru Miscellaneous Paper No. 56.

Simpson, J. [1988]: *The Economics of Livestock Systems in Developing Countries.* Boulder, CO, Westview Press.

Sims, M. [1981]: *United States Doctoral Dissertations in Third World Studies, 1869-1978.* Wal-tham, MA, Brandeis Univ., Crossroads Press.

Sims, M., and A. Kagan [1976]: *American and Canadian Doctoral Dissertations and Master's Theses on Africa, 1886-1974.* Waltham, MA, African Studies Association.

Singer, H., J. Wood, and T. Jennings [1987]: *Food Aid: The Challenge and Opportunity.* Ox-ford, England, Oxford University Press.

Singh, I. J. [1976]: *A Note on the Economics of Agricultural Mechanization.* Washington, World Bank.

———— [1977]: "Appropriate Technologies in Tanzanian Agriculture: Some Empirical and Policy Considerations." *In Tanzania Basic Economic Report, Annex VII.* Washington, World Bank.

Singh, R. [1988]: *Economics of the Family and Farming Systems in Sub-Saharan Africa: Devel-opment Perspectives.* Boulder, CO, Westview Press.

Sisaye, S., and E. Stommes [1980]: "Agricultural Development in Ethiopia: Government Budgeting and Development Assistance in the Pre and Post 1975 Periods." *J. Developing Areas* 16(2):156-185.

Skarstein, R., and S. Wangwe [1986]: *Industrial Development in Tanzania: Some Critical Is-sues.* Uppsala, Sweden, SIAS.

Skinner, E. [1965]: "Labour Migration among the Mossi of Upper Volta." In Kuper, ed., 1965, pp. 60-84.

Sklar, R. L. [1976]: "Post-Imperialism: A Class Analysis of Multinational Corporate Ex-pansion." *Comparative Politics* 9(1):75-92.

———— [1988]: "Beyond Capitalism and Socialism in Africa." J. Modern African Studies 26(1):1-21.

Smith, A. K., and C. E. Welch, Jr., eds. [1978]: *Peasants in Africa.* Waltham, MA, Cross-roads Press.

Smith, C. A., ed. [1976a]: *Regional Analysis. Vol. 1. Economic Systems.* New York, Aca-demic Press.

———— [1976b]: "Regional Economic Systems: Linking Geographical Models and Socio-economic Problems." In C. A. Smith, ed., 1976a, pp. 3-63.

Smith, J. A. [1978]: *The Development of Large-Scale Integrated Sugar Schemes in Western Kenya*. Nairobi, Univ. of Nairobi, IDS Working Paper No. 343.

Smith, M. [1984]: "Nutrition and Farming Systems Research and Extension." In Butler, ed., 1984.

Smith, M. E., ed. [1977]: *Those Who Live from the Sea: A Study in Maritime Anthropology*. St. Paul, MN, West Publishing for the American Ethnological Society, Monograph No. 62.

Smith, R. T., ed. [1978]: *Market-Place Trade — Periodic Markets, Hawkers and Traders in Africa, Asia and Latin America*. Vancouver, Univ. of British Columbia, Center for Transportation Studies.

Smith, S. [1980]: "The Ideas of Samir Amin: Theory or Tautology?" *J. Dev. Studies* 17(1):5-21.

Smith, T. [1979]: "The Underdevelopment of Development Literature: The Case of Dependency Theory." *World Politics* 31(2):247-288.

Smith, V. E. [1975]: *Efficient Resource Use for Tropical Nutrition: Nigeria*. East Lansing, MSU, Graduate School of Business Administration.

Smith, V. E., S. Lynch, W. Whelan, J. Strauss, and D. Baker [1979]: *Household Food Consumption in Rural Sierra Leone*. East Lansing, MSU, DAE, Rural Development Working Paper No. 7.

Smith, V. E., J. Strauss, and P. Schmidt [1981]: *Single-Equation Estimation of Food Consumption Choices in Rural Sierra Leone*. East Lansing, MSU, DAE, Rural Development Working Paper No. 13.

Smith, V. E., J. Strauss, P. Schmidt, and W. Whelan [1980]: *Non-Price Factors Affecting Household Food Consumption in Sierra Leone*. East Lansing, MSU, DAE, Rural Development Working Paper No. 12.

Smith, W. E. S. [1971]: *We Must Run While They Walk*. New York, Random House.

Smock, D. R., and A. C. Smock [1972]: *Cultural and Political Aspects of Rural Transformation: A Case Study of Eastern Nigeria*. New York, Praeger.

Soligo, R. [1973]: *Factor Intensity of Consumption Patterns, Income Distribution, and Employment Growth in Pakistan*. Houston, TX, Rice Univ., Program of Development Studies, Paper No. 44.

Songre, A. [1973]: "Mass Emigration from Upper Volta: The Facts and Implications." In ILO, 1973, pp. 199-215.

Sorenson, L., J. R. Pedersen, and N. C. Ives [1975]: *Maize Marketing in Zaire*. Manhattan, Kansas State Univ., Food and Feed Grain Institute, Storage, Processing, and Marketing Report No. 51.

Sorenson, M. P. K. [1967]: *Land Reform in the Kikuyu Country: A Study in Government Policy*. Nairobi, Oxford Univ. Press.

Southern African Development Coordination Conference [SADCC, 1984]: *Agricultural Research Conference, 1987*. Sebele, Botswana, Department of Agriculture.

Southworth, V. R., W. O. Jones, and S. R. Pearson [1979]: "Food Crop Marketing in Atebuba District, Ghana." *Food Research Institute Studies* 17:157-195.

Sparling, E. W. [1981]: "A Survey and Analysis of *Ex Post* Cost-Benefit Studies of Sahelian Irrigation Projects." Ft. Collins, Colorado State Univ., Department of Economics, draft.

Spencer, D. S. C. [1972]: *Micro-Level Farm Management and Production Economics Research among Traditional African Farmers: Lessons from Sierra Leone*. East Lansing, MSU, DAE, African Rural Employment Paper No. 3.

——— [1976]: *African Women in Agricultural Development: A Case Study in Sierra Leone.* Washington, ACE, OLC.

——— [1986]: "Agricultural Research in Sub-Saharan Africa: Using the Lessons of the Past to Develop a Strategy for the Future." In R. J. Berg and Whitaker, eds., 1986, pp. 215-241.

Spencer, D. S. C., and D. Byerlee [1976]: "Technical Change, Labor Use, and Small Farmer Development: Evidence from Sierra Leone." *Amer. J. Agr. Econ.* 58:874-880.

——— [1977]: *Small Farms in West Africa: A Descriptive Analysis of Employment, Incomes and Productivity in Sierra Leone.* East Lansing, MSU, DAE, ARE Working Paper No. 19.

Spencer, D. S. C., D. Byerlee, and S. Franzel [1979]: *Annual Costs, Returns and Seasonal Labour Requirements for Selected Farm and Nonfarm Enterprises in Rural Sierra Leone.* East Lansing, MSU, DAE, ARE Working Paper No. 27.

Spencer, D. S. C., I. I. May-Parker, and F. S. Rose [1976]: *Employment, Efficiency and Income in the Rice Processing Industry of Sierra Leone.* East Lansing, MSU, DAE, ARE Paper No. 15.

Spencer, D. S. C., and V. Nyateng [1987]: "The Prospects for Increased Rice Production in West Africa." Paper presented at an *IFPRI/ISRA Conference on the Dynamics of Cereal Production and Consumption in West Africa, Dakar, July 15-17.*

Spencer, D. S. C., and M. V. K. Sivakumar [1987]: "Pearl Millet in African Agriculture." In ICRISAT, 1987, pp. 19-31.

Spencer, P. [1973]: *Nomads in Alliance: Symbiosis and Growth among the Rendille and Samburu of Kenya.* London, Oxford Univ. Press.

Spencer, W. P., D. L. Pfost, and J. R. Pedersen [1975]: *Grain Storage and Preservation in Senegal.* Manhattan, Kansas State Univ., Food and Feed Grain Institute, Storage, Processing, and Marketing Report No. 54.

Srinivasan, T. N. [1985]: "Neoclassical Political Economy, the State and Economic Development." *Asian Dev. Rev.* 3(2):38-58.

Staatz, J. [1979]: *The Economics of Cattle and Meat Marketing in the Ivory Coast.* Ann Arbor, Univ. of Michigan, CRED.

——— [1980]: "Meat Supply in Ivory Coast, 1967-1985." In Ariza-Nino and Steedman, eds., 1980, pp. 1-210.

——— [1988]: *Food Supply and Demand in Sub-Saharan Africa: Material prepared for the 1988 World Food Day Packet.* East Lansing, MSU, DAE, Staff Paper 88-51.

Staatz, J., and C. K. Eicher [1984]: "Agricultural Development in Historical Perspective." In Eicher and Staatz, eds., 1984, pp. 3-30.

Staatz, J., J. Dione, and N. Nango Dembele [1989]: "Cereals Market Liberalization in Mali." *World Dev.* 7(5):703-718.

Stanning, J. [1988]: "Policy Implications of Household Grain Marketing and Storage Decisions in Zimbabwe." In Rukuni and Bernsten, eds., 1988, pp. 329-358.

Staudt, K. [1975]: "Women Farmers and Inequalities in Agricultural Services." *Rural Africana* 29:81-94.

——— [1978-79]: *Tracing Sex Differentiation in Donor Agricultural Programs.* Washington, AID Office of Women in Development.

Stavis, B. [1979]: *Agricultural Extension for Small Farmers.* East Lansing, MSU, DAE, Rural Development Working Paper No. 3.

Steedman, C., et al. [1976]: *Mali-Agricultural Sector Assessment. Final Report.* Ann Arbor, Univ. of Michigan, CRED.

Steel, W. F. [1977]: *Small-Scale Employment and Production in Developing Countries: Evidence from Ghana.* New York, Praeger.

_____ [1979]: "Development of the Urban Artisanal Sector in Ghana and Cameroun." *J. Modern African Studies* 17:271-284.

Steeves, J. S. [1978]: "Class Analysis and Rural Africa: The Kenya Tea Development Authority." *J. Modern African Studies* 16:123-132.

Stein, L. [1979]: *The Growth of East African Exports and their Effect on Economic Development.* London, Croom Helm.

Stenning, D. J. [1957]: "Transhumance, Migratory Drift, Migration: Patterns of Pastoral Nomadism." *J. Royal Anthropological Institute* 87:57-73.

_____ [1959]: *Savannah Nomads: A Study of the WoDaaBe Pastoral Fulani of Western Bornu Province, Northern Region, Nigeria.* London, Oxford Univ. Press.

Stern, N. H. [1972]: "Experience with the Use of the Little-Mirrlees Method for an Appraisal of Small-Holder Tea in Kenya." *Oxford Bull. Econ. and Stat.* 34(1):93-123.

Stern, R. [1965]: "The Determinants of Cocoa Supply in West Africa." In I. G. Stewart and Ord, eds., 1965, pp. 65-82.

Stevens, C. [1979]: *Food Aid and the Developing World—Four African Case Studies.* New York, St. Martin's Press.

Stevens, P., Jr., ed. [1978]: *The Social Sciences and African Development Planning.* Waltham, MA, Crossroads Press.

Stewart, F. [1979]: "Employment and the Choice of Technique: Two Case Studies in Kenya." In Ghai and Godfrey, eds., 1979, pp. 47-74.

Stewart, I. G., R. C. Ogley, and W. D. C. Wright [1962]: *Nigeria: Determinants of Projected Level of Demand, Supply and Imports of Farm Products in 1965 and 1975.* Washington, USDA, ERS-Foreign 32.

Stewart, I. G., and H. W. Ord, eds. [1965]: *African Primary Products and International Trade.* Edinburgh, Edinburgh Univ. Press.

Stillman, C. W., ed. [1955]: *Africa in the Modern World.* Univ. of Chicago Press.

Stolper, W. F. [1969]: *Planning without Facts: Lessons in Resource Allocation from Nigeria's Development.* Cambridge, MA, Harvard Univ. Press.

Stoltz, D. [1979]: "Smallholder Dairy Development in Past, Present and Future in Kenya." Unpublished Ph.D. dissertation, Univ. of Hohenheim, West Germany.

Stoop, W. A. [1987]: "Adaptation of Sorghum/Maize and Sorghum/Pearl Millet Intercrop Systems to the Toposequence Land Types in the North Sudanian Zone of the West African Savanna." *Field Crops Research* 16:255-272.

Storm, R. [1976]: "Government Cooperative Groundnut Marketing in Senegal and Gambia." *J. Rural Cooperation* 5(1):29-42.

Stout, B. A., and C. M. Downing [1976]: "Agricultural Mechanization Policy." *International Labour Rev.* 113:171-187.

Strauss, J. [1981]: "Determinants of Household Food Consumption in Sierra Leone: Estimation of a Household-Firm Model with Application of the Quadratic Expenditure System." Unpublished Ph.D. dissertation, MSU, East Lansing.

Strauss, J. [1984]: "Joint Determination of Food Consumption and Production in Rural Sierra Leone: Estimates of a Household-Firm Model." *J. Dev. Econ.* 14:77-103.

Streeten, P. [1972a]: *The Frontier of Development Studies.* New York, Wiley.

_____ [1972b]: "Single Barrier Theories of Development." In Streeten, 1972a, pp. 13-20.

Stryker, J. D. [1974a]: "The Malian Cattle Industry: Opportunity and Dilemma." *J. Modern African Studies* 12:441-457.

_____ [1974b]: "Exports and Growth in the Ivory Coast: Timber, Cocoa, and Coffee." In Pearson and Cownie, 1974, pp. 11-66.

Sudan, Democratic Republic of ___ [1977]: *Food Investment Strategy, 1977-1985*. Khartoum, Ministry of Agriculture, Food, and Natural Resources.

Sudarkasa, N. [1973]: *Where Women Work: A Study of Yoruba Women in the Marketplace and in the Home*. Ann Arbor, Univ. of Michigan, Museum of Anthropology, Paper No. 53.

Sukhatme, P. V. [1970]: "Incidence of Protein Deficiency in Relation to Different Diets in India." *British J. Nutrition* 24:477-487.

Sundrum, R. M. [1987]: *Growth and Income Distribution in India*. New Delhi, Sage Publications.

Sundstrom, L. [1974]: *The Exchange Economy of Pre-Colonial Tropical Africa*. London, Hurst.

Sünkel, O. [1973]: "Transnational Capitalism and National Disintegration in Latin America." *Social and Economic Studies* 22:132-176.

Surarerks, V. [1986]: *Historical Development and Management of Irrigation Systems in Northern Thailand*. Thailand, Chiang Mai Univ., Department of Geography.

Suret-Canale, J. [1971]: *French Colonialism in Tropical Africa, 1900-1945*, translation by T. Gottheiner. New York, Crea Press.

Swamy, G. [1981]: *International Migrant Workers' Remittances: Issues and Prospects*. Washington, World Bank, Staff Working Paper No. 481.

Swanson, R. A. [1980]: "Development Interventions and Self-Realization among the Gourma (Upper Volta)." In Brokensha, Warren, and Werner, eds., 1980, pp. 67-92.

_____ [1981]: *Household Composition, Rainfall and Household Labor Time Allocation for Planting and Weeding: Some Oservations and Recommendations*. Ouagadougou, Burkina Faso, SAFGRAD, Farming Systems Unit, Document No. 4.

Swift, J. [1975]: "Pastoral Nomadism as a Form of Land-Use: The Twareg of the Adrar in Iforas (Mali)." In Monod, ed., 1975, pp. 443-454.

_____ [1977]: "Sahelian Pastoralist: Underdevelopment, Desertification and Famine." *Annual Rev. Anthropology* 6:457-478.

_____ [1981]: "Seeds of Famine: A Review." *Science* 211:473-474.

Swindale, L. D. [1984]: "ICRISAT, its Programmes and its Work in Africa." In *SADCC, 1984*, pp. 73-87.

Swindell, K. [1978]: "Family Farms and Migrant Labour: The Strange Farmers of The Gambia." *Canadian J. African Studies* 12(1):3-17.

Szal, R. J. [1979]: *Income Inequality and Fiscal Policies in Botswana*. Geneva, ILO, Income Distribution and Employment Programme Working Paper No. 73.

Szereszewski, R. [1961]: *Structural Change in the Economy of Ghana, 1891-1911*. London, Weidenfield and Nicholson.

Talbot, L. M. [1972]: "Ecological Consequences of Rangeland Development in Masailand, East Africa." In Farvar and Milton, eds., 1972, pp. 694-711.

Tapsoba, E. [1981]: "An Economic and Institutional Analysis of Formal and Informal Credit in Eastern Upper Volta: Empirical Evidence and Policy Implications." Unpublished Ph.D. dissertation, MSU, East Lansing.

Taylor, B. R., and T. B. Bailey [1979]: "Response of Maize Varieties to Environment in West Africa." *Tropical Agriculture* 56(2):89-97.

Technical Advisory Committee [TAC, 1982]: *Report of the Quinquennial Review Mission to ILCA*. Rome, FAO, TAC Secretariat.

_____ [TAC, 1988]: *Sustainable Agricultural Production: Implications for International Agricultural Research*. Rome, FAO, TAC Secretariat.

Tecle, T. [1975]: *The Evolution of Alternative Rural Development Strategies in Ethiopia: Implications for Employment and Income Distribution*. East Lansing, MSU, DAE, ARE Paper No. 12.

Tempelman, A. [1972]: "The Economics of Rice Milling in the Western State of Nigeria." Ibadan, Nigeria, NISER, mimeograph.

Temu, P. [1975]: "Marketing Board Pricing and Storage Policy with Particular Reference to Maize in Tanzania." Unpublished Ph.D. dissertation, Stanford Univ., Stanford, CA.

Tench, A. B. [1975]: *Socio-Economic Factors Influencing Agricultural Output—with Special Reference to Zambia.* Saarbrucken, West Germany, Verlag des SSIP-Schriften.

Terray, E. [1972]: *Marxism and Primitive Societies. Part 2. Historical Materialism and Segmentary Linear-Based Societies,* translation by M. Klopper. New York, Monthly Review Press, pp. 93-186.

Terry, E., K. Oduro, and F. Caveness, eds. [1981]: *Tropical Root Crops: Research Strategies for the 1980s.* Ottawa, Canada, IDRC.

Thenevin, P. [1978]: *L'investigation en milieu rural et la pratique du développement (Cadre d'integration et approche systémique).* Paris, INSEE, AMIRA note de travail No. 18.

_____ [1980]: "L'aide alimentaire en cereales dans les pays Sahéliens." Paris, Ministère de la Coopération, Service des Études et Questiones Internationales.

Thodey, A. R. [1969]: *Marketing of Grains and Pulses in Ethiopia.* Menlo Park, CA, Stanford Research Institute.

Thompson, V., and R. Adloff [1975]: "French Economic Policy in Tropical Africa." In Duignan and Gann, eds., 1975a, pp. 127-164.

Thorner, D., B. Kerblay, and R. E. F. Smith, eds. [1966]: *The Theory of Peasant Economy.* Homewood, IL, Irwin.

Thornton, D. S. [1973]: *Agriculture in South East Ghana.* Vol. I. *Summary Report.* Reading, England, Univ. of Reading, DAE and Management.

Thornton, D. S., and R. F. Wynn [1968]: "An Economic Assessment of the Sudan's Khasm el Girba Scheme." *Eastern Africa J. Rural Dev.* 1(2):1-21.

Tiffen, M. [1976]: *The Enterprising Peasant: Economic Development in Gombe Emirate, North Eastern State, Nigeria, 1900-1968.* London, HMSO, Ministry of Overseas Development.

Timmer, C. P. [1970]: "On Measuring Technical Efficiency." *Food Research Institute Studies* 9:99-171.

_____ [1972]: "Employment Aspects of Investment in Rice Marketing in Indonesia." *Food Research Institute Studies* 11:59-88.

_____ [1986]: *Private Decisions and Public Policy: The Price Dilemma in Food Systems of Developing Countries.* East Lansing, DAE, MSU International Development Papers, No. 7.

Timmer, C. P., W. P. Falcon, and S. R. Pearson [1983]: *Food Policy Analysis.* Baltimore, Johns Hopkins Univ. Press.

Tinker, I. [1976]: "The Adverse Impact of Development on Women." In Tinker, Bramsen, and M. Buvinic, eds., 1976, pp. 22-34.

Tinker, I., M. B. Bramsen, and M. Buvinic, eds., [1976]: *Women and World Development.* New York, Praeger.

Todaro, M. P. [1969]: "A Model of Labor Migration and Urban Unemployment in Less Developed Countries." *Amer. Econ. Rev.* 59:138-148.

_____ [1976]: *Internal Migration in Developing Countries: A Review of Theory, Evidence, Methodology and Research Priorities.* Geneva, ILO.

_____ [1980]: "Internal Migration in Developing Countries: A Survey." In Easterlin, ed., 1980, pp. 361-402.

Toksoz, S. [1981]: *An Accelerated Irrigation and Land Reclamation Program for Kenya: Dimensions and Issues.* Cambridge, MA, Harvard Univ., HIID, Discussion Paper No. 114.

Tollens, E. F. [1975]: *Problems of Microeconomic Data Collection on Farms in Northern Zaire.* East Lansing, MSU, DAE.

Toupet, C. [1977]: *La sédentarisation des nomades en Mauritanie centrale Sahélienne.* Paris, Librairie Honore Champion.

Tourte, R. [1974]: "Réflexions sur les voies et moyens d'intensification de l'agriculture en Afrique de l'Ouest." *L'Agronomie Tropicale* 24(9):917-946.

Tourte, R., *et al.* [1971]: "Thémes legers— thèmes lourds: systèmes intensifs: voies différentes ouvertes au développement agricole du Sénégal." *L'Agronomie Tropicale* 26:632-671.

Traore, B. [1980]: "Socioeconomic Field Assessment of Prospective Technologies in the SAT of Mali: A Case Study in an OACV Zone." In ICRISAT, 1980a, pp. 125-135.

Traore, N., and M. Toure [1978]: "Le machinisme agricole au Mali." *Journées d'études technico-économiques sur le tracteur Swazi Tiukabi.* Bamako, Mali, CEPI et DMA.

Tripp, R. [1978]: "Economic Strategies and Nutritional Status in a Compound Farming Settlement of Northern Ghana." Unpublished Ph.D. dissertation, Columbia Univ., New York.

Truong Binh, J. Pichot, and P. Beunard [1978]: "Caractérisation et comparison des phosphates naturels tricalciques d'Afrique de l'Ouest en vue de leur utilisation directe en agriculture." *L'Agronomie Tropicale* 33:136-145.

Tryfos, P. [1974]: "Canadian Supply Functions for Livestock and Meat." *Amer. J. Agr. Econ.* 56:107-113.

Turok, B., ed. [1979]: *Development in Zambia: A Reader.* London, Zed Press.

Tyc [1975]: *L'élevage en Haute Volta: analyse et proposition d'orientation.* Paris, SEDES.

Uchendu, V. C. [1967]: "Some Issues in African Land Tenure." *Tropical Agriculture* 44(2):91-101.

—————— [1968]: "Socioeconomic and Cultural Determinants of Rural Change in East and West Africa." *Food Research Institute Studies* 8:225-242.

Uchendu, V. C., and K. R. Anthony [1975a]: *Agricultural Change in Kisii District, Kenya.* Nairobi, East African Literature Bureau.

—————— [1975b]: *Agricultural Change in Teso District, Uganda.* Nairobi, East African Literature Bureau.

Uhlig, S. J., and B. A. Bhat [1979]: *Choice of Technique in Maize Milling.* Edinburg, Scottish Academic Press.

UN [1981]: *World Population Prospects as Assesssed in 1980.* New York, United Nations Secretariat, Population Division.

UNDP and IBRD [1985]: *Somalia: Report of a Joint Technical Cooperation Assessment Mission.* New York and Washington.

UN/ECA [1971]: *Africa's Strategy for Development in the 1970s: First Conference of Ministers of the Economic Commission for Africa.* New York.

UN/ECA/FAO [1964]: *Report of the FAO/ECA Expert Meeting on Government Measures to Promote the Transition from Subsistence to Market Agriculture in Africa, Addis Ababa, April 27-May 7, 1964.* Rome, FAO.

—————— [1966]: *African Agricultural Development: Reflections on the Major Lines of Advance and the Barriers to Progress.* New York.

UNESCO [1988]: *UNESCO Statistical Yearbook, 1988.* UNESCO, New York.

Unité d'Évaluation [1978]: *Évaluation de l'operation arachide et cultures vivrières. Étude agro-économique de 32 exploitations agricoles en zone OACV.* Bamako, Mali, IER.

Univ. of Nairobi [1975]: *Second Overall Evaluation of the Special Rural Development Programme.* Nairobi, IDS, Occasional Paper No. 12.

Upton, M. [1967]: *Agriculture in South-West Nigeria: A Study of the Relationship between Production and Social Characteristics in Selected Villages.* Reading, England, Univ. of Reading, Development Studies Paper No. 3.

———— [1970]: "The Influence of Management on a Sample of Nigerian Farms." *Farm Economist* 11:526-536.

———— [1973]: *Farm Management in Africa: The Principles of Production and Planning.* London, Oxford Univ. Press.

———— [1979]: "The Unproductive Production Function." *J. Agr. Econ.* 30:179-194.

———— [1985]: "Returns from Small Ruminant Production in South West Nigeria." *Agr. Systems* 17:65-83.

———— [1986]: "Production Policies for Pastoralists: The Botswana Case." *Agr. Systems* 20:17-35.

———— [1987]: *African Farm Management.* Cambridge, England, Cambridge Univ. Press.

Upton, M., and H. Casey [1974]: "Risk and Some Pitfalls in the Use of Averages in Farm Planning." *J. Agr. Econ.* 25:147-152.

Upton, M., and D. H. Petu [1964]: "An Economic Study of Farming in Two Villages in Illorin Emirate." *Bull. Rural Econ. and Sociology* 1(1):7-30.

Urdang, S. [1980]: *Fighting Two Colonialisms: Women in Guinea-Bissau.* New York, Monthly Review Press.

USDA, ERS [1981]: *Food Problems and Prospects in Sub-Saharan Africa: The Decade of the 1980's.* Washington, FAER No. 166.

Vadlamundi, Y. R., and H. U. Thimm [1974]: "Economic Analysis of Fertilizer Trials Conducted in Kenya on Maize." *East African Agr. and Forestry J.* 40(2):189-201.

Vail, D. J. [1973]: "Induced Farm Innovation and Derived Scientific Research Strategy." *Eastern Africa J. Rural Dev.* 6(1):

Valdés, A., ed. [1981]: *Food Security for Developing Countries.* Boulder CO, Westview Press.

Van Apeldoorn, G. J. [1971]: *Markets in Ghana—A Census and Some Comments.* Vol. 1. *Northern and Upper Regions.* Legon, Univ. of Ghana, ISSER, Technical Publication No. 17.

Van Binsbergen, W. M. J., and H. A. Meilink, eds. [1978]: *Migration and the Transformation of Modern African Society.* Leiden, Netherlands, Afrika-Studiecentrum, African Perspectives, 1978/a.

Van der Hoeven, R. [1977]: *Zambia's Income Distribution during the Early Seventies.* Geneva, ILO, Income Distribution and Employment Programme, Working Paper No. 54.

Van der Wiel, A. C. A. [1977]: *Migratory Wage Labour: Its Role in the Economy of Lesotho.* Mazenod, Lesotho, Book Center.

Van Ginneken, W. [1976]: *Rural and Urban Income Inequalities in Indonesia Mexico, Pakistan, Tanzania and Tunisia.* Geneva, ILO.

Van Hekken, N., and H. U. E. T. van Velzen [1972]: *Land Scarcity and Rural Inequality in Tanzania: Some Case Studies from Rungwe District.* The Hague, Netherlands, Mouton.

Van Raay, H. G. T. [1975]: *Rural Planning in a Savanna Region.* Rotterdam, Rotterdam Univ. Press.

Van Rensburg, P. [1978]: *The Serowe Brigades: Alternative Education in Botswana.* London, Macmillan.

Vasthoff, J. [1968]: *Small Farm Credit and Development: Some Experiences in East Africa with Special Reference to Kenya.* Munich, West Germany, Weltforum Verlag.

Venema, L. B. [1978]: *The Wolof of Saloum: Social Structure and Rural Development in Senegal.* Wageningen, Netherlands, Agricultural Research Report No. 871.

Vengroff, R. [1977]: "Dependency and Underdevelopment in Black Africa: An Empirical Test." *J. Modern African Studies* 15:613-630.

Vermeer, D. E. [1970]: "Population Pressures and Crop Rotational Changes among the Tiv of Nigeria." *Annals, Assoc. of Amer. Geographers* 60:299-314.

———— [1979]: "The Tradition of Experimentation in Swidden Agriculture among the Tiv of Nigeria." *Applied Geography Conferences,* Vol. 2, J. M. Frazier and B. J. Epstein, eds., SUNY-Binghamton.

———— [1981]: "Collision of Climate, Cattle and Culture in Mauritania during the 1970s." *Geographical Rev.* 71:281-297.

Vinay, B. [1968]: *L'Afrique commerce avec l'Afrique: problèmes et impératifs Africains de coopération économique et monétaire.* Paris, Presses Universitaires de France.

Von Braun, J., and D. Puetz [1987]: "An African Fertilizer Crisis: Origin and Economic Effects in The Gambia." *Food Policy* 12:337-48.

Von Freyhold, M. [1979]: *Ujamaa Villages in Tanzania: Analysis of a Social Experiment.* New York, Monthly Review Press.

Von Kaufmann, R. [1976]: "The Development of the Range Land Areas." In Heyer, Maitha, and Senga, eds., 1976, pp. 225-287.

Von Pischke, J. D. [1980]: "The Political Economy of Specialized Farm Credit Institutions." In Howell, ed., 1980, pp. 81-103. Also available as *World Bank Working Paper No. 446,* 1981, Washington.

Vyas, V., and D. Casley [1988]: *Stimulating Agricultural Growth and Rural Development in Sub-Saharan Africa.* Washington, World Bank, PPR Working Paper.

Waelti, J. [1988]: *Indigenous Social Science and Economic Development in Kenya.* St. Paul, Univ. of Minnesota, Econ. Dev. Center, Research Report No. 5.

Wallace, T. [1981]: "The Kano River Project, Nigeria: The Impact of an Irrigation Scheme on Productivity and Welfare." In Heyer, Roberts and Williams, eds., 1981, pp. 281-305.

WARDA [1984]: "Programme Achievement, Contribution to and Impact on Rice Development in West Africa." Monrovia, Liberia 84/STC-14/17.

———— [1988]: *WARDA: Strategic Plan, 1990-2000.* Monrovia, Liberia.

Wardle, C. [1979]: *Promoting Cattle Fattening among Peasants in Niger.* London, ODI, Pastoral Network Paper No. 8c.

Warren, A., and J. K. Maizels [1977]: "Ecological Change and Desertification." In *UN Conference on Desertification: Desertification: its Causes and Consequences.* London, Pergamon Press, pp. 169-260.

Watts, E. R. [1969]: "Bureaucracy and Extension." *East African J.* 6(8):37-40.

Weber, A., and T. T. Hartmann [1977]: "A Comparative Study of Economic Integration with Special Reference to Agricultural Policy in the East African Community." In Dams and Hunt, eds., 1977, pp. 379-395.

Weber, M., J. Staatz, J. Holtzman, E. Crawford, and R. Bernsten [1988]: "Informing Food Security Decisions in Africa: Empirical Analysis and Policy Dialogue." Paper presented at the Annual Meeting of the AAEA, Knoxville, TN, July 31-August 3.

Weil, P. [1970]: "The Introduction of the Ox Plow in Central Gambia." In McLoughlin, ed., 1970, pp. 229-263.

———— [1980]: "Land Use, Labor and Intensification among the Mandinka of Eastern Gambia." Paper presented at the Annual Meeting of the African Studies Association, Philadelphia, October 15-18.

Weiler, E. [1979]: "Social Cost-Benefit Analysis of the Nianga Project, Senegal." Unpublished M.S. thesis, Purdue Univ., West Lafayette, IN.

Weinrich, A. K. H. [1975]: *African Farmers in Rhodesia: Old and New Peasant Communities in Karangaland*. New York, Oxford Univ. Press.

Wells, J. C. [1974]: *Agricultural Policy and Economic Growth in Nigeria, 1962-68*. Ibadan, Nigeria, Oxford Univ. Press.

Welsch, D. E. [1965]: "Response to Economic Incentives by Abakaliki Rice Farmers in Eastern Nigeria." *J. Farm Econ.* 47:900-914.

_____ [1966]: "Rice Marketing in Eastern Nigeria." *Food Research Institute Studies* 6:329-352.

Westley, S., and B. F. Johnston, eds. [1975]: *Proceedings of a Workshop on Farm Equipment Innovations for Agricultural Development and Rural Industrialization*. Nairobi, Univ. of Nairobi, IDS Occasional Paper No. 16.

Westley, S., B. F. Johnston, and M. David [1975]: *Summary Report of a Workshop on a Food and Nutrition Strategy for Kenya*. Nairobi, Univ. of Nairobi, IDS Occasional Paper No. 14.

Whetham, E. H. [1972]: *Agricultural Marketing in Africa*. London, Oxford Univ. Press.

Whetham, E. H., and J. I. Currie, eds. [1967]: *Readings in the Applied Economics of Africa*, 2 Vols. London, Cambridge Univ. Press.

Whitney, A. [1968]: *Marketing of Staple Foods in Eastern Nigeria*. East Lansing, MSU, DAE, Agricultural Economics Report No. 114.

Whitney, T. R. [1981]: "Changing Patterns of Labor Utilization, Productivity and Income: The Effects of Draft Animal Technology on Small Farms in Southeastern Mali." Unpublished M.S. Thesis, Purdue Univ., West Lafayette, IN.

WHO [1976]: *Methodology of Nutritional Surveillance, Report of a Joint FAO/UNICEF/WHO Expert Committee*. Geneva.

Whyte, W. F. [1981]: *Participatory Approaches to Agricultural Research and Development: A State-of-the-Art Paper*. Ithaca, NY, Cornell Univ., Rural Development Committee.

Widstrand, C. G., ed. [1971]: *Co-operatives and Rural Development in East Africa*. New York, Holmes and Meier.

_____ [1972]: *African Co-operatives and Efficiency*. Uppsala, Sweden, SIAS.

_____ [1975]: *Multinational Firms in Africa*. Stockholm, Sweden, Almquist and Wiksell.

Wilber, C., ed. [1973]: *The Political Economy of Development and Underdevelopment*. New York, Random House.

Wilcock, D. C. [1978]: *The Political Economy of Grain Marketing and Storage in the Sahel*. East Lansing, MSU, DAE, ARE Working Paper No. 24.

_____ [1981]: *Small-Scale, Non-Farm Enterprises in Eastern Upper Volta: Survey Results*. East Lansing, MSU, DAE, ARE Working Paper No. 36.

Williams, G. [1981]: "The World Bank and the Peasant Problem." In Heyer, Roberts, and Williams, eds., 1981, pp. 16-51.

Williamson, O. [1985]: *The Economic Institutions of Capitalism*. New York, Free Press.

Wilson, F. [1972]: *Labour in the South African Gold Mines, 1911-1969*. Cambridge, England, Cambridge Univ. Press.

Winch, F. [1976]: "Costs and Returns of Alternative Rice Production Systems in Northern Ghana: Implications for Output, Employment and Income Distribution." Unpublished Ph.D. dissertation, MSU, East Lansing.

Winter, G. [1970]: *Méthodologie des enquêtes niveau de vie en milieu rural Africain*. Paris, ORSTOM, Initiatives et documents techniques No. 15.

_____ [1975]: *Le point de vue du planificateur sur le problème de l'amélioration des méthodes d'investigation en milieu rural Africain*. Paris, INSEE, AMIRA note de travail No. 2.

_____ [1978]: *Réflexion sur les enquêtes-menages à fins multiples dans les pays en voie de développement*. Paris, INSEE, AMIRA note de travail no. 21.

Wolf, E. [1969]: *Peasant Wars of the Twentieth Century*. New York, Harper and Row.

Wolgin, J. M. [1975]: "Resource Allocation and Risk: A Case Study of Smallholder Agriculture in Kenya." *Amer. J. Agr. Econ.* 57:622-630.

_____ [1979]: "Resource Allocation and Risk: A Case Study of Smallholder Agriculture in Kenya: A Reply." *Amer. J. Agr. Econ.* 61:114-115.

Wolpe, H. [1972]: "Capitalism and Cheap Labour Power in South Africa: From Segregation to Apartheid." *Economy and Society* 1:425-456.

Wood, L. J. [1974]: *Market Origins and Development in East Africa*. Kampala, Makerere Univ., Department of Geography, Occasional Paper No. 57.

World Bank [1980]: *World Development Report, 1980*. Washington.

_____ [1981a]: *World Development Report, 1981*. Washington.

_____ [1981b]: *Accelerated Development in Sub-Saharan Africa: An Agenda for Action*. Washington.

_____ [1984]: *Annual Report, 1984*. Washington.

_____ [1985a]: *Ensuring Food Security in the Developing World: Issues and Options*. Washington.

_____ [1985b]: *The Smallholder Dimension of Livestock Development: A Review of World Bank Experience*, 2 Vols. Washington, Report No. 5979.

_____ [1986]: *Poverty and Hunger: Issues and Options for Food Security in Developing Countries*. Washington.

_____ [1987a]: *The World Bank Atlas, 1987*. Washington.

_____ [1987b]: *World Development Report, 1987*. Washington.

_____ [1987c]: *The Twelfth Annual Review of Project Performance Results*. Washington.

_____ [1988a]: *Food Security for Africa*. Washington, draft.

_____ [1988b]: *World Development Report, 1988*. Washington.

_____ [1988c]: *Rural Development: World Bank Experience, 1965-86*. Washington.

_____ [1988d]: *Cotton Development Programs in Burkina Faso, Côte d'Ivoire and Togo*. Washington.

_____ [1988e]: *Strengthening Agricultural Research in Sub-Saharan Africa: A Proposed Strategy*. Washington.

_____ [1991]: *World Development Report, 1991*. Washington.

World Commission on Environment and Development [1987]: *Our Common Future*. London, Oxford Univ. Press.

Wrigley, C. C. [1965]: "Kenya's Pattern of Economic Life, 1902-45." In Harlow and Chilver, eds. 1965, pp. 209-264.

Wuyts, M. [1981]: "The Mechanization of Present-Day Mozambican Agriculture." *Dev. and Change* 12:1-27.

Yansane, A. Y. [1976]: "An Evaluation of Socialist Experiments in African States of French Colonial Legacy." *Nigerian J. Econ. Soc. Studies* 18(3):363-407.

Yap, L. Y. L. [1977]: "The Attraction of Cities: A Review of the Migration Literature." *J. Dev. Econ.* 4:239-264.

Young, D. L. [1979]: "Resource Allocation and Risk: A Case Study of Smallholder Agriculture in Kenya: Comment." *Amer. J. Agr. Econ.* 61:111-113.

Young, M. D. [1979]: "Influencing Land Use in Pastoral Australia." *J. Arid Environments* 2:279-288.

Young, S. [1977]: "Fertility and Famine: Women's Agricultural History in Southern Mozambique." In Palmer and Parsons, eds., 1977, pp. 66-81.

Youngs, A. J. [1972]: "Wheat Flour and Bread Consumption in West Africa: A Review with Special Reference to Ghana." *Tropical Science* 14(3):235-244.

Yudelman, M. [1975]: "Imperialism and the Transfer of Agricultural Techniques." In Duignan and Gann, eds., 1975a, pp. 329-359.

Zachariah, K. C., and J. Conde [1980]: *Migration in West Africa: Demographic Aspects*. New York, Oxford Univ. Press for the World Bank and OECD.

Zalla, T. [1981]: "Political, Technical and Economic Aspects of Smallholder Milk Production in Northern Tanzania." Unpublished Ph.D. dissertation, MSU, East Lansing.

Zalla, T., D. J. Campbell, J. Holtzman, L. Lev, and D. Trechter [1981]: *Agricultural Production Potential in the Mandara Mountains in Northern Cameroon*. East Lansing, MSU, DAE, Rural Development Working Paper No. 17.

Zalla, T., R. B. Diamond, and M. S. Mudahar [1977]: *Economic and Technical Aspects of Fertilizer Production and Use in West Africa*. East Lansing, MSU, DAE, ARE Working Paper No. 22 and Muscle Shoals, AL, IFDC.

Zartman, I. W., and C. L. Delgado, eds. [1984]: *The Political Economy of the Ivory Coast*. New York, Praeger.

Zerbo, D., and M. Le Moigne [1977]: "Problèmes posés par la mécanisation dans les pays membres du CILSS." Antony, France, CEEMAT.

Zimbabwe, Government of the Republic of ___ [1980]: *Zimbabwe Research Index, 1979*. Salisbury, Government Printer.

_____ [1981]: *Growth with Equity: An Economic Policy Statement*. Salisbury, Government Printer.

_____ [1982]: *Report of the Commission of Inquiry into the Agricultural Industry*. Harare.

Zolberg, A. A. [1967]: "The Political Use of Economic Planning in Mali." In Harry G. Johnson, ed., 1967b, pp. 98-123.

Zuckerman, P. S. [1977]: "Different Smallholder Types and Their Development Needs." *J. Agr. Econ.* 28:119-128.

_____ [1979a]: *A Micro-Level Farm Management Study in Western Nigeria: Field Work Design and Application*. Ibadan, Nigeria, IITA, Discussion Paper No. 7/79.

_____ [1979b]: *A Micro-Level Farm Management Study in Western Nigeria: Some Results and Experiences with Questionnaires*. Ibadan, Nigeria, IITA, Discussion Paper No. 8/79.

_____ [1979c]: "Simulating the Decision Making of a Nigerian Smallholder." *Canadian J. Agr. Econ.* 27(2):17-26.

Zurek, E., ed. [1985]: *Integrated Rural Development— Research Results and Programs Implementation, Bonn Conference*. Hamburg, Verlag Weltarchiv.

PART TWO. Agriculture in Economic Development:
Theories, Findings, and Challenges in an Asian Context

Agriculture in Economic Development: Theories, Findings, and Challenges in an Asian Context

John W. Mellor and Mohinder S. Mudahar

Chapter I. Introduction

Since World War II and the end of colonialism in Asia, there has been a substantial evolution in thought about the role of agriculture in economic development and the processes by which agriculture develops. That evolution has been reflected not only in the substance of the literature but also in the relative weight given to different areas of analysis.

Also, the proportion of research and literature from the developing world has increased. Western economists had a large, perhaps even dominant, influence on published thought about agriculture and development in the early postwar period. The remains of the colonial legacy and the flow of foreign capital and technical assistance from the West contributed to that influence. With the sharp decline of U.S. foreign assistance and increase in the number of economists and research institutions in Asian countries, that influence has declined. As a result, the perceptions that shape research and its ultimate use have changed. A relatively smaller proportion of research on agricultural development in Asia is reported or even reflected in Western journals today. And American and Asian scholars interact less, so that knowledge of agricultural development has declined in America.

Modern thought about what is now called economic development began in the 1930s and 1940s, when Europeans were concerned about what they perceived to be the backwardness of Eastern and Southeastern Europe. They generally emphasized industrialization [e.g., Rosenstein-Rodan, 1943], an emphasis that has been consistently reflected in the mainstream of thought on economic development.

It was recognized early that the shift of labor from rural to urban areas was an important aspect of industrialization. Initially, the literature on that subject was concerned with the use of surplus rural labor to facilitate capital formation in a nonagricultural sector which, with the transfer of labor to industry along with the food that it was already consuming, made the social cost of labor almost zero. Most of the approaches of that time did not call for the diversion of scarce resources to develop agriculture.

Agriculture was given a positive role in development in Johnston and Mellor [1961]. They used a labor surplus approach, and recognized that industrialization was essential to modernization, a view later elaborated in Mellor [1966, 1976], Johnston and Kilby [1975], and an update by Mellor and Johnston [1984]. In addition to agriculture's supply of labor, Johnston and Mellor [1961] emphasized the need to increase agricultural production to supply wage goods in support of labor transfers, the ability of agriculture to provide industrial capital through foreign trade, and the stimulus that demand for industrial goods emanating from rising rural incomes would give to growth of the nonagricultural sector.

The ideas of Johnston and Mellor [1961] were not in the mainstream of thought on economic development in Asia at that time, however. The idea was more widely accepted that the basic limitation to economic growth was capital and that resources should, therefore, be concentrated in the capital goods industry. This implied that resource allocations to consumer goods industries, including the principal one, agriculture, should be minimized. That strategy was carefully delineated and quantified over the following decade. It culminated in highly sophisticated, multisectoral, mathematical growth models based solely on capital, giving a dynamic role neither to agriculture nor to labor [Chakravarty, 1969]. The results were similar to the results of import displacement models used for Latin America at the same time.

By the late 1960s, it was recognized that growth from the development strategy was slow and that its benefits were distributed narrowly. The latter problem was expected and was to have been solved by rapid growth that would increase consumption and distribute benefits more broadly. Slow growth was an unpleasant surprise. It led thought in the 1970s in two directions: to supplementary programs that would try to abate poverty directly, and to alternative strategies of growth that emphasized agriculture and developed growth linkages between agriculture and other sectors of the economy.

Alleviating poverty through social welfare schemes was consistent with an orientation toward capital and industrialization. Many adherents of these schemes began to view agriculture as a place to hold the poor and underemployed. They recognized the need for social welfare programs to mitigate rural poverty, to retain the poor in rural areas, and to avoid disturbing the urban economic and political processes as industrialization and urbanization proceeded. Others envisaged social welfare in a basically rural society and rejected an emphasis on industrialization. During the 1970s, the basic human needs approach found a wide following in western countries. But most leaders of Asian developing countries presumed that the benefits of such an approach would be wiped out by population growth; that it would not benefit large, politically important urban constituencies and large farmers; and that it would leave their countries in permanently weak positions in international power politics. Nevertheless, much thought continues to be given to such strategies.

A contrasting approach that emphasized agricultural production as a way to raise incomes in agriculture through vigorous development processes and as a way of fostering linkage and multiplier effects elsewhere in the economy became attractive in the 1980s as the deficiencies of capital-intensive approaches became apparent. Mellor [1976, 1986] and Mellor and Johnston [1984] provided an integrated statement of the strategy as a means of accelerating growth and modernization. It is an approach, however, which is not yet fully conceptualized or described using sound empirical research. During the 1980s, progress was made in both conceptualizing and quantifying the key relationships. Not surprisingly, as that process proceeded, increasing emphasis was given to expanded rural infrastructure investment as the means of bringing about the rural specialization and integration essential to modernization, technological change, and rising productivity [Ahmed and Hossain, 1987]. This interest in rural linkages was fostered by recognition of the failure of older approaches to industrialization, and social welfare to provide for the rising expectations of a rapidly growing rural population.

Thought about how to develop agriculture has evolved more consistently than thought about the role of agriculture. In the 1950s, the view was widely held that farmers in developing countries were ignorant, inefficient, and exploited. It was believed that if exploitation could be stopped by removing the rapacious landlord and money lender, if leadership could be provided through local government bodies, and if ignorance could be removed through extension and community development programs, then agriculture would grow and prosper.

It became apparent, however, that while these measures were important to political development, they had little effect on production. A view developed, with T. W. Schultz [1964] as the most articulate spokesman, that farmers were indeed intelligent and sensible optimizers but that their environment and lack of incentives discouraged increases in output. Specifically, many believed that fertilizer, water, pesticides, and credit were not reaching the farmer. Development programs were redesigned to provide these inputs. Again, success was only modest.

The next step in the evolution of agricultural development came with the reminder that modern agriculture in developed countries had grown and developed largely because of the application of science. By the mid-1960s, many recognized that in developing countries, production increases and greater use of inputs were impeded by limitations of technology. Subsequently, much effort was devoted to increasing the ability to develop new high-yielding crop varieties and associated practices [Dalrymple 1986a, b]. Signs of accelerated growth then became apparent. But progress from the new high-yield technologies tended to be more rapid where there had already been a major effort to inform farmers, to develop extension services, and to develop input distribution systems. New technology was indeed the key to the puzzle, but many other parts had to be put in place as well. A slowly building body of analysis came to recognize the critical role for physical infrastructure investment—a matter of major importance because of massive resource requirements.

Growth in the knowledge of how to develop agriculture has increased the attention given to problems of implementation. This is an important new direction that leads past the traditional boundaries of economic thought and empirical techniques. Increasing agricultural production has also brought greater attention to other development questions such as what are the determinants of effective demand for agricultural output, how resources to agricultural research and modern inputs are to be allocated, what patterns of income distribution are emerging, and what processes broaden participation in economic growth. It is recognized that small farmers and the landless have difficulty manipulating and, in turn, benefiting from, the institutions that are critical to growth and development. Concern with broadening participation in such institutions is growing.

Following the outline of the preceding overview, we begin with a brief description of the characteristics and role of Asian agriculture, followed by a review of agriculture's place in the dominant theories of economic growth. In these discussions the bias is, of course, towards the Asian context. The theme is that of a major, positive role for agriculture in the process of economic development. Given that context, we proceed to discuss the process of modernizing agriculture so as to contribute to overall economic growth. The discussion is followed by further treatment of two particularly important aspects of modernization—technology and market development, leading to a treatment of the thorny issues of income distribution and welfare in the context of agricultural growth. The story then reverts to the issue of overall economic growth—how contribution to real national income arising from technological change in agriculture can stimulate growth in other sectors. This is a view that sees overall economic growth rising from the multipliers that result from agricultural growth. A brief treatment of the vast subject of trade and aid is presented before a discussion of implementation issues is taken up. The question of how to achieve implementation is presented in the earlier sections.

The subject is vast, the literature immense. We have included only a small fraction of an increasingly diverse and specialized literature, selecting those studies that fit into the progression outlined above. Many important works are not mentioned in the text, but are included in the references. We no doubt have missed important contributions—in the interest of keeping at least the vestiges of a theme and due to space constraints. This problem of a massive body of literature is much more substantial for the literature on Asia than for Africa and Latin America. For the latter two, the literature is large but at least a pretense can be made at a comprehensive survey. For Asia, India alone has an overwhelming literature. Our solution is to have a theme we believe is particularly relevant to contemporary problems in Asia, to stay with it, and to keep the total presentation brief.

Chapter II. Agriculture's Characteristics and Role

1. Stages of Agricultural Development

Following Rostow [1960], the growth of agriculture in developing countries can be divided into three stages: traditional (static), transitional, and modern (dynamic).[1] These stages and their major attributes are summarized in Table 1. The contribution of agriculture to economic development increases as it develops from the static stage to the dynamic.[2] The value of this classification into stages is severely limited by the lack of characteristics unique to any one stage and clearcut demarcations between stages. Nevertheless, it still delineates the changing characteristics of agriculture and the implications of these changes for agriculture's relationships with other sectors in the economy. It also emphasizes the changing objectives and instruments of agricultural development. These insights are essential for policymakers to understand not only the role of agriculture in economic development but also the processes by which agriculture develops.

2. Special Characteristics of Asian Agriculture

The role of agriculture varies from one stage of economic development to another and from one country to another. The importance of agriculture in Asia's economic development comes from its relative and absolute size, as shown by the comparative agricultural development indicators (Table 2). There are, however, large variations within Asia. In 1985, the contribution of agriculture to gross domestic product ranged from 25 percent (Pakistan) to 62 percent (Nepal) in South Asia; 17 percent (Thailand) to 48 percent (Burma) in Southeast Asia; and 3 percent (Japan) to 33 percent (China) in East Asia (Table 3). More strikingly, the proportion of the labor force employed in agriculture in 1980 ranged from 53 percent (Sri Lanka) to 93 percent (Nepal) in South Asia; 42 percent (Malaysia) to 71 per-

Table 1. Summary of major characteristics of agricultural development from stage I
through stage II and into stage III

General characteristic	Stage I (static)	Stage II (transitional)	Stage III (dynamic)
1. Values, attitudes, motivations	Negative or resistant (does *not* imply non-national)		Positive or receptive
2. Goals of production	Family consumption and survival		Income and net profit
3. Technology or state of arts	Static or traditional with no or slow innovation		Dynamic or rapid innovation
4. Degree of commercialization of farm production	Subsistence or semisubsistence		Commercial
5. Degree of commercialization of farm inputs	Family labor and farm produced		Commercial
6. Factor proportions and rates of return	High labor/capital ratio, low labor return		Low labor/capital ratio, high labor return
7. Institutions affecting or serving agricultural and rural areas	Deficient and imperfect		Efficient and well developed
8. Availability of unused agricultural resources	Available		Unavailable
9. Share of agricultural sector in total economy	Large		Small

Source: Hayami and Ruttan [1971, 1985], originally from Wharton [1963a].

cent (Thailand) in Southeast Asia; and 11 percent (Japan) to 74 percent (China) in East Asia.

In noting that the relative size of the agricultural sector suggests major attention to that sector, two important points should be kept in mind. First, even at best, the maximum growth rates in agriculture tend to be low relative to those achievable in nonagriculture. Thus, a nonagricultural sector may achieve, even for rather sustained periods of time, growth rates of 10–15 percent. In agriculture, one is doing well to get above the range of 3–6 percent. This leads to the implication that agriculture's potential can only be realized if the development process occurs broadly throughout the agricultural sector.

Table 2. Selected comparative agricultural development indicators for Asia, Africa, Latin America, and the rest of the world[a]

	Year[b]	Unit	World	Percent share in			
				Asia[c]	Africa[c]	Latin America[c]	The rest of the world
Population	1984–86	Million	4,838.8	58	11	8	22
Agricultural population	1984–86	Million	2,220.0	74	16	5	5
Arable land and permanent crops	1983–85	Million ha	1,475.0	31	12	12	45
Irrigated land	1983–85	Million ha	217.8	63	4	7	26
Nitrogen consumption	1983/84–1985/86	Million mt	69.2	39	3	4	54
Phosphate consumption	1983/84–1985/86	Million mt	33.2	27	4	7	63
Potassium consumption	1983/84–1985/86	Million mt	25.6	12	2	6	80
Total nutrient consumption	1983/84–1985/86	Million mt	128.1	30	3	5	62
Agricultural tractors	1983–85	Million	24.0	18	2	6	74
Harvesters/threshers	1983–85	Million	3.8	30	1	3	65
Total cereal production	1984–86	Million mt	1,839.3	42	4	6	48
Paddy rice production	1984–86	Million mt	473.5	92	2	4	2
Wheat production	1984–86	Million mt	519.4	35	2	4	59
Maize production	1984–86	Million mt	473.9	21	6	11	62
Other cereal production[d]	1984–86	Million mt	372.5	16	8	4	72
Total pulses production	1984–86	Million mt	51.7	46	12	10	32
Total roots and tubers production	1984–86	Million mt	590.3	38	17	8	37
Fruit production[e]	1984–86	Million mt	315.6	30	12	20	38
Vegetable production[f]	1984–86	Million mt	404.6	55	7	5	33
Meat production	1984–86	Million mt	150.2	24	5	11	60
Milk production[g]	1984–86	Million mt	511.8	16	3	7	73
Egg production[h]	1984–86	Million mt	31.1	33	4	10	53
Tea production	1984–86	Million mt	2.3	78	12	3	7
Coffee (green) production	1984–86	Million mt	5.4	13	23	63	1
Sugarcane production	1984–86	Million mt	928.0	38	8	49	6
Seed cotton production	1984–86	Million mt	50.1	49	7	10	33

[a] Derived from data obtained from the annual publications (and various previous issues) of FAO [1989a] and FAO [1989b]. The percent share is approximate due to rounding.
[b] Three-year average for the years shown.
[c] Refers to FAO's definition of continental Asia (including China, Israel, and Japan); continental Africa (including South Africa); and Latin America.
[d] Includes barley, rye, oats, millet, sorghum, and other minor cereals.
[e] Excludes melons.
[f] Includes melons.
[g] Includes milk from cows, buffaloes, sheep, and goats.
[h] Includes hen eggs and other eggs.

The broadly participatory growth in the agricultural sector means not only bringing in the bulk of geographic regions, but the bulk of the people within those regions. This latter process has two dimensions. The first, particularly in the context of smallholder agriculture, is that one must have institutions that bring most of the small farmers into the process. This requires not only complex and intricate institutional development, but also full geographic participation. This is the major argument for a well-developed rural infrastructure. Urban-based officials may have the impression that all rural people have good roads because that is the only kind they know. However, the fact of the matter is that in developing countries road systems are much less developed than they were at the same stage of development in modern developed countries; and it follows that a high proportion of rural people are located too far from roads to enter fully into the exchange and specialization economy. One of the important benefits of physical infrastructure is to distribute the educated people who run modern institutions broadly throughout the economy. Pioneering work by Raisuddin Ahmed and Mahabub Hossain has brought out these relationships [Ahmed and Hossain, 1987].

It can also be seen from Table 3 that South, Southeast, and East Asia (excluding China) represent successively higher levels of economic growth. Our review is biased towards discussion of the earlier stages of growth and problems of take-off in growth, both for agriculture and the economy generally. Thus, the emphasis is largely on South and Southeast Asia. We draw heavily on the historical literature for East Asia.

Agriculture also demands attention in economic development because of the peculiar nature of its production conditions. It is seasonal and heterogenous in nature; subject to large agroclimatic, environmental, and economic risks; highly unorganized and nonunionized; and involves a large number of decision-makers. Limited awareness of knowledge about the transformation of agriculture, and the complexity of technology and technological change in it, has baffled many development economists and administrators and turned them away from agriculture as an engine of employment-oriented economic growth.

Agriculture is the only sector of the economy that comes close to meeting the conditions of atomistic competition. Yet, agriculture has often been regarded as an impenetrable mystery, not yielding to the tools of economic analysis and incapable of being integrated with other sectors of the economy. Furthermore, the common view that the farmer is bounded by tradition, irrational, and unresponsive to economic stimuli has prevented agriculture from receiving adequate attention and resources from planners and policymakers.

This erroneous view of the farmer arose from the failure to understand the complex relations between the farmer's business world and his household life; the economic implications of the high risks farmers face; the effect of the heterogeneity of physical, economic, and institutional conditions on innovations; and the burden that the limited land base places on the ability of technological change to increase agricultural production. In the complex decisionmaking environment of

Table 3. Comparative economic development indicators for selected countries in Asia

Region/country	Population mid-1985 (millions)	GNP per capita 1985 (US$)	Average annual real growth rate (1980–85) (percent)				Agriculture as percent of GDP (1985)[b]	Percent of total expenditure on agriculture (1981)	Infant mortality rate per thousand of live births (1985)
			Population	Gross domestic product (GDP)	GDP per capita	Gross agricultural production			
South Asia									
Bangladesh	100.6	150	2.6	3.6	1.0	2.8	50	12[d]	123
India	765.1	270	2.2	5.2	3.0	2.7	31	7[c]	89
Nepal	16.5	160	2.4	3.4	1.0	—	62[c]	18	133
Pakistan	96.2	380	3.1	6.0	2.9	2.1	25	2[c]	115
Sri Lanka	15.8	380	1.4	5.1	3.7	4.0	27	8[f]	36
Southeast Asia									
Burma	36.9	190	2.0	5.5	3.5	5.4	48	24[c]	66
Indonesia	162.2	530	2.1	3.5	1.4	3.1	24	10[c]	96
Malaysia	15.6	2000	2.5	5.5	3.0	3.0	—	7[g]	28
Philippines	54.7	580	2.5	−0.5	−3.0	1.7	27	6[c]	48
Thailand	51.7	800	2.1	5.1	3.0	3.4	17	10	43
East Asia									
Japan	120.8	11300	0.7	3.8	3.1	1.6	3	—	6
South Korea	41.1	2150	1.5	7.9	6.4	6.3	14	6[c]	27
China[a]	1040.3	310	1.2	9.8	8.6	9.4	33	—	35

Region/Country	Years of life expected at birth (1985)	Population per square kilometer of agricultural area (1980)	Rural population as a percent of total (1985)[h]	Percent of labor force in agriculture (1980)	Percent of national income received by lowest 20% (1981)	Percent of calorie requirements supplied per capita (1980)	Percent of adults who are literate (1981)
South Asia							
Bangladesh	51	908	82	75	7[i]	84	26[f]
India	56	373	75	70	7[j]	88	36
Nepal	47	356	93	93	5[f]	86	19[c]
Pakistan	51	324	71	55	8[k]	106	24[g]
Sri Lanka	70	570	79	53	7[i]	102	85[g]
Southeast Asia							
Burma	59	321	76	53	8[m]	113	66[c]
Indonesia	55	461	75	57	7[n]	110	62[d]
Malaysia	68	320	62	42	4[l]	121	60[c]
Philippines	63	442	61	52	5[k]	116	75[g]
Thailand	64	257	82	71	6[n]	105	86[c]
East Asia							
Japan	77	2,139	24	11	8[o]	124	99[c]
South Korea	69	1,702	36	36	6[n]	128	93[d]
China[a]	69	308	78	74	7[g]	107	69[p]

Source: Compiled from World Bank [1983a, 1987c].

a Excluding Taiwan.
b At current prices.
c 1984.
d 1978.
e 1980.
f 1977.
g 1979.
h 100 minus urban population as a percent of total.
i 1974.
j 1975.
k 1970.
l 1973.
m 1972.
n 1976.
o 1969.
p 1982.

the farmer, the welfare of his family must be improved if innovations are to raise agricultural output.

3. Contributions of Agriculture

According to Kuznets [1961], agriculture makes product, market, and factor contributions to economic development. According to Johnston and Mellor [1961], agriculture increases food supplies, enlarges agricultural exports, transfers manpower, forms capital, and stimulates industrialization through increased rural net cash income. These two approaches have been synthesized in Figure 1 to describe the contribution of agriculture to economic development. The capital-oriented development strategies attach little importance to these contributions, and so, either ignore agriculture entirely or include it only marginally. According to the World Bank [1982b], economic growth has been rapid in virtually all those countries where agricultural development has been strong. Faster agricultural growth in low-income countries can also reduce rural poverty since over 90 percent of the absolute poor are rural people. The different elements of this powerful role of agriculture are discussed below.

4. Structural Change and Growth Patterns

The process of economic growth consists of growth in economic variables and structural change in the economy. Of particular importance is the secular growth in the absolute size of agriculture, its secular decline in relative importance, and the concurrent increase in the relative importance of the industrial sector.[3] In an empirical study of fifty-one countries, Chenery [1960] showed that the share of industrial output increased from 17 percent when per capita income was $100 to 38 percent when it was $1,000 while the share of primary production (agriculture) declined from 45 percent to 15 percent, the share of transportation and communication doubled, and the share of other services did not change.

Goreux [1959] and Houthakker [1957] demonstrated that as development takes place, the composition of demand changes in favor of industrial goods, and the share of food in the household budget declines. This is consistent with the well-known Engel effects. Chenery [1960], on the other hand, argued that the factor supply conditions lead to a systematic change in the structure of industrial growth as incomes rise.

In noting the inelastic demand for most basic food staples, one should not ignore the fact that there is a substantial set of agricultural commodities, including livestock and horticultural products, for which the demand is elastic. These commodities, particularly livestock, have a significant initial base and a rapid growth in effective demand as incomes rise. Indeed, this set of factors explains why agriculture can play quite an important role even in relatively late stages of devel-

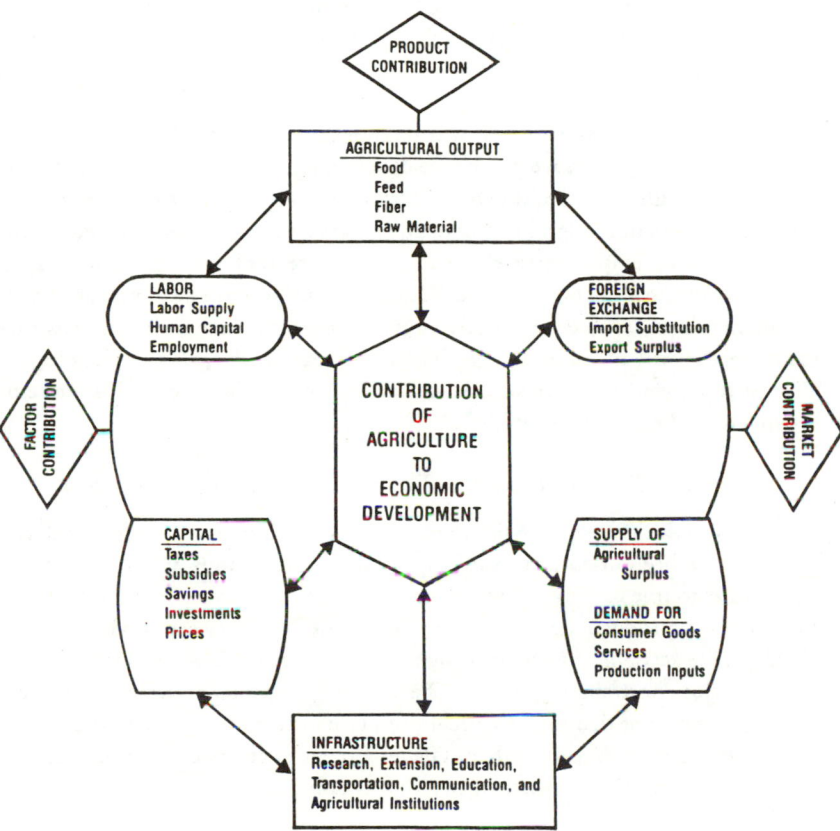

Figure 1. Sectoral linkages and contribution of agriculture to economic development.

opment. The income-elastic sectors become more and more important, and because they are not constrained as much by land area as the basic staples, they can provide a more substantial supply response as well. Thus, one has an important interaction between urbanization and growth of urban incomes, on the one hand, and acceleration of the growth rate in the agricultural sector, on the other hand, as demand swings toward the income-elastic commodities for which land is less constraining.

Agriculture has a key role, which diminishes over time, in fostering these structural changes which are part of the development process. Recent technological breakthroughs in agriculture have enhanced its ability to foster these changes through changes in foodgrains production, marketable surpluses, income distribution patterns, employment patterns, and intersectoral resource transfers. In particular, changes in income distribution determine farm and nonfarm consumption patterns, subsequently affecting the supply of wage goods and raw materials to the rest of the economy. These linkages, triggered by technological changes in agriculture, have significant implications for labor participation, employment, and economic growth.[4]

Chapter III. Agriculture and Theories of Economic Development

Particularly with respect to Asia, most theories and models of economic development have been oriented toward capital and capital goods and so have given agriculture no role or only a passive one.[5] They have not taken labor to be a limiting factor of production, have not specifically dealt with wage goods, and have failed to take account of features of agriculture that cause food production to become a major constraint to growth. The following sections describe the five economic relationships that define agriculture's role in economic growth and review the models of economic growth and development in terms of their relevance to low-income countries in which agriculture is dominant.

1. Key Economic Relationships

LAND AND LAND-AUGMENTING TECHNOLOGICAL CHANGE

Recognizing land as a factor of production leads to diminishing returns to variable inputs because land is assumed to be fixed in supply. Diminishing returns ultimately lead an economic system into a stationary state as was amply demonstrated by Ricardo. Although this paper deals with post-World War II literature, it should be noted that the classical economists, e.g., Petty, Smith, and Ricardo, had much to say about agriculture's role in economic development and anticipated much of the contemporary discussion.

Most Asian countries encompass a high proportion of the population in areas of high density, a substantial fraction of the population is underemployed or employed at extraordinarily low levels of productivity, and agriculture—for which land accounts for a major factor share—contributes significantly to GNP. As a result, the effect of diminishing returns on national output can be large. Land-augmenting technological change can offset or delay that effect by increasing the effective supply of land in agricultural production. Hence, discussion of the literature on technological change has a central place in this review.

INELASTIC AGGREGATE SUPPLY OF FOODGRAINS

The more inelastic the aggregate, as opposed to individual crops, supply of foodgrains is with respect to prices, the more important technological change is to the growth of crop yields, agricultural production, and labor demand. According to Herdt [1970], the long-run aggregate supply elasticity for Punjab agriculture is 0.1 to 0.2. According to Barnum [1973], the aggregate supply elasticity for foodgrains in India is about 0.1. Based on the profit-function approach but utilizing pooled time-series and cross-section data for the nine regions of Philippines from 1948-74, Quizon [1981] found a statistically significant estimate of 0.104 for the short-run price elasticity of aggregate agricultural (crops and live-stock) supply. Such elasticities appear to be typical for countries in which most of the cultivatable land is already being cultivated. Foodgrain supplies for such countries can be increased by imports or by land-augmenting technological change. Since the inelasticity of supply arises from the fixity of the land base, it follows that the supply of agricultural commodities that use little land will not be inelastic. These commodities include fruits, vegetables, livestock products, and other agricultural products whose demand is highly elastic; but they do not include foodgrains, the basic calorie source for most Asian populations.

Long-run supply elasticity may be defined to include an induced response in public investments and institutional development [Hayami and Ruttan 1971, 1985]. Similarly, Mundlak [1988] describes price response in terms of capital accumulation and includes all forms of capital such as human capital and the institutions of technological change. In this view, long-term price response may be quite elastic. However, Mundlak is explicit that such response is conditional on the realization of a substantial pace of technological change. In this review, we treat the processes of technological change separately in later sections as more appropriately dealt with by explicit public policy. This is because of the long lags in developing such measures and the consequent need to abstract from short-run processes of market price formation.

HIGH MARGINAL PROPENSITY OF LABORERS TO CONSUME FOODGRAINS

The low elasticity of the aggregate supply of foodgrains will not limit additional employment if labor's marginal propensity to consume foodgrain is low.

However, empirical evidence has shown that low-income consumers spend most of any additional income on food, a major part of it going to foodgrains and other staples. For India, the lower 20 percent of the income distribution which can be defined as the laboring class, spend 59 percent of their incremental income on foodgrains alone and 79 percent on all food commodities [Mellor and Lele, 1973; Mellor, 1978].

ELASTIC SUPPLY OF LABOR

In most low-income Asian countries, the supply of labor to the nonagricultural sector seems to be highly elastic. This is because population growth is rapid and employment conditions in the agricultural sector are poor. Increasing employment opportunities in the nonagricultural sector, with only a small increase in the real wage rates, will elicit a large increase in labor supply to that sector. The elasticity of aggregate labor supply from agriculture depends on the relative size of the agricultural sector, on technical conditions of agricultural production, and on the family labor-leisure choice function.[6] Much work has been done in this area to determine whether or not the marginal product of labor in agriculture is zero.

The marginal product of labor in agriculture need not be zero to provide a highly elastic labor supply [Mellor, 1963]. It is likely that agricultural labor is fully employed seasonally and that small expenditures on selective mechanization or on reorganizing production could save large amounts of labor at seasonal peaks and make the labor supply elastic. The empirical evidence is generally consistent that the supply of labor to the nonagricultural sector is highly elastic with little increase in real wage rate [Nabi, 1984; T. H. Lee, 1971; Ohkawa and Rosovsky, 1960; Umemura, 1969].

Thus, despite a complex set of determinants, it appears that the labor supply can increase rapidly if jobs and wage goods are made available. Note that a constraint on wage goods is effectively a constraint on labor supply. Lele and Mellor [1981] underline this connection with a two sector model which presents and achieves a general equilibrium with two separate, interacting markets, for labor and for wage goods.

LESS-THAN-PERFECT SUBSTITUTABILITY OF CAPITAL FOR LABOR

To the extent that capital and labor are less than perfect substitutes, restraints on the labor supply (or wage goods supply) will require increasingly high rates of savings simply to maintain a given rate of growth. Although growth models may assume fixed factor proportions, in reality, there is always more than one process to produce the same commodity, each with a different capital-labor ratio. Depending on the factor-price ratio, capital and labor can be combined in different proportions by using different production techniques. And, at an aggregate level with trade, the choices of production can change capital-labor ratios substantially. Changes in income distribution may change the structure of demand and consequently the average capital-labor ratio. The technological data indicate that the possibilities of capital-labor substitution vary considerably among sectors, are greater in agriculture than in industry, and are particularly so in rural activities for which rural people have a high marginal propensity to consume. As a result, it is possible to increase employment without a corresponding increase in capital.

2. Aggregate Theories of Economic Growth

The aggregate economic growth models that form the basis of modern growth theory were developed in Harrod [1948] and Domar [1957]. The Harrod-Domar model provides a conceptual framework for analyzing the growth process by focusing on a few crucial economic variables and their relations. One of the major assumptions of the Harrod-Domar formulation is a production function with fixed factor proportions. The steady-state growth rate is determined by the average productivity of capital and the average propensity to save. This approach is elegant in its simplicity. The models of Solow [1956] and Swan [1956] eliminate the knife-edge property of the Harrod-Domar model by allowing substitution between capital and labor and introducing technological change.

However, investment plays a dominant role in determining economic growth in these models. While the frameworks were designed essentially for high-income countries, they form the intellectual basis for a wide range of models that are applied to developing countries. They are particularly important in understanding inattention to agriculture. In particular, aggregate models are inappropriate for analyzing the sectoral linkages, structural changes, and market feedbacks in the process of economic growth. Since these models generally have ignored technological change, growth is achieved mainly by increasing the use of capital, which is relatively scarce in most low-income countries. Furthermore, these models may also be inappropriate when labor supply is highly elastic and the supply of wage goods seriously constrains employment.[7]

3. Multisector Models of Economic Growth

The neoclassical two-sector model is an elaboration of the single-sector neoclassical growth model.[8] But despite all the efforts made to make the two-sector models more realistic, "they do not represent any great advance in realism over one-sector models," according to Hahn and Matthews [1965, p. 39]. Two-sector models appear to disregard low-income economies that have elastic supplies of labor and in which wage goods constrain both employment and the rate of growth. Because these models assume full employment they are even less suitable for analyzing the growth process in labor surplus economies. However, the concepts behind such models have been important to development theory and practice in Asia and explain much of the early inattention to agriculture.

In the two-sector model of Mahalanobis [1953, 1955], which guided Indian planning for a decade or two commencing in the mid-1960s, labor is assumed to be perfectly elastic and the growth rate of the system is determined asymptotically by the proportion of investment going to the capital-goods sector and the ratio of incremental income to investment in that sector. Mahalanobis maintained that the main constraint to employment is the scarcity of capital goods. In general, however, Mahalanobis's framework ignored demand considerations for both consumer and capital goods. The main paradox of the Mahalanobis model in fact is that income will eventually be higher if a large proportion of investment goes to the capital-goods sector even though capital may be more productive if invested in the consumption-goods sector.

The Fel'dman [1957] model, developed for the Soviet Union in the 1920s, is similar to the Mahalanobis model. It emphasizes investment in heavy industry as a source of economic growth. Agriculture, as a source of wage goods and as a source of employment and economic growth, is virtually ignored in both the Mahalanobis and the Fel'dman models. Despite obvious shortcomings, the Mahalanobis model formed the basis of the Second Five-Year Plan in India.

Chinese leaders, including Mao, explicitly recognized agriculture's role as a source of wage goods and eventually, through bitter experience, the central importance to overall economic growth of a dynamic, healthy agricultural sector. They did not, however, appreciate the investment requirements necessary to achieve this dynamism on an appropriate scale, and thus China's sectoral allocation of state investment bore striking similarity to that of India's in its overwhelming emphasis on heavy industry [Tang and Stone, 1980; Lardy, 1983a; Stone, 1985], while various mechanisms were employed to channel rural savings to the greatest extent possible through the public sector [Stone, 1988a; Ishikawa, 1982, 1988].

The multisector model developed in Chakravarty and Lefeber [1965] is an intertemporal optimizing planning model. It assumes that the labor supply is perfectly elastic and that capital is the only bottleneck to economic growth. The objective of the model is to maximize the discounted sum of composite consumption bundles over the planning period subject to technological, market, and political constraints and to the terminal conditions. Land, labor, and natural resources are assumed to have no limits and are not included in the model. As Srinivasan [1965] pointed out, the model has many shortcomings. Since technology is assumed to be constant, the projections of supply and demand of inputs and hence, the investment patterns, are not realistic. Despite surplus labor, the objective of the model can only be achieved if capital-intensive techniques are adopted. Eckaus and Parikh [1968] tried to make this model more realistic by adding sectors, by incorporating nonlinear relationships and technological change, and by dividing the economy into different regions.

The consistency models developed in Manne and Rudra [1965] and Bergsman and Manne [1966] are multisectoral, intertemporal, and use dynamic Leontief input-output frameworks, but treat agriculture superficially. They are designed to analyze the implications that alternative paths of economic growth have for the balance of trade by postulating different aggregate growth targets and import substitution targets. They were tested by using data from India. Their main conclusions are that investment in import substitution and export promotion of capital goods industries relieves foreign exchange bottlenecks in the long run. Von Neumann's [1945-46] general equilibrium multisector model of a uniform expanding economy was highly abstract but was a major step in the development of multisector planning models. Specifically, it allowed incorporation of labor and wage goods into the formulation.

None of the models described above had a place for a dynamic agriculture in economic growth. Nor did they have a place for accelerating growth by mobilizing labor. They led, in practice, to a capital-intensive, low-employment strategy, with little investment in agriculture. The theory behind the Mahalanobis plan for India was simple, internally consistent, and suitable for an economy presumed to have poor prospects for growth of agriculture and exports.

4. Dualistic Development Models

The dualistic models focus on the transfer of labor from the agricultural sector to more productive employment in the industrial sector and the supply of wage goods as a constraint to economic growth.

POPULATION GROWTH AND LABOR TRANSFER

W. Arthur Lewis [1954], in an original formulation of a dualistic model, emphasizes the potential for capital formation of a transfer of low-productivity labor from the agricultural to the industrial sector without a corresponding decline in agricultural output. Lewis's model is termed "classical" because it assumes that some agricultural labor is redundant and that labor is paid a constant, institutionally determined wage. The Lewis model was more formally elaborated and generalized in Ranis and Fei [1961]. They assumed no technological change and showed that the agricultural sector passes through three distinct phases as labor is transferred to the industrial sector.

In phase I, some labor in the agricultural sector is redundant. The average product of labor in the agricultural sector is assumed to equal wages in the industrial sector. Hence, labor can be transferred from the agricultural to the industrial sector without affecting either agricultural output or the terms of trade between the two sectors. In phase II, the marginal product of labor in agriculture is positive but less than industry's institutionally determined wage. As a result, labor cannot be transferred to industry without decreasing agricultural output. This implies that the terms of trade go against industry and that the industrial real wage measured in industrial goods rises. When all agricultural labor with a marginal product less than industry's wage has been transferred, the agricultural sector enters into phase III where labor becomes a scarce factor and wages are no longer fixed. Both the agricultural and industrial sectors become competitive and wages are determined by marginal labor productivity. In this phase, the agricultural surplus falls rapidly because total output falls and the wage rate rises. In phases II and III, the industrial sector faces a rising labor supply curve.

In both the Lewis and Ranis-Fei models, the demand for labor in the industrial sector was determined by capital accumulation. In their later work, Fei and Ranis [1963, 1964] made labor absorption in the industrial sector a function of capital, innovational intensity and its labor-using bias, growth in real wages, and the elasticity of labor demand. On the other hand, Jorgenson [1961] assumed that agricultural labor has a positive marginal product, that labor leaves the agricul tural sector when the wage rate in the nonagricultural sector provides an income equal to the average income in agriculture, and that the wage rate in the nonagricultural sector is equal to the marginal product of its labor.[9] Population growth is determined endogenously in the Jorgenson model and exogenously in the Lewis and Ranis-Fei models. None of these models incorporates other factors that determine labor supply. If intersectoral labor transfers and changes in labor participation rates are an important part of growth in low-income countries, then models must deal with the real complexities of the labor markets in such economies. Mellor and Stevens [1956], Mellor [1963, 1976], A. K. Sen [1966], Todaro [1969], and Hymer and Resnick [1969] shed some light on labor supply questions.

WAGE GOODS PRODUCTION

Because of the high marginal propensity of laborers to spend on food, the supply of a marketable surplus of wage goods is important in determining labor transfer and capital accumulation in the industrial sector. In classical dualistic models, since land is fixed and capital is not included, productivity can only be raised through improved production techniques. Fei and Ranis [1964] assume that technical change is neutral, which raises the marginal productivity of labor and hence total agricultural output. Since wages are fixed, an increase in total output increases the marketable surplus which, in turn, changes the terms of trade against the agricultural sector and lowers the labor supply curve in the industrial sector.

A. K. Dixit [1969] and Hornby [1968] examined the question of how to increase the production of wage goods in response to a change in relative prices and a favorable public investment policy using a classical framework. In the absence of technological change, this approach inferred that returns would diminish and the prices of wage goods would rise, which would increase labor costs and cause capital to be substituted for labor. In the Jorgenson model, since capital is excluded, land is fixed, and labor is fully employed, production can only be maintained and labor transferred through technological change in the agricultural sector.

Since technological change is critical to the growth of agricultural production and tends to be biased, normally land-augmenting, it is important to incorporate these elements in a dualistic model. Lele and Mellor [1981] presented a dualistic model that examines the relationships between increased foodgrain production, achieved through alternative technologies, and the rates of growth of nonagricultural employment, the nonagricultural sector capital-labor ratios, relationships of the prices of agricultural and nonagricultural commodities, and changes in the per capita incomes of the labor force. The model clarifies that land-augmenting technological change plays a critical role in release of labor to the industrial sector; is accompanied by declining terms of trade for agriculture; and that factor bias is an important determinant of the pace of labor transfer. In a further extension, Mellor and Ranade [1988] show how, under the conditions of developing countries, technological change in agriculture does not depress agricultural income.

WAGE GOODS TRANSFER

Dualistic development models have assumed that the amount and pattern of consumption per capita does not change and that the potential marketable surplus will automatically be transferred to the industrial sector without cost. These are highly simplifying assumptions. In fact, because of leakages in the system, not all the marketable surplus gets transferred. In the classical model [W. A. Lewis, 1954] the capitalist landlord, responsive institutions, and the government make sure that all the potential surplus is properly channeled to the industrial sector. The "neoclassical" models have not really addressed this question.

The Lewis, Ranis-Fei, and Jorgensen models assume that consumption per capita does not change as growth rates in the economy rise. Consequently, these models ignore the effects of both price and income on the level and pattern of consumption. Kelley, Williamson, and Cheetham [1972], however, analyzed consumption using a linear expenditure system. This added a realistic component to the dualistic models. Zarembka [1970] also focused on the problem of marketable surplus by assuming that income and price elasticities of demand are not zero. Lele and Mellor [1981] explored this relationship in the context of technological change and changed distribution of income, showing that it is the factor bias of technology which is key to a substantial transfer of wage goods to support a high growth rate of nonagricultural employment which, in turn, requires either accelerated capital formation or lower capital-labor ratios. The factor bias towards land distributes incremental income to relatively higher income people who market additional agricultural commodities to purchase nonagricultural goods and services. The factor bias not only supplies wage goods, but provides demand for the goods produced by added employment. The added employment, of course, requires at least some added capital to combine with the added labor.

CAPITAL FORMATION AND TRANSFER

Except for the Kelley-Williamson-Cheetham model, the "classical" and "neo-classical" dualistic models incorporate capital only in the nonagricultural sector. This is a significant shortcoming, especially since technological change is often the main source of increases in labor productivity in the agricultural sector and since technological change normally is either embodied in new capital goods or requires a large amount of working capital to purchase modern farm inputs produced in the nonagricultural sector. Furthermore, technological change itself is determined by capital formation because it requires large initial investments [Hayami and Ruttan 1971, 1985; Mundlak, 1988]. On the other hand, employment and hence, growth in the nonagricultural sector, depends on capital accumulation. Mellor [1974] delineated the attributes of a model for a country with a dominant and dynamic agricultural sector. The appropriate model would have much in common with the models of A. K. Sen [1968] and Von Neumann [1945/1946], particularly with reference to the role of wage goods.

5. Market Orientation

We should not leave this discussion on theories of economic development without discussing market orientation, though briefly. The essence of a market orientation is substantial decentralization of decisionmaking, with the guidance for that decisionmaking coming from market prices. As economies become more and more complex, it becomes less and less possible for governments to allocate resources in an effective way. Thus, as development proceeds, it drives toward a market orientation that becomes stronger and stronger. It should be noted that, in an agriculturally oriented development strategy, market orientation is particularly important. On the one hand, a myriad of small farmers can hardly be managed by fiat from a central government; on the other hand, the activities generated by the rising incomes of small farmers tend themselves to be labor-intensive and hence, suited to small-scale production, again leading to large numbers of entrepreneurs pursuing diverse activities. Thus, a market orientation is critical to an agriculturally led development strategy.

Having made the basic case for a market orientation, one should recognize that in the early stages of development when relatively labor-intensive activities are most productive and efficient, there is a substantial need for public sector activities in support of the small firms that tend to go along with a high degree of labor intensity. Thus, agriculture itself requires substantial support from public sector research and educational institutions; it may require an initial impetus even in areas like fertilizer distribution and credit. Thus, striking the correct balance between public and private sector activities and, within each of those, between larger-scale centralized activities and small-scale decentralized activities is one of the most important sets of decisions in an agriculture-oriented development strategy.

In the 1980s, a large number of developing countries undertook reforms that involved substantial increases in market orientation. These changes were required in part because of major structural maladjustments, which in many cases were postponed by taking on massive amounts of debt. The maladjustments rose significantly from the large increases in oil import bills which represented a decline in real incomes for the large importers, the effect of which was hidden by large borrowings. But constant pressure to increase public expenditure for developmental and welfare reasons, combined with fixed exchange rates, also created major and growing distortions. The transition from the one set of policies to the other tends to be politically difficult and often disruptive. Foreign assistance can play a major role in such processes, but its instability has often added to the problems rather than decreasing them [Lele and Nabi, eds., 1990].

Chapter IV. Agricultural Production Behavior, Technology, and Policy

1. Technology and Agricultural Productivity

GREEN REVOLUTION AND SOURCES OF GROWTH

Agricultural production is affected by farm inputs, weather, government programs, and technological change. In many Asian countries, however, the most important factor in the substantial increase of agricultural productivity of recent years has been the introduction of land-augmenting farm technology. These farm innovations have set into motion events which have had important multiplier effects.

Since land-augmenting agricultural technology (often referred to as the "Green Revolution") was introduced in Asian countries, many studies have appeared describing its various implications for growth and income distribution.[10] These studies have acknowledged that the Green Revolution has contributed positively to agricultural growth. But they have also warned about its negative effects, at least on the distribution of income.

Rice is a staple food crop for much of Asia's population. The results reported in Table 4 indicate that, except for Thailand, the major source of increases in rice production in major Asian countries has been increases in yield per acre. Expansion of area with controlled irrigation accounted for a significant part of the increase in production attributed to the growth in area planted to rice. Increased

Table 4. Estimated relative contributions to growth in rice production in selected Asian countries at the height of the Green Revolution

| | | annual | Percentage of total increased production | | | | | |
| | Year[a] | growth (in %) | attributed to area | | | attributed to yield | | |
			Irrigated land	Rainfed & upland	Total	Fertilizer[b]	Others[c]	Total
Burma	1965-73	0.8	35.8	-23.3	12.5	47.8	39.7	87.5
India	1965-70	3.2	19.2	5.8	25.0	47.3	27.7	75.0
Indonesia	1965-72	4.8	46.4	-6.8	39.6	25.2	35.2	60.4
Philippines	1965-73	3.4	33.1	-7.7	25.4	44.5	30.1	74.6
Sri Lanka	1960-68	4.8	34.7	11.1	45.8	31.9	22.3	54.2
Thailand	1965-72	2.1	10.8	82.2	93.0	13.6	-6.6	7.0

Source: Asian Development Bank [1977].
[a] Five-year average centered on the years sown.
[b] One additional kilogram of nutrients (N + P$_2$O$_5$ + K$_2$O) is assumed to produce ten kilograms of paddy.
[c] Includes increased factor productivity resulting from new technology.

yield accounted for approximately 75 percent of the increase of rice production in India and the Philippines. The performance of high-yielding varieties of wheat was even more spectacular.[11] This is partly because the growing season of wheat and the geographical regions suitable for wheat cultivation are drier and hence subject to fewer environmental hazards than those for rice [Barker and Mangahas 1971].

A comparative analysis of the adoption of new rice technology and changes in rice farming for selected Asian countries is available in IRRI [1975, 1978b]. This study presents summary results based on farm-level research from 1971 to 1973 in thirty-six villages, fourteen study areas in six countries in Asia.[12] According to Duff [1978], there is little evidence to indicate a causal relationship between the adoption of modern rice varieties and mechanization, particularly tractors. However, water control, through the establishment of pumping units, has increased the use of modern rice varieties. According to Parthasarathy and Prasad [1978], there was significant association between farm size and adoption of modern rice varieties in both wet and dry seasons in Andhra Pradesh, an important rice growing province in India. Barker, Herdt, and Rose [1985] provided a classic survey of The Rice Economy of Asia, with a particularly full picture of technological change.

Wheat has also been an important component in the Green Revolution. The increase in wheat production illustrates what an outstanding research breakthrough can do when applied in a locale where there is an impressive experimental system, room to expand irrigated area rapidly, a well-developed set of institutions and facilities that can efficiently transmit knowledge, and a marketing system that can deliver production inputs and outputs. Wheat production grew rapidly primarily because of a sharp increase in yields following the widespread application of dwarf wheat varieties. This growth increased the profitability of irrigated wheat production, which in turn accelerated investment to expand cropped area. The rate of growth in irrigated area sown with wheat was dramatic. This is attributable to large, cheaply developed groundwater resources which facilitated development of private tube wells. The investment in tube wells became highly profitable after the new varieties were introduced. The rapid expansion of rural electrification further aided this growth in irrigated area. It is also significant that the infrastructure, including transportation and communication systems, were highly developed in the wheat regions of Asia; and that the wheat area had a more developed infrastructure of public services than many other regions of India.[13]

The belief that technological change helps to expand agricultural production is further corroborated by the empirical results reported in Hayami and Ruttan [1971, 1985]. These results are based on a cross-sectional analysis of thirty-eight developed and developing countries. Chemical, biological, and mechanical innovations, the use of which is determined by domestic resource endowments and factor price ratios, are clearly important in determining growth in agricultural productivity. The study also found that the contribution of general and technical education, as a major part of the embodiment of human capital in labor, was quite large.

CONTRIBUTION AND DEMAND OF FERTILIZER

The input analysis reported in Table 5 emphasizes the dramatic change in the sources of growth of foodgrain production in India. The increased contribution of fertilizer to foodgrain production has been startling. From 1949/50 and 1960/61, the use of additional fertilizer accounted for less than 10 percent of increased foodgrain output in India, while from 1960/61 to 1973/74 it was responsible for 53 percent. Much of this increase came after fertilizer responsive crop varieties were introduced in the late 1960s and began to increase the productivity of fertilizer.[14] Increased use of fertilizer has been the single most important indicator of technological change in agriculture. It reflects increases in irrigation and the development of new crop varieties, since they raise the productivity and the profitability of using greater quantities of fertilizer.[15] However, yield response to applied fertilizer varies across soils and crops, and under different technological, management, and climatic conditions. The rice yield response to applied nitrogen, for example, varies considerably from one location to another and over time, but generally it is higher in dry season than in wet season because of greater solar energy and lower damage due to weather and insect infestation.

In India, fertilizer use is concentrated in a few districts. In northern India, fertilizer use grew at a compound annual growth rate of 38 percent from 1960-61 to 1970-71. The rapid rise in fertilizer consumption in the region in the early 1960s was associated with improvements in price ratio between fertilizer and wheat, but increases in the adoption of high-yielding wheat varieties and in controlled irrigation were probably far more important.[16] Increases in domestic fertilizer production require large capital outlays and place a heavy burden on foreign-exchange supplies, thus restricting the expansion of fertilizer supply. This and other fertilizer related policy issues and their national and farm-level implications in developing countries are discussed in Mudahar and Pinstrup-Andersen [1977] and Mudahar [1978].

The available cross-country empirical evidence on price elasticity of fertilizer demand is presented in Table 6. The large variation in elasticity estimates is partly due to differences in data, time period, methodology, definition of fertilizer, and domestic policies related to fertilizer prices. However, the tremendous variability in elasticities in Table 6 suggests highly imperfect knowledge of the relationship between fertilizer price and use. The elasticities computed vary so much because farmers typically use much less fertilizer than would be profitable in perfectly functioning markets. Hence, price may easily pick up a move toward

Table 5. Estimated contribution of agricultural inputs as a percentage of total and incremental foodgrain production in India for selected years[a]

Crop years	Unirrigated land and labor on it		Irrigated land and labor on it		Intensification of labor		Inorganic fertilizer	
	Total	Incremental	Total	Incremental	Total	Incremental	Total	Incremental
1956–57	73	–	22	–	4	–	1	–
1960–61	68	20	21	14	9	55	2	11
1964–65	63	6	21	21	11	34	5	38
1970–71	51	–5	22	29	12	16	15	59
1973–74	47	–5	22	22	13	30	18	53
1978–79[b]	38	10	19	6	13	10	30	76
1983–84[b]	31	0	17	10	12	9	40	79

Source: Mellor [1976, Appendix Table 9].

[a] The percentages of incremental contribution have been taken from the increments to foodgrain production from one selected year to the next. The 1978–79 and 1983–84 data are extrapolations. New technology played a key role in determining the levels of factor productivity, particularly for fertilizer, that lie behind the numbers presented.

[b] Extrapolations.

Table 6. Summary of fertilizer demand studies and price elasticity of demand for fertilizer[a]

Country/region	Fertilizer	Time period	Elasticity of demand		Adjustment coefficient	Source
			Short run	Long run		
Asia						
India	N	1953/54–1967/68	−0.31[b]	−0.34	0.92	M. S. Rao [1974]
India	N	1953/54–1967/68	−0.53[c]	−6.63	0.08	M. S. Rao [1974]
India	N	1958/59–1963/64	−1.20[b]	−2.50	0.50	Parikh [1966]
Japan	NPK	1883–1937	–	−0.74[d]	–	Hayami [1964]
Korea	NPK	1960–72	−0.17	−0.88	0.20	Sung, Dahl, and Shim [1973]
Korea	NPK	1971	−0.70[d]	–	–	Shim, Dahl, and Sung [1974]
Pakistan	N	1959/60–1972/73	−0.52[d]	–	–	Salam [1975]
Philippines	N	1958–72	−0.59[d]	–	–	Rodriguez [1974]
Taiwan	N	1950–66	−0.55[c]	–	–	Hsu [1972]
Taiwan	N	1950–66	−2.03[d]	−2.99	0.68	Hsu [1972]
Thailand	NPK	1954–72	−0.29[c]	–	–	Puapanichaya [1976]
Thailand	NPK	1954–72	−0.27[c]	−0.37	0.72	Puapanichaya [1976]
Other countries						
Brazil	NPK	1949–71	−1.12[d]	–	–	Larson and Cibautos [1974]
Brazil	NPK	1949–71	−0.33	−1.94	0.17	Larson and Cibautos [1974]
United States	NPK	1911–56	−0.53	−2.99	0.23	Griliches [1958a]

[a] Many of these studies have also been summarized in Mudahar [1978] and Timmer [1974c]. Short-run elasticity = adjustment coefficient × long-run elasticity.

[b] Denotes significance between 0.7 and 0.8.

[c] Denotes significance between 0.8 and 0.9.

[d] Denotes significance at 0.9 or higher.

"equilibrium" involved in improved policies, distribution networks or even technology, depending on the specification of the model.

ROLE OF IRRIGATION AND WATER MANAGEMENT

A comprehensive list of publications dealing with social and economic aspects of irrigation systems in Asia was compiled by IRRI [1976]. This bibliography covers materials up to mid-1970s and includes a total of 689 publications. It is divided into eight categories: economic analysis of design and construction of irrigation and drainage systems; operation and management of irrigation systems; irrigation policy and planning; economic analysis of irrigation performance; water rates; social and institutional factors in irrigation; interaction of irrigation systems with their environments; and selected technical issues in the design and operation of irrigation systems. About 75 percent of the publications covered in this bibliography dealt with three topics: irrigation policy and planning, economic analysis of irrigation performance, and social and institutional factors in irrigation.

Irrigation and water management reduce risk and increase agricultural production. Irrigation increases crop yields at existing levels of inputs, allows application of higher levels of inputs, increases cropping intensity through multiple cropping and makes it possible to grow modern crop varieties and high value cash crops. According to Bagi [1981c], irrigated farms used much larger amounts of labor per unit of land as compared to unirrigated farms in crop production in Haryana, India. The availability of irrigation and efficient water management can substantially increase agricultural production, farm employment and farm income. Increased farm income, in turn, increases farm investment as well as backward and forward linkages between production and consumption.

Irrigation and water management have contributed significantly to agricultural development in Asia by increasing crop production and reducing production risks. However, the analysis of irrigation projects, allocation of water and water use efficiency, water pricing policies, and equity effects of irrigation has received growing attention from economists only in the last few years. According to Kikuchi and Hayami [1978a], the efforts to develop irrigation systems in the Philippines were induced largely by an increase in the social rates of return to investment in irrigation [also see Feeny 1983a]. The increase in social profitability to investment in irrigation was due to the introduction of modern rice varieties, changes in the world price of rice, and a rise in the cost of bringing new land under cultivation. The desire to become self-sufficient in food production and to improve food security has motivated many governments in Asia to make large investments in irrigation.

A research seminar on irrigation systems in Southeast Asia was held at IRRI during 1976 and the papers are published in the proceedings [IRRI, 1978c].[17]

Some of the key conclusions of this seminar were: the development of irrigation systems in areas where farmers are small and poor would promote equity; since large-scale irrigation projects are always subsidized, landowners receive windfall gains through an increased value of their lands; and higher water charges alone may not provide incentives for more efficient use of irrigation water.

A series of studies commissioned by IFPRI, however, shows that participation in the benefits from irrigation has a broad base and includes small farmers and landless laborers. Much of the gain from the increased value of land accrues directly to small owner-operators [Mongkolsmai, 1985; Paris and Pascual, 1984; Prabowo, 1985; Sriswasdilek and Wattanutchariya, 1985].

The relative success of a number of Asian countries in achieving at least short-term self-sufficiency in rice, combined with increasing investment costs for irrigation and declining real rice prices, has rekindled debate on the efficiency of irrigation investment. In a number of countries, there has been a substantial shift of resources from new construction to rehabilitation and upgrading of infrastructure and management of existing systems. There is much debate about the benefits of these investments. Chambers [1987] suggests that the payoffs to improved main system management in South Asia could be very large. However, other recent analyses indicate that the returns to rehabilitation and management improvement may be lower in general than the returns to investment in small and medium scale new irrigation systems in the Philippines [Rosegrant, 1985; Rosegrant, Gonzales, et al., 1987] and in Indonesia [Rosegrant, Kasryno, et al., 1987]. Weaver [in Mellor, Weaver, Lele, and Simon, 1968] in an earlier work reports similarly on the failure to develop tertiary channels and management as deriving from basic flaws in management and design upstream.

No one denies the large direct and indirect contributions of irrigation and water management to agricultural development in Asia. However, this should not lead planners and policymakers to ignore other investment opportunities in agriculture which may have the potential of even higher rates of return. For example, modern rice technology is most suitable for irrigated rice, yet a large share of rice in Asia is still rainfed. Barker and Herdt [1979, p. 29] concluded:

> Asian governments, particularly among the rice-importing countries, are likely to emphasize irrigation as a fairly sure but costly means of increasing rice production and achieving price and political stability. However, if the estimates of the potential for yield increase in rainfed rice are correct, the total benefits for the Asian economies may be greater if more emphasis is given to research on rainfed rice problems.

Clearly, this calls for greater efforts in determining economic returns from alternative investment opportunities for different technology regimes for a specific crop.

CONSTRAINTS TO HIGHER CROP YIELDS

The Green Revolution has been concentrated in wheat and rice areas with well developed irrigation systems and institutional networks. The arid and semiarid areas in Asia are still experiencing little technological change in agriculture. This is due to limited technological possibilities, high risk, and serious socioeconomic constraints [ICRISAT, 1980]. Binswanger, Jodha, and Barah [1980], based on a survey of sample farms in the semiarid tropics of India, concluded that income risk is high and it is primarily due to production rather than price risk, virtually all farmers are risk-averse; high-risk and risk-averse attitudes of farmers lead to underinvestment in agriculture in the semiarid tropics. However, the extent of underinvestment relative to socially optimal levels may prove to be quite small.

Despite large increases in average rice yields in Asia, the gap between the potential and actual rice yields is still quite large. In recognition of this yield gap, IRRI initiated a yield constraints project in 1974 to determine the role of biological and socioeconomic constraints in explaining yield gaps.[18] The results of these country studies are summarized in IRRI [1977], and IRRI [1979a]. According to Barker [1979], the maximum rice yields on experiment stations were between 4.5 to 5.5 mt/ha in wet season and between 5.5 to 6.5 mt/ha in dry season. The actual national average rice yield of about 2 mt/ha was much lower than potential yield. Lack of control over water, low levels of fertilizer use and high risk appear to be important constraints to expanded rice yields. Herdt [1979] concluded that, given various constraints, the available technology is being used to its potential. For future growth, the development of technology must be accompanied by institutional reforms that make current technology more attractive to farmers. This was further emphasized by Ruttan [1978].

The policy and institutionally related questions of technological advance were addressed in the "Stanford Project on the Political Economy of Rice in Asia." The purpose of this project was to trace the history of rice policies in several Asian countries and to understand the causal mechanisms in formulating national rice policy. These comparative rice policy studies go beyond the narrow economic factors in understanding the formulation and implementation of national rice policies and in understanding the behavior of policymakers. The overall methodology and results are summarized in Timmer [1975a, c]; and the results of individual country studies are summarized in Timmer [1975b] for Indonesia; Siamwalla [1975] for Thailand; R. H. Goldman [1975] for Malaysia; Mangahas [1975] for Philippines; Hayami [1975a] for Japan; Moon [1975] for South Korea; and Chen, Hsu, and Mao [1975] for Taiwan. A related set of studies on comparative advantage, government policies, and international trade in rice, covering

Indonesia, Philippines, Taiwan, and Thailand, is reported in Pearson, Akrasanee, and Nelson [1976], and Monke, Pearson, and Akrasanee [1976]. A more recent set of papers on rice price policy in China, Indonesia, Nepal, Philippines, South Korea, and Thailand can be found in Sicular [1989a].

VARIABILITY IN CEREAL PRODUCTION

As has been discussed in the previous sections, the introduction of modern agricultural technology (including modern seeds, fertilizer, and irrigation) has resulted in impressive growth in food production, particularly cereal production, in many developing countries, especially Asian countries. However, as cereal production has grown so has year-to-year variability in cereal production. The primary source of cereal production variability has been variability in cereal yields. These issues have been addressed in a series of studies carried out at IFPRI which include J. R. Anderson, Hazell, and Evans [1987], J. R. Anderson and Hazell, eds. [1989], Hazell [1982, 1984, 1985, 1986], and Mehra [1981]. The variability in cereal yields has been attributed to biological, climatic, and economic factors. Any strategy designed to minimize fluctuations and variability in crop yields has important implications for agricultural research and agricultural policies such as crop insurance, crop diversification, marketing, and buffer stock arrangements.

Several other follow-up studies, including Sahn, ed. [1989] and Sahn and von Braun [1987], examine the relationship between food production and consumption variability. An analysis of data from thirty-eight countries by Sahn and von Braun [1987] indicates that increased production variability does translate into increased variability in consumption; year-to-year consumption variability has declined during the past twenty-five years, mainly due to stocking operations and trade practices; and food insecurity, as measured in terms of fluctuations around trend levels of consumption, does remain a problem, especially for the poor. Consumption variability can be reduced through appropriate technology, trade, storage, and pricing policies.

2. Farm Size, Productivity, and Resource Allocation

INVERSE RELATIONSHIP BETWEEN FARM SIZE AND PRODUCTIVITY

The empirical relationship between farm size and productivity in land scarce Asian countries has important policy implications for agricultural development strategy, land reforms, and agricultural taxation.

Studies based mainly on "Farm Management Data" collected in the 1950s from selected districts in India show that farm size and productivity are inversely related in traditional agriculture.[19] This relationship has important implications for policies affecting land reform and farm organization, as it implies that output could be increased simply by dividing large farms into small ones. In other words, land reform can improve both equity and efficiency. The controversy on the farm-size issue has been argued at great lengths for India and that controversy illuminates the issue well, hence we report it in some detail.

A. K. Sen [1962, 1964] hypothesized that high labor input and low labor costs were responsible for higher productivity on small farms. This explanation was later questioned by P. K. Bardhan [1973] on the grounds that dual labor markets cannot exist.[20] However, empirical support for A. K. Sen's explanation can still be found by explicitly incorporating differences between the quality of hired and family labor. Family, as compared to hired, labor is relatively more productive, works harder and longer, and requires less supervision.

Khusro [1964] hypothesized that the inverse relationship between farm size and productivity is explained by differences in soil fer tility. Small farms may be more fertile either because the soil is managed better or because the quality of land is better [Bardhan, 1973]. According to Khusro [1964] for India and Roumasset [1976] for rice in the Philippines, if land quality is accounted for, the inverse relationship disappears. It might also be true that higher irrigation and cropping intensities on small farms make them more productive than large ones.

Rudra [1968a, b], on the other hand, questioned the statistical validity of the inverse relationship. According to Rudra, this relationship is partly the result of aggregation and might disappear if ungrouped data were used. Most of the Indian studies, which verify the inverse relationship statistically, use data from the 1950s. At that time, Indian agriculture was characterized by the absence of technological change and little use of modern inputs. On the other hand, during the 1960s and 1970s, the major sources of growth in agricultural production were capital and such modern farm inputs as high-yielding crop varieties, fertilizer, insecticides, and controlled water.

The inverse relationship between farm size and productivity does not exist for crops characterized by improved production technologies [Ghosh, 1986; and A. Sen, 1981]. In modern agriculture, the role of purchased farm inputs becomes crucial, with limited possibilities of substitution between labor and purchased farm inputs. It is well established that the small farmers, as compared to large farmers, have relatively limited access to credit and purchased inputs. As a result, the small farmers may not be able to apply optimal amounts of purchased modern farm inputs and the relationship between farm size and productivity may eventually become positive. In a detailed survey of small farmers in South Asia, I. J. Singh [1988a] has summarized the empirical evidence which shows that the inverse relationship between farm size and productivity held during the pre-HYV period but does not hold during the post-HYV period.

Based on a large sample of farmers in Punjab, S. S. Sidhu [1974a] found that the technical efficiency of small and large farmers producing wheat was about the same. Johl [1973a], on the other hand, observed a positive relationship between farm size and productivity in Punjab, as did Utami and Ihalauw [1973] in a sample survey of rice farmers in central Java, Indonesia (where farm size is small even by Asian standards) during 1971. This positive relationship can be explained by the better access large farmers have to scientific information, modern farm technology, and financial institutions or by their ability to use fertilizers and mechanization to overcome land quality and labor constraints.

It has been suggested that small farmers achieve higher crop yields at higher cost of production per unit of land by using relatively more labor and animal draft power. Consequently, economic efficiency on small farms is likely to be lower than on large farms. In the absence of gainful off-farm employment for family labor, however, the opportunity cost is lower than the market wage rate paid to hired labor [Bagi, 1983a; and A. K. Sen, 1962, 1964, 1966]. As a result, the use of market wage rate tends to overestimate labor and hence production costs. Yotopoulos and Lau [1973] estimated the relative economic efficiency of small and large farms in India using the concept of restricted profit function and concluded that small farms had higher relative economic efficiency than that of large farms. S. S. Sidhu [1974a] used farm level data for wheat production in Punjab to estimate a similar model and concluded that there was no difference in the economic efficiency of small and large farms.

Economic efficiency is a combination of technical and allocative efficiency [Yotopoulos and Lau, 1973]. Technical efficiency refers to the ability to produce maximum output from a given set of inputs; whereas, allocative efficiency refers to the ability to choose an optimal combination of inputs for a given set of input and output prices. More recent analysis based on farm level data in India found no definite superiority in economic efficiency, or its technical and allocative efficiency components, for either group of farms [Huang and Bagi, 1984; and Huang, Tang, and Bagi, 1986]. These results indicate that there is no convincing economic case for land redistribution or land ceiling, except for socio-political considerations.

The major policy question remains unsettled. Can rural unemployment and mass poverty be reduced by redistributing land and reducing farm size? Ladejinsky [1972] observed that an emphasis on a land ceiling in India can lead a government to neglect land reform measures such as increasing security for tenants, regulating land rents and farm wages, and consolidating land. C. H. H. Rao [1970] emphasized the need for more responsive credit institutions to serve small farmers in their efforts to modernize agriculture. C. H. H. Rao and Subbarao [1976] concluded that market imperfections do not put small farmers in India at as great a disadvantage in marketing rice as is generally believed. Instead, the lack of a well-developed infrastructure reduces the marketing efficiency of both small and large farmers.

Two other important questions need to be addressed. What farm size is optimal, taking both efficiency and equity into consideration? And, is there a relationship between farm size and the adoption of modern farm technology, including farm machinery? According to Schultz [1964, p. 111], "The size of farms may change as a consequence of the transformation, . . . but changes in size are not the source of the economic growth to be had from the modernization process." Barraclough [1967, p. 264] added that "when a society's institutional parameters are in flux and the kinds of activities carried out by the farm unit are themselves changing, there is no possibility of identifying optimum-size farms."

A related concept is the relationship between farm size and returns to scale in agricultural production. In order to formulate appropriate policies on farm organization, information on the economies and diseconomies of scale is needed. Empirical investigations carried out in Asia, most of which used the standard Cobb-Douglas production function in log-linear form, indicate that agricultural production is generally ruled by constant returns to scale.[21] The obvious inconsistency between the inverse relationship of farm size and productivity and constant returns to scale can be explained by the law of variable proportions.

In real farm situations, not all inputs increase in the same proportion. In determining optimum farm size, "an appeal to the concept of 'returns to scale' is, as a rule, barren because the transformation of traditional agriculture always entails the introduction of one or more new agricultural factors" [Schultz, 1964, p. 111]. In most Asian countries, agricultural land is becoming a serious constraint on the expansion of agricultural production. This implies that if agriculture is to be modernized, some inputs are needed more than others, new inputs should be brought into the production process, and input substitution possibilities should be explored.

EFFICIENCY OF RESOURCE ALLOCATION IN AGRICULTURE

To transform traditional agriculture and to expand the contributions of agriculture to economic development, there is a need both to determine the efficiency of resource allocation in agriculture and to formulate public policies that would remove inefficiencies in agricultural production. Schultz [1964, p. 27], in his classic book *Transforming Traditional Agriculture*, maintained that "there are comparatively few significant inefficiencies in the allocation of the factors of production in traditional agriculture." Consequently, agriculture cannot be modernized merely by altering the prevailing pattern of resource allocation in the absence of technological change.

In Asia, a high proportion of the empirical studies evaluating the efficiency of resource allocation in agriculture have been conducted in India.[22] The major impetus for most of these studies came from the "Farm Management Data" collected in different states of India during the 1950s. Resource allocation was evaluated by estimating agricultural production functions (usually of the Cobb-Douglas type) and then comparing the computed marginal value products for different factors with their corresponding market prices. Equality between the marginal value product and market price of different inputs implied that resources were allocated efficiently.

The main conclusions of aggregate studies using "Farm Management Data" are that the variable factors of production are allocated and used efficiently but that bullocks, fixed factors, and the main source of draft power, are not. Bullocks are used uneconomically mainly because of resource fixity and because of diseconomies of scale. This puts into question the relevance of an approach that, in analyzing the efficiency of resource allocation for a particular enterprise, measures capital input as a flow. Econometric studies indicate that resources in India are allocated with no significant inefficiencies. On the other hand, D. K. Desai [1963], in a linear programming study of resource use in Maharashtra, showed that resources are allocated inefficiently, as gaps between actual and potential agriculture output indicate.[23] This raises a question about the relevance and ability of different methodological approaches to analyze the efficiency of resource allocation. Also, one can always question the appropriateness of prevailing factor prices in evaluating the efficiency of resource allocation in agriculture, using an aggregate production function.

Most of these studies deal with the period before the Green Revolution in India. Not only has the use of modern factors of production, especially fertilizer, become popular but bullocks are being replaced gradually by tractors. The Government of India [1976] study concludes that "the studies available so far could not be considered adequate enough in coverage to provide guidance in policy formulation" for tractorization [vol. I, p. 445]. Similarly, A. K. Sen [1975a, p. 164] observed that "the factual picture is unclear, e.g., the extent of the yield impact of tractorization has not yet been isolated from variations in other factors not complementary to tractor use for a sufficiently large number of cases." Because tractors are indivisible, used for several purposes, and used in all enterprises, how to measure their economic efficiency and their contribution to agricultural production presents serious analytical questions. Furthermore, a farmer does not have to own a tractor to make use of it. In many Asian countries, farmers hire the services of tractors in order to perform specific farm operations on time and more efficiently.

Finally, most of the studies discussed were based on the standard neoclassical approach. Day and I. J. Singh [1977, p. ix] argued that agricultural development theory based on neoclassical economics "underplays the complexity of technology, overplays the rationality and information content of decisions, and exaggerates the equilibrium and efficiency of market." Consequently, where reality contradicts the underlying key assumptions of studies, their results may be of little relevance (or may even be misleading) to policymakers.

3. Tenancy, Productivity, and Resource Allocation

TENANCY ARRANGEMENT

Most of the studies that used "Farm Management Data" deal with the owner-operated farms, yet tenancy is prevalent all over Asia, including India. According to traditional theory, resources are allocated inefficiently under share tenancy mainly because of a lack of incentives.[24] However, Cheung [1969, pp. 3–4], based on Taiwan's experience, rejected the inefficiency argument and concluded that

> resource allocation under private property rights is the same whether the landowner cultivates the land himself, hires farm hands to do the tilling, leases his holding on a fixed-rent basis, or shares the actual yield with his tenant. In other words, different contractual arrangements do not imply different efficiencies of resource use as long as these arrangements are themselves aspects of private property rights.

Sharecropping exists for several reasons. First, sharecropping helps tenants share the risk and uncertainty in crop production, since tenants may not have risk-bearing ability to rent-in land for cash. Second, tenants may not have the necessary cash or ability to borrow in order to rent-in land and purchase modern farm inputs. Third, tenants may be able to borrow from landlords and may even be able to share the cost of inputs. Fourth, landowners will feel safe to lend to tenants since they will be able to recover their loan with interest at the time of harvest. Fifth, landowners have the opportunity to participate in decisions related to farm operations, crop selection, and input use. Sixth, if tenants are faced with the challenge of making a subsistence living from sharecropping, they will have the incentive to produce the maximum possible output from available inputs and do a better job at managing farm operations. In this case, landowners may be able to share higher crop output due to tenant's hard work.

The sharecropping contracts, however, may vary in different parts of the world and even within small geographical areas. The neo-marxian explanation for this variation is based on the degree of economic dominance of the landowner relative to the economic deprivation of the tenant. Since the relative bargaining power of the two parties varies in individual cases, the prevailing sharecropping contracts can also vary. On the other hand, the neo-classical explanation for this variation is based on the degree of contribution of landowner relative to tenant. According to this view, the output and input shares of the two parties may seem to be uniform in sharecropping contracts in a small geographical area. The actual contribution and reciprocal obligations, however, may vary in every case.

According to the Marxian view, landowners exploit the tenant and the extent of this exploitation will depend on the relative economic and hence the bargaining power of the two parties [Bhaduri, 1983]. On the other hand, the neo-classical view denies the possibility of exploitation since the actual shares in crop output represents the return to monetary as well as nonmonetary favors and reciprocal obligations. In many densely populated developing Asian countries, there are large number of small and marginal farmers who must depend on sharecropping for their subsistence. Land available for sharecropping is limited and it gives enormous bargaining advantage to landowners. This can create a strong temptation for the landowners to exploit tenants. Whether such exploitation actually happens may vary from one case to another.

RELATIVE EFFICIENCY OF SHARECROPPING

There has been a debate on the relative efficiency of sharecropping as compared to owner-operated and other forms of tenancy arrangements [Byres, 1983]. In case landowners exploit tenants, tenants will not have the incentive to apply optimal level of inputs and perform various farm operations efficiently. As a result, suboptimal application of inputs will lead to economic inefficiency. This view is represented by P. K. Bardhan and Srinivasan [1971]. On the other hand, when off-farm gainful employment opportunities are limited and there is excess demand for rented land, sharecropping can be as efficient as other forms of tenancy, even if tenants receive less than their fair share. When there is competition for limited rental land, landowners have the opportunity to choose the most efficient sharecroppers. The challenge of making a living and competition among sharecroppers forces the tenants to apply optimal levels of farm inputs and be efficient producers [D. G. Johnson, 1950].

On the other hand, if the neo-classical view is correct and the output shares received by the landowner and the tenant truly reflect their contributions, the efficiency on sharecropped land may not be any different from that on cash rented and owner-operated land. Furthermore, if owner-operated or cash rented farms do not have necessary resources to apply optimal levels of inputs, whereas the sharecropping farms are able to do so, the efficiency of sharecropping may actually be higher than owner-operator and other forms of tenancy. According to Bagi [1981b], there was no significant difference in the technical efficiency of the owner-operated and sharecropped farms when both irrigated and unirrigated farms were aggregated. However, when irrigated farms were analyzed separately, technical efficiency was significantly higher on sharecropped farms than on owner-operated farms. Furthermore, sharecropped farms made more intensive use of labor than owner-operated farms, thereby having positive implications for employment in labor-abundant Asian countries.

Sharecropping may be more prevalent in traditional agriculture when agricultural production is primarily nonmechanized and highly labor intensive. As the agricultural sector modernizes, it may become possible for most of the large landowners to purchase farm machinery and cultivate the entire land by themselves [A. Sen, 1981]. The supply and use of modern farm inputs such as disease-resistant high-yielding crop varieties, fertilizers, pesticides, herbicides, and irrigation increase crop output; increase net farm income per unit of land; and reduce risk in crop production. All these factors tend to reduce the need for sharecropping arrangements. The incidence of sharecropping may decline with general economic development of a region or country, but it may not totally disappear, as long as there are sound economic reasons for its existence.

In analyzing fertilizer use behavior, Minhas and Srinivasan [1966] assumed that the greater the share of the crop a share tenant has, the more fertilizer he will use. C. H. H. Rao [1971] argued that sharecropping is more prevalent for those crops and areas with little entrepreneurship, little substitution between crops, and a negligible amount of uncertainty. According to C. H. H. Rao, fixed contractual arrangements may be more prevalent where uncertainty is high. He also speculated that fixed contractual arrangements may be preferred to crop-sharing arrangements after modern farm technology is adopted. P. K. Bardhan and Srinivasan [1971] also rejected Cheung's argument that resource allocation under share tenancy is efficient.

More recently, Reid [1976, p. 576], using theoretical analysis, argued that "gain from the joining of tenant and landlord interests, not gain from the dispersion of agriculture risk, is the impetus to share tenancy, and that sharecropping is chosen for its efficiency, not in spite of its inefficiency or efficiency." As C. H. H. Rao [1971] and Day [1967] argued, the efficiency of a sharecropping arrangement depends on the farm technology used and on how risk and management tasks are shared. Most of these issues were discussed in detail in Roumasset [1976]. He also argued, in contrast to the prevailing view, that risk aversion does not inhibit the use of modern farm inputs. This conclusion was based on a sample of Filipino rice farmers and his analysis of the decision to use nitrogen fertilizer.

The empirical evidence on the efficiency of share tenancy remains inconclusive. If share tenancy is as efficient as owner-operator arrangements, and if efficiency is the only objective, then the case for land reforms is weakened.[25] Since existing agricultural institutions tend to be geared to serve the interests of landlords, there is a need to analyze the access tenants have to these services and its implications for resource use and agricultural output.[26]

4. Rationality, Incentives, and Price Policy

SUPPLY RESPONSE AND PRICE POLICY

Relative prices have a significant role to play in transforming a traditional agricultural sector into a modern and dynamic sector. That, of course, presumes that farmers and others respond to changing price incentives. The rationality of farmers in traditional agriculture was once a controversial issue. In view of this, a large number of supply response studies were undertaken with a principal aim of demonstrating a farmer response to price as a means of demonstrating farmer rationality. Tables 7 and 8 report results from a substantial number of those studies. They clearly show response and contribute to T. W. Schultz's [1964] finding of farmer rationality.

Having once established farmer rationality, more complex problems remain. The term supply response has often been used vaguely. There are important empirical distinctions between the response to price changes of area sown with individual crops, of total cropped area, of crop yields, and of aggregate agricultural production. Furthermore, the price in question could be the price of a specific output, the price of one output relative to another, or the relative price of inputs and outputs. Most empirical supply response studies, however, deal with acreage (as opposed to production) response of individual crops to changes in relative output prices.[27]

The differences in the size of the price elasticities, even for the same crop in the same region, are large [Tables 7 and 8]. These differences are caused by differences in the time period and length of time series data; the nature of dependent variables: for instance, area can be irrigated, unirrigated, total, standardized, change over time, or a ratio; the price variable used, such as the specific output price, the price index, the relative output price, and the competing crop used to obtain relative output price; and the nature of the model, its specification, and the techniques used to estimate it.[28]

The voluminous literature on acreage response has left several questions unsettled. There are still large gaps in our knowledge about aggregate production behavior, though the limited evidence available, for example, from Herdt [1970], Barnum [1973], and Bapna [1980], indicates that the price elasticity of aggregate agricultural output is low at between 0.1 and 0.2. The price elasticity for cash crops is presumed to be higher than for food crops, perhaps because they tend to occupy a smaller area. It has become routine to estimate short-term elasticity, long-term elasticity, and coefficient of adjustment. Little is known, however, about what determines a coefficient of adjustment. The meaning and relevance to policymakers of long-term elasticity remain uncertain. We still know little of how the farmer allocates nonland production resources in response to price policy and what effect price policy has on technological change in agriculture.[29] Mellor and Ahmed, eds. [1988], in a compendium of papers, specifically address price policy in the context of technological change. They conclude that technological change introduces substantial problems of instability and potential secular decline in prices that price policy must address but it does so in a favorable environment of declining costs of production.

Table 7. Summary of selected acreage response studies for food crops in Asia

Crop	Country/region	Time period	Price elasticity[a]		Dependent variable	Source
			Short run	Long run		
Rice	India–Pakistan/Punjab	1914/15–1945/46	0.31	0.59	Standard irrigation area	R. Krishna [1963]
	India/Punjab	1951-64	0.24	0.40	Area	Kaul [1967]
	India/Punjab	1948/49–1965/66	0.33	0.38	Standard irrigated area	Maji, et al. [1971]
	Bangladesh (9 districts)[b]	1948/49–1962/63	0.12	–	Rice area relative to rice and jute area	Hussain [1964]
	Philippines/Central Luzon	1953/54–1963/64	0.13–0.27	0.62–2.15	Area	Mangahas, et al. [1966]
	Indonesia	1951-62	0.30	–	Area	Fletcher and Mubyarto [1966]
	Thailand	1940-64	0.18	0.31	Area	Behrman [1968]
Wheat	India–Pakistan/Punjab	1914/15–1943/44	0.08	0.14	Standard irrigated area	R. Krishna [1963]
	India/Punjab	1951-64	0.08	0.09	Area	Kaul [1967]
	India/Punjab	1948/49–1965/66	0.67	0.67	Standard irrigated area	Maji, et al. [1971]
	India/Uttar Pradesh	1950/51–1962/63	0.21	0.64	Area	J. Krishna and M.S. Rao [1967]
	Pakistan (7 districts)	1933/34–1958/59	0.10–0.20	–	Percent change in irrigated area	Falcon [1964]
Corn[c]	India–Pakistan/Punjab	1914/15–1943/44	0.23	0.56	Standard irrigated area	R. Krishna [1963]
	India/Punjab	1948/49–1965/66	0.49	0.54	Standard irrigated area	Maji, et al. [1971]
	Philippines	1946/47–1963/64	0.07	0.42	Area	Mangahas, et al. [1966]
	Thailand (8 corn regions)	1950-63	1.03	2.29	Area	Behrman [1968]
Sorghum	India–Pakistan/Punjab	1914/15–1943/44	–	-0.58	Unirrigated area	R. Krishna [1963]
	India/Tamil Nadu	1947-65	0.20	0.28	Area	Madhavan [1972]
Bajra[d]	India–Pakistan/Punjab	1914/15–1945/46	0.09	0.36	Unirrigated area	R. Krishna [1963]
	India/Punjab	1951-64	0.05	0.06	Area	Kaul [1967]
	India/Tamil Nadu	1942-66	0.03	0.15	Area	Madhavan [1972]
Barley	India–Pakistan/Punjab	1914/15–1945/46	0.39	0.50	Unirrigated area	R. Krishna [1963]
	India/Punjab	1951-64	0.53	0.60	Area	Kaul [1967]
Gram[e]	India–Pakistan/Punjab	1914/15–1945/46	–	-0.33	Unirrigated area	R. Krishna [1963]
	India/Punjab	1951-64	-0.30	-0.65	Area	Kaul [1967]

[a] Short-run elasticity = coefficient of adjustment × long-run elasticity.
[b] Summer rice only.
[c] Maize.
[d] Millet.
[e] Chickpeas.

Table 8. Summary of selected acreage response studies for cash crops in Asia

Crop	Country/region	Time period	Price elasticity[a]		Dependent variable	Source
			Short run	Long run		
Cotton	India–Pakistan/Punjab (American cotton)	1922/23–1941/42	0.72	1.62	Standard irrigated area	R. Krishna [1963]
	India–Punjab (American cotton)	1951–64	0.34	2.84	Area	Kaul [1967]
	India–Pakistan/Punjab (Desi cotton)	1922/23–1943/44	0.59	1.08	Standard irrigated area	R. Krishna [1963]
	India/Punjab (Desi cotton)	1951–64	0.29	1.19	Area	Kaul [1967]
	Pakistan (8 districts)	1933/34–1958/59	0.41	–	Percent change in area	Falcon [1964]
Sugarcane	India–Pakistan/Punjab	1915/16–1943/44	0.34	0.60	Standard irrigated area	R. Krishna [1963]
	India/Punjab	1951–64	0.09	0.73	Area	Kaul [1967]
	India/North Bihar	1950/51–1964/65	0.66	0.79	Area	D. Jha [1970]
	India/Tamil Nadu	1947–65	0.63	0.76	Area	Madhavan [1972]
Groundnuts	India/Tamil Nadu	1947–65	0.34	0.65	Area	Madhavan [1972]
	India/Andhra Pradesh	1930/31–1943/43	0.76	–	Area	Reddy [1970]
Jute	India–Pakistan	1911–38	0.46	0.73	Area	Venkataramanan [1958]
	Bangladesh (9 districts)	1948/49–1962/63	0.42	–	Jute area relative to rice and jute area	Hussain [1964]

[a] Short-run elasticity = coefficient of adjustment × long-run elasticity.

The role of price and nonprice factors in raising agricultural output has also been reviewed by Chhibber [1988]. The available evidence indicates that in developing countries the long-run aggregate supply elasticity of agriculture with respect to price is in the range of 0.3 to 0.9. The elasticity is not greater than 1.0, as is sometimes claimed by those who ascribe primacy to price policy. On the other hand, elasticity is not as low as zero, as claimed by those who view price policy effects as insignificant. The price elasticity is in the range of 0.6 and 0.9 in relatively advanced and land-abundant countries, and around 0.2 to 0.5 in developing countries with inadequate infrastructure. The supply elasticity with respect to nonprice factors (public goods and services) tends to be much higher. It is around 1.0 in countries with inadequate infrastructure, imperfect markets, limited capital, and lack of private research organization. On the other hand, the supply elasticity to nonprice factors in developing countries with better developed infrastructure is smaller.

The available empirical evidence indicates that there is a need for a judicious blend of improvements in price incentives and nonprice factors such as infrastructure, technology, delivery systems, and services. These results have also been supported by a study conducted by Binswanger, Khandker, and Rosenzweig [1989] on the impact of infrastructure and financial institutions on agricultural output and investment in India. The study covers 1960/61 to 1980/81 period and is based on panel data from eighty-five randomly selected districts from thirteen states in India. Based on detailed empirical analysis, they conclude that prices really do matter but so do infrastructure, markets, and banks.

From a policy point of view, it is important to recognize the conflicting roles of relative prices in allocating resources, in distributing incomes, and in capital formation [Mellor, 1968, 1969a, 1978]. These conflicts limit the scope for public policy in manipulating prices and raise complex welfare questions.[30] Finally, although a less researched issue in Asia, the effect of a broad macro policy on exchange rates and hence on relative prices of agricultural tradeables to nontradeables is still significant [Bautista, 1987]. In particular, pro-industry trade policies may result in unfavorable price relations for agriculture.

The distortions in price policy and lack of necessary incentives are quite common in most developing countries [Schultz, ed. 1978]. The empirical evidence, however, does confirm that prices play a significant role in achieving specific policy goals in economic development. Varying relative crop prices can induce a shift of area from one crop to another. But with technology stagnant and with few purchased inputs, such a shift cannot increase the aggregate foodgrain supply, as foodgrains already occupy the bulk of cultivated area. Agricultural prices influence the balance between agricultural and industrial development through labor and capital flows across sectors. But initial disequilibrium and many other forces are at work. Prices are important in distributing the benefits and losses of growth. Yet, there are large gaps in our knowledge about the effects of price policy on irrigation and the use of fertilizers and about the relative merits of output price support and input subsidy programs.[31]

CROP PRICE SUPPORT AND INPUT SUBSIDY

The discussion related to crop price support and input subsidy programs is summarized in Barker and Hayami [1976], Mudahar and Pinstrup-Andersen [1977], Mudahar [1978], R. Ahmed [1978, 1979], Bagi [1984], Timmer [1986a], and Mellor and R. Ahmed, eds. [1988]. Both crop price support and input subsidy programs can increase the output of a specific crop but initially the main beneficiaries are those farmers who have positive marketable surplus of that crop. Input subsidies, on the other hand, can increase the output of all crops that use the subsidized input, provided there is enough supply to satisfy increased demand for that input due to the price subsidy. In a regime of input scarcity, however, large and influential farmers will be able to purchase more and the smaller farmers will be squeezed out of the market. Input subsidy programs targeted at a single crop often do not work due to leakages.

Crop price support and input subsidy programs are non-neutral to farm size because the gains to the producers from these policies are more or less directly related to the size of crop output sold or quantity of inputs purchased. Economic incentives provided through crop price supports and/or input price subsidies are essential to adopt modern technology and increase agricultural output. However, economic incentives must be accompanied by investment in infrastructure, agricultural research, irrigation systems, and technology transfer in order to have greater social rates of return and higher agricultural growth.

Both crop price support and input subsidies tend to be biased in favor of large farmers. The small and marginal farmer with no marketable surplus may actually end up spending more on consumption, if they are net buyers of that commodity. In the absence of technological change and increases in area for the target crop, an increase in output will be at a higher unit cost. Crop price support in such a case may benefit the producer but not the consumer of that commodity. Domestic production at a unit cost slightly higher than international price may be justified if it creates additional employment in the production and/or processing of that commodity and saves scarce foreign exchange which may have a higher opportunity cost and pressing national priorities elsewhere.

On the other hand, input subsidy policy assumes that farmers are applying sub-optimal levels of farm inputs due to depressed output prices. Some farmers, however, may not have the necessary resources to purchase adequate amount of inputs. Input subsidies can increase crop output with no increase in crop prices for the consumers. The small and marginal farmers also benefit from input subsidy, provided they have equal access to those subsidies. The supplies of agricultural credit, fertilizer, chemicals, gasoline, and diesel fuel in most developing countries are generally scarce. Implementation of input subsidy programs generally increase demand. This results in an increase in market price and the small farmers, who may have only limited access to subsidized inputs, may end up paying even higher prices. Consequently, any input subsidy program must be accompanied by programs designed to increase input supply and remove marketing constraints.

According to Barker and Hayami [1976], the use of fertilizer subsidy is preferable to rice price support if the objective is to achieve rice self-sufficiency in Philippines. R. Ahmed [1979] also examined the relative efficiency of price support and fertilizer subsidy policies in increasing rice production by half a million tons in Bangladesh. The results indicate that fertilizer subsidy policy is superior to price support policy. Mudahar [1978] identifies and analyzes the information and economic analysis needed to design fertilizer subsidy and price policy in developing countries. Garcia [1981] provides a detailed Colombian case study which also has relevance to Asian countries. Based on the analysis of farm-level data from India, Bagi [1984] concluded that both crop price support and input subsidies can increase crop output. The former increases output of the targeted crop while the latter increases the output of all crops that use the subsidized input. The choice between those two policies will depend on the objective and their economic implications with respect to efficiency, equity, and cost to achieve that objective.

According to Rosegrant et al. [1987] fertilizer subsidy has been employed as a key instrument to stimulate rice production in Indonesia. It has led to rapid growth in fertilizer use, rice production, and subsidy expenditure. Any reduction or elimination of fertilizer subsidy would achieve significant financial gains for the government. However, the results indicate that fertilizer subsidy remains a powerful policy instrument for accelerating domestic agricultural production. Farmers remain highly responsive to changes in fertilizer price. Complete elimination of fertilizer subsidy has a large negative impact on production because net imports, hence import costs, increase significantly. Depending on rice price strategies, removal of subsidies causes either consumers or producers to suffer significant welfare losses. As a result, fertilizer subsidy should be reduced only in pace with improvements in quality of irrigation, expansion of credit program, dissemination of improved crop technology, improved pest management practices, higher efficiency in fertilizer use, and improved fertilizer production and distribution systems.

One of the main reason for fertilizer subsidies is the prevalence of inefficient fertilizer production and distribution systems in many developing countries [Mudahar, 1978]. The issue of fertilizer production pricing policies was discussed in a seminar organized by the World Bank and the papers are published in the conference proceedings by Segura, Shetty, and Nishimizu [1986]. According to them, economically optimal fertilizer pricing policy—whether determined by the free market or by an official agency—must perform the following functions: provision of stimulus to mobilize and allocate adequate resources for fertilizer capacity expansion; optimal choice of production processes; effective control of feedstock and other operating costs; satisfactory level of capacity utilization; and when necessary, closure of obsolete, high cost fertilizer plants.

ECONOMIC INCENTIVES AND AGRICULTURAL POLICIES

Agricultural growth and farmer incentives are influenced not only by direct sectoral agricultural policies, but also by developments in other sectors of the economy, particularly trade, exchange rate, and other macroeconomic policies. There are at least four stylized facts about the agricultural policies of developing countries, the interactions among which are not fully appreciated or analyzed. These are: promotion of industry through policies of import substitution and protection against imports competing with domestic production; maintenance of overvalued exchange rates through exchange-control regimes and import licensing mechanisms; suppression of producer prices of agricultural commodities through government procurement policies, agricultural marketing boards, export taxes, and/or export quotas; and compensation of various disincentive effects on producers through subsidization of agricultural inputs and capital investment in irrigation and other inputs. The net effect of these direct and indirect policies leads to a tax on agriculture and transfer of substantial resources from agricultural to nonagricultural sectors.

Several years ago, the World Bank initiated a research project entitled "A Comparative Study of the Political Economy of Agricultural Pricing Policies." The purpose of the project was to provide a detailed history of pricing policies; to measure the degree of intervention affecting agriculture; and to analyze their effects on output, consumption, trade, the budget, intersectoral resource transfers, and income distribution. In other words, the study provides a systematic comparative analysis of the impact of government intervention and measures discrimination against agriculture. The study was carried out for eighteen countries for the 1975-84 period. All the country studies used a common methodology which facilitated a comparative analysis. The initial results of these studies are reported in Krueger, Schiff, and Valdes [1988]. The detailed findings of country studies and synthesis will be published in four volumes; the country studies from Asia—which includes Korea, Malaysia, Pakistan, the Philippines, Sri Lanka, and Thailand—are reported in Krueger, Schiff, Valdes, eds. [1990].

The study measures the impact of sector-specific (direct) and economy-wide (indirect) policies on agricultural incentives for both exportable and importable agricultural commodities. The direct effect is measured by the proportional difference between the producer price and the border price with appropriate adjustments for distribution, storage, transportation, and other marketing costs. The indirect effect has two components. The first is the impact of the unsustainable portion of the current account deficit and of industrial protection policies on the real exchange rate, and thus on the price of agricultural commodities relative to nonagricultural nontradeables. The second is the impact of industrial protection policies on the relative price of agricultural commodities to that of nonagricultural tradeable goods. The initial results for two time periods (1975-79 and 1980-84) for selected exportable and importable agricultural commodities are reported in Table 9. These results provide estimates for the direct and total (direct and indirect) impact of government policies on agriculture.

The results show that for exportables there is a high degree of discrimination; in many cases, discrimination increased over time and the indirect discrimination is much stronger than the direct discrimination. On the other hand, importables are subsidized directly but are taxed indirectly; the net effect being discrimination on importables also. In other words, the findings indicate that in almost all cases the direct effect is equivalent to a tax on exportables (-11 percent on average) and to a subsidy on importables (20 percent on average); the indirect effect also taxes agriculture (-27 percent on average) and dominates the direct effect. There are, of course, variations across different countries depending on their respective stages of development. Two other impacts of these policies are annual transfer of substantial resources from agriculture to government and nonagricultural sectors in most countries, and stabilization of domestic producer prices for both exportables and importables. However, these intervention policies may not be the most appropriate effective mechanisms to achieve price stabilization.

The World Bank also analyzed trade and pricing policies in world agriculture in *1986 World Development Report* [World Bank, 1986a]. One of the conclusions was that economic growth and stability would be greatly enhanced if pricing and trade policies were improved in developing countries. Unfavorable economic policies have discouraged agricultural production and hindered agricultural development in many of these countries. Some of these policies include overvalued exchange rates, protection of industrial activities, taxation of agricultural exports, and import-competing food crops. In addition, public policies designed to subsidize consumers and farm inputs and policies designed to stabilize consumer and producer prices have led to significant losses in the real national income of developing countries.

The available evidence indicates that the developing countries discriminate against their farmers, even though agriculture accounts for a larger share of gross domestic product, employment, and export earnings. On the other hand, the industrial countries provide subsidies to farmers, even though agriculture accounts for a small share of gross domestic product and employment. Elimination

Table 9. Direct and total nominal rate of protection for export and import commodities in selected developing countries[a]

Region/country	Commodity[b]	Exports 1975/79 Direct (percent)	Exports 1975/79 Total	Exports 1980/84 Direct	Exports 1980/84 Total	Commodity[b]	Imports 1975/79 Direct (percent)	Imports 1975/79 Total	Imports 1980/84 Direct	Imports 1980/84 Total
Asia										
Korea	—	—	—	—	—	Rice	91	73	86	74
Malaysia	Rubber	-25	-29	-18	-28	Rice	38	34	68	58
Pakistan	Cotton	-12	-60	-7	-42	Wheat	-13	-61	-21	-56
Philippines	Copra	-11	-38	-26	-54	Corn	18	-9	26	-2
Sri Lanka	Rubber	-29	-64	-31	-62	Rice	18	-17	11	-20
Thailand	Rice	-28	-43	-15	-34	—	—	—	—	—
Africa										
Côte d'Ivoire	Cocoa	-31	-64	-21	-47	Rice	8	-25	16	-10
Egypt	Cotton	-36	-54	-22	-36	Wheat	-19	-37	-21	-35
Ghana	Cocoa	26	-40	34	-55	Rice	79	13	118	29
Morocco	—	—	—	—	—	Wheat	-7	-19	0	-8
Zambia	Tobacco	1	-41	7	-50	Corn	-13	-55	-9	-66
Latin America										
Argentina	Wheat	-25	-41	-13	-50	—	—	—	—	—
Brazil	Soybeans	-8	-40	-19	-33	Wheat	35	3	-7	-21
Chile	Grapes	1	23	0	-7	Wheat	11	33	9	2
Colombia	Coffee	-7	-32	-5	-39	Wheat	5	-20	9	-25
Dominican Republic	Coffee	-15	-33	-32	-51	Rice	20	2	26	7
Others										
Portugal	Tomatoes	17	12	17	4	Wheat	15	10	26	13
Turkey	Tobacco	2	-38	-28	-63	Wheat	28	-12	-3	-38
Average (simple unweighted)		-11	-36	-11	-40		20	-5	21	-6

Source: Krueger, Schiff, and Valdes [1988]

[a] The direct nominal protection rate is defined as the difference between the total and the indirect nominal protection rates, or equivalently, as the ratio of (1) the difference between the relative producer price and the relative border price, and (2) the relative adjusted border price measured at the equilibrium exchange rate and in the absence of all trade policies.

[b] The commodities for which the results are reported are considered fairly representative of government policy toward exportables or import-competing food crops.

of these distortions can result in large potential gains to the world economy. A summary of national and international aspects of agricultural policies in developing countries and their potential economic consequences are also provided in A. Ray [1988].

The economic impact of various agricultural, tax, and trade policies can be analyzed through various programming, simulation, and econometric models. Braverman and Hammer [1988] have provided an overview of multi-market simulation models which have been designed and applied to evaluate the economic consequences of proposed agricultural policy reforms in several developing countries in Asia, Africa, and Latin America. Bautista [1987] has analyzed the effects of trade and exchange rate policies on production incentives in Philippine agriculture for the thirty-year period from 1950 to 1980. The empirical findings indicate an existence of persistent and significant bias in relative incentives against agricultural export production and in favor of nontraditional (mainly industrial) exports, and most strongly in favor of import-competing industrial consumer goods. The use of such policies has been, and still continues to be, widespread in most developing countries in Asia.

PRICE STABILIZATION AND INTERVENTION POLICIES

Producer and consumer price stability for agricultural commodities remains one of the important national objectives of policymakers in most Asian countries. Price stability can be achieved indirectly through policies affecting demand and/or supply or directly through price controls. Different policies, individually or as a policy package, have different economic implications with respect to agricultural growth and welfare. An important recent study by R. Ahmed and Bernard [1989] examines rice price fluctuations and an approach to price stabilization in Bangladesh.

Rice is a staple commodity in Bangladesh and other Asian countries. The rice sector is also subject to various kinds of government intervention policies. According to Ahmed and Bernard [1989], actual rice price fluctuations have increased in the post-independence period (1970s onward) in Bangladesh. In the past, government used ration distribution, open-market sales, and post-harvest procurement to stabilize prices. These policy interventions were based on quantity targets. In the absence of a consistent approach based on price targets, the government's efforts to stabilize rice price have been relatively ineffective and wasteful.

Ahmed and Bernard [1989] propose an alternative framework for rice price stabilization which involves a shift from quantity-targeted approach to price-targeted approach. Policy framework includes specification of price band in annual prices, linking price band to seasonal prices, coordinating ration prices for priority groups with market prices, maintaining optimal stocks in the public sector, developing flexible procurement and open-market sales programs, and careful monitoring of crop production levels. The framework accommodates private traders as principal actors in consonance with a complementary public intervention in foodgrain markets. The results based on a simulation model indicate that the proposed alternative program would increase rice price stability, both annual and seasonal, and reduce pressure on the rationing system.

Two other major studies dealing with price policy in selected Asian countries were initiated in the mid-1980s. The first study deals with a comparative analysis of food price policy and government intervention programs in China, Indonesia, Philippines, Nepal, Republic of Korea, and Thailand [Sicular, ed., 1989a]. The second study deals with evaluation of rice market intervention policies in Bangladesh, India, Republic of Korea, and Malaysia [ADB, 1988]. Timmer [1988a] has examined four country case studies commissioned by the ADB and has drawn lessons which may be of particular relevance to policymakers in other Asian countries. First, it is important to make a clear distinction between short-run and long-run objectives and the analyst must provide a bridge between expediency and long-run efficiency. Second, global cost of production information does not provide an appropriate reference for determining domestic support prices. Third, there is frequently a wide gap between the first-best policy prescription based on border price paradigm and the outcome of events in reality. Fourth, certain type of interventions may be impractical when borders are permeable to unauthorized trade.

5. Rural Credit

Rural credit[32] is an extraordinarily complex and contentious subject that includes the broad issues of financial market development, overall economic growth, agricultural growth, and equity. The basic controversy at the macro level is whether intervention in financial markets, including direct credit allocation and interest rate regulation, accelerates development or misallocates resources and slows development. [Fry, 1988]. McKinnon [1988], generally counted on the "liberalization" side of the controversy, has delineated a number of problems of liberalization, particularly a tendency to indiscriminate lending at high rates with inadequate assessment of risk, leading to a policy recommendation of some regulation of interest rates and to some use of credit quotas [McKinnon, 1988; see also de Macedo, 1988]. A similar debate has occurred as to whether credit "leads" supply-creating activities, playing either a proactive or a neutral role. Patrick [1966] reviews these arguments and Mellor [1966, 1976] suggests a synthesis with particular reference to agriculture.

A debate parallel to that in the broad macro area has occurred in the area of rural credit. This debate is summarized in relation to expansion of rural banking as well as interest rates and their determination, with the latter leading back into issues of scale economies and expansion of the credit system.

In developing countries, the rural sector—and even the smaller commercialized portion of that sector—includes a major portion of the economy's real and financial resources. The fragmentation of the rural sector due to its spatial dispersion, poor communication system, and consequent high transaction costs results initially in a low degree of integration of rural financial markets into the national and global systems. Although opening of rural credit branches is discussed largely, if not exclusively, in terms of credit extension, it is of perhaps greater importance in deposit mobilization and market integration. In general, deposit mobilization of rural branches of the commercial credit system are far larger than credit extended, and those net deposits are readily moved to other regions and sectors in response to opening of dispersed rural branches.

In countries, such as India, that have expanded rural banking rapidly, a dilemma has been exposed. The same political forces that have led to rapid expansion of the system for deposit mobilization have brought high overdues on loans and high bad debt ratios. Deposit mobilization is facilitated by convenience, which requires a multiplicity of branches and loss of scale economies. Such expansion tends to occur only slowly without political pressure. New branches inevitably make initial losses while scale economies are built up. The fact of expansion under political pressure may bring the fact of political pressure on lending and the excuse as well for poor loan recoveries and gross overdues.

The issue of poor loan recovery in the context of rapid expansion of rural credit systems has led to a major controversy on the role of institutional credit—indeed, whether it even has an important role—and the issue of subsidized interest rates. See Adams [1980], and Adams, Graham, and von Pischke, eds., [1984] for an exposition of this issue. The interest rate issue is complex because of the role of interest rates in influencing the savings rate, mobilization of savings, and investment rates. The "high interest rate" school emphasizes the elasticity of savings and deposit mobilization with respect to interest rates and the misallocation of resources attendant on low interest rates (see, for example, Adams [1980] and Adams, Graham, and von Pischke, eds. [1984]).

The empirical record on interest rate elasticities is sparse and provides mixed results, but the weight is more on the side of relatively lower elasticity of savings than of investment with respect to interest rates. This is not surprising, since rural saving may be heavily weighted by a drive to save specific sums for specifically calculated future events, including the possibility of major economic reverses. This issue is even more complex because deposits mobilized by higher interest rates are drawn not only from hoarding in nonproductive forms but also from investment in productive activities—especially in a dispersed small-scale sector such as agriculture and allied activities. Such a conversion into financial assets from directly productive instruments may be, in net, production- and employ-ment-depressing, given the high employment content of rural activities.

Much of the interest rate controversy revolves around the issue of lending margins, itself a complex issue. Liberalization is often interpreted to mean inter-est rate spreads between borrowing and lending costs that cover margins. But start-up margins are always high in new branches and should be viewed in part as a capital cost to be spread over several years of income—not just the year in which the cost is incurred. Besides, scale economies are substantial in rural branches and proximity is important to enlarging business. Thus, initial large spreads between borrowing and lending interest rates may themselves lead to low volume, and conversely, initial apparent subsidies may lead to a rise in volume that will eventually reduce margins. These are complex processes in which po-litical forces may well push rates too low.

A further complication with respect to margins and efficiency of rural banking operations arises from the interaction of the dispersed nature of the rural sector, the need for convenience and hence proximity, and scale economies. All of these argue for single institutional monopolies in a particular area, while other effi-ciency-based arguments favor proliferation of institutions to provide competi-tion. Competition between a cooperative system and commercial branch bank-ing, plus access to urban-based finances through low-cost transport, represents a solution to this dilemma by offering different systems that can tap somewhat dif-ferent markets through somewhat different modes of operations and still overlap sufficiently to provide competition.

The problem of massive overdues is the overriding issue in rapidly expanding rural credit systems. In this complex area, several shibboleths need to be disposed of: overdues are generally not greater with small borrowers than large ones, in poverty-oriented programs than others, in one type of institution compared with others, or now compared to several years ago; overdues are not the same as bad debts (inappropriate dates for including repayment and lack of certainty about reborrowing in the future are due to inappropriate definition and accounting practices); the issue is not simple, to be dealt with by a rule or two (rather it is more like line losses in electricity distribution, to be brought down slowly by many complex steps).

Finally, there is an issue of whether special banks should be created for the poor. The arguments against such actions are of course clear: reduction of scale economies by fractionating the institutional credit market; high costs incident to small-scale borrowers with little opportunity for cross-subsidization; the political attraction of large-scale subsidies when they can be targeted to such institutions; and the general argument that markets work well and should not be tampered with. In practice, a number of examples exist of successful credit programs tar-geted to low-income people. The Grameen Bank of Bangladesh is a prime example [Hossain, 1988a]. Its borrowing rates were subsidized, but margins were kept within reasonable bounds and lending rates were high enough to cover them. Loans are to the poor and particularly to poor rural women.

Chapter V. Agricultural Research and Transfer of Technology

1. Investment in Agricultural Research

Modern agricultural technology, the result of agricultural research, has been the major source of agricultural growth gained through increases in the efficiency of resource use. Despite this contribution, developing countries usually underinvest in agricultural research.[33]

Many developing countries still accord low priority to agricultural research as reflected by budget allocation to agricultural research. According to Yadav [1987], agricultural research expenditure as a proportion of the agricultural budget in Nepal declined from 32 percent in 1970-71 to 14 percent in 1980-81. The real expenditure on agricultural research increased by about 4 percent per year from 1970-71 to 1980-81 as compared to expenditures on agricultural extension and agricultural support services which increased by 7 percent and 14 percent per year, respectively. Even what is allocated to agricultural research is not really used for this purpose. For example, annual agricultural research expenditure in Nepal during 1979/80 (based on 3-year average) was Rs 35.4 million. Out of this, only 35 percent was spent on actual research and the remaining 65 percent was spent on nonresearch activities.

Estimated expenditures on agricultural research and extension are reported in Table 10. Asia, which has 72 percent of the world's agricultural population, 57 percent of its total population, and 32 percent of its arable land, contributed only 17 percent of the world's expenditure on agricultural research and 20 percent of the world's expenditures on agricultural extension.[34]

Despite the importance of agricultural research, there has been only limited economic analysis of the optimum amount of investment in agricultural research. Nor has much work been done to provide guidelines for the allocation of research resources among projects, crops, and regions.[35] Most of the studies on agricultural research have been limited to measuring the rates of return to investment in agricultural research.

A systematic conceptualization of the process of allocating resources to agricultural research was developed in Pinstrup-Andersen and Franklin [1977]. Binswanger and Ryan [1977] proposed that agricultural research resources should be allocated according to the share a particular crop has in the gross value of agricultural output. Research resources need to be allocated on the basis of efficiency and equity. In labor-surplus agrarian economies, it is also important to evaluate the impact of research resource allocation on employment. Research resource al

Table 10. Estimated land, population, and expenditure on agricultural research and extension in Asia

Region	Expenditure on agricultural				Ratio of research to extension expenditure	Arable land 1975 (%)	Total population, 1975 (%)	Agricultural population, 1975 (%)
	Research, 1974		Extension, 1974					
	Amount	Percent	Amount	Percent				
	(US$ millions, constant 1971)	(US$ millions, constant 1971)						
North America and Oceania	1,289	34	288	22	4.5	21	7	1
Asia	646	17	259	20	2.5	32	57	72
World	3,841	100	1,326	100	2.9	100	100	100

Source: Boyce and Evenson [1975] for expenditure and FAO [1989a] for land and population estimates.

location designed to help the poorest consumers should emphasize commodities with low income and price elasticities of demand [Pinstrup-Anderson, de Londono, and Hoover, 1976].

As argued in Mellor [1977], several factors make market prices inappropriate for allocating agricultural research resources optimally. The relationship between research expenditure and economic returns are poorly understood. Market prices might not include externalities of agricultural research, such as its effect on health and nutritional status, and thus understate their true value to society. Finally, new technology affects the distribution of income by changing both the relative returns to owners of productive resources and the prices of goods consumed in unequal proportions by different income classes. Research resources could be allocated by taking into consideration their effects on the supply of wages goods, on the demand for labor, on the nutritional composition of food supply, and on the size of and variation in producers' net income. But, because little is known about them, the allocation of research resources remains ripe for research. Hayami and Ruttan [1971, 1985] explore the complex institutional aspects of the question.

2. Contribution of Agricultural Research

Pinstrup-Andersen and Franklin [1977] name three contributions that are the direct result of agricultural research. These are increases in the technical efficiency of at least one resource; changes in the characteristics and composition of existing products and development of new ones, and reductions in production risk. These contributions result in changes in the composition, quantity and quality of agricultural supply, in aggregate resource demand, and in domestic farm income.

Estimates—from selected empirical studies—of the internal rates of return on investment in agricultural research for a number of commodities are given in Table 11.[36] The estimated rates of return in Asian countries are 35 to 63 percent for aggregate output, 25 to 75 percent for rice in Japan, 32 to 78 percent for rice in Asia, 25 percent for rubber in Malaysia, and 60 percent for sugarcane in India.[37] The evidence is limited to only a few crops and a few countries in Asia. However, the internal rates of return estimated for other crops and other regions of the world confirm its general order of magnitude. Agricultural research in Asia has made a major contribution to agricultural growth by relaxing constraints imposed by the inelastic supply of cultivable land.[38]

3. The Transfer of Agricultural Technology and Research

Agricultural research requires large investments, particularly in human capital, which, in developing and developed countries, requires time and economic resources. And, the agricultural problems for each set of agroclimatic and socioeconomic conditions are unique; this situation limits the transfer of research results. Therefore, the developing countries need to emphasize adaptive research so

Table 11. Estimated annual internal rates of return on investment in agricultural research for different commodities[a]

Country/region	Commodity	Time-period	Rate of return (%)	Source
Asia				
Japan	Aggregate	1880-1938	35	Tang [1963]
India	Aggregate	1953-71	40	Evenson and Jha [1973]
India	Aggregate	1960/61-1972/73	63	Kahlon et al. [1977]
Japan	Rice	1915-50	25-27	Hayami and Akino [1977]
Japan	Rice	1930-61	73-75	Hayami and Akino [1977]
Philippines	Rice	1966-75	27	Flores-Moya, Evenson, and Hayami [1978]
Asia	Rice	1950-65	32-39	Evenson and Flores [1978]
Asia	Rice	1966-75	73-78	Evenson and Flores [1978]
Tropics	Rice	1966-75	46-71	Flores-Moya, Evenson, and Hayami [1978]
Bangladesh	Rice and wheat	1961-77	30-35	Pray [1979]
Malaysia	Rubber	1932-73	25	Pee [1977]
India	Sugarcane	1945-58	60	Evenson [1976]
Other Countries				
United States	Aggregate	1949-59	35-40	Griliches [1964]
United States	Aggregate	1949-59	47	Evenson [1968]
United States	Hybrid corn	1940-55	35-40	Griliches [1958b]
United States	Hybrid sorghum	1940-57	20	Griliches [1958b]
Mexico	Wheat	1943-63	90	Ardito-Barletta [1970]
Mexico	Maize	1943-63	35	Ardito-Barletta [1970]
Peru	Maize	1954-67	35-40	Hines [1972]
Colombia	Rice	1957-72	60-82	Hertford et al. [1977]
Colombia	Rice	1957-74	79-96	Scobie and Posada-Torres [1978]
Brazil	Cotton	1924-67	77+	Ayer [1970]
Brazil	Cotton	1924-67	77-110	Ayer and Schuh [1972]
Colombia	Soybeans	1960-71	79-96	Hertford et al. [1977]
Canada	Rapeseed	1960-75	95-110	Nagy and Furtan [1978]
South Africa	Sugarcane	1945-62	40	Evenson [1976]
Australia	Sugarcane	1945-58	50	Evenson [1976]
Australia	Pastures	1948-69	58-68	Duncan [1972]
United States	Poultry	1915-60	21-25	Peterson [1967]

[a] Many of these results have also been summarized in Arndt, Dalrymple, and Ruttan, eds. [1977]; Boyce and Evenson [1975]; Evenson, Waggoner, and Ruttan [1979]; Pinstrup-Andersen [1982]; and Ruttan [1982].

that they can maximize the benefits of research results by increasing the transferability of technology from whatever source.

A particular need is for research on how to increase the effectiveness of agricultural research systems. It is clear that returns to research success are high and that success requires an effective national research system. The national research systems in many countries are relatively ineffective.[39]

That constraint can be erased by borrowing research results. The ability of high-income countries to generate the bases for technological change is well developed. Hence, that capacity is one of the most coveted elements that these countries have to offer the low-income countries. Substantial analysis is needed on the appropriate technology, on the way education and training helps or hinders technology transfer, and on how the transfer of technology relates to national systems. Technology transfer requires highly effective national systems.

An excellent overview of the performance and prospects of international transfer of agricultural technology is provided by Evenson [1988]. There are four channels of technology transfer, including direct, adaptive or indirect, pretechnology science, and capacity. Based on an examination of large volumes of data, Evenson concludes that direct transfer between countries is very limited for most technology fields. Even within most countries, direct transfer between regions for most technology areas is limited. Indirect transfer is probably more important. However, indirect transfer does not take place without research capacity in the destination country. Pretechnology science transfer has become an increasingly important form of transfer as research systems throughout the world have expanded capacity. Capacity transfer has also been important. United States' assistance and training have facilitated the development of this capacity, especially in several Asian countries.

The development of domestic technology transfer capacity depends on at least three factors, including investment in agricultural research, investment in agricultural extension, and creation of scientific manpower. The available evidence, as reflected by the above three factors, shows a significant growth in technology transfer capacity in the last twenty years in the developing countries, particularly in Asia (Table 12). In the 1950s and 1960s, most Asian countries (except East Asia) were spending more on agricultural extension than on agricultural research. This strategy was reversed in the 1970s and 1980s. A rapid growth in spending in agricultural research, combined with little or no growth in expenditure on agricultural extension, produced roughly equal spending intensities (public expenditure in research or extension as percent of the value of agricultural product) in research and extension in most developing countries, particularly in Asia.

The World Bank (through its lending) and CGIAR (through its agricultural research programs) provided major stimulus for the development of national agricultural research and extension capacity. For example, the World Bank initiated and funded the development of Training and Visit System (T&V) of agricultural extension in the 1970s [Benor and Harrison, 1977; and Benor and Baxter, 1984]. The T&V system emphasizes simplicity, flexibility, and continuous feedback from farmers to extension and research. Central aspect of the system is the role of

Table 12. Expenditure on agricultural research and extension and estimated agricultural scientific manpower in world regions

Region	Agr. research expenditure, 1980[a]		Public expenditure as % of the value of agr. product						Scientific manpower[c]	
			Agricultural research			Agricultural extension			Number	
	Mill. US$	Growth[b]	1959	1970	1980	1959	1970	1980	1980	Growth[d]
Northern Europe	410	4.3	0.55	1.05	1.60	0.65	0.85	0.84	8,027	4.4
Central Europe	871	6.2	0.39	1.20	1.54	0.29	0.42	0.45	8,827	3.1
Southern Europe	209	5.3	0.24	0.61	0.74	0.11	0.35	0.28	2,686	1.7
Eastern Europe	553	2.8	0.50	0.81	0.78	0.32	0.36	0.40	20,220	3.5
USSR	939	2.5	0.43	0.73	0.70	0.28	0.32	0.35	31,394	2.6
Oceania	387	4.2	0.99	2.24	2.83	0.42	0.76	0.98	3,302	1.9
North America	1336	2.0	0.84	1.27	1.09	0.42	0.53	0.56	10,305	1.5
Temperate S. America	80	2.6	0.39	0.64	0.70	0.07	0.50	0.43	1,527	4.2
Tropical S. America	269	7.7	0.25	0.67	0.98	0.34	0.71	1.19	4,840	8.5
Caribbean & C. America	113	8.3	0.15	0.22	0.63	0.09	0.18	0.33	2,167	4.4
North Africa	62	3.0	0.31	0.62	0.59	1.27	2.21	1.71	2,340	4.4
West Africa	206	4.6	0.37	0.61	1.19	0.58	1.24	1.13	2,466	6.0
East Africa	71	5.9	0.19	0.53	0.81	0.67	0.88	1.16	1,632	7.4
Southern Africa	82	2.0	1.13	1.10	1.23	1.64	0.67	0.46	1,650	2.4
West Asia	125	5.1	0.18	0.37	0.47	0.25	0.57	0.51	2,329	5.1
South Asia	191	6.0	0.12	0.19	0.43	0.20	0.23	0.20	5,691	4.0
Southeast Asia	103	11.4	0.10	0.28	0.52	0.24	0.37	0.36	4,102	9.3
East Asia	735	5.2	0.69	2.01	2.44	0.19	0.67	0.85	17,262	2.2
China	644	11.9	0.09	0.68	0.56	n.a.	n.a.	n.a.	17,272	13.8

Source: Evenson [1988]; originally from Judd, Boyce, and Evenson [1986].

n.a. = not available.

[a] In constant 1980 U.S. dollars

[b] Growth expressed as a ratio of expenditure in 1980 over 1959.

[c] In terms of scientist man-years.

[d] Growth expressed as a ratio of scientific manpower in 1980 over 1959.

contract farmers and subject matter specialists and the primacy of field work. Similarly, the CGIAR system had a significant impact on the development of national agricultural research systems, allocation of needed resources, and the training of scientific manpower.

Evenson [1988] also provided an overview of emerging agricultural technologies based on both the conventional methods and biotechnology. The development of biotechnology has been stimulated by institutional developments enabling private firms to capture more returns from their research. According to Evenson, the 1990s will be dominated by technologies emerging from the conventional systems, with some shift from public sector to private sector. By year 2000, the biotechnology sector will produce a number of significant technologies but the developments will be volatile, including failures, bankruptcies, and mergers. United States, Japan, and Western Europe will play dominant roles and most of these technologies will emerge from the private sector. Direct transfer and multinational firms will play increasingly important roles in the development and transfer of agricultural technology.

4. International Agricultural Research Centers

In this context, the international agricultural research centers (IARCs), under the auspices of Consultative Group on International Agricultural Research (CGIAR), have made significant contributions to agricultural growth in Asia not only through research results but also by facilitating the transfer of appropriate agricultural technology (Table 13). All of these programs will affect the nature and level of employment generated by transfers of technology and the competitiveness of domestic production employing imported technology.[40] The initial emphasis of the IARCs was on individual crops mandated for respective centers. However, over time farming systems research (FSR) became an increasingly important component of research programs of the IARCs as well as national research programs and agricultural development projects. A comprehensive review of FSR is provided by Simmonds [1985].

CGIAR, which was established in 1971, is an association of countries, international and regional organizations, and private foundations dedicated to supporting a system of international agricultural research centers around the world. The number of both CGIAR members and international agricultural research centers has grown over time. During 1986, CGIAR had fifty-one members, of which thirty-nine were donors [CGIAR, 1987b]. Total contributions have grown from US$12.0 million in 1972 to US$235.5 million in 1986.

In 1987, W. David Hopper, then CGIAR chairman, outlined two broad challenges for the CGIAR [CGIAR, 1987b]. First, policies and programs: how the CGIAR can best contribute to the enhancement of income and the enhancement of food availability for poor people throughout the developing world. Second,

Table 13. International agricultural research centers with implications for agricultural development in Asia

Center	Location	Primary research mandate	Year of establishment
IRRI (International Rice Research Institute)	Philippines	Rice, farming systems	1960
CIMMYT (International Maize and Wheat Improvement Center)	Mexico	Wheat, maize, and farming systems	1966
CIAT (Centro Internacional de Agricultura Tropical)	Colombia	Cassava and beans	1967
AVRDC (Asian Vegetable Research and Development Center)	Taiwan	Vegetables	1971
CIP (International Potato Center)	Peru	Potatoes	1972
ICRISAT (International Crops Research Institute for the Semi-Arid Tropics)	India	Sorghum, pearl millet, pigeon peas, chickpeas, groundnuts, and farming systems	1972
IBPGR (International Board for Plant Genetic Resources)	Italy	Plant genetic material	1973
IFDC (International Fertilizer Development Center)	United States	Fertilizer	1975
IFPRI (International Food Policy Research Institute)	United States	Food policy	1975
ICARDA (International Center for Agricultural Research in the Dry Areas)	Syria	Barley, wheat, broad beans, lentils, and farming systems	1976
ISNAR (International Service for National Agricultural Research)	Netherlands	National agricultural research systems	1980
IIMI (International Irrigation Management Institute)	Sri Lanka	Irrigation	1984

technologies: how best to maintain the research drive to find and exploit new technologies of producing basic food materials for growing world demand, especially in the light of the accelerating revolution in biological research fundings and methodologies.

The CGIAR was initially an agricultural production research system. In the mid-1970s, a need was felt to increase the effectiveness of production science research both in increasing production and distributing benefits to the poor by adding a policy research center. In that context, the International Food Policy Research Institute (IFPRI) was formed to generate new knowledge of policy, to stimulate the building of similar capacity in developing countries, and finally to have a policy impact in collaborations with sister institutions in the national systems of developing countries. One result of this activity has been to give a substantial boost to the attention paid to policy-oriented research relating to agricultural development, with a particular emphasis on field research that elicits facts at the farm and family level and pyramids those to policy conclusions.

Chapter VI. Marketable Surplus and Marketing Behavior

The generation of a marketable surplus and its transfer from the agricultural to the nonagricultural sector are crucial to the achievement of self-sustaining economic growth in dualistic development models. With the exception of A. K. Dixit [1969], Hornby [1968], Lele and Mellor [1981], and Zarembka [1970], these models do not deal with changes in marketable surplus caused by changes in technology, prices, and output. These issues have been analyzed in theoretical and empirical studies that are specifically about marketable surplus. Knowledge about how a marketable surplus responds to changes in price, output, and the size of land holdings has important implications for the design of agricultural policy.

1. Price, Output, and Marketable Surplus

Knowledge of the price elasticity of a marketable surplus is of paramount importance as a guide to the formulation of appropriate price policy for output. Most of the studies on marketable surplus have estimated the price elasticities of marketable surpluses for different crops from different regions at different time periods. Mathur and Ezekiel [1961] argued that the cash obligations of farmers in subsistence agriculture are relatively fixed. Consequently, a rise in output prices would result in a fall in the marketable surplus, implying that the price elasticity is negative. This would lead to an increase in the amount of crop produce retained for home consumption, which is determined as a residual.[41] But this study assumed that the income elasticity of demand for nonfoodgrains was zero and that there were no substitution effects. Neither of these assumptions is supported by the empirical evidence.[42]

Based on data from 1959/60 to 1962/63, T. N. Krishnan [1965] estimated the price elasticity of marketable surplus for foodgrains in India to be -0.03. The underlying logic is that as price goes up, real income also goes up, and the consumption demand for farm produce increases, which leads to a reduction of the marketable surplus. The whole analysis assumed that farm output is fixed at a certain level. If the effect of price on output is explicitly incorporated, however, the price elasticity of the marketable surplus may become positive.

R. Krishna [1962] developed a conceptual framework to determine the relationship between the price and marketable surplus of a single crop by incorporating positive price elasticities of both output and consumption. He found the price elasticity of marketable surplus for subsistence crops in India to be positive. According to Nowshirvani [1967b], however, R. Krishna [1962] ignored the income effect of a change in the value of initial consumption caused by changes in price. Once this is incorporated, a negative price elasticity of the marketable surplus must be considered a possibility. Khusro [1967] was more affirmative, saying that farmers would retain more of the foodgrains they produce if the market price was lower. This implies that the price elasticity for marketable surplus is positive.

K. Bardhan [1970] estimated the price and output elasticities of marketable surplus of foodgrains in North India.[43] The study was based on a cross-sectional survey of twenty-seven villages in Punjab and Uttar Pradesh. The price elasticity of marketable surplus was estimated to be -0.6 and the output elasticity was estimated to be 1.8. In Thamarajakshi [1971], the price elasticity for marketable surplus of foodgrains was -0.6 and the output elasticity 1.01. These two studies clearly imply that if the marketable surplus is to be expanded, then foodgrain prices should be reduced and foodgrain output expanded through technological change. But, since the price elasticity of output is usually positive, a reduction in the price of foodgrains alone will reduce the marketable surplus.

Many of the empirical analyses of marketable surplus have dealt with India, but a few have analyzed the behavior of marketable surplus in other Asian countries.[44] Mangahas, Recto, and Ruttan [1966], using a time series analysis for Philippines, found the price elasticities of marketable surplus for rice and corn generally to be positive, with the price elasticities for rice higher than for corn. Furthermore, the price elasticities for marketable surplus were higher than the corresponding elasticities for production. The price elasticities of marketable surplus for rice were found to be positive in Taiwan by Chinn [1976], in Indonesia by Mubyarto [1965] and Mubyarto and Fletcher [1966], and in Thailand by Behrman [1966]. Using a 1972-74 sample survey of rice farmers in central and south Luzon in the Philippines, Toquero, Duff, Lacsina, and Hayami [1975] estimated the price elasticities to be 0.0, partial output elasticity to be 1.37, and the total price elasticity to be 0.41. Output has a strong, positive effect on marketable surplus. Price influences marketable surplus positively through its indirect effect on output. Therefore, the findings were consistent with the indirect inference of others that price elasticities are positive.

2. Farm Size and Marketable Surplus

Knowledge of the relationship between farm size and marketable surplus is relevant to the design of land tenure policies. The most detailed analysis of the distribution of the marketable surplus of agricultural production was done by Narain [1961] who found that there is a nonlinear relationship between farm size and marketed surplus. Marketed surplus as a proportion of the value of agricultural output first drops from 20.7 percent for 0.5 acre size group to 9.7 percent for 10–15 acre size group and then rises rapidly thereafter. According to Narain, the drop in the proportion of marketable surplus is due to distress sales.

Using twenty-three samples drawn from eight states in India, R. Krishna [1965a] found a linear relationship with a negative intercept between output and marketable surplus.[45] The output elasticities of marketable surplus were between 1.04 and 1.60 for wheat and between 1.04 and 1.36 for rice. These results imply that land reform policy should try to reduce fragmentation and increase farm size if its objective is to increase marketable surplus. This analysis dealt with a single crop. The linear relationship observed for a single crop may not hold for aggregate agricultural production.

Muthiah [1964], using a sample survey of six villages in Punjab and western Uttar Pradesh in India, found a positive relationship between marketable surplus and the size of agricultural holdings. Dandekar [1964] provided empirical evidence that the relationships between farm size and the marketable surpluses of wheat, jowar, and other cereals are positive.[46] Other studies that show positive relationships between farm size and marketable surplus, and between production and marketable surplus, are Parthasarathy and Subbarao [1964] for rice in south India; Bhargawa and Rustogi [1972] for rice in West Bengal; and Dayal [1963] for cereals in Uttar Pradesh.[47] These studies indicated that the marketable surplus is concentrated in the hands of large farmers who control large parts of total production and sell larger shares of their produce than smaller farmers because their income elasticity for foodgrain consumption is lower. There appear to be no studies that analyze the effect of farm size on marketable surplus after the Green Revolution.

3. Sectoral Terms of Trade and Marketable Surplus

In addition to its positive impact on foodgrain production and marketable surplus, an increase in foodgrain prices also determines changes in the distribution of income and the transfer of resources between sectors, the distribution of income between classes, and the growth of the economy.[48] As reported in Parthasarathy and Mudahar [1976], policies based on the observed empirical relationship often ignore the unfavorable consequences that higher foodgrain prices have for economic growth and the distribution of income, effects that sometimes outweigh the positive effects of higher foodgrain prices.

Changes in the terms-of-trade in favor of foodgrains relative to manufactured goods were found to reduce aggregate savings in India through their effect on wage rates and nonagricultural sector profits [Parthasarathy and Mudahar, 1976; Mellor, 1976]. Shifting the terms of trade in favor of foodgrains is only one instrument available to policymakers. Its efficacy when compared to other instruments, such as a direct public investment in agriculture, continues to be in doubt.[49] However, major distortions in relative prices arising from macro policy, including public investment policy, and their effects on exchange rates have been shown in other parts of the world to have a substantial deleterious effect on agriculture [see, for example, Garcia 1981]. Bautista [1987] documents these exchange rate relationships for the Philippines, showing a strong negative effect on agriculture of a whole set of macro policies.

4. Agricultural Marketing and Its Efficiency

In regions experiencing the Green Revolution, marketed surplus has increased faster than production; this increase has resulted in postharvest gluts and wastage where marketing facilities are inadequate. As reported in Gill [1972] and Mudahar [1974], the proportion of total wheat production marketed increased from 33 percent in 1967/68 to 57 percent in 1970/71 in the Indian Punjab. Furthermore, the postharvest (April 15 to July 15) market arrivals, as a percentage of the total marketed surplus for wheat, increased from 55 percent in 1967/68 to 84 percent in 1970/71. Markets were not able to cope with this sudden increase in arrivals because of deficiencies in storage, communication, and transportation facilities.

In India, policies for the system of marketing agricultural commodities have been particularly complex. It was widely believed that private trade was exploitative and discouraged production, there was concern with the free market's inequity in allocating food, and there was a stated belief that cooperative structure would have social advantages. Lele [1971], however, showed that the shortcomings in the efficiency and productivity of the private marketing system stem largely from deficiencies in transportation, communications, and information.[50] The study concluded that these inadequacies were aggravated by haphazard government policies that were usually made in reaction to short-term crises without consideration of long-term production and welfare. The Indian experience is typical of the experience of many developing Asian countries.

It is difficult to argue that cooperatives have social advantages as long as they are controlled by the wealthier rural people. In periods of scarcity and rising prices, market forces reduce the consumption of the poor disproportionately because their low incomes and the high proportion of those incomes spent on food offer few alternatives. It is doubtful that cooperative marketing can alleviate a problem rooted in income inequality and scarcity. Lacking viable solutions to this politically explosive problem, politicians usually resort to displacing private trade.

The environment for the development of cooperatives improves as technological change increases the demand for new forms of purchased inputs and market outlets and as the infrastructure develops. In the long-run, trade organized by cooperatives may have a useful social and economic role to play. For a full discussion of these issues as they relate to equity, see Lele [1974, 1981].

Chapter VII. Modernizing Agriculture and Rural Welfare

1. Food, Nutrition, and Consumption Patterns

INCIDENCE OF MALNUTRITION

The literature on the economic aspects of human nutrition has attempted to estimate the incidence of malnutrition by determining the size and importance of deficiencies in calorie and protein consumption, and to understand how the Green Revolution affects nutrition by affecting food supply.

It was generally believed in the 1960s that there existed a serious protein gap that was responsible for malnutrition in developing countries. But, empirical evidence suggests that protein deficiency is less important than calorie deficiency.[51] Sukhatme [1977, p. 1] argued that there is "no evidence to support that our diet is seriously deficient in protein. . . . The protein deficiency undoubtedly prevails but it appears to be the indirect result of inadequate energy in the diet. It is only right therefore that research workers should turn their attention to estimating the size of the calorie gap and its incidence in the population."

It is important to use an appropriate procedure to determine energy requirements and the number of the malnourished. Dandekar and Rath [1971a, b] estimated the incidence of poverty in India and Reutlinger and Selowsky [1976] estimated it in Brazil by using the "average" amount of energy for the "reference" individual required to lead a healthy active life. Sukhatme [1977], however, argued that this procedure overestimates the incidence of poverty because "average" requirements are higher than the "minimum" requirements. Using National Sample Survey data for 1971/72, Sukhatme [1977] estimated that the incidence of malnutrition is 25 percent in urban India and 20 percent in rural India.[52]

The incidence of energy-deficient diets in eighty-seven developing countries (excluding China) during 1980 is reported in Table 14. About 16 percent of the population (340 million) did not have adequate calories to prevent stunted growth and serious health risk. Out of 340 millions of affected population, 65

Table 14. Prevalence of energy-deficient diets in eighty-seven developing countries, 1980[a]

Region	No. of countries	Not enough calories for an active working life[b]		Not enough calories to prevent stunted growth and serious health risks[c]	
		Share in population (%)	Population (million)	Share in population (%)	Population (million)
South Asia	7	50	470	21	200
East Asia and Pacific	8	14	40	7	20
Sub-Saharan Africa	37	44	150	25	90
Middle East and North Africa	11	10	20	4	10
Latin America and the Caribbean	24	13	50	6	20
Developing countries	87	34	730	16	340

Source: World Bank [1986b].

[a] The eighty-seven countries accounted for 92 percent of the population in developing countries in 1980, excluding China.
[b] Below 90 percent of FAO/WHO requirements. This category includes population described in footnote C.
[c] Below 80 percent of FAO/WHO requirements.

percent was in Asia—59 percent in South Asia alone. On the other hand, 34 percent of the population (730 million) did not have adequate calories for an active working life. Out of 730 million affected population, 70 percent was in Asia—64 percent in South Asia alone. Over time, there has been a small reduction in the relative shares of affected population, but not in the absolute numbers.

According to the World Bank [1980c], the number of people in absolute poverty in eighty-seven developing countries (excluding China and other centrally planned economies) is estimated at about 780 million. Absolute poverty refers to a condition of life characterized by malnutrition, illiteracy, and disease as to be beneath any reasonable definition of human decency. The problem is relatively more serious in Asia. These malnutrition problems exist even when the world has ample food. The growth in global food production has been faster than population growth in the last forty years [World Bank, 1986b]. Despite this, many poor countries and millions of poor people do not share in this abundance. They suffer from both chronic and transitory food insecurity, caused mainly by lack of purchasing power. Based on research in India, Binswanger and Quizon [1984] concluded that an increase in domestic food production will not reduce food insecurity unless it reduces food prices.

GREEN REVOLUTION, NUTRITION, AND PRICES

The studies that implicitly assumed a serious protein gap exists in developing countries concluded that the Green Revolution has reinforced this problem by replacing area under pulses, which are rich in protein, with cereals. However, Ryan and Asokan [1977], using empirical data from the wheat region of India, argued that the yield-oriented strategy for breeding wheat did not adversely affect nutrition. Furthermore, one could argue that a yield-oriented strategy and the Green Revolution increased the food supply, and thus reduced the calorie gap, increased the income of farmers, and even increased the real income of consumers by reducing prices. Lele [1972] reported a positive relationship between the Green Revolution, income distribution, and nutrition. Pinstrup-Andersen, et al. [1976] argued that an increase in the total supply of nutrients is a poor indicator of relative nutritional impact because both the wastage of nutrients and adjustments by consumers of their food consumption are functions of the commodity from which the additional food nutrients are obtained.

In analyzing the impact of CGIAR, J. R. Anderson, Herdt, and Scobie [1988] provide an overview of the impact of Green Revolution on food production, nutrition, and prices. Since food accounts for the largest share of expenditure by low-income groups, any reduction in food prices leads to about twice the relative increase in real income for poor households than for rich households (Table 15). Modern crop varieties have helped to keep real food prices from rising. If modern varieties of rice and wheat had not replaced traditional varieties (with other inputs

Table 15. Estimated impact of a decrease in price of food on the real income of low-income and high-income consumer groups

Country	Increase in real per capita income due to 10% decrease in the price of food (in %)		Source
	Lowest 10%	Highest 10%	
Egypt	5.6	1.0	Alderman and von Braun [1984]
India[a]	5.3[b]	1.2[c]	Mellor [1978]
India	7.3	2.9	Murty [1983]
Sri Lanka	8.5	4.1	Sahn [1988a, b]
Thailand	6.0	2.0	Trairatvorakul [1984]

[a] Foodgrains only.
[b] Lowest 20 percent.
[c] Highest 5 percent.

unchanged) in the early 1980s, annual rice output would have been ten to thirty million tons less and wheat output ten to twenty million tons less. Modern varieties of other crops have added at least three to five million tons to available food supplies.

However, consumption gains of the poor may be limited due to several factors. First, increased food output may be absorbed by the rich if income growth favors them and the poor lack the purchasing power. Second, added domestic food production may merely displace food imports. Third, in response to a slow increase in prices of staples, employers may hold wages down and the real purchasing power of the poor may not improve much. However, the available evidence indicates that introduction of modern varieties, by moderating food prices, has been the main factor in improving the nutrition of the poor in developing countries. In most of Asia, modern varieties of rice and wheat have prevented mass starvation, and improved nutrition of urban residents, farm households, *and* landless rural poor.

A. K. Sen [1981a, b, 1987] has provided very perceptive insights into the ingredients of famine analysis and the role of food production, poverty and entitlements in famines. According to him, decline in food production and availability is not necessarily the primary cause of famine, rather a series of factors may converge to reduce the exchange entitlement of households, precipitating reduced consumption. These factors include redistribution of available food, inflation, reduction in income due to unemployment and lower farm profits. Governments often pay excessive attention to aggregate food supplies, failing to recognize the other key elements that result in the decline of food consumption of population segments which, at the extreme, result in famine. The policymakers must also be cognizant of the fact that the proportion of poor in a population may not change over time but this does not mean that it is the same households [Srinivasan, 1985]. It is important to generate such information since it is essential in designing and implementing effective poverty alleviation and nutrition programs.

TARGETED NUTRITION AND SUBSIDY PROGRAMS

A series of studies undertaken by Behrman [1988], Behrman and Deolalikar [1987], and Behrman, Deolalikar, and Wolfe [1988] in several developing countries, including India, analyze the influence of prices, income, and schooling on the nutrient intake and the effect of nutrition on health, productivity, wages, and fertility. The studies suggest strong impact of food prices on nutrition, particularly for the poor; growth in income may be less likely to improve nutrition than has been suggested by others; women's schooling is important in improving nutrition; nutrition exerts positive influence on productivity, wages, and fertility; and direct association between nutrition and health is not shown.

Such information is essential in designing effective nutrition and food subsidy programs as a mechanism to improve nutrition and keep the cost of food subsidy as low as possible. Food subsidy programs are popular in many developing countries, particularly in Asia [Pinstrup-Andersen, ed., 1988]. The blanket subsidy program in Sri Lanka became too expensive for government to sustain and was replaced in 1979 with targeted food stamp scheme [Edirisinghe, 1987]. In the long run, it is essential to promote welfare through policies designed to promote economic growth, as was argued for Sri Lanka by S. S. Bhalla and Glewwe [1986]. However, these policies need to be complemented with targeted nutrition and food subsidy programs in order to improve the welfare of low-income households who may not benefit from economic growth.

As has been shown by Garcia and Pinstrup-Andersen [1987] for a sample survey of a pilot project area in the Philippines, targeted food subsidy schemes could be very effective in reducing malnutrition. The pilot project was implemented in three provinces for one year during 1983/84. Prior to the implementation of pilot program, the survey showed that the distribution of food within household was biased in favor of adults, about one-fourth of preschoolers were malnourished, and malnutrition decreased with an increase in income. Food (rice and cooking oil) subsidies had positive impact on household food expenditure, calories acquired and consumed, and average weight of preschoolers. It was shown that consumers are more likely to increase food consumption if foods are subsidized than if incomes are raised directly. The total cost of the scheme was attributed to food subsidy (84 percent), administrative cost (9 percent), and incentive payments to private retailers to assure efficient distribution of subsidized food (7 per-

cent). The annual cost of eliminating calorie deficiencies in the sample was estimated to be about twenty-five dollars per adult equivalent unit. The food subsidy
scheme was based on geographical targeting rather than targeting based on
household income—which is relatively more difficult to administer, implement,
and monitor. However, there was a strong relationship between poverty and
malnutrition, indicating that any long-term solution to the nutrition problem
must focus on poverty alleviation. Overall, the scheme was considered efficient
and cost effective. Blanket subsidy programs, like urban food subsidies in China,
become too expensive to sustain in the long run and even lead to wastage of food
by the high-income households.

INCOME AND CONSUMPTION PATTERNS

There is a direct relationship between income and consumption behavior. The
Green Revolution has been responsible for increases in income levels and for
changes in the distribution of income. As a result, the Green Revolution has important implications for food consumption patterns.[53] Mellor and Lele [1973] argued that, particularly for the peasant farming classes, the incremental share of
budget spent on foodgrains declines with the income increases associated with
improved agricultural technology; and the incremental share spent on goods and
services produced by nonfoodgrains agriculture and industrial sectors increases.
This can increase economic growth by increasing the marketable surplus of foodgrains to the nonagricultural sector; increasing the incremental demand for the
consumption of those consumer goods that can be produced by labor-intensive
techniques; increasing labor demand and employment; and increasing demand
for foodgrains, which keeps foodgrain prices from falling below the incentive
price levels that are important for continuing growth in agricultural production.

There are sharp contrasts between rural and urban expenditure patterns [Gov-
ernment of India, 1986] in Asia as well as in other developing and even developed
countries. However, all these differences cannot be explained by differences in
income alone. This suggests that short-term development patterns in different
sectors are dissimilar. The expenditure elasticities for foodgrains, milk, and milk
products are consistently greater in rural than in urban areas [Harrison, Hitchings, and Wall, 1981]. In rural compared to urban areas, expenditure elasticities
for clothing are higher in the low-expenditure groups and fall much less rapidly
as expenditures rise but are lower in the upper expenditure groups. C. H. H. Rao
[1969] showed that foodgrains may be as much as 19 percent cheaper and urban
goods 34 percent more expensive in rural than in urban areas. Apparent elasticities may change substantially as rural markets are integrated more fully into the
national centers. Consumers' tastes may also change over time, especially as income increases rapidly, markets grow and become integrated, and income distribution changes. In order to shift the structure of domestic production, so that

employment is increased, taxes and subsidies may be used to channel consumption toward more labor-intensive commodities.

2. Labor, Employment, and Wages

THE NATURE OF RURAL LABOR SUPPLY

Surprisingly little is known about the rural labor supply.[54] The rural labor force in much of South Asia appears to be overemployed at seasonal peaks and underemployed in slack seasons. Yet contrary to commonly held views that labor is chronically in surplus, it seems likely that there is an active and pervasive labor market that adjusts to interseasonal equilibria.[55] The labor market is probably in equilibrium at all seasons, but with sharply different wage rates. During slack seasons, the supply may be so elastic that a modest increase in wages will draw a large increase in labor supply. The labor supply may be highly inelastic during seasonal peaks. Farm operators would then try to reduce labor requirements during peak season by mechanizing, that is, by substituting more elastic supplies of capital. Thus, a combina tion of inelastic supply of labor during seasonal peaks and an inability to mechanize may reduce employment in slack seasons. Most mechanization in India seems to have been suitable for resolving this problem.[56]

Low-income households provide disproportionately large numbers of women in the rural labor force. According to Mellor [1976], among landless and near landless families in India, 43 percent of the workers are female, compared to 33 percent of the workers among families with holdings of over five acres. This suggests that women withdraw from the rural labor force as incomes rise. This result is corroborated for India in Agarwal [1984], for Indonesia in B. White [1984], and for rural Egypt in B. Hansen [1969]. Child labor responds similarly to rising incomes. There is also a high degree of substitution between women's and children's labor [Yotopoulos and Mergos, 1986].

There are about three times as many households of landowners as there are of landless laborers in India. While most landowners do farm work, little is known of their need for labor. It seems logical, given the low wage rate, that well-to-do landowners would hire much of the labor needed for arduous physical work and spend most of their own time on supervisory, managerial, and marketing activities. On the other hand, it is possible that the introduction of machinery, rising wage rates, and declining seasonality of employment raise direct labor participation rates of landowners.[57]

Rapid population growth prevents the increased demand for labor from tightening the labor market and raising per capita employment. No group has a greater stake in reducing the rates of population growth than the landless laborers.[58] Unfortunately, there is probably no better example of a divergence between individual and group interests. If the poor participate in economic growth, however, then mortality rates decline, the potential income of the poor goes up, and the opportunity to invest in education expands. A strategy of growth that increases the demand for labor may reduce the long-run supply of labor by reducing fertility, thereby improving employment and the wage rates of the laboring class.

THE DEMAND FOR RURAL LABOR

The major forces affecting the demand for agricultural labor are modern farm technology, mechanization, changes in crop acreage, structure of demand for agricultural products, and the pattern of cultivation. Again, surprisingly little is known about the effects these forces have on the demand for labor. In the few areas that have undergone major technological changes, it is clear that wages, employment, and mechanization have all increased substantially.[59] These developments are, however, the product of complex interactions between unusually successful, geographically localized innovations.[60]

Based on 1967/68 sample data for the Indian Punjab, S. S. Sidhu [1974b] estimated that the per acre labor demand function increased by 25 percent as a result of a shift to modern wheat varieties. For Rajasthan, Acharya [1973] estimated that labor demand increased by 32 percent as a result of the same kind of shift. Barker and Cordova [1978] reported that for rice production in the Philippines the introduction of modern varieties of rice increased labor input per hectare, but decreased labor input per ton of rice production.

Johl [1973b], in a careful analysis of agriculture in the Indian Punjab, concluded that mechanization, along with elements of improved farm technology, has directly and indirectly increased not only employment and labor productivity but also the returns to various factors of production. On the other hand, Billings and A. Singh [1969] reported that labor demand declines marginally as a farm shifts completely from traditional to mechanized cultivation methods. Kahlon and Grewal [1972] concluded that the use of tractors in the agriculture of the Indian Punjab decreased labor demand slightly but increased the productivity of land and off-farm employment. Donovan [1974], using a linear programming model for a group of villages in the state of Mysore (now Karnataka) in India, showed that mechanization can break labor bottlenecks and increase production.[61]

Binswanger [1978] in a comprehensive review for South Asia concludes that there is no convincing evidence that tractors are responsible for substantial increases in intensity, yields, timeliness, and gross returns. There is no evidence that tractors have high benefit-cost ratios in semiarid zones or even in the eastern rice belt of the subcontinent. A comparative historical perspective of agricultural mechanization is provided by Binswanger [1984]. Policy issues and options related to agricultural mechanization in the context of economic development are discussed in detail by Binswanger and Donovan [1987].

A concern for the laboring classes should not lead automatically to the simplistic solution of not mechanizing. And there is not a dichotomy between labor-saving and land-saving mechanization. Mechanization, which pays for itself in increased crop yields and larger supply of wage goods necessary to increase employment, may also displace labor that has little alternative use in agriculture and for which provision has not been made elsewhere in the economy.[62] Given the complexity and variability of these relationships, one can well argue that such decisions about mechanization are necessarily best left to the marketplace which in turn requires that there be no direct or indirect (e.g., through credit) subsidies.

The difficulty of the employment problem and the natural desire for simple answers lend substantial support to increased use of rural public works programs.[63] While the returns to rural infrastructure are potentially high, however, rural public works created solely for employment creation with inadequate complementary resources may be less useful for long-term poverty reduction than alternative programs. In those areas where there is little apparent prospect for future development, there is a difficult tradeoff between public works as a means of distributing welfare to such regions, or to invest elsewhere and expect migration to solve the problem.[64] Finally, one of the most important determinants of the real wages of the poor is foodgrain prices [Mellor and Desai, 1985]. If food production increases, the poor can benefit whether or not rural public works projects are effective. If such projects do not increase food production, the poor will lose much of their gain from higher employment through higher prices.[65]

3. Poverty, Equity, and Technology

POVERTY PROBLEM

Any definition of poverty is bound to be arbitrary. Dandekar and Rath [1971a, b] set the poverty line for India at 20 rupees or $4.19 per person per month according to 1960/61 prices and exchange rates, on the ground that persons with less than that income do not obtain enough calories for normal health and activity. According to this definition, 40 percent of India's rural population and 50 percent of its urban population were below the poverty line during 1960/61. Since urban employment has grown consistently faster than the natural increase in urban population, poverty in the cities largely reflects rural poverty and the migration of the unemployed to urban areas.

Although there is controversy on this point, the real incomes of the people in the lower 40 percent of the income distribution scale probably have not increased in the past two decades.[66] A particularly important finding by Narain, reported in Mellor and Desai, eds. [1985], is that in India underlying structural changes have reduced poverty secularly with that effect more than balanced by rising food prices and declining food availability per capita. The implication is that insufficient Green Revolution and population growth are the culprits behind lack of progress in reducing poverty. These results are corroborated by Alamgir [1975a, b] for Bangladesh, and T. Alauddin [1975] for Pakistan.[67]

According to a World Bank [1975] study, 40 percent of the Asian population was below the poverty line set at fifty dollars per capita annual income during 1969, with 85 percent of the Asian poor in the rural areas. Time series data show large swings in the proportion of the Indian rural population in poverty—from 40 percent to 60 percent. Some of these swings lasted for several years. The nature of these swings and their causes are discussed by several authors in Mellor and Desai, eds. [1985]. The authors put special emphasis on food prices and food production.

The income of landless or almost landless laborers depends on the demand for their labor by the owners of land and capital. Virtually all of them fall into the Dandekar and Rath [1971a, b] poverty class. Only a massive increase in employment can change this. The rural landless labor class is larger in India than in most other low-income countries, which makes the problem more difficult, economically and politically. India, however, offers particularly timely and instructive lessons because population growth and the diminishing availability of uncultivated lands are bringing increases in the size of the landless class in many other Asian countries.

TECHNOLOGY, GROWTH, AND EQUITY

The possible trade-offs between growth and equity in agricultural development and strategies for their improvement were discussed at the Eighteenth International Conference of Agricultural Economists, held in Jakarta, Indonesia, August 24 to September 2, 1982. The deliberations were published in Maunder and Ohkawa, eds. [1983] and Greenshields and Bellamy, eds. [1983]. According to Vyas [1983a, b], the pattern of agricultural growth in South and Southeast Asian countries during the 1970s did not favor the small farmers and the landless laborers who continued to be marginal producers and consumers. Hayami [1983], however, argued that technology does not promote inequity in the rural sector, and concludes that there is no trade-off between growth and equity in the long run. Narain and Roy [1980] showed the very powerful effects of irrigation, as a proxy for technology, on increasing employment.

Though it is ironic that a solution to the problem of rural poverty is provided by the spread of yield-increasing technological innovations that may raise the incomes of the landowning classes markedly, increased food supplies are essential if the welfare of the poor is to be improved. The nature of the new technology, the extent of its application, and the physical environment within which it is applied determine the size of the increase in yield, the quantity of inputs purchased, and the amount of employment created. These forces, in combination with the distribution of production resources, are responsible for the initial allocation of benefits among different income classes.

Although the use of modern technology has certainly increased average per capita income, there is controversy about how it affects the distribution of income. Some claim that the income distribution is becoming much more inequitable.[68] Using a sample of 126 representative farm holdings during 1967/68 to 1969/70, K. Singh [1973] argued that farm income inequality declined in the Aligarh district of Uttar Pradesh in India from 1963/64 to 1968/69. Based on an empirical study in Gujarat (India), Schluter [1974] concluded that much of the incremental income from high-yielding varieties is a residual return to landowners; only a small proportion comes from a greater use of labor.

In a farm-level analysis of rice farms in selected Asian countries, Barker and Herdt [1978] concluded that the adoption of modern rice varieties has been accompanied by an increase in labor use per hectare, small farmers have lagged significantly behind large farmers in the adoption of labor-saving innovations but not in the adoption of technology that would increase yields and income, and in the survey villages in India, Indonesia, and Pakistan, large farmers consistently used higher levels of fertilizer and obtained higher yields.

The proportion of the additional income paid to labor is nevertheless smaller than the proportion paid to each of the other inputs. This has confused many appraisers of the Green Revolution. According to R. S. Dixit and P. P. Singh [1970] actual imputed payments to labor in Uttar Pradesh (India) absorbed only 10 percent of the increased income from high-yielding varieties of wheat, while other purchased inputs absorbed 23 percent, leaving 67 percent as a reimbursement to the owners of land and capital. Similarly, in thirteen of the fifteen cases in Table 16, the proportion of increased output attributable to labor was between 5 and 15 percent. The percentage gain in income to labor was large, more than 25 percent in seven of the fifteen cases, because of the large rise in production.

In a study based on survey data of rice farmers for 1966, 1970, and 1974 for Laguna and Central Luzon/Laguna in the Philippines, Ranade and Herdt [1978] concluded that even though relative share of total labor declines, and because employment of hired labor increases, hired laborers became relatively "better-off" and new technologies were not landlord biased. Commenting on this study, Sinaga and Sinaga [1978] indicated that in Indonesia the benefit from the use of modern varieties of rice went to the operators and landlords, despite the fact that the labor requirement for rice production did not decline. The decline in labor share was because of a decline in real wages in the rural areas of Indonesia.

In a comprehensive study of crop diversification, Schuh and Barghouti [1988] argue that the rice industry in Asia in recent years has the found itself in a para-

Table 16. Division of increased agricultural production between labor and other farm inputs in Asia

Location	Increase in gross value of output		Increase in labor "payments"		Percent of increased output to labor[b]	Percent of increased output to other Inputs[c]	Source
	Rupees per acre[a]	Percent increase	Rupees per acre[a]	Percent increase			
Wheat							
Aligarh, U.P., India[d]	462	71	46	58	10	90	Dixit and Singh [1970]
Varanasi, U.P., India[d]	620	65	11	15	2	98	Misra and Shukla [1969]
Udaipur, Rajasthan, India	343	43	18	13	5	95	Acharya [1969]
Punjab, India	450	100	56	42	12	88	Government of India [1967/68]
Kharif Paddy							
West Godavari, Andhra Pradesh, India	269	38	32	17	12	88	Agro-Economic Research Center [1968/69]
East Godavari, Andhra Pradesh, India	216	33	20	13	10	90	Agro-Economic Research Center [1968/69]
Uttar Pradesh, India	1,100	200	67	92	6	94	Misra and Shukla [1969]
Tamil Nadu, India	550	100	33	20	6	94	Government of India [1968/69]
Laguna, Philippines	374	72	3	3	1	99	Crisostomo et al. [1971]
Sambulpur, Orissa, India	404	95	36	28	11	89	Tripathy and Samal [1969]
Rabi Paddy							
West Godavari, Andhra Pradesh, India	562	86	39	16	7	93	Agro-Economic Research Center [1968/69]
East Godavari, Andhra Pradesh, India	761	153	39	30	5	95	Agro-Economic Research Center [1968/69]
Tamil Nadu, India	625	100	46	21	7	93	Government of India [1968/69]
Gumai Bil, Bangladesh	948	208	302	125	32	68	Masud and Underwood [1970]
Bajra[e]							
Kaira, Gujarat, India	300	85	39	27	13	87	B. M. Desai and M. D. Desai [1969]

Source: Mellor and Lele [1973].
[a] Peso in the case of Philippines and Taka in the case of Bangladesh.
[b] Labor "payment" is defined as physical labor input (family and hired) in man-days at a constant wage.
[c] Other inputs "payments" are defined as gross value of output minus share to labor.
[d] U.P. refers to state of Uttar Pradesh in India.
[e] Millet.

doxical situation of "immiserizing growth," under which rice producers and workers are actually worse off than they were before the new rice production technology was adopted. Rice, which is a staple food in Asia, has limited international trade prospects. Domestic demand for rice has relatively low income and price elasticity. Improved varieties and use of modern inputs has increased rice supply faster than demand, pushing the price of rice down and leading to lower income for rice producers. Public policies that facilitate the transfer of resources to other productive activities can alleviate the problem. As a result, there is a need for rational agricultural diversification out of rice, and development of nonagricultural activities that will shift labor out of the agricultural sector. Some of these issues have also been addressed by Gonzales [1987] in the context of rice production and regional crop diversification in the Philippines.

Quizon and Binswanger [1986] analyzed the impact of agricultural growth and government policy on income distribution in India. The study is based on a limited general equilibrium model and deals with all-India data for a twenty-year period from 1961 to 1981. Based on an empirical analysis, they conclude that income gains from the Green Revolution initially accrued to the wealthier rural groups, but after 1972/73 they were transferred to urban consumers. By 1980/81 the per capita income of poor and wealthier rural groups alike were barely above their respective 1960/61 levels. They propose a reduction in population growth and an increase in nonagricultural employment and income to convert agricultural growth into reduced poverty.

POVERTY ALLEVIATION STRATEGY

Programs for alleviating rural poverty logically should emphasize helping low-income small farmers gain access to new production technology and inducing expenditures from the increased incomes of the landowning classes that would stimulate the demand for labor.[69] If access to goods and services is to be broadened, the incomes of low-income families must be increased. Most of such increases must occur by increasing employment and labor productivity, particularly in countries with the lowest per capita incomes.

Experience shows, however, that development and investment strategies that raise production do not necessarily cause a commensurate increase in the employment or incomes of low-income people. An increase in foodgrain production may not provide enough direct increase in employment to create adequate demand to maintain price ratios, even though grain is an important staple of low-income families in many countries. Conversely, a large increase in the income of the lower income majority caused by an increase in employment cannot be sustained in low-income countries unless there is a commensurate increase in the production of food and other goods to meet the greater demand allowed by higher incomes. A practical strategy that increases the productivity and incomes of low-income families is complex and must be adapted to the different factor endowments and levels of development of countries.

The critical point to keep in mind in assessing the effect of new technology on poverty is that growth in population and hence of the labor force is a major force increasing poverty. The principal offset is migration to nonfarm employment. The combination of the high rate of population growth and little growth in non-farm employment is too powerful a negative factor for technological change to offset. But if the multiplier effects of new agricultural technology on nonfarm employment are strong, a major reduction in poverty can occur. It is for policy to encourage those nonfarm employment multipliers.

Major inequalities in the ownership or control of land and other productive assets impose significant obstacles to economic development. Programs that expand production in such an environment tend to exclude many of the poor from production and consumption. This holds back growth, does little or nothing to alleviate poverty, and aggravates economic and social inequalities.

To enable the poor to participate in the growth process, low-income countries may need either to redirect public expenditures or to redistribute land. Either choice is likely to be viewed as an unacceptable short-run sacrifice by powerful urban and rural elite groups. Appeals to a long-run concern for the existing system may not be able to overcome or modify these views.

A major set of problems facing the landless and the poor is the lack of their own capital and their inability to obtain institutional credit for any productive activities. Banks often discriminate against them because they do not have land or any other assets for collateral. In this context, the experience of the Grameen Bank in Bangladesh—which provides credit to alleviate rural poverty for the landless—is very instructive for other developing countries as well as Bangladesh [Hossain, 1988a]. The Grameen Bank was established as a specialized financial institution in 1983 in order to provide credit to the rural poor. In 1987, there were 298 branches, and women accounted for 74 percent of total membership. Loans are given for noncrop activities. Out of all the loans granted in 1986, 46 percent were for livestock and poultry, 25 percent were for processing and manufacturing, and 23 percent were for training and shopkeeping. The loan recovery rate was excellent since less than 1 percent of the loans were overdue. The Grameen Bank has significantly contributed to the generation of employment (especially for women) and the alleviation of rural poverty. However, the cost of operations is considered rather high due to the low rate of interest and high cost of intense supervision. The cost may be high compared with commercial bank operations, but, for an objective comparative analysis this cost needs to be compared instead with the cost of other national programs that are designed to provide employment and alleviate rural poverty, particularly for rural women.

The World Bank has provided leadership in financing projects designed to alleviate rural poverty. At present, one of the major thrusts of the World Bank is gradually to alleviate poverty in low-income developing countries. The World Bank experience in rural development from 1965 to 1986 is summarized in World Bank [1988b]. The Bank defined rural development projects as those projects in the agricultural sector where 50 percent or more of the direct benefits were intended to go to poverty target groups. The operational goals of rural development projects were to improve productivity, employment, and income for the target groups as well as to provide a minimum acceptable levels of food, shelter, education, and health. In broader terms, the rural development strategy was appropriate and effective. Millions of rural people benefited from investment in rural infrastructure and food production. However, many often-ambitious targets have not always been met. Many valuable lessons have been learned and applied to later operations. The program continues to evolve and the challenge to alleviate rural poverty in developing countries remains undiminished.

4. Health and Education

IMPROVED HEALTH

Health epitomizes the complex interaction between production, employment, consumption, and human well-being.[70] But, the surplus labor in many countries lessens the importance of the argument that improving the health of the labor force would make it larger and more productive.

Wyon and Gordon [1971] found the death rate nearly twice as high among low-caste agricultural laborers as among high-caste owner-cultivators in a set of villages in relatively prosperous Indian Punjab. Levinson [1974], in another Punjab study—this one of infants aged six to twenty-four months—frequently found significant malnutrition in all income and caste groups, reflecting a lack of knowledge as well as resources. But malnutrition was much more widespread among the laboring classes than among the landowning classes. According to Chowdhury, Alauddin, and Chen [1977], during the 1971 and 1974 famines in Bangladesh, the population which was affected more than others included the young, elderly, poor, and the disadvantaged.

The children of the poor frequently suffer from poor health and high mortality rates because they lack proper care. Minkler [1970] noted that in both rural and urban India the need to supplement low incomes forces the majority of women from low-income households to seek jobs outside the home. As a result, their children may not receive proper attention. This leads to high infant mortality, which encourages a large number of births [Kocher, 1973]. Research by S. Kumar [1979] indicates that large-scale food subsidies improve substantially the food consumption of low-income households and the health of infants. Recent research on this subject has corroborated these early results [Kennedy and Alderman, 1987]. The research results suggest various ways to reduce the cost of food subsidies, including targeting them towards the poorest households, inferior commodities, and low-income regions.

There is increasing evidence that public activities, including education and public health measures, are necessary if improved health is to follow quickly after rising incomes and increased food consumption. There is a particular need for research on the factors that affect the relationship between increased food intake, nutrition, and health; on the extent to which these factors are subject to public intervention; and on the extent to which private actions may be encouraged by increased education. On the education factor, there is evidence that, at the very lowest levels of incomes, education has little effect on nutritional status and health [Bouis, 1990]. That may be because, at the lowest income level, there is simply not the resource flexibility to take advantage of improved education with respect to health and nutrition. However, as incomes rise, the impact of education on nutrition also rises.

RURAL EDUCATION

No effective poverty program can ignore education. Education increases access to jobs, production resources, and power. By improving labor productivity and the efficiency with which resources are used, it contributes to higher national income. As a rule, the poor are the least educated and get the fewest benefits from their schooling. They also fail to attend school, particularly in rural areas, largely because jobs requiring education are not readily available and the opportunity cost of the labor hours of family members is relatively high even in very poor families.[71] A review of literature in the past twenty years and discussion of issues dealing with education and development is provided by Psacharopoulos [1988a].

The relative costs of education are much higher for the needy because the income foregone for education is more important to them than for the upper-income classes. A number of studies show that the rates of return from investment in primary education are considerably higher than from investment in other levels of schooling.[72] These studies, however, are based largely on urban areas, where a much larger percentage of children attend primary schools that are substantially superior to those in rural districts.

It is likely, as analyses of educational processes proceed, that the returns to secondary education in the dynamic context of a modernized agriculture will turn out to be quite high. We already see in studies of management-intensive modernization practices, such as use of cross-bred dairy animals, that the returns to upper levels of education are high [Alderman, 1987]. One could make a reasonable generalization that, as technology advances, it becomes more complex and therefore it requires higher levels of education. It stands to reason that in developing countries where at any given level of education, instruction is relatively poor, it will take more years of such education to achieve a given result. In these circumstances, primary education may largely provide the tools for further learning, and it is at the secondary school level that one acquires the ability to make the judgments necessary for modernizing agriculture.

The contrasts in the enrollment of different economic classes in universities are similarly associated with greater costs and lower rewards for the poor. Full participation of the poor in education cannot be achieved simply by adding primary school facilities in rural areas suffering from economic stagnation and extreme poverty. Fundamental economic and, possibly, social changes are needed.[73]

5. Environmental Considerations

As agriculture modernizes, incomes rise and rural welfare increases. There is scope for improving the environment in which rural people live. It must be kept in mind that the bulk of rural environmental destruction, including deforestation and destruction of perennial grasses in arid lands, takes place as a result of growing populations of increasingly poor people [Mellor, 1988c]. Increasing poverty forces people onto land of lower and lower productivity.

S. Kumar and Hotchkiss [1988] analyze the consequences of deforestation for women's time allocation, agricultural production, and nutrition in hill areas of Nepal. The main causes of deforestation in Nepal are found to be the need for more land to grow food to sustain growing population, and fuelwood consumption. The main consequences of deforestation are found to be low agricultural productivity on existing cultivated land since more time is spent on collecting essential forest products, and rapid environmental degradation. Clearly, any strategy to deal with environmental concerns needs to be analyzed in the context of growth in population and poverty.

It is also important to keep in mind, when looking at environmental issues in the context of modernization and rural welfare, that values with respect to the environment probably differ little between low and high income people, whether located in developed or developing countries. Most people are quite risk-averse with respect to ill-understood changes in the circumstances in which they live. This leads to an environmental conservatism. Most poeple are concerned with intergenerational income transfers. They are concerned about their children and their grandchildren. This concern leads them away from practicing a discounted rate of return in their day-to-day life. Some of these issues, in the context of population policy and individual choice, are analyzed by Nerlove, Razin, and Sadka [1987].

There are three basic reasons why destruction of environmental resources is more rapid in rural areas of developing countries. First, intense poverty, which as indicated above, drives people onto marginal resources and leads to destructive land use patterns. Second, the low level of education in rural areas makes it difficult for people to apply complex modern practices. For example, there is a clear close relationship between adoption of integrated pest management with its requirement for careful insect population counts, and for quick response to changes in insect populations. And fertilizer-conserving practices include scientific soil testing, variation in fertilizer use from year to year according to the findings of those tests, and precise timing and placement in the application of fertilizer. All of these practices, which are important from an environmental preservation point of view, require widespread education through the secondary-school level.

Finally, an intricate grid of rural infrastructure is essential to sound environmental practices. This is because land and water resources vary in the rural sector over small geographic areas. Optimal use of these resources calls for variations in agricultural production systems, which may not be tuned precisely to the narrow demand in a small region defined by lack of infrastructure. To put the case simply, in an area of subsistence agriculture with a small amount of flat valley land and a considerable amount of steeply sloping land, annual grain crops such as maize are likely to be grown on both. As the potential opens up for specialization in trade, which accompanies good infrastructure, the level lands in the valley may be used for annual crops, but the hillsides can be planted to perennial crops, including fruit trees and grasses. This steep land may even be used quite intensively in combination with the flat valley lands. That is possible, however, only if the surplus production for the local area can be exported and other goods brought in.

Thus, we can expect environmental destruction to occur as long as rural populations are exceedingly poor, ill-educated, and without basic means of transportation and communication. Modernization of agriculture increases rural welfare generally and has substantial externalities with respect to the environment.

These theories of environment and sustainable agriculture, which propound that farming practices can and should be designed to maintain optimal crop yields indefinitely, has received a great deal of attention from scholars and policymakers in the last few years [Bunting, ed., 1987; Conable, 1989; FAO, 1988b; Grimshaw, 1989; Mellor, 1988c; Oram, 1988; Pimentel et al., 1987; Tisdell, 1988; York, 1988; Warford and Partow, 1989; T. J. Davis and Schirmer, eds., 1987; and World Commission on Environment and Development, 1987a, b]. However, the concept of sustainability is not precise and has different meanings for different people. As has been emphasized by Oram [1988], the causes and effects of major environmental concerns—such as deforestation, acid rain, eutrophication of water resources, ozone depletion, and the greenhouse effect—transcend economic, social, and geographic boundaries. Agriculture is both a contributor to the erosion of sustainability and a victim of other environmental abuses caused by industry, rapid population growth, and urbanization.

Agricultural resource management, especially of agricultural land and water, is extremely important and must be included in any strategy of agricultural development. This is particularly important for developing Asian countries with high population pressure and limited supplies of agricultural land and water. According to the results reported in Pimentel *et al.* [1987], in the United States, for example, soil erosion averages about eighteen tons per hectare per year. About half of the forty-five million tons of fertilizer applied annually in the United States are replacing the soil nutrients lost by erosion. Pimentel *et al.* [1987] find that soil erosion and associated water runoff cost the United States about $43.5 billion dollars annually in direct and indirect effects. The long-term environmental and social costs may be several times this level. Clearly, it would pay society to invest in soil and water conservation. These problems in tropical Asia are even more serious, and the poor developing countries can ill afford such physical and economic losses. Both conventional and unconventional strategies for soil and water conservation need to be promoted. For example, according to Grimshaw [1989], the use of vetiver grass has proved to be an effective vegetative alternative for reducing soil erosion and preserving soil moisture. It is also environmentally sound, less expensive, and has the potential to make a large contribution to sustainable agriculture.

6. Women in Development

The influence of women is, of course, pervasive in the household and rural development context. Women provide a major part of the labor force in the rural sector, participate in a wide range of household decision-making activities, and play a particularly critical role in family welfare circumstances. Presumably, no aspect of rural development fails to have a female component. For that reason, instead of drawing attention to women's roles as we proceed, we refer the reader to a particularly perceptive discussion on women and structural transformation by Lele [1986]. Specific guidelines to help policymakers and development institutions bring women into the economic mainstream are outlined in Herz [1989].

Having emphasized the pervasiveness of the role of women in the modernization process, one should pay specific attention to critical aspects of the process from which they may be excluded, either accidentally or by design. For example, the importance of education increases in the process of modernization. Education rapidly becomes the most important asset, particularly for lower income people. If women are systematically excluded from formal education, they are excluded from ownership of one of the key assets in the development process. It is not uncommon for the opportunity cost of young girls' time to be higher than that of boys, primarily because of their role in child rearing of siblings, and therefore, girls are often withdrawn from school at an early age. This has major repercussions not only for their own development, but also in generational terms because of the critical role of women in molding the next generation.

Women may also have problems with respect to ownership and control of assets, including land. This again may inhibit women, particularly those who are single either by choice or due to the deaths of male family members. Because of the critical role of women in the family and household, they may be more restricted in their geographic movement. This means that, if women and young girls are to be included in the processes of education and of development generally, particular attention may have to be given to how they can be integrated into complex institutions while still fulfilling heavy household responsibilities. Such considerations are particularly important because unthinking male domination of institutional structures may in fact exclude women, even when movement out of the household and related factors are not at work. Thus, it is often necessary to consider the role of women explicitly.

Chapter VIII. Growth Linkages and Agricultural Development

Poor performance of agriculture is widely regarded as a retardant to economic growth, but the fact that rural areas can stimulate growth is rarely recognized. According to Hirschman [1958, pp. 109-110]: "Agriculture certainly stands convicted on the count of its lack of direct stimulus to the setting up of new activities through linkage effects—the superiority of manufacturing in this respect is crushing."[74] Traditional agriculture may have diminishing returns and increasing costs, but the conclusion that increased agricultural production must become a drain on the productivity of other sectors ignores the potential and the implications of modernizing agriculture more rapidly through technological change. Increases in the efficiency of technologically advanced farming make possible large net increases in national income, which provide growth with positive consumption as well as production multiplier and linkage effects.

1. Sectoral Linkages and Development Strategy

Hirschman [1958] operationalized the linkage approach to economic development. In Hirschman's framework, investment dominated economic development. The economic development strategy he proposed relied primarily on linkages to induce and facilitate investment. More recently, Hirschman [1977, p. 80] provided a broader definition of linkages: "development is essentially the record of how one thing leads to another, and the linkages are that record, from a specific point of view. They focus on certain characteristics inherent in the productive activities already in process at a certain time. These ongoing activities, because of their characteristics, push, or more modestly, invite some operators to take up new activities. Whenever that is the case, a linkage exists between the ongoing and the new activity."[75]

The linkage theory of economic development has been operationalized to measure the forward and backward linkages, and provide guidelines to policy-makers to induce investment in those industries with strong linkages and remove obstacles from industries with potentially strong but existing weak linkages. The Leontief input-output framework is the basis for measuring alternative linkage indexes and sectoral interdependence.[76] The direct backward linkage index measures the amount of intermediate inputs required from different sectors in the economy to produce one unit of output in any one sector. The direct forward linkage index measures the amount of output from any one sector supplied for intermediate use to different sectors of the economy as a proportion of its total demand.

Hirschman [1958] argued that backward linkages are better guides to the design of economic development strategies than forward linkages, since increased demand for intermediate inputs is a better stimulus than increased supply. In Hirschman's framework, sectors with high forward *and* backward or high backward linkages should be preferred to agriculture, which, according to Chenery and Watanabe [1958], possesses high forward and low backward linkages. Modernized agriculture, however, promises higher potential linkages than traditional agriculture. Furthermore, the empirical evidence generated by Yotopoulos and Nugent [1973] from cross-country time-series data does not support the extreme version of the linkage hypothesis, according to which countries that give high priority to high-linkage industries have higher rates of growth than those that give low priority to them. According to Mudahar [1982], however, since the linkage indexes are calculated from input-output tables based on highly aggregated data, one should be very careful in interpreting these indexes and using them to guide sectoral planning.

2. Growth Linkages and Agricultural Development

Mudahar [1982] classified sectoral growth linkages in agriculture into four categories: production linkages, consumption linkages, investment linkages, and employment linkages.[77] The Asian Development Bank [1977] estimated production linkage coefficients for India and found that crop and animal husbandry has "weak" backward and "medium-strong" forward linkages, and that agroprocessing sector has "strong" backward and "medium-weak" forward linkages. Experience since the Green Revolution demonstrates that modernizing agriculture possesses strong backward and forward linkages.[78] This is corroborated empirically in an econometric model of the Indian economy by Rangarajan [1982]. Critical to the assessment of agricultural sector linkages is the role of technological change in making a *net* addition to income which may provide its stimulus through increased *consumption* expenditure.

It is common to think of a pattern of demand either as fixed or as malleable only as a welfare measure. The contrasting view, presented in Mellor [1976], is that demand can be manipulated to increase production—for example, by investment policy, including emphasis on peasant agriculture and by taxation policy. That is, it can be manipulated to accommodate capital shortages (by structural changes towards a labor-intensive product mix) and other obstacles to expansion, to increase the productive use of land, and to increase the consumption of foodgrains enough to sustain the price in the face of an increase in agricultural production. Crucial to this view of growth is the potential for a major increase in the national income through efficiency-increasing technological change in agriculture and the mobilization of underutilized labor by the expansion of effective demand, particularly in the service and small-scale manufacturing sectors. The growth of such activities has important locational implications—what activities occur where and in what order—which a facilitating policy needs to grasp [Wanmali, 1983, 1985].

To plan effectively to meet increased demand and to correct imbalances between the production and consumption of different goods, the patterns of expenditure of additional income must be known. Increased expenditures on production inputs and capital goods for agriculture may have strong domestic growth linkages. Such investments are largely for fertilizer, pesticides, improved seeds, irrigation works, and labor-saving machinery. Fertilizer and pesticides tend to have a large import content. Many mechanical items can be produced in small-scale enterprises with low capital-labor ratios.[79] In the long run, however, 60 to 80 percent of additional rural income is spent on consumer goods. A comparative analysis by Hazell and Röell [1983] brings out the strong linkages of middle peasants in Asia with growth in other sectors. The nature of new foodgrain technologies and the availability of resources determine the distribution of the additional income.

Increases in per capita income and growing disparities in income distribution have important implications for households in different expenditure classes and for growth linkages. They cause marketable surplus of the higher expenditure classes to increase, which increases the proportion of incremental income they spend on nonfoodgrains, processed foods, and nonagricultural goods and services.[80] The production processes of most of these goods and services are highly labor intensive, and so can create large increases in the demand for labor.

The linkages from increased foodgrain production cannot have their full stimulative effect on growth unless restraints on production in the domestic consumer goods sector are removed. Growth in industrial production may be constrained by institutional barriers, particularly within capital, input, and output markets.

Public policy must analyze these barriers and make appropriate adjustments. The nonfoodgrain agricultural sector (dairy, vegetable, and fruit production) is the most important beneficiary of demand increased by rising incomes. It is in many respects the most attractive sector for expansion because it is labor intensive and has a geographically dispersed pattern of demand and production. As the process of economic development gathers momentum, the rapid growth in demand for this subsector of agriculture and its lack of land constraint allows an acceleration in growth well beyond what the land-bound basic food staples sector could sustain—four to six percent growth rates, rather than two to three percent.

While demand generated by agricultural expansion allows shifts in the composition of industry, shifts that reduce the capital required for each employee, and other elements of the process, can greatly increase savings and investment. Added savings may not only finance much of the larger capital needs of agriculture but may also finance part of an expansion of the nonagricultural sector.[81] The extent to which agriculture supports investment in other sectors depends on the net capital requirements for agriculture's own increased production, the development of institutions for transferring agricultural savings within agriculture or to the other sectors in the economy, the economic returns from capital in those sectors, and the form of institutional growth in those sectors.

The basic components of the rural-led employment-oriented strategy of growth were developed fully in Mellor [1976]. The most important principle in the argument was that the supply of foodgrains as wage goods is a major constraint on employment growth, but that attainable increases of foodgrain production can allow employment to grow significantly faster than in the past. The analytical framework for the strategy was developed in Lele and Mellor [1981]. The test of key quantitative relationships of rural-led growth strategy was provided in Mellor and Mudahar [1974a, b]. The simulation confirms that foodgrain production is a constraint to employment, that a demand-derived expansion of nonfoodgrain agriculture is important, and that the choice of technology is significant in determining these forces.

Chapter IX. International Trade and Resource Transfers

1. Resources for Economic Development

The ability of low-income countries to develop rapidly with broad participation in that development is strongly influenced by their international trade performance, the flow of capital resources, and other elements of the international environment. Trade based on comparative advantage makes productivity increases possible through specialization. In low-income countries, this tends to increase employment opportunities by encouraging them to produce labor-intensive goods and to import capital-intensive goods. Specialization according to comparative advantage conserves the scarce capital resources of low-income countries, making them available for employment-oriented production. For the developed countries, it expands markets for capital-intensive goods and raw materials. Thus trade policies interact with other elements of the development strategy, including an emphasis on agriculture, the structure of industry, and the growth of employment. All too often, developing countries have misallocated resources to capital-intensive industries, even to the point of subsidizing substantial exports of such commodities [Mellor and Johnston, 1984; Mellor and Lele, 1975].

Foreign resource transfers may facilitate a participatory growth strategy by allowing an increase of the rate of investment without a commensurate reduction in consumption [Mellor, 1976]. Similarly, foreign transfers may facilitate development programs that increase employment and raise the consumption of necessities by low-income people. Simultaneously, they add to the amount of foreign exchange available and thus facilitate the adjustments most developing countries have to make in their internal structures before they can realize their trade potential. On the other hand, developed countries that are unable to make the adjustment needed to allow an increased volume of low-income countries' imports in their markets, can transfer resources which, because they add foreign exchange available to low-income countries, are often a convenient, if second best, alternative.

There are significant differences in the nature and mix of the resource transfers appropriate or possible for low-income countries. If trade and resource transfers are to succeed in influencing the choice and pattern of development, the nature of the relationships, the relative advantages of trade and different resource transfer instruments, and how these instruments can be used to achieve particular objectives must be clearly understood.

2. Agricultural Trade and Economic Development

Low-income countries are emphasizing their trade needs even more than resource transfers, because trade is far more important than foreign aid as a source of foreign exchange, and because foreign exchange obtained through trade carries fewer of the conditions commonly associated with aid or private capital. This increases the relevance of a number of trade issues for research and development policy. Most important, trade and development strategies are so closely related that any consideration of aid when choosing strategies must also incude a look at trade.

AGRICULTURAL TRADE PATTERNS AND GROWTH STRATEGY

Tolley and Gwyer [1967] reported that the shares of agricultural products in the exports of developing countries are higher than those of developed countries.[82] Furthermore, in some countries one or two crops account for most of the earnings from agricultural exports. India's exports in the 1950s were weighted heavily toward such agricultural commodities as jute and tea [Mellor, 1976]. The belief that the prospects for the growth of these commodities were poor led to an import displacement policy, in spite of the rising capital intensity of such a strategy. Initially displaced imports may be produced domestically by labor-intensive techniques,[83] but over time the imported goods to be displaced are increasingly capital intensive. However, in contrast to India's first two plans, subsequent plans emphasized exports explicitly.[84] Generally, however, export promotion is still done in the context of a foreign exchange regime that discriminates strongly against agriculture [see, for example, Bautista, 1987].

The large demand for imports of raw materials and capital goods and the uncertainties of foreign aid for financing them indefinitely led to many policy changes in developing countries, but the progress towards more liberal trade regimes has been slow. For example, in the 1960s, India instituted a variety of export subsidies and licensing preferences, the complexity and inefficiency of which precipitated the 1966 devaluation. Following that, exports reversed the declining trend of the 1950s. However overall, during the 1950s, India's imports grew at a rate comparable to Taiwan, the Philippines, and Hong Kong. The sharp contrast between the imports of India and the other countries occurred in the 1960s. While most countries rapidly increased the growth rate of their imports in the 1960s, India's actually declined. This represented a loss of comparative advantage, reflected in poorer growth and poorer export performance [Mellor, 1976].

To provide the means of payment for imports, exports must be at the core of a strategy to support increased employment with imports of necessary capital-intensive goods. In India, one of the primary arguments for the capital-intensive approach to development is based on pessimism about export prospects. According to Bhagwati and P. Desai [1970], this is usually grounded on the expectations that demand for primary commodities will grow poorly and that barriers against imports of manufactured goods by high-income countries will rise.[85] According to Mellor [1976], however, the evidence, particularly of the 1960s, does not support the gloomy view of exports of low-income countries generally or of India specifically. More recently, Bhagwati [1988] discusses issues and provides evidence in support of "export-promotion" as opposed to "import-substituting" strategy.

The efforts of less developed countries to industrialize seem to have paid off by rapid advancement of exports of manufactured goods. The emphasis on manufacturing may also have resulted in temporary neglect of the primary commodity categories and a consequent loss of productive output and export potential. This neglect has probably been most evident in agriculture and has taken the form of underinvestment in cost-reducing technological change.[86] Most importantly, the trade policies favoring capital-intensive industry resulted in an over-valued exchange rate that was deleterious to agriculture.

A set of research studies which were conducted under the umbrella of the "Stanford Project on the Political Economy of Rice in Asia" dealt specifically with the comparative advantage of rice production in Thailand, Philippines, and Taiwan.[87] The results are summarized in Monke, Pearson, and Akrasanee [1976]. Akrasanee and Wattanaukit [1976] concluded that Thailand had a strong comparative advantage in rice production but the taxation system discriminated against its expansion. Herdt and Lacsina [1976] concluded that in the Philippines rice production would be preferable to importation if the long-term price of rice is $600 per mt and input prices remain at their 1974 level. On the other hand, if the long-run price of rice is below $280 per mt and the input prices remain at their 1974 level, the country would be better off to import additional rice requirements. Similarly, Wu and Mao [1976] concluded that for Taiwan, self-sufficiency in rice may be justified if the world price of rice remains high. K. Anderson and Ahn [1984] found that South Korea's advantage in food production is declining, and the cost of protection policy is increasing. K. Anderson and Hayami [1986] provide a comprehensive review of protectionist agricultural policies in East Asia, emphasizing the very high levels of agricultural protection and its relation to declining comparative advantage in agriculture as industrial productivity rises very rapidly.

The lessons for India on exports are similar to those of most Asian, African, and Latin American countries. Trade relations with a richer and more powerful partner had to be laboriously changed before trade could have a vigorous role in development. According to Hecksher [1919] and Ohlin [1933], trade between countries with unlike proportions of factors of production reflects one of the more plausible patterns of trade. According to this view, however, India would produce labor-intensive commodities. In practice, however, India's pattern of industrial growth has been highly capital intensive.[88] There is also a tendency for industries with greater capital intensity to expand exports faster, although the average increase in the capital intensity of exports was somewhat less than for the economy as a whole.[89]

Agricultural trade and protection policies in Asia are reviewed by DeRosa [1988]. During 1985, the Asian countries accounted for about 15 percent of world trade in agricultural commodities (Table 17). China, India, Indonesia, and Malaysia account for the major share of the region's agricultural exports. Overall, Asia's imports of agricultural commodities are much more than exports. Japan alone accounts for over 40 percent of the region's agricultural imports. Other major importers are Hong Kong, the Republic of Korea, Singapore, and Taiwan. The existing patterns of trade indicate that Asia's strongest trading relationship in agriculture is with industrial countries outside the region.

Table 17. Agricultural trade of Asian countries by commodity groups, 1985[a]

Commodity division	World exports	Asian countries		Direction of trade (in %)[b]		
		Exports	Imports	Japan	Other Asia	Other industrial countries[c]
		Billion US$				
Food	146.6	19.7	24.3	11	31	46
Meats, fish, dairy products	42.3	5.5	7.7	24	29	40
Cereals	32.5	2.2	6.9	2	24	47
Fruits, vegetables	25.2	3.6	2.5	13	45	41
Sugar	5.8	0.8	0.9	8	52	26
Seeds, oils	19.8	4.4	4.6	5	39	47
Beverages, spices	21.1	3.3	1.9	4	31	31
Raw materials	49.2	9.1	13.4	15	30	46
Tobacco	3.8	0.3	0.7	2	6	79
Rubber	6.7	3.7	1.7	12	49	31
Wood	14.4	2.9	4.5	27	33	34
Natural fibers	11.1	1.9	3.7	12	33	46
Others (hides, pulp)	13.1	0.3	2.8	8	26	60
All commodities	195.8	28.8	37.8	13	31	46

Source: DeRosa [1988]; originally from World Bank's Trade, Analysis, and Reporting System.
[a] The data cover the trade of fifteen major Asian economies, including China and Taiwan Province of China. The underlying data for China are for 1984.
[b] Average percentage share in Asian countries' exports and imports.
[c] Consists of Australia, Canada, New Zealand, the United States, and the industrial countries of western Europe.

AGRICULTURAL PROTECTIONISM AND ECONOMIC DEVELOPMENT

According to DeRosa [1988], *ad valorem* tariffs are the most common form of restriction imposed against imports of food commodities and agricultural raw materials. Hong Kong and Singapore impose virtually no tariffs. On the other hand, the highest tariffs in the region are imposed by South Asian countries. Furthermore, tariff rates are generally higher for food than for agricultural raw materials, indicating a strong desire for national food self-sufficiency.

However, agricultural protection policies are much more prevalent in the developed industrialized countries of the world. This area has been extensively researched, and the literature is immense. Selected studies that deal with Asia or have trade implications for Asian countries include K. Anderson [1981, 1983a, b, 1989b]; K. Anderson and Ahn [1984]; K. Anderson and Tyers [1985a, b, c, 1987b]; K. Anderson and Warr [1987]; K. Anderson, Hayami, *et al.* [1986]; Binswanger and Scandizzo [1983]; Fitchett [1988]; Koester [1985]; T. C. Miller [1986]; Tyers and K. Anderson [1986, 1987, 1990]; Winglee [1989]; and the World Bank [1987c].

The empirical evidence developed by Binswanger and Scandizzo [1983] and reviewed by D. Gale Johnson [1988] indicates that there is a systematic relationship between the level of protection (positive, which implies subsidy; or negative, which implies tax) and the characteristics of countries, commodities, and the structure of farming. The protection coefficients are positively related to the level of per capita income, and commodities produced on the larger or more specialized farms. On the other hand, the protection coefficients are negatively related to the percentage of the countries' labor force engaged in agriculture or, alternatively, the percentage of GNP produced by agriculture; the amount of agricultural land per capita; the value of agricultural exports per capita; and the products classified as tropical beverages. Demand for agricultural protection increases as the share of agriculture in GNP declines and per capita income increases. However, at high levels of per capita income, agricultural protection becomes affordable. Asia, Japan, and now the Republic of Korea fall in this category. On the other hand, at low levels of per capita income, agriculture is an important source of revenue for government.

A survey of agricultural trade and protection policies in Japan is available in Fitchett [1988]. Japan has emerged as one of the major importers of agricultural commodities in the last two decades. Japanese farmers have benefited from various protection policies. The reasons cited for these protection policies include food security, rural-urban income parity, and smoothing of the sectoral adjustment process. The nominal rates of protection (as measured by the nominal protection coefficient, NPC) for seven selected agricultural commodities in Japan, EC-10, and EFTA member countries are provided in Table 18. The results indicate that during 1980-82 the weighted average rates of protection were 133 per-

Table 18. Nominal rates of protection for selected agricultural commodities in Japan, EC-10, and EFTA, 1980–82 (in %)

Commodity	Nominal rate of protection		
	Japan	EC-10	EFTA[a]
Rice	235	40	0
Wheat	290	40	65
Coarse grains	330	40	55
Beef and lamb	180	25	130
Pork and chicken	50	25	40
Dairy products	190	75	145
Sugar	200	50	55
Weighted average	133	55	90

Source: Fitchett [1988]; originally from K. Anderson and Tyers [1987b].
[a] European Free Trade Association member countries.

cent, 90 percent, and 55 percent in Japan, EFTA, and EC-10 member countries, respectively. Earlier, K. Anderson [1983a] reported that during the period 1960–64 to 1980–82 the weighted average level of NPC for seven principal agricultural commodities in Japan increased from 68 percent to 151 percent. On the other hand, the average rate of protection for these commodities in the United States was about 16 percent [Tyers and K. Anderson, 1986].

Such protection policies discourage structural transformation of the economy into more efficient production systems. Furthermore, the financial and economic costs of agricultural protection policies are enormous both to taxpayers (in the form of budgetary subsidies to farmers) and to consumers (in the form of higher food prices). The average annual cost of agricultural support and protection policies for selected industrial countries during 1984–85 are summarized in Table 19. The direct costs to taxpayers are the highest in the United States (US$ 49 billion) and the European Community (US$ 25 billion). On the other hand, direct costs to consumers are highest in the European Community (US$ 42 billion) and Japan (US$ 35 billion). Direct costs to taxpayers and consumers of agricultural support and protection policies were about US$ 185 billion per year in 1984–86, which is equivalent to about 40 percent of the gross value to agricultural producers. These results indicate that in the industrialized, developed countries there is a transfer of substantial resources from the government and consumers to producers. On the other hand, in the developing countries of the world, there is a transfer from producers to the government and to consumers [Krueger, Schiff, and Valdes, 1988].

Table 19. Costs of agricultural support policies for selected industrial countries (annual average for 1984-86)

Country	Direct cost to			Producer subsidy equivalent (in %)[b]
	Taxpayers[a]	Consumers	Total	
	BillionUS$			
United States	49.1	17.1	66.3	28.3
Canada	3.0	2.7	5.7	39.1
Australia	0.6	0.7	1.3	14.5
New Zealand	0.4	0.1	0.5	22.5
Japan	7.4	34.9	42.3	68.9
Austria[c]	0.6	1.0	1.6	35.3
European community	25.2	42.2	67.2	40.1
Total	86.3	98.5	184.9	38.4

Source: Winglee [1989]. Originally from M. Kelly *et al.* [1988], based on data from the OECD.
[a] Net of budgetary receipts from tariffs.
[b] The subsidy that would be required to maintain producers' income at the current level if all support policies were removed; measured as a percent of the gross value to agricultural producers.
[c] Refers to 1984-85.

A comprehensive analysis of international trade issues, trade policy reforms, and protectionism in the context of industrialization is provided by the World Bank [1987c]. According to the World Bank [1988a], protectionism broadly declined up to 1974 as tariffs were cut under successive agreements of the GATT. Average import tariffs on manufactures declined from about 40 percent in the early 1950s to less than 10 percent in 1974. However, agricultural products and textiles—two major exports from the developing countries—remained the biggest exception to the trend towards more liberal trade. Furthermore, liberal trade has been seriously threatened since the mid-1970s.

TRADE BARRIERS AND INTERNATIONAL TRADE NEGOTIATIONS

There is great potential for exporting labor-intensive manufactured commodities from low-income, labor-surplus countries. But, realization of this potential depends on the effects the trade policies have on the efforts of industrialized countries, the efforts of low-income countries to expand labor-intensive exports, and on the attitude of industrialized countries toward low-income countries' policies to expand manufactured exports by using subsidies. Limits placed by industrialized countries on labor-intensive exports by low-income countries obviously have adverse implications for employment in low-income countries, for an employment-oriented strategy of growth, and hence on the domestic demand for food.

The concerns of producers in high-income countries should be addressed in ways that take into account low-income countries' interests. The objective is to obtain economic efficiency by encouraging low-income countries to adopt outward-looking growth strategies that support participatory development. The problem was less urgent for high-income countries when only a few low-income countries chose an export-oriented growth strategy. Now that more low-income countries are adopting such a strategy, it is important that the full implications be analyzed.[90]

Most high-income countries have a generalized system of trade preferences. In the multilateral trade negotiations, special attention is being paid to tropical products of interest to low-income countries. High-income countries have also made political commitments to insure that tariff-cutting formulas and other negotiating schemes to reduce nontariff barriers cover products of interest to low-income countries. It is important to analyze the effects such policies have on trade and to determine how effective these measures would be in promoting low-income countries' exports of labor-intensive manufactures, especially since many low-income countries' exports are produced by capital-intensive processes.[91] As rural modernization occurs, the comparative advantage of developing countries in labor-intensive agricultural commodities is bound to increase. Thus it is important that GATT negotiations in the late 1980s and 1990s open trade to these commodities.

As developed countries increase factor productivity in their agriculture and cease to expand domestic demand for cereals, the rapidly expanding export markets to developing countries become increasingly attractive. If those markets are to expand, the developed countries must be open to labor-intensive imports, including labor-intensive agricultural commodities.[92]

A basic instrument of the export promotion policies of low-income countries is subsidization of export industries. There is evidence that indiscriminate use of export subsidies tends to be inefficient. On the other hand, undiscriminating opposition by developed countries overlooks the legitimate needs for subsidies to offset the costs of entry into foreign markets, particularly when a low-income country is diversifying the legitimate needs for subsidies to offset the costs of entry into foreign markets, particularly when diversifying exports. The problem is how to devise international guidelines that discourage economically wasteful subsidies and encourage policies promoting labor-intensive exports from low-income countries.

Tariff and nontariff barriers to agricultural imports and possibilities for bilateral trade negotiation in Asia are summarized in Table 20. Tariff rates are among the highest in South Asia. Nontariff barriers, generally regarded as more trade-distorting than tariffs, are applied widely but more selectively by most of the Asian countries. The nontariff barriers include restrictive licensing, prohibitions, state trading, quotas and entry regulations on health and product standards. There are possibilities for achieving economic gains from liberalizing agricultural trade in Asia through bilateral or multilateral trade negotiations. The World Bank [1986a] has also suggested that liberalization of trade should be a high priority for international action in agriculture.

According to Valdes [1988] and Zietz and Valdes [1988], much of the trade in agriculture is not covered by the spirit or letter of the Generalized Agreement on Tariffs and Trade (GATT) rules. Nontariff barriers proliferate, export subsidies increasingly substitute for a natural competitive advantage and trade wars erupt with frightening regularity. There is some hope that the Uruguay round of multilateral trade negotiations can achieve more for agricultural trade than prior GATT rounds. Sharply rising budgetary costs of support and protection policies have made the United States, the European Community, and other industrial countries more open to the idea of agricultural trade reform than ever before. Lack of a reasonable agreement in the Uruguay round could be detrimental to agriculture in developing countries.

Finger and Olechowski, eds. [1987] have published a handbook on multilateral trade negotiations for the Uruguay round that provides needed information base for participating developing countries. Jalali, ed. [1989] has completed a

Table 20. Barriers to agricultural imports and possibilities for bilateral trade negotiations in Asia

Region/country	Commodity division	Tariffs[a]			Nontariff barriers					
		Frequency	Av. level (percent)	Bilateral negotiation possibilities[b]	Quantitative restrictions				State trading	Entry regulations
					Restrictive licensing	Quotas	Prohibitions	Bilateral negotiation possibilities[b]		
Japan (JA)[c]	All	–	11		24	12	3	KO	–	–
	Foods		15		31	15	2			
	Raw materials		7		1		6			
South Asia										
Bangladesh (BA)	All	97	62	PA, IO, MA, PH, TH	36		37	PA, KO, SI	1	2
	Foods	98	66		44		30		1	3
	Raw materials	94	57		18		56		3	
India (IN)	All	99	106	MA, PH, TH	40		71	IO, PH	19	3
	Foods	99	119		47		65		20	4
	Raw materials	98	92		26		83		18	
Pakistan (PA)	All	93	59	BA, MA, PH, TH	43	2	39	BA, MA	5	
	Foods	93	75		33	3	55		7	
	Raw materials	94	43		70		5			
Sri Lanka (SR)	All	95	34	IO, MA	9			KO	4	
	Foods	95	48		9				4	
	Raw materials	94	21		9				1	
Southeast Asia										
Indonesia (IO)	All	80	14	BA, SR, KO	26	23	50	IN, MA	2	4
	Foods	75	19		8	29	63		3	5
	Raw materials	100	10		98					
Malaysia (MA)	All	77	9	BA, IN, PA, SR	8		1	PA, IO, PH, TH	2	20
	Foods	68	8		8		2		3	29
	Raw materials	97	9		10					1

(continued on next page)

Table 20. Barriers to agricultural imports and possibilities for bilateral trade negotiations in Asia [continued]

Region/country	Commodity division	Tariffs[a]			Nontariff barriers					
					Quantitative restrictions				State trading	Entry regulations
		Frequency	Av. level	Bilateral negotiation possibilities[b]	Restrictive licensing	Quotas	Prohibitions	Bilateral negotiation possibilities[b]		
		(percent)								
Philippines (PH)	All	100	28	BA, IN, PA, KO	49	6	9	IN, MA, TH	3	71
	Foods	100	35		53	4	13		5	96
	Raw materials	100	21		39	10				11
Thailand (TH)	All	88	29	BA, IN, PA, KO	35	1	24	MA, PH	1	20
	Foods	86	36		40	0	32		1	24
	Raw materials	95	22		23	1				7
East Asia										
Hong Kong (HK)	All				8	1		KO		3
	Foods				8	2				4
	Raw materials				7					
Korea (KO)	All	100	21	IO, TH	30			JA, BA, SR, HK		
	Foods	100	29		38					
	Raw materials	100	13		7					
Singapore (SI)[d]	All				22			BA		9
	Foods				26					12
	Raw materials				13					

Sources: DeRosa [1988]; originally from United Nations Conference on Trade and Development's Trade Information System and the World Bank.

Notes: The data refer to the following years: 1985 (Pakistan, Thailand), 1986 (Bangladesh, Japan, Korea, Singapore, Sri Lanka), and 1987 (Hong Kong, India, Indonesia, Malaysia, the Philippines).

[a] The data refer to general or statutory ad valorem tariff rates.

[b] Possibilities for bilateral negotiations for main agricultural commodities are based on correlations relating indices of comparative advantage to average tariff levels and on frequency ratios of quantitative restrictions between all pairs of countries.

[c] The tariff data for Japan refer to Tokyo Round-bound rates. The nontariff barriers data do not include information about Japan's state trading and entry regulations.

[d] Singapore is grouped with the East Asian countries for analytical purposes.

second volume to follow the volume published in 1988 on research inventory for the multilateral trade negotiations for the Uruguay round. This research inventory provides a guide to recent and ongoing economics research relevant to the issues under multilateral trade negotiations.[93]

COMMODITY PRICES, MARKET INSTABILITY, AND TRADE

An employment-oriented strategy of growth can expand total production and exports steadily if labor supply is elastic. But, such a strategy implies substantial increases of imports under most circumstances. Therefore, low-income countries may become more vulnerable to fluctuations in export earnings caused by cyclical fluctuations in high-income countries or by other factors, especially since the low-income countries' exports tend to be concentrated by commodity or by market. Fluctuations in the demand of high-income countries tend to be reflected in export commodity prices and in the instability of low-income countries' net foreign exchange earnings. In the more developed low-income countries, instability is also reflected in changes in demand, volume, and earnings of manufactured exports.[94] The efforts of developed countries to stabilize their domestic food markets increases instability in international markets [Josling, 1980; Koester, 1982]. And, this is in the context of increasing food production instability in both developed and developing countries [Mellor, 1981; Hazell, 1982].

According to the World Bank [1986a], international market prices of major agricultural products vary more than the prices of industrial products (Table 21). The price instability indices for major agricultural products were over 10 and as high as 90 for sugar during 1964–84. On the other hand, the price instability indices for the majority of manufactured products were lower than 10 for the same period. The high variability in agricultural commodity prices explains, to some extent, why developing countries adopt various kinds of stabilization schemes to protect farmers from large price falls and consumers from large price increases.

In a recent study, Grilli and Yang [1988] find that from 1900 to 1986 relative prices of all primary commodities, relative to those of traded manufactured goods, declined on trend by 0.5 percent a year and those of nonfuel primary commodities by 0.6 percent a year. Morrison and Wattleworth [1988] analyze the relative contribution of supply and demand factors to sharp declines in the prices of primary commodities. The results indicate that rising supplies of food and larger production capacity of agricultural raw materials were the major factors responsible for depressing primary commodity markets in the 1980s, and particularly during 1984–86. Relatively low economic growth during this period in the industrial countries was another factor.

The effects of the instability of export earnings on the economic growth of low-income countries and on employment have often been explored. Yet there is little understanding of how instability of export or import prices and significant

Table 21. Price instability indices for world market prices of major agricultural commodities[a]

Commodity	International price instability index (%)	
	1964–84	1974–84
Sugar	90.8	51.5
Cocoa	37.3	34.1
Rice	33.0	21.9
Coffee	32.0	37.7
Palm kernels	27.5	32.5
Wheat	24.3	16.9
Tea	21.7	23.6
Jute	21.2	26.8
Soybeans	20.8	9.9
Beef	16.7	11.3
Corn	16.6	15.6
Rubber	16.1	14.0
Sorghum	15.6	13.6
Cotton	14.3	10.7

Source: World Bank [1986a].
[a] Price instability index measures the average deviation from the price trend in any particular year. Prices are mainly from the London and New York markets, and they are deflated by the manufacturing unit value (MUV) index (1984 = 100).

shifts in a country's terms of trade are transmitted within low-income countries; or of how they affect aggregate employment, employment in specific sectors, economic growth, and diversification.[95] These questions are of special importance when low-income countries try to increase employment and improve income distribution through concerted rural development efforts. Heavy dependence on earnings from commodity exports and major fluctuations in commodity prices may have major long-run effects on the ability of the country to mount and pursue a participatory strategy. These problems are particularly severe for thinly traded commodities such as rice [Siamwalla and Haykin, 1983].

Recent strong fluctuations in the prices of the export commodities of low-income countries and of crucial raw material and food imports, such as fertilizer and grains, dramatize the effect that commodity market fluctuations have on the development objectives of low-income countries. Rising food and fertilizer prices and scarcity have reduced the availability of food, especially in countries dependent on food aid. Wildly fluctuating fertilizer prices have adversely affected incentives for food production and posed complex problems of food pricing for the producers and consumers of low-income countries. But, one must be aware that many developing countries have tariff policies to protect their fertilizer industry [Mudahar, 1978]. Such policies can raise fertilizer prices and thus slow the process of agricultural modernization. M. S. Rao [1974], in a detailed analysis of the implications that implicit tariffs have for fertilizer in India, estimated that the cost of protection was large both in the form of additional production costs and in the form of lost agricultural output.[96]

TRADE ISSUES, INFORMATION NEEDS, AND POLICY

Analytical work on the role of agricultural trade in economic development and the interaction between trade and development strategy needs to focus on the following issues: first, internal adjustment mechanisms and low-income country policies, including policies on reserve holdings of food and commodities or financial assets, that can effectively cushion the effects of external shocks on internal growth and employment objectives; second, international actions such as buffer stocks that promote stability of trade in commodities and foodstuffs, and support policies of low-income countries to diversify production and adjust in response to abrupt changes in external demand and supply; third, domestic and international efforts to increase demand for the commodity exports of low-income countries, through appropriate institutional and promotional arrangements; fourth, existing market structures from the producer to the retailer to determine how to improve their performance; fifth, the implications that exploiting the resources of the sea have for internationally traded commodities of interest to low-income countries; finally, determination of the trade patterns for agricultural products.[97]

In addition, some of the concerns of high-income countries now receiving attention need further analysis of their implications for low-income countries. These include access to low-cost supply, the effects of instability in commodity prices on domestic consumers, and the usefulness of commodity arrangements. Several proposed schemes for international investment can also have important effects on supply, price, and aggregate export earnings. The policies high-income countries adopt to solve these problems will influence the prospects of low-income countries by affecting commodity export earnings and the rate and method of resource exploitation.

3. National and International Resource Transfers

FOREIGN ASSISTANCE AND DEVELOPMENT

Agriculture-based, high-employment strategies of growth are becoming more widely accepted. But, the changes needed in policy often require political decisions that can be taken more rapidly if some of the economic costs are absorbed by foreign resource transfer.[98] Foreign assistance includes food aid, loans, and grants from both bilateral and multilateral agencies. Resource transfers may also provide important investment resources to those sectors or activities that are responsible for developing crucial elements of the participatory strategy or that help meet basic human needs while domestic capabilities are being developed.

There is a need to analyze the most suitable amount, form, and terms of assistance.

It can be argued that foreign assistance given in support of a participatory development strategy should improve the long-term ability for raising domestic resources in the recipient country, encourage improvements in resource allocation, increase the employment and income of the lower-income majority, and help expand the provision of basic services to improve human well-being. Previous analyses of the volume of resource transfers needed to attain the objectives of low-income countries used the aggregate approach of the two-gap models.[99] These models suffer from serious deficiencies when used to forecast the long-term needs of low-income countries for resource transfers to achieve self-sustaining participatory growth.

Badly needed is an alternative to the two-gap analysis that would focus more on the distributive aspects of growth. Such an approach would need to take into account the investment needs of participatory strategies, the savings potential of a given distribution of income, the constraints to providing a posited amount of consumption by the low-income majority, and the effects aid and trade have on national savings under alternative international monetary and trade-policy regimes. This core of the proper methodology could then allow long-term resource needs to be quantified on the basis of an improved analytical framework.[100]

The source of resource transfers, their use, and the terms and conditions under which they are provided all bear on their effectiveness. Transfers have implications for employment growth and a country's long-term balance-of-payments position, and can directly influence the ability of the country to pursue its development strategy. Given the large resources available to private firms and the private capital market, questions must be asked about how they can contribute to development. These include how they affect overall growth, employment, the balance of payments and income distribution; how to obtain development resources; and what relationship such sources of finance have to the stage and strategy of development. Analysis is needed of the institutional and policy questions of access to private equity or bank capital and of those operations of the international monetary system that influence the supply of private capital to low-income countries.

TRENDS IN MULTILATERAL ASSISTANCE

The amount of assistance given as loans and grants, the relationship of present assistance to the servicing of past debt, and the relationship of these to the current stage of development of the recipient are changing rapidly and require current analysis. The effects of current assistance practices that tie procurement, commodity, or generalized sector loans to the development objectives of low-income countries should also be investigated. It is especially important to identify and

Table 22. Total World Bank lending for agriculture and rural development

Purpose	Allocation by Purpose	
	1975-79	1980-85
	(percent)	
Agricultural credit	14.2	17.5
Agricultural sector loan	1.4	6.2
Area development	25.2	20.4
Irrigation	32.1	30.6
Research and extension	5.1	4.3
Others (forestry)	21.9	20.9
Total	100.0	100.0
	(Billion US$)	
Total bank lending	38.02	81.17
Lending for agriculture	11.58	21.22
	(percent)	
Lending for agriculture	30.5	26.1

Source: World Bank [1986a].

analyze the distortions of development objectives introduced by assistance practices and to explore means of reducing or eliminating such distortions.

The World Bank is the principal donor and plays a leading role in providing development assistance to agriculture in developing countries. According to Lipton and Paarlberg [1989], the World Bank provided 29 percent of official resource flows to agriculture and rural development in 1980-83, and 36 percent in 1985, only slightly less than total bilateral resources for this purpose. During 1979-83, only 16 percent of bilateral aid reached agriculture as against 30 percent of Bank flows. In this context, the Bank provides a much needed corrective thrust against urban bias in national public investment and policy decisions prevalent in most developing countries, including Asia.

Agriculture and rural development has been, and continues to be, an important objective of the World Bank. World Bank lending for agriculture and rural development averaged about 27.5 percent of total Bank lending during 1975-85 (Table 22). The major focus of World Bank lending has been projects dealing with irrigation, drainage, area development, and agricultural credit. The Bank finances only part of the project costs. According to the World Bank [1986a], the $33 billion it lent for agriculture during 1975-85 has helped finance a total investment of about $87 billion. The Bank's experience has demonstrated that economic rates of return for agricultural projects are comparable to projects in other sectors. Furthermore, agricultural projects have been successful in raising agricultural productivity and food production, increasing rural employment, and improving income of the rural poor. This has been particularly true for the Asian region which also accounts for the largest share of total lending for agriculture (Table 23). However, this is much smaller than Asia's share in total population, agricultural population, rural poor, and poverty [Lipton and Paarlberg, 1989].

In addition to project lending, the Bank has also been involved in supporting sectoral and structural adjustment programs since 1979. The sector adjustment loans (SECALs) and structural adjustment loans (SALs) are designed to address sector-specific policy issues and broad economy-wide policy reforms, respectively. The share of agricultural sector loans in total agricultural lending has increased from 1.4 percent in 1975-79 to 6.2 percent in 1980-85, and is even higher at present. The initial impact of SALs and SECALs has been positive. One must realize that restructuring the economies and reforming existing agricultural and economic policies can be a long and difficult process. However, it is essential to carry out needed economic and policy reforms in order to improve the overall economic environment for investment. Otherwise, investment in agricultural projects cannot be effective in achieving stated national goals. The policy issues that different SALs and SECALs address include reforming various pricing, subsidy, exchange rate, and trade policies as well as reduction in government intervention in various production, marketing, and trade activities. An evaluation of the Bank's adjustment lending is provided in World Bank [1988c].

As shown in Tables 22 and 23, the relative share of Bank lending for agriculture and rural development has declined over time. This declining trend in the share of agricultural lending, which does not appear to be deliberate, has become even more pronounced in recent years. However, the degree of observed decline in the relative share of lending for agriculture depends on the definition used in estimating agricultural and rural development lending. The issue of lending for agriculture—sources of decline, amount, share, components, and future allocations—is currently being debated within the Bank. As has been discussed earlier, the agricultural sector makes a substantial contribution to economic development in most developing countries, including Asia. Over three-fourths of the population lives in rural areas. Hence, any reduction in agricultural lending, unless it is compensated by other donors or by the national governments, can have an adverse impact on agricultural development and rural welfare. In order to address effectively the Bank's current initiatives of poverty alleviation, women in development, and the environment, there is need to reverse the declining trend in the relative share and increase Bank lending for agriculture and rural development.

FOOD AID TRENDS AND IMPACT

Food aid represents a particularly important basic issue. It provides real resources and public sector revenues and can particularly help employment growth

Table 23. Amount and regional shares of World Bank lending to agriculture[a]

	Bank lending[b] (millions US$)			Agriculture's share in total lending (in %)		Regional shares in narrow definition of agricultural lending (in %)				
	Lending for agriculture									
Fiscal year	Total	Broad definition[c]	Narrow definition[c]	Broad definition	Narrow definition	Africa	Asia	EMENA[d]	Latin America	Past borrowers
1970	2186.1	426.4	426.4	18.9	18.3	13.4	38.1	16.4	22.0	10.2
1971	2505.2	419.2	391.4	17.0	16.3	13.9	46.6	15.3	15.7	8.5
1972	2965.9	436.3	411.0	19.6	17.3	19.4	35.0	15.1	20.8	9.7
1973	3408.0	937.1	766.6	21.5	19.2	21.9	35.5	16.4	18.8	7.4
1974	4313.9	955.9	922.9	27.1	24.9	23.5	29.6	15.8	23.3	7.9
1975	5895.9	1857.6	1823.6	26.1	25.2	19.3	36.8	18.3	20.6	5.0
1976	6632.4	1927.6	1540.1	29.6	28.1	18.1	40.2	19.5	19.2	3.0
1977	7066.8	2307.9	2126.9	32.0	29.5	13.5	46.0	21.0	17.7	1.8
1978	8410.7	3269.7	2962.7	32.2	29.6	14.2	45.2	20.8	18.6	1.2
1979	10010.5	2521.8	2362.3	31.4	28.5	12.9	49.3	21.1	16.0	0.7
1980	11481.7	3468.4	3054.4	28.6	26.2	15.4	46.9	18.7	18.6	0.4
1981	12291.0	3763.0	3495.4	28.1	25.7	16.0	46.6	16.4	20.9	–
1982	13015.9	3078.4	2889.4	26.6	23.7	15.8	43.5	18.1	22.6	–
1983	14477.0	3698.3	2944.5	23.9	20.3	16.2	45.6	18.0	20.1	–
1984	15522.3	3472.9	2848.8	24.7	19.9	14.3	50.3	19.0	16.3	–
1985	14384.4	3749.3	3003.2	25.9	20.2	13.4	50.0	18.9	17.7	–
1986	16318.7	4777.4	3497.4	24.0	19.3	12.7	44.2	17.1	25.9	–
1987	17674.0	2930.3	2736.0	23.1	17.5	15.1	43.7	12.7	28.5	–

Source: Lipton and Paarlberg [1989]; originally from Agriculture and Rural Development Department of the World Bank.
[a] Agriculture here refers to agriculture and rural development.
[b] Three-year moving average, centered on the year shown.
[c] Broad definition corresponds to the World Bank definition of loans for agriculture. Narrow definition excludes both sector loans in agriculture and loans for agroindustry.
[d] Europe, Middle East, and North Africa region of the World Bank.

by assuring the supply of wage goods. The disincentive effects of food aid seem not to be significant for most countries.[101] Food aid represents a very low-cost source of resource transfer if it is seen as a means of price discrimination among markets with differing price elasticities [Mellor, 1983].

Food aid plays an important role in the economies of developing countries. During 1981-83, total food aid to developing countries was about $2.5 billion a year and accounted for about 9.4 percent of all official development assistance [Mellor, 1987]. The trends in food aid and cereal imports for developing world regions are summarized in Table 24. The results indicate that while cereal imports by developing countries have increased dramatically over the last twenty years, food aid has declined both absolutely and on a per capita basis. Furthermore, the share of food aid in total cereal imports has also declined over time. The decline has been particularly pronounced in Asia. Two other trends in food aid have emerged: sources of food aid have diversified, and over 25 percent of food aid is now being channeled through international agencies like the World Food Programme.

According to H. Ezekiel [1988], food aid can be classified into four categories: program food aid, project food aid, emergency food aid, and adjustment food aid. However, most of the analytical discussions on food aid do not make a clear distinction between different types of food aid. There appears to be a consensus that food aid has made an important contribution to food security, nutrition, employment, and economic growth in the developing world. Food aid can, and does, help provide the means needed to protect (and raise) the consumption status and labor productivity of the poor. This is particularly relevant in Asia. Using an applied general-equilibrium model, Srinivasan [1989] has shown that a well-designed and efficiently implemented food-for-work program can virtually eliminate abject poverty in India at a modest cost.

In many of the developing countries, especially in Asia, national food stocks tend to be very large and expensive because of food security concerns, partly due to the random occurrence of poor crop years and the potential for a sequence of bad years. Reutlinger and Bigman [1981] have estimated that a six million MT domestic food stock could cost between $59 and $82 million a year to operate. In this context, free trade and food aid represent a far more cost-effective approach to food security than large food stocking arrangements. Meeting such food security concerns of developing countries was the basic principle behind the creation in 1981 of a Cereal Import Facility at the International Monetary Fund (IMF). This facility was designed to provide financing to countries facing short-term problems of domestic food production shortfalls or high international prices. Between 1981 and 1985, a total of only seven developing nations made use of this facility. It has been suggested that, first, the facility needs to be broadened

Table 24. Cereal imports and food aid receipts by ninety-nine developing countries by region over time[a]

Region	Year	Aggregate (million metric tons)			Per capita (kg)	
		Commercial cereal imports	Food aid[b]	Total cereal imports	Food aid[b]	Total cereal imports
Asia[c]	1961-63	11.4	5.7	17.1	3.82	11.54
	1976-78	22.2	4.2	26.4	2.06	12.98
	1981-83	36.9	2.7	39.6	1.18	17.14
Latin America	1961-63	3.7	1.9	5.6	8.31	25.00
	1976-78	14.2	0.4	14.6	1.17	43.26
	1981-83	21.6	0.9	22.5	2.30	60.80
North Africa/Middle East	1961-63	1.9	3.9	5.7	24.13	35.81
	1976-78	14.6	2.5	17.1	10.22	70.96
	1981-83	27.6	2.7	30.3	10.19	112.72
Sub-Saharan Africa	1961-63	1.5	0.1	1.6	0.62	7.87
	1976-78	4.1	0.9	4.9	2.89	16.21
	1981-83	6.4	2.1	8.5	5.85	23.29
Total developing countries	1961-63	18.5	11.6	30.0	5.59	14.49
	1976-78	55.1	8.0	63.0	2.74	21.59
	1981-83	92.5	8.4	100.9	2.55	30.50

Source: Mellor [1987]; originally from Huddleston [1984] and FAO [1985]. Information from FAO [1988a] and World Bank [1984b] was also used to obtain cereal imports and food aid estimates.

[a] The ninety-nine developing countries include those covered by the Huddleston study. Out of these ninety-nine developing countries, nineteen were in Asia, twenty-four in Latin America, seventeen in North Africa/Middle East, and thirty-nine in Sub-Saharan Africa.

[b] Food aid total for 1976–78 does not include approximately 0.7 million metric ton reported by FAO, most of which went to Indochina and Portugal.

[c] Including China.

to include noncereals, and, second, the rules regarding drawings from the facility need to be liberalized so as to make it accessible to more countries.

The potential benefits and costs of food aid programs are summarized by Srinivasan [1989]. Food aid "can" (not necessarily "will") further economic development through several channels: it adds resources that can be used for current consumption and accumulation; it provides balance-of-payments support by reducing the foreign exchange spent on imports; it augments the domestic availability of food; to the extent it is targeted to the poor, it can alleviate poverty and improve health and nutritional status of the poor; it promotes development if it is tied to development-oriented projects that would not have been undertaken otherwise; and to the extent it can be credibly tied to the initiation of growth-promoting policies and reform of policies detrimental to growth, it can promote development. Clearly, food aid has the potential to improve food security, nutrition, employment, and economic growth in the recipient countries. The potential costs of food aid include the following: it may provide disincentives to domestic food production and hence, increase the probability of long-run dependency on food aid; and by alleviating food shortages, it enables the regime in power to postpone politically costly economic reforms.

Any food aid program must accentuate the positive and eliminate the negative effects. In other words, for food aid to make the maximum contribution to economic development, a donor must provide reliable amounts of food aid so that long-term development programs can be designed and implemented, provide large amounts of food aid in order to make a significant impact on employment, and recognize the conditions of effective food aid use. On the other hand, the recipient country must give priority to agricultural development in order to minimize the disincentive effects of food aid, and pursue policies that spread capital supplies as evenly as possible over the labor force in order to maximize employment growth. Ultimately, the effectiveness of food aid depends on flexibility in its use, coordination of donor objectives with development objectives, and whether domestic, economic, political, and institutional environments in recipient countries are conductive to efficient utilization of food aid as a development tool.

Chapter X. Implementation and Assessment of Agricultural Strategy

1. Implementation of Agricultural Development Strategy

There is, at best, a fine line between analysis of the policies, programs, and projects that should be pursued and analysis of ways to implement those decisions. Each interacts with the other. Implementation deserves special attention because it has not been emphasized enough and is particularly poorly understood in the context of participatory approaches to development, which take place under highly heterogenous conditions.

Problems of implementation arise both in poor countries and in developed ones. Policies or programs designed to obtain desired objectives may not be feasible because of sociopolitical constraints on the governments of developing countries and on the international institutions trying to assist them. Analysis of these problems is important in determining the policy and program mix that developing countries should adopt.

In developed and developing countries alike, the process of implementing policies and programs designed to use resources efficiently interacts with policies and programs designed to increase the availability of resources. This is especially true for financial resources. In the context of rural development in low-income countries, it is even more important for particularly scarce indigenous resources — personnel and institutions. The need to conserve and simultaneously expand the supply of these resources supports the need for inquiry to determine appropriate forms of implementation, including traditional institutions and private organizations.

Implementation must be analyzed with the methodology and perspective of other disciplines as well as economics. Research projects that integrate several disciplines and aspects of the problem are appropriate for analysis. Projects of separate disciplines are also appropriate, with policy advisors integrating the results of several of these projects into an action program. Little is known about how to conduct such research although it is especially necessary now. Because it is necessary to formulate sound policy and operations, a research program must be designed to address policy and operational questions and problems. There must be an explicit concern not only with how to use resources effectively to meet given objectives, but also on how to increase the supply of resources effectively.

Two aspects of implementation in poor countries deserve special attention — the political and social processes that affect the ability of the poor to participate in development, and the processes that determine how institutions function and grow.[102] Each set of processes develops from a particular cultural and historical framework that requires programs and analyses to be carefully adapted. Generalizing presents difficulties analogous to those presented by economic research on agricultural production policy. Each takes place in highly heterogeneous physical, social, political, and economic environments.

Analysis of political processes should explore the range of means by which bureaucratic and other institutions are related to their clienteles.[103] These clienteles may include elected local and national legislators, traditional leaders, and even members of the bureaucracy itself. Such analysis could fruitfully focus on the competition of groups and interests and the implications that competition has for the participation of low-income people in development. Analysis of the political determinants of success for policies or programs is also desirable. It must be recognized that the purpose of such analysis would not be just to provide insights into how to channel goods, services, and income to the lower income groups but also to explore the means of communication and feedback mechanisms that are essential for developing and operating the relevant institutions effectively.[104]

Analysis of institutional dynamics should address the problems of using resources effectively and of increasing the quality and quantity of financial, physical, human, and institutional resources. A range of questions about the interaction between formal technical knowledge and informal local, intuitive knowledge needs to be examined. Questions about the roles of traditional and commercial structures and of their desired degree of autonomy from the central government must also be examined.

The more research done on operational problems and on aspects of implementation, the more it merges into the very processes of project development and program evolution, and the more the distinctions between research, monitoring, and evaluation become blurred. However, because these processes are now in such an early stage of development, research can, if not generalize about these processes, then at least provide a sense of how they vary among countries and cultures. This would help create more flexible approaches to projects and improve techniques for project identification, development, implementation, monitoring, and evaluation.

One promising way to analyze the institutional aspects of project implementation and evaluation is to develop better ways of structuring and using participant observer micro research. Because such research focuses on the social organization of production and resource allocation—that is, on the institutional environment in which producers actually make decisions—it can be a useful source of information for project management.

The major challenges in pursuing this kind of micro research more effectively are: to develop more efficient ways of establishing how representative particular communities and situations are, to develop better ways of using micro research to identify reliable indicators that projects do or do not achieve their goals, and to develop more standardized methodologies to facilitate comparison and generalization from selected case studies.

2. Evaluation of Agricultural Development Strategy

One of the purposes of measuring progress is to judge the degree of national commitment to a particular strategy of growth. Such a judgment puts different weights on economic and social sectors and on the reallocation of national budgetary resources required to pursue a participatory growth strategy. An important element in such analyses is the distribution of revenues and expenditures. An equally important element is an examination of a country's fiscal and monetary policies to determine how much they support and are consistent with a particular strategy. Such an analysis is difficult and technical. To be useful it must be comparative, since there are few absolute criteria. To be effective, such an analysis must take into account differences in budgeting procedures and practices and differences in how resources are allocated.

Consistent and effective assessment will often require that data on public finance be consolidated by the central government and, at times, at lower levels of government. A preliminary effort would require the evaluation of current sources of information, recommendations to improve information flows, and standards for judging performance. Experimentation is needed to see whether comparative analysis can help judge the commitment to particular strategies and perhaps to learn the size of resource commitments particular approaches imply.

Study is needed of how much improvement can be achieved on procedures for providing data to support policymaking and analysis and for improving the effectiveness of projects and programs. Most Asian countries are better off than other developing countries in their abilities to collect, disseminate, and analyze data relevant for food and agricultural analysis and planning [FAO, 1987]. Even in these countries, there are two problems: some countries do not have adequate and appropriate micro and macro data to carry out policy analysis, and/or data may be available but the country does not have capability for data processing and policy analysis. At a minimum, assessment is required of the data for a recursive approach to project design, implementation, and evaluation under a strategy of broadly participatory growth, and of the content and size of special surveys to measure the effects of programs and projects, the causes of those effects, and the means of improving programs and projects.[105]

Chapter XI. Summary, Conclusions, and Research Agenda

Conceptualization of the role of agriculture in economic development and the means of achieving agricultural development has improved immensely over the past few decades. Empirical knowledge of economic relationships has grown even faster. The number of studies carried out under different conditions and on a wide range of topics has grown at an accelerating rate. As a result, documentation has been achieved for a wide range of diverse conditions. Of course, there is also increasing difficulty in substantiating generalizations, but perhaps a diminishing need to do so. This growth in research is the product of public concern about the need to improve development policy and rapid growth in the stock of trained research personnel and the institutional capacity of developing countries.

The research needs of the future include refinement and replication of past research. A simple comparison of the number of agricultural economists in each state in the United States, and the wide range of replicated studies done with the number of economists and studies in individual developing countries drives home the point that there is much further to go in expanding these capabilities. Similarly, as circumstances are constantly changing over time, research needs to be brought up-to-date. The research needs of the future require expansion of research capacity and will tax the ability of the scholarly community to digest such knowledge and to use it effectively.

Four areas of inquiry of particular importance are still inadequately conceptualized and lack a solid empirical base. These four areas are: the linkages by which a large, dynamic agriculture multiplies its effects on total economic growth through the other sectors of the economy—the role of technology, infrastructure, and education in those processes; the processes by which policy measures may be implemented quickly and efficiently; the links between improved income, food intake, nutrition and human well-being; and the interactions among the Third World countries, as their growth accelerates, as well as between the developing countries and the already industrialized countries.

The rationale for a strategy of growth based on agriculture is that a technologically dynamic agriculture stimulates accelerated growth in other sectors. That growth is not only faster and more broadly participatory than growth from other strategies of development, but it encourages a widely dispersed pattern of urbanization. The megalopolises that we see in many developing countries are quite contrary to the pattern of urbanization in present-day developed countries. That pattern is the product of growth without a broad rural base. It is the linkages from broadly based rural development that encourage development of a wide base of small urban centers that eventually, of course, may achieve a life of their own and grow to quite a large size.

Such linkages work through the demand for consumption goods and services, a demand which derives from the higher incomes that result from efficiency-increasing technological change in agriculture. But, the precise dimensions of these linkages and how public policy can encourage them is still poorly understood. Similarly, it is clear that these linkages are part of a commercialization process that depends on a large and expensive infrastructure of roads, other means of communication, and electrification. The size and composition of the needed investment and the principles by which it should be allocated have received little attention.

There is a continuous tendency for the academic community to understate the importance of rural education in these processes of growth. This underrating of the importance of education by the academic and intellectual communities follows from an inadequate understanding of the role that formal education plays in preparing people to deal with complex situations. The strong relation between the level of formal education and the acceptance of integrated pest management, of high levels of feeding of livestock, and of efficient use of fertilizer, all three of which involve complex management interactions, is an illustration of these processes. Of course, rural people themselves understand the importance of such education and try to drive their political systems to provide it.

This statement of the importance of physical infrastructure and education to the process by which agricultural development drives accelerated development in other sectors of the economy underlines another *lacunae* in agricultural development research. Rural physical infrastructure and rural education both require vast quantities of resources. Raising these resources at the national government level, with the inevitable detachment of revenue raising from expenditure patterns, is apt to produce deleterious effects on incentives as marginal tax rates are pushed to high levels. It can be argued that the same taxation for local purposes that are fully understood and desired by the people will have a less strong negative incentive effect. Thus, we must confront a complex set of questions relating to the development of local government.

It is notable that local government is weaker, generally speaking, in developing countries than was the case in developed countries at a similar stage of development. This contrast may arise from the nature of colonial regimes and the freedom movements that ended colonialism. Both tended to be urban based and to favor centralized power structures. We need to understand these processes of development of local government and the role that local government plays in raising resources and allocating those resources effectively for broad-based rural development. This moves us into the complex areas of politics and political economy as well as economics.

Increasingly, agricultural practitioners have been frustrated by problems in implementing agreed upon policies. There has been little comparative analysis of development projects—analysis to determine what needs to be done to implement particular policies.

Growth in agricultural production has sometimes been accompanied by declining prices, rising stocks, and decreased imports or increased exports while the per capita consumption of food-deficient people has failed to increase. This problem has been partly met in the long run by effective linkages, and by stimulating employment and the incomes of low-income people with high income elasticities of demand for food. But, there is also a short-run problem that can be effectively dealt with by employment and food subsidies. There is still a lack of knowledge of the effect such policies have on public finance, food consumption, and nutrition.

A discussion of policies also relates to the complex areas of politics, political economy, and economics. There is an increasing recognition that this inadequate knowledge base must inevitably lead to more substantial research into health issues and particularly public health problems. It has become more and more clear that increased productivity in agriculture does lead to rather broad-based participation in the income benefits of that increased productivity. The conversion of rising incomes into increased food consumption on the part of low-income people seems quite efficient. There is clear evidence, however, of a much lower level of efficiency in converting increased food consumption into improved health. A major expansion of research is necessary in order to understand these health problems. Undoubtedly, that research will lead back to issues of public expenditure and the development of rural local government. It seems likely that public health measures are important in these processes.

Even more, the importance of private aspects of health, including individual home sanitary systems, may require widespread acceptance before there is a measurable impact on the health of the population generally. Thus, economists, nutritionists, and public health practitioners must come together in order to understand these problems and to move toward a solution. It is also clear that the role of women in these processes is particularly important. A substantial body of research shows that income controlled by women has a more important effect on food consumption and nutritional status than income controlled by men. As understanding of health issues broadens, it seems highly likely that considerations of the role of women and women's education will expand.

Finally, large populations in Third World countries will enter periods of accelerated growth over the next few decades. We know little about the effects such growth will have on the aggregate demand for food and international trade, or on aid and food security relationships and the ways they interact with development strategy, breadth of political participation, and political systems. Yet we know that major changes in these relationships are likely, that they will affect all countries profoundly, and that the benefits from economically sound, long-run policies will be immense.

This quest for knowledge, with respect to relations among developing countries and between developing countries and developed countries, must, as a matter of course, include increasingly sophisticated trade analysis. We understand fully that open trading regimes can be favorable to economic growth. Given the restrictionism endemic in developing countries, the first round of policy from that knowledge has been to push toward freeing trade and drawing back from the stultifying effects of government interference. However, it is becoming increasingly clear that there are substantial scale economies, particularly for a wide range of agricultural commodities, and substantial institutional requirements that in effect act like scale economies. Thus, an effective horticultural export program requires research systems tuned to the specific horticultural commodities being exported and institutions specifically oriented to marketing those commodities. For most developing countries, such capacity cannot be built for a large number of commodities, and hence, the scale issues become important. This, in turn, requires careful analysis of commodities to be emphasized, the result of that analysis in effect driving comparative advantage in an important way.

Thus, we face an exciting future world, growing out of nearly half a century of political and economic change in Asia. Agricultural economists have played a major role in providing the basic knowledge that has allowed these processes to move far more rapidly than they moved when the present-day developed countries were progressing through similar stages of development. This rapid growth is possible because increases in trained people and institutional capacity generate more new knowledge and that knowledge base can be applied more rapidly. There is a danger that we could lose our sense of strategy in the myriad details of process. Thus, we end our review on the note that it is a sense of strategy that has led to an emphasis on agriculture, allowing it to play a driving role in essentially all Asian countries where accelerated overall economic growth has been achieved. It is this sense of strategy that has allowed efficient allocation of both development resources and the analytical resources of our profession, thus forwarding accelerated growth and broad participation in those processes of growth.

Notes

1. For an early effort to categorize stages of agricultural development, see Johnston and Mellor [1961] and Mellor [1962a]; for a historical perspective on growth stage theories and their relevance for agricultural development, see Wharton [1963a, b], Ruttan [1965], and Hayami and Ruttan [1971, 1985].

2. Seminal efforts to articulate the process of agricultural development and strategies for economic development, especially in the Asian context, include Day [1963], Schultz [1964], Mellor [1966, 1976], Ishikawa [1967a], Hayami and Ruttan [1971, 1985], Johnston and Kilby [1975], and Binswanger and Ruttan, *et al.* [1978]. However, as has been argued by Kamarck [1976], most theories of economic development or agricultural development do not take into account the peculiar conditions faced by countries in the tropics.

3. For detailed discussion on structural change and patterns of development in relation to the agricultural sector, see Schultz [1953], Clark [1957], Chenery [1960, 1979], Kuznets [1971], Chenery and Syrquin [1975], and Chenery and Watanabe [1958].

4. These relations are spelled out more fully in Lele and Mellor [1981], Johnston and Kilby [1975], Mellor and Lele [1973], Mellor and Mudahar [1974a, b], Mellor [1976], and Mudahar [1982].

5. For analyses of the positive role of agriculture in economic development, see Johnston and Mellor [1961], Mellor and Johnston [1984], Kuznets [1961], Nicholls [1961, 1963, 1964], Eicher and Witt [1964], Witt [1965], Mellor [1966, 1967, 1974, 1976], Southworth and Johnston [1967], Thorbecke [1969], Hayami and Ruttan [1971, 1985], the World Bank [1982b], Hwa [1983], Ghatak and Ingersent [1984], and Eicher and Staatz, eds. [1984].

6. For further discussion and conceptualization of labor-leisure choice and its impact on labor supply, see Mellor [1962a, 1963], Nakajima [1969], and A. K. Sen [1966]; and, more recently, in the context of agricultural household models, see Barnum and Squire [1979b], and I. J. Singh, Squire, and Strauss, eds. [1986].

7. Unlike single-sector models where labor supply is exogenous, the models developed by Solow [1956], Buttrick [1958], Leibenstein [1957], and R. R. Nelson [1956] determine labor supply endogenously.

8. Two-sector models were developed by Meade [1962], Solow [1961], Stiglitz [1969], Takayama [1963], Uzawa [1961, 1963], and others.

9. The Jorgenson model is termed "neoclassical" in the literature. The "classical" and "neoclassical" theories of dualistic development were tested and contrasted in Jorgenson [1966, 1967] and A. K. Dixit [1970]. For a complete statement on generalized dualistic development models, see Kelley, Williamson, and Cheetham [1972]. The role of agriculture in dualistic development models has also been reviewed and summarized in Ghatak and Ingersent [1984].

10. Selected studies dealing with different aspects of the Green Revolution include L. R. Brown [1968], Wharton [1969b], Johnston and Cownie [1969], Ladejinsky [1969a, b, 1970], Falcon [1970], Griffin [1972], Lele and Mellor [1972], Collier, Soentoro, Wiradi, and Makali [1974], Evenson [1974], Mudahar [1974], Randhawa [1974], C. H. H. Rao [1975], M. H. Khan [1975], Mellor [1976], Day and I. J. Singh [1977], Narain [1977], Ruttan [1977], Farmer [1979], Farmer, ed. [1977], Hayami, Kikuchi, *et al.* [1978], Dahlberg, ed. [1979], Hayami and Kikuchi [1981], Feder and O'Mara [1981], IRRI [1978a], ICRISAT [1980], Pearse [1980], Chaudhury [1982], Barker and Herdt [1985], Mellor and Desai, eds. [1985], Lipton and Longhurst [1985, 1989], and Hossain [1988b].

11. For further discussion on the introduction and adoption of modern crop varieties, see Streeten [1969]; for wheat in India, Dalrymple [1974, 1986a, b]; for rice and wheat in developing countries, Bernsten, Siwi, and Beachell [1982]; for rice in Indonesia, Herdt and Capule [1983]; and Barker and Herdt [1985] for rice in Asia. More recently, CGIAR has provided a comprehensive survey of the development and transfer of modern crop technology and its impact on agricultural development, J. R. Anderson, Herdt, and Scobie [1988].

12. These six countries include India and Pakistan from South Asia and Indonesia, Malaysia, Philippines, and Thailand from Southeast Asia.

13. For a detailed discussion of the sources of productivity growth in Indian Punjab, the showcase of the Green Revolution in India, see I. J. Singh [1971], Mudahar [1974], Johl and Mudahar [1974], and Day and Singh [1977]. In addition to the factors mentioned, land consolidation was an important factor in expanding irrigated area by facilitating profitable capital investment in private tube wells operated by diesel engines or electric motors. The relative economics of prospective technologies for semiarid tropics (unlike wheat and rice varieties which were suited primarily for areas with assured irrigation) in India is discussed in Ryan, Sarin, and Pereira [1980]. A comparative analysis of the sources, nature and impact of Green Revolution in Bangladesh is provided by Hossain [1988b].

14. The role of modern rice varieties in raising fertilizer productivity was demonstrated by Herdt and Mellor [1964], and the complementarities among irrigation, fertilizer, and modern rice varieties in the Philippines are analyzed in Wickham, Barker, and Rosegrant [1978] and Herdt, Te, and Barker [1977/78, 1980]. Attribution of a large share of production growth to fertilizer arises partly by subsuming returns to various complementary inputs to fertilizer [Herdt and Capule, 1983]. However, such relatively raw analyses properly place the public policy focus on the need for developing infrastructure, distribution systems and incentive policies for massive increases in fertilizer use.

15. For a detailed analysis of fertilizer in agricultural development of India, see G. M. Desai [1969, 1973, 1978, 1979, 1982]; G. M. Desai and G. Singh [1973]; and G. M. Desai, Chary, and Bandyopadhyay [1972]. The role of fertilizer in the Asian rice economy is dis-

cussed in David [1976, 1978]; Barker [1978]; David and Barker [1978]; Herdt, Te, and Barker [1977/78]; Wickham, Barker, and Rosegrant [1978]; Pitt [1983b]; and Barker, Herdt, and Rose [1985]. Methodology to measure the contribution of fertilizer to food production and its application to estimate fertilizer's contribution to wheat and rice production in India is demonstrated in Mudahar [1987]. A comprehensive study dealing with the role of fertilizer in economic development in the United States was carried out by Sahota [1968b].

16. According to Hayami [1964], 70 percent of the increase in commercial fertilizer input in Japan is explained by technological progress in agriculture and 30 percent by decrease in price of fertilizer between 1883 and 1937. This conclusion is corroborated by the recent experience of rice growing in Asian countries by David [1976, 1978], Sidhu and Baanante [1981], and Rosegrant, Kasryno, Gonzales, Rasahan, and Saefudin [1987].

17. IRRI published the proceedings, *Irrigation Policy and Management in Southeast Asia*, during 1978. Among others, the proceedings include papers by Wickham and Valera [1978], D. C. Taylor [1978], Hafid and Hayami [1978], Dozina, Kikuchi, and Hayami [1978], Tagarino and Torres [1978], and Trung [1978].

18. The project was implemented in six Asian countries: Bangladesh, Indonesia, Philippines, Sri Lanka, Taiwan, and Thailand. The basic methodology used is summarized in De Datta, *et al.* [1978]. All the country studies used a common methodology which made it possible to compare the results across different countries. Furthermore, these studies were carried out by interdisciplinary teams of researchers both at the national and international levels.

19. The districts, from which farm management data were collected, were from selected Indian states, including Uttar Pradesh, Madhya Pradesh, Andhra Pradesh, Punjab, West Bengal, Bombay (now Maharashtra), and Madras (now Tamil Nadu). The empirical evidence on an inverse relationship between farm size and productivity was provided in P. K. Bardhan [1973]; Bharadwaj [1974]; Khusro [1964, 1969]; Lau and Yotopoulos [1971]; Mazumdar [1963, 1965]; A. P. Rao [1967]; C. H. H. Rao [1963, 1966]; Saini [1971]; A. K. Sen [1962, 1964]; and Yotopoulos and Lau [1973]. Some of these studies were summarized in Bachman and Christensen [1967], Kanel [1967], Bhagwati and Chakravarty [1969], and I. J. Singh [1988a]. More recently, the empirical evidence in support of an inverse relationship between farm size and productivity is provided in Bagi [1981a, 1983a, b, 1984]; Huang and Bagi [1984]; Deolalikar [1981]; Huang, Tang and Bagi [1986]; V. Rao and Chotigeat [1981]; and A. Sen [1981].

20. Based on the survey of six villages in South India (semiarid tropical area) Ryan, Ghodake, and Sarin [1980] found (a) no consistent evidence that small farmers use more labor per hectare than large farmers and (b) little evidence for the existence of dual labor markets.

21. The empirical evidence for India is provided in Khusro [1964], R. Krishna [1964], C. H. H. Rao [1965a], Saini [1969], and Yotopoulos and Lau [1973]. These studies are also based on "Farm Management Data" for the 1950s. Similar studies for other Asian countries include: Yotopoulos and Lau [1979] for the methodology and summary; Lau, Lin, and Yotopoulos [1979] for Taiwan; Kuroda [1979] for Japan; Adulavidhaya, *et. al* [1979] for Thailand; and Tamin [1979] for Malaysia. All these studies, except for Japan, confirm the hypothesis of constant returns to scale; Japanese agriculture has been found to experience increasing returns to scale.

22. These studies include P. K. Bardhan [1973], Chennareddy [1967], D. K. Desai [1963], Day and I. J. Singh [1977], Hopper [1965], Khusro [1964], R. Krishna [1964], C. H. H. Rao [1965a], Sahota [1968a], Saini [1969], S. S. Sidhu [1974a], Lau and Yotopoulos

[1971], and Yotopoulos and Lau [1973]. Some of these studies have been summarized in Bhagwati and Chakravarty [1969] and in Government of India [1976].

23. Based on the results of comprehensive linear programming model of agriculture in the Pakistan Punjab, Gotsch and Falcon [1975] concluded that there was scope to increase net revenue through optimal cropping patterns and cropping intensities. Further details of the model and its empirical analysis are available in Gotsch, B. Ahmed, et al. [1975]. A brief survey of literature on farm planning models with a focus on agriculture (both crop and livestock production) in the developed world is available in Glen [1987]. The survey concludes that although these models are used as research tools or teaching aids, few of them are used directly by farmers to improve their decisionmaking.

24. Among the studies that hold this view are Heady [1947], Georgescu-Roegen [1960], Issawi [1957], and Schickele [1941]. The notable exception is D. G. Johnson [1950].

25. Vyas [1970] and Ladejinsky [1972] argued that policy prescriptions which emphasize abolishing tenancy and setting ceilings on land ownership might overlook reforms that would make existing tenancy arrangements more effective and equitable. The empirical analysis carried out in three provinces in Thailand shows a statistically significant effect of ownership security on land prices [Chalamwong and Feder, 1988]. The risk of eviction on untitled land and the advantages in access to credit associated with titled land are shown to account for higher price of titled land. As a result, the authors conclude that granting full legal ownership to squatters can be a socially beneficial policy.

26. For related studies on the economic aspects of tenancy and land reforms, see Abdullah [1976] and Zaman [1973] for Bangladesh and Mangahas [1974] for the Philippines. Newbery [1977] and Newbery and Stiglitz [1979] discuss the theoretical issues; T. C. Smith [1959] and Feeny [1983a, b] discuss a broad range of issues related to tenancy.

27. A detailed survey of supply response studies is available in Askari and Cummings [1976].

28. The most popular model used for empirical acreage response studies has been the Nerlovian adjustment model, Nerlove [1958]. Multiple regression models, even though less elegant, are also widely used. Narain [1965], in his detailed and particularly insightful study on acreage response in India, used graphical analysis.

29. According to R. Krishna [1982], since elasticity of output with respect to technological change appears to be higher than price elasticity, a balanced policy should stress technology policy more than the price policy while maintaining price incentives.

30. For further discussion on price policy and economic development, see R. Krishna [1967a] and Mellor [1966, 1968] and for the price policy debate in India, see Dantwala [1967, 1972] and Lele [1969]. More recent literature on agricultural price policy includes ADB [1988], Bale and Lutz [1978, 1981], Bertrand [1980], Braverman, Ahn, and Hammer [1983], Byerlee and Sain [1986], Cheong and D'Silva [1984], de Janvry and Subbarao [1986], Krishna and Raychaudhuri [1980, 1981], Krueger, Schiff, and Valdes [1988], Meier, ed. [1983], Mellor and Ahmed, eds. [1988], Pinstrup-Andersen, ed. [1988], Scandizzo and Bruce [1980], Schultz, ed. [1978], Sicular, ed. [1989a], Timmer [1986a, b, c], and Timmer, Falcon, and Pearson [1983]. The evidence provided by various studies indicates that (1) the agricultural sector in developing countries is heavily taxed while that in the developed countries received substantial subsidies, and (2) there are large income transfers from the rural to the urban sector in developing countries and from the urban to the rural sector in developed countries.

31. The issue of choice between crop price support and input subsidy has become extremely important for policymakers. Both crop price support and input subsidy policies are widespread in developing countries. The issue is far from settled and the initial analysis is

provided in Barker and Hayami [1976], Mudahar [1978], R. Ahmed [1978, 1979, 1981], Bagi [1984], Timmer [1986a], and Mellor and R. Ahmed [1988].

32. We are particularly grateful for the input of Bupendra M. Desai in the development of this section.

33. For the evidence and detailed analysis of international data on agricultural research and extension, see Boyce and Evenson [1975]; Evenson and Kislev [1973, 1975a, b]; Evenson [1978a, 1986b, 1987, 1988]; Evenson, Pray, and Scobie [1985], Judd, Boyce, and Evenson [1986]; Oram and Bindlish [1981]; Oram [1985]; Pray [1979, 1983]; Ruttan [1982]; and Ruttan and Pray, eds. [1987].

34. According to Boyce and Evenson [1975], in Asia during 1974, 1.9 percent of the value of agricultural production was spent on research and 0.9 percent was spent on extension. Also, the expenditure on agricultural research was only about 26 percent of the total annual expenditure for research in Asia. More recent experience indicates that there has been a slight shift in public resource allocation in favor of agricultural research relative to agricultural extension.

35. These issues and problems were addressed carefully in many papers contained in Fishel, ed. [1971] and in Arndt, Dalrymple, and Ruttan, eds. [1977]. For a review of the models and methods used to allocate resources in agricultural research, see Shumway [1977]. Barker and Herdt [1979] analyzed different aspects of resource allocation for rice research in Asia and concluded that rainfed lowland rice needs to be given research priority in South and Southeast Asia since modern rice technology has largely bypassed the rainfed rice. The efficiency and equity issues in allocation of research resources and design of agricultural technology in developing countries are discussed in Binswanger and Ryan [1977] and Ryan [1984]. Resource allocation, structure, and incentives for agricultural research and its contribution to agricultural development in Nepal are analyzed by Yadav [1987].

36. Note the pioneering work using cost-benefit analysis to estimate the contribution of agricultural research by Griliches [1958b]—in the context of hybrid corn and related innovations.

37. For a description of the Indian agricultural research system and its contribution to agricultural production, see Mohan, Jha, and Evenson [1973]; Evenson and Jha [1973]; and Indian Society of Agricultural Economics [ISAE, 1977a]. Resource allocation, structure, and its contribution to agricultural development in Nepal are analyzed by Yadav [1987].

38. There is a growing realization of the need to estimate returns from investment in agricultural research. Pinstrup-Andersen [1982] deals with contribution of agricultural research to economic development; Ruttan [1982] and Ruttan and Pray, eds. [1987] deal with agricultural research policy. Other selected studies dealing with agricultural research are Andrew and Hildebrand [1982], Busch and Lacy [1983], Evenson [1978a], Evenson and Kislev [1973, 1975a, b], Evenson, Pray, and Quizon [1986], Evenson, Putnam, and Pray [1983], Evenson, Waggoner, and Ruttan [1979], Khan and Akbari [1976], Norton and Davis [1981], Pray [1979, 1983], Ruttan [1986a, 1986b, 1986d], and the World Bank [1985e].

39. For the problems and major issues in the development of agricultural research systems in developing countries, see Moseman [1970]. The influence of international research on national agricultural research systems is discussed in Evenson [1986b, 1987, 1988] and Evenson, Pray, and Scobie [1985]. Neglected dimensions and emerging alternatives in agricultural research are discussed in Dahlberg, ed. [1985].

40. Hillman and Monke [1983], after reviewing the literature on international transfer of agricultural technology, concluded that the greatest successes in technology transfer have been with technologies which are neutral with respect to the economic, biological, and in-

stitutional environment into which they are transferred. Technological opportunities and international technology transfer in agriculture are discussed in detail by Evenson [1988]. Evenson, Putnam, and Pray [1983] have analyzed the effects of international transfer of agricultural technology on the competitiveness of U.S. agriculture.

41. But in subsistence agriculture, household consumption requirements are more important and production patterns are adjusted according to consumption constraints. For detailed analysis along these lines, see Day and I. J. Singh [1977], I. J. Singh [1971], Mudahar [1973], and Mudahar and Day [1978]. More recently, household consumption and its implications for production decisions and related relationships in developing countries are evaluated in the context of agricultural household models by Barnum and Squire [1979a, b]; Binswanger, Evenson, Florencio, and White, eds. [1980]; I. J. Singh, Squire, and Strauss [1986a]; and I. J. Singh, Squire, and Strauss [1986b]. The agricultural households combine two fundamental units of microeconomic analysis: the household and the firm. The so-called new theory of agricultural households combines the producer and consumer behavior. These models have been applied in several Asian countries, including India, Indonesia, Japan, Korea, and Malaysia.

42. Khatkhate [1962] similarly argued that the marketable surplus will increase as the price falls because the farmer is interested in maintaining his money income at the same level. Dubey [1963] and Dandekar [1964] challenged these conclusions. For further discussion of the fixed-cash requirement theory of marketable surplus, see Nowshirvani [1967a].

43. For related studies on marketable surplus and marketing of cereals and foodgrains in India, see P. K. Bardhan and K. Bardhan [1969, 1971], Lele [1971], and Moore, Johl, and Khusro [1972].

44. Two other studies on marketable surplus of agricultural produce in South Asia include A. R. Khan and Chowdhury [1962] and Zaman [1966] for Pakistan.

45. A critical review and detailed comments on this study are available in Majumdar [1965], C. H. H. Rao [1965b], and R. Krishna [1965b].

46. Dandekar [1964], based on empirical evidence for jowar, wheat, and other cereals, also rejected the fixed-cash requirement theory of marketable surplus put forth by Mathur and H. Ezekiel [1961].

47. Additional studies on marketable surplus in other parts of India include Kahlon and Vashishtha [1968], Muthiah [1964], and Vyas and Maharaja [1966].

48. Mellor [1973a] has analyzed the impact of accelerated growth in agricultural production on intersectoral transfer of resources; and Mellor [1978] has discussed the impact of food price policy on income distribution in low-income countries, with particular reference to India.

49. T. H. Lee [1971] has provided a detailed empirical analysis of intersectoral resource transfers in Taiwan. Analyses of trends in foodgrain prices and the terms of trade in India and economic consequences for foodgrain production and economic growth are available in Mellor and Dar [1968] and Thamarajakshi [1969] for the period between 1952/53 and 1964/65, and Parthasarathy and Mudahar [1976] for the period between 1952/53 and 1973/74.

50. Related studies dealing with agricultural marketing and marketing efficiency in Asia and developing countries include J. C. Abbott [1962], Lele [1967, 1971, 1974]; Jasdanwalla [1966]; R. W. Cummings, Jr. [1968]; Farruk [1970]; Moore, Johl, and Khusro [1972]; Ruttan [1969]; Timmer [1972, 1974a]; Wharton [1962]; Harriss [1979, 1986]; von Oppen, Raju, and Bapna [1980]; and World Bank [1988d].

51. The empirical evidence from India that protein deficiency is not as widespread as has been believed, and that the calorie gap is more serious than the protein gap, is contained in the proceedings of the Indian Society of Agricultural Economics [ISAE, 1977b]. A de-

tailed assessment of energy and protein requirements conducted by a joint FAO/WHO Ad Hoc Expert Committee is available in FAO and WHO [1973]. The findings of this report also reject the simplistic view that the protein gap is widespread.

52. Further discussion on incidence of malnutrition and related issues is available in Berg [1973, 1981, 1987], Berg, Scrimshaw, and Call, eds. [1973], Caliendo [1979], Kalirajan [1976], Mitra [1973], L. Taylor [1977], FAO [1977], FAO and WHO [1973], Piazza [1983, 1986], Pitt [1983a], Knudsen and Scandizzo [1979], Poleman [1981], Srinivasan [1981], and World Bank [1986b]. Poleman [1981] points out disagreements among organizations, such as FAO, World Bank, and USDA, on the nature and extent of world food and nutrition problems, and hence their perceptions about appropriate remedial actions.

53. Consumer expenditure patterns and the implications of income growth are analyzed in Azizur Rahman [1963], M. I. Khan [1963], B. M. Desai [1972], Mellor and Lele [1973], and Mellor [1978]. The implications of industrialization for the demand for food in low-income countries are discussed in Stevens [1963]. The relation of income, expenditure patterns and food subsidies is extensively analyzed in R. Ahmed [1979, 1981]; J. M. Davis [1977]; George [1979]; Sarma, Roy, and George [1979]; Gavan and Chandrasekera [1979]; R. Krishna and Chhibber [1983]; Mateus [1983]; Trairatvorakul [1984]; Bienen and Gersovitz [1986]; and Pinstrup-Andersen, ed. [1988].

54. A critical review of issues dealing with contractual arrangements, employment, and wages in rural labor markets is available in P. K. Bardhan [1978, 1979, 1980, 1984a, b], P.K. Bardhan and Rudra [1978, 1981, 1983], and Binswanger and Rosenzweig, eds. [1984]. An excellent survey of labor market performance in developing countries was carried out by A. Berry and Sabot [1978].

55. Kao, Anschel, and Eicher [1964] found no consistent evidence that disguised unemployment exists in agriculture; also see I. J. Singh [1971] and Mudahar [1973]. The marginal productivity of labor may not be zero but Visaria and Visaria [1973] found that labor productivity in rural India was very low.

56. See, for example, C. H. H. Rao [1974], I. J. Singh [1971], Mudahar [1974], Johl [1973b], and Day and Singh [1977] for analysis dealing with Indian Punjab.

57. Theoretical discussion on some of these issues is available in N. Islam [1964], Mellor [1963], and Mellor and Stevens [1956].

58. In an excellent survey of landless poor in South Asia, I. J. Singh [1983] concludes that in the long run a reduction in population growth, an increase in agricultural growth and an increase in opportunities in the nonagricultural sector can benefit the landless through increased employment and can eradicate poverty. The relationship between the Green Revolution, prices, and poverty is discussed in several papers published in Mellor and Desai, eds. [1985], and in more recent papers by I. J. Singh [1988a, b, c].

59. Several case studies which discuss these issues are Billings and A. Singh [1970]; Kahlon [1976]; Lal [1976]; Mudahar [1974]; Johl and Mudahar [1974]; C. H. H. Rao [1972]; Visaria [1972]; Kahlon, Gupta, and Sondhi [1971]; Roy and Blase [1978]; and Barker and Cordova [1978].

60. Collier, Soentoro, Wiradi, and Makali [1974], in analyzing the impact of modern technology on institutional change in Java, concluded that there appears to be a significant relationship between the spread of modern rice varieties and the expansion of *tebasan* (a traditional method of selling a crop just before harvest), which is responsible for a reduction in employment opportunities for harvest labor. This is corroborated by Utami and Ihalauw [1978].

61. Other studies that analyze the interaction between modern farm technology, mechanization, and employment in agriculture include Acharya [1973], Agarwal [1980, 1984b],

Johnston and Cownie [1969], Mehra [1976], A. K. Sen [1975a, b], B. Ahmed [1975], Binswanger [1978, 1984], and Binswanger and Donovan [1987].

62. Johnston and Mellor [1961], Mellor [1966, 1976], Mellor and Johnston [1984], Mellor and Mudahar [1974a, b] argued that the major constraint to creating and sustaining increased employment in developing countries is wage goods supply. The Green Revolution and modernizing agricultural sectors have the potential for relaxing this constraint.

63. Rural works programs provide an alternative means of employment in areas facing large unemployment. This may be important since the proportion of landless laborers has gone up in the last decade or so. Several case studies, for example, include Abdullah, Hossain, and Nations [1976]; Booth and Sundrum [1976]; Raj [1976]; and Rath [1974].

64. Examples of rural works programs in India are provided in Apte [1973], Donovan [1973], and Dantwala [1975]. However, Dantwala [1975] seemed to conclude that the rural works program does not really benefit the poor. R. Krishna [1973, 1982] concluded that radical politicization of the poorest groups will be necessary if they are to obtain the share allocated to them.

65. Most of the studies on employment deal with the implications of modern farm technology on employment in the production process. Timmer [1972], however, analyzed the implications for employment of investment in rice marketing in Indonesia.

66. This controversy is reflected in several studies dealing with poverty in India, such as Dandekar and Rath [1971a, b], P. K. Bardhan [1970], Minhas [1970], and Vyas [1972]. A review of changes in income distribution and poverty in India is available in D. Kumar [1974].

67. Cross-country data on income distribution and incidence of poverty in the world are available in S. Jain [1975] and World Bank [1975].

68. See, for example, Frankel [1971], C. H. H. Rao [1975], Swenson [1976], and Shah [1976] for India, Hossain [1988b] for Bangladesh and Griffin [1974] for several developing countries. For a balanced view of the practices and attempts to increase equity in the context of agricultural growth, see Sarma [1981].

69. For further discussion on rural income distribution, especially the effects of technological change in agriculture, see Gotsch [1972], Hayami and Herdt [1977], Barker and Herdt [1978], and Barker, Herdt, and Rose [1985].

70. Some of the complex interactions between nutrition and health are discussed by Selowsky and L. Taylor [1973], C. E. Taylor [1976], McCord [1977], and Selowsky [1981b]; and the linkage between nutrition and labor productivity is discussed by Leibenstein [1957] and Deolalikar [1988]. Related studies that analyze nutrition, health, and labor productivity interactions include Behrman [1988], and Behrman, Deolalikar, and Wolfe [1988].

71. For discussion on school participation rates in India, see Blaug, Layard, and Woodhall [1969] and Shortlidge [1976].

72. See, for example, Harberger [1965] and Kothari [1967].

73. The importance of education in economic development and as an investment to develop human capital was well articulated in a pioneering effort by T. W. Schultz [1963]. The importance of investment in rural education and its effect on agricultural development was discussed in an excellent paper by Welch [1978].

74. In 1977, Hirschman clarified his earlier position on linkages in agriculture: "Perhaps the principal reason why it is difficult to establish backward and forward linkage industries around the staples is not so much that, as I argued originally, there are fewer linkage effects in agriculture than in industry, but that they largely point to industries whose technologies are alien to the grower of the staple" [p. 78].

75. According to Hirschman, this is an all-encompassing definition of linkages which includes the well known forward and backward linkages.

76. See Leontief [1951]. For definitions of alternative linkage indexes and their measurement, see Chenery and Watanabe [1958], Hirschman [1958], Rasmussen [1956], Yotopoulos and Nugent [1973], and Mudahar [1982].

77. The implications growth linkages have for agriculture development can best be analyzed by agricultural sector models that incorporate these linkages explicitly. Thorbecke [1973] discussed alternative approaches to agricultural sector analysis in developing countries. For a recursive programming approach, see Day [1963], Day and I. J. Singh [1977], Day and Cigno, eds. [1978], Mudahar and Day [1978], and Mudahar [1973]. For a simulation approach, see Manetsch et al. [1971] and Mellor and Mudahar [1974a, b].

78. A comprehensive analysis of modernizing agriculture and structural transformation was made in Johnston [1970], Johnston and Kilby [1975], and Mellor [1976]. The growth linkages of new foodgrain technologies in Asia were discussed in Mellor and Lele [1973] and Mudahar [1982]. Also see Flanders [1969], Hazell and Röell [1983], Krueger [1962], Lipton [1968a], and Nicholls [1963] for different aspects of linkages between agricultural and industrial sectors.

79. The growth of small-scale industry in response to agricultural development was discussed in Johl and Mudahar [1974] for the Indian Punjab and in van der Veen [1973] for Gujarat state in India.

80. According to D. W. Adams, Canh, and Chin [1975]: "Rural purchasing power in Taiwan during the past two decades (1952-72) has provided a major market for goods produced in the nonagricultural sector. These final demand linkages were especially important in the 1950s when rural residents made up a large part of the total population and when nonagricultural exports were relatively small. . . . The underutilized 'industrial islands' surrounded by seas of rural poverty in Northeast Brazil, Colombia, and Pakistan, for example, are vivid contrasts to the way development has evolved in Taiwan" [p. 141]. T. C. Smith [1959] sheds light on these important issues for pre-Meiji Japan.

81. Mellor [1973a] has reviewed this controversy and provided an elaboration of the basic case for net resource transfers from agriculture. For detailed analysis of intersectoral resource transfer in Taiwan, see T. H. Lee [1971].

82. Despite this, there are few analytical studies dealing with patterns of agricultural trade and the implications of alternative trade policies in determining these patterns. For discussion of agricultural trade issues and their interaction with economic development, see Ojala [1969], Sisler [1971], Mellor and Lele [1975], Tolley and Zadrozny, eds. [1975], Sorenson [1975], Mellor [1976], and Nagle [1976]. Taiwan was analyzed in Tang and Liang [1975] and Pakistan in S. R. Lewis, Jr. [1968].

83. This confirms the policy recommended for most of Latin America by Prebisch [1964].

84. A detailed account of India's trade policies is available in Bhagwati and P. Desai [1970]. Other studies that deal with different aspects of trade in India are R. Bharadwaj [1962], Mellor and Lele [1975], and M. Singh [1964].

85. The implications of agricultural protectiveness on distribution of gains from the dissemination of technology were analyzed by Josling [1975, 1980].

86. According to Evenson [1975], "There has been a significant decrease in the comparative productivity of cereal grain production in less developed regions of the world . . . the decline in relative productivity in most of these countries has diminished their export performance as well." Evenson attributed this to a lack of investment in agricultural research. For discussion on related issues and the interaction of technological change and in-

ternational trade, see Hayami and Ruttan [1971, 1985]; Evenson, Houck, and Ruttan [1970]; and Vernon, ed. [1970].

87. These studies used a domestic resource cost approach to determine the comparative advantage in rice production [Pearson, Akrasanee, and Nelson, 1976].

88. See Mellor and Lele [1975]. Also see R. Bharadwaj [1962], which maintains that the capital intensity of exports *vis à vis* import replacement rose in India in 1958/59 and 1953/ 54.

89. According to Mellor and Lele [1975], India's rank correlation coefficients between capital intensity and export growth from 1964 to 1969 were statistically significant at the 90 percent level.

90. D. Gale Johnson [1975] analyzed the effects of a hypothetical worldwide shift to free trade of agricultural products on the outputs and prices of those products. Tolley and Za- drozny, eds. [1975] paraphrased Johnson's conclusion: "that even though free trade would permit a more efficient allocation of resources in the long run, the immediate effect would be to lower food production for a decade, with the distribution of food consumption shift- ing toward high-income countries at the expense of the less developed countries" [1975]. Note the more recent work reaching similar conclusions by K. Anderson and Ahn [1984] and Hayami [1983]. The prospects for world demand for the agricultural exports of devel- oping countries were analyzed in Rojko and Mackie [1970]. The structure, conduct, and performance of the international rice market and patterns of rice trade were analyzed in Fal- con and Monke [1979/80], and Siamwalla and Haykin [1983].

91. A comprehensive discussion of some of these issues is available in H. G. Johnson [1967], which continued the work of the 1964 United Nations Conference on Trade and Development and provided background for the 1967 Conference. This work points out that the GATT system is discriminatory against developing countries. Also, see Perez and Benedick, eds. [1978] which deals with some of these issues in the context of multilateral trade negotiations. A number of more recent studies have analyzed the impact of developed country trade practices on developing country agriculture and the potential implications of trade liberalization. See, for example, Josling [1980], Koester [1982], Koester and Valdes [1984], OECD [1987], Tyers and Anderson [1986, 1987], and Valdes and Zietz [1980].

92. A number of studies indicate that rapid rates of agricultural growth in developing countries, facilitated by open trade regimes, produce a rapid growth in those countries' food imports, particularly cereals. See, for example, Bachman and Paulino [1979], de Janvry and Sadoulet [1986a, b], Houck [1986], and J. E. Lee, Jr. and Shane [1985].

93. The multilateral trade negotiations (MTN) for the Uruguay round are divided into fifteen negotiating groups, out of which fourteen are for goods and one for services. These fifteen negotiations groups are: (1) tariffs, (2) nontariff measures, (3) natural resource-based products, (4) textiles and clothing, (5) agriculture, (6) tropical products, (7) GATT articles, (8) MTN agreements and arrangements, (9) safeguards, (10) subsidies and countervailing measures, (11) trade-related aspects of intellectual property rights, including trade in coun- terfeit goods, (12) trade-related investment measures, (13) dispute settlement, (14) func- tioning of the GATT system, and (15) services.

94. Export earnings also depend on the terms of trade. It has been argued that there is a long-run tendency for the terms of trade to turn against primary commodities, the major source of export earnings for many of the developing countries. However, Morgan [1959, 1963] found no strong evidence of a consistent pattern to support this general conclusion.

95. Hayami and Ruttan [1971, 1985] indicated that a major source of disequilibrium in world agriculture since World War II has been agricultural protectionism in the developing

countries. They also analyzed the implications of the Green Revolution for agricultural trade.

96. Balassa *et al.* [1971] have provided a comprehensive theoretical and empirical discussion of the structure of protection in developing countries, including case studies for Malaysia, Pakistan, and the Philippines.

97. More recently, Timmer and Falcon [1975a, b] analyzed the determinants of rice trade in Asia and concluded that "differences in rice and fertilizer prices across countries (and over time) are an important determinant of levels and patterns of international trade in rice" [1975b, p. 89].

98. The political aspects of foreign aid, in the context of India, are discussed in Eldridge [1969].

99. A general discussion of the two-gap model approach is available in Chenery and Strout [1966] and Chenery and MacEwan [1966].

100. See Mellor [1976] for a general discussion on foreign aid along these lines. The needed volume of foreign assistance can be viewed as the difference between domestic financial resource requirements and their availability. Domestic resource availability can be manipulated through policies on taxes, savings, and international resource transfers. For discussion of agricultural taxation, see Gandhi [1966], Johl [1972], and Pathak and Patel [1970]; for savings, see Mikesell and Zinser [1973] and Pannikar [1961]; and for intersectoral resource transfers, see T. H. Lee [1971], Mellor [1973a], and C. H. H. Rao [1969].

101. These issues are summarized in Bhagwati and Chakravarty [1969] and are addressed in more detail in Clay and Singer [1984], Dantwala [1967], M. Ezekiel [1958], Fisher [1963], Huddleston [1984], Isenman and Singer [1977], Khatkhate [1962], J. S. Mann [1967], Mathur and H. Ezekiel [1961], Pinstrup-Andersen and Tweeten [1971], and S. R. Sen [1960]. Maxwell and Singer [1979] provide a survey of the issues surrounding the use of food aid. More recent work on food aid includes H. Ezekiel [1988], Hopkins [1984], Mellor [1987], Srinivasan [1989], Wallerstein [1980], and World Food Programme [1985, 1987].

102. Russell and Nicholson, eds. [1981] discuss a range of such issues related to collective action.

103. For example, it is known that large areas under irrigation in Pakistan are experiencing soil salinity problem due to water-logging and poor drainage. Yet, according to S. H. Johnson III [1982b], the government has not been able to find a long-term satisfactory solution due to lack of success in implementing salinity control and reclamation projects, and lack of positive response from the bureaucracy due to project-related conflicts between provincial and central government organizations.

104. A major effort at comparative analysis of agricultural development projects in order to elucidate elements of implementational strategy was made by Mosher [1966] and Lele [1975]. Because broad access to data and project knowledge is difficult except in large agencies such as the World Bank, the sponsor of the Lele study, such work has not been duplicated. Birowo [1983] discussed the problems associated with implementing rural development strategies in Indonesia.

105. Based on the evaluation of the Companiganj project on health, nutrition, and family planning in Bangladesh, McCord [1977] concluded that for successful implementation of a project there is need for a realistic assessment of what is likely to work outside the pilot project and evidence that the program will work before it is implemented nationally.

References

Abbie, L., J. Q. Harrison, and J. W. Wall [1982]: *Economic Returns to Investment in Irrigation in India*. Washington, World Bank Staff Working Paper No. 536.

Abbott, J. C. [1962]: "The Role of Marketing in the Development of Backward Agricultural Economies." *J. Farm Econ.* 44:349-362.

Abbott, P. C. [1979]: "Modeling International Grain Trade with Government Controlled Markets." *Amer. J. Agr. Econ.* 61:22-31.

Abdullah, A. [1976]: "Land Reform and Agrarian Change in Bangladesh." *Bangladesh Dev. Studies* 4:67-114.

Abdullah, A., M. Hossain, and R. Nations [1976]: "Agrarian Structure and the IRDP—Preliminary Considerations." *Bangladesh Dev. Studies* 4:209-222.

Abdullah, T. A., and S. Ziedenstein [1982]: *Village Women in Bangladesh: Prospects for Change*. Oxford, England, Pergamon Press.

Abel, M. E. [1975]: "The Impact of U.S. Agricultural Policies on Trade of the Developing Countries." In Tolley and Zadrozny, eds., 1975, pp. 21-56.

Abercrombie, K. [1969]: "Population Growth and Agricultural Development." *Monthly Bull. Agr. Econ. and Stat.* 18(4):3-9.

Acharya, S. S. [1969]: "Comparative Efficiency of High Yielding Varieties Programme: Case Study of Udaipur District." *Econ. and Polit. Weekly* 4(44):1755-1757.

———— [1973]: "Green Revolution and Farm Employment." *Indian J. Agr. Econ.* 28(1):30-45.

———— [1982]: *Green Revolution: Impact on Farm Employment and Incomes*. Jaipur, India, Sanghi Prakashan.

Adams, D. W. [1980]: "Recent Performance of Rural Financial Markets." In Howell, ed., 1980, pp. 15-34.

Adams, D. W., T. Q. Canh, and L. A. Chin [1975]: "Changes in Rural Purchasing Power in Taiwan, 1952-72." *Food Research Institute Studies* 14:127-145.

Adams, D. W., and D. H. Graham [1981]: "A Critique of Traditional Agricultural Credit Projects and Policies." *J. Dev. Econ.* 8:347-366.

Adams, D. W., D. H. Graham, and J. D. von Pischke, eds. [1984]: *Undermining Rural Development with Cheap Credit*. Boulder, CO, Westview Press.

Adams, D. W., and R. C. Vogel [1986]: "Rural Financial Markets in Low-Income Countries: Recent Controversies and Lessons." *World Dev.* 14:477-487.

Adams, D. W., J. D. von Pischke, and G. Donald, eds. [1984]: *Rural Financial Markets in Developing Countries: Their Use and Abuse*. Baltimore, Johns Hopkins Univ. Press.

Adams, M. [1983]: *Economic Development and Change in East and Southeast Asia*. Canberra, Australian Government Publishing Service.

Adas, M. [1974]: *The Burma Delta Economic Development and Social Change on an Asian Rice Frontier, 1850-1941*. Madison, Univ. of Wisconsin Press.

Adelman, I., and C. T. Morris [1968]: "An Econometric Model of SocioEconomic and Political Change in Underdeveloped Countries." *Amer. Econ. Rev.* 58:1184-1218.

———— [1973]: *Economic Growth and Social Equity in Developing Countries*. Stanford, CA, Stanford Univ. Press.

Adelman, I., and D. Sunding [1987]: "Economic Policy and Income Distribution in China." *J. Comparative Econ.* 11(3):441-461.

Adelman, I., and E. Thorbecke, eds. [1966]: *The Theory and Design of Economic Development*. Baltimore, Johns Hopkins Univ. Press.

Adulavidhaya, K., Y. Kuroda, L. J. Lau, P. Lerttamrab, and P. A. Yotopoulos [1979]: "A Microeconomic Analysis of the Agriculture of Thailand." *Food Research Institute Studies* 17: 79-86.

Afiff, S., W. P. Falcon, and C. P. Timmer [1980]: "Elements of a Food and Nutrition Policy in Indonesia." In Papanek, ed., 1980, pp. 406-428.

Afiff, S. and C. P. Timmer [1971]: "Rice Policy in Indonesia." *Food Research Institute Studies* 10:131-159.

Agarwal, B. [1980]: "Tractorization, Productivity, and Employment: A Reassessment." *J. Dev. Studies* 16:375-386.

_____ [1984a]: "Rural Women and High Yielding Variety Technology." *Econ. and Polit. Weekly* 19(13):A39-A52.

_____ [1984b]: "Tractors, Tubewells, and Cropping Intensity in the Indian Punjab." *J. Dev. Studies* 20:290-302.

Agarwala, R. [1983]: *Price Distortions and Growth in Developing Countries.* Washington, World Bank Staff Working Paper No. 575.

Aghevli, B. B., and J. Marquez-Ruarte [1985]: *A Case of Successful Adjustment: Korea's Experience during 1980-84.* Washington, IMF, Occasional Paper No. 39.

Agricultural Projects Services Center [APROSC, 1982]: *Nepal: Foodgrain Marketing and Price Policy Study.* Kathmandu, Nepal.

_____ [APROSC, 1984]: *A Study on Agricultural Productivity in Nepal.* Kathmandu, Nepal.

Agro-Economic Research Center [1968-69]: *Report on the Study of High-Yielding Varieties Programme.* Waltair, Andhra Pradesh, Andhra Univ.

Ahammed, C. S., and R. W. Herdt [1983]: "Farm Mechanization in a Semi-closed Input-Output Model: The Philippines." *Amer. J. Agr. Econ.* 65:516-525.

_____ [1984]: "Measuring the Impact of Consumption Linkages on the Employment Effects of Mechanisation in Philippine Rice Production." *J. Dev. Studies* 20:242-255.

Ahluwalia, M. S. [1976]: "Inequality, Poverty, and Development." *J. Dev. Econ.* 3:307-342.

_____ [1978]: "Rural Poverty and Agricultural Performance." *J. Dev. Studies* 14:298-323.

Ahmad, N. [1976]: *A New Economic Geography of Bangladesh.* New Delhi, Vikas Publishing House.

Ahmed, B. [1975]: "The Economics of Tractor Mechanization in the Pakistan Punjab." *Food Research Institute Studies* 14:47-64.

Ahmed, I. [1974]: "Green Revolution with or without Tractors—The Case of Sri Lanka." *Marga Quarterly J.* 2(3):64-74.

_____ [1976]: "Technical Change and Labour Utilisation in Rice Cultivation: Bangladesh." *Bangladesh Dev. Studies* 5(3):359-366.

_____ [1978]: "Unemployment and Underemployment in Bangladesh Agriculture." *World Dev.* 6:1281-1296.

_____ [1981]: "Farm Size and Labour Use: Some Alternative Explanations." *Oxford Bull. of Econ. and Stat.* 43:73-88.

Ahmed, I., and V. W. Ruttan [1987]: *Generation and Diffusion of Agricultural Innovations: The Role of Institutional Factors.* Brookfield, VT, Gower.

Ahmed, R. [1977]: *Food Production in Bangladesh: An Analysis of Growth, Its Sources and Related Policies.* Dhaka, Bangladesh, BARC, Agricultural Economics and Rural Social Science Papers No. 2.

_____ [1978]: "Price Support *versus* Fertilizer Subsidy for Increasing Rice Production in Bangladesh." *Bangladesh Dev. Studies* 6:119-138.

_____ [1979]: *Foodgrain Supply, Distribution, and Consumption Policies within a Dual Pricing Mechanism: A Case Study of Bangladesh.* Washington, IFPRI, Research Report No. 8.

_____ [1981]: *Agricultural Price Policies under Complex Socioeconomic and Natural Constraints: The Case of Bangladesh.* Washington, IFPRI, Research Report No. 27.

_____ [1987a]: "Structure and Dynamics of Fertilizer Subsidy in Bangladesh." *Food Policy* 12:63–76.

_____ [1987b]: "A Structural Perspective of Farm and Non-Farm Households in Bangladesh." *Bangladesh Dev. Studies* 15(2):87–112.

_____ [1988]: "Structure, Costs, and Benefits of Food Subsidies in Bangladesh." In P. Pinstrup-Andersen, ed., 1988, pp. 219–228.

Ahmed, R., and A. Bernard [1989]: *Rice Price Fluctuation and An Approach to Price Stabilization in Bangladesh.* Washington, IFPRI, Research Report No. 72.

Ahmed, R., and M. Hossain [1987]: *Infrastructure and Development of a Rural Economy: A Case Study of Bangladesh.* Washington, IFPRI and Dhaka, BIDS, mimeo.

Ahmed, R., and N. Rustagi [1987]: "Marketing and Price Incentives in Asian and African Countries: A Comparison." In Elz, ed., 1987, pp. 104–118.

Ahn, C. Y., I. Singh, and L. Squire [1981]: "A Model of an Agricultural Household on a Multi-Crop Economy: The Case of Korea." *Rev. Econ. and Stat.* 63:520–525.

Akino, M., and Y. Hayami [1974]: "Sources of Agricultural Growth in Japan, 1880-1965." *Quart. J. Econ.* 83:454–479.

_____ [1975]: "Efficiency and Equity in Public Research: Rice Breeding in Japan's Economic Development." *Amer. J. Agr. Econ.* 57:1–10.

Akrasanee, N., and A. Wattananukit [1976]: "Comparative Advantage in Rice Production in Thailand." *Food Research Institute Studies* 15:177–212.

Alagh, Y. K., and P. S. Sharma [1980]: "Growth of Crop Production: 1960-61 to 1978-79—Is it Decelerating?" *Indian J. Agr. Econ.* 35(2):104–118.

Alam, M. F. [1981]: "Cost of Credit from Institutional Sources in Bangladesh." *Bangladesh J. Agr. Econ.* 4:51–61.

Alamgir, M. [1975a]: "Poverty, Inequality, and Social Welfare: Measurements, Evidence, and Policies." *Bangladesh Dev. Studies* 3:153–180.

_____ [1975b]: "Some Analysis of Distribution of Income, Consumption, Savings, and Poverty in Bangladesh." *Bangladesh Dev. Studies* 3:737–818.

_____ [1980]: *Famine in South Asia: Political Economy of Mass Starvation.* Cambridge, MA, Oelgeschlager, Gun and Hain.

Alauddin, M. [1982]: "Inputs and Returns to Agricultural Research in Bangladesh." *J. Management, Business, and Economics* 8(2):130–146.

Alauddin, M., and C. Tisdell [1987]: "Trends and Projections for Bangladesh Food Production: An Alternative Viewpoint." *Food Policy* 12:332–336.

Alauddin, T. [1975]: "Mass Poverty in Pakistan: A Further Study." *Pakistan Dev. Rev.* 14:431–450.

Alderman, H. [1984]: "Attributing Technological Bias to Public Goods." *J. Dev. Econ.* 14:375–393.

_____ [1987]: *Cooperative Dairy Development in Karnataka, India: An Assessment.* Washington, IFPRI, Research Report No. 64.

Alderman, H., and C. P. Timmer [1980a]: "Food Policy and Food Demand in Indonesia." *Bull. Indonesian Econ. Studies* 16(3):83–93.

_____ [1980b]: "Consumption Parameters for Sri Lanka Food Policy Analysis." *Sri Lanka J. of Agrarian Studies* 1(2):1–12.

Alderman, H., and J. von Braun [1984]: *The Effects of the Egyptian Food Ration and Subsidy System on Income Distribution and Consumption.* Washington, IFPRI, Research Report No. 45.

Ali, I. [1986]: *Rice in Indonesia: Price Policy and Comparative Advantage*. Manila, Philippines, ADB, Economic Staff Paper No. 29.

Ali, I., B. M. Desai, R. Radhakrishna, and V. S. Vyas [1981]: "Indian Agriculture at 2000 — Strategies for Equity." *Econ. and Polit. Weekly* 16(10, 11, 12):409-424.

Alim, A. [1974]: *An Introduction to Bangladesh Agriculture*. Dhaka, M. Alim.

Allen, F. [1985]: "On the Fixed Nature of Sharecropping Contracts." *Econ. J.* 95:30-48.

Amranand, P. [1983]: "Impact of the Price Support Programme for Rice." In *The Siam Project in Macro Economic Management of the Thai Economy*. Washington, World Bank, and Bangkok, NESDB.

Amranand, P., and W. Grais [1984]: *Macroeconomic and Distributional Implications of Sectoral Policy Interventions: An Application to Thailand*. Washington, World Bank Staff Working Paper No. 627.

Anderson, C. A., and M. J. Bowman, eds. [1965]: *Education and Economic Development*. Chicago, Aldine.

Anderson, D., and F. Khambata [1985]: "Financing Small-Scale Industry and Agriculture in Developing Countries: The Merits and Limitations of 'Commercial' Policies." *Econ. Dev. and Cultural Change* 33:349-371.

Anderson, D., and M. W. Leiserson [1980]: "Rural Nonfarm Employment in Developing Countries." *Econ. Dev. and Cultural Change* 28:227-248.

Anderson, J. R., J. L. Dillon, and J. B. Hardaker [1977]: *Agricultural Decision Analysis*. Ames, Iowa State Univ. Press.

Anderson, J. R., and K. B. Hamal [1983]: "Risk and Rice Technology in Nepal." *Indian J. Agr. Econ.* 38(2):217-222.

Anderson, J. R., and P. B. R. Hazell, eds. [1989]: *Variability in Grain Yields: Implication for Agricultural Research and Policy in Developing Countries*. Baltimore, Johns Hopkins Univ. Press for IFPRI.

Anderson, J. R., P. B. R. Hazell, and L. T. Evans [1987]: "Variability of Cereal Yields: Sources of Change and Implications for Agricultural Research and Policy." *Food Policy* 12:199-212.

Anderson, J. R., R. W. Herdt, and G. M. Scobie [1988]: *Science and Food: The CGIAR and its Partners*. Washington, World Bank.

Anderson, J. R., R. W. Herdt, G. M. Scobie, C. E. Pray, and H. E. Jahnke [1987]: *International Agricultural Research Centers: A Study of Achievements and Potential*. Armidale, South Wales, Australia, Univ. of New England, Dept. of Ag. Econ. and Bus. Mgmt., microfiche.

Anderson, K. [1981]: *Northeast Asian Agricultural Protection in Historical and Comparative Perspective: The Case of South Korea*. Canberra, ANU, Australia-Japan Research Center, Research Report No. 82.

———— [1983a]: "Growth of Agricultural Protection in East Asia." *Food Policy* 8:327-336.

———— [1983b]: "Economic Growth, Comparative Advantage and Agricultural Trade of Pacific Rim Countries." *Rev. Marketing and Agr. Econ.* 51:231-248.

———— [1983c]: "Fertilizer Policy in Korea." *J. Rural Dev.* 6(1):43-57.

———— [1983d]: "The Peculiar Rationality of Beef Import Quotas in Japan." *Amer. J. Agr. Econ.* 65:108-112.

———— [1987]: "On Why Agriculture Declines with Economic Growth." *Agr. Econ.* 1:195-207.

———— [1989a]: *Agriculture and Economic Growth in China: An Intersectoral and International Perspective*. In Longworth, ed., 1989, pp. 242-252.

_____ [1989b]: "Korea: A Case of Agricultural Protection." In Sicular, ed., 1989a, pp. 109-153.

Anderson, K., and I. C. Ahn [1984]: "Protection Policy and Changing Comparative Advantage in Korean Agriculture." *Food Research Institute Studies* 19:139-151.

Anderson, K., C. Findlay, and P. Drysdale [1987]: *China's Trade and Pacific Economic Growth.* Canberra, ANU, Australia-Japan Research Centre, Research Paper.

Anderson, K., and Y. Hayami, with A. George, M. Honma, K. Otsuka, E. Saxon, S. Shei, and R. Tyers [1986]: *The Political Economy of Agricultural Protection: East Asia in International Perspective.* Sydney, Allen and Unwin.

Anderson, K., and R. Tyers [1985a]: "Price, Trade, and Welfare Effects of Agricultural Protection: The Case of East Asia." *Rev. Marketing and Agr. Econ.* 53:113-140.

_____ [1985b]: "China's Economic Growth, Structural Transformation, and Food Trade." *Australian J. Chinese Affairs* 14:65-83.

_____ [1985c]: *Economic Growth and Agricultural Protection in East and Southeast Asia: Implications for International Grain and Meat Trade.* Canberra, ANU, ASEAN-Australian Economic Paper No. 21.

_____ [1987a]: "Economic Growth and Market Liberalisation in China: Implications for Agricultural Trade." *Developing Economies* 25(2):124-151.

_____ [1987b]: "Japan's Agricultural Policy in International Perspective." *J. Japanese and International Econ.* 1(1):131-146.

Anderson, K., and P. G. Warr [1987]: "General Equilibrium Effects of Agricultural Distortions: A Simple Model for Korea." *Food Research Institute Studies* 20:245-263.

Anderson, R. S., P. R. Brass, E. Levy, and B. M. Morris, eds. [1982]: *Science, Politics and the Agricultural Revolution in Asia.* Boulder, CO, Westview Press.

Andrew, C. O., and P. E. Hildebrand [1982]: *Planning and Conducting Applied Agricultural Research.* Boulder, CO, Westview Press.

Antholt, C., and H. E. B. Wennergren [1985]: "Agricultural Credit Reform in Bangladesh: Lessons Learned from a Pilot Project." *Bangladesh J. Agr. Econ.* 8(1):35-44.

Antle, J. M. [1984]: "Human Capital, Infrastructure, and the Productivity of Indian Rice Farmers." *J. Dev. Econ.* 14:163-181.

Antonelli, G., and A. Quadrio-Curzio [1988]: *The Agro-Technological System towards 2000: A European Perspective.* Amsterdam, North-Holland.

Apte, D. P. [1973]: "Crash Scheme for Rural Employment: Evaluation of the Programme in a District in Maharashtra." *Econ. and Polit. Weekly* 8(12):595-600.

Ardito-Barletta, N. [1970]: "Costs and Social Benefits of Agricultural Research in Mexico." Unpublished Ph.D. dissertation, Univ. of Chicago.

Armas, A., and D. J. Cryde [1981]: "Economic Incentives, Wage Policy, and Comparative Advantage in Philippine Agriculture." Quezon City, Philippines, Council on Asian Manpower Studies, Discussion Paper Series 81-03.

Arndt, T. M., D. G. Dalrymple, and V. W. Ruttan, eds. [1977]: *Resource Allocation and Productivity in National and International Agricultural Research.* Minneapolis, Univ. of Minnesota Press.

Ash, R. F. [1988]: "The Evolution of Agricultural Policy." *China Quarterly* (Special Issue on Food and Agriculture in China During the Post-Mao Era) 116:529-555.

Asian Development Bank [ADB, 1968]: *Asian Agricultural Survey.* Manila.

_____ [ADB, 1977]: *Asian Agricultural Survey, 1976: Rural Asia—Challenge and Opportunity.* New York, Praeger.

_____ [ADB, 1978]: *Rural Asia: Challenge and Opportunity. Second Asian Agricultural Survey.* Supplementary Papers, Vol. 1, Manila.

_____ [ADB, 1982]: *Nepal Agricultural Sector Strategy Study.* Vol. I and II, Manila. Also published in Kathmandu, Nepal.

_____ [ADB, 1985]: *Agriculture in Asia: Its Performance and Prospects, The Role of ADB in its Development.* Manila.

_____ [ADB, 1987]: *Rural Employment Creation in Asia and the Pacific.* Manila.

_____ [ADB, 1988]: *Evaluating Rice Market Intervention Policies: Some Asian Examples.* Manila.

Asian Productivity Organization [APO, 1975]: *Impact of Fertilizer Shortage: Focus on Asia—Report of the Symposium on Inter-relationship between Agricultural Input Industry and Agriculture.* Tokyo.

_____ [APO, 1979a]: *Fertilizer Distribution in Selected Asian Countries.* Tokyo.

_____ [APO, 1979b]: *Food Grain Distribution in Selected Asian Countries.* Tokyo.

_____ [APO, 1982]: *Grain Legumes Production in Asia.* Tokyo.

_____ [APO, 1983a]: *Farm Mechanization in Asia.* Tokyo.

_____ [APO, 1983b]: *Agricultural Research Management in Asia.* Tokyo.

_____ [APO, 1984]: *Agricultural Mechanization in Selected Asian Countries.* Tokyo.

_____ , ed. [APO, 1987]: *Productivity Measurement and Analysis: Asian Agriculture.* Tokyo.

Askari, H., and J. T. Cummings [1976]: *Agricultural Supply Response: A Survey of the Econometric Evidence.* New York, Praeger.

Aubert, C. [1987]: *Rural Capitalism versus Socialist Economics? Rural-Urban Relations and the Agricultural Reforms in China.* Communication for the Eighth International Conference on Soviet and East European Agriculture. Berkeley, Univ. of California.

Ayer, H. W. [1970]: "The Costs, Returns, and Effects of Agricultural Research in a Developing Country—The Case of Cotton Seed Research in São Paulo, Brazil." Unpublished Ph.D. dissertation, Purdue Univ., West Lafayette, IN.

Ayer, H. W., and G. E. Schuh [1972]: "Social Rates of Return and Other Aspects of Agricultural Research—The Case of Cotton Research in São Paulo, Brazil." *Amer. J. Agr. Econ.* 54:557-569.

Axinn, N. W., and G. H. Axinn [1983]: *Small Farms in Nepal.* Kathmandu, Rural Life Associates.

Azam, K. M. [1973]: "The Future of the Green Revolution in West Pakistan: A Choice of Strategy." *International J. Agrarian Affairs* 5:404-429.

Azizur Rahman, A. N. M. [1963]: "Expenditure Elasticities in Rural West Pakistan." *Pakistan Dev. Rev.* 3:232-249.

Bachman, K. L., and R. P. Christensen [1967]: "The Economics of Farm Size." In Southworth and Johnston, eds., 1967, pp. 234-257.

Bachman, K. L., and L. A. Paulino [1979]: *Rapid Food Production Growth in Selected Developing Countries: A Comparative Analysis of Underlying Trends, 1961-76.* Washington, IFPRI, Research Report No. 11.

Badan Urusan Logistik [BULOG, 1983]: *Main Operations of BULOG.* Jakarta, Indonesia.

Bagi, F. S. [1981a]: "Relationship between Farm Size and Economic Efficiency: An Analysis of Farm-Level Data from Haryana (India)." *Canadian J. Agr. Econ.* 29:317-326.

_____ [1981b]: "Economic Efficiency of Sharecropping in Indian Agriculture." *Malayan Econ. Rev.* 26:15-24.

_____ [1981c]: "Economics of Irrigation in Crop Production in Haryana." *Indian J. Agr. Econ.* 36(3):15-26.

_____ [1983a]: "Relationship between Farm Size, Productivity and Returns to Scale in Haryana (India) Agriculture." *Rural Systems* 1:189-212.

————— [1983b]: "Relationship between Farm Size, Productivity, Input Demand and Production Cost." *Artha Vijnana* 25:231-245.

————— [1984]: "Economic Evaluation of Some Agrarian Policies: An Analysis of Farm-Level Data from Haryana, India." *Rural Systems* 2:27-40.

Bajracharya, D. [1983]: "Fuel, Food, or Forest? Dilemmas in a Nepali Village." *World Dev.* 11:1057-1074.

Balassa, B. [1982]: "Structural Adjustment Policies in Developing Economies." *World Dev.* 10:23-38.

————— [1987]: "China's Economic Reforms in a Comparative Perspective." *J. Comparative Econ.* 11(3):410-426.

Balassa, B., and Associates [1971]: *The Structure of Protection in Developing Countries.* Baltimore, Johns Hopkins Univ. Press for the World Bank.

Bale, M. D. [1985]: *Agricultural Trade and Food Policy: The Experience of Five Developing Countries.* World Bank Staff Working Paper No. 724.

Bale, M. D., and R. C. Duncan [1983]: *Prospects for Food Production and Consumption in Developing Countries.* Washington, World Bank Staff Working Paper No. 596.

Bale, M. D., and B. L. Greenshields [1978]: "Japanese Agricultural Distortions and their Welfare Value." *Amer. J. Agr. Econ.* 60:59-64.

Bale, M. D., and E. Lutz [1978]: *Trade Restrictions and International Price Instability.* Washington, World Bank Staff Working Paper No. 303.

————— [1981]: "Price Distortions in Agriculture and their Effects: An International Comparison." *Amer. J. Agr. Econ.* 63:8-22.

Ban, S. H. [1981]: "The Growth of Agricultural Output and Productivity in Korea, 1918 to 1978." *J. Rural Dev.* 4(1):1-18.

Ban, S. H., P. Y. Moon, and D. H. Perkins [1980]: *Rural Development Studies in the Modernization of the Republic of Korea: 1945-1975.* Cambridge, MA, Harvard Univ. Press, Harvard East Asian Monograph No. 89.

Bandaranaike, R. O. [1984]: *Tea Production in Sri Lanka: Future Outlook and Mechanisms for Enhancing Sectoral Performance.* Colombo, Central Bank of Ceylon, Occasional Paper No. 7.

Banister, J. [1984]: "An Analysis of Recent Data on the Population of China." *Population and Dev. Rev.* 10:241-271.

Bansil, P. C. [1966]: "Impact of Food Policy on Agricultural Development in Ceylon." *Indian J. Agr. Econ.* 21(1):238-245.

————— [1971]: *Ceylon Agriculture: A Perspective.* New Delhi, Dhanpat Rai and Sons.

————— [1975]: *Agricultural Problems in India.* 2nd rev. and enlarged edition. New Delhi, Vikas Publishing.

Banta, G. R. [1982]: *Asian Cropping Systems Research: Microeconomic Evaluation Procedures.* Ottawa, IDRC.

Banta, S., and C. V. Mendoza, eds. [1984]: *Organic Matter and Rice.* Manila, IRRI.

Bapna, S. L. [1980]: *Aggregate Supply Response of Crops in a Developing Region.* New Delhi, Sultan and Chand.

Bardhan, K. [1970]: "Price and Output Response of Marketed Surplus of Foodgrains: A Cross-Sectional Study of Some North Indian Villages." *Amer. J. Agr. Econ.* 52:51-61.

Bardhan, P. K. [1969]: "Agriculture in China and India: Output, Input, and Prices." *Econ. and Polit. Weekly* 4(1,2):53-65.

————— [1970]: "On the Minimum Level of Living and the Rural Poor." *Indian Econ. Rev.* NS5:129-136.

_____ [1973]: "Size, Productivity, and Returns to Scale: An Analysis of Farm Level Data in Indian Agriculture." *J. Polit. Economy* 81:1370-1386.

_____ [1978]: "Labor Absorption in Rice Agriculture in South Asia." In Bardhan, Viadyanathan, *et al.*, eds., 1978.

_____ [1979]: "Labor Supply Functions in a Poor Agrarian Economy." *Amer. Econ. Rev* 69:73-83.

_____ [1980]: "Interlocking Factor Markets and Agrarian Development: A Review of Issues." *Oxford Econ. Papers* NS32:82-98.

_____ [1982]: "Agrarian Class Formation in India." *J. Peasant Studies* 10(1):73-94.

_____ [1983]: "Regional Variations in the Rural Economy of India." *Econ. and Polit. Weekly* 18(30):1319-1334.

_____ [1984a]: *Land, Labor, and Rural Poverty: Essays in Development Economics.* New York, Columbia Univ. Press.

_____ [1984b]: "Determinants of Supply and Demand for Labour in a Poor Agrarian Economy: An Analysis of Household Survey Data from Rural West Bengal." In Binswanger and Rosenzweig, eds., 1984, pp. 242-262.

Bardhan, P. K., and K. Bardhan [1969]: "The Problem of Marketed Surplus of Cereals." *Econ. and Polit. Weekly* 4(26):A103-A110.

_____ [1971]: "Price Response of Marketed Surplus of Foodgrains." *Oxford Econ. Papers* NS23:255-267.

Bardhan, P. K., and A. Rudra [1978]: "Interlinkage of Land, Labor, and Credit Relations: An Analysis of Village Survey Data in East India." *Econ. and Polit. Weekly* 13(6,7):367-384.

_____ [1981]: "Terms and Conditions of Labour Contracts in Agriculture: Results of a Survey in West Bengal." *Oxford Bull. Econ. and Stat.* 43(1):89-111.

_____ [1983]: *Agrarian Relations in West Bengal.* Bombay, India, Somaiya.

_____ [1986]: "Labour Mobility and the Boundaries of the Village Moral Economy." *J. Peasant Studies* 13(3):90-115.

Bardhan, P. K., and T. N. Srinivasan [1971]: "Cropsharing Tenancy in Agriculture: A Theoretical and Empirical Analysis." *Amer. Econ. Rev.* 61:48-64.

_____ [1987]: *Land, Labor, and Rural Poverty: Essays in Development Economics.* New York, Columbia Univ. Press.

Bardhan, P. K., A. Viadyanathan, Y. Alagh, G. S. Bhalla, and A. L. Bhadem, eds. [1978]: *Labour Absorption in Indian Agriculture, Some Exploratory Investigations.* Bangkok, ILO-ARTEP.

Barker, R. [1966]: "The Response of Production to a Change in Rice Price." *Philippine Econ. J.* 5(2):260-276.

_____ [1978]: "Yield and Fertilizer Input." In IRRI, 1978b, pp. 35-66.

_____ [1979]: "Adoption and Production Impact of New Rice Technology: The Yield Constraints Problem." In IRRI, 1979a, pp. 1-26.

_____ [1982]: "Recent Trends in Labor Utilization and Productivity in Philippine Agriculture with Comparisons to Other Asian Experiences." In Hainsworth, ed., 1983, pp. 141-172.

_____ [1985]: *An Overview of Research in Irrigation Management in Asia.* Kandy, Sri Lanka, IIMI, Research Paper No. 1.

_____ [1987]: "A Perspective on Studies of Productivity Growth in Asian Agriculture." In APO, ed., 1987, pp. 93-101.

Barker, R., and V. G. Cordova [1978]: "Labor Utilization in Rice Production." In IRRI, 1978a, pp. 113-136.

Barker, R., E. W. Coward, Jr., G. Levine, and L. E. Small [1984]: *Irrigation Development in Asia: Past Trends and Future Directions*. Ithaca, NY, Cornell Univ. Studies in Irrigation Series, No. 1.

Barker, R., and Y. Hayami [1976]: "Price Support vs. Input Subsidy for Food Self-Sufficiency in Developing Countries." *Amer. J. Agr. Econ.* 58:617-628.

Barker, R., and R. W. Herdt [1977]: "Small-Farmers and Changing Rice Technology." *Ekonomi Dan Keuangan Indonesia* 25(2):157-179.

_____ [1978]: "Equity Implications of Technology Changes." In IRRI, 1978b, pp. 83-108.

_____ [1979]: "Rainfed Lowland Rice as a Research Priority: An Economist's View." In IRRI, 1979b, pp. 3-50. Also in IRRI Research Paper Series 26, 1979.

_____ [1982]: "Setting Priorities for Rice Research in Asia." In R. S. Anderson, Brass, Levy, and Morrison, eds., 1981, pp. 427-462.

Barker, R., and R. W. Herdt, with B. Rose [1985]: *The Rice Economy of Asia*. Washington, RFF.

Barker, R., and M. Mangahas [1971]: "Environmental and Other Factors Influencing the Performance of New High Yielding Varieties of Wheat and Rice in Asia." In *IAAE*, 1971, pp. 397-408; also in *ISAE*, 1972, pp. 223-236.

Barker, R., and R. Sinha, with B. Rose [1982]: *The Chinese Agricultural Economy*. Boulder, CO, Westview Press.

Barlow, C. [1978]: *The National Rubber Industry, its Development Technology, and Economy in Malaysia*. New York, Oxford Univ. Press.

Barlow, C., S. Jayasuriya, and E. C. Price [1982]: *Evaluating Technology for New Farming Systems: Case Studies from Philippine Rice Farms*. Los Baños, Philippines, IRRI.

Barnett, A. D. [1981]: *China's Economy in Global Perspective*. Washington, Brookings.

Barnett, A. D., and R. Clough, eds. [1986]: *Modernizing China: Post-Mao Reform and Development*. Boulder, CO, Westview Press.

Barnum, H. [1973]: *A Model of the Market for Foodgrains in India, 1948-1964*. Berkeley, Univ. of California, Institute of International Studies, Technical Report No. 23.

Barnum, H. N., and L. Squire [1979a]: "An Econometric Application of the Theory of the Farm-Household." *J. Dev. Econ.* 6:79-102.

_____ [1979b]: *A Model of an Agricultural Household: Theory and Evidence*. Baltimore, Johns Hopkins Univ. Press.

_____ [1980]: "Predicting Agricultural Output Response." *Oxford Econ. Papers* 32:284-295.

Barraclough, S. L. [1967]: "Comment on the Economics of Farm Size." In Southworth and Johnston, eds., 1967, pp. 263-267.

Bartsch, W. H. [1977]: *Employment and Technology Choice in Asian Agriculture*. New York, Praeger.

Bateman, D. I. [1976]: "Agricultural Marketing: A Review of the Literature of Marketing Theory and Selected Applications." *J. Agr. Econ.* 27:171-225.

Bauer, P. T. [1972]: *Dissent on Development: Studies and Debates in Development Economics*. Cambridge, MA, Harvard Univ. Press.

Bauer, P. T., and B. S. Yamey [1959]: "A Case Study of Response to Price in an Under-Developed Country." *Econ. J.* 69:800-805.

Bautista, R. M. [1987]: *Production Incentives in Philippine Agriculture: Effects of Trade and Exchange Rate Policies*. Washington, IFPRI, Research Report No. 59.

Bautista, R. M., and S. Naya, eds. [1984]: *Energy and Structural Change in the Asia-Pacific Region: Papers and Proceedings of the Thirteenth Pacific Trade and Development Conference*. Manila, PIDS and ADB.

Beaton, G. [1982]: "Evaluation of Nutrition Interventions: Methodologic Considerations." *Amer. J. Clinical Nutrition* 35:1280-1289.

Beaton, G. H., and H. Ghassemi [1982]: "Supplementary Feeding Programs for Young Children in Developing Countries." *Amer. J. Clinical Nutrition* 35:864-916.

Behrman, J. R. [1966]: "Price Elasticity of the Marketed Surplus of a Subsistence Crop." *J. Farm Econ.* 48:875-893.

———— [1968]: *Supply Response in Underdeveloped Agriculture: A Case Study of Four Major Annual Crops in Thailand, 1937-63.* Amsterdam, North-Holland.

———— [1988]: "Nutrition and Health and their Relations to Economic Growth, Poverty Alleviation, and General Development." In Psacharopoulos, ed., 1988b.

Behrman, J. R., and A. B. Deolalikar [1987]: "Will Developing Country Nutrition Improve with Income? A Case Study for Rural South India." *J. Polit. Economy* 95:492-507.

Behrman, J. R., A. B. Deolalikar, and B. L. Wolfe [1988]: "Nutrients: Impacts and Determinants." *World Bank Econ. Rev.* 2:299-320.

Behrman, J. R., and K. N. Murty [1985]: "Market Impacts of Technological Change for Sorghum in Indian Near-Subsistence Agriculture." *Amer. J. Agr. Econ.* 67:539-549.

Bell, C. L. G., and P. B. R. Hazell [1980]: "Measuring the Indirect Effects of an Agricultural Investment Project on its Surrounding Region." *Amer. J. Agr. Econ.* 62:75-86.

Bell, C. L. G., P. B. R. Hazell, and R. Slade [1982a]: *Project Evaluation in Regional Perspective: A Study of an Irrigation Project in Northwest Malaysia.* Baltimore, Johns Hopkins Univ. Press.

———— [1982b]: "The Prospects for Growth and Change in the Muda Project Region, Malaysia." In Hainsworth, ed., 1982, pp. 255-276.

Bellamy, M., and B. Greenshields, eds. [1987]: *Agriculture and Economic Instability.* Brookfield, VT, Gower Publishing, IAAE Occasional Paper No. 4.

Beneria, L., ed. [1982]: *Women and Development: The Sexual Division of Labor in Rural Societies.* New York, Praeger.

Benjamin, M. P. [1981]: *Investment Projects in Agriculture: Principles and Case Studies.* Harlow, Essex, Longman.

Bennagen, M. E. C. [1982]: *Staple Food Consumption in the Philippines.* Washington, IFPRI, IFDC and IRRI, Working Paper No. 5.

Benor, D., and M. Baxter [1984]: *Training and Visit Extension.* Washington, World Bank.

Benor, D., and J. Q. Harrison [1977]: *Agricultural Extension: The Training and Visit System.* Washington, World Bank.

Benus, J., J. Kmenta, and H. Shapiro [1976]: "The Dynamics of Household Budget Allocation to Food Expenditures." *Rev. Econ. and Stat.* 58:129-138.

Berg, A. D. [1973]: *The Nutrition Factor: Its Role in National Development.* Washington, Brookings.

———— [1981]: *Malnourished People: A Policy View.* Washington, World Bank.

———— [1987]: *Malnutrition: What Can be Done? Lessons from World Bank Experience.* Baltimore, Johns Hopkins Univ. Press.

Berg, A. D., N. S. Scrimshaw, and D. L. Call, eds. [1973]: *Nutrition, National Development, and Planning.* Cambridge, MA, MIT Press.

Bergmann, H., and J. M. Boussard [1976]: *Guide to the Economic Evaluation of Irrigation Projects.* Paris, OECD.

Bergsman, J., and A. S. Manne [1966]: "An Almost Consistent Intertemporal Model for India's Fourth and Fifth Plans." In Adelman and Thorbecke, eds., 1966, pp. 239-256.

Bernsten, R. H., B. H. Siwi, and H. M. Beachell [1982]: *The Development and Diffusion of Rice Varieties in Indonesia.* Los Baños, Philippines, IRRI, Research Paper No. 71.

Berrill, K., ed. [1964]: *Economic Development with Special Reference to East Asia*. London, Macmillan.

Berry, A., and R. H. Sabot [1978]: "Labor Market Performance in Developing Countries: A Survey." *World Dev.* 6:1199-1242.

Berry, D. A., and W. R. Cline [1979]: *Agrarian Structure and Productivity in Developing Countries*. Baltimore, Johns Hopkins Univ. Press.

Berry, R. A. [1971]: "Land Reform and the Agricultural Income Distribution." *Pakistan Dev. Rev.* 11:30-44.

Bertrand, T. [1980]: *Thailand: Case Study of Agricultural Input and Output Pricing*. Washington, World Bank, Staff Paper No. 385.

Bhaduri, A. [1973]: "Agricultural Backwardness under Semi-Feudalism." *Econ. J.* 83:120-137.

_____ [1983]: *The Economic Structure of Agricultural Backwardness*. London, Academic Press.

Bhagwati, J. N. [1962]: "The Theory of Comparative Advantage in the Context of Underdevelopment and Growth." *Pakistan Dev. Rev.* 2:339-353.

_____ [1988]: "Export-Promoting Trade Strategy: Issues and Evidence." *World Bank Research Observer* 3:27-57.

Bhagwati, J. N., and S. Chakravarty [1969]: "Surveys of National Economic Policy Issues and Policy Research: Contribution to Indian Economic Analysis—A Survey." *Amer. Econ. Rev.* 59(4, Part 2):1-73.

Bhagwati, J. N., and P. Desai [1970]: *India: Planning for Industrialization—Industrialization and Trade Policies since 1951*. London, Oxford Univ. Press for the OECD Development Centre.

Bhagwati, J. N., and T. N. Srinivasan [1975]: *Foreign Trade Regimes and Economic Development: India*. New York, Columbia Univ. Press for NBER.

Bhalla, G. S. [1983]: *Green Revolution and the Small Peasant: A Study of Income Distribution among Punjab Cultivators*. New Delhi, India, Concept Publishing.

Bhalla, G. S., and Y. K. Alagh [1979]: *Performance of Indian Agriculture: A Districtwise Study*. New Delhi, Sterling.

Bhalla, S. [1977]: "Agriculture Growth: Roles of Institutional and Infrastructural Factors." *Econ. and Polit. Weekly* 12(45,46):1898-1904.

_____ [1978]: "The Role of Sources of Income and Investment Opportunities in Rural Savings." *J. Dev. Econ.* 5:259-281.

_____ [1980]: "The Measurement of Permanent Income and its Application to Savings Behavior." *J. Polit. Economy* 88:722-744.

Bhalla, S. S., and P. Glewwe [1986]: "Growth and Equity in Developing Countries: A Reinterpretation of the Sri Lankan Experience." *Econ. Rev.* 1(1):35-64.

Bharadwaj, K. [1974]: *Production Conditions in Indian Agriculture: A Study Based on Farm Management Surveys*. Cambridge, England, Cambridge Univ. Press.

Bharadwaj, R. [1962]: *Structural Basis of India's Foreign Trade: A Study Suggested by the Input-Output Analysis*. Univ. of Bombay, Series in Monetary and International Economics No. 6.

Bhargawa, P. N., and V. S. Rustogi [1972]: "Study of Marketable Surplus of Paddy in Burdwan District." *Indian J. Agr. Econ.* 27(3):63-68.

Bhooshan, B. S. [1979]: *The Development Experience of Nepal*. New Delhi, Concept Publishing.

Bienen, H. S., and M. Gersovitz [1986]: *Consumer Subsidy Costs, Violence, and Political Stability*. Princeton, NJ, Princeton Univ. Press.

Bigman, D. [1982]: *Coping with Hunger: Toward a System of Food Security and Price Stabilization.* Cambridge, MA, Ballinger.

———— [1985]: *Food Policies and Food Security Under Instability: Modeling and Analysis.* Lexington, MA, Lexington Books.

Bigman, D., and S. Reutlinger [1979]: "Food Price and Supply Stabilization: National Buffer Stocks and Trade Policies." *Amer. J. Agr. Econ.* 61:657-667.

Billings, M. H., and A. Singh [1969]: "Labour and the Green Revolution: The Experience in Punjab." *Econ. and Polit. Weekly* 4(52):A221-A224.

———— [1970]: "Mechanization and Rural Employment, with Some Implications for Rural Income Distribution." *Econ. and Polit. Weekly* 5(26):A61-A72.

Bindlish, V., R. Barker, and T. D. Mount [1989]: "Can Yield Variability Be Offset by Improved Information? The Case of Rice in India." In J. R. Anderson and Hazell, eds., 1989, pp. 287-300.

Binswanger, H. P. [1978]: *The Economics of Tractors in South Asia: An Analytic Review.* New York, ADC and ICRISAT.

———— [1980]: "Attitudes towards Risk: Experimental Measurement in Rural India." *Amer. J. Agr. Econ.* 62:395-407.

———— [1981]: "Attitudes toward Risk: Theoretical Implications of an Experiment in Rural India." *Econ. J.* 91:867-890.

———— [1984]: *Agricultural Mechanization: A Comparative Historical Perspective.* Washington, World Bank Staff Working Paper No. 673.

———— [1986]: "Behavioral and Material Determinants of Production Relations in Agriculture." *J. Dev. Studies* 22:503-539.

Binswanger, H. P., S. L. Bapna, and J. B. Quizon [1984]: "Systems of Output Supply and Factor Demand Equations for Semi-Arid Tropical India." *Indian J. Agr. Econ.* 39(2):179-202.

Binswanger, H. P., and G. Donovan [1987]: *Agricultural Mechanization: Issues and Options.* Washington, World Bank Policy Study.

Binswanger, H. P., R. E. Evenson, C. A. Florencio, and B. N. White, eds. [1980]: *Rural Household Studies in Asia.* Singapore, Singapore Univ. Press.

Binswanger, H. P., N. S. Jodha, and B. C. Barah [1980]: "The Nature and Significance of Risk in the Semi-Arid Tropics." In ICRISAT, 1980, pp. 303-316.

Binswanger, H. P., S. R. Khandker, and M. R. Rosenzweig [1989]: *The Impact of Infrastructure and Financial Institutions on Agricultural Output and Investment in India.* Washington, World Bank, Policy, Planning and Research Working Paper Series No. 163.

Binswanger, H. P., and J. Quizon [1984]: *Distributional Consequences of Alternative Food Policies in India.* Washington, World Bank, Agricultural Research Unit Report No. 20.

———— [1986]: *What Can Agriculture Do for the Poorest Rural Groups.* Washington, World Bank, Agricultural Research Unit, Discussion Paper No. 57.

Binswanger, H. P., J. B. Quizon, and G. Swamy [1984]: "The Demand for Food and Foodgrain Quality in India." *Indian Econ. J.* 31(4):72-96.

Binswanger, H. P., and M. R. Rosenzweig, eds. [1984]: *Contractual Arrangements, Employment, and Wages in Rural Labor Markets in Asia.* New Haven, CT, Yale Univ. Press.

Binswanger, H. P., and V. W. Ruttan, with U. Ben-Zion, A. de Janvry, R. E. Evenson, Y. Hayami, T. L. Roe, J. H. Sanders, W. W. Wade, A. Weber, and P. Yeung [1978]: *Induced Innovation: Technology, Institutions, and Development.* Baltimore, Johns Hopkins Univ. Press.

Binswanger, H. P., and J. G. Ryan [1977]: "Efficiency and Equity Issues in Ex Ante Allocation of Research Resources." *Indian J. Agr. Econ.* 32(3):217-231.

Binswanger, H. P., and P. L. Scandizzo [1983]: *Patterns of Agricultural Protection.* Washington, World Bank, Agricultural Research Unit Report No. 15.

Bird, R. M. [1974]: *Taxing Agricultural Land in Developing Countries.* Cambridge, Mass., Harvard Univ. Press.

Birowo, A. T. [1983]: "Rural Development Planning and Implementation." In Maunder and Ohkawa, eds., 1983, pp. 417–426.

Biswas, A. K., D. K. Zuo, J. E. Nickum, and C. M. Liu [1983]: *Long Distance Water Transfer: A Chinese Case Study and International Experiences.* Dublin, Ireland, Tycooly International Publishing.

Blaug, M., R. Layard, and M. Woodhall [1969]: *The Causes of Graduate Unemployment in India.* London, Allen Lane.

Bliss, C. J., and N. H. Stern [1982]: *Palanpur: The Economy of an Indian Village.* Oxford, England, Oxford Univ. Press.

Blyn, G. [1966]: *Agricultural Trends in India, 1891-1947: Output, Availability, and Productivity.* Philadelphia, Univ. of Pennsylvania Press.

———— [1973]: "Price Series Correlation as a Measure of Market Integration." *Indian J. Agr. Econ.* 28(2):56–59.

———— [1983a]: "The Green Revolution Revisited." *Econ. Dev. and Cultural Change* 31:705–726.

———— [1983b]: "Income Distribution among Haryana and Punjab Cultivators, 1968/69–1975/76." *Indian Econ. Rev.* 18:199–224.

Bogahawatte, C. [1983]: "Simulating the Impact of the Changes in Wheat Flour Imports and World Price of Rice on the Supply and Demand for Rice in Sri Lanka." *Indian J. Agr. Econ.* 38(1):15–26.

Booth, A. [1977]: "Irrigation in Indonesia Part I." *Bull. Indonesian Econ. Studies* 13:33–74.

Booth, A., C. C. David, S. Baharsjah, S. Meyanathan, G. Sivalingam, F. Chan, D. Mongkolsmai, A. G. Bennett, and R. G. Mauldon [1986]: *Food Trade and Food Security in ASEAN and Australia.* Kuala Lumpur, Malaysia, ASEAN-Australian Joint Research Project.

Booth, A., and R. M. Sundrum [1976]: "The 1973 Agricultural Census." *Bull. Indonesian Econ. Studies* 12(2):90–105.

———— [1985]: *Labour Absorption in Agriculture: Theoretical Analysis and Empirical Investigation.* New York, Oxford Univ. Press.

Borton, R. E., ed. [1967]: *Case Studies to Accompany "Getting Agriculture Moving'.* New York, ADC.

Bose, S. R., and E. H. Clark II [1969]: "Some Basic Considerations on Agricultural Mechanization in West Pakistan." *Pakistan Dev. Rev.* 9(1):273–308.

Boserup, E. [1965]: *The Conditions of Agricultural Growth: The Economics of Agrarian Change under Population Pressure.* Chicago, Aldine.

———— [1981]: *Population and Technological Change: A Study of Long-Term Trends.* Chicago, Univ. of Chicago Press.

Bottomley, A. [1963a]: "The Cost of Administering Private Loans in Underdeveloped Rural Areas." *Oxford Econ. Papers* NS15:159–162.

———— [1963b]: "The Premium for Risk as a Determinant of Interest Rates in Underdeveloped Rural Areas." *Quart. J. Econ.* 77:637–647.

———— [1964a]: "Monopoly Profit as a Determinant of Interest Rates in Underdeveloped Rural Areas." *Oxford Econ. Papers* NS16:431–437.

———— [1964b]: "The Determinants of Pure Rates of Interest in Underdeveloped Rural Areas." *Rev. Econ. and Stat.* 46:301–304.

_____ [1964c]: "The Structure of Interest Rates in Underdeveloped Rural Areas." *J. Farm Econ.* 46:313-322.

Bouis, H. [1990]: *The Effects of Commercialization on Land Tenure, Household Resources, and Nutrition in the Philippines.* Washington, IFPRI, Research Report No. 79.

Bouis, H., and R. W. Herdt [1982]: "Evaluating Trade-offs and Complementarities among Public Investments in the Rice Sectors of Asian Countries." In Chisholm and Tyers, eds., 1982, pp. 263-280.

Boussard, J. M., and M. Petit [1967]: "Representation of Farmers' Behavior under Uncertainty with a Focus-Loss Constraint." *J. Farm Econ.* 49:869-880.

Boyce, J. K. [1986]: "Water Control and Agricultural Performance in Bangladesh." *Bangladesh Dev. Studies* 14:1-35.

Boyce, J. K., and R. E. Evenson [1975]: *National and International Agricultural Research and Extension Programs.* New York, ADC.

Braverman, A., C. H. Ahn, and J. S. Hammer [1983]: *Alternative Agricultural Pricing Policies in the Republic of Korea: Their Implications for Government Deficits, Income Distribution, and Balance of Payments.* Washington, World Bank Staff Working Paper No. 621.

Braverman, A., and J. L. Guasch [1986]: "Rural Credit Markets and Institutions in Developing Countries: Lessons of Policy Analysis from Practice and Modern Theory." *World Dev.* 14:1253-1267.

Braverman, A., and J. S. Hammer [1988]: "Computer Models for Agricultural Policy Analysis." *Finance and Dev.* 25(2):34-37.

Braverman, A., and T. N. Srinivasan [1981]: "Credit and Sharecropping in Agrarian Societies." *J. Dev. Econ.* 9:289-312.

Braverman, A., and J. E. Stiglitz [1982]: "Sharecropping and the Interlinking of Agrarian Markets." *Amer. Econ. Rev.* 72:695-715.

Bray, F. [1986]: *Rice Societies: Technology and Development in Asian Societies.* Oxford, Basil Blackwell.

Breimyer, H. F. [1973]: "The Economics of Agricultural Marketing: A Survey." *Rev. Marketing and Agr. Econ.* 41:115-165.

Breman, J., and S. Mundle, eds. [1990]: *Rural Transformation in Asia.* New Delhi, Oxford Univ. Press.

Brown, D. D. [1971]: *Agricultural Development in India's Districts.* Cambridge, MA, Harvard Univ. Press.

Brown, G. T. [1973]: *Korean Pricing Policies and Economic Development in the 1960s.* Baltimore, Johns Hopkins Univ. Press.

Brown, L. R. [1968]: "The Agricultural Revolution in Asia." *Foreign Affairs* 46:688-698.

_____ [1970]: *Seeds of Change: The Green Revolution and Development in the 1970s.* New York, Praeger for ODC.

Brown, M. L. [1979]: *Farm Budgets: From Farm Income Analysis to Agricultural Project Analysis.* Baltimore, Johns Hopkins Univ. Press for the World Bank.

Brun, E., and J. Hersh [1976]: *Socialist Korea: A Case Study in the Strategy of Economic Development.* New York, Monthly Review Press.

Buck, J. L. [1956]: *Land Utilization in China.* New York, ADC.

Bunting, A. H., ed. [1987]: *Agricultural Environments: Characterization, Classification and Mapping.* Wallingford, UK, Commonwealth Agricultural Bureaux International.

Burmeister, L. L. [1987]: "The South Korean Green Revolution: Induced or Directed." *Econ. Dev. and Cultural Change* 35(4):767-790.

Busch, L., and W. B. Lacy [1983]: *Science, Agriculture and Politics of Research.* Boulder, CO, Westview Press.

Bussink, W. C. F. [1970]: "A Complete Set of Consumption Coefficients for West Pakistan." *Pakistan Dev. Rev.* 10(2):193-231.

Buttrick, J. [1958]: "A Note on Professor Solow's Growth Model." *Quart. J. Econ.* 72:633-636.

Buvinic, M., and M. A. Lycette, eds. [1983]: *Women and Poverty in the Third World.* Baltimore, Johns Hopkins Univ. Press.

Buvinic, M., M. A. Lycette, and W. P. McGreevey, eds. [1979]: *Women and Poverty.* Washington, International Center for Research on Women.

Byerlee, D., and G. Sain [1986]: "Food Pricing Policy in Developing Countries: Bias against Agriculture or for Urban Consumers." *Amer. J. Agr. Econ.* 68:961-969.

Byres, T. J., ed. [1983]: *Sharecropping and Sharecroppers.* London, Cass Publishers.

Cabanilla, L. S. [1983]: "Economic Incentives and Comparative Advantage in the Livestock Industry." Manila, PIDS 83-07.

Cain, M. [1983]: "Landlessness in India and Bangladesh: A Critical Review of National Data Sources." *Econ. Dev. and Cultural Change* 32:149-168.

Caliendo, M. A. [1979]: *Nutrition and the World Food Crisis.* New York, Macmillan.

Calkins, P. H. [1980]: *The New Decision Making Environment in Chinese Agriculture.* Ames, ISU, International Studies in Economics Monograph No. 14.

_____ [1982]: "Why Development Fails: The Evolution Gap in Nepal's Subsistence Agriculture." *World Dev.* 10:397-411.

Carloni, A. S. [1981]: "Sex Disparities in the Distribution of Food Within Rural Households." *Food and Nutrition* 7:3-12.

Carruthers, I. D., and C. Clark [1981]: *The Economics of Irrigation.* Liverpool, England, Liverpool Univ. Press.

Carruthers, I. D., and R. Stoner [1981]: *Economic Aspects and Policy Issues in Groundwater Development.* Washington, World Bank Staff Working Paper No. 496.

Carter, C. A., and F. N. Zhong [1988]: *China's Grain Production and Trade: An Economic Analysis.* Boulder, CO, Westview Press.

Carter, H., ed. [1981]: *Food Security in a Hungry World: Proceedings of the International Food Policy Conference.* Davis, Univ. of California Press.

Castillo, G. T. [1975]: *All in a Grain of Rice.* Laguna, Philippines, SEARCA.

Castle, E., and K. Hemmi, eds. [1982]: *U.S.-Japanese Agricultural Trade Relations.* Washington, RFF.

Centro Internacional de Agricultura Tropical, ed. [CIAT, 1986]: *Cassava in Asia, its Potential and Research Development Needs.* Cali, Colombia.

Chakravarty, S. [1969]: *Capital and Development Planning.* Cambridge, MA, MIT Press.

Chakravarty, S., and L. Lefeber [1965]: "An Optimizing Planning Model." *Econ. Weekly* 17(5, 6, 7):237-252.

Chalamwong, Y., and G. Feder [1988]: "The Impact of Land Ownership Security: Theory and Evidence from Thailand." *Econ. Rev.* 2:187-204.

Chambers, R. [1975]: *Water Management and Paddy Production in the Dry Zone of Sri Lanka.* Colombo, ARTI, Occasional Publication No. 8.

_____ [1978]: "Project Selection for Poverty-Focused Rural Development: Simple is Optimal." *World Dev.* 6:209-219.

_____ [1981]: "Rural Poverty Unperceived: Problems and Remedies." *World Dev.* 9:1-19.

_____ [1983]: *Rural Development: Putting the Last First.* London, Longman.

_____ [1987]: *Managing Canal Irrigation: Practical Analysis from South Asia.* New Delhi, Oxford Univ. Press; and Cambridge, England, Cambridge Univ. Press.

Chand, R., and J. L. Kaul [1986]: "A Note on the Use of the Cobb-Douglas Profit Function." *Amer. J. Agr. Econ* 68:162–164.

Chandler, W. U. [1986]: *The Changing Role of the Market in National Economies*. Washington, Worldwatch Institute Paper No. 72.

Chandraratna, M. F. [1976]: "Key Factors in Increasing Agricultural Production." *Marga Quarterly J.* 3(1):67–76.

Chang, K. Y., ed. [1985]: *Perspectives on Development in Mainland China*. Boulder, CO, Westview Press.

Chao, K. [1970]: *Agricultural Production in Communist China, 1949-1965*. Madison, Univ. of Wisconsin Press.

Chaudhry, M. G. [1982]: "Green Revolution and Redistribution of Rural Incomes: Pakistan's Experience." *Pakistan Dev. Rev.* 21:173–205.

Chaudhry, M. G., M. A. Gill, and G. M. Chaudhry [1985]: "Size-Productivity Relationship in Pakistan's Agriculture in the Seventies." *Pakistan Dev. Rev.* 24:349–361.

Chayanov, A. V. [1966]: *The Theory of Peasant Economy*. Homewood, Ill., Richard Irwin. Edited by D. Thorner, B. Kerblay and R.E.F. Smith. The American Economic Association Translation Series.

Chen, H. Y., W. F. Hsu, and Y. K. Mao [1975]: "Rice Policies of Taiwan." *Food Research Institute Studies* 14:403–417.

Chen, L. C., E. Huq, and S. D'Souza [1981]: "Sex Bias in the Family Allocation of Food and Health Care in Rural Bangladesh." *Population and Dev. Rev.* 7:55–70.

Chen, N. R., and W. Galenson [1969]: *The Chinese Economy under Communism*. Chicago, Aldine.

Chen, W. H., and J. T. Scott, Jr. [1982]: *The Simulation of Grain Supply, Producer Income and Consumer Price Policy: The Case of Rice in Taiwan*. Urbana-Champaign, Univ. of IL, DAE, No. 181.

Chenery, H. B. [1960]: "Patterns of Industrial Growth." *Amer. Econ. Rev.* 50:624–654.

————— [1979]: *Structural Change and Development Policy*. New York, Oxford Univ. Press.

Chenery, H. B., M. S. Ahluwalia, C. L. G. Bell, J. H. Duloy, and R. Jolly [1974]: *Redistribution with Growth*. London, Oxford Univ. Press for the World Bank.

Chenery, H. B., and A. MacEwan [1966]: "Optimal Patterns of Growth and Aid: The Case of Pakistan." In Adelman and Thorbecke, eds., 1966, pp. 149–178.

Chenery, H. B., and T. N. Srinivasan, eds. [1988]: *Handbook on Development Economics*, Vol. I. New York, Elsevier.

Chenery, H. B., and A. Strout [1966]: "Foreign Assistance and Economic Development." *Amer. Econ. Rev.* 56:679–733.

Chenery, H. B., and M. Syrquin [1975]: *Patterns of Development, 1950-1970*. London, Oxford Univ. Press.

Chenery, H. B., and T. Watanabe [1958]: "International Comparisons of the Structure of Production." *Econometrica* 26:487–521.

Cheng, C. Y. [1982]: *China's Economic Development: Growth and Structural Change*. Boulder, CO, Westview Press.

Cheng, S. H. [1968]: *The Rice Industry of Burma, 1852-1940*. Singapore, Univ. of Malaya Press.

————— [1973]: *The Rice Trade of Malaya*. Singapore, University Education Press.

Chennareddy, V. [1967]: "Production Efficiency in South Indian Agriculture." *J. Farm Econ.* 49:816–820.

Cheong, K. C., and E. H. D'Silva [1984]: *Prices, Terms of Trade, and the Role of Government in Pakistan's Agriculture*. Washington, World Bank Staff Working Paper No. 643.

Chernichovsky, D., and O. A. Meesook [1984]: *Patterns of Food Consumption and Nutrition in Indonesia*. Washington, World Bank Staff Working Paper No. 670.

Cheung, S. N. S. [1969]: *The Theory of Share Tenancy with Special Application to Asian Agriculture and the First Phase of Taiwan Land Reform*. Univ. of Chicago Press.

Chhibber, A. [1988]: "Raising Agricultural Output: Price and Nonprice Factors." *Finance and Dev.* 25(2):44–47.

Chichilnisky, G., and L. Taylor [1980]: "Agriculture and the Rest of the Economy: Macro Connections and Policy Constraints." *Amer. J. Agr. Econ.* 62:303–309.

Chinn, D. L. [1976]: "The Marketed Surplus of a Subsistence Crop: Paddy Rice in Taiwan." *Amer. J. Agr. Econ.* 58:583–587.

———— [1978]: "Income Distribution in a Chinese Commune." *J. Comparative Econ.* 2:246–265.

Chisholm, A. H., and R. Tyers, eds. [1982]: *Food Security: Theory, Policy, and Perspectives from Asia and the Pacific Rim*. Lexington, MA, Lexington Books.

Chopra, R. N. [1981]: *Evolution of Food Policy in India*. Delhi, Macmillan.

———— [1985]: *Green Revolution in India: The Relevance of Administrative Support for its Success*. New Delhi, Intellectual Publishing.

Chow, G. C. [1987]: *The Chinese Economy*. New York, Harper and Row.

Chowdhury, A., K. M. Alauddin and L. C. Chen [1977]: "The Interaction of Nutrition, Infection, and Mortality during Recent Food Crises in Bangladesh." *Food Research Institute Studies* 16:47–61.

Chowdhury, A., and G. S. Ram [1978]: "Price Response of a Perennial Crop: A Case Study of Indian Tea." *Indian J. Agr. Econ.* 33(3):74–83.

Choudhry, Y. A., and Z. Sattar [1984]: "National Marketing Agencies for Rural South Asia: The Case for Public Enterprise with Private Management." *Econ. Bull. Asia and the Pacific* 35(2):36–44.

Chung, J. S.-H. [1974]: *The North Korean Economy: Structure and Development*. Stanford, CA, Hoover Institution Press.

Clarete, R. L., and J. A. Roumasset [1983]: *An Analysis of the Economic Policies Affecting the Philippine Coconut Industry*. Manila, PIDS Working Paper No. 83–08.

Clark, C. [1957]: *The Conditions of Economic Progress*, 3rd ed. London, Macmillan.

Clark, C., and M. Haswell [1964, 1966, 1967]: *The Economics of Subsistence Agriculture*. 1st, 2nd, 3rd editions. London, Macmillan.

Claudon, M. P., ed. [1986]: *World Debt Crisis: International Lending on Trial*. Cambridge, MA, Ballinger.

Clay, E. J. [1975]: "Equity and Productivity Effects of a Package of Technological Innovations and Changes in Social Institutions: Tubewells, Tractors, and High-Yielding Varieties." *Indian J. Agr. Econ.* 30(4)74–87.

———— [1976]: "Institutional Change and Agricultural Wages in Bangladesh." *Bangladesh Dev. Studies* 4:423–440.

Clay, E. J., and J. Shaw, eds. [1987]: *Poverty, Development and Food*. New York, Macmillan.

Clay, E. J., and H. W. Singer [1984]: *Food Aid and Development — Issues and Evidence: A Survey of the Literature since 1977 on the Role and Impact of Food Aid in Developing Countries*. Rome, World Food Programme, Occasional Paper No. 3.

Cleaver, H. M., Jr. [1972]: "The Contradictions of the Green Revolution." *Amer. Econ. Rev., Papers and Proceedings* 62:177–186.

Collier, W. L., J. C. Colter, Sinarhadi, and R. Shaw [1974]: "Choice of Techniques in Rice Milling on Java: A Note." *Bull. Indonesian Econ. Studies* 10(1):106–120.

Collier, W. L., Soentoro, K. Hidayat, and Y. Yuliali [1982]: *The Acceleration of Rural Development on Java: From Village Studies to a National Perspective.* Bogor, Indonesia, Agro-Economic Survey, Occasional Paper No. 6.

Collier, W. L., Soentoro, G. Wiradi, and Makali [1974]: "Agricultural Technology and Institutional Change in Java." *Food Research Institute Studies* 13:169-194.

Collier, W. L., G. Wiradi, and Soentoro [1973]: "Recent Changes in Rice Harvesting Methods." *Bull. Indonesian Econ. Studies* 9(2):36-45.

Colombo, U., D. Gale Johnson, and T. Shishido [1978]: *Reducing Malnutrition in Developing Countries: Increasing Rice Production in South and Southeast Asia.* New York, Trilateral Commission, Trilateral Papers No. 16.

Conable, B. [1989]: "Development and the Environment: A Global Balance." *Finance and Dev.* 26(4):2-4.

Consultative Group on International Agricultural Research [CGIAR, 1987a]: *Sustainable Agricultural Production: Implications for International Agricultural Research.* Rome, FAO, TAC Secretariat.

——— [CGIAR, 1987b]: *1986-87 Annual Report.* Washington, CGIAR Secretariat.

Conway, G. I., I. Manwan, and D. S. McCarley [1984]: *Sustainability of Agricultural Intensification in Indonesia.* Jakarta, Indonesia, Agency for Agricultural Research and Development, Workshop Report.

Conway, P. J., and M. Bale [1988]: "Approximating the Effective Protection Coefficient without Reference to Technological Data." *World Bank Econ. Rev.* 2:349-363.

Cowan, C. D., ed. [1964]: *The Economic Development of Southeast Asia—Studies in Economic History and Political Economy.* New York, Praeger.

Coward, E. W., Jr., ed. [1980]: *Irrigation and Agricultural Development in Asia.* Ithaca, NY, Cornell Univ. Press.

Cownie, J., B. F. Johnston, and B. Duff [1970]: "The Quantitative Impact of the Seed-Fertilizer Revolution in West Pakistan: An Exploratory Study." *Food Research Institute Studies* 9:57-95.

Coyle, W. T. [1981]: *Japan's Rice Policy.* Washington, USDA, ESS, FAER No. 164.

Coyle, W. T. [1983]: *Japan's Feed—Livestock Economy.* Washington, USDA, ESS, FAER No. 17.

Crisostomo, C. M., W. H. Meyers, T. B. Paris Jr., B. Duff, and R. Barker [1971]: "The New Rice Technology and Labor Absorption in Philippine Agriculture." *Malayan Econ. Rev.* 16:117-158.

Croll, E. [1979]: *Women in Rural Development—The People's Republic of China.* Geneva, ILO.

——— [1982]: "The Sexual Division of Labor in Rural China." In Beneria, ed., 1982, pp. 223-247.

——— [1985]: *Women and Rural Development in China: Production and Reproduction.* Geneva, ILO, Women, Work and Development No. 11.

Crosson, P. R. [1975]: "Institutional Obstacles to Expansion of World Food Production." *Science* 188:519-524, 9 May.

Culter, P. [1985]: "Detecting Food Emergencies: Lessons from the 1979 Bangladesh Crisis." *Food Policy* 10:207-224.

Cummings, R. W., Jr. [1968]: "Effectiveness of Pricing in an Indian Wheat Market: A Case Study of Khanna, Punjab." *Amer. J. Agr. Econ.* 50:687-701.

Cummings, R. W., Jr., and S. Baharsjah [1980]: "The Role of Agricultural Research in National Development—Implications for Research Strategies in a Region." *Indonesian Agr. Research and Dev. J.* 2(2).

Cummings, R. W., Jr., R. W. Herdt, and S. K. Ray [1979]: *Policy Planning for Agricultural Development*. New Delhi, Tata McGraw-Hill.

Cummings, R. W., Sr., W. P. Falcon, R. Revelle, and R. Tyers [1978]: *Research Issues Affecting Agricultural Development Policy*. Islamabad, Government of Pakistan, Planning Commission.

Currie, J. M. [1981]: *The Economic Theory of Agricultural Land Tenure*. Cambridge, England, Cambridge Univ. Press.

Dagli, V., ed. [1974]: *A Regional Profile of Indian Agriculture*. Bombay, Vora.

Dahlberg, K. A., ed. [1979]: *Beyond the Green Revolution: The Ecology and Politics of Global Agricultural Development*. New York, Plenum Press.

_____ , ed. [1985]: *New Directions for Agriculture and Agricultural Research: Neglected Dimensions and Emerging Alternatives*. Totowa, NJ, Rowman and Allanheld.

Dalrymple, D. G. [1971]: *Survey of Multiple Cropping in Less Developed Nations*. Washington, USDA, ERS, FAER No. 12.

_____ [1974]: *Development and Spread of High-Yielding Varieties of Wheat and Rice in the Less Developed Nations*. Washington, USDA, ERS, FAER No. 95.

_____ [1985]: "The Development and Adoption of High-Yielding Varieties of Wheat and Rice in Developing Countries." *Amer. J. Agr. Econ.* 67:1067-1073.

_____ [1986a]: *Development and Spread of High-Yielding Rice Varieties in Developing Countries*. Washington, AID, Bureau of Science and Technology.

_____ [1986b]: *Development and Spread of High-Yielding Wheat Varieties in Developing Countries*. Washington, AID, Bureau of Science and Technology.

Dandekar, V. M. [1964]: "Prices, Production, and Marketed Supply of Foodgrains." *Indian J. Agr. Econ.* 19(3, 4):186-195.

Dandekar, V. M., and N. Rath [1971a]: "Poverty in India, Part I—Dimensions and Trends." *Econ. and Polit. Weekly* 6(1):29-48.

_____ [1971b]: "Poverty in India, Part II—Policies and Programmes." *Econ. and Polit. Weekly* 6(2):106-146.

Dando, W. A. [1980]: *The Geography of Famine*. New York, Wiley.

Dantwala, M. L. [1967]: "Incentives and Disincentives in Indian Agriculture." *Indian J. Agr. Econ.* 22(2):1-25.

_____ [1972]: "Foreword." In ISAE, 1972, pp. 7-10.

_____ [1975]: "A Profile of Poverty and Unemployment in 12 Villages." *Indian J. Agr. Econ.* 30(2):1-17.

_____ [1976]: "Agricultural Policy in India since Independence." *Indian J. Agr. Econ.* 31(4):31-53.

Dasgupta B. [1980]: *The New Agrarian Technology and India*. New Delhi, Macmillan.

David, C. C. [1976]: "Fertilizer Demand in the Asian Rice Economy." *Food Research Institute Studies* 15:109-124.

_____ [1978]: "Factors Affecting Fertilizer Consumption." In IRRI, 1978b, pp. 67-81.

_____ [1982]: *Credit and Price Policies in Philippine Agriculture*. Manila, PIDS, Staff Paper Series No. 82-02.

_____ [1983]: "Economic Policies and Philippine Agriculture." Manila, PIDS, Working Paper No. 83-02.

_____ [1989]: "Philippines: Rice Policy in Transition." In Sicular, ed., 1989a, pp. 154-182.

David, C. C., and A. R. Balicasan [1981]: "An Analysis of Fertilizer Policies in the Philippines." *J. Philippine Dev.* 3(1/2):21-37.

David, C. C., and R. Barker [1978]: "Modern Rice Varieties and Fertilizer Consumption." In IRRI, 1978a, pp. 175-211.

————— [1982]: "Labour Demand in the Philippine Rice Sector." In Gooneratne, ed. 1982, pp. 119-157.

David, C. C., R. Barker, and A. Palacpac [1987]: "The Nature of Productivity Growth in Philippine Agriculture, 1948-1982." In APO, ed., 1987, pp. 409-438.

Davie, J. L. [1986]: "China's International Trade and Finance." In U.S. Congress, ed., 1986, Vol. 2, pp. 311-344.

Davis, J. M. [1977]: "The Fiscal Role of Food Subsidy Programs." *IMF Staff Papers* 24:100-127.

Davis, T. J., and I. A. Schirmer, eds. [1987]: *Sustainability Issues in Agricultural Development: Proceeding of the 7th Agriculture Sector Symposium.* Washington, World Bank.

Dawson, O. L. [1970]: *Communist China's Agriculture: Its Development and Future Potential.* New York, Praeger.

Day, R. H. [1963]: *Recursive Programming and Production Response.* Amsterdam, North-Holland.

————— [1965]: "Probability Distribution of Field Crop Yields." *J. Farm Econ.* 47:713-741.

————— [1967]: "The Economics of Technological Change and the Demise of the Share Cropper." *Amer. Econ. Rev.* 57:427-449.

Day, R. H., and A. Cigno, eds. [1978]: *Modelling Economic Change: The Recursive Programming Approach.* Amsterdam, North-Holland.

Day, R. H., and I. J. Singh [1977]: *Economic Development as an Adaptive Process: The Green Revolution in the Indian Punjab.* London, Cambridge Univ. Press.

Dayal, R. [1963]: "Some Aspects of the Marketable Surplus of Foodgrains." *Agricultural Situation in India* 18(1):23-25.

Deaton, A., and J. Muellbauer [1980]: *Economics and Consumer Behavior.* New York, Cambridge Univ. Press.

Deboeck, G., and R. Ng [1980]: *Monitoring Rural Development in East Asia.* Washington, World Bank Staff Working Paper No. 439.

De Datta, S. K., K. A. Gomez, R. W. Herdt, and R. Barker [1978]: *A Handbook on the Methodology for an Integrated Experiment: Survey on Rice Yield Constraints.* Los Baños, Philippines, IRRI.

De Janvry, A. [1978]: "Social Structure and Biased Technical Change in Argentine Agriculture." In Binswanger, Ruttan, *et al.*, 1978, pp. 297-323.

De Janvry, A., and J. J. Dethier [1985]: *Technological Innovation in Agriculture: The Political Economy of its Rate and Bias.* Washington, World Bank, CGIAR Paper No. 1.

De Janvry, A., and E. Sadoulet [1986a]: *Agricultural Growth in Developing Countries and Agricultural Imports: Econometric and General Equilibrium Analysis.* Berkeley, Univ. of California, Dept. of Agr. and Resource Econ., Working Paper No. 424.

————— [1986b]: "The Conditions for Harmony between Third World Agricultural Development and U.S. Farm Exports." *Amer. J. Agr. Econ.* 68:1340-1346.

De Janvry, A., and K. Subbarao [1986]: *Agricultural Price Policy and Income Distribution in India.* New York, Oxford Univ. Press.

De Kosinsky, V., and M. de Somer, eds. [1985]: *Water Resources for Rural Areas and their Communities: Proceedings of the 5th World Congress on Water Resources.* Brussels, International Water Resources Association.

De Leon, M. S. J. [1982]: *Intersectoral Capital Flows and Balanced Agro-Industrial Development in the Philippines.* Manila, PIDS, Working Study Paper No. 84-01.

————— [1983]: *Government Expenditures and Agricultural Policies in the Philippines, 1955-1980.* Manila, PIDS, Working Paper No. 83-06.

De Macedo, J. [1988]: "Comments on 'Financial Liberalization in Retrospect: Interest Rate Policies in LDCs.' " In Ranis and T. P. Schultz, eds., 1988, pp. 411–415.

Deolalikar, A. B. [1981]: "The Inverse Relationship between Productivity and Farm Size: A Test Using Regional Data from India." *Amer. J. Agr. Econ.* 63:275–279.

———— [1988]: "Nutrition and Labor Productivity in Agriculture: Estimates for Rural South India." *Rev. Econ. and Stat.* 70:406–413.

Dernberger, R. F., ed. [1980]: *China's Development Experience in Comparative Perspective.* Cambridge, MA, Harvard Univ. Press.

DeRosa, D. A. [1988]: "Agricultural Trade and Protection in Asia." *Finance and Dev.* 25(4):50–52.

Desai, B. M. [1972]: *Analysis of Consumer Expenditure Patterns in India.* Ithaca, NY, Cornell Univ., DAE, Occasional Paper No. 54.

Desai, B. M., and M. D. Desai [1969]: *The New Strategy of Agricultural Development in Operation: A Case Study of the Kaira District in Gujarat.* Bombay, Thacker.

Desai, D. K. [1963]: *Increasing Income and Production in Indian Farming: Possibilities with Existing Resource Supplies on Individual Farms. Application of Linear Programming Technique.* Bombay, ISAE.

Desai, G. M. [1969]: *Growth of Fertiliser Use in Indian Agriculture: Past Trends and Future Demand.* Ithaca, NY, Cornell Univ., International Agricultural Development Bulletin 18.

———— [1970]: "Factors Determining Demand for Pesticides." *Econ. and Polit. Weekly* 5(52):A181–A183.

———— [1971]: "Some Observations on Economics of Cultivating High-Yielding Varieties of Rice in India." *Artha-Vikas* 7(2):1–19.

———— [1973]: *Nitrogen Use and Foodgrain Production, India, 1973-74, 1978-79, and 1983-84.* Ithaca, NY, Cornell Univ., DAE Occasional Paper No. 55.

———— [1978]: "A Critical Review of Fertilizer Consumption after 1974–75 and Prospects for Future Growth." *Fertilizer News* 23:7–18.

———— [1979]: "Fertilisers in India's Agricultural Development: Problems and Policies." In Vakil and Shah, eds., 1979, pp. 377–426.

———— [1982]: *Sustaining Rapid Growth in India's Fertilizer Consumption: A Perspective Based on Composition of Use.* Washington, IFPRI, Research Report No. 31.

———— [1985a]: *Policy Issues for Long-term Growth of Fertilizer Use in Bangladesh.* Ahmedabad, Indian Institute of Management, Working Paper No. 578.

———— [1985b]: "Market Channels and Growth of Fertilizer Use in Rainfed Agriculture: Conceptual Considerations and Experience in India." In ICRISAT, 1985, pp. 41–52.

———— [1986]: "Fertilizer Use in India: The Next Stage in Policy." *Indian J. Agr. Econ.* 41(3):248–270.

Desai, G. M., P. N. Chary, and S. C. Bandyopadhyay [1972]: *Dynamics of Growth in Fertiliser Use at Micro Level.* Ahmedabad, Indian Institute of Management, Center for Management in Agriculture Monograph No. 32.

Desai, G. M., and J. W. Mellor [1969]: "Changing Basis of Demand for Fertiliser in Indian Agriculture." *Econ. and Polit. Weekly* 4(39):A175–A188.

Desai, G. M., and N. V. Namboodiri [1983]: "The Deceleration Hypothesis and Yield Increasing Input in Indian Agriculture." *Indian J. Agr. Econ.* 38(4):497–508.

Desai, G. M., and G. Singh [1973]: *Growth of Fertilizer Use in Districts of India: Performance and Policy Implications.* Ahmedabad, Indian Institute of Management, Center for Management in Agriculture, Monograph No. 41.

Dhar, P. N., and H. F. Lydall [1961]: *The Role of Small Enterprises in Indian Economic Development.* New York, Asia Publishing.

Dillon, J. L., and J. R. Anderson [1971]: "Allocative Efficiency, Traditional Agriculture, and Risk." *Amer. J. Agr. Econ.* 53:26-32.

Dixit, A. K. [1969]: "Marketable Surplus and Dual Development." *J. Econ. Theory* 1:203-219.

———— [1970]: "Growth Patterns in a Dual Economy." *Oxford Econ. Papers* NS 22:229-234.

Dixit, R. S., and P. P. Singh [1970]: "Impact of High Yielding Varieties on Human Labour Inputs." *Agricultural Situation in India* 24:1081-1089.

Dixon, J. A. [1982]: *Food Consumption Patterns and Related Demand Parameters in Indonesia: A Review of Available Evidence.* Washington, IFPRI, IFDC and IRRI.

Dixon, R. B. [1982]: "Mobilizing Women for Rural Employment in South Asia." *Econ. Dev. and Cultural Change* 30:373-390.

Dixon, T. A., and R. Tyers [1982]: "India's Food Security: Supply, Demand, and the Signs of Success." In Chisholm and Tyers, eds. 1982, pp. 191-216.

Dobb, M., ed. [1967a]: *Papers on Capitalism, Development, and Planning.* New York, International Publishers.

———— [1967b]: "Some Problems of Industrialization in Agricultural Centers." In Dobb, ed., 1967a, pp. 71-88.

Domar, E. D. [1957]: *Essays in the Theory of Economic Growth.* New York, Oxford Univ. Press.

Dong, F. [1988]: *Rural Reform, Nonfarm Development, and Rural Modernization in China.* Washington, World Bank EDI Seminar Paper No. 38.

Donnithorne, A. [1967]: *China's Economic System.* New York, Praeger.

Donovan, W. G. [1973]: *Rural Works and Employment: Description and Preliminary Analysis of a Land Army Project in Mysore State, India.* Ithaca, NY, Cornell Univ., DAE, Occasional Paper No. 60.

———— [1974]: *Employment Generation in Agriculture: A Study in Mandya District, South India.* Ithaca, NY, Cornell Univ., DAE, Occasional Paper No. 71.

Dorner, P. [1972]: *Land Reform and Economic Development.* Baltimore, Penguin Books.

Douglas, G. K., ed. [1984]: *Agricultural Sustainability in a Changing World Order.* Boulder, CO, Westview Press.

Dovring, F. [1959]: "The Share of Agriculture in a Growing Population." *Monthly Bull. Agr. Econ. and Stat.* 8(8,9):3-11.

Dozina, G., Jr., M. Kikuchi, and Y. Hayami [1978]: "Mobilizing Local Resources for Irrigation Development: A Communal System in Central Luzon, Philippines." In IRRI, 1978c, pp. 135-142.

Dubey, V. [1963]: "The Marketed Agricultural Surplus and Economic Growth in Underdeveloped Countries." *Econ. J.* 73:689-702.

Duff, B. [1978]: "Mechanization and Use of Modern Rice Varieties." In IRRI, 1978a, pp. 145-164.

———— [1986]: *Some Consequences of Agricultural Modernization in the Philippines, Thailand, and Indonesia.* Los Baños, Philippines, IRRI, Research Paper Series No. 120.

Duncan, R. C. [1972]: "Evaluating Returns to Research in Pasture Improvement." *Australian J. Agr. Econ.* 16(3):153-168.

Dutt, A. K., and M. Geib [1987]: *An Atlas of South Asia.* Boulder, CO, Westview Press.

Easter, K. W., ed. [1986]: *Irrigation Investment, Technology, and Management Strategies for Development.* Boulder, CO, Westview Press.

Easter, K. W., M. E. Abel, and G. Norton [1977]: "Regional Differences in Agricultural Productivity in Selected Areas of India." *Amer. J. Agr. Econ.* 59:257-265.

Easter, K. W., J. A. Dixon, and M. M. Hufschmidt, eds. [1986]: *Watershed Resource Management: An Integrated Framework with Studies from Asia and the Pacific.* Boulder, CO, Westview Press.

Eaton, D. J., and W. S. Steele, eds. [1976]: *Analyses of Grain Reserves: A Proceedings.* Washington, USDA ERS, and NSF, ERS Report No. 634.

Eckaus, R. S., and K. S. Parikh [1968]: *Planning for Growth: Multisectoral, Intertemporal Models Applied to India.* Cambridge, MA, MIT Press.

Eckstein, A. [1975]: *China's Economic Development.* Ann Arbor, Univ. of Michigan Press.

_____ [1977]: *China's Economic Revolution.* New York, Praeger.

_____ [1980]: *Quantitative Measures of China's Economic Output.* Ann Arbor, Univ. of Michigan Press.

Eckstein, A., W. Galenson, and T. C. Liu, eds. [1968]: *Economic Trends in Communist China.* Chicago, Aldine.

Eckstein, O. [1958]: *Water Resource Development: The Economics of Project Evaluation.* Cambridge, MA, Harvard Univ. Press.

Edirisinghe, N. [1987]: *The Food Stamp Scheme in Sri Lanka: Costs, Benefits, and Options for Modification.* Washington, IFPRI, Research Report No. 58.

Edirisinghe, N., and T. T. Poleman [1976]: "Implications of Government Intervention in the Rice Economy of Sri Lanka." Ithaca, NY, Cornell Univ., International Agr. Paper No. 48.

_____ [1979]: "Welfare or Growth? Sri Lanka's Problem in Peasant Agriculture." Ithaca, NY, Cornell Univ., DAE, Staff Paper 79-18.

Eicher, C. K., and J. M. Staatz, eds. [1984]: *Agricultural Development in the Third World.* Baltimore, Johns Hopkins Univ. Press.

Eicher, C. K., and L. W. Witt, eds. [1964]: *Agriculture in Economic Development.* New York, McGraw-Hill.

Eldridge, P. J. [1969]: *The Politics of Foreign Aid in India.* London, Widenfeld and Nicholson.

Ellis, H. S., and L. A. Metzler, eds. [1950]: *Readings in the Theory of International Trade.* Philadelphia, Blakiston.

Elvin, M. [1973]: *The Pattern of the Chinese Past: A Social and Economic Interpretation.* London, Eyre Methuen.

Elz, D., ed. [1984]: *The Planning and Management of Agricultural Research.* Washington, World Bank, Proceedings of a World Bank and ISNAR Symposium.

_____ [1987]: *Agricultural Marketing Strategy and Pricing Policy.* Washington, World Bank.

Encarnacion, J. R. [1976]: *Philippine Economic Problems in Perspective.* Manila, Univ. of the Philippines, Institute of Economic Development and Research.

Ender, G. [1983]: *Food Security Policies of Six Asian Countries.* Washington, USDA, ERS, FAER No. 190.

Enke, S. [1962]: "Industrialization through Greater Productivity in Agriculture." *Rev. Econ. and Stat.* 44:88-91.

Ensminger, D. [1962]: "Overcoming the Obstacles to Farm Economic Development in the Less Developed Countries." *J. Farm Econ.* 44:1367-1387.

Evans, J. A. [1986]: *Indonesia: An Export Market Profile.* Washington, USDA, ERS 86-6.

Evenson, R. E. [1968]: "The Contribution of Agricultural Research and Extension to Agricultural Production." Unpublished Ph.D. dissertation, Univ. of Chicago.

_____ [1974]: "The 'Green Revolution' in Recent Development Experience." *Amer. J. Agr. Econ.* 56:387-394.

_____ [1975]: "Agricultural Trade and Shifting Comparative Advantage." In Tolley and Zadrozny, eds., 1975, pp. 181-200.

_____ [1976]: "International Transmission of Technology in the Production of Sugarcane." *J. Dev. Studies* 12:208-231.

_____ [1978a]: "Agricultural Research and Education in Asia." *Second Asian Agriculture Survey*. ADB, Survey Supplementary Papers, Vol. 2.

_____ [1978b]: "Time Allocation in Rural Philippine Households." *Amer. J. Agr. Econ.* 60:322-330.

_____ [1983a]: "The Allocation of Women's Time: An International Comparison." *Behavior Science Research* 17(3,4):196-215.

_____ [1983b]: "Economics of Agricultural Growth: The Case of Northern India." In Nobe and Sampath, eds., 1983, pp. 145-192.

_____ [1984]: "Estimating Labor Demand Functions for Indian Agriculture." In Binswanger and Rosenzweig, eds., 1984, pp. 263-279.

_____ [1986a]: "Infrastructure, Output Supply and Input Demand in Philippine Agriculture: Provisional Estimates." *J. Philippine Dev.* 13:62-76.

_____ [1986b]: *The CGIAR Centers: Measures of Impact on National Research, Extension and Productivity*. Washington, World Bank.

_____ [1987]: *The International Agricultural Research Centers: Their Impact on Spending for National Research and Extension*. Washington, World Bank, CGIAR Study Paper No. 22.

_____ [1988]: "Technological Opportunities and International Technology Transfer in Agriculture." In Antonelli and Quadrio-Curzio, 1988, pp. 133-172.

Evenson, R. E., and P. Flores [1978]: "Social Returns to Rice Research." In IRRI, 1978a, pp. 243-265.

Evenson, R. E., C. F. Habito, A. R. Quisumbing, and C. S. Bastilon [1986]: "Methods for Agricultural Policy Analysis: An Overview." *J. Philippine Dev.* 13:1-39.

Evenson, R. E., J. P. Houck, and V. W. Ruttan [1970]: "Technical Change and International Trade: Three Examples—Sugarcane, Bananas, and Rice." In Vernon, ed., 1970, pp. 415-480.

Evenson, R. E., and D. Jha [1973]: "The Contribution of the Agricultural Research Systems to Agricultural Production in India." *Indian J. Agr. Econ.* 28(4):212-230.

Evenson, R. E., and Y. Kislev [1973]: "Research and Productivity in Wheat and Maize." *J. Polit. Economy* 81:1309-1329.

_____ [1975a]: "Investment in Agricultural Research and Extension: A Survey of International Data." *Econ. Dev. and Cultural Change* 23:507-522.

_____ [1975b]: *Agricultural Research and Productivity*. New Haven, CT, Yale Univ. Press.

Evenson, R. E., C. Pray, and J. Quizon [1986]: *Research and Extension Productivity and Income in Asian Agriculture*. Ithaca, NY, Cornell Univ. Press.

Evenson, R. E., C. Pray, and G. M. Scobie [1985]: "The Influence of International Research on National Agricultural Research Systems." *Amer. J. Agr. Econ.* 67:1074-1079.

Evenson, R. E., J. Putnam, and C. Pray [1983]: "Effects of International Transfer of Agricultural Technologies on the Competitiveness of U.S. Agriculture." Prepared for the OTA Report on *U.S. Competitiveness in Agricultural Trade*. Washington.

Evenson, R. E., and J. Roumasset [1986]: "Markets, Institutions and Family Size in Rural Philippine Households." *J. Philippine Dev.* 13:141-162.

Evenson, R. E., and L. Sardido [1986]: "Regional Total Factor Productivity Change in Philippine Agriculture." *J. Philippine Dev.* 13:40-61.

Evenson, R. E., P. E. Waggoner, and V. W. Ruttan [1979]: "Economic Benefits from Research: An Example from Agriculture." *Science* 205:1101-1107, 14 Sept.

Ezekiel, H. [1988]: "An Approach to a Food Aid Strategy." *World Dev.* 16:1377-1387.

Ezekiel, M. [1958]: "Apparent Results in Using Surplus Food for Financing Economic Development." *J. Farm Econ.* 40:915-923.

Falcon, W. P. [1964]: "Farmers' Response to Price in a Subsistence Economy: The Case of West Pakistan." *Amer. Econ. Rev., Papers and Proceedings* 54(3):580-591.

———— [1970]: "The Green Revolution: Generations of Problems." *Amer. J. Agr. Econ.* 52:698-710. Also in ISAE, 1972, pp. 295-317.

———— [1978]: "Food Self-Sufficiency: Lessons from Asia." In *International Food Policy Issues, A Conference.* Washington, USDA, ERS, FAER No. 143.

———— [1980]: "Elements of a Food and Nutrition Policy in Indonesia." In Papanek, ed., 1980, pp. 406-428.

———— [1984]: "Recent Food Policy Lessons from Developing Countries." *Amer. J. Agr. Econ.* 66:180-187.

———— [1987]: "Aid, Food Policy Reform, and U.S. Agricultural Interests in the Third World." *Amer. J. Agr. Econ.* 69(5):929-935.

Falcon, W. P., W. O. Jones, S. R. Pearson, J. A. Dixon, G. C. Nelson, F. R. Roche, and L. J. Unneveher [1984]: *The Cassava Economy of Java.* Stanford, CA, Stanford Univ. Press.

Falcon, W. P., L. A. Mears, M. M. Hastings, and S. R. Pearson [1985]: *Rice Policy in Indonesia, 1985-1990: The Problems of Success.* Jakarta, Indonesia, BULOG.

Falcon, W. P., and E. A. Monke [1979-80]: "International Trade in Rice." *Food Research Institute Studies* 18:279-306.

Falcon, W. P., and G. F. Papanek, eds. [1971]: *Development Policy: The Pakistan Experience.* Cambridge, MA, Harvard Univ. Press.

Falcon, W. P., and C. P. Timmer [1975]: "The Political Economy of Rice Production and Trade in Asia." In Reynolds, ed., 1975, pp. 373-408.

Farmer, B. H., ed. [1977]: *Green Revolution? Technology and Change in Rice-Growing Areas of Tamil Nadu and Sri Lanka.* Boulder, CO, Westview Press.

———— [1979]: "The 'Green Revolution' in South Asian Ricefields: Environment and Production." *J. Dev. Studies* 15:304-319.

Farrington, J., F. Abeyratne, and G. J. Gill [1984]: *Farm Power and Employment in Asia.* Bangkok, ADC.

Farruk, M. O. [1970]: *The Structure and Performance of the Rice Marketing System in East Pakistan.* Ithaca, NY, Cornell Univ. DAE, Occasional Paper No. 31.

Feder, G. [1980]: "Farm Size, Risk Aversion and the Adoption of New Technology Under Uncertainty." *Oxford Econ. Papers* 32:263-283.

———— [1982]: "Adoption of Interrelated Agricultural Innovations: Complementarity of Risk, Scale and Credit." *Amer. J. Agr. Econ.* 64:94-101.

———— [1987]: "Land Ownership Security and Farm Productivity: Evidence from Thailand." *J. Dev. Studies* 24:16-30.

Feder, G., R. Just, and D. Zilberman [1982]: *Adoption of Agricultural Innovation in Developing Countries—A Survey.* Washington, World Bank Staff Working Paper No. 542.

Feder, G., and G. T. O'Mara [1981]: "Farm Size and the Diffusion of Green Revolution Technology." *Econ. Dev. and Cultural Change* 59:59-76.

Feeny, D. [1982]: *The Political Economy of Productivity: Thai Agricultural Development, 1880-1975.* Vancouver, Univ. of British Columbia Press.

———— [1983a]: "Extensive *versus* Intensive Agricultural Development: Induced Public Investments in Southeast Asia, 1900-1940." *J. Econ. History* 43:687-704.

———— [1983b]: "The Moral or the Rational Peasant? Competing Hypotheses of Collective Action." *J. Asian Studies* 42:769-789.

Fei, J. C. H., and G. Ranis [1963]: "Innovation, Capital Accumulation, and Economic Development." *Amer. Econ. Rev.* 53:283-313.

———— [1964]: *Development of the Labor Surplus Economy: Theory and Policy*. Homewood, IL, Irwin.

———— [1966]: "Agrarianism, Dualism, and Economic Development." In Adelman and Thorbecke, eds., 1966, pp. 3-46.

———— [1975]: "A Model of Growth and Employment in the Open Dualistic Economy: The Cases of Korea and Taiwan." *J. Dev. Studies* 11(2):32-63.

Fei, J. C. H., G. Ranis, Y. Y. Bien, S. H. Jo, and S. Nigata [1982]: "A Comparative Analaysis of the Role of the Agricultural Sector in the Development Process: Japan, Taiwan, and South Korea." In Hou and Yu, eds., 1982, pp. 783-840.

Fei, J. C. H., G. Ranis, and S. W. Y. Kuo [1979]: *Growth with Equity—The Taiwan Case*. New York, Oxford Univ. Press.

Fei, J. and B. Reynolds [1987]: "A Tentative Plan for the Rational Sequencing of Overall Reform in China's Economic System." *J. Comparative Econ.* 11:490-502.

Fel'dman, B. A. [1957]: "A Soviet Model of Growth." In Domar, 1957, pp. 223-261.

Feuerwerker, A. [1968]: *The Chinese Economy, 1912-1949*. Ann Arbor, Univ. of Michigan, Papers in Chinese Studies.

———— [1969]: *The Chinese Economy, 1870-1911*. Ann Arbor, Univ. of Michigan, Papers in Chinese Studies.

Fforde, A. [1983]: "The Historical Background to Agricultural Collectivization in North Vietnam: The Changing Role of Corporate Economic Power." Univ. of London, Department of Economics, Birkbeck Discussion Paper No. 148.

Fine, J. C., and R. G. Lattimore, eds. [1982]: *Livestock in Asia—Issues and Policies*. Ottawa, Canada, IDRC.

Finger, J. M., and A. Olechowski, eds. [1987]: *The Uruguay Round: A Handbook on the Multilateral Trade Negotiations*. Washington, World Bank.

Fishel, W., ed. [1971]: *Resource Allocation in Agricultural Research*. Minneapolis, Univ. of Minnesota Press.

Fisher, F. M. [1963]: "A Theoretical Analysis of the Impact of Food Surplus Disposal on Agricultural Production in Recipient Countries." *J. Farm Econ.* 45:863-875.

Fitchett, D. A. [1988]: *Agricultural Trade Protectionism in Japan: A Survey*. Washington, World Bank Discussion Paper No. 28.

Flanders, M. J. [1969]: "Agriculture *vs.* Industry in Development Policy: The Planners' Dilemma Re-examined." *J. Dev. Studies* 5:171-189.

Fletcher, L., and Mubyarto [1966]: "Supply and Market Surplus Relationships for Rice in Indonesia." Paper presented at the ADC/Univ. of Minnesota Conference on Supply and Market Relationships in Peasant Agriculture, Minneapolis.

Flinn, J., K. Kalirajan, and L. Castillo [1982]: "Supply Responsiveness of Rice Farmers in Laguna, Philippines." *Australian J. Agr. Econ.* 26:39-48.

Flinn, J. C., and B. Duff [1985]: *Energy Analysis, Rice Production Systems and Rice Research*. Los Baños, Philippines, IRRI, Research Paper Series No. 114.

Flinn, J. C., and D. P. Garrity [1986]: *Yield Stability and Modern Rice Technology*." Los Baños, Philippines, IRRI, Research Paper Series No. 122.

Flinn, J. C., and P. B. R. Hazell [1988]: "Production Instability and Modern Rice Technology: A Philippine Case Study." *Developing Economies* 26(1):34-50.

Flinn, J. C., and L. J. Unnevehr [1985]: *Contributions of Modern Rice Varieties to Nutrition in Asia*. Los Baños, Philippines, IRRI, Research Paper Series No. 110.

Flores-Moya, P., R. E. Evenson, and Y. Hayami [1978]: "Social Returns to Rice Research in the Philippines: Domestic Benefits and Foreign Spillover." *Econ. Dev. and Cultural Change* 26:591-607.

Folbre, N. [1984]: "Household Production in the Philippines: A Non-Neoclassical Approach." *Econ. Dev. and Cultural Change* 32:303-330.

Food and Agriculture Organization [FAO, 1977]: *The Fourth World Food Survey*. Food and Nutrition Series No. 10. Rome.

_____ [FAO, 1981]: *Agriculture: Towards 2000*. Rome, Economic and Social Development Series No. 23.

_____ [FAO, 1985]: *Food Aid in Figures, 1984*. Rome.

_____ [FAO, 1987]: *Expert Consultation on Data Needs for Food and Agricultural Analysis and Planning in Developing Countries*. Rome.

_____ [FAO, 1988a]: *Trade Yearbook, 1987* (and other earlier issues). Rome.

_____ [FAO, 1988b]: *Sustainable Agricultural Production: Implications for International Agricultural Research*. Rome, TAC Secretariat.

_____ [FAO, 1989a]: *Production Yearbook, 1988*, Vol. 42 (and other earlier issues). Rome.

_____ [FAO, 1989b]: *Fertilizer Yearbook, 1988*, Vol. 38 (and other earlier issues). Rome.

_____ [FAO, 1989c]: *State of Food and Agriculture, 1987-88*. Rome.

_____ [FAO and WHO, 1973]: *Energy and Protein Requirements, Report of a Joint FAO/WHO 'Ad Hoc' Expert Committee*. FAO Food and Nutrition Series No. 7, Rome, WHO Technical Report Series No. 522, Geneva.

Frank, C. R., K. S. Kim, and L. Westphal [1975]: *Foreign Trade Regimes and Economic Development: South Korea*. New York, Columbia Univ. Press.

Frankel, F. R. [1969]: "India's New Strategy of Agricultural Development: Political Costs of Agrarian Modernization." *J. Asian Studies* 28:693-710.

_____ [1971]: *India's Green Revolution: Economic Gains and Political Costs*. Princeton, NJ, Princeton Univ. Press.

Fry, M. J. [1988]: *Money, Interest, and Banking in Economic Development*. Baltimore, Johns Hopkins Univ. Press.

Gandhi, V. P. [1966]: *Tax Burden on Indian Agriculture*. Cambridge, MA, Harvard Univ. Law School.

Gans, O., ed. [1986]: *Appropriate Techniques for Development Planning*. Ft. Lauderdale, FL, Verlag Brietenbach.

Garcia, J. G. [1981]: *The Effects of Exchange Rates and Commercial Policy on Agricultural Incentives in Colombia*. Washington, IFPRI, Research Report No. 24.

Garcia, M., and P. Pinstrup-Anderson [1987]: *The Pilot Food Price Subsidy Scheme in the Philippines: Its Impact on Income, Food Consumption and Nutritional Status*. Washington, IFPRI, Research Report No. 61.

Gardner, B. L. [1979a]: *Optimal Stockpiling of Grains*. Lexington, MA, Lexington Books.

_____ [1979b]: "Robust Stabilization Policies for International Commodity Agreements." *Amer. Econ. Rev.* 69:169-172.

Garg, V. K. [1980]: *State in Foodgrain Trade in India: A Study of Policies and Practices of the Public Distribution System*. New Delhi, Vision Books.

Garnaut, R., ed. [1980]: *ASEAN in a Changing Pacific and World Economy*. Canberra, ANU Press.

Gasser, W. R. [1981]: *Survey of Irrigation in Eight Asian Nations: India, Pakistan, Indonesia, Thailand, Bangladesh, South Korea, Philippines, and Sri Lanka*. Washington, USDA, ERS, FAER No. 165.

Gavan, J. D., and I. S. Chandrasekera [1979]: *The Impact of Public Foodgrain Distribution on Food Consumption and Welfare in Sri Lanka*. Washington, IFPRI, Research Report No. 13.

Geertz, C. [1963]: *Agricultural Involution: The Process of Ecological Change in Indonesia*. Berkeley, Univ. of California Press.

Gemmill, G. [1985]: "Forward Contracts or International Buffer Stocks? A Study of Their Relative Efficiencies in Stabilizing Commodity Export Earnings." *Econ. J.* 95:400–417.

George, P. S. [1979]: *Public Distribution of Foodgrains in Kerala—Income Distribution Implications and Effectiveness*. Washington, IFPRI, Research Report No. 7.

—————— [1985]: *Some Aspects of Procurement and Distribution of Foodgrains in India*. Washington, IFPRI, Working Paper No. 1.

Georgescu-Roegen, N. [1960]: "Economic Theory and Agrarian Economics." *Oxford Econ. Papers* NS12:1–40. Also in Eicher and Witt, eds., 1964, pp. 144–169.

Ghatak, S., and K. Ingersent [1984]: *Agriculture and Economic Development*. Baltimore, Johns Hopkins Univ. Press.

Ghodake, R. D., and J. G. Ryan [1981a]: "Human Labour Availability and Employment in Semi-Arid Tropical India." *Indian J. Agr. Econ.* 36(4):31–44.

Ghodake, R. D., J. G. Ryan, and R. Sarin [1981b]: "Human Labour Use with Existing and Prospective Technologies of the Semi-Arid Tropics of South India." *J. Dev. Studies* 18(1):25–46.

Ghose, A. K. [1980]: "Wages and Employment in Indian Agriculture." *World Dev.* 8:413–428.

—————— [1982]: "Food Supply and Starvation: A Study of Famines with Reference to the Indian Sub-continent." *Oxford Econ. Papers* 34:368–389.

Ghosh, M. [1986]: "Farm-Size Productivity Nexus under Alternative Technology." *Indian J. Agr. Econ.* 41(1):17–28.

Gill, G. J. [1983]: "Mechanized Land Preparation, Productivity, and Employment in Bangladesh." *J. Dev. Studies* 19:329–348.

Gill, K. S. [1972]: *Wheat Market Behavior in Punjab and Haryana, Post-Harvest Period, 1968/69 to 1970/71: Emerging Problems of Wheat Marketing*. Ludhiana, India, Punjab Agricultural Univ. Press.

Gilland, B. [1983]: "Considerations on World Population and Food Supply." *Population and Dev. Rev.* 9:203–211.

Gittinger, J. P. [1982]: *Economic Analysis of Agricultural Projects*, 2nd ed. Baltimore, Johns Hopkins Univ. Press for the World Bank.

Gittinger, J. P., J. Leslie, and C. Hoisington, eds. [1987]: *Food Policy: Integrating Supply, Distribution, and Consumption*. Baltimore, Johns Hopkins University Press.

Glaeser, B., ed. [1987]: *The Green Revolution Revisited: Critique and Alternatives*. New York, St. Martin's Press.

Glassburner, B. [1986]: "Survey of Recent Developments." *Bull. Indonesian Econ. Studies* 22(1):1–133.

Glen, J. J. [1987]: "Mathematical Models in Farm Planning." *Operations Research* 35:641–666.

Goldberg, R. A., ed. [1981]: *Research in Domestic and International Agribusiness Management*, Vol. II. Greenwich, CT, JAI Press.

Goldman, H. W., and C. G. Ranade [1977]: "Analysis of Income Effect on Food Consumption in Rural and Urban Philippines." *J. Agr. Econ. and Dev.* 2(2):150–165.

Goldman, R. H. [1974]: "Seasonal Rice Prices in Indonesia, 1953-69: An Anticipatory Price Analysis." *Food Research Institute Studies* 13:99–143.

_____ [1975]: "Staple Food Self-Sufficiency and the Distributive Impact of Malaysian Rice Policy." *Food Research Institute Studies* 14(3):251-293.

Goldman, R. H., and L. Squire [1982]: "Technical Change, Labor Use and Income Distribution in the Muda Irrigation Project." *Econ. Dev. and Cultural Change* 30:753-775.

Gomez, A. A. [1986]: *Philippines and the CGIAR Centers: A Study of Their Collaboration in Agricultural Research.* Washington, World Bank, CGIAR Study Paper No. 15.

Gonzales, L. A. [1987]: "Rice Production and Regional Crop Diversification in the Philippines: Economic Issues." *Philippines Rev. Econ. and Business* 24:125-148.

Goodell, G. E. [1984]: "Bugs, Bunds, Banks, and Bottleneck: Organizational Contradictions in the New Rice Technology." *Econ. Dev. and Cultural Change* 33:23-41.

Gooneratne, W., ed. [1982]: *Labour Absorption in Rice-Based Agriculture.* Bangkok, ILO, ARTEP.

Gordon-Ashworth, F. [1984]: *International Commodity Control: A Contemporary History and Appraisal.* London, Croom Helm.

Goreux, L. M. [1959]: *Income Elasticity of Demand for Food.* Rome, FAO.

_____ [1960]: "Income and Food Consumption." *Monthly Bull. Agr. Econ. and Stat.* 9(10):1-12.

Gorou, P. [1945]: *Land Utilization in French Indochina.* Washington, Institute of Pacific Relations.

Gotsch, C. H. [1972]: "Technical Change and the Distribution of Income in Rural Areas." *Amer. J. Agr. Econ.* 54:326-341.

_____ [1975a]: "Traditional Agriculture in the Pakistan Punjab: The Basic Model." *Food Research Institute Studies* 14:7-25.

_____ [1975b]: "Linear Programming and Agricultural Policy: Summary and Suggestions for Further Work." *Food Research Institute Studies* 14:99-105.

Gotsch, C. H., B. Ahmed, W. P. Falcon, M. Naseem, and S. Yusuf [1975]: "Linear Programming and Agricultural Policy: Micro Studies in the Pakistan Punjab." *Food Research Institute Studies* 14:3-107.

Gotsch, C. H., and G. Brown [1980]: *Taxes and Subsidies in Pakistan Agriculture, 1966-76.* Washington, World Bank Staff Working Paper No. 387.

Gotsch, C. H., and W. P. Falcon [1975]: "The Green Revolution and the Economics of Punjab Agriculture." *Food Research Institute Studies* 14:27-46.

Gotsch, C. H., and S. Yusuf [1975]: "Technical Indivisibility and the Distribution of Income: A Mixed Integer Programming Model of Punjab Agriculture." *Food Research Institute Studies* 14:81-98.

Government of India, Department of Statistics [1986]: "A Note on the Second Quinquennial Survey on Consumer Expenditure." *Sarvekshana: J. of the National Sample Survey Organization* 9(3):17-23.

Government of India, Planning Commission, Program Evaluation Organization [1967/68]: *Evaluation Study of the High Yielding Varieties Programme.* New Delhi.

_____ [1968/69]: *Evaluation Study of the High Yielding Varieties Programme.* New Delhi.

Government of India, Planning Commission, Ministry of Agriculture and Irrigation [1976]: *Report of the National Commission on Agriculture.* New Delhi.

Govind, N. [1985]: *Regional Perspectives in Agricultural Development: A Case Study of Wheat and Rice in Selected Regions of India.* New Delhi, Concept Publishing.

Grabowski, R. [1979]: "The Implications of an Induced Innovation Model." *Econ. Dev. and Cultural Change* 27:723-734.

_____ [1981]: "Induced Innovation, Green Revolution, and Income Distribution: Reply." *Econ. Dev. and Cultural Change* 30:177-181.

Greenshields, B. L., and M. A. Bellamy, eds. [1983]: *Rural Development: Growth and Equity.* IAAE Occasional Paper No. 3, Brookfield, VT, Gower.

Grewal, S. S., and B. S. Bhullar [1982]: "Impact of Green Revolution on the Cultivation of Pulses in Punjab." *Indian J. Agr. Econ.* 37(3):406-406.

Grewal, S. S., and P. S. Rangi [1976]: "Impact of Input Prices on Production and Profitability of Wheat and Paddy in the Punjab." *Indian J. Agr. Econ.* 31(3):94-100.

Griffin, K. B. [1972]: *The Green Revolution: An Economic Analysis.* Geneva, UNRISD, Report No. 72.6.

_____ [1974]: *The Political Economy of Agrarian Change: An Essay on the Green Revolution.* Cambridge, MA, Harvard Univ. Press.

_____ , ed. [1984]: *Institutional Reform and Economic Development in the Chinese Countryside.* Hong Kong, Macmillan.

Griffin, K., and A. Saith [1981]: *Growth and Equality in Rural China.* Geneva, ILO.

Grigg, D. B. [1970]: *The Harsh Lands: A Study in Agricultural Development.* London, St. Martin's Press.

Griliches, Z. [1957]: "Hybrid Corn: An Exploration in the Economics of Technological Change." *Econometrica* 25:501-522.

_____ [1958a]: "The Demand for Fertilizer: An Economic Interpretation of Technological Change." *J. Farm Econ.* 40:591-606.

_____ [1958b]: "Research Costs and Social Returns: Hybrid Corn and Related Innovations." *J. Polit. Economy* 66:419-431. Also in Eicher and Witt, eds., 1964, pp. 369-386.

_____ [1959]: "Distributed Lags, Disaggregation, and Regional Demand Functions for Fertilizer." *J. Farm Econ.* 41:90-102.

_____ [1960]: "Measuring Inputs in Agriculture: A Survey." *J. Farm Econ.* 42:1411-1427.

_____ [1963]: "The Sources of Measured Productivity Growth: United States Agriculture, 1940-1960." *J. Polit. Economy* 71:331-346.

_____ [1964]: "Research Expenditures, Education, and the Aggregate Agricultural Production Function." *Amer. Econ. Rev.* 54:961-974.

Grilli, E. R., and M. C. Yang [1988]: "Primary Commodity Prices, Manufactured Goods Prices, and the Terms of Trade of Developing Countries: What the Long Run Shows." *Econ. Rev.* 2:1-47.

Grimshaw, R. G. [1989]: "A Review of Existing Soil Conservation Technologies and A Proposed Method of Soil Conservation Using Contour Farming Practices Backed by Vetiver Grass Hedge Barriers." In Meyers, ed., 1989, pp. 81-93.

Gulbrandsen, O., and A. Lindbeck [1973]: *The Economics of the Agricultural Sector.* Stockholm, Almqvist and Wicksell.

Gupta, A. P. [1975]: *Marketing of Agricultural Produce in India.* Bombay, Vora.

Haessel, W. [1975]: "The Price and Income Elasticities of Home Consumption and Marketed Surplus of Foodgrains." *Amer. J. Agr. Econ.* 57:111-115.

Hafid, A., and Y. Hayami [1978]: "Mobilizing Local Resources for Irrigation Development: 'The *Subsidi desa*' Case of Indonesia." In IRRI, 1978c, pp. 123-133.

Hahn, F. H., and R. C. O. Matthews [1965]: "The Theory of Economic Growth: A Survey." *Econ. J.* 74:779-902. Also in *Survey of Economic Theory* 2, 1965, New York, St. Martin's Press:1-124.

Hainsworth, G. B., ed. [1982]: *Village-Level Modernization in Southeast Asia: The Political Economy of Rice and Water.* Vancouver, Univ. of British Colombia Press.

Hallett, A. J. H. [1981]: "Data Analysis as a Sufficient Condition for Economic Planning: The Case of Sri Lanka's Development Plans." *Rev. Public Data Use* 9:283-300.

_____ [1983]: "Employment, Investment, and Production in Sri Lanka, 1959-80: Reflection on What the Figures Reveal." *Marga Quarterly J.* 7(1):78-100.

Hameed, N. D. A., N. Amerasinghe, B. L. Panditharatna, G. D. L. Gunasekera, J. Selvadurai, and S. Selvanayaham, eds. [1977]: *Rice Revolution in Sri Lanka.* Geneva, UNRISD.

Hanrahan, C. E., F. S. Urban, and J. L. Deaton [1987]: *Long-Run Changes in World Food Supply and Demand.* Washington, USDA.

Hansen, B. [1969]: "Employment and Wages in Rural Egypt." *Amer. Econ. Rev.* 59:298-313.

Hansen, G. E., ed. [1981]: *Agricultural and Rural Development in Indonesia.* Boulder, CO, Westview Press.

Harberger, A. C. [1964]: "The Measurement of Waste." *Amer. Econ. Rev.* 54(2):58-76.

_____ [1965]: "Investment in Men *versus* Investment in Machines: The Case of India." In C. A. Anderson and Bowman, eds., 1965, pp. 11-50.

Harberger, A. C., ed. [1985]: *World Economic Growth: Case Studies of Developed and Developing Nations.* San Francisco, Institute for Contemporary Studies.

Hardaker, J. B. [1979]: "A Review of Some Farm Management Research Methods for Small Farm Development in LDCs." *J. Agr. Econ.* 30:315-328.

Hargrove, T. R., V. L. Cabanilla, and W. R. Coffman [1985]: *Changes in Rice Breeding in 10 Asian Countries: 1965-84.* Los Baños, Philippines, IRRI, Research Paper Series No. 111.

Harling, K. [1983]: "Agricultural Protectionism in Developed Countries: Analysis of Systems of Intervention." *European Rev. Agr. Econ.* 10:223-247.

Harrison, J. Q., J. A. Hitchings, and J. W. Wall [1981]: "Demand Projections for India." In *India: Demand and Supply Prospects for Agriculture.* Washington, World Bank, Staff Working Paper No. 500.

Harriss, B. [1979]: "There is Method in My Madness: Or Is It Vice Versa? Measuring Agricultural Market Performance." *Food Research Institute Studies* 17:197-218.

_____ [1986]: "Implementation of Food Distribution Policies: A Case Study in South India." In Kaynak, ed., 1986, pp. 223-241.

Harrod, R. F. [1948]: *Towards a Dynamic Economics: Some Recent Developments of Economic Theory and their Application to Policy.* London, Macmillan.

Harrod, R. F., and D. C. Hague, eds. [1963]: *International Trade Theory in a Developing World.* New York, St. Martin's Press.

Hartley, M., M. Nerlove, and R. Peters [1984]: *The Supply Response for Rubber in Sri Lanka.* Washington, World Bank, Staff Working Paper No. 657.

Hasan, P. [1976]: *Korea: Problems and Issues in a Rapidly Growing Economy.* Baltimore, Johns Hopkins Univ. Press for the World Bank.

_____ [1984]: "Adjustment to External Shocks: Why East Asian Countries have Fared Better Than Other LDC's." *Finance and Dev.* 21(4):14-17.

Haseyama, T., A. Hirata, and T. Yaragihara, eds. [1983]: *Two Decades of Asian Development and Outlook for the 1980s.* Tokyo, Institute of Developing Economies, Symposium Proceedings No. 8.

Hassan, N., and K. Ahmad [1984]: "Intra-Familial Distribution of Food in Rural Bangladesh." *Food and Nutrition Bull.* 6(4):34-42.

Hayami, Y. [1964]: "Demand for Fertilizer in the Course of Japanese Agricultural Development." *J. Farm Econ.* 46:766-779.

_____ [1969]: "Resource Endowments and Technological Change in Agriculture: U.S. and Japanese Experiences in International Perspective." *Amer. J. Agr. Econ.* 51:1293-1303.

—— [1972]: "Rice Policy in Japan's Economic Development." *Amer. J. Agr. Econ.* 54:19–31.

—— [1974]: "Conditions for the Diffusion of Agricultural Technology: An Asian Perspective." *J. Econ. Hist.* 34:131–148.

—— [1975]: "Japan's Rice Policy in Historical Perspective." *Food Research Institute Studies* 14:359–380.

—— [1981a]: "Agrarian Problems of India: An East and Southeast Asian Perspective." *Econ. and Polit. Weekly* 16(16):707–712.

—— [1981b]: "Induced Innovation, Green Revolution and Income Distribution: Comment." *Econ. Dev. and Cultural Change* 30:169–176.

—— [1983]: "Growth and Equity—Is There a Trade-Off?" In Maunder and Ohkawa, eds., 1983, pp. 109–119.

—— [1986]: "Poverty and Beyond: The Forces Shaping the Future in Asia." In Maunder and Renborg, eds., 1986, pp. 37–48.

—— [1988]: *Japanese Agriculture under Siege: The Political Economy of Agricultural Policies.* New York, St. Martins Press.

Hayami, Y., and M. Akino [1977]: "Organization and Productivity of Agricultural Research Systems in Japan." In Arndt, Dalrymple and Ruttan, eds., 1977, pp. 29–59.

Hayami, Y., in association with M. Akino, M. Shintani, and S. Yamada [1975]: *A Century of Agricultural Growth in Japan: Its Relevance to Asian Development.* Tokyo, Univ. of Tokyo Press.

Hayami, Y., E. Bennagen, and R. Barker [1977]: "Price Incentive versus Irrigation Investment to Achieve Food Self-Sufficiency in the Philippines." *Amer. J. Agr. Econ.* 59:717–721.

Hayami, Y., C. C. David, P. Flores, and M. Kikuchi [1976]: "Agricultural Growth against a Land Resource Constraint: The Philippine Experience." *Australian J. Agr. Econ.* 20:144–159.

Hayami, Y., and A. Hafid [1979]: "Rice Harvesting and Welfare in Rural Java." *Bull. Indonesian Econ. Studies* 15(2):94–112.

Hayami, Y., and R. W. Herdt [1977]: "Market Price Effects of Technological Change on Income Distribution in Semisubsistence Agriculture." *Amer. J. Agr. Econ.* 59:245–256.

Hayami, Y., T. Kawagoe, Y. Morooka, and M. Siregar [1988]: *Agricultural Marketing and Processing in Upland Java: A Perspective from a Sunda Village.* Bogor, Indonesia, ESCAP-CGPRT Centre.

Hayami, Y., and M. Kikuchi [1978]: "Investment Inducements to Public Infrastructure: Irrigation in the Philippines." *Rev. Econ. and Stat.* 60:70–77.

—— [1981]: *Asian Village Economy at the Crossroads: An Economic Approach to Institutional Change.* Tokyo, Univ. of Tokyo Press. Published in 1982 by Johns Hopkins Univ. Press, Baltimore.

—— [1985]: "Agricultural Technology and Income Distribution: Two Indonesian Villages Viewed from the Japanese Experience." In Ohkawa and Ranis, eds., 1985, pp. 91–109.

Hayami, Y., M. Kikuchi, P. F. Moya, L. M. Bambo, and E. B. Marciano [1978]: *Anatomy of a Peasant Economy: A Rice Village in the Philippines.* Los Baños, Philippines, IRRI.

Hayami, Y., P. F. Moya, L. Maligalia, and M. Kikuchi [1980]: "The Economic Accounts of Households in a Philippine Village." In Ohkawa and Key, eds., 1980, pp. 29–62.

Hayami, Y., A. Quisumbing, and L. S. Adriano [1987]: *In Search of a Land Reform Design for the Philippines.* Univ. of the Philippines at Los Baños, Agricultural Research Program, Monograph Series No. 1.

Hayami, Y., and V. W. Ruttan [1970a]: "Factor Prices and Technical Change in Agricultural Development: The United States and Japan, 1880-1960." *J. Polit. Economy* 78:1115-1141.

_____ [1970b]: "Agricultural Productivity Differences among Countries." *Amer. Econ. Rev.* 60:895-911.

_____ [1970c]: "Korean Rice, Taiwan Rice, and Japanese Agricultural Stagnation: An Economic Consequence of Colonialism." *Quart. J. Econ.* 84:562-589.

_____ [1971, 1985, rev. and enl. ed.]: *Agricultural Development: An International Perspective.* Baltimore, Johns Hopkins Univ. Press.

_____ [1984]: "The Green Revolution: Inducement and Distribution." *Pakistan Dev. Rev.* 23:37-63.

Hayami, Y., V. W. Ruttan, and H. M. Southworth, eds. [1979b]: *Agricultural Growth in Taiwan, Korea, and the Philippines.* Honolulu, Univ. Press of Hawaii for the East-West Center.

Hayami, Y., K. Subbarao, and K. Otsuka [1982]: "Efficiency and Equity in the Producer Levy of India." *Amer. J. Agr. Econ.* 64:655-663.

Hayami, Y., and S. Yamada [1975]: "Agricultural Research Organization in Economic Development: A Review of the Japanese Experience." In Reynolds, ed., 1975, pp. 224-249.

Hazell, P. B. R. [1970]: "Game Theory — An Extension of its Application to Farm Planning under Uncertainty." *J. Agr. Econ.* 21:239-252.

_____ [1982]: *Instability in Indian Foodgrain Production.* Washington, IFPRI, Research Report No. 30.

_____ [1984]: "Sources of Increased Instability in India and U.S. Cereal Production." *Amer. J. Agr. Econ.* 36:145-159.

_____ [1985]: "Sources of Increased Variability in World Cereal Production Since the 1960s." *J. Agr. Econ.* 36:145-159.

_____ [1986]: *Summary Proceedings of a Workshop on Cereal Yield Variability.* Washington, IFPRI.

Hazell, P. B. R., and C. Ramasamy, eds. [1990]: *Green Revolution Reconsidered: The Impact of High-Yielding Rice Varieties in South India.* Baltimore, Johns Hopkins Univ. Press for IFPRI.

Hazell, P. B. R., and A. Röell [1983]: *Rural Growth Linkages: Household Expenditure Patterns in Malaysia and Nigeria.* Washington, IFPRI, Research Report No. 41.

He, G. T., A. Te, G. Zhu, S. L. Travers, F. Lai, and R. W. Herdt [1984]: "The Economics of Hybrid Rice Production in China." Los Baños, Philippines, IRRI, Research Paper Series No. 101.

Heady, E. O. [1947]: "Economics of Farm Leasing Systems." *J. Farm Econ.* 29:659-678.

Hecksher, E. [1919]: "The Effect of Foreign Trade on the Distribution of Income." *Economisk Tidskrift* 21:497-512. Also in Ellis and Metzler, eds., 1950, pp. 270-300.

Hedley, D. A. [1979]: *Rice Buffer Stocks for Indonesia: A First Approximation.* Washington, IFPRI, IFDC and IRRI. Working Paper No. 2.

Helsel, Z. R., ed. [1987]: *Energy in Plant Nutrition and Pest Control. Vol. 2, Energy in World Agriculture.* Amsterdam, The Netherlands, Elsevier Science.

Herath, H. M., J. B. Hardaker, and J. R. Anderson [1982]: "Choice of Varieties by Sri Lanka Rice Farmers: Comparing Alternative Decision Models." *Amer. J. Agr. Econ.* 64:87-93.

Herdt, R. W. [1970]: "A Disaggregate Approach to Aggregate Supply." *Amer. J. Agr. Econ.* 52:512-520.

_____ [1979]: "An Overview of the Constraints Project Results." In IRRI, 1979a, pp. 395-411.

_____ [1980]: "Studies in Water Management Economics at IRRI." In IRRI, 1980a, pp. 115-138.

_____ [1987]: "A Retrospective View of Technological and Other Changes in Philippine Rice Farming, 1965-1982." *Econ. Dev. and Cultural Change* 35:329-350.

Herdt, R. W., and R. Barker [1979]: "Sources of Growth in Asian Food Production and an Approach to Identification of Constraining Factors." In UN Univ. and IRRI, eds., 1979, pp. 21-44.

Herdt, R. W., and C. Capule [1983]: *Adoption, Spread, and Production Impact of Modern Rice Varieties in Asia.* Los Baños, Philippines, IRRI.

Herdt, R. W., and T. A. Lacsina [1976]: "The Domestic Resource Cost of Increasing Philippine Rice Production." *Food Research Institute Studies* 15:213-231.

Herdt, R. W., and A. M. Mandac [1981]: "Modern Technology and Economic Efficiency of Philippine Rice Farmers." *Econ. Dev. and Cultural Change.* 29:375-399.

Herdt, R. W., and J. W. Mellor [1964]: "The Contrasting Response of Rice to Nitrogen: India and the United States." *J. Farm Econ.* 46:150-160.

Herdt, R. W., and M. Rosegrant [1978]: "The Impact of Price and Income Support Policies on Small Rice Farmers in the Philippines." *Philippine Rev. Bus. and Econ.* 15(4):1-36.

_____ [1981]: "Simulating the Impacts of Credit Policy and Fertilizer Subsidy on Central Luzon Rice Farms, The Philippines." *Amer. J. Agr. Econ.* 63:655-665.

Herdt, R. W., and P. J. Stangel [1984]: "Population, Rice Production, and Fertilizer Outlook." In IRRI, 1984, *Organic Matter and Rice.* Los Baños, Philippines.

Herdt, R. W., A. Te, and R. Barker [1977/78]: "The Prospects for Asian Rice Production." *Food Research Institute Studies* 16: 183-203.

_____ [1980]: "Prospects for Rice Production and Consumption in the Philippines—A Projection." *J. Agr. Econ. and Dev.* 10(1):1-28.

Herdt, R. W., and J. Y. S. Tsai [1977]: "Economic Changes in Typical Rice Farms in Taiwan, 1895-1976." *Ekonomi Dan Keuangan Indonesia* 25(3):243-268.

Herdt, R. W., and T. H. Wickham [1975]: "Exploring the Gap between Potential and Actual Rice Yield in the Philippines." *Food Research Institute Studies* 14:163-181.

Hertford, R., J. Ardila, A. Rocha, and C. Trujillo [1977]: "Productivity of Agricultural Research in Colombia." In Arndt, Dalrymple and Ruttan, eds., 1977, pp. 86-123.

Herz, B. [1989]: "Bringing Women into the Economic Mainstream." *Finance and Dev.* 26(4):22-25.

Hillel, D. [1987]: *The Efficient Use of Water in Irrigation: Principles and Practices for Improving Irrigation in Arid and Semiarid Regions.* Washington, World Bank Technical Paper No. 64.

Hillman, J., and E. A. Monke [1983]: "International Transfer of Agricultural Technology." In Maunder and Ohkawa, eds., 1983, pp. 519-528.

Hines, J. [1972]: "The Utilization of Research for Development—Two Case Studies in Rural Modernization and Agriculture in Peru." Unpublished Ph.D. dissertation, Princeton Univ., Princeton, NJ.

Hirashima, S. [1977]: *Hired Labor in Rural Asia.* Tokyo, Institute of Developing Economies.

Hirashima, S., and M. Muqtada [1986]: *Hired Labour and Rural Labour Markets in Asia.* New Delhi, ILO, ARTEP.

Hirschman, A. O. [1958]: *A Strategy of Economic Development.* New Haven, CT, Yale Univ. Press.

_____ [1977]: "A Generalized Linkage Approach to Development, With Special Reference to Staples." In Nash, ed., 1977, pp. 67-98.

Hoefer, J. A., and P. J. Tsuchitani, eds. [1980]: *Animal Agriculture in China*. Washington, National Academy Press.

Hong, W., and L. Krause [1981]: *Trade and Growth of the Advanced Developing Countries in the Pacific Basin*. Seoul, Korea Development Institute Press.

Hoogstraten, H. J. [1985]: *Irrigation and Social Organization in West Malaysia*. Wageningen, Holland, Agricultural University.

Hooley, R., and V. W. Ruttan [1969]: "The Philippines." In Shand, ed., 1969, pp. 215-250.

Hoon, K. L. [1969]: *Land Utilization and Rural Economy in Korea*. New York, Greenwood Press.

Hopkins, R. [1984]: "The Evolution of Food Aid." *Food Policy* 9:343-363.

Hopkins, R., D. Puchala, and R. Talbot, eds. [1979]: *Food, Politics and Agricultural Development: Case Studies in Public Policy of Rural Modernization*. Boulder, CO, Westview Press.

Hopper, W. D. [1965]: "Allocative Efficiency in a Traditional Indian Agriculture." *J. Farm Econ.* 47:611-624.

Hoque, M. Z. [1984]: *Cropping Systems in Asia: On-Farm Research and Management*. Los Baños, Philippines, IRRI.

Horii, K. [1981]: *Rice Economy and Land Tenure in West Malaysia: A Comparative Study of Eight Villages*. Tokyo, IDE, Occasional Paper Series No. 18.

Hornby, J. M. [1968]: "Investment and Trade Policy in a Dual Economy." *Econ. J.* 78:96-107.

Hossain, M. [1980]: "Foodgrain Production in Bangladesh: Performance, Potential and Constraints." *Bangladesh Dev. Studies* 8(1/2): 39-70.

———— [1984]: "Agricultural Development in Bangladesh: A Historical Perspective." *Bangladesh Dev. Studies* 12(4):29-57.

———— [1986]: "A Note on the Trend of Landlessness in Bangladesh." *Bangladesh Dev. Studies* 14:93-100.

———— [1987]: "Agricultural Growth Linkages—The Bangladesh Case." *Bangladesh Dev. Studies* 15:1-30.

———— [1988a]: *Credit for Alleviation of Rural Poverty: The Experience of Grameen Bank in Bangladesh*. Washington, IFPRI and BIDS, Research Report No. 65.

———— [1988b]: *Nature and Impact of the Green Revolution in Bangladesh*. Washington, IFPRI and BIDS, Research Report No. 67.

Hou, C. M., and T. S. Yu, eds. [1982]: *Agricultural Development in China, Japan, and Korea*. Taipei, Taiwan, Academica Sinica.

Houck, J. P. [1986]: "A Note on the Link between Agricultural Development and Agricultural Imports." St. Paul, Univ. of Minnesota, Department of Agricultural and Applied Economics, Staff Paper No. 86-26.

Houthakker, H. S. [1957]: "An International Comparison of Household Expenditure Patterns, Commemorating the Centenary of Engel's Law." *Econometrica* 25:532-551.

Howell, J., ed. [1980]: *Borrowers and Lenders: Rural Financial Markets and Institutions*. London, ODI.

Howell, J., ed. [1985]: *Recurrent Costs and Agricultural Development*. London, ODI.

Hsieh, S. C., and T. H. Lee [1966]: *Agricultural Development and its Contributions to Economic Growth in Taiwan*. Taipei, Taiwan, JCRR, Economic Digest Series No. 17.

Hsieh, S. C., and V. W. Ruttan [1967]: "Environmental, Technological, and Institutional Factors in the Growth of Rice Production: Philippines, Thailand, and Taiwan." *Food Research Institute Studies* 7:307-341.

Hsu, R. C. [1972]: "The Demand for Fertilizer in a Developing Country: The Case of Taiwan, 1950-1966." *Econ. Dev. and Cultural Change* 20:299-309.

———— [1982]: *Food for One Billion: China's Agriculture since 1949.* Boulder, CO, Westview Press.

Huang, C. J., and F. S. Bagi [1984]: "Technical Efficiency on Individual Farms in Northwest India." *Southern Econ. J.* 51:108-115.

Huang, C. J., A. M. Tang, and F. S. Bagi [1986]: "Two Views of Efficiency in Indian Agriculture." *Canadian J. Agr. Econ.* 34:209-226.

Huddleston, B. [1984]: *Closing the Cereals Gap with Trade and Food Aid.* Washington, IFPRI, Research Report No. 43.

Huddleston, B., D. Gale Johnson, S. Reutlinger, and A. Valdes [1984]: *International Finance Arrangements for Food Security.* Baltimore, Johns Hopkins Univ. Press.

Hughes, H., ed. [1988]: *Achieving Industrialization in East Asia.* Cambridge, England, Cambridge Univ. Press.

Hung, G. N. T. [1977]: *Economic Development of Socialist Vietnam, 1955-80.* New York, Praeger.

Hunter, G. [1969]: *Modernizing Peasant Societies.* New York, Oxford Univ. Press for Institute of Race Relations.

———— [1973]: "Agricultural Administration and Institutions." *Food Research Institute Studies* 12:233-251.

Hussain, S. M. [1964]: "A Note on Farmer Response to Price in East Pakistan." *Pakistan Dev. Rev.* 4:93-106.

Hwa, E. C. [1983]: *The Contribution of Agriculture to Economic Growth: Some Empirical Evidence.* Washington, World Bank Staff Working Paper No. 619.

Hymer, S., and R. Resnick [1969]: "A Model of an Agrarian Economy with Nonagricultural Activities." *Amer. Econ. Rev.* 59:493-506.

Ichimura, S. [1986]: *Development Strategies and Productivity Issues in Asia.* Tokyo, APO.

Indian Society of Agricultural Economics [ISAE, 1972]: *Comparative Experience of Agricultural Development in Developing Countries of Asia and South-East since World War II.* Bombay.

———— [ISAE, 1977a]: "Returns from Investment on Agricultural Research." *Indian J. Agr. Econ.* 32(3):181-247.

———— [ISAE, 1977b]: "Economics of Nutrition." *Indian J. Agr. Econ.* 32(3):1-112.

Ingram, J. C. [1971]: *Economic Change in Thailand, 1850-1970.* Stanford, CA, Stanford Univ. Press.

International Association of Agricultural Economists [IAAE, 1971]: *Policies, Planning, and Management in Agricultural Development: Papers and Reports, 14th ICAE, August 23-September 2, 1970, Minsk, Russia.* Oxford, England, Oxford Institute of Agrarian Affairs.

International Crops Research Institute for the Semi-Arid Tropics [ICRISAT, 1980]: *Socioeconomic Constraints to Development of Semi-Arid Tropical Agriculture.* Patancheru, India.

———— [ICRISAT, 1985]: *Agricultural Markets in the Semi-Arid Tropics.* Patancheru, India.

International Food Policy Research Institute [IFPRI, 1976]: *Meeting Food Needs in the Developing World: Location and Magnitude of the Task in the Next Decade.* Washington, Research Report No. 1.

International Fertilizer Development Center [IFDC, 1982]: *Bangladesh Policy Options for the Development of the Fertilizer Sector.* Muscle Shoals, AL.

International Institute for Environment and Development and the World Resources Institute [1987]: *World Resources, 1987.* New York, Basic Books.

International Labour Organization [ILO, 1973]: *Mechanization and Employment in Agriculture: Case Studies from Four Continents.* Geneva.

_____ [ILO, 1974]: *Sharing in Development: A Program of Employment, Equity, and Growth for the Philippines.* Geneva.

_____ [ILO, 1977, 1979]: *Poverty and Landlessness in Rural Asia.* Geneva.

_____ [ILO, 1978]: *Labour Absorption in Agriculture; The East Asian Experience.* Bangkok, ARTEP.

_____ [ILO, 1980]: *Employment Expansion in Asian Agriculture: A Comparative Analysis of South Asian Countries.* Bangkok, ARTEP.

_____ [ILO, 1981]: *Employment Expansion through Local Resource Mobilisation.* Bangkok, ARTEP.

_____ [ILO, 1986]: *Employment and Development in the Domestic Food Crop Sector, Sri Lanka: Report of the ILO-ARTEP Rural Employment Mission to Sri Lanka.* New Delhi, Asian Employment Programme.

International Monetary Fund [IMF, 1988]: *Primary Commodities: Market Developments and Outlook.* World Economic and Financial Surveys, Washington.

International Rice Research Institute [IRRI, 1975]: *Changes in Rice Farming in Selected Areas of Asia.* Los Baños, Philippines.

_____ [IRRI, 1976]: *Bibliography on Socio-Economic Aspects of Irrigation in Asia.* Los Baños, Philippines and New York, ADC.

_____ [IRRI, 1977]: *Constraints to High Yields on Asian Rice Farms: An Interim Report.* Los Baños, Philippines.

_____ [IRRI, 1978a]: *Economic Consequences of the New Rice Technology.* Los Baños, Philippines.

_____ [IRRI, 1978b]: *Interpretive Analysis of Selected Papers from "Changes in Rice Farming in Selected Areas of Asia."* Los Baños, Philippines.

_____ [IRRI, 1978c]: *Irrigation Policy and Management in Southeast Asia.* Los Baños, Philippines.

_____ [IRRI, 1979a]: *Farm-Level Constraints to High Rice Yields in Asia: 1974-77.* Los Baños, Philippines.

_____ [IRRI, 1979b]: *Rainfed Lowland Rice—Selected Papers from the 1978 IRRI Conference.* Los Baños, Philippines.

_____ [IRRI, 1980]: *Report of a Planning Workshop on Irrigation Water Management.* Los Baños, Philippines, IRRI.

_____ [IRRI, 1982a]: *Report of a Workshop on Cropping Systems Research in Asia.* Los Baños, Philippines.

_____ [IRRI, 1982b]: *Rice Research Strategies for the Future.* Los Baños, Philippines.

_____ [IRRI, 1982c]: *Report of a Workshop on Cropping Systems Research in Asia.* Los Baños, Philippines.

_____ [IRRI, 1983]: *Consequences of Small-Farm Mechanization.* Los Baños, Philippines.

_____ [IRRI, 1985]: *Women in Rice Farming.* Brookfield, VT, Gower.

_____ [IRRI, 1988]: *Green Manure in Rice Farming—Proceedings of a Symposium on Sustainable Agriculture.* Los Baños, Philippines.

International Rice Research Institute and Chinese Academy of Agricultural Sciences [IRRI, 1980]: *Rice Improvement in China and Other Asian Countries.* Los Baños, Philippines.

International Service for National Agricultural Research and the Rockefeller Foundation [ISNAR, 1985]: *Women and Agricultural Technology: Relevance for Research: Report from the CGIAR Inter-Center Seminar on Women and Agricultural Technology.* New York, Rockefeller Foundation, 2 Vols.

Iqbal, B. A., and M. Talha [1986]: "Sustaining Growth in Indian Agriculture." *Food Policy* 11:94-98.

Isarangkura, R. [1986]: *Thailand and the CGIAR Centers: A Study of their Collaboration in International Agricultural Research*. Washington, World Bank, CGIAR Study Paper No. 16.

Isenman, P. [1980]: "Basic Needs: The Case of Sri Lanka." *World Dev.* 8:237-258.

Isenman, P., and H. W. Singer [1977]: "Food Aid: Disincentive Effects and their Policy Implications." *Econ. Dev. and Cultural Change* 25:205-237.

Ishii, Y., ed. [1978]: *Thailand: A Rice Growing Society*. Honolulu, Univ. of Hawaii Press.

Ishikawa, S. [1967a]: *Economic Development in Asian Perspective*. Tokyo, Kinokuniya Bookstore.

———— [1967b]: "The Resource Flow between Agriculture and Industry: The Chinese Experience." *Developing Economies* 5(1):3-49.

———— [1970]: *Agricultural Development Strategies in Asia — Case Studies of the Philippines and Thailand*. Manila, ADB.

———— [1971]: "Changes in the Structure of Agricultural Production in Mainland China." In Jackson, ed., 1971, pp. 346-377.

———— [1975]: "Peasant Families and Agrarian Community in the Process of Economic Development." In Reynolds, ed., 1975, pp. 451-496.

———— [1977]: "China's Food and Agriculture — A Turning Point." *Food Policy* 2:90-102.

———— [1978]: *Labour Absorption in Asian Agriculture. An "Issues" Paper*. Bangkok, ILO, ARTEP.

———— [1980]: "East Asian Experience in Labour Absorption in Agriculture: A Summary of East Asian Papers and Some Further Thought." In ILO, 1980, pp. 1-58.

———— [1981]: *Essays on Technology, Employment, and Institutions in Economic Development: Comparative Asian Experience*. Tokyo, Kinokuniya Bookstores.

———— [1982]: "Labour Absorption in China's Agriculture." In Ishikawa, Yamada and Hirashima, eds., 1982, pp. 1-25.

———— [1983]: "China's Economic Growth since 1949 — An Assessment." *China Quarterly* 94:242-281.

———— [1988]: "Patterns and Processes of Intersectoral Resource Flows: Comparisons of Cases in Asia." In Ranis and T. P. Schultz, eds., 1988, pp. 283-338.

Ishikawa, S., S. Yamada, and S. Hirashima, eds. [1982]: *Labour Absorption and Growth in Agriculture: China and Japan*. Bangkok, ILO, ARTEP.

Islam, B. [1978]: "Price, Income, and Foreign Exchange Elasticity of Asian Rice Imports." *Amer. J. Agr. Econ.* 60:532-535.

Islam, N. [1964]: "Concepts and Measurement of Unemployment and Underemployment in Developing Economies." *International Labour Rev.* 89:240-267.

———— [1966]: *Studies in Consumer Demand*. Vols. 1 and 2. Karachi, Pakistan, Oxford Univ. Press.

———— , ed. [1974]: *Agricultural Policy in Developing Countries*. New York, Macmillan.

———— [1978]: *Development Strategy of Bangladesh*. New York, Pergamon Press.

Islam, R., ed. [1985]: *Strategies for Alleviating Poverty in Rural Asia*. Bangkok, ILO, ARTEP, Asian Employment Programme and Dhaka, BIDS.

Islam, R., A. R. Khan, and E. Lee [1982]: *Employment and Development in Nepal*. Bangkok, ILO, ARTEP.

Issawi, C. [1957]: "Farm Output under Fixed Rents and Share Tenancy." *Land Econ.* 33:74-77.

Jackson, W. A. D., ed. [1971]: *Agrarian Policies and Problems in Communist and Non-Communist Countries*. Seattle, Univ. of Washington Press.

Jain, H. C. [1971]: "Growth and Recent Trends in the Institutional Credit in India." *Indian J. Agr. Econ.* 26(4):555-556.

Jain, S. [1975]: *Size Distribution of Income: A Compilation of Data.* Washington, World Bank.

Jaiswal, P. L., ed. [1985]: *Rice Research in India.* New Delhi, ICAR.

Jalali, J., ed. [1989]: *A Research Inventory for the Multilateral Trade Negotiations, 1989.* Washington, World Bank Papers for the Uruguay Round, Second Annual Compilation.

James, W. E. [1978]: *An Economic Analysis of Land Settlement Alternatives in the Philippines.* Los Baños, Philippines, IRRI, Agricultural Economics Paper No. 78-30.

———— [1982]: *Asian Agriculture and Economic Development.* Manila, ADB, Economic Staff Paper No. 5.

———— [1983]: *Asian Agriculture in Transition: Key Policy Issues.* Manila, ADB, Economic Staff Paper No. 19.

Jamison, D. T., and L. J. Lau [1981]: *Farmer Education and Farm Efficiency.* Baltimore, Johns Hopkins Univ. Press for the World Bank.

Jannuzi, F. T., and J. T. Peach [1980]: *The Agrarian Structure of Bangladesh: Impediments to Development.* Boulder, CO, Westview Press.

Jasdanwalla, A. [1966]: *Marketing Efficiency in Indian Agriculture.* Bombay, India, Allied Publishers.

Jayasuriya, S., C. Barlow, and R. Shand [1981]: "Farmers' Long-Term Investment Decisions: A Study of Sri Lanka Rubber Smallholders." *J. Dev. Studies* 18:47-67.

Jayasuriya, S. K., A. Te, and R. W. Herdt [1986]: "Mechanisation and Economics of Machinery Use in Low-Wage Economies." *J. Dev. Studies* 22:328-335.

Jha, D. [1970]: "Acreage Response of Sugarcane in Factory Areas of North Bihar." *Indian J. Agr. Econ.* 25(1):79-91.

Jha, S. C. [1987]: "Rural Development in Asia: Issues and Perspectives." *Asian Dev. Rev.* 5:83-99.

Jodha, N. S. [1971]: "Land-Based Credit Policies and Investment Prospects for Small Farmers." *Econ. and Polit. Weekly* 6(39):A143-A148.

———— [1985]: "Population Growth and Decline of Common Property Resources in Rajasthan, India." *Population and Dev. Rev.* 11:247-264.

———— [1986]: "Common Property Resources and Rural Poor in Dry Regions of India." *Econ. and Polit. Weekly* 21(26):1169-1186.

Jodha, N. S., and R. P. Singh [1982]: "Factors Constraining Growth of Coarse Grain Crops in Semi-Arid Tropical India." *Indian J. Agr. Econ.* 37(3):346-354.

Johl, S. S. [1972]: "Agricultural Taxation in a Developing Economy: A Case of India." *Indian J. Agr. Econ.* 27(3):1-19.

———— [1973a]: "Farm Size, Economic Efficiency and Social Justice (A Case of Punjab)." *Agricultural Mechanization in Asia* 4(1):56-61.

———— [1973b]: "Mechanisation, Labour Use, and Productivity in Agriculture." *Agricultural Situation in India* 28:3-16.

Johl, S. S., and M. S. Mudahar [1966]: "Distribution of Agricultural Processing and Supply Industries in Punjab." *Indian J. Agr. Econ.* 21(4):47-54.

———— [1974]: *The Dynamics of Institutional Change and Rural Development in Punjab, India.* Ithaca, NY, Cornell Univ., Center for International Studies, Rural Development Committee, Monograph No. 5. Also Chapter 4 in Uphoff, ed., 1982, Vol. 1.

Johnson, D. Gale [1950]: "Resource Allocation Under Share Contracts." *J. Polit. Economy* 58:111-123.

———— [1973]: *World Agriculture in Disarray.* London, Macmillan.

———— [1975]: "Free Trade in Agricultural Products: Possible Effects on Total Output, Prices, and the International Distribution of Output." In Tolley and Zadrozny, eds., 1975, pp. 3-20.

———— [1976]: "Increased Stability of Grain Supplies in Developing Countries: Optimal Carryover and Insurance." *World Dev.* 4:977-987.

———— [1982]: *Progress of Economic Reform in the People's Republic of China.* Washington, American Enterprise Institute.

———— [1988]: "Constraints on Price Adjustments: Structural, Institutional and Financial Rigidities." In Antonelli and Quadrio-Curzio, 1988, pp. 81-102.

Johnson, D. Gale, K. Hemmi, and P. Lardinois [1985]: *Agricultural Policy and Trade: Adjusting Domestic Programs in an International Framework.* New York, New York Univ. Press.

Johnson, Harry G. [1967]: *Economic Policies toward Less Developed Countries.* Washington, Brookings Institution.

Johnson, M., W. L. Parish, and E. Lin [1987]: "Chinese Women, Rural Society, and External Markets." *Econ. Dev. and Cultural Change* 35(2):257-278.

Johnson, S. E., and J. W. Couston [1970]: "High-Yielding Varieties in the Strategy of Development." *Monthly Bull. Agr. Econ. and Stat.* 19:1-8.

Johnson, S. H. III [1982a]: "The Effects of Major Dam Construction: The Nam Pong Project in Thailand." In MacAndrews and Sien, eds., 1982, pp. 172-207.

———— [1982b]: "Large-Scale Irrigation and Drainage Schemes in Pakistan: A Study of Rigidities in Public Decision-Making." *Food Research Institute Studies* 18:149-180.

———— [1985]: "Economic, Social, and Technical Considerations Determining Investments in Groundwater in Bangladesh." In de Kosinsky and de Somer, eds., 1985.

———— [1986a]: "Social and Economic Impacts of Investments in Groundwater: Lessons for Pakistan and Bangladesh." In Nobe and Sampath, eds., 1986, pp. 179-216.

———— [1986b]: "Agricultural Intensification in Thailand: Complementary Role of Infrastructure and Agricultural Policy." In Easter, ed., 1986, pp. 111-127.

Johnson, S. H. III, and T. Charoenwatana [1981]: "Economics of Rainfed Cropping Systems: Northeast Thailand." *Water Resources Research* 17:462-468.

Johnson, S. H. III, and J. O. Reuss [1984]: "Economics of Changes in Irrigation Management in Pakistan: An Integrative Modeling Approach." *Water International* 9:66-71.

Johnston, B. F. [1951]: "Agricultural Productivity and Economic Development in Japan." *J. Polit. Economy* 59:498-513.

———— [1962]: "Agricultural Development and Economic Transformation: A Comparative Study of the Japanese Experience." *Food Research Institute Studies* 3:223-276.

———— [1966]: "Agriculture and Economic Development: The Relevance of the Japanese Experience." *Food Research Institute Studies* 6:251-312.

———— [1970]: "Agricultural and Structural Transformation in Developing Countries: A Survey of Research." *J. Econ. Literature* 8:369-404.

———— [1977]: "Food, Health, and Population in Development." *J. Econ. Literature* 15:879-907.

———— [1989]: "The Political Economy of Agricultural Development in the Soviet Union and China." *Food Research Institute Studies* 21:97-137.

Johnston, B. F., and W. C. Clark [1982]: *Redesigning Rural Development: A Strategic Perspective.* Baltimore, Johns Hopkins Univ. Press.

Johnston, B. F., and J. Cownie [1969]: "The Seed-Fertilizer Revolution and Labor Force Absorption." *Amer. Econ. Rev.* 59:569-582.

Johnston, B. F., and H. Kaneda, eds. [1969]: *Agriculture and Economic Growth: Japan's Experience.* Tokyo, Univ. of Tokyo Press.

Johnston, B. F., and P. Kilby [1975]: *Agricultural and Structural Transformation*. London, Oxford Univ. Press.

Johnston, B. F., and R. Martorell [1977]: "Interrelationships among Nutrition, Health, Population, and Development." *Food Research Institute Studies* 16:1-19.

Johnston, B. F., and J. W. Mellor [1960]: "The Nature of Agriculture's Contributions to Economic Development." *Food Research Institute Studies* 1:335-356.

_____ [1961]: "The Role of Agriculture in Economic Development." *Amer. Econ. Rev.* 51:566-593.

Johnston, B. F., and S. T. Nielson [1966]: "Agriculture and Structural Transformation in a Developing Economy." *Econ. Dev. and Cultural Change* 14:279-301.

Johnston, B. F., J. M. Page, and P. Warr [1972]: "Criteria for the Design of Agricultural Development Strategies." *Food Research Institute Studies* 11:27-58.

Johnston, B. F., and G. S. Tolley [1965]: "Strategy for Agricultural Development." *J. Farm Econ.* 47:365-379.

Johnston, B. F., and T. P. Tomich [1985]: "Agricultural Strategies and Agrarian Structure." *Asian Dev. Rev.* 3(2):1-37.

Jones, J. R., ed. [1986]: *East-West Agricultural Trade*. Boulder, CO, Westview Press.

Jorgenson, D. W. [1961]: "The Development of a Dual Economy." *Econ. J.* 71:309-334.

_____ [1966]: "Testing Alternative Theories of the Development of a Dual Economy." In Adelman and Thorbecke, eds., 1966, pp. 45-60.

_____ [1967]: "Surplus Agricultural Labor and the Development of a Dual Economy." *Oxford Econ. Papers* NS19:288-312.

_____ [1969]: "The Role of Agriculture in Economic Development: Classical *versus* Neoclassical Models of Economic Growth." In Wharton, ed., 1969a, pp. 320-348.

Joshi, P. C. [1975]: *Land Reforms in India: Trends and Perspective*. Bombay, Allied Publishers.

Josling, T. [1975]: "Agricultural Trade: Implications for the Distribution of Gains from Technical Progress." In Tolley and Zadrozny, eds., 1975, pp. 169-180.

_____ [1980]: *Developed Country Agricultural Policies and Developing Country Food Supplies: The Case of Wheat*. Washington, IFPRI, Research Report No. 14.

Judd, M. A., J. K. Boyce, and R. E. Evenson [1986]: "Investing in Agricultural Supply: The Determinants of Agricultural Research and Extension." *Econ. Dev. and Cultural Change* 35:77-114.

Kahlon, A. S. [1976]: "Impact of Mechanisation on Punjab Agriculture with Special Reference to Tractorisation." *Indian J. Agr. Econ.* 31(4):54-70.

Kahlon, A. S., and S. S. Grewal [1972]: "Farm Mechanisation in a Labor-Abundant Economy: A Comment." *Econ. and Polit. Weekly* 7(20):991-992.

Kahlon, A. S., J. R. Gupta, and R. K. Sondhi [1971]: "Impact of New Farm Technology on Farm Labor Use in Punjab, 1966-67 to 1969-70." *Agricultural Situation in India* 26:629-636.

Kahlon, A. S., and M. S. Mudahar [1969]: "Rationale of Maximum Credit Limit Based on Farm Production Plans." In Reserve Bank of India, ed., 1969, pp. 190-201.

Kahlon, A. S., P. N. Saxena, H. K. Bal, and D. Jha [1977]: "Returns to Investment in Agricultural Research in India." In Arndt, Dalrymple and Ruttan, eds., 1977, pp. 124-167.

Kahlon, A. S., and K. Singh [1984a]: "Eastern Region: Potential, Constraints, and Development Strategies." *Econ. and Polit. Weekly* 19(4):174-180.

_____ [1984b]: *Managing Agricultural Finance—Theory and Practices*. New Delhi, Allied Publishers.

Kahlon, A. S., and D. S. Tyagi [1983]: *Agricultural Price Policy in India*. New Delhi, Allied Publishers.

Kahlon, A. S., and S. D. Vashishtha [1968]: "A Study of the Factors Governing the Flow of Marketable Surplus of Major Crops in Ludhiana District." *Agricultural Situation in India* 23:815-823.

Kalirajan, K. [1976]: "Calorie Intakes of Food Comparisons across States and Classes." *Indian J. Agr. Econ.* 31(2):53-63.

—— [1980]: "The Contribution of Location Specific Research to Agricultural Productivity." *Indian J. Agr. Econ.* 35(4):8-16.

—— [1981]: "The Economic Efficiency of Farmers Growing High-Yielding, Irrigated Rice in India." *Amer. J. Agr. Econ* 63(3):566-570.

Kalirajan, K., and J. C. Flinn [1981]: "Comparative Technical Efficiency in Rice Production." *Philippine Econ. J.* 46:31-43.

Kalirajan, K., and R. T. Shand [1982]: "Location Specific Research: Rice Technology in India." *Land Economics* 58:537-546.

Kamarck, A. M. [1976]: *The Tropics and Economic Development*. Baltimore, Johns Hopkins Univ. Press.

Kaneda, H. [1968]: "Substitution of Labor and Nonlabor Inputs and Technical Change in Japanese Agriculture." *Rev. Econ. and Stat.* 47:162-176.

—— [1969]: "Economic Implications of the 'Green Revolution' and the Strategy of Agricultural Development in West Pakistan." *Pakistan Dev. Rev.* 9:111-143.

Kaneda, M. [1968]: "Long-Term Changes in Food Consumption Patterns In Japan, 1878-1964." *Food Research Institute Studies* 8:1-32.

Kanel, D. [1967]: "Size of Farm and Economic Development." *Indian J. Agr. Econ.* 22(2):26-44.

Kanwar, J. S., and M. S. Mudahar [1983]: *Fertilizer Sulfur and Food Production: Research and Policy Implications for Tropical Countries—Executive Brief*. Muscle Shoals, AL, IFDC, Technical Bulletin T-27.

—— [1986]: *Fertilizer Sulfur and Food Production*. Dordrecht, Netherlands, Martinus Nijhoff Publishers.

Kao, C. H. C., K. R. Anschel, and C. K. Eicher [1964]: "Disguised Unemployment in Agriculture: A Survey." In Eicher and Witt, eds., 1964, pp. 129-144.

Kaul, J. L. [1967]: "A Study of Supply Responses to Price of Punjab Crops." *Indian J. Econ.* 48:25-39.

Kaul, J. L., and D. S. Sidhu [1971]: "Acreage Response to Prices for Major Crops in Punjab: An Econometric Study." *Indian J. Econ. Studies* 26(4):427-434.

Kawagoe, T., and Y. Hayami [1985]: "An Intercountry Comparison of Agricultural Production Efficiency." *Amer. J. Agr. Econ.* 67:87-92.

Kawagoe, T., Y. Hayami, and V. W. Ruttan [1985]: "The Intercountry Agricultural Production Function and Productivity Differences among Countries." *J. Dev. Econ.* 19:113-132.

—— [1988]: "The Intercountry Agricultural Production Function and Productivity Differences among Countries: Reply." *J. Dev. Econ.* 28:125-126.

Kawagoe, T., K. Otsuka, and Y. Hayami [1986]: "Induced Bias of Technical Change in Agriculture: The United States and Japan, 1880-1980." *J. Polit. Economy* 94:523-544.

Kaynak, E., ed. [1986]: *World Food Marketing Systems*. London, Butterworths.

Kelley, A. C., and J. G. Williamson [1974]: *Lessons from Japanese Development*. Univ. of Chicago Press.

Kelley, A. C., J. G. Williamson, and R. J. Cheetham [1972]: *Dualistic Economic Development*. Univ. of Chicago Press.

Kelly, M., N. Kirmani, M. Xafa, C. Boonekamp, and P. Winglee [1988]: *Issues and Developments in International Trade Policy*. Washington, IMF, Occasional Paper No. 63.

Kennedy, E. T., and H. H. Alderman [1987]: *Comparative Analyses of Nutritional Effectiveness of Food Subsidies and Other Food Related Interventions*. Washington, IFPRI.

Khadka, N. [1985]: "The Political Economy of the Food Crisis in Nepal." *Asian Survey* 25(9):943-962.

Khan, A. R. [1978]: "Taxation, Procurement and Collective Incentives in Chinese Agriculture." *World Dev.* 6:827-836.

Khan, A. R., and A. H. M. N. Chowdhury [1962]: "Marketable Surplus Function: A Study of the Behavior of West Pakistan Farmers." *Pakistan Dev. Rev.* 2:354-376.

Khan, M. H. [1966]: *The Role of Agriculture in Economic Development: A Case Study of Pakistan*. Wageningen, Holland, Centre for Agricultural Publications and Documentation.

——— [1975]: *The Economics of the Green Revolution in Pakistan*. New York, Praeger.

Khan, M. H., and A. H. Akbari [1976]: "Impact of Agricultural Research and Extension on Crop Productivity in Pakistan: A Production Function Approach." *World Dev.* 14:757-762.

Khan, M. H., and D. R. Maki [1980]: "Relative Efficiency by Farm Size and the Green Revolution in Pakistan." *Pakistan Dev. Rev.* 19:51-64.

Khan, M. I. [1963]: "A Note on Consumption Patterns in the Rural Areas of East Pakistan." *Pakistan Dev. Rev.* 3:399-413.

Khatkhate, D. R. [1962]: "Some Notes on the Real Effects of Foreign Surplus Disposal in Underdeveloped Economies." *Quart. J. Econ.* 76:186-196.

Khusro, A. M. [1964]: "Returns to Scale in Indian Agriculture." *Indian J. Agr. Econ.* 19(3,4):51-80.

——— [1967]: "The Pricing of Food in India." *Quart. J. Econ.* 81:270-285.

——— [1969]: "Farm Size and Land Tenure in India." *Indian Econ. Rev.* NS4:123-145.

Kikuchi, M., G. Dozina, Jr., and Y. Hayami [1978]: "Economics of Community Work Programs: A Communal Irrigation Project in the Philippines." *Econ. Dev. and Cultural Change* 26:211-225.

Kikuchi, M., A. Hafid, C. Saleh, S. Hartoyo, and Y. Hayami [1980]: "Class Differentiation, Labor Employment and Income Distribution in a West Java Village." *Developing Economies* 18(1):45-64.

Kikuchi, M., and Y. Hayami [1978a]: "New Rice Technology and National Irrigation Development Policy." In IRRI, 1978a, pp. 315-335.

——— [1978b]: "Agricultural Growth against a Land Resource Constraint: A Comparative History of Japan, Taiwan, and the Philippines." *J. Econ. Hist.* 38:839-864.

——— [1980a]: "Technology and Labor Contract: Two Systems of Rice Harvesting in the Philippines." *J. Comparative Econ.* 4:357-377.

——— [1980b]: "Inducements to Institutional Innovations in an Agrarian Community." *Econ. Dev. and Cultural Change* 29:21-36.

——— [1983]: "New Rice Technology, Intrarural Migration and Institutional Innovation in the Philippines." *Population and Dev. Rev.* 9:247-257.

Kim, D. I. [1980]: "A Profile of Korean Rural Villages, Farmers and their Changing Quality of Life." *J. Rural Dev.* 4(1):1-18.

Kim, K. S., and M. Roemer [1980]: *Growth and Structural Transformation Studies in the Modernization of the Republic of Korea: 1945-1975*. Cambridge, MA, Harvard Univ, Harvard East Asian Monograph No. 86.

King, E., and R. E. Evenson [1979]: "Time Allocation and Home Production in Philippine Rural Households." In Buvinic, Lycette, and McGreevey, eds., 1979, pp. 35-61.

Kirk, D. [1972]: "Prospects for Reducing Birth Rates in Developing Countries: The Interplay of Population and Agricultural Policies." *Food Research Institute Studies* 11:3-26.

Klein, L., and K. Ohkawa [1968]: *Economic Growth, the Japanese Experience since the Meiji Era. Proceedings of the Japan Economic Research Center.* Homewood, IL, Irwin.

Kluemper, S. A. [1986]: "Towards an Approach to Food Policy Planning: The Case of Indonesia." In Gans, ed., 1986, pp. 73-117.

Knapp, K. C. [1982]: "Optimal Grain Carryovers in Open Economies: A Graphical Analysis." *Amer. J. Agr. Econ.* 64:197-204.

Knight, H. [1954]: *Food Administration in India, 1939-47.* Stanford, CA, Stanford Univ. Press.

Knudsen, O. K., and P. L. Scandizzo [1979]: *Nutrition and Food Needs in Developing Countries.* Washington, World Bank Staff Working Paper No. 328.

Kocher, J. E. [1973]: *Rural Development, Income Distribution, and Fertility Decline.* New York, Population Council.

Koester, U. [1982]: *Policy Options for the Grain Economy of the European Community: Implications for Developing Countries.* Washington, IFPRI, Research Report No. 35.

———— [1985]: "Agricultural Market Intervention and International Trade." *European Rev. Agr. Econ.* 12:87-103.

Koester, U., and M. D. Bale [1984]: *The Common Agricultural Policy of the European Community: A Blessing or a Curse for Developing Countries.* Washington, World Bank Staff Working Paper No. 630.

Koester, U., and A. Valdes [1984]: "The EC's Potential Role in Food Security for LDC's: Adjustment in its STABEX and Stock Policies." *European Rev. Agr. Econ.* 11:415-437.

Kothari, V. N. [1967]: "Returns to Education in India." In B. Singh, ed., 1967, pp. 127-156.

Krauss, M. B. [1983]: *Development without Aid: Growth, Poverty and Government.* New York, McGraw-Hill.

Krishna, J., and M. S. Rao [1967]: "Dynamics of Acreage Allocation for Wheat in Uttar Pradesh: A Study in Supply Response." *Indian J. Agr. Econ.* 22(1):37-52.

Krishna, R. [1962]: "A Note on the Elasticity of the Marketable Surplus." *Indian J. Agr. Econ.* 17(2):79-84.

———— [1963]: "Farm Supply Response in India-Pakistan: A Case Study of the Punjab Region." *Econ. J.* 73:477-487.

———— [1964]: "Some Production Functions for the Punjab." *Indian J. Agr. Econ.* 19(3,4):87-97.

———— [1965a]: "The Marketable Surplus Function for a Subsistence Crop: An Analysis with Indian Data." *Econ. Weekly* 17(5,6,7):309-324.

———— [1965b]: "The Marketable Surplus for a Subsistence Crop: Reply to Comments." *Econ. Weekly* 17(44,45):1665-1668.

———— [1967a]: "Agricultural Price Policy and Economic Development." In Southworth and Johnston, eds., 1967, pp. 497-540.

———— [1967b]: "Government Operations in Foodgrains." *Econ. and Polit. Weekly* 2(37):1695-1706.

———— [1973]: "Unemployment in India." *Indian J. Agr. Econ.* 28(1):1-23.

———— [1982]: "Some Aspects of Agricultural Growth, Price Policy, and Equity in Developing Countries." *Food Research Institute Studies* 18:219-260.

Krishna, R., and A. Chhibber [1983]: *Policy Modeling of a Dual Grain Market: The Case of Wheat in India.* Washington, IFPRI, Research Report No. 38.

Krishna, R., and G. S. Raychaudhuri [1980]: *Some Aspects of Wheat and Rice Price Policy in India*. Washington, World Bank Staff Working Paper No. 381.

———— [1981]: "Agricultural Price Policy in India—The Case Study of Rice." *Indian Econ. J.* 28:16-34.

———— [1982]: "Trends in Rural Savings and Capital Formation in India, 1950-51 to 1973-74." *Econ. Dev. and Cultural Change* 30:271-298.

Krishnan, T. N. [1965]: "The Marketable Surplus of Foodgrains: Is It Inversely Related to Price?" *Econ. Weekly* 17(5,6,7):325-328.

Krueger, A. O. [1962]: "Interrelationships between Industry and Agriculture in a Dual Economy." *Indian Econ. J.* 10:1-13.

———— [1979]: *The Development Role of the Foreign Trade Sector. Studies in the Modernization of the Republic of Korea: 1945-1975*. Cambridge, MA, Harvard Univ. Press, Harvard East Asian Monograph No. 87.

———— [1986]: "The Impact of External Economic Assistance to Korea." *J. Rural Dev.* 9:219-256.

Krueger, A. O., M. Schiff, and A. Valdes [1988]: "Agricultural Incentives in Developing Countries: Measuring the Effect of Sectoral and Economy-wide Policies." *Econ. Rev.* 2:255-272.

————, eds. [1990]: *A Comparative Study of the Political Economy of Agricultural Pricing Policies, Vol. 2, Asia*. New York, Oxford Univ. Press.

Kueh, Y. Y. [1983]: "Weather, Technology, and Peasant's Organization as Factors in China's Food Grain Production, 1952-1981." *Econ. Bull. for Asia and the Pacific* 34:15-26.

———— [1984a]: "Fertilizer Supplies and Foodgrain Production in China, 1952-1982." *Food Policy* 9:219-231.

———— [1984b]: "A Weather Index for Analyzing Grain Yield Instability in China, 1952-81." *China Quarterly* 97:68-83.

———— [1986]: "Weather Cycles and Agricultural Instability in China." *J. Agr. Econ.* 37:101-104.

Kumar, D. [1974]: "Changes in Income Distribution and Poverty in India: A Review of the Literature." *World Dev.* 2:31-41.

Kumar, S. [1979]: *Impact of Subsidized Rice on Food Consumption and Nutrition in Kerala*. Washington, IFPRI, Research Report No. 5.

Kumar, S., and D. Hotchkiss [1988]: *Consequences of Deforestation for Women's Time Allocation, Agricultural Production, and Nutrition in Hill Areas of Nepal*. Washington, IFPRI, Research Report No. 69.

Kuo, L. T. C. [1972]: *The Technical Transformation of Agriculture in Communist China*. New York, Praeger.

Kuroda, Y. [1979]: "Production Behavior of the Japanese Farm Households in the Mid-1960s." *Food Research Institute Studies* 17:67-78.

Kuroda, Y. [1987]: "The Production Structure and the Demand for Labor in Postwar Japanese Agriculture, 1952-82." *Amer. J. Agr. Econ.* 69:328-337.

Kuroda, Y., and P. A. Yotopoulos [1978]: "A Microeconomic Analysis of Production Behavior of the Farm Household in Japan—A Profit Function Approach." *Econ. Rev.* (Japan) 29(2):116-129.

———— [1980]: "A Study of Consumption Behavior of the Farm Household in Japan—An Application of the Linear Logarithmic Expenditure System." *Econ. Rev.* (Japan) 31(1):1-15.

Kuznets, S. [1961]: "Economic Growth and the Contribution of Agriculture: Notes on Measurement." *International J. Agrarian Affairs* 3:59-75. Also in Eicher and Witt, eds., 1964, pp. 102-119.

———— [1971]: *Economic Growth of Nations: Total Output and Production Structure.* Cambridge, MA, Harvard Univ. Press.

Ladejinsky, W. [1964]: "Agrarian Reform in Asia." *Foreign Affairs* 42:445-460.

———— [1969a]: "The Green Revolution in Punjab: A Field Trip." *Econ. and Polit. Weekly* 4(26):A73-A82.

———— [1969b]: "Green Revolution in Bihar, the Kosi Area: A Field Trip." *Econ. and Polit. Weekly* 4(39):A147-A162.

———— [1970]: "Ironies of India's Green Revolution." *Foreign Affairs* 48:758-768.

———— [1972]: "Land Ceilings and Land Reform." *Econ. and Polit. Weekly* 7(5, 6, 7):401-407.

Lal, D. [1976]: "Agricultural Growth, Real Wages, and the Rural Poor in India." *Econ. and Polit. Weekly* 11(26):A47-A61.

Lamer, M. [1957]: *The World Fertilizer Economy.* Stanford, Calif. Stanford Univ. Press.

Landau, D. [1986]: "Government and Economic Growth in the Less Developed Countries: An Empirical Study for 1960-80." *Economic Dev. and Cultural Change* 35:35-75.

Langham, M. R., and R. H. Retzlaff, eds. [1982]: *Agricultural Sector Analysis in Asia.* Singapore, Univ. of Singapore Press for ADC.

Lardy, N. R. [1978]: *Economic Growth and Distribution in China.* Cambridge, England, Cambridge Univ. Press.

———— [1983a]: *Agriculture in China's Modern Economic Development.* New York, Cambridge Univ. Press.

———— [1983b]: *Agricultural Prices in China.* Washington, World Bank, Staff Working Paper No. 606.

———— [1984]: "Prices, Markets and the Chinese Peasant." In Eicher and Staatz, eds., 1984, pp. 420-435.

———— [1986a]: "Prospects and Some Policy Problems of Agricultural Development in China." *Amer. J. Agr. Econ.* 68:451-457.

———— [1986b]: "Agricultural Reforms in China." *J. International Affairs* 39(2):91-104.

———— [1988]: "Dilemmas in the Pattern of Resource Allocation in China: 1978-85." In Nee and Stark, eds., 1988.

Larson, D. W., and J. S. Cibantos [1979]: *The Demand for Fertilizer in Southern Brazil, 1948-71.* Microfiche in AID, Washington.

Lau, L. J., W. L. Lin, and P. A. Yotopoulos [1978]: "The Linear Logarithmic Expenditure System: An Application to Consumption-Leisure Choice." *Econometrica* 46:843-868.

———— [1979]: "Efficiency and Technological Change in Taiwan Agriculture." *Food Research Institute Studies* 17:23-50.

Lau, L. J., and P. A. Yotopoulos [1971]: "A Test for Relative Economic Efficiency and Application to Indian Agriculture." *Amer. Econ. Rev.* 61:94-109.

———— [1988]: "Do Country Idiosyncrasies Matter in Estimating a Production Function for World Agriculture." *J. Dev. Econ.* 13:7-19.

Leaf, M. J. [1983]: "The Green Revolution and Cultural Change in a Punjab Village, 1965-1978." *Econ. Dev. and Cultural Change* 31:227-270.

Lecaillon J., C. Morrisson, H. Schneider, and E. Thorbecke [1987]: *Economic Policies and Agricultural Performance of Low-Income Countries.* Paris, OECD.

Ledesma, A. J., P. Q. Makil, and V. A. Miralao, eds. [1983]: *Second View from the Paddy.* Manila, Ateneo de Manila Univ., Institute for Philippine Culture.

Lee, D. H. K. [1957]: *Climate and Economic Development in the Tropics*. New York, Harper for the Council on Foreign Relations.

Lee, E. H. L. [1977]: "Rural Poverty in Sri Lanka, 1963-73." In ILO, 1977, pp. 161-184.

Lee, J. E., Jr., and M. Shane [1987]: *United States Agriculture and Third World Economic Development: Critical Interdependency*. Washington, National Planning Association.

Lee, R. [1986]: *Population, Food, and Rural Development*. Oxford, England, Oxford Univ. Press.

Lee, T. H. [1971]: *Intersectoral Capital Flows in the Economic Development of Taiwan, 1895-1960*. Ithaca, NY, Cornell Univ. Press.

Lee, T. H., H. H. Chan, and Y. E. Chen [1978]: "Labour Absorption in Taiwan Agriculture." In ILO, 1978, pp. 167-236.

Leibenstein, H. [1957]: *Economic Backwardness and Economic Growth*. New York, Wiley.

Lele, U. J. [1967]: "Market Integration: A Study of Sorghum Prices in Western India." *J. Farm Econ.* 49:147-159.

_____ [1969]: "Agricultural Price Policy." *Econ. and Polit. Weekly* 4(35):1413-1419.

_____ [1970]: "Modernization of the Rice Milling Industry: Lessons from Past Performance." *Econ. and Polit. Weekly* 5(28): 1081-1090.

_____ [1971]: *Food Grain Marketing in India: Private Performance and Public Policy*. Ithaca, NY, Cornell Univ. Press.

_____ [1972]: "The Green Revolution: Income Distribution and Nutrition." In P. L. White, ed., 1972, pp. 20-25.

_____ [1974]: "The Roles of Credit and Marketing in Agricultural Development." In N. Islam, ed., 1974, pp. 413-441.

_____ [1975]: *The Design of Rural Development: Lessons from Africa*. Baltimore, Johns Hopkins Univ. Press for the World Bank.

_____ [1981]: "Cooperatives and the Poor: A Comparative Perspective." *World Dev.* 9:55-72.

_____ [1986]: "Women and Structural Transformation." *Econ. Dev. and Cultural Change* 34:195-221.

Lele, U. J., and J. W. Mellor [1972]: "Jobs, Poverty, and the 'Green Revolution.'" *International Affairs* 48:20-32.

_____ [1981]: "Technological Change, Distributive Bias, and Labour Transfer in a Two Sector Economy." *Oxford Econ. Papers* NS33:426-441.

Lele, U., and I. Nabi, eds. [1990]: *Transitions in Development: The Role of Concessionary and Commercial Capital Flows*. San Francisco, CA., Institute for Contemporary Studies.

Leontief, W. W. [1951]: *The Structure of American Economy, 1919-1939*. 2nd ed. New York, Oxford Univ. Press.

Levinson, F. J. [1974]: *Morinda: An Economic Analysis of Malnutrition among Young Children in Rural India*. Cambridge, MA, Cornell-MIT International Nutrition Policy Series.

Lewis, J. P. [1972]: "Public Works Approach to Low-End Poverty Problems: The New Potentialities of an Old Answer." *J. Dev. Planning* 5:85-114.

Lewis, J. P., and V. Kallab, eds. [1986]: *Development Strategies Reconsidered: U.S.-Third World Policy Perspectives No. 5*. New Brunswick, NJ, Transaction Books for the ODC.

Lewis, S. R., Jr. [1968]: "Effects of Trade Policy on Domestic Relative Prices: Pakistan, 1951-64." *Amer. Econ. Rev.* 58:60-78.

_____ [1973]: "Agricultural Taxation and Intersectoral Resource Transfers." *Food Research Institute Studies* 12:93-114.

Lewis, W. A. [1954]: "Economic Development with Unlimited Supplies of Labour." *Manchester School Econ. and Soc. Studies* 22:139-191.

———— [1958]: "Unlimited Labor: Further Notes." *Manchester School Econ. and Soc. Studies* 26:1–32.

———— [1984]: "The State of Development Theory." *Amer. Econ. Rev.* 74:1–10.

Lim, D. [1973]: *Economic Growth and Development in West Malaysia*. Kuala Lumpur, Oxford Univ. Press.

————, ed. [1983]: *Further Readings in Malaysian Economic Development*. Kuala Lumpur, Oxford Univ. Press.

Lin, Y. F. [1987]: "The Household Responsibility Reform System in China: A Peasant's Institutional Choice." *Amer. J. Agr. Econ.* 69:410–415.

———— [1988]: "Rural Factor Markets in China After the Household Responsibility Reform." *Econ. Dev. and Cultural Change* 36:S199–S224 (Supplement).

Lipton, M. [1968a]: "Strategy for Agriculture: Urban Bias and Rural Planning." In Streeten and Lipton, eds., 1968, pp. 83–148.

———— [1968b]: "The Theory of the Optimising Peasant." *J. Dev. Studies* 4:327–351.

———— [1977]: *Why Poor People Stay Poor: Urban Bias in World Development*. Cambridge, MA, Harvard Univ. Press.

———— [1978]: "Inter-farm, Inter-regional and Farm–Nonfarm Income Distribution: The Impact of the New Cereal Varieties." *World Dev.* 6:319–337.

———— [1983]: *Poverty, Undernutrition, and Hunger*. Washington, World Bank Staff Working Paper No. 597.

———— [1984]: "Conditions of Poverty Groups and Impact on Indian Economic Development and Cultural Change: The Role of Labour." *Dev. and Change* 15:473–493.

———— [1989]: "Agriculture, Rural People, the State and the Surplus in Some Asian Countries. Thoughts on Some Implications of Three Recent Approaches in Social Science." *World Dev.* 17:1551–1571.

Lipton, M., J. Connell, B. Dasgupta, and R. Laishley [1977]: *Migration from Rural Areas: The Evidence from Village Studies*. London, Oxford Univ. Press.

Lipton, M., and R. Longhurst [1985]: *Modern Varieties, International Agricultural Research and the Poor*. Washington, World Bank, CGIAR Study Paper No. 2.

———— [1989]: *New Seeds and Poor People*. London, Unwin.

Lipton, M., and R. Paarlberg [1989]: *The Role of the World Bank in Agricultural Development in the 1990s*. Washington, IFPRI, mimeo.

Liu, J. C. [1970]: *China's Fertilizer Economy*. Chicago, Aldine.

Lo, Fu-chen, ed. [1987]: *Asia by the Year 2000*. Kuala Lumpur, Malaysia, Asian Pacific Development Center.

Lockheed, M. E., D. T. Jamison, and L. J. Lau [1980]: "Farm Education and Farm Efficiency: A Survey." *Econ. Dev. and Cultural Change* 29:37–76.

Lockwood, B., P. K. Mukherjee, and R. T. Shand [1971]: *The High-Yielding Varieties Programme in India*, 2 Vols. Canberra, ANU and New Delhi, Government of India, Planning Commission.

Lockwood, W. W., ed. [1954]: *The Economic Development of Japan: Growth and Structural Change, 1868–1938*. Princeton, NJ, Princeton Univ. Press.

————, ed. [1965]: *The State of Economic Enterprise in Japan*. Princeton, NJ, Princeton Univ. Press.

Long, F. A., and A. Oleson, eds. [1980]: *Appropriate Technology and Social Values: A Critical Appraisal*. Cambridge, MA, Ballinger.

Long, M. [1968]: "Interest Rates and the Structure of Agricultural Credit Markets." *Oxford Econ. Papers* NS20:273–288.

Longworth, J. W., ed. [1989]: *China's Rural Development Miracle: With International Comparisons*. St. Lucia, Australia, Univ. of Queensland Press.

Luu, N. N. [1979]: *The Technological Development of Agriculture in the People's Republic of China*. The Hague, Netherlands, Institute of Social Studies, Research Report Series No. 5.

Lynch, F., ed. [1977]: "View from the Paddy: Empirical Studies of Philippine Rice Farming and Tenancy." *Philippine Sociological Rev.* 20(1,2):1-274.

MacAndrews, C., and C. L. Sien, eds. [1982]: *Too Rapid Rural Development*. Athens, Ohio Univ. Press.

MacBean, A. I. [1966]: *Export Instability and Economic Development*. Cambridge, MA, Harvard Univ. Press.

Madhavan, M. C. [1972]: "Acreage Response of Indian Farmers: A Case Study of Tamil Nadu." *Indian J. Agr. Econ.* 27(1):67-86.

Mahalanobis, P. C. [1953]: "Some Observations on the Process of Growth of National Income." *Sankhya* 12:307-312.

―――― [1955]: "The Approach of Operational Research to Planning in India." *Sankhya* 16:3-130.

Mahalingasivam, R. [1978]: "Food Subsidy in Sri Lanka." *Sri Lanka J. Social Sciences* 1(1):75-94.

Mahapatra, I. C., D. J. Bhumbla, and S. D. Bokil [1986]: *India and ICRISAT: A Study of their Collaboration in Agricultural Research*. Washington, World Bank, CGIAR Study Paper No. 18.

Mahmud, W., ed. [1981]: *Development Issues in an Agrarian Economy—Bangladesh*. Bangladesh, Dhaka Univ., Centre for Administrative Studies.

Maji, C. C., D. Jha, and L. S. Venkataraman [1971]: "Dynamic Supply and Demand Models for Better Estimations and Projections: An Econometric Study for Major Foodgrains in the Punjab Region." *Indian J. Agr. Econ.* 26(1):21-34.

Majumdar, I. M. [1965]: "Marketable Surplus Function for a Subsistence Crop: Further Comments." *Econ. Weekly* 17(20):820-823.

Malenbaum, W. [1959]: "India and China: Contrasts in Development Performance." *Amer. Econ. Rev.* 49:284-309.

―――― [1962]: *Prospects for Indian Development*. London, Allen and Unwin.

―――― [1982]: "Modern Economic Growth in India and China: The Comparison Revisited, 1950-1980." *Econ. Dev. and Cultural Change* 31:45-84.

Mandal, G. C., and M. G. Ghosh [1976]: *Economics of the Green Revolution: A Study in East India*. New Delhi, Asia Publishers.

Manetsch, T., J., M. L. Hayenga, A. N. Halter, T. W. Carroll, M. H. Abkin, D. R. Byerlee, K. Y. Chong, G. Page, E. Kellogg, and Glenn L. Johnson [1971]: *A Generalized Simulation Approach to Agricultural Sector Analysis with Special Reference to Nigeria*. East Lansing, MSU.

Mangahas, M. [1974]: "Economic Aspects of Agrarian Reform under the New Society." *Philippines Rev. Business and Econ.* 11(2):175-187.

―――― [1975]: "The Political Economy of Rice in the New Society." *Food Research Institute Studies* 14:295-309.

―――― [1979]: "Planning for Improved Equity in ASEAN, Hong Kong, and the Republic of Korea." *Econ. Bull. for Asia and the Pacific* 30(20):1-19.

―――― [1984]: "The Relevance of Poverty Measurement to Food Security Policy." *J. Philippine Dev.* 11(2):191-202.

Mangahas, M., A. E. Recto, and V. W. Ruttan [1966]: "Price and Market Relationships for Rice and Corn in the Philippines." *J. Farm Econ.* 48:685–703.

Mann, C. K., and B. Huddleston [1986]: *Food Policy: Framework for Analysis and Action.* Bloomington, Indiana Univ. Press.

Mann, J. S. [1967]: "The Impact of Public Law 480 Imports on Prices and Domestic Supply of Cereals in India." *J. Farm Econ.* 49:131–146.

Manne, A. S., and A. Rudra [1965]: "A Consistency Model of India's Fourth Plan." *Sankhya* Series B27:57–144.

Marga Institute [1982]: "Food." *Marga Quart. J.* 6(4):1–108.

Marten, G. G., ed. [1986]: *Traditional Agriculture in Southeast Asia: A Human Ecology Perspective.* Boulder, CO, Westview Press.

Martin, M. V., and J. A. McDonald [1986]: "Food Grain Policy in the Republic of Korea: The Economic Costs of Self-Sufficiency." *Econ. Dev. and Cultural Change* 34:315–332.

Marzouk, G. A. [1972]: *Economic Development and Policies: Case Study of Thailand.* Rotterdam, Holland, Rotterdam Univ. Press.

Masud, S. M., and F. L. Underwood [1970]: *Gumai Bil Boro Paddy Profits and Losses, 1967/68 Season.* Mymensingh, East Pakistan Agricultural Univ., Research Report No. 5.

Mateus, A. [1983]: *Targetting Food Subsidies for the Needy.* Washington, World Bank Staff Working Paper No. 617.

Mathur, P. N., and H. Ezekiel [1961]: "Marketable Surplus of Food and Price Fluctuations in a Developing Economy." *Kyklos* 14:396–408.

Maunder, A., and K. Ohkawa, eds. [1983]: *Growth and Equity in Agricultural Development: Proceedings, 18th ICAE, Jakarta, Indonesia, 24 August-2 September, 1982.* Brookfield, VT, Gower.

Maunder, A., and U. Renborg, eds. [1986]: *Agriculture in a Turbulent World Economy: Proceedings of the 19th ICAE, 26 Aug- 4 Sept, 1985.* Brookfield, VT, Gower.

Maxwell, S. J., and H. W. Singer [1979]: "Food Aid to Developing Countries: A Survey." *World Dev.* 7:225–247.

Mazumdar, D. [1963]: "On the Economics of Relative Efficiency of Small Farmers." *Econ. Weekly* 5(28,29,30):1259–1263.

———— [1965]: "Size of Farm and Productivity: A Problem of Indian Peasant Agriculture." *Economica* NS32:161–173.

McCalla, A. F. [1969]: "Protectionism in International Agricultural Trade, 1850–1968." *Agr. Hist.* 43:329–344.

McCalla, A. F., and T. E. Josling, eds. [1981]: *Imperfect Markets in Agricultural Trade.* Montclair, NJ, Allanheld and Osmen.

McCalla, A. F., and T. E. Josling [1985]: *Agricultural Policies and World Markets.* New York, Macmillan.

McCarthy, F. D., and L. Taylor [1980]: "Macro Food Policy Planning: A General Equilibrium Model for Pakistan." *Rev. Econ. and Stat.* 62:107–121.

McCord, C. [1977]: "Integration of Health, Nutrition, and Family Planning: The Companiganj Project in Bangladesh." *Food Research Institute Studies* 16:91–105.

McCraw, T. K., ed. [1986]: *America versus Japan.* Boston, MA, Harvard Business School Press.

McCrone, G. [1962]: *The Economics of Subsidising Agriculture: A Study of British Policy.* London, Allen and Unwin.

McKinnon, R. I. [1988]: "Financial Liberalization in Retrospect: Interest Rate Policies in LDCs." In Ranis and T. P. Schultz, eds., 1988, pp. 386–410.

Meade, J. E. [1962]: *A Neo-Classical Theory of Economic Growth*, 2nd ed. London, Unwin.

Mears, L. A. [1961, 1975]: *Rice Marketing in the Republic of Indonesia*. Djakarta, Univ. of Indonesia, School of Economics.

_____ [1981]: *The New Rice Economy of Indonesia*. Yogyakarta, Indonesia, Gadjah Mada Univ. Press.

Mears, L. A., M. H. Agabin, T. L. Anden, and R. C. Marquez [1974]: *The Rice Economy of the Philippines*. Quezon City, Univ. of the Philippines Press.

Meenakshi, J. V., R. Sharma, and T. T. Poleman [1986]: *The Impact of India's Green Revolution on the Pulses and Oilseeds*. Ithaca, NY, Cornell Univ., International Agricultural Economics Study, A. E. Res. 86-22.

Mehra, S. [1976]: "Some Aspects of Labor Use in Indian Agriculture." *Indian J. Agr. Econ.* 31(4):95-121.

_____ [1981]: *Instability in Indian Agriculture in the Context of the New Technology*. Washington, IFPRI, Research Report No. 25.

Meier, G. M., ed. [1983]: *Pricing Policy for Development Management*. Baltimore, Johns Hopkins Univ. Press for the World Bank.

_____ , ed. [1987]: *Pioneers in Development*, 2nd Series. New York, Oxford Univ. Press.

Meier, G. M., and D. Seers, eds. [1984]: *Pioneers in Development*. New York, Oxford Univ. Press for the World Bank.

Meinzen-Dick, R. S. [1984]: *Local Management of Tank Irrigation in South India: Organization and Operation*. Ithaca, NY, Cornell Univ., Studies in Irrigation No. 3.

Mellor, J. W. [1962a]: "Increasing Agricultural Production in Early Stages of Economic Development—Relationships, Problems, and Prospects." *Indian J. Agr. Econ.* 17(2):29-46.

_____ [1962b]: "The Process of Agricultural Development in Low-Income Countries." *J. Farm Econ.* 44:700-716.

_____ [1963]: "The Use and Productivity of Farm Family Labor in the Early Stages of Agricultural Development." *J. Farm Econ.* 45:517-543.

_____ [1966]: *The Economics of Agricultural Development*. Ithaca, NY, Cornell Univ. Press.

_____ [1967]: "Towards A Theory of Agricultural Development." In Southworth and Johnston, eds., 1967, pp. 21-60. Also in ISAE, 1972, pp. 23-37.

_____ [1968]: "The Functions of Agricultural Prices in Economic Development." *Indian J. Agr. Econ.* 23(1):23-37.

_____ [1969a]: "Agricultural Price Policy in the Context of Economic Development." *Amer. J. Agr. Econ* 51:1413-1420.

_____ [1969b]: "The Subsistence Farmer in Traditional Economies." In Wharton, ed., 1969a, pp. 209-227.

_____ [1973a]: "Accelerated Growth in Agricultural Production and the Intersectoral Transfer of Resources." *Econ. Dev. and Cultural Change* 22:1-16.

_____ [1973b]: "Nutrition and Economic Growth." In A. D. Berg, Scrimshaw and Call, eds., 1973, pp. 70-73.

_____ [1974]: "Models of Economic Growth and Land-Augmenting Technological Change." In N. Islam, ed., 1974, pp. 3-40.

_____ [1976]: *The New Economics of Growth: A Strategy for India and the Developing World*. Ithaca, NY, Cornell Univ. Press.

_____ [1977]: "Relating Research Resource Allocation to Multiple Goals." In Arndt, Dalrymple and Ruttan, eds., 1977, pp. 478-497.

_____ [1978]: "Food Price Policy and Income Distribution in Low-Income Nations." *Econ. Dev. and Cultural Change* 27:1-26.

_____ , ed. [1979]: *India: A Rising Middle Power*. Boulder, CO, Westview Press.

_____ [1980]: "Food Aid and Nutrition." *Amer. J. Agr. Econ.* 62:979-983.

_____ [1981]: "Global Dynamics of the World Food Situation." In H. Carter, ed., 1981, pp. 71-83.

_____ [1983]: "The Utilization of Food Aid for Equitable Growth." In *Report of the World Food Programme/Government of the Netherlands Seminar on Food Aid*. Rome, World Food Programme/Government of the Netherlands Publication.

_____ [1986]: "Agriculture on the Road to Industrialization." In J. P. Lewis, and Kallab, eds., 1986, pp. 67-90.

_____ [1987]: "Food Aid for Food Security and Economic Development." In Clay and Shaw, eds., 1987, pp. 173-191.

_____ [1988a]: "Global Food Balances and Food Security." *World Dev.* 16:997-1011.

_____ [1988b]: *Lectures on Agricultural Growth and Employment: An Equitable Growth Strategy and its Knowledge Needs*. Islamabad, Pakistan, PIDE, Lectures in Development Economics No. 7.

_____ [1988c]: "The Intertwining of Environmental Problems and Poverty." *Environment* 30(9):8-13 and 28-30.

_____ [1988d]: "Food and Development: The Critical Nexus between Developing and Developed Countries." In Antonelli and Quadrio-Curzio, eds., 1988, pp. 175-184.

_____ [1989]: "Agricultural Development in the Third World: The Food, Poverty, Aid, Trade Nexus." *Choices* 1Q:4-8.

Mellor, J. W., and R. Ahmed, eds. [1988]: *Agricultural Price Policy for Developing Countries*. Baltimore, Johns Hopkins Univ. Press for IFPRI.

Mellor, J. W., and A. K. Dar [1968]: "Determinants and Development Implications of Foodgrain Prices in India, 1949-64." *Amer. J. Agr. Econ.* 50:962-974.

Mellor, J. W., and B. de Ponteves [1964]: "The Effect of Growth in Demand for Milk on the Demand for Concentrate Feeds, India, 1951-1976." *Indian J. Agr. Econ.* 19(3,4):131-146.

Mellor, J. W., and G.M. Desai, eds. [1985]: *Agricultural Change and Rural Poverty: Variations on a Theme by Dharm Narain*. Baltimore, John Hopkins Univ. Press for IFPRI.

Mellor, J. W., and S. Gavian [1987]: "Famine: Causes, Preventions, and Relief." *Science* 235:539-545.

Mellor, J. W., and B. F. Johnston [1984]: "The World Food Equation: Interrelations among Development, Employment, and Food Consumption." *J. Econ. Literature* 22:531-574.

Mellor, J. W., and U. J. Lele [1965]: "Alternative Estimates of the Trend in Indian Foodgrains Production during the First Two Plans." *Econ. Dev. and Cultural Change* 13:217-235.

_____ [1973]: "Growth Linkages of the New Foodgrain Technologies." *Indian J. Agr. Econ.* 28(1):35-55.

_____ [1975]: "An Interaction of Growth Strategy, Agriculture, and Foreign Trade—The Case of India." In Tolley and Zadrozny, eds., 1975, pp. 93-113.

Mellor, J. W., and W. A. Masters [1990]: "The Changing Roles of Multilateral and Bilateral Foreign Assistance." In Lele and Nabi, eds., 1990.

Mellor, J. W., and T. V. Moorti [1971]: "Dilemma of State Tubewells." *Econ. and Polit. Weekly* 6(13):A37-A45.

Mellor, J. W., and M. S. Mudahar [1974a]: *Modernizing Agriculture, Employment, and Economic Growth: A Simulation Model*. Ithaca, NY, Cornell Univ., DAE, Occasional Paper No. 75.

———— [1974b]: *Simulating an Economy with Modernizing Agricultural Sector: Implications for Employment and Economic Growth in India.* Ithaca, NY, Cornell Univ., DAE, Occasional Paper No. 76.

Mellor, J. W., and C. G. Ranade [1988]: "Technological Change in a Low Labor Productivity, Land Surplus Economy: The African Development Problem." Washington, IFPRI, mimeo.

Mellor, J. W., and M. Schluter [1972]: "New Seed Varieties and the Small Farm." *Econ. and Polit. Weekly* 7(13):A31-A38.

Mellor, J. W., and R. D. Stevens [1956]: "The Average and Marginal Product of Farm Labor in Underdeveloped Countries." *J. Farm Econ.* 38:780-791.

Mellor, J. W., T. F. Weaver, U. J. Lele, and S. R. Simon [1968]: *Developing Rural India: Plan and Practice.* Ithaca, NY, Cornell Univ. Press.

Mergos, G., and R. Slade [1987]: *Dairy Development and Milk Cooperatives: The Effects of a Dairy Project in India.* Washington, World Bank Discussion Paper No. 15.

Merrey, D. J., and J. M. Wolf [1986]: *Irrigation Management in Pakistan: Four Papers.* Sri Lanka, IIMI Research Paper No. 4.

Meyers, L. R., ed. [1989]: *Innovation in Resource Management: Proceedings of the Ninth Agricultural Sector Symposium.* Washington, World Bank.

Mikesell, R. F., and J. E. Zinser [1973]: "The Nature of the Savings Function in Developing Countries: A Survey of the Theoretical and Empirical Literature." *J. Econ. Literature* 11:1-26.

Millar, J. R. [1970]: "A Reformulation of A. V. Chayonov's Theory of the Peasant Economy." *Econ. Dev. and Cultural Change* 18:219-229.

Miller, T. C. [1986]: *Explaining Agricultural Price Policy across Countries and across Commodities Using Political Interest Group Theory.* Unpublished Ph.D. dissertation, Univ. of Chicago.

Minhas, B. S. [1970]: "Rural Poverty, Land Redistribution, and Development Strategy: Facts and Policy." *Indian Econ. Rev.* NS5:97-128.

Minhas, B. S., and T. N. Srinivasan [1966]: "New Agricultural Strategy Analyzed." *Yojana* 10(1):20-24.

Minkler, M. [1970]: "Fertility and Female Labor Force Participation in India: A Survey of Workers in Old Delhi Area." *J. Family Welfare* 17:31-43.

Misra, J. P., and B. D. Shukla [1969]: "A Study on the Economics of High Yielding Varieties Programme." *Agricultural Situation in India* 24:107-114.

Mitchell, D. O., and R. C. Duncan [1987]: "Market Behavior of Grain Exporters." *World Bank Research Observer* 2(1):3-21.

Mitra, A. [1973]: "The Nutrition Movement in India." In A. D. Berg, Scrimshaw and Call, eds., 1973, pp. 357-365.

Mohan, R., and R. E. Evenson [1974]: "The Intensive Agricultural District Program in India: A New Evaluation." *J. Dev. Studies* 11:135-154.

Mohan, R., D. Jha, and R. Evenson [1973]: "The Indian Agricultural Research System." *Econ. and Polit. Weekly* 8(13):A21-A26.

Moll, P. [1988]: "The Intercountry Agricultural Production Function and Productivity Differences among Countries: Comment." *J. Dev. Econ.* 28:121-124.

Monke, E. A., S. R. Pearson, and N. Akrasanee [1976]: "Comparative Advantage, Government Policies, and International Trade in Rice." *Food Research Institute Studies* 15: 257-283.

Monke, E. A., and S. S. Salam [1986]: "Trade Policies and Variability in International Grain Markets." *Food Policy* 11:238-252.

Mongkolsmai, D. [1983]: *Status and Performance of Irrigation in Thailand*. Washington, IF-PRI, IFDC and IRRI, Working Paper No. 8.

―――― [1985]: *The Distributional Impacts of Irrigation in Thailand*. Washington, IFPRI, mimeo.

Mongkolsmai, D., and M. W. Rosegrant [1989]: "The Effect of Irrigation on Seasonal Rice Prices, Farm Income, and Labor Demand in Thailand." In Sahn, ed., 1989, pp. 246-263.

Moon, P. Y. [1975]: "The Evolution of Rice Policy in Korea." *Food Research Institute Studies* 14:381-402.

Moore, J. R., S. S. Johl, and A. M. Khusro [1972]: *Indian Foodgrain Marketing*. New Delhi, Prentice-Hall.

Moorti, T. V. [1971]: *A Comparative Study of Well Irrigation in Aligarh District, India*. Ithaca, NY, Cornell Univ., International Agricultural Development, Bulletin No. 19.

Morgan, T. [1959]: "The Long-Run Terms of Trade between Agriculture and Manufacturing." *Econ. Dev. and Cultural Change* 8:1-23.

―――― [1963]: "Trends in Terms of Trade and their Repercussions on Primary Producers." In Harrod and Hague, eds., 1963, pp. 52-95.

Morris, D. [1974]: "What is a Famine?" *Econ. and Polit. Weekly* 9:1855-1864.

Morris, M. D. [1979]: *Measuring the Conditions of the World's Poor: The Physical Quality of Life Index*. New York, Pergamon Press for ODC.

Morrison, T., and M. Wattleworth [1988]: "Causes of the 1984-86 Commodity Price Decline." *Finance and Dev.* 25(2):31-33.

Moseman, A. H. [1970]: *Building Agricultural Research Systems in the Developing Nations*. New York, ADC.

Mosher, A. T. [1966]: *Getting Agriculture Moving: Essentials for Development and Modernization*. New York, Praeger for ADC.

―――― [1969]: *Creating a Progressive Rural Structure*. New York, ADC.

―――― [1971]: *To Create a Modern Agriculture*. New York, ADC.

―――― [1981]: *Three Ways to Spur Agricultural Growth*. New York, IADS.

Motooka, I. [1971]: *Agricultural Development in Thailand, I, II, III, IV*. Kyoto Univ., Center for Southeast Asian Studies, Discussion Papers 26, 27, 28 and 29.

Mubyarto [1965]: "The Elasticity of the Marketable Surplus of Rice in Indonesia: A Study in Java-Madura." Unpublished Ph.D. dissertation, ISU, Ames.

―――― , ed. [1982]: *Growth and Equity in Indonesian Agricultural Development*. Jakarta, Indonesia, Yayasan Agro Ekonomika.

Mubyarto and L. B. Fletcher [1966]: *The Marketable Surplus of Rice in Indonesia: A Study in Java-Madura*. Ames, ISU, Dept. of Economics, International Studies in Economics, Monograph No. 4.

Mudahar, M. S. [1973]: *Dynamic Models of Agricultural Development with Demand Linkages*. Ithaca, NY, Cornell Univ., DAE, Occasional Paper No. 59.

―――― [1974]: "Dynamic Analysis of Direct and Indirect Implications of Technological Change in Agriculture: The Case of Punjab, India." *Proceedings: 47th Annual Meeting, Western Agricultural Economic Association*. pp. 160-172. Also published as Cornell Univ., DAE, Occasional Paper No. 79.

―――― [1978]: "Needed Information and Economic Analysis for Fertilizer Policy Formulation." *Indian J. Agr. Econ.* 33(3):40-67. Also available as ADC Teaching and Research Reprint No. 24, 1980.

―――― [1982]: "Backward and Forward Linkages in Agricultural Sector Analysis Models." In Langham and Retzlaff, eds., 1982, pp. 255-288.

_____ [1983]: *Monitoring Fertilizer Price, Availability, and Quality in Developing Countries: The Case of Bangladesh*. Muscle Shoals, AL, IFDC.

_____ [1984]: *Fertilizer Price Deregulation and Public Policy: The Case of Bangladesh*. Muscle Shoals, AL, IFDC.

_____ [1987]: "Measuring the Contribution of Fertilizer to Food Production." *Indian J. Quantitative Econ.* 3(2):1-19.

Mudahar, M. S., and R. H. Day [1978]: "A Generalized Cobweb Model of Punjab Agriculture." In Day and Cigno, eds., 1978, pp. 249-264.

Mudahar, M. S., and T. P. Hignett [1982]: *Energy and Fertilizer: Policy Implications and Options for Developing Countries*. Muscle Shoals, AL, IFDC, Technical Bulletin IFDC-T-20.

_____ [1985]: "Energy Efficiency in Nitrogen Fertilizer Production." *Energy in Agriculture* 4:159-177.

_____ [1987a]: "Fertilizer and Energy Use." In Helsel, ed., 1987, pp. 1-23.

_____ [1987b]: "Energy Requirements, Technology, and Resources in the Fertilizer Sector." In Helsel, ed., 1987, pp. 25-61.

_____ [1987c]: "Energy Efficiency, Economics, and Policy in the Fertilizer Sector." In Helsel, ed., 1987, pp. 133-164.

Mudahar, M. S., and E. C. Kapusta [1987]: *Fertilizer Marketing Systems and Policies in the Developing World*. Muscle Shoals, AL, IFDC, Technical Bulletin T-33.

Mudahar, M. S., and P. Pinstrup-Andersen [1977]: "Fertilizer Policy Issues and Implications in Developing Countries." *Proceedings of Fertilizer Seminar on Trends in Consumption and Production*. New Delhi, Fertiliser Association of India, PS/2:1-20.

Mukerjee, K. K. [1985]: "Paddy-Rice Marketing in India." In Jaiswal, ed., 1985, pp. 692-703.

Mundlak, Y. [1979]: *International Factor Mobility and Agricultural Growth*. Washington, IFPRI, Research Report No. 6.

_____ [1988]: "Capital Accumulation, the Choice of Techniques, and Agricultural Output." In Mellor and Ahmed, eds., 1988, pp. 171-189.

Muqtada, M., and M. M. Alam [1983]: *Hired Labour and Rural Labour Market in Bangladesh*. Bangkok, ILO, ARTEP.

Murty, K. N. [1983]: *Consumption and Nutrition Patterns of ICRISAT Mandate Crops in India*. Hyderabad, ICRISAT, Economics Program Progress Report 53.

Murty, K. N., and M. von Oppen [1985]: "Nutrient Distribution and Consumer Policies in India with Policy Implications." In ICRISAT, 1985, pp. 179-200.

Muthiah, C. [1964]: "Marketed Surplus of Foodgrains by Size, Level of Holdings and Income." *Agricultural Situation in India* 19(2):95-98.

Myrdal, G. [1968]: *Asian Drama: An Enquiry into the Poverty of Nations*, 3 vols. New York, Twentieth Century Fund.

Nabi, I. [1984]: "Village-End Considerations in Rural-Urban Migration." *J. Dev. Econ.* 14:129-145.

_____ [1985]: "Rural Factor Market Imperfections and the Incidence of Tenancy in Agriculture." *Oxford Econ. Papers* 37:319-329.

Nabi, I., N. Hamid, and S. Zahid [1986]: *The Agrarian Economy of Pakistan*. Karachi, Oxford Univ. Press.

Nadiri, M. I. [1970]: "Some Approaches to the Theory and Measurement of Total Factor Productivity: A Survey." *J. Econ. Literature* 8:1137-1177.

Nadkarni, M. V. [1986]: " 'Backward' Crops in Indian Agriculture: Economy of Coarse Cereals and Pulses." *Econ. and Polit. Weekly* 21(38, 39):A113-A115, A117-A118.

Nagai, I. [1959]: *Japonica Rice, Its Breeding and Culture*. Tokyo, Yokendo.

Nagle, J. C. [1976]: *Agricultural Trade Policies*. Lexington, MA, Lexington Books.

Nagy, J. G., and W. H. Furtan [1978]: "Economic Costs and Returns from Crop Development Research: The Case of Rapeseed Breeding in Canada." *Canadian J. Agr. Econ.* 26(1):1-14.

Nair, K. [1979]: *In Defense of the Irrational Peasant*. Univ. of Chicago Press.

Nakajima, C. [1969]: "Subsistence and Commercial Family Farm: Some Theoretical Models of Subjective Equilibrium." In Wharton, ed., 1969a, pp. 165-185.

Nakamura, J. I. [1966]: *Agricultural Production and the Economic Development of Japan, 1873-1922*. Princeton, NJ, Princeton Univ. Press.

Narain, D. [1961]: *Distribution of the Marketed Surplus of Agricultural Produce by Size-Level of Holdings in India, 1950/51*. Bombay, Asia Publishing.

———— [1965]: *The Impact of Price Movements on Areas under Selected Crops in India, 1900-1939*. Cambridge, England, Cambridge Univ. Press.

———— [1977]: "Growth of Productivity in Indian Agriculture." *Indian J. Agr. Econ.* 32(1):1-44.

Narain, D., and P. C. Joshi [1969]: "Magnitude of Agricultural Tenancy." *Econ. and Polit. Weekly* 4(39):A139-A142.

Narain, D., and S. Roy [1980]: *Impact of Irrigation and Labor Availability on Multiple Cropping: A Case Study of India*. Washington, IFPRI, Research Report No. 20.

Narkswasdi, U., and S. Selvadurai [1968]: *Economic Survey of Padi Production in West Malaysia*. Kuala Lumpur, Malaysia, Ministry of Agriculture and Cooperatives, Bulletin No. 120.

Naseem, M. [1975]: "Credit Availability and the Growth of Small Farms in the Pakistan Punjab." *Food Research Institute Studies* 14:65-80.

Nash, M., ed. [1977]: *Essays in Economic Development and Cultural Change, in Honor of Bert F. Hoselitz*. Univ. of Chicago Press.

National Council of Applied Economic Research [NCAER, 1975]: *Changes in Rural Income in India, 1968-71*. New Delhi.

Nee, V., and D. Stark, eds. [1988]: *The Politics of Markets in Reforming Socialist Countries*. Stanford, CA, Stanford Univ. Press.

Nehen, I. K., and I. R. Wills [1986]: *Land Preparation in West Java: Benefits and Costs of Alternative Techniques*. Bogor, Indonesia, Agency for Agricultural Research and Development; Los Baños, Philippines, IRRI.

Nelson, G. C. [1988]: *Agricultural Price Policy in Nepal*. Manila, Philippines, ADB.

Nelson, G., and M. Agcaoili [1983]: *Impact of Government Policies on Philippine Sugar*. Manila, PIDS, Working Paper No. 83-04.

Nelson, R. R. [1956]: "A Theory of the Low-Level Equilibrium Trap in Underdeveloped Economies." *Amer. Econ. Rev.* 46:894-908.

Nepal, Ministry of Food and Agriculture [1972]: *Rice Marketing in Nepal*. Kathmandu.

———— [1971]: *Farm Management Study in the Selected Regions of Nepal, 1968/69*. Kathmandu.

———— [1981]: *Nepal's Experience in Hill Agricultural Development*. Kathmandu.

Nepal, National Planning Commission [1977]: *A Survey of Employment, Income Distribution and Consumption Patterns in Nepal*. Kathmandu.

Nepal Rastra Bank [1982]: *An Evaluation of Small Farmers Development Projects*. Kathmandu.

Nerlove, M. [1958]: *The Dynamics of Supply: Estimation of Farmers Response to Price*. Baltimore, Johns Hopkins Univ. Press.

Nerlove, M., A. Razin, and E. Sadka [1987]: *Population Policy and Individual Choice: A Theoretical Investigation*. Washington, IFPRI, Research Report No. 60.

Nestel, B., ed. [1983]: *Agricultural Research for Development Potentials and Challenges in Asia.* The Hague, Netherlands, ISNAR.

———— [1985]: *Indonesia and the CGIAR Centers: A Study of Their Collaboration in Agricultural Research.* Washington, World Bank, CGIAR Study Paper No. 10.

Newbery, D. M. G. [1975]: "The Choice of Rental Contract in Peasant Agriculture." In Reynolds, ed., 1975, pp. 109-137.

———— [1977]: "Risk Sharing, Sharecropping, and Uncertain Labor Markets." *Rev. Econ. Studies* 44:585-594.

Newbery, D. M. G., and N. Stern, eds. [1987]: *The Theory of Taxation for Developing Countries.* New York, Oxford Univ. Press for the World Bank.

Newbery, D. M. G., and J. E. Stiglitz [1979]: "Sharecropping, Risk Sharing, and Importance of Imperfect Information." In Roumasset, Boussard and Singh, eds., 1979, pp. 311-339.

———— [1981]: *The Theory of Commodity Price Stabilization: A Study in the Economics of Risk.* Oxford, England, Clarendon Press.

———— [1982]: "Optimal Commodity Stock-Piling Rules." *Oxford Econ. Papers* 34:403-427.

Newman, P., ed. [1968]: *Readings in Mathematical Economics, Vol. II: Capital and Growth.* Baltimore, Johns Hopkins Univ. Press.

Nghiep, L. T., and Y. Hayami [1979]: "Mobilizing Slack Resources for Economic Development: The Summer-Fall Rearing Technology of Sericulture in Japan." *Explorations in Economics History* 16:163-181.

Nicholls, W. H. [1961]: "Industrialization, Factor Markets, and Agricultural Development." *J. Polit. Economy* 69:319-340.

———— [1963]: "An 'Agricultural Surplus' as a Factor in Economic Development." *J. Polit. Economy* 71:1-29.

———— [1964]: "Agricultural Policy: The Place of Agriculture in Economic Development." In Berrill, ed., 1964, pp. 336-371. Also in Eicher and Witt, eds., 1964, pp. 11-44.

———— [1969]: "Development in Agrarian Economies: The Role of Agricultural Surplus, Population Pressures, and System of Land Tenure." In Wharton, ed., 1969a, pp. 269-319.

Nicholson, N. K. [1968]: "Political Aspects of Indian Food Policy." *Pacific Affairs* 41:34-50.

———— [1984]: "Landholding, Agricultural Modernization and Local Institutions in India." *Econ. Dev. and Cultural Change* 32:569-592.

Nicol, K. J., S. Sriplung, and E. O. Heady, eds. [1982]: *Agricultural Development Planning in Thailand.* Ames, ISU Press.

Niu, R., and P. H. Calkins [1986]: "Towards an Agricultural Economy for China in a New Age: Progress, Problems, Response, and Prospects." *Amer. J. Agr. Econ.* 68:445-450.

Nobe, K. C., and R. K. Sampath, eds. [1983]: *Issues in Third World Development.* Boulder, CO, Westview.

———— [1986]: *Irrigation Management in Developing Countries: Current Issues and Approaches.* Boulder, CO, Westview Press.

Nolan, P. [1983a]: *Growth Processes and Distributional Change in a South Chinese Province: The Case of Guangdong.* Univ. of London, Contemporary China Institute, School of Oriental and African Studies.

———— [1983b]: "De-collectivisation of Agriculture in China, 1979-82: A Long-Term Perspective." *Cambridge J. Econ.* 7:381-403.

Norlund, I. [1986]: "Social and Economic Studies on Vietnam: An Overview." In Norlund *et al.*, eds., 1986, pp. 176-202.

Norlund, I., S. Cederroth, and I. Gerdin, eds. [1986]: *Rice Societies: Asian Problems and Prospects.* London, Curzon Press.

Norton, G. W., and J. S. Davis [1981]: "Evaluating Returns to Agricultural Research: A Review." *Amer. J. Agr. Econ* 63:685-699.

Nowshirvani, V. F. [1967a]: "A Note on the 'Fixed Cash Requirement' Theory of Marketed Surplus in Subsistence Agriculture." *Kyklos* 20:772-773.

——— [1967b]: "A Note on the Elasticity of the Marketable Surplus: A Comment." *Indian J. Agr. Econ.* 22(1):110-114.

Nulty, L. [1972]: *The Green Revolution in West Pakistan: Implications for Technological Change.* New York, Praeger.

Nurkse, R. [1953]: *Problems of Capital Formation in Underdeveloped Countries.* Oxford, England, Oxford Univ. Press.

Nyberg, A. J., and D. Prabowo [1982]: *Status and Performance of Irrigation in Indonesia and the Prospects to 1990 and 2000.* Washington, IFPRI, IFDC and IRRI, Working Paper No. 4.

Oasa, E. K. [1981]: "The International Rice Research Institute and the Green Revolution: A Case Study in the Politics of Agricultural Research." Unpublished Ph.D. dissertation, Honolulu, Univ. of Hawaii.

——— [1987]: "The Political Economy of International Agricultural Research: A Review of the CGIAR's Response to Criticisms of the 'Green Revolution.' " In Glaeser, ed., 1987, pp. 13-55.

Oasa, E. K., and B. H. Jennings [1982]: "Science and Authority in International Agricultural Research." *Bull. Concerned Asian Scholars* 14(4):30-44.

Oberlander, L. [1986]: *A Cost-Benefit Approach to Agricultural Price Policy: The Case of Thailand,* tr. D. A. Valencia. Ft. Lauderdale, FL, Brietenbach.

Ogura, T. [1980]: *Can Japanese Agriculture Survive? A Historical and Comparative Approach.* Tokyo, Agricultural Policy Research Center.

Ohkawa, K. [1972]: *Differential Structure and Agriculture: Essays on Dualistic Growth.* Tokyo, Kinokuniya Bookstores for Hitotsubashi Univ., Institute of Economics Research, Research Series, Vol. 13.

Ohkawa, K., B. F. Johnston, and H. Kaneda, eds. [1970]: *Agriculture and Economic Growth: Japan's Experience.* Princeton, NJ, Princeton Univ.

Ohkawa, K., and B. Key, eds. [1980]: *Asian Socio-Economic Development.* Tokyo, Univ. of Tokyo Press.

Ohkawa, K., and G. Ranis, eds. [1985]: *Japan and the Developing Countries.* Oxford, England, Basil and Blackwell.

Ohkawa, K., and H. Rosovsky [1960]: "The Role of Agriculture in Modern Japanese Economic Development." *Econ. Dev. and Cultural Change* 9, Part 2:43-68. Also in Eicher and Witt, eds., 1964, pp. 45-69.

Ohlin, B. [1933]: *Interregional and International Trade.* Cambridge, MA, Harvard Univ. Press.

Oi, J. [1986]: "Peasant Grain Marketing and State Procurement: China's Grain Contracting System." *China Quarterly* 106:272-290.

Ojala, E. M. [1952a]: *Agriculture and Economic Progress.* London, Oxford Univ. Press.

——— [1952b]: *Problems and Issues in a Rapidly Growing Economy.* Baltimore, Johns Hopkins Univ. Press.

——— [1969]: "The Pattern and Potential of Asian Agricultural Trade." *Monthly Bull. Agr. Econ. and Stat.* 18(9):1-11.

Ongkingco, P. S., J. A. Galvez, and M. W. Rosegrant [1982]: *Irrigation and Rice Production in the Philippines: Status and Projections*. Washington, IFPRI, IFDC and IRRI, Working Paper No. 3.

Oomen, M. A. [1972]: *Small Industry in Indian Economic Growth: A Case Study of Kerala*. Delhi, Mudralaya Press.

Oram, P. A. [1985]: "Agricultural Research and Extension: Issues of Public Expenditure." In Howell, ed., 1985, pp. 59-82.

_____ [1988]: "Building the Agroecological Framework." *Environment* 30(9):14-17 and 30-36.

Oram, P. A., and V. Bindlish [1981]: *Resource Allocation to National Agricultural Research: Trends in the 1970s (A Review of Third World Systems)*. The Hague, Netherlands, ISNAR; and Washington, IFPRI.

Oram, P., J. Japata, G. Alibaruho, and S. Roy [1979]: *Investment and Input Requirements for Accelerating Food Production in Low-Income Countries by 1990*. Washington, IFPRI, Research Report No. 10.

Organization for Economic Cooperation and Development [OECD, 1985]: *Agriculture in China: Prospects for Production and Trade*. Paris.

_____ [OECD, 1987]: *National Policies and Agricultural Trade*. Paris.

Oshima, H. T. [1965]: "Improving the Statistics of National Accounts for Development Planning with Special Emphasis on Southeast Asia." *Philippine Econ. J.* 4:249-283.

_____ [1972]: "Income Inequality and Economic Growth: The Postwar Experience of Asian Countries." *Malayan Econ. Rev.* 15(2):7-41.

_____ [1986]: "The Transition from an Agricultural to an Industrial Economy in East Asia." *Econ. Dev. and Cultural Change* 34:783-810.

_____ [1987a]: *Economic Growth in Monsoon Asia: A Comparative Study*. Tokyo, APO.

_____ [1987b]: "Food and Agriculture in Asia by the Year 2000." In Fu-chen Lo, ed., 1987.

_____ [1988]: "Diversified Agricultural Development in Philippine Development Strategy." *Philippine Rev. Econ. and Bus.* (1).

Otsuka, K., and Y. Hayami [1985]: "Goals and Consequences of Rice Policy in Japan, 1965-80." *Amer. J. Agr. Econ.* 67:529-538.

_____ [1988]: "Theories of Share Tenancy: A Critical Survey." *Econ. Dev. and Cultural Change* 37:31-68.

Owen, N. G. [1971]: "The Rice Industry of Mainland Southeast Asia, 1850-1914." *J. Siam Society* 59(Part 2):75-143.

Palmer, I. [1975]: *The New Rice in the Philippines*. Geneva, UNRISD.

_____ [1977]: *The New Rice in Indonesia*. Geneva, UNRISD.

Panayotou, T. [1985]: *Food Policy Analysis in Thailand*. Bangkok, ADC.

_____ [1989]: "Thailand: The Experience of a Food Exporter." In Sicular, ed., 1989a, pp. 65-108.

Panchamukhi, V. R. [1975]: "Linkages in Industrialization." *J. Dev. Planning (UN)* 8:121-165.

Pannikar, P. G. K. [1961]: "Rural Savings in India." *Econ. Dev. and Cultural Change* 10:64-85.

Papanek, G., ed. [1980]: *The Indonesian Economy*. New York, Praeger.

Parikh, A. K. [1966]: "Consumption of Nitrogenous Fertilizer: A Continuous Cross-Section Study and Covariance Analysis." *Indian Econ. J.* 14:258-274.

Paris, T. B., Jr., and N. P. Pascual [1984]: *Impact of Irrigation on Income Distribution in Selected Rice Farming Areas of the Philippines*. Washington, IFPRI, mimeo.

Parthasarathy, G., and M. S. Mudahar [1976]: "Foodgrain Prices and Economic Growth." *Indian J. Agr. Econ.* 31(2):16-30.

Parthasarathy, G., and D. S. Prasad [1978]: "Response to the Impact of the New Rice Technology by Farm Size and Tenure—Andhra Pradesh, India." In IRRI, 1978b, pp. 111-127.

Parthasarathy, G., and B. V. Subbarao [1964]: "Production and Marketed Surplus of Rice in the Deltas of the South." *Agricultural Situation in India* 19:721-726.

Pathak, M. I., and A. S. Patel [1970]: *Agricultural Taxation in Gujarat.* New York, Asia Publishing House.

Patrick, H. T. [1966]: "Financial Development and Economic Growth in Underdeveloped Countries." *Econ. Dev. and Cultural Change* 14:174-189.

Pauker, G. J. [1968]: "Political Consequences of Rural Development Programs in Indonesia." *Pacific Affairs* 41:386-402.

Paulino, L. A. [1986]: *Food in the Third World: Past Trends and Projections to 2000.* Washington, IFPRI, Research Report No. 52.

Pearse, A. [1980]: *Seeds of Plenty, Seeds of Want: Social and Economic Implications of the Green Revolution.* Oxford, England, Clarendon Press.

Pearson, S. R., N. Akrasanee, and G. C. Nelson [1976]: "Comparative Advantage in Rice Production: A Methodological Introduction." *Food Research Institute Studies* 15:127-137.

Pearson, S. R., T. Josling, and W. Falcon [1986]: *Food Self-Reliance and Food Self-Sufficiency: Evaluating the Policy Options.* Washington, Aurora Associates.

Pee, T. Y. [1977]: "Social Returns from Rubber Research in Peninsular Malaysia." Unpublished Ph.D. dissertation, MSU, East Lansing.

Perez, L. L., and G. P. Benedick, eds. [1978]: *Trade Policies toward Developing Countries: The Multilateral Trade Negotiations.* Washington, AID, Bureau of Intragovernmental and International Affairs.

Perkins, D. H. [1964]: "Centralization and Decentralization in Mainland China's Agriculture, 1949-1962." *Quart. J. Econ.* 78:208-237.

——— [1966]: *Market Control and Planning in Communist China.* Cambridge, MA, Harvard Univ. Press.

——— [1967]: "Economic Growth in China and the Cultural Revolution (1960- April 1967)." *China Quarterly* 30:33-48.

——— [1969]: *Agricultural Development in China, 1368-1968.* Chicago, Aldine.

———, ed. [1975]: *China's Modern Economy in Historical Perspective.* Stanford, CA, Stanford Univ. Press.

——— [1980]: "China's Experience with Rural Small-Scale Industry." In F. A. Long and Oleson, eds., 1980, pp. 177-192.

——— [1983]: "Research on the Economy of the People's Republic of China: A Survey of the Field." *J. Asian Studies* 2:345-372.

——— [1986]: *China: Asia's Next Economic Giant?* Seattle, Univ. of Washington Press.

——— [1988]: "Reforming China's Economic System." *J. Econ. Literature* 26:601-645.

Perkins, D. H., A. De Angelis, R. Dernberger, S. Hallford, A. Khan, O. Livingston, W. Parish, T. Rawski, K. Simmons, A. Stinchcombe, P. Timmer, and L. van Slyke [1977]: *Rural Small-Scale Industry in the People's Republic of China.* Berkeley, Univ. of California Press.

Perkins, D. H., and S. Yusuf [1984]: *Rural Development in China.* Baltimore, Johns Hopkins Univ. Press for the World Bank.

Perrin, R. K., and G. M. Scobie [1981]: "Market Intervention Policies for Increasing the Consumption of Nutrients by Low-Income Households." *Amer. J. Agr. Econ.* 63:73-82.

Perrin, R., and D. Winkelmann [1976]: "Impediments to Technical Progress on Small versus Large Farms." *Amer. J. Agr. Econ.* 58:888-894.

Perrin, R. K., D. L. Winkelmann, E. R. Moscardi, and J. R. Anderson [1976]: *From Agronomic Data to Farmer Recommendations: An Economics Training Manual.* Mexico City, CIMMYT, Information Bulletin 27.

Peterson, W. L. [1967]: "Returns to Poultry Research in the United States." *J. Farm Econ.* 49:656-669.

_____ [1979]: "International Farm Prices and the Social Cost of Cheap Food Policies." *Amer. J. Agr. Econ.* 61:12-26.

Petit, M. J. [1988]: "Presidential Address." Twentieth International Conference of Agricultural Economists, Buenos Aires, Argentina, August 24-31, 1988.

Philippine Council for Agriculture and Resources Research and Development [PCARRD, 1983]: *Agriculture and Resources Research Manpower Development in South and Southeast Asia.* Los Baños, Philippines.

Philippines, National Economic and Development Authority, ed. [PNEDA, 1983]: *The Consequences of Small Rice Farm Mechanization in the Philippines.* Tagaytay City, Philippines, Development Academy of the Philippines.

Piazza, A. [1983]: *Trends in Food and Nutrient Availability in China, 1950-81.* Washington, World Bank, Staff Working Paper No. 607.

_____ [1986]: *Food Consumption and Nutritional Status in the PRC.* Boulder, CO, Westview Press.

Pimentel, D., J. Allen, A. Beers, L. Guinand, R. Linder, P. McLaughlin, B. Meer, D. Musonda, D. Perdue, S. Poisson, S. Siebert, K. Stoner, R. Salazar, and A. Hawkins [1987]: "World Agriculture and Soil Erosion." *BioScience* 37(4):277-283.

Pinckney, T. C. [1986]: "Stabilizing Pakistan's Supply of Wheat: Issues in the Optimization of Storage and Trade Policies." *Pakistan Dev. Rev.* 25:451-466.

_____ [1989]: *The Demand for Public Storage of Wheat in Pakistan.* Washington, IFPRI, Research Report No. 77.

Pinstrup-Andersen, P. [1981]: *Nutritional Consequences of Agricultural Projects: Conceptual Relationships and Assessment Approaches.* Washington, World Bank Staff Working Paper No. 456.

_____ [1982]: *Agricultural Research and Technology in Economic Development.* New York, Longman.

_____ [1985]: "Food Prices and the Poor in Developing Countries." *European Rev. Agr. Econ.* 12:69-81.

_____ , ed. [1988]: *Food Subsidies in Developing Countries: Costs, Benefits, and Policy Options.* Baltimore, Johns Hopkins Univ. Press for IFPRI.

Pinstrup-Andersen, P., N. R. de Londono, and E. Hoover [1976]: "The Impact of Increasing Food Supply on Human Nutrition: Implications for Commodity Priorities in Agricultural Research and Policy." *Amer. J. Agr. Econ.* 58:131-142.

Pinstrup-Andersen, P., and D. Franklin [1977]: "A Systems Approach to Agricultural Research Resource Allocation in Developing Countries." In Arndt, Dalrymple and Ruttan, eds., 1977, pp. 416-435.

Pinstrup-Andersen, P., and P. B. R. Hazell [1985]: "The Impact of the Green Revolution and Prospects for the Future." *Food Reviews International* 1(1):1-25.

Pinstrup-Andersen, P., and L. G. Tweeten [1971]: "The Value, Cost, and Efficiency of American Food Aid." *Amer. J. Agr. Econ.* 53:431-440.

Pitt, M. [1977]: "Economic Policy and Agricultural Development in Indonesia." Unpublished Ph.D. dissertation, Berkeley, Univ. of California.

―――― [1983a]: "Food Preferences and Nutrition in Rural Bangladesh." *Rev. Econ. and Stat.* 65:105-114.

―――― [1983b]: "Farm-Level Fertilizer Demand in Java: A Meta Production Function Approach." *Amer. J. Agr. Econ.* 65:502-508.

Plusquellec, H. L., and T. Wickham [1985]: *Irrigation Design and Management: Experience in Thailand and its General Applicability.* Washington, World Bank Technical Paper No. 40.

Poleman, T. T. [1972]: "Employment, Population, and Food: The New Hierarchy of Development Problems." *Food Research Institute Studies* 11:11-26.

―――― [1981]: "Quantifying the Nutrition Situation in Developing Countries." *Food Research Institute Studies* 18:1-58.

Poleman, T. T., and D. K. Freebairn, eds. [1973]: *Food, Population, and Employment: The Impact of the Green Revolution.* Praeger, New York.

Popkin, S. L. [1979]: *The Rational Peasant: The Political Economy of Rural Society in Vietnam.* Berkeley, CA, Univ. of California Press.

Prabowo, D. [1983]: "Demand for the Supply of Basic Food Products in the ASEAN Countries." In Maunder and Ohkawa, eds., 1983, pp. 267-276.

―――― [1985]: "The Impact of Irrigation on Income Distribution in Central Java, Indonesia." Washington, IFPRI, mimeo.

Prais, S. J., and M. S. Houthakker [1971]: *The Analysis of Family Budgets.* Cambridge, England, Cambridge Univ. Press.

Pray, C. E. [1979]: "The Economics of Agricultural Research in Bangladesh." *Bangladesh J. Agr. Econ.* 2(2):1-34.

―――― [1983]: "Underinvestment and the Demand for Agricultural Research: A Case Study of the Punjab." *Food Research Institute Studies* 19:51-79.

Pray, C. E., and J. R. Anderson [1985]: *Bangladesh and the CGIAR Centers: A Study of their Collaboration in Agricultural Research.* Washington, World Bank, CGIAR Study Paper No. 8.

Prebisch, R. [1964]: *Toward a New Trade Policy for Development.* New York, United Nations.

Psacharopoulos, G. [1988a]: "Education and Development: A Review." *World Bank Research Observer* 3(1):99-116.

―――― , ed. [1988b]: *The Political Economy of Poverty, Equity, and Growth: Country Perspectives.* Washington, World Bank.

Puapanichaya, K. [1976]: "Analysis of Demand for Fertilizer in Thailand, 1965-1972." Unpublished Ph.D. thesis, Univ. of the Philippines, Los Baños.

Pudasaini, S. P. [1979]: *Farm Mechanization, Employment, and Income in Nepal: Traditional and Mechanical Farming in Bara District.* Manila, Philippines, IRRI, Research Paper Series 38.

―――― [1983]: "The Effect of Education in Agriculture: Evidence from Nepal." *Amer. J. Agr. Econ.* 65:505-515.

Quiggin, J., and B. L. Anh [1984]: "The Use of Cross-Sectional Estimates of Profit Functions for Tests of Relative Efficiency: A Critical Review." *Australian J. Agr. Econ.* 28:44-55.

Quisumbing, M. A. R. [1986]: "The Effects of Food, Price, and Income Policies on the Nutrition of Low-Income Groups: A Philippine Case Study." *Food and Nutrition Bull.* 8(2):24-49.

Quizon, J. B. [1981]: "Factor Input Demand and Output Supply Elasticities in Philippine Agriculture." *Philippine Econ. J.* 20(2):103-126.

Quizon, J. B., and H. P. Binswanger [1983]: "Income Distribution in Agriculture: A Unified Approach." *Amer. J. Agr. Econ.* 65:526-37.

_____ [1986]: "Modeling the Impact of Agricultural Growth and Government Policy on Income Distribution in India." *Econ. Rev.* 1:103-148.

Rahim, A. M. A. [1985]: "Agricultural Credit Institutions and Policies in Bangladesh." *Bangladesh J. Agr. Econ.* 8:17-34.

Rahman, R. I. [1980]: "New Technology in Bangladesh Agriculture: Adoption and its Impact on Rural Labor Market." Bangkok, ILO-ARTEP Asian Employment Programme Working Paper.

Rahman, S. H. [1986]: "Supply Response in Bangladesh Agriculture." *Bangladesh Dev. Studies* 14:57-100.

Raj, K. N. [1975]: "Linkages in Industrialization and Development Strategy: Some Basic Issues." *J. Dev. Planning (UN)* 8:105-119.

_____ [1976]: "Trends in Rural Unemployment in India, An Analysis with Reference to Conceptual and Measurement Problems." *Econ. and Polit. Weekly* 11(31,32,33):1281-1292.

Rana, S., and T. R. Joshi [1968]: "Nepal's Food Grain Surplus and Deficit Regions." *National Geographic J. India* 14(2,3):165-175.

Ranade, C. G. [1980]: "Impact of Cropping Pattern on Agricultural Production." *Indian J. Agr. Econ.* 35(2):85-93.

_____ [1986]: "Growth of Productivity in Indian Agriculture: Some Unfinished Components of Dharm Narain's Work." *Econ. and Polit. Weekly* 21(25, 26):A75-A80.

Ranade, C. G., and R. W. Herdt [1978]: "Shares of Farm Earnings from Rice Production." In IRRI, 1978a, pp. 87-104.

Ranade, C. G., K. H. Rao, and D. C. Sah [1982]: *Groundnut Marketing in India: A Study of Cooperative and Private Trade Channels*. Ahmedabad, Indian Institute of Management, CMA Monograph 82.

Randhawa, M. S. [1974]: *Green Revolution: A Case Study of Punjab*. Delhi, Vikas Publishing House.

Rangarajan, C. [1982]: *Agricultural Growth and Industrial Performance in India*. Washington, IFPRI, Research Report No. 33.

Ranis, G. [1984]: "The Dual Economy Framework: Its Relevance to Asian Development." *Asian Dev. Rev.* 2(1):39-51.

_____ [1988]: "Towards a Model of Development for the Natural Resources Poor Economy." In L. Krause, ed., 1988.

Ranis, G., and J. C. H. Fei [1961]: "The Theory of Economic Development." *Amer. Econ. Rev.* 51:535-565. Also in Eicher and Witt, eds., 1964, pp. 181-194.

Ranis, G., and T. P. Schultz, eds. [1988]: *The State of Development Economics: Progress and Perspectives*. United Kingdom, Basil Blackwell Ltd.

Rao, A. P. [1967]: "Size of Holding and Productivity." *Econ. and Polit. Weekly* 2(44):1989-1991.

Rao, C. H. H. [1963]: "Farm Size and Economies of Scale." *Econ. Weekly* 14(50):2041-2044.

_____ [1965a]: *Agricultural Production Functions, Costs, and Returns in India*. Bombay, Asia Publishing House.

_____ [1965b]: "The Marketable Surplus Function for a Subsistence Crop: Comments." *Econ. Weekly* 17(16):677-678.

_____ [1966]: "Alternative Explanations of the Inverse Relationship between Farm Size and Output per Acre in India." *Indian Econ. Rev.* NS1:1-12.

_____ [1969]: "Resource Prospects for the Rural Sector: The Case of Indirect Taxes." *Econ. and Polit. Weekly* 4(13):A52-A58.

_____ [1970]: "Farm Size and Credit Policy." *Econ. and Polit. Weekly* 5(52):A157-A162.

―――― [1971]: "Uncertainty, Entrepreneurship, and Sharecropping in India." *J. Polit. Economy* 79:578-595.

―――― [1972]: "Farm Mechanization in a Labor-Abundant Economy." *Econ. and Polit. Weekly* (5,6,7):393-400.

―――― [1974]: "Employment Implications of the Green Revolution and Mechanization: A Case Study of the Punjab." In N. Islam, ed., 1974, pp. 340-350.

―――― [1975]: *Technological Change and Distribution of Gains in Indian Agriculture*. Delhi, Macmillan.

―――― [1986]: "Science and Technology Policy: An Overall View and Broader Implications." *Indian J. Agr. Econ.* 41(3):229-233.

Rao, C. H. H., S. K. Ray, and K. Subbarao [1988]: *Unstable Agriculture and Drought: Implications for Policy*. New Delhi, Vikas.

Rao, C. H. H., and K. Subbarao [1976]: "Marketing of Rice in India: An Analysis of the Impact of Producers' Price on Small Farmers." *Indian J. Agr. Econ.* 31(2):1-15.

Rao, J. M. [1986]: "Agriculture in Recent Development Theory." *J. Dev. Econ.* 22:41-86.

Rao, K. P. C., and V. K. Pandey [1976]: "Supply Response of Paddy in Andhra Pradesh." *Indian J. Agr. Econ.* 31(2):46-53.

Rao, M. S. [1974]: "Protection of Fertilizer Industry and its Impact on Indian Agriculture." Unpublished Ph.D. dissertation, Univ. of Chicago.

Rao, V., and T. Chotigeat [1981]: "The Inverse Relationship between Size of Land Holdings and Agricultural Productivity." *Amer. J. Agr. Econ.* 63:571-574.

Rao, V. K. R. V. [1974]: *Growth with Justice in Asian Agriculture: An Exercise in Policy Formulation*. Geneva, UNRISD.

Rao, V. M. [1972]: "Agricultural Wages in India—A Reliability Analysis." *Indian J. Agr. Econ.* 27(3):38-68.

Rao, V. M., and R. S. Deshpande [1986]: "Agricultural Growth in India: A Review of Experiences and Prospects." *Econ. and Polit. Weekly* 21(38):A101-103, A105-109, A111-112 and 21(39):20-27.

Rasmussen, P. N. [1956]: *Studies in Inter-Sectoral Relations*. Amsterdam, North-Holland.

Rath, N. [1974]: "Has Rural Unemployment Declined?" *Indian J. Agr. Econ.* 29(4):20-31.

―――― [1985]: "Prices, Costs of Production and Terms-of-Trade of Indian Agriculture." *Indian J. Agr. Econ.* 40(4):451-481.

Rawski, T. G. [1979]: *Economic Growth and Employment in China*. New York, Oxford Univ. Press for the World Bank.

Ray, A. [1988]: "Agricultural Policies in Developing Countries: National and International Aspects." In Antonelli and Quadrio-Curzio, 1988, pp. 33-52.

Ray, S. K., R. W. Cummings, Jr., and R. W. Herdt [1979]: *Policy Planning for Agricultural Development*. New Delhi, Tata McGraw Hill.

Raychaudhuri, G. S., and R. Krishna [1979]: "Some Aspects of Wheat Price Policy in India." *Indian Econ. Rev.* 14:101-125.

Reddy, J. M. [1970]: "Estimation of Farmers' Supply Response—A Case Study of Groundnut." *Indian J. Agr. Econ.* 25(4):57-63.

Regmi, M. C. [1976]: *Land Ownership in Nepal*. Berkeley, Univ. of California Press.

Reich, M., Y. Endo, and C. P. Timmer [1986]: "The Political Economy of Structural Change: Conflict between Japanese and United States Agricultural Policy." In McCraw, ed., 1986, pp. 151-192.

Reid, J. D., Jr. [1976]: "Sharecropping and Agricultural Uncertainty." *Econ. Dev. and Cultural Change* 24:549-576.

Renaud, B. M., and P. Suphaphiphat [1971]: "The Effect of the Rice Export Tax on the Domestic Rice Price Level in Thailand." *Malayan Econ. Rev.* 16:84-102.

Reserve Bank of India, ed. [1969]: *Financing Agriculture by Commercial Banks.* Bombay.

Reutlinger, S. [1977]: *Food Insecurity: Magnitude and Remedies.* Washington, World Bank Staff Working Paper No. 267.

Reutlinger, S., and D. Bigman [1981]: "Feasibility, Effectiveness, and Cost of Food Security Alternatives in Developing Countries." In Valdes, ed., 1981, pp. 185-212.

Reutlinger, S., and J. Katona-Apte [1984]: "The Nutritiional Impact of Food Aid: Criteria for the Selection of Cost-Effective Foods." *Nutrition Today* 19(2):20-28.

Reutlinger, S., and M. Selowsky [1976]: *Malnutrition and Poverty: Magnitude and Policy Options.* Baltimore, Johns Hopkins Univ. Press for the World Bank.

Reynolds, L. G., ed. [1975]: *Agriculture in Development Theory.* New Haven, CT, Yale Univ. Press.

Reynolds, L. G. [1985]: *Economic Growth in the Third World, 1850-1980.* New Haven, CT, Yale Univ. Press.

Rice, E. B., and S. Bunyasi, eds. [1986]: *Agricultural Pricing and Trade Policy, Seminar Background Readings.* Washington, World Bank, EDI.

Richards, P., and W. Gooneratne [1980]: *Basic Needs, Poverty and Government Policies in Sri Lanka.* Geneva, ILO, WEP.

Richter, H. [1976]: *Burma's Rice Surpluses: Accounting for the Decline.* Canberra, ANU, Development Studies Centre, Working Paper No. 3.

Rijk, A. G., and C. L. J. van der Meer [1984]: *Thailand Agricultural Sector Assessment.* Manila, ADB.

Ritson, C. [1973]: "A Framework for Analyzing the Contribution of the Agricultural Sector to Economic Development." *J. Agr. Econ.* 79:57-79.

Robequain, C. [1944]: *The Economic Development of French Indochina.* New York, Oxford Univ. Press.

Rodriguez, G. [1974]: "The Demand for Fertilizer in the Philippines." Unpublished M. A. dissertation, Ateneo de Manila Univ., Manila, Philippines.

Rojko, A. S., and A. B. Mackie [1970]: *World Demand Prospects for Agricultural Exports of Less Developed Countries in 1980.* Washington, USDA, ERS, FAER No. 60.

Rose, B. [1982]: *An Overview of the Indonesian Rice Economy.* Ithaca, NY, Cornell Univ., DAE Res. 82-44.

Rosegrant, M. W. [1985]: *The Production and Income Effects of Water Distribution Methods and Canal Maintenance in Diversion Irrigation Systems in the Philippines.* Report submitted to the ADB. Washington, IFPRI.

———— [1990]: "Production and Income Effects of Rotational Irrigation and Rehabilitation of Irrigation Systems in the Philippines." In Sampath and Young, eds., 1990.

Rosegrant, M. W., L. A. Gonzales, H. E. Bouis, and J. F. Sison [1987]: *Price and Investment Policies for Food Crop Sector Growth in the Philippines.* Report Submitted to the ADB. Washington, IFPRI.

Rosegrant, M. W., and R. W. Herdt [1978]: "The Impact of Price and Income Support Policies on Small Rice Farmers in the Philippines." *Philippine Rev. Bus. and Econ.* 15(4):1-36.

———— [1981]: "Simulating the Impacts of Credit Policy and Fertilizer Subsidy in Central Luzon Rice Farms, The Philippines." *Amer. J. Agr. Econ.* 63:655-665.

Rosegrant, M. W., F. Kasryno, L. A. Gonzales, C. Rasahan, and Y. Saefudin [1987]: *Price and Investment Policies in the Indonesian Food Crop Sector.* Washington, IFPRI; Bogor, Indonesia, Center for Agro Economic Research.

Rosegrant, M. W., and J. A. Roumasset [1985]: "The Effect of Fertilizer on Risk: A Hete-roscedastic Production Function with Measurable Stochastic Inputs." *Australian J. Agri. Econ.* 29:107-121.

Rosegrant, M. W., J. A. Roumasset, and A. M. Balisacan [1985]: "Biological Technology and Agricultural Policy: An Assessment of Azolla in Philippine Rice Production." *Amer. J. Agr. Econ.* 67:726-732.

Rosegrant, M. W., and A. Siamwalla [1988]: "Government Credit Programs: Justification, Benefits, and Costs." In Mellor and Ahmed, eds., 1988, pp. 219-238.

Rosen, G. [1975]: *Peasant Society in a Changing Economy: Comparative Development in South-east Asia and India.* Urbana, Univ. of Illinois Press.

Rosenberg, D. A., and J. G. Rosenberg [1978]: *Landless Peasants and Rural Poverty in Selected Asian Countries.* Ithaca, NY, Cornell Univ., Center for International Studies, Rural Development Committee, LNL No. 2.

Rosenberg, J. G., and D. A. Rosenberg [1980]: *Landless Peasants and Rural Poverty in Indo-nesia and the Philippines.* Ithaca, NY, Cornell Univ., Center for International Studies, Rural Development Committee, LNL No. 3.

Rosenstein-Rodan, P. N. [1943]: "Problems of Industrialization of Eastern and South-East-ern Europe." *Econ. J.* 53:202-211.

Rosenzweig, M. R. [1978]: "Rural Wages, Labor Supply, and Land Reform: A Theoretical and Empirical Analysis." *Amer. Econ. Rev.* 68:847-861.

———— [1980]: "Neoclassical Theory and the Optimizing Peasant: An Econometric Anal-ysis of Market Family Labor Supply in a Developing Country." *Quart. J. Econ.* 94:31-55.

———— [1982]: "Educational Subsidy, Agricultural Development, and Fertility Change." *Quart. J. Econ.* 97:67-88.

———— [1984]: "Determinants of Wage Rates and Labor Supply Behavior in the Rural Sec-tor of a Developing Country." In Binswanger and Rosenzweig, eds., 1984, pp. 211-241.

Rostow, W. W. [1960]: *The Stages of Economic Growth: A Non-Communist Manifesto.* Cam-bridge, England, Cambridge Univ. Press.

Roumasset, J. A. [1976]: *Rice and Risk: Decision-Making among Low-Income Farmers.* Amster-dam, North-Holland.

Roumasset, J. A., and A. M. Balisacan [1983]: "The Political Economy of Rice Policy and Trade in the Asian-Pacific Region." Honolulu, HI, East-West Center Resource Systems Institute.

Roumasset, J. A., J. M. Boussard, and I. J. Singh, eds. [1979]: *Risk, Uncertainty, and Agri-cultural Development.* Laguna, Philippines, Southeast Asian Regional Center for Gradu-ate Study and Research in Agriculture; New York, ADC.

Roumasset, J. A., M. W. Rosegrant, U. N. Chakravorty, and J. R. Anderson [1989]: "Fer-tilizer and Yield Variability—A Review." In J. R. Anderson and Hazell, eds., 1989, pp. 223-233.

Roumasset, J. A., and S. Setboonsarang [1988]: "Second-best Agricultural Policy: Getting the Price of Thai Rice Right." *J. Dev. Econ.* 28:323-340.

Roumasset, J. A., and J. Smith [1981]: "Population, Technological Change, and the Evo-lution of Labor Markets." *Population and Dev. Rev.* 7:401-419.

Roumasset, J. A., and G. Thapa [1983]: "Explaining Tractorization in Nepal: An Alterna-tive to the "Consequences Approach"." *J. Dev. Econ.* 12:377-395.

Roy, S., and M. G. Blase [1978]: "Farm Tractorisation, Productivity and Labour Employ-ment: A Case Study of Indian Punjab." *J. Dev. Studies* 14:193-209.

Rudra, A. [1968a]: "Farm Size and Yield Per Acre." *Econ. and Polit. Weekly* 3(26,27,28):1041-1044.

_____ [1968b]: "More on Returns to Scale in Indian Agriculture." *Econ. and Polit. Weekly* 3(43):A33-A38.

Rudra, A., and A. K. Sen [1980]: "Farm Size and Labour Use: Analysis and Policy." *Econ. and Polit. Weekly* 15 (Annual Number):391-394.

Russell, C. S., and N. K. Nicholson, eds. [1981]: *Public Choice and Rural Development.* Washington, RFF.

Ruthenberg, H. [1980]: *Farming Systems in the Tropics.* Oxford, England, Oxford Univ. Press.

Ruttan, V. W. [1960]: "Positive Policy in the Fertilizer Industry." *J. Polit. Economy* 68:634.

_____ [1964]: "Equity and Productivity Objectives in Agrarian Legislation: Perspectives on the New Philippine Land Reform Code." *Indian J. Agr. Econ.* 19(3,4):114-130.

_____ [1965]: "Growth Stage Theories and Agricultural Development Policy." *Australian J. Agr. Econ.* 9:17-32.

_____ [1966]: "Technological and Environmental Factors in the Growth of Rice Production in the Philippines and Thailand." *Rural Econ. Problems* 3(1):63-107.

_____ [1969]: "Agricultural Product and Factor Markets in Southeast Asia." *Econ. Dev. and Cultural Change* 17:501-519.

_____ [1970]: *Agricultural Revolution in Southeast Asia: Impact on Grain Production and Trade.* New York, The Asia Society.

_____ [1972]: "Planning Technological Advance in Agriculture: The Case of Rice Production in Taiwan, Thailand, and the Philippines." In Solo and Rogers, eds., 1972, pp. 52-77.

_____ [1975a]: "Technological Transfer, Institutional Transfer, and Induced Technical and Institutional Change in Agricultural Development." In Reynolds, ed., 1975, pp. 165-191.

_____ [1975b]: "Integrated Rural Development Programs: A Skeptical Perspective." *International Dev. Rev.* 17(4):9-16.

_____ [1977]: "The Green Revolution: Seven Generalizations." *International Dev. Rev.* 19(4):16-23.

_____ [1978]: "New Rice Technology and Agricultural Development Policy." In IRRI, 1978a, pp. 367-381.

_____ [1982]: *Agricultural Research Policy.* Minneapolis, MN, Univ. of Minnesota Press.

_____ [1984a]: "Perspectives on Population and Development." *Indian J. Agr. Econ.* 39(4)630-638.

_____ [1984b]: "Integrated Rural Development Programmes: A Historical Perspective." *World Dev.* 12:393-401.

_____ [1985a]: "The Intercountry Agricultural Production Function and Productivity Differences among Countries." *J. Dev. Studies* 19:113-132.

_____ [1985b]: "Toward a Global Agricultural Research System." In *Agricultural Research Policy and Organization in Small Countries: Report of a Workshop.* The Hague, Netherlands, Wageningen Agricultural Univ.

_____ [1986a]: "Some Concerns about Agricultural Research Policy in Developing Countries." *Quart. J. International Agriculture* 25:318-321.

_____ [1986b]: "Toward a Global Agricultural Research System: A Personal View." *Research Policy* 15:307-327.

_____ [1986c]: "Induced Institutional Innovation." In Schuh, ed., 1986, pp. 119-140.

_____ [1986d]: "Development of Agricultural Research Capacity: Some Perspectives from Asian Experience." In Schuh and McCoy, eds., 1986, pp. 206-222.

Ruttan, V. W., and Y. Hayami [1972]: "Strategies for Agricultural Development." *Food Research Institute Studies* 11:129-148.

_____ [1984]: "Toward a Theory of Induced Institutional Innovations." *J. Dev. Studies* 20:203-223.

Ruttan, V. W., and C. Pray, eds. [1987]: *Policy for Agricultural Research.* Boulder, CO, Westview Press.

Ruttan, V. W., and G. Traxler [1986]: "Assistance for Water Resource Development in Pakistan: Some Lessons from Experience." *Pakistan J. Agr. Social Sciences* 1:72-91.

Ryan, J. G. [1984]: "Efficiency and Equity Considerations in the Design of Agricultural Technology in Developing Countries." *Australian J. Agr. Econ.* 28(2, 3):109-135.

_____ [1987]: *Building on Success: Agricultural Research, Technology, and Policy for Development. Proceedings of the ACIAR Policy Symposium.* Canberra, Australia.

Ryan, J. G., and M. Asokan [1977]: "Effect of Green Revolution in Wheat on Production of Pulses and Nutrition in India." *Indian J. Agr. Econ.* 32(3):8-15.

Ryan, J. G., R. D. Ghodake, and R. Sarin [1980]: "Labor Use and Labor Markets in Semi-Arid Tropical Rural Villages of Peninsular India." In ICRISAT, 1980, pp. 357-379.

Ryan, J. G., R. Sarin, and M. Pereira [1980]: "Assessment of Prospective Soil-, Water-, and Crop-Management Technologies for the Semi-Arid Tropics of Peninsular India." In ICRISAT, 1980, pp. 52-72.

Sahn, D. E. [1988a]: "Changes in the Living Standards of the Poor in Sri Lanka during a Period of Macroeconomic Restructuring." *World Dev.* 15:809-830.

_____ [1988b]: "The Effect of Price and Income Changes on Food-Energy Intake in Sri Lanka." *Econ. Dev. and Cultural Change* 36:315-340.

_____ , ed. [1989]: *Seasonal Variability in Third World Agriculture: The Consequences for Food Security.* Baltimore, Johns Hopkins Univ. Press for IFPRI.

Sahn, D. E., and J. von Braun [1987]: "The Relationship between Food Production and Consumption Variability: Policy Implications for Developing Countries." *J. Agr. Econ.* 38:315-327.

Sahota, G. S. [1968a]: "Efficiency in Resource Allocation in Indian Agriculture." *Amer. J. Agr. Econ.* 50:584-605.

_____ [1968b]: *Fertilizer in Economic Development: An Econometric Analysis.* New York, Praeger.

Saini, G. R. [1969]: "Resource-Use Efficiency in Agriculture." *Indian J. Agr. Econ.* 24(2):1-18.

_____ [1971]: "Holding Size, Productivity, and Some Related Aspects of Indian Agriculture." *Econ. and Polit. Weekly* 6(26):A79-A85.

Saito, K. A., and D. P. Villanueva [1981]: "Transaction Costs of Credit to the Small-Scale Sector in the Philippines." *Econ. Dev. and Cultural Change* 29:631-640.

Salam, A. [1975]: "Economic Analysis of Fertilizer Application in Punjab Pakistan." Unpublished Ph.D. dissertation, Univ. of Hawaii, Honolulu.

_____ A. [1981]: "Farm Tractorisation, Fertiliser Use and Productivity, and Mexican Wheat in Pakistan." *Pakistan Dev. Rev.* 20:323-345.

Salita, D. C., and D. Z. Rossel [1980]: *Economic Geography of the Philippines.* Bicutan, Philippines, National Research Council of the Philippines.

Sampath, R. K., and R. A. Young, eds. [1990]: *Social, Institutional and Economic Issues in Irrigation Management in Developing Countries.* Boulder, CO, Westview Press.

Sanders, J. H., and R. C. Hoyt [1970]: "The World Food Problem: Four Recent Empirical Studies." *Amer. J. Agr. Econ.* 52:132-135.

Sanderson, F. [1978]: *Japan's Food Prospects and Policies.* Washington, Brookings Institution.

_____ [1984]: *Global Demand for Food and Fiber through 2000.* Washington, RFF.

Sanderson, F., and S. Roy [1979]: *Food Trends and Prospects in India.* Washington, Brookings Institution.

Santos, C. L. G. [1987]: *Identifying Nutritionally Vulnerable Households in the Philippines.* Ithaca, NY, Cornell Univ., DAE, International Agriculture Mimeograph No. 114.

Sanyal, S. K. [1972]: "Has There Been a Decline in Agricultural Tenancy?" *Econ. and Polit. Weekly* 7(19):943-945.

Sarma, J. S. [1981]: *Growth and Equity: Policies and Implementation in Indian Agriculture.* Washington, IFPRI, Research Report No. 28.

_____ [1982]: *Agricultural Policy in India: Growth with Equity.* Ottawa, IDRC.

_____ [1986]: *Cereal Feed Use in the Third World: Past Trends and Projections to 1990 and 2000.* Washington, IFPRI, Research Report No. 57.

Sarma, J. S., S. Roy, and P. S. George [1979]: *Two Analyses of Indian Foodgrain Production and Consumption Data.* Washington, IFPRI, Research Report No. 12.

Sarma, J. S., and P. Yeung [1985]: *Livestock Products in the Third World: Past Trends and Projections to 1990 and 2000.* Washington, IFPRI, Research Report No. 49.

Sarris, A. H. [1980]: "Grain Imports and Food Security in an Unstable International Market." *J. Dev. Econ.* 7:489-504.

Sathirathai, S., and A. Siamwalla [1987]: "GATT Law, Agricultural Trade and Developing Countries: Lessons from Two Case Studies." *World Bank Econ. Rev.* 1:595-618.

Sawant, S. [1983]: "Investigation of the Hypothesis of Deceleration in Indian Agriculture." *Indian J. Agr. Econ.* 38(3):475-496.

Saxon, E., and K. Anderson [1983]: *Japanese Agricultural Protection in Historical Perspective.* Canberra, ANU, AJRC, Pacific Economic Paper No. 92.

Scandizzo, P. L. [1984]: *The Consequences of Price Stabilization Policies: Theoretical Problems and Empirical Measurements.* Rome, FAO.

Scandizzo, P. L., and C. Bruce [1980]: *Methodologies for Measuring Agricultural Price Intervention Effects.* Washington, World Bank Staff Working Paper No. 394.

Scandizzo, P. L., P. B. R. Hazell, and J. R. Anderson [1984]: *Risky Agricultural Markets: Price Forecasting and the Need for Intervention Policies.* Boulder, CO, Westview Press.

Scandizzo, P. L., and O. K. Knudsen [1980]: "The Evaluation of the Benefits of Basic Needs Policies." *Amer. J. Agr. Econ.* 62:46-57.

Scandizzo, P. L., and G. Swamy [1982]: *Benefits and Costs of Food Distribution Policies: The Indian Case.* Washington, World Bank Staff Working Paper No. 509.

Schickele, R. [1941]: "Effect of Tenure System on Agricultural Efficiency." *J. Farm Econ.* 23:185-207.

Schiff, M. W. [1985]: *An Econometric Analysis of the World Wheat Market and Simulation of Alternative Policies, 1960-80.* Washington, USDA, ERS, Staff Report No. AGES-850827.

Schluter, M. G. G. [1971]: *Differential Rates of Adoption of the New Seed Varieties in India: The Problem of the Small Farm.* Ithaca, NY, Cornell Univ., DAE, Occasional Paper No. 47.

_____ [1974]: *The Interaction of Credit and Uncertainty in Determining Resource Allocation and Incomes on Small Farms, Surat District, India.* Ithaca, NY, Cornell Univ., DAE, Occasional Paper No. 68.

Schluter, M. G. G., and G. O. Parikh [1974]: "The Interaction of Cooperative Credit and Uncertainty in Small Farmer Adoption of the New Cereal Varieties." *Artha-Vikas* 11(2):31-48.

Schran, P. [1970]: *The Development of Chinese Agriculture: 1950-1959*. Urbana, Univ. of Illinois Press.

Schuh, G. E. [1968]: "Effects of Some General Economic Development Policies on Agricultural Development." *Amer. J. Agr. Econ.* 50:1283-1293.

⸺ [1976]: "The New Macroeconomics of Agriculture." *Amer. J. Agr. Econ.* 58:802-811.

⸺ [1984]: "Future Directions for Food and Agricultural Trade Policy." *Amer. J. Agr. Econ.* 66:242-247.

⸺ , ed. [1986]: *Technology, Human Capital, and the World Food Problem*. St. Paul, MN, Univ. of Minnesota Press.

Schuh, G. E., and S. Barghouti [1988]: "Agricultural Diversification in Asia." *Finance and Dev.* 25(2):41-44.

Schuh, G. E., and J. McCoy, eds. [1986]: *Food, Agriculture and Development in the Pacific Basin*. Boulder, CO, Westview Press.

Schuh, G. E., and H. Tollini [1979]: *Costs and Benefits of Agricultural Research: The State of the Arts*. Washington, World Bank Staff Working Paper No. 360.

Schultz, T. W. [1953]: *The Economic Organization of Agriculture*. New York, McGraw-Hill.

⸺ [1960]: "Value of Farm Surpluses to Underdeveloped Countries." *J. Farm Econ.* 42:1019-1030.

⸺ [1963]: *The Economic Value of Education*. New York, Columbia Univ. Press.

⸺ [1964]: *Transforming Traditional Agriculture*. New Haven, CT, Yale Univ. Press.

⸺ [1968]: "Institutions and the Rising Economic Value of Man." *Amer. J. Agr. Econ.* 50:1113-1122.

Schultz, T. W., ed. [1978]: *Distortions of Agricultural Incentives*. Bloomington, Indiana Univ. Press.

⸺ [1979]: *The Economics of Being Poor*. Stockholm, Sweden, Carolinske Institute, Nobel Lecture. Also published in *J. Polit. Economy* 88:639-651, 1980.

⸺ [1980]: "Effects of the International Donor Community on Farm People." *Amer. J. Agr. Econ.* 62:873-878.

Schutjer, W. A., and C. S. Stokes, eds. [1984]: *Rural Development and Human Fertility*. New York, Macmillan.

Scobie, G. M., and R. Posada-Torres [1978]: "The Impact of Technical Change on Income Distribution: The Case of Rice in Colombia." *Amer. J. Agr. Econ.* 60:85-92.

Scott, J. C. [1976]: *The Moral Economy of the Peasant: Rebellion and Subsistence in Southeast Asia*. New Haven, CT, Yale Univ. Press.

Seddon, D., S. Westwood, and P. Blaike [1987]: *Nepal—A State of Poverty: The Political Economy, Population Growth and Social Deprivation*. New Delhi, Vikas.

Segura, E. L., Y. T. Shetty, and M. Nishimizu [1986]: *Fertilizer Producer Pricing in Developing Countries—Issues and Approaches*. Washington, World Bank, Industry and Finance Series No. 11.

Selowsky, M. [1979]: "Target Group-Oriented Food Programs: Cost Effective Comparisons." *Amer. J. Agr. Econ.* 61:988-994.

⸺ [1981a]: "Income Distribution, Basic Needs, and Tradeoffs with Growth: The Case of Semi-industrialized Latin American Countries." *World Dev.* 9:73-92.

⸺ [1981b]: "Nutrition, Health, and Education: The Economic Significance of Complementarities at an Early Age." *J. Dev. Econ.* 9:331-346.

Selowsky, M., and L. Taylor [1973]: "The Economics of Malnourished Children: An Example of Disinvestment in Human Capital." *Econ. Dev. and Cultural Change* 22:17-30.

Sen, A. [1981]: "Market Failure and Control of Labour Power: Towards an Explanation of 'Structure' and Change in Indian Agriculture, Part I." *Cambridge J. Econ.* 5:201-228.

Sen, A. K. [1962]: "An Aspect of Indian Agriculture." *Econ. Weekly* 14(4,5,6):243-246.

_____ [1964]: "Size of Holdings and Productivity." *Econ. Weekly* 16(5,6,7):323-326.

_____ [1966]: "Peasants and Dualism with or without Surplus Labor." *J. Polit. Economy* 74:425-450.

_____ [1968]: *Choice of Techniques*, 3rd ed. New York, A. M. Kelley.

_____ [1975a]: *Employment, Technology, and Development*. London, Oxford Univ. Press.

_____ [1975b]: "Employment, Institutions, and Technology: Some Policy Issues." *International Labor Rev.* 112:45-73.

_____ [1981a]: "Ingredients of Famine Analysis: Availability and Entitlements." *Quart. J. Econ.* 96:433-464.

_____ [1981b]: *Poverty and Famines: An Essay on Entitlement and Deprivation*. Oxford, England, Oxford Univ. Press.

_____ [1987]: *Food, Economics and Entitlements*. Helsinki, United Nations Univ., World Institute for Development Economics Research.

Sen, B. [1974]: *The Green Revolution in India—A Perspective*. New York, Wiley.

Sen, S. R. [1960]: "Impact and Implications of Foreign Surplus Disposal on Underdeveloped Economies—The Indian Perspective." *J. Farm Econ.* 42:1031-1042.

Setboonsarng, S. [1985]: *The Vegetable Oil and Animal Feed Model for Thailand*. Bangkok, Thailand Development Research Institute, Research Report.

Shaban, R. A. [1987]: "Testing between Competing Models of Sharecropping." *J. Polit. Economy* 95:893-920.

Shah, C. H. [1976]: "Growth and Inequality in Agriculture." *Indian J. Agr. Econ.* 31(4):71-94.

_____ , ed. [1979]: *Agricultural Development in India: Policy and Problems*, R. P. Nevatia Felicitation Volume. Bombay, Orient Longman.

Shahabuddin, Q. [1985]: "Testing of Cobb-Douglas Myths: An Analysis with Disaggregated Production Functions in Bangladesh Agriculture." *Bangladesh Dev. Studies* 13:88-97.

Shand, R. T., ed. [1969]: *Agricultural Development in Asia*. Berkeley, Univ. of California Press.

_____ , ed. [1973]: *Technical Change in Asian Agriculture*. Canberra, ANU Press.

_____ [1987]: "Income Distribution in a Dynamic Rural Sector: Some Evidence from Malaysia." *Econ. Dev. and Cultural Change* 36:35-50.

Shari, I. [1979]: "The Impact of Public Policies on Income Distribution in Peninsular Malaysia." *Economic Bull. for Asia and the Pacific* 30(2):20-45.

Sharma, D. P., and V. V. Desai [1980]: *Rural Economy of India*. New Delhi, Vikas Publishing.

Sharma, R. P., and J. R. Anderson [1985]: *Nepal and the CGIAR Centers: A Study of their Collaboration in Agricultural Research*. Washington, World Bank, CGIAR Study Paper No. 7.

Shen, T. H. [1964]: *Agricultural Development in Taiwan since World War II*. Ithaca, NY, Cornell Univ. Press.

_____ [1974]: *Agriculture's Place in the Strategy of Development: The Taiwan Experience*. Taipei, Taiwan, JCRR.

Shetty, S. L. [1971]: "An Inter-Sectoral Analysis of Taxable Capacity and Tax Burden." *Indian J. Agr. Econ.* 26(3):216-246.

Shim, Y. K., D. C. Dahl, and B. Y. Sung [1974]: "Estimation of Fertilizer Consumption in Korea, 1975-1985." *Biological and Agricultural Series*, Vol. 3. Seoul National Univ., Faculty Papers, pp. 64-83.

Short, D. E., and J. C. Jackson [1971]: "The Origins of Irrigation Policy in Malaysia." *J. Malaysian Branch of the Royal Asiatic Society* 44:78-103.

Shortlidge, R. L. [1975]: "University Training for 'Gram Sevaks' in India." *Econ. Dev. and Cultural Change* 24:139-153.

———— [1976]: *A Socioeconomic Model of School Attendance in Rural India.* Ithaca, NY, Cornell University, DAE, Occasional Paper No. 86.

Shukla, V. P. [1971]: *Interaction of Technological Change and Irrigation in Determining Farm Resource Use, Jabalpur District, India, 1967-68.* Ithaca, NY, Cornell Univ., International Agricultural Development Bulletin No. 20.

Shumway, C. R. [1977]: "Models and Methods to Allocate Resources in Agricultural Research: A Critical View." In Arndt, Dalrymple and Ruttan, eds., 1977, pp. 436-457.

Siamwalla, A. [1975]: "A History of Rice Policies in Thailand." *Food Research Institute Studies* 14:233-249.

———— [1987]: *Issues in Thai Agricultural Development.* Bangkok, Thailand Development Research Institute Report.

Siamwalla, A., and S. Haykin [1983]: *The World Rice Market: Structure, Conduct, and Performance.* Washington, IFPRI, Research Report No. 39.

Siamwalla, A., and S. Setboonsarng [1987]: *Agricultural Pricing Policies in Thailand, 1960-1984.* Bangkok, Thailand Development Research Institute Report.

Siamwalla, A., and A. Valdes [1980]: "Food Insecurity in Developing Countries." *Food Policy* 5:258-272.

Sicular, T. [1986a]: "Recent Agricultural Price Policies and their Effects: The Case of Shandong." In U.S. Congress, Joint Economic Committee, 1986, Vol. 1, pp. 407-430.

———— [1986b]: "Agricultural Planning in China: The Case of Lee Willow Team No. 4." *Food Research Institute Studies* 20:1-24.

———— [1986c]: "Prospects and Some Policy Problems of Agricultural Development in China: Discussion." *Amer. J. Agr. Econ.* 68:458-460.

———— [1988a]: "Plan and Market in China's Agricultural Commerce." *J. Polit. Economy* 96:283-307.

———— [1988b]: "Agricultural Planning and Pricing in the Post-Mao Period." *China Quarterly* (Special Issue on Food and Agriculture in China During the Post-Mao Era) 116:671-705.

————, ed. [1989a]: *Food Price Policy in Asia: A Comparative Study.* Ithaca, NY, Cornell Univ. Press.

———— [1989b]: "China: Food Pricing under Socialism." In Sicular, ed., 1989a, pp. 243-288.

Sidhu, D. S. [1979]: *Price Policy for Wheat in India: An Economic Analysis for Production and Marketing Problems.* New Delhi, Chand.

Sidhu, S. S. [1974a]: "Relative Efficiency in Wheat Production in the Indian Punjab." *Amer. Econ. Rev.* 64:742-751.

———— [1974b]: "Economics of Technical Change in Wheat Production in the Indian Punjab." *Amer. J. Agr. Econ.* 56:217-226.

Sidhu, S. S., and C. A. Baanante [1981]: "Estimating Farm-Level Input Demand and Wheat Supply in the Punjab Using a Translog Profit Function." *Amer. J. Agr. Econ.* 63:237-246.

Sigurdson, J. [1977]: *Rural Industrialization in China.* Cambridge, MA, Harvard Univ. Press.

Silcock, T. H. [1967]: *Thailand: Social and Economic Studies in Development.* Canberra, ANU Press.

———— [1970]: *The Economic Development of Thai Agriculture.* Ithaca, NY, Cornell Univ. Press.

Simmonds, N. W. [1985]: *Farming Systems Research: A Review.* Washington, World Bank Technical Paper No. 43.

Sinaga, R. S., and B. M. Sinaga [1978]: "Comments on Shares of Farm Earnings from Rice Production." In IRRI, 1978a, pp. 105-109.

Singh, B., ed. [1967]: *Education as Investment.* Meerut, India, Meenakshi Prakashan.

Singh, G., and H. S. Sandhu [1971]: "Income Distribution by Farm Size." *Agricultural Situation in India* 26:193-199.

Singh, I. J. [1971]: "The Transformation of Traditional Agriculture: A Case Study of the Punjab, India." *Amer. J. Agr. Econ.* 53:275-284.

———— [1979]: *Small Farmers and the Landless in South Asia.* Washington, World Bank, Staff Working Paper No. 320.

———— [1983]: "The Landless Poor in South Asia." In Maunder and Ohkawa, eds., 1983, pp. 379-400.

———— [1988a]: *Small Farmers in South Asia: Their Characteristics, Productivity, and Efficiency.* Washington, World Bank Discussion Paper No. 31.

———— [1988b]: *Tenancy in South Asia.* Washington, World Bank Discussion Paper No. 32.

———— [1988c]: *Land and Labor in South Asia.* Washington, World Bank Discussion Paper No. 33.

Singh, I. J., and R. H. Day [1975]: "A Microeconomic Chronicle of the Green Revolution." *Econ. Dev. and Cultural Change* 23:661-686.

Singh, I. J., L. Squire, and J. Strauss [1986a]: "A Survey of Agricultural Household Models: Recent Findings and Policy Implications." *World Bank Econ. Rev.* 1:149-179.

————, eds. [1986b]: *Agricultural Household Models: Extensions, Applications, and Policy.* Baltimore, Johns Hopkins Univ. Press for the World Bank.

Singh, K. [1973]: "The Impact of New Agricultural Technology on Farm Income Distribution in the Aligarh District of Uttar Pradesh." *Indian J. Agr. Econ.* 28(2):1-11.

Singh, M. [1964]: *India's Export Trends and the Prospects for Self-Sustained Growth.* London, Oxford Univ. Press.

Sinha, R. [1975]: "Chinese Agriculture: A Quantitative Look." *J. Dev. Studies* 11:202-223.

Sinha, R., and G. Drabek, eds. [1978]: *The World Food Problem: Consensus and Conflict.* Oxford, England, Pergamon Press.

Sisler, D. G. [1971]: "International Trade Policies and Agriculture." In IAAE, 1971, pp. 259-275.

Smith, J., and G. Umali [1985]: *Production, Risk and Optimal Fertilizer Rates: An Application of the Random Coefficient Model.* Manila, Philippines, IRRI Research Paper Series No. 115.

Smith, T. C. [1959]: *The Agrarian Origins of Modern Japan.* Stanford, CA, Stanford Univ. Press.

Snodgrass, D. R. [1980]: *Inequality and Economic Development in Malaysia.* Kuala Lumpur, Oxford Univ. Press.

Soejono, I. [1976]: "Growth and Distributional Change of Income in Paddy Farms in Central Luzon Java, 1968-74." *Bull. Indonesian Econ. Studies* 12(2):80-89.

Solo, R. A., and E. M. Rogers, eds. [1972]: *Inducing Technological Change for Economic Growth and Development*. East Lansing, Michigan State Univ. Press.

Solow, R. M. [1956]: "A Contribution to the Theory of Economic Growth." *Quart. J. Econ.* 70:65-94. Also in Stiglitz and Uzawa, eds., 1969, pp. 58-87; and in Newman, ed., 1968, pp. 142-171.

_____ [1961]: "Note on Uzawa's Two Sector Model of Economic Growth." *Rev. Econ. Studies* 29:48-50.

Sorenson, V. L. [1975]: *International Trade Policy: Agriculture and Development*. East Lansing, MSU, International Business and Economic Studies.

Southworth, H. M., ed. [1972]: *Farm Mechanization in East Asia*. New York, ADC.

Southworth, H. M., and M. Barnett, eds. [1974]: *Experience in Farm Mechanization in Southeast Asia*. New York, ADC.

Southworth, H. M., and B. F. Johnston, eds. [1967]: *Agricultural Development and Economic Growth*. Ithaca, NY, Cornell Univ. Press.

Squire, L. [1981]: *Employment Policy in Developing Countries: A Survey of Issues and Evidence*. New York, Oxford Univ. Press for the World Bank.

Srinivasan, T. N. [1965]: "A Critique of the Optimizing Planning Model." *Econ. Weekly* 17(5,6,7):255-264.

_____ [1981]: "Malnutrition: Some Measurements and Policy Issues." *J. Dev. Econ.* 8:3-19.

_____ [1985]: "Agricultural Production, Relative Prices, Entitlements, and Poverty." In Mellor and Desai, eds., 1985, pp. 41-53.

_____ [1989]: "Food Aid: A Cause of Development Failure or An Instrument for Success." *Economic Review* 3(1):39-65.

Srinivasan, T. N., and P. K. Bardhan, eds. [1975]: *Poverty and Income Distribution in India*. Calcutta, Statistics Publishing Society.

Srivastava, A. K. [1975]: "International Trade and Regulation of Agricultural Commodities: A Review." *Indian J. Econ.* 56(221):165-194.

Srivastava, U., E. O. Heady, K. D. Rogers, and L. V. Mayer [1975]: *Food Aid and International Economic Growth*. Ames, ISU Press.

Sriwasdilek, J., and S. Wattanutchiraya [1985]: *Income Distribution Effects of Small and Medium Scale Irrigation in Thailand*. Washington, IFPRI, mimeo.

Stavis, B. [1974]: *Making Green Revolution: The Politics of Agricultural Development in China*. Ithaca, NY, Cornell Univ., Rural Development Committee, Rural Development Monograph No. 1.

_____ [1978]: *The Politics of Agricultural Mechanization in China*. Ithaca, NY, Cornell Univ. Press.

Steinberg, D. I. [1981]: *Burma's Road to Development: Growth and Ideology under Military Rule*. Boulder, CO, Westview Press.

Stevens, R. D. [1963]: "The Influence of Industrialization on the Income Elasticity of Demand for Retail Food in Low-Income Countries." *J. Farm Econ.* 45:1495-1499.

Stevens, R. D., H. Alavi, and P. Bertocci, eds. [1976]: *Rural Development in Bangladesh and Pakistan*. Honolulu, Univ. Press of Hawaii.

Stiglitz, J. E. [1969]: "Allocation of Heterogenous Capital in a Two-Sector Economy." *International Econ. Rev.* 10:373-390.

_____ [1974]: "Incentives and Risk Sharing in Share-Cropping." *Rev. Econ. Studies* 41:219-255.

——— [1987]: "Some Theoretical Aspects of Agricultural Policies." *World Bank Research Observer* 2:43–60.

Stiglitz, J. E., and H. Uzawa, eds. [1969]: *Readings in the Modern Theory of Economic Growth.* Cambridge, MA, MIT Press.

Stone, B. [1980]: *A Review of Chinese Agricultural Statistics, 1949-79.* Washington, IFPRI, Research Report No. 16.

——— [1985]: "The Basis for Chinese Agricultural Growth in the 1980s and 1990s: A Comment on Document No. 1, 1984." *China Quarterly* 101:114–121.

——— [1986]: "China's Fertilizer Application in the 1980s and 1990s: Issues of Growth, Balance, Allocation, Efficiency and Response." In U.S. Congress, Joint Economic Committee, 1986, Vol. 1, pp. 453–496.

——— [1988a]: "Relative Foodgrain Prices in the People's Republic of China: Rural Taxation through Public Monopoly." In Mellor and Ahmed, eds., 1988, pp. 124–154.

——— [1988b]: "Developments in Agricultural Technology." *China Quarterly* (Special Issue on Food and Agriculture in China During the Post-Mao Era) 116:767–822.

Streeten, C. P. [1969]: *A Partnership to Improve Food Production in India.* New York, Rockefeller Foundation.

Streeten, P., and M. Lipton, eds. [1968]: *The Crisis of Indian Planning in the 1960s.* London, Oxford Univ. Press.

Streeten, P., S. J. Burki, M. ul Haq, N. Hicks, and F. Stewart [1981]: *First Things First: Meeting Basic Human Needs in the Developing Countries.* New York, Oxford Univ. Press for the World Bank.

Stross, R. E. [1986]: *The Stubborn Earth: American Agriculturalists on Chinese Soil, 1898-1937.* Berkeley, Univ. of California Press.

Subbarao, K. [1985]: "State Policies and Regional Disparity in Indian Agriculture." *Dev. and Change* 16:523–546.

Suh, S. C. [1978]: *Growth and Structural Changes in the Korean Economy, 1910-1940.* Cambridge, MA, Harvard Univ. Press for the Council on East Asian Studies.

Sukhatme, P. V. [1962]: "The Food and Nutrition Situation in India." *Indian J. Agr. Econ.* 17(2):1–28 and 17(3):1–34.

——— [1971]: *Three Papers on Food and Nutrition: The Problem and the Means of its Solution.* Brighton, England, Univ. of Sussex, IDS.

——— [1977]: "Incidence of Undernutrition." *Indian J. Agr. Econ.* 32(3):1–8.

——— [1983]: "Farm Prices in India and Abroad: Implications for Production." *Econ. Dev. and Cultural Change* 32:169–182.

Sung, B. Y., D. C. Dahl, and Y. K. Shim [1973]: "Projection of the Demand for Fertilizer: Time-Series Data Analysis." *J. Agr. Econ. (Korea)* 15.

Svejnar, J., and E. Thorbecke [1986]: *Economic Policies and Agricultural Performance: The Case of Nepal, 1960-1982.* Paris, OECD.

Swamy, G., and H. P. Binswanger [1983]: "Flexible Consumer Demand Systems and Linear Estimation: Food in Asia." *Amer. J. Agr. Econ.* 65:675–684.

Swan, T. W. [1956]: "Economic Growth and Capital Accumulation." *Econ. Record* 32:334–361. Also in Stiglitz and Uzawa, eds., 1969, pp. 88–115; and in Newman, ed., 1968, pp. 172–199.

Swenson, C. G. [1976]: "The Distribution of Benefits from Increased Rice Production in Thanjavur District, South India." *Indian J. Agr. Econ.* 31(1):1–12.

Tacke, E. F. [1983]: "Performance and Development Needs in Asian Agriculture." *Asian Dev. Rev.* 1(1):63–85.

Tagarino, R. N., and R. D. Torres [1978]: "The Pricing of Irrigation Water: A Case Study of the Philippines' Upper Pampanga River Project." In IRRI, 1978c, pp. 143-150.

Takase, K. [1984]: "Irrigation Development and Cereal Production in Asia." *Asian Dev. Rev.* 2(2):80-91.

Takase, K., and T. Wickham [1978]: "Irrigation Management and Agricultural Development in Asia." In ADB, 1978, pp. 123-157.

Takayama, A. [1963]: "On a Two-Sector Model of Economic Growth: A Comparative Static Analysis." *Rev. Econ. Studies* 30:95-104.

Tamin, M. [1979]: "Microeconomic Analysis of Production Behavior of Malaysian Farms: Lessons from Muda." *Food Research Institute Studies* 17:87-98.

Tang, A. [1963]: "Research and Education in Japanese Agricultural Development." *Econ. Studies Quarterly* 13:27-41 and 91-99.

––––––– [1984]: *An Analytical and Empirical Investigation of Agriculture in Mainland China, 1952-1980.* Taipei, Taiwan, Chung-Hua Institute for Economic Research.

Tang, A., and K. S. Liang [1975]: "Agricultural Trade in the Economic Development of Taiwan." In Tolley and Zadrozny, eds., 1975, pp. 115-146.

Tang, A., and B. Stone [1980]: *Food Production in the People's Republic of China.* Washington, IFPRI, Research Report No. 15.

Tarrant, J. R. [1982]: "Food Policy Conflicts in Bangladesh." *World Dev.* 10:103-113.

Taylor, C. E. [1976]: "Nutrition and Population in Health Sector Planning." *Food Research Institute Studies* 16:77-90.

Taylor, D. C. [1978]: "Financing Irrigation Services in the Pekalen Sampean Irrigation Project, East Java, Indonesia." In IRRI, 1978c, pp. 111-122.

––––––– [1981]: *The Economics of Malaysian Paddy Production and Irrigation.* Bangkok, ADC.

Taylor, D. C., K. M. Noh, and M. A. Hussein [1979]: *An Economic Analysis of Irrigation Development in Malaysia.* Washington, IFPRI, IFDC and IRRI, Working Paper No. 1.

Taylor, D. C., and T. H. Wickham, eds. [1979]: *Irrigation Policy and the Management of Irrigation Systems in Southeast Asia.* Bangkok, ADC. Also published by IRRI, 1978.

Taylor, L. [1977]: "Research Directions in Income Distribution, Nutrition, and the Economics of Food." *Food Research Institute Studies* 16:29-45.

Te, A. [1982]: *An Economic Analysis of a Reserve Stock Program for Rice in the Philippines.* Washington, IFPRI, IFDC, and IRRI, Working Paper No. 7.

Tendler, J. [1975]: *Inside Foreign Aid.* Baltimore, Johns Hopkins Univ. Press.

Terjung, W. H., J. T. Hayes, H-Y. Ji, P. E. Todhunter, and P. A. O'Rourke [1985]: "Potential Paddy Rice Yields for Rainfed and Irrigated Agriculture in China and Korea." *Annals, Assoc. Amer. Geographers* 75:83-101.

Terjung, W. H., H-Y. Ji, J. T. Hayes, P. A. O'Rourke, and P. E. Todhunter [1984a]: "Actual and Potential Yield for Rainfed and Irrigated Maize in China." *International J. Biometerology* 28:115-135.

––––––– [1984b]: "Actual and Potential Yield for Rainfed and Irrigated Wheat in China." *Agricultural and Forest Meterology* 31:1-23.

Thakur, D. S. [1974]: "Foodgrain Marketing Efficiency: A Case Study of Gujarat." *Indian J. Agr. Econ.* 29(4):61-74.

Thamarajakshi, R. [1969]: "Intersectoral Terms of Trade and Marketed Surplus of Agricultural Produce, 1951-52 to 1965-66." *Econ. and Polit. Weekly* 4(26):A91-A102. Also in ISAE, 1972, pp. 141-157.

––––––– [1971]: "Prices, Production, and Marketed Surplus of Foodgrain in the Indian Economy, 1951-52 to 1965-66." *Agricultural Situation in India* 25:1047-1052.

Thandee, D. [1986]: "Socioeconomic Factors and Small-Scale Farmers in Southeast Asia." In Marten, ed., 1986, pp. 159-170.

Thiam, T. B., and S. Ong, eds. [1979]: *Readings in Asian Farm Management*. Singapore, Univ. of Singapore Press.

Thomas, V. [1985]: *Linking Macroeconomic and Agricultural Policies for Adjustment with Growth: The Colombian Experience*. Baltimore, Johns Hopkins Univ. Press for the World Bank.

Thorbecke, E., ed. [1969]: *The Role of Agriculture in Economic Development*. New York, Columbia Univ. Press for the NBER.

———— [1973]: "Sector Analysis and Models of Agriculture in Developing Countries." *Food Research Institute Studies* 12:73-89.

Thorbecke, E., and J. Svejnar [1987]: *Economic Policies and Agricultural Performance in Sri Lanka, 1960-1984*. Paris, OECD.

Thorner, D. [1962]: *Land and Labour in India*. Bombay, Asia Publishing.

Timmer, C. P. [1970]: "On Measuring Technical Efficiency." *Food Research Institute Studies* 9:99-171.

———— [1972]: "Employment Aspects of Investment in Rice Marketing in Indonesia." *Food Research Institute Studies* 11:59-88.

———— [1973]: "Choice of Techniques in Rice Milling in Java." *Bull. Indonesian Econ. Studies* 9(2):57-76.

———— [1974a]: "A Model of Rice Marketing Margins in Indonesia." *Food Research Institute Studies* 13:145-167.

———— [1974b]: "Choice of Techniques in Rice Milling in Java: A Reply." *Bull. Indonesian Econ. Studies* 10(1):121-126.

———— [1974c]: "The Demand for Fertilizer in Developing Countries." *Food Research Institute Studies* 13:197-224.

———— [1975a]: "The Political Economy of Rice in Asia: A Methodological Introduction." *Food Research Institute Studies* 14:191-196.

———— [1975b]: "The Political Economy of Rice in Asia: Indonesia." *Food Research Institute Studies* 14:197-231.

———— [1975c]: "The Political Economy of Rice in Asia: Lessons and Implications." *Food Research Institute Studies* 14:419-432.

———— [1975d]: "Interaction of Energy and Food Prices in Less Developed Countries." *Amer. J. Agr. Econ.* 57:219-224.

———— [1976]: "Food Policy in China." *Food Research Institute Studies* 15:53-69.

———— [1981]: "China and the World Food System." In Goldberg, ed., 1981, Vol. II, pp. 75-118.

———— [1984]: "Energy and Structural Change in the Asia-Pacific Region: The Agricultural Sector." In Bautista and Naya, eds., 1984, pp. 51-72.

———— [1986a]: *Getting Prices Right: The Scope and Limits of Agricultural Price Policy*. Ithaca, NY, Cornell Univ. Press.

———— [1986b]: "The Role of Price Policy in Rice Production in Indonesia." In Rice and Bunyasi, eds., 1986, pp. 168-241.

———— [1986c]: "Rice Price Policy in Indonesia: Keeping Prices Right." In Rice and Bunyasi, eds., 1986, pp. 144-167.

———— [1986d]: "Prospects and Some Policy Problems of Agricultural Development in China: Discussion." *Amer. J. Agr. Econ.* 68: 461-463.

———— , ed. [1987a]: *The Corn Economy of Indonesia*. Ithaca, NY, Cornell Univ. Press.

———— [1987b]: *Crop Diversification in Rice-Based Agricultural Economies: Conceptual and Policy Issues*. Cambridge, MA, HIID, Development Discussion Paper No. 352AFP.

———— [1987c]: *Food Price Policy in Indonesia*. Cambridge, MA, HIID, Development Discussion Paper No. 250AFP.

———— [1988a]: *Analyzing Rice Market Interventions in Asia: Principles, Issues, Themes, and Lessons*. In ADB, 1988, pp. 323–368.

———— [1988b]: "The Agricultural Transformation." In Chenery and Srinivasan, eds., 1988, pp. 275–331.

———— [1989]: "Indonesia: Transition from Food Importer to Exporter." In Sicular, ed., 1989a, pp. 22–64.

Timmer, C. P., and H. Alderman [1979]: "Estimating Consumption Parameters for Food Policy Analysis." *Amer. J. Agr. Econ.* 61:982-987.

Timmer, C. P., and W. P. Falcon [1975a]: "The Political Economy of Rice Production and Trade in Asia." In Reynolds, ed., 1975, pp. 373–408.

———— [1975b]: "The Impact of Price on Rice Trade in Asia." In Tolley and Zadrozny, eds., 1975, pp. 57–89.

Timmer, C. P., W. P. Falcon, and G. Nelson [1979]: "China's Food Policy: Incentives and Mechanisms; How Reserves Build Confidence; Posed on a Knife Edge." *CERES* 12(2):25-30; 12(3):36-38; 12(4):41-45.

Timmer, C. P., W. P. Falcon, and S. R. Pearson [1983]: *Food Policy Analysis*. Baltimore, Johns Hopkins Univ. Press.

Timmer, C. P., and M. Guerreiro [1981]: "Food Aid and Development Policy." In *Food Aid and Development*. New York, ADC, pp. 13-30.

Timmer, C. P., and J. R. Jones [1986]: "China: An Enigma in the World Grain Trade." In J. R. Jones, ed., 1986, pp. 153-180.

Tisdell, C. [1988]: "Sustainable Development: Differing Perspectives of Ecologists and Economists." *World Dev.* 16:373.

Todaro, M. P. [1969]: "A Model of Labor Migration and Urban Unemployment in Less Developed Countries." *Amer. Econ. Rev.* 59:138-148.

Tolley, G. S., and G. D. Gwyer [1967]: "International Trade in Agricultural Products in Relation to Economic Development." In Southworth and Johnston, eds., 1967, pp. 403-447.

Tolley, G. S., V. Thomas, and C. M. Wong [1982]: *Agricultural Price Policies and the Developing Countries*. Baltimore, Johns Hopkins Univ. Press.

Tolley, G. S., and P. A. Zadrozny, eds. [1975]: *Trade, Agriculture, and Development*. Cambridge, MA, Ballinger.

Tomek, W. G., and K. L. Robinson [1972]: *Agricultural Product Prices*. Ithaca, NY, Cornell Univ. Press.

Toquero, Z., B. Duff, T. A. Lacsina, and Y. Hayami [1975]: "Marketable Surplus Functions for a Subsistence Crop: Rice in the Philippines." *Amer. J. Agr. Econ.* 57:705-709.

Trairatvorakul, P. [1984]: *The Effect on Income Distribution and Nutrition of Alternative Rice Price Policies in Thailand*. Washington, IFPRI, Research Report No. 46.

Travers, L. [1984]: "Post-1978 Rural Economic Policy and Peasant Income in China." *China Quarterly* 98:241-259.

Tripathy, R. N., and B. Samal [1969]: "Economics of High Yielding Varieties in IADP: A Study of Sambalpur In Orissa." *Econ. and Polit. Weekly* 4(43):1719-1724.

Trung, N. Q. [1978]: "Economic Analysis of Irrigation Development in Deltaic Regions of Asia: The Case of Central Thailand." In IRRI, 1978c, pp. 155-164.

Tsuchiya, K. [1976]: *Productivity and Technological Progress in Japanese Agriculture*. Univ. of Tokyo Press.

Tsuji, H. [1977a]: "An Economic and Institutional Analysis of the Rice Export Policy of Thailand with Special Reference to the Rice Premium Policy." *Developing Economies* 15(2):202-220.

―――― [1977b]: "Rice Economy and Rice Policy in South Vietnam up to 1974: An Economic and Statistical Analysis." *Southeast Asian Studies* 15(3):263-294.

―――― [1982]: "A Quantitative Model of the International Rice Market and Analysis of National Rice Policies, with Special Reference to Thailand, Indonesia, Japan, and the United States." In Langham and Retzlaff, eds., 1982, pp. 291-321.

Tuan, F. C. [1986]: *China's Agricultural Output in 1990 and 2000*. Washington, USDA, ERS 86-8, pp. 34-35.

Tuan, F. C., and F. W. Crook [1983]: *Planning and Statistical Systems in China's Agriculture*. Washington, USDA, ERS, FAER No. 181.

Turvey, R., and E. Cook [1976]: "Government Procurement and Price Support of Agricultural Commodities: A Case Study of Pakistan." *Oxford Econ. Papers* 28:102-117.

Tyagi, D. S. [1987]: "Domestic Terms of Trade and their Effect on Supply and Demand of Agricultural Sector." *Econ. and Polit. Weekly* 22(13):A30-A36.

Tyers, R. [1982]: "Food Security in ASEAN: Potential Impacts of a Pacific Economic Community." *Malayan Econ. Rev.* 27:40-60.

Tyers, R., and K. Anderson [1986]: "Distortions in World Food Markets: A Quantitative Assessment." Paper prepared for *World Development Report, 1986*. Washington, World Bank.

―――― [1987]: *Liberalizing OECD Agricultural Policies in the Uruguay Round: Effects on Trade and Welfare*. Canberra, ANU, Working Papers in Trade and Development No. 87/10.

―――― [1990]: *Distortions in World Food Markets*. Cambridge, England, Cambridge Univ. Press.

Umemura, M. [1970]: "Agriculture and Labor Supply in the Meiji Era." In Ohkawa, Johnston and Kaneda, eds., 1970, pp. 175-198.

Unnevehr, L. J. [1982]: "The Impact of Philippine Government Intervention in Rice Markets." Los Baños, Philippines, IRRI, Agricultural Economics Paper Series 82-24.

―――― [1983]: *The Effect and Cost of Philippine Government Intervention in Rice Markets*. Washington, IFPRI, IFDC, and IRRI, Working Paper No. 9.

―――― [1985]: "The Costs of Squeezing Marketing Margins: Philippine Government Intervention in Rice Markets." *Developing Economies* 23(2):158-172.

―――― [1986]: "Changing Comparative Advantage in Philippine Rice Production: 1966 to 1982." *Food Research Institute Studies* 20:43-70.

Unnevehr, L. J., B. O. Juliano, C. M. Perez, and E. B. Marciano [1985]: *Consumer Demand for Rice Grain Quality in Thailand, Indonesia, and the Philippines*. Los Baños, Philippines, IRRI, Research Paper Series No. 116.

U. N. University and International Rice Research Institute [UN Univ. and IRRI, 1979]: *Interfaces between Agriculture, Nutrition, and Food Science*. Los Baños, Philippines.

United States Agency for International Development [AID, 1982]. *Sri Lanka: The Impact of PL-480 Title I Food Assistance*. Washington, Project Evaluation Report No. 39.

Uphoff, N. T., ed. [1982]: *Rural Development and Local Organization in Asia*. Vol. I of III. Delhi, Macmillan.

U.S. Congress, Joint Economic Committee [1978]: *Chinese Economy—Post-Mao Policy Performance*. Vol. 1. Washington, GPO.

―――― [1982]: *China under the Four Modernizations*. Part I. Washington, GPO.

_____ [1986]: *China's Economy Looks toward the Year 2000*. Vols. 1, 2. Washington, GPO.

Utami, W., and J. Ihalauw [1973]: "Some Consequences of Small Farm Size." *Bull. Indonesian Econ. Studies* 9(2):46-56.

_____ [1978]: "The Relation of Farm Size to Production, Land Tenure, Marketing, and Social Structure—Central Java, Indonesia." In IRRI, 1978b, pp. 127-139.

Uzawa, H. [1961]: "On a Two Sector Model of Economic Growth: I." *Rev. Econ. Studies* 29:40-47.

_____ [1963]: "On a Two Sector Model of Economic Growth: II." *Rev. Econ. Studies* 30:105-118. Also in Stiglitz and Uzawa, eds., 1969, pp. 411-424.

Vaidyanathan, A. [1986]: "Labor Use in Rural India—A Study of Spatial and Temporal Variations." *Econ. and Polit. Weekly* 21(52):A130-A146.

Vakil, C. N., and P. R. Brahmanand [1956]: *Planning for an Expanding Economy: Accumulation, Employment, and Technical Progress in Underdeveloped Countries*. Bombay, Vora.

Vakil, C. N., and C. H. Shah, eds. [1979]: *Agricultural Development of India: Policy and Problems*. Bombay, Orient Longman.

Valdes, A., ed. [1981]: *Food Security for the Developing Countries*. Boulder, CO, Westview Press.

_____ [1988]: "Constraints on Adjustments through International Trade: An Analysis of the Relationship between Industrial and Developing Countries." In Antonelli and Quadrio-Curzio, 1988, pp. 123-132.

Valdes, A., and J. Zietz [1980]: *Agricultural Protection in OECD Countries: Its Cost to Less Developed Countries*. Washington, IFPRI, Research Report No. 21.

Van der Veen, J. [1973]: *A Study of Small-Scale Industries in Gujarat State, India*. Ithaca, NY, Cornell Univ., DAE, Occasional Paper No. 65.

Venkataramanan, L. S. [1958]: "A Statistical Study of Indian Jute Production and Marketing with Special Reference to Foreign Demand." Unpublished Ph.D. dissertation, Univ. of Chicago.

Verma, B. N., and D. W. Bromley [1987]: "The Political Economy of Farm Size in India: The Elusive Quest." *Econ. Dev. and Cultural Change* 35:791-808.

Vermeer, E. B. [1977]: *Water Conservancy and Irrigation in China*. Leiden, Leiden Univ. Press.

_____ [1982]: "Income Differentials in Rural China." *China Quarterly* 89:1-33.

Vernon, R., ed. [1970]: *The Technology Factor in International Trade*. New York, Columbia Univ. Press for the NBER.

Virmani, A. [1982]: *The Nature of Credit Markets in Developing Countries: A Framework for Policy Analysis*. Washington, World Bank Staff Working Paper No. 524.

_____ [1985]: *Government Policy and the Development of Financial Markets: The Case of Korea*. Washington, World Bank Staff Working Paper No. 747.

Visaria, P. [1972]: "Rural Employment: Rapporteur Report." *Indian J. Agr. Econ.* 27(4):179-189.

_____ [1981]: "Poverty and Unemployment in India—An Analysis of Recent Evidence." *World Dev.* 9:277-300.

Visaria, P., and L. Visaria [1973]: "Employment Planning for the Weaker Sections in Rural India." *Econ. and Polit. Weekly* 8(4,5,6):A269-A276.

Von Bremen, L., Y. H. Chuang, R. Diamond, A. Martinez, and M. S. Mudahar [1981]: *Economic Evaluation of Fertilizer Supply Strategies for the ASEAN Region: Linear Programming Approach*. Muscle Shoals, AL, IFDC Technical Bulletin IFDC-T-21.

Von Neumann, J. [1945/46]: "A Model of General Economic Equili rium." *Rev. Econ. Studies* 13:1-9. Also in Newman, ed., 1968, pp. 221-229.

Von Oppen, M., R. P. Parthasarathy, and K. V. Subbarao [1985]: "Impact of Market Access on Agricultural Production in India." In ICRISAT, 1985, pp. 159-168.

Von Oppen, M., V. T. Raju, and S. L. Bapna [1980]: "Foodgrain Marketing and Agricultural Development in India." In ICRISAT, 1980, pp. 173-192.

Von Oppen, M., and K. V. Subbarao [1980a]: *Tank Irrigation in Semi-arid Tropical India, Part I: Historical Development and Spatial Distribution.* Patancheru, India, ICRISAT Economics Program Report No. 5.

———— [1980b]: *Tank Irrigation in Semi-Arid Tropical India, Part II: Technical Features and Economic Performance.* Patancheru, India, ICRISAT Economics Program Report No. 8.

———— [1987]: *Tank Irrigation in Semi-Arid Tropical India, Economic Evaluation and Alternatives for Improvement.* Patancheru, India, ICRISAT Research Bulletin No. 10.

Von Oppen, M., and J. G. Ryan [1985]: "Research Resource Allocation: Determining Regional Priorities." *Food Policy* 10:253-264.

Vyas, V. S. [1970]: "Tenancy in a Dynamic Setting." *Econ. and Polit. Weekly* 5(26):A72-A80.

———— [1972]: "Institutional Change, Agricultural Production, and Rural Poverty: The Experience of Two Decades." *Commerce* 125(3198):40-43.

———— [1973]: "Rural Works in Indian Development." *Dev. Digest* 11:62-66.

———— [1976]: "Structural Change in Agriculture and Small Farm Sector." *Econ. and Polit. Weekly* 11(1,2):24-32.

———— [1983a]: "Growth and Equity in Asian Agriculture: A Synoptic View." In Maunder and Ohkawa, eds., 1983, pp. 52-59.

———— [1983b]: "Asian Agriculture: Achievements and Challenges." *Asian Dev. Rev.* 1(2):27-44.

Vyas, V. S., and M. H. Maharaja [1966]: "Factors Affecting Marketable Surplus and Marketed Supplies: A Case Study in Two Regions of Gujarat and Rajasthan." *Artha-Vikas* 2(1):52-78.

Wald, H. P. [1959]: *Taxation of Agricultural Land in Underdeveloped Countries: A Survey and Guide to Policy.* Cambridge, MA, Harvard Univ. Press.

Walinsky, L. J., ed. [1977]: *Agrarian Reform as Unfinished Business — The Selected Papers of Wolf Ladejinsky.* London, Oxford Univ. Press for the World Bank.

Walker, K. R. [1984]: *Food Grain Procurement and Consumption in China.* London, Cambridge Univ. Press.

Walker, T. S., and K. G. Kshirsagar [1985]: "The Village Impact of Machine Threshing and Implications for Technology Development in the Semi-Arid Tropics of Peninsular India." *J. Dev. Studies* 21:215-231.

Wall, J. [1978]: "Foodgrain Management: Pricing, Procurement, Distribution, Import, and Storage Policy." In *India: Occasional Papers.* Washington, World Bank Staff Working Paper No. 279.

Wallace, M. B. [1986]: "Fertilizer Price Policy in Nepal." Kathmandu, Nepal, Winrock Research and Planning Paper Series No. 6.

———— [1989]: "Nepal: Food Pricing with an Open Border." In Sicular, ed., 1989a, pp. 183-242.

Wallerstein, M. [1980]: *Food for War, Food for Peace.* Cambridge, Mass., MIT Press.

Wanmali, S. [1983]: *Service Provision and Rural Development in India: A Study of Miryalguda Taluka.* Washington, IFPRI, Research Report No. 37.

———— [1985]: *Rural Household Use of Services: A Study of Miryalguda Taluka, India.* Washington, IFPRI, Research Report No. 48.

Ward, W. B. [1985]: *Science and Rice in Indonesia.* Boston, MA, AID Science and Technology Development Series, Oelgeschlager, Gunn and Hain.

Warford, J., and Z. Partow [1989]: "Evolution of the World Bank's Environmental Policy." *Finance and Dev.* 26(4):5-8.

Warriner, D. [1973]: "Results of Land Reforms in Asian and Latin American Countries." *Food Research Institute Studies* 12:115-131.

Welch, F. [1978]: "The Role of Investments in Human Capital in Agriculture." In T. W. Schultz, ed., 1978, pp. 259-281.

Wells, J. C. [1989]: "On the Agricultural Performance of Developing Nations, 1950-85." *Food Research Institute Studies* 21:165-191.

Wen, S. M. [1986]: *National and Regional Trends in Production and Productivity for Post-1978 Chinese Agriculture.* Ithaca, NY, Cornell Univ., Cornell International Agriculture Bulletin No. 111.

Wennergren, E. B., C. Antholt, and M. D. Whitaker, eds. [1984]: *Agricultural Development in Bangladesh: Prospects for the Future.* Boulder, CO, Westview Press.

Wharton, C. R., Jr. [1962]: "Marketing, Merchandizing, and Money Lending: A Note on Middleman Monopsony in Malaya." *Malayan Econ. Rev.* 7(2):24-44.

———— [1963a]: "Research on Agricultural Development in Southeast Asia." *J. Farm Econ.* 45:1161-1174.

———— [1963b]: "The Economic Meaning of Subsistence." *Malayan Econ. Rev.* 8(2):46-58.

————, ed. [1969a]: *Subsistence Agriculture and Economic Development.* Chicago, Aldine.

———— [1969b]: "The Green Revolution: Cornucopia or Pandora's Box?" *Foreign Affairs* 47:464-476.

White, B. [1976]: "Population, Involution, and Employment in Rural Java." *Dev. and Change* 7:267-290.

———— [1984]: "Measuring Time Allocation Decision-Making and Agrarian Changes Affecting Rural Women: Examples from Recent Research in Indonesia." *IDS Bulletin* 15(1):18-33.

White, P. L., ed. [1972]: *Proceedings, Western Hemisphere Nutrition Congress.* Mt. Kisco, NY, Futura.

Whyte, R. O. [1974]: *Rural Nutrition in Monsoon Asia.* Kuala Lumpur, Malaysia, Oxford Univ. Press.

Whyte, R. O., and P. Whyte [1982]: *The Women of Rural Asia.* Boulder, CO, Westview Press.

Wickham, T. H., R. Barker, and M. Rosegrant [1978]: "Complementarities among Irrigation, Fertilizer, and Modern Rice Varieties." In IRRI, 1978a, pp. 221-232.

Wickham, T. H., and A. Valera [1978]: "Practices and Accountability for Better Water Management." In IRRI, 1978c, pp. 61-76.

Wickizer, V. D., and M. K. Bennett [1941]: *The Rice Economy of Monsoon Asia.* Stanford Univ., CA, Food Research Institute.

Wiens, T. B. [1978]: "The Evolution Policy and Capabilities in China's Agricultural Technology." In U.S. Congress, Joint Economic Committee, 1978, pp. 671-703.

———— [1982]: "The Limits to Agricultural Intensification: The Suzhou Experience." In U.S. Congress, 1982, pp. 462-474.

———— [1983]: "Chinese Economic Reforms: Price Adjustments, the Responsibility System and Agricultural Productivity." *Amer. Econ. Rev.* 73:319-324.

Winglee, P. [1989]: "Agricultural Trade Policies of Industrial Countries." *Finance and Dev.* 26(1):9-11.

Wipf, L. J. [1971]: "Tariff, Nontariff Distortions, and Effective Protection in U.S. Agriculture." *Amer. J. Agr. Econ.* 53:423-430.

Witt, L. W. [1965]: "Role of Agriculture in Economic Development: A Review." *J. Farm Econ.* 47:120-131.

Wittfogel, K. A. [1957]: *Oriental Despotism: A Comparative Study of Total Power.* New Haven, CT, Yale Univ. Press.

Wong, C. M. [1978]: "A Model for Evaluating the Effects of Thai Government Taxation of Rice Exports on Trade and Welfare." *Amer. J. Agr. Econ.* 60:65-73.

Wong, J., ed. [1979]: *Group Farming in Asia: Experience and Potentials.* Singapore, Singapore Univ. Press.

World Bank [1975]: *The Assault on World Poverty: Problems of Rural Development, Education, and Health.* Baltimore, Johns Hopkins Univ. Press.

_____ [1979a]: *Nepal Agricultural Sector Review.* Washington, Report No. 2205-NEP.

_____ [1979b]: *Philippines Sector Study: Grain Production Policy Review.* Washington, Report No. 2192A-PH.

_____ [1980a]: *Sri Lanka: Key Development Issues in the 1980s,* 2 vols. Washington, Report No. 2955-CE.

_____ [1980b]: *Thailand—Toward a Development Strategy of Full Participation.* Washington.

_____ [1980c]: *Poverty and Human Development.* New York, Oxford Univ. Press.

_____ [1982a]: *Indonesia: Policy Options and Strategies for Major Food Crops.* Washington, Report No. 3686B.

_____ [1982b]: *World Development Report, 1982.* New York, Oxford Univ. Press.

_____ [1983a]: *World Tables, 3rd ed., Vol. I, Economic Data; Vol, II, Social Data.* Baltimore, Johns Hopkins Univ. Press.

_____ [1983b]: *Thailand: Rural Growth and Employment.* Washington.

_____ [1983c]: *China: Socialist Economic Development.* Washington.

_____ [1983d]: *Wages and Employment in Indonesia.* Washington, Report No. 3586.

_____ [1983e]: *World Development Report, 1983.* New York, Oxford Univ. Press.

_____ [1984a]: *Republic of Korea—Agriculture Sector Survey.* Washington, Report No. 4709-KO.

_____ [1984b]: *World Development Report, 1984.* New York, Oxford Univ. Press.

_____ [1985a]: *Thailand: Pricing and Marketing Policy for Intensification of Rice Agriculture.* Washington.

_____ [1985b]: *China: Long-term Issues and Options.* New York, Oxford Univ. Press, Report No. 5206-CHA.

_____ [1985c]: *China: Agriculture to the Year 2000. Annex 2.* Washington.

_____ [1985d]: *World Development Report, 1985.* New York, Oxford Univ. Press.

_____ [1985e]: *Agricultural Research and Extension: An Evaluation of the World Bank's Experience.* Washington.

_____ [1986a]: *World Development Report, 1986.* New York, Oxford Univ. Press.

_____ [1986b]: *Poverty and Hunger: Issues and Options for Food Security in Developing Countries.* Washington.

_____ [1987a]: *Philippines, A Framework for Recovery.* Washington.

_____ [1987b]: *China—The Livestock Sector.* Washington.

_____ [1987c]: *World Development Report, 1987.* New York, Oxford Univ. Press.

_____ [1988a]: *World Development Report, 1988.* New York, Oxford Univ. Press.

_____ [1988b]: *Rural Development: World Bank Experience, 1965-86.* Washington.

_____ [1988c]: *Adjustment Lending: An Evaluation of Ten Years of Experience.* Washington, World Bank Policy and Research Series No. 1.

_____ [1988d]: *Agricultural Marketing: World Bank's Experience.* Washington, Operations Evaluation Department Report No. 7353.

———— [1988e]: *Conditionality in World Bank Lending: Its Relation to Agricultural Pricing Policies*. Washington, Operations Evaluation Department Report No. 7357.

———— [1989]: *World Development Report, 1989*. New York, Oxford Univ. Press.

———— [1990]: *World Development Report, 1990*. New York, Oxford Univ. Press.

World Commission on Environment and Development [1987a]: *Food 2000: Global Policies for Sustainable Agriculture*. London, Zed Books.

———— [1987b]: *Our Common Future*. New York, Oxford Univ. Press.

World Food Programme [1985]: *Review of Food Aid Policies and Programmes*. WFP/CFA, 19/5, Rome.

———— [1987]: *Roles of Food Aid in Structural and Sectoral Adjustment*. WFP/CFA, 23/5, Rome.

Wu, Craig C. [1977]: "Education in Farm Production: The Case of Taiwan." *Amer. J. Agr. Econ.* 59:699-709.

Wu, Carson K.-H., and Y. Mao [1976]: "Interregional Comparative Advantage of Rice Production in Taiwan." *Food Research Institute Studies* 15:233-256.

Wyon, J. B., and J. E. Gordon [1971]: *The Khanna Study: Population Problems in the Rural Punjab*. Cambridge, MA, Harvard Univ. Press.

Yadav, R. P. [1987]: *Agricultural Research in Nepal: Resource Allocation, Structure, and Incentives*. Washington, IFPRI, Research Report No. 62.

Yamada, S. [1975]: *A Comparative Analysis of Asian Agricultural Productivity and Growth Patterns*. Tokyo, APO, Productivity Series No. 10.

Yoneo, I. [1978]: *Thailand: A Rice Growing Society*. Honolulu, Univ. of Hawaii Press.

York, E. T., Jr. [1988]: "Improving Sustainability with Agricultural Research." *Environment* 30(9):18-20 and 36-40.

Yoshimura, H., M. P. Perera, and P. J. Gunawardena [1975]: *Some Aspects of Paddy and Rice Marketing in Sri Lanka*. Colombo, Sri Lanka, Agrarian Research and Training Institute, Occasional Publication Series No. 10.

Yotopoulos, P. A. [1983a]: "A Micro Economic-Demographic Model of the Agricultural Household in the Philippines." *Food Research Institute Studies* 19:1-24.

———— [1983b]: "The Interface of Poverty and Hunger with Equity and Growth: Global Results and East and Southeast Asian Perspective." In Haseyama, Hirata and Yaragihara, eds., 1983, pp. 61-75.

———— [1985]: "Middle-Income Classes and Food Crises: The 'New' Food-Feed Competition." *Econ. Dev. and Cultural Change* 33:463-483.

Yotopoulos, P. A., and K. Adulavidhaya [1977]: "The Green Revolution in Thailand: With a Bang or with a Whimper." *J. Econ. Dev.* 2(1):7-30.

Yotopoulos, P. A., and L. J. Lau [1973]: "A Test for Relative Economic Efficiency: Some Further Results." *Amer. Econ. Rev.* 63:214-223.

————, eds. [1979]: "Resource Use in Agriculture: Application of the Profit Function to Selected Countries." *Food Research Institute Studies* 17:1-119.

Yotopoulos, P. A., L. J. Lau, and W. L. Lin [1976]: "Microeconomic Output Supply and Factor Demand Functions in the Agriculture of the Province of Taiwan." *Amer. J. Agr. Econ.* 58:333-340.

Yotopoulos, P. A., and G. J. Mergos [1986]: "Family Labor Allocation in the Agricultural Household." *Food Research Institute Studies* 20:87-104.

Yotopoulos, P. A., and J. B. Nugent [1973]: "The Balanced-Growth Version of the Linkage Hypothesis: A Test." *Quart. J. Econ.* 87:157-171.

———— [1976a]: *Economics of Development: Empirical Investigations*. New York, Harper and Row. Also published in 1984 by Keio-tsusin Co., Tokyo.

_____ [1976b]: "In Defense of a Test of the Linkage Hypothesis." *Quart. J. Econ.* 90:334-343.

Young, S., ed. [1985]: *Trade Policy Issues in the Pacific Basin.* Seoul, Development Institute Press.

Yudelman, M., G. Butler, and R. Banerji [1971]: *Technological Change in Agriculture and Employment in Developing Countries.* Paris, OECD.

Zaman, M. R. [1966]: "Marketed Surplus Function of Major Agricultural Commodities in Pakistan." *Pakistan Dev. Rev.* 6(3):376-394.

_____ [1973]: "Sharecropping and Efficiency in Bangladesh." *Bangladesh Econ. Rev.* 1:149-172.

Zarembka, P. [1970]: "Marketable Surplus and Growth in the Dual Economy." *J. Econ. Theory* 2:107-121.

Zietz, J., and A. Valdes [1988]: *Agriculture in the GATT: An Analysis of Alternative Approaches to Reform.* Washington, IFPRI, Research Report No. 70.

Statistical References

1. General Statistical References

Asian Development Bank Economics Office [ADB]: *Key Indicators of Developing Member Countries of ADB.* Annual.

Central Intelligence Agency [CIA, 1969]: *Agricultural Acreage in Communist China, 1949-68: A Statistical Compilation.* Washington.

Food and Agriculture Organization [FAO, 1978]: *National Methods of Collecting Agricultural Statistics: Asia and the Far East.* Bangkok.

_____ [FAO, 1986]: *World Rice Situation and Outlook 1985/86.* International Rice Commission 35(1):69-72, also other years.

_____ [FAO]: *Food Balance Sheets, and per Capita Food Supplies* Rome. Annual on tape.

_____ [FAO]: *Fertilizer Yearbook.* Rome. Annual.

_____ [FAO]: *Production Yearbook.* Rome. Annual.

_____ [FAO]: *Trade Yearbook.* Rome. Annual.

International Agricultural Development Service, now Winrock International [IADS, year]: *Agricultural Development Indicators—A Statistical Handbook.* New York. Periodic.

International Maize and Wheat Improvement Center [CIMMYT]: *World Wheat Facts and Trends.* Mexico. Periodic.

_____ [CIMMYT, year]: *World Maize Facts and Trends.* Mexico. Periodic.

International Monetary Fund [IMF]: *International Financial Statistics Yearbook.* Annual.

International Rice Research Institute [IRRI, 1986]: *World Rice Statistics, 1985.* Los Baños, Philippines. Periodic.

Rose, B. [1985]: *Appendix to the Rice Economy of Asia.* Washington. RFF.

United Nations: *Statistical Yearbook for Asia and the Pacific.* Bangkok. Annual.

United States Department of Agriculture [USDA]: *Agricultural Statistics.* Washington. Annual.

USDA, ERS: *Agricultural Situation—Asia.* Washington. Annual.

_____ *Agricultural Situation—Review of and Outlook for People's Republic of China.* Washington. Annual.

_____ *East Asia and Oceania Situation and Outlook Report.* Washington. Annual.

_____ *South Asia Situation and Outlook Report.* Washington. Annual.

_____ *Southeast Asia Situation and Outlook Report.* Washington. Annual.

_____ *Grains*. Washington. Also other commodities such as cotton, oilseeds, livestock, etc.

_____ *Reference Tables on Rice Supply*. Washington. Annual.

_____ *Reference Tables on Wheat, Corn, and Total Coarse Grain Supply*. Washington. Annual.

_____ *Rice Outlook and Situation*. Washington. Monthly.

World Bank: *World Development Report*. New York, Oxford Univ. Press. Annual.

_____ *World Tables*. Baltimore, Johns Hopkins Univ. Press. Annual.

_____ *The World Bank Atlas*. Washington. Annual.

_____ *World Debt Tables*. Washington. Annual.

_____ *Price Prospects for Major Primary Commodities*. Washington, Biennial.

_____ *Commodity Trade and Price Trends*. Baltimore, Johns Hopkins Univ. Press. Annual.

2. Statistical References for Specific Asian Countries

Bangladesh

Bangladesh, Bureau of Statistics: *Monthly Statistical Bulletin of Bangladesh*. Dhaka. Annual.

_____ . *Statistical Digest of Bangladesh*. Dhaka. Annual.

_____ . *Statistical Yearbook of Bangladesh*. Dhaka. Annual.

_____ . *The Yearbook of Agricultural Statistics of Bangladesh*. Dhaka. Annual.

_____ . Ministry of Agriculture [1974]: *Bangladesh Agriculture in Statistics*. Dhaka.

Burma

Statistical Yearbook of Burma. Rangoon. No longer published.

_____ , Central Statistical and Economics Department: *Quarterly Bulletin of Statistics*. Rangoon. No longer published.

_____ , Ministry of Planning and Finance: *Report to the Pyithu Hluttaw on the Financial, Economic and Social Conditions of the Socialist Republic of the Union of Burma*. Rangoon. Annual.

China

Chen, N. R. [1967]: *Chinese Economic Statistics: A Handbook for Mainland China*. Chicago. Aldine.

China, State Statistical Bureau [1987a]: *China Trade and Price Statistics*. New World Press, China Statistical Information and Consultancy Service Centre. Annual

_____ [1987b]: *China Agriculture Yearbook, 1987*. Beijing, Agricultural Publishing House. Annual.

_____ [1987c]: *Yearbook of Rural, Social, and Economic Statistics of China, 1986*. Beijing, China Reconstructs and China Statistical Information and Consultancy Service Center. Annual.

_____ [1988]. *Statistical Yearbook of China, 1988*. Beijing, China Statistical Information and Consultancy Service Centre. Annual.

Li, C. M. [1962]: *The Statistical System of Communist China*. Berkeley. Univ. of California Press.

Stone, B. [1982]: "The Use of Agricultural Statistics: Some National Aggregate Examples and Current State of the Art." In Barker, Sinha, and Rose, 1982.

United States, Census Bureau [1988]: *China: Consumer Demand Statistical Update*. Washington.

USDA, ERS [1988]: *Agricultural Statistics of the People's Republic of China, 1949-1985*. Washington. USDA, ERS. On diskettes, also annual by year on paper.

India

Elhance, D. N. [1962]: *Economic Statistics of India since Independence.* New Delhi, Kitab
    Mahal Private.
Fertilizer Association of India [FAI]: *Fertilizer Statistics.* Delhi. Annual.
Government of India, Central Statistical Organization: *Statistical Abstract of India.* Delhi.
    Annual.
———— , Department of Agriculture and Cooperatives: *Indian Agriculture in Brief.* Delhi.
    Annual.
———— . *Abstract of Agricultural Statistics.* Delhi, Manager of Publications. Annual.
———— . *Agricultural Situation in India.* Delhi. Annual.
———— , Directorate of Economics and Statistics: *Agricultural Prices in India.* Delhi. Annual.
———— . *Bulletin on Food Statistics:* Delhi, Controller of Publications. Annual.
———— . *Estimates of Area, Production, and Yield of Principal Crops in India.* Delhi. Annual.
———— . *Farm Harvest Prices in India.* Delhi. Annual.
———— , Ministry of Finance: *Economic Survey.* Delhi. Annual.
Kulkarni, V. G. [1968]: *Statistical Outline of Indian Economy.* Bombay, Vora and Co.

Indonesia

Central Bureau of Statistics: *National Socioeconomic Expenditure Survey (Susenas).* Jakarta.
    Periodic since 1964.
———— . *Statistical Pocketbook of Indonesia.* Jakarta. Annual.
———— . *Statistical Yearbook of Indonesia.* Jakarta. Annual.

Japan

Bank of Japan, Bureau of Statistics [1966]: *Hundred Year Statistics of the Japanese Economy.*
    Tokyo.
Bureau of Statistics: *Monthly Statistics of Japan.* Tokyo, Prime Minister's Office.
———— . *Monthly Statistics of Japan.* Tokyo, Prime Minister's Office.
———— . *Statistical Handbook of Japan.* Tokyo.
Japan, Ministry of Agriculture, Forestry, and Fisheries: *Abstract of Statistics on Agriculture,
    Forestry and Fisheries.* Tokyo. Annual.
———— . *Statistical Yearbook of Ministry of Agriculture, Forestry and Fisheries.* Tokyo. Annual.
Japan, Institute of Developing Economies [1969]: *One Hundred Years of Agricultural Statistics
    in Japan.* Tokyo, Kabushiki Kaisha Sangyo Tokei Kenkyusha.
Kikuchi, M., K. Mochida, and Y. Hayami [1975]: *Rice Statistics in Japan.* Los Baños, Phil-
    ippines, IRRI.
Ohkawa, K., M. Shinohara, and M. Umemura, eds. [1965]: *Estimates of Long-Term Eco-
    nomic Statistics of Japan since 1868.* Vol. 9, *Agriculture and Forestry*, Tokyo, Toyo, Keizai
    Shinposha.

Korea

Bank of Korea: *Economic Statistics Yearbook.* Seoul. Annual.
Korea, Ministry of Agriculture and Forestry: *Food Crop Statistics.* Seoul. Annual.
———— . *Yearbook of Agriculture and Forestry Statistics.* Seoul. Annual.

Malaysia

Malaysia, Department of Statistics: *Annual Bulletin of Statistics.* Kuala Lumpur. Annual.
———— . *Statistical Handbook of Peninsular Malaysia.* Kuala Lumpur. Annual.

Malaysia, Sabah

Sabah, Department of Agriculture: *Agricultural Statistics of Sabah*. Kota Kinabalu, Agricultural Information Division.
Sabah, Department of Statistics: *Annual Bulletin of Statistics, Sabah*. Kota Kinabalu.
———— , *Monthly Statistics, Sabah*. Kota Kinabalu.
———— , *Statistical Handbook of Sabah*. Kota Kinabalu. Annual.

Malaysia, Sarawak

Sarawak, Department of Agriculture: *Agricultural Statistics of Sarawak*. Kuching.
———— . *Annual Statistical Bulletin*. Kuching.
———— . *Quarterly Bulletin of Statistics*. Kuching.
Sarawak, Department of Statistics: *Statistical Handbook of Sarawak*. Kuching. Annual.

Nepal

Agriculture Inputs Corporation [AIC, 1983]: *Basic Statistics of Agricultural Inputs in Nepal*.
Nepal, Central Bureau of Statistics [1985]: *National Sample Census of Agriculture, 1981/82*. Kathmandu.
———— . *Foreign Trade Statistics*. Kathmandu. Annual.
———— . *Monthly Statistical Bulletin*. Kathmandu.
———— . *Statistical Pocketbook of Nepal*. Kathmandu. Annual.
———— . *Yield Data for Principal Crops*. Kathmandu. Annual.
Nepal, Department of Food and Agricultural Marketing Services [DFMAS]: *Agricultural Statistics of Nepal*. Kathmandu. Periodic.
———— [DFMAS, 1982]: *Food Statistics of Nepal, 1981*. Kathmandu.
Nepal Rastra Bank [NRB, 1985]: *Some Important Statistics in Agriculture, Nepal*. Kathmandu, Agricultural Credit Division.

Pakistan

Pakistan, Central Statistics Office [CSO, 1972]: *25 Years of Pakistan in Statistics*. Karachi.
Pakistan, Federal Bureau of Statistics: *Monthly Statistical Bulletin*. Karachi.
———— : *Pakistan Statistical Yearbook*. Karachi.
Pakistan, Ministry of Finance, Planning, and Provincial Coordination: *Statistical Pocketbook of Pakistan*. Karachi, Manager of Publications. Annual.
Pakistan, Ministry of Food, Agriculture, and Development: *Yearbook of Agricultural Statistics*. Islamabad. Annual.
Rab, A. [1961]: *Acreage, Production, and Prices of Major Agricultural Crops of West Pakistan: 1931-59*. Karachi, Institute of Development Economics, Statistical Papers No. 1.

Philippines

Anden, T., and A. Palacpac [1981]: *Data Series on Rice Statistics in the Philippines*. Los Baños, Philippines, PCARR.
Philippines, Bureau of Agricultural Economics: *Crop, Livestock, and Natural Resources Statistics*. Manila. Annual.
Philippines, National Census and Statistics Office: *Foreign Trade Statistics*. Manila. Annual.
———— . *Philippine Household Surveys*. Periodic.
Philippines, National Economic and Development Authority [NEDA]: *Philippine Statistical Yearbook*. Manila. Annual.
Philippines, National Science Development Board, Food and Nutrition Research Institute: *Nutrition Surveys*. Periodic.

Philippine Council for Agriculture and Resources Research [PCARR, 1981]: *Data Series on Corn and Sorghum Statistics in the Philippines*. Los Baños.

Sri Lanka

Central Bank of Ceylon: *Economic and Social Statistics of Sri Lanka*. Colombo.
_____ [1980]: *Review of the Economy*. Colombo.
Sri Lanka, Department of Agriculture [1976]: *Agricultural Statistical Information*. Agricultural Economics Publication No. 4. Colombo.
Sri Lanka, Department of Census and Statistics: *Quarterly Bulletin of Statistics*. Colombo.
_____ . *Sri Lanka Yearbook*. Colombo.
_____ . *Statistical Abstract of Sri Lanka*. Colombo. Annual.
_____ . *Statistical Pocketbook of Sri Lanka*. Colombo. Annual.
Sri Lanka, Ministry of Agricultural Development and Research [1981]: *Agricultural Statistics of Sri Lanka: 1951/52-1980/81*. Colombo.
Sri Lanka, Statistics Department: *Economic and Social Statistics of Sri Lanka*. Colombo. Occasional.

Taiwan Province

Taiwan, Department of Agriculture and Forestry: *Taiwan Agricultural Prices Monthly*. Taipei.
_____ . *Taiwan Agricultural Yearbook*. Taipei. Annual.
Taiwan, Department of Budget, Accounting, and Statistics: *Statistical Abstract of Taiwan*. Taipei. Annual.
_____ . *Statistical Yearbook of Taiwan*. Taipei.
Taiwan, Food Bureau: *Taiwan Food Statistics Book*. Taipei. Annual.
Taiwan, Governor-General, Directorate of Statistics [1946]: *Summary of Statistics for 51 Years*. Taipei.
Taiwan, Joint Commission on Rural Reconstruction [JCRR, 1966]: *Taiwan Agricultural Statistics, 1901-1965*. Economic Digest Series No. 18. Taipei.
_____ [1977]: *Taiwan Agricultural Statistics, 1961-1975*. Economic Digest Series No. 22, Taipei.

Thailand

Gaesuwan, Y., A. Siamwalla, and D. Welsch [1974]: *Thai Rice Production and Consumption Data, 1947-74*. Bangkok, Thammassat Univ.
Pookkachatikul, J. S., S. Tongpan, and D. Welsch [1974]: *Thai Rice Price Data*. Bangkok, Kasetsart Univ., DAE, Staff Paper No. 14.
Thailand, Department of Customs: *Foreign Statistics of Thailand*. Bangkok. Annual.
Thailand, Ministry of Agriculture, Center for Agricultural Statistics: *Agricultural Statistics of Thailand*. Bangkok. Annual.
Thailand, Ministry of Agriculture, National Statistical Office: *Quarterly Bulletin of Statistics*. Bangkok.
_____ . *Statistical Handbook of Thailand*. Bangkok. Annual.
_____ . *Statistical Yearbook of Thailand*. Bangkok. Annual.

PART THREE. The Theory, Empirical Evidence, and Debates
on Agricultural Development Issues in Latin America:
A Selective Survey

This enterprise began over a decade ago when Dr. Schuh was on the faculty of Purdue University. It was then taken to the University of Minnesota, where he served as Head of the Department of Agricultural and Applied Economics from 1979 through 1984. Antonio Salazar joined the endeavor in 1983, when he began a two-year period as visiting professor at the University of Minnesota. The project was then taken to the World Bank, where Dr. Schuh served as Director of Agriculture and Rural Development for three years. The final revision was made at the University of Minnesota.

We are indebted to many people for assistance in completing this survey. An important debt is owed to the reviewers of an earlier draft, who by calling our attention to literature we had neglected, spotting errors, suggesting rearrangements of material, and making incisive comments have improved the manuscript considerably. These reviewers include Arthur J. Coutu, Jaime M. Fernandez, Delbert Fitchett, Alain de Janvry, Richard A. King, José Olivares, Vernon W. Ruttan, Robert D. Stevens, and Alberto Valdés. Dale W. Adams and William C. Thiesenhusen gave us special assistance on credit and land reform issues, respectively, as did Avishay Braverman on contractual arrangements among factors of production. Vernon Ruttan read the manuscript a second time and gave us many helpful suggestions. Remaining deficiencies in the paper are our responsibility alone, for we didn't always follow the advice given.

We also owe a special debt to a number of people. In addition to making helpful comments on various versions and doing yeoman's duty in checking the references, Lee Martin was more patient in tolerating our delays in completing the manuscript than we had a right to expect. Maria Ignez Angeli Schuh did a great deal of library work for us, and also permitted her home to be cluttered with stacks and boxes for far too long. Only one who has done such a paper can appreciate the enormous work involved in locating and verifying references. Lee and Maria Ignez have an enormous number of hours invested in this paper and we are very grateful to them both.

Our debt to Alain de Janvry is also very great. He commented on almost every page of the original text, and repeatedly asked good questions and challenged us to penetrate more in our analysis. Ruy Miller Paiva also gave us detailed and extensive comments on an earlier version, with similar valuable comments and questions.

The librarians in the World Bank's Sectoral Library and the IMF–World Bank's Joint Library deserve a special note of thanks. These include Sue Dyer, Karen Eggert, Jane Keneskey, and Chris Windhausen from the Bank's Sectoral Library, Beverly Tait from the Joint Library, and Louise Letnes from Waite Library of the Department of Agriculture and Applied Economics of the University of Minnesota.

For financial support, without which this project would not have been possible, we are indebted to Purdue University, the University of Minnesota, the Ford Foundation, Emprêsa Brasileira de Pesquisa Agropecuária (EMBRAPA), and the World Bank. We are also grateful to the Escola de Pós-Graduação em Economia (EPGE), Fundação Getulio Vargas (FGV), for releasing Dr. Brandão to be a visiting professor at the University of Minnesota.

For typing the various drafts we are grateful to Laura Heiberg, Carol Hansen, Cheryl Johnson, Elizabeth Wynand-Green, Fernando Leobons, and Sonia Moral. Cheryl Johnson and Jodie Kaden ably did the final version.

Finally, we are grateful to the American Agricultural Economics Association for commissioning the survey, and to successive Boards of the Association for their patience with what must have seemed like interminable delays.

The Theory, Empirical Evidence, and Debates on Agricultural Development Issues in Latin America: A Selective Survey

G. Edward Schuh and Antonio Salazar P. Brandão

Preface

This survey makes no pretense at being complete. The literature is too vast and published in too diverse a form to make any pretention of comprehensiveness. Some twenty countries are involved (the Caribbean countries, with the exception of Cuba, are excluded), at least three languages, and a rather large scholarly community. Historians, sociologists, and anthropologists have had as much to say about agricultural development as have economists.

The literature on Latin America is quite relevant to academic communities and students of agriculture worldwide. The problems of the region are important and challenging, and the perspectives on these problems of both the Latin American and North American scholarly communities (which tend to be emphasized in the survey) are unique and valuable. Of special interest, in our view, is the extent to which disciplinary distinctions are blurred among Latin American scholars, with many of them being some combination of historian, sociologist, economist, and political scientist. This often provides a particularly rich and, especially for North American scholars, rewarding insight into the various problems.

We set for ourselves an ambitious goal when we began this undertaking—to relate advances in development thought to empirical research and to development policy. At the same time we wanted to assess whether developments in the theory evolved from problems of the region, whether empirical research responded to or contributed to developments in the theory, and whether empirical research helped clarify policy issues. Hence, we focus on theory, on empirical research, and on policy.

Our bias in doing the survey is towards the English language literature and to that from Brazil. Without blurring the uniqueness of the Brazilian literature, we believe it is at least representative of the literature coming from Latin America. After all, this country covers nearly half of South America and contains about half of the South American population. Even in the case of the Brazilian and English language literature, however, we have been forced to be more selective than we really would have liked. For example, our review of the literature and its interpretation admittedly reflect our own personal biases. Similarly, we recognize that we have not done complete justice to the literature on structuralism and dependency theory.

Important journals in which the Spanish literature can be found include *Estudios Rurales Latinoamericanos*, published in Colombia; *El Trimestre Económico*, published in Mexico City but with a regional emphasis; *Cuadernos de Economía* from the Catholic University in Chile and *Estudios de Economía* of the University of Chile, as well as the research papers series at Escolatina in Santiago, Chile; *Económica* in Buenos Aires and the journal of the faculty of economics at Cuyo (Mendoza); *Economía y Demografía* for México; and the *CEPAL Review*, formerly known as the *Economic Bulletin for Latin America*. We draw on some of the literature from these sources but our coverage is by no means complete. Important Portuguese language journals include *Revista Brasileira de Economia*, Fundação Getulio Vargas; *Pesquisa e Planejamento*, Ministry of Planning; *Estudos Econômicos*, University of São Paulo; *Revista de Economia Rural*, published by the Brazilian Society of Rural Economics (SOBER); and *Relatório de Pesquisa*, published by the São Paulo State Secretariat of Agriculture and Supply. Sources for much of the literature on structuralism and dependency theory can be found in de Janvry [1981a] and in other sources cited later in this paper.

Our apologies are due to those we have neglected. We hope our goal of bringing some of the stimulating and original ideas from Latin America to those less familiar with the region is sufficient reward for any personal oversight or neglect. More importantly, perhaps, our biases and omissions will motivate others to help fill in the gaps, set the record straight, and provide different perspectives.

Chapter I. Introduction

1. Historical Background on the Region[1]

Latin America is similar to the United States, Canada, and the Caribbean in that all are cultural off-shoots of Europe. However, Latin America differs from other parts of the hemisphere in that it grew from a different branch of Europe, in a very different manner, and with little contact with the other parts until this century.

Latin America as we now know it was visited, conquered, and settled by Europeans in the century *before* the first landing at Plymouth Rock. The Europeans who went to Latin America went in the name of the kings of Spain and Portugal, and took with them the mental and social structures of monarchy and feudal organization. Moreover, they established colonial regimes that were politically, economically, culturally, and religiously dependent upon the mother countries. Colonization was a state enterprise, carried out directly in the case of the Spanish territories and by delegation in the case of the Portuguese ones.

In contrast, the first colonists on the eastern seaboard of what is now the United States came from England and northern Europe. Many came in protest against their mother countries, and often to escape religious persecution. Rather than Catholicism they brought the Protestant ethic, which emphasized individual responsibility, work, and material success as signs of divine predestination.

Despite the later colonization of what is now the United States, the colonies located there obtained their political independence in the late eighteenth century. In contrast, the revolutions of independence in Latin America did not come until the early nineteenth century. Moreover, the structures of economic and cultural dominance by Spain and Portugal left a heavy mark, although change did take place through trade; through large British and French (and later on, German) commercial, financial, and cultural influences; and through significant transfers of land assets from people born in Spain to people born in the colonies.

The new political constitutions for the emerging Latin American countries were very much in the spirit of the French Revolution, and many of them were modeled after the U.S. Constitution. The extent to which countries in the region have been governed democratically has varied a great deal over time. In the 1970s, for instance, there was a large number of authoritarian or military governments. In the 1980s, there has been a shift back to democratic governments.

The conquerors of Latin America miscegenated extensively with the black slave population (particularly in Brazil, Cuba, and the Dominican Republic). This also was in marked contrast to the United States, where the power of the state was used to maintain segregation and with it a greater racial purity. Exceptions in Latin America include Argentina and Costa Rica. Miscegenation with the indigenous Indian population was also extensive in the region, with the exception of Argentina and to a lesser extent Uruguay and Chile.

Colonialist dependence continued in Latin America even after political independence was obtained. The new countries participated in the international capitalist system by selling raw materials and importing manufactured goods. But the exchange later shifted from Spain and Portugal toward England and France, with England in particular taking measures to preserve this pattern of trade. Still later, the trade pattern would shift toward the United States, with many Latin American observers viewing these new trade patterns as continued dependency relationships.

McGrath [1973] noted that during the early stage of independence in Latin America, when there was little contact between the United States and the nations to the south, Latin American political and literary leaders voiced strong admiration for the new democracy to the north that had shown the way to freedom. He noted that the Monroe Doctrine of 1832 was well received by them, and that they appreciated the promise of help against European intervention. In fact, their disappointment was that help was not given in the face of concrete interventions.

Perspectives eventually changed, however, due in part to different patterns of development and in part to differences in relationships between the United States and the countries of Latin America. In the early 1800s there was not much difference in the levels of material well-being between the two regions. Cultural and artistic manifestations were richer in Latin America. The first university in the New World was located in Latin America, not the United States.

But in contrast to Latin America, the United States benefited from the Industrial Revolution between 1820 and 1910. Per capita income in the United States was soon double that of Latin America.[2] By 1970 it was eighteen times greater, with the gap growing wider.

Eventually, the United States began to appear as a predatory threat to the countries of Latin America. In addition to the growing economic disparities, specific actions by the United States dramatically changed its image. The Mexican War and 1846–48 land grab had a profound effect on both Mexico and other countries in the region. Growing U.S. interest in intervention in the Caribbean burst into the Spanish-American War. There was the "taking of Panama" by means of an imposed treaty, and the "big stick" policy of "dollar diplomacy" and armed intervention which marked relations with Mexico and the Caribbean. A famous Mexican *dicho* (saying) describes the new perspectives quite well: "Poor little Mexico; born so close to the United States and so far from God!"

New processes slowly began to work in Latin America, however. The application of modern medical technology toward the end of the nineteenth century led to a phenomenal population explosion which is still evident today in most countries in the region. The Industrial Revolution began to have an effect in major urban areas at about the same time, and massive rural-urban migrations began. Political concerns focused increasingly on economic matters, with an emerging concern for economic and social justice both within individual countries and between the countries of the region and the advanced countries, especially the United States.

In the last decade of the 19th century and the early decades of the 20th century there were the beginnings of labor unions and of Socialist parties of the Left. After the Russian Revolution there was a growing Marxist-Communist influence in the region—an influence which has had a greater impact on socio-economic thought in the region than it has had in the United States. Universities grew in numbers and size and became the nerve centers of political awareness and agitation. And of course there was the Mexican Revolution, which has had a significant influence at least on the rhetoric of the region.

Franklin Roosevelt launched the "Good Neighbor Policy" in the 1930s in an attempt to reverse the growing alienation between the United States and Latin America. United States involvement in World War II caused its interests to be focused elsewhere, however, and this policy was soon forgotten. But World War II marked the emergence of a serious concern on the part of U.S. scholars with understanding the culture and the socioeconomic problems of the region, a concern that was aided and abetted by the emergence of the United States as a world power and its growing economic involvement abroad.

There followed a period of neglect of Latin America on the part of the United States, as it first engaged in the reconstruction of Western Europe and its wartime enemies, and later became caught up in a Cold War with the Soviet Bloc. Fidel Castro's takeover of Cuba in 1959 and the imposition of a Communist government some ninety miles off the coast of Florida quickly refocused U.S. attention on the region. The ambitious Alliance for Progress, whose terms were set at Punta del Este in 1961, was the result. Two important assessments include Perloff [1969] and Levinson and Onis [1970]. But the United States soon became distracted by another war in another part of the world, and both the idealism and any semblance of a cohesive economic and political policy towards the region were eventually abandoned. Only the coming to power of another leftward leaning government in Nicaragua, together with the emergence of major international debt crises in Latin America in the early 1980s, has brought the region back to the attention of policy makers, academics and lay people.

2. Economic Background of the Region

Summary data which characterize the countries of Latin America some fifteen years into the approximately forty-year period covered by this summary are presented in Table 1. (The goal was to strike an approximate midpoint in the period.) The individual countries vary a great deal in their stage of development, in their resource endowment, and in the development problems they face. Despite their common Latin heritage, they also vary a great deal in their character or personality. And despite the common image of the region as Spanish-speaking, Portuguese-speaking Brazil has approximately half the population of South America. The population of Haiti speaks French. This great diversity in the region makes easy generalizations rather hard to come by.

Table 1. Area, population, per capita income, and GDP of
Latin American countries in the 1960s

Country	Area (km²)	Population in 1967 (000)	Population densities (per km²)	Per capita income[a] 1960 ($US)	GDP[b] 1960 ($mil.)
Argentina	2,766,656	23,031	8.3	868	17,947
Bolivia	1,098,581	3,801	3.5	165	609
Brazil	8,511,965	86,580	10.2	289	20,305
Chile	741,767	9,010	12.1	658	5,128
Colombia	1,138,338	19,215	16.8	336	5,203
Costa Rica	50,900	1,536	29.7	471	568
Cuba	114,524	8,033	70.2	–	–
Dominican Republic	48,442	3,889	81.0	253	766
Ecuador	270,670	5,508	20.3	304	1,312
El Salvador	20,935	3,151	147.3	280	698
Guatemala	108,889	4,717	43.0	291	1,094
Haiti	27,750	4,515	165.0	94	390
Honduras	112,088	2,445	21.8	208	406
Mexico	1,969,300	45,611	23.6	518	18,688
Panama	75,650	1,329	17.5	474	484
Nicaragua	130,000	1,778	14.0	243	359
Paraguay	406,752	2,121	5.1	255	450
Peru	1,280,219	12,385	9.6	338	3,387
Uruguay	186,926	2,783	14.6	853	2,124
Venezuela	898,805	9,352	10.4	809	5,933
Latin America	20,019,000	250,950	12.0	–	–

Sources: Economic Commission for Latin America [ECLA, 1967]; Instituto
Interamericano de Estadísticas [1969]. With permission.

[a] An average of two income estimates in equivalent purchasing power, with one
estimate based on average relative prices in Latin America and the other based on
relative U.S. prices.

[b] Exchange rates were the same as those used in constructing the estimates of per capita
income.

The countries in the region range from giant Brazil, which is larger than the
continental United States by a Texas and covers roughly 43 percent of the land
area encompassed by the twenty Latin American countries, to tiny El Salvador.
Brazil also has the largest population, almost 35 percent of the total, with Panama
having the smallest. Population densities also vary a great deal, ranging from
sparsely populated Bolivia and Paraguay to densely populated El Salvador and
Haiti. Large areas of Brazil are also sparsely populated and as yet unopened to
human settlement.

The three largest economies of the region as measured by gross domestic
product (GDP) are those of Brazil, Mexico, and Argentina, in that order. The

total GDP of these countries are roughly three times the total of the three countries having the next largest gross domestic products—Venezuela, Colombia, and Chile.

The level of development, as measured by per capita income, varies almost as much as the relative size of the countries. The high-income countries in 1960 were Argentina, Uruguay, and Venezuela, followed by Chile, Mexico, Panama, and Costa Rica. The poorest country by far was Haiti, followed by Bolivia and Honduras.

The use of geopolitical boundaries is a bit misleading in these comparisons, however. The poverty-stricken Northeast of Brazil, with its some 30 million people, had a per capita income of only slightly over US $100 in 1960. The level of poverty was of the same order as Haiti; and if the region were treated as a country, only Mexico and the rest of Brazil would be larger in population. Similarly, the per capita income of the state of São Paulo in the South is substantially above that of even other states in the South of that country.

Some countries such as Argentina, Brazil, and Mexico are classified as semi-industrialized, and prior to their difficulties in the 1980s Brazil and Mexico were usually classified among the Newly Industrialized Countries (NICs), the handful of countries that had pulled themselves up by their economic bootstraps during the 1960s and 1970s [McMullen, 1982]. The state of São Paulo in Brazil in particular has the largest industrial park in Latin America, and has all the appearances of an advanced country. Changes in economic fortunes have also changed how countries rank in per capita terms. Oil-rich Venezuela prospered greatly in the 1970s, and Brazil benefited from an extraordinary economic boom during the years 1968-74. Argentina and Uruguay have stagnated, on the other hand, as have some other countries in the region. The international debt crisis and the severe world economic recession of the early 1980s hit Latin American countries especially hard. As of 1984 per capita income levels were approximately what they had been in 1975, indicating almost a decade of economic stagnation.

The population of Latin America, which now exceeds 300 million, represents about 7 percent of the world total. In 1950-60 the annual growth rate of this region's population was by far the highest for any regional grouping in the world:

Region	Annual percentage	Region	Annual percentage
Latin America	2.8	U.S. & Canada	1.8
Africa	2.1	Soviet Union	1.7
Asia	1.9	Europe	0.8

As always, the averages conceal a great deal. Growth rates for individual countries ranged from 1.5 to 3.5. In 1960-70 the population growth rate of Latin

America increased to 2.9, still the highest rate in the world. However, some seven countries experienced declines in their population growth rate during this decade. In the period 1973-84, the population growth rate for Latin America had declined to 2.4 percent, while that for Africa had increased to 2.9. The growth rate for Asia was 2.0 percent in this period, but if China is excluded it was 2.4 percent.

More recent data on the Latin American countries are presented in Table 2. In taking stock of the stage of development of the region it is important to note that as of 1985 only Haiti was classified by the World Bank among the thirty-five low-income countries in its classification. All the rest were classified among middle-income countries, with the last seven countries in the table (starting with Chile) classified among the upper middle income countries and all the rest among lower middle income countries.

The share that agriculture makes up of GDP was less than 30 percent in all cases, with the share often being relatively small. The share that the agricultural labor force makes up of the total labor force is significantly larger, with quite a number around 50 percent or greater. The comparison of these two shares suggests quite low labor productivity for agriculture. It should be noted in making this comparison, however, that the estimates of agriculture's share of GDP tend to be biased downward due to the discrimination against agriculture (with farmers' prices below world market levels) and the high protection afforded the manufacturing sectors (prices of manufactured products above world levels). Taking this measurement bias into account still leaves labor productivity in agriculture quite low.

The share that the agricultural labor force made up of the total labor force declined significantly in most countries of the region from 1965 to 1981, despite slow rates of economic growth in the 1970s. In a similar fashion, the urbanization of the regional economies proceeded at a rapid rate.

It is interesting to note that food production per capita declined in nine of the twenty countries between 1974-76 and 1981-83. This poor performance of food production tended to be associated with the low-income countries of the region, in which population growth rates tended to be high.

Finally, the data in Table 2 also show that agricultural and forest products play a significant role in the total merchandise exports of the region. However, as of 1982, the share that food imports made up of total merchandise imports tended to be quite low, and was for the most part declining over time.

Economic activity in Latin America has historically been oriented very heavily to its natural resource base.[3] Almost half the population has been engaged in the primary sector, earning its living by farming, mining, forestry, or fishing. The importance of natural resources is even more evident in the trade mix of the region with the rest of the world. Ever since the Spanish Conquest the region has

Table 2. Selected data on Latin America[a]

Country	GNP per capita Average annual growth rate (%) Dollars 1983	GNP per capita Average annual growth rate (%) 1965-83	Average annual growth rate in agriculture 1965-73	Average annual growth rate in agriculture 1973-83	Share agriculture is of GDP 1965	Share agriculture is of GDP 1983	Share agric. labor force is of total 1965	Share agric. labor force is of total 1981	Fertilizer consumption (in 100 g of plant nutrients/ha arable land) 1970	Fertilizer consumption 1982	Ave. index of food production per cap. (1974-76=100) 1981-83	Share of other primary commodities of total exports[b] (%) 1965	Share of other primary commodities 1982	Food imports, share of total imports (%) 1965	Food imports 1982	Average annual growth of population (percent) 1973-83	Urban population as share of total population 1965	Urban population 1983
Haiti	300	1.1	-0.3	0.7	-	-	77	74	4	51	90	-	-	-	26	1.8	18	27
Bolivia	510	0.6	3.5	1.5	21	23	58	50	13	8	87	3	-	20	12	2.6	26	43
Honduras	670	0.6	2.4	3.3	40	27	68	63	160	137	107	90	87	12	10	3.5	26	38
El Salvador	710	-0.2	3.7	4.0	36	27	59	50	1048	830	91	81	55	16	18	3.0	39	42
Nicaragua	880	-1.8	2.8	1.4	25	22	57	39	184	186	74	90	91	13	12	3.9	43	55
Costa Rica	1020	2.1	7.0	1.7	24	23	47	29	1086	1134	88	84	71	9	9	2.4	38	45
Peru	1040	0.1	2.0	0.9	15	8	50	40	297	266	82	54	17	17	18	2.4	52	67
Guatemala	1120	2.1	5.8	2.3	-	-	64	55	224	498	102	86	69	11	6	3.1	34	40
Dom. Rep.	1370	3.9	5.9	3.2	26	17	64	49	354	353	95	68	82	25	16	2.4	35	54
Paraguay	1410	4.5	2.7	6.0	37	26	55	49	58	39	109	92	-	4	13	2.5	36	41
Ecuador	1420	4.6	3.9	1.9	27	14	54	52	123	277	92	96	33	10	5	2.6	37	46
Colombia	1430	3.2	4.0	3.7	30	20	45	26	310	538	106	75	68	8	11	1.9	54	66
Cuba	-	-	-	-	-	-	35	23	1539	1726	127	92	-	29	-	0.8	58	70
Chile	1070	-0.1	-1.1	3.7	9	10	26	19	317	189	102	7	27	20	12	1.7	72	82
Brazil	1880	5.0	3.8	4.2	19	12	49	30	169	365	113	83	43	20	8	2.3	51	71
Argentina	2070	0.5	-0.1	1.5	17	12	18	13	24	31	112	93	67	7	4	1.6	76	84
Panama	2120	2.9	3.4	1.4	18	-	46	33	391	469	102	-	64	-	9	2.3	44	50
Mexico	2240	3.2	5.4	3.5	14	8	50	36	246	778	106	62	10	5	10	2.9	55	69
Uruguay	2490	2.0	0.4	1.5	15	12	18	11	392	376	106	95	67	10	7	0.5	81	85
Venezuela	3840	1.5	4.5	2.6	7	7	30	18	165	408	91	1	()	12	17	3.5	72	85

Source: World Bank, 1985a. With permission.

Notes: [a] Most data for 1983 are preliminary estimates.

[b] Other primary commodities excludes fuels, minerals, and metals.

been an exporter of primary goods: foodstuffs, fibers, forest derivatives, metals, and other minerals. Raw materials were essentially the only exports until well into the 20th century, and only in the last century has the share fallen below 95 percent of total regional exports [Grunwald and Musgrove, 1970, p. 1].

General economic development in the post-World War II period has been rather uneven both across countries and over time within the same country. Brazil and Mexico, for example, are as noted above usually included among the NICs. Yet in the 1980s both countries have experienced serious economic difficulties. For the region as a whole, the geometric growth rate for GNP was 5.1 percent in 1950-60, 5.5 percent in 1960-70, 6.6 percent in 1967-70 [ECLA, 1970a], and 4.15 percent in the period 1973-83 [World Bank, 1985a]. The data for the 1967-70 period were heavily influenced by the extraordinary growth rate of Brazil in this period. Except for the 1973-83 period, these growth rates may appear to be satisfactory. But the average masks the unevenness of the growth process in individual countries, and also falls far short of what might be expected from countries "catching up," and certainly far short of the experiences of Japan and the NICs.

Economic growth in the postwar period took place under the shadow of an almost continuous sequence of balance of payments crises.[4] The region has also been subject to high and unstable rates of inflation, with countries such as Argentina, Bolivia, Brazil, Chile, and Peru having chronic and almost pathological problems. We will see below that the balance of payments problems and the chronic inflation have had important influences on the shape of economic thought in the region, with important influences on economic policy relative to agriculture.

In addition to having low levels of per capita income and rather uneven rates of growth, most Latin American countries have rather highly skewed distributions of incomes, one of the important debates in the region that will be discussed in more detail later in the paper. An important aspect of the income distribution problem in most of Latin America is the very wide gap that separates the rural from the urban population. Some summary data from a study by the Economic Commission for Latin America [ECLA, 1970b] are presented in Table 3. These data are not entirely consistent across countries, nor are the concepts on which they are based exactly what one would want. However, the relationships shown appear to be adequate for present purposes.

The sharp differences in income between the agricultural and nonagricultural sectors are clearly demonstrated. Urban or nonagricultural incomes tend to be more than double the rural or agricultural average. The only exception is Argentina, where nonagricultural incomes are only slightly higher than agricultural incomes. But with the possible exception of Uruguay, Argentina is a unique case in Latin America. Land in that country was largely occupied by large *haciendas*, and

Table 3. Latin America: Rural and urban, agricultural and nonagricultural incomes in selected countries

Country and sector	Index of average income (rural average = 100)	Share of all income units	Income groups				
			Lowest 20%	30% below the median	30% above the median	15% below the top 5 percent	Top 5%
	(percent)		(percent)				
Venezuela[a]							
Rural	100	40.8	72.9	48.6	28.7	16.3	12.2
Urban	250	59.2	27.1	51.4	71.3	83.7	87.8
Large cities	274	45.2	10.9	34.7	59.3	73.2	76.2
Small cities	176	14.0	16.2	16.7	12.0	10.5	11.6
Mexico[a]							
Rural	100	44.2	68.7	54.7	34.5	21.5	10.7
Urban	231	55.8	31.3	45.3	65.5	78.5	89.3
Mexico							
Agricultural	100	43.7	68.2	56.3	26.6	26.4	20.7
Nonagricultural	198	56.3	31.8	43.7	73.4	73.6	79.3
Brazil							
Agricultural	100	45.4	62.2	65.1	34.5	17.3	12.1
Nonagricultural	273	54.6	37.8	34.9	65.5	82.7	87.9
Costa Rica							
Agricultural	100	50.0	76.4	80.3	23.8	16.5	19.6
Nonagricultural	184	50.0	23.6	19.7	76.2	83.5	80.4
El Salvador							
Agricultural	100	60.2	100.0	87.9	30.4	23.3	18.8
Nonagricultural	229	39.8	–	12.1	69.6	76.7	81.2
Argentina							
Agricultural	100	14.8	21.9	20.0	6.9	12.2	14.9
Nonagricultural	115	85.2	78.1	80.0	93.1	87.8	85.1

Source: ECLA, 1970b, p. 402. With permission.
[a] The Venezuelan and Mexican classifications differ somewhat. In Mexico the urban category includes all cities with 2,500 or more inhabitants. In Venezuela the dividing line is 5,000 inhabitants. About 3 percent of the population in Venezuela resides in towns of from 2,500 to 5,000 inhabitants. Small cities in Venezuela are those with 5,000 to 25,000 inhabitants, and large cities those with 25,000 or more.

Peron's prolabor policies contributed to early mechanization. Consequently, less than 15 percent of the total labor force in Argentina is engaged in agricultural activities, and within agriculture the man/land ratio is favorable.

Unfortunately, to the best of our knowledge, there are no careful studies of the differences in cost of living between rural and urban areas in Latin America, at least in terms of comparing a carefully standardized market basket of goods. However, our judgment would be that, if anything, the measured income differences understate the true income differences due to the disparity in public services and public inputs between the two sectors and to the unavailability of many goods and services in rural areas that are accessible to urban groups.

In any case, despite the unsightly *favelas* (slums) that dominate the large urban centers in Latin America, the bulk of poverty in most countries of the region is located among the rural population.[5] This fact of economic life is not yet fully reflected in economic and social policy.

The agricultural sector, given the discrimination it received at the hands of policy makers, has not been a dynamic force for development, but neither has it been the constraint to development it is often alleged to be. In the aggregate, per capita production of food products has tended to rise over the period (Table 2). Two points from the output data are of particular interest. First, total agricultural output increased only at about the same rate as population growth because of relatively slow rates of growth in such nonfood products as cotton, coffee, wool, and tobacco. Second, the growth rates for agriculture vary a great deal among countries, ranging from 0.7 percent per year for Haiti to 6.0 percent per year for Paraguay in the 1973-83 period.

By far the major part of the increase in agricultural output has come from extension of the area planted rather than from improvements in yields, especially in the first half of the post-World War II period. New areas were brought into cultivation, the agricultural labor force was still increasing absolutely in most countries, and inputs of capital still took the form of conventional capital such as buildings and machinery and equipment. Modern inputs were used on a relatively limited scale compared to other parts of the world (excluding Africa), and the so-called Green Revolution—of such force in other regions—had had a relatively limited impact in the aggregate.[6] To the contrary, yields and other measures of agricultural productivity were quite low by international standards. Although the situation was changing rapidly in some countries such as Argentina, agriculture was still very much a natural resource-based activity in most of the region, with biological technology in particular still playing a relatively small role.[7]

Valdés and Muchnik de R. [1984] have provided an excellent overview of trends in consumption, production, and trade of agricultural products in Latin America over the last two decades. Over that period, they found that food consumption in the region grew at an annual rate of 2.8 percent, a rate similar to its population growth rate. During the same period, total animal feed use of grain grew at a rate close to 5.4 percent due to the rapid increase in the consumption and exports of meat and dairy products.

Some writers maintain that the nutritional status of the lowest income groups in Latin America has worsened [Caballero and Maletta, 1983, for example]. Valdés and Muchnik de R. [1984] note that the average caloric intake in Latin America rose moderately from the early 1960s through the late 1970s, from 2,432 calories per day in 1961/65 to 2,591 per day in 1979/81, with calories originating from animal sources increasing from 403 to 455 calories per capita per day. They also note that the rate of protein intake in the region had been quite stable, and that the average protein supply per capita had been greater than the minimum recommended level in each of the countries in the region. The regional average was close to the world average, although considerably less than that of developed nations.

Valdés and Muchnik de R. cite studies by Mohan, Garcia and Wagner [1981], Urrutia [1981], and Castenado [1984] which provide evidence that makes it difficult to accept the argument that the nutritional status of the lowest income groups in the countries those authors studied had worsened. In fact, the evidence these authors provide is to the contrary. As Valdés and Muchnik de R. note, however, the fact that malnutrition seems to be diminishing does not imply that it has disappeared.

These authors also note that there have been significant changes in the composition of the Latin American diet, which is gradually coming closer to the food patterns of more developed countries. These changes include:

> An increase in per capita consumption of wheat and rice, and a significant decline in the consumption of maize and other cereals typical of the traditional regional diet, with cereals as a whole continuing to account for approximately 40 percent of total calories.
> A great acceleration in the per capita consumption of vegetable oils, with consumption of fruits and vegetables increasing somewhat.
> A large decline in the consumption per capita of roots and tubers (cassava, potatoes) and dry legumes (beans), typical staples of the traditional Latin American diet.
> An increase in the per capita consumption of meats (especially poultry), eggs, and dairy products. This caused an increase in use of feed grains such as maize.

These changes in consumption patterns are attributed to five factors: increases in per capita income and the effect of Engel's Law; urbanization, which favors the consumption of more storable and processed foods with a lower cost of preparation (i.e., wheat derivatives, rice, and vegetables), but not of typical foods such as cassava, potatoes, *quinoa* (in Andean countries), and dry legumes; growing participation of women in formal labor markets, which raises the opportunity cost of preparation time in the household; the modification of relative prices as a result of technological changes, with poultry and rice being important examples; and price policies which have distorted price relatives, as in the case of wheat in Brazil.

With respect to trends in food and agricultural production, Valdés and Muchnik de R. [1984] note the following:

> Between 1961 and the middle of the 1970s, food production in Latin America grew at an annual rate of 3.2 percent, which is 0.5 percent faster than its population growth. Latin America in this period had the fastest growing food production of all the developing world.
>
> This situation changed radically in the second half of the 1970s, when food production in the Third World as a whole accelerated while in Latin America it diminished sharply from 4.2 percent annually for 1961/70 to only 1.7 percent in 1971/80.
>
> The main difference between the source of the food production increase in the 1960s and the 1970s was in the expansion in cultivated area. During the 1960s it expanded at an annual rate of 2.7 percent, while yields increased 1.5 percent. In the 1970s, the increase in cultivated areas diminished to 0.6 percent, and the rise in yields went down slightly to around 1 percent annually.
>
> On an average, gross value of agricultural production per capita in Latin America went up 0.8 percent annually in this twenty-year period. Four countries (Brazil, Colombia, Guatemala, and Paraguay) had annual growth rates in the Gross National Farm Product (GNFP) that were above 4 percent. At the other extreme, six nations (Chile, Haiti, Honduras, Panama, Peru, and Uruguay) had annual growth rates lower than 2 percent in their GNFP.
>
> Livestock production rose at a faster pace than crop production (around 3.3 percent annually), and poultry and egg production were the most dynamic (9.3 and 5.1 percent, respectively). Beef production had the slowest growth rate (2.1 percent annually), lower than the population growth rate.

With respect to agricultural trade:

> Agricultural exports still accounted for more than 50 percent of total income from goods and services exports in Argentina, Brazil, Colombia, Costa Rica, El Salvador, Guatemala, and Dominican Republic.

More than 50 percent of the most dynamic agricultural exports of the region go to industrialized nations, which are by far their main markets. More than 70 percent of all Latin American farm exports are sold to industrialized countries, and only 7 to 9 percent are exported to other nations of the region.

In 1973/77, 80 percent of food exports originated in the Southern Cone (Argentina, Chile, Paraguay, and Uruguay), another 14 percent in Tropical South America, and only 5 percent in Mexico and Central America.

Approximately 70 percent of total agricultural and livestock imports in Latin America came from industrialized nations, and another 26 to 28 percent came from the region itself.

Cereals have been a dominant group in total regional imports, with wheat rating first, followed by maize and cereal preparations.

In concluding this section it should be noted that Latin America still has vast, almost untouched hinterlands. These are made up of jungle and tropical forests as well as open areas such as the *campo cerrado* in Brazil and the *llanos* in Colombia. Given the present state of technology, however, the immediate value of these abundant lands for agriculture appears to be doubtful, although the expansion of soybean production on the *cerrado* soils of Brazil this last decade shows that conditions can change. In general, however, even if the problem of access could be resolved and an adequate road network and other infrastructure created, much technical progress would still be required before many of these lands could be utilized for agricultural production at reasonable costs. Hence, technological change will have to be the major source of growth in the future, with a shift in the policy emphasis away from land to price, technology, trade, and exchange rates.

3. Previous Surveys of the Literature on Agricultural Development in Latin America

There has been a number of partial surveys of the literature on Latin American agricultural development, and there are two important contemporary bibliographic sources. Perhaps the most notable of the latter is the Land Tenure Center (LTC) Library at the University of Wisconsin. This library for some years made a serious attempt to assemble all the bibliographic material on Latin American agriculture. It divulged information on new source material in the form of periodic accession lists and assembled bibliographies on special topics such as "Agrarian Reform in Brazil" and related subjects. The Center also published current papers on Latin American agriculture.

In the mid-1970s the International Center for Tropical Agriculture (CIAT) in Cali, Colombia, published annotated bibliographies on the Latin American literature. For an example, see Centro de Documentación Económica para América

Latina [CEDEAL, 1976]. This material was assembled by CEDEAL. This service was soon discontinued, and the program of the LTC has been scaled down significantly (and its effort shifted to Africa) due to lack of funding. Computerized bibliographies and data banks have replaced analytical documentation centers.

An important continuing source of bibliographic material on Latin America is the *Latin American Research Review* (published by the Latin American Studies Association), which contains many synthesis and review articles on specific topics. Other continuous journals in the English language include *Journal of Latin American Studies*, published by the Cambridge University Press in England; *Inter-American Economic Affairs*, published by Inter-American Affairs Press in Washington; and *Journal of Interamerican Studies and World Affairs*, published by Sage Publications, Inc., University of Miami. For insights into work by scholars of the Soviet Union on Latin America see Blasier [1981] and Hough [1981].

Another general bibliographic compilation was made by Utah State University [Daines, Le Baron, and Whitaker, 1971; and Le Baron et al., 1973a, b, c] under a contract with the United States Agency for International Development (AID).[8] This reference gives major emphasis to irrigation and water management, but provides a good coverage of the economics of agriculture as well. By design this collection excluded material on land tenure/agrarian reform and agricultural credit. On the other hand it has a comprehensive listing of source materials, covering the various censuses, agricultural statistics, demographic and population information, national accounts, and other bibliographies—all organized on a country basis.

In the 1960s the Agricultural Development Council (ADC) sponsored a series of seven surveys of research on agricultural development in different parts of the world. One of these dealt with Central America and another with Brazil [Lombardo, 1969; Schuh, 1970a]. These surveys are somewhat more analytical than the others mentioned, and attempt to relate the literature to the major problems of the respective agricultural sectors they cover and to evaluate in a qualitative way the research produced up to the time of the surveys.

These surveys also provide some notion of the volume of literature available on Latin American agriculture, while at the same time giving some notion of the magnitude (and impossibility!) of our present task. The bibliography by Daines, Le Baron and Whitaker [1971] has over 3,000 entries, and that by Schuh [1970a] approximately 1,000. The annotations published by CEDEAL [1976] filled almost 400 pages of fine print in the first volume. In a rapid search the Agricultural Sector Library of the World Bank located 10,000 titles on agriculture in Latin America.

Two other features of the situation in Latin America make it difficult to access the literature. First, in some countries a great deal of the research is financed by government organs or semiprivate research institutions. The literature that results is typically of limited distribution. Second, libraries in Latin America have received very modest financial support. Consequently, their collections are not kept up to date and for the most part they are not very useful as sources of material.

A description of the research can be obtained from the two studies by Lombardo [1969] and Schuh [1970a]. The Lombardo study for Central America covered primarily the period 1953 to 1966, and included 230 entries. The distribution of entries by subject matter and country is presented in Table 4.

The original classification scheme for the Lombardo study listed fifteen major subject matter titles. However, no entries were recorded for "international agricultural trade," "values, attitudes, and motivations," and "rural industry." Other categories such as "agricultural finance," "population and agricultural labor," "agricultural mechanization," and "agricultural statistics and research methodology" had only a few items each.

Over two-fifths of the items were on "farm management and farm organization," and on "land and water use," with heavy emphasis on cost-of-production studies, and on land tenure, land reform, and settlement. One-third of the projects were dedicated to "economic and agricultural development" and to "supply and demand for agricultural commodities," with the majority discussing agricultural development or industry or crop studies. Seventy-six percent of all the items were in these four major categories.

The classification by countries also shows important differences. Thirty-seven percent of the items listed came from Costa Rica and 24 percent from Guatemala. A longer tradition in agricultural economics research and a larger number of agricultural economists in these countries were offered as explanations for this concentration.

Lombardo argued that research in Central America could be regrouped into two general strata judging by the projects' objectives and content. Although in his view most studies were totally lacking or deficient in economic analysis, he argued that they represented serious efforts to conduct organized, systematic research. Others, particularly those dealing with land reform issues, were in his view pursued primarily to present the authors' views and opinions on the matter. Nevertheless, he felt that a satisfactory base had been established upon which subsequent research could be built.

Lombardo also found a relation between research emphases and policy interests. For example, cost studies had been made in response to the need for information to implement price support programs. Similarly, after the Alliance for Progress was initiated in 1961, all of the countries had announced one or more five-year plans for developing agriculture; land tenure and its reform remained a

Table 4. Number of research items in agricultural economics and development for Central American countries as of July, 1966

Classification	Costa Rica	El Salvador	Guatemala	Honduras	Nicaragua	Panama	Other Central America	Totals	
								Number	%
Land and water use	12	3	11	4	4	5	6	45	20
Farm management and farm organization	38	1	3	–	1	6	1	50	22
Agricultural finance	2	1	1	–	–	–	–	4	2
Supply and demand for agricultural commodities	5	4	21	1	–	3	3	36	16
Agricultural marketing	1	3	3	1	–	3	2	13	6
Rural life and organization	2	2	7	2	–	–	–	13	6
Agricultural policy	3	–	2	2	–	–	2	9	4
Population and agricultural labor	1	–	–	–	–	1	–	2	*
Agricultural mechanization	–	–	–	–	–	1	–	1	*
Agricultural statistics and research methodology	1	1	–	–	–	–	–	2	*
Extension, education, and innovation	6	1	–	2	–	2	–	11	5
Economic and agricultural development	13	3	7	5	5	5	4	42	18
Total	84	19	55	16	10	26	18	228	100
Percent	37	8	24	8	4	11	8	100	

Source: Lombardo, 1969.
* Less than 1 percent.

burning political, social, and economic issue. Hence, land-use studies were relatively large in number.

Exceptions to this responsiveness of the research establishment occurred in the cases of "agricultural finance," "agricultural marketing," and "extension, education, and innovation." Lombardo noted that credit programs had been in operation for decades and extension services began in the early 1950s, yet there had been little research on these topics. Similarly, there was a growing awareness of the importance of orderly marketing and of the need to improve present structures and practices, but little research was available to serve as a guide to either private or public policy. Lombardo was especially concerned about the lack of studies on world and domestic trends and their implications for coffee, sugar cane, cotton, and bananas.

Schuh's comparable classification for Brazil is presented in Table 5. This classification was somewhat different than that of Lombardo since a given piece of research was classified under more than one subject matter title if it made a significant contribution in more than one area. The reason for using this approach was to give a somewhat better picture of the total research "mix" than is obtained if each item is classified according to only one subject matter area.

Schuh expressed surprise at the volume of research and research reports he found on the Brazilian agricultural sector. However, similar to Lombardo, he cautioned that it would be difficult to call many of the items included in the inventory "research," in the usual definition of the word. A large portion of the entries were rather superficial studies, with little depth and little analysis or hypothesis testing.

Schuh attempted to provide perspective on the literature by including a chapter which described the profession of agricultural economics in Brazil and the major institutions that served as its basis. He noted that until 1968 there was not one indigenous agricultural economist in Brazil with a Ph.D., and that as late as 1960 there were fewer than a dozen agricultural economists with the M.S., with most of those obtained some time earlier.

This situation was fairly typical in Latin America at that time. Moreover, general economic training in Brazil and the rest of Latin America had been little influenced by neoclassical economics, resting instead on the influence of the French distributionist tradition and a strong bent for institutionalism. The consequence was that much of what we would call agricultural economics research was done by historians, sociologists, and general social scientists. Much of the training in economics faculties was provided by lawyers and accountants. Many of the courses in economics faculties consisted of accounting and management of firms. Lawyers were knowledgeable not only about firms but about institutions.

Table 5. Distribution of research among subjects and regions, Brazil

	North	Northeast	Minas Gerais	Guanabara & Rio de Janeiro	Goiás & Mato Grosso	São Paulo	Paraná & St. Catarina	Rio Grande do Sul	National	Others	Total	%
Land and water use	–	20	9	–	4	9	–	6	44	4	96	7.1
Farm management and farm organization	2	19	85	5	12	65	3	29	21	11	252	18.6
Agricultural finance	–	10	10	–	–	6	–	6	7	3	42	3.1
Supply and demand for agricultural commodities	2	30	17	6	–	53	2	10	125	14	259	19.1
Agricultural marketing	–	23	24	5	1	33	–	–	34	11	131	9.6
International agric. trade	–	2	–	1	–	1	–	–	35	–	38	2.8
Rural life & organization	3	17	27	1	7	18	1	3	22	–	99	7.3
Agricultural policy	2	14	3	1	–	–	–	3	26	–	49	3.6
Population & agric. labor	–	10	4	1	7	10	1	3	45	2	83	6.1
Values, attitude, & motivation	–	1	4	–	–	1	–	5	–	–	12	0.9
Agricultural mechanization	–	–	–	–	–	2	–	–	3	–	5	0.4
Rural industry	1	3	3	–	–	1	–	–	7	–	15	1.1
Agricultural statistics and research methodology	–	3	5	–	2	5	–	3	9	–	27	2.0
Extension, education and innovation	–	3	17	2	1	7	1	15	8	–	54	4.0
Economic & agric. dev.	2	31	6	3	6	7	4	13	84	6	162	11.9
General background on agriculture, economy and people	2	11	1	–	1	1	1	2	13	–	32	2.4
Sum	14	197	215	24	42	219	13	98	483	51	1356	100.0
Percent	1.0	14.5	15.8	1.8	3.1	16.2	1.0	7.2	35.6	3.8	100.0	
Number of studies	11	140	154	21	33	155	10	61	356	36	977	
Percent	1.1	14.3	15.8	2.1	3.4	15.9	1.0	6.2	36.4	3.8	100.0	

Source: Schuh, 1970a.

During the decade of the 1960s and the early 1970s there were major institutional development efforts in Latin America in both economics and agricultural economics. The Ford Foundation and AID played major roles in these efforts, together with selected U.S. universities. The impetus for these developments came in part from the Alliance for Progress with its emphasis on plans and planning.

The magnitude of the development effort can be seen from the case of Brazil.[9] That country currently has some six M.S.-level graduate programs in agricultural economics patterned after U.S. training programs and that are reasonably well institutionalized, plus a number of less well-developed programs. It also has the only Ph.D. program in Latin America that is similar to a U.S. program, although this program is still in its incipient stages. The M.S. programs have trained well over 500 people at the M.S. level, and over twenty Ph.D.s have been trained domestically. Some eighty or more Ph.D.s have been trained abroad, with more still in the pipeline, albeit at a decreasing rate due to the debt crisis, the reduction in foreign aid to Brazil, and the ever higher cost of training abroad.

Compared to the stock of human capital in the mid-1960s, these are quantum increases. But Brazil is a large and diverse country, has a rapidly growing population of over 140 million people, and faces complex and serious economic problems. The number of Ph.D.s in the country is no more than double what would be found in a single agricultural economics department in many Land Grant Universities in the United States.

Schuh was also surprised at the limited amount of research done on the Brazilian agricultural economy by foreigners, despite the economic importance of that country, and the relative ease with which foreigners could do research there. Brazil's economic "miracle" of 1968-74 attracted considerable attention by foreign *general* economists, and gave rise to what the Brazilian press labeled a new generation of "Brazilianists." But U.S. *agricultural* economists have given very little attention either to that important and interesting country or to the other countries of the region—part of their general neglect of world agriculture.[10]

Turning to the classification of research, Schuh found that the geographic focus of the research reflected the location of the stronger research centers in the country. The Northeast, and the states of Minas Gerais and São Paulo received the greatest research attention, with the state of Rio Grande do Sul following in fourth place, but with substantially less research than the other three regions. A serious problem at the time of Schuh's writing was the lack of research on the important frontier areas which were producing an ever larger share of total agricultural output, and the lack of research on the vast Amazon basin.

By far the greatest research emphasis had been given to problems of a national scope, but Schuh cautioned that a large share of these studies were rather superficial in quality and contained neither the rigor nor analytical content of studies that focused on smaller areas. There were few high quality sectoral studies at that time.

When viewed from a subject matter standpoint, two areas stood out in the Brazil inventory: supply and demand analyses, and farm management studies. Schuh noted that much of the research classified as demand and supply analysis was for the most part situation reports or industry studies that tended to be basically descriptive in nature. A number of reasonably good supply studies had been made with the good quality data on the state of São Paulo, and the state of knowledge with respect to demand was progressing, despite the work still needed before a complete set of knowledge of these important behavioral parameters was available. The next grouping of studies in terms of relative importance was on agricultural development, marketing, and land and water use, with each in part reflecting important regional emphases.

The regional differences in research emphasis were interesting for what they revealed about the responsiveness of the research process to local problems. For example, two subject matter areas had received major emphasis in the Northeast of Brazil: supply and demand studies, and agricultural and economic development. The supply and demand studies reflected the important contribution to the demand literature by the Bank of the Northeast, plus the natural interest of that bank in industry studies as a basis for credit policy. The emphasis on agricultural development reflected the interest of the bank as a development institution.

Lesser, but still significant, emphasis had been given to agricultural marketing, land and water use, and farm management and farm organization. Schuh noted the almost complete lack of work on the economics of water use—in a region that had serious drought problems, and a lack of research on labor or on rural industry—in an area which had serious problems of excess labor on the land. Finally, there had been little or no research on education, extension, or innovation—in an area with very low educational levels, and which had stagnated at very low levels of technical efficiency.

The state of Minas Gerais had one of the most unbalanced research efforts of any of the regions considered. Almost 40 percent of the literature collected was on farm management and farm organization, perhaps a consequence of a large share of it being done at the Rural University of Viçosa (now the Federal University of Viçosa) with its natural emphasis on farming. Considerably less, but still a sizeable amount, was directed to marketing, rural life and organization, extension, education and innovation, and supply and demand studies.

An interesting comment on the research on the agriculture of this state was that it rated relatively high on rigor and strict adherence to the scientific method—a reflection of the new graduate program at the Federal University of Viçosa—but seldom tackled the important problems of the state's agriculture. This was probably a logical outgrowth of the research being in large part a product of theses in the M.S. program, which had both teaching and research objectives.

Research on the state of São Paulo was better balanced as well as more responsive to the problems of the state. Major attention had been given to farm management and to supply and demand studies, in that order. Somewhat less attention had been given to marketing. Beyond these three subject matter areas, the concentration was rather slim on any one subject matter area, but rather widely distributed.

São Paulo was the region which had received the most attention in terms of marketing research, perhaps a reflection of the city of São Paulo being a major marketing center for the country as a whole, and the state itself being by far the most developed state in the union. On the other hand, very little attention had been given to problems of international trade, although a major portion of Brazilian agricultural exports moved through the port of that state, nor was much attention given to price policy.

Research in the state of Rio Grande do Sul also had a heavy emphasis on farm management. However, in total, the effort was reasonably well-balanced. Secondary areas of emphasis were supply and demand studies, extension, education and innovation, and economic and agricultural development.

In providing an overall evaluation of the research at the time of his inventory, Schuh noted the general failure of supply and demand studies to estimate underlying behavioral parameters, and the general tendency of the research on agricultural development to be descriptive in nature, with little or no analysis or conceptualization. Research on land use and land tenure tended to be based on census data and usually described the nature of the size distribution of land holdings with little in-depth analysis. Similarly, research on international trade tended to be little more than market surveys in other countries (generally for coffee). There had been little evaluation of the impact on the agricultural sector of exchange rate policy, protective tariffs, or export licensing, and only a few penetrating studies of agricultural price policies. These were the marks of a nascent profession, in which the tools of modern economics and quantitative analysis were only beginning to be introduced and support for social science research was quite limited.

Chapter II. Agriculture in Theories of Economic Development

Most Latin American countries have either neglected their agriculture and/or discriminated against it rather severely during most of the post-World War II period. This was not the case during the liberal period of Latin America's history (1850-1930). Moreover, in the late 1970s and in the 1980s, many countries in the region were giving increased attention to their agriculture, first as a means of promoting economic growth, second as a consequence of stabilization and adjustment policies which shifted domestic terms of trade in favor of agriculture, and third as a means of earning foreign exchange to service international debt.

The discrimination against agriculture which prevailed for so long appears to have been rooted in the history of the region, in particular economic events and what they were perceived as foretelling for the future, and in the particular evolution of economic theory in the region. Whether economic events shaped economic theory, as Wesley C. Mitchell [1949] and the Marxists would argue, or whether economic theory shaped economic policy and in turn the particular development path taken by the region, we are not prepared to say, although our inclinations take us to the side of Wesley Mitchell.[11] What does seem apparent is that there has been a counterpoint or dialectic between the two, and that one cannot be understood in isolation from the other. Hence, what follows in this section is a cursory appraisal of the historical features of the development of the region that appear to have influenced economic thought as it has evolved in the postwar period. A review of trade policies and patterns of trade is presented in a later section.

At least two phenomena stand out in the historical evolution of the region as factors influencing not only the character and nature of development, but the development of economic thought as well. The first is the particular distribution of land holdings and institutional arrangements that was a result of, and evolved out of, the original colonization of the countries in the region. Contrary to the case of the United States, access to land was not provided on a broad basis in most Latin American countries. Rather, land was granted in large blocks to a relative few who were already privileged members of society.

This concentration of control over a major economic resource, plus the economic, social and political institutions which evolved out of the Spanish and Portuguese cultures, created—and has sustained over time—rigid stratification systems and bipolar class structures. To oversimplify a bit, at the one extreme are the elitist groups (typically referred to as the oligarchy) who are relatively few in number, but who have maintained a dominance over political, economic, and social power. At the other extreme are the large masses of small freeholders, sharecroppers, landless workers, members of Indian communities, tenants, and others who have little economic power and have for the most part wielded little political power. The middle class, which is the great stabilizer in most advanced countries, is relatively small and unimportant (Table 3, p. 18). Argentina, Chile, and Costa Rica currently have significant middle classes; and the middle class is growing rapidly in Brazil in response to economic development.

One does not have to be an institutional determinist to recognize the importance of institutions that determine the access to income streams and to the control over political power. Clearly, institutions do change in response to changing economic conditions [T. W. Schultz, 1968], and those same institutions can be altered on a piecemeal basis with something less than revolutionary means. But it is equally as clear that, in most Latin American countries, economic conditions have not been such as to induce significant and broad-based changes in institutions, nor have the appropriate piecemeal changes been accomplished by reformists at a sufficient rate to broaden significantly the distribution of political and economic power. This probably explains in part the emphasis given by Latin American social scientists to redistributive, as opposed to efficiency, issues.

We will review some of the literature which deals with this problem later. The important point for now is that the socioeconomic and political institutions in Latin America are very different from those in countries such as the United States, and that these institutions very much affect the access to income streams and the incentives to produce particular kinds of income streams in the respective countries. Moreover, the problems that have arisen as a result of these particular institutional arrangements have had an important and more general influence on economic and social science thought.

The second major factor affecting the nature of Latin American development in the postwar period, and the evolution of economic thought, was the series of major shocks imposed on the Latin American economies in the first half of the twentieth century.[12] Latin America came out of its colonial period with fundamentally an export economy or, more properly, a group of export economies, each oriented to Spain or Portugal and exchanging a small number of resource-based products for a relatively wide range of consumption and luxury goods. During the first century of its independence, and especially up to World War I, the countries of Latin America became more fully integrated into world markets. The trade sector expanded and underwent considerable transformation. The European countries began to specialize in producing manufactures for export, and required large imports of foodstuffs and raw materials. Latin America became an attractive source of supply, particularly as the demand for products such as sugar, coffee, and cocoa (previously luxury products) expanded in response to rising incomes. Products that had not previously figured in trade became important: meat and grains from the River Plate countries and bananas from Central America.

There was a major sorting out of comparative advantage in this period, in part in response to the free trade policies then prevailing, with important shifts in location of production and degree of specialization. The post-World War II dependence of certain countries on particular products—Brazil on coffee, Central America on coffee and bananas, Cuba on sugar, Argentina on meat and grain— was formed in the second half of the 19th century. In summary, the Industrial Revolution produced a much more specialized world economy into which Latin America was drawn as a supplier of large amounts of foodstuffs and raw materials.

This phase of development through integration into world markets and specialization in production was brought to an abrupt end with the Great Depression of the 1930s. Between 1930 and 1945 the region suffered a severe crisis as a result of the large shocks imposed by the Great Depression and World War II. Although these were not the first shocks as a result of fluctuations in industrial activities in the advanced countries, their magnitude was certainly different, as were the reactions they produced.

The trade crisis of the 1930s was especially serious. Demand for all kinds of raw materials fell drastically, with consequent declines in the price or quantity, or both, of almost every one of the region's products. Ten of the commodities included in the Grunwald-Musgrove study [1970], for example, experienced an average decline in real price of about 60 percent between the peak year of the late 1920s and the low point of the Great Depression. These declines provoked a massive external disequilibrium, and before balance was restored at a much lower level of trade, gold reserves had been exhausted and convertibility of currencies was suspended. The sharpest contraction in trade any place in the world took place in Chile, which suffered a decline of 85 percent in the value of its total trade (imports plus exports) between 1929 and 1932. The contraction exceeded 65 percent for seven other Latin American countries. For countries that were dependent on exports, these were enormous shocks to their economies.

The Second World War, which followed on the heels of the Great Depression, was not especially helpful. Demand for raw materials rose sharply, and although most prices were controlled during the war, there was a substantial recovery of export value. But the income generated by exports could not find an outlet. Only high priority goods such as fuel and capital equipment could be imported, and even these were largely limited to the needs of the export sector. Shortages of fuel were so severe that several million tons of coffee and grain were burned in Brazil and Argentina, respectively. (In the case of coffee in Brazil, this burning of coffee beans had certain price-enhancing values as well.) Thus, only one side of the reduced trade balance of the 1930s was restored, and the Latin American countries were, in effect, forced to enjoy a trade surplus and to accumulate sizeable reserves. These conditions contributed to a significant amount of import-substituting industrialization.

It is little wonder that as World War II terminated, few countries in the region were willing to revert to the old system of integrated world trade with specialization in the production of foodstuffs and raw materials. The roller coaster— comprised of the economic boom of the 1920s, the Great Depression, the Second World War, and the distortions brought about by the rapid shifts in the fortunes of individual countries—persuaded many Latin American countries that autarchic development and independence from the world capitalistic system would be a more appropriate path to follow. They were reinforced in this belief by the considerable amount of industrialization which they were able to accomplish on their own, particularly during World War II when foreign sources of supply were cut off.

It is interesting to note that economic shocks from the international economy in the early 1980s may in a similar way be inducing many countries in the region to shift to more broad-based development policies which give more attention to agriculture, substituting export promotion for import-substituting industrialization, and thus reintegrating themselves back into the international economy, in part through agriculture. The motivations for these changes in basic policy include the severe international debt crises many of these countries have experienced, huge realignments in exchange rates on the international scene,[13] especially in the value of the U.S. dollar, and the unprecedentedly high interest rates that occurred in the early 1980s.

Returning to the factors that gave rise to the drive for import-substituting industrialization, we note that economic factors were not the only driving force. Raúl Prebisch gave policy-makers both an economic model and an ideology that rationalized industrialization as the prescription for economic growth. Associated strains of economic thought influencing or reflecting the role of agriculture in economic development include the economics of ECLA, structuralism, dependency theory, and the monetarist-structuralist controversy. It is to these subjects that we now turn.

1. The Prebisch Thesis

With the exception of Keynes, seldom has one economist dominated the economic thought and policy of a region as has Prebisch in the post-World War II period of Latin American development. His influence also extended far beyond Latin America. Prebisch was an Argentine economist-*cum*-diplomat who was Executive Secretary of ECLA, and later Secretary-General of UNCTAD. Both positions gave him a ready forum for his ideas, but the economic conditions of the times and the stage of development of Latin America caused his ideas to fall on fertile soil in that region.[14]

The "Prebisch Thesis" has to do with a supposed decline in the external terms of trade of low-income countries and the deleterious consequences of that decline. Prebisch argued that "peripheral" countries had experienced (and in his view probably would continue to experience) long-run deterioration in their terms of trade with the center, that this deterioration led to their exploitation by the center countries, and that countries of the periphery should take measures to prevent this exploitation.[15] The Prebisch thesis also has to do with one of the oldest quarrels in economics: the quarrel between free traders and protectionists. In post-World War II Latin America, the Prebisch thesis was the main intellectual basis for import-substituting industrialization and the general neglect of agriculture by policy makers.

Prebisch first advanced his [1950] ideas on the basis of a computation of the British terms of trade in the period from 1870 up to the Second World War.[16] Those computations showed a decline in the terms of trade for those countries trading with Britain. The discussions which followed from this original seminal paper are some of the most confusing and, in some sense, unjoined discussions in the economic literature. There have been arguments over, among other things, what Prebisch really meant, how many Prebisch theses there really are, what the terms of trade really show, and whether the issue had to do with the balance of payments or the welfare of a country. The Prebisch thesis also spawned a new set of concepts that have dominated the North-South debate of recent years over the necessity of income transfers from the advanced to the low-income countries.

To gain a "flavor" of Prebisch's argument and its apparent plausibility—a factor which has led to the persistence of his ideas—first consider the original rationale he had for his thesis. In his 1950 paper, Prebisch pointed to Latin America as a region where "reality is undermining the outdated schema of the international division of labor" [p. 1]. The basis for his argument was his view that technical progress is more intensive in industrial than in primary production.[17] Therefore, in his view, the price of primary products should remain higher than those of industrial products. Consequently, the price relatives between industrial and primary commodities *should* show a steady decrease in favor of underdeveloped countries (in Prebisch's words, in favor of the *periphery* and against the *center* or advanced countries).

> Had this happened, the phenomenon would have been of profound significance. The countries of the periphery would have benefited from the fall in price of finished industrial products to the same extent as the countries of the center. The benefits of technical progress would, thus, have been distributed alike throughout the world, in accordance with the implicit premise of the schema of the international division of labor, and Latin America would have no economic advantage in industrializing. On the contrary, the region would have suffered a definite loss, until it had achieved the same productivity efficiency as the industrial countries. [p. 8]

Drawing on data from a 1949 United Nations study, Prebisch argued that these predictions of the "theory" did not correspond to real world facts. Instead, the terms of trade "moved against the periphery, contrary to what should have happened had prices fallen as cost decreased as a result of high productivity" [p. 9]. In his view, the reason for this disparity between the "theory" and the "facts" was that the increases in the incomes of entrepreneurs and productive factors in the industrial countries were greater than the decreases in costs resulting from higher productivity, while the reverse occurred in the peripheral countries. Consequently, even for such technical progress as would occur in the peripheral countries, a share of it would be transferred to the industrial countries [p. 10]. With this assertion, the basis was laid for later predatory and imperialist theories of trade and international relations.

Over the years Prebisch offered a number of different explanations for how this mechanism worked. In his original paper, he argued that the process could be understood by means of the trade cycles, "since the cycle is the characteristic form of growth of capitalistic economy, and increased productivity is one of the main factors of that growth" [p. 12]. He argued that during the upswing of the cycle, prices of primary products increased faster than prices of industrial products, while during the downswing they fell more. But in industrial nations, where competition is keener and laborers are better organized, part of the profits obtained from upswing price increases are absorbed by wage increases. On the downswing, the downward rigidity of wages impedes a price contraction in the advanced countries consistent with the real gain in productivity during the period. On the other hand, "the characteristic lack of organization among the workers employed in primary production (in the periphery) prevents them from obtaining wage increases comparable to those of the industrial countries and from maintaining the increases to the same extent" [p. 13]. The result, in his view, was that as price adjustments took place through international trade, the need for adjustment in the center was transferred to the periphery. The final result was that prices of primary products would decrease more sharply as a result of the contraction of demand, with the result that over time there would be a secular decline in the terms of trade.

In a 1954 comment on a paper by Gunnar Myrdal [1961], Prebisch [1961b] developed a somewhat different argument—one that is familiar to U.S. agricultural economists—and combined it with a protectionist argument. He argued on this occasion that "imports of primary products in developed countries do not increase generally in a degree compatible with a satisfactory rate of growth in underdeveloped countries" [pp. 277-78]. This he attributed partly to the low income elasticity of demand for primary products in developed countries, and partly to the increasing protection extended by these countries to their agricultural producers. Therefore, "imports of primary products in advanced countries tend to grow with much less intensity than the demand for industrial goods in underdeveloped countries" [p. 278]. For Prebisch, the only way to correct the problems arising from this disparity in the income elasticities of demand was to promote industrial production in the underdeveloped countries.

In his 1959a article, Prebisch presented a somewhat more formal treatment of the reasons for the so-called decline in the terms of trade which integrated the previous two explanations. He considered two countries, one mainly industrial

(A) and the other mainly primary-producing (B). A's industrial productivity is assumed to be three times that of B, and B's primary productivity three times that of A. In other respects (wage rate, trade, marginal productivity, population, and per capita income growth), both countries are assumed to be in equal conditions.

If there are no differences in income elasticities of demand between primary and industrial commodities, there will be no deterioration in the terms of trade against primary producers. But now assume that the income elasticity of demand for industrial commodities is higher than for primary commodities. In this case, Prebisch argued that if there is not perfect mobility of labor among countries, country B would have to transfer part of its labor force from a high productivity primary sector to a low productivity industrial sector. As a result, country B would, in his view, experience the pressure of surplus manpower, forcing unemployment and a decline in wage rates. As a result, B's export prices would fall, transferring income to country A.

Country A, of course, would experience an opposite effect. Therefore, in country B, wages would increase less than the increase in in-country productivity, while in country A they would increase more than in-country productivity. Furthermore, Prebisch argued that if industrial productivity in B is much below that of A, the level of wages in B will drop even more.

Thus, in Prebisch's view [1959a], the difference in wage created by differences in income elasticities of demand and further increased by differences in productivity would result in a transfer of real income from primary producing countries to industrial countries. To quote him:

> Therefore, the combination of disparities of income elasticities of demand and in technological densities put the periphery in a weaker position *vis-à-vis* the center, as regards the terms of trade. The center is in a better position to retain the fruits of its general increase in productivity because the increment in manpower does not need, as in the periphery, to press on occupations with a lower productivity ratio to the detriment of the wage level. In other words, general improvements in productivity tend to be fully reflected in the increment of the wage rate at the center, while at the periphery a part of the fruits of these improvements is transferred through the fall of export prices and the corresponding deterioration in the terms of trade. [p. 262]

To close his argument, Prebisch states his "working hypothesis":

> A higher rate of increase of productivity in export than in domestic activities, coupled with a rather weak industrialization

> process, may in the past have been powerful forces
> contributing to the deterioration in the terms of trade for some
> products. Further deterioration may occur in the future if
> efforts are concentrated on technical improvements in primary
> production without a vigorous development of industries and
> their technical advance, accompanied by a cautious policy of
> interference with the free play of international market forces
> to support the prices of important primary commodities.
> [pp. 263-264]

In this context it is useful to note the extent to which Prebisch's recommendations were a prescription for the path that development policy took in Latin America. Investments in the modernization of agriculture were to be avoided for they would only lead to a deterioration in the terms of trade.[18] The emphasis was to be on industrialization, accompanied by protectionist measures. Moreover, one can already see antecedents of the later demands on the part of the developing countries for commodity agreements to support the prices of primary commodities.

In a report prepared for the Secretary-General of the United Nations for the first United Nations Conference on Trade and Development [UNCTAD, 1964], Prebisch again dealt with these issues at length. Two new factors were introduced, however. Dwelling again on the importance of technical progress in the advanced countries, he pointed out that this leads to the substitution of synthetics for natural products, further weakening the market for primary products from the periphery and having a further deleterious effect on the terms of trade. This emphasis on technology led to a concern with technological dependence in the developing countries, and eventually to demands for access to the technology from the advanced countries.

UNCTAD I led to UNCTAD II, III, and IV. The U.S. and other industrialized countries got much the worse of UNCTAD I as Prebisch dominated the proceedings both intellectually and diplomatically. Later, Prebisch eventually molded the developing countries into a political force. The seminal ideas of Prebisch permeated the Integrated Commodities Program that came out of the Manila Conference in 1976. Harry G. Johnson's penetrating response [1967b] to UNCTAD I notwithstanding, the U.S. and other countries of the industrialized West have been on the intellectual defensive ever since.

The Prebisch thesis has given rise to a lengthy and extensive intellectual exchange, and has been challenged on both empirical and theoretical grounds. A useful selection from among this voluminous literature would include Baer [1962], Ellsworth [1956], Fink [1955], Flanders [1964a], Haberler [1968], Hyde [1963], Harry G. Johnson [1967b], Kindleberger, van der Tak, and Vanek [1956], and Rogge [1956]. For a collection of his work in Spanish, see Gurrieri, ed.

[1982]. The appraisal by Flanders is perhaps the most comprehensive and penetrating, so we draw on it here to help focus on the key issues.[19]

The first thing to note is that there is no such thing as *a* Prebisch model; rather, there are various Prebisch models—a factor that has led to some of the difficulty in joining the issues. But in each of the models there are two basic elements: an "instrument of exploitation" and an "instrument of correction." The instrument of exploitation is the secular decline in the terms of trade between primary products (export commodities of the LDCs) and the industrial products (export commodities of the advanced countries). The instrument of correction is industrialization of the low-income countries, aided and abetted by protectionism.

A synthesis of Prebisch's writings would suggest that the tendency for the prices of primary products to decline relative to the prices of industrial products can be explained by at least three factors:

(1) differences in market structures between the LDCs and the DCs;

(2) differences in technological densities (which we will define below); and

(3) failure of the periphery to exploit their presumed monopoly power in primary tropical commodities.

Clearly there are two sets of issues here. The first is the empirical question of how to measure the terms of trade, whether they have, in fact, declined for an individual country or group of countries, and what they reflect given that they have changed. The second is the theoretical validity of the explanations given for the decline if it has in fact occurred.

Let's consider first the empirical questions. A first point to note is that the terminal years chosen to measure the change in terms of trade is important in determining whether they rise or fall. This point seems obvious, but failure to recognize its importance has led to a great deal of mischief, even if unintended, on the part of proponents of both sides of the controversy.

A second issue is the coverage of the respective price indexes, and in particular, whether services should be considered. Prebisch and his followers have tended to include only the prices of goods. Kindleberger *et al.* [1956], however, argued that the terms of trade for goods and services would be a superior concept to that for goods alone. They cited the case of Norway, which at the time of their writing earned 40 percent of its total credits on current accounts from shipping, as a case in point, and noted that if the price of shipping behaves differently from the price of commodity exports, the merchandise terms of trade would be misleading as an indication of the basis for trade with other countries. The importance of tourism to a country like Mexico would cause the same point to be relevant.

Other complications under this same heading include the fact that changes in the composition of products traded can cause a change in the measured price index, even if the prices of none of the products included in the index have changed. Similarly, a rise in the quality of a product over time might be reflected in a higher price. Clearly, there may be no loss in welfare to the importing country from a rise in price under these circumstances.

Neither of these problems can be handled adequately by fixed-base index numbers with base-year weights. An index with current weights would overcome some of the difficulties. However, comparisons of two years that do not include the base year also involve quantity changes and therefore statements with respect to the pure movement in prices are not feasible.

It should also be noted that changes in the composition of trade and changes in the quality of the products traded are pertinent only when one is attempting to establish a link between the terms of trade and the welfare of the country. The effect of a shift in the external terms of trade on the balance of payments of a country would be the same, independently of the source of the change in price. It should also be noted in passing that this is another source of confusion in the literature on the Prebisch thesis. Arguments are often advanced on both sides of the controversy without explicit reference to whether it is a balance of payments problem that is being considered or a problem of national welfare. They are not the same thing, of course.

Prebisch's reaction to these discussions about whether the terms of trade did actually decline was that the situation of the periphery *vis-à-vis* the center is parallel to that of agriculture *vis-à-vis* industry within an economy, and indeed, within the great centers themselves. "There, too, relative prices are following a downward trend." But, "no one has attempted to deny the existence of this trend, or to belittle its importance, on the grounds that have been adduced in connection with the corresponding deterioration in the peripheral regions, namely, the statistics do not reflect the improvement in the quality of industrial products, or that the ratio cannot be accurately calculated on the basis of price indices." The measures taken to defend agricultural relative prices on the home market "have the merit of affording a concrete demonstration of the importance attached by the world centers to the decline in the prices of primary commodities, and at the same time indicate a possible way of solving the problem at the international level" [Prebisch, 1963, pp. 78–88].

Let's now consider the various arguments for the declines in the terms of trade. The first is the differences in market structure between the periphery and the center. Prebisch argued that in the center the gains from increases in productivity are not reflected in lower prices of goods since they are absorbed by the workers in the form of higher wages. This he attributed to the power of labor unions which cause a downward stickiness of both wages and prices. The markets in the periphery, on the other hand, were alleged to be more competitive in nature, so the downward stickiness does not exist.

Fink [1955, pp. 71-72] argued that the solution to this problem was to devalue the currency:

> Whether in the industrial centers (assuming productivity increases) money wages are stable and prices fall, or prices are stable and money wages rise should make no difference in the terms of trade except perhaps in the very short run. . . . Since foreign exchange values are inversely related to the domestic currency costs of production of internationally traded goods and services, then lower price levels in the industrial centers would have required further adjustments of the peripheral countries' currencies—in short, more devaluation.

We cite this reference because it illustrates one of the frequent confusions in the literature between a balance of payments problem and a welfare problem. Under rather general conditions (which may not prevail in the individual instance!), a devaluation would eventually improve the balance of payments. But a devaluation involves a loss in national welfare, since the devaluing country has to export more of its domestic resources to import a given amount of imports.

Flanders [1964a, pp. 310-311] is more penetrating:

> (Prebisch's) argument seems to be based on the factor-price equalization theorem, but it is a fallacious and naive interpretation . . . as he does elsewhere, Prebisch seems to be identifying wages with personal income, that is, assuming that there is only one factor of production in the world, labour. If that is the case, however, the factor-price equalization theorem is not relevant and the equalization of income throughout the world is not an . . . "implicit premise of the schema of the international division of labour." Alternatively, if there are two factors of production, the theorem is valid only when both countries produce both commodities. This is a difficult assumption to make when one country is the periphery and the other is the centre.

Flanders [1964a, pp. 312-313] further notes:

> There is a peculiar asymmetry here. The rigidity of wages in the periphery, if it existed, would result in a decline in employment. . . . Thus, income would be decreased in the periphery by means of unemployment rather than by means of lower prices and real wages. This reasoning is unobjectionable, but surely it should be applied also to the centre.

Prebisch's argument in relation to the differences in technological densities is confusing, in part because he did not make explicit his assumptions relative to production. Flanders [1964a, pp. 313-315] points out that he was making use of Graham's "notion of a list of products, marked in order of the degree of comparative advantage." Graham's results assumed the existence of a unique factor of production (labor), and production proportional to the quantity of labor utilized. Under these assumptions, the products that have a comparative advantage will be chosen for which the ratio of the average products of labor (average product in developing country/average product in developed country) is greater than or equal to the wage ratio. Therefore, an increase in productivity in the developing countries will cause a transfer of income to the rest of the world by causing the wage ratio (and, consequently, the terms of trade) to decline. The transfer of income is measured by the differences between the wage ratio and the productivity ratios of exported products. Moreover, the transfer will be higher the higher are the differences in the productivity ratios in the periphery. If the technology is dense (that is, the productivity ratios are close to each other), the decline in the terms of trade (and the associated income transfer) will be small.

The third component of the Prebisch models deals with the differences in income elasticities of demand for industrial and primary products. Since in his view the income elasticity of demand for primary products is lower than for industrial products, the desires of developing countries to grow at the same rate as the developed countries will create a balance of payments problem for the developing countries. To correct for such a disparity in elasticities, the classical mechanism — and Prebisch recognizes this — would be devaluation of the currency. However, in his view developing countries should take advantage of the "monopoly" position which he believes they have, or could have, in world markets instead of letting competitive forces drive the price of their exports down to a competitive equilibrium. Moreover, he argued that if the developing countries let the market mechanisms work freely, they will experience an income transfer to the rest of the world that is greater than what is socially optimum. This social optimum, in Prebisch's view, is the income transfer that would result if the developing countries were to take advantage of their presumed monopoly position.

The policy idealized by Prebisch and his followers to correct the misfortune to which the periphery was submitted was protectionism. In fact, Prebisch advanced three reasons for protectionism: to retain scarce foreign exchange; to countervail the decline in the terms of trade; and as a means of halting the "excessive" expansion of the export sector, as described above. The heavy taxation of agriculture by means of implicit and explicit export taxes, as well as by export embargoes and complicated export licensing procedures, is consistent with this general perspective.

The argument in favor of rationing foreign exchange was first proposed in Prebisch [1950] and was repeated again in Prebisch [1959a]. In the first, he was concerned with the dollar shortage, which dominated discussions of international trade at that time. Later, he was concerned with the low import coefficient of the United States, and advocated protection as a means of rationing foreign exchange that was scarce due to that low coefficient. Given the importance he attached to the disparities in income elasticities, he believed a balance of payments equilibrium could be achieved either by slower growth or protection. In his view, of course, the latter was the least unpleasant.

It should be noted that in today's world, perceptions about relative import coefficients are reversed. Observers in the United States are concerned with the high import coefficients for that country and the low import coefficients of its trading partners. Obviously, the value of real exchange rates and the level of protectionist measures have an influence on what these coefficients are.

Both Haberler [1968] and Flanders [1964a] have pointed out that Engel's Law applies to food and not to raw materials, and that Prebisch's assertions about relative income elasticities were questionable. Baer [1962], on the other hand, believed that the available evidence supported the Prebisch position for many important parts of the world.

Flanders [1964a] challenged Baer's argument and argued that there were two components to the Prebisch argument. One had to do with relative income elasticities of demand, and these gave rise to balance of payments problems that could only be resolved as explained above. The second had to do with the price elasticity of demand of the center for the imports from the periphery. This elasticity presumably is low due to the "monopoly" positions that the periphery countries have in these markets.[20] The result is that the free market, responding to existing relative prices, misallocates resources in the periphery between export industries and import-competing industries so that aggregate real income in the periphery is not maximized. This, of course, carries a rather heavy load of assumptions about the structure of markets.

Flanders developed a "minimal model" that provides an integrated explanation for the presumed deterioration in the terms of trade and its relationship to welfare and balance of payments. This model is useful in that it abstracts from the many asides that Prebisch introduces, and also provides a useful means of drawing the present discussion to a close.

Consider two countries that are identical with respect to population growth, rate of increase in per capita income, and technology. Assume further that one of the countries (say Country I) exports only primary goods, while the other country exports only industrial goods. (This difference in comparative advantage may be due to differences in resource endowments.) Differences in the income elasticities of demand imply that the rate of increase in demand for industrial goods will be greater than the rate of increase in demand for primary goods. That, in turn, implies that there will be a relative increase in the resources employed in the urban sector of both countries. However, in order for the industrial goods produced by Country I to compete in its internal market with imports of industrial goods from Country II, Country I will have to accept a deterioration in its terms of trade, either through a devaluation of its currency or through a decrease in the wage rate.

Counteracting forces will develop as growth proceeds, however. First, with a larger share of the population in Country I employed in the urban sector, the growth rate in demand for Country II's industrial goods may decline somewhat. Similarly, there will be a tendency for the growth rate of supply of primary products in Country I to decline, further counteracting the adverse trend in the terms of trade.

The other force at work is the decline in real income caused by the decline in the terms of trade. That will reduce the rate of growth of demand for Country II's industrial products.

However, in discussing Prebisch [1959a], Flanders [1964a] notes that, even in the presence of such counteracting tendencies, the following holds:

> [A]s a result, the periphery's growth rate will be less than it would be if the demand for its exports were more income elastic (or less price elastic) or the demand for imports less income elastic. This, indeed, is the heart of the "Prebisch thesis" and, we repeat, it stems from the assumption of different income elasticities of demand, not from alleged differences in market structures and wage-price mechanisms. [p. 322]

Flanders summarizes the Prebisch [1959a] model as follows:

> We can, therefore, characterize the basic Prebisch thesis as consisting of two components: (1) A "balance-of-payments" problem, with demand for imports in Country I tending to grow faster than import demand in Country II, so that equilibrium can be achieved only if Country I grows more slowly than Country II. This problem arises from Country II's inelastic demand (for imports from Country I) with respect to income. (2) A "real income problem." This is frequently stated as Country II's price inelastic demand for imports from Country I, but more correctly should be attributed to Country I's monopolistic position in the world market, which causes the demand for her exports to be less than infinitely elastic with

respect to price. The result of this is that the "free market,"
responding to existing relative prices, misallocates resources in
Country I between export industries and import competing
industries, so that real income in Country I is not maximized.
(It should be remembered, however, that this misallocation is
the result of two assumptions about market structure, both of
which are necessary: (1) the monopolistic position of Country I
in the world market and *either* (2) perfect competition among
producers of the export goods in each I-Country or (3)
perfectly competitive behavior on the part of decision makers
in each Country I, each assuming that output of its export
goods remain constant in the other countries of the periphery.)
[pp. 323-324]

Attempts to test the Prebisch-Singer thesis with empirical data have received
renewed attention in recent years. Spraos [1980] presents a general and quite
comprehensive modern view of the statistical issues involved in measuring
changes in the terms of trade. Brandão [1978], Brandão and Schuh [1980], Gonçalves and Barros [1982], and Souza [1984] analyze the empirical evidence for
Brazil.

Brandão [1978] and Brandão and Schuh [1980] used a variety of price indices
to calculate the terms of trade for Brazil during the period 1953-75. Their indices
indicate a clear declining trend for this period as a whole. However, a great deal
of the trend was due to the very high level of the terms of trade in the 1950s—a
consequence in part of the commodity boom associated with the Korean War. For
the decade of the 1960s, there was no trend in the data, although there were some
very sharp year-to-year fluctuations which undoubtedly created balance of payments problems and made the management of macroeconomic policy more difficult.

Among the different criticisms made of empirical measures of the terms of
trade, perhaps one of the more important is the difficulty of taking into account
changes in the composition of trade over time. Brandão and Schuh [1980] attempted to identify those effects separately on the export, import, and terms of
trade indices. (The method is due to Lipsey [1963], and consists of comparing
Laspeyres and Paasch indexes.) They found that in the case of Brazil the composition of both exports and imports changed over time in the direction of lower-
priced products. However, the net effect for the terms of trade was not deter-
minable.

The studies by Gonçalves and Barros [1982] and Souza [1984] referred to the
period 1850-1979, with Gonçalves and Barros using simple statistical procedures
and Souza using more sophisticated time series methods. Their evidence is consistent with the hypothesis that there has been no longer-term tendency for the

terms of trade to deteriorate against Brazil. Not surprisingly, their results indicate that in some periods (1887-1940 being the best example) the evidence is consistent with the Prebisch thesis, while in other periods it isn't.

An important problem with these studies is the difficulty in taking account of changes in the quality of industrial goods. The quality of primary commodities (for the most part the exports) has not changed over time while the quality of manufactured products (the imports) has improved significantly. To the extent improvements in quality are reflected in higher prices, the indexes of the terms of trade will be biased to show a downward trend when in fact, in terms of constant quality products, there has been no trend.

The prices need to be adjusted to reflect changes in quality if meaningful welfare inferences are to be made from changes in the terms of trade.[21] Moreover, one has to decide whether what is to be tested is the tendency of the terms of trade of primary *versus* industrial commodities, or of products exported and imported by developing countries.[22] Clearly, each of these will require special handling of the data since primary commodities are exported by both developed and developing countries, while developing countries engage in trade with both developed and other developing countries.

To conclude, in general it seems fair to say that there has been a tendency of proponents of the declining terms-of-trade hypothesis to overgeneralize about their data. The terms of trade are country-specific, and fluctuate rather widely from one year to another. Difficulties in taking account of changes in the quality of nonagricultural commodities are legion. And finally, the fact that countries like the United States are major exporters of primary commodities themselves, while many developing countries such as Brazil and Mexico now export manufactured commodities on a large scale, further complicates the issue. (Prebisch, if he were alive today, would undoubtedly be pleased with this last development.)

Another important issue associated with the Prebisch thesis and the behavior of the terms of trade involves the welfare inferences made by Prebisch [1959a] and others from changes in the terms of trade. Brandão and Schuh [1979] have addressed this issue by obtaining a welfare indicator for a nation engaged in international trade and making use of international capital markets. They show that the problem has not been properly posed in most of the literature dealing with this issue, especially for those countries for which foreign capital is important.

Their model shows that when a country participates in both trade and borrowing and lending internationally, its national welfare is a rather complicated function of a large number of parameters. Moreover, when capital accounts are taken into account, there is no direct relationship between changes in the terms of trade and changes in welfare. The failure to recognize this complication is part of the general tendency to ignore what takes place on the capital account in attempting to understand international economic relations, even in the contemporary world. Prebisch has in his defense the fact that international capital markets were largely atrophied at the time of his early writing. Hence, what happened on the trade accounts in that period was for the most part dominant. That excuse is less acceptable in his and others' later writings.

Brandão and Schuh [1979] show, based on well-known results from trade theory, that before a connection can be established between the terms of trade and welfare it is necessary to establish the relationship between the terms of trade and the rate of return to foreign investment. This rate determines the remittances made to foreigners on account of their investment and is therefore a component of the change in welfare. It may be, for example, that an improvement in the terms of trade will lead to an increase in that rate of return (*via* a Stolper-Samuelson relation), which in turn may offset (or even reverse) the favorable effects on the trade account. Welfare inferences under these circumstances will not be unambiguous, and much more specific hypotheses have to be made about the quantitative effects of changes in the terms of trade on the rate of return on investment before precise welfare inferences can be drawn.

Brandão [1978] investigated this issue empirically for Brazil by assuming plausible values for the key parameters. Although his results need to be taken with some caution, they indicate that Brazil's protectionist policies for the manufacturing sector have had a negative welfare effect, thus making Brazil worse off. Moreover, he found that in some years when the external terms of trade moved strongly in Brazil's favor, the effects on the rate of return on foreign investment were not favorable, while in other years a decline in the terms of trade had a favorable effect on the rate of return. In those cases in which the rate of return effects were large enough to more than offset the other effects, national welfare would move in an opposite direction to the change in the terms of trade.

There is a final point that seems to be neglected in much of this literature. Leaving the capital account aside, the effects on national welfare of shifts in the terms of trade will depend importantly on the relative size of exports and domestic utilization, of imports and domestic production, and of the relative size of the elasticity of demand. Consider the case, for example, of a country that consumes only a small proportion of its total production of a given commodity and in which consumption of that commodity makes up only a small proportion of total consumption (coffee in Brazil, for example). If the total demand elasticity (domestic demand plus foreign demand) for that commodity is less than one, the loss in welfare to producers from a decline in the commodity price would probably swamp the benefits on the consumption side. If the total demand elasticity were greater than one, however, total revenue to producers would increase as product price declined, and total welfare would likely increase from the decline in price.

Similarly, consider the case in which the country is only a marginal exporter of the commodity and the consumption items are a relatively large component of the total consumption bundle. The relative demand elasticities would again have a significant effect on the size and sign of the net welfare effects. Similar cases can be worked out on the import side. The size of the net welfare gain is again determined in part by whether the domestic production sector is a small or large component of total domestic utilization.

Finally, it should be noted that if the terms of trade are shifting against a country because of technological changes elsewhere, one defense a country has is to improve its own technology. In this way, gains in productivity may offset the decline in price in terms of the domestic economy and may also make it possible to sustain foreign exchange earnings.

To conclude this section, it is worth emphasizing five points. First, the Prebisch thesis has had an extraordinary influence on economic thought in Latin America. Second, the trade pessimism implied by the thesis was generalized to what one might call an overall market pessimism (or market failure assumption) which involved assumptions about the price inelasticity of agricultural supply, low income elasticities of demand for agricultural output, and inelasticities in other sectors of the economy. This caused adherents of the Prebisch thesis to want to substitute national planning for the unfettered play of market forces. Third, agriculture was viewed as an impediment to economic development because it provided an important means by which periphery countries were exploited by the center through international trade. Fourth, the thesis gave rise to a structuralist and associated dependency school of thought on economic development and economic policy. And fifth, the thesis also produced a perspective on economic development and policy which gave great weight to what was happening in the international economy rather than to what was happening in the domestic economy alone.

In the next section we turn to a discussion of structuralism and dependency theory and the institution which served as their intellectual headquarters, ECLA.

2. The Economic Commission for Latin America (ECLA), Structuralism and Dependency Theory

Prebisch and his work have been strongly associated with ECLA, headquartered in Santiago, Chile. The work of that Commission was focused on the terms of trade and other aspects of the Prebisch thesis, on the structuralist and dependency schools of economic development, and on the structuralist theory of inflation. In this section we consider the ECLA and its evolution over time and its associated development models of structuralism and dependency theory. The structuralist theory of inflation is considered in the next section in the context of the monetarist–structuralist controversy.

Before proceeding, however, it is worth synthesizing the main policy prescriptions that evolved out of the Prebisch thesis and the structuralist–dependency schools of thought. These prescriptions were essentially directed at isolating national economies from the international trading system and giving a greater role to planning. Autarchic development was achieved by high levels of protection for the industrial sector and severe discrimination against agriculture. The goal of protecting the industrial sector was to promote import-substituting industrialization. The discrimination against agriculture was not so much by design as by neglect. It was severe and persistent, however, and some would debate whether it was just by neglect.

The most common and important means of discrimination against agriculture was by means of an overvalued currency, but explicit export taxes, export quotas and embargoes, and other limitations on exports were used extensively. Most countries in the region have persisted with an overvalued currency for long periods of time (and with large degrees of overvaluation). For a summary of the distortions in the case of Brazil, see Schuh [1978, pp. 311-313]. This distortion is in effect an implicit tax on exports (and an implicit subsidy on imports) and has been an important means of transferring the agricultural surplus to the urban sector of the economy. However, discrimination took other forms as well. High tariffs and other restrictions on imports of modern inputs such as fertilizers, machinery, and pesticides contributed to their increased cost. And use of price ceilings at the retail level also redounded negatively to agriculture. Ironically, by means of the high protection of the industrial sector and the discrimination against agriculture, policymakers shifted the domestic terms of trade against agriculture in the same sense that proponents of the Prebisch thesis alleged that the *external* terms of trade were shifted against the primary sector.

ECLA is better known in Latin America by its *sigla* in Spanish and Portuguese, CEPAL (Comisión Económica Para la América Latina), and adherents to its views are known as *Cepalistas* (Portuguese) or *Cepalinos* (Spanish). ECLA has been a unique institutional forum that has served as a distinctive Third World voice in international forums. Its history and thought have been essentially those of Raúl Prebisch, and in fact ECLA was Prebisch's principal theoretical and ideological vehicle from its inception. Through the years ECLA has been the headquarters of the structuralist theory of economic development, and many of Latin America's prominent "structuralists," such as Celso Furtado, have been associated with it. Guzmán [1976] is a detailed and well documented history of Latin American development and ECLA.

Joseph Love's paper [1980], "Raúl Prebisch and the Origins of the Doctrine of Unequal Exchange," provides an important and perceptive historical background on both Prebisch and the ECLA and the evolution of the ideas they promulgated. (We draw on his article for what follows.) Love pointed out that both the ideas of Prebisch and the intellectual position of the ECLA were rooted in the economic history of Argentina. In the latter half of the 19th Century and the first thirty years of the 20th, Argentina had made the theory of comparative advantage a near-sacrosanct doctrine as it reaped the benefits of export-led growth and an international division of labor. Moreover, as Diaz-Alejandro [1970] has noted: "From 1860 to 1930 Argentina grew at a rate that has few parallels in economic history, perhaps comparable only to the performance during the same period of other countries of recent settlement" [p. 2]. It was widely believed that Argentina had prospered during this period according to the theory of comparative advantage, and protection of the industrial sector was vigorously opposed in Latin America as late as the 1920s [Love, 1980, p. 48].

The 1920s, however, was a period of disequilibrium as well as expansion in world trade, and though Argentina continued to prosper, the country experienced the same problems as a number of other primary-producing nations in the final years before the October 1929 crash—falling prices, rising stocks, and debt payment difficulties. The global economic collapse of the 1930s worked a tremendous hardship on Argentina, and began the period of strong government intervention in the Argentine economy, especially by means of trade controls.

With sharply restricted export earnings throughout the Great Depression, self-sufficiency of industry was a necessity. Manufacturing in Argentina grew impressively in the 1930s and early 1940s. But this industrial development was common to southern South America, where Brazil and Chile found themselves in similar situations with the collapse of their export sales [Love, 1980, p. 51].

Prebisch was an active participant in the government of Argentina throughout this period and already in the 1930s was formulating and writing his ideas on unequal exchange [Love, 1980, p. 54]. His interest in industrialization as a solution to Latin America's economic problems originally arose from a desire to make Argentina less vulnerable economically, a vulnerability it experienced during the entire 1930-45 period.

By 1948 Prebisch was attacking the theory of comparative advantage. He noted that its precepts were repeatedly violated by the industrialized nations, whose economists nonetheless used classical trade theory as an ideological weapon against the developing countries. It was in February of that year that the establishment of ECLA was approved by the U.N. Economic and Social Council (UNESCO), and the first meeting of ECLA was held in Santiago, Chile, in that same year. Prebisch was already dominating economic thought in the region, so much so that the chief outcome of the first meeting was a resolution which called for a study of Latin America's terms of trade. The need for Latin America to industrialize was also stressed at this first meeting.

Prebisch was asked to head the ECLA at the very beginning, but he was reluctant to do so. However, in May 1949, the Commission distributed in Spanish (in mimeographed form) his famous *The Economic Development of Latin America and Its Principal Problems*. The United Nations published it in 1950 and it was reprinted in full in *Economic Bulletin for Latin America* in 1962 [see Bath and James, 1976]. Hirschman later dubbed this paper the Latin American Manifesto [Love, 1980, p. 75].

ECLA became a permanent arm of the United Nations in 1951. Love [1980, p. 57] notes that Prebisch's personality, his theses, and his programs so dominated the Commission in its formative years that it stood in sharp contrast to the Economic Commission for Asia (established in 1947), and the Economic Commission for Africa (1958), agencies which tended to have a more technical orientation.

ECLA has from its inception challenged the basic assumption of the theory of comparative advantage, largely on the grounds that growth in productivity in the center's industrial sector was not reflected in lower prices, which would have spread the benefits of this technical progress throughout the system, but rather was captured by higher wages in the center. As noted above, Prebisch and his colleagues believed that technical progress in the center was faster in the industrial sector than in agriculture and hence that agricultural prices should have *risen* relative to the price of manufactured products. That they didn't was viewed as a defect in the international system and of the tenets of comparative advantage.

ECLA's work focused on the trade cycle, on the import coefficient among countries, on the structure of labor markets, and on the disparities of income elasticities of demand among countries. Monopolistic pricing in the center was another important topic, as was national planning.

ECLA's attack on the international division of labor entailed a call for rapid industrialization in the periphery. Policies to induce this import substitution generally involved high protection of the manufacturing sector, large subsidies to the manufacturing sector (usually through cheap credit—which induced a capital-intensive manufacturing sector), overvalued currency and rationing of foreign exchange, and taxation of agriculture by means of trade and exchange rate policies, explicit export taxes, and ceilings on retail food prices.

ECLA's "manifesto" also called for international agreements that would provide price protection for primary products during the downswing of the trade cycle. Love [1980, pp. 59-60] notes that this call followed an earlier U.N. attempt to establish an International Trade Organization (ITO). The idea of commodity price stabilization at the international level had been discussed at the same Bretton Woods Conference (1944) which created the International Bank for Reconstruction and Development (better known today as The World Bank) and the International Monetary Fund. In 1947 a special U.N. conference at Havana established the ITO, whose principles included intergovernmental action to prevent violent price swings in primary commodities. In the end, the U.S. Congress failed to ratify the treaty establishing the ITO and the less ambitious General Agreement on Tariffs and Trade (GATT) was created in its place.

In effect, then, ECLA proposed a program of price stabilization for primary commodities which, as Love notes [1980, p. 60], implicitly included the idea of countermonopoly against the industrialized countries. This proposal was taken up by UNCTAD when Prebisch became its first Director in 1963 and has been part of the policy proposals of that organization ever since.

Let us now turn to the schools of thought that evolved out of the Prebisch thesis and which were developed in large part at ECLA's headquarters in Santiago, Chile. Prebisch's thesis about the declining external terms of trade is in an important sense a theory of economic stagnation for the developing or peripheral countries. An important contribution it made to the analysis of developing country experience was the attention it gave to conditions in the international economy as factors affecting economic development in those countries. In the case of Latin America, in particular, it pointed to the relationship of the United States to the countries in the region. This international perspective is in sharp contrast with both the growth-stage and dual economy models popular in other parts of the world at that time. Those models looked within national economies for the timing and transformation which was expected to lead to economic development.

However, the Prebisch thesis was at one and the same time a theory of stagnation and a theory of development in that it explained how the center was able to grow and why the periphery stagnated. The concept of *underdevelopment* thus became an important concept in this literature, and development and underdevelopment were viewed as two components of one unified system. Another important feature of the model was an implicit assumption that one part of the world cannot become better off without making another part worse off as long as they are part of the same system. The Prebisch thesis also at one and the same time implied both a *structural* and a *dependency* theory of development. Dependence took the form of exporting raw commodities with low income elasticities of demand and importing goods with high income elasticities. Moreover, labor unions and oligopolistic markets in developed countries were postulated to capture productivity gains in the form of higher wages and profits. On the other hand, surplus labor and foreign ownership made it impossible for the raw-commodity exporting country to retain all its potential gains.

Stated in this simple and stylized way, one sees a counterpoint between structure and dependence, and there is a tendency in the literature to refer to proponents of this view of the world as either structuralists or dependency theorists, or to use the two concepts interchangeably. Our inclination, however, is to refer to structuralists as those who give more emphasis to the structures of markets (including the elasticity of demand), and to refer to dependency theorists or *dependistas* as those who introduce political theory into the model and thus become political economists. Dependency theorists have tended to introduce conspiracy or volition into their models, culminating in theories of imperialism. Prebisch himself, and some of his followers, were much less conspiracy-minded and gave more emphasis to the structure of the economy, including the distribution of income, as the causal factors. It is for this reason that we believe it makes sense analytically to make a distinction between the structuralists and the *dependistas*.

W. Arthur Lewis [1984, p. 124] has one of the most succinct descriptions of dependency theory we have found and it is worth quoting. After noting that the dependency argument is about power and its cumulative accretion, his summary description is as follows:

> A peripheral country that begins to export agricultural commodities becomes paralyzed in ways that preclude an industrial takeoff. Its trade and all that goes with it—shipping, banking, insurance, port facilities—fall into the hands of a few foreigners, with or without association with a few rich local families. The profits of this trade are transferred overseas instead of being invested in the country. The best jobs are reserved to foreigners, so that local talent is untrained and unable either to compete in the old trades or to start new ones. The talented young become frustrated, lose confidence in their abilities, emigrate, or lower their horizons. Domestic industries are destroyed by imports. The foreign countries are interested in foreign trade and, if they can, will block attempts to create new industries that might diminish their trade or render it more costly. Mass advertising teaches the people to prefer imported consumer goods to their own products, thereby raising the propensity to import foreign brands or materials or machinery in place of local resources. This trend imperils the balance of payments, makes it harder to provide jobs, and pushes displaced workers back into the subsistence sector.

Lewis believed this to be a reasonable description of what was happening in most developing colonies in the first half of the twentieth century, though he believed it exaggerated the share accruing overseas. He also found it difficult to understand why independent countries should fall into such a trap. Thus, he doesn't believe it was a good description of Brazil around 1880, which had already begun to build its own industrial bourgeoisie, or of Argentina, which in his view was bossed by its great landowners rather than by foreign capital. He further noted that as early as 1962 he had argued that dependency theory seemed important to him for study of the second half of the nineteenth century, but not the second half of the twentieth century when independent governments were engaged in restructuring the place of foreigners in the country. Moreover, as a contribution to deciding whether the small farms should be encouraged to plant more tea or rubber, it seemed to him unhelpful [Lewis, 1965].

Andre Gundar Frank [1967] has been the intellectual (and radical) leader of the dependency school, although F. H. Cardoso and Faletto [1969] is generally given equal weight as an original source of dependency thought. For a perceptive description and synthesis of the various strands in this school of thought, see F. H. Cardoso [1977]. Other useful references include Amin [1974]; Baran [1957], a classical treatment of the underdevelopment perspective; F. H. Cardoso [1972]; Chilcote [1974, 1978]; Chilcote and Edelstein [1974]; Cockcroft, Frank, and D. L. Johnson [1972]; T. dos Santos [1970a,b]; Duvall [1978]; Emmanuel [1972]; Furtado [1970a]; Jaguaribe *et al.* [1970]; Pinto [1974]; Pinto and Knakal [1973]; Stavenhagen [1974]; Stern [1988]; Sünkel [1973]; W. B. Taylor [1985] and Wallerstein [1974]. Frank [1977] is an interesting response to his critics.

It is generally recognized that dependency analysis developed out of Marxism and Latin American structuralism. Love [1990] delineates and assesses the relative importance of these two intellectual traditions in the initial articulation of dependency in the mid- and latter 1960s.

Bath and James [1976] provide a general overview of the dependency analysis of Latin America. They note that dependency analysis, despite its long tradition in Latin America, emerged from the relative obscurity of Latin American writers to be used by U.S. scholars when it was first adopted by a group of "radicals" in the United States partially as a reaction to U.S. involvement in Vietnam. Earlier in this volume Eicher and Baker refer to dependency theory as a Latin American export to the other developing countries. The U.S. radicals also used it as part of a general attack on the capitalist system, the military-industrial complex, and U.S. imperialism.[23]

Bath and James noted that the *dependistas* had not been subjected to the critical scrutiny they deserve and that systematic criticism of the dependency approach had been lacking.[24] They attributed this to the lack of academic respectability of the radicals, and to the different channels through which their publications appeared.

Alain de Janvry's [1975] interpretation of rural development in Latin America is an important example of dependency theory applied to agricultural and rural development.[25] Consequently, we sketch out the main elements of his argument to illustrate this perspective when directly applied to agriculture.[26]

De Janvry's starting point is a criticism of T. W. Schultz's *Transforming Traditional Agriculture* [1964] on the grounds that in Schultz's perspective the origins of agricultural poverty are dissociated from the dynamics of development in other sectors of agriculture, in other economic activities, and in the world economic system. Instead, Schultz argued that to help farmers escape from the misery of traditional farming, all that is needed is to provide new technological alternatives.

De Janvry believes that to deal with rural poverty in this way is an historical inconsistency that seriously impairs our ability to delineate and interpret the origins and dynamics of poverty and to identity the means by which it can be attacked. His interpretation is that underdevelopment cannot be treated apart from development if backward areas or countries are related by the market to the advanced areas or countries. In fact, he argues that within the world capitalist system, a theory of underdevelopment and rural poverty needs to be a theory of economic space which can explain how the contradictions of development in certain areas (the developed countries) transform traditional societies in other areas (the developing countries) into underdeveloped ones.

De Janvry [1975] rejects the notion that rural poverty can be analyzed from the framework of a theory of traditional societies. Instead, the analysis has to be done, in his view, against a background of historical events that have created structural dualism and marginalization of large sections of society. Because agriculture serves as a natural refuge for surplus populations, rural poverty should be analyzed in the framework of marginality rather than of traditional culture.

Defining the concepts of periphery and marginality are de Janvry's first task. In his view these concepts are essential for the construction of a theory of underdevelopment. He defines the periphery as that portion of economic space which is characterized by backward technology with consequent low levels of remuneration of the labor force, and/or by advanced technology with little capacity to absorb the mass of the population into the modern sector. These excess human masses are created by the very process of economic growth. They can be found in all sectors of the economy and are functionally related to the modern sector that needs them in order to confront the conditions under which growth occurs in the periphery. In agriculture they are the farmers who lose control of the means of production because they cannot withstand the competitive pressure of the modern sector, or the farmers who see their economic condition deteriorate as a consequence of their retention of traditional production techniques. These groups join the ranks of the marginals as *minifundistas*[27] and subsistence farmers.

The economies of the center countries also have marginal populations. But these tend to be small groups who have been bypassed by economic development and are widely dispersed in the economy. In contrast, in peripheral economies the marginals involve large masses who have been "objectively" created by the dynamics of accumulation.

Historically, de Janvry says that the transformation of Latin America into a periphery of the world capitalistic system occurred first through colonialization, which integrated the region into commercial markets, and later through efforts, essentially by England, to destroy the Iberian colonial empire in order to establish free trade with the rest of the world. It was this system which encouraged the export of primary products from the periphery and the export of manufactured consumption goods from the center. This trade provided the basis for class alliances between traditional landed elites and British capital to maintain the internal social *status quo*.[28] Hence, from this perspective there was a certain economic rationality to the transformation of Latin American countries into peripheries, for such transformation enabled the landed elites to capture part of the large surplus of the agroexporter sector and hence to enjoy consumption patterns similar to those of developed countries while at the same time retaining all the advantages conferred by their social position in underdeveloped economies. Under this model, unequal exchange to the benefit of the center is obtained on a "voluntary" basis by co-opting the traditional elites into sharing in the surpluses extracted. Maintenance of low agricultural wages through precapitalist relations of production, imposed by the elites in order to tie labor to the land and alienate it from its own opportunity cost on the labor market, made it possible to set the external terms of trade to the benefit of the center.

De Janvry [1975] argues that the second industrial revolution of the 20th century changed the mechanism by which this process worked. For a while, import substitution policies provided a new dynamic for the industrial sector. But after the 1930s, the terms of trade in the exchange between raw materials and capital goods deteriorated "coercively" against the periphery. Given that there was a reasonably well integrated world market for capital, the cost of unequal exchange had to be transferred to what for de Janvry was the only nontradeable factor— unskilled labor. Continued industrial development in the periphery was now conditioned by the ability of nations to reduce their labor cost.

De Janvry says that two means can be used to cheapen labor. One is to impose repressive labor policies—a mechanism that clearly has been extensively used but which is limited by the organization and insurgency potential of the working classes. The second means is to lower the cost of those mass consumption items that constitute the bulk of laborers' budgets and hence determine labor cost. Food, textiles, and popular construction are the producing sectors most negatively affected by the consequent commercial distortions that governments use to depress the internal terms of trade against wage goods. Consequently, profitable investment ventures are confined to that subset of the industrial sector which is oriented to satisfying demand from the upper classes—demand which is intensive in advanced technology and in imported capital goods. In the development of Latin America these industries were transformed into large enclaves of modernization that were unable to generate enough employment to absorb available labor.

Under conditions of a highly unequal distribution of income, de Janvry argues that the effect of import substitution policies was to replace the initial sectoral linkage (characteristic of preindustrial peripheries) between the export sector of primary goods and the import sector of *luxury consumption items* by a new linkage between the export sector of primary products and the import sector of *capital goods* for the production of luxury consumption items. By way of contrast, de Janvry notes that in the economies of the center, the fundamental sectoral linkage is between the production sector of capital goods and the production sector of mass consumption goods. It is in this contrast between the industrial structures of the center and peripheral economies that lies the economic rationality of labor incorporation into the economies of the center and of marginality of labor in peripheral areas.

De Janvry [1975] draws other contrasts between the economies of the center and those of the periphery. In the economies of the center, labor constitutes both a cost and a benefit to capital. It represents a cost since wages are subtracted from profits, but it is also a benefit since wages serve to generate the demand for wage goods that will allow further accumulation. This symbiotic relationship leads to a "social contract" in the advanced countries between capital and labor, under the auspices of the state. It is this social contract which allows real wages to relate effectively to increases in labor productivity. (De Janvry is clearly in the Prebisch tradition.)

Such a social contract is prohibited in peripheral economies in de Janvry's view by their distinct sectoral linkage (between export of primary products and import of capital goods for the production of luxury items). In those economies labor constitutes only a cost to capital and is not simultaneously a benefit. This is because industrial production is oriented not toward mass consumption but toward consumption by the upper income classes or towards exports under conditions of surplus labor.

The dynamic of peripheral accumulation in the context of unequal exchange is based on continued dominance of capital over cheap labor. In fact, it is by this very process of labor submission to modern capital that societies make the transition from traditional to peripheral. We thus have the interesting contrast in which incorporation of labor is, in the countries of the center, a condition for growth, whereas marginality is in the periphery a contradiction of growth. Moreover, in the periphery, structural dualism will in de Janvry's view deepen as development proceeds since he believes it a necessary condition for growth of the modern sector.

De Janvry then turns to the discussion of agricultural stagnation and rural poverty based on his theory of marginalization. He argues that in the 20th century, agriculture and industrial interests have integrated themselves in Latin America, generally with the traditional landed elites extending their sectoral control of land in the *latifúndio*. Although these interests were integrated, he refers to Stavenhagen [1968] to argue that the agricultural interests were increasingly subject to the dominance of the urban interests.

Since the type of industrialization pursued in Latin America requires low wages and overvalued exchange rates to cheapen capital imports, commercial and market distortions result in domestic terms of trade that are dramatically unfavorable to agriculture. This has led to a long period of stagnation in Latin American agriculture. He argues that an exception was in Mexico, where land reform eliminated the dominance of the traditional elites and opened the way for ambitious infrastructure investments and the diffusion of land-saving technological change. But beyond examples such as that, development of capitalist farming in food crops has been blocked. In his view only the *latifúndios* have been maintained as viable economic units, and that has occurred through institutional control by the dominating elites who monopolize institutional services (institutional credit, technology, information, and so on) and derive from them economic compensation for the unfavorable terms of trade. Part of the cost of stagnation has also been shifted to labor through miserable wages.

Within agriculture, the rise of what de Janvry describes as rampant marginality changes the social relations of production in such a way as to further weaken labor. The social relations of production have been gradually redefined from precapitalist (where the workers' subsistence plots were located within the *latifúndio*) to a functional dualism between the commercial sector and the *minifúndio* (where the subsistence sector is now external to a capitalistic agricultural sector). In his view, precapitalist relations were needed to cheapen labor by tying it to the land and alienating it from its own opportunity cost. Once marginality became widespread, labor cost could be further reduced by making use of wage labor instead of labor partially paid in land privileges, since that permits landlords to recover the land previously given to workers and to tailor the hiring of labor to fluctuating seasonal and annual needs. Development of a functional dualism between the commercial sector and subsistence agriculture is fully consistent with the needs for growth in the periphery where industry is oriented to luxury goods and exports. In this structure the subsistence sector produces cheap labor for the commercial sector which, in turn, can produce cheap food for the market.

Generalized rural poverty in rural Latin America is hence the logical outcome of a three-level chain of exploitive relations. The first is at the international level between dominant centers and dependent peripheries in the context of unequal exchange between raw materials and industrial capital goods. The second is at the sectoral level between modern industry, which produces commodities for the upper classes and the external market, and the sectors that produce mass consumption items in the context of the need for cheap wages and the consequent deterioration of the domestic terms of trade. The third is at the social level between landlords and agricultural labor and marginal populations in the context of transmission to labor of the costs of unequal exchange at the international and sectoral level, as well as exclusion of the mass of population from modern sector employment. It is only when the food sector is itself a major component of the export sector that this chain is reduced to a two-level set of relations, as unequal international exchange is brought to bear directly on the terms of trade for agriculture.

De Janvry [1975] then argues that specific contradictions that jeopardize economic growth and social stability in the periphery correspond to each of these three levels of exploitive relations. At the international level, the structural deficit in the balance of payments blocks expansion of the modern industrial enclave in the periphery. At the sectoral level, agricultural stagnation raises labor costs, unleashes inflationary pressures, and worsens the balance of payments deficit. At the social level, miserable wages build up revolutionary pressures that are reinforced by ecological degradation and demographic explosion in the subsistence sector. In this context, de Janvry argues that rural development programs in the regions do not have redistributive goals determined by the need to increase market size for the modern industrial sector, as one might naively expect. Instead, those programs generally pursue two simultaneous objectives whose relative importance is determined by particular historical circumstances. The first of these is a production goal designed to alleviate deficits in the balance of payments, contain inflationary pressures, and cheapen wages. The second is a distributive goal which is fundamentally political and designed to promote social integration of the potentially revolutionary strata of the peasantry and their eventual incorporation into social groups that will favor maintenance of the social *status quo*.

De Janvry then concludes his paper with a discussion of the political economy of rural development projects, the popular form of agricultural development during the 1970s. He tends to view these projects as a means of preserving the *status quo*, rather than as a means of undertaking true reform and economic development. He refers to Mexico, Cuba, and the new (at that time) Latin American nationalism of Peru as exceptions to this rule. In these cases, since the luxury goods bias in the industrial sector was presumably eliminated, the state was set for planned autocentered agricultural and industrial development. In these countries industrial production can be oriented towards mass consumption items, and redistributive policies acquire a logic in terms of market expansion. As in the economies of the center, the social management of a balance between the capacity of the system to produce and to consume wage goods becomes a necessary condition for growth. Terms of trade for agriculture as well as institutional processes aimed at land-saving technological change and at infrastructure investment can be managed consistently.

Schuh's critique [1984b] of de Janvry's paper took issue with its central propositions, and stressed the following: that de Janvry has a theory of income distribution and not a theory of growth or development; that the external terms of trade can shift against a country for natural economic reasons and need not shift "coercively"; that the income losses caused by a decline in the terms of trade will be borne by *both* land and labor and not just labor; that center countries benefit from growth and development in the periphery in terms of expanded markets and thus do not necessarily have a vested interest in continued poverty in the periphery; that modern industrial sectors require skilled labor, not unskilled labor, and that consequently the large masses of unskilled workers are largely irrelevant to them; that foreign exchange constraints in the region are largely due to misguided economic policies and hence are self-imposed; and that labor costs can be cheapened by raising productivity rather than by lowering wage rates. These points are elaborated in Schuh's paper.

Let us now return to policy issues, with the focus on the structuralist model as it evolved from the Prebisch thesis. Bath and James [1976, p. 6] noted that by the end of the 1950s Prebisch's policy recommendations could be synthesized in four propositions: push import-substituting industrialization through protective measures so as to absorb surplus labor and capture productivity gains in industry within the periphery; encourage economic integration among periphery countries to increase market sizes and capture productivity gains within the periphery as a whole; negotiate a scheme of reciprocity for transferring resources from the center to the periphery to compensate for the productivity gains that are leaked from the periphery through falling raw commodity prices; and organize raw commodity control schemes in order to reduce price fluctuations and to raise farm prices above what they would otherwise be.

It is important to note that there are two forms of self-sufficiency in the structuralist model which lead to autarchic development. The first is on the industrial side, with protectionist measures put in place to substitute domestic production for imports. The original expectation was that production of consumption goods would be internalized and imports of consumption goods would be replaced with imports of capital goods.

Self-sufficiency on the agricultural side had two dimensions. On the one hand there was a concern with increasing food supplies by means of land reform (see the next section on the structuralist-monetarist controversy), which had both efficiency and equity goals. But an equally important dimension was the tendency to channel domestic production away from foreign markets to the domestic market. This was done primarily by means of overvalued currencies, designed primarily to make critical imports needed for development of the industrial sector cheap in terms of domestic resources, and the other discriminatory trade policies described above.

In effect, export policy for agricultural commodities was based on a "vent for surplus" export model. Exports were permitted only after the domestic market had been satisfied. See Leff [1968] for a description of these policies in the case of Brazil.

Finally, it should be noted that the structuralist-dependency models later evolved in two other important directions. At the international level increased attention was given to the *technological* dependence of the peripheral countries on the center, with subsequent demands for the transfer of technology from the center to the periphery. Wionczek has been a frequent writer on this theme [1969-70, 1971a, b, and 1972], but Prebisch [1971] also recognized it. Second, dependency theory was applied internally to the developing countries, with exploitation of marginalized groups in society, including those in agriculture, by the domestic oligarchy [Goodman, 1976].

We conclude this section with a summary and an assessment of ECLA and the two schools of thought it evolved. The experience of the 1930s and World War II, together with the popularity of the Prebisch model, caused development policy in Latin America to focus on import-substituting industrialization. High protection of the manufacturing sector, overvalued currencies, explicit export taxes, export embargoes, and complicated export licensing procedures shifted the domestic terms of trade against agriculture and made it difficult for that sector to compete in foreign markets. This caused agricultural export performance to be less than dynamic, and made it difficult for agriculture to contribute as much to general economic development as it might have.

In the view of those who have more confidence in domestic and international markets, this set of policies to promote economic development was responsible for the poor performance of the trade sector during the period following World War II. In their view, the agricultural sector was heavily penalized by the policies pursued by national governments and thus the sometimes weak performance of exports and food production should not be attributed to "structural" factors. The explanation for the facts lies in the set of policies pursued, and the wounds were largely self-inflicted. Moreover, the social costs of these policies were quite high.[29] In addition, the experiences of some countries in the late 1960s and early 1970s show a clear, positive response of exports in general, and agricultural exports in particular, to incentives provided through a proper mix of policies, as we shall see in a later section.

Prebisch in UNCTAD [1964] himself later went a long way in retracting his own narrow position on import substitution. As data and studies of the experience in the 1950s began to accumulate, it became clear that import-substituting industrialization was not a panacea.[30] Countries that pursued such policies in the 1950s and early 1960s found that the high protection called for by his perspectives created an enclave economy, that the employment-generating performance of their industrial sector was weak, and that in point of fact they did not reduce their dependence on imports. The principal effect of these policies was to change the import mix, not to reduce the import quantum. Moreover, vested interests built up in the protected industries, and the trade liberalization that would have been consistent with infant-industry arguments never came about. Prebisch argued that more attention had to be given to agriculture and to exports in development policy, and that ways had to be found to improve the employment performance of the economy.

ECLA almost completely neglected exchange rate issues, as well as the effects of changes in relative prices and what they could accomplish in reallocating resources. They neglected agriculture as the basis of economic growth because they believed this sector inherently lacked dynamism. One of the reasons they perceived so little dynamism in agriculture was that they tended to view it as isolated from the rest of the economy. This also trapped them into believing there was no aggregate supply response.

Another important feature of the dependency literature is the lack of symmetry in the postulated relationship between the periphery and the center. The periphery countries are supposedly dependent on the center. But if the relationship is as described by the *dependistas*, surely the center is just as dependent on the periphery for its growth.[31] This must provide a basis for bargaining and negotiations which could lead to a more equitable distribution of the benefits of development. Prebisch's efforts to mobilize the developing countries into an effective political force was of course pointed in that direction. But the failure to note the symmetrical nature of the dependency relationship in this literature is striking.

Today, ECLA seriously lags intellectually behind recent developments in development theory and policy in other parts of the world. The structuralist-dependency perspective tends to be very inward-looking in terms of development policy. However, the outstanding successes in development these last ten to twenty years have been in Asia and in countries that have promoted export-led growth. As countries in Latin America increasingly turn outward in their attempt to deal with their debt problems, the relevance of ECLA becomes less and less.

Perhaps the most serious indictment of the structuralist-dependency schools, however, has been the failure of their models to correctly predict the pattern of development which has emerged in the post-World War II period. As de Janvry and Crouch [1980a] note, the periphery has outperformed the center in much of this period.

3. The Monetarist-Structuralist Controversy

The monetarist-structuralist controversy has been about the causes of inflation in the inflation-racked economies of Latin America. Its relevance to a review of the literature on agricultural development in Latin America is that the structuralists believe the causes of inflation are rooted at least in part in an agriculture that fails to respond to price incentives. Moreover, adherents of this view believe that reform of the agricultural sector, especially a land reform that changes the size distribution of land holdings and breaks up the *latifúndio-minifúndio* complex, is an important means of reducing the high rates of inflation. The connection to the structuralist school of development-underdevelopment should be obvious.

This subject has generated such a lively debate in the Latin American literature in part because Latin America has been characterized by high, chronic, and unstable rates of inflation. Harberger [1981] identified twelve cases of what he called chronic and acute inflation in the world in the period, 1952-76. Not surprisingly, ten of these were in Latin America. As measured by the annual percentage increase in consumer prices, inflation in Argentina averaged 27 percent in the 1949-74 period and then increased to an average annual rate of 293 percent from 1974 to 1976, and to a rate of almost 700 percent in 1985 compared to 1984. Inflation in Brazil averaged 35 percent from 1957 to 1976, but in early 1986 was at an annual rate over 700 percent per year. Inflation in Bolivia averaged 117 per year from 1952 to 1959; Chile averaged 273 percent per year from 1971 to 1976, and also was hyperinflating by 1985. In no other part of the world has inflation been so endemic.

Sünkel [1958] provides an early and classic structuralist interpretation of Chilean inflation. In this section we focus on the core of the debate. Those interested in more detail should see the Baer [1967] survey article, Baer and Kerstenetzky [1964], Grunwald [1964], and Wachter [1976].

Understanding the context in which this controversy arose is quite important. From one perspective, it arose in the broader debate on the development alternatives for the region. With ECLA providing much of the intellectual leadership for that debate, it is not surprising that market failures make up a central part of at least one side of the controversy.

From another perspective, some of the most influential participants in the debate were, at one time or another, involved in policy making. Some of them, in fact, have held important policy positions, as was the case of Roberto Campos (one-time President of Brazil's National Bank of Economic Development and later Minister of Planning), Celso Furtado (one-time architect of development policy for Brazil's Northeast, and later Planning Minister), Raúl Prebisch (described above), and Aldo Ferrer, (some-time policy maker in Argentina). This explains why on occasion some of the authors seem to be using inflation as a means of reinforcing points already made in other contexts, or to be suggesting policy measures that would have a scope much greater than the one in question. The most convincing example of the latter perspective is the structuralists' quest for land reform to halt inflation. As Olivera [1964, p. 331] put it: "It is as if believers in the 'Pigou effect', being persuaded that price and wage reductions can prevent involuntary unemployment, recommended long-term structural changes favorable to price and wage flexibility as a practical way of correcting a slump."

MONETARISM

Latin American monetarism shares important similarities with monetarism in the United States in that each has some form of the quantity theory of money as a basic theoretical construct. However, the question of the effectiveness of financing government deficits by monetary expansion *vis-à-vis* the creation of debt has not been of any significance for Latin American monetarists. All deficit financing in these countries has until recently (and up to now in many of them) been done by means of expansion of the monetary base. Consequently, the monetary expansion *versus* creation of debt issue, which has been the subject of a heated debate among U.S. economists, has not played a significant role in the Latin American debate.

One of the most fundamental elements of the monetarist position in Latin America is the claim that a stabilization program is a precondition for economic growth. In other words, monetarists condemn the use of inflation as a means of financing the progress of economic growth. This position is mainly due to the distortions they associate with such a policy; as they argue, any kind of growth so financed will be short-lived. This can be seen in the following statements:

> A strong inflation creates distortions in the economy, which may be regarded as comparable to the undesirable incentives induced by unsatisfactory forms of taxation. [Dorrance, 1964, p. 40]

> Inflation diminishes the value of resources available for domestic investment. Community saving is reduced, and a considerable part of this saving is channeled to foreign rather than domestic investment, while the flow of capital from abroad is discouraged. A substantial investment is diverted to uses which are not of the highest social priority. . . . The apparent profitability of certain short-lived investments lead to distortions in the productive structure which make the economy less adaptable. Balance of payments difficulties are symptoms of the underlying stresses. To reduce the foreign deficits the authorities are almost forced to resort to contracts, which in most cases, protect uneconomic production . . . However, if the economic system has been allowed to get out of hand, the authorities must decide to stabilize or not to stabilize. There is no doubt that the process of stabilization is difficult, but difficult or not, it is a prerequisite to rapid economic growth. [Dorrance, 1964, p. 68]

> Though inflation is alleged to be the means of bringing about

full utilization of the available resources, this is a self-defeating proposition. Inflation itself generates distortions and disequilibrium which make it more difficult to ever reach balanced complementarity in the use of resources. An increased money supply may speed up economic activity in an economy suffering from cyclical unemployment. But Latin American economic systems, vulnerable as they are in many ways, are allergic to unemployment. [Gudin, 1962, p. 346]

The basic content of the stabilization programs proposed by monetarists is the reorganization of government finances. This includes a reduction in the budget deficits by diminishing the level of subsidies granted to the private sector, an appropriate price policy for public goods and public utilities, and a modernization of the tax system (tax increases if necessary) in order to increase government revenue [Dorrance, 1964; and Gudin, 1962].

In addition, strong opposition to any kind of price control is apparent in their position. The well-known distortions caused by price controls are frequently mentioned as the basic reason for avoiding them. Moreover, they attribute the low level of modernization and/or efficiency of some sectors, especially agriculture, to these policies, which tend to distort relative prices in an undesirable way. Exchange rate controls, of course, are a crucial part of the problem.

However, given that most governments in the region do in fact use some kind of price controls, an important element of any stabilization plan is believed to be the liberalization of prices. This, they point out, might delay the decline in inflation, but it is extremely important in establishing the basis for the economic growth which is expected to follow. The inflation which results when prices are deregulated is often referred to as "corrective" inflation.

Although an extreme monetarist position would deny any effect to changes in wages on the rate of inflation, some monetarists have considered them, together with devaluations of the exchange rate, as important factors in the dynamics of inflation. Harberger [1963], for example, considered the change in the minimum wage as an explanatory variable in his empirical study of Chilean inflation, and later [1964] developed a model to account for the effects of devaluation on the general price level. Delfim Netto et al. [1965], in their pioneer empirical study of inflation in Brazil, also emphasized the movements of wages and exchange rates as important autonomous determinants of the rate of inflation in Brazil.[32] Pastore [1973a], in a later study, included the rate of change in legally-fixed minimum wages and in the exchange rate in the equations he estimated and obtained significant results.

It is worth quoting Eugênio Gudin [1962, p. 352], the Brazilian monetarist *par excellence*, on the question of wages:

I should like to add that two other important factors distort the economic system, although these are not directly related to the investment patterns. One is the annual or biennial adjustment of 'minimum wages' by government decree with the consequent disruption of the price system. Apart from the wage readjustments between employers and employees, often after long and hard disputes, if not strikes, minimum wages are often fixed under pressure from vociferous demagogues, at levels well above those of real wages. This, naturally, has a big effect on the upper part of the wage scale in all industries and progressively reduces differentials between skilled and unskilled labor. This is the exact opposite of what would have happened if the labor market had been left free.

Summing up, the monetarist position seems to accept the view that wages are capable of exerting autonomous pressure on the rate of inflation. However, it requires a significant leap of the imagination to attribute to them any position in favor of wage controls. Despite this, Baer [1967] in his survey of this literature, points out that wage controls are a policy recommendation of monetarists. In his words:

> The rationale for a control of wage increases . . . is twofold. On the supply side, it is argued that with credit controls, declines in real wages will increase the profit rate of the capitalist, who needs increased profits to replace credit, now in short supply, as a source of short-term working capital and as a source of long-term expansion. On the demand side, wage restraint will lower the excess demand which is the basic cause of inflation. [p. 7]

It seems to us, however, that this position cannot be characterized as monetarist. On the supply side, one can dismiss the point by noting that it would not be consistent with the monetarist view of the price system. Wage controls would only add an additional distortion to the economy. Moreover, it would be very difficult to defend wage controls on the above grounds and, at the same time, defend less exchange rate control, a policy which even Baer would admit the monetarists favor. On the demand side, the statement has a typical Keynesian flavor which is not compatible with the fundamentals of the monetarist model. Monetarists are, of course, opposed to strong labor unions and to minimum wage laws which push wages above market-clearing levels, but that is another issue.

The point of this discussion is not to argue that stabilization plans actually implemented were free of wage controls. Clearly, governments have distributive biases and these are reflected in their actions. Our point is that there is nothing in the monetarist model that would say that only wages should be controlled and no other price in the economy.

Dornbusch [1982], writing on the developing countries, has recently formulated a model which stresses the role of the real exchange rate and government deficits in the determination of internal and external equilibrium as well as the rate of inflation. In his model both inflation and external disequilibrium have the same cause: budget deficits. Once these are reduced, the economy is able to find its long-run equilibrium with a lower rate of inflation. Budget deficits generated by parastatals (public sector corporations or companies that permit private ownership of stock) have been an important cause of increases in the money supply and in turn inflation in Latin America in the early 1980s.

STRUCTURALISM

A synthesis of the essence of the structuralist model is more difficult than for the monetarist model, due to the wide variety of such models. However, structuralist theories generally rest on hypothesized slow increases in agricultural productivity, administered industrial prices, and a passive monetary policy. The slow increases in agricultural productivity are usually attributed to "poor" land tenure patterns [Sünkel, 1966; Olivera, 1964].

Individual authors stress different types of structural imbalance, influenced by their experience in particular countries. This variety of imbalance is also due to the more intricate relationship which proponents of these models established between economic growth and inflation as compared to the clear-cut monetarist position.

The core of the inflationary process in the structuralist model is postulated to lie in imbalances in the economy. Changes in the structure of the economy are viewed as necessary if the process of development is to be sustained without a parallel process of inflation.

> Economic development calls for constant changes in the form of production, in the economic and social structure and in patterns of income distribution. Failure to make these changes in time or to undertake them partially and incompletely leads to these maladjustments and stresses which release the ever latent and powerful inflationary forces in the Latin American economy.
>
> This should not be construed as meaning that inflation is inevitable in our countries. Far from it. To avoid inflation, however, there must be a rational and far-sighted policy of economic development and social betterment, in other words,

an essentially new approach in which an answer other than inflation is sought to these maladjustments and stresses arising from development. [Prebisch, 1961a, p. 1]

Different authors classify the causes of inflation in the structuralist model as to whether they are structural, circumstantial, or propagation factors [Sünkel, 1958; Grunwald, 1961]. The most frequently cited and important of the structural factors are a presumed inelasticity of agricultural supply, a low and unstable capacity to import, the distribution of income, and deficiencies in the tax systems. In the paragraphs which follow we discuss each of these structural factors. A brief discussion of circumstantial and propagation factors follows, plus a discussion of some general issues.

The inelasticity of agricultural supply is usually assumed to be rooted in the land tenure system that prevails in most of the region. This system is characterized by the so-called *latifúndio-minifúndio* complex, the coexistence side by side of very large farms with a low intensity of land use and extremely small farms (with a very weak potential for income generation) that are intensively cultivated. This system is assumed to have a weak capacity to respond to changes in demand. On the one hand the *latifúndio* owners are assumed to have access to an infinitely elastic supply of labor and thus have no incentive for modernization,[33] which many adherents to this view associate with agricultural mechanization. On the other hand, the argument goes, the prevailing technology is not profitable for the small farmers.[34] As a result, the growth in demand associated with industrialization leads to substantial increases in food prices but little increase in output.

Some writers also emphasize cost elements as factors explaining the high prices of food. Prebisch [1961a, p. 5] points out that the costs of modern inputs might contribute to high food prices, with these high costs being attributed to import-substituting industrialization. Similarly, he notes [1961a, p. 3] that in the "agricultural sector wage increases arising from higher productivity in other branches of the economy . . . the answer may be found to the question why in some countries costs and prices relative to agricultural production have risen with a marked effect on mass consumption."

Most of this literature in our view focuses excessively on the *latifúndio-minifúndio* complex. In most of Latin America, what matters in terms of supply is the medium to large-sized farms that have a capitalistic organization and which respond to profit incentives. These farms account for a large share of total output.

The second structural factor is *the limited and unstable capacity of a country to import*, an element obviously related to the export performance of the economy. A number of factors are postulated to be behind this low capacity to import. First, the dependency on primary goods, which are assumed to have limited potential in world markets due to the low income elasticity of demand for these goods as well as the appearance of synthetic substitutes for them.[35] Second, the cyclical downturns of world trade which are assumed to affect especially primary commodities, which are thus subjected to higher degrees of price instability. Third, an unsatisfactory import-substituting industrialization which does not give appropriate consideration to the growth of exports. Prebisch [1961a] stressed the latter point.

Facing these limitations on the import capacity of the economy, it was believed that governments would have to make constant devaluations of the domestic currency in order to re-establish equilibrium. As a consequence, cost pressures would develop following the increase in import prices and this in turn would bring about increases in the general price level.[36] Any positive effects on exports from the devaluations were largely ignored, probably a reflection of the export pessimism of adherents to this view and their belief that agriculture could not respond even if incentives were provided.

The third structural factor is *the distribution of income*, which it was believed could give rise to inflationary pressures in at least three distinct ways. First, there was the low savings coefficient in the region, viewed as a natural outcome of the highly concentrated distribution of income in most countries. The naive Keynesian consumption function has been an essential element of the structuralist model. The Friedman perspective that consumers tend to save a constant proportion of their permanent income does not seem to have been accepted by authors writing in the structural tradition.

This low savings coefficient, in turn, was assumed to contribute to inflationary pressures as the need to increase the investment level above that established by the low saving coefficient appeared. The mechanism was credit expansion which generated forced savings in a manner similar to that described long ago by Wicksell. It is worth noting that seldom was it recognized that low savings behavior might have something to do with the incentives to save. Moreover, institutional reforms of capital markets to eliminate usury laws and to provide viable savings instruments were seldom mentioned, even though Brazil's economic "miracle" of the late 1960s and early 1970s was driven in part by reforms of this nature.

The second way in which income distribution was assumed to contribute to inflationary forces was through the patterns of demand it induced. It was argued that the highly skewed distribution of income diverts investment and import substitution from social overhead investment as well as from mass consumption goods, two kinds of investments that would give more flexibility to the economy [Grunwald, 1961], and channels it instead to luxury goods and investments with low social productivity, such as the construction of apartments and the holding of idle land.

The third way in which the distribution of income was assumed to affect inflation involves the agricultural sector and price incentives. Since most people receive very low incomes, policy makers find it politically difficult to provide proper incentives to agricultural producers. Higher prices would be tantamount to a regressive redistribution of income and consequently are avoided. The result is a further disincentive to agricultural production [Grunwald, 1961]. From this perspective, it isn't that agriculture doesn't respond; it isn't given the *incentive* to respond.

Finally, the fourth structural factor responsible for inflationary pressures is assumed to be *the tax system*. The key issue here is the lack of efficiency in collecting direct taxes (especially land taxes) and the regressivity in taxation that ultimately results as governments turn to indirect taxes. With inflexible current expenditures on the part of the government the budget deficit could only be eliminated by cuts in investment. Since this is undesirable in terms of sustaining growth, the deficit must be financed by inflation unless a change in the tax structure were to occur, including a more progressive schedule [Prebisch, 1961a, p. 21].

Circumstantial factors as structural causes of inflation are presumed to be exogenous, and include episodic events such as increases in import prices due to external factors, general wage increases due to union activities, catastrophes,[37] wars, and the like. *Propagation factors*, defined by Grunwald [1961] and Sünkel [1958], are those elements which, although not primary sources of inflation, provide feedback into the process. They are the result of adjustments induced by the economic agents in their struggle to retain their share of output, and include wage earners trying to keep their relative wages, entrepreneurs trying to maintain profit margins, and the government trying to increase nominal expenditures. In the latter case, the level of real government expenditures has been typically assumed to be more or less fixed because of the need for infrastructure, investment, and the need for the government to be an employer of last resort. Because of this "fixity," it was generally believed that government expenditures could not be cut.

E. A. Cardoso's [1980b, 1981] model, developed in a paper with the title "Food Supply and Inflation," is an example of a structuralist model in which wage behavior plays a central role. The inflationary process is closely linked to inconsistent claims for shares in income, with real desired wages assumed to be rigid. She shows that the structuralist hypotheses are capable of generating a disequilibrium that can lead to constantly rising prices without any mechanism to offset them. Steady inflation can persist only if workers allow their real wage to lag. If the workers try to catch up, the model explodes with accelerating inflation.

From a structuralist perspective there is no such a thing as an anti-inflationary policy, at least as generally conceived as short-term stabilization policy. All that can be done is to devise an appropriate *development* strategy that would eliminate the structural imbalances.

It is interesting to note that the one proposition the monetarists and structuralists agree on is the need for a reform of the tax system, with the reform leading to a more progressive system. This agreement on such a controversial issue is remarkable, especially since a change in the tax system would in almost all cases pose a major political challenge. Beyond that, however, the structuralists argue that the monetarists' stabilization plans are directed only toward the propagation factors and consequently will only reduce growth in the economy without dealing with the fundamental problems.

The asymmetry in the two approaches should be noted. The monetarists believe that markets work and that the emphasis should be on short-term stabilization so as to make long-term growth possible. They give little attention to structural issues or to institutional reform. The structuralists, for their part, are pessimistic about short-term stabilization policy since it is believed to have only deleterious consequences without addressing the fundamental problems. The only solution they envisage is the reform of the rural land tenure system, a redistribution of income and with it a restructuring of demand, a change in the tax system, and in some cases the provision of incentives to the export sectors. Each of these, of course, involves a significant change in the social structure of the economy.

We would make three points in concluding this section. First, as the above discussion implies, the structuralists have been the reformers in Latin America. Second, the policy mix that follows from the structuralist perspective leads to significant redistributions of income. Third, neither the monetarists nor the structuralists gave much attention to factors that would alter the long-term productivity of the society such as investments in the capacity to produce new technology and in other forms of human capital. The reforms proposed by the structuralists *imply* a potential rise in productivity.[38] But they are fundamentally redistributionist in character.

EMPIRICAL TESTS OF THE STRUCTURALIST AND MONETARIST MODELS

There has been a number of tests of the structuralist and monetarist explanations for inflation in Latin America. A useful starting point in discussing these models is with Harberger's [1963] study of Chilean inflation in the period 1939-58. His model contains strictly monetarist variables, but a wage variable was included as well on the grounds that the labor market was characterized by market failures and that the government's wage policy gave wages a special degree of autonomy. At least one monetary variable was significant in all equations estimated, thereby confirming in his view the monetarist case. However, the coefficient of the wage variable was significant in some equations.

Vogel [1974] used data from sixteen Latin American countries for the period 1950-69. His goal was to see how well a monetarist model would do in explaining inflation and thus no structural variables were included. His results confirm Harberger's model, with the caveat that he did not include a wage variable in his equations.

Pastore [1974] used a monetarist model to explain Brazil's inflation in the post-World War II period, but like Harberger he included structuralist variables such as minimum wages and devaluations of the national currency. He found a role for both the monetarist and structuralist variables.

Wachter [1976], contrary to the authors referred to above, was explicitly concerned with testing the structuralist hypothesis. She used data from Chile, Argentina, Brazil, and Mexico, and used the rate of change in the relative price of food as a structural variable. The results for Chile confirm those obtained by Harberger. For the other countries, the monetary variables had statistically significant coefficients, but so does the relative price of food in the cases of Brazil and Mexico.

F. de H. Barbosa [1983, chapter 3] undertook comprehensive tests of both the monetarist and structuralist hypotheses to explain inflation in Brazil in the post-World War II period. His results indicate that monetarist variables are important. When both the monetarist and structuralist variables (oil and agriculture) were included, both sets of variables had significant coefficients. However, the effects of the monetarist variables were far stronger than the other variables in the equation. He concluded that one cannot account for the high rate of inflation in Brazil in recent years solely on the basis of supply shocks.

The strong conclusion that emerges from this research is that monetary variables are an important factor explaining the rate of inflation in Latin America.[39] The results support the monetarists, although not the extreme monetarists. All of the studies except for Vogel's found that a structuralist variable had some role in explaining the dynamics of inflation. We should note, however, that there is some ambiguity among some of the authors as to whether wage rates and exchange rates are true structuralist variables. These variables, as noted above, are included in empirical models that are purportedly monetarist in design. Wachter [1976], on the other hand, is explicit in emphasizing that she has a structuralist variable, namely, the rate of change of the relative price of food. Similarly, F. de H. Barbosa [1983] included explicit structuralist variables in his models.

CONCLUDING COMMENTS

The structuralist-monetarist controversy in the Latin American literature parallels in a very real sense the cost-push *versus* demand-pull controversy in the U.S. economic literature. One difference, however, is that among the structuralists, the price of food is a key variable, although in many models the wage rate is used as the structuralist variable. In the American literature the analysis seldom presses back beyond the wage rate *per se*. Another difference is that proponents of cost-push inflation in the United States seldom have reformist ambitions as strong as those of the structuralists in Latin America. In the latter case the reforms would include a major restructuring of agriculture and major redistributions of income in both agriculture and the economy as a whole.

There are other links to agriculture in the Latin American literature. For example, Dornbusch's [1982] formulation of the monetarist model shows how the failure to implement balanced budgets leads to overvaluation of the currency. An overvalued currency is a tax on exports and a subsidy on imports. Either way, it discriminates against agriculture to the extent that it is a trade sector, and in Latin America it generally is.

The structuralist models make a case for reform of the agricultural sector, but this is to focus on long-term measures as the basis of short-term stabilization policy. The reform measures proposed, moreover, have seldom gone beyond land or agrarian reform, with its strong emphasis on the redistribution of income and assets. The political difficulties in bringing about such reforms have meant that little or nothing has been done to stimulate agricultural development. Interestingly enough, however, the international debt crisis many Latin American countries faced in the early 1980s led to short-term reforms of trade and exchange rate policies and this is having a significant effect on agriculture. Argentina, Brazil, and to a lesser extent, Mexico are important examples.

More generally, the structuralists posited a link between economic development and its effect on the structure of the economy as the key to solving the inflation problem. Reforms which ease the import constraint, improve the distribution of income, or improve the efficiency of the tax system or which increase its progressivity are all seen as solving the inflation problems at the same time they promote economic development.

Finally, modern authors in Latin America still differ and disagree on the most desirable policy for stabilization. However, the old propositions about the utility of structural reform as a means of controlling or reducing inflation have clearly been abandoned. Longer-term reform issues are now properly a part of development economics. Recent discussions have focused on cost-push elements and on monetary variables and purchasing power parity as factors explaining inflation. The stabilization programs of Chile, Argentina, Uruguay, and, more recently, Brazil, have been reflections of the new monetarism in the region. These new stabilization programs are discussed in a later section.

4. Summary

Latin America was at one time closely integrated into the international capitalist system. A series of large external shocks, including World War I, the Great

Depression of the 1930s, and World War II, to say nothing of recurrent trade cycles, caused both intellectuals and policymakers to see a brighter future for their countries if they reduced this dependence. Raúl Prebisch provided the ideas and the leadership to promote the concept of inward-looking economic development. Since Latin America exported mainly primary commodities, and the external terms of trade were viewed as shifting against these commodities, industrialization was viewed as the engine of growth and agriculture was neglected.

Within national economies in the region, inflation has been a pervasive and persistent problem. The structuralists viewed agriculture as one important source of this problem, and proposed reform of the land tenure system as the means to resolve the problems. Kahil [1963] provided a comprehensive criticism of the structuralist position on the role of the agricultural sector in Brazil. Hence we see that throughout much of the post-World War II period agriculture was generally viewed as a problem rather than an opportunity. Over the last ten years or so, however, as import-substituting industrialization policies became discredited and as exports from any sector were needed to service large foreign debt, agriculture has come back closer to center stage. Dependency theory has declined as an academic interest and as a basis for economic policy, as has structuralism as the basis for explaining inflation.

Chapter III. Theories of Agricultural Development

Hayami and Ruttan [1985, chapter 3] identify five models or theories of agricultural development: the resource exploitation model; the conservation model; the location model (also referred to as the urban-industrial impact model); the diffusion model; and the high pay-off input model. We should add to that list their own model, which integrates many of the other models and provides essentially a theory of technological and institutional change.

This chapter provides a selective review of the Latin American literature which bears on theories of agricultural development. After a brief summary of the Hayami-Ruttan model, we review Schuh's eclectic interpretation of Brazil's agricultural modernization, which uses the Hayami-Ruttan perspective but draws explicitly on Schultz' urban-industrial impact model and the high-payoff input model (references provided below). This is followed by other applications of the Hayami-Ruttan model to a better understanding of Brazil's agricultural development. Paiva's model of technological dualism is then considered. This model is important because it shows how technological progress can generate, through developments in the product markets, economic forces which bring that progress to a halt. Paiva's work is also important in that it exemplifies the general pessimism that Latin Americans tend to have about agricultural development.

De Janvry's interpretation of agricultural development in Argentina is then considered. This interpretation is based on a more operational specification of the Hayami-Ruttan model, and provides a test of monetarist and structuralist explanations for agricultural development. Tests of the urban-industrial impact model (implicit in the Hayami-Ruttan model) are then reviewed with data from Brazil. Katzman's research [1974] is especially important in this regard since he introduces von Thuenen's theory of agricultural rent and land use into the model. This gives it a more rigorous formulation.

The remainder of the chapter refers to literature that is not in the Hayami-Ruttan tradition, but which is important in understanding agricultural development in Latin America. This includes a section on resource-based development and the frontier, and Celso Furtado's structuralist model. The resource-based development models address the role of the frontier in the development of Latin American agriculture, and especially hollow-frontier models. In the final section we call attention to some of the literature on various socialist experiments in the region.

This survey obviously neglects a lot of work that has been done on agricultural development. Our goal is to provide a flavor of some of the more comprehensive models, and to relate that work to some of the larger streams of thinking in the international community. Moreover, many of the studies discussed in the following chapters bear on elements of agricultural development theories.

There has been a number of estimates of the rates of return to investments in agricultural research (the high pay-off input model), plus some work on the diffusion of technology in Latin America (the diffusion model). Discussion of these studies is deferred to a later chapter.

1. The Hayami-Ruttan Model

The Hayami-Ruttan [1985] model of induced innovation is probably the most widely accepted model used today both for understanding agricultural growth processes and as the basis for agricultural development policy. The wide acceptance of their model is due in part to the central role it gives to developing the capacity to produce and distribute new production technology as the core of the agricultural development problem, a perspective now widely accepted among policy makers and analysts alike. Hayami and Ruttan relate this core process to institutional innovation and development, and to the nonfarm sector as a source of cheap inputs that substitute for the primary factors of land and labor within agriculture.

Hayami and Ruttan identify the capacity to develop a technology consistent with environmental and economic conditions as the single most important variable to explain the growth of agricultural productivity. They postulate a theory of induced development that gives a major role to factor prices in explaining the innovative activities of public sector research institutions in producing the basis for technical change. Successful achievement of continued productivity growth over time involves dynamic institutional arrangements consistent with the new growth potentials, and must be complemented by investments in general education and in production education for farmers if the full productive potential of the new knowledge and the new inputs is to be realized.

This model is a natural evolution and synthesis of what Hayami and Ruttan identify [1985, chapter 3] as the conservation model, which is capable of producing modest rates of agricultural growth by means of on-farm innovations [see Boserup, 1965]; T. W. Schultz's urban-industrial impact model [1953], which hypothesizes a strong link between urban-industrial development and agricultural development; the diffusion model, which explains how innovations are spread once they are produced; and the high pay-off input model, which shows that the pay-off to investments in the production and distribution of new production technology is quite high. The authors are critical of most previous work of development economists concerned with agricultural development in the post-World War II period who, with their growth stage and dual economy models, had concentrated on identifying the *role* of agriculture in the total development process, to the neglect of attempting to understand the agricultural development process itself.

This same criticism applies to much of the literature on agricultural development in Latin America up through the 1960s, with one important caveat. As noted earlier, the prevalence of agricultural stagnation caused the literature on Latin American agricultural development to concentrate on explaining stagnation not growth. This emphasis caused many Latin American writers to focus on what were perceived as existing institutional constraints such as land tenure arrangements and the size distribution of land holdings, rather than to perceive the potential role and importance of nonexistent or ineffective agricultural research institutes.

The Hayami-Ruttan model has some limitations as a complete theory of agricultural development, although the analytical and empirical perspective they provide in their 1985 edition is significantly broader and richer than what appeared in their original 1971 edition. Their model is about the factor-saving bias that technical change takes in easing constraints to development, and about the institutional arrangements needed to keep technological advances on an efficient growth path. Their analysis extends beyond this, however, to include such things as international transfer of technology, equity issues associated with rapid technological change, disequilibrium in global agriculture, and mobilizing agricultural growth for overall development.

To provide a somewhat broader perspective on some of the models and interpretations of agricultural development to be provided below, it is useful to sketch out a somewhat broader framework than that encompassed by the Hayami-

Ruttan model. A broader analytical framework might start with the concept of an agricultural (production) "surplus" as elaborated by Nicholls [1963b]. Although only a one-sector model, Nicholls makes it clear that it is a *production* surplus (production greater than is needed to feed the agricultural population) that is important and not a labor surplus, and explains how that surplus can be used for more general economic development. Nicholls's perspective was rooted in his extensive experience in Latin America.

The second component of a more general model would be the Hayami-Ruttan model, which would explain how the production "surplus" can be most efficiently generated. The modern theory of agricultural factor markets, with its more rigorous treatment of interlinkages among them (see Binswanger and Rosenzweig, [1984a, b] and the section below), complemented with Hirschmann's [1977] generalized linkages perspective, would constitute a third component. The demand for agricultural output would be the fourth component, with this demand including foreign demand. An explanation of how institutional arrangements for product and factor markets evolve would also be needed. And the model would be completed with a theory of the state and public institutions to explain how government intervenes in the process and how public institutions evolve. This more comprehensive model would explain agricultural output and the prices of that output, the level of resource use in agriculture and the returns to those factors, the disposition of the agricultural surplus, what policy makers do, and the evolution of institutional arrangements.

We now consider various models and interpretations of agricultural development in Latin America.

2. Schuh's Eclectic Interpretation of Brazil's Agricultural Modernization [Schuh, 1975c]

Up until the second half of the 1960s, Brazil's agriculture was characterized by low levels of productivity by international standards, and a rather slow rate of modernization. At the same time, the relatively small state of São Paulo, which not many years before that period had accounted for 30 percent of Brazil's agricultural output, had developed a relatively modern agriculture, with the productivity of land and labor growing throughout the post-World War II period. In addition, the use of modern inputs for Brazil as a whole had been growing at a rapid rate from 1967 up to the early 1970s.

Schuh [1975c] provided an interpretation of why the agricultural sector had grown in this unbalanced way, and an analysis of the factors that influenced the rate of modernization. He rested his analysis on the Hayami-Ruttan model [1971], Professor T. W. Schultz's urban-industrial impact model [1953], and the interactions between economic policy and technical change. We review his synthesis here in part because it draws in important ways on some of the significant work done on Brazilian agricultural development.[40]

Brazil's economic history has been characterized by a long succession of cycles in basic products, among them sugar, rubber, coffee, gold, cattle, cocoa, and cotton.[41] With one exception, the booms in each of these products eventually died out. The exception, coffee, gave rise to self-sustaining development and constituted the basis of the rapid industrialization of São Paulo and of the central, industrialized region of Brazil. Associated with the industrialization, there was a successful transformation of agriculture in the region that sustained and contributed to the development of the economy despite a severely discriminatory economic policy.

Schuh [1975c] suggested that there were four enigmas to the modernization of Brazilian agriculture in the post-World War II period: Why was there in the aggregate so little modernization in this period, as reflected in low levels of productivity, only small tendencies for this productivity to rise—with some important exceptions—and relatively low levels of use of modern inputs? Why did the agriculture of São Paulo experience substantial modernization, despite the lack of modernization in the rest of the country? Why, even in the state of São Paulo, did modernization not occur in the food or subsistence crops? And, how does one explain the rapid increase in the use of modern inputs starting in 1967?

These four enigmas constituted parts of a larger picture, and could in Schuh's view be explained by available economic theory. For evidence in support of his interpretation he drew on the research of others. His starting point was Hayami and Ruttan's theory of induced technical change. In addition to the need for technical change to ease the constraint implied by land, the primary factor of production that is relatively inelastic in supply, the Hayami-Ruttan model has two additional components. The first is the need for a *local capacity* for agricultural research and development (R&D) to generate new production technology since biological research tends to be highly location specific. The second is the need to develop an industrial capacity to produce the modern inputs needed to capitalize on the innovations produced by the research process, or sufficient growth in exports so these modern inputs can be imported.

Even in this simplified form, Schuh argued that the Hayami-Ruttan model was useful in explaining the lack of modernization of Brazilian agriculture. In the first place, it was possible to bring additional land into production at greater distances from the urban centers with very little increase in the supply price of food. This was possible in part because a modern road system was built to accompany the expansion of the frontier, and in part because there was a rapid modernization of the transportation system so that transportation costs declined, offsetting the increasing distances. At the same time, there was practically no increase in the real supply price of labor to agriculture. With these two important conditions in the markets for the primary inputs, there was little tendency for the real price of agricultural products to rise, and hence little incentive to invest in agricultural research. This last statement is not part of the Hayami-Ruttan model, of course.

These conditions were reinforced by two other developments. First, the incorporation of the relatively fertile lands of Paraná and of the South of Goiás and Mato Grosso was primarily a phenomenon of the post-World War II period and tended to coincide with the rapid rural-urban migration and expansion of the urban-industrial economies of São Paulo and Rio de Janeiro. Second, by damping off exports, economic policy, which was highly discriminatory against agriculture in most of the post-World War II period, caused total demand for agricultural output to be less than it otherwise would have been, and thus further reduced any incentive to invest in agricultural research. Distortions in trade and exchange rate policy caused an even larger share of a modestly expanding agricultural output to be channeled to the domestic economy.

But there were still other factors at work. Import-substitution industrialization policy concentrated on developing local industries for products that were previously imported. Since modern agricultural inputs were used at low levels at the time this process started, no local capacity to supply these inputs had been developed. Later, when these industries *were* established, relatively high levels of protection caused the prices of these inputs to be higher than they otherwise would have been and thus reduced the incentives to use these inputs. This same industrialization policy had a strong anti-employment bias, with the result that labor left agriculture at a slower rate than it otherwise would have, helping to keep agricultural wage rates relatively low, despite the rapid pace of industrialization and a significant flow of rural-urban migration.

Finally, high and unstable rates of inflation created a great deal of price uncertainty for agriculture, making farmers less likely to incur technological risk. These high and unstable rates of inflation also contributed to significant imperfections in capital markets and discriminated against small producers, with the result that such technological innovations as were available were adopted only by the large producers. Moreover, inflation caused asset owners to shift their portfolios in favor of real assets, with the expectation being a gain in capital value, providing little incentive to raise the productivity of these assets.

Thus, Schuh [1975c] saw the lack of modernization in Brazilian agriculture as a whole to be a function of a resource endowment that provided little incentive for land or labor-saving technical change and economic policy that reduced price incentives for agriculture. Thus there was little incentive to invest in agricultural research for the nation as a whole.

How, then, does one explain the paradox that the state of São Paulo experienced a very rapid rate of modernization in its agriculture? The immediate answer is that the state of São Paulo did, in contrast to the rest of the country, invest in agricultural research and extension and in the education of its rural population. The data are truly remarkable. As of 1965, for example, São Paulo had more agricultural researchers than the rest of the country as a whole. Its number of extension workers was only slightly less than in the rest of the country. And in many years the budget for São Paulo's State Secretariat of Agriculture was larger than the budget for the national Ministry of Agriculture. Similarly, in some years towards the end of the 1930s and at the beginning of the 1940s, the state of São Paulo invested more in breeding improved seeds for cotton than did the combined private and public sectors in the United States in the development of hybrid seed corn.[42]

These relative proportions showed the low level of investment in agricultural R&D for the nation as a whole, and also why the technological performance differed so much between São Paulo and the rest of the country. But an important question still remained. Why did São Paulo invest so much in the modernization of its agriculture and the rest of the country so little? Schuh explained this by means of the urban–industrial impact model,[43] together with an analysis of who benefited and who bore the costs of Brazil's development programs.

T. W. Schultz's [1953] urban–industrial impact model (see below) has to be understood both in large and in small contexts. Putting the analysis in a somewhat larger context shows the close interaction between agricultural and industrial development and the complementarity between the two in attaining self-sustained economic development.

Schuh [1975c] takes as his starting point the fact that Brazil's economic history had been dominated by a long sequence of commodity cycles. The last of these booms, coffee in São Paulo, was transformed into self-sustaining development because its agricultural "surplus" could be converted into a broader-based industrialization and development. Dean [1969], in an unusually perceptive analysis, shows how the coffee economy was different than the economies of previous basic commodities and how its boom culminated in industrialization. Nicholls [1969a] had a similar analysis, but with less detail than Dean's.

The important points are that the industrialization of São Paulo was based fundamentally on the coffee boom and that a large part of the capital and the entrepreneurial talent for industrialization came from agriculture. At this level one interprets the urban–industrial impact model in the large. The combination of agricultural and industrial development resulted in a diversified economy that was strong and grew rapidly. When the coffee sector began to decline, the strong local economy kept the decline from spreading greatly beyond the sector because it was diversified. Perhaps the strong point was the ability to sustain a strong agricultural research effort, even though agriculture was in a down cycle.

The best example of this robustness was the ability of the research system to develop alternatives to coffee as it declined. One of the outstanding alternatives was cotton, and the advantage of cotton was that it supplied fiber for a rapidly growing textile industry. This reinforced the complementarity between industry and agriculture and caused the industrial sector to be willing to support agricultural research. Moreover, the production of cotton grew so rapidly that the country became a major exporter of cotton.

At this point it is useful to understand why the research program focused on export or cash crops at the expense of food crops. Schuh argued that this could be explained in large part by identifying the beneficiaries of the research and by understanding the distribution of political power. Power in the state legislature at that time was largely in the hands of the landowners-*cum*-industrialists. The benefits of research on food or subsistence crops, given their low price elasticity of demand, would have been distributed in large part to consumers in the form of lower prices. The benefits of research on export crops, on the other hand, went in large part to the landowners in the form of economic rents which redounded to the land and to entrepreneurial skills, both of which were in relatively inelastic supply [Ayer and Schuh, 1972].

Landless workers may have gained in a relative sense because cotton was probably more labor-intensive than other alternatives, but probably not as labor-intensive as coffee. Moreover, in general it should be noted that strong development efforts directed to subsistence crops tend to produce immiserizing growth for the agricultural sector itself. Both the price and income elasticities of demand for these commodities tend to be relatively low. If productivity rises significantly, there will in general be a need to transfer resources out of the sector and a decline in land values, other things being equal. For these reasons, the frequent arguments that development efforts should be focused on small producers of subsistence crops seem misguided, especially when it is argued as being in *their* interest. Society as a whole, of course, might benefit handsomely from such efforts, especially consumers and nonfarm workers. The small producer would benefit only if there were room for import substitution, or if the off-farm market for labor were particularly strong.

The urban-industrial impact model interpreted in a smaller context explains why the development in São Paulo became self-sustaining. Industrialization created strong and more efficient capital and labor markets, led to the development of the strongest infrastructure in Brazil, and so on. These features benefited agriculture, but a strong agriculture also benefited the industrial sector as well. Moreover, increases in per capita incomes led to diversification of agriculture toward fruits and vegetables and other high-value crops such as sugarcane, which further strengthened agriculture. The fact that there was a continuous flow of migrants into Paulista agriculture from other parts of Brazil helped make these labor-intensive crops economically viable.

The urban-industrial impact model in these two senses helps explain how the process of agricultural modernization in Brazil got started in São Paulo. However, Schuh [1975c] argued that this didn't tell the whole story. The remainder of the story rested with the interactions between economic policy and technological change.

The analytical framework for understanding this interaction was taken from de Janvry's [1973] elaboration of S. Ahmad's [1966] model of induced technical change (see below). This model permits both factor prices and product prices to become endogenous as a function of technical change. Thus, for example, if the commodity is tradeable and therefore faces a relatively—if not perfectly—elastic demand, the benefits of biological innovations will be capitalized in land values, thus giving ever stronger incentives for biological innovations. This process drives technological innovation in the direction of an ever larger land-saving bias (Cochrane's technological treadmill) [Cochrane, 1958, chapter 5]. If, on the other hand, the demand for the commodity has a relatively low price elasticity of demand, the declining price for the commodity may lead to a decline in the price of land, and at best a more neutral resource-saving effect of the new technology. Moreover, the decline in product price may drive a faster rate of adoption of new production technology in order for farmers to stay alive economically [see Schuh, 1974, for example].

Given that São Paulo already had a well established research system at the end of World War II, and the rest of the country did not, the discriminatory economic policies towards agriculture associated with import-substituting industrialization had a significant and differential effect between São Paulo and the rest of the country. With that research program already focused on export crops, there were strong pressures generated for biological, land-saving production technology. Although the rapid expansion of the industrial sector pulled a great deal of labor out of São Paulo agriculture, this labor was continuously replaced by migrants from other parts of the country. In effect, agriculture faced a perfectly elastic supply curve of labor. In this setting, discriminatory economic policy which kept product prices lower than they would otherwise have been only accelerated the process of technical change since a steady stream of new technology was being made available.

For most of the rest of Brazil, however, the situation was different.[44] The agricultural research capability was quite limited and thus there was only a limited flow of new production technology. The consequence of the discriminatory economic policies was to push labor and other mobile resources out of agriculture, leading to a more extensive form of organization and sustaining pressures to expand the frontier. Economic policies thus exaggerated a duality in Brazilian agriculture that a geographic differential in research capacity and a geographically differentiated configuration of political interests had started.

In conclusion, Schuh provides an explanation for the broad lines of agricultural development in Brazil that is very much along standard neoclassical lines, extended by a consideration of political-economic interests and how these interests affected the flow of public resources. No conspiracy or dependency thesis is required. In fact, the emphasis on export crops may have been in the best interests of landless workers and small producers, since export crops were more labor-intensive and created a greater demand for labor than would have been the case had the research process focused on subsistence crops. Whether this was in the best interests of the country as a whole, however, is an open question.

3. Other Brazilian Applications of the Hayami-Ruttan Model

Alves and Pastore [1985][45] analyzed the impact of Brazil's general economic development and agricultural policy on the modernization of agriculture by means of the induced innovation hypothesis. They noted, as have others, that Brazil's agriculture had grown at a rate consistent with the expansion of the general economy as a whole in the post-World War II period, not being a brake on economic development but also not being a driving force. The main issue was that agriculture had expanded on the extensive frontier, with only modest gains in productivity overall.

The authors noted that there have been two explanations offered for the low indices of productivity in Brazilian agriculture. The first, very popular in the 1960s, attributed the stagnation to the agrarian structure and its inability to respond to price incentives (the structuralist interpretation). In their view, however, Pastore's [1973b] study of supply response (see below), which showed significant supply response to changes in prices, demonstrated that Brazilian farmers did in fact respond to prices and that structuralist interpretations were not relevant.

The structuralist argument, in the view of Alves and Pastore [1985], rested on a number of propositions: imperfections in the capital market, created in part by high and unstable rates of inflation, lead to imperfections in the market for land; land is a store of value in Brazil and is acquired more for this value than as a factor of production; imperfections in the capital market also limit access to land, and the desire of the large land-owners to hold their land as a store of value reduces still further the available supply of land; high population growth rates and a limited labor absorptive capacity of the nonfarm sector (imperfections in the labor market) lead to subdivision of small and medium-sized farms; and as a final result, a dual structure of labor use is created: the small farms, which use family labor intensively, and the large farms, which use relatively little labor in the factor proportions sense.

Alves and Pastore [1985] refer to Cline's [1970] study of the agrarian structure, in which he argued that agriculture is operating at a point interior to the production possibility curve. They argue that there is nothing in a land reform that will deal with the basic market imperfections that created the problem, so they see little in land reform that would create a self-sustaining agriculture.

The other explanation of the stagnation in agricultural productivity in Brazil is offered by people who are in the tradition of Professor T. W. Schultz [1964]. These authors, whom Alves and Pastore identify as Schuh [1968], Nicholls [1973], and G. W. Smith [1969], attributed the stagnation to discrimination against agriculture by means of economic policy as part of an import-substituting industrialization policy; insufficient investment in agricultural research and in the education of the rural population; the failure to develop efficient industries to supply modern inputs; and general "cheap food" policies which reduce the incentives for investment in agriculture. These authors do not see the agrarian structure as the main obstacle to modernization, but each sees a place for localized modest land reforms.

Alves and Pastore noted that two routes have been taken to stimulate agriculture in Brazil. The first was directed at expanding the area in farming, while the second focused on raising productivity *per se*. On the first count, they used the Hayami-Ruttan model to argue that incentives never developed in Brazil to provide the motivation for investments in agricultural research. Land was an abundant factor, and the policy of the government in the post-World War II period was to "fill up" the interior. Incentives to this end were created by the construction of a new capital, Brasília, in the middle of the country, and a massive road-building program.

By the same token, the population growth rate was high throughout this period, and the import-substituting industrialization policy, with its anti-employment bias, caused labor to remain dammed up in agriculture, thereby creating little incentive for mechanization. As they put it, the growth in the agricultural labor force, which didn't begin to turn down until the 1970s, essentially matched the growing demand for food and fiber.

Finally, economic policy was designed, in their view, to minimize the use of capital in agriculture. The objective was to channel as much capital as possible to the industrialization drive.

Interestingly enough, Alves and Pastore [1985] argued that this policy was perfectly rational up to the mid-1960s, when the economic "miracle" of the late 1960s and early 1970s led to rapid economic growth and rising prices of food. In their view, using the relatively abundant land and labor was an appropriate policy until this spurt in growth occurred.

This conclusion is somewhat surprising in light of the high rates of return that agricultural research has been shown to have, even in Brazil,[46] and the other benefits that productivity growth could have contributed. Productivity growth would have undoubtedly created more capital for industrial expansion, and also earned more foreign exchange, even with discriminatory trade and exchange rate

Table 6. Annual growth rates of crop production and its sources,
Brazil and regions, 1950–68

Component	Region	Period (in %)	
		1950–60	1960–68
Change in production per area:	Brazil	1.77	2.03
$\dfrac{\Delta(Y/A)}{Y/A}$	São Paulo	3.76	4.79
	Central-South	1.55	2.09
	Northeast	0.48	0.62
Change in area/man relations:	Brazil	0.54	1.96
$\dfrac{\Delta(A/N)}{A/N}$	São Paulo	−0.54	0.62
	Central-South	3.39	1.68
	Northeast	4.39	1.65
Change in rural labor:	Brazil	3.53	1.36
$\dfrac{\Delta N}{N}$	São Paulo	1.21	−1.32
	Central-South	3.39	1.68
	Northeast	4.39	1.65
Change in agricultural production:	Brazil	5.84	5.35
$\dfrac{\Delta Y}{Y}$	São Paulo	4.92	4.09
	Central-South	6.56	5.76
	Northeast	5.03	5.40

Source: Pastore, Alves, and Rizzieri, 1976. With permission.

policies. These could have powered a more rapid rate of industrial growth, and alleviated the persistent balance of payments problems that characterized this period, at the same time creating more employment.

Pastore, Alves, and Rizzieri [1976] later developed evidence on the induced innovation hypothesis in Brazil in what is a careful and important analysis of the agricultural development process in that country. The authors first calculated trends in agricultural production and their sources of growth for three important regions of Brazil and for Brazil as a whole (Table 6). The three regions included the poverty-stricken Northeast, which includes roughly a third of Brazil's population; the Central-South—Brazil's agricultural and industrial heartland—excluding São Paulo; and São Paulo itself. This division in effect divides the country into low, middle, and high-income regions, with the level of agricultural modernization classified approximately the same.

In the 1950s Brazilian crop production increased principally by means of an increase in the area cultivated (accounting for 70 percent of the total increase). Thirty percent of the increase was accounted for by an increase in production per hectare. Viewed from the perspective of labor, an increase in production per worker accounted for 40 percent of the total increase in output, with the increase in employment accounting for the remaining 60 percent.

The regional differences were very great in that decade. In the state of São Paulo, an increase in production per hectare accounted for almost all of the increase in output. In the Central-South, productivity accounted for a somewhat smaller share, but in the Northeast, increases in land and labor productivity were essentially nonexistent.

These patterns changed significantly in the 1960s. The contribution of land productivity increased significantly in the Central-South, São Paulo, and Brazil as a whole, and the absorption of labor declined very significantly. The growth in area per worker was significant in all cases, although less in the Northeast than in the other two regions.

The average annual percentage growth rates in land and labor productivity in the period 1955-65 were as follows:

	Labor productivity	Land productivity
	Y/N	Y/A
Brazil	4.0	2.0
Central-South	4.1	2.1
São Paulo	5.4	4.8
Northeast	3.8	0.6

These data suggest that the technological path in Brazil was closer to that of the United States than to that of Japan.

Pastore, Alves, and Rizzieri [1976] assembled the limited data available on changes in factor prices. In the case of São Paulo, they found that the price of labor rose slowly relative to the price of agricultural output, while the price of capital goods declined. This, combined with the rapid increase in area under cultivation, was inducing mechanization in a manner consistent with the Hayami-Ruttan model. In other regions, the price of labor rose very slowly, but the price of capital goods also declined. This was sufficient to induce some mechanization. In the case of land, the price was rising on a more localized base. São Paulo, of course, with the large rise in its urban population and industrial sector, was experiencing larger increases. This helps to explain the more rapid rise in the productivity of land in that state.

The authors concurred with the previous study by Alves and Pastore [1978] that the slow growth in productivity in Brazilian agriculture was consistent with the Hayami-Ruttan model and that, moreover, this extensive development model was rational for Brazil. Little mention was made, however, of the significant protection of both the fertilizer industry and the machinery and equipment industries, or that economic policy limited the export market for agricultural products and thus sacrificed growth potential in agriculture, or that increases in productivity are the basis for increasing per capita incomes both in the agricultural and nonfarm sectors. They referred to the high social rates of return Ayer and Schuh [1972] found for cotton research, but said nothing about the economic growth sacrificed when such high rates of return were foregone.

4. Paiva's Model of Technological Dualism

Ruy Miller Paiva is one of the leading agricultural economists in Latin America and has long been a student of agricultural development both in Brazil and elsewhere. Although very much in the neoclassical tradition, he is not optimistic about the potential for modernization or the development of agriculture. Perhaps his first articulation of a model to explain his pessimism was in a paper on technological dualism, which was published in 1971.[47]

Paiva's analysis had to do with the limits of modernization within the agriculture of a given country. He suggested that there is a mechanism of self-control (or endogenous control, if you will) to the process of modernization, by means of which agriculture will reach what he calls an "adequate degree of modernization." Market forces inhibit growth in agricultural productivity beyond this level of modernization, thus "braking" it at a low level. This "adequate degree of modernization" is characterized by the simultaneous existence of modern and traditional techniques of production.

Paiva [1971] argued that a technological innovation, once it is adopted by producers, tends to generate or bring with it large increases in the physical productivity of the modern factors of production, with a consequent increase in production. Due to the low elasticity (price and income) of demand for agricultural products, prices of these products experience a substantial decline, creating an attenuation in the incentive to adopt the new technique. As the dissemination of the modern technique proceeds, there will also be a decline in the prices of the traditional inputs. This purportedly raises the relative economic advantage of the traditional techniques, due principally to the fact that the cost of production with these techniques depends more on the prices of the traditional factors (land and labor). This inverts the situation in part. The economic advantage of the modern technique relative to the traditional technique will decline more, or even become lower than that of the modern technique, as the diffusion process proceeds, thus reducing the number of producers disposed to adopt the modern technology. Thus, one is left not only with a technological dualism in agriculture, but a situation in which the level of modernization is limited by the expansion of *the economy as a whole* since such expansion will be necessary to make the product market strong and thus keep the product price from declining, as well as to keep the traditional factors of production relatively scarce. The model emphasizes the growth of the nonfarm sector of the economy as an important element in the process of agricultural modernization in a market economy.

Paiva's model offers some important insights, although we don't believe the pessimism which follows from his model is fully warranted. One of the insights is a reminder that the degree of modernization within agriculture is seldom uniform, and instead varies a great deal from one farm to another and from one region to another. Similarly, it seems clear that induced changes in product prices and in the prices of the primary factors can influence the adoption of modern inputs and techniques, and thus the process of modernization can be self-limiting if factor markets are imperfect, and if agriculture makes up a large share of the total economy. By focusing on the product price as an endogenous variable that can affect the rate of adoption of new production technology, Paiva extends the Hayami-Ruttan model with the proposition that extensive diffusion of a new production technology can eventually drive down the prices of primary factors of production (land and labor) and thus reduce the incentive to adopt the new technology. This is an important general equilibrium effect.

However, Paiva tends to neglect some of the positive effects of modernization. In the first place, his basic model is that of a closed economy. If trade is admitted and the country is a small country in a trade sense, there may be no negative effect on the price of the commodity from the adoption of new production technology. Moreover, under these conditions the value of biological technology will tend to be capitalized into land values, thus driving the process of innovation and adoption *forward*, rather than braking it, by increasing the price of land relative to the price of fertilizer.

But consider the closed economy case. The important point is that the new production technology tends to be an important source of new income streams, first to early adopters of the technology and later to consumers as economic rents are competed away and product prices decline. If the country is quite poor, the income elasticity of demand for agricultural products (especially food) will be relatively high. The decline in product price will distribute the benefits of the new technology progressively in favor of the poor. The increases in income for those affected by the decline in product price will increase demand. How strong this increase will be is determined by the stage of development and other characteristics of the economy. The increase in demand can be fairly large at low levels of development.

There are similar questions about the effects of product price declines on factor markets. Paiva seems to assume either that the markets for agricultural land and labor are completely isolated from other land and labor markets in the rest of the economy, or are highly imperfect. In point of fact, the agricultural land and labor markets are generally linked to those in the rest of the economy, with the result that the price of land and labor tend to be determined as much by conditions in the rest of the economy as within agriculture alone.

Of course, these intersectoral input markets are often imperfect. Consequently, rapid rates of technical change which under other conditions would drive labor out of agriculture, may occur under conditions in which labor is unable to migrate, or in which only a small proportion of the labor force can migrate. Under these conditions, there could actually be a decline in the real wage in agriculture. But the issue then becomes what to do about the imperfections in the land and labor markets. Government policy should be directed at improving these markets so modernization can proceed, rather than accepting the self-induced constraint.

It is well known that agricultural labor experiences serious adjustment problems in an economy undergoing rapid technological change. Moreover, it is well known that this adjustment problem must be dealt with if both agriculture and the economy as a whole are to capitalize fully on the modernization process [T. W. Schultz, 1961b].

Paiva also fails to recognize that an increase in productivity implies a decline in the average cost of production. Hence, the product price can well decline from the increase in output without excessively reducing production. Clearly, those who have not adopted the new technology will suffer economic pressure and eventually be pushed out of production. But presumably much of that land will be channeled into the hands of those who are willing and able to adopt the new technology.

Producers located in areas where the new production technology is not adapted can experience serious adjustment problems and eventually be pushed either out of production entirely or into the production of another commodity. The solution to their problem is to have a locally adapted technology developed, or to shift to other crops to which their resources are adapted. The infrastructure and marketing arrangements to handle these new commodities would have to be developed.

Finally, Paiva leaves little room for the diversification of agriculture as modernization proceeds. As per capita incomes rise in the country, perhaps driven by the process of agricultural modernization, demand will shift towards commodities such as livestock and livestock products, fruits, and vegetables. This shift in resources in response to this shift in demand will reduce the market pressure on the commodity experiencing the technological change.

Paiva [1979] later expanded his ideas in a book on the limitations of agricultural development as a dynamic factor in economic development. Here he extends his pessimism, and almost concludes that there is no potential for the modernization of agriculture and that the source of economic development must be sought elsewhere. This larger analysis is almost as interesting as his more narrow model.

Paiva started his analysis with a critique of three leading proponents of modern agricultural development theory: T. W. Schultz [1964], Hayami and Ruttan [1971, 1985], and Mellor [1966]. His criticism can be summarized in the one phrase he used to characterize these studies: "An Excess of Optimism." The three reasons he believed these authors are excessively optimistic were as follows: lack of land that is agronomically suited; the limitations of research; and the lack of economic opportunity and the small size of consumer markets. He believes that students of agricultural development have not given enough attention to these issues, and dedicates a chapter of his book to each limitation.

On the issue of land, Paiva was essentially a fundamentalist. He seemed to believe that land productivity is based inherently on its natural state, and little can be done to improve it. He made his case on the basis of Brazilian data, and went to great lengths to show that only a small proportion of the available land is now capable of producing adequate yields and to show how low those yields are. Paiva failed to recognize that the role of knowledge and new production technology is precisely to raise land productivity, and that a large share of the production potential of land in countries that now have a modern agriculture is man-made, and not inherent in the "original" properties of the soil.

Paiva's concerns with the limitations of agricultural research focused on three issues: poor soils that do not react to the application of fertilizers and lime; soils that are rocky or have steep slopes and thus do not lend themselves to mechanization; and poor climates that rule out regular harvests and satisfactory yields. Again, the emphasis was on the land, and the basis for the analysis is the present *status quo*. He focused on applied research, and expressed doubts over whether such research can in fact solve the difficult production problems. This position fails to recognize the extent to which science and technology *can* remove these constraints, and the extent to which it has, in fact, done so in the past, including a reduction in the role of climate as a production constraint. Experience with the modern rices and wheats illustrates well the potential of applied research programs.

Finally, Paiva discusses market limitations. A key part to this analysis is his model of self-control, as described above. In terms of the domestic market he considers low prices and income elasticities of demand, especially for subsistence crops, and market imperfections. These were discussed above. On the trade side, Paiva emphasizes the advantages the developed countries have, and the weak market outlook for trade in agricultural commodities. An important point in this context is that developed countries, with their well-developed physical infrastructure and marketing arrangements, can better capitalize on new production technology than can the developing countries. He believes this gives them a new permanent advantage.

Brazil has in a short period of time taken away a large share of the international soybean market once dominated by the U.S., has become the world's leading exporter of orange juice, and also one of the largest exporters of poultry. Moreover, as noted above, Brazil, like many other Latin American countries, has suffered self-inflicted wounds in their trade in agricultural commodities. Hence, the degree of trade pessimism which Paiva expresses also seems unwarranted.

To conclude, Paiva's model of self or endogenous control has some important insights for development theory and policy. Moreover, his perspective is useful as an illustration of how one of the leading thinkers about agricultural development in Latin America views the problem. His general pessimism about the potential for agricultural development is not atypical.

5. De Janvry's Model of Argentine Agricultural Development

De Janvry's analysis [1973] of Argentine agricultural development was based on a more operational specification of the Hayami-Ruttan model of induced technical change, but it was very much in that tradition. His contribution to the theory was to specify more rigorously the decision processes underlying the generation of agricultural innovations by the public sector and the adoption of new technologies by individual entrepreneurs. In both of these decision processes, an attempt was made to show the dynamic interplay of economic stress *versus* economic incentive. In doing this, he introduced social elements into the economic theory of inducement of innovations.

The extent to which de Janvry [1973] extended and enriched the Hayami-Ruttan model can be seen in his own statement of the contribution of his paper:

> Of prime concern is the prevalence of agricultural underdevelopment that is conceptualized through a lag in the dynamic adjustment path toward potential technological and institutional innovations. For this purpose, the generation of innovations is specified in a demand and supply framework, and differences between a latent and an actual demand are distinguished. Latent demand for innovations is that which, when met by supply, leads agricultural development to an optimum consistent with prevailing economic and scientific conditions. If these conditions are socially optimum, so will the latent demand be. Actual demand, which guides the course of current public sector innovations, is conditioned by government and by socially and politically dominant farm interests, and generally will diverge from latent demand, thus

creating lags in the generation of socially optimum innovations. Hence, of major importance for understanding agricultural underdevelopment is the specification of shifters in latent and actual demands for and supply of innovations.

The socioeconomic model of induced innovations presented in this paper is aimed at understanding the stagnation of the Argentine agricultural sector and, consequently, also of its national economy. That stagnation is due in great part to the unavailability to farmers of land-saving, yield-increasing technology is shown by two case studies. This unavailability is traced to a lag in the course of technological and institutional innovations, and this lag is in turn explained by the interplay of shifters in latent and actual demands and supply for innovations in that country.

Turning to the diffusion process of new technologies in Argentina, we analyze the very particular dynamic properties of a land market-induced treadmill that acts as a coercive drive for the adoption of new production techniques. It is shown that the long-run nature of this mechanism calls upon an activation of the land market through land taxation and more flexible land rental laws in order to accelerate the agricultural technification process. Finally, the extractive mechanisms of agricultural surplus are contrasted to those of Owen's "Mill-Marshallian" model to show that in this case no income transfer occurs naturally through market forces and that taxation schemes need to be implemented to redistribute among sectors the welfare gains from technological change.

De Janvry's model follows from Ahmad [1966], and makes use of the *ex ante* concept of a historical Innovation Possibility Curve (IPC), unit isoquants in a factor-factor quadrant, and a unit price line. Factor and output prices are taken to be endogenous and the innovation process can be shown to have specific resource-saving effects. Moreover, the consequences of the demand for output being price elastic or inelastic can be taken into account, and a *latent* demand for technological innovations is part of the analytical apparatus.

De Janvry [1973] points out that in the Hayami-Ruttan model, the generation of innovations is studied at the reduced form, supply-and-demand level and not at the structural form level of the decision functions of government, farmers and researchers. He views their approach as satisfactory because they work in a comparative statics framework and are not overly concerned with a specification of the dynamic adjustment path generated by the successive interplays of individual adoption decisions and of public innovation decisions.

In his view the first step in specifying a socioeconomic model of induced innovations consists of distinguishing between latent and actual demands for innovations. The second step consists of specifying the decision processes that underlie the *actual demand* for innovations. He believes this demand will materialize in essentially two forms: the budget allocated for research, and a flow of information from producers to the Agricultural Experiment Station. The crucial question then is to determine *whose* demands affect the size and allocation of money to research and constitute sources of information for the researchers as to the type of innovations currently needed. The answer is found by identifying the economic agents to whom the welfare returns from technological changes accrue. In Argentina, with a highly elastic long-run demand for exportable agricultural products, the direct welfare gains from technological change accrue wholly to the agricultural sector under the form of higher producer surpluses. Because the increase in gross income results from higher exportable surpluses, welfare gains also accrue indirectly to the whole economy through particularly high import multipliers.

Two demands for technological innovations correspond to these two destinations of welfare gains—the demand from the agricultural sector that is voiced through the dominant farm organizations, and the demand from the public at large that is voiced through the government. Because there exists a variety of forms of technological change and because they can affect producer and exportable surpluses differentially, the two demands may be in conflict.

De Janvry [1973] argues that almost universally the demand for innovations that has dominated the course of agricultural research originates in the agricultural sector. He asserts that the actual demand in Argentina for innovations originates only in the agricultural sector due to what he describes as the total absence of government policies toward technological change in agriculture. He attributes power to the large landowners, who tend to dominate agricultural interests. Hence, he postulated that the actual demand for public innovations resulted from the maximization of the utility function of the dominant farm interests.

De Janvry identifies several schools of thought which have provided interpretations of the stagnation of Argentina agriculture. The "monetarist" position is conceptualized in his model as a shift in latent demand away from its socially optimum position due to the level and variability of relative prices. This is the interpretation provided by Lucio Reca [1967] in his Ph.D. thesis.

The traditional "structuralist" interpretation of stagnation, on the other hand, attributes the inelasticity of aggregate supply response to the lack of entrepreneurial behavior in the agricultural sector. De Janvry cites Ferrer [1967] as an example of this perspective, and quotes him as follows:

> Large landowners do not seem to have followed the pattern of

behavior characteristic of the entrepreneur in the capitalist
system. Frequently, land is held for prestige or social status and
as a hedge against inflation, rather than as capital that should
be turned to yielding a maximum profit through the use of
manpower and investment. The system of large landowners
and tenant farms for the most part explains the continued low
yields per hectare of the main products of the Pampean region.
It also explains the failure of price incentive policies followed
after 1950 for the purpose of raising agricultural output in the
Pampean region. [pp. 106-107]

A third interpretation of stagnation that calls upon both monetarist and structuralist arguments is that it results from the unavailability of yield-increasing new
technologies for adoption by individual farmers. In de Janvry's model this technological barrier was explained by the lag between actual and latent demand for
innovations (structuralist factors) and by the shift of latent demand away from a
social optimum (monetarist factors). This proposition coincides with Diaz-
Alejandro's diagnosis [1970] of the performance of Argentine agriculture, which
will be referred to later.

In de Janvry's view the traditional structuralist position is easily refuted
through the analysis of adoption patterns of available profitable technologies and
through the high supply responsiveness of individual products obtained through
reallocation of land among activities. By contrast, the lack of aggregate supply
response indicates either the unprofitableness or risk of technical change or the unavailability of yield-increasing technologies for adoption by individual farmers.

To show the absolute and relative contributions of monetarist and structuralist
factors to stagnation, de Janvry [1973] made an analysis of the actual and potential
processes of technological change in Argentine exportable agricultural products,
specifically in corn, wheat, and beef cattle, in an attempt to interpret their stagnation within his model of induced innovations.

He viewed his task [p. 430] as showing that:

(1) current price ratios generate a *latent demand* for land-saving, yield-
   increasing technologies.
(2) this latent demand is eventually in substantial divergence from its
   socially optimal position, mainly for monetarist reasons.
(3) the new land-saving technologies are incongruent with the current
   factor ratios of dominant farm interests. Hence, because there is no
   stress, the *actual demand* for these innovations to the public sector is
   weak, with a consequent lack of funds for research on those
   technologies and an orientation of research away from them.

(4) lack of these technologies is due to their real *unavailability* for adoption by individual farmers because of lack of technical and economic research and information.

De Janvry found that both seeds and mechanical improvements spread as rapidly in the past in Argentina as they did in the United States. In fact, mechanization spread so rapidly that it transformed Argentine agriculture into a labor-surplus economy, but had little effect on output levels and exportable surpluses. Interestingly enough, there was a similar lack of yield effect for the seeds, since they were designed to work in the traditional production contexts and not within *packages* of new techniques and modern inputs.

De Janvry then investigated the low levels of fertilizer used in Argentina by estimating the parameters of production functions. These results were published in de Janvry [1972]. He found that there was a latent demand for the fertilizer technology even at prevailing prices, and still more at potential prices that would be lower. However, he found that large farms with high land/capital ratios have no potential for fertilizer use until their land is cropped more intensively and the fertilization frontier shifted to the right. For this reason, he argued that there was no actual demand for the fertilizer technology deriving from the dominant farm interests. Consequently, systematic research on fertilizer was only initiated by the national research organization in 1962 and given low priority. Moreover, he found that a high percentage of the fertilizer trials were on the wrong side of the fertilization frontier, mainly because it was easier for scientists to work with the educated, large farmers who have contacts with the research stations and also have land to spare for experiments.

In conclusion, lack of adoption of fertilizer technology was not due solely to unfavorable prices or to perverse individual attitudes toward profit maximization or change, but mainly to the real unavailability of techniques at the farm level—a lack of agronomic and economic research and information. This was due at least in part to the dominance by large farmers of the decision-making process which determined the allocation of research funds.

Similar results were found when de Janvry analyzed technical change in the traditional area of beef breeding, dominated by large absentee landowners. Pounds of beef produced per acre had remained constant since 1935, and the use of yield-increasing, management-intensive breeding techniques—such as permanent pastures, forage reserves, pregnancy tests, etc., was low. By contrast, the region was technologically advanced with respect to the genetic improvement of breeds, and new breeds had been rapidly diffused in the region.

Only in recent years had public research been initiated in intensive breeding techniques. De Janvry found that economic analyses of the limited experimental data available all coincided in showing the high profitability of those techniques, especially when combined appropriately into packages. Hence, a latent demand for the technology existed.

The analysis of profit functions, however, showed that while the adoption of new technology and management efforts were profitable on small farms, this was not the case on large farms. Moreover, highest profits per acre were still obtained by large absentee and traditional landowners, even though they didn't adopt the new technology. As a result, there was not, nor had there been, a strong demand for research on these technologies.

In de Janvry's view, lack of adoption of the new breeding techniques by small and medium-sized farmers reflected in part the simple lack of technical and economic information on their uses. It also reflected, in his view, the existence of a strong regressive institutional bias against small farmers in their access to credit, information, and education, all of which are strong determinants of adoption of new technologies.

The important point, according to de Janvry, is that the lack of technology comes from the lack of profitability for large farmers and from institutional biases against small farmers, reflecting in all cases the unavailability of these techniques for individual adoption and not perverse economic behavior. This refutes the traditional structuralist arguments according to which absenteeism and large scale are barriers to technical change and lead to stagnation. He found, in contrast, that causality runs from the lack of profitable techniques available for adoption to absenteeism and large scale as the most rational behavior.

De Janvry's paper and model are important on a number of counts. First, they show how powerful the Hayami-Ruttan perspective can be—a perspective that breaks down innovations into particular classes and gives them particular roles to play in easing the constraints to development. Second, his paper is insightful in its own right and broke important new ground. Third, trying to identify who benefits and who pays the costs of technical change can be a powerful factor in understanding why the political process—as well as the economic process—does some things and doesn't do others.

Finally, one might ask why, if the large farmers had the power to influence the research process, they did not also influence economic policy so that it was not so discriminatory against agriculture. The probable answer is that agricultural research decisions were in effect more nearly within the agricultural sector, while the overvalued currency and trade distortions involved more powerful groups in the rest of the economy.

De Janvry has published extensively on Latin American agricultural development. In addition to the articles and books referred to elsewhere in this paper, see A. de Janvry and C. Garramon [1977a, b, c]; Deere, C. D., and A. de Janvry [1979]; A. de Janvry and L. Crouch [1980b]; A. de Janvry [1981b]; A. de Janvry and E. Sadoulet [1986, 1988]; A. de Janvry et al. [1987]; A. de Janvry [1978]; A.

de Janvry [1987a, b]; and A. de Janvry [1989]. De Janvry's perspective is also reflected in other sources discussed in more detail elsewhere in this paper.

6. Tests of the Urban-Industrial Impact Model

The urban-industrial impact model is the result of work by T. W. Schultz [1953], who was concerned with explaining the failure of agricultural production and price policy to remove the substantial regional disparities in the rate and level of development in U.S. agriculture. His hypothesis for explaining this unevenness in development and in per capita incomes was formulated in three propositions: "(1) Economic development occurs in a specific locational matrix. . . . (2) These locational matrices are primarily industrial-urban in composition. . . . (3) The existing economic organization works best at or near the center of a particular matrix of economic development and it also works best in those parts of agriculture which are situated favorably in relation to such a center" [p. 147].

Schultz's rationale for the urban-industrial impact hypothesis was that facor and product markets functioned more efficiently in areas of rapid urban-industrial development than in areas where the urban economy had not made a transition to the industrial stage. Major attention was given to structural imperfections in labor and capital markets. The role of the urban-industrial sector as a source of new and more productive inputs was also stressed.

This hypothesis stimulated a number of studies in the United States designed to test its main empirical propositions. Among these were studies by Ruttan [1955], Sisler [1959], Tang [1958], Nicholls [1961], and Hathaway [1964]. Results of these studies have generally supported Schultz's propositions with respect to the impact of urban-industrial growth on geographic differentials in per capita (or per farm worker) farm income. Tests of the factor and product market rationale have been less conclusive.

The urban-industrial impact hypothesis has been tested in three studies in Brazil: William Nicholls [1969a] with data from the state of São Paulo; Petrônio Rios [1969] with data from the state of Minas Gerais; and finally, Martin Katzman [1974, 1975b] with data from Goiás. The first two tests were broadly supportive of the hypothesis, with Nicholls's study in particular showing that by the 1950s urban-industrial development was influencing productivity in agriculture by facilitating the flow of capital into agriculture and the flow of labor out of agriculture. He noted that the impact of urban-industrial development was limited, however, because of the locational impact of resource-based opportunities for development and the failure of the Brazilian government to invest in the research capacity and the agricultural services necessary to permit the agricultural sector to respond to growth in the urban-industrial sector.

In commenting on the Nicholls's paper, Schuh [1969] raised two issues. First, he questioned the efficacy of considering urban-industrialization as the basis of a policy for agricultural development by posing the question, "What would you do with the next million dollars to promote agricultural development in São Paulo?" Obviously, the answer would not be to promote urban industrialization, although this does not deny the fact that urban-industrial concentrations do have an impact on agriculture. The second question he raised was whether the causality did not run at least in part the other way—with successful agricultural development in the state of São Paulo contributing to the urban-industrialization of São Paulo. Interestingly enough, Rios [1969] later pointed out that the agricultural data Nicholls used predated the urban-industrialization data by a decade, suggesting that causality did in fact go the other way.

Katzman's [1974] starting point was to note that von Thuenen's [1842, 1966] theory of agricultural rent and land use had surprisingly little impact even on those facets of the agricultural development literature where it would seem to have most relevance. He noted that this neglect was certainly not due to lack of scholarly interest in spatial variations in agriculture since the literature on economic development is replete with such concepts as regional dualism, the center versus the periphery, growth rates, and lagging regions. He referred in particular to the urban-industrial impact model which, with its emphasis on the efficiency of markets, had up to the time of his writing been considered an institutional model, *sui generis*, with no intellectual roots in the von Thuenen tradition.

Katzman [1974] developed a neoclassical formulation of the von Thuenen paradigm as a basis for reinterpreting the urban-industrial impact model and to identify both similarities and contradictions between the two. He then tested the two models with data both from the United States and for a frontier region in Brazil. He believed his to be the first numerical presentation of the complete system of von Thuenen *agricultural intensity gradients*. The von Thuenen paradigm has been widely used in urban economics.

The von Thuenen paradigm comprises a theory of crop substitution and a theory of factor substitution or agricultural intensity. The crop theory predicts around a market center a formation of rings that specialize in the production of different staples. Katzman focused on the intensity theory that describes either factor substitution or discrete changes in systems of farming within a given ring. He noted that this latter theory would be applicable to a region characterized by the production of a single staple which is exported through a single market center.

Katzman also noted that while the two paradigms are concerned with somewhat different aspects of reality, each has something to say about spatial variations in factor proportions, labor productivity, and yields. He captures the essence of the two models by contrasting their explanations of the use of fertilizer. The von Thuenen paradigm would explain its use in areas more accessible to urban-industrial centers by higher location rents, while the urban-industrial impact paradigm would explain it by more perfect current input markets or perhaps by greater exposure to the center of innovational diffusion and modern ideas.

The contrast between the two paradigms can be further contrasted by quoting from Katzman [1974, p. 694]:

> The von Thuenen and industrial-urban models are independently formulated explanations of regional variations in agricultural structure and income. Ostensibly, the explicand in the von Thuenen model is access to commodity markets, in the urban-industrial impact model, access to urban-industrial complexes. To the extent that spatial variations in commodity and factor prices reflect transfer costs, the urban-industrial impact model logically reduces to a von Thuenen model. The urban-industrial impact model genuinely differs from the von Thuenen model in its emphasis on market imperfections associated with monopoly and monopsony in rural areas dominated by small towns.

Katzman found that in the Brazilian frontier state of Goiás the spatial structure of agriculture conforms well to a von Thuenen model in which there is limited land-labor factor substitution. Predictions of the capital intensity of farming were improved when local industrial development was taken into account. He concluded from the evidence he reviewed, from both the United States and Brazil, that a synthesis of the two paradigms provided powerful insights into the relationship between urban and rural development.

In his 1975b article, Katzman explicitly examined the role of growth poles and development highways on the development of the frontier state of Goiás. Goiás has been one of the fastest growing states in Brazil, yet its per capita income is only half that of the national average. As the only state with below-average income attracting net migration, it is a striking anomaly. Moreover, associated with net in-migration has been a slight gain in relative per capita income, again an exception.

The explanation for these anomalies is that considerable public investment was poured into the state starting in about 1940, and especially from 1960 to 1975. This was part of Brazil's drive to "fill up its frontier" and, in this period, involved the construction within the state of the new national capital, Brasília, and massive road-building programs. Earlier, in the late 1930s, a new state capital had been constructed at Goiânia.

Katzman used the von Thuenen and the central place paradigm as the theoretical basis for his analysis. In the former, the city is taken as a given or as exogenous, and its impact on the agricultural frontier is mediated by transportation costs. In the latter, agricultural development is taken as an exogenous influence on urbanization, an impact again mediated by transportation costs.

Katzman [1975b] found that the major impact of Brasília on the state of Goiás was through the agricultural sector. Moreover, the direct impact of transportation had been swamped, in the case of Goiás, by the impact of rural growth itself on urbanization. A rural population which is expanding and increasing in income will demand a whole range of services which support the rural export base (e.g., consumer services, finance, repair, and retailing). As the rural market becomes larger, the "threshold" for import-substituting a wider range of goods and services is attained, and becomes a spur to urbanization.

In conclusion, Katzman's work on Brazil expanded the perspective of Schultz's urban-industrial impact model by introducing locational factors as implied formally by von Thuenen's theory of agricultural rent and land use. This gives the model a more rigorous theoretical base, reduces its institutional flavor, and gives a smaller role to market imperfections. His work also makes clear that at early stages of development the direction of causality probably goes from agricultural development to urban-industrial development, rather than the other way around. His expanded model of urban-industrial impacts thus probably has its greatest relevance at later stages of economic development, when the need is to transfer resources out of agriculture. It was at that stage of development that Schultz's model found its greatest use in the United States.

7. Resource-Based Development and the Frontier

Throughout most of history, expansion in the areas cultivated or grazed has been the main means of increasing agricultural output. Latin America constitutes an important region of the world where there are still large areas of unsettled land, where there are still frontiers, and where policy makers still have the choice of whether to expand their agriculture on the extensive or intensive frontier.

A substantial literature in economic history and in development economics has attempted to interpret the implications of agricultural development in newly settled regions. Hayami and Ruttan [1985] identified two bodies of literature which deal with these issues. The first is the "staple" model developed by the Canadian economic historian, Innis [1927, 1933, 1940] to explain the rapid growth of commodity production and exports in the newly settled areas of North America.[48] The staple model of development focuses on the crucial importance of export staples such as furs, fish, timber, and grain as factors shaping economic development.

The second body of literature is the "vent-for-surplus" model, which emerged from Myint's [1958] efforts to explain the rapid growth of production and trade in a number of tropical countries during the nineteenth century. His explanation was that surplus land and labor enabled peasant producers, even though facing relatively fixed technical coefficients, to expand production rapidly under the stimulus of new markets opened up by the reduction of transportation costs. Caves [1968] provides a review of the vent-for-surplus model and a comparison with the staple theories.

Hayami and Ruttan [1985] agree that in the past, exploitation of natural resources along the lines suggested by the staple and vent-for-surplus models has been a major source of agricultural and economic development. But they believe there are relatively few remaining areas of the world where development along these lines constitutes an efficient source of growth. Their argument is based on two premises [pp. 44–45]. First, the opening of new lands in Latin America and Africa awaits the development of technologies for control of pests and diseases in the African plains infested with tsetse flies and for dealing with the problem soils of the Brazilian *campo cerrado* and the Venezuelan and Colombian *llanos*. Second, investment in new production technology for use on lands already settled constitutes such an efficient source of growth that governments will not incur the investments in infrastructure needed to open up new lands.

We believe that three caveats should be placed on these arguments. First, there are large areas of quality land in Latin America that can be settled with prevailing technology. Second, rapid population growth and the difficulty of absorbing labor in the manufacturing sector gives governments more incentive to open these lands as a means of alleviating the mass migration of people to urban centers. And in the case of Brazil, the political drive to occupy the national territory provides ample incentives to invest in the needed infrastructure—as successive governments have done throughout the post-World War II period. Occupying the national territory is seen as a national security issue.

In fact, Brazil's recent success in bringing *cerrado* lands into production provides an important example of what can be done. EMBRAPA, the national research organization, has helped develop new production technology for these regions. A decline in the real price of fertilizers and lime make it economical to "correct" the soils. And the construction of penetration roads has effectively brought regions of *cerrado* soils into the market economy.

M. Nelson [1973] has made the most ambitious attempt to assess the potential of new lands in Latin America. He notes [p. 1]:

> In some quarters it is an article of faith that the great forested heartland of South America can and must be utilized if Latin America is to realize its development goals. This belief rests squarely on the premise that the mere existence of unused forest and land resources is sufficient reason to warrant the investment of capital and labor in their exploitation. Such an approach runs counter to the widely held economic doctrine

> that natural resource endowment is far less vital to
> development than the rate of sociotechnical change that extends
> resources through substitution, alters the location of economic
> activity, and provides a climate for the adoption of new
> techniques or the application of existing ones.

Nelson examined the economic bases underlying investment and policy for the development of new lands for agriculture and forestry. The area considered is limited to the humid tropical lowlands and uplands and the semiarid Chaco—a region that consists of parts of Argentina, Bolivia, and Paraguay. All told this area covers approximately 12 million square kilometers or 60 percent of Latin America.

Nelson reviewed the current theory and practice of humid tropical land development in Latin America, evaluated a wide range of projects under way at the time of his research, and derived implications for both investment policy and project design. His conclusions are largely negative on the potential of these natural resources for the development of the region. Among other factors, he believes that the emphasis on these development prospects causes nations to divert resources and energy from the really crucial problems to be addressed for sustained economic development. He argued that expansion of the agricultural frontier will inevitably result in the destruction of natural resources and that the weight of evidence is on the side of those who would restrain expansion. The prudent approach in his view is a gradual one designed to attract spontaneous settlers, together with—subject to caveats he prescribes—private investment and enterprise.

Katzman [1975a] was motivated by Frederick Jackson Turner's [1920] paradigm of the role of the frontier in American history to examine the Brazilian frontier in comparative perspective.[49] He noted that while there have been numerous articles in both English and Portuguese on the Brazilian frontier (see his article for references), these studies were largely descriptive and have not been systematically integrated into any conceptual framework, much less the comparative tradition inspired by the Turner thesis. He tested the frontier paradigm against four experiences from Brazil: the São Paulo coffee frontier; the Paraná coffee frontier; the original Amazonian rubber frontier; and the new Amazônia, 1970 to the present.

Katzman focused on agricultural settlement to address three economic questions: What determines which of various empty or semiempty regions are settled, if at all? To what extent does the physical environment, as opposed to demographic, technological, or political factors, determine the social relations of production (e.g., family farms vs. slave plantations)? To what extent has the existence of the frontier increased wages in the settled regions of the nation, a necessary condition for the functioning of the labor safety valve?

Katzman [1975a] presented data to show that of the four continental countries (Australia, Canada, and the United States are the other three), only Brazil continued to experience a vigorous expansion in its cropland. This included new physiographic zones: wheatlands on the southern prairie; upland rice cultivation on the savannah of the Central West highlands (around Brasília); and planned colonies in the equatorial forests of Amazônia. He further noted that frontier expansion cannot be explained by either the sheer availability of vacant land or by the growth in demand for food, as exemplified by the fact that land in farms in the United States and Canada stabilized at 65 and 7 percent of total land area, respectively, and that cropland in the United States had been declining towards its 1910 cropland area.

Katzman [1975a] made a distinction in understanding the spread of a frontier between *subsistence* economies and *export-propelled* economies, thus adding a category to the export-propelled explanation of both the staple and vent-for-surplus theorists. He further noted that while both types may be open to foreign immigration, they differ in their degree of participation in international trade. In the former, the value of land for resettlement is largely determined by *sita* factors (climate, fertility, or topography). In the latter, the *situation* of land with respect to roads or ports also affects the value of land.

In further elaboration, Katzman noted that a subsistence economy is generally one that is too distant from the market to engage in interregional trade or whose production possibilities are too similar to those of its potential trading partners. If land for such production is so abundant as to be virtually rent-free, agriculture will be undertaken on an extensive basis, "wasteful" of land and with low yields on a per acre basis but with maximization of returns to labor. This may stimulate a rapid increase in "natural" population growth due to improved nutrition, which increases fertility and reduces mortality. Children are sent off to occupy new virgin lands (when they are of sufficient age) and thus to continue the occupation of the frontier.

Katzman noted that these conditions held in southern Brazil in the early nineteenth century, when modern European immigration began from Italy and Germany. But while the subsistence frontier was expanding north from Rio Grande do Sul and Santa Catarina, an export-propelled frontier was radiating in all directions from metropolitan São Paulo.

A crucial factor in export-propelled settlement, however, is that increased distance from the market requires greater transportation outlays in exporting the staple and importing consumer goods and farm inputs. This may either justify massive investments in railroads, highways or other infrastructure, or it may impose a limit on the export-propelled expansion. A subsistence-based expansion, on the other hand, could continue without such constraints.

Another issue is what happens to wage rates. If population were fixed, expansion of the frontier would increase the supply of land per worker and thus raise labor productivity and wages, a necessary condition for Turner's labor safety valve to work. If natural population growth and/or immigration from abroad accompanied the expansion of the frontier, the labor safety valve would not work. Katzman noted that in the case of São Paulo, massive immigration from Italy dampened the wage-enhancing effect of frontier expansion. Consequently, the opening up of São Paulo provided more of a safety valve for Italy than for the Northeast of Brazil.

Katzman argued that each of the four cases of frontier expansion he examined in Brazil were export-propelled. He has perceptive observations on factors determining the social relations of production, i.e., whether farming is organized on the basis of independent family farms or plantations with dependency relations, and argues that the key to the pattern of subdivision of lands in export-propelled frontiers is the existence of a wide range of financial investments which provide landowners with alternative investments.

To conclude, Katzman provides—both in his tests of the urban-industrial impact model and in his discussion of expansion on the frontier—very perceptive and penetrating analyses on the location and spatial dimensions of agricultural development. Contrary to others, we suspect that both subsistence-based frontier expansion and export-propelled development will continue to play a role in understanding agricultural development in Latin America, at least until industrialization becomes more effective in absorbing the rapidly growing labor forces in the region.

Another version of the frontier model is that of the "hollow frontier," which Katzman [1975a, p. 267] attributes to Preston Jones. In the tropics, the practice of shifting cultivation or slash-and-burn agriculture leads to the abandonment of a cultivated plot for as long as a generation pending regrowth of the forest. It is for this reason that frontier expansion may not increase the amount of land devoted to agriculture. New land brought into cultivation may simply offset in area land abandoned in settled regions.

Mandell [1969] provides an overview of the hollow frontier models. G. L. da S. Dias [1976] has developed an explanation for how and why this process continues up until the present time in Brazil, and perhaps in other countries. His model also explains how land on the frontier tends to accumulate in large holdings.

There are five components to Dias's model [p. 21]: successive occupation of new lands by farmers who migrate from one place to another; the use of a traditional system of soil recovery, based on fallow, which is land extensive; the life cycle of the small farmer and his capacity for capital accumulation; the availability of modern techniques for soil recovery which leads to a more intensive system of land utilization; and discrimination against the small farmer in the capital market.

Dias noted that a key characteristic of the process by which new fertile lands are settled or colonized is its low requirement for working capital, which consists for the most part of providing food for the farmer's family. The investment in fixed capital on the other hand, involves cutting and burning the forest, removing the stumps, and planting and caring for cash crops (annual and/or perennial). Subsistence crops are planted to guarantee the survival of the farmer's family, but not all of the time available will be dedicated to this activity. The "surplus" labor is allocated to the investment activities of clearing and planting cash crops. When all of the available area (however determined) is totally occupied with cash crops, the farmer's income will probably be at its maximum.

Since only traditional cultivation methods are generally applied in these frontier colonizations, the fertility of the soil will eventually decline and in addition a proliferation of weeds and brush will emerge. Under these conditions more and more labor will be required to do the cultivation and the area planted will start to decline. Costs will also rise because of declining yields and thus income will also decline. Migration to new fertile land will eventually be the only economic alternative for the family.

Dias noted [1976, p. 144], insightfully, that the process described above would take place independently of the particular tenure arrangement that prevails. Moreover, small as well as large farmers are subjected to the same process. In the case of large landowners, however, *colonos* would have dedicated their surplus labor to investment activities on the landlord's land; in the case of the smallholder, the investment would have been on their own plot. The income distribution consequences might be very different in the two cases, depending on the nature of the labor market.

Dias also noted that under the traditional technology of soil recovery, which consists of alternating periods of cultivation with periods of fallow, the large farmers have an obvious advantage over the small farmers. The pressure on the soils of the smallholder may thus be much stronger and the decline in productivity may occur faster on their land than on that of the large producer. This will cause these producers to migrate at an earlier date, with their plots either abandoned or sold to larger producers. By whichever means, their plots are eventually reaggregated into larger farms. Dias also noted that the smallholder may be subject to capital rationing, both internal and external. This will make the adoption of modern techniques of soil conservation more difficult for them, thus reinforcing the general tendency of decline and out-migration.

Dias [1976] obviously was referring to export-propelled settlement. What we see in his model is a process by which the land may be settled by smallholders, but in which either the lack of proper technology, pressures of the life cycle, or imperfect capital markets lead to eventual out-migration and aggregation into larger holdings. Depending on the availability of techniques to sustain productivity, land may either go out of production permanently or into fallow. If it goes into fallow, the presumption is that land productivity in the aggregate will decline compared to an earlier period. Hence, the "hollow frontier" effect may occur on either or both of the extensive and/or the intensive margins.

It should be noted that when processes associated with the hollow frontier are important in a nation's agriculture, as in Brazil, aggregate data on yields can be very misleading as an indicator of the development process. The opening of new lands, with their higher yields, may offset the yield declines that are occurring in traditional areas.

Nicholls's [1969c] description and analysis of the settlement of the state of Paraná is another important contribution to the literature on the frontier. He contrasted the settlement of this frontier with the settlement of the frontier in the United States, giving special attention to the fact that despite its proximity to urban centers such as Rio de Janeiro and São Paulo this region was bypassed for a long period of time. Its development came only after São Paulo had set the stage by establishing an efficient system of production and marketing of a highly profitable export product eminently suited to the soils of the region, and by extending its railway and highway network in such a way as to make both foreign and domestic markets readily accessible to farmers of the new region. Of particular interest in the colonization of this region was the creation of a private land company which—in sharp contrast with Brazil's always erratic public colonization efforts—brought about orderly, efficient, and rapid land settlement in the region.

Nicholls noted that the agricultural frontier in Brazil long played only a minor developmental role, in contrast to the frontier in U.S. history. He attributes this to the fact that Brazil's Northeast, which he describes as similar to the U.S. southern Tidewater, was dominant politically, socially, and economically in Brazil. Scarce development resources were thus channeled, for a long time, to the Northeast, rather than to the frontier. Early U.S. economic development, in contrast, was far more balanced, being based on nearly simultaneous development of southern agricultural exports, New England shipping and manufactures, and cheap midwestern food. The contrasting influence on democratization and defense of special economic interests between the two countries was in Nicholls's view striking.

Nicholls [1969c] noted that later agricultural development in the Brazilian South was able to avoid many of the unfavorable social effects of a slave-based agrarian society, which the Northeast and East of Brazil could not avoid. As in the United States, these agricultural frontiers of the Brazilian South served to permit many migrants to enjoy a considerable degree of upward social mobility, preventing Brazil—particularly the South—from becoming the closed society which it is so commonly alleged to be.

At the same time, Nicholls noted that, as in the eastern United States a century earlier, the existence of vast empty spaces in Brazil has not been conducive to land and forest conservation. The availability of new lands makes it possible for the nation to avoid facing the problems of soil maintenance through rational systems of crop rotation and soil conservation, the use of organic and commercial fertilizers, and generally improved cultural practices. Brazil's political drive to occupy all its land causes it to continue to build huge penetration roads which open up new frontier areas, which continue to be farmed in an exploitive fashion and with serious environmental consequences. It also reduces the incentives to invest in agricultural research.

Bunker's recent book [1985] on the Amazon is important, as is Katzman's [1987] review article. Katzman [1977] on *Cities and Frontiers in Brazil* is also important, as is Mandell's analysis of the expansion of the rice industry on the Brazilian frontier [1971, 1972]. Mueller [1980] examines frontier-based agricultural expansion in the Brazilian state of Rondônia. Foweracker [1981] provides historical perspective, in a political economy, of the struggle for land on the frontier of Brazil. And Willems [1972] examines the rise of a middle class in the frontier society of Brazil.

Binswanger and McIntire [1987] examine the production relations in land-abundant tropical agriculture. They define production relations as:

> the relations of people to products and factors of production in terms of their rights of ownership and use and the corresponding relationships of people among each other as buyers and sellers, as factor owners and renters, as landlords, tenants, workers, employees, creditors and debtors. [p. 73]

These authors set themselves two objectives: to explain the major institutions and customary features of production relations in three agroclimatic subzones of the land-abundant tropics that have simple technology and high transportation costs, and to provide predictions of how these institutions and features will change in response to increases in population densities and the opening of substance-oriented systems via external migration and interregional or international trade [Binswanger and McIntire, p. 73]. Although their paper has basically an African perspective, it is relevant to Latin America as well.

To conclude, the frontier has played an important role in many Latin American countries, with equally important influences on the trajectory of agricultural development in the older regions of the individual countries. The frontier will probably continue to play a major role in countries as disparate as Brazil, Peru, Bolivia, and Mexico. The political pressures of burgeoning population pressures in urban centers and the drive to occupy the nation on national security grounds will drive the development process in these directions despite the well known high social rates of return from investing in the intensification of agriculture by means of creating and diffusing new production technology.

8. Furtado's Structuralist Model of Agricultural Stagnation

Some years ago Mellor and Johnston [1967] perceptively pointed out that a complete theory of agricultural development would have to explain why agriculture stagnated as well as how and why it developed. In this context, Hayami and Ruttan's book [1971, 1985], undoubtedly the most influential source today on agricultural development, is by scholars who came from the United States and Japan, two unusually successful cases of agricultural development. In contrast, much of the writing on agricultural development in Latin America has concentrated on explaining why agriculture in that region has stagnated for so long. Structuralist thought, which focuses on the identification of bottlenecks rooted in land tenure arrangements, bonded labor, and the distribution of income and its effect on the structure of demand, was relatively popular in the 1960s and early 1970s, when Latin American agricultural output did little more than increase at the same rate as increases in demand, rather than becoming a driving force for general economic development by growing at a faster rate. As agriculture contributed more to economic development in response to the commodity boom of the mid and late 1970s, structuralist models and interpretations became less popular.

Celso Furtado is in the structuralist tradition. We discuss his model because he went beyond many analysts in developing a reasonably systematic explanation of why agriculture stagnated at a low level of productivity. Furtado's analysis is also important because he is one of the few structuralists who presented a model sufficiently comprehensive to permit land reform to be a means of increasing the level of income of the rural population in general, and of the rural worker in particular.[50]

Furtado writes as an economic historian and an important characteristic of his work is the sweeping generalizations he makes about Brazil and Latin America's economic history. The central themes of Furtado's interpretation of the underdevelopment of the region are the structural rigidities of its economies and their dependency on the U.S. economy (and, to some extent, on other developed economies). In these respects, his work is in the tradition of the ECLA and of its influential economist, Raúl Prebisch. His book *A Economia Brasileira* [1954] is dedicated to Prebisch. Nonetheless, his original contributions to this literature are significant.

The central core of Furtado's model can be found in a chapter entitled "The

Agrarian Structure in Brazilian Underdevelopment" in Furtado [1972], which gives a good flavor of its content. Three propositions dominate this essay:

(1) The style of the colonization of Brazil had deep implications for the formation of the agrarian structure (land tenure system) and this, in turn, is a major element in the explanation of the low level of modernization of agriculture.

(2) The agricultural entrepreneur in Brazil has always been a profit maximizer who responds (quickly) to economic incentives. This is a significant departure from other writers in the structuralist tradition who emphasize the lack of profit-maximizing behavior in conjunction with alleged inelasticity of agricultural supply.

(3) In the context of the agrarian structure that exists in Brazil, the *latifúndio-minifúndio* complex is a system that provides conditions for the growth of agriculture in a situation in which natural resources (especially the fertility of the soil) are being depleted.

Furtado [1972] argued that the colonization of Portuguese America was different from the colonization of the rest of the Americas, and that Brazil was the only country in the Americas that was organized, from the beginning, on the basis of commercial capitalism in the form of an *empresa agro-mercantil* (an agromercantile firm, or perhaps what is better described as a crop plantation). Hispanic America, on the other hand, was colonized on the basis of the *Conquista*, which also handed out land in large holdings but which permitted an initial "accumulation" on the basis of exploitation of a pre-existent population (the American Indians). The essential difference between the two is thus in the source and conditions of labor supply, not in the structure of land holdings *per se*.

The structural inheritance of Brazil is based, in Furtado's view, on the following characteristics: there was an abundance of land in tropical or subtropical conditions; a local labor supply was practically nonexistent in the initial phases of colonization; and with exports being the *raison d'être* of territorial occupation, the form of organization that was most economical for *this* activity tended to predominate. This was essentially a plantation form of organization, and one that in Furtado's view has predominated down to present times.

Furtado argued that both Portuguese and Anglo-Saxon America were essentially creations of European commercial expansion. In both cases the initial "accumulation" was made in part through the plunder of Africa by the importation of slaves. However, there was an important difference. In Brazil, the institution of slavery left an indelible imprint on the fabric of national institutions. In the U.S., on the other hand, the institutions of New England prevailed where, at the side of an agriculture of small and medium-sized farms, there was a commercial bourgeoisie of considerable autonomy.[51]

Contrary to this fragmentation of land holdings and economic power in the U.S., in Brazil the original organization was on the basis of large crop-producing plantations (*empresas agro-mercantil*) that in Furtado's view left their decisive marks on the structure of the economy and on the society that formed the country. The ruling class was from the beginning made up of economically powerful men. They controlled access to the land, and in a country that was primarily agricultural this meant, in Furtado's view, that they reduced the nonslave population to what were *essentially* slaves. As a result, the system was able to pass through the abolition of slavery virtually unchanged.

For those not familiar with Latin America, some comments on the *latifúndio-minifúndio* complex may be pertinent. This is a complex social arrangement in which large (plantation-type) farms are surrounded by and associated with a large number of small farmers and rural workers who provide a ready source of labor. In some cases the small farmers are landowners. In other cases they are sharecroppers who also supply labor to the larger landowners' plots. And in still other cases they may be permanent workers who have access to a small plot of land the landowner has given to them on which to grow subsistence crops (these are called *colonos*). Because the *colonos* have access to this land they receive a wage less than the going wage for their work on the landowner's farm. This system is widely referred to as "exploitation." In the Northeast of Brazil, where the *latifúndio-minifúndio* complex tends to be more prevalent, one can find farmers-workers who are various combinations of small owners-sharecroppers and sharecroppers-*colonos*.

The crop-producing plantations and, later on, the large cattle farms are essential elements in Furtado's interpretation. The former is, in his view, the most important, because it was based in large part on three and one-half centuries of slavery which in his view enabled the *empresas* to frustrate any other form of agricultural organization. Furtado believes the institution of slavery has dominated rural life in Brazil, and that it imparted an authoritarian character still present today in many parts of Brazil (see note 51). It is a paradox of the system, in Furtado's view, that labor is simultaneously scarce and cheap for the crop-producing plantations. This apparent anomaly is rooted in Furtado's assumption that labor is exploited by monopsony hiring of labor by landowners who monopolize the control of land. This causes labor to be cheap. At the same time, however, labor is scarce relative to land in the aggregate. But when the demand for labor increases in response to a particular commodity boom, the supply of labor to that sector is perfectly elastic, since migrants will come from other parts of agriculture. Hence, the combination of presumed local monopolies on land holdings and the availability of abundant supplies of labor from elsewhere keep labor costs down to individual producers, despite the fact that labor is scarce relative to land.

In Furtado's view [1972], the scarcity of labor in the aggregate is what causes the extensive use of land in Brazil. It perpetuates the practice of shifting field cultivation (the widespread use of fallow) and an itinerant agriculture (the hollow frontier referred to above). Such a system requires that the *empresa* have at its disposition large quantities of land that it "underutilizes." But it also requires that the *empresa* attempt to assure itself positions on new agricultural frontiers, since a loss of fertility manifests itself with both temporary and permanent crops.

Furtado did not explain why the scarcity of labor in Brazil did not lead to mechanization, as it did in the United States. However, presumably this was because the large land owners never faced a competitive labor market, or one in which wages were rising. The concentration of land ownership in Furtado's view keeps labor cheap, induces an extensive use of the land, and provides little incentive to raise land productivity. Presumably this is a rational use of resources from the perspective of the land owner. Furtado recognized it to be a Schultz-type, low-level equilibrium trap.

Furtado believed that this system of land tenure and exploitation of the land also leads to plundering of the nation's natural resources. As land becomes more scarce as a result of population pressures, the fallow period is shortened, with a consequent acceleration in the degradation of the soils. Eventually the land is converted into low-productivity pastures. In effect, society is "mining" or using up a nonreproducible natural resource.

In Furtado's view, it was difficult for agriculture in other parts of Brazil to compete with this predatory agriculture since he believed a more capital-intensive agriculture was a high-cost agriculture. Thus the "technification" or modernization of agriculture is precluded by the system of land holdings. Moreover, since only subsistence wages are paid, there is little incentive to invest in education and training, and these investments are needed for the assimilation of technical progress in agriculture.

Furtado's solution to this problem was to distribute land to those who do not have it, or to the *minifundistas*, thereby breaking the monopoly that the *empresas* have on the land and creating a more competitive labor market. He believed the *minifúndio* (the small landowner) receives implicit wages as low as those of the lowest paid workers, and is underemployed because of the small size of his farm. Moreover, the small size of his plot makes it difficult for him to evolve technically. If, in contrast, the increase in land is made available to the *empresa*, the monopsony hire of labor will continue, the *empresa* will attract additional labor from the readily available supply, and the same level of technology will be used.

Land given to the *minifundistas*, however, opens to the worker the possibility of raising the value of his own labor, which eventually should increase the level of technology he employs. Moreover, if this increase in the value of the labor of the small owner affects the supply price of rural labor generally, the *empresa* will have to increase the productivity of the labor it uses also in order to compete on the labor market. This will cause agriculture to become more capital intensive, and the traditional practices of cultivation would have to be abandoned progressively. Clearly this transformation would take place only if the cost of labor increased substantially, which could occur only if a substantial part of the rural "mass" had the possibility of working for its own account under conditions much more favorable than those currently encountered on the *minifúndios* and in the marginal lands of the moving frontier. Furtado believes that as long as the *empresa agromercantil* is the principal source of employment of this rural mass, there will be few possibilities of consolidating other forms of agricultural organization.

Furtado believed, however, that this redistribution of income within agriculture would not be sufficient to have sustained agricultural development. Redistribution policies would be needed more generally in the economy in order to create sufficient demand to absorb the increased agricultural output. Thus stagnation in agriculture is seen to be rooted in structuralist issues in both the farm and nonfarm sector.

Rezende [1975] noted the fundamental inconsistency in Furtado's argument that labor is both cheap and scarce. This can be reconciled, however, if the landowning class does in fact exercise a monopoly on land and colludes to hold down wages. The plausibility of Furtado's model is rooted in the failure of an expanding industrial sector to pull labor out of agriculture at a significant rate until the 1970s; in the significant segmentation of the labor market due to the high incidence of illiteracy among the rural labor force and its consequent lack of marketable skills; in the prevalence of large farms with a significant amount of idle land; in the *patrão*-authoritarian regime which still exists in much of rural Brazil; and in the apparent lack of employment alternatives for workers in the interior.

The theory of large land owners colluding to hold down rural wages is not the only theory that will interpret the data, however, nor is the *a priori* evidence all that strong in its favor. In the first place, it implies a control over the labor force which does not in practice appear to exist, even though as noted the *patrão*-authoritarian regime still predominates in some parts of Brazil. This caveat aside, A. W. Johnson [1971] found that there is considerable shifting of labor among employers in a traditional region of the Northeast; for more detail on this labor market, see Teixeira [1976]. Furtado himself implied a reasonably competitive labor market when he said that wage increases from local increases in demand are rapidly competed away from an influx of workers. This influx would not be possible if the workers were kept in bondage to the extent that Furtado implied. Moreover, the labor force in Brazil is notoriously mobile, which argues against exploitation based on a local monopoly of land.

An important part of Furtado's argument was that land is withheld from production, presumably as a guise to keep it from the hands of the laborer, and thereby keep him subservient. The hollow frontier model of development, described above, will explain this process equally as well.

Finally, the extent to which Furtado's interpretation applies in today's Brazil is a major issue. He implied that it refers to all of Brazilian agriculture, yet surely this is to ignore the important differences that exist among regions in the degree of modernization. He also implied that there has been practically no modernization of the agricultural sector in Brazil. Yet, agriculture in São Paulo and the south of Brazil has modernized rapidly in the post-World War II period, and there have been technical breakthroughs in individual crops in other regions. Moreover, fertilizer consumption in Brazil has increased rapidly in recent years in response to a decline in its relative price.

Perhaps Furtado's greatest contribution is his sensitivity to the political and institutional consequences of land reform. Many of the criticisms of Furtado have focused on the (potentially negative) consequences of reform. However, one cannot do a full evaluation of the effect of land reform without considering its impact in creating a political environment more favorable to the interests of small scale agriculture, as evidenced by the land reforms in Japan, Korea, and Taiwan. The issue then reduces to whether the longer-term positive benefits outweigh the short-term negative disruptions, and to whether economic forces can generate a decline in average farm size without the disruption of a reform.

9. Socialist Agriculture

There have been three socialist or centrally planned experiences in Latin America, if one does not consider the *ejido* system in Mexico a socialist system [see Venezian and Gamble, 1969, for a description of the *ejido* system; in addition, see comments later in this section]. The most significant socialist transformation was in Cuba. Lesser transformations took place in Chile under Allende and Nicaragua under the Sandinistas. In this section we review some of the literature on Cuba and Chile, and comment briefly on the Nicaraguan and Mexican cases.

The modern history of Cuba is dominated by three important themes: the persistent difficulties of its agricultural sector, and the challenges of reforming that sector along socialist lines; the strong interdependence between the Cuban and U.S. economies up to the time of the Revolution and the international political issues that have dominated economic policy in the period since the Revolution; and the importance Castro has given to education as the means to bring about a cultural and economic transformation of Cuban society. Some of the key issues and authors who have written on these issues are reviewed in the following paragraphs.

The history of Cuban agriculture has been the history of tobacco and sugar, with sugar playing a predominant role. As will be noted below, however, tobacco and sugar involve different organizational forms of agriculture. The sugar sector has distinguished itself over the years by being a major source of foreign exchange earnings. The problems of plantation agriculture in this sector are complicated by the importance of sugar mills, which bring their own special problems of industrial organization. Because of the problems inherent in plantation agriculture, and Castro's desire to shift income from the urban to the rural sector, agrarian reform was a critical component of the Revolution.

A useful place for the uninitiated to start reading about Cuba and its revolution is with Dumont [1970], the French agricultural planner, which is a translation of the original French version [Dumont, 1964]. The author referred to Cuba as an economy with unbelievable richness of soil and weather which permitted year-round cropping. He describes Castro's early attempts at reform (1959-60) as agrarian reform amid revolutionary and romantic anarchy [chapter 2], which was later compounded when it became bureaucratized [chapters 3 and 4]. Dumont writes with a perceptive, critical eye, while still being sympathetic with the goals of the revolution.

Ortiz [1947] describes the counterpoint between tobacco and sugar in Cuba's history. Abstracts from this book can be found in Ortiz [1966, pp. 168-175]. Tobacco, with its potential for a free-holder agriculture, created a middle class, a free bourgeoisie. Sugar, in contrast, in his view created two extremes, slaves and masters, or the proletariat and the rich. In addition, tobacco was always under the control of the home government, while sugar tended to be under foreign control superimposed on the island's government [Ortiz, 1966, p. 173]. Mintz [1964], in a foreword to Guerra y Sánchez, discusses the industrialization of sugar production and its relationship to social and economic change (abstracts reproduced in Mintz, 1966, pp. 176-186).

Bianchi provides a careful statistical analysis of agriculture in the prerevolutionary period [1964a] and to the extent the availability of data permits, an assessment of its early postrevolutionary development [1964b]. The year of agrarian reform was 1960, and a massive effort to diversify and increase farm production began in 1961. His analysis focuses on the disappointments of this initiative, which led to the eventual return to an emphasis on sugar production.

MacEwan [1981b] provides one of the most, if not the most, comprehensive treatments of the agricultural sector, of the successive agrarian reforms, and of their roles in the development of the Cuban economy [also, see MacEwan, 1982]. The discussion of the agricultural sector is placed in the context of the broader socialist developments taking place in the economy. Particular attention is given to the development of an agriculture-based development strategy. The final part of the book draws the lessons of agriculture and development in Cuba, including an analysis of the weakness of agriculture as a leading sector and the importance and role of external (Soviet) assistance.

O'Connor [1970] has two extensive chapters on agriculture. The first [chapter 4, pp. 55–89] is titled "The Mismanaged Economy: Prerevolutionary Agriculture." The author concentrated on the political economy of the agricultural sector, and argued that many of the problems of the sector were rooted in the extent to which the economy was made up not of individuals, but of organized groups. Some of the groups were private, while others were public. Taken in their entirety, they constituted economic baronies that were well organized and special-interest minded, and amounted to a system of corporate economy, each component of which rejected the ideology of economic liberalism. The second chapter, "The Agrarian Revolution" [chapter 5, pp. 90-134], describes in some detail the *process* of the agrarian revolution.

Forster [1982] provides a comparison of agricultural productivity between state and private farms in Cuba. Forster and Handelman [1984] assess the impact of the Revolution on food production and distribution. Sanches and Scobie [1986] examine the collaboration between Cuba and the International Agricultural Research Centers. S. Eckstein [1981, 1983] analyzes the domestic and international constraints on agricultural development.

Three papers are useful for a more detailed analysis of sugar policy. Hagelberg [1979] provides a synoptic overview of Cuba's sugar policy. Brunner [1977] is a more detailed analysis of sugar policy in the period 1963-70. Roca [1976] analyzes the failure of the famous 10 million ton sugar harvest. On the latter issue, Horowitz [1972, p. 7] perceptively notes:

> [A] single crop economy labeled socialist is no more effective in getting the crops out and the cane cut than was a single crop capitalism. . . . And, while the problem is clearly, in some sense, diversifying crops and using modern agricultural technology, it is also a problem of incentives — or, more simply, the problem of material rewards and commodity goods for hard labor.

To conclude, Castro was no more successful in modernizing agriculture by socialist means than was the Soviet Union, Eastern European, or other socialist countries. Cuba's agrarian reform was a serious attempt to redistribute income from the urban to the rural economy. That effort was at least partially successful. But self-sustaining agricultural development failed to emerge.

Educational reform was the second component of the Cuban revolution. Fagen [1969] provides a comprehensive discussion of the role of education, including the literacy campaign, in transforming the political culture of Cuba. Jolly [1964] makes a thoroughgoing analysis from the prerevolutionary background through the literacy campaign and including school and university education. Jolly draws effectively on the available data. Carnoy and Werthein [1979] discuss educational reform in the context of economic change, also with ample use of the available data.

Turning to more general reference on the Cuban economy, Mesa-Lago is one of the more prominent authors in the English language [see 1972, 1978—a revision of a 1974 edition of the same title, and 1981]. Mesa-Lago was the editor of a wide-ranging collection of essays on Cuba [1971]. Ritter [1974] is a comprehensive analysis of the macroeconomic performance of the Cuban economy. Brundenius [1984] and Pérez-López [1987] are more recent attempts to assess quantitatively the performance of the economy, with significantly disparate results. Other general references include Dominguez [1971] and MacEwan [1981b].

For still more general treatments, the reader may be interested in Huberman and Sweezy [1969], Seers [1964], and Horowitz [1981, fourth edition]. Horowitz is a collection of papers that has expanded with each successive edition. An important Cuban journal on development issues is *Economia y Desarrollo*. An English-language journal that has material on Cuba is *Latin American Perspectives*.

To conclude, it should be noted that the volume of scholarship on the Cuban economy, and especially on the agricultural sector is rather modest. Data problems and deficiencies make quantitative analysis especially difficult, and these have been compounded by the political barriers created by both sides to limit the mobility of researchers.

Turning to the case of Chile, the socialist experiment during the Allende regime did not go much beyond land reform and associated measures, and this land reform itself did not go at a significantly faster pace than it did under the preceding government. Valdés [1974] provides a useful description and evaluation of Allende's reforms and the problems associated with it. Chonchol [1977], a participant in the reform movement, describes the social and economic organization of the Chilean reformed sector during the Popular Unity government. Jarvis [1985] describes and analyzes Chilean agriculture under military rule.

According to Valdés [1974], an important feature of the reforms associated with the Allende government was significant distortions in the price system and in the administrative allocation of modern inputs. In the latter case, a significant share of available inputs was channeled to the *asentamientos*, while larger (privately owned) farms received little of such inputs. This resulted in a gross distortion of factor proportions between the two sectors, with the expected effects on output. Valdés also noted that the reforms did not have the expected effect on the distribution of income, since much of the land was distributed to wage earners in agriculture. These workers constituted a lower middle class in Chilean agriculture, and were for the most part better off than small farm owner-operators.

Chonchol [1977] provides an objective description and analysis of the great difficulties faced by the Allende government in implementing its reform measures. Among these were excessive paternalism on the part of CORA, which developed the yearly plans for working the farms and kept their accounts, with little participation by the *campesinos*; remunerations which were excessively egalitarian and did not serve to stimulate quality and effort in the farm work; conflicts on the *asentamientos* because of differences in status, with resistance to new *campesinos* out of fear that newcomers would reduce the amount of land available to those already on the *asentamientos*; *asentamientos* that were inadequate in size; and irresponsibility in the use of funds provided by CORA.

Chonchol also described the political resistance to change by the opposition. He also provided brief descriptions of the Agrarian Reform Centers and the Production Centers that were eventually to be the centerpieces of the reformed agricultural sector, and the policy that was to be followed in assigning land to the *campesinos*. There are valuable lessons in this perceptively done piece for anyone attempting to carry out a major land reform in agriculture.

Jarvis [1985] has provided a valuable postscript to the Allende reforms by analyzing developments under the military government which followed. He questioned the government's data on agricultural output, and analyzed the impact of their policies on income and employment in agriculture. His analysis showed that at the time of his analysis agriculture had performed less well than government data showed. He also found that the impact of policy changes had been serious on employment and wages in agriculture. The structure of agriculture had been shifted towards the small farmer, but the government was not yet providing them the services they needed to develop into a productive sector of the economy.

Other useful sources on Chilean agriculture include Lehman [1972, 1973], Roxborough [1974], Kay [1974], and Stanfield [1976].

Nicaragua's reform of the rural sector, like Chile's, did not go beyond land redistribution and significant distortions in the price system. Mexico's land reform and the creation of the *ejido* system could be interpreted as a socialist system since the state retains ultimate ownership of the land distributed in the form of *ejidos*. However, this land remains in the family and can be passed on from one generation to another so long as land is farmed by the family. Moreover, a very vital system of privately-owned land holdings exists side by side with the *ejido* system. This private system dominates Mexican agriculture.

In conclusion, the socialization of agriculture has produced disappointing results almost every place around the world it has been tried, with Latin America being no exception. The Cuban experience has been the most significant of the socialist experiments in Latin America, and probably the only one from which longer term lessons might be drawn. It does not provide a great deal of expectation that socializing agriculture offers an attractive means of developing agriculture. The collection by Dorner, ed. [1977] provides perspective on global experiences with collective farming approaches of various kinds.

10. Concluding Comments

Research on agricultural development in Latin America has produced some valuable applications and tests of generally accepted agricultural development models, such as the Hayami–Ruttan model; it has extended some models, such as de Janvry's extension of the Hayami–Ruttan model and Katzman's extension of the urban-industrial impact model; and it has generated a rather original but modest interpretation of the frontier, as well as Paiva's model which helps explain the pace of agricultural development. One of the paradoxes that arises from the attempts to apply the Hayami–Ruttan model is the failure of mechanization to proceed, until recently, at a faster pace than it has in countries such as Brazil where land is abundant relative to labor. The structuralists such as Furtado have an explanation for this which is rooted in dominance of the land market by large landowners and the effect this has on agricultural wages. However, the problem can be explained equally well by other more plausible models.

As long as Latin American agriculture was relatively stagnant and not contributing significantly to the more general development of the economy, the literature tended to concentrate on explaining that agricultural stagnation. When agriculture took off in response to the boom in international commodity markets starting in the early 1970s, the literature shifted to attempts to explain that development with models developed elsewhere. In a later section we will note that systematic discrimination against agriculture by means of trade and exchange rate policies, an integral component of import-substituting industrialization, can explain the relative stagnation of the 1960s and early 1970s, without appeal to structuralist factors. This monetarist interpretation of agricultural development in our view provides a more robust explanation of past development than does the structuralist interpretation, and provides a sounder model to guide agricultural development for the future.

Chapter IV. Supply Response, Marketing, Consumption, and Production: Empirical Studies

This chapter provides an overview of some of the empirical research on Latin American agriculture. The literature on supply response is given more ample treatment than is the literature on agricultural marketing, consumption, and production because of its relevance to policy issues, to the monetarist-structuralist debate, and because some of this research has made important contributions to the literature.

1. The Evidence on Supply Response

The central motivation for studies of agricultural supply response in Latin America derives from at least two sets of issues: questions related to the economic efficiency and rationality of farmers, and the structuralist concern with the low elasticity of supply of agricultural output. The first of these questions was motivated in large part by T. W. Schultz's [1964] seminal book in which he argued against the idea of noneconomic behavior on the part of farmers in low-income countries. The second question, although obviously related to the first, has different *a prioris*. Structuralists claim that farmers have a low response to price because of the nature of their assets, their resource endowments, and the character of the labor market they face. In effect, they argue that farmers maximize returns from their portfolio, but the nature of choice is such that a low response to agricultural prices is the optimal decision.

The policy implications of whether product supply response is positive or virtually inelastic are quite great. If the response is positive, then resource allocation can be influenced by price, trade, and exchange rate policies and the so-called monetarist perspective prevails. Proper incentives will elicit desired responses and thus help to obtain agricultural development. Discrimination against agriculture by means of price, trade, and exchange rate policies, on the other hand, will lead to sluggish growth and stagnation.

If supply response is extremely low, on the other hand, then the structuralist models and interpretations are vindicated. Economic policies that operate through changes in relative prices will be ineffective, and development can be obtained only through policies which affect the "structure" of the economy. Such policies include on the production side land reform, which affects the size distribution or structure of land holdings. On the demand side, these policies include more general redistribution of income, which affects the structure of demand.

Supply response studies can be divided into two broad classes: supply response within the agricultural sector, and aggregate or intersectoral supply response. Each of these has a bearing on different policy questions. Knowledge on the within-agriculture supply response is useful information for shaping policies designed to influence the composition of agricultural output (say, between food and nonfood, traded and nontraded goods), and for designing policies that affect individual subsectors of agriculture such as maize or beef. Knowledge on the aggregate supply response, on the other hand, is more relevant for policies that deal with the relationships between agriculture and the rest of the economy. The structuralist analysis of the inflationary process in Latin America, for example, is concerned with the (assumed) low aggregate elasticity of agricultural supply and its role as a factor explaining chronic inflation in the region.

Besides these considerations, knowledge of the aggregate supply elasticities is important information for policy purposes, especially to the extent that it may indicate limitations of price policies in achieving certain objectives. For example, if the long-run aggregate elasticity is of the order of 0.4 and if one desires to keep output growing at a rate of 3 percent per year (16 percent in five years), this would require a 7.5 percent per year increase in the relative prices of agricultural goods (44 percent over five years). This, in many instances, may not be politically feasible [Krishna, 1982]. However, even if feasible, it would impose very high costs on low income groups, costs which might be undesirable on income distribution grounds. If this were the true elasticity of supply, one can appreciate the practical (as opposed to theoretical) constraints to the implementation of price policies that it implies.[52]

Another dimension of the dynamics of supply that applies to less developed countries is the question of how to incorporate the effects of the process of modernization. Nerlove [1979, p. 883] stresses this point and some of its implications:

> In modern agriculture, or in an agricultural sector in the course of modernization, constant changes are occurring. These changes are typically large, frequently discontinuous, and require major reallocation of resources both within the agricultural sector and between agriculture and the rest of the economy. Moreover, more often than not, these changes are not reflected in visible prices although in market oriented economies, major shifts in the demand for various agricultural commodities or in the supply of inputs used in agricultural production do not take the form of price changes. In the supply response studies discussed earlier in this paper, and indeed in my discussion of models of response based on dynamic optimization, I tacitly assumed that visible prices convey all of the information to which farmers find it necessary to respond. This is certainly not true even in recent times in a highly developed economy such as we have in the United States.

In less developed countries the major sources of shocks referred to by Nerlove generally involve investments in infrastructure (which as a rule lead to the development of new markets), the process of technical change, demographic factors, changes in the labor market, and the intervention of the government. In commenting on his own basic supply response model, which is now widely used, Nerlove [1979, p. 886] notes:

> It is inadequate, despite the many ingenious modifications and additions others have made to it, either to model dynamic optimization in response to changing prices or to understand the nature of dynamic supply response in the context of a developing economy.

Despite these important limitations, a review of existing studies is pertinent to a review of the literature on Latin American agricultural development. The available studies constitute the existing knowledge on agricultural supply response in the region and it is from these basic studies that a stronger base must be built for the future. Similarly, although somewhat limited in their scope, these studies constitute a source of empirical evidence for the policy debates that take place in the region and, as such, provide a (perhaps limited) base for policy recommendations.

This section is divided into three major parts. In the next section, we examine the studies that deal with annual crops. In the following section, we deal with perennial crops, and following that, we deal with livestock supply response studies. In the final section, we make some concluding comments and attempt to derive the major implications of these studies for policy purposes.

We make no attempt to evaluate the econometric procedures underlying the various studies, since this would take us too far afield. Our main goal is to provide a flavor of the work that has been done and to extract some (hopefully) robust lessons for economic policy.

Prior to reviewing the literature that has focused on Latin America, Mundlak [1988] and J. M. Rao [1989] have two general studies that are worth noting. Mundlak [1988] is a broad analytical perspective on aggregate agricultural supply, with some attempt to draw on previous empirical studies. Rao [1989] is a (selective) survey of previous studies of agricultural supply response from around the world. Rao concludes, among other things, that cross-country estimates exaggerate aggregate supply responsiveness to prices while time-series studies underestimate the response somewhat. Rao believes a tentative range of 0.4 to 0.5 seems plausible for developing countries. He concludes that major shifts in the terms of trade will alter resource allocation between sectors much less than the distribution of incomes. Rao draws other important policy inferences from his survey of empirical results.

STUDIES OF ANNUAL CROPS

The Nerlovian supply-response model has had a significant influence on most of the supply response studies we reviewed. See Nerlove [1956, 1958] and Askari and Cummings [1976]. We draw importantly on the last two references. For the studies of annual crops, the great majority of the estimated models were specified as follows:

$$A_t^* = a_0 + a_1 P_t^* + a_2 Z_t + M_t \qquad (1)$$
$$A_t - A_{t-1} = \delta (A_t^* - A_{t-1}) \qquad (2)$$
$$P_t^* - P_t = \beta (P_{t-1} - P_{t-1}^*) \qquad (3)$$

where:

A_t^* is the actual area under cultivation in period t;

P_t^* is the actual price of the crop in period t;

A_t is the desired area to be under cultivation in period t;

P_t is the expected "normal" price in year t for subsequent periods;

Z_t are other observable exogenous variables in period t;

β is the coefficient of adaptation;

δ is the coefficient of adjustment;

M_t is a random variable.

Most of the studies reviewed assume that either β or δ is equal to one.[53] Sometimes a yield equation is included in the analysis or the adaptive expectation equation is modified by a term that reflects the difference between expected and actual yields.[54]

In synthesizing the empirical results, the cross elasticities have been suppressed since estimates of them have been far less frequent than the estimates of own elasticities. In particular policy situations, knowledge of these parameters might play a decisive role in determining government action. However, given that the studies reviewed tend to have drastic simplifications with respect to other (nonmarket) effects on supply (say technological change), we choose to neglect them for purposes of this review, although we comment on this and other features of the models below.

The elasticities are organized by commodity in Tables 7 through 12. Table 12 contains a miscellaneous grouping. We are aware that there is a fairly large number of estimates of supply elasticities in graduate theses from Latin American graduate programs. To the extent that at least some of these would use more recent data than those reported here, estimated elasticities might tend to be a bit larger as a consequence of the more widespread use of modern inputs in recent years.

Generalization about these results is not easy, although a few patterns do stand out. First, short-run supply elasticities tend to be less than one, but definitely positive, with a statistically significant response to price in most cases. Second, given this caveat about a relatively low response, the elasticities tend to vary a great deal even for the same commodity. This undoubtedly reflects the different production alternatives producers face in different parts of the region, different policy regimes, different specifications of the model, and the use of different statistical procedures. More generally, a low short-term supply response is consistent with the low level of use until recent years of purchased inputs such as fertilizer.

Third, the long-run elasticities tend to be larger than the short-term elasticities, and in some cases significantly so. This is expected, and has important policy implications.

Fourth, we found very few estimates of aggregate or sectoral supply elastici-

Table 7. Estimates of wheat supply elasticities

Author	Period	Country	Province/state	Short-run elasticity	Long-run elasticity
Friere [1966]	–	Argentina	–	0.57	–
Reca [1967]	1924–44	Argentina	Buenos Aires	not significant	–
Reca [1967]	1945–65	Argentina	Buenos Aires	not significant	–
Reca [1967]	1924–44	Argentina	Córdoba	0.32	
Reca [1967]	1945–65	Argentina	Córdoba	not significant	–
Reca [1967]	1924–44	Argentina	Santa Fé	0.31	
Reca [1967]	1945–65	Argentina	Santa Fé	not significant	
Reca [1967]	1924–44	Argentina	Entre Rios	0.83	
Reca [1967]	1945–65	Argentina	Entre Rios	not significant	
Maffucci [1969]	1925–45	Argentina	Buenos Aires	–	0.50
Maffucci [1969]	1946–65	Argentina	Buenos Aires	–	not significant
Maffucci [1969]	1925–45	Argentina	Córdoba	–	0.17
Maffucci [1969]	1946–65	Argentina	Córdoba	–	0.50
Maffucci [1969]	1925–45	Argentina	Santa Fé	–	0.61
Maffucci [1969]	1946–65	Argentina	Santa Fé	–	0.31
Maffucci [1969]	1925–45	Argentina	Entre Rios	–	0.75
Maffucci [1969]	1946–65	Argentina	Entre Rios	–	0.75
Swift [1969]	1942–64	Chile	Coquimbo	0.74	0.92
Swift [1969]	1942–64	Chile	O'Higgins	1.30	2.00
Swift [1969]	1942–64	Chile	Cautin	0.50	1.20
Swift [1969]	1942–64	Chile	–	0.37	3.65
Fitchett (I) [1971][a]	–	Chile	–	0.14	–
Fitchett (II) [1971]	–	Chile	–	0.155	0.246

[a] These results refer to a model estimated without lagged area as an explanatory variable.

Table 8. Estimates of corn supply elasticities

Author	Period	Country	Province/state	Short-run elasticity	Long-run elasticity
Reca [1967]	1924-44	Argentina	Buenos Aires	0.23	—
Reca [1967]	1945-65	Argentina	Buenos Aires	not significant	—
Reca [1967]	1924-44	Argentina	Córdoba	0.30	
Reca [1967]	1945-65	Argentina	Córdoba	not significant	—
Reca [1967]	1924-44	Argentina	Santa Fé	0.25	
Reca [1967]	1945-65	Argentina	Santa Fé	0.10	
Reca [1967]	1924-44	Argentina	Entre Rios	0.35	
Reca [1967]	1945-65	Argentina	Entre Rios	not significant	—
Maffucci [1969]	1925-45	Argentina	Buenos Aires	—	not significant
Maffucci [1969]	1946-65	Argentina	Buenos Aires	—	0.39
Maffucci [1969]	1925-45	Argentina	Córdoba	—	not significant
Maffucci [1969]	1946-65	Argentina	Córdoba	—	not significant
Maffucci [1969]	1925-45	Argentina	Santa Fé	—	not significant
Maffucci [1969]	1946-65	Argentina	Santa Fé	—	0.28
Maffucci [1969]	1925-45	Argentina	Entre Rios	—	0.75
Maffucci [1969]	1946-65	Argentina	Entre Rios	—	not significant
Fitchett (I) [1971][a]	—	Chile	—	not significant	—
Fitchett (II) [1971]	—	Chile	—	0.349	5.229
Pastore [1973b]	1945-65	Brazil	—	0.148	0.571
Pastore (KW)[b] [1973b]	1945-65	Brazil	—	0.193	0.395
Pastore [1973b]	1945-65	Brazil	Northeast[c]	0.10	0.15
Toyama-Pescarin [1970]	—	Brazil	São Paulo	0.83	3.32
Brandt et al. [1968]	—	Brazil	São Paulo	0.45	2.55
Tallone-Rosso [1965]	1944-62	Brazil	Minas Gerais	0.830	1.430
Santos, L. F. dos [1971]	1947-69	Brazil	Minas Gerais	0.07	0.09
Santos, D. B. dos [1974]	1947-70	Brazil	Pernambuco	0.179	0.279
Ribeiro [1975]	1947-70	Brazil	Piauí	0.25	1.04
Ayres [1976]	—	Brazil	Espírito Santo	1.08	—
Iamaguchi [1982]	1948-80	Brazil	São Paulo	0.013	0.018

[a] These results refer to a model estimated without lagged area as an explanatory variable.
[b] Pastore (KW) means the estimates were corrected for autocorrelation utilizing the method of Kenneth-Wallis.
[c] The Northeast region comprises the states from Maranhão and Piauí (middle north) all the way down to the state of Bahia in the East.

Table 9. Estimates of oats supply elasticities

Author	Period	Country	Province/state	Short-run elasticity	Long-run elasticity
Reca [1967]	1924-44	Argentina	Buenos Aires	0.08	–
Reca [1967]	1945-65	Argentina	Buenos Aires	0.01	–
Reca [1967][a]	1924-44	Argentina	Buenos Aires	–	0.24
Reca [1967][a]	1945-65	Argentina	Buenos Aires	–	0.15
Fitchett (I) [1971][b]	–	Chile	–	not significant	
Fitchett (II) [1971]	–	Chile	–	0.120	0.247

[a] These models did not include lagged area as an explanatory variable. We interpret them as long run, contrary to Fitchett's [1971] interpretation.

[b] These results refer to a model estimated without lagged area as an explanatory variable.

Table 10. Estimates of flaxseed supply elasticities

Author	Period	Country	Province/state	Short-run elasticity	Long-run elasticity
Friere [1966]	–	Argentina	–	1.10	–
Reca [1967]	1924-44	Argentina	Buenos Aires	0.41	–
Reca [1967]	1945-65	Argentina	Buenos Aires	not significant	
Reca [1967][a]	1924-44	Argentina	Buenos Aires	–	0.48
Reca [1967][a]	1945-65	Argentina	Buenos Aires	–	0.89

[a] Models from which these estimates are taken did not include lagged area as an explanatory variable. We interpret them as long-run, contrary to Fitchett's [1971] interpretation.

Table 11. Estimates of rice supply elasticities

Author	Period	Country	Province/state	Short-run elasticity	Long-run elasticity
Fitchett (I) [1971]	–	Chile	–	2.038	–
Fitchett (II) [1971]	–	Chile	–	not significant	–
Merrill [1967]	1945–64	Peru	–	–	0.50
Pastore [1973b]	1945–65	Brazil	–	0.305	1.169
Pastore (KW) [1973b][a]	1945–65	Brazil	–	0.294	0.786
Pastore [1973b]	1945–65	Brazil	Center South[b]	0.23	0.49
Pastore [1973b]	1949–66	Brazil	São Paulo	0.61	1.96
Pastore [1973b]	–	Brazil	São Paulo	0.62	4.10
Brandt et al. [1968]	1948–68	Brazil	Goiás	0.30	2.34
Vilas [1975]	–	Brazil	São Paulo	0.42	0.69
Toyama-Pescarin [1970]	–	Colombia	–	–	0.235
Gutierrez and Hertford [1974]	1947–69	Brazil	Minas Gerais	0.093	0.405
Santos, L. F. dos [1971]	1960–71	Brazil	Santarem	0.315	1.507
Santos, D. B. dos [1974]	1951–70	Brazil	Pará	0.289	0.518
Rebello [1973]	1947–70	Brazil	Pernambuco	0.320	0.478
Santos [1974]	1947–70	Brazil	Piauí	0.11	0.30
Ribeiro [1975]	–	Brazil	Espírito Santo	0.41	–
Ayres [1976]	1948–80	Brazil	São Paulo	0.309	0.366
Iamaguchi [1982]					

[a] Pastore (KW) refers to estimates that are corrected for autocorrelation by the means of the Kenneth-Wallis method.
[b] Center South region comprises the states of Minas Gerais, Espírito Santo, Mato Grosso, Goiás, Paraná, Santa Catarina, and Rio Grande do Sul.

Table 12. Estimates of miscellaneous crop supply elasticities

Author	Crop	Period	Country	Province/state	Short-run elasticity	Long-run elasticity
Fitchett (I) [1971][a]	Barley	–	Chile	–	0.658	0.66
Fitchett (II) [1971]	Barley	–	Chile	–	0.091	1.857
Fitchett (I) [1971][a]	Rye	–	Chile	–	0.887	–
Fitchett (II) [1971]	Rye	–	Chile	–	0.881	2.351
Fitchett (I) [1971][a]	Potatoes	–	Chile	–	0.321	–
Fitchett (II) [1971]	Potatoes	–	Chile	–	not significant	–
Belli [1970][b]	Cotton	–	Nicaragua	–	–	0.66
Belli [1970][c]	Cotton	–	Nicaragua	–	–	0.42
Belli [1970][d]	Cotton	–	Nicaragua	–	–	1.37
Pastore [1973b]	Cotton	1945–65	Brazil	–	0.192	0.627
Pastore (KW) [1973b][c]	Cotton	1945–65	Brazil	Center South[f]	0.213	0.213
Pastore [1973b]	Cotton	1945–65	Brazil	São Paulo	0.230	0.230
Pastore [1973b]	Cotton	1945–65	Brazil	São Paulo	1.22	2.03
Brandt et al. [1968]	Cotton	–	Brazil	São Paulo	0.37	–
Ayer & Schuh [1972]	Cotton	–	Brazil	São Paulo	0.69	1.57
Pastore [1973b]	Peanuts	1945–65	Brazil	–	–	0.944
Pastore (KW) [1973b][c]	Peanuts	1945–65	Brazil	–	0.719	1.546
Pastore [1973b]	Peanuts	1945–65	Brazil	–	0.736	1.658
Pastore [1973b]	Peanuts	1945–65	Brazil	Center South[f]	1.47	1.47
Pastore [1973b]	Peanuts	1945–65	Brazil	São Paulo	0.47	1.02
Pastore [1973b]	Tobacco	1945–65	Brazil	–	0.109	0.201
Pastore (KW) [1973b][c]	Tobacco	1945–65	Brazil	–	0.105	0.211
Pastore [1973b]	Tobacco	1945–65	Brazil	Northeast[g]	0.52	0.70
Pastore [1973b]	Castor beans	1945–65	Brazil	–	0.199	0.677
Pastore (KW) [1973b][c]	Castor beans	1945–65	Brazil	–	0.199	0.648
Pastore [1973b]	Castor beans	1945–65	Brazil	Center South[f]	0.23	2.09
Pastore [1973b]	Castor beans	1949–66	Brazil	São Paulo	0.39	0.77
Pastore [1973]	Manioc	1945–65	Brazil	–	0.106	0.955
Pastore (KW) [1973b][c]	Manioc	1945–65	Brazil	–	0.068	0.216
Pastore [1973b]	Manioc	1945–65	Brazil	Northeast[g]	0.12	0.18
Pastore [1973b]	Manioc	1945–65	Brazil	Center South[f]	0.09	0.90
Pastore [1973b]	Manioc	1949–66	Brazil	São Paulo	0.26	0.47
Pastore [1973b]	Potatoes	1949–66	Brazil	São Paulo	0.29	0.29
Pastore [1973b]	Sugar cane	1945–65	Brazil	–	0.156	–
Pastore [1973b]	Sugar cane	1945–65	Brazil	Center South[f]	0.26	0.26
Pastore [1973b]	Sugar cane	1949–66	Brazil	São Paulo	0.12	0.12
Pastore [1973b]	Food[h]	1945–65	Brazil	–	0.138	0.274
Pastore [1973b]	Nonfood[h]	1945–65	Brazil	–	0.411	0.533
Pastore [1973b]	Domestic consumption[h]	1945–65	Brazil	–	0.147	0.212
Pastore [1973b]	Edible beans	1949–65	Brazil	São Paulo	0.37	0.37
Pastore [1973b]	Soybeans	1949–66	Brazil	São Paulo	2.63	15.47

[continued]

Table 12. Estimates of miscellaneous crop supply elasticities [Continued]

Author	Crop	Period	Country	Province/state	Short-run elasticity	Long-run elasticity
Toyama-Pescarin [1970]	Sugar cane	—	Brazil	São Paulo	0.27	0.39
Toyama-Pescarin [1970]	Edible beans	—	Brazil	São Paulo	0.31	0.43
Brandt et al. [1968]	Edible beans	—	Brazil	São Paulo	0.10	0.31
Garcia [1981]	Food[i](1)	1950-76	Colombia	—	—	0.4478
Garcia [1981]	Food[j]	1953-76	Colombia	—	—	0.5282
Garcia [1981]	Food (2)	1953-76	Colombia	—	—	0.445
Garcia [1981]	Food (1)	1953-67	Colombia	—	—	0.8298
Garcia [1981]	Food (2)	1953-67	Colombia	—	—	0.8530
Reca [1967]	Agricultural production	1950-74	Argentina	—	0.21-0.35	0.92-0.78
Gemma [1983]	Soybeans	1971-77	Brazil[k]	—	2.53	4.17
Santos, D. B. dos [1974]	Manioc	1947-70	Brazil	Pernambuco	0.187	0.235
Santos, D. B. dos [1974]	Edible beans	1942-70	Brazil	Pernambuco	0.140	1.085
Santos, D. B. dos [1974]	Potatoes	1947-70	Brazil	Pernambuco	0.088	0.199
Santos, D. B. dos [1974]	Cotton	1947-70	Brazil	Pernambuco	0.145	0.215
Hemerly [1975]	Peanuts	1950-72	Brazil	—	0.631	1.022
Leite [1975]	Soybeans	1951-71	Brazil	—	0.61	1.62
Ribeiro [1975]	Manioc	1947-70	Brazil	Piauí	0.15	0.31
Ribeiro [1975]	Edible beans	1947-70	Brazil	Piauí	0.19	1.63
Ribeiro [1975]	Cotton	1947-70	Brazil	Piauí	0.25	1.76
Ribeiro [1975]	Aggregate[l]	1947-70	Brazil	Piauí	0.12	0.42
Ayres [1976]	Edible beans	—	Brazil	Espírito Santo	0.56	—
Iamaguchi [1982]	Manioc	1948-80	Brazil	São Paulo	0.065	0.128
Iamaguchi [1982]	Edible beans	1948-80	Brazil	São Paulo	0.199	0.279

a These results refer to a model estimated without lagged areas as an explanatory variable.
b Estimate based on ordinary least squares applied to the production function.
c Estimates based on indirect least squares.
d Estimates based on relative shares method.
e Pastore (KW) refers to estimates that were corrected for autocorrelation by means of the Kenneth–Wallis method.
f The center south region comprises the states of Minas Gerais, Espírito Santo, Mato Grosso, Goiás, Paraná, Santa Catarina, and Rio Grande do Sul.
g The Northeast region comprises the states from Maranhão and Piauí (middle North) all the way down to the state of Bahia in the East.
h The author has aggregated his data on the following categories: food, nonfood domestically consumed, and total domestic consumption, which is the aggregate of the first two.
i Prices deflated by the total implicit price deflator of GNP.
j Prices deflated by the implicit deflator of the nonagricultural sector.
k This study utilizes a cross-section time series methodology. The cross-section includes the states of São Paulo, Paraná, and Rio Grande do Sul.
l This includes the following crops: rice, edible beans, cotton, manioc, corn, sugar cane, garlic, castor beans, onions, sweet potatoes, and tomatoes.

ties. Pastore [1973b] found a long-run elasticity of food supply of 0.274 for Brazil, and a long-run elasticity of nonfood agricultural supply of 0.533. For Colombia, J. G. Garcia [1981] obtained estimates of aggregate food supply elasticities in the range of 0.445–0.9424, and Reca [1967] for Argentina obtained estimates of the long-run elasticity of agricultural supply in the range of 0.42–0.78. These values are in broad agreement with most other studies of aggregate response done throughout the world as summarized by Scandizzo and Bruce [1980] and also by Krishna [1982].

Peterson [1979], in a paper that has attracted a great deal of attention, obtained estimates of the long-run aggregate supply elasticity by using cross-country data from a sample of fifty-three countries. He obtained an estimate of two for this important parameter. Binswanger et al. [1985] also estimated the aggregate supply response parameter, with their estimates based on cross-section/time-series data (annual observations for fifty-eight countries in the period 1969-78). Their principal results [p. 4] were:

> A weak positive supply response was obtained from the variations over time for the individual countries (within-country variations). A negative supply response is obtained from the between-country variations. The shifters, as a group, account for most of the variations in supply in the within-country and between-country analysis. These findings are in contrast to the result obtained by Peterson. In view of the importance of the issue we analyze the reason for the difference in the two studies and conclude that Peterson's results were obtained from FAO price data that since have been revised by FAO. Our analysis is based on the revised data.

It is difficult to believe that most of the differences between Binswanger et al. [1985] and Peterson [1979] can be accounted for by revisions in the FAO data (see Peterson [1983] for later comments). Peterson's results clearly incorporated the effects of both short-run response to price along the static short-run supply function as well as induced shifts in the supply function itself. In view of this it is not surprising to find supply elasticities above one. The statistical results reported by Binswanger et al. may have suffered from problems of multicollinearity.

To summarize, the statistical results show relatively small but significant supply responses in the short term for annual crops, with large responses as longer term adjustments are taken into account. Aggregate supply responses tend to be less than 1 also, but also statistically significant. Estimates of the parameters of the metasupply function with cross-country data tend to show a relatively large response.

STUDIES OF PERENNIAL CROPS

Similar to the case of annual crops, the Nerlovian model underlies most of the work with perennials. However, given the special character of these crops, it is with them that most of the creative research has been done. We focus on coffee and cocoa, the two crops for which significant research has been reported, but with special emphasis on coffee.

The first model (and one of the most interesting) we examine was developed by Arak [1968].[55] She studied the price responsiveness of coffee growers by considering the changes in both the number of trees and yields. To do that, it was necessary to analyze planting, abandonment, and removal decisions.

The first decision, *planting*, is postulated to depend on the expected price of coffee, P_t; the percentage of old trees (ten years or more), D_t; and the area previously planted with coffee,

$$\sum_{j=0}^{t-1} N_j,$$

which, by assumption, cannot be used to plant coffee again in view of soil depletion. Here N_j is the area planted with coffee in year j and the sum is taken from an initial period dated zero.

An interesting aspect of this decision is the role of labor costs. The author noted that labor is a major constraint to São Paulo coffee growers and therefore the planting of new trees will depend on these costs [Arak, 1968, p. 14]:

> The relative number of old trees is used as a proxy for short-
> run labor cost expectations. When a large percent of the tree
> stock is old, it is likely that abandonment of old trees will
> make labor relatively plentiful in the near future; farmers
> would, therefore, plant desired areas more quickly when the
> tree stock is relatively old. Conversely, with a relatively young
> stock of coffee trees, expectations of short-run labor cost
> would be less favorable and farmers would be likely to
> postpone planting desired trees.

The equation which reflects the decision process is estimated as follows:

$$N_t = \delta_0 + \delta_1 D_t + \delta_2 (D_t P_t) + \delta_3 \left(D_t \sum_{j=0}^{t-1} N_j\right)$$

where the variables are as defined as above, and δ_1 and δ_2 are the parameters to be estimated.

The *removal* decision is postulated to be dependent on the expected price for that year and on physical yields. Thus,

$$R_t = (d_0 + d_1 P_t + d_2 F_{t-1}) T_{t-1}^E$$

where R_t is the percentage of trees to be removed in year t, F_{t-1} is a dummy variable equal to 1 if there was a frost in year $t-1$, and zero otherwise and T_{t-1}^E is the number of trees in the age group for which removal is one alternative. The parameters to be estimated are d_0, d_1, and d_2.

The *abandonment* decision is considered separately for the small farmer (facing a land constraint) and the large farmer. For the small farmer it is postulated that abandonment depends negatively on expected prices, while for the large farmer the relationship is hypothesized to be positive. Arak [1968, p. 217, emphasis added] argued as follows:

> Provided that a farmer plans to maximize the present
> discounted value of future earnings from coffee over an infinite
> horizon (constrained by labor, *but not by land*), he may find it
> profitable to abandon trees at an earlier age the higher are the
> real coffee prices: when coffee price is high, it pays to incur
> planting costs more frequently in order to reap only the highest
> yields on each tree. The farmer will maintain each tree for a
> longer period when prices fall, providing, of course, that they
> do not fall so low as to convince him to forget future coffee
> cultivation. For these farmers, there may well be a range of
> coffee prices for which the relationship between optimal
> abandonment age and prices is negative:
>
> $$Z_t = g\ (P_t)\ \frac{dZ_t}{dP_t} \le 0, \text{ where } Z_t$$
>
> is the optimal abandonment age of trees in year t.

A consideration of the abandonment decision for young trees and an assumption about linear relations completes the model to be estimated.

The parameters of the planting equation were estimated with data for the period 1930-55 (excluding observations for 1937, 1942, and 1943), while those for the other two equations were estimated with data for the period 1933-50. The short-run price elasticity of annual planting was estimated to be 2.28.[56]

Arak [1968] also estimated the price responsiveness of coffee growers for the states of Minas Gerais and Espírito Santo. Since there were no separate data on new planting and removal of the trees for these states and, as before, these decisions depend on different factors, a model was developed to explain the desired change in the stock of trees [p. 19]. The removal decision was also considered and it was separated into removal of "desired" and "undesired" trees.

Maximum likelihood estimates of the parameters were obtained with data for the period 1927-59 for the states of Minas Gerais and Espírito Santo. The long-run elasticities of change in the desired area for coffee with respect to expected price were estimated to be 0.54 for the state of Minas Gerais, and 0.28 for the state of Espírito Santo. Short-run elasticities were 0.08 and 0.20, respectively.

Arak [1967, cited by Askari and Cummings, 1976, p. 242] also estimated the supply response of coffee growers in the state of Paraná. For that analysis the following equation was utilized:

$$\frac{N_t}{L_{t-1}} = a_0 + \frac{a_1}{P_t} + \frac{a_2 \tilde{R}_t}{L_{t-1}} + M_t$$

where L_{t-1} is the cultivated agricultural land in period t-1 and $\tilde{R}_t$ is a measure of the replacement of frost-damaged trees in period t. The equation was estimated by ordinary least squares with data for the period 1945-62. The price elasticity of the desired tree stock was found to be 0.96.

Bacha [1968] undertook a study of the world coffee market and included in his empirical work estimates of price responsiveness for the states of Paraná and São Paulo in Brazil, and for Colombia, Latin America and Africa. Bacha criticized Arak's model for Paraná by pointing out that although she had correctly discussed the coffee sector of that state with a frontier-type of model, her statistical analysis did not take this rather unique situation into account. He, in turn, estimated the supply of coffee as a function of price (lagged four years) and one-year lagged tree stock. The Nerlovian adjustment coefficient was made a function of the adult tree stock. The estimate of the long-run price elasticity was 0.71 [p. 96].

For the case of São Paulo, explicit consideration was given to the fact that coffee has a two-year production cycle. It was assumed that the relation between output and area was different for odd and even years. The estimated equation was:

$$Q_t = a_0 + a_1 D_t + a_2 P_{t-4} + a_3 D_t P_{t-4} + a_4 Q_{t-2}$$

where:

 D_t is a dummy variable equal to zero in even years and equal to one otherwise;

 P_{t-4} is the coffee price lagged four years; and

 Q_{t-2} is production lagged four years.

The estimated short-and long-run elasticities were 0.23 and 1.00, respectively [p. 74].

A simpler version of the above equation with only P_{t-4} and Q_{t-1} included as independent variables was estimated for Colombia and for Latin America. The dependent variable for Latin America was exports [p. 70]. The results were as indicated in Table 13.

Still another attempt to estimate the supply response of São Paulo coffee growers was made by Saylor [1974]. He tested four specifications of the supply

Table 13. Supply elasticities for coffee

	Short run	Long run
Colombia, 1936–64	0.07	0.453
Latin America, 1943–60	0.276	0.518

Source: Bacha, 1968, p. 74. Used by permission of Frank Cass & Co. Ltd.

function: a purely Nerlovian equation; a modification of the pure Nerlovian model to include an intercept shift dummy variable to account for the rapid changes that occurred during the coffee eradication period (1962-67); another modification of the initial equation to include an intercept and slope change dummy variable for the period 1948-62 and 1963-70 in order to identify possible shifts in structure; and a model of irreversible supply in which it was hypothesized that in years of price increases there would be a positive response, while in years of declining prices there would be none or, at best, a much smaller response.

The results considered by Saylor to be his best gave rise to the following short-run elasticities: 0.117, 0.173, and 0.101. The corresponding long-run elasticities were 0.625, 0.733, and 0.605. These results show a fairly high degree of consistency in the elasticities obtained, with the long-run elasticity significantly greater than the short-run elasticity.

Behrman and Klein [1970] also estimated a supply function for coffee as part of their estimation of a multiple equation model for the Brazilian economy. The basic description of the model is as follows [Askari and Cummings, 1976, p. 251]:

> Acreage was represented as a function of the ratio of the domestic coffee price to a gross domestic production deflator, lagged two years, and coffee acreage lagged one year; yield was explained in terms of domestic coffee prices and rainfall, both lagged one year.

The estimated price elasticities were as follows:

> Short-run elasticity of acreage, 0.10
> Long-run elasticity of acreage, 0.11
> Short-run elasticity of yield, 0.15

Finally, Bateman [1969] estimated a supply function for Colombian coffee. He found that the elasticity of new acreage planted was between 0.47 for the period 1947-65, and 0.87 for the period 1952-65.

Behrman [1968b] estimated supply functions for cocoa for a number of Latin American countries. The structural model was of the Nerlovian type. Estimates of the price elasticities obtained are summarized in Table 14.

Table 14. Supply elasticities for cocoa

	Short run	Long run
Brazil	0.53	0.95
Ecuador	–	0.28
Dominican Republic	0.03	0.15
Venezuela	0.12	0.38

Source: Behrman, 1968b. With permission.

STUDIES OF THE LIVESTOCK SECTOR

A number of studies of livestock supply response, especially for cattle, has been developed for Latin American countries. Most of these studies are based on microeconomic models of the behavior of cattle producers. The methods, despite the inclusion of time, are essentially static in nature and do not make a distinction between the long-run and the short-run in the specification of the model. Consequently the estimated elasticities are a mixture of short-run and long-run responses.

A different approach was taken by Mascolo [1979, 1980], however, inspired by a previous study by Carvalho [1972]. Mascolo and Carvalho used methods that are essentially dynamic in nature and their models open the way for an entirely new approach to the study of the livestock sector and, for that matter, perennial crops [on this latter issue see Nerlove, 1979].

Reca's study [1967] of the Argentine cattle sector is important in historical terms and is indicative of the prevailing methodology at that time. His model was essentially Nerlovian, and consisted of an equation which explains the desired size of the herd, another which postulates the partial adjustment of the herd, still another which postulates that production depends on the size and on the average growth of the herd, and an identity which constrains the change in the herd. Estimates of the elasticities obtained were as follows:

> *Period, 1923-47.* (a) *Slaughter:* There were eight estimated price elasticities with values in the range −0.43 to −0.16.[57] The equation with the best fit yields an elasticity of −0.36. (b) *Desired herd:* As above, there were eight estimated price elasticities in the range of 0.32 to 1.03. The equation with the best fit yielded an elasticity of 0.38.
> *Period, 1948-65.* (a) *Slaughter:* Price elasticities were estimated in the range −0.026 to −0.19. (b) *Desired herd:* Price elasticities were estimated in the range 0.26 to 0.28.

These results show the usual result that the short-run slaughter response to changes in price is negative, rather than positive as theory generally predicts. This is because the livestock sector involves a special kind of inventory investment phenomenon. We take up this issue next.

Studies of livestock supply response took an important change in direction after seminal studies by Yver [1971] and Jarvis [1969, 1974], and later by Carvalho [1972]. The studies by Yver and Jarvis were the first to incorporate specifically the investment behavior of cattle producers. The decision, they argue, can be thought of as deciding how much capital to own and the composition of that capital.

Jarvis [1974] is a good published source for this material. He develops microeconomic capital-theory models which explain the optimum slaughter age and feed input for steers, and their market prices, as well as similar models for the cows in the herd. The models for cows are particularly complex since they can be slaughtered, fattened, and/or be maintained as a capital good or "growing machines."

More specifically, consider a very simple model in which the only input is the steer itself. Let

$$V(\Theta) = w(\Theta)e^{-r\Theta}$$

where $V(\Theta)$ is the present value of an animal that lives to age Θ, $w(\Theta)$ is the weight of an animal of age Θ and r is the interest rate. It is assumed that

$$w' > 0 \text{ and } w'' < 0.$$

Then the optimal slaughter age is the value of $\hat{\Theta}$ that satisfies

$$V'(\Theta) = w'(\Theta) e^{-r\Theta} - rw(\Theta) e^{-r\Theta} = 0$$

or

$$\frac{w'(\Theta)}{w(\Theta)} = r$$

The slaughter age is determined by the condition that the rate of growth in the weight of the animal be equal to the interest rate. This is a true maximum since one can easily verify that $V''(\Theta) < 0$. It can also be easily verified that

$$\frac{d\hat{\Theta}}{dr} < 0$$

where $\hat{\Theta}$ is the optimal slaughter age. In other words, the higher the interest rate, the lower the slaughter age. The interest rate has a fundamental role in the decision process because there is no other cost to cattle production in this simple model than the interest foregone.

The model can be made somewhat more realistic by considering the existence of other inputs in the production process. In this case, the producer will maximize the present value of the fattening process, i.e.,

$$\max_{\Theta,i} \pi (\Theta,i) = p(i,\Theta) \ w(i,\Theta) - C_i\!\int_0^c e^{-rt}dt$$

where p is the price of the steer (assumed equal to one in the previous model), i is a fixed bundle of inputs utilized to feed the steer, and C is the cost of this bundle.

Jarvis [1974] goes on to derive the first-order necessary conditions for an interior maximum and some comparative static results. He finds [p. 493] that:

> the rate of weight gain plus the rate of price change due to
> aging is equal to the interest rate plus the cost per day of
> feeding the animal as a percentage of its total value. Similarly,
> at î the present discounted value of the marginal net weight
> gain and price increase corresponding to a higher stream of
> inputs throughout the steer's life, less the present discounted
> cost of feeding the animal these inputs, must be zero.

The comparative static results were of indeterminate sign. The data from Argentina then were used to determine the signs of the relevant partial derivatives. It is notable that in the short run a negative slaughter response was obtained. This follows because higher prices induce higher slaughter age; producers withhold the animals from the market in order to feed them to a higher weight.

The model was then extended to consider female cattle. This is done by including in the profit function the expected value of calves born in addition to the revenue from the sale of the animal. The qualitative nature of the analysis is the same, but considerably more complex.

The next stop was the specification of an econometric model for the purpose of explaining changes in herd size, in the number and weight of the animals slaughtered, in domestic consumption, in exports, and in the price of beef. The advantage of this model is that it gives a clear rationale for the perceived negative slaughter response in the short run, and also provides a rationale for expecting to find different supply responses between males and females and between the long run and short run. Jarvis [1974, pp. 517-518] summarized his results as follows:

> I believe the results obtained should abolish doubt as to
> whether Argentine producers respond to prices. These results,
> which have been obtained using herd data which I believe are
> much improved over those previously available, are statistically
> quite significant and identify the slaughter response of different
> animal categories. The results show that producers
> systematically reallocate their portfolios in the expected manner
> when the recursive effect of their decisions is strongly evident,
> that is, with continuously operating markets for disposable
> productive assets. And because much of the indicated response

is an interactivity shift within the agricultural sector between
grains and livestock, the price response shown by cattle
producers implies a response by field crop producers as well.

Lattimore [1974] undertook a similar study of the Brazilian cattle industry in
the period 1947-71. His study was unique in that it included a behavioral equation for the interventions of policy makers in the economy. He estimated an
econometric model of the sector considering that Brazil is a marginal exporter in
the world market (small country) facing an infinitely elastic demand for beef. In
addition, he tried to explain the level of government intervention, defining intervention as the difference between the world beef price converted into cruzeiros
by the free trade exchange rate minus the domestic beef price.

The model consists of eight equations: the policy intervention equation, a
price equation which links the domestic economy to the world market, slaughter
equations for male and female cattle, a domestic demand for beef, investment
functions for males and females, and an identity which defines the exports of
beef. The policy intervention equation is the major novelty of this model. It is
expected, according to Lattimore [1974] and Lattimore and Schuh [1976], that
higher levels of inflation should induce stronger government intervention to
maintain lower beef prices in the domestic economy. Similarly, when there is a
favorable situation in the balance of payments, the government increases its level
of intervention since it can afford the loss of foreign exchange that will be induced
by the desired intervention.

Parameters of the model were estimated and the supply elasticities obtained
were as follows [Lattimore, 1974; or Lattimore and Schuh, 1976]:

> Price elasticities of the cattle stocks: (a) short run (one-period impact):
> females 0.046, males 0.078; (b) long run: females 0.788, males 1.775.
> Price elasticities of slaughter: (a) short run (one-period impact): females
> −0.575, males −0.113; (b) long run: females 1.538, males 1.596.

Both the relative size and signs of these elasticities are as predicted by the models developed by Jarvis [1974] and Yver [1971]. Cattle producers were again
found to respond significantly to price, and in ways predicted by the theory. Statistically significant coefficients were also obtained for the variables in the government behavior equation. Although *ad hoc* in nature, this behavioral model anticipated more formal modeling of government behavior that later appeared in
the U.S. literature [see Rausser, Lichtenberg, and Lattimore, 1982].

As we noted earlier, Nerlove in his recent appraisal of supply response studies
[1979] stressed the need for the development of truly dynamic models of the livestock (and perennial crops) sectors. Carvalho's work [1972] on the U.S. cattle industry was a first successful attempt to model the decision making of profit-maximizing agents in this sector along the lines suggested by Nerlove.

Mascolo [1979, 1980] developed a similar model for the cattle industry in Brazil. The model was designed to explain the whole set of decisions agents have to take. It includes, for males and females, the options of keeping them as reproductive stock, keeping them as fattening stock, or slaughtering them. This is illustrated by Figure 1, which was taken from Nerlove, Grether, and Carvalho [1979, p. 238]. Also see Mascolo [1979, 1980].

The model actually utilized by Mascolo was a somewhat simplified version of the above. In Brazil, producers are not allowed to slaughter females before their first calving. Thus, there are no data recorded for these animals and therefore this alternative has been excluded from the model.

The solution to the supply problem can be specified as a dynamic programming model in which the return function in period t is the following: total revenue from current sales, plus the present value of animals born in period t, $t-1$ and $t-2$, plus the present value of the females born in $t-3$[58] minus the cost of maintaining herd numbers, minus the aging cost of the capital stock, minus the cost of maintaining the males in the fattening process, minus the aging costs of the male stock while in the fattening process, minus the costs of maintaining the animals born in t, $t-1$ and $t-2$. All of the cost functions are assumed to be quadratic. The author solves Bellman's equation by backward substitution subject to the biological constraints on mortality and natality.

The elasticities obtained with the model are price elasticities of the demand of the cattle producer for his own stock (fattening and reproduction). These are of a different nature than the supply elasticities presented above. However, Mascolo's [1979, 1980] results are as follows:

Elasticity of the fattening (male) stock with respect to current beef price: -0.054

Elasticity of the fattening (male) stock with respect to next year's expected price: 0.045

Elasticity of the reproductive stock with respect to long-run price (defined as an average of this year's and next year's expected price): 0.015

These elasticities are very low. The author was unhappy with the results and attributed them to the poor quality of the data he had. It is important to note that the stock data he used were estimates he developed on the basis of scattered information available, as well as estimates developed by other analysts.

To close this section, we note the study of Witherell [1969] of the supply elasticity of wool in Argentina and Uruguay. The model was Nerlovian and the price elasticities are as presented in Table 15.

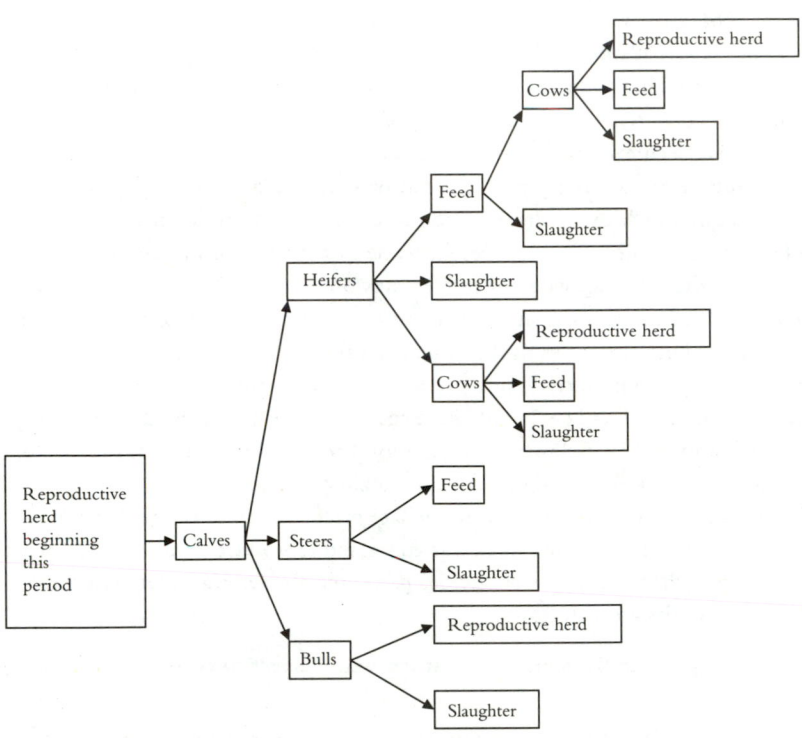

Figure 1. Flow Chart of Decisions in Cattle Production
(*Source:* Nerlove, Grether, and Carvalho, 1979. By permission.)

Table 15. Wool supply elasticities

	Short run	Long run
Argentina	0.042	0.202
Uruguay	0.212	0.481

Source: Witherell, 1969. With permission.

To conclude this section a number of comments are in order. First, there is little literature on the role of livestock in agricultural development, despite the obviously large amount of resources committed to this sector, especially in Latin America. Second, the supply response studies for Latin America help a great deal in understanding the nature of the investment process in this sector, an important first step in understanding the role of the sector in the development process. Third, the estimates of the parameters obtained from empirical research, especially their negative supply coefficients for cattle in the short run, indicate that policy makers need to be especially careful in intervening with price policy so as not to create longer term consequences contrary to their policy goals.

Much needs to be done to better understand the livestock sector. In large countries with heterogeneous ecological conditions there are important regional interactions. Similarly, technical progress in the livestock sector is poorly understood, as is the role it might play in the development process. Thus, important as the contributions reported above are to understanding the cattle and other sectors, they do little more than scratch the surface.

CONCLUDING COMMENTS

Generalizing about the overall results on supply is not easy, although a number of comments seem to be in order. First, the analysis of supply response has ranged from a relatively large number of studies that use fairly simple models, often Nerlove-type distributed lag models, to a more limited number of studies that tend to be innovative and to use fairly sophisticated methodologies. These latter studies tend to focus on the livestock and perennial (tree crops) sectors.

Unfortunately, with the exception of the livestock studies and some of the tree crop studies, very little effort has been made to dig beneath the econometric results to understand what they mean, or to understand the underlying factors affecting supply. It may be that the ease with which plausible econometric results can be obtained by using simple Nerlove-type models has been a barrier to more in-depth analysis of underlying supply behavior.

Second, there is ample evidence that producers do respond to prices. To argue that farmers are irrational, or overcome by inertia, is just not supported by the statistical results. An important issue, of course, is the size of the response. The structuralists posit that the intersectoral supply response is quite low, approaching zero. The statistical results summarized above, however, indicate that the elasticity is positive, but smaller than one.

Two points seem pertinent. First, the absolute change in output from a change in price can be fairly large if the change is on a sufficiently large base, even though the supply elasticity itself is relatively small. Second, we know from studying resource flows among sectors that the long-term supply response can be quite large. The perceived premature and large out-migration from agriculture in country after country is ample evidence of that, as is the large and rapid intersectoral flows of capital in response to changes in the domestic terms of trade.

We believe the estimated responses summarized above are of sufficient magnitude to cast considerable doubt on the structuralist model. The export (excess) supply elasticities consistent with these estimates of the direct supply elasticities obtained are sufficiently large to produce a significant response to changes in policy. Moreover, one does not have to turn to irrationality or to problems of the "structure" of agriculture to rationalize the results obtained. The elasticities reported are consistent with those obtained in other parts of the world, including the United States, where "structure" is apparently not a problem and where presumably producers are quite rational.

A third point is that, with a few exceptions, it was only in the 1970s that there began to be significant technical change in Latin American agriculture as reflected in improved varieties, increased use of modern inputs, and more sophisticated management and production techniques. These developments tend to cause supply elasticities to be larger, and may explain why structuralist interpretations of Latin American economic performance have declined in importance over the last decade. This suggests that structuralist models may have been more relevant for the 1950s and 1960, but that monetarist models have come into their own in the late 1970s and early 1980s.

We don't believe the data support that interpretation, however. We will show in a later section that the agriculture of countries like Argentina was quite dynamic in the pre-World War II period. The stagnation and sluggish growth of the 1950s and 1960s was a response to serious discrimination against agriculture by economic policies that promoted import-substituting industrialization. Producers were responding to discriminatory economic policies in very rational ways, and it had little to do with structural factors.

Fourth, there has been little tendency for output to outpace demand in the region, nor have there been many attempts to understand the sources of supply or to account systematically for changes in output over time. We are aware of only three studies which represent serious efforts at growth accounting other than those associated with tests of the Hayami-Ruttan model. The first was a study of São Paulo agriculture in Brazil [Thomazinni-Ettori, 1964], which showed that total factor productivity in that state grew at about the same rate in the 1950s as it did in the U.S. in the same period. Later, Hertford [1971] did a series of growth analyses for Mexico which showed significant growth in total factor productivity, but with output growth still largely a function of growth in conventional inputs. Finally, G. F. Patrick [1975] used shift-share analysis to understand output growth in the crop sector of Brazil. The distinguishing feature of his results was the role given to geographic shifts in location and shifts in commodity mix, both of which led to increased output or to production of high value crops. More recently, Graham, Gauthier, and Mendonça de Barros [1987] disaggregated the increase in output of Brazilian agriculture in the period 1950-80, finding that increases in area cultivated were still the major factor explaining increases in output compared to increases in yields.

The global study by Judd, Boyce, and Evenson [1986] is the kind of research needed to understand longer term supply and growth issues. This article is useful in the present context in that it puts the Latin American data on investments in research and extension in perspective with similar investments in other parts of the world. It is of value both for the original data and the comparative perspective.

Finally, it should be noted that recent developments in economic theory have important implications for future studies of supply response in Latin American agriculture. The theory of rational expectations, and its important lessons with respect to econometric practice, should make us aware of one other possible weakness of available estimates of supply response. These estimates fail to account fully for the knowledge that economic agents have (and acquire) with respect to the policy regimes in which they make their decisions. Furthermore, they completely fail to account for the changes in behavior induced by changes in these policy regimes.[59]

To illustrate this point we review briefly the theoretical model of supply response as developed by Z. Eckstein [1981, 1984]. The point of departure for the formulation of the model is the observation that the dynamic element in the Nerlovian model, as presented above, is given by *ad hoc* disequilibrium behavior supposedly induced by adjustment costs.[60] From this observation, a model that accounts for the equilibrium behavior of the agent can be formulated taking into account the restrictions faced and the maximization of the discounted expected value of profits. That is, this agent is assumed to maximize:

$$E_{-1} \lim_{N \to \infty} \sum_{t=0}^{N} \beta^t (X_{1t} + \frac{P_{2t}}{P_{1t}} X_{2t}) \tag{1}$$

subject to the following constraints:

$$A_{1t} + A_{2t} = \bar{A}_t \tag{2}$$

$$X_{1t} = \{(f_1 + a_{1t} - \frac{d_0}{2} A_{1t}) + \frac{d_1}{\bar{A}_t}(\bar{A}_t - A_{1(t-1)} - A_{1t})\} A_{1t} \quad (3)$$

$$X_{2t} = (f_2 + a_{2t}) A_{2t} \quad (4)$$

Where:

X_{it} is the production of crop i at time t;

P_{it} is the relative price of crop i in year t;

$\bar{A}_t$ is the total cultivated land available in year t;

A_{it} is the land allocated to crop i in year t;

$0 < \beta < 1$ is the discount factor;

a_{it} is the shock to production of crop i in year t;

S_t is a vector on $n-3$ exogenous variables at time t, such as taxes, tariffs, and other variables that contain information on the prices A_t's and a_{it}'s;

f_1, d_0, d_1, are positive parameters of the production functions;

E_t (X) is the mathematical expectation of variable X conditioned on the information available at time t.

The constraints on the above problem are the availability of land, given by (2), and the production functions for the two crops, (3) and (4). Z. Eckstein [1981, pp. 28-29] explains equation 3 as follows:

> The production function of crop 1 . . . includes a linear shock to productivity, a_{1t}, which is uncontrollable and random and a dynamic term, $d_1 \{\bar{A}_t - A_{1(t-1)}\}$, which is meant to approximate the deterioration of productivity due to successive cultivation of crop on the land. The last term implies that, on the average, land productivity at time t increases proportionately to the quantity of current soil which has not been used for crop 1 in the previous period.
>
> The deterioration of land yields is due to exhaustion of the soil and accumulation of crop-specific insects and worms, arising from growing the same crop in successive periods. This element is captured by the term $-d_1 A_{1t-1} \ldots$, but we argue that if more land is available for cultivation ($\bar{A}_t$ is increasing), farmers are more flexible and can avoid deterioration of the average product. Notice that this term introduces a dynamic element into the production function. In what follows, we show that a positive d_1 gives rise to a land allocation process that can be regarded as a crop rotation.

The vector S_t did not appear explicitly in the above formulation. However, it is a part of the information that the agents have at the time they make decisions. As such, it is included in the process of calculating the expected value and plays a fundamental role in the derivation of the decision rule of agents. This, in fact, is the essence of the rational expectations hypothesis, i.e., that the agents know the stochastic processes that generate S_t, a_{1t}, and $P_t(f_2 + a_{2t})$, which are the exogenous variables of the models.

It would be beyond the scope of this paper to go through the somewhat intricate process of solving the above maximization problem, even with simple stochastic structures for the above process. Instead, we simply present some of Z. Eckstein's [1981, pp. 41–46] general comments on the decision rules obtained in his simplified version of the model which is used to make a comparison with the Nerlovian, or supply response model, as he calls it [pp. 42-43]:

> . The main difference in interpreting the land allocation response
> to a once-and-for-all change can be summarized as follows.
> The elasticity in the supply response model depends mostly on
> the values of the correlation between lag prices and current
> land. Further, the long-run elasticity is greater than the short-
> run elasticity. The elasticities on the above model depend on
> d_1, the dynamic parameter in the objective function and the
> serial correlation in prices. The short-run elasticity is
> *independent* of the cross-correlation between lag prices and
> current land allocation and in general may be higher or lower
> than the long run elasticity.

In concluding the comparison between the two models, he states [p. 46]:

> We can conclude that even though the model in this section
> and Nerlove's supply response model may give rise to similar
> equations for estimation, and both use time series observations,
> they will interpret the data in a completely different way.
> Furthermore, any pattern of serial and cross-correlation
> between land allocations and prices that has been documented
> by the supply response models can be *explained* via the models
> of optimizing agents who know the true (equilibrium)
> distribution of prices.

The significance of the rational expectations approach in Latin America may ultimately be in the fact that it takes policy regimes into account. Policy changes in Latin America are legion. If producers understand these changes and adjust accordingly, empirical work that fails to take this into account has little theoretical or empirical base. In implementing rational expectations models to understand Latin American development better, the lack of data will be a serious problem.

But that is no reason to neglect this important body of work totally and the econometric procedures it implies.

2. Marketing Studies

The most ambitious study of agricultural marketing institutions in Latin America was that carried out by the Latin American Market Planning Center at Michigan State University. This research involved a series of diagnostic studies that extended over a period of some ten years, and was financed by AID and by local government agencies. The studies in each case were carried out in collaboration with local professional personnel.

The maintained hypothesis of these studies was that improvements in the marketing system that links producers to consumers could contribute both to economic and to agricultural development. The basic premises of the study and the perspective they take are expressed well in the following statements taken from the summary report [Harrison et al., 1974]:

> The long-held belief that effective marketing systems will evolve automatically is at best dubious. Since it is widely recognized that farmers and industrialists must be educated, motivated, assisted and sometimes subsidized to encourage the necessary innovation to promote development, there is no apparent reason to expect market intermediaries (or more accurately, marketing systems firms) to be any different. In fact, our evidence suggests that at some stage public agency efforts to stimulate the development of effective internal markets may become crucial to development. [pp. 2-3]
>
> For our purposes we have found it convenient to regard the "marketing system" as a primary mechanism for coordinating production, distribution and consumption activities. When viewed in this manner, marketing would include the exchange activities associated with the transfer of property rights to commodities, the physical handling of products and the institutional arrangements for facilitating these activities. . . .
>
> The "systems orientation" emphasizes interdependence of related activities and is concerned with the coordination of economic activities as a system. Thus, production and distribution of farm inputs, farm production and food distribution, and production and distribution of consumer goods are viewed as a system because they are interdependent. Small increases in productivity in one part of the system may greatly improve the potential for the whole system. Similarly

failure at any functional level may cause stagnation in the entire system. [pp. 4, 5]

The study resulted in six major research reports. Their respective titles indicate the geographical focus of the studies, plus the particular perspective taken: *Food Marketing in the Economic Development of Puerto Rico* [Riley et al., 1970a] (the first of the studies); *Market Processes in the Recife Area of Northeast Brazil* [Slater et al., 1969b]; *Market Processes in La Paz, Bolivia* [Slater et al., 1969a]; *Market Coordination in the Development of the Cauca Valley Region—Colombia* [Riley et al., 1970b]; *Agricultural Marketing in Economic Development, An Annotated Bibliography* [Torrealba, 1971]; and *Fomenting Improvements in Food Marketing in Costa Rica* [PIMA, 1973]. The title of the summary report is *Improving Food Marketing Systems in Developing Countries: Experiences from Latin America* [Harrison et al., 1974]. A list of occasional papers, graduate theses, and technical reports can be obtained from the summary report.

It is difficult to do justice to these studies in any systematic way. We would emphasize that they do, however, represent the most ambitious effort to do empirical work on food and agricultural marketing systems in Latin America. A comprehensive systems approach was taken in the studies and the focus for the most part was on identifying marketing processes and evaluating the efficiency of the system. Some of the studies included the marketing system for agricultural inputs as well as that which connects producers to the consumers.

In general the different studies identify inefficiencies in the marketing systems and argue that reducing these inefficiencies can contribute to economic development. Another uniform finding is that the food marketing system in each of the areas studied is clearly deficient in supplying low-income consumers at a reasonable price level. In all cases, the poor paid more for food items than did higher income groups. The authors attributed these higher prices to problems of store location. The problem is exacerbated in some cases by the fact that lower income consumers spend a substantial share of their food budget on transport to and from the main market areas.

Another general finding was the existence of the wholesale retailer, a link in the channel that has disappeared from modern food distribution systems. This additional link causes an additional cost and margin in small retailer systems, which are a major outlet for low and middle income consumers.

More generally, the authors found from the La Paz [Slater et al., 1969a] and Recife [Slater et al., 1969b] studies that gross margins at wholesale are comparable to those at retail. They found this surprising, since in their judgment one would expect lower margins at the wholesale level, where volume would compensate for reduced unit margins. They believe these data suggest a lack of competitiveness and/or high uncertainty costs and spoilage at the wholesale level.

As part of their conclusions the authors identified seven categories of marketing problems in each of the countries where they conducted commodity subsystem studies, problems which keep rural marketing costs and coordination uncertainties high [see Harrison *et al.*, 1974, pp. 78–82]. The first is the relative lack of regional specialization and relatively small-scale farming units in the production of individual commodities. Farmers tend to produce several products, often in small quantities. This results in small production quantities scattered over large geographic areas with subsequent high assembly costs. This is a special problem in milk, and fruit and vegetable production.

Second, partly as a result of the lack of geographical concentration of agricultural production, there are frequently very few rural traders available to purchase an individual farmer's produce. The result is some tendency toward spatial monopsonies or oligopsonies in assembly markets. Nicholls [1969a] has observed in the case of Brazil, however, that independent truckers do a great deal to keep these local monopsonies and oligopsonies from becoming too powerful.

Third, there is an overwhelming prevalence of crude and inefficient handling, packaging, storage, and product preservation practices and little grading.

Fourth, price distortions and uncertainties are always a problem and are sometimes ruinous. Markets are "thin" and stocks are limited. Government programs designed to stabilize prices are often mismanaged, thus accentuating fluctuations and instabilities.

Fifth, there is a shortage of both short- and long-term credit for financing commercial activities at reasonable interest rates. This is in part because in most countries the capital market is heavily regulated, with the result that credit is either directed elsewhere or the limited amount provided for this purpose is absorbed by a few large, financially sound firms.

Sixth, there is a pervasive failure of the traditional assembly trader to perform the communications function linking the farmer to potential markets. We must note that there has also been a failure of governments to provide effective market information systems.

Finally, the seventh category of commodity subsystem problems is the traditional physical facilities bias inherent in most development planning. Enthusiasm for a US$50 million project to build a network of publicly owned storage facilities can be easily generated, but hardly anyone is interested in a US$2 million supervised credit and training program designed to improve managerial competence among marketing cooperatives and private intermediaries.

The reader is directed to chapters 5 and 6 of the summary report [Harrison *et al.*, 1974] for an overview of more general conclusions and for the recommendations of the authors on how marketing systems can be improved.

A more recent collection of papers of interest is in the proceedings of an international workshop on the domestic marketing of foods held in Cali, Colombia, in July 1984 [Scott and Costello, eds., 1985]. The focus of this workshop was on the processes and mechanisms involved in delivering foodstuffs from the producer to the rural and urban consumer in Latin America. Researchers from the fields of rural and agricultural economics, urban sociology, anthropology, and geography who have worked on topics such as small producer marketing, rural markets and intermediaries, urban food marketing and distribution systems, and food marketing policies for development presented the results of their research projects and discussed questions of research methodology and policy development to improve food marketing systems in the region. This volume also contains a bibliography of over 130 substantial writings on food marketing in Latin America.

The editors of this volume note in their abstract [Scott and Costello, eds., 1985] that to date very little applied research has been undertaken on the marketing system in Latin America. They also note that the results of the workshop clearly indicate the need for the development of more sophisticated research methodologies.

Two studies of how markets and marketing arrangements evolved on the Central-West frontier of Brazil are also of particular interest. G. W. Smith's Ph.D. dissertation [1965] on agricultural marketing and economic development is especially perceptive both on the role of markets and on how they evolve. He found that rather efficient marketing arrangements evolved in a fairly short period of time. Mandell's Ph.D. dissertation [1969] was equally perceptive, as was his later study [1971] of the evolution of the rice sector on the frontier. G. W. Smith's later paper [1969] is also a perceptive review and analysis of the tendency of the Brazilian government to repeatedly attribute periodic supply crises to imperfections in agricultural markets and to respond with large investments in silos and warehouses.

Scott's [1985a,b] papers on the marketing of potatoes in Lima are also valuable. He marshals substantial evidence to argue that in this sector, at least, the middleman does not have the exploitive power he is alleged to have.

An important feature of the marketing scene in Latin America is the prevalence of centralized marketing institutions that are either a part of the government or are government owned. These are present in many countries and reflect either the general lack of regionwide confidence in the capacity of the private sector to perform marketing functions efficiently, or attempts to distribute food to poor people at less cost than it is delivered to other income groups. An important example of such an institutional arrangement is CONASUPO in Mexico, which has pervasive effects in the food and agricultural sector in that country.

Many of these organizations are given monopoly control of imports for some or all commodities. There have been few empirical studies or evaluations of these institutions. An important exception is Quezada's [1981] study of INESPRE in the Dominican Republic.

Parallel to these government food distribution systems there are often completely unrelated price setting organizations to protect consumers at the retail level and to set prices for farmers. In the case of Brazil, SUNAB sets the retail prices for consumers. G. W. Smith [1969] provides insights as to how this organization has worked over time. Often such organizations intervene to fix margins, which can be particularly disruptive to marketing processes.

At the farm level in Brazil, a program of minimum prices has long been in place, operated by the CFP. The CFP manages an extensive stocks program which results in frequent intervention in the markets. SUNAB also intervenes, often by bringing in imports to cushion price increases. Both institutions create a great deal of uncertainty in markets, probably lowering overall resource and marketing efficiency rather than improving it. Aside from Quezada's [1981] study of INESPRE in the Dominican Republic and G. W. Smith's [1969] evaluation of SUNAB, we know of no serious evaluation of these centralized marketing institutions.

To conclude, we would emphasize that this survey of empirical research on food and agricultural markets is far from complete. However, we would also note that aside from those studies cited, there are not many studies which penetrate in any depth. Many of them are highly descriptive, or consider only a narrow part of the system. An interesting example of such a descriptive study is that by Weekes-Vagliani [1985] for Peru; this study also has references to the Peru literature. One of the anomalies of this situation is that the middleman is an all too familiar bogeyman in Latin America; agricultural markets are widely perceived to be imperfect and dominated by monopolists, oligopolists, monopsonists, and oligopsonists. Moreover, governments often intervene because of the widespread belief that these imperfections are pervasive. The empirical evidence to support the existence of these imperfections, or to justify the government interventions, is generally sadly lacking, however.

3. Consumption Studies

If consumption studies are defined to include studies of demand and the estimation of demand parameters, there is a fairly large number of such studies in the Latin American literature. To pull them all together and make some sense out of them would be a major undertaking in itself and we abandoned such an enterprise early on.

We would like to identify a number of key contributions to the literature for the reader, however. The first is the set of studies conducted or commissioned by the USDA's Economic Research Service in the late 1960s to make long-run projections of demand and supply for agricultural output. Such studies were made for a number of Latin American countries [see L. R. Martin, ed., 1977, pp. 320–325]. They are largely of historical interest at this point, but they did pull together a great deal of empirical data, often from primary sources.

Another significant study of demand and consumption was carried out under the auspices of SUDENE, the development agency for Brazil's Northeast. These involved a large number of rather carefully done studies of urban consumption patterns. Five of these dealt with the consumption of food in urban areas, and were eventually pulled together by Moura [1968]. This study gives a rather complete set of demand elasticities for that region of Brazil. A similar set of elasticities is available for Argentina [de Janvry, Bieri, and Nunez, 1972].

The most ambitious study of consumption in Latin America was the ECIEL study of household income and consumption in urban Latin America. The ECIEL study of household budget data was the result of unique collaboration among a large number of investigators in member institutions in ten South American countries and a coordinating staff at the Brookings Institution in Washington, and extended over a decade. This was part of a larger Program of Joint Studies on Latin American Economic Integration. For background and some history of the program, see Musgrove [1982].

The larger program of which the consumption study was a part was founded in 1963, with the participation of three Latin American institutions — The Instituto Torcato di Tella of Argentina, the Fundação Getulio Vargas of Brazil, and the Universidad de Chile — and its component studies were coordinated by the Brookings Institution. After eleven years under this arrangement, during which the program expanded to include more than two dozen Latin American institutions and to undertake four major research projects, the coordination was transferred to a newly created, autonomous institution (with the name ECIEL) in Rio de Janeiro, Brazil, in 1974 [Musgrove, 1982]. This was an unusually successful example of international collaboration among social scientists, and at one time or another involved many if not most of the best social scientists working on the region's problems.

During a period of almost ten years, the ECIEL Program conducted, as one of its projects, a study of household income and consumption. This study was based on original surveys carried out during six years in eighteen cities in ten South American countries. This was the first time anywhere in the world that parallel national studies of this nature were conducted through the collaboration of a number of independent statistical and research institutions, with a common purpose, organization, and set of procedures. Musgrove describes and evaluates that experience in order to provide an analytical history of the study for the extensive LSMS, which was launched by the World Bank in the early 1980s. His analytical history of ECIEL's study is fascinating, and should be of value to researchers and research institutions more generally.

The substantive results of the ECIEL consumption study, as it is usually described, have been presented in a series of books, articles, monographs, and working papers. These are listed in the bibliography of Musgrove's paper, and are not discussed here. It is interesting to note, however, how the ECIEL came to do the consumption study in the first place. Musgrove (we draw on him directly for what follows) notes that the ECIEL Program undertook as its first project a study of optimum locations of several industries on the assumptions of cost minimization and free trade among the countries of LAFTA, discussed below. Work on this project required estimates and projections of demand for the products of those industries, and incidentally revealed how little was known about the structure of consumption in Latin America.

As a result of discussions begun in 1968, the ECIEL Program decided to include a detailed study of consumption in its next round of research rather than to continue to stress investment. Since it quickly became evident that data of the sort required were available in few if any countries, and consequently that new household surveys would have to be conducted, the project dropped from consideration all nonhousehold components of final demand. A further restriction to major urban areas was imposed by the anticipated costs and difficulties of budget surveys in rural areas. Rural income and consumption surveys have since been conducted in several countries of the region, however.

Interestingly, the project soon took on a life and justification of its own. It detached itself from the initial industrial-integration orientation of ECIEL as well as from a study of price comparisons to which it was originally linked, although the consumption data continued to supply the weights for price indices. Moreover, as the project evolved, the structure of consumption itself became less important as a topic of study, and attention shifted steadily to the determinants of consumption levels and living standards, and particularly to the level and distribution of household incomes. The distribution of income is an important problem in most of Latin America, and the household data provided the means of doing more work on it. Some of these studies are referred to in a later section on the distribution of income.

A book edited by Robert Ferber [1980] is a convenient source for some of the consumption studies. Of particular interest in that volume are papers by Cline [1980], Crockett and Friend [1980], A. C. Meyer [1980], and Musgrove [1980c]. Other consumption studies based on the ECIEL data include Chaigneau and Szalachman [1977], Junta del Acuerdo de Cartagena [1975], Hill [1978], Howe [1974], Howe and Musgrove [1977], and Musgrove [1974, 1977a, b, 1978a, b, 1980a, b].

Gray [1982] has made an unusually careful and perceptive analysis of household consumption and expenditure data on food for Brazil. Her study combines specific analysis of Brazilian food and nutrition issues with a more general model for the study of the nutritional effects of a variety of governmental policies on both agricultural and nonagricultural sectors of the economy. The main objective of the study was to estimate the responses of malnourished Brazilians to changes in incomes and relative prices. A secondary objective was to shed some light on differences between malnourished and well-nourished groups and on differences across income strata. A third objective was to explore the trade-off between quantity and quality in the additional purchases made possible by changes in the real income of different groups. Finally, her study illustrates how the empirical results can be applied in policy analysis.

Some of the results of this study are of particular interest. For example, according to anthropometric measurements, more than half of the children in Brazil were malnourished to some degree in 1975. This was true despite the fact that food balance data show that, on the average, consumption levels of food were sufficient to provide an adequate diet. In addition, although the lowest income groups were concentrated in the rural areas of Brazil, the malnourished population was centered in urban areas, probably because the cost of many nonfood necessities are significantly higher in cities. Classifying consumers by calorie consumption rather than by income thus provides a more accurate basis for studying the effects of food and nutrition policies on the malnourished.

Calorie consumption functions were estimated for individual commodities and for total calorie intake, yielding a matrix of own-price, cross-price, and income elasticities for commodity-specific and aggregate calorie intake. The resulting elasticity estimates indicate that caloric consumption is highly correlated with income and that the malnourished readily adjust their eating patterns in response to changes in relative prices. These results point out the potentially powerful effects of income and pricing policies on the nutritional status of the population.

Gray's [1982] study also indicated strongly, however, that quality and variety in the diet are as important to the poor and malnourished as they are to the more affluent. Even those with large daily calorie deficits may choose to spend additional income on smaller quantities of higher quality food or nonfood amenities rather than on larger amounts of calories from less "desirable" foods such as cassava. She also found that income transfers are a rather expensive means of tackling the malnutrition problem. It is estimated that it would have taken at least the cruzeiro equivalent of US $5 billion in 1980 to raise the per capita calorie consumption of the malnourished to 2,000 calories per day through income transfers alone.

Gray also examined the efficacy of several alternative food subsidies in increasing the calorie consumption of the malnourished. She specifically considered subsidies on wheat bread, milk, and rice. Her results indicate that subsidizing wheat bread may actually add to the problem of calorie deficiency because bread is a poor substitute for higher calorie foods. A subsidy on milk, in contrast, can be expected to have positive nutritional effects, but is a very expensive way to tackle malnutrition because the greatest benefits can be expected to go to higher income individuals (unless the subsidy is targeted). She found that rice is the best food to subsidize for nutritional purposes, especially the lower quality rice that is not generally consumed by the well-to-do.

Gray also analyzed the Brazilian government's effort to promote domestic production of alcohol from sugarcane and other raw materials as a substitute for imported petroleum. This program potentially has significant nutritional effects because of its enormous size and agricultural focus. She found that the most harmful effects on low-income and malnourished segments of the population could be expected if the program were based on mechanized sugarcane planta- tions in the intensively farmed southeastern region of the country. The effects would be especially serious if rice production were displaced and rice imports were not increased in proportion. In contrast, the most beneficial nutritional re- sults could be expected from labor-intensive, cassava-based alcohol production concentrated in marginal agricultural areas in the northeastern interior. If based on cassava grown on small farms, crop displacement would be minimized while providing more employment and higher incomes in areas of concentrated poverty.

We suspect that many of the basic relationships found by Gray [1982] have more general applicability in Latin America. Other studies of malnutrition in Latin America include those by Mohan, Garcia, and Wagner [1981], Solimano and Taylor [1980], studies referred to in Berg [1987], and also, Pinstrup- Andersen, ed. [1988].

Another study of general interest, especially in the context of agricultural modernization and development, is by Sanint and ERS [1983]. The author sought to understand the effects on consumption patterns and imports of wheat and maize of the dramatic increases over the past twenty-five years in rice pro- duction in Colombia and Venezuela (partly due to the adoption of high-yielding varieties). The study was limited to an analysis of the changes in availability and consumption of staple carbohydrate foods. Wheat, corn, potatoes, cassava, and plantains were thought to be the most likely consumption items for which rice would be a substitute. Beef and edible beans were expected to be less important as substitutes, but were included to make the analysis more comprehensive.

The study found that consumers adjusted their dietary patterns in response to the greater availability and lower price of rice. In Colombia, the degree of direct substitution of rice for other commodities was rather small. Instead, the increase in rice consumption appears to have been the result of its own declining price and a large propensity on the part of Colombian consumers to spend additional in- come on rice consumption. The direct substitution of rice for other commodities was stronger in Venezuela, but for locally produced potatoes and plaintains rather than wheat and corn.

For exporters of wheat and corn, the most significant finding was that the direct substitution of rice for wheat and corn was either minimal (as in Colombia) or nonexistent (as in Venezuela). The evidence from this study suggests that a growing import demand for food grains such as wheat and corn can exist side by side with rapid expansion in domestic production of another grain such as rice which has traditionally been regarded as a close substitute for imported grains in consumers' diet. It should be noted that in both countries the real price of rice declined relative to other staples and that per capita consumption more than doubled between 1956 and 1977.

The substitution effect that accompanied lower rice prices was greater in Venezuela than in Colombia and tended to have a greater effect on the locally produced foods—potatoes, cassava, and plantains. In Colombia, this effect was felt directly on the consumption of plantain and corn, and indirectly (due to government rationing of wheat imports) on wheat.

The effect of increases in per capita incomes also differed between the two countries. Starchy foods (potatoes and cassava) are superior commodities in Colombia but not in Venezuela. Corn, on the other hand, is a superior and quite important staple in Venezuela, while it is considered to be an inferior good in Colombia. The income effect was in the same direction for rice, wheat, and plantains in both countries. These results suggest that measured parameters can vary a great deal from one country to another, based on stage of development, distribution of income, and prevailing consumption patterns.

To conclude this section, there is a great deal to be done in understanding basic consumption patterns and trends over time, and in developing a complete set of quantitative relationships between consumption and relative prices and per capita income. Up to now a basic problem has been the lack of reliable data on the major variables. For example, there has been little research on the permanent income hypothesis in Latin America. The availability of the ECIEL data made such research possible, and Musgrove has used it to that end [1974]. But such studies are few in number. Recent efforts to generate more basic data, such as through the World Bank's LSMS [Altimir and Sourrouille, 1980] should improve this situation in the near future.

4. Production Studies

There is a large number of production studies in Latin America and, as in the case of the studies of consumption, it would require a major effort to pull them together and make sense out of them. Again, we decided early in the review process that such a review would be beyond our resources. We can provide clues as to where such studies might be found, however. In the first place, the graduate programs at places like the Catholic University in Santiago, Chile; Chapingo, outside of Mexico City; and at Fortaleza, Viçosa, Piracicaba, and Porto Alegre in

Brazil are excellent sources. In each case an effort was made early in the establishment of the programs to have students collect field data. These data and their analysis are reported in what is now a large number of M.S. theses.

There is also a large number of "cost of production studies" in Latin America. These have been motivated by the tendency to use cost of production as the basis for agricultural price policy. These cost studies have not always been well grounded in production theory and suffer from other limitations as well [Schuh, 1976b].

In the case of Brazil, the Institute of Agricultural Economics (Instituto de Economia Agrícola) in the Secretariat of Agriculture of the state of São Paulo has long maintained a sample for collecting longitudinal farm level data, while also undertaking special studies by means of well-designed samples. These efforts for the state of São Paulo have provided what is perhaps the most complete and accurate set of data on the agriculture of any place in Latin America. The results of these studies are published in *Agricultura em São Paulo*, and in other publications of the Institute.

The Nicholls and Paiva studies, *Ninety-Nine Fazendas: Structure and Productivity of Brazilian Agriculture* [1963/1973], are another rich source of production data on Brazilian agriculture. Although this sample was small, it was collected with care and the insights and perceptions from the analysis of the data are remarkable.

For Latin America as a whole, there is a fairly large number of peasant studies. This literature is diverse, and does not lend itself to easy generalization. There are many micro-level field studies done by anthropologists which offer both analysis and insight. References to much of this anthropological literature can be found in publications of the University of Wisconsin Land Tenure Center. Other peasant studies border on political tracts, with little use of data and much exhortation. Isaacmen and Sunseri [no date] is a selected annotated bibliography of peasant studies.

A final reference that may be of value to those not familiar with Latin America is *The Latin American Peasant*, by Pearse [1975]. Pearse describes the various peasant groups and the institutional arrangements that link them to the rest of the economy.

To conclude this section, there is a very high payoff to the person or persons willing to collate the many production studies from the region and to analyze them to identify patterns of resources use, production systems, and enterprise combinations. We hope somebody takes up this challenge in the near future.

5. Concluding Comments

There has been a great deal of empirical work on various aspects of agricultural development in Latin America. How one views this literature depends a great deal on whether one views the cup as half full or half empty. Certainly there have been some ambitious efforts, such as the MSU AID attempt to better understand the marketing system in Latin America and ECIEL's attempt to develop data on consumption patterns. There has also been some creative and original research designed to understand supply response in tree crops and livestock, especially cattle. The various supply studies, when added up, throw considerable light on the structuralist-monetarist debate.

On the other hand, much of the work on supply has been fairly mechanical, with little effort to dig below the surface. Lack of data is a serious problem. And there have been few attempts to pull together the results from disparate isolated studies to see what larger stories could be told, or to assess the limitations of the theory and the methodological approaches used in the past. There should be a high payoff from such synthesis, analysis, and evaluation.

Chapter V. The Distribution of Income and Empirical Studies

Latin American countries tend to be distinguished by very unequal distributions of income. Detailed measurements are not required to perceive this, since the disparities of income are visible even to the most casual observer. Slums located side by side with the condominiums of the affluent are facts of life that strike one's eye in any of the major cities in the region. Less visible, but perhaps even more important, is the pervasive poverty of rural Latin America, especially among the landless workers.

These disparities probably help explain the political instability of many countries in the region, for they create social tensions and provide fertile soil for political activists. They may also be behind some of the long periods of dictatorships and authoritarian governments that many countries have experienced. Political disturbances lead to unstable economic conditions, with the result that the interests of the military or powerful political leaders frequently converge with the interests of the private sector in attempting to create a more stable political-economic environment.

We focus on the economic dimensions of the income distribution problem. Nonetheless, as is apparent from a perusal of the literature, discussions of the income distribution issue among social scientists (including economists) can become extremely political, often sounding as if it were a debate between the government and the opposition. In the case of Brazil, which is a central concern of this chapter, there was also a clear distinction between the "official" (government) economists and the critics. Moreover, as one reads the literature one often finds between the lines a voicing of deep concerns about "the other side's" treatment of the problem, especially when manipulations of large sets of data are involved.

Table 16. Gross domestic product and rate of growth
in gross domestic product per capita, Brazil, 1960-70

	GDP per capita in 1978 cruzeiros	Rate of growth of GDP per capita
1960	12,996.3	–
1961	13,807.4	6.24
1962	14,211.5	2.93
1963	14,003.4	−1.46
1964	14,015.5	0.09
1965	14,259.7	1.74
1966	14,404.5	1.02
1967	14,709.0	2.11
1968	15,930.3	8.30
1969	17,064.0	7.12
1970	18,098.3	6.03

Source: Conjuntura Econômica, various issues.

It would be unfair to say that the debate over changes in Brazil's distribution of income has been useless. As we review the various studies, it becomes clear that there is agreement on some issues, even though most of the authors insist on emphasizing disagreement at the expense of taking more constructive views of their empirical results.

A major part of this chapter is devoted to the extensive debate about changes over time in the distribution of income in Brazil. Other literature on income distribution issues will be considered, however. The chapter is divided into five parts. The first part provides a comprehensive review of the Brazil debate. The second and third parts discuss, in a less comprehensive way, the literature on the income distribution issue in Colombia and Mexico, respectively. The fourth section reviews studies that bear on the income distribution issue in Latin America. The fifth and final section presents some concluding comments.

1. The Income Distribution Debate in Brazil

This section is divided into three parts. The first part provides the empirical evidence on changes in the distribution of income in Brazil during the decade of the 1960s, the period about which there was the most debate. In the second part we review the interpretations given to these changes. The third part reviews more recent studies and provides concluding comments on the Brazil debate.

CHANGES IN BRAZIL'S DISTRIBUTION OF INCOME IN THE 1960s

Data on the growth of GDP and GDP per capita in Brazil during the 1960s provide useful background on the discussion which is to follow (Table 16). Brazil experienced very low growth rates in GDP per capita during the 1962-67 period. Growth resumed in 1968, and at a very rapid rate extending into the 1970s. This

Table 17. Comparison of income distribution, Brazil, 1960-70

Population deciles		Percentage of income			Average income [a]		
		1960	1970	% change	1960	1970	% change
Lowest	10	1.17	1.11	− 5.13	25	32	+28.00
	10	2.32	2.05	−11.64	48	58	+20.83
	10	3.42	2.97	−13.16	71	84	+18.31
	10	4.65	3.88	−16.55	96	110	+14.58
	10	6.15	4.90	−20.32	127	139	+ 9.45
	10	7.66	5.91	−22.75	158	168	+ 6.33
	10	9.41	7.37	−21.68	195	210	+ 7.69
	10	10.85	9.57	−11.80	225	272	+20.89
	10	14.69	14.45	− 1.64	305	411	+34.75
Highest	10	39.66	47.79	+20.50	815	1,360	+66.87
Highest	5	27.69	34.86	+25.90	1,131	1,984	+75.42
Highest	1	12.11	14.57	+20.32	2,389	4,147	+73.59
Lowest	40	11.57	10.00	−13.57	60	71	+18.33
	20	13.81	10.81	−21.73	142	153	+ 7.74
Highest	40	74.62	79.19	+ 6.13	385	563	+46.23
Total		100.00	100.00	−	206	282	+36.89

Source: Langoni, 1973, p. 64.
[a] In 1970 cruzeiros.

rapid growth led to frequent references to an "economic miracle." It is important to keep these data in mind since it is possible that the worsening in Brazil's distribution of income during this decade was a consequence of the rapid growth experienced at the end of the decade. This expectation that the distribution of income would become more skewed with rapid economic growth is referred to as the Kuznets effect [see Kuznets, 1955]. Unfortunately, some participants in the debate seem to assume that the whole decade was a period of rapid economic growth, and to ignore the fact that the more unequal distribution by the end of the 1960s may have been due at least in part to six years of sluggish growth in the middle of the decade.

The characterization of the distribution of income in 1960 and 1970 was done essentially by Fishlow [1972a], Langoni [1973], and Fox [1983]. Table 17 is taken from Langoni and provides a characterization of the distribution of income in 1960 and 1970 by deciles and other groupings. The controversy started with these data, and in particular on such issues as the methodologies used to match the income classes of the 1970 census with those of the 1960 census; on whether individual or family income should be the variable of concern; and on the treatment given to open-ended income classes and inclusion or not of individuals declaring zero income. However, we take them as the basic characterization of the income distribution in 1960 and 1970.

It is clear, as Langoni and others have noticed, that these data show a deteri-

Table 18. Variation in the concentration indices, Brazil, 1960-70

Indices	Excludes zero income persons			Includes zero income persons		
	Gini	Var. logs	Theil	Gini	Var. logs	Theil
1960	0.4999	0.8971	0.4699	0.5570	1.69	0.5802
1970	0.5684	0.9763	0.6442	0.6049	1.71	0.7267
% change	13.7	8.82	37.09	8.60	1.18	25.25

Source: Langoni, 1973, p. 67.

oration in the relative distribution of income during the 1960s. All but the highest decile group experienced decreases in their income share during the decade. The highest decile group, however, experienced a 20 percent *increase* in its share.

When one considers average incomes, the same phenomenon is observed. The increases in average incomes were much higher for the upper classes. Only the upper decile group experienced an increase in average income that exceeded the overall growth of income, which was 37 percent.

Another characterization of this same process is by means of concentration indices, of which Langoni computed the Gini, the Theil and the variance of the logarithms of income. His estimates of these parameters are presented in Table 18, including results which both include and exclude those which declared zero income. The distribution worsened during the decade according to each of these indices, although the magnitude of the changes varies depending on which index one chooses. Note that the inclusion of the zero income earners reduces the changes in all the indices.

To characterize the changes in the income distribution further, Langoni went beyond these aggregate measures to consider other dimensions of the problem. For example, Table 19 summarizes data on the regional and sectoral changes in the distribution of income during this period. The main inferences from this table are as follows:

(1) The primary sector's share in the economically active population (henceforth referred to as EAP) declined during the decade, as did its share of income, which was reduced by 32 percent.

(2) The secondary sector experienced the largest increases in both its shares of the EAP and of income.

(3) There was very little change in either of these shares for the tertiary sector.

(4) The average income of all sectors increased during the decade, with that of the secondary sector experiencing the largest increase, followed by the tertiary and primary sectors, in that order.

(5) In 1960, the average income of the urban sector was approximately

Table 19. Sectoral and regional comparisons of the income distribution, Brazil, 1960-70

Sector region	Share in the EAP (%)			Share in income (%)			(cr$/month 1970)			Gini			Var. of logs			Theil		
	1960	1970	1970/60	1960	1970	1970/60	1960	1970	1970/60	1960	1970	1970/60	1960	1970	1970/60	1960	1970	1970/60
P	46.56	40.05	-13.98	29.13	19.64	-32.58	121	138	+14.05	0.4290	0.4418	+ 2.98	0.5110	0.5474	+ 7.12	0.3746	0.4302	+14.84
S	15.24	19.74	+29.52	18.89	25.18	+33.30	256	359	+40.23	0.4174	0.5010	+20.03	0.5580	0.7411	+32.81	0.3386	0.5169	+52.66
T	38.20	40.21	+ 5.26	51.86	55.18	+ 6.40	280	387	+38.21	0.5030	0.5726	+13.84	0.8852	1.1725	+32.46	0.4516	0.6051	+33.99
U	53.44	59.95	+12.18	70.78	80.36	+13.53	273	378	+38.46	0.4816	0.5514	+14.49	0.7924	1.0326	+30.31	0.4229	0.5779	+36.65
I	10.36	10.58	+ 2.12	16.78	16.28	- 2.98	334	448	+34.13	0.4540	0.5297	+16.67	0.7027	0.8423	+19.87	0.3654	0.5166	+41.38
II	20.86	22.78	+ 9.20	28.60	34.42	+20.35	283	426	+50.53	0.4366	0.5429	+24.35	0.6346	0.8663	+36.15	0.3529	0.5637	+59.73
III	14.72	16.77	+13.93	16.25	16.14	- 0.68	228	271	+18.86	0.4061	0.5012	+23.42	0.5304	0.7329	+38.18	0.3197	0.5057	+58.18
IV	15.95	13.51	-15.30	13.02	9.83	-24.50	169	205	+21.30	0.5261	0.5484	+ 4.12	0.7484	0.8962	+19.75	0.5378	0.6119	+13.78
V	30.66	27.63	- 9.88	17.43	15.42	-11.53	117	157	+34.19	0.4895	0.5565	+13.69	0.5296	0.8128	+53.47	0.4931	0.6796	+37.82
VI	7.44	8.72	+17.20	7.78	7.36	- 5.40	216	238	+10.18	0.4416	0.4864	+10.14	0.5887	0.6385	+ 8.46	0.4005	0.5003	+24.92
Brazil							206	282	+36.89	0.4999	0.5684	+13.70	0.7647	0.9893	+29.37	0.4699	0.6442	+37.09

Source: Langoni, 1973, p. 81.

Notes: P = Primary

S = Secondary

T = Tertiary

U = Urban (S + T)

I = Guanabara and Rio de Janeiro.

II = São Paulo.

III = Paraná, Santa Catarina, and Rio Grande do Sul.

IV = Minas Gerais and Espírito Santo.

V = Maranhão, Piauí, Ceará, Rio Grande do Norte, Paraíba, Pernambuco, Alagoas, Sergipe, and Bahia.

VI = Roraima, Acre, Amazonas, Rondônia, Pará, Amapá, Mato Grosso, Goiás, and Distrito Federal.

EAP = Economically active population.

2.2 times higher than that of the primary sector, and in 1970, that figure went up to approximately 2.7.

(6) All three concentration indices show positive increases in 1970 as compared with 1960; the increases were higher for the secondary sector, followed by the tertiary and the primary sectors, in that order.

(7) The degree of concentration, according to these indices, was higher in the urban than in the primary sector.

(8) In 1970, the degree of concentration in the secondary sector was higher, according to all indices, than it was in the primary sector; in 1960, however, the Gini and the Theil indices showed a higher concentration in the primary sector.

(9) The income share in each region declined except region II (the state of São Paulo).

(10) Average income increased in all regions, with that for region II well above the others.

(11) An increase in the degree of concentration occurred in all regions.

The roles of individual (personal) characteristics in this process of concentration were also analyzed by Langoni [1973]. Although he provided data for age, sex, and education in 1960 and 1970, we present only the data for education (Table 20) since this variable played a key role in the subsequent debate over the causes of the worsening of the income distribution over the decade.

The most important points that emerged from Table 20 are as follows:

(1) There was a substantial reduction in the share of illiterates in the EAP.

(2) The groups with Secondary I, II, or higher educational achievement experienced increases in their income shares.

(3) The change in the income share of each group was positively correlated with that group's level of education; that is, the highest positive increase occurred for those with college education and the highest negative change took place for illiterates.

(4) The average income of illiterates was practically the same in 1960 and 1970.

(5) All groups other than the illiterate class experienced an increase in average income even though, except for the group with a college education, that increase was smaller than the average change shown in Table 17.

(6) Except for the illiterate class, each of the classes experienced an increase in the concentration within classes over the decade according to all three indices.

Table 20. Comparisons of the distribution of income by level of educational attainment, Brazil, 1960-70

Educational attainment[1]	Share in the EAP (%)			Share in income (%)			Average income (1970 cr$/month)			Gini			Var. of logs			Theil		
	1960	1970	change	1960	1970	change	1960	1970	change	1960	1970	change	1960	1970	change	1960	1970	change
Illiterate	39.05	29.75	−23.81	21.12	11.79	−44.18	111	112	+ 0.90	0.4162	0.3886	− 6.63	0.4755	0.5304	+11.55	0.2963	0.2395	−19.17
Primary	51.71	54.47	+ 5.34	53.17	46.46	−12.62	211	240	+13.74	0.4183	0.4614	+10.30	0.6262	0.7282	+16.29	0.3122	0.3905	+25.08
Secondary I	5.16	8.03	+55.62	11.06	13.74	+24.23	440	482	+ 9.54	0.4387	0.5134	+17.03	0.6084	0.8525	+40.12	0.3150	0.4483	+42.32
Secondary II	2.67	5.24	+96.26	6.97	12.79	+83.50	536	688	+28.36	0.4247	0.5007	+17.89	0.5167	0.7406	+43.33	0.2788	0.3945	+41.50
College	1.40	2.51	+79.28	7.66	15.21	+98.56	1,123	1,706	+51.91	0.4590	0.4596	+ 1.31	0.5916	0.8572	+44.90	0.2463	0.2904	+17.91

Source: Langoni, 1973, p. 98.

[1] Secondary schooling in Brazil consists of seven years altogether. In this table, Secondary I corresponds to the first four years of this period and Secondary II corresponds to the three additional years. Primary schooling covers four years. Hence, the total of primary and secondary schooling is eleven years.

(7) Two of the concentration indices (Gini and Theil) show that the degree of concentration increased less for the college educated than for others; the variance of the logarithms, however, shows that the increase in concentration in that group was the highest of all.[61]

Concerned not only with the characteristics of the income distribution but also with the characteristics of poverty, Fishlow [1972a] earlier had examined the 1960 census data from another perspective. He defined poverty as follows:

> The real minimum wage for 1960 in the Northeast, the poorest region, is taken as the lower limit of acceptable income for a family of 4.3 persons. For rural Brazil, the wage prevailing in the rural areas of the Northeast is taken; for the urban Northeast, the standard of the medium-sized *município* is applied; and for all other urban residents, the Northeast level, incremented by 15 percent to allow for higher relative prices, is applied. [p. 393]

This poverty measure was then corrected for family size, with the results shown in Table 21. According to Fishlow, by this criterion, 31 percent of all Brazilian families were below the poverty line. He summarized the results from Table 21 as follows:

> The differentiating characteristics of poverty emerge clearly in Table 2 [21]: low levels of education; concentration in agricultural activities; location in, and nonmigration from, rural areas; limited number of workers per family; residence in the Northeast; larger than average family size and number of children; and relatively smaller opportunities for education of those children. . . . The Brazilian problem is more one of low levels of productivity within the mainstream of the rural economy.
>
> The policies appropriate to dealing with poverty are correspondingly differentiated. Negative income taxes, subsidies, and welfare programs have a lesser role to play in Brazil than efforts directed at disseminating modern techniques in agriculture and accelerating growth more generally. Note that policies designed to tie the population to agriculture by making urban conditions less satisfactory will not help; the poor are not to be especially found as migrants engaged in marginal activities in urban areas. [p. 394]

Fox [1983] later performed a similar set of calculations to characterize the changes in the characteristics of poverty during the 1960s. Utilizing the same

Table 21. The profile of Brazilian poverty, 1960 (in %)

	Poor families	All other
Sex of head of household		
Male	83	92
Female	17	8
Age of head of household		
14–29	17	20
30–60	70	66
61 +	13	14
Education of head of household		
None	64	35
Primary	35	55
Lower secondary	1	5
Upper secondary	–	2
University	–	2
Head of household economically active	83	92
Sectoral distribution of economically active		
Agriculture and extractive industry	68	49
Industry	10	15
Commerce	5	11
Services	9	8
Transport and communications	5	8
Liberal professions, government, administrative, etc.	2	8
Position in occupation of head of household		
Employer	1	4
Self-employed	51	45
Employee in private sector	37	38
Employee in public sector	3	9
Sharecropper	8	4
Number of workers per family		
0	11	3
1	62	59
2	15	21
3 +	12	17
Migratory status of head of household		
Migrant from rural area	14	14
Migrant from urban area	19	37
Nonmigrant	67	49
Location of family		
Urban	40	54
Rural	60	46
Region of family		
Northeast	43	15
East	40	38
South	17	47
Family size		
1	4	6
2–3	18	33
4–5	27	32
6 +	51	29
Number of children, 0-14		
0	15	35
1-2	29	39
3-4	29	39
5 +	27	7
Number of children in school		
0	67	67
1	13	16
2	10	8
3 +	10	8

Source: Fishlow, 1972a, p. 393. With permission.

Table 22. Brazil: characteristics of poor, nonpoor households in 1960, 1970

	1960			1970		
	Poor households	Nonpoor households	Relative incidence of poverty[a]	Poor households	Nonpoor households	Relative incidence of poverty[a]
Percent of households	27	73		23.7	76.3	
Percent of population	34	66		30.5	69.5	
Average family size	6.1	4.5		6.4	4.6	
		(percent of households)			(percent of households)	
Region						
Northeast	40.8	17.1	1.73	48.1	23.1	1.65
East	41.1	37.8	1.06	21.4	24.8	0.89
South	18.2	45.1	0.48	20.9	43.2	0.54
Frontier[b]				9.6	8.6	1.05
Location						
Urban and suburban	35.7	54.3	0.78	49.4	61.2	0.85
Rural	64.3	45.7	1.26	50.6	38.8	1.18
Economic activity of head						
Working or looking for work	82.2	91.6	0.92	79.7	87.0	0.93
Not working	17.8	8.4	1.62	20.3	13.0	1.37
Sex of head						
Male	82.9	91.5	0.93	82.0	89.4	0.94
Female	17.1	8.5	1.57	18.0	10.6	1.45
Age of head						
Under 30	14.0	21.3	0.73	12.2	19.8	0.67
30–39	28.2	28.2	1.01	29.6	26.0	1.10
40–49	27.4	21.4	1.19	22.7	22.7	1.16
50–59	15.4	15.6	0.99	16.8	17.0	0.99
60+	14.4	13.5	1.04	13.6	14.5	0.95
Education of head						
None	65.0	36.0	1.48	67.8	34.1	1.53
Some primary	34.2	54.3	0.70	33.4	43.9	0.81
Some lower secondary	0.6	5.2	0.16	3.3	11.0	0.35
Some upper secondary	0.1	2.3	0.06	0.5	7.7	0.08
Some university	0.1	2.1	0.05	0.1	3.3	0.02
Migratory status of head						
Migrant from rural areas	13.4	14.6	0.94	6.0	5.5	1.14
Migrant from urban areas	19.2	35.9	0.61	66.4	55.6	1.06
Nonmigrant	67.4	49.5	1.24	27.6	38.9	0.76
Sector of employment of head						
Agriculture and extraction	72.9	48.7	1.22	66.4	41.6	1.38
Industry	8.4	15.7	0.56	14.4	21.4	0.70
Services	8.3	8.4	0.91	5.5	11.5	0.52

[continued]

Table 22. Brazil: characteristics of poor, nonpoor households in 1960, 1970 [Continued]

	1960			1970		
	Poor households	Nonpoor households	Relative incidence of poverty[a]	Poor households	Nonpoor households	Relative incidence of poverty[a]
Commerce	4.3	11.4	0.41	6.5	6.1	1.03
Transport and communications	4.2	7.9	0.55	3.1	6.8	0.51
Government, liberal arts, others	2.0	7.9	0.29	4.1	12.6	0.37
Position in occupation of head						
Public sector employee	2.4	9.2	0.30	4.8	11.3	0.45
Private sector employee	35.9	38.5	0.88	35.9	40.8	0.84
Self-employed	52.4	44.6	1.04	45.2	38.9	1.05
Sharecropper or family worker	8.6	3.6	1.66	13.6	5.6	1.72
Employer	0.6	4.1	0.17	0.5	3.4	0.18
Number of economically active						
0	11.4	3.0	2.16	13.7	5.1	1.92
1	54.9	62.2	0.91	56.9	58.4	0.98
2	16.5	20.6	0.85	15.8	21.7	0.78
3 or more	17.2	14.3	1.14	3.2	14.8	0.94
Number of children under 14						
0	15.0	33.4	0.53	19.8	37.2	0.59
1-2	28.7	37.7	0.81	19.5	36.0	0.61
3-4	26.5	19.6	1.23	27.7	18.6	1.34
5 or more	29.7	9.4	1.99	33.0	8.3	2.33
Number of children under 14 in school						
0	65.6	67.3	0.98	43.8	47.1	1.13
1-2	23.0	25.4	0.93	34.9	39.5	1.09
3 or more	11.4	7.3	1.36	21.3	13.4	1.63

Source: Fox, 1983. With permission.
[a] Incidence of poor in category as a multiple of the overall incidence of poverty (26.6 percent).
[b] Fishlow's 1960 sample excluded from the Frontier area.

definition of poverty as Fishlow (updating values to 1970), she compared the percentages of poor and nonpoor households and their characteristics in 1960 and 1970.[62] Her results are shown in Table 22, about which the following two points should be noted. First, the share of the poor living in urban areas grew from 35.7 percent in 1960 to 49.4 percent in 1970. Second, the corresponding change in the number of poor living in rural areas was a reduction from 64.3 percent in 1960 to 50.6 percent in 1970.

The above studies present characterizations of both the distribution of income in Brazil and of poverty. The logical next question is: What were the causes of these changes? We examine how this question has been addressed in the next section.

THE CAUSES OF THE WORSENING OF THE INCOME DISTRIBUTION
IN THE 1960s

From a general point of view, the debate over the causes of the worsening (more skewed) distribution of income in Brazil can be divided into two groups of arguments: those which emphasize market responses to demand and supply shifts; and those which emphasize the nature of the government policies (for example, wage policy) adopted after 1964.

Proponents of the first set of arguments emphasized that the appearance of many disequilibria was a "natural" consequence of the process of growth experienced by Brazil. Emphasis was given to the labor market, in which it was hypothesized that there were larger shifts in the demand for skilled labor as compared to the demand for unskilled labor. Consequently, the relative wages of skilled (sometimes identified with educated) persons went up in the process.

Proponents of the second set of arguments, on the contrary, focused on government policy, especially with respect to wages, and used their analysis and empirical results to score points against the government in power. After 1964, a substantial freezing of wage rates was enforced by the government, which periodically readjusted the minimum wages at a rate less than the rate of inflation. Other elements of policy frequently mentioned were tax rebates for special projects, and subsidized credit, especially for agriculture. This issue is discussed in a later section.

To examine the evidence that was raised to support the different positions, consider the decomposition of the variance of the logarithms of income undertaken by Langoni [1973, pp. 92-97] and by Bacha and Taylor [1978, especially pp. 279-282]. Langoni decomposed the variance of the logs both among the five educational groups that appear in Table 20 and within each group. Table 23 presents a summary of his findings with respect to education as a whole, without showing his within-group analysis. Changes in relative income, in the educational level of the work force, and in education within the individual groups account for 23 per-

Table 23. Decomposition of the change in the variance of the logarithms of income,
Brazil, 1960-70: Education

Share of the changes in the variance explained by	
Changes in relative income	23.15%
Changes in the educational composition	34.78%
Changes within groups	42.07%

Source: Langoni, 1973, p. 92.

cent, 35 percent, and 42 percent, respectively, of the change in the total variance
of the logarithms.

To study the effects of other variables (sex, age, region, and sector), Langoni
adopted a different strategy. Instead of decomposing the variance of the loga-
rithm of income, he performed an analysis of variance by means of regression
techniques and then identified the marginal contribution of each variable in explaining the total variance in the logarithms of income (Table 24). These results
show that the marginal contribution of education to the total variance of the logarithms of income increased from 31 percent to 41 percent. The marginal contributions of both region and sex to the total variance of income, however, were
smaller in 1970 than in 1960. Langoni emphasized the fact that this reduction,
especially when one is considering regions, shows that all other things constant,
regional differences in income had declined.[63] Finally, age becomes more important in 1970 reflecting, according to Langoni, the influx of youngsters into the
economically active population.

Langoni continued his analysis by developing simulations of the income distribution. Given that the "disequilibrium" induced by the very process of growth
was one of the basic features of Langoni's explanation for the increased concentration in Brazil's distribution of income, he set out to test that hypothesis. To
perform such a test, he calculated the estimated income distribution in 1970 (using the 1970 regression coefficients) with the independent variables set at their
1960 level, and then compared this estimated distribution with that obtained
when the independent variables were set at their 1970 levels. The differences in
the two were then attributed to changes in the composition of the EAP, or a "scale
effect," as he called it.

We summarize some of his main findings [from 1973, pp. 117-121]. First, the
relative share of the 40 percent at the bottom of the distribution would have been
22 percent if there had been no change in the composition of the EAP, while their
actual share in 1970 was 15 percent. Thus, the "scale effect" was responsible for a
substantial loss in income to this group. Second, for the 40 percent at the top of
the distribution, the picture was reversed. Their actual share in 1970 was 70 percent, while it would have been only 61 percent had the EAP retained the same

Table 24. Comparison of income distributions in 1960 and 1970:
marginal contribution of each variable, Brazil

	1960		1970	
	Marginal contribution[a]	Share	Marginal contribution[a]	Share
Education	9.98	30.79	15.43	41.01
Region	7.57	23.36	4.86	12.92
Age	6.05	18.67	7.74	20.57
Sex	4.85	14.96	4.99	13.26
Activity	3.96	12.22	4.61	12.25
Total	32.41	100.00	37.62	100.00
$\bar{R}^2$	50.74	–	59.28	–
Multicollinearity[b]	18.33		21.66	

Source: Langoni, 1973, p. 112.

[a] The marginal contribution of a variable is measured by $\bar{R}^2 - \bar{R}^2_i$ where $\bar{R}^2$ is the corrected R^2 of the regression with all variables and $\bar{R}^2_i$ is the corrected R^2 for the regression with ith variable(s) omitted.

[b] Multicollinearity ith is measured as $\bar{R}^2 - \sum_i (\bar{R}^2 - \bar{R}^2_i)$.

composition. Considering only the top 10, 5, and 1 percent, the actual and estimated income shares in 1970 were as follows:

	Actual	Estimated
10 percent	33.88	25.18
5 percent	23.50	16.44
1 percent	8.15	5.90

Second, when comparing the "scale effect" with the "income effect" (which amounts to changes in relative incomes with constant composition of the EAP), Langoni notes that scale effects are usually much larger than income effects, a finding which gives support to his general hypothesis about the process of concentration of income.

Using this same methodology, Bacha and Taylor [1978] examined Langoni's data by expanding the analysis to include all five variables (education, sex, sector, age, region) considered by Langoni. Their results are reproduced in Table 25. Their interpretation of the results was as follows:

> Both our findings and parallel results reported by Fishlow [1973, 1977] are rather surprising in view of Langoni's insistence upon the importance of compositional changes in explaining the increase in the variance of the log income in the period. In the

Table 25. Brazil: Decomposition of the changes in the variance of logs of incomes between 1960 and 1970

Classification (no. of groups)	Estimated change in variance of logs	Explained change in variance of logs	Proportional contributions to explained change of:		
			Composition changes	Relative income changes	Within groups variance changes
Education (5)					
1960 weights	0.2916	0.3296	14.6	56.2	29.2
1970 weights	0.2916	0.2534	3.9	52.7	43.4
Sex (2)					
1960 weights	0.2218	0.2162	2.4	−0.2	97.8
1970 weights	0.2218	0.2270	4.7	−0.4	95.7
Sector (3)					
1960 weights	0.2573	0.2596	1.8	42.3	59.5
1970 weights	0.2573	0.2547	2.9	37.6	62.3
Age (8)					
1960 weights	0.2642	0.2592	1.2	31.8	67.0
1970 weights	0.2642	0.2704	3.1	33.7	63.2
Region (6)					
1960 weights	0.2232	0.2269	−4.5	13.4	91.1
1970 weights	0.2232	0.2199	−6.2	13.4	92.8

Source: Bacha and Taylor, 1978, p. 281. With permission.

first place, population movements between regions and between sectors of activity had an *equalizing* influence on the overall variance, in spite of the fact that migrations from the rural sector to urban activities and from poorer to richer regions were substantial between 1960 to 1970. This result seems to contradict the original version of Kuznets' hypothesis for the Brazilian case. On two other population dimensions, sex and age, the compositional changes contributed to increase the overall variance, but only slightly so. No more than 1.2 to 3.1 percent of the total change in variance is explained by shifts in age composition of the population, and only 2.4 to 4.7 percent by changing sex composition.

The results in Table 3 [25] also indicate that changes in the educational composition of the labor force account for at most 14.6 percent of the increase in the variance of log incomes (with 1960 weights), but also that this contribution could be as little as 3.9 percent, with 1970 weights. This finding clashes with Langoni's contention, apparently based upon the same data and methodology, that "the changes in relative income explain 23 percent of the total increase observed in the period, whereas *changes in composition explain 35 percent*, and the increased inequality within groups represents 42 percent" [Langoni, 1973, p. 93, emphasis added]. According to the results reported in Table 3 [25] relative income changes were responsible for no less than 52.7 percent of the explained change in variance, with within group variance accounting for from 29.2 (with 1970 weights) to 43.4 percent (with 1970 weights) of the total. We convinced ourselves that our results were consistent with Langoni's basic data (except as noted in note 6), but could not find the reasons why our final results differ from his. [pp. 280-282]

The reference to note 6 in the above had to do with the fact that these authors used arithmetic means to compute the w^i (weighting) factors since they could not obtain the geometric means.

The conflicting empirical results of Langoni and of Bacha and Taylor indicate that there are still unanswered questions in this debate. This is unfortunate in view of the importance of this issue for the evaluation of past policies and the speculation with respect to future policies. It is even more unfortunate in that there seems to be little interest in pursuing the matter further. To our knowledge, for example, nobody has yet come up with an explanation for the difference in results obtained by Langoni [1973] and by Bacha and Taylor [1978] with respect to education. Moreover, Bacha and Taylor are apparently satisfied with their findings, as the following suggests:

We convinced ourselves that our results were consistent with Langoni's basic data [except as noted in 6], but could not find the reasons why our final results differ from his. Similar scepticism by Malan and Wells [1973] and Fishlow [1973, 1977] leads us to the conclusion that Kuznet's effect did not importantly shape trends in the Brazilian size distribution in the 1960s. [p. 282]

Despite this rather laconic conclusion, however, there have been other tests of the various hypotheses underlying the causes of the worsening of the income distribution in Brazil. We now turn to a review of some of this literature.

Fishlow [1972a] performed a decomposition of the Theil information index in an attempt to understand the 1960 census data. This index can be disaggregated in much the same way as the variance of the logarithms to show the importance of various groups in the determination of the overall inequality.

Table 26. Decomposition of inequality coefficients[a], Brazil, 1960

	Corrected	Uncorrected
Total	0.57	0.72
Within	0.25	0.29
Between[b]	0.32	0.43
Education	0.20 (0.11)	0.25 (0.11)
Sector	0.12 (0.03)	0.19 (0.05)
Age	0.09 (0.09)	0.13 (0.11)
Region	0.04 (0.03)	0.05 (0.03)
Interactions		
E-S	−.10	−.15
E-A	−.01	−.02
E-R	−.03	−.04
S-A	−.02	−.04
S-R	−.02	−.03
A-R	.00	.00
E-S-A	+.02	+.04
E-S-R	+.02	+.03
S-A-R	.00	.00
E-A-R	.00	+.01

Source: Fishlow, 1972a, p. 396. With permission.

[a] Measured in natural log units.

[b] The three regions are the census-defined Northeast, East, and South. The seven ages are 10-14, 15-19, 20-29, 30-39, 40-49, 50-59, and 60+. The five sectors are agriculture and extractive; industry and construction; services; merchandise commerce, transport, and communications; and financial services, independent professions, and public administration. The six educational categories are none, primary incomplete, primary complete, lower secondary, upper secondary, and university.

Fishlow's results are reproduced in Table 26. It is interesting to observe that the four characteristics used to classify the data explain more than 50 percent of the total inequality of both the corrected and the uncorrected distribution.[64] Notice also that regional differences accounted for a very small percentage of the observed inequality. Education, sector, and age had an important role in the explanation of the income inequality in 1960. Nonetheless, Fishlow [1972a, p. 398] warns that "it is well to remember not only how much of inequality is explained by education, but also how little. Age and education together do not account for more than a third of the variation in individual incomes."

Fishlow also examined data on the 1970 distribution of income. Unfortunately, he did not present any decomposition of the inequality in 1970 or, for that matter, of the change between 1960 and 1970 (apparently for lack of data). Despite this lack of more detailed analyses, he attributed the worsening of the distribution in the 1960s to government policies, emphasizing the minimum wage policy in the post-1964 stabilization years and also what he calls the liberal concession made by fiscal policy in the form of tax incentives. In reference to education, he pointed out that the distribution of educational opportunities has not been in favor of more equality. He further added that:

> Some increases in variance and inequality may virtually be
> inevitable owing to the age structure of the labor force and the
> prior lack of education, but there is clearly scope for a policy
> that emphasizes to a greater extent extension of educational
> opportunities to the underprivileged and various calculations of
> the rate of return to elementary schooling suggest it is a highly
> profitable strategy as well [Lerner, 1970; Levy, 1970]. Thus, an
> educational policy that succeeded in elimination of illiteracy
> among the young between 1960 and 1970 could have
> simultaneously increased the average level of educational
> attainment more, while reducing the variance less than the
> pattern actually occurring. Current plans, however, seem to
> favor continuing emphasis upon secondary and university
> enrollment without sensibility to the distributional implications
> of such a structure. [pp. 401-402]

Finally, Fishlow [1972a] concluded his article by stating:

> There is no necessary inconsistency between greater equality
> and expanding output. Brazilian poverty is directly linked to
> low levels of productivity, particularly rural, that are subject to
> attack. Policies can be developed. But, first there must be
> recognition of an accounting system that reckons and applauds
> not only increases in aggregate output, but also tabulates the
> differential gains in welfare that are reflected in the distribution
> of income. [p. 402]

There are three remaining issues we need to examine: the hypothesis that the demand for skilled labor shifted more than the demand for nonskilled labor and related issues; the wage squeeze hypothesis; the hypothesis that the absolute income of the poor grew much more than the absolute income of the nonpoor. Our discussion of the first two issues is based on the excellent summary presented by Bacha and Taylor [1978, especially pp. 282-295].

The assertion made by the proponents of the skill differential hypothesis is that growth has led to large shifts in the demands for skilled labor *vis-à-vis* the demand for nonskilled labor. This, in turn, is presumed to have produced a widening of the wage differentials among these groups. Two important assumptions with respect to the labor market are essential to this argument. The first is that the supply of unskilled labor is infinitely elastic and that the supply of skilled labor is less than infinitely elastic. In fact, it generally is assumed that the higher the skill level, the lower is the elasticity. The second assumption is that the elasticity of substitution between any two given skill levels is low.

Table 27. Income distribution comparisons by educational levels, Brazil, 1960-70

Educational level	Population share (in %)		Average income (in 1970 cr$/month)		Proportional change 1960-70 (in %)	
	1960[a]	1970	1960	1970	Income	Population
			(in 1970 cruzeiros/mo.)			
Illiterates	39.05	29.75	111	112	0.1	2.2
Primary	51.71	54.47	211	240	13.7	41.3
Secondary I	5.16	8.03	440	482	9.5	108.9
Secondary II	2.68	5.24	536	688	28.4	163.4
College	1.40	2.51	1.123	1.706	51.9	140.8

Source: Bacha and Taylor, 1978, p. 283. With permission. Taken from Langoni, 1973, Table 4-2, p. 86.
[a] 1960 population figures were obtained from Fishlow and Mesook [1972].

There have been various attempts to test this hypothesis. Table 27, which shows data on population shares and average income per educational level together with estimates of the percentage changes in population per educational level, is a useful starting place for the discussion. This is Table 4 from Bacha and Taylor's article [1978], and they make the following comments:

> Table 4 [27] shows that the population share of college graduates increased by 80 percent, while their per capita income increase (51.9) percent was greater than any other educational group. Contrariwise, the population share of illiterates fell by a quarter, and their per capita income stayed constant. Any sort of competitive labor market theory would suggest that average payments to the college educated should have fallen under this sort of shift while those of the increasingly scarce illiterates should have gone up. Widening skill differentials in the face of increasing numbers of the skilled during the 1960's make up the real puzzle in Table 4 [27]. It is not resolved by observations about labor force composition in 1960. [p. 283]

Most of the research to test the hypothesis of unequal demand shifts has been based on macroeconomic models [Morley and Williamson, 1975; E. A. Cardoso, 1980a; and Lysy and Taylor, 1980]. Morley and Williamson's analysis was based on a model developed earlier [1974] for the purpose of providing a link between labor absorption rates and changes in the size distribution of income and in the distribution of earnings. In the more recent paper [1975], they estimated that the demand for the highest skill group (eight skill levels were considered) grew 4.8 percent annually, while for the lowest skill group, demand increased by 3.2 percent annually between 1960 and 1970. They went further to estimate whether this growth pattern was sufficient to explain the observed wage differentials. To do that, they (arbitrarily) assumed values for the growth in supply of each skill (2.6 percent per year for all skill levels) as well as for the demand and supply elasticities for each skill level. From these assumptions, they concluded that wage differentials could be explained by demand and supply factors.

As might be expected, Morley and Williamson were criticized for the arbitrary nature of their elasticities [Bacha and Taylor, 1978, p. 283]. Unfortunately, no alternative analysis has been forthcoming from either the critics or from Morley and Williamson. Readers of those papers are left with the uncomfortable feeling of not knowing the sensitivity of the results to the stated assumptions.

Fishlow [1973, 1977] chose to test the labor skill hypothesis by means of the following equation:

$$\Delta \log \frac{L_{it}}{L_{ot}} = \sigma_i \log \frac{w_{it}}{w_{ot}} + R_i \, \Delta t$$

where: L_{it} = demand for skill i;
        w_{it} = wage of skill i;
        L_{ot} = demand for unskilled workers;
        w_{ot} = wage of unskilled workers;
        R_i = percentage difference between the earnings rates of labor using technical progress for skill class i and unskilled workers;
        σ = elasticity of substitution
        t = time
        Δ = difference operator

He estimated the R_i's in the above equation, assuming that demand equals supply and assumed elasticities of substitution between one and four. He found that the R's should be between 6.1 and 11.9 percent for the higher skills and between 4.0 and 5.2 percent per year for the medium skills. When compared with per capita GDP growth in Brazil during the 1960s, about 3 percent per year, this value seems to be very high.

Working with a macroeconomic model in the Keynesian tradition, E. A. Cardoso [1980a] reached results that differ from those of Morley and Williamson. In her words:

> Our results differ from Morley and Williamson's because our model is closed with respect to savings. Like them, we show that an increase in aggregate demand leads to more employment, which may well benefit the poor. The increased demand, however, has to be "financed" by more saving, and in any kind of widow's curse model, this shifts the income distribution toward the rich. This effect occurs in the identity

> model and not in that of Morley-Williamson. Their results are
> optimistic because they have conveniently left out the
> nonfavorable distribution mechanisms. Wage repression
> combined with stimulation of effective demand is not the sort
> of policy that shifts income away from rentiers and other high-
> saving classes. The poor almost always pay for rapid economic
> growth, and Brazilian policy in the 1960s (hypothesis B) made
> their burden more onerous. [pp. 87-88]

Finally, Lysy and Taylor [1980], using a multisector version of the kind of model used by E. A. Cardoso [1980a], arrived at conclusions that tended to confirm her findings.

We turn our attention now to the wage squeeze hypothesis, which argues that the wage controls exercised during the post-1964 period in Brazil were an important element behind the increased skewness in the distribution of income. Bacha and Taylor [1978] analyzed various criticisms of this hypothesis, such as whether the share of those receiving minimum wages declined substantially and whether an important part of the population was outside the formal labor market and, as such, not affected by the minimum wage.

These observations are undoubtedly important if one wants to appraise this argument carefully. However, the evidence available does not make it possible to evaluate quantitatively the importance of those elements. Bacha and Taylor estimated some simple regressions of the median wages in the manufacturing sector of Rio de Janeiro against the minimum wage, the cost of living index in Rio, and the GDP per capita, and found significant coefficients for the minimum wage [p. 288]. For example, elasticities of 0.48 and 0.57 were estimated in two cases. Nonetheless, this appears to be a very crude way of testing the relationship between these variables, especially in the absence of an explicit model that justified the links specified in their regressions.

We now consider a final issue, Gary Fields's [1977] rather surprising (to most observers) conclusion that the poor in Brazil experienced a much larger increase in absolute income in the 1960s than did the nonpoor. Recall the data from Table 16 and compare them with the title of Field's paper, "Who Benefits from Economic Development? A Re-examination of Brazilian Growth in the 1960s." The so-called economic miracle started only in 1968; most of the decade was characterized by very sluggish economic growth.

Fields was concerned with changes in absolute incomes during the decade. Thus, it was necessary to have the population distributed among the same income brackets in 1960 and 1970. To obtain homogeneous brackets, he took the data from Fishlow's [1972a] paper and used 1960 income brackets as bases. To convert the 1970 brackets into comparable 1960 brackets, he began with the ratio of the mean incomes in 1970 and 1960 calculated in constant 1960 dollar terms

Table 28. Brazilian size distribution of income, adjusted to make 1960
and 1970 income brackets comparable

Monthly income in 1960 cruzeiros	Percentage of population		Cumulative percentage of population	
	1960	1970	1960	1970
None	14.7	11.7	14.7	11.7
0- 2.1	22.3	23.8	37.0	35.5
2.1- 3.3	14.4	12.2	51.4	47.7
3.3- 4.5	10.5	11.0	61.9	58.6
4.5- 6.0	13.1	14.5	75.0	73.1
6.0-10.0	13.8	9.4	88.8	82.5
10.0-20.0	.2	10.9	97.0	93.4
20.0-50.0	2.6	5.0	99.6	98.4
Over 50.0	0.5	1.6	100.1	100.0

Source: Fields, 1977, p. 571. With permission.

(US\$679/US\$513). These same means, expressed in current new cruzeiros were 5.52 and 258.1. Thus, the ratio of the real means was 1.32 and of the nominal means 46.76. The ratio of these, 35.32, is then an inflation factor which can be used to deflate the 1970 brackets. For example, the first positive bracket in 1970 runs from 0 to 2.8 new cruzeiros per month. Then applying a linear approximation to the population frequency within each bracket, 2.1/2.8 of the population in the 0-2.8 category was assigned to the 0-2.1 category, and the remaining 0.7/2.8 was assigned to the next higher category. An analogous procedure was followed for the other brackets [Fields, 1977, note 6]. The resulting distribution for 1970 is shown in Table 28, together with the 1960 distribution.

Fields's next step was to define a poverty level. He noted that in 1960, 31 percent of the population was poor according to Fishlow's criteria which, as both Fishlow and Fields argued, was the Brazilian definition. Since there were no means by which these families could be identified, he took as poor those families with incomes less than Cr\$2.1 per month; that is, 37 percent of the population in 1960 and 35.5 percent in 1970.

Based on these data, Fields estimated the growth of the mean income of the poor and of the nonpoor in 1960 and 1970 by solving the following equations. These are transcriptions from Fields's paper [p. 574].

$$(37\%)\,(P^{60})\,(\bar{y}_p^{60}) + (63\%)\,(P^{60})\,(\bar{y}_n^{60}) = (P^{60})\,(5.52)$$
$$(37\%)\,(P^{60})\,(\bar{y}_p^{60}) = (5.2\%)\,(P^{60})\,(5.52)$$
$$(35.5\%)\,(P^{70})\,(\bar{y}_p^{70}) + (64.5\%)\,(P^{70})\,(\bar{y}_n^{70}) = (P^{70})\,(258.1/35.32)$$
$$(35.5\%)\,(P^{70})\,(\bar{y}_p^{70}) = (8\%)(P^{70})\,(2.1/2.8)\,(258.1/35.32)$$

where:

P^j is the population in year j; j = 1960, 1970;

$\bar{y}_i^j$ is the mean income of class i in year j; i = p (poor) and n (nonpoor); quotients in parentheses are the mean incomes in 1960 cruzeiros: 5.2 percent is the income share of the poor in 1960; and
(8.0%) (2.1/2.8) is the income share of the poor in 1970.

From these equations he concluded that the mean income of the poor grew by 63 percent during the decade while the mean income of the nonpoor grew only 28 percent.

Fields was heavily criticized by all of his commentators. Ahluwalia, Duloy, Pyatt, and Srinivasan [1980] pointed to a logical mistake in Fields's procedure in that it implies the contradiction that the mean income of 7.9 percent of the population with incomes between Cr$2.1 and Cr$2.8 is Cr$1.84. This obviously invalidates the conclusion that the mean incomes of the poor grew by 63 percent between 1960 and 1970. These authors pursued the subject further and concluded that Fields's data were not suitable for the kind of analysis for which he used them.

Beckerman and Coes [1980] showed that Fields's procedure is sensitive to the type of income deflation utilized. They reproduced Fields's methodology using the change in the Cost of Living Index for São Paulo between 1960 and 1970 as the deflator for the income brackets. Their results differ from those obtained by Fields. For example, they found the percentage of the poor in the population to be 37.3 percent in 1970, compared to Fields's estimate of 35.5 percent. The rate of growth of the mean income of the poor between 1960 and 1970 was 38 percent, in contrast to the 63 percent calculated by Fields. And finally, they found the rate of growth in the mean income of the nonpoor between 1960 and 1970 to be 22 percent, as opposed to 28 percent estimated by Fields.

Fishlow [1980] pointed to difficulties with the data used by Fields to reach his conclusions, especially with respect to the definition of the poverty line. Fishlow argued that:

> Such a procedure is both critical to the findings and quite illegitimate. It ignores the difference between the distributions among individuals and families; it obscures the fact that the original analysis corrected for income in kind and regional price differences; and it takes too light the problem posed by a large group of individual zero-income recipients. [p. 250]

Fishlow then examined the data on family incomes in 1960 and 1970. As he notes, this was not done in his earlier research since he did not have the data at that time. Table 29 presents the distribution in 1960 and 1970.

After noting the clear picture which emerges from this table, and which reinforces his earlier findings with respect to the concentration of income in Brazil, Fishlow took the minimum wage in the rural Northeast of Brazil in 1970

Table 29. Decile distribution of family income, Brazil, 1960 and 1970

1960		1970	
Percent of families	Percent of income	Percent of families	Percent of income
10−	0.5	10−	0.65
10	2.2	10	1.7
10	2.9	10	2.4
10	3.8	10	3.3
10	4.9	10	4.3
10	6.0	10	5.5
10	8.1	10	7.5
10	11.0	10	10.6
10	17.0	10	17.9
10+	43.6	10+	46.2

Source: Fishlow, 1980, p. 253. With permission.

(Cr$125) as a definition of the poverty line and concluded that 32.5 percent of the families (and 39 percent of the population) were still below this line in 1970. Acknowledging the deflation problem raised by Beckerman and Coes [1980], Fishlow showed that under the most favorable deflator (the GDP implicit deflator), the number of families below the poverty line was 29.7 percent, a reduction of 9 percent in relation to 1960.

Furthermore, the average income of nonpoor families grew 27 percent in the period while those of the poor families grew only 8 percent in this period. These results obviously clash with Fields's findings.

While recognizing some of the failings of his procedures, Fields [1980a] put up a sturdy defense of what he had done, while at the same time taking issue with his critics both on procedural and analytical grounds. While expressing some ambivalence about what the data really showed, he defended his approach, which focused on what happened to the absolute income of the poor rather than what happened to the relative share of the poor.

The experience of the 1970s probably provides a more valid test of the effect of economic growth on the distribution of income. In fact, the availability of the 1980 census data, plus data from household surveys, have painted a rather different picture about what was happening to poverty and the distribution of income in Brazil. The decade of the 1970s was a period of fairly rapid and sustained economic growth in Brazil, in fact, one of the most rapid, if not *the* most rapid of any country in the world. A combination of export promotion and import-substituting industrialization policies was pursued, with the emphasis on export promotion. The latter policies tended to benefit agriculture, and agricultural exports grew rapidly in the period (see below). But continued expansion of the manufacturing sector led to a sustained and rapid out-migration from agriculture in this decade.

Analyses of the 1980 census data show that the distribution of income improved during this decade for the country as a whole, although *within* agriculture, the distribution became more skewed. Of special note was a reduction in the income gap between agriculture and the rest of the economy and a reduction in regional income disparities. To quote Denslow and Tyler [1983]:

> Despite persistent poverty, during the 1970s substantial progress in reducing poverty and improving living standards occurred. While there is some limited and weak evidence of overall increased income concentration, growing overall relative inequality, if it did occur, was minor and could not be fully discerned by our means. In addition, there were observed reductions in income inequality among regions and among sectors. The agricultural sector in particular, while witnessing growing income inequality within it, was characterized by rapidly growing average incomes which seemed to reduce the gap between agricultural and urban-based occupational incomes. [p.3]

Denslow and Tyler drew a number of conclusions from their analysis of the data [pp. 29-30]. First, there was substantial progress in improving living standards in the 1970s. Second, their nonincome measures of poverty attest to considerable progress, despite continued and pressing problems of poverty. Third, average real incomes also increased substantially, even among the poorest 40 percent of the economically active population.

At the same time, however, overall income inequality did not undergo significant change between 1970 and 1980. This relative stability of a high degree of inequality was due in their judgment to two offsetting changes. On the one hand, there was a large rise in rural incomes relative to urban incomes. On the other hand, however, there was an increase in the inequality of incomes within agriculture.

Finally, they believed their evidence pointed to a reduction in absolute poverty during the 1970s. The improvement in agricultural incomes and the reduction in regional disparities were important components of that improvement. These occurred in a setting of rapid economic growth and an associated generation of a large number of employment opportunities.

Hoffman and Kageyama [1984] also analyzed the 1980 census data and their results on the relative distribution of income are broadly similar to those of Denslow and Tyler, but with some added details. For example, as measured by the distribution of income per economically active person, inequality increased in the decade of the 1970s, but it decreased when the distribution of income by family was considered.

They found a reduction in sectoral and regional income disparities similar to those found by Denslow and Tyler [1983]. But their analysis of the regional data showed a reversal of concentration patterns in the 1970s. Contrary to the 1960s, the North and Central West, followed by the Northeast, experienced an increase in the concentration of income, while in the South and the Southeast, there was virtually no change in the concentration of incomes in the 1970s.

Hoffman [1984] also analyzed the data on absolute poverty. He, too, found that there was a decrease in absolute poverty by almost all measures considered, and that other measures of the "quality of life" also showed improvements. Hoffman and Kageyama [1985] investigated the relationship between the modernization of agriculture and the distribution of income in Brazilian agriculture. They concluded that modernization was associated with an increase in average incomes in the sector and a reduction in absolute poverty. But it was also associated with an increase in the concentration of income among upper income groups in agriculture. They concluded that the solution to this problem is not to slow down the process of modernization, but instead to extend the political rights of the agricultural working class so that political measures would be taken to assure their greater participation in the benefits of technical progress.

Pfefferman and Webb [1983] made a careful analysis of the data on poverty and income distribution in Brazil. Their analyses are perceptive, and their insights are penetrating. Among other things, they were concerned about the many oversimplifications about the income distribution and poverty in Brazil. To quote:

> The population is frequently reduced to a few categories that fit theories and paradigms, such as industrial labor, the Northeast landless, and senior executives. But factory workers and the Northeast landless together account for only 12 percent of the Brazilian labor force, while the total income of senior managers is also a relatively small proportion of total personal income. The great bulk of income and employment is left out in discussions that center on those categories. Within the Northeast, for example, there are as many owners and self-employed in nonfarm business as there are landless families; the former are growing much faster in number, and their average

income is over three times that of the landless. There are
almost as many domestic servants as factory workers. In 1974,
family employment on small and medium farms in the South
region alone (States of Paraná, Santa Catarina, and Rio Grande
do Sul) was as large as that in all factories with 50 or more
workers in Brazil, and total income received by each group
was similar. In 1976 there were twice as many workers in
government and welfare services as in factories. Market and
institutional forces affecting employment and incomes in these
other, generally neglected components of the economy are
thus, by far, the biggest part of the story. [p. 102]

Pfefferman and Webb [1983] were also skeptical that *any* conclusion could be
drawn as to whether the distribution of income in Brazil has changed over time,
on the grounds that the margin for error in the data is very large. To quote:

In conclusion the data on distribution all convey a picture of
skewedness. They do not, however, lend themselves to
comparisons of the degree of skewedness in time. No strong
conclusion on "improving" or "worsening" of inequality can
be drawn from the available data. [p. 123]

These authors made a detailed analysis of poverty and how it has changed over
time, with comparisons to the research of others. They used the ENDEF data,
which are taken from a well-designed household sample and in their judgment a
better set of data than can be obtained from the various censuses. Defining the
poverty line at US$260 per capita in metropolitan areas and making an arbitrary
approximation to regional differences in the cost of living, they found that in
1974, 27 percent of the population was defined as poor. In addition, 61 percent of
the poor were rural and one-half in the Northeast. One of their findings was that
almost three quarters of the urban poor were in smaller cities and towns rather
than in metropolitan areas [p. 103].

Another major finding of their study was that the income levels of the poor
have been underestimated in previous studies [pp. 103-105]. Among other comparisons, they compare their results with those of Fishlow and Mesook [1972] for
1960, using census data, which they consider to be the most thorough previous
measurement of poverty levels. Fishlow and Mesook's poverty level was US$130
in 1974 prices, equal to a Northeast minimum wage. After adjusting for nonmonetary incomes, Fishlow and Mesook estimated that 31 percent of all families
fell below that level in 1960. Applying the same poverty level to the ENDEF
data, Pfefferman and Webb found that the share of families that were poor in 1974
was on the order of 15 percent, one-half of the 1960 level. Based on their assumption that the underestimation of the incomes of the poor would have to be im-

plausibly large to reconcile the difference, they concluded that there was a real reduction in poverty over the period [p. 105].

Pfefferman and Webb's [1983] results on occupational and regional differences are also important:

> One point to note is the big differential between farm and nonfarm manual labor. In all regions, the landless laborer doubles his income by moving to urban manual employment within his own region. Allowances for urban-rural cost of living differences would still leave increases of well over 50 percent since the landless buy much of their food, while at least half of nonfarm manual employment is in small cities and towns where cost of living differences with rural areas are not as large as in metropolitan areas. The rural-Northeast to urban-Rio move roughly triples income, while the family income of a São Paulo manual worker is 4.7 times that of a Northeast farm laborer. [pp. 105-107]

The authors had similar important insights on the relation between regional productivity and incomes in agriculture (small farmers *versus* farm laborer households):

> The São Paulo farm wage earner's household income is higher than in the Northeast only partly because the regional wage rate is higher. Over half the higher family income of the São Paulo farm laborers must be explained in other ways. One possibility is more days worked per year, which may in turn be related to more developed labor markets (better information, quicker transport, and more efficient intermediation). Another is more and better-paid work opportunities outside agriculture—in towns and cities during agricultural slack seasons. These greater employment opportunities apply as well to secondary earners. Finally, since many wage-earners have access to some land, they may enjoy higher productivity on those plots than their counterparts in the Northeast.
>
> What is most interesting about the cross-section relationship between small farmer and wage-earner incomes and regional productivity is that it suggests that the gains from productivity growth and improved terms of trade have been widely shared within the agricultural sector. [Pfefferman and Webb, 1983, p. 108]

Their analysis of trends in wage rates are similarly important for students of growth:

> The single most striking and important feature of those series is the marked increase after 1970 of real wage rates for casual rural farm laborers. These wage rates went up by 75 percent in real terms between 1970 and 1980. [Pfefferman and Webb, 1983, p. 108]
>
> For comparison, per capita income went up by 77 percent during the period 1970-80. Thus an important group of workers who are among the poorest in the country experienced a substantial absolute improvement in their daily earnings. The convergence of rural and urban wages for unskilled labor as well as the convergence between wages in different parts of Brazil suggest the emergence of an increasingly homogeneous labor market. Convergence has meant progress, especially for the poorest groups. [p. 108]

The authors noted later [Pfefferman and Webb, 1983, p. 112] that the gap between urban and rural unskilled wages fell from 56 percent in 1968 to 16 percent in 1977. They further noted that an important feature of what was occurring was a substantial move out of agriculture. The absolute number of farm households increased by only 11 percent over 16 years, and the rate of increase appeared to be slowing. Of special significance in this analysis is that the differential between agricultural and nonagricultural mean household incomes remained virtually constant in the period between 1974-75 and 1980. This suggests to them that the enormous absorption of rural-urban migrants by the urban areas had occurred without a flooding of the lower income categories [Pfefferman and Webb, 1983, p. 114].

Turning to a more specific study of rural poverty that used micro data, researchers at Purdue collaborated with a number of Brazilian researchers and institutions to collect and analyze data from three poverty-stricken areas in Brazil [R. D. Singh, Kehrberg, Patrick, and Schuh, 1979]. One of these was in the Northeast (poverty in a poor region), one in Minas Gerais (poverty in a region of intermediate incomes), and one in São Paulo (poverty in a high-income region). This study was the first to collect intensive household data, including the use of time.

The *municípios* from which these samples of data were drawn were selected because they were known to be regions of extreme poverty. One of the interesting findings from the study was the extent to which each region had historically experienced extended and large out-migrations. On the basis of the simple neoclassical labor migration model, one would have expected some equilibrium to have taken place. However, that was not the case, and for obvious reasons once one examined the data. Geographic migration tends to be highly selective [Greenwood, 1975], and these three regions were no exception. The populations that remained in each of the regions were made up mostly of the aged, the very young, or those disadvantaged for a variety of reasons. The more well educated, the middle-aged and young, and the entrepreneurial had all left. Thus, the region had lost its high quality human capital, and the migrants had taken much of their physical capital with them. Remittances did little to restore a balance in incomes.

Schuh [1982] used these data, the results of earlier studies on the United States, and brain drain models to argue that out-migration, contrary to the simple neo-classical model of migration, tends to impose negative externalities on a region by virtue of this exodus of human capital, and at the same time bestows this human capital virtually as a free good on the higher-income recipient region. The extent of the income transfer will depend in part on the extent to which lower-level education is financed locally. In any case, from this perspective, geographic migration is not necessarily an effective means of narrowing regional income differences. Although the regional differences may eventually narrow and disappear, it generally takes a long time. The some 100 years it took for the regional differences in per capita income between the South and the rest of the United States just to narrow significantly is a case in point. Out-migration from the Northeast to the rest of Brazil has also been proceeding for a long time, also with very little narrowing of the gap. The policy implications of this finding will be discussed in a later section.

Regional disparities in per capita income also often have an important policy dimension to them. As Baer [1964a] and later M. A. Martin [1976] have shown, import-substituting industrialization policies have contributed importantly to the poverty problem in the Northeast of Brazil, and to the persistence of the regional disparity in per capita incomes. Much of the industrial sector in Brazil is located in the South, especially in Rio de Janeiro and São Paulo. The Northeast is primarily agricultural. High levels of protection for the industrial sector and policies that discriminate against agriculture shift the domestic terms of trade against agriculture. But this in effect shifts the terms of trade against the Northeast as well. Both Baer [1964a] and M. A. Martin [1976] estimated the interregional income flows from this terms of trade effect. They found them to be substantial, and for a significant period of time they more than offset the direct resource transfers back to the Northeast through fiscal and other means designed to promote more rapid economic development in the region.

Finally, the reader is referred to the essays in Taylor, Bacha, Cardoso, and Lysy [1980] for important uses of quantitative models, general equilibrium considerations and simulations to understand the relationship between economic growth and income distribution in Brazil. Some of these studies have been referred to above, but the general reference is important in its own right. Sahota and Rocca [1985] also use quantitative models to understand the factors shaping the distribution of income in Brazil. The same authors earlier [1981] had used micro-level farm data to make similar analyses for the agricultural sector in Brazil.

CONCLUDING COMMENTS ON THE INCOME DISTRIBUTION DEBATE IN BRAZIL

The debate over changes in the distribution of income in Brazil centered around two distinct sets of arguments: the role of markets and market responses, and the role of government policies, regulations, and controls. The evidence produced in the extensive debate over these issues, based on a comparison of the 1960 and 1970 census data, did not provide a clear-cut conclusion to the relative importance of each set of factors. Unfortunately, neither did it give sufficient emphasis to the precarious nature of the data on which the debate rested, in particular the sorely deficient 1960 census. Perhaps even more unfortunately, studies of income distribution in Brazil became a means of making political statements about the regime in power at the time. This did not lead to careful, rigorous analyses, and contributed to the unresolved nature of the debate.

The data from the 1960 and 1970 censuses did consistently show, however, the extent to which poverty in that country is essentially a problem of generalized low productivity in the agricultural sector. With the exception of Argentina, this is generally the case in the region. This finding was first documented for Brazil by Fishlow [1972a], but was consistently supported by the other studies. Had development policy given more attention to agriculture, economic growth might have been even faster than it was, and with a more equitable distribution of the fruits of that development. As it was, rapid economic growth at the end of the 1960s did little to alter the situation. In fact, average incomes in the primary sector lagged even further behind those in the secondary and tertiary sectors by 1970.

Also worthy of note is the extent to which much of the debate on these earlier Brazil data focused on the *relative* distribution of income and neglected what happened to the absolute income of the poor. Fields [1977] was an exception, of course, but his analysis has not stood the test of professional scrutiny. From the standpoint of economic development *per se*, it may be that what happens to the absolute income of the poor is more important. In the context of the relative distribution of income issue, it is important to note the extent to which frequent pleas for redistribution of land in Latin America are based on equity considerations (see below). In this regard, it is interesting to note that in 1970, the Brazil data showed that the income distribution was more highly skewed in the secondary sector than in agriculture.

More recent data from Brazil and more recent analyses of these data painted a more positive view of the impact of development on the alleviation of poverty and the distribution of income. While the skewness of the income distribution may not have changed much, the absolute income of the poor appears to have improved dramatically, at least until the international debt crisis of the 1980s struck. Moreover, sectoral as well as regional differentials declined, perhaps in part due to the increased attention given to agriculture, the pursuit of a more rational exchange rate policy, and the addition of, and a greater attention to, export-promoting economic policies.

2. Colombia

Colombia experienced substantial economic growth during the 1970s and associated with that growth there was, as in Brazil, a considerable debate over who received the benefits. The weight of opinion for some time was that the growth in the 1970s did not benefit the poor or improve the distribution of income. See, for example, Ranis [1980], Bejarano [1980], and R. A. Berry and Soligo [1980]. If these authors are correct, their findings are particularly significant since two governments in the 1970s followed policies specifically designed to improve the income distribution and the democratic nature of Colombia's political institutions should have helped eliminate biases toward income concentration [Urrutia, 1985].

Urrutia's careful study [1985] of the data challenged the previous studies and showed that a complete analysis of all the existing statistical data revealed that the income distribution did not worsen in the 1970s, and that the real incomes of the poor improved significantly, especially in the latter half of the decade. He attributed the differences in findings to a number of factors. First, redistribution policies in the fields of education, health, nutrition, foreign trade policy, financial policy, and fiscal policy cannot yield results in the short run. Hence, he believed that assessing policies put into place in 1974 on the basis of 1975 data—the most recent statistics available to R. A. Berry and Soligo when they published their study—may not be methodologically correct.

Second, the most commonly available statistics suggest a process of income concentration. National income data on salaries and independent statistics on real industrial wages showed little improvement in an economy with rapidly growing income per capita. This suggests a worsening of the relative income position of labor. He noted that a simple comparison of income distributions derived from labor force surveys suggest the same thing. Ranis [1980] uses such data for his analysis.

Urrutia argued that the income data used by Ranis and others were incomplete and of highly variable quality. Thus the estimates of income distribution were not comparable and therefore could not be used to estimate changes in the indices of concentration over time. Urrutia made a detailed analysis of the data to establish comparability over time. He also broadened the base from the usual concentration on wages in the manufacturing sector. Among other things, he found that the real wages of the very poor — the landless agricultural worker — increased rapidly in the decade of the 1970s, as did the wages of various categories of unskilled urban workers. In addition, he found that all the data pointed to little growth in real income for the poor in the first part of the decade, but to rapid progress in their standard of living in the second part of the decade, after economic policy started to be consciously designed with distributional goals in mind.

Finally, Urrutia noted that income surveys and wage series show average conditions and changes for different population categories. They did not, however, show what has happened to the real incomes and to the economic welfare of individuals and families over time. To address this issue he attempted to follow the fortunes of a group of poor families in the city of Cali through the entire decade. He found that the real incomes for this sample of poor families increased by about 100 percent in the decade.

Urrutia analyzed (in chapter 6) a number of hypotheses for why the distribution of income improved in the 1970s. He found that one of the most significant changes in Colombia in the 1960s was a cessation in the growth of the rural labor force. As the demand for agricultural output continued to grow, rural labor productivity began to grow markedly, reducing the degree of underemployment in agriculture. In the 1970s, the real income of rural workers began to grow because of an increase in agricultural prices and a continued increase in labor productivity. Since landless workers were the poorest group in Colombian society, their improved income led to a decrease in the proportion of families below the poverty line and to an improvement in the income distribution. This tight labor market in the countryside not only reduced wage differentials between agricultural laborers and unskilled urban workers, it also contributed to an increase in the real earnings of unskilled urban workers.

Interestingly enough, labor demand did not decline in the rural areas despite the gains in labor productivity in agriculture. The principal reason for this appears to be an expansion in agricultural exports, especially coffee. (We note later that Colombia has pursued less discriminatory agricultural trade policies than other Latin American countries.) The export of flowers also became an important activity at this time. But other factors were important as well. First, part of the increases in world coffee prices in this period were passed on to producers, and most coffee production was still in the hands of small and medium-sized producers. The supply of nonagricultural imported products was also allowed to increase to avoid the increases in prices of consumer goods that would have occurred had rural incomes increased and import growth been restricted.

Urrutia [1985] considered other factors as well. For example, the supply of education increased significantly in the 1960s and 1970s, attenuating wage increases for white collar workers. Worsening of the income distribution in the 1950s coincided with the period of most rapid import substitution. The gradual shift away from import-substituting policies in the later period promoted some labor-intensive exports and decreased monopoly rents previously generated by quantitative import controls. This decreased the earnings of the top 5 percent of income recipients. Similarly, the fiscal system seems to have had some positive effects on the distribution of incomes.

To conclude, a number of inferences can be drawn from the Colombian studies. The first is the need to make a careful analysis of the data to put them on a comparable basis, and the need to find some means of following the income of individuals or families over time. Second, Urrutia's study shows that rapid economic growth need not necessarily lead to a concentration of income. Third, trade policies can have an important influence on the distribution of income. And finally, general economic policies which attempt to channel income to or away from disadvantaged groups can have a significant impact on the distribution of income, for good or bad.

3. Mexico

Mexico is another country with a highly skewed distribution of income, with the bulk of the poverty rooted in agriculture and rural areas [C. W. Reynolds, 1970]. Mexico is well known for its land reform and the creation of the *ejido* system of small farms. Venezian and Gamble's [1969] analysis of the data suggest that by tying people to the land, the *ejidátario* system probably slowed rural-urban migration, perhaps to the disadvantage of those remaining behind. Villa-Issa [1976], however, has shown that part-time farming with off-farm employment is important in Mexico.

Bergsman [1980] made a fairly detailed analysis of poverty and the distribution of income in Mexico. His analyses were based on household survey data, not on census data. He found that the poorest 40 percent of households received between 8 and 12 percent of total income. This is lower than in most of the sixteen other less developed countries for which data were presented in the World Bank's *World Development Report, 1979* [1979]. Only in Honduras, Peru, and Brazil—all Latin American countries—were the shares clearly lower than in Mexico. The ratio of the share of the highest 20 percent to the share of the lowest 20 percent was similarly higher than in most other countries. Moreover, there was virtually no change in inequality between 1963 and 1977.

Taking the 1977 minimum wage as a poverty line, Bergsman found that the percentage, and even the absolute number of households whose incomes fell below that line, were decreasing. As of 1977, only 20-30 percent of all households earned less than the minimum wage [Bergsman, 1980, p. 19].

Based on data for 1975, Bergsman found that 52 percent of poor families were in agriculture. Of this group, 33 percent were self-employed (probably *ejidatários*), and 18.5 percent were salaried workers. This low-income group (the 52 percent) included 76 percent of all families in agriculture in Mexico. Thus a major share of poverty in Mexico is in agriculture, and an overwhelming share of agricultural families are poor.

Bergsman [1980] also found that the sectoral structure of Mexican poverty had not changed much between 1963 and 1975. The percentage distribution by sector of "poor" families was virtually identical in 1975 to what it had been in 1963, when agriculture accounted for 54.5 percent of the total. He found that in 1963 there were proportionately more landless workers and fewer of those who owned land. He concluded that if these data were correct, they reflected a failure of Mexico's land redistributions during the period to improve the incomes of the so-called beneficiaries [p. 22].

Bergsman's analysis identified education of the head of the household as the variable most closely associated with income differences. Thus inequality in educational opportunities was a major source of inequality in the distribution of income. Other causes of poverty and inequality were identified as follows: rapid population growth, especially since about 1950, with a resulting high dependency ratio and a rapid growth of the labor force; neglect of the production capacity of many agricultural laborers and owners of poor land (including many *ejidatários*) who at the time of his analysis comprised over 2 million of Mexico's 11 million families and some three-fourths of all Mexicans employed in agriculture; and policies that made capital equipment cheaper and labor more expensive than would otherwise have been the case.

4. Other Studies

M. de R. Lopes [1977] and Matus-Gardea [1981] studied the impact of trade policy on the distribution of income in Brazilian and Mexican agriculture, respectively. They found that trade and exchange rate policies which discriminated against agriculture had more serious consequences for the rural poor (see below). The incidence of these implicit export taxes tend to fall on small producers since they cannot escape the tax, as do larger producers, by shifting to more extensive systems of production.

V. Thomas [1987] made one of the few attempts to estimate regional differences in living standards. Using data from the ENDEF consumption survey in Brazil, referred to above, he found that although costs of living adjustments narrow spatial differences, large regional disparities remain in the living standards in that country, particularly upon comparing the Northeast and the Southeast. He also found that the application of price indices reduces urban-rural variations more than it does regional differences. In particular, the urban-rural gap in food consumption is drastically narrowed. Nevertheless, he found poverty much more concentrated in the rural areas than in the urban areas, although differences in its incidence are more striking than the urban-rural divergences.

Epstein's [1975] analysis of Peron's role in redistributing income in Argentina is interesting. He linked the major redistribution of income in Argentina to the continuing political power of Peron. In a short period of time, 1943–49, Peron raised the income of labor dramatically. Real wages increased 81 percent, primarily after he became president in 1946. This was done largely through the power of the General Confederation of Labor (CGT), and accomplished largely through increases in nominal wages. But price controls on basic food items and other economic policies were also important. Peron capitalized on economic difficulties of the rural-urban migrants, who were thrown out of work by the collapse of the agricultural export market during the 1930 depression, by unionizing them.

Upon Peron's ouster, the pursuit of orthodox economic policies discriminated severely against the workers and in favor of the upper income groups. Between 1953 and 1959 wage and salary earners had their share of all national income decline from 52.3 to 44.9 percent, a drop of 7.4 percentage points. By 1969 real wages had been pushed down to the economic level of 1946.

A potpourri of empirical studies which use the ECIEL data on selected urban centers in Latin America can be found in *Consumption and Income Distribution in Latin America*, edited by Robert Ferber [1980]. The reader is referred especially to the papers by Cline [1980], Figueroa and Weisskoff [1980], Mantel and Martirena-Mantel [1980], and Lubbert [1980]. The latter two papers consider economic integration and the distribution of income. Musgrove [1974] used the ECIEL data to investigate the determination and distribution of permanent household income in urban South America, one of the few uses of the permanent income hypothesis with data from the region. See also his paper [1980c] in the volume edited by Ferber. Hill [1978] used the same data to analyze the impact of imperfect capital markets on life cycle consumption.

For those interested in the problems involved in measuring levels of living and poverty with particular reference to Latin American concerns, practices and experience, Altimir and Sourrouille [1980] is an excellent reference. Both conceptual and data problems in the measurement of levels of living and of poverty are illustrated with quantitative evidence from the data base accumulated by the joint ECLA-World Bank Project on the Measurement and Analysis of Income Distribution. The authors have endeavored to stimulate further discussion by summarizing quantitative exercises carried out by the project, stating personal opinions, and leaving questions open.

Table 30. Estimates of the incidence of poverty
in Latin American countries around 1970

Country	Households below the poverty line (in %)			Households below the destitution line (in %)[a]		
	Urban	Rural	National	Urban	Rural	National
Argentina	3	19	8	1	1	1
Brazil	35	73	49	15	42	25
Colombia	38	54	45	14	23	18
Costa Rica	15	30	24	5	7	6
Chile	12	25	17	3	11	6
Honduras	40	75	65	15	57	45
Mexico	20	49	34	6	18	12
Peru	28	68	50	8	39	25
Uruguay	10	–	–	4	–	–
Venezuela	20	36	25	6	19	10
Latin America	26	62	40	10	34	19

Source: Altimir, 1982, p. 82. With permission.
[a] That is, whose income is not sufficient to buy even the minimum diet.

Altimir [1982] made a fairly comprehensive analysis of the extent of poverty in Latin America. This paper is useful, among other reasons, for its extended discussion of the conceptual issues involved in the definition of a poverty line. He based his poverty lines on diet, estimating the cost of a food basket to cover minimum nutritional needs adequately. The poverty lines he used correspond to a figure that is double that minimum food cost, on the grounds that such a sum would cover the value at current prices of the goods required to satisfy the basic needs which are usually satisfied in these societies through private consumption expenditure [p. 40].

Tables 30 and 31 are taken from Altimir's study. These data show the extent to which poverty is concentrated in the rural sector throughout the region. They also show the extent to which poverty is pervasive in the agricultural or rural sectors — Argentina, Chile, and Costa Rica being the important exceptions. Altimir is optimistic that in the majority of countries in Latin America the poverty problem is manageable, at least as regards the magnitude of the resources involved and bearing in mind the considerable additional resources that would be required to provide basic public services [Altimir, 1982, p. 94].

The reader may also have some interest in four more general references on income distribution issues. The first is by Hirschman [1973], a prolific writer on general development issues in Latin America. Although addressing the generic income distribution problem, his analysis helps explain why such a highly skewed distribution of income has been tolerated in Latin America in a period of fairly rapid increases in per capita incomes. He argues that in the early stages of

Table 31. Estimates of relative poverty in Latin American countries around 1970

	Households below the line of relative poverty (in %)	
	Urban	Rural
Argentina	27	28
Brazil	52	54
Colombia	43	48
Costa Rica	34	36
Chile	38	39
Honduras	40	58
Mexico	44	48
Peru	34	48
Uruguay	25	–
Venezuela	37	38

Source: Altimir, 1982, p. 96. With permission.

rapid economic development, when inequalities in the distribution of income may even increase sharply, a society's tolerance for such disparities may be substantial. This is because of what he refers to as a tunnel effect (from his analogy with parallel lines of cars stuck in a tunnel), whereby members of society (those in one lane) are content with the fact that others (those in other lanes) get ahead in the world, on the grounds that it suggests that the members of society who do not progress will eventually have their day. This is no reason for complacency in his view, however, since if the income disparities do not eventually narrow, trouble and perhaps disaster may occur.

Another important source is the impressive collection of empirical material in *Redistribution with Growth* by Chenery, Ahluwalia, Bell, Duloy, and Jolly [1974]. This volume contains an impressive array of facts on inequality in most developing countries, some interesting and pertinent country studies, and significant analyses of how poverty and inequality problems might be addressed. The basic descriptive data shows that poverty is pervasive in Asia and Africa (with over half the poorest people in the world in India), while inequality, as distinct from absolute poverty, is more a Latin American problem. Pyatt's [1977] perceptive review of their volume is also of interest.

Thurow [1971] argued that the distribution of income can be viewed as a public good. This implies that redistribution may be a prerequisite for Pareto-efficiency. Brown, Fane, and Medoff [1973] demonstrated that Thurow's analysis was erroneous and present what they believe to be a correct solution to the question of what constitutes a Pareto-efficient distribution of income when the distribution of income itself affects utility. Thurow's reply [1973] is also worth reading.

Williamson's [1977] discussion of strategic wage goods, prices, and inequality is also useful. The relationship of wage goods to food and agricultural development is obvious, although often neglected.

5. Concluding Comments

Carefully drawn data on the distribution of income in Latin America are scarce, and longitudinal data even more scarce. With the exception of Argentina, Chile, and Costa Rica, however, it is clear that the distribution of income in the region is highly skewed. The academic community has given a great deal of attention to this relative income distribution issue, even though their data are precarious. When the issue becomes highly politicized, as it did in the case of Brazil, the analytical and empirical results can often be questionable.

Three main conclusions seem to surface from the above review. First, failure to direct resources to agricultural development will result in a continuation of this skewed income distribution, despite rapid rural-urban migration. Second, export promotion policies, as eventually occurred in the case of Brazil and Colombia, can do a great deal to reduce absolute poverty, especially in rural areas where it tends to be concentrated. Third, census data can be quite misleading as a source of income distribution data. There appears to be no substitute for well-conceived household surveys designed to collect budget and expenditure data for this purpose.

One aspect that has been badly neglected in the use of crude data on income distribution is the implication of the life-cycle hypothesis and the well-known age pattern of income flows. The life pattern of earnings takes on a well-known form related to the accumulation of human capital. Rapid population growth has resulted in rapid influxes of young people into the labor force. This by itself would cause an increase in the proportion of the population receiving low incomes.

The World Bank's evaluation [1986a] of the impact of the crisis of the early 1980s on poverty in Latin America is perhaps the best way to end this section on income distribution. The background to this episode is that between World War II and the depression of the 1980s Latin America was one of the fastest growing areas in the developing world. Between 1960 and 1980 only a few developing countries, including the successful Southeast Asian economies, expanded at a faster rate. The gap in per capita GDP between Latin America and the United States narrowed in this period, and with growth there was considerable social improvement. Between 1960 and the late 1970s life expectancy at birth, perhaps the most important indicator of welfare, increased from fifty-six to sixty-four years. Clean water had been made accessible to more than two-thirds of the growing population, and the share of the labor force in agriculture had declined from one-half to one-third. By the late 1970s most children attended primary school, and while the distribution of income remained highly skewed in all but a few countries, most of the poor had seen their standard of living improve at about the same pace as the entire population.

This growth was driven by the domestic mobilization and allocation of resources, but it was further propelled by powerful external forces. Among these were a rapidly expanding world economy, a reasonably strong demand for primary commodities, a substantial transfer of financial resources from the industrialized countries to Latin America, and low real interest rates in the world's capital markets.

The bubble burst in the wake of the second oil shock in 1979. The demand for Latin American exports collapsed as the industrialized countries were thrown into recession. Real interest rates soared and the supply of capital from the rest of the world all but dried up. Interest payments on debt accumulated during the 1970s became a tremendous drain on export earnings and domestic savings. Starting in 1982, the region began to transfer resources to the rest of the world on a massive scale that represented between 25 and 30 percent of the exports of goods and services not including capital flight [World Bank, 1986a]. What was a recession in the industrialized countries became a depression in Latin America, with the exception of Brazil and Colombia.

Governments in the region made considerable efforts to adjust to the new realities. Domestic currencies had been depreciated by 25–40 percent in real terms since 1980-82 in Argentina, Brazil, Chile, Colombia, Ecuador, Mexico, and other countries. Public sector deficits had been reduced, and export and efficient import-substitution projects had been undertaken. Adjustment had been primarily through import cuts rather than through employment-creating export expansion, however, partly because world demand, except in 1984, had remained sluggish. Except in Brazil and Colombia, Latin America's development efforts had been set back by more than a decade.

Any assessment of the impact of this depression on the poor must recognize that in most Latin American countries the majority of the poorest people live in the rural areas. The report [World Bank, 1986a] noted, for example, that in Mexico the three bottom deciles of the income distribution scale were entirely rural and that in Brazil 70 percent of the lowest four deciles were rural households in the mid-1970s. Even in urban Chile, three-quarters of the rural population belong to the two poorest deciles, although the absolute number of urban poor is twice that of the rural poor. Among the rural population in Latin America, moreover, the poorest are usually the landless workers who purchase all or a large part of their food.

The report documented the serious decline in employment, despite continued population growth, the rise in unemployment, a serious decline in real wages, and except in Chile, where a successful attempt was made to focus government social spending on the poorest segments of the population, social services and social indicators declined. Capital flight had benefited a relatively small segment of the population, while average incomes had declined, which suggests that the distribution of income had tended to become even worse.

The report showed the impact on the rural-urban balances to be mixed. Agricultural employment suffered less than urban employment, in part because of a good performance of agriculture relative to other sectors. This good performance of agriculture was due at least in part to the adjustment policies themselves, especially the sharp devaluations and a relaxation of agricultural price controls designed to keep food prices down in the cities. The domestic terms of trade improved for farmers and worsened for the urban population. Farmers found themselves relatively better off.

For farm laborers, however, the situation deteriorated. This very poorest group in most Latin American countries, who must purchase most or all of the food their families consume, experienced declines in real wages as real food prices increased (by over 30 percent in Mexico and Brazil). Farm laborers alone accounted for 15 percent of all Brazilian households and nearly 40 percent of all agricultural households [World Bank, 1986a, p. 17].

The report noted that much had been written during the high-growth years of the 1960s and 1970s about the alleged negative effects of Latin American development on the poor. Citing studies reviewed above [Urrutia, 1985; Pfefferman and Webb, 1983], plus the study on Mexico by Gregory [1986], the World Bank report further noted that these earlier studies were found to be largely without foundation. In their view the depression of the 1980s showed, unhappily, that growth had been beneficial to the poor and that it is economic stagnation and decline which have worsened their already precarious livelihood.

These results show the important role the food and agricultural sector in Latin America has to play as the countries of the region adjust to the new international realities. They also show the importance of the income distribution dimension to that adjustment. Despite weak international commodity markets, agriculture can expect to benefit from policies that seek to give more attention to external markets and that give greater attention to comparative advantage. Whether rural laborers will eventually share in this benefit will probably depend on how long it takes for the economy as a whole to recover.

The significance of food as a wage good clearly comes to the fore as a distributional issue. The World Bank report noted that in dealing with its adjustment problem Argentina had seen the need for a large-scale National Food Program (PAN), which periodically distributed food packages and various social services. As of 1986 this program covered about 5.5 million of the country's 30 million people and involved the daily packaging and distribution of about 1,000 tons of food, covering about 30 percent of the needs of an average family of four. The report noted that the sheer number of recipients in one of the world's best endowed agricultural countries conveyed a dramatic image of the social cost of the 1981-82 crisis.

Looking to the future, Latin American policy makers will need to develop similar targeted feeding programs if they are to pursue policies which capitalize on their comparative advantage and shift the domestic terms of trade in favor of agriculture. Williamson [1977], in a study of the post-World War II history of the United States, found that even in such a high income country, where food is a relatively small portion of the consumer budgets, the price of food as a wage good, and its changes over time, had a significant effect on the distribution of income.

Chapter VI. Trade, Exchange Rates, and Trade Performance: A Policy Perspective

Policy issues are discussed in this and the following chapters. The present chapter is divided into four main parts: trade and exchange rate policies and agricultural trade performance; wheat imports—indirect discrimination against traditional commodities; policy reform in the Southern Cone; and summary, conclusions, and some related literature. The literature on common markets, customs unions, and free trade associations is reviewed in the next chapter.

Much of the material surveyed in this and the following chapters is empirical in nature. The distinguishing feature which caused it to be reviewed separately was its relevance to or emphasis on policy issues.

1. Trade and Exchange Rate Policies and Trade Performance: Selected Countries and Issues

The objectives of this section are to provide a flavor of trade and exchange rate policies used in the region, some perspective on the linkages between trade and exchange rate policies and the trade performance of agriculture, and a review of some of the literature on these issues.[65] The countries considered are either economically important or distinguished by the economic experiences they have undergone. Descriptive data are provided to make the literature more meaningful in terms of the selected countries. A recurrent theme in the survey is that agricultural trade performance and, in turn agricultural development performance, have been less than each might have been in much of the post-World War II period, in large part because trade and exchange rate policies discriminated against agriculture. This was a consequence of the widespread use of import-substituting industrialization policies to promote general economic development and to break the so-called "binds" of dependency. It was also often a consequence of export barriers designed to channel output to the domestic economy to combat inflation or increases in the cost of living (*custo de vida*).

Coffee is by far the principal export of Latin America, with the region including major exporters such as Brazil and Colombia, plus numerous other coffee-exporting countries.[66] Coffee is followed (by a considerable margin) by cereals, fiber, fruits and vegetables, and meats. Trends in exports show an interesting dichotomy. The region's exports of oilseeds, vegetable oils, cocoa, animal feed, fruits and vegetables, and beverages (wines) have been growing at a fast rate in recent years, while exports of cotton, wool, hides, and vegetable fibers have been declining at a similarly fast rate.

Latin American trade patterns are oriented strongly towards the industrialized countries (70 percent in 1973-79). In this same period only 9 percent of the region's exports went to other Latin American countries (up only slightly from 7 percent in 1962-64). Developing countries outside Latin America received very few exports from the region. Trade with North Africa and the Middle East was significant only for sugar (12 percent) and vegetable oils (13 percent). Trade with Asia was significant only for vegetable oils (19 percent) and fiber (11 percent).

Valdés [1984a] argued that the potential for Latin America to increase or maintain its share of the rapid-growth export markets is quite favorable. Fast-growth commodities like coffee, oilseeds, vegetable oils, tobacco, beverages, and fruits and vegetables are important export products for the region. On the negative side, exports of sugar and meats (the latter especially since 1974) face, in his judgment, slow-growing foreign markets.

ARGENTINA

Export taxes have been one of the most important forms of (explicit) taxation of agriculture in Argentina, while quantitative controls of exports have been, as a rule, less important.[67] Quotas have been the most common form of intervention on the import side. In some cases, such as nitrogen fertilizer, there was an outright prohibition of imports for many years [Reca, 1980, p. 11].

Argentina has followed a fixed exchange rate policy during a substantial part of the post-World War II period. Given that domestic price inflation has been relatively high and unstable during this period, there have thus been rather large swings (cycles) in the real exchange rate (Table 32).[68] The size of the shocks involved in these swings can be seen from the table. The effect on agriculture of these swings was attenuated by the use of explicit export taxes. The basic policy stance towards agriculture was to extract resources from the sector through trade and exchange rate policy. In effect, the policy instruments were orchestrated so that export taxation by means of an overvalued exchange rate was alternated with explicit export taxes. At the time of a devaluation, the export tax would be raised. Then as the peso became increasingly overvalued, the explicit tax would be reduced. In effect, policy makers were trying to balance the conflicting goals

Table 32. Argentina: Nominal and real exchange rates, 1950–81

Period	Average nominal exchange rate	Average real exchange rate[a]
1950–54	0.10	34.86
1955–60	0.58	60.90
1961–70	2.38	29.41
1971–74	5.00	18.80
1975–78	484.10	44.87
1979–81	3,619.67	21.03
1950–81[b]	401.35	35.99

Source: IMF, International Financial Statistics, various years.

[a] The real exchange rate is defined in purchasing power terms and is calculated as follows:

$$\text{Nominal exchange rate} \times \frac{\text{Wholesale price index of the U.S.}}{\text{Wholesale price index of Argentina}}$$

[b] The standard deviations of the nominal and real exchange rate over the entire period were 1,336.96 and 19.20, respectively.

Base year: 1975.

of low food prices, the need for foreign exchange and tax revenues for development purposes, and the proper degree of incentives for producers.

Before turning to the more recent literature on Argentine trade, which for the most part refers to the period since 1950, it is useful to provide a bit of the background for this history and some of the early studies. The period 1862-1940 in Argentina was a period of liberal commercial policies which Diaz-Alejandro [1970, chapter 3] referred to as "The Great Expansion" (also known as the liberal period). The year 1862 inaugurated a period of relative political stability during which the government became committed to close cooperation with foreign capital and to foreign markets. A spectacular rural expansion took place that lasted up until 1929. This expansion was so remarkable that little conflict arose over the division of rural production between exports and domestic consumption. For many goods such as wool, linseed, and maize, domestic consumption even in 1929 represented a small fraction of output. For wheat and beef the share of domestic absorption was higher, oscillating between 40 and 50 percent of output. During 1920-29, 49 percent of all rural output went to domestic consumption and 51 percent to exports.

The period 1930-63 was just the opposite—a period of stagnation in the rural sector.[69] During this period rural output grew at an annual rate of around 1 percent per year, well below the population growth rate. Associated with this decline in output growth was a dramatic decline in Argentina's share of world exports from 1934-38 to 1959-62. Some of this decline Diaz-Alejandro ascribed to external factors such as the Great Depression of the 1930s and World War II, when transportation difficulties made European markets difficult to access.

Table 33. Argentina: Trade in agricultural products, 1955-80

Period	Average value of exports	Average value of imports (US$000)	Average trade balance[a]
1955-60	923,567	184,833	738,734
1961-70	1,291,087	127,513	1,163,574
1971-80	3,307,344	282,994	3,024,357
1955-80	1,981,758	200,541	1,781,217

Source: FAO, Trade Yearbook, various years.
[a] Average trade balance of agricultural products, i.e., value of agricultural exports minus value of agricultural imports.

After this period, however, he ascribed the continued weakness of trade to internal supply difficulties rather than world demand conditions. Indeed, it was not uncommon during this period for the Argentine government to limit by decree the exports of certain commodities to assure an adequate supply of exportables for domestic consumption at "reasonable" prices, even at the expense of cutting back deliveries to traditional overseas markets.

The interesting question, of course, is whether an inadequate supply capability caused the export stagnation, or whether trade policy was so discriminatory against agriculture that it caused supplies not to be forthcoming. Although recognizing both alternatives, Diaz–Alejandro favored a domestic supply-induced (structuralist) explanation for the stagnation in exports. He based this conclusion on his empirical analysis which suggested that the short-run elasticity of supply for crops was quite low, as was domestic elasticity of demand for these commodities, while the short-run supply response of beef was perverse due to the inventory nature of this production activity. He generalized by arguing that with low elasticities on both the demand and supply side of the market, a devaluation would have little effect.

While recognizing that it is excess supply that matters, Diaz–Alejandro did not seem to recognize that both demand and supply elasticities can be low and the excess-supply elasticity still be large. This is because the relative importance of trade plays a significant role in determining size of the excess-supply elasticity. When Argentina was exporting over one-half of its domestic production, the excess-supply elasticity would have tended to be low. But as exports declined as a share of total output, the export-supply elasticity would tend to increase.

The evolution of Argentina's agricultural trade in the 1955-80 period is shown in Table 33. Exports grew almost steadily throughout the period covered by the table, with an increase of 270 percent between the first and last periods. Agricultural imports, on the other hand, experienced a different pattern. There was a declining trend up through the end of the 1960s, and an increasing trend thereafter.

Table 34. Argentina: Effective protection coefficients, 1950-74

Period	Wheat	Corn	Sorghum	Beef	Rice	Cotton	Wool
1950-54	0.29	0.27	–	–	–	–	–
1955-59	0.62	0.69	–	–	–	–	–
1960-64	0.72	0.79	0.54	0.84	–	0.81	0.79
1965-69	0.80	0.76	0.60	0.84	1.001[a]	0.72	0.80
1970-74	0.60	0.64	0.52	0.64	1.40	1.14	0.72

Source: Reca, 1980. With permission.
[a] Based only on 1968-69 and 1969-70.
Note: Coefficients of effective protection recognize that the inputs used in producing a good may experience protection as well as the good itself and estimates the net effect of both protections.

Despite the importance of agricultural trade to the Argentine economy, and despite the importance of trade policy to the agricultural sector, there have been few really penetrating studies of these issues. Perhaps the first was by Diaz-Alejandro [1965], who considered agriculture as part of his larger study of the Argentine economy. Reca's study [1967] of supply response in Argentine agriculture was indirectly related to these issues, but did not consider the trade issues directly. (Estimates of the supply parameters from this study were considered in the section on supply response.)

One of the more comprehensive studies of Argentine trade policy and its effect on the performance of agriculture was a later study by Reca [1980] for the World Bank. His results showed that trade policy affected subsectors of agriculture differently, with more severe discrimination against grains relative to beef cattle, wool, rice, and cotton. This study suffers from the deficiency of not taking into account distortions in the exchange rate. If the large overvaluations of the peso which occurred with frequency were taken into account, the measures of effective protection shown in Table 34 would undoubtedly have at times been negative, and by a large margin. This probably would not alter in a major way the basic proposition about the *relative* discrimination among subsectors of the agricultural economy. Moreover, Reca evaluated the bias inherent in his procedures in an appendix, and used an estimate of the equilibrium exchange rate in his welfare analysis.

Reca [1980, p. 50] also calculated coefficients of domestic resource costs (DRC), which are defined as the ratio of the opportunity cost of domestic resources and the international value added. The DRC, in effect, gives an indication of whether the potential benefits of a project to expand the exports of a sector are positive. If the coefficient is less than one, this indicates the desirability of increasing the exports of that sector.[70]

A summary of Reca's findings is presented in Table 35. For the majority of the commodities and years he considered, the coefficient is below one. This sug-

Table 35. Argentina: Domestic resource costs,
selected agricultural commodities, 1960–74

Product	Period		
	1960–64	1965–69	1970–74
Wheat	0.46	0.54	0.69
Corn	0.58	0.63	0.45
Sorghum	1.27	0.97	0.63
Beef	1.30	1.22	0.86
Wool	0.48	0.74	0.60
Rice	–	1.47	0.88
Cotton	0.75	0.74	0.83

Source: Reca, 1980. With permission.

gests that there was potential for an increase in exchange earnings from the expansion of exports; the domestic cost was smaller than the international value added.

Reca's interpretation [1980] of these results is as follows:

> They indicate a comparative advantage [*sic*] in all seven commodities studied in 1970–74 when international agricultural prices were high. Even when world prices were lower throughout the 1960s, Argentina appears to have had a comparative advantage [*sic*] in wheat, corn, grain sorghum, wool, and cotton at the official exchange rate. . . .
>
> The evolution of DRC coefficients from 1960 to 1974 showed a relative loss in the case of wheat largely explained by increases in beef prices in the early 1970s, and a strong increase in the comparative advantage [*sic*] of corn and grain sorghum, attributable to gains in productivity in both crops in the recent past. The DRC coefficients for beef cattle production stayed somewhat above the critical level of one, except in the period 1970–74 when beef prices were at a record high. However, because estimates of DRC coefficients for beef cattle production during the 1960s can range from 0.67 to 1.88, depending entirely on the opportunity cost assigned to land, it is very difficult to come to a final conclusion about the overall comparative advantage [*sic*] of beef production. [Reca, 1980, p. 50]

It should be noted that Reca used the official exchange rate for this analysis also. This fails to take account of the chronic tendency to overvalue the Argentine peso. If this distortion were taken into account, Argentine agriculture would have been shown to be inherently much more competitive in international markets throughout this period than it actually was. Thus agriculture could have earned much more foreign exchange to support a higher rate of economic growth were it not for restrictive trade and exchange rate policies.

A significant recent study by Cavallo and Mundlak [1982] provides insights into the consequences of these policies, and at the same time illustrates the political dilemmas faced by policy makers. These authors estimated the parameters of a two-sector growth model and used them to simulate the evolution of the economy under alternative policy regimes.

The first simulation of interest is that which results from the elimination of all export taxes and import tariffs. As background, it should be noted that the estimated effect of export taxes on agriculture as calculated by the authors consists of two components: the direct or explicit tax on exports and the indirect or implicit tax represented by the shift in domestic terms of trade against the sector.[71] The total extraction of resources from the sector is the sum of the two components. On average, Cavallo and Mundlak [1982] found the direct tax to be only 15 percent while the indirect tax was 34 percent.

The total extraction from agriculture oscillated substantially, especially in the 1940s when a peak of 148 percent was attained in 1947. The extraction is measured as a proportion of agricultural output measured at factor cost, with output and exports expressed in per capita terms. This explains how the extraction can be greater than 100 percent. The total tax increased again in the early 1970s, reaching a level above 50 percent in 1973 as the domestic economy was isolated from the large rise in prices in international commodity markets.

Cavallo and Mundlak [1982] also calculated the *total* flow of funds from agriculture during the period, a flow which included the taxes as well as the net flow of savings out of agriculture [p. 62]. (The net flow of savings is defined as total savings from agricultural income minus investment in agriculture.) Their results show very high levels of extraction in the 1940s (almost 200 percent in 1947). The total extraction declined to around 25 percent in 1953 and 1954, then rose to around 70 percent from 1956 through 1962, and then declined to around 50 percent by the early 1970s. Clearly, the extraction from agriculture was large and significant.

Finally, the authors calculated the international and domestic terms of trade [Cavallo and Mundlak, 1982, p. 65]. In view of the trade and exchange rate policies that prevailed in the 1941-71 period considered, it comes as no surprise that the domestic terms of trade were significantly lower than the international terms of trade, with the disparity especially wide during the 1940s.

Cavallo and Mundlak's [1982] first simulation considered a complete liberalization of trade, with export taxes and import tariffs set equal to zero and, to avoid the effects of changes in government revenue, internal taxes set at levels which guaranteed that the ratio of tax collections to output was maintained at the observed level. The simulation was started in 1950 "after the main abnormalities of the war and postwar period disappeared and world trade was developing rapidly" [p. 63].

To consider the effects of this simulation, we start with the real exchange rate (which is defined in this article as the nominal exchange rate deflated by the domestic price of nonagricultural products produced domestically), a concept that determines the terms of trade between the traded and nontraded sectors in the nonagricultural sector [Cavallo and Mundlak, 1982, pp. 29-30, 64]. A reduction in export taxes, according to Cavallo and Mundlak,

> increases the domestic price of the exported commodity, increases the incentive for producing exportables, reduces their domestic consumption, and thereby leads to an increase in exports. This change requires a shift of resources to the exporting sector. Resources mobilized from the production of nontradeables decrease their output, thus leading to excess demand, an increase in their prices, and a *decrease in the real exchange rate*. [p. 137, emphasis added]

On the other hand, a reduction in tariffs will have an opposite effect on the real exchange rate:

> A reduction in the import tariff (t_m) will lower the domestic price of the imported good. Following the preceding argument with respect to a decline in t_x, this change will result in excess supply of nontradeables, decrease their price, and consequently *increase the real exchange rate*. [Cavallo and Mundlak, 1982, p. 137, emphasis added]

In the empirical analysis of the trade liberalization scheme, it turned out that the effect of reducing export taxes was predominant and therefore the real exchange rate declined when trade policy was liberalized. As a consequence, the increase in the domestic price of exported goods was smaller than the reduction in the export tax and the reduction in the domestic price of imported goods was larger than the tariff reduction.[72] The effects of these changes in prices on agricultural production and food consumption were summarized by Cavallo and Mundlak [1982, p. 64] as follows:

> The economy responded to those changes in prices.
> Agricultural production increased gradually and reached the largest deviation from the base run in 1963 with a growth of 4.7 percent. . . . At the same time, there was a reduction in the

consumption of food. This decline reflects the increase in food prices caused by the elimination of t_x and, to some extent, by the compensatory increase in the domestic tax.

Subsequently, Cavallo and Mundlak [1982, pp. 64–65] commented also on the effects of this policy change in other sectors of the economy. Nonagricultural output declined sharply right after the policy change (in view of the price decline that was experienced at that time) and this decline in per capita output continued for four years, after which a recuperation started. Nonetheless, by the end of the period in 1971, income measured in terms of nonagricultural goods was still less than the historical value by approximately 8 percent. Another "unpleasant" feature of these results was a decline in nonagricultural wages in terms of food (that is, nominal wage in nonagriculture deflated by the market price of agricultural goods), which in 1971, the last year of the simulation, was only 87 percent of the historical value.

The authors of this study attributed some of the "unpleasant" results they obtained in this simulation to the behavior of the real exchange rate. The decline in that variable following the liberalization of trade and exchange rate policy eventually reduced the effects of the liberalization on domestic prices of imported and exported goods (as noted above). To investigate this issue further, they repeated the exercise, with the real exchange rate not allowed to decline. Their description of some of the results of this simulation was as follows:

> The agricultural price increases considerably, whereas the price of the imported good is less than its base-run value, as it should be, although the difference is relatively small. This change in prices results in a strong effect on agricultural output. . . . At the same time, the per capita output of nonagriculture also increased continuously and never declined below the value of the base run. . . . At the end of the 10-year period of this experiment, per capita income increased around 33 percent and consumption and investment changed similarly. The economy accumulated foreign assets, which by the end of the period represented 18 percent of the new augmented income. This accumulation of foreign assets reflects the developments in exports and imports. . . . At the end of the period, exports were 44 percent greater than in the base run. Imports also rose considerably, becoming 60 percent greater than in the base run and declining to 40 percent greater at the end of the period. [Cavallo and Mundlak, 1982, pp. 68, 74]

Despite all these positive features, this simulation also resulted in a decline in real wages in terms of food. Cavallo and Mundlak [1982, p. 74] argued that such policies would be very difficult to implement in a country such as Argentina in which policy makers have always been concerned with the real wages in terms of food. However, this simulation shows the extent to which economic growth was sacrificed as a consequence of pursuing cheap food policies.

Another experiment was developed to see whether it would be possible to use part of this foregone growth to implement a policy of keeping real wages at their historical values. The authors' summary comments on this simulation were as follows:

> It is thus possible to liberalize trade while protecting the food wage. This policy is, however, not costless. The performance of the economy under this policy is inferior to that without the food subsidy. . . . After 21 years, per capita income reached, under the latter policy, 811.2 pesos as compared to 920.2 without food subsidy. This loss of 13 percent in income resulted in a decline in the per capita consumption of the agricultural product from 42.2 pesos to 38.9, a decline of 8 percent. So, even though such a policy may be appealing and acceptable for other reasons, less food and fewer nonagricultural goods are consumed under it. [Cavallo and Mundlak, 1982, p. 81]

A number of comments are pertinent in the context of these findings. First, keeping food prices low relative to wages has been a predominant objective of policy makers in Latin America. In the case of Argentina, strong labor unions have been a driving political force in this direction, especially during and after the Peron period [see Epstein, 1975].

Second, a cheap food policy is consistent with an import-substituting industrialization policy. It makes it possible to keep nominal wages low to the industrial sector, increasing the profitability of this sector and providing incentives for expansion. Workers benefit as food prices decline, even though nominal wages may not increase. In effect they receive implicit income transfers from the agricultural sector.

Third, if agriculture were not a traded sector, or if it were sufficiently large so that its output had an effect on international prices, investments in agricultural research would be a solution to this problem since the adoption of new production technology would lower agricultural and food prices. However, with prices determined in international markets, producers would reap most of the direct benefits of technical change. This probably explains why Argentina (and other Latin American countries as well) have not invested heavily in the capacity to produce new agricultural technology and opted instead for trade and exchange rate distortions to siphon resources from agriculture for the development of the nonfarm sector. The belief has been that resources could be extracted from agriculture without affecting domestic food supplies.[73]

The key issue in the modernization and thus development of agriculture in Latin America is therefore whether a greater investment in the production of new technology for agriculture would have earned sufficient foreign exchange to generate a higher rate of growth—a rate that would have promoted rapid industrialization and thus an increase in the food wage equal to that which was obtained. A simulation by Cavallo and Mundlak which examined this proposition would have been most enlightening.

BRAZIL

The analysis of Brazilian trade policy after World War II reveals much the same set of interventions as in the case of Argentina and most of the other Latin American countries. Brazil has relied heavily on implicit taxation of its agricultural sector by means of an overvalued exchange rate and the widespread use of quantitative restrictions, both on the export and import sides (especially fertilizers). In addition, imported inputs (fertilizers are again an important example) have been subjected to relatively high tariffs in order to protect a nascent domestic industry.[74]

Explicit taxes on agricultural exports have also been used extensively. An important example of this is coffee, which has persistently been subjected to a *confisco cambial* (foreign exchange confiscation). This tax in the early 1960s amounted to more than 50 percent of the total FOB value of coffee exports, but has declined somewhat since then [see M. A. Martin, 1976, for example]. Explicit export taxes have also been extensively used on beef over the years, and more recently on soybeans, in the latter case primarily to channel the raw beans to the domestic milling industry.

An overview of the basic trade statistics for Brazil again provides a background for the interpretation of the literature. It is useful to start by noting that Brazil has been very successful in diversifying its exports. For example, in 1953, coffee, sugar, and cocoa made up 77 percent of the value of Brazil's exports. This share had declined to 26 percent in 1975 [Brandão, 1978, p. 81]. Furthermore, the share that agricultural exports as a whole make up of total exports declined substantially during this period. In 1979, this share was around 50 percent.

Another aspect of Brazil's agricultural trade experience has been its ability to gain new export markets. Starting in the late 1960s and extending up through the end of the 1970s, Brazil came to account for almost half of the total world trade in soybeans—a market which the U.S. previously dominated. During the 1970s and the early 1980s Brazil came to be the leading exporter of frozen orange juice and a major exporter of poultry. In the latter two cases this penetration of foreign

Table 36. Brazil: Nominal and real exchange rates, 1950–81

Period	Average nominal exchange rate	Average real exchange rate[a]
1950–57	0.050	6.68
1958–64	0.544	11.28
1965–71	3.701	8.76
1972–79	15.096	9.15
1950–81[b]	10.755	8.97

Source: IMF, International Financial Statistics, various years.

[a] The real exchange rate is calculated as follows:

$$\text{Nominal exchange rate} \times \frac{\text{Wholesale price index of the U.S.}}{\text{Wholesale price index of Brazil}}$$

[b] The standard deviations on the annual data on nominal and real exchange rates are 25.34 and 2.53, respectively, for the period as a whole.

Base year: 1975.

markets was obtained by use of export subsidies designed to promote the development of value-added industries.

Data on the nominal and estimated real exchange rates are shown in Table 36. Brazil followed a fixed exchange rate rule until 1968. In that year, a major change took place with the introduction of a crawling peg exchange rate policy, which was pursued with increasing sophistication over time.

The averages in the table mask a great deal of the instability in the exchange rate in this period as indicated by the standard deviations. For example, the real rate was 3.80 in 1953, then increased to 10.87 in 1954, declined to 7.13 in 1956, and peaked at 13.00 in 1959. It was greater than ten except for one year, 1963, in the period 1958 through 1965. Then a period of relative stability followed from 1966 through 1978, when it ranged between 7.73 and 9.00. In 1979, with the second large increase in petroleum prices, the real exchange rate surged to 13.18 — the highest level in the whole period, but then declined to 10.48 and 11.80 in 1980 and 1981, respectively. Overall, the standard deviation for the real exchange rate was 2.53 for the period as a whole.

Brazilian exchange rate policy in the immediate post-World War II period illustrates how external factors can influence economic policy. At the end of World War II Brazil was the dominant coffee producer in the international economy. The evidence [Veiga, 1974] is that Brazil overvalued its currency in this period at least in part as a means of exploiting the downward sloping demand curve for coffee it faced in international markets. It has persistently tried to exploit that dominant position by exchange rate policy and export taxes ever since.

Unfortunately, Brazil did not have a dominant position in the markets for its other exports. As it gradually recognized this, it shifted to a multiple exchange rate system. This system was also extensively used to allocate resources on the

Table 37. Brazil: Trade in agricultural products, 1951-79

Periods	Average value of exports	Average value of imports (US$000)	Average trade balance[a]
1951-57	1,423,914	312,929	1,110,986
1958-64	1,159,129	269,857	771,371
1965-71	1,588,075	301,430	1,286,645
1972-79	5,266,842	1,138,361	4,378,481
1951-79	2,459,744	527,462	1,932,282

Source: FAO, Trade Yearbook, various years.
[a] Trade balance of agricultural products, i.e., value of agricultural exports minus value of agricultural imports.

import side, and eventually grew to include twenty-one different rates for imports and some seven for exports. Veiga describes and analyzes the various policy "epochs" of Brazilian trade and exchange rate policy, with estimates of the extent to which policy was discriminatory towards agriculture.

Table 37 shows the evolution of the value of agricultural exports, imports, and the balance of agricultural trade for the period 1951-79. The period can be divided into approximately four epochs: 1951-57, 1958-64, 1965-71, and 1972-79. Exports were lower in the second period than they were in the first, largely because of a strong, overvalued cruzeiro (see Table 36). They began to increase with the reforms (devaluations) of 1964 and following years and increased significantly through 1971. Starting in 1966-67, Brazil added an export-promoting policy to its import-substituting industrialization policy. By 1972 the distortion in the exchange rate had been virtually eliminated and agricultural exports responded accordingly, aided and abetted by the global commodity boom of the 1970s. Exports peaked at US$7.6 billion in 1977.

Imports of agricultural commodities approximately paralleled the performance on the export side—an important finding in its own right. The performance of the agricultural trade balance is even more important, however, since this is a measure of agriculture's contribution to economic growth by supplying foreign exchange. The trade balance was significantly lower in the 1958-64 period than it was in the earlier period (Table 37). The real exchange rate in this period averaged 11.28 (Table 36), compared to 6.68 in the earlier period. In the periods 1965-71 and 1972-79, the real exchange rate averaged 8.76 and 9.15, respectively.

Clearly, the real exchange rate has a significant effect on trade performance. However, a global commodity boom, combined with a rational exchange rate policy, in some sense makes for the best of all possible worlds. Contrary to many developing countries, Brazil took advantage of the international markets of the 1970s, and its growth performance reflected that favorable policy stance. It is easy to see how Brazil optimistically took on a great deal of foreign debt in this period, not foreseeing the difficulties that were to follow in the 1980s. However, it also failed to keep its exchange rate in line. The real exchange rate surged from 8.66 in 1977 to 13.18 in 1979, and the balance on its agricultural trade accounts declined from US$6.6 billion (the peak) in 1977 to US$4.7 billion in 1979.

The performance of Brazil's agricultural trade sector has been analyzed in some detail in the literature. Some of the main research dealing with the decades of the 1950s and 1960s are referred to in note 74. The essential message of these studies is the discrimination suffered by agriculture by means of trade and exchange rate policy, and the identification of the instruments utilized by the policy makers to tax the sector. A second important lesson is the instability of trade and exchange rate policy throughout this period, and the repeated shocks this imposed on agriculture.

There are two other dimensions of trade and exchange rate policy that are quite important but which seldom receive the attention they deserve. These refer to the distribution effects of trade policy, effects which occur on at least two levels in the case of agriculture. First, recall that the basic trade and exchange rate policy for many Latin American countries—evolving from the import-substituting industrialization model—involves high levels of protection for the industrial sector and serious discrimination against agriculture. The fact that individual regions of a country differ greatly in their sector-mixes thus causes sectoral discrimination to become inherently regional discrimination.

Baer [1964a] perceptively noted this effect in the case of Brazil. Trade and exchange rate policy in Brazil significantly discriminated against the "agricultural" Northeast (the largest collection of poverty in the Western hemisphere) and favored the industrialized South. The resource flow induced by the regional distortion in the terms of trade was quite sizeable. In fact, Baer found that this flow was larger than the fiscal transfer to SUDENE (the regional development authority for the Northeast) from the South at an important stage of SUDENE's program. For those interested, M. A. Martin [1976] updated this analysis.

The analysis of the impact of trade and exchange rates by size of farm was equally revealing [M. Lopes, 1977; M. Lopes and Schuh, 1979]. Large farms were able to escape the effects of export taxes (which lower agricultural prices below what they otherwise would be) by shifting to more extensive forms of production. They reduced their use of labor and capital and increased their use of land in a relative sense. The small producer, facing rather different conditions in the factor markets because of market imperfections, especially in the labor markets, did not share that opportunity, at least not to the same extent. As a result, these small producers bore a larger share of the burden of the tax.

Lopes's study has a number of important implications. First, it explains why rural-urban migration in Brazil and other countries has been so large and, in many respects, premature. Labor is being pushed out of agriculture by means of trade and exchange rate policies which severely shift the domestic terms of trade against agriculture. Second, it explains (at least in part) why land is used so extensively on large farms. Third, it explains, in part, the persistent pattern of low land productivity on large farms and high land productivity on small farms. And finally, it explains why extracting resources from agriculture by means of trade and exchange rate policy can be so inefficient and counterproductive. Such policies have strong disincentive effects and thus extract resources from the very sector that may have a strong comparative advantage. Lopes shows that a land tax which extracted approximately the same quantity of resources from agriculture would have none of these deleterious consequences.

Despite the apparent high costs of these policies, they were consistent with other policy objectives of the government. It has been a long-term goal of Brazilian policy makers to "fill up the interior"—to push the population that has been concentrated along the coast into the interior. A great deal of the migration in Brazil has been rural to rural [Schuh, 1975c]. Hence, labor pushed off farms in traditional areas has been channeled to frontier areas, where the construction of penetration roads has opened ever larger areas to colonization. There has also been a great deal of migration from the rural areas of the East and Northeast to the rural areas of São Paulo state.

The decade of the 1970s was a rather unusual period in Brazilian post-World War II history since it was a period in which the domestic terms of trade shifted in favor of agriculture. There were two components to this shift. First, Brazil made remarkable progress in the late 1960s and 1970s in getting its economic house in order. The rate of domestic inflation was brought down significantly and the frequent minidevaluations caused the exchange rate to approach equilibrium levels by 1972. In fact, equilibrium was so close that Brazil for the first time in a long time *revalued* its currency slightly at the time of the second devaluation of the U.S. dollar (1973).

What followed, however, was in large part a function of the changed economic environment created by the bloc-floating exchange rate system that emerged after the United States floated the dollar. The U.S. dollar proceeded to decline by successive steps during the 1970s, reaching a trough or low point in 1979. This decline contributed importantly to the export boom in commodity markets of that decade. Brazil, of course, benefited from this decline in the value of the dollar because it continued to sustain a relatively constant purchasing power parity of the cruzeiro *vis-à-vis* the dollar by means of its crawling-peg exchange rate policy.

The conditions thus created are what Schuh [1984a] refers to as the "third-country" effects of exchange rate realignments in a bloc-floating exchange rate system, and help explain a great deal of Brazil's economic history in the 1970s and early 1980s. Many observers believed that Brazil should have devalued at the time of the 1973 increase in petroleum prices, since at that time it imported a major share of its petroleum. However, it didn't devalue any more in nominal terms than the exchange rate policy it was pursuing at that time indicated it should. As it turned out, that didn't matter a great deal. The value of the dollar fell and the cruzeiro was tied to it in purchasing power terms. Brazil thus benefited from the fall in the value of the dollar, and gained a competitive advantage relative to "third" countries. In part as a result, Brazil experienced one of the highest rates of economic growth in the world in the 1970s, despite its dependence on high-priced petroleum.[75]

The experience of the 1980s was just the opposite. The second petroleum shock occurred in 1979. This time, however, the United States reversed its domestic policies compared to the earlier period. Starting in October 1979, the Federal Reserve declined to monetize the burgeoning Federal deficit. The result was a shift from high negative real rates of interest in that country in the 1970s to unprecedentedly high real rates of interest by early 1980. These high interest rates caused a shift into dollar assets and started the dollar on an unprecedented rise. The dollar was further strengthened by President Reagan's deregulation of the domestic petroleum industry shortly after he took office in 1981, which in effect virtually eliminated the implicit import subsidy for petroleum.

The cruzeiro being tied to the dollar was no longer beneficial to Brazil. The cruzeiro was already significantly overvalued in 1979 as a result of the acceleration of domestic inflation, thus necessitating a maxidevaluation in that year. However, the effects of that devaluation were soon lost due to a failure to follow up with proper complementary monetary and fiscal policies. Once Brazil's foreign debt crisis erupted in 1982, it had essentially no choice but to accelerate its rate of minidevaluations.

The experience of the 1970s also gave rise to another set of problems and controversy. The shift in domestic terms of trade in this decade caused domestic food prices to rise in relative terms and critics became increasingly outspoken against what was perceived as an export promotion policy. Mendonça de Barros and Graham [1978] made one of the first empirical studies of this issue. Their hypothesis was that the more export-oriented (outward looking) trade policy implemented by the government in the final years of the 1960s, together with the commodity boom in international markets in the 1970s, had accentuated a dichotomy between the export and domestic or subsistence food subsectors of agriculture. The remarkable response of agriculture in the early 1970s was attributed to the crawling-peg exchange rate system and to the increased participation of processed agricultural products in the value of exports.[76]

This same policy, on the other hand, was said to be responsible for the poorer performance of the domestic or subsistence food subsector.[77] The authors showed a substantial decline in the rate of growth of production of the domestic crops *vis-à-vis* the exportables. This was attributed to the fact that, given the rate at which demand for these latter goods was expanding, domestic crops were not able to compete effectively for the inputs necessary for production.

A word of caution is in order with respect to these results, those to be presented below, and those by authors not cited in this study. It is true that exports did respond quite well to the incentives provided. But the whole story of competition between exports and the production of domestic food crops has not, to the best of our knowledge, been told. In the first place, the classification of the commodities has been along traditional lines and this is misleading in some cases. For example, sugarcane has traditionally been classified as an export crop and the hectares planted to it grew very rapidly over the past decade. But that growth was largely to produce alcohol as a fuel for automobiles and in displacement of imported petroleum (see below). That obviously is a trade issue since it essentially involves the use of agriculture as an import-substitution industry. But for policy purposes, clarity would be added by identifying it as a part of Brazil's *energy* policy, rather than to attribute it to export policy.

Similarly, there has been no attempt to our knowledge to trace direct shifts in hectares between export crops and domestic food crops to verify whether there has been a direct displacement. To illustrate the significance of this, in the beginning of the soybean boom a great deal of the soybean crop was planted in conjunction with wheat, resulting in a double crop. Later, soybeans expanded onto land that was previously used extensively as pasture. It was only later that it competed directly with other crops, and then often with other export crops (coffee, for example).

These issues aside, the remarkable response of exports[78] to policy changes brings us again to the question of comparative advantage in the agricultural sector. Calculations of the DRC for agricultural products and some of their transformations have been made by Mendonça de Barros [1974] and are presented in Table 38.[79] Comparison of the data in Table 38 with the exchange rates prevailing at the beginning of the 1970s (the period of Mendonça de Barros's study) shows clearly that there was a potential benefit in increasing the exports of most of the products considered. The exceptions were soybean oil and meal and wheat.

In a later study, Mendonça de Barros et al. [1975] went still further and calculated DRCs for other sectors of the economy. The results confirmed the favorable ranking of agriculture as an export sector. In fact, in the opinion of the authors, the three sectors in which comparative advantage was weak were some intermediate products, mining and energy, and machinery. Although not presenting a detailed statistical analysis, they concluded that the Brazilian system of export subsidies appeared to be negatively correlated with comparative advantage.

The soybean sector is of special interest in Brazil's agricultural trade policy. The general perception has been that it was highly subsidized from an export perspective. This is because both the milling industry and the exports of its two

Table 38. Domestic resource costs for selected commodities,
Brazil, 1971-72 crop year

Commodity	Domestic resource cost
Cotton (mechanized)	4.34
Cotton (animal traction)	4.10
Peanuts (animal traction)	2.41
Cassava flour	3.46
Corn (mechanized)	5.31
Corn (animal traction)	5.27
Soybeans (mechanized)	5.78
Soybeans (animal traction)	6.24
Cassava ("raspa")	2.17
Wheat	12.60
Peanut oil	4.65
Peanut meal	4.65
Soybean oil	7.53
Soybean meal	7.53

Source: Mendonça de Barros, 1974.

main products—soybean oil and meal—were highly subsidized. However, the soybean sector itself was treated very differently than the two processed commodities. Raw soybeans suffered from the significant overvaluation of the cruzeiro during this period, together with periodic explicit export taxes and export embargoes designed to channel production to the domestic market.

Carlos Santana [1984] analyzed the effects of these policies by means of the model of effective protection. Some of his estimates are presented in Table 39. Rather than being subsidized, as commonly believed, soybean production was severely taxed during most of the period, and a large part of this taxation was done by means of an overvalued exchange rate.

The development of Brazil's soybean sector has been of particular interest to U.S. producers and policy makers. Of particular interest has been whether U.S. embargoes provided the incentive for expansion of Brazilian soybean production. Faminous and Hillman [1986] explore this issue, and their report is a rich source of data on the soybean sector in that country.

A second interesting trade case is wheat, for which Brazil has long provided significant subsidies. This commodity has over the years been the most important agricultural import in Brazil. Self-sufficiency in wheat has been a goal of successive Brazilian governments since World War II. In order to achieve that goal, various schemes for subsidizing producers have been implemented. The data in Table 38 (for wheat) indicate a very high social cost for these subsidies, a result which is supported by other studies [Knight, 1971; Contador, 1974b].

A comparison of the earlier results by Knight [1971] with the results from more recent studies shows an improvement in the situation, however. This im-

Table 39. Estimates of effective rates of protection for soybean production, Brazil, 1977-78 through 1982-83 (in %)

Crop year	Effective rate of protection[a]	Net effective rate of protection[b]
1977-78	−30.32	−52.56
1978-79	−26.34	−43.92
1979-80	−27.83	−38.19
1980-81	− 1.99	−40.33
1981-82	43.13	−25.74
1982-83	− 0.24	23.54

Source: Santana, 1984, p. 204. With permission.
[a] Estimated with the official exchange rate. A negative sign indicates the sector was taxed; the larger the number the larger the tax.
[b] Estimated with an equilibrium exchange rate. Negative sign has same implication as above.

provement, according to Mendonça de Barros [1974], can be attributed to two factors: the increase in yields experienced in recent years, and the introduction of soybeans, which decreases the capital costs and land costs through an increase in the number of days these resources are used per year. As noted earlier, soybeans are often double-cropped with wheat.

Brazil's wheat policy illustrates just how far economic policy can go astray, as well as the importance of digging beneath the surface to understand all dimensions of policy. For example, although wheat prices were often set as much as twice border price levels when evaluated at official exchange rates, the fact that the cruzeiro was significantly overvalued also has to be taken into account. Of equal interest is the fact that at the time of the large rise in international commodity prices in the early 1970s, Brazil implemented large consumer subsidies for wheat. Calegar [1984] and Calegar and Schuh [1988] have analyzed these policies. Once established, the consumption subsidies became a policy fixture. Moreover, they burgeoned—reaching approximately US$1 billion a year in some years. This is a large subsidy for a low income country that has ample consumption alternatives. It also contributed to the rapid growth of imports in the late 1970s and early 1980s. Another dimension of these policies is that it was largely middle- and upper-income groups that benefited from the subsidy. Low income groups consumed much smaller quantities of wheat products and the consumption subsidies were of little value to them.

To elaborate further on the question of comparative advantage, we draw on the results of Homen de Melo's study [1981] of the domestic and exportable subsectors of Brazilian agriculture. Starting with the same classifying principle that Mendonça de Barros and Graham [1978] used, Homen de Melo included the following goods in each category:

Domestic goods: Rice, edible beans, cassava, corn, potatoes, and onions

Exportable goods: Soybeans, oranges, sugar, tobacco, cocoa, coffee, peanuts, and cotton

Nominal protection coefficients were calculated for some of these commodities plus wheat. The coefficients (presented in Table 40) show the following:

(1) For the domestic goods, the ratios of internal to international prices were consistently greater than one, revealing that this has been a protected sector (in nominal terms).

(2) In more recent years, the level of nominal protection for the domestic goods declined, but the ratios were still greater than one for many products.

(3) For the exportable goods, the ratio was consistently less than one for all products.

Homen de Melo [1981] also calculated these coefficients of nominal protection taking into account the overvaluation of the cruzeiro during 1948-65. When this distortion is taken into account, the earlier conclusion that rice, potatoes and onions have been protected persists even though the level of protection is lower. As was true before, the data for edible beans do not show a very clear pattern. The most noticeable change occurs with respect to corn, which becomes a product that is taxed instead of protected.

There is an important point that follows from de Melo's study. That is the value of disaggregating the agricultural sector into traded and nontraded components. The importance of making this distinction, of course, will depend on the mobility of resources among the subsectors. If resources are highly mobile among subsectors within agriculture, discrimination against one or a limited number of sectors will spill over into the remaining subsectors.

Another important aspect of understanding the effects of trade policies on agricultural sectors is the significant degree of natural protection that distance from consumption sectors gives the agriculture of countries such as Brazil and Argentina, especially for "bulky" commodities such as maize. For such commodities, the disparity between border prices evaluated at CIF and FOB prices is quite large.

Another area in which research is needed is in understanding the mechanisms of price formation and the plausibility of the classification used by both Mendonça de Barros and Graham [1978], and Homen de Melo [1981]. For example, corn is a product in which Brazil would appear to have considerable potential in world markets. Yet it has not been exported in the past in large part due to the discriminatory effects of government policy [see, for example, Thompson and

Table 40. Nominal protection coefficients for selected commodities in Brazilian agriculture, 1948-77

Year	Rice	Edible beans	Potatoes	Onions	Corn	Peanuts	Sugar cane	Cotton	Soybeans	Wheat	Coffee
1948	1.72	–	2.02	–	0.74	0.76	0.62	0.75	0.77	1.31	–
1949	2.13	–	1.69	–	1.08	0.73	0.74	0.80	1.05	1.45	–
1950	1.71	0.74	3.14	–	0.86	1.31	0.70	0.85	1.31	1.99	–
1951	1.47	0.64	2.40	–	0.92	0.93	0.69	1.00	1.15	2.24	–
1952	2.35	1.55	2.16	–	1.17	0.88	0.78	0.89	1.37	2.35	–
1953	3.59	2.37	3.70	–	1.59	0.88	0.78	0.75	1.49	2.67	–
1954	3.06	2.18	3.52	–	1.02	1.02	0.74	1.82	1.08	2.92	–
1955	2.74	1.55	2.04	–	1.41	0.66	0.58	1.80	1.23	2.90	–
1956	2.69	1.57	1.56	–	1.29	0.78	0.77	0.82	1.08	2.60	–
1957	2.59	1.61	2.19	2.92	1.24	0.87	0.52	0.93	1.27	2.53	–
1958	2.28	1.36	1.74	5.51	0.99	0.65	0.45	0.74	1.02	2.37	–
1959	1.58	0.92	1.61	–	0.93	0.62	0.45	0.77	1.04	2.22	–
1960	1.65	0.66	0.95	1.92	0.68	0.76	0.47	0.66	0.95	1.69	–
1961	1.40	0.58	1.33	2.65	0.99	0.70	0.38	0.70	0.79	1.59	–
1962	2.50	1.01	1.30	1.31	0.97	0.62	0.42	0.67	0.90	1.48	–
1963	2.67	1.11	1.48	2.22	0.70	0.65	0.53	0.70	0.81	1.67	–
1964	1.73	0.68	1.05	2.69	0.87	1.03	0.50	0.66	0.84	1.76	–
1965	0.94	0.52	1.18	2.57	0.58	0.70	0.61	0.64	0.66	1.73	–
1966	1.79	1.48	2.08	1.10	0.62	0.78	0.47	0.61	0.85	1.73	0.30
1967	1.67	0.78	1.07	1.28	0.79	0.80	0.45	0.66	0.71	1.59	0.36
1968	1.36	0.84	0.81	1.87	0.62	0.89	0.43	0.65	0.79	1.57	0.39
1969	1.31	1.80	1.57	1.55	0.74	0.76	0.45	0.64	0.85	1.57	0.58
1970	1.26	1.04	1.06	0.76	0.73	0.68	0.41	0.70	0.89	1.70	0.33
1971	2.36	1.05	1.06	1.55	0.71	0.75	0.38	0.76	0.88	1.52	0.51
1972	2.28	1.22	1.38	1.43	0.83	0.61	0.35	0.71	0.71	1.41	0.65
1973	1.41	2.09	1.70	1.67	0.83	0.89	0.31	0.87	0.81	1.05	0.69
1974	1.16	0.93	1.47	1.06	0.63	0.56	0.21	0.79	0.72	0.90	0.63
1975	1.62	1.30	1.02	1.41	0.72	0.61	0.20	0.77	0.73	1.21	1.00
1976	1.24	2.17	0.95	1.15	0.75	0.63	0.34	1.24	0.76	1.20	1.34
1977	1.32	1.46	1.13	–	0.70	0.71	0.47	0.81	0.67	1.45	0.71

Source: Homen de Melo, 1981. With permission.

Schuh, 1978; and Matsunaga, 1983]. Inquiry into the nature of the tradeable *versus* nontradeable subsectors of the agricultural sector would make an important contribution to understanding the long-run choices open to Brazilian policy makers, as well as policy makers in other Latin American countries.

One important issue in this discussion of trade has to do with the relative elasticity of export supply. A significant aspect of the trade-pessimistic perspective discussed earlier in this paper is a belief that the export supply elasticity is low. Consequently, even when conditions for trade become more favorable, the trade pessimists believe the economy will not be able to respond. Thompson and Schuh [1978] examined this issue in their evaluation of trade policy and corn in Brazil. The point they make is that the supply elasticity for the domestic sector should not be confused with the elasticity of export supply, which is an excess supply concept. In the case of corn in Brazil they find that a domestic supply elasticity of 0.25 is consistent with an export supply elasticity greater than twelve, a function, of course, of the small proportion of domestic corn production that Brazil exported at the time. This implies that relatively modest shifts in the domestic terms of trade can produce a relatively large increase in the exportable surplus.

Consideration of two additional studies and of Brazil's gasohol program bring this section on Brazil to a close. First, it is useful to consider another dimension of the Santana study [1984] of the Brazilian soybean industry, referred to above. This case is of interest because Brazil has discriminated against the production and raw soybean sector in order to favor the development of a processing and milling industry at home. The objectives of this policy were twofold: to contribute to the development of the rapidly growing feed and ration industry; and to be able to internalize more value added at home, rather than to export only raw materials.

A complex set of policies was followed to attain these objectives. Part of the economic environment, as noted above, was a significantly overvalued exchange rate, which was a tax on exports. But explicit taxes on exports (*confiscos*, or confiscations) were also used on the raw soybeans, while export subsidies were used for the processed products of soybean oil and meal. Institutional credit for the production of soybeans was highly subsidized, but the effect of this was offset at least in part by other policy instruments.

An interesting question, of course, is whether Brazil gained more by exporting the processed products than it would have gained by exporting the raw materials. Santana [1984] tried to answer that question, after taking into account the fact that Brazil could no longer be considered a small country from a trade standpoint in the export of this commodity. The results were mixed. When second-order effects were not taken into account, it appears that foreign exchange earnings are smaller to the overall soybean complex than they would have been if trade and exchange rate policy had been liberalized and Brazil had exported raw beans rather than processed products. When second-order effects were taken into account, however, it appears that Brazil increased its foreign exchange earnings by pursuing the policies it did. A more robust answer to this question is still needed.

Santana made no attempt to estimate the alternative effects on employment of the two strategies. It isn't clear which way these would go in light of the availability of additional land to increase soybean production.

The second concluding study is by J. do C. Oliveira [1984], who has provided a fairly comprehensive analysis of the incidence of Brazil's implicit taxation of agriculture in the period from 1950 to 1974. To make consistent use of the small-country assumption, the author omitted coffee from the analysis. Implicit in this procedure is the presumption that coffee policy would be the same, independently of the changes made *vis-à-vis* other commodities. Moreover, to capture the effect of quantitative controls on trade and to exclude redundant protection in the estimate of the shadow exchange rate, implicit rates were considered instead of the legal explicit rates (nominal rates) for import tariffs and export taxes [p. 403]. All the crops considered were assumed to be tradeable [p. 403]. Thus, the domestically produced share of agricultural inputs that were in part imported were considered in terms of value added (the return to the primary factors) to be equally protected (or discriminated against) by the implicit tax structure. This choice reflects the objective of minimizing any upward bias in the estimation of the effective incidence of the tax.

For purposes of his study, J. do C. Oliveira [1984] assumed that the exchange rate consistent with free trade would be appropriate. To obtain estimates of this exchange rate he used the same methodology proposed by Bacha and Taylor [1971]. In this context an equilibrium rate is defined as the exchange rate that maintains the current account (merchandise and services) of the balance of payments equal to zero.

Besides the equilibrium exchange rate, the author also calculated its difference from prevailing rates for both exports and imports and obtained a rather surprising result. During the greater part of the 1950s the exchange rate paid by the majority of the importers was undervalued. Recall that Brazil used a complex set of multiple exchange rates in this period, but the traditional presumption has been that they were all overvalued, differing only in degree. Moreover, during the 1950s the taxation of exports was quite high relative to the import tariffs. Thus, he concludes that during the 1950s the market for foreign exchange, which was completely monopolized by the government, taxed imports as well as exports, but with the tax on exports being much higher in this period than they were in the later period.

During the 1960s the export tax reflected in the overvalued exchange rate was reduced. But as a consequence of the high import tariffs (tariffs are an implicit tax on the export sector), the implicit global tax on the export sector (which was almost completely agricultural) remained at the same high level as in the 1950s. However, following a significant reduction in tariffs in the second half of the 1960s, the average implicit tax on exports due to overvaluation of the exchange rate declined. But associated with an increase in average export taxes and a growing deficit in the current accounts, the rate of overvaluation of the exchange rate accelerated again in the first half of the 1970s.

These results suggest that the extent of discriminatory trade policies against agriculture during the period of industrialization of the Brazilian economy was more serious than has usually been recognized. Relative to the 1950s, resources were extracted from the sector by means of a product price distortion, not only on exported products, but on products produced for the domestic market as well. The discriminatory policy was particularly intense in the case of imported foods. The domestic prices for these commodities did not follow the same trends as did world prices (for example, for onions and potatoes). The divergence of prices was especially large in the final years of the 1960s and the beginning of the 1970s. This was a consequence of strong domestic price controls and the rapid increase in world prices. These results for domestic commodities are contrary to those reported above by Homen de Melo [1981], but they came from a more comprehensive model.

Up to the mid-1960s, imports of food were normally admitted at a subsidized exchange rate (*custo de câmbio*). This, together with the agreement with the United States on PL-480 wheat, starting in 1954, probably had a depressing effect on domestic prices. (An analysis of the PL-480 program in Brazil can be found in Hall, 1980.) This was not only due to the fact that aggregate supply was increased by the imports, but also because the imports were made available at average dollar import costs that were significantly reduced and under conditions of loan payments that were especially favorable. This facilitated the subsidies that were being provided to the consumers.

Thus it does not appear that Brazil really pursued a consistent policy of import substitution for wheat. This is the case despite the fact that imports were limited in quantity both by terms of the AID agreement and by government control in response to organized pressures from producer groups. Moreover, J. do C. Oliveira's results [1984] are consistent with the hypothesis that the function of agriculture during the period of import-substituting industrialization is to supply cheap foreign exchange materials that made it possible for the industrial sector to avoid increases in nominal wages and in the share of labor in total factor costs. In any case, the evolution of the aggregate relative tax was such that its incidence on agricultural output after the end of the 1950s was substantially higher (about 25 percentage points, on the average) than at the end of the 1940s and the beginning of the 1950s.

Oliveira's results also make it possible to identify at least three clear phases in the postwar history of indirect implicit taxation of Brazilian agriculture. The first refers to the period extending from 1953 through 1961-64, when the foreign exchange tax on the sector grew consistently and significantly, ending the period approximately 40 percentage points higher at the end than at the beginning of the period. The second period extended from 1965 to 1970, and was characterized by a decline in the rate of taxation, it being approximately half at the end of the period of what it was at the beginning. The third period, extending from 1971 through 1974, was a period of rising tax rates again.

J. do C. Oliveira [1984] also found that the relative implicit taxation on the agricultural export sector and on the domestic market sector moved in a very similar manner during the period. The tax on the sector producing for the domestic market was pushed upwards by the influence of the imported products (wheat, milk, onions, and potatoes), for which the average relative tax was in general higher than for the other commodities for the domestic market. Thus nonimported agricultural commodities for the domestic market were in general taxed less than those exported and imported. These results confirm the role of trade and exchange rate policies as the main means of extracting resources from agriculture.

Finally, there is Brazil's gasohol (proálcool) program which, aside from its wheat policy, is one of the few examples of import-substituting programs which have favored agriculture in the region. This program was a much larger and far more ambitious import-substituting program than that for wheat. Its objective has been to replace petroleum imports with a fuel produced from domestic biomass sources (almost completely sugarcane). Given that such biofuel is not competitive with petroleum-based fuels, the subsidies inherent in the program have been quite large. Brazil's alcohol fuel program has been the largest and most extensive in the world, and has involved extensive credit subsidies to both the farm and nonfarm sector, the development of specialized motors that use pure alcohol rather than a mixture, and the actual export of alcohol as a fuel to countries such as the United States.

A government publication [Brasil, 1984] sets out the goals for this program. From 1973 to 1983, alcohol's share of national consumption of primary energy grew from 8 to 12.4 percent. [See Tourinho et al., 1985. This source has many references to other Brazilian studies of the alcohol program.] Production of alcohol in the 1983/84 crop year was 7.86 billion liters, and in 1983 alcohol accounted for approximately 17 percent of the fuels used for road transportation. With projects actually contracted for, production was expected to reach nine billion liters in the 1984/85 cropyear [Tourinho et al., 1985]. Under a goal of self-sufficiency, in 1984, the government projected that it would require a production of 19.3 billion liters by 1993, with alcohol also used as a fuel for hauling freight [Brasil, 1984].

This program has been evaluated extensively. All the analyses showed the cost of the program to be quite high. Homen de Melo and Pelin [1984], for example, estimated the social cost of alcohol to range between US$79 and US$90 per barrel equivalent. Tourinho et al. [1985] also estimated the marginal social cost of the alcohol to be high. Mendonça de Barros et al. [1983] made an evaluation of the costs of raw material for the production of alcohol. Serôa da Motta [1985, 1986] has evaluated the costs and benefits of the program.

Barzelay and Pearson [1982] analyzed the program from a social cost-social benefit perspective, specifically seeking to determine the increase in the price of imported petroleum that would make alcohol fuel economically efficient. Among other things, they found the social opportunity costs of alcohol production in the base case (US$0.21) were about twice as high as the social returns (US$0.10) [p. 140]. For alcohol to break even in terms of social profitability in 1990, the annual rate of increase in real oil prices would have had to be 5 percent on a 1981 base. They note that even with assumptions very favorable to the production of alcohol, it will take a long time to offset accumulated social losses.

Tourinho et al. [1985], in one of the more comprehensive analyses of the impact of the alcohol program, reached rather contradictory conclusions. For example, they concluded [p. 88] that the greater or lesser expansion of sugarcane production, depending on alternative scenarios, does not appear to affect the performance of the principal crops for domestic consumption and export, all of which remain on practically the same growth trend line. However, they concluded that the prodlcool program induces substantial domestic costs in the form of higher prices for agricultural commodities, as well as for the alcohol itself.

Tourinho et al. [1985] concluded that the impact of the program on the trade balance is positive, ranging from US$700 million in 1983/84 to almost US$1 billion annually in 1993/94. They note that these numbers are not larger in large part because of the rapid growth of agricultural imports that will be a consequence of the program. They further note that these imports would grow rapidly whether prodlcool was in place or not.

Brazil's alcohol program obviously has very high costs to society and sacrifices a great deal of economic growth. The program was put in place by a military government, and carried large elements of national security as its justification. Policy makers, and Brazilian citizens as well, will need to decide soon whether the program justifies its costs in terms of sacrificed economic growth, and whether substituting petroleum imports with domestically produced biofuels is the most efficient way to generate the foreign exchange needed to service its foreign debt and to help finance its further economic growth. It already has itself "locked in" with a large distillery sector which has few alternative uses, and a large fleet of automobiles specialized to consume alcohol.

It is precisely in this context that Serôa da Motta and Ferreira [1988] made their recent economic reappraisal of the alcohol program. They show that the social viability of ethanol production will only become a reality with considerable increases in productivity and with international oil prices above US$30, conditions which in their view are only likely to come about at the end of the 1990s. However, they estimate that if one accepts that the investments already undertaken in the sector are considered to be sunk costs, then the prospect of viability is not so remote, and perfectly feasible with oil prices in the US$15-20 range. Completely abandoning the program would in their view be a disaster, since it would result in losses on fixed assets in production capacity and sugarcane already planted, in the cost of converting alcohol cars to gasoline, in unemployment, and in the shifts required in the labor force. Thus, they argue that the program should be continued, but with the production of alcohol limited to the sector's current capacity.

To conclude this section, it should be noted that Brazilian trade policy was dominated during the 1950s and 1960s, and to a lesser extent even in the more recent period, with a "vent for surplus" export policy [Leff, 1968], in which exports were permitted only after the domestic market had been satisfied. For a long period of time the criterion used was whether domestic prices were rising or not. With high rates of inflation, the signal to export was seldom given. Thus, one sees the extent to which concern for the cost of living came to dominate agricultural policy, and to result in serious discrimination against the sector. The impressive thing is that agriculture did as well as it did given the level and persistence of the discrimination it suffered. It was only with the massive alcohol program in the 1970s and 1980s that government policy shifted favorably toward agriculture, and even then, not on a consistent basis nor with the main objective to benefit agriculture.

COLOMBIA

The case of Colombia is in many aspects similar to the cases of Brazil and Argentina. It, too, relied heavily on the classical trade instruments—exchange rate policy, tariffs, export taxes, subsidies, and quantitative controls—to implement an import-substituting industrialization policy. An important difference in the case of Colombian economic policy, however, was the existence of an explicit goal of food self-sufficiency which led to outright prohibition of food imports.[80] In other respects, moreover, there was a more positive attitude towards the exports of some commodities, called minor exports, than in the other two countries. These commodities benefited from various incentives in the form of preferential exchange rates and, at times, fiscal incentives.

Table 41. Colombia: Nominal and real exchange rates, 1950–82

Periods	Average nominal exchange rate	Average real exchange rate[a]
1950–58	3.37	21.77
1959–66	9.23	28.62
1967–73	19.81	33.53
1974–82	41.42	28.75
1950–82[b]	17.94	27.80

Source: IMF, International Financial Statistics, various years.

[a] The real exchange rate is calculated as follows:

$$\text{Nominal exchange rate} \times \frac{\text{Wholesale price index of the U.S.}}{\text{Wholesale price index of Columbia}}$$

[b] The standard deviations of the nominal and real exchange rates are 15.84 and 6.02 respectively.

Base year: 1975.

Up to the end of the 1960s, however, the country had followed a rather discriminatory policy against its major exports (especially coffee). For example, in calculating the effective purchasing power parity exchange rate for coffee and minor exports for various subperiods of the decades of the 1950s and 1960s, Diaz-Alejandro [1976, p. 17] found that minor exports persistently benefited from a more favorable exchange rate, with the difference sometimes being on the order of 100 percent or more.

In 1967, however, Colombia undertook a major change in its policy towards exports. This basic change in attitude was implemented by means of the adoption of a crawling-peg exchange rate, similar to the Brazilian system, by means of which the exchange rate was devalued at very small and unpredictable intervals. Data on the evolution of the exchange rate for selected time periods are presented in Table 41. Although not shown in the table, the purchasing power parity or real exchange rate showed a pattern during the 1950s and 1960s that is characteristic of a fixed exchange rate system. The currency appreciated in the years following a devaluation, eventually reaching a peak, at which the the next devaluation was undertaken. However, the fluctuations in the real exchange rate for Colombia were much less pronounced than for the Argentine or Brazilian cases, a function of the Colombian authorities doing a better job of controlling domestic inflation. The real exchange rate actually declined during the 1950s, and to a lesser extent in the 1960s. After 1967, with the introduction of the crawling peg, the behavior of the real rate was much more stable. It stayed approximately constant initially, and then experienced a slight depreciation towards the end of the 1970s.

Data on Colombia's agricultural trade are presented in Table 42. The underlying data on which the table is based show virtually no trend up through 1967,

Table 42. Colombia: Trade in agricultural products, 1952-80

Period	Average value of exports	Average value of imports	Average trade balance[a]
		(US$000)	
1952-58	460,829	80,557	380,271
1959-66	378,458	64,402	314,056
1967-73	579,594	89,494	490,099
1974-80	1,899,106	284,429	1,614,677
1952-80	813,944	127,468	686,475

Source: FAO, Trade Yearbook, various years.
[a] Trade balance of agricultural products, i.e., value of agricultural exports minus value of agricultural imports.

with a significant growth in the period thereafter. Exports in the first period oscillated somewhat, with the oscillation roughly coinciding with fluctuations in coffee prices. This commodity's share of the total value of exports was still around 60 or 70 percent by the end of the decade of the 1970s.

The remarkable growth of exports after 1967 is even more impressive if coffee and noncoffee agricultural exports are considered separately. For the latter group, 1967 was apparently a benchmark year which set the conditions for a substantial growth up through 1980. Although noncoffee exports included the minor exports[81] which benefited from special treatment in terms of trade and exchange rate policies, Diaz-Alejandro [1976] attributed the remarkable performance of these commodities, in particular bananas, cotton, sugar, and tobacco (BCST), not only to external policy, but also to domestic agricultural policies. In particular, he noted:

> Another characteristic of the BCST group is that besides being subjected to influences emanating from foreign trade policy, it has benefited from special agricultural policies which regulate its internal prices and provide subsidized credit. The case of cotton is perhaps the most dramatic example of the pay-off to such ad hoc, crop-specific programs . . . during the postwar years Colombia passed from being a net importer to a net exporter of that commodity within a short period of time. During the 1950s, cotton growers (mainly large-scale growers, it may be noted) received generous tax concessions as well as credit and price support from an institute designed exclusively to promote that crop. Such policies have continued, raising not only output, but also yields. (While Colombia became an important cotton exporter during the 1960s, competing exports from some traditional sources, such as Mexico, stagnated.)

> Sugar and bananas have also benefited greatly from special
> government credit programs. [pp. 44–45]

Diaz-Alejandro [1976, pp. 63–71] attempted to quantify, within a simplified regression framework, the importance of the exchange rate policy as well as the other policies just mentioned. A major conclusion of the analysis was that the rate of growth of the supply of exports was sensitive to changes in the exchange rate, as well as to a measure of its instability. He also found evidence to support his claim that other policies had significant effects on the performance of these minor exports.

Despite this positive aspect of its trade policy, Colombia basically followed an import-substitution industrialization policy. As usual with such a development policy, protectionist policies were followed which tended to focus the concerns of policy makers on the nonagricultural sector. An analysis of the biases in policies followed by the government was developed in an important study by J. G. Garcia [1981], who states his criterion as follows:

> The ratio of import and export exchange rates can be a
> measure of the bias. In determining the exchange rate for
> imports, all taxes and other surcharges on imports should be
> included and import subsidies and other measures that reduce
> the cost of importing should be deducted. The exchange rate
> for exports should be net of taxes and include all subsidies,
> direct or indirect, that in one way or another can raise the price
> received by exporters. While the ratio between the two sets
> measures bias, it tells nothing about the incidence of the
> discrimination created by any particular protective structure.
> [p. 40]

The results obtained by J. G. Garcia lead to the following comments based on the author's observations:

(1) Import exchange rates have, for most of the period, been more favorable than export rates. There is an antiexport bias in the set of trade policies followed by Colombia. These have caused the prices of importable goods relative to exportable goods to increase by 15–35 percent.

(2) Coffee exports have, during the whole period, been penalized by a lower than average exchange rate, as was noted earlier.

(3) Industrial and nonagricultural exports were the most favored by exchange rate policy.

J. G. Garcia [1981, p. 42] alerts the reader to possible shortcomings in this kind of analysis. He notes that some of these coefficients may be misleading and sug-

Table 43. Colombia: Summary of import tariffs and export subsidies, 1956–78 (in %)

Premium, tariff or subsidy	1956–67	1967–78
Import premium	37 to 54	–
Nominal tariff	16	20
"True" import tariffs	53 to 70	20
Export subsidies		
Agriculture		
Coffee	−15	−16
Other	33	16
Industry	43	30
Net subsidy		
Agriculture		
Coffee	−68 to −85	−36
Other	−20 to −37	− 4
Industry	−10 to −27	10

Source: J. G. Garcia, 1981. With permission.
Note: Negative signs indicate taxation.

gests the need to go beyond these measures to have a clearer picture of the structure of bias in Colombian trade policy. He was especially concerned that the costs of imports might be higher than what he obtained because of the existence of quantitative restrictions. This would also affect exports, as Garcia explains:

> The nominal cost of importing only tells part of the story. As will be shown below, an import substitution bias in some industrial sectors discriminated against the agricultural and the industrial export sectors and other import-competing industrial sectors. The nominal cost of importing is low because 70 percent of imports are of capital goods, raw materials, and intermediate inputs used by industry. Duties on these imports were low, but they were subject to severe quantity rationing. Most imports used as inputs in agriculture were severely restricted and had high import duties because they were domestically produced.
>
> When a move to promote all minor exports was made during the 1953–60 period, noncoffee agricultural exports increased, but industrial exports did not grow rapidly until after the export promotion of 1959 and 1960. Starting in 1960, industrial exports were promoted more than noncoffee agricultural exports. [J. G. Garcia, 1981, pp. 42–43]

Garcia's results are shown in Table 43. The import premium is defined as the gain importers obtain in view of the quantitative restrictions. In other words, the same level of imports would prevail if one unit of foreign exchange was priced from 37 to 54 percent more. This estimation is based on import demand elasticities calculated by Garcia.

The following observations can be made:

(1) There was a substantial reduction in the level of protection after 1967.
(2) All exports were taxed in the period 1956–67 and all but industrial exports were taxed in the period 1967–78.
(3) Coffee was the most heavily taxed commodity during the whole period.
(4) Export taxes were substantially smaller in the more recent period.

It is important to recognize the important link that Garcia's work provides in understanding the effects of trade policy. As we noted at the beginning of this discussion of the Colombia case, there was a substantial number of special programs designed to provide incentives for minor exports. These programs were indeed effective in many instances and thus other agricultural exports have been much less penalized than coffee exports. The fact remains, however, that all exports were penalized, so policy was basically discriminatory towards agriculture and the export sector. This suggests that partial analyses of policies should always be understood as first approximations which, in many instances, may well give a very distorted picture of the effects of any particular measure.

The final aspect of this analysis focuses on food production. As mentioned earlier, self-sufficiency has been an explicit policy goal of Colombia and has in fact been achieved for most crops. Garcia's measures of nominal protection [Garcia, 1981, p. 50] are presented in Table 44. It is clear that a considerable degree of protection for these commodities indeed existed. The level of protection, however, tended to decline in more recent years, with some products even being taxed during the decade of the 1970s (sorghum, soybeans, sugar, barley, and rice).

Garcia and Llamas [1988] analyze the effects of variations in the external terms of trade (specifically the coffee boom of the 1970s) and in government expenditures on relative product prices, real agricultural wages, and the distribution of income between labor and nonlabor factors of production in Colombia during the period 1967–83. Their careful analytical and empirical analysis includes a comparison of the various policies that might alleviate the eventual adverse effects on other sectors of a boom in one sector (the so-called Dutch disease). They find that agricultural output does respond to price changes, which implies that spurring the economy by increasing government expenditure may not be the answer to agricultural growth because it may only serve to reduce relative agricultural prices.

V. Thomas *et al.* [1984] made a rather comprehensive study of the adjustment problems faced by Colombia in the early 1980s. Among other things, these au-

Table 44. Colombia: Nominal protection in the food sector, 1953-78

Year	Producer/international prices							Wholesale/international prices		Consumer/international prices	
	Sorghum	Soybeans	Milk	Wheat	Corn	Sugar	Barley	Vegetable oil	Rice	Meat	Cotton fiber[a]
1953	na	na	na	2.24	1.48	1.79	na	na	1.77	na	1.14
1954	na	na	na	2.58	2.09	1.91	na	na	1.97	na	1.25
1955	na	na	na	2.53	2.00	1.78	na	na	1.39	3.03	0.73
1956	na	na	na	2.12	1.77	1.39	na	na	1.50	2.87	0.57
1957	na	na	na	2.49	1.95	1.24	na	na	1.58	2.59	0.77
1958	na	1.45	2.37	2.37	1.31	1.82	na	na	1.13	1.98	0.52
1959	na	1.78	1.85	2.59	1.56	2.33	na	na	1.20	2.22	0.68
1960	na	1.35	2.17	2.34	1.69	2.13	1.51	na	1.52	2.28	1.26
1961	na	0.99	2.35	2.15	1.77	2.12	1.62	na	1.99	2.18	0.93
1962	na	1.05	1.95	1.74	1.20	2.10	na	na	1.05	2.23	0.90
1963	na	1.17	1.65	1.74	1.55	0.82	1.54	2.73	1.04	2.12	1.00
1964	na	1.52	2.08	2.15	1.95	1.24	1.71	2.99	1.41	1.83	1.11
1965	1.27	1.17	1.93	2.05	1.32	2.50	0.97	2.00	1.55	1.42	0.83
1966	1.25	1.14	2.60	2.17	1.45	3.12	1.32	2.98	1.66	1.97	1.17
1967	1.05	1.13	2.37	1.76	1.58	2.43	1.28	2.08	1.37	1.92	1.02
1968	1.54	1.21	2.95	1.85	1.56	2.55	1.22	2.56	1.26	1.55	0.93
1969	1.23	1.25	2.86	1.88	1.31	1.35	1.18	2.63	1.09	1.54	0.89
1970	1.12	1.29	2.69	1.81	1.32	1.54	1.12	2.56	1.06	1.20	0.93
1971	1.07	1.11	1.76	1.30	1.33	0.96	1.12	2.64	1.05	1.01	0.84
1972	1.34	0.94	1.44	1.47	1.58	0.59	0.57	2.56	0.83	0.89	0.74
1973	0.94	0.54	1.88	0.74	1.25	0.47	0.65	1.53	0.52	0.80	0.76
1974	0.77	0.74	1.57	0.85	0.87	0.17	0.68	1.52	0.54	0.73	0.71
1975	0.90	0.97	1.25	1.34	1.06	0.30	1.06	1.49	0.65	1.63	0.95
1976	1.02	1.07	1.41	1.40	1.21	0.63	0.98	1.88	0.83	1.79	1.00
1977	1.65	1.22	1.93	2.03	2.30	1.37	na	1.61	1.13	1.95	1.02
1978	na	na	1.67	na	1.82	na	na	na	0.95	na	0.95

Source: J. G. Garcia, 1981. With permission.

Notes: The international prices CIF, except where noted; na = not available.

[a] The figures from 1953 to 1959 represent the ratio of the domestic price to the import CIF price. The figures from 1960 to 1978 represent the ratio of the domestic price to the export fob price.

thors present updated estimates of nominal protection rates for importable goods and of welfare costs of the wheat policy.

The agricultural sector in Colombia is well integrated with macroeconomic developments in the economy as a whole. The performance of agriculture—especially its external sector—strongly influences the economic aggregates and macroeconomic policies. These, in turn, affect the developments in agriculture as well. V. Thomas *et al.* [1984] explain these links:

> This enables us to see that the country has exhibited symptoms of the "*booming-sector syndrome*" experienced elsewhere in the world, where the rapid growth of a few primary exports and the resulting inflow of foreign exchange have been coupled with a real appreciation of the exchange rate causing major changes in labor deployment and resource use which have hurt other productive sectors by drawing resources into the booming sectors. In Colombia, the coffee and drug export windfalls in the second half of the 1970s produced these adverse effects on noncoffee agriculture and the rest of the economy and the impact has persisted even after the end of the commodity boom period. Growth in noncoffee exports (agricultural and other) has been decelerating since the mid-1970s, first as the relative producer prices in the domestic markets for these commodities worsened, and later as international export conditions became increasingly unfavorable. [p. iii]
>
> The main link between the macroeconomy and agriculture that is established in this report operates through the effect of coffee production and exports; an important impact may have been experienced by illegal drugs also. These commodities have contributed significantly towards raising the aggregate supply of output in Colombia. Less obvious, but equally important, has been their impact on increasing the money supply, aggregate demand and inflation, in causing an appreciation of the real exchange rate, and in hurting the *performance of noncoffee production sectors*. Increases in the price of coffee resulted in higher disposable incomes in Colombia and an increase in the demand for all goods; since the domestic price of tradeables is to a significant degree determined by their world prices and the exchange rate, this rise in incomes tended to raise aggregate demand, and consequently, the relative prices of *nontradeables*. This shift in relative prices was reinforced by

Table 45. Colombia: Nominal rates of protection for selected importable crops, 1980–82 (in %)

	1980	1981	1982
Wheat	36	45	91
Corn	87	67	79
Soybeans	37	46	85
Sorghum	67	57	110
Beans	42	186	106

Source: V. Thomas *et al.*, 1984, p. 61. With permission.

the impact of accelerated foreign exchange inflows and of the domestic supply to the higher inflation rates (average over 25%) witnessed in the second half of the 1970s. [pp. iv–v]

The differential between domestic and external inflation during 1975–83 was high, with domestic prices measured at the official exchange rate rising 115 percent, compared to a 42 percent increase in one index of external prices. Partly as a result, the producer prices of noncoffee tradeables (which, as mentioned earlier, are strongly influenced by international prices) have been falling relative to the price of domestic goods and services in this period. Since agricultural output has a higher share of tradeables than the rest of the economy, this fall in relative producer prices has been especially adverse for the sector. Unfortunately, the shift of incentives in favor of nontradeables has not produced any significant output response from this domestic sector (for example, services) as a whole, so that there has been little offset to the production and employment losses which resulted from slower growth in the tradeable goods sector. [p.v]

The updated estimates of nominal protection were calculated as the percentage differences between farmgate and international CIF prices and are shown in Table 45. It should be noted that this ratio measures both the protection created by government policy and the "natural" protection given by transportation costs and port charges (since data are not available to estimate those).

Lack of data prevented the authors from calculating nominal protection rates for exportables. But they noted:

Nominal rates of protection for exportables are expected to be significantly lower than for importables and may be negative at least for several crops in several years. [p. 61]

Lack of data also precluded the calculation of effective rates of protection (ERPs), but the authors noted:

> Since agricultural inputs in Colombia receive much lower protection than outputs, especially importables, the ERP would exceed the nominal rates given earlier. Therefore, importables unambiguously receive high rates of protection. The ERP for exportables, however, is more difficult to determine for the same reasons mentioned in the case of nominal rates. [p. 61]

Finally, the authors estimated the welfare effects of reducing the level of protection of wheat. In making these estimates, the level of protection was not reduced to zero, however, because it was recognized that there were distortions in the rest of the economy. Thus, the second-best policy would be the reduction of protection to the average level of the economy. Due to lack of reliability of the data (the breakdown of GDP into tradeables and nontradeables and the levels of protection of tradeables), sensitivity analysis was performed by varying the level of average distortion in the economy. The base case considered a 15 percent level of distortion, with 10 and 20 percent levels also considered. The results were explained as follows:[82]

> The distortion level is based on the assumption that the resources diverted from wheat production and consumption are used to purchase or produce an average bundle composed of exportables, importables and nontraded goods. The range for the distortion from 10% to 20% depends on plausible parameters for various contributions of the distribution of importables, exportables and nontraded goods coupled with average levels of import tariffs and export subsidies. Based on analysis of producer and consumer surplus, the results show that it is the consumer who would stand to gain most from the policy, whereas the government and efficiency gain varies from Col$163 to Col$226 million depending on the distortions in the rest of the economy. The net efficiency loss is modest compared to the value of consumption (less than 1%). Even when the elasticity of supply was increased from a short-run elasticity of 1.2, the efficiency gain as a percentage of GDP increases only slightly over 1% in the 10% distortion case. [V. Thomas et al., 1984, pp. 62-62]

To conclude, just as in the case of Argentina, Brazil, Mexico (see below), and other Latin American countries, the pattern of government intervention in Colombian agriculture has been complex and multifaceted. Some sectors, primarily the export sectors, have experienced serious discrimination, while others, generally those producing for domestic production, have been protected. Given the high levels of protection provided the industrial sectors, it is likely that agriculture as a whole experienced serious discrimination. Of special interest in the case of Colombia, however, has been the drug trade. Production of these commodities has been an important source of income for the agricultural sector. Equally as important, the foreign exchange earnings from this trade has helped to sustain the value of the Colombian currency, to the detriment of the more conventional parts of agriculture.[83] To the authors' knowledge, Colombia is the only case where black market rates of the currency have periodically been below the official exchange rates.

The level and character of intervention in the agricultural sector by means of trade and exchange rate policies, as in the case of other countries—and especially Brazil—have varied a great deal over time. This makes it difficult to generalize about the general thrust of policy. In the case of Colombia, policy towards agriculture has probably been more favorable in the post-World War II period than is the case in most other Latin American countries. As noted earlier, this had a positive effect on the distribution of income. Colombia also had a more effective policy with respect to technological change [see Scobie and Posada-Torres, 1977]. This more positive attitude toward agriculture is probably a reflection of there being more well-organized commodity associations in Colombia. The power of these lobbies is a distinctive feature of Colombian agriculture policy.

CHILE

Trade policy in Chile has also been characterized by exchange controls and various other forms of interventions in trade flows. Import-substituting industrializaton has been an important goal of policy makers. However, two important and distinctive aspects of the Chilean economy should be recognized. The first is the relatively small share that agricultural exports have made up, until the post-Allende years, of total exports; the main export has been copper. The second is that prior to World War II, Chile was a net exporter of agricultural products [Valdés, 1973]. Since that period, however, the situation has changed and the trade balance in agricultural products has always been negative, at least up through 1980 (Table 46).

Chilean economic policy has been characterized by a great deal of instability in the postwar period. This has been reflected in part in large swings in the real exchange rate. (Averages for selected periods are presented in Table 47.) But other aspects of trade policy were unstable as well.[84] For example, the 1931-55 period was dominated by substantial quantitative interventions in trade. In 1956-58, a

Table 46. Chile: Trade in agricultural products, 1962-80, US$000

Period	Average value of exports	Average value of imports	Average trade balance[a]
1962-73	38,412	190,259	-151,847
1974-80	212,283	537,298	-325,015
1962-80	102,470	318,116	-215,646

Source: FAO, Trade Yearbook, various years.

[a] Trade balance of agricultural products, i.e., value of agricultural exports minus value of agricultural imports.

Table 47. Chile: Nominal and real exchange rates, 1955-81

Period	Average nominal exchange rate	Average real exchange rate[a]
1955-59	0.001	0.35
1960-64	0.002	0.19
1965-69	0.006	2.03
1970-73	0.103	3.68
1974-81	25.840	5.53
1955-81[b]	7.673	2.66

Source: IMF, International Financial Statistics, various years.

[a] Real exchange rate calculated as follows:

$$\text{Nominal exchange rate} \times \frac{\text{Wholesale price index of the U.S.}}{\text{Wholesale price index of Chile}}$$

[b] The standard deviation for the nominal and real exchange rates is 14.260 and 2.617, respectively.

stabilization and liberalization plan was implemented, designed by the Klein-Saks consulting firm. In 1959-61, President Jorge Alessandri continued the liberalization policy of the previous period, but with more emphasis on liberalization and less emphasis on stabilization. However, in 1961, the Alessandri government had to make a sharp turn from its liberalization policy in response to a rapidly deteriorating situation for foreign reserves and foreign debt. Again, during 1965-70, the country went back to a stabilization and liberalization plan under the Presidency of Eduardo Frei. It was in this period that Chile shifted from a policy of fixed exchange rate to a crawling-peg system.

During 1971-73, the period of Salvador Allende's presidency, there was again a return to a more interventionist policy. The exchange rate was initially frozen, but the government was forced to devalue in 1971 and again in 1972. Exchange controls became very restrictive in this period [Behrman, 1976].

After 1973, the political regime of General Pinochet adopted more orthodox economic policies. However, instability in the implementation of policies was a

Table 48. Chile: Effective protection coefficients, 1947-65

	1947-50	1951-55	1956-60	1961-65
Import competing activity				
Wheat	−0.43	−0.25	−0.16	−0.67
Beef	−0.45	−0.29	−0.39	−0.25
Export activity				
Barley	−0.30	a	b	c
Lamb	−0.33	−0.27	−0.39	−0.28
Wool	−0.65	−0.45	−0.42	−0.34

Source: Valdés, 1973. With permission.
a 1951-53: −0.23 average; 1954-55: 0.25 average.
b 1956-60: −0.18, −0.4, 0.09, 0.11.
c 1961-65:   0.42, −0.01, −0.02, 0.14, 0.08.

characteristic of this period as well. In the early years, a more liberal path was taken: the exchange rate was devalued and an overall reduction in tariffs was executed. But, as the rate of inflation accelerated in the second half of the decade, stabilization again became the main goal and a frozen exchange rate became an important component of policy.[85] This led again to overvaluation of the currency.

In an attempt to understand the change in the position of Chile from being a net exporter overall to being a net importer of agricultural goods, Valdés [1973] estimated effective protection rates in the period 1945-65 for five agricultural commodities: wheat, beef, wool, barley, and lamb. The first two commodities represented the import-competing sector and accounted for a substantial share of all agricultural imports (approximately 50 percent between 1946 and 1960). Wool and barley were traditional exports, and lamb was a marginal export. Together these commodities made up 30 percent of the total value of agricultural exports during the period 1946-60 [p. 160].

Valdés's results are presented in Table 48. All of the products experienced negative protection or taxation. The only exception was barley, which benefited from (positive) protection in selected years.

The findings of Valdés should be contrasted with those of Behrman [1976]. Although Behrman did not provide a disaggregated measure, he found the following for effective protection and domestic resources costs (DRC) for the aggregate of agriculture and forestry:

	1961	1967	1968
Effective protection rates (EPR)	50 [145]*	−7	15
Domestic resource costs (DRC)	250	–	111

* There are two estimates for 1961. The one in brackets is referred to later on as EPR2.

Table 49. Mexico: Trade in agricultural products, 1962-80

Period	Average value of exports	Average value of imports (US$000)	Average trade balance[a]
1962-64	506,967	102,867	404,100
1965-69	651,926	115,552	536,374
1970-74	861,695	462,639	399,055
1975-80	1,512,348	1,262,772	249,576
1962-80	955,952	567,168	388,784

Source: FAO, Trade Yearbook, various years.
[a] Trade balance of agricultural products, i.e., value of agricultural exports minus value of agricultural imports.

These results suggest that substantial protection was provided to this aggregate sector in 1961.[86] However, this result was due to Behrman's failure to take into account distortions in the exchange rate.

In ranking the various sectors with regard to preferential treatment, Behrman's [1976] calculations show that industry has had a preference over agriculture which, in turn, has had a preference over mining. Behrman refers to this situation, and also to the issue of the comparative advantage of agriculture, in the following way:

> On the sectoral level, protection has been highest for industry, lowest for mining, and in between for agriculture. However, in 1961, agriculture had a relatively high EPR (EPR2) and DRC. This result raises questions about the widespread assumption that agriculture is a low-DRC sector against which substantial discrimination has occurred. [p. 141]

This latter statement was based on improperly calculated protection coefficients, however.

As shown in Table 46, exports increased significantly in the 1970s. This corresponded with the massive devaluation of the Chilean currency, the implementation of a unique set of macroeconomic policies, and the liberalization measures taken in the Pinochet period. Agricultural imports grew even faster in this same period, despite the large devaluation of the currency. This was a consequence of the liberalization of import policies and the recovery of economic growth.

MEXICO

During most of the period since the end of World War II, agriculture has been a major source of foreign exchange earnings for Mexico [Table 49]. In 1960, for example, it accounted for 57 percent of total exports. It has not been able to sustain that performance, however. By 1975, agriculture's share of the total was

Table 50. Mexico: Nominal and real exchange rates, 1950-81

Period	Average nominal exchange rate	Average real exchange rate[a]
1950-54	9.42	15.29
1955-59	12.50	15.52
1960-64	12.50	13.87
1965-69	12.50	13.32
1970-74	12.50	13.02
1975-81	21.47	13.51
1950-81[b]	13.98	14.05

Source: IMF, International Financial Statistics, various years.

[a] The real exchange rate is calculated as follows:

$$\text{Nominal exchange rate} \times \frac{\text{Wholesale price index of the U.S.}}{\text{Wholesale price index of Mexico}}$$

[b] The standard deviations for the nominal and real exchange rate were 4.64 and 1.44, respectively.

down to 28.5 percent, and by 1980 it was down to only 10 percent, as the export boom in petroleum—which started in 1974—accelerated in the late 1970s. By 1980, petroleum exports accounted for 64.5 percent of total exports, and by 1982, 76.9 percent. Moreover, Mexico in 1980 was running a large deficit on its agricultural trade account (US$1.2 billion) as the value of agricultural imports burgeoned from US$563 million in 1976 to US$2.9 billion in 1982. The only previous deficit on the agricultural trade account in the period since 1962 was in 1974, when it was a modest US$34 million. Data on the nominal and real exchange rates of the Mexican peso are presented in Table 50. For all intents and purposes Mexico pursued a fixed exchange rate policy up to 1976. Given that domestic inflation was managed quite well in that period, and given that capital markets were managed rather tightly, no great shocks were imposed on agriculture from this source up through 1975. Since that date, however, the fluctuations in both nominal and real exchange rates have been relatively large, especially into the 1980s.

The authors of three rather comprehensive studies of Mexican agriculture [Venezian and Gamble, 1969; C. W. Reynolds, 1970; and Bueno, 1971] concluded that agricultural policies (and by implication trade and exchange rate policies) were reasonably efficient during the 1950s and most of the 1960s. However, during the late 1960s and up through the early 1980s, agriculture appears to have suffered discrimination in economic policy. Prices for most agricultural commodities rose less than the general price level during this period [Pardo, 1984], and Mexico began to import two of its most basic commodities—corn and wheat—in significant quantities. As recently as 1969-71 Mexico was a net exporter of corn, while in 1964-67 exports of wheat represented approximately 20 percent of production [Pardo, 1984]. For example, in 1980 corn imports represented more than 23 percent of total domestic utilization. It was to offset this growing external dependency, driven in part by rising exports of petroleum and a bad case of the Dutch disease, that the government in the early 1980s established the SAM (Mexican Food Supply System)[87] program to promote self-sufficiency in basic commodities, a program which was terminated after little more than a year due to budget difficulties.

The government has long played a strong interventionist role in Mexican agriculture. The Mexican land reform is widely known; the government for many years maintained a dual price system for corn, with small producers receiving prices significantly above the prices paid to larger commercial producers; a government agency, CONASUPO,[88] has played a major role in the food distribution system; and the exchange rate has been strongly managed in the presence of a highly interventionist trade policy.

Despite this strong intervention, there are relatively few studies that have attempted to measure the net impact of government intervention on the agricultural sector. Both Bueno [1971] and R. Wallace and Ten Kate [1979] have made comprehensive sectoral analyses of the impact of selected economic policies on the Mexican economy. Both studies were based on aggregate input-output tables for 1960 and 1970. Bueno estimated that in 1960 the nominal tariff and the implicit tariff were 6.7 and 6.5 percent for the primary sector (6.2 and 4.7 percent for crops, respectively). Within agriculture, he found that subsidies to small producers raised the price of corn by 42 percent and the price of wheat by 14 percent above world (U.S.) prices for that group. The manufacturing sector, on the other hand, was receiving 33 percent nominal tariff protection plus 25 percent protection by implicit means as measured by the level of nominal protection.

In terms of effective rates of protection, Bueno [1971] found that the primary activities received between 1 and 2 percent protection (4.4 and 1.7 percent for crops), while the manufacturing sector was receiving protection on the order of 72 percent as measured by the effective tariff using the Balassa criterion. When distortions in the exchange rate were taken into account, Bueno found the nominal tariff and the implicit tariff both to be −3 percent for the primary sector versus 22 and 15 percent, respectively, for manufacturing. Thus, domestic and trade policies were distorting the domestic terms of trade against the resources employed in the agricultural sector.

Bueno's [1971] results are reinforced by the findings of R. Wallace and Ten Kate [1979] for 1970. These authors estimated the net nominal protection for primary activities to be 0.7 percent (2.3 percent for crops), which is very low compared to the estimate of 15.3 percent they found for the average of all sectors of the economy. When distortions in the exchange rate were taken into account, the nominal protection of primary activities was found to be −12.6 percent (−11.2 percent for crops).

The effective rates of protection for the primary sectors were found to be more unfavorable than were the nominal rates. Measured by this criterion, the primary sector experienced negative protection in the amount of 2.5 percent, while intermediate goods and durable and capital goods received positive protection in the amounts of 14.5 and 52.8 percent, respectively. When distortions in the exchange rate were taken into account, R. Wallace and Ten Kate [1979] estimated the effective protection to be −12.5 for crops, −12.9 for stockraising, and −14.0 for forestry. Again, the primary sector was relatively the less protected sector compared with the rest of the economy, and in point of fact was rather heavily taxed by means of trade and exchange rate policy.

All of the above studies were sectoral analyses of the economy, with little attempt to disaggregate to the individual commodity level. Bueno [1971] provided some results for corn and other commodities, but did not analyze the impact of the policies in any detail.

Another global or comprehensive analysis of the agricultural sector was made by means of the so-called CHAC Model [Norton and Solis, 1983], which was a study of the agricultural sector using a linear programming model and a partial equilibrium framework. This study included an analysis of the efficiency of resource use in the sector, and an analysis of the impact of the overvaluation of the currency. This study encompassed 33 major crops based on data for 1968.

Except for the CHAC Model, the authors found no rigorous detailed studies of the consequences of government intervention in particular agricultural commodity sectors, except for the study by Pardo [1984], discussed below. This is somewhat surprising in light of the fact that government policies for the most part have been directed to individual commodities, with a fairly high degree of differentiation in policies among commodity sectors.

Using an effective protection model which made it possible to consider the effects of both domestic and trade and exchange rate policies, Pardo [1984] made a comprehensive analysis for the corn and wheat sectors in Mexico. The aggregate results (using three alternative measures) for the wheat sector are presented in Table 51. Those on the left-hand side of the table are evaluated at the official exchange rate, while those on the right-hand side are evaluated using estimates of the equilibrium exchange rate. A comparison of the two sets makes it possible to isolate the effects of distortions in the foreign exchange market.

The importance of the overvaluation of the peso as a source of distortion in the wheat sector can be clearly seen. Throughout the period the size of the distortions were more discriminatory against wheat producers when output and inputs were evaluated at the shadow exchange rates. In net terms, price policy discriminated against wheat producers throughout the period except in 1971, 1972, 1977, and

Table 51. Rates of distortion in the wheat sector, 1970–82 (in %)

	NRP	ERP	ERS	NNRP	NERP	NERS
1970	− 6.34	−15.08	−13.99	−17.72	−25.02	−24.06
1971	23.70	23.04	25.21	8.78	8.61	10.51
1972	13.60	14.01	16.32	1.23	1.97	4.02
1973	−16.38	−18.09	−16.13	−27.09	−28.27	−26.57
1974	−29.41	−22.00	−21.34	−40.26	−34.36	−33.15
1975	−17.50	−12.99	−11.55	−30.89	−26.88	−25.68
1976	−12.40	−12.21	−10.47	−22.76	−22.35	−20.87
1977	19.18	40.01	44.00	14.58	35.13	38.97
1978	1.49	8.77	12.41	− 7.71	− 0.69	2.61
1979	− 6.36	3.58	7.14	−18.24	− 9.00	− 5.92
1980	− 1.78	14.53	31.27	−19.72	− 5.33	− 8.11
1981	20.98	48.78	87.69	6.09	17.30	48.63
1982	−30.07	−31.37	−12.29	−25.26	−26.90	− 6.29

Source: Pardo, 1984, p. 137. With permission.
NRP: Nominal rate of protection.
ERP: Effective rate of protection.
ERS: Effective rate of subsidy.
NNRP: Net nominal rate of protection.
NERP: Net effective rate of protection.
NERS: Net effective rate of subsidy.

1981. However, positive incentives were given to producers through both the tradeable and nontradeable inputs, thus raising value added. This can be observed by comparing the values of ERP and ERS *versus* NRP and the values of NERP and NERS *versus* NNRP.

The discrimination against producers appears to have been greatest during the 1973-75 period, with a tax of between 27 and 40 percent in terms of the NNRP, between 27 and 34 percent in terms of the NERP, and between 26 and 33 percent in terms of the NERS. In other words, domestic value added was being diminished by up to 33 percent, compared to a situation of evaluating value added at world prices. Put somewhat differently, returns to primary factors used in the production of wheat were diminished by this proportion. These levels of taxation occurred during the commodity boom of the 1970s. Hence, Mexican producers were not permitted to participate in the strong international markets of that period.

The various measures of distortions move in a parallel fashion both in size and sign from 1970 through 1976. The main difference arises from taking into account distortions in the value of the peso. From 1977 through 1982, however, differences in both the size and the sign of the various indicators became significant. This suggests that policies towards both tradeable and nontradeable inputs became more important in this latter period.

Table 52. Nominal and effective rates of distortions in the corn sector, 1970-82 (in %)

Year	NNRP	NERP5	NERP4	NERP3	NERS5	NERS4	NERS3
1970	−17.73	−26.82	−31.76	−33.93	−25.60	−31.30	−33.51
1971	−45.00	−52.87	−56.73	−58.00	−52.06	−56.40	−57.69
1972	−23.08	−28.93	−32.38	−34.73	−27.55	−31.81	−34.20
1973	−30.12	−31.92	−33.50	−34.75	−30.46	−32.81	−33.70
1974	−37.48	−26.44	−20.90	−18.58	−24.27	−19.10	−16.75
1975	−19.44	−10.03	−5.36	−8.06	−7.62	−3.35	−6.15
1976	6.14	14.27	19.07	6.94	17.67	21.66	9.21
1977	6.51	18.25	25.99	11.40	22.78	28.10	13.24
1978	−10.31	−4.15	−0.92	−10.83	0.59	0.68	−9.44
1979	−11.40	−0.18	9.03	−5.17	7.48	13.05	−0.90
1980	5.45	31.74	52.50	23.65	55.84	60.93	30.32
1981	13.60	48.92	81.03	34.64	97.70	95.37	44.85
1982	5.07	16.15	22.60	11.77	49.67	33.26	20.33

Source: Pardo, 1984, p. 162. With permission.
NNRP: Net nominal rate or protection.
NERP5:   Net effective rate of protection in technology 5.
NERS5:   Net effective rate of subsidy in technology 5.
NERP4:   Net effective rate of protection in technology 4.
NERS4:   Net effective rate of subsidy in technology 4.
NERP3:   Net effective rate of protection intechnology 3.
NERP3:   Net effective rate of subsidy in technology 3.

Finally, when policies toward nontradeable inputs are taken into account by calculating the net effective rate of subsidy, overall policy is seen to be even more favorable to wheat producers. Policies *vis-à-vis* nontradeable inputs more than offset the discrimination through the other two markets in 1978, 1980, and 1981. They made the negative protection smaller and the positive protection larger in the remaining years. The importance of subsidies provided through nontradeable inputs is an important finding of the Pardo study [1984]. An important part of the subsidy reflects the failure to adjust charges for irrigation water at a sufficiently rapid rate in a highly inflationary environment.

The comparable data for the corn sector are presented in Table 52. The corn sector in Mexico involves more diversified technologies and production is more geographically dispersed than is the case for wheat. Consequently, Pardo [1984] carried out the analysis for three different levels of technology: one under irrigation and two technologies for rainfed areas associated with different mixes of modern inputs. These three technologies caused price policies that affect the production and input markets to have different effects on the sector. The bulk of corn is produced in rainfed areas, but the production of corn in irrigated areas accounted for up to 24 percent of total production in the early 1980s.

The results presented in Table 52 are those using estimates of the equilibrium exchange rate. With the exception of the net nominal rate of protection, they are presented by technology. Technology 5 refers to the technology with irrigated water, technology 4 to the technology in rainfed areas with use of tractors, and technology 3 to production in rainfed areas without use of tractors.

Distortions in product price, measured by NNRP, indicate that corn producers experienced serious discrimination during the first half of the 1970s. From 1970 to 1975, real guaranteed prices in Mexico were declining while world prices reached quite high levels in 1974 and 1975. In addition, the overvaluation of the peso was estimated to be greater than 20 percent in those two years.

Corn producers experienced less discrimination in terms of product price in 1978 (−10 percent) and 1979 (−11 percent), and except for those two years the sector was favored after 1976. The large devaluation of 1976, a revision of support prices, the establishment of the SAM program in the 1980s, and the large devaluations in 1982 at the time the foreign debt crisis broke out were factors which tended to favor the corn sector. In 1981, the level of nominal protection was 14 percent as a consequence of an increase in the domestic price of 16 percent and a simultaneous reduction in the world price of 5 percent. This level of nominal protection was provided despite the fact that the overvaluation of the peso was estimated to be over 35 percent at that time.

When evaluated in terms of net effective protection (NERP), the protection and discrimination were in the same direction as when only nominal protection was considered, but the degree of protection or discrimination was in a different direction. In the first four years of the 1970s, the discrimination calculated by the NERP was higher than that when product market effects alone were considered. This suggests that policies vis-à-vis tradeable inputs (i.e., fertilizer) were also discriminating against producers, reducing the value added for one group by up to 58 percent.

Price policies tended to favor producers using technology 4, while producers in rainfed areas without tractors (NERPS, the more traditional producers) experienced the lowest rates of protection. In part this is explained by the small share of tradeable inputs in their costs of production. Overall, however, policy interventions were very favorable for corn producers during the early 1980s.

The net effective rate of subsidy (NERS) measures the combined impact of policies affecting the product market and the tradeable and nontradeable inputs. The fact that it is close in magnitude to the NERP indicates that interventions relative to nontradeable inputs were insignificant in this period, so the key role was played by policies affecting the markets for product and tradeable inputs.

To conclude, the corn and wheat sectors have been treated very differently in Mexico. Corn is the traditional food grain, and although economic policy was discriminatory against the sector in the first half of the 1970s, it provided positive incentives in the latter half of the 1970s and into the early 1980s. This was undoubtedly an attempt to reduce external dependence on a commodity so important in the diets of most Mexicans.

Policy in Mexico has also had the avowed intent to favor small producers relative to large commercial producers, a policy that is almost unique in Latin America. This probably explains the differential treatment afforded to the corn and wheat sectors. However, within the corn sector, policy tended to be more favorable to the larger, more commercial producer. This probably reflects the difficulty of reaching small subsistence producers with effective policy instruments.

In an earlier study Matus-Gardea [1981], using a methodology similar to that used by Mauro Lopes [1977] in the case of Brazil, estimated the impact of overvaluations of the Mexican peso on different sized farms (large private farms, small private farms, and *ejidal* production units). He found that the positive effects of the domestic policies referred to above were offset, or at least counterbalanced, by the negative effects of the implicit tax represented by the overvalued currency. As in the case of Brazil [M. de R. Lopes, 1977], large Mexican farmers are able to escape this tax by reorganizing their resources. The incidence of the tax thus tended to be on small producers and hired labor.

Balassa [1983] provides an overview of Mexican trade policy from 1956 through the early 1980s. The period was divided into three subperiods. The 1956-71 years were described as the period of stabilizing development. Inflation was low, although higher than in the United States, with the result that the purchasing power parity exchange rate appreciated over the period. The current account deficit in Mexico went from 1.4 percent of GDP in 1956 to 2.3 percent in 1971. Tariffs and quantitative restrictions were gradually increased throughout the period, and with them the level of effective protection. By 1970, agriculture, by his estimate, was being taxed at a rate of -1.4 percent as measured by the effective rate of protection. Agricultural exports further suffered, in his view, from the effects of the price policies introduced under the aegis of CONASUPO, established in 1961, which fixed the prices of certain domestically consumed goods above world market levels. With exports sold at world market prices, there was increased discrimination against export crops, in particular cotton. Mexican cotton exports declined between 1955-57 and 1970-72. The share of Mexico in world exports declined from 11.4 percent to 4.9 percent in a period in which world exports were growing.

Balassa [1983] noted that Mexico's share in world exports of beef and coffee also declined. Its share of world exports of cattle and sugar, on the other hand, increased. The increase in cattle exports was explained by the rise in U.S. demand, and the increase in sugar exports by an increase in the U.S. quota allotment.

Overall, Balassa found that the value of Mexican primary exports increased by only 2 percent during the 1960s. In this same period, primary exports rose by 16 percent in Korea and 18 percent in Taiwan, two countries which followed out-ward-oriented economic policies.

Balassa describes the 1972-76 period as the Echeverria expansion. This period was one of stimulative fiscal policy, accelerating inflation, and a growing deficit on current account. Tariffs were raised, and import controls tightened. Some ex-port promotion mechanisms were put in place, but primarily for the manufac-turing sector. Primary exports as a whole declined in absolute terms in this period as the growing overvaluation of the peso was not offset by export incentives. The decline extended to most of Mexico's major primary export commodities, while new primary exports failed to develop.

The Lopez-Portillo period (1977-82) was a period of growing imbalance. The deficit in the public sector grew rapidly and as a result the peso appreciated sig-nificantly in real terms, more than offsetting the devaluations of 1976 and 1977. By the end of 1981, the extent of appreciation of the real value of the peso was 29 percent compared to 1977. In Balassa's view the appreciation of the real exchange rate adversely affected primary exports, with the introduction of the SAM further discriminating against exports with its food self-sufficiency objec-tives. Under SAM, support prices were raised on crops destined for domestic consumption, thereby giving further impetus to the shift of irrigated area from higher-value export crops to lower-value domestic crops. The SAM program may have been a proper response to a petroleum-induced rise in the real exchange rate. However, the peso was rising in real terms at least in part because of do-mestic fiscal policy imbalances.

2. Wheat Imports: Indirect Discrimination Against Traditional Commodities

Byerlee [1986], in a very perceptive paper, shows how policy interventions can have significant indirect effects on agriculture and thus, how complex policy analysis often needs to be. Writing specifically about the political economy of wheat imports by developing countries, he shows that policies which subsidize the consumption of one commodity can have significant discriminatory effects on other parts of agriculture. Moreover, he shows how a constellation of vested interests in both the domestic economy and in the international economy can converge to bring this about.

Wheat has special significance in the analysis of food policy and food imports in developing countries. Cereals constitute the bulk of food imports by these countries, and among cereals, wheat is by far the dominant food grain import. In 1980, for example, wheat accounted for an estimated 86 percent of their food grain imports. Moreover, these countries account for two-thirds of total wheat imports. Wheat exports, on the other hand, are dominated by developed countries which produce about two-thirds of the world's wheat and account for about 95 percent of total exports.

During the 1960s and the 1970s, wheat showed a remarkably rapid and widespread increase in its contribution to diets in developing countries. To a large extent this represented a significant substitution for so-called inferior food staples — coarse grains and roots and tubers. This increased consumption occurred even in countries that did not produce wheat. An estimated 80 percent of the increase in total wheat consumption during the 1970s occurred in the developing world, where wheat consumption grew at a rate of 5.4 percent annually. Aside from the largest wheat producers (China, India, Pakistan, and Turkey), increases in consumption were largely met by imports. As a result, import dependence increased in almost all developing countries to reach high levels by 1982.

Interestingly enough, Byerlee and Sain [1986] argue that pricing policies in developing countries have not necessarily been biased against producers of wheat, contrary to the case with export and other crops, and to the evidence from other studies.[89] To the contrary, for a sample of 32 countries and data from the 1980s, they found that in 36 percent of the countries wheat producers were taxed, in 39 percent they received no significant tax or subsidy, and in 26 percent of the countries the producers actually received subsidies. *Consumption* subsidies, on the other hand, were found to be extensive, with 81 percent of the countries subsidizing the consumption of bread, only 4 percent providing no tax or subsidy, and 15 percent actually taxing bread.

Ten countries from Latin America were included in Byerlee and Sain's sample. The results are somewhat different than for the sample as a whole, but not greatly so. Brazil and Colombia subsidized their producers; Argentina, Mexico, Peru, and Uruguay taxed theirs significantly; and four — Bolivia, Chile, Ecuador, and Paraguay — neither significantly taxed nor subsidized theirs, with the exception of Ecuador where the tax was around 15 percent. But only Colombia taxed consumption, with all others providing significant subsidies (data were not available for Argentina).

The case of Brazil stood out clearly, with both producers and consumers receiving large subsidies. Mexico, a significant wheat producer, heavily subsidized the consumption of wheat, with a nominal protection of 0.34. Only Argentina, of these ten countries, was a major exporter.

Byerlee [1986] argued that the growth in imports in the period covered by his analysis was a result of a combination of regular economic forces and government interventions that gave rise to vested interests. For example, increased wheat consumption in the developing world has to a large extent occurred in urban areas, and especially in those located at port cities or close to port cities. The regular economic forces at work are the rise in per capita incomes, the rise in opportunity cost of time associated with urbanization and the increased participation of women in the labor force—each of which increases the demand for convenience foods—and the low transportation costs involved in bringing wheat or other food staples to these cities.

But there are other economic issues and interests at work as well. For example, the consumers of wheat are middle- and upper-income groups, and these are generally able to influence consumer price policy and strategies for urban food supply. But interests of wheat exporting countries have also played a significant role, as have interests of the domestic wheat processing industry in developing countries, which often have strong linkages to exporting and processing firms in exporting countries. Wheat-exporting countries have played an important role by influencing prices and consumer exposure and tastes through subsidized exports, food aid, and market promotion campaigns. These include export subsidies, subsidized credit for imports, and assistance in milling and baking.

But similarly important is the vested interest of the processing industry for wheat. Bread and other wheat products are highly processed. Byerlee [1986] notes that some of the largest milling establishments are in countries that produce virtually no wheat. Once these processing facilities are installed, the political pressure to keep them operating is great. Moreover, with domestic transportation infrastructure weak and wheat-producing areas often a long distance from urban centers, the economics of the supply situation favors bringing the wheat in from abroad.

Calegar [1984], in a study referred to earlier, notes that Brazilian wheat policy has surely discriminated against the producers of other food commodities. Byerlee [1986] cites a personal communication from Lucio Reca in which Reca estimates that wheat subsidies in Brazil reduced the demand for maize by 10–20 percent and that this led to a reduction of maize prices by 11–15 percent. Maize is probably the most widespread crop in Brazil and it is grown extensively by small producers. Hence, although the wheat sector itself is small in Brazil, both in terms of the resources used and the share of food expenditures on wheat, the effect of the subsidies was quite widespread in the economy.

The international debt crisis of the 1980s and the conditionality (policy reforms as a condition for loans) imposed by the IMF and the World Bank are inducing countries to shift away from such heavy use of subsidies. However, the dilemmas posed for the developing countries are still quite serious. To the extent these countries have merely lagged in developing their capacity to diffuse a modern production technology in the wheat sector, the proper pricing policy is still a challenge. To foster development, policy makers in the short term may find it advantageous to take advantage of cheap food supplies from abroad. However, this may impede the development of the domestic industry in the longer term. In dealing with this dilemma, elimination of the large consumption subsidies (and converting them to targeted feeding programs) would at least save on foreign exchange.

For countries that have no inherent comparative advantage in wheat production, elimination of consumption subsidies for wheat would benefit producers of competing commodities. But over the longer term, there is probably little alternative to taking advantage of lower production costs elsewhere, and increasing exports of other goods and services to earn the foreign exchange to pay for imported wheat. This is called specialization and the international division of labor, and there are important benefits from it. However, Byerlee [1986] raises an interesting question at the end of his paper. Imports of several other food commodities, especially feed grains and dairy products, have also risen rapidly in the past decade in many developing countries. Since many of the same interest groups, especially urban middle-income consumers and farmers of exporting countries, benefit from these imports, he concludes that a similar set of forces resulting from government interventions also underlies the growing import dependence on these commodities, with some of the same deleterious consequences to longer-term development.

Finally, it is worth noting that the direct evidence that external food aid has a direct disincentive effect on agriculture in the recipient countries in Latin America is rather mixed. In the Andean region, domestic wheat production declined in response to reduced producer prices for wheat during the 1960s when most wheat was imported as food aid [Dudley and Sandilands, 1975; Valderama, 1979]. In Brazil, however, where much of the wheat imported was until recently on concessional terms in the form of food aid, there was little or no effect on producer incentives. By pursuing a two-price system, Brazil was able to use the food aid to subsidize consumption while at the same time subsidizing producers [Hall, 1980].

3. Policy Reform in the Southern Cone

During the 1974–83 period the three countries of the Southern Cone—Chile, Argentina, and Uruguay—implemented innovative reform packages in an attempt to rescue themselves from the severe macroeconomic disequilibria in which they found themselves as a consequence of their own past inward-looking policies and rather unique shocks from the international economy. These reforms are relevant to agriculture and to agricultural development because they represent attempts to open their respective economies to a greater play of domestic and international economic forces, and thus implicitly to give agriculture a greater role in the development of their respective economies. It is for this reason that we refer to the literature on these reforms and discuss the reforms and their consequences.[90]

These three countries, similar to most others in the region, had pursued inward-looking development strategies which relied heavily on government intervention and which discriminated severely against agriculture. Their economies were characterized by anti-export biases, high spreads in protection across sectors, and heavily controlled financial sectors. They also suffered from recurrent balance-of-payments crises and slow growth.

The reforms of the 1970s essentially involved attempts to switch from import-substituting industrialization to a more neutral policy. Markets were liberalized to improve resource allocation, commodity price controls were practically eliminated, trade barriers were reduced, interest rates were decontrolled, constraints on capital flows were reduced (as were constraints on labor markets), and government deficits were reduced and/or eliminated. In short, the reforms involved a combination of short-term stabilization measures and long-term policies to remove government intervention progressively across product and factor markets.

Interestingly enough, all three countries experienced initial success with the early stages of their reform packages. But each also eventually encountered a boom-bust cycle, large increases in external indebtedness, and major internal financial crises.

Two of the countries were experiencing enormous macroeconomic imbalances at the time the reforms were implemented. Annual inflation rates were approaching 1,000 percent in Chile (September 1973) and 2,300 percent in Argentina (March 1976). Government deficits as a share of GDP were 12.0 percent in Argentina and 16.1 percent in Chile. These imbalances were less severe in Uruguay, with inflation only 97 percent in 1973 and the government deficit averaging only 3.2 percent of GDP in the year prior to the reform (1973). But Uruguay had also experienced almost zero growth for twenty years.

Two kinds of reforms were implemented in commodity markets: deregulation of domestic prices, and reduction of tariff and nontariff barriers to trade. According to Corbo, J. de Melo, and Tybout [1986]—hereafter referred to as CMT—Chile went farthest on both counts. A synthesis of the reforms in that country will illustrate just how extensive they were.

According to CMT, in the early stages of implementation in Chile, prices in the domestic commodity market were deregulated, and subsidies were practically eliminated. Most nontariff restrictions on trade were eliminated, and commodity markets were indirectly deregulated by privatizing over 500 enterprises that had been seized or nationalized during the Allende years. The multiple exchange rate system was unified with one initial large devaluation. All nontariff restrictions were lifted and the tariff structure was also reformed. Over a five-year period starting in 1974, the average tariff was brought down from 90 percent to a uniform 10 percent.

Financial reforms in each country sought to reduce or eliminate nonprice allocation of credit and highly negative real rates of interest. Each country also tried to open its economy to international capital flows, but the speed and extent of this varied across countries. Liberalization of labor markets was in practice relatively minor, but the weakening of trade union power was a part of the reform in each country.

It was in the stabilization policies that the most innovation took place. In the first phase of these policies, anti-inflationary policy was based on major reductions in monetary growth and fiscal deficits—rather orthodox measures. As complements to these efforts, and to eliminate balance of payments crises, each country also made major attempts at expenditure switching. In Chile and Argentina, this was achieved through large real devaluations; in Uruguay, switching included a combination of real devaluation and promotion of nontraditional exports. Then, to avoid repetition of external crises, these initial adjustments were followed with a "passive" crawling-peg exchange rate regime aimed toward maintaining purchasing power parity.

These initial policies successfully eliminated balance of payments crises. But inflation remained at disturbingly high levels. This motivated a shift in stabilization tactics that involved using the exchange rate as an anti-inflationary tool, in recognition of the fact that the economies were now more open to trade and capital flows. The policies that evolved were designed to take into account the presumption that expectations about inflation and devaluation were important in determining the dynamics of stabilization. Thus it was assumed that *preannounced* exchange rate targets would break inflationary expectations. The rate of devaluation into the future (some six months) was set in a preannounced schedule (known as the *tablita*) at less than the existing difference between domestic and world inflation. This policy was labeled by CMT as an *active* crawling peg.

The expectation was that deregulation of the product and factor markets would cause both purchasing power parity and interest parity to obtain fairly rapidly. Corbo, J. de Melo, and Tybout [1986] indicate that the Rodriguez [1982] model came closest to describing how the preannouncements were supposed to work—at least in Argentina and Uruguay. They note:

> In this model, interest-rate parity obtains continuously because of the absence of controls on capital flows and the assumption of perfect asset substitutability. The law of one price holds for tradeables, and the rate of change in nontradeable goods prices is a function of inflationary expectations—which are assumed to form adaptively—and of excess demand for nontradeable goods. The model predicts that the implementation of a *tablita* should immediately reduce nominal interest rates and, to a lesser extent, inflation. The decline in real interest rates should first stimulate demand, creating an excess demand for nontradeable goods, thereby inducing a temporary appreciation

(that is, a fall) of the real exchange rate. As inflation falls, both the real interest rate and the real exchange rate should increase, approaching their long-run equilibrium from below. *The economy should stabilize without undergoing the recession associated with traditional contractionary measures.* The avoidance of a recession associated with this new approach was quite attractive politically. [p. 10, emphasis added]

CMT [1986] reviewed each case in some detail. They found that there were significant gains in productivity and export performance early in the reform process, especially in Chile and Uruguay. Agricultural exports grew at a rapid pace in the case of Chile. But these efficiency gains were ultimately overshadowed by problems with inconsistencies in the policies, difficulties in implementation, and overloaded market frictions (which are detailed in their paper). These factors generated a sustained appreciation of the real exchange rate, and a large spread between the cost of dollar-denominated and peso-denominated loans as domestic interest rates rose to very high levels. In turn, the appreciation and interest rate spreads created protracted opportunities for arbitrage that distracted firms from the business of production. Eventually, firms became deep enough in debt that, as expectations of a major devaluation developed, they were forced into crisis borrowing to cover soaring interest costs. Collapse was the eventual result.

CMT [1986] drew a number of lessons from their review of the three cases. First, they noted that rarely do countries carry out reform packages as profound as those in the Southern Cone. The reform packages spanned all markets, and were implemented in a short time, starting from a crisis situation and heavily distorted markets.

Second, the three countries started with conditions much further from the reform targets than did other countries that had liberalized successfully. Import-substituting industrialization policies had been in place since the Great Depression of the 1930s. Interest rates had been controlled for decades. Price controls had been in place since the early 1950s. Labor markets had been subject to innumerable regulations. Political interest groups were deeply entrenched. Expectations were deeply imbedded. Trade regimes had traditionally discriminated against exporting activities and resulted in much redundant protection for import-competing sectors. And macro policies were frequently constrained by balance of payments crises.

Third, the preannounced devaluations, combined with the removal of restrictions on capital flows, led to boom-bust cycles. With regulated interest rates, setting the devaluation schedules at much less than the differential between internal and external inflation led to large short-term capital inflows, increased dollar indebtedness, and booms. But the booms caused strong real appreciation of the peso, which caused a loss of competitiveness for tradeables and generated doubts about the sustainability of the *tablita*. Capital flight followed and the economies went bust, with each country having accumulated a huge foreign debt in the process. The obvious lesson is that use of the exchange rate as an anti-inflationary tool leads to a major macroeconomic disturbance, and this disturbance eventually undermines the stabilization effort.

CMT [1986] also argued that adjustment delays are more likely to affect exportables adversely than importables during a period of real exchange rate appreciation. Consequently, exporters are particularly vulnerable to movements in the real exchange rate.

These three important social experiments provide valuable lessons for others attempting reform. They make it clear that reform is difficult and complex and that there are no easy routes, especially given the magnitude of the distortions and the degree to which past policy paradigms are imbedded in expectations. The previously cited experiences of Brazil, Chile, Argentina, and Colombia suggest that agriculture can respond significantly on the export side when the right incentives are provided. The difficulty is in sustaining the reforms, because the redistributions of income involved are quite great, given the magnitude of the distortions. Past protection measures have created concentrations of monopoly and monopsony power and other vested interests.

More generally, key determinants of successful devaluations are the elasticities of export supply and demand, the mobility of resources between tradeable and nontradeable sectors, and the elasticity of substitution in production between domestic and imported capital goods and intermediate products. These parameters have to be evaluated in the context of the individual economy. Expectations in a political or policy sense are also important in determining investment response. If the reforms appear to be infeasible, or the political regime precarious, it is not likely that the expected investment response will be forthcoming.

4. Summary, Conclusions and Some Related Literature

The "Abstract" from Corbo's [1986] excellent review of problems, development theory and strategies of Latin America provides an excellent overview of the major swings over time in economic policy perspectives in the region:

> In this paper I examine the influence of both economic
> problems and development theories on the growth strategies
> followed by Latin American countries in the last 70 years. Up
> to the Great Depression, most countries in the region were
> fairly open to trade, and comparative advantage was the basis
> for the leading development theory. From the Great Depression
> to World War II, most countries practiced crisis management as
> a reaction to the sharp external shocks coming from the

Depression. Toward the end of the forties, the first attack on free trade orthodoxy was launched by the structuralist school, headed by Raúl Prebisch. It championed import-substitution industrialization. By the early sixties, however, disappointed with the prospect for growth, Brazil and Colombia and to a lower degree Chile, started reforms aimed at reducing the antiexport bias of their import-substitution policies. In Brazil and Colombia, the supply response of exports to the new economic incentives was dramatic: they grew even faster than did world exports in the golden decade of world trade. Later on, a second frontal attack on structuralism was launched, this time by the southern cone countries, which recently instituted liberalization attempts. Had it not been for ill-fated stabilization programs and the external shocks of the seventies and early eighties, the reform measures would have lifted the southern cone countries out of their economic stagnation. However, from these recent experiences, we have also learned that the coordination of macroeconomic policies with the liberalization policies is fundamental to achieve the expected gain from a less distorted economy. [Corbo, 1986]

Corbo [1986, p. 40] also notes that in the early stages of the import-substitution strategy articulated by ECLA and its leader, Raúl Prebisch, the anti-export bias was small and the Latin American economies performed quite well. But by the late 1950s and the early 1960s the costs of the resulting distortions became all too explicit. The annual rate of increase for Latin American exports between 1951 and 1960 was 1.4 percent, as against a worldwide rate of 4.0 percent [Corbo, 1986, p. 38]. The increasingly distorted trade regimes which created a strong anti-export bias constrained overall growth.

The anti-export bias was inherently anti-agriculture. As liberalization of these policies proceeded in the late 1960s and through the 1970s, the discrimination against agriculture declined and as the surveys and analyses above showed, agricultural exports responded accordingly. The added dimension, which Corbo does not cover, is that continued protection of the manufacturing sector even after the anti-export bias declined was a continuing discrimination against agriculture. There also was the spillover between the traded and nontraded sector of agriculture.

Balassa [1985] has an excellent discussion of outward oriented trade policies, with a comparison of Latin American experiences with those of countries from other parts of the developing world. He, too, stresses that with the protection of the manufacturing sector, inward orientation also involves a bias against primary, in particular agricultural, products that is not found under outward orientation.

Balassa also provides evidence that an outward orientation has given superior results both in times of world boom and in periods of world recession, as well as evidence that application of an outward-oriented strategy by an increasing number of developing countries would not encounter serious foreign market constraints.

Much was made in the discussion of the individual country experiences (earlier in this section) of the role of distortions in exchange rates as a factor affecting the performance of agricultural sector as a whole. Valdés [1985b] presents a framework to estimate the combined effects of trade and exchange rate policies on the structure of relative prices for agriculture and then applies it to selected developing countries. He puts emphasis on defining and measuring the implicit protection or taxation of agriculture which results from the linkages between it and the rest of the economy. The real exchange rate, defined as the ratio of the price of tradeables to nontradeables, is portrayed as playing a central role in the profitability of tradeables in agriculture—import-competing and exportable; and it is through the real exchange rate that the macromanagement of the economy affects agriculture.

The essence of Valdés's [1985b] perspective is as follows:

> It is a well-accepted argument in theory that a tariff on imports also taxes exports, and that a subsidy for exports also subsidizes imports. A policy that protects industry directly raises the cost of importable inputs such as fertilizers, machinery, and other materials used by farmers. More importantly, indirectly, through its effects on the real exchange rate, such a policy affects the relative profitability of other tradeables. The exchange rate that maintains a balance in the external account at a "higher" rate of protection to industry is below the rate at lower rates of protection. The result is that the domestic prices of tradeable goods from agriculture are lower relative to the prices of protected tradeable goods from industry and of nontradeable goods. This drives up the prices of labor and other inputs to agriculture relative to the output prices, reducing the profitability of producing tradeables in agriculture. [p. 2]

Four of the seven countries which Valdés considered were from Latin America (Argentina, Chile, Colombia, and Peru). His findings were as follows:

> The results suggest a high degree of substitution between home goods and importables. A clear implication from these results is that at least one-half of the burden of protection is born by

exportables. Since the exports of many LDC's are predominantly agricultural, an import-substitution strategy taxes agriculture substantially more than a comparison of the nominal rates of protection would suggest. For example, the values. . . . for Chile and Argentina indicate that a uniform tariff on imports of 20 percent—which is not high by LDC standards—represents an implicit tax on exports of approximately 10 percent. If exports are taxed directly, say at a rate of 15 percent (as beef exports in Argentina were in the past), the total tax on exports is 25 percent. Similarly, only part of the tariff is a tax on consumers of importables and protection to producers of import-competing goods. The rest is an implicit tax on producers of exportables (and of import-competing activities with lower protection, as food), and an implicit subsidy to consumers of exportables and of those importables (like food). The implications of these results for economic policy are strong. [pp. 8-9]

To capture the same type of incidence parameter for subsectors of agriculture, Valdés [1985b] applied the same approach to data from Peru[91] for the period 1949-83. His results are equally revealing:

[I]f the uniform tariff on nonagricultural importables is raised by 10 percent and tariffs on agricultural goods did not change, an implicit tax of 5.6 percent (with respect to home goods) is imposed on import-competing agricultural activities (such as rice), and an implicit tax of 6.6 percent is imposed on importable agricultural goods (such as cotton and sugar). When prices are compared to the prices of nonagricultural importables, the implicit tax on both types of agricultural goods is 10 percent. In contrast, similar calculations made with respect to an increase in protection of agricultural importables resulted in a much lower incidence on the price of home goods. That is, during the same period in Peru, changes in the prices of nonagricultural importables had a much greater effect on the prices of home goods than changes in the prices of agricultural importables. . . . Similar computations for agricultural exportables indicate that an increase of 10 percent in the price of agricultural exportables raises the value of home goods 2.6 percent, compared to 0.6 percent resulting from a rise in the price of nonagricultural exportables. [pp. 9-10]

Finally Valdés [1985b] noted that as part of an industrialization strategy through protection, the real exchange rate falls consistently through time. That is, the higher average tariff implied a fall in the equilibrium exchange rate. He noted that as the Peruvian economy became more closed during the 1960s and 1970s with increases in restrictions on trade, the real exchange rate underwent a major and persistent decline after the 1960s, reducing the profitability of producing tradeables *vis-à-vis* nontradeables.

Valdés also noted that such declines in the long-run real exchange rate have been particularly harmful for the production of agricultural tradeables in LDCs, slowing their production and speeding up increases in the domestic consumption of tradeables (imported cereals and exportables), reducing the contribution of agriculture to growth and to the balance of payments, and making LDC's more dependent on imported food. He notes that a falling real exchange rate is not necessarily a sign for a devaluation, since protection of the domestic industry or larger inflows of capital, including foreign assistance, could cause the external accounts to be in equilibrium at a low exchange rate.

A result would be that agriculture, together with exportables in general, would be taxed implicitly. This penalty on agriculture is inherent and lasts as long as industry is highly protected. It cannot be eliminated by better management in other areas of economic policies.

Thus we see historically the important role of the exchange rate in the taxation of agriculture in Latin America. Moreover, the "strong" currencies may be as much, or more, a reflection of distortions in other parts of the economy, such as high protection of the industrial sector, as of direct interventions in the foreign exchange markets.

It seems clear that much of the poor trade performance of Latin American agriculture in the 1950s and 1960s was due to self-inflicted wounds imposed by national trade and exchange rate policies. There is ample evidence that Latin American agriculture can penetrate foreign markets if they have proper trade and exchange rate policies and if they keep their production technology up to date. The experiences of Argentina, Brazil, Chile, and Colombia all support this proposition.

These conclusions also have important implications for the emphasis given by writers on Latin America to the role of the external terms of trade. Agriculture did not become unprofitable because of what was happening to the external terms of trade, but because of what was happening to domestic trade and exchange rate policies. Although it is true that declines in the external terms of trade can create balance of payments problems, the only rational response to this is to become more efficient by raising productivity, not to impose resource distortions on the economy. The real decline in the price of primary commodities, which constitutes part of the decline in the external terms of trade, can be a source of income growth for consumers if policy makers make it possible for consumers to share in them in a rational way.

Two issues remain. The first is the protectionism and the predatory policies of the developed countries, especially the European Community, the United States, and Japan. Valdés and Zietz [1980] analyzed the impact of OECD protectionist policies on less developed countries as a whole. Valdés [1983] analyzed the effects on Latin American countries as part of the larger study. An important consequence of present protectionist policies is to cause international prices for protected products to be lower than they otherwise would be. Consequently, liberalization could impose welfare losses on consumers in the developing countries. Moreover, the lower prices associated with protectionist policies can be viewed as a source of exchange savings for the developing countries.

Valdés [1983] or Valdés and Zietz [1980] estimated that a 50 percent reduction in protectionist barriers by the OECD countries would cause a gross loss in real income of US$116 million for Latin America, but a gain of US$558 million due to additional exports. On balance the gain is positive. De Rubenstein and Budge C. [1988] make a careful analysis of North-South grain policies and their effects on international trade, with special reference to Chile.

The World Bank's *World Development Report, 1986* [1986c] contains a comprehensive analysis of distortions in agricultural trade by both the developing and developed countries. It, too, includes an analysis of the effects of trade liberalization on the part of both the developed and developing countries. It shows the gains to be modest but significant.

Under the leadership of Anne Krueger (aided and abetted by Maurice Schiff and Alberto Valdés) the World Bank has undertaken a large research project on the political economy of agricultural pricing policies. The objective of this study has been to examine the effects of the multifarious general and sectoral policies affecting agriculture through a consistent analytical framework, and to attempt to understand why governments acted as they did. Studies of some eighteen countries have been included in the project, including five (Argentina, Brazil, Chile, Colombia, and the Dominican Republic) from Latin America. Krueger, Schiff, and Valdés [1988] provides an overview of the analytical framework used and a synthesis of some of the empirical results. As this volume goes to press, the individual country studies are being published and a two-volume collection of summaries of the country studies is in the works. Authors of the country studies may be found in Krueger, Schiff, and Valdés [1988]. This project is the most comprehensive analysis to date of policy distortions in these countries.

K. Anderson and Tyers [1986] have made detailed analyses of liberalization in restrictive trade policies. These were drawn on in the *World Development Report, 1986*. Zietz and Valdés [1986] update the Valdés and Zietz study [1980].

In assessing these various empirical studies it is important to note that the impact of the protectionist policies by the OECD countries, especially the U.S. and the EC, is not just in terms of the markets that are foregone. In 1985, 1986, and 1987 both the U.S. and the EC made extensive use of export subsidies, the U.S. especially with the 1985 farm bill. In the case of the United States, an important part of the export subsidy is in implicit form due to its extensive use of deficiency payments. The use of export subsidies makes the world prices even lower than they would be from the protectionist measures themselves.

One of the implications of the results of the various studies of protectionist measures is the importance of a greater participation of developing countries, and especially those of Latin America, in multilateral trade negotiations. In the past, participation of the developing countries has been marginal. Although this has been in part due to the stance of the developed countries, in the past many of the developing countries were not even members of the GATT. With agricultural trade given increased attention in the current trade negotiations, there is added incentive for the developing countries to participate. For an excellent analysis of alternative approaches to the reform of agricultural trade policies in the GATT, see Zietz and Valdés [1988]. Valdés [1987] discusses the potentially most feasible approaches for developing countries to obtain some measure of liberalization in the present Uruguay Round.

Balassa and Michalopoulos [1985] demonstrate that multilateral trade liberalization is in the mutual interests of both the developed and developing countries, not only in terms of promoting economic growth but also in terms of increasing market possibilities for both sets of countries. In addition to proposing that trade liberalization be undertaken in the framework of a new round of negotiations, they make recommendations on the process of liberalization in agriculture, manufacturing, and services as well as on the establishment of an appropriate safeguard mechanism, dispute settlement procedures, and surveillance by the GATT.

Finally, there is the issue of just how strong an export-promotion strategy individual countries should take from the point of view of their own welfare. Adelman [1984] has argued for an agriculture-first development model for developing countries, but as a means of broadening the domestic base for economic development rather than as the basis for an export drive. This broadening of the domestic base by strengthening agriculture would, in her view, reduce the need for imports in many countries, at the same time providing a broader market for manufactured products and thus making it possible to realize scale economies in that sector without launching an export drive. Mellor [1976] also made a strong argument for agriculture-first development policies.

Adelman's perspective may in part be misguided. Reducing the discrimination against agriculture, especially that through trade and exchange rate policies, can lead to broad-based increases in per capita income. This can increase demand for agricultural output as fast or faster than the supply side can respond, especially if it leads to a significant upgrading of diets. The demand for imports of some agricultural commodities may thus increase. The individual country may thus need to increase its exports to pay for these imports. This is the essence of the international division of labor and the gains from trade. Moreover, for those countries with large foreign debts, there seems to be little alternative to increasing exports from agriculture or elsewhere.

The related issue, of course, is that a fundamental cause of the concern about the development of agriculture is that per capita incomes are so low in that sector relative to those in the rest of the economy. This problem will be resolved only with the transfer of large numbers of the agricultural labor force to the nonfarm sector of the economy. This requires not only balanced expansion of the economy as a whole, but improvements in economic policy to increase the labor-absorption capacity of the nonfarm sector of the economy.

Chapter VII. Common Markets, Customs Unions, and Free Trade Associations

Corbo [1986, p. 22] notes that the crisis in import-substituting industrialization at the national level—a crisis which was emerging in the late 1950s—gave rise to a second stage of import substitution by means of regional integration. (The following paragraphs draw from Corbo.) ECLA was aware that growth through import substitution at the national level was becoming increasingly costly. Prebisch himself [1959a, b] concluded that further import substitution would have to be undertaken at a regional level. Prebisch [1959a] wrote:

> Trade between Latin American countries forms only 10 percent of their total foreign trade, and industrial exports are relatively very small by contrast with countries such as Italy, Japan, and others with similar income levels. All this has resulted in the splitting of the industrialization process into as many watertight compartments as there are countries, without the advantages of specialization and the economies of scale. [pp. 267-268]

Instead of concluding from the evidence at hand that the system of protection should be rationalized to reduce its anti-export bias, Prebisch [1959a] recommended that

> the response to this should be the enlargement of national markets through the gradual establishment of a common market. . . . Without the common market, there will be a continued tendency by each country to try to produce

everything—say, from automobiles to machinery—under the sheltering wing of very high protection. [p. 268]

The perceived success of regional integration in other parts of the world was also a motivating factor in the drive for Latin American integration.

This chapter is divided into four parts. Successive sections which address the Latin American Free Trade Association (LAFTA), the Andean Common Market, and the Central America Common Market (CACM) are followed by a section of concluding comments.

1. Latin American Free Trade Association (LAFTA)

For reasons noted above, ECLA in the sixties became the champion of regional integration. With ECLA's intellectual leadership and the support of the United States, a Latin American Free Trade Association (LAFTA) was created in 1961. The reduction of trade barriers within the region was to be negotiated commodity by commodity, and the industrialists in the highly protected manufacturing sectors were to play a central role as members of the country negotiating teams. Not surprisingly, it proved very difficult to reach agreement on tariff reductions except in the case of a small number of insignificant commodities.

The original expectation was that LAFTA would eventually evolve to be a Latin American Common Market. In addition to the difficulties noted above, however, the wide disparity in level of development of member countries proved to be a serious stumbling block. Axline's [1981] overview of Latin American regional integration was rather pessimistic about such integration without socioeconomic changes which would fundamentally alter the political and social structure of Latin American countries [pp. 184-185].

Krause and Mathis [1968] made a brief evaluation of this customs union early in its history. They found, significantly, that computation of concessions on a percentage basis promised to affect those countries most dependent on trade, which were disproportionately the lesser-developed countries. In the absence of considerable diversification during the interim, they expected that these countries, generally in a weak competitive position anyway, stood to experience an adverse impact from integration. Recognition of this prospect was cited as a justification to delay application of reductions, extending beyond 1973, on certain agricultural products of dominant importance to the lesser-developed countries [pp. 12-13].

In reviewing previous experiences, Krause and Mathis [1968] found that the index for intraregional trade rose from 100 to 222 between 1961 and 1966, while that for extrazonal trade during the same period rose from 100 to 126. Intraregional trade rose from 6 percent to 10.5 percent during the period. They also found that the distribution of intraregional trade changed little during the period. The more developed countries (especially Argentina and Brazil) dominated intraregional trade, while the contribution of some of the lesser developed countries (for example, Ecuador and Bolivia) actually decreased [p. 14]. In looking to the future, the authors argued that the LAFTA would have to pay special attention to the special status of the lesser developed countries if it wanted to evolve further.

2. Andean Common Market

In 1969, with the LAFTA initiative going nowhere, a subset of middle-sized LAFTA members formally approved an Andean Common Market Pact (the Cartagena Agreement). This initiative was first launched in 1966 when Chile, Colombia, Ecuador, Peru, and Venezuela signed the Declaration of Bogotá.

There is a lack of good analytical studies of either LAFTA or the Andean Common Market. Morawetz [1974], Ferris [1979], and Vaitsos [1978] are useful sources for descriptive material and some analysis, however. Corbo [1986] noted that in designing its rules of operation, members of the Andean Pact took into account many of the lessons learned from the LAFTA initiatives:

> Tariffs and nontariff barriers were to be fully eliminated among member countries by the end of 1980; Chile and Colombia had advocated an even faster decline. [Diaz-Alejandro, 1973]
>
> Instead of proceeding commodity by commodity, tariffs were to be reduced each year by 10 percent of the minimum *ad valorem* tariff then existing in Colombia, Chile and Peru which in no case was to exceed 100 percent. Thus reduction of the tariffs was going to be automatic. The less developed members (Ecuador and Bolivia) were, however, given more favorable terms.
>
> Parallel with the general rule of automatic reductions, the Pact called for the allocation of new manufacturing activities to individual countries to avoid duplication and to reap benefits from economies to scale. The result would be import substitution at a regional level. The countries were also to negotiate a common external tariff. [Corbo, 1986, p. 24]

Corbo [1986, p. 24] judged that the Andean market was a definite improvement over continued import substitution at the country level in that it allowed countries to carry out intra-industry specialization and to create trade. However, to be sustainable a mildly protective common external tariff had to be implemented. That goal was never achieved. Corbo [p. 25] noted that the politics of import substitution at a regional level was more difficult than that within a country, and the Andean Pact lost its dynamism in the second half of the 1970s. He noted that the final blow came when Chile, which had played a central role in the creation of the Pact, withdrew after failing to obtain agreement on its proposals for sharply reducing the common external tariff and for lifting the Pact's restriction on direct foreign investment. Ffrench-Davis [1977] had noted earlier that a less nationalistic attitude was essential if the Pact was to be successful.

3. Central American Common Market (CACM)

The most ambitious and successful of these efforts at regional integration was the Central American Common Market (CACM). Agreement to establish the common market was reached in 1960, and the General Treaty on Central American Integration of 1961 sought to accelerate the process of integration by providing for free trade immediately on virtually all products. Only 5 percent of the items traded were excluded, but even these were subject to interim regulations that were in turn subject to automatic liberalization within a five-year period. Other dimensions to the proposed integration included: a movement towards a common external tariff; the introduction of a zonal payments system; and support for special "integration industries." The goal was to have complete uniformity in the external tariff by 1970.

Rapid progress was made in the beginning, with some 80 percent of Central American tariffs established at uniform levels by 1967. There were already recurring problems, however, in selecting a common level of protection acceptable to all five countries. Moreover, as of 1968 a number of important trade items had continued untackled: transportation equipment, electrical appliances, petroleum products, certain agricultural products, etc. [Krause and Mathis, 1968, p. 10].

A book edited by Cline and Delgado [1978] brings together analyses of the costs, benefits, successes, and failures of the CACM. Delgado noted that there were two phases to the evolution of the CACM. The first phase was under the tutelage of the ECLA, and integration efforts were to center on a prospective scheme of "integration industries" that would be allocated among member countries [p. 9]. By 1960, however, the year the market was established, the focus of integration had turned to free trade, prompted by the positions of the more industrialized countries (Guatemala and El Salvador) and encouraged by U.S. policy makers, who feared the ECLA approach would establish a series of regional monopolies [p. 9].

The formation of the CACM in 1960 was followed by early success as interregional trade mushroomed to unexpected levels. Krause and Mathis [1968] noted that such trade increased by 300 percent between 1960 and 1965. The resulting atmosphere of euphoria gave way to one of contention by the mid-1960s, however, as the poorest member country, Honduras, and to some extent Nicaragua and Costa Rica, protested the concentration of the benefits of integration in the other countries. It then became clear that the modality of free trade as the chief theme of integration had omitted any effective mechanism for redistributing benefits among partners. Honduras eventually left the market in 1969, after the so-called football war with El Salvador, posing a significant institutional crisis.

The traditional components in analyses of customs unions have been trade creation and trade diversion. In chapter 3 of Cline and Delgado [1978], Cline [1978] considers these two facets as well as several other economic effects not conventionally treated in analyses of economic integration. He found that the total net economic benefits of integration were large, amounting to 3 or 4 percent of regional product by 1972. Expressed as a single present discounted value for all future years, the decision to integrate was worth US$3 billion. The most important economic benefits arose from foreign exchange savings and from the CACM's stimulus to investment. Economies of scale and employment effects were also important, but less so than the other two effects. Economic benefits from structural transformation were minimal, even though the shift in sectoral composition from agriculture to industry was considerable.

Interestingly enough, Cline found that the *traditional* effects of a common market, trade creation and trade diversion, had extremely small benefits and costs. The absence of high costs of trade diversion throws important light on one controversial aspect of integration. A common criticism of the CACM was that it established highly inefficient industries behind protective tariffs. But Cline's data show that CACM product costs were reasonable compared to world prices and that if anything, average tariffs *fell* instead of rising as member countries moved to the common external tariff.

Cline found that the distribution of net benefits of integration across countries was relatively equitable. By 1982 the percentage shares of member countries in net benefits resembled their shares in regional population. The only clear case of low relative benefits was that of Honduras. Even for that country, however, net benefits were positive, indicating that Honduras lost economic opportunities by dropping out.

An important finding from Cline's analysis was that the benefits tended to even out over time. As of 1968, for example, there was a high concentration of net benefits in Guatemala and El Salvador. By 1972, the distribution was much more uniform among member countries. This is what might have been expected, with countries having the largest industrial sector tending to dominate regional trade at the beginning, and countries with a smaller industrial base at the beginning eventually catching up as the new regional sales horizons stimulated investments. And indeed, an investment survey conducted by Cline found that it was precisely in the two economies with the smallest industrial bases, Honduras and Nicaragua, that the greatest proportional stimulus to investment occurred as a result of the CACM.

Agricultural issues were examined in chapter 7 of Cline and Delgado [1978] by Cappi, Fletcher, Norton, Pomerada, and Wainer [1978]. The emphasis was on trade and potential trade in basic grains within Central America. Unlike trade in industrial products, trade in agricultural products had, since the beginning of the CACM, been subject to numerous restrictions. Individual countries pursued self-sufficiency in grains, ignoring the cheaper sources of supply available through imports from partner countries. The authors concluded that even a partial liberalization of this trade would provide considerable welfare benefits in the form of cheaper grains. Production and exports of basic grains would increase in Honduras and Nicaragua, imports by Costa Rica would increase, and there would be little change in trade for Guatemala and El Salvador. There would be corresponding benefits for producers in Honduras and consumers in Costa Rica, with erosion in the incomes of producers, particularly of large farmers in Costa Rica, and substantially increased incomes of farm owners in Honduras.

A final set of analyses of agricultural trade explored the comparative advantage of agriculture with respect to the rest of the world. The results showed a diversity of specialization possibilities among the five countries. No single country or subgroup of countries was found to have a monopoly of comparative advantages in all export crops. Instead, each country showed favorable specialization prospects in specific alternative products.

In its 1976 appeal for the reactivation of the CACM, ECLA argued the need for political integration [CEPAL, 1976]. In particular, it argued that a common external economic policy was the only means by which Central America could decisively impose its negotiating capacity relative to other countries and groups of countries, especially in terms of the Generalized System of Preferences and the Multilateral Trade Negotiations (MTNs) sponsored by GATT [p. 40]. To strengthen agriculture as the base for a reactivated Common Market they suggested the need for improved fertilizer supply and a regional dimension to irrigation programs. They expected this to increase the supply both of basic foods and exports [p. 66]. On the supply of fertilizers, they argued that there was a marked inverse relation between the volume purchased and price, which suggests that joint purchases could reduce prices for the region by approximately 15 percent [p. 67].

Two authors have addressed the specific problems of agriculture and the Central American Common Market. Quiros-Guardia [1973b] argued that regional integration is not a *sine qua non* condition for agricultural development in this region because in his view there was a lack of complementarity in the structure of agricultural production. Moreover,

> Despite these innovations, this theory (customs unions) was
> never intended to apply to an agricultural development
> situation characterized by economic and technological dualism,

extreme specialization in export production, rapid population growth, and widespread unemployment. [p. 5]

Quiros-Guardia discussed the role of agriculture in economic integration from two perspectives: the potential benefits (losses) that integration may create; and the extent to which integration and development depend on agriculture. The benefits for agriculture in his view include: expansion of market demand due to income growth and urbanization, more stable markets, and possibilities for import substitution; regional specialization and improved production efficiency due to more scope for taking advantage of comparative advantage; external economies through such things as integrated agricultural research; and export diversification and an improved bargaining position.

Possible detrimental effects, in his view, would arise if import substitution entailed unduly high prices for capital and manufactured goods purchased by the farm sector. Agricultural development would be depressed either through reduced effective demand or through reduced incentives to invest in productivity-raising capital goods.

> Although this will tend to redress the balance of trade terms [sic] in favor of agriculture, longer term productivity gains in agriculture may offset the income effects of adverse terms of trade. This, however, may create serious distributive effects on development gains for the agricultural sector unless industrial efficiency and productivity is expanded to keep a certain balance of relative prices in the regional market. [Quiros-Guardia, 1973b, p. 12]

Quiros-Guardia [1973b] made an interesting comment in a footnote to this statement:

> The effects of deteriorating internal terms of trade will follow essentially the same pattern as the one described by the argument concerning development gains from international trade between "center" and "periphery" countries in the international economy. [p. 12]

This statement suffers from the same failure, noted above, to distinguish between the effects of changes in the terms of trade on the balance of payments and its effects on national welfare.

Quiros-Guardia [1973b] also argued that successful economic integration of the region would depend on successful integration of agriculture; agriculture is the most important sector of the regional economy and failure to integrate it may cause serious external problems to arise, leading to difficulties of integration. He reviewed the various difficulties in obtaining regional integration of agriculture, and the several attempts to facilitate it. Among the latter, the 1965 Limón Protocol (effective 1967) was designed to establish and regulate free trade in rice, beans, corn and sorghum within the Common Market. It also involved coordinating production and supply policies, coordinating marketing and price support programs, and limiting the volume of basic grains imports from nonmember countries by means of quotas. To administer and enforce such agreements, the Protocol created a Marketing and Price Stabilization Commission (CCMEP) under the Central American Economic Council. He noted, however, that the absence of a regional agency with the responsibility and authority to coordinate agricultural planning, together with the lack of an adequate enforcement mechanism, made the coordination of production policies virtually impossible. Moreover, regional coordination of marketing policies was hampered by the failure of the CCMEP to design a harmonious system of price supports, although progress was made in establishing a regional grain storage system with funds allocated by the Central American Bank of Integration.

In addition to the institutional limitations, the CACM was also affected by other problems. For instance, the emergence of a rice industry in El Salvador threatened to disrupt the internal markets in Costa Rica and Nicaragua, resulting in the interruption of free trade and inducing retaliation measures on the part of El Salvador. The development of the poultry industry in that same country adversely affected Honduras. Similar conflicts of interest affected trade flows between Nicaragua and Costa Rica [Quiros-Guardia, 1973b, p. 53].

In conclusion, Quiros-Guardia attributed the lack of an adequate institutional structure for the design and implementation of a regional agricultural policy to several factors. One of these was the fact that the CACM entailed only free trade and tariff equalization. Consequently, a common agricultural policy was limited to those measures that would insure product mobility among member nations. There were no mechanisms for dealing with adjustment problems, which could be severe in the case of agriculture. At the national level, he believed that the general weakness and diffuseness of domestic market policy for agriculture in member nations inhibited the design of a more comprehensive regional policy. Similarly, the large number of agencies with responsibility for the formulation and implementation of agricultural policy made it difficult to establish a more complete and functioning organization at the regional level [Quiros-Guardia, 1973b, pp. 54-55].

The second study of agriculture in the CACM also noted that agriculture has been a barrier to full integration [Lizano-Fait, 1975b]. This author argued, however, that agriculture had received different treatment than the manufacturing sector, with the objective being to protect the interests of important social groups. The objective of his study was to answer these questions: Is it necessary to include agricultural products in the process of integration? What arguments have been used to exclude agricultural products from the process of integration? Is it necessary to give a different treatment to agricultural production than to industrial output? What has been the experience of the CACM in this respect? And, what are the prerequisites necessary to include agricultural products in the process of integration?

His answer to the first question was in the affirmative, on the grounds that agriculture produces raw materials for important parts of the manufacturing sector of member countries (milk, sugar, oilseeds, hides, etc.) and that different prices in different countries would affect the relative competitive advantage of industries using these raw materials, some of which are important sources of employment in the region. The arguments given for *excluding* agriculture from integration had included: there are no economies of scale in agriculture, thus there will be no reduction in production costs from integration; production patterns are the same in the member countries so there is nothing to be gained from integration; self-sufficiency in agriculture is a desirable policy goal; and free trade would have important employment repercussions. He effectively rebutted each of the first three arguments and, while recognizing the validity of the fourth, argued that special measures should be implemented to deal with it.

His response to the third question was also in the affirmative, on the grounds that the geography of production (location specificity) was different for agriculture than for manufactured products, with many more workers involved and with the potential for greater political pressure. In particular, he noted that prevailing policy interventions in the food and agriculture sectors were diverse and needed to be reconciled; small farmers might be harmed by competition and find it difficult to adjust; certain price stabilization measures would be needed as integration proceeded; and import substitution behind protective barriers for manufactured products would shift the domestic terms of trade against agriculture, thus prejudicing this sector, especially in light of the general political support for industrialization.

Finally, despite the *Protocolo de Limón* (also referred to as the *Protocolo Especial Sobre Granos*) referred to earlier, integration of the agricultural sector was a failure, and Lizano-Fait [1975b] discussed the dimensions of this failure and the reasons for it. In concluding his paper he noted that the process of integration requires free trade in agricultural products; it takes longer to establish free trade in agricultural products because of its effects on employment and land values; integration of agriculture requires a great deal of coordination of national policies as well as of financing and infrastructure: discrimination against agriculture must be reduced; and regional production patterns must be made more efficient so that consumers do not have to pay permanent subsidies to producers in such regions.

To conclude this review of literature on the CACM, three other studies are of interest. Wilford and Christon [1973] noted that in addition to the increase in trade associated with the elimination of internal tariff barriers, the composition of intraregional trade was also dramatically altered. By 1971, food products and raw materials constituted a far smaller proportion of regional trade, while regional manufacturing and chemical exports increased dramatically. Moreover, Guatemala and El Salvador posted substantial regional balance of trade surpluses over the 1960-71 period, while Honduras, Nicaragua, and Costa Rica experienced trade deficits. The retirement of Honduras from the CACM was the climax to a steady decline in that country's trade benefits from the market.

Willmore [1976] analyzed the patterns of trade creation and trade diversion for the market, the traditional components in analyses of customs unions. He found that there was substantial trade diversion of nondurable consumer goods, an important commodity category that accounted for 52 percent of interregional trade by the end of the period he studied. External trade creation in intermediate goods compensated for trade diversion in final goods, and accounted for the observed absence of trade diversion in aggregate studies. He found that there was considerable compensating trade creation in consumer nondurables, but on balance the CACM appears to have been a trade-diverting customs union for nondurable consumer goods.

Finally, Brada and Mendez [1983] used econometric techniques to compare the effects of integration for developing and developed countries. The areas studied were the EEC, EFTA, CACM, LAFTA, and the Andean Pact. Their conclusions:

> In sum we have found that economic integration among developing countries can have the same beneficial effect on intramember trade that it has among developed countries. Trade among CACM countries was found to be augmented by a factor not different from that estimated for EFTA and the EEC. On the other hand, LAFTA and the Andean Pact have had no positive effect on intramember trade, even when the greater distance and lower levels of development are taken into account.
>
> We have also found that the effects of trade preferences on integration are influenced by the level of development of the integrating countries and by the distance between them. Fortunately, from the standpoint of developing countries, the level of per capita incomes encountered in most of these countries is not so low as to represent an absolute barrier to the creation of effective integration schemes. Distance, on the other hand, has a powerful depressing effect on the ability of preference schemes to increase intramember trade. Thus, integration of very large, or distant developing countries is

unlikely to be effective in promoting intramember trade.
[pp. 599-600]

4. Concluding Comments

Common markets, customs unions and free trade associations emerged in Latin American areas as the basis of a second stage of import substitution for the region. These efforts at regional integration arose in response to disappointment and frustration with national input-substituting policies, and in recognition of the success with the European common market. The first, general effort at such integration, the creation of the LAFTA, essentially met with failure. Later efforts, the Andean Common Market and especially the CACM, experienced success, at least for a period of time. Eventually, they, too, reached limits beyond which it was not possible to go under existing institutional arrangements.

A number of factors appear to have put limits on the extent of integration obtained. One important factor was the difficulty in integrating countries that have wide disparities in stages of development and in per capita incomes. A second factor was the lack of institutional means for sharing in the benefits and costs of the integration. Still a third was the distance in terms of transportation costs which separated the countries, which probably helps explain why only the CACM was a success while the LAFTA got virtually nowhere.

Other important reasons for the failure to attain fuller economic integration were the inability to deal with the important adjustment costs that arise from economic integration, and the failure to deal with the special problems of agriculture. Although certainly not to be emulated by others, an important feature of European economic integration has been its Common Agricultural Policy (CAP). By providing common (high) protection to its agricultural producers the Common Agricultural Policy minimized (or at least reduced) the adjustments needed by the agricultural sector. Nothing comparable was provided in any of the Latin American initiatives.

Future endeavors at economic integration will need to give more attention to the special adjustment problems faced by agriculture. It is hoped that these problems will be dealt with in a positive way, and not by restrictive protectionist measures. Further endeavors will also need to give more attention to the development of regional institutional arrangements, especially those needed to provide for a more equitable sharing of the costs and benefits of integration.

Chapter VIII. Domestic Commodity Markets, Modern Input Markets, and Traditional Factor Markets: A Policy Perspective

This chapter is divided into seven sections. The first reviews the literature dealing with domestic commodity and modern input markets and policies. The second section covers the literature on contractual arrangements among factors of production. This is followed by sections on land tenancy reform—general issues, and on the Latin American land reform literature. The literature on agricultural credit and credit policies and on labor markets and migration is discussed in the following sections. The final section presents some concluding comments.

1. Domestic Food Policies, Agricultural Policies, and Modern Input Policies

In this section we discuss the literature on domestic policies affecting the food and agricultural sector, including the literature on modern inputs. The coverage includes food subsidies and food policies more generally, domestic commodity policies (such as minimum price policies) and marketing boards and parastatal marketing institutions. Some literature on the modeling of commodity markets is also included.

It is very difficult to generalize about these policies, for contrary to the trade and exchange rate policies, they have varied a great deal from one country to another and also over time in individual countries. Moreover, it is fair to say that there have not been very many penetrating analyses of these policies. An important exception to the rule was G. W. Smith's [1969] perceptive analysis of Brazilian agricultural policy in the 1950-67 period.

In an important and influential study, Echeverria [1969] investigated the effects of agricultural price policies on intersectoral income transfers in Chile during the period 1959 to 1967. This study is important for a number of reasons. First, shifts in the domestic terms of trade have played a central role in two-sector development models. Moreover, many policies affect agriculture by changing the domestic terms of trade. Second, the shifts in income expected to result from changes in the domestic terms of trade are assumed to influence the availability of capital for investment in different productive sectors, as well as the welfare of different groups of people. Third, Echeverria's analytical framework is useful for general analyses of agricultural policies. In summary, his study is valuable for the background it provides on the policies affecting agricultural development in Chile, for the methodology it develops, and for its empirical results.

Echeverria's maintained hypothesis was that the analysis of income distribution effects has to be conducted at a disaggregated level, in terms of productive sectors, income strata, and groups of people, if it is to be useful to policy makers. Moreover, the criteria used to make this disaggregation are important. The income groups he used in analyzing income transfers included large producers, small producers, tenant laborers, and agricultural workers for the agricultural sector, and nonagricultural producers, nonagricultural workers, and the rest of the world for the nonfarm sector.

For the empirical analysis, Echeverria divided the 1959 to 1967 period into three subperiods to reflect three distinct policy regimes. He found, as expected, that the relative income shifts among the different groups varied a great deal in each of the three periods, influenced in part by the particular policy instruments used in each period. Echeverria also noted that in drawing inferences about relative income shifts it is important to keep in mind the relative proportion of income of the particular group actually generated by agricultural activities. Hence, nonagricultural producers may benefit from relatively large transfers of income in an absolute sense, but this may involve a modest shift in relative terms because agricultural activities make up only a small share of the gross income of that group. Similarly, he stressed that it is important in making such analyses to consider all prices, and especially the price of labor, which he considers to be the most important of all.

In addition to his specific analyses of particular policies, Echeverria drew three general conclusions of interest. First, identification of the groups for analysis, as well as the type of price policies and their orientation, depend fundamentally on each country's historical and political reality. Hence, it is difficult to generalize across countries about particular policies. Second, he reminds the reader that the price system has both distributive and resource allocation effects and these are often in conflict. In this case, he points particularly to wage policy. These dual functions must be taken into account in the formulation of a strategy of economic development. Third, distributive goals can, in his view, probably be more effectively attained by redistributing assets, especially land in the case of Chile, and by reform of the fiscal system, than by manipulating price policy. He recognized, however, the political difficulties in implementing land reform and in reforming the fiscal system.

Turning to the more general literature on agricultural policies, a useful place to obtain a "feel" for the diversity of agricultural policies in Latin America is from the so-called "bench-mark" studies sponsored by the Ford Foundation in the second half of the 1960s. Volumes were published for Argentina [Fienup, Brannon, and Fender, 1969]; Uruguay [Brannon, 1968]; Brazil [Schuh—in collaboration with Alves, 1970b, 1971]; Peru [Coutu and King, 1969]; Venezuela [Heaton, 1969]; and Mexico [Venezian and Gamble, 1969]. This series of studies was motivated by a concern the Foundation had that economic policy research was being neglected in the region. The bench-mark studies were designed to take stock of what was known about agricultural policy in the region, and led to a concentrated effort to develop the capacity to train agricultural economists for policy analysis and to develop centers of policy research. The Foundation eventually supported such programs in Mexico, Colombia, Peru, Chile, Argentina, and Brazil under the leadership of Lowell Hardin, who at that time was the Foundation's program officer for agriculture in Latin America. Hardin, together with

F. F. Hill, International Vice President for the Foundation, conceived the bench-mark studies in the first place.

The graduate and policy research programs were based on the premise that, while the capacity to advance production technology was viewed as necessary for agricultural development, this capability in itself constituted only part of the total package needed for rapid economic growth. Structural, institutional, policy, and organizational changes were expected to be equally important. The studies focused on policy and institutional issues, including import policies, price policies, marketing, land tenure, extension, and research. An overview of some of the preliminary results of these studies can be found in Fletcher and Merrill [1968], which contains short papers on each of the seven country studies originally commissioned, plus a couple of synthesis pieces.

Another important volume on agricultural policy in the region reports the proceedings of a conference sponsored by the Interamerican Development Bank on agricultural policy as a limiting factor in the development process [Alexander, ed., 1975]. This volume has some excellent papers in it, and also gives the reader a flavor of the complexity and diversity of agricultural policies in the regions.

Modern input policies in the region have been importantly influenced by trade and exchange rate policies, by subsidized credit policies, and by import-substituting industrialization policies. Overvalued currencies constitute import subsidies and thus, on the surface, the use of modern inputs tended to be subsidized in the region. But countries such as Brazil, which tried to develop modern fertilizer and machinery and equipment industries behind protectionist barriers, typically more than offset these import subsidies with protectionist measures, thus discriminating against producers. These policies in turn were offset (in part or totally) by subsidized credit.

The World Bank's policy paper on mechanization [1986b] argues that general economic policies such as credit, trade, and exchange rate policies tend to have a far greater effect on the pace of mechanization in most developing countries than do mechanization policies *per se*. (Schuh had argued the importance of general economic policies as factors affecting agricultural development in two earlier papers, 1968 and 1975d.) For an excellent study of the role of policy in influencing choice of technology, see Sanders and Ruttan [1978]. For more general analyses of mechanization in Latin America, see Sanders's Ph.D. thesis [1973] and Thirsk [1972, 1980, and 1985].

Brazil's minimum price policy has been the subject of a great deal of analysis and empirical work. Some key references include G. W. Smith [1969], Paniago [1969], Rollemberg-Mollo [1983], Coelho [1979], and J. D. de Araujo [1980]. The last three references were conducted by Brazil's Company for the Financing of Production (CFP), which in recent years has had an unusually good analytical capability and good leadership. Other excellent papers published by the CFP include a careful analysis of the relations between Brazil's domestic prices and border prices [G. L. da S. Dias and I. G. V. Lopes, 1983; also, see CFP, 1983]; an excellent analysis of speculation in agricultural markets and producer incomes [M. de R. Lopes, 1983]; an analysis of government intervention in Brazilian agricultural markets [M. de R. Lopes, 1986]; and an analysis of alternative prices for stabilizing agricultural prices in Chile.

Per Pinstrup-Andersen has pioneered in the analysis of the effects of food prices on the poor, and the impact of food production on nutrition. Some useful references include Pinstrup-Andersen [1977, 1984, 1985]; Pinstrup-Andersen and Calcedo [1978]; and Pinstrup-Andersen, de Londono, and Hoover [1976]. The latter paper was particularly innovative.

Another important source is the collection by Solimano and Taylor [1980], which examines the role of food price policies on nutrition in Latin America. In addition to a couple of useful analytical papers, this volume reports on some seven case studies and contains three sets of policy recommendations.

The econometric modeling of commodity markets and the use of these models for policy analysis has been an important growth industry for economists in recent decades. Walter Labys and Jere Behrman and their colleagues have pioneered in this work, and have contributed importantly by bringing together the work of others. Some important basic references include Labys [1973, 1975], F. G. Adams and Behrman, ed. [1978], Labys, Nadiri, and del Arco [1980], and Labys and Pollak [1984]. Labys, Nadiri, and del Arco focused on commodity markets and Latin American development. Each of these sources has extensive references to bibliographic material. Labys, Nadiri, and del Arco contains analyses of commodity agreements. Tropical tree crops, such as coffee and cocoa, have received considerable attention from modelers. Unfortunately, despite the numerous attempts to evaluate sectoral policies with these models, it is difficult to generalize about the results.

An important part of domestic food and agriculture policy in Latin America is implemented by means of parastatal and domestic price fixing organizations. Examples of the parastatal food agencies include CONASUPO in Mexico and INESPRE in the Dominican Republic. Often these agencies are predicated on the assumption that domestic markets are imperfect and exploit both consumers and producers. While initiated in an attempt to provide competition for perceived monopolists and monopsonists, these parastatals typically have grown to take over large parts of the marketing sector.

Another institutional arrangement important in domestic food and agriculture policy is domestic price fixing organs such as SUNAB in Brazil. In inflationary Latin America, it is not surprising that policy makers have a keen interest in the *custo de vida* (cost of living) and that they see price fixing as the means to control it. These agencies probably do more to create risk and uncertainty than they do to affect the level of prices, for if prices are effectively lowered supplies eventually disappear from the markets and policy makers have to relent in their price fixing.

Unfortunately, there have been very few attempts to evaluate the parastatals, either in terms of their internal efficiency or in terms of their impact on the markets. An important exception is Quezada's [1981] study of INESPRE in the Dominican Republic.

Risk is an important characteristic of agricultural production systems and many countries have ambitious crop insurance programs to assist farmers in coping with risks. Moreover, problems associated with risks are one of the reasons governments intervene in their agricultural product and factor markets. Hazell, Pomerada, and Valdés [1986] have edited a collection of papers addressing crop insurance as a means of dealing with risk and as a means of promoting agricultural development. They try to put crop insurance in perspective as one means of managing agricultural risks. They note that the opportunity costs of resources used for such programs are quite high, and that in practice, multiple-risk crop insurance has proved disappointing. Papers in the volume address alternative means of attaining risk reduction. The experiences of Brazil, Mexico, Costa Rica, and Panama are considered.

Finally, recent years have witnessed some attempts at assessing the commitment of government to agriculture through its expenditure on the sector. Schuh and Thompson [1980] suggested the use of government expenditures as the means of assessing this commitment, and illustrated the difficulties of doing so with data from Brazil. Elias [1981] has made a careful analysis of the data for Latin America.

To conclude, domestic commodity policies have taken a back seat to trade and exchange rate policies in Latin America, especially in terms of their impact on the agriculture's domestic terms of trade. Government interventions in the domestic economy are often pervasive, however, especially in the form of parastatal food distribution and marketing agencies, and in terms of domestic price-fixing agencies. These institutions typically have either distributional or inflation-control objectives. Unfortunately, we know all too little about their impact on the food and agriculture sector.

2. Contractual Arrangements Among Factors of Production

The size distribution of land holdings in most Latin American countries is highly skewed. The poor in these countries tend to be concentrated in agriculture and in rural areas. Rather unique politico-socio-economic arrangements have grown up around what is called the *latifúndio-minifúndio* complex, with a common perception being that there is a dependency relationship between the two kinds of agricultural operations and that the *latifúndio* exploits the *minifúndio*.

Moreover, research persistently finds that the productivity of land is higher on small farms than it is on large farms.

This conjuncture of stylized facts about Latin American agriculture has generated a strong interest in the size distribution of land holdings, and a persistent interest in land reform on the part of intellectuals concerned about agricultural and more general economic development. This interest is reflected in pleas for land reform that would distribute land to the landless workers, attacks on sharecropping as a tenure arrangement, criticisms of the contractual arrangements by which sharecroppers have a portion of their labor services tied to the farm activities of their landowner at less than the prevailing wage, and a general condemnation of the *latifúndio-minifúndio* complex. There are many unsettled issues, however, not the least of which is whether small landowners are really exploited, or whether it is the landless worker who is worse off.

In this section we provide a general discussion of contractual arrangements among factors of production. Although these arrangements have received scant attention in Latin American literature, a discussion of the issues provides a general setting for the sections which follow.

In the literature on Asian agricultural development, a vital literature on tenure and land, credit, and labor contractual arrangements has emerged these last twenty years [see Binswanger and Rosenzweig, eds., 1984a, for references to much of this literature]. This literature seeks to evaluate these contractual arrangements to ascertain their efficiency and their effects on resource use and the distribution of income. We draw on Binswanger and Rosenzweig [1984b] to set the stage for a more detailed discussion of some of this literature.

Development thought in the 1950s and 1960s was dominated by macroeconomic models of the economic development process. To the extent these models focused on the labor market, agricultural employment, and agricultural wages, they did so in the context of viewing agriculture as a source of labor for nonagricultural sectors of the economy. Agriculture was viewed as a source of "surplus" labor that was to be mobilized for the expansion of the nonfarm sector. Put somewhat differently, industrialization was viewed as the means to provide employment for that surplus labor, which in many models, and especially those pioneered by Lewis [1954], was viewed as having zero marginal productivity, and thus generating economic development.

The models of intersectoral labor markets, such as they were, were highly abstract simplifications, with little attempt to relate them specifically to empirical reality. Recent developments in the theory, however, have attempted to get at the microeconomic roots of the markets and to understand the behavior of the various participants in the market. In this sense, these developments have paralleled the work of Hayami and Ruttan [1985] in understanding the microeconomics of the process of technical change, including their concern with institutional arrangements.

A central theme that arises in the Binswanger-Rosenzweig [1984b] review of the literature is that an understanding of institutional arrangements or imperfections in any one market (for example, the labor market) requires attention to the imperfections in or constraints on other markets (for example, the land or credit markets). This perspective would appear to be of particular value in Latin America, where tenure arrangements so frequently involve apparent "tie" contracts between the use of land and labor [see Teixeira, 1976].

Binswanger and Rosenzweig noted that two lines of thought characterize the surplus labor model of rural agriculture. According to the first, rural labor can be withdrawn from the agricultural sector because there are large pools of unemployed or underemployed rural workers. The theoretical problem is to reconcile large scale unemployment or underutilization of laborers with a nonzero wage for labor. The second line of thought seeks to explain the assumed insensitivity of agricultural output to the number of available laborers by distinguishing between that number and total labor supplied. This approach focuses on the labor supply behavior of the peasant household as well as on the structure of the labor market.

According to Binswanger and Rosenzweig [1984b], the most influential surplus labor model that explains the coexistence of idleness and constant wages is the nutritionally based efficiency wage hypothesis, elaborated first by Leibenstein [1957] and later by Mazumdar [1959], Mirrlees [1975], and Stiglitz [1976]. This model assumes that at low levels of income there is a technically determined, positive relationship between nutritional level and labor effort per unit of time (or per worker). Binswanger and Rosenzweig note that direct empirical tests of this relationship are extremely rare. The most rigorous experimental study they found was of sugar cane workers laboring under actual field conditions in Guatemala [Immink and Viteri, 1981]. That study found no relationship between energy availability and supply of work units.

Binswanger and Rosenzweig find this model implausible because of the apparent lack of any relationship between energy intake and effort, especially in the short run, and because workers can be easily hired on a daily basis. The alternative route to rationalizing the possibility that agricultural workers will have a zero marginal productivity requires a distinction between labor time supplied and laborers. Substitutability within the household then makes it possible to remove one member of the family without affecting output.

This perspective leads to duality in the agricultural sector, with some farms maximizing profits and hiring workers whereas other, smaller farms use only family labor and participate in the agricultural labor market only as suppliers of labor. But this duality in the labor market implies a duality in the land market.

Moreover, Binswanger and Rosenzweig note how mobile in employment farm people really are, which makes it unlikely they can be easily exploited.

Reference to duality in the land markets takes one to the subject of tenancy. In the rural areas of most Latin American countries some labor is combined with land by means of contracts which link the delivery of labor to access to land. Thus, the terms and arrangements associated with the market for land have a significant effect on the earnings of rural households and on aggregate production processes. Moreover, the interconnectedness of the agricultural factor markets is one of the important findings that has come out of the new land tenure literature. With such interconnectedness, one cannot understand one market without understanding what is happening in the others. More specifically, one cannot understand what is happening to agricultural wage rates and the earnings of labor without understanding the rental value of land and what is going on in the land market.

Binswanger and Rosenzweig note that the absence of a sales market for land is not sufficient to force the use of tenancy since landowners could still hire all cooperating factors in quantities that are optimal for their own land. However, the institution of tenancy and the market for tenancies do substitute for the sales market. When there are no scale economies, at least one other factor market must be absent before the temporary rental of land becomes a necessary means to achieving the most efficient factor ratios for all factors of production and all ratios. This issue is discussed in more detail in the next section.

3. Land Tenancy Reform — General Issues

The issue of sharecropping has long been of interest to general economists and agricultural economists alike. Hayami and Ruttan [1985, p. 389] note that "tenure reform has been viewed as essential to the mobilization of labor resources and the generation of productivity growth in both liberal and Marxist development perspectives." By the early 1950s there was a consensus that from the standpoint of productivity alone, an owner-operator agricultural system was best.

This consensus led to major emphasis on land reform in the technical and economical assistance efforts of a number of national and international development assistance agencies after World War II. The U.S. commitment to land reform as an instrument of political and economic policy began with reforms carried out in Germany and Japan. The success of the Japanese land reform was important in U.S. support for land reform under the Alliance for Progress in Latin America [see L. J. Walinsky, ed., 1977; the other main thrust of the Alliance was in economic planning, as noted above].

Reviews of the land reform efforts and accomplishments of the 1950s and 1960s in Latin America can be found in E. Feder [1965], Raup [1967], and Dorner and Kanel [1971]. Other important sources for that early period include United Nations [1951] and AID [1970].

As evidence accumulated over time, it became clear that the relationships among farm size, tenure, and productivity were not as expected. Share tenants frequently achieved higher yields than owner-operators, especially in the smaller size ranges. Even in the larger size classes, owner-operators seldom exhibited any clear-cut productivity differentials relative to other classes. In many cases the highest levels of productivity were achieved by owner-tenants—typically small landowners who cultivated rented land in addition to their own land.

These anomalies gave rise to a growing literature which has tried to explain the occurrence of sharecropping. As Quibria and Rashid [1986] note, "For almost two decades sharecropping has been an institution in search of a generally agreed upon theoretical basis." These same authors [1984] provide a survey of this literature.

There is a number of phenomena that need to be explained by a comprehensive theory of contractual arrangements in agriculture. The first is the existence of share tenancy as a contractual arrangement. The second is the existence of fixed rental and fixed wage contracts, which often coexist in the same area with sharecropping. And the third is the spatial and temporal variation of each form of contractual arrangement, and thus of their links with the socioeconomic environment. Past experience suggests that the dominant contractual form can vary with the crop, the prevailing technology, the extent of market development, and other characteristics of the economic and social environment.

A related set of issues is whether the interlinked contracts, referred to above, are in fact a form of exploitation of less powerful agents by more powerful agents. Bhaduri [1973, 1977] has taken this view, as has much of the populist literature on tenure issues in Latin America.

Eswaran and Kotwal [1985, p. 352], drawing on Binswanger and Rosenzweig [1984b], note that three types of explanation have been offered for the existence of different technical contracts: a tradeoff between risk sharing and transaction costs; screening of workers of different qualities; and market imperfections for inputs besides land. Cheung [1969] began the recent interest in share tenancy by challenging the prevailing view that share arrangements were less efficient than land ownership or fixed rents. His perspective was anticipated as early as 1950 by D. Gale Johnson, who questioned the presumed inefficiency of share arrangements in U.S. agriculture.

Cheung developed a general equilibrium model in which the terms of contract such as share rate, farm size, and input levels are variables, determined through negotiations between landlords and tenants, while the wage rate is determined in the market. If private property rights are well defined and the enforcement of contractual terms is costless, Cheung concluded that "the implied resource allocation under private property rights is the same whether the landlord cultivates the land himself, hires farm hands to do the tilling, leases his holdings on a fixed-rent basis, or shares the actual yield with his tenant" [Cheung, 1969, p. 4]. He further postulated that sharecropping offers the advantage of risk sharing, while the other two contracts involve lower transaction costs. The optimal tradeoff in a given set of circumstances would then determine the dominant contractual form.

There have been two challenges to the Cheung perspective. First, the argument that risk sharing is the main motivation behind sharecropping lacks empirical support [Pant, 1981; C. H. H. Rao, 1971; K. Chao, 1983]. C. H. H. Rao in particular argued that both the enforcement cost and the return to the allocative ability of tenants can be very high for the crops characterized by dynamic changes in their production and demand functions. For such crops the leasehold contract would be preferred even if the risk were higher.

The second challenge to the Cheung perspective is that the tradeoff argument lacks credibility since there is no reason to believe that sharecropping involves greater transaction costs than a wage contract if transactions costs include supervision costs, as they should. Further, Newbery [1975], Reid [1976], and Stiglitz [1974] argued that risks cannot be a decisive factor in explaining the choice of share contract because the same degree of risk sharing can be achieved by combining leasehold and fixed-wage contracts. The indivisibility of the tenant's entrepreneurial and managerial ability, however, would make it inefficient to allocate his time between wage employment and leasehold farming on a smaller scale. [See Jaynes, 1984, for a detailed critique of Cheung's argument.]

The second type of explanation for the existence of different tenurial contracts is what are characterized as screening models. This perspective is associated with Hallagan [1978], and Newbery and Stiglitz [1979] when referred to tenants, and to Allen [1982], who used a screening model similar to that of the other authors cited to incorporate the unobservability of the quality of land in addition to the unobservability of the tenant's ability.

These authors argued that it is the asymmetry of information about tenants' abilities (and about the quality of land in the case of Allen) that warrant sharecropping. Different contracts are needed to sort tenants out by their abilities. Eswaran and Kotwal [1985], however, argued that the assumption of ignorance on the part of landlords about tenants' abilities is quite inappropriate for most rural communities. In their view there is typically little mobility among these groups and thus information about abilities and assets is easily available. They also note that screening models cannot explain why a certain contractual form might predominate in one area while quite another form predominates elsewhere, nor can they explain the change in the contractual structure that has been observed to result from a change in technology or the development of markets. They note, for example, that tenancy contracts changed to wage contracts in India after the introduction of new technology in the 1960s [C. H. H. Rao, 1977, ch. 12], and that the sharecropping contracts in the post-bellum South of the United States changed to wage contracts with the advent of mechanization [Day, 1967].

The recent literature on sharecropping is converging on market imperfections for inputs other than land as the explanation for variations in contractual arrangements. From this perspective, tenancy substitutes for the absence of, or imperfections of, a market for some factor input besides land. Eswaran and Kotwal [1985] noted that the absence or incompleteness of markets can typically result from the high costs of quality enforcement. They noted the extent to which recent literature has identified factors for which markets are highly imperfect: technical know-how [Reid, 1976], managerial ability [Bell and Zusman, 1979], bullocks [Bliss and Stern, 1982], credit [Jaynes, 1984], and family labor [Pant, 1983].

Eswaran and Kotwal [1985] developed a theory of contractual structure in agriculture based on this perspective. They put it well:

> An effective way of gaining access to such a factor (for which the market is imperfect or absent) is to offer a self-monitoring (incentive) contract to the factor owner, involving him in the production process. The factor input is thus available only as a package deal with the factor owner's time. However, the self-monitoring contract does not have to be a share contract. The landlord could gain access to the tenant's supervision ability or to his bullocks by offering him a fixed rental contract. Why then does sharecropping exist?
>
> Following Reid [1977] we envisage the landlord and tenant as both contributing unmarketed resources in a sharecropping arrangement. We view sharecropping as a partnership in which both agents have incentives to self-monitor. Such a contract arises to mitigate morally hazardous behaviour on the part of both agents — a phenomenon as yet unexplored in the literature. If all the monitoring of input quality is undertaken by a single agent, he becomes the sole residual claimant; in a wage contract it is the landlord, and in a fixed rental contract it is the tenant. The different contracts thus reflect different techniques of combining unmarketed productive inputs. The choice of technique depends on exogenous parameters such as the endowment distribution across the classes of factor owners and the prevailing production technology. The equilibrium contractual structure emerges from the optimizing decisions of both landlord and tenant in a given environment. In order to

> facilitate the derivation of the testable implications of this view, we shall abstract from all considerations of risk in this paper. This abstraction enables us to make a comparison of the hypothesis that sharecropping exists to pool unmarketed resources with the implications of the conventional wisdom that it exists to share risk. [Eswaran and Kotwal, 1985, p. 353]

Eswaran and Kotwal [1985] noted that Bliss and Stern [1982, p. 309] also viewed sharecropping as an arrangement that involves the pooling of managerial and cultivating skills. They also noted that their proposition that the different contracts reflect different techniques of combining unmarketed productive inputs is implicit in Reid's work on sharecropping [1976, 1977], which Reid later makes explicit [1979]. What Eswaran and Kotwal did was to formalize the idea and relate it to the exogenous parameters of the social and economic environment.

Eswaran and Kotwal [1985] developed a model of agricultural production that enables them to endogenize the type of contractual arrangement (fixed rent, fixed rental, or sharecropping) that will prevail in a given section. They concluded that their view of sharecropping as a *partnership* is consistent with three significant empirical observations made in the literature [p. 364]:

> First, the yields on farms cultivated under sharecropping are sometimes found to be higher than on farms alternatively cultivated, despite the moral hazard inherent in the noncooperative nature of the share contract. In their model this may result from the advantage of being able to pool unmarketed resources. Second, the hypothesis that the greater the production risk, the greater the prevalence of sharecropping, has received little support in recent empirical investigations [C. H. H. Rao, 1971; Pant, 1981]. . . . Third, the relative insensitivity of the share to the variation in technology and market characteristics across different regions has also been shown to be consistent with our model. [Eswaran and Kotwal, 1985, p. 364]

Quibria and Rashid [1986] took a narrower view to the imperfect factor market hypothesis and rested their model on the existence of dual labor markets. These authors argued that there are two assumptions which have frequently been recognized in descriptive accounts of sharecropping economies, yet inadequately explored in the theoretical literature: the existence of a dual labor market; and the fact that sharecroppers are not always landless peasants but rather are themselves small landowners. They then point to some seven facts which they believe are consistent with dual labor markets [p. 95]. After finding that their analytical results are consistent with these facts, they remind the reader that one reason why sharecropping has appeared to provide such a disparate series of results is the in-herent complexity of the issues. A general model which encompasses all the choices available becomes unwieldy and analytically intractable.

There are a variety of other important papers in this tradition of the interlock-ing markets perspective—papers that bear on the sharecropping issue. Braver-man and Srinivasan [1981] model the linkages among land, labor, and credit transactions in the context of sharecropping. They show that regardless of the presence or absence of linkage, or any other control by the landlord, in equilib-rium contracts a tenant's utility under sharecropping will be the same as that which he could have obtained as a full-time wage laborer, so long as the landlord can vary the size of the plot given to a tenant and there are enough potential ten-ants. This result implies that policies other than land reform will not affect the welfare of tenants. For example, government subsidization of tenant's credit re-sults only in the subsidization of landlords.

Braverman and Stiglitz [1982] addressed the issue of whether interlinked mar-kets are a means of exploiting less powerful agents by more powerful agents. They concluded that the presence of interlinkages need not be taken as evidence that agrarian markets in less developed countries are noncompetitive, although it is clear that such linkages have distributive as well as allocative effects. More im-portantly, they concluded that in many situations, competitive and noncompeti-tive markets may look quite similar. Thus, distinguishing among the various possibilities may in their view require greater subtlety than is frequently em-ployed in empirical and policy work in this area.

Braverman and Guasch [1984] viewed the linkage of tenancy and credit con-tracts as a screening device in an environment characterized by a heterogeneous labor pool and imperfect information. Braverman and Stiglitz [1986] found that cost-sharing contracts have a decided advantage over contracts which specify the level of inputs whenever there are asymmetries of information regarding produc-tion technology between the landlord and the tenant.

Finally, the important Braverman and Stiglitz [1986] paper addressed the issue of linkages and the adoption of new production technology. They noted that there is a long-standing belief that landlords and capitalists have used their control over the means of production to direct the development and adoption of technol-ogies which have increased their welfare at the expense of the workers. They fur-ther noted that there is also a widespread belief that the interlinkage between credit and tenancy markets provides further impetus to the resistance to innova-tions. Innovations which make tenants better off presumably reduce their de-mand for loans, and thus make landlords (*qua* creditors) worse off. Bhaduri [1973, 1977] has made such an argument.

These contentions have typically been dismissed out of hand by standard wel-fare economics arguments, which insist that incentives always exist to adopt the technology. Braverman and Stiglitz [1986] pointed out, however, that the rural environment of most developing countries may not be adequately described by the standard economic model on which this welfare analysis is based. In particular, in developing countries, where sharecropping contracts are widely used, there is evidence of widespread unemployment, and the full set of risk and capital markets required by the competitive paradigm is lacking.

The Braverman and Stiglitz [1986] paper shows that under quite general conditions the institutional structure of the economy may indeed be an important determinant of whether a particular innovation will or will not be adopted. In particular, they show that landlords may wish to—and can—resist innovations which unambiguously increase production whenever sharecropping contrasts are employed; conversely, landlords may adopt innovations which not only lower the welfare of workers, but even lower net national product; and the pressure of interlinkage may, indeed, affect the adoption of a new technology. The reason it does, however, is only partly related to the effect of the innovations on tenants' borrowing. To the contrary, they show that innovations may increase as well as decrease the tenants' demand for borrowing. They also refer to the possibility that the mere presence of unemployment (even under a pure wage system) may be sufficient to generate resistance by landlords and employers to the adoption of superior technological innovations.

Some other general references on sharecropping and interlinkages include Bardhan [1977, 1979, 1980], Mitra [1983], Srinivasan [1979], Bardhan and Srinivasan [1971], Lucas [1979], Mazumdar [1975] and Quibria and Rashid [1984].

To conclude this section we want to make four points. First, in the environments of rural areas in many developing countries, the conditions for competitive allocative efficiency simply do not prevail. Many markets are imperfect; others necessary for the competitive paradigm to apply do not exist; and the lack of information is a serious limitation. In the final analysis, whether markets are efficient is an empirical question, not something to be taken on faith.

Second, the new land tenure economics of the last twenty years represents an antithesis to the paradigm that governed land reform programs in the post-World War II period. It demolished the traditional view that share tenancy is always inefficient and thereby removed the rationale for its replacement by leasehold tenancy or owner-operatorship as a prerequisite to rapid productivity growth in agriculture [Hayami and Ruttan, 1985, p. 395]. On the contrary, the new theory demonstrates that under certain circumstances share tenure can be a more efficient method of sharing risk and minimizing transactions costs, compared to other leasing arrangements. Clearly, the efficiency of alternative contractual arrangements is critically dependent on the level of technology which affects the capacity to supervise labor. However, all these statements are contingent upon a given property rights allocation. They do not negate the reason often given for prefer-

ring land reform, i.e., that giving land to the tillers will eliminate some inefficiencies by allowing the operator of the land to obtain his full marginal productivity.

This leads to our third point, that successful land reform must be designed
with due recognition for the underlying conditions in the factor markets and their
interlinkages. Therefore, it often has to be complemented with reforms in the
other markets.

Finally, the observant reader will have noticed that much of the literature on
the new tenure economics was generated by those concerned with tenure issues in
Asia. Hayami and Ruttan [1985, p. 396-397] attribute this to the fact that land
reforms in Asia were carried out with the expectation that reforming labor and
tenure relationships through elimination of intermediaries, regulation of rental arrangements, and transfer of ownership rights to the tiller would achieve both equity and productivity objectives.

Paradoxically, the literature on interlinking markets has had very little impact
on the Latin American literature. Hayami and Ruttan [1985, p. 397] attribute this
to the fact that neither the practice nor the theory of land tenure reform in Latin
America has given much weight to the achievement of productivity objectives
through the transfer of land ownership to the tiller. It has concentrated more on
equity and political objectives instead.

This possible explanation for the neglect of the new tenure economics in the
Latin American literature strikes us as only partially correct. In the first place, the
new perspective provides a useful framework for analyzing both equity and productivity issues. In the second place, many of the justifications for land reform
measures are based on the empirical observation that land productivity is higher
on small farms than it is on large farms. The implication for many observers has
been that breaking up the large farms into small farms will raise overall land productivity.

4. The Latin American Land Reform Literature[92]

Land reform has been a popular and recurrent issue among intellectuals concerned with Latin America. One motivation for this interest was noted early in
this paper—the way the land was originally distributed at colonization, in large
blocs, and the highly skewed distribution of land holdings that still prevails in
most countries today. Another motivation is the social relations among land
owners and workers in the region, which is widely viewed as a means whereby
the economically powerful exploit the economically less powerful.

Carroll [1961] has noted that the main features of the agrarian structure in
Latin America are: the importance of *latifúndios*, or very large farms; the large
number of *minifúndios*, or very small farms; the special situation of the *comuni-
dades*, or communal holdings; and the peculiar form of farm labor known as the

colono system. The *latifúndios* not only account for the major share of the land in most countries, they also are assumed to include land of the higher qualities, and are viewed as exercising monopoly power. Carroll [1961] argued that the worst feature of land concentration is the resulting concentration of power which it carries with it. He further noted that it is against this concentration of power that most of the fury of popular land reforms has been directed [p. 165].

The *minifúndios* are much larger in number, but account for a very small part of total land holdings. They are typically closely interrelated with the *latifúndios* in very intricate ways. They supply seasonal labor to the *latifúndios* and in many ways contribute to the maintenance of the system.

The third major type of land holding in Latin America is the *communidad*, which dates back to the Incas, Mayas, and Aztecs, all of whom held land in collective fashion. The system survives today in areas of native Indian populations, mostly in the Andean areas. The base of the *communidad* is the extended family, who together have claim over a specific land area. The Mexican *ejido* is a throwback to the *communidades*.

The last major feature of the Latin American tenure system is the *colono*. These workers have the status of tenant laborers. They are paid in the temporary or traditional usufruct of a parcel of land and certain other privileges. In return, the *colono* must serve a specific number of days on the estate and fulfill other customary obligations, such as making available members of his family for certain tasks in the field or in the owner's household. This system is often combined with sharing or with tenancy on a cash rent basis.

> The concept of *colono* varies from one part of Latin America to another, and often from one region to another within the same country. In the states of Paraná, São Paulo, and Minas Gerais in Brazil, for example, the *colono* is responsible for the care of a given area of coffee (a number of trees) or sugar cane. In the states of Rio Grande do Sul and Santa Catarina, further to the south, the *colono* tends instead to be a small farmer who is a descendant of immigrants and who owns his own land. The most common concept in Brazil, however, is that of a worker paid in wages (*assalariado*), hired on a "permanent," monthly, or daily basis. [Paiva, 1988]

From the North American side, intellectual leadership on Latin American land reform issues has been closely connected with the Land Tenure Center at the University of Wisconsin. Perhaps the best representation of the views of this Center can be found in *Land Reform in Latin America: Issues and Cases*, edited by Peter Dorner, ed. [1971] and dedicated to Kenneth H. Parsons and Raymond J. Penn.

The Foreword to this book contains a brief synopsis of the intellectual antecedents of the Center and we draw on it here. It notes that one of Wisconsin's
early economists, Richard T. Ely, held a continuing interest in the ownership of
resources and how ownership affected land use. In 1914 he published his *Property
and Contract*. Ely brought to Wisconsin John R. Commons, who worked on labor
problems. Ely, Commons, and their associates, among other things, included in
their basic approach both interdisciplinary cooperation and comprehension of
working rules or property interests involved.

Professor Hibbard, a student and associate of Professor Ely, directed attention
to land use problems, particularly those in rural areas. *A History of the Public Land
Policies* by Hibbard [1924] is a scholarly source of information on the land policies
of the United States. *Land Economics* by Ely and George Wehrwein [1940] was for
many years the primary text for students interested in this field.

International land tenure problems were a part of the regular seminars at Wisconsin as early as the 1920s. Ely, Hibbard, and later Wehrwein devoted every
other year of their Land Tenure Problems seminar to the study of these issues in
other countries. Later, Professor Kenneth H. Parsons and his students developed
further this work in land tenure.

In 1950, Hugh H. Bennett, the first chief of the Technical Cooperation Administration (TCA, a predecessor of the present AID), spoke to the annual meeting of the Land Grant College Association. Mr. Bennett asked what the universities could best do to improve United States foreign policy and to assist
developing countries. The University of Wisconsin decided to attempt a new
contribution in the field of land tenure. Professor Raymond J. Penn of Agricultural Economics and William Small of Rural Sociology were asked to begin the
preparation of a proposal to TCA. The proposal, prepared with assistance from
Professor Kenneth H. Parsons of Agricultural Economics, suggested a Wisconsin
Conference on World Land Tenure Problems to be held in the Fall of 1951.

Three representatives from each of forty countries were invited to the conferences. The conference met for six weeks in Madison and traveled one week
through the midwest and southeast United States, ending in Washington. The
proceedings of the conference were published by the University of Wisconsin
Press in 1956 in a book entitled *Land Tenure*, edited by Parsons, Penn, and Raup.

The steering committee for the conference, which excluded representation
from the University of Wisconsin and the U.S. government, recommended the
establishment of a land tenure center at Wisconsin. In 1961, when Congress permitted some of the foreign assistance appropriation to be used for research, the
University proposed the creation of the Land Tenure Center, which was established in 1962.

True to its intellectual antecedents, the Land Tenure Center has focused on the
institutional arrangements and on political power, to the relative neglect of effi-

ciency issues. The first paragraph of the collection edited by Dorner [1971] states the perspective very clearly:

> In rural areas land ownership or other secure forms of tenure which assure the farmer of some control over the returns from his labor and the land he works is the real and practically the only means of participation in the political and economic life of the country. This *is* the access route to economic and political citizenship and to a share in the sovereign power of the nation state. [p. xvii]

and later:

> Latin America needs increased total agricultural output, increased employment, and increased productivity per worker. The contribution of all these is unlikely to be achieved without appropriation and reorganization of many of these large farms. [p. xvii]

This emphasis on institutional arrangements and political power, and the significant influence of the LTC on research on land tenure issues in Latin America, probably explains why the issue of interlinked markets and the efficiency questions of tenure arrangements have received so little attention in the Latin American literature. This, of course, takes nothing away from the importance of the Center's emphasis on institutional (read microinstitutional) arrangements. The Dorner [1971] volume gives a good flavor of this perspective, with a combination of conceptual papers and case studies.

Another important source on agrarian reform in Latin America is de Janvry's [1981a] *The Agrarian Question and Reformism in Latin America*. Although heavy reading for one not steeped in the concepts of the dependency literature, and made more complicated by his own introduction of new concepts, de Janvry's scholarship is massive as he draws together a great deal of the literature and the disparate lines of thoughts.

De Janvry claims that he was motivated to write his book by two frustrations that he believes are widely shared by others. We quote from his preface:

> One is that practitioners of economic development . . . lack a global interpretative framework that would allow them to give consistency to the variety of actions they undertake. As a result the practice of economic development has been reduced to the evaluation and implementation of disconnected projects. Underlying these projects is a dramatic paucity of understanding of how they fit into the broader economy and

what their political implications are. Most importantly, there is
a general incomprehension (or neglect) of the state as a social
phenomenon and of its logic, role, and limits in the
management of reforms. In terms of the practice of
development, a crying need thus exists for a unifying
framework that is both sufficiently comprehensive to explain
the multidimensional facets of underdevelopment and
sufficiently simple to provide a broad set of guidelines that can
in turn be made more specific in particular historical,
geographical, and ideological contexts to serve as a basis for
policy foundation and political action.

The other frustration . . . is that the global interpretations
that have been developed . . . have remained at very general
and abstract levels. This, while pleasing to academic scholars of
world economic systems and armchair revolutionaries, has left
the practitioners of economic development, from policy makers
to political activists, in a vacuum: the global frameworks have
provided only sweeping directives that can hardly be translated
into pragmatic programs, and specific projects can be fitted
into the global frameworks only in a rather distant and artificial
fashion. In this book, I consequently attempt to show that this
vacuum need not exist; a global interpretative framework is
indeed essential for policy and political action, and it can serve
to give very pragmatic directives for the definition and
implementation of "what is to be done." . . .

Consequently, this book is also written for scholars and
students of economic development in an attempt to further our
understanding of the *global* phenomenon of underdevelopment
in *the world economic systems* and to show how this knowledge
can be used in the analysis and design of specific projects.
[de Janvry, 1981, pp. xi, xii; emphasis added]

One thus sees that de Janvry in this volume is in the tradition of dependency theory, although he would prefer to describe it as political economy.

Chapter headings in de Janvry's book include "Laws of Motion in the Center-Periphery Structure," "Transformation of the Agrarian Structure and the Peasantry," "Disarticulated Accumulation and Agrarian Crisis," and "The Political Economy of Reformism." There is one chapter directed to land reform, and another to the strategy of integrated rural development.

In the chapter on land reforms, de Janvry reviewed the sixty year history of land reform in Latin America, starting with the Mexican Revolution. There have been seventeen land reforms in ten countries of the region, including three in

Mexico, three in Chile, and two in Guatemala. None of these has been extensive in scope, or, with the exception of the Mexican reforms, has had significant effects on the size distribution of land holdings.

De Janvry developed a typology of land reforms using the concepts of mode of production and social class, and then classified the various reforms. This permitted him to contrast reforms that sought to redistribute land to peasants within a precapitalist or a capitalist agriculture, induce a transition from precapitalist to capitalist agriculture along either the "junker" or the farmer road of development, or promote a shift from "junker" road to farmer road. De Janvry concluded that

> while these reforms have been successful in inducing a rapid
> elimination of precapitalist social relations in agriculture, and in
> some cases in eliminating the landed elites from control of the
> state, their impact has been severely limited by the constraints
> of disarticulated accumulation. The permanence of these
> constraints, which reproduce the objective and subjective
> dimensions of the agrarian crisis, implies that the solution to
> the agrarian crisis increasingly lies outside the domain of
> agrarian reformism. [de Janvry, 1981a, p. 5]

De Janvry noted [1981a, p. 5] that with the end of the land reforms in the early 1970s, rural development and basic needs programs assumed a key position in the strategy of agrarian reformism. He devoted a chapter to an analysis of these programs.

De Janvry wrote with authority from the perspective of dependency theory. Somewhat surprisingly, in the context of his analysis of the agrarian "crisis" in Latin America, he is not a revolutionary. For his policy recommendation, he advanced an economic program for equitable *growth*, not revolution. Specifically, he argued that the social articulation of economic systems, whereby the final-goods market for the modern sector is located in the expenditure of wage and peasant incomes, should be the key objective of structural transformation, and that the rapidly emerging rural semiproletariat and landless labor are the fundamental progressive political forces in agriculture, even though they are fraught with all the ideological contradictions that their peasant roots and continued linkages to subsistence agriculture imply [de Janvry, 1981a, pp. 5-6]. The interested reader will want to read Barraclough's [1984] review article of de Janvry's book.

Hayami and Ruttan [1985] have some interesting observations on land reform in Latin America:

> In land reforms initiated under leftist regimes, such as Cuba in
> the 1960s and Chile in the late 1960s, distrust of the peasantry

> has often resulted in attempts to establish group farming
> enterprises guided or controlled by a state agency. Similar
> transformations have also been attempted by a number of
> populist or reformist governments, such as Peru in the early
> 1970s and El Salvador in the late 1970s. But regardless of
> whether the reforms have been inspired by radical, populist, or
> reformist ideologies, the political leadership has sought to
> retain control over the organization of agricultural production.
> The group farming enterprises resulting from such reforms
> have more often become a burden on development than an
> efficient source of food and fiber. And they have typically
> failed to satisfy the land hunger of the peasants. [p. 397]

Valdés's [1974] analysis of land reform under Allende is also of interest, and for two reasons. The first was his observation that the reform did not benefit the really poor in agriculture, and instead benefited middle income groups. The really poor in Chilean agriculture were the small landowners, who did not receive additional land. The landless workers, on the other hand, did receive land but they were already part of the middle class.

The second aspect of the Chilean reform which Valdés noted was the significant distortion in resource use which the reform induced. The worker was permitted to retain his private plot while working on the group farm. The workers would take the highly subsidized modern inputs intended for the reform sector and use them, together with their own labor, on their privately-owned plots. The reform land thus received very limited nonland inputs per unit of land, while the privately-owned land received the bulk of the nonland inputs.

Hayami and Ruttan [1985] were concerned about a somewhat different dimension of this same problem. They noted [p. 397] that in many cases improvements in the production support system, including agricultural research and extension and the development of input markets and credit institutions, were often focused on the modern sector and neglected the small-scale reform sector.

Two studies stand out in the literature in the extent to which they used systematic research procedures to evaluate the benefits and costs of proposed reform programs. The first of these was Cline's [1970] analysis of the economic consequences of land reform in Brazil, which considered a number of issues. First, Cline examined whether economies of large scale production exist in Brazilian agriculture. Second, he considered the relationship of land utilization to farm size. This is important because the data from many countries in the region indicate that small farms use their land more intensively than do large farms. Seven major influences are examined to explain this phenomenon. Cline noted [p. xiv], "The low utilization of land by large farms, in the face of rural unemployment and underemployment on small farms, is the main reason to expect land redistribution to increase agricultural production." This is a common justification for land redistribution in the Latin American literature.

Much of the empirical work in Cline's study is based on sample survey data for some 1000 farms in seven Brazilian states. The parameters of Cobb-Douglas type production functions were estimated for each of seventeen state-product sectors. Statistical tests were made for whether the degree of homogeneity of these production functions differs significantly from one. Statistical tests were also made of the hypothesis that land-use intensity declines as farm size rises. The results of both sets of tests suggest there are efficiency gains to be had from land redistribution.

Estimates of the underlying production functions and the input-size relationships are then used to make estimates of the impact of land redistribution on agricultural production. One estimate is made for a total reform in which all farms are reorganized and another set for "political reform" in which farms under 300 hectares, and farms with below average land price, are excluded from expropriation. Appendices deal with colonization, and especially with the feasibility of colonization in frontier areas.

This study is far from a complete analysis of the problem. The analysis is partial equilibrium in nature, and the general equilibrium effects of such a comprehensive reform would be great. Moreover, in estimating the effects of the reform on production no estimates were made of the costs of carrying out the reform, of the long-run effects of postreform savings behavior and adoption of new techniques, or of short-run changes in output due to the disruption of reform. The important contribution of the study is its serious attempt to provide empirical evidence which bears on the reform question, and the attempts to estimate the static efficiency gains from the reforms.

Another serious and comprehensive attempt to examine empirically the effects of land redistribution is in Kutcher and Scandizzo's study [1981] of the agricultural economy of Northeast Brazil. The analysis of this study is based on a large, well-designed sample of data drawn from the northeast as a whole. In addition to a number of analyses of specific dimensions to the reform issue, Kutcher and Scandizzo used a large programming model to evaluate general land redistribution in the region. They also considered the effects of other policy options for the region.

Like Cline [1970], Kutcher and Scandizzo [1981] concluded that their analyses support land redistribution as a policy option. Similarly, an important contribution of the study, whether one agrees with the conclusion or not, is the ambitious attempt to produce data that would bear on the policy choice, an important deficiency in most of the appeals for land reform, based as they often are on political considerations rather than on careful analyses of the underlying data. An important missing question in all analyses we have seen, for example, is an evaluation of redistributing assets by means of education and other forms of human capital rather than through redistribution of land.

The Cline and Kutcher and Scandizzo studies both provide *ex ante* assessments of possible land reform programs. Thiesenhusen [1987b] has provided an *ex post* assessment of agrarian reform in Panama. From 1969 to 1973 about 16 percent of the farmland in Panama was converted into land-reform settlements. Thiesenhusen analyzes primary economic data from forty-three agrarian reform settlements, or *asentamientos*, in Panama to determine the level and sources of beneficiary incomes (and a comparison to opportunity costs), the relationship of possible independent variables to settlement performance, and the current return from various collective projects. The study documents the generally unimpressive economic performance of agrarian reform settlements — performance that was hinted at by former studies on Panamanian agriculture.

Griffin's [1974] essay on the political economy of agrarian change, focused primarily on the Green Revolution, has a chapter dedicated to Colombia and Mexico. He argued that biased technical change, market imperfections, and government policy combine to insure that the benefits of the Green Revolution accrue largely to the more prosperous regions and the more prosperous landowners. Thus, in his view technical change in agriculture results in greater income inequality and a polarization of social classes. His analysis is based on the major hypothesis that economic and political power are concentrated in the hands of a small group and as a result factor markets are highly imperfect. Consequently, many members of the rural community have restricted access to the means of production, and this affects the methods of cultivation that are used and the efficiency of the system.

The evidence from the early experience with the Green Revolution suggested that this technology did appear to concentrate income and assets [Falcon, 1970]. Later research showed these effects not to be as great as originally expected, however. See Hayami and Ruttan [1985] and Scobie and Posada-Torres [1978].

Another important paper is Thiesenhusen's review essay on rural development questions in Latin America [1987a]. This rewarding essay covers twelve books published in the period 1982 to 1985, synthesizing the different perspectives they offer on Latin American agrarian issues. R. A. Berry and Cline [1979] is a general analysis of agrarian structure and productivity in developing countries.

The literature on land reform in Latin America is quite large. The reader interested in this subject will want to take advantage of the two-volume, 621-page annotated bibliography prepared by the Land Tenure Center at the University of Wisconsin [LTC, 1974]. This three-year effort covered most of the holdings of the Center at that time, which the then Center Director, William Thiesenhusen, described as the largest collection on agrarian reform in the world.

Beyond the above references, the following are some of the significant references the reader may want to consider. Horowitz, ed. [1970] is a collection of papers, *Masses in Latin America*. Solon Barraclough has a paper on land reform in this collection. Stavenhagen's collection [1970] of papers on *Agrarian Problems and Peasant Movements in Latin America* is important for the variety of positions presented on reform, with emphasis on the political dimensions to the problem, and for the papers it includes by important Latin American reformists. Important among these are papers by Flores and Chonchol.

Some important general references on land reform in Latin America include "Agrarian Structure in Seven Latin American Countries," by Barraclough and Domike [1966]; *Agrarian Structure in Latin America*, by Barraclough [1973]; *Land Reform and Economic Development*, by Dorner [1972]; *Cooperative and Commune*, edited by Dorner [1977]; *Land and Labour in Latin America*, edited by Duncan and Rutledge [1977]; *Land Reform in Latin America: Bolivia, Chile, Mexico, Peru, and Venezuela*, by S. Eckstein *et al.* [1978]; *Land Concentration and Rural Poverty*, by Griffin [1976]; *Land Reform: A World Survey*, by Roger King [1977]; *International Dimensions of Land Reform*, edited by Montgomery [1984]; *State Policies and Migration*, edited by Peek and Standing [1982]; "Latin America's Employment Problem," by Thiesenhusen [1971]; and "A Suggested Policy for Industrial Reinvigoration in Latin America," Thiesenhusen [1972].

Tannenbaum's study [1929] of the Mexican Revolution is a classic. Other important studies of the Mexican land reform include "Land Reform and Productivity in Mexico," by Dovring [1970]; "El Macro Económico del Problema Agrario Mexicano," by S. Eckstein [1969]; *Transformation of Mexican Agriculture*, by Sanderson [1986]; and *Mexico's Agricultural Dilemma*, by Yates [1981]. Other pertinent references include *Agrarian Reform and Rural Poverty: A Case Study of Peru*, by Alberts [1983]; "Achievements and Contradictions of the Peruvian Agrarian Reform," by Kay [1982]; *Agrarian Reform and Peasant Organization on the Ecuadorian Coast*, by Redclift [1977]; "Chile's Experiments in Agrarian Reform: Four Colonization Projects Revisited," by Thiesenhusen [1974]; and "Agrarian Reform (in Nicaragua)," by Thome and Kaimowitz [1985]. Teixeira's [1976] study of the sharecropping system in a *município* of the Northeast of Brazil is also of interest.

We conclude this section by making four points. First, land reform has been a dominant political issue in Latin America. Second, the literature on land reform is very large, but a great deal of it is polemical in nature and not very analytical. For example, there are very few studies of the land market *per se*, despite its obvious importance to the issues above. An exception is Brandão [1986], who used a portfolio model to test a number of hypotheses about factors affecting the price of land. This research showed that in the aggregate the land market worked well.

Land was shown to be very attractive as an asset. Capital gains were shown to be the most important source of the profitability of owning land.

The third point we would make is that those trained abroad in neoclassical economics seem to be reluctant to undertake the difficult analytical and empirical research needed to provide a sound basis for economic policy. And fourth, the Land Tenure Center at the University of Wisconsin has had a significant influence on the Latin American literature, with an important and proper emphasis on institutional arrangements.

5. Agricultural Credit and Credit Policies[93]

Credit programs have been an important component of agricultural policies in many Latin American countries. Governments and international agencies have relied, sometimes heavily, on credit as a means to foster the growth of agriculture. Moreover, in some instances, as we shall see later, credit was believed to be a means of improving the distribution of income.

The literature dealing with these issues can be usefully classified into two groups. The first is the earlier literature (material published prior to the mid-1970s), which emphasized the problems and difficulties encountered in the implementation of credit programs. This literature had a flavor of institutionalism to it, in the sense that the problems encountered were usually attributed to "flaws" in administration of the programs or to random causes (such as a bad crop justifying default), and made little contribution to evaluating the farm-level impacts of the programs. The second group involves the more recent literature, which concentrates on the functioning of the financial markets and how these markets may (adversely) affect credit policies in attaining their targets. Our analysis of the literature is thus carried out under two headings: the institutional approach, and the financial market approach.

The literature on rural credit is fairly large, and we will not discuss it in detail. Instead, we chose to focus on a few key issues and to call the reader's attention to some of the important sources.

Two major institutions have generated a significant part of the literature on credit: the capital formation project at the Ohio State University [summarized in D. W. Adams et al., 1975], under the leadership of Dale Adams and his colleagues; and the Spring Review sponsored by U.S. AID [AID, 1973].

Sources not cited below, but which the reader may find of interest, include D. W. Adams [1978], on mobilizing household savings through rural financial markets; D. W. Adams [1971], a critical review of external funding policy for agricultural credit in Latin America; D. W. Adams and Graham [1981], a critical review of traditional agricultural credit projects and policies; P. F. C. de Araujo and Meyer [1977], on agricultural credit policy in Brazil; P. F. C. de Araujo [1967], on factors affecting the demand for credit at the farm level; Peres [1976], the derived demand for credit under conditions of risk; W. S. Becker [1970], agricultural credit and Colombia's economic development; Colyer and Jiménez [1971], supervised credit as a tool in agricultural development; Costa Rego and Wright [1981], an analysis of the distribution of rural credit in Brazil; Gonzales-Vega [1977], interest rate restrictions and income distribution; Ladman and Tinnermeier [1981], the political economy of agricultural credit in Bolivia; B. P. Rao [1973], the economics of agricultural credit use in southern Brazil; G. Singh [1974], farm land determinants of credit allocation and use in southern Brazil; I. J. Singh and Ahn [1978], a dynamic multicommodity model of the agricultural sector; Soares [1977], the Northeast of Brazil; Tendler [1962, 1970], a penetrating overview of agricultural credit in Brazil; and White [1975], another overview of the Brazil case. A more general discussion of credit markets and credit policy can be found in Virmani [1985].

This section is divided into four parts. The next part examines the earlier literature which discusses the fundamentals of credit policies as viewed by international agencies. We also review here some of the justifications that were attributed to governments as the basis for the programs. This is followed by consideration of the more recent emphasis on financial markets. The next part deals briefly with the issue of informal credit markets, and the section ends with some concluding comments.

THE INSTITUTIONAL APPROACH: CREDIT POLICIES—SOME JUSTIFICATIONS

AID has been one of the more important suppliers of funds for agricultural credit programs. This agency undertook a major review of such programs in the AID Spring Review of 1972/73. That seminar provided a summary of the state of the arts at that time, as well as the more frequent justifications for such programs [AID, 1973]. The interested reader should also see Donald [1976], which was a follow-up to the Review and, according to the author, conceived as a means of distilling and disseminating its results [p. viii].

The emphasis at that time was on agricultural credit for small farmers. This raises at least three questions: Why agriculture? Why credit? And why small farmers? The answers to these questions lie at the core of the justification for these programs and Donald [1976] answers them in the following quotations:

> It has become apparent that the lagging development of agriculture is a much more important constraint on national growth than had been realized, whether growth is defined as the general welfare of the population or even in narrow terms of industrial achievement. Agricultural contributions to exports, to domestic food and raw materials supply, and to the purchasing power of a national market for industry are being

given increasing priority; and the harmful effects on agriculture of the usual import substitution methods are more widely appreciated. [p. 12]

Given that a relatively high priority should be accorded to increasing agricultural production and employment in future development strategies, a focus on small farmers is: (1) inevitable in some degree, since they are the largest body of producers and have most of the underemployed labor resources to be mobilized; and (2) more productive where land is becoming the scarce resource and labor more plentiful, since they are more likely than large farmers to get high yields per hectare. . . . It is possible, though far from assured, that if efforts to achieve increases of agricultural output are concentrated on small rather than large farmers, there might be a relative savings in the use of scarce capital and industrial inputs in obtaining a given increment in production, due to the greater labor intensity with which they are applied. [p. 13]

So one cannot say without qualification that the smallholder sector is the most promising area for improvements in efficiency or in production levels—indeed, there is a short-run probability in the opposite direction. What can be said is that a comprehensive, more enduring effort to raise the level of agricultural output and productivity, and to utilize underemployed rural resources generally will require far greater attention to small farmers than has been usual in the past. The labor intensity of smallholder farming will more effectively diffuse the productive employment and incomes, and the potential for higher yields and for economies in uses of scarce inputs should not be neglected. And since small farmers usually represent the largest and least "developed" part of the agricultural sector, their production is the area of greatest potential for improvement in the long run. [pp. 13-14]

With respect to small farmers, two of the leading forms of expression of this desire for equity will have been the initiation of special credit programs for small farmers, and the promotion of cooperatives among them. The efficiency of these activities will be dissected at length below; for the present, it will suffice to note that a concern for equity, as distinct from development *per se*, has been an important reason why such programs have come into existence and found the political support that is necessary to obtain government funds. While much of the discussion that follows will be couched in terms of economic growth as the assumed objective of developing countries, it will be well to remember that a concern with equity and its political ramifications will often be the driving force that brings small farmer problems to the attention of governments. [pp. 14-15]

If priority attention is given to small farmers with growth potential, what reasons are there for supposing that credit assistance is a pertinent vehicle for their development? We may assume that additional resources are needed by the target groups. Financial credit is the most universal and flexible transferable form of economic resource: with cash obtained via credit, one can buy anything that is for sale. While goods and services could also be transferred to desired parties by administrative allocation, the transfer can be more easily affected by credit and with much greater freedom of choice and efficiency . . . For those who believe in the efficiency of the market forces and individual rationality to serve social ends, the efficiency of transfer via credit is self-evident. For those who have some doubts about this efficiency, conditional and guided credit is still likely to be more efficient than price allocation attempts. [p. 17]

The question "why credit?" can be answered in summary terms by noting that it has a number of advantages as a means of transferring resources to a relatively neglected target group, that it can lead to productive gains when conditions are right, and that its absence is sometimes a consequential bottleneck to increased production. [Donald, 1976, pp. 17-18]

In summary, from the perspective of the Spring Review, credit was seen as the most effective means of transferring resources to the small farmer which, in turn, was assumed to have the highest potential to induce output growth. Small farmers were also viewed as the segment of society in greatest need of an income transfer on equity grounds.

This earlier literature failed to distinguish between two aspects of credit: its availability and its cost. In fact, there appears to have been a contradictory view in the above arguments with respect to the interest rate to be charged. From one side, viewed as the most important source of growth for agricultural output, small farmers should be able to pay market rates for credit. From another side, viewed as discriminated against and in need of an income transfer, small farmers should receive concessionary rates. Needless to say, this last view was predominant in most credit programs.

However, the failure to actually distinguish between the two dimensions of the problem may have led to the failure to recognize more valid arguments for increasing credit availability. Many arguments presented in the development literature provide a basis for extending more credit to agriculture, but not necessarily more credit at concessionary rates. Krueger and Ruttan [1983] summarize the usual arguments in favor of credit; we note that only the last two would, perhaps, imply concessionary rates:

> This emphasis on credit is based on five perspectives. First is the Schumpeterian view, which identifies innovation as the critical element in economic development and credit as the principal instrument that allows the innovator to bid resources away from other activities. A second perspective is based on a view similar to that of market reform. The farmer obtains credit and sells his output to the same middleman and is thought to be exploited in each transaction. A third perspective, closely related to the second, views public credit institutions as providing part of the supervised education and credit package designed to induce traditional farmers to adopt modern inputs. A fourth perspective views credit as an income transfer mechanism to lessen inequities in income distribution in rural areas. A fifth perspective views subsidized credit as an incentive to farmers to expand production in spite of disincentives resulting from market interventions or exchange rate distortions that discriminate against farmers in product markets. [p. 52]

All of the above justifications allegedly have some theoretical or ethical basis. In some cases this is clearer than in others, as will be seen in the discussion below of the literature on financial markets. Nonetheless, governments may have other, more pragmatic reasons to rely on credit policies. Compared with price policies, for example, governments may see credit subsidies as less uncertain and quicker. That is, the credit subsidy "gets there immediately" while the effect of a price policy depends on farmers' receiving and understanding the price signals [Sayad, 1977a]. In Latin American countries in which lack of confidence in markets has been so widespread among policy makers, this may very well have been a strong reason behind these kinds of policies.

A second reason for extensive government use of credit policies is perhaps that they can be more easily financed through monetary expansion, which usually does not require a great deal of political discussion. By financing the subsidy via monetary expansion, governments avoid the need to deal with legislative bodies and interest groups.[94]

THE INSTITUTIONAL APPROACH: EVALUATIONS

The institutional evaluations tended to be of a general nature. They discussed the major problems faced in the implementation of the projects, but very little analysis was provided of problems at the farm or micro level.[95] Donald [1976] provided a list of the conditions expected for a credit program to meet its objectives:

> To summarize the conditions necessary for success in a small farmer credit program that have been mentioned so far: if a new technology or crop is available with adequate market potential, and lack of capital is, in fact, a constraint on its adoption, if the farmers' motivations direct them to borrow for productive purposes and repay loans, and if the credit institutions can be structured to serve small rather than large farmers, then favorable results are possible. [p. 31]

The evaluations from the Spring Review [AID, 1973] showed problems with many programs. Among the most frequently mentioned difficulties were repayment problems and the fact that most of the loans were made to large farmers. With respect to production, the Latin American evidence apparently differed from that of other regions. While in other regions there had been no substantial increases in output, in Latin America, based on the experience in Brazil and Colombia, there was [Donald, 1976, p. 29].

More detailed comments with respect to the failure of the programs to reach small farmers are deferred until later when the issue of financial markets and their relation to credit programs are examined. At that point the default question will also be addressed.

David and Meyer [1980] summarized the evidence on the impacts of credit programs at the farm level, classifying the existing studies into three classes: descriptive, econometric, and mathematical programing. Table 53 presents some results from these descriptive studies which refer to Latin American countries. All of the studies, except one from Colombia, were based on cross-sectional data of borrowers and nonborrowers.

Based on the evidence from Table 53, one can see that borrowers tended to have substantially larger farms, higher operating expenses and higher investment and production per hectare. There are a few studies for Latin America that report net farm income per hectare; they show substantially smaller differences between borrowers and nonborrowers than do the studies referred to in the table. David and Meyer [1980, p. 207] note that in the Guatemala study [Daines, 1975], differences in the value of production between borrowers and nonborrowers were decomposed into price, yield, crop mix, and crop area effects. This decompos:-

Table 53. Percentage differences in selected measures between borrowers and nonborrowers, selected countries

Countries	Number of observations	Farm size (ha)	Percentage differences in:			
			Operating expenses per ha	Invest-ment per ha	Pro-duction per ha	Net farm income per ha
Brazil, 1965	132	78	112	NA	30[a]	2
Southern Brazil, 1965	954	94	127	80	62[a]	NA
Southern Brazil, 1969	732	68	281	338	133[a]	NA
Colombia, 1968	52	74	104	NA	6	NA
Colombia, 1965/68[b]	25	30	56	NA	35	NA
Guatemala, 1975	1600	5	39	NA	−3	2[c]

Source: David and Meyer, 1980, and their references. With permission.
[a] Gross farm income per hectare.
[b] Comparison of borrowers in the credit program before 1965 and after 1968.
[c] Based on lower 75 percent of farms in terms of size.
NA = Not available.

tion showed that an important source of the expansion in crop area (one of the principal sources of growth of production) was credit.

The econometric studies surveyed by David and Meyer [1980] were classified as follows: production function, input demand functions, and efficiency-gap functions. Studies for Latin American countries were reported only for the production function approach and were based on data from Brazil and Colombia. Commenting on the methodologies utilized in those studies, David and Meyer noted:

> The credit variable was specified in several ways (Table 2). The Colombian studies treated credit as a separate input. The later Colombian study further hypothesized that borrowers have a completely different production technology so separate production functions were estimated for borrowers, nonborrowers, and borrowers prior to the supervised credit program. A modified Cobb-Douglas production model was used in the Brazilian model where credit was assumed to shift production coefficients for operating expenses, modern inputs, and machinery, but not for land, labor or animal power.
> [David and Meyer, 1980, p. 211]

The estimates for Brazil and Colombia (from Table 2 of David and Meyer) are presented in Table 54. David and Meyer's views of these studies were as follows:

Table 54. Estimates of the effects of borrowing on the Cobb-Douglas
production function, selected countries

		Colombia			Brazil
Item	Colombia 1960	Borrower[a] 1965	1968	Non-borrower 1968	1971/72
Log a		1.174	2.899	0.740	1.514
Land	0.303 (1.620)	0.379* (1.560)	0.777 (3.964)	0.418 (1.742)	0.293 (4.420)
Labour	–	0.396 (1.472)	0.049 (0.383)	0.456* (2.505)	0.009 (0.880)
Farm Equipment	−0.103 (−1.873)	0.144 (1.043)	0.048 (0.533)	0.034 (0.354)	0.045* (1.340)
Livestock	–	–	–	–	0.009* (1.830)
Operating Expense	0.115 (1.885)	0.314 (1.377)	0.279* (1.898)	0.405 (3.092)	0.246* (4.300)
Modern Varieties	–	–	–	– (5.020)	0.356*
Credit	0.6410 (3.705)	0.064 (0.877)	−0.084 (−1.000)	0.104* (1.825)	–
Credit × Operating Expense	–	–	–	–	0.0001* (1.970)
Credit × Modern Inputs	–	–	–	–	−0.00003 (−0.370)
R^2	0.89	0.57	0.90	0.80	0.96
Number of Observations	17	27	27	25	129

Source: David and Meyer, 1980, and their references. With permission.

[a] Borrowers are participants in supervised credit programs. Nonborrowers are
nonparticipants including farmers borrowing from nonformal sources.

[b] Figures in parentheses are t–values.

* Indicates statistical significance at 10 percent or better confidence interval.

First, specifying credit as a separate production input presents a
conceptual problem because loans may permit purchasing
optimal input levels, but do not directly generate output.
Double counting of inputs may also occur with credit as a
separate variable. An example exists with the Colombian
results where a higher production coefficient for credit was

found in the earlier study. In this study, the credit variable in effect captures the contribution of labor and other variables explicitly specified in the later model.

Second, attributing differences in production functions between borrowers and nonborrowers to borrowing implicitly assumes a relationship between source of liquidity and production function. A slight difference exists between borrowers and nonborrowers in the operating expense coefficient in the Colombian and Brazilian studies, but not in coefficients for modern inputs also expected to be influenced by loans. The direction of the differences, however, is inconsistent. [David and Meyer, 1980, pp. 211-214]

Turning now to the mathematical programming studies, there has been a wide diversity of modelling approaches used to address the problem. The most common approach is single-period linear programming with a representative model of a farm maximizing profits under constraints specified by resource endowments and technology. Other approaches include multiperiod models, as well as recursive models. A flavor of the general results is given by David and Meyer [1980] in the following paragraph:

Several similar results emerge from these studies. Technological change, adoption of new varieties and cropping systems, mechanization and farm income are frequently found to be constrained by current formal loan supplies. Borrowing limits must be relaxed to obtain socially desired changes in these variables. Likewise, evaluations of credit programs conclude that formal loans have resulted in desirable farm changes. Furthermore, productive alternatives exist so farmers could pay substantially higher interest rates with limited reduction in borrowings. Small farmers appear particularly insensitive to interest rates. [p. 288]

A comprehensive study of a Colombian project of supervised credit was made by D. W. Adams, Peña, and Giles [1966]. Data for this study were collected from five areas: Tolima, Antioquia, Boyaca, Valle del Cauca, and Caldas, with a concentration in the first four areas. In each area the (random) sample included about 10 to 15 percent of the borrowers who were engaged in the program for at least one year. A brief summary of the most relevant findings of this study is as follows:

(1) The borrowers in the program were better educated than the average farm operator of small farms in the areas in which the research was performed.

(2) It was noted during the course of the study that several important changes in farm organization were being induced by SC (supervised credit). In a few cases, the size of the farm was increased as a direct result of SC. In a number of cases the farm operation had become more diversified as farmers specialized their farm production. In almost all cases, SC had helped orient the borrowers, operations more toward the money market. [p. 317]

(3) There was an increase in the intensity of crop cultivation as well as an increase in annual crop production for those in the SC program. More use of modern inputs also was observed for some borrowers.

(4) One of the most noticeable changes which SC has introduced on the farms of borrowers is a substantial increase in the number of animal units owned. Many of the SC farmers now have several head of additional cattle, more pigs, and more chickens because of SC. This has generally helped farmers better utilize pasture, as well as helping to spread the farm risks among several different types of enterprises. [p. 139]

The authors of this study made several recommendations to improve the supervised credit program. Their suggestion with respect to interest rates seems especially important:

> Another important point relates to the interest rate charged by INCORA for the loans to farmers. The rate now charged by INCORA (five percent, plus one percent insurance) is somewhat lower than what the *Caja Agraria* or the IDB/Fondo are charging. The INCORA loans are larger and the terms of repayment more favorable than the *Caja Agraria*'s, and since INCORA provides a good deal more supervision, the authors feel that INCORA's rate should be raised. Colombia's annual inflationary rate is substantially higher than 10 percent. Thus, loans let at rates of interest below this inflationary rate have negative rather than positive rates of real interest. Furthermore, the field studies indicated that borrowers are not very responsive to interest rates. Borrowers pay much more attention to amount loaned, service, and terms of repayment. The interest rate should be raised to a basic six percent, with one percent additional for insurance, and another one percent to cover the handling charges of the *Caja Agraria*. [D. W. Adams et al., 1966, p. 141]

We conclude this section by noting that the institutional approach to credit gives a somewhat distorted view of the role of credit. The studies surveyed by

David and Meyer [1980] showed, in general, positive output and investment responses, but they failed to answer a number of important questions, including the critical one of what it would have been possible to achieve under the given circumstances as opposed to what actually was achieved. The financial market approach provides a broader approach for analyzing more general questions.

THE FINANCIAL MARKET APPROACH

The more recent literature on agricultural credit views these programs from the perspective of general financial markets and relates the most common problems that were discussed in the earlier literature, such as default and the perverse income distribution effects, to the intrinsic nature of this market. Von Pischke and Adams [1980] is an important paper in this tradition. See also the papers in Von Pischke, Adams, and Donald, eds., [1983]. We discuss these issues with the following two questions in mind: how much, under the normal operating conditions of the financial system, should an additional unit of subsidized credit increase the total amount of external funds in the agricultural sector (this is an efficiency question); and what are the distributive consequences of these policies. To provide a framework to examine these two questions we draw on Sayad [1977a], whose model provides a means of discussing the current debate and, as well, furnishes empirical measures of efficiency.

Following Sayad [1977a] and Brandão and Magalhães [1982], assume there are two assets, A and I, with rates of return r_a and r_i, respectively. These rates are given and are independent of the way the portfolios are financed. The demand for these assets can be specified as follows:

$$A = A(r_a, r_i, r_c)$$
$$I = I(r_a, r_i, r_c)$$

where r_c is the market rate of interest. It is assumed that:

$$\frac{\delta A}{\delta r_a} > 0; \frac{\delta A}{\delta r_i} < 0; \frac{\delta A}{\delta r_c} < 0$$

$$\frac{\delta I}{\delta r_a} < 0; \frac{\delta I}{\delta r_i} > 0; \frac{\delta I}{\delta r_c} < 0$$

This is a simplified specification, since other variables could have been included on the right-hand side, the most obvious examples being the covariance between the returns of A and I and the individual's wealth. However, for our purposes, the simplified model is sufficient.

To maintain the assets A and I, the firms in this economy can either finance them with own capital, E, and/or borrowed capital, K. It is assumed that E depends positively on r_c; that is, the higher the interest rate the greater the amount

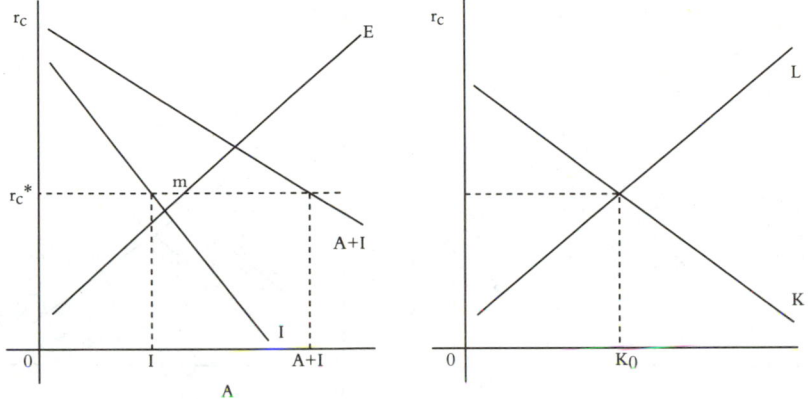

Figure 2. Financial market equilibrium and allocation of funds among alternate investments (adapted from Sayad, 1977a, with permission)

of own capital, given the individual's wealth and the rates of return to A and I. More simply

$$E = E(r_c), E' > 0$$

Notice now that the identity

$$A + I \equiv E + K$$

implies that $K = K(r_c)$, with $K' < 0$.

To close the model, specify the supply of loans by the banking system as an increasing function of the interest rate, *ceteris paribus*. That is,

$$L = L(r_c), L' > 0$$

This also is a simplification since other variables could be included, but this specification is again sufficient for our purpose.

Assume next that the ratio A/I is independent of r_c. That is, given r_a and r_i, this relationship is entirely determined. The equilibrium of the model can be represented with the help of the diagrams in Figure 2. The demand for and the supply of loans is represented in the diagram on the right, with the interest rate, r_c^*, determined by the intersection of the two curves. In the diagram on the left, the volume of own resources (corresponding to the distance mr_c^*) and the investment in A + I are determined.

Assume now that the government decides to increase the supply of loans by

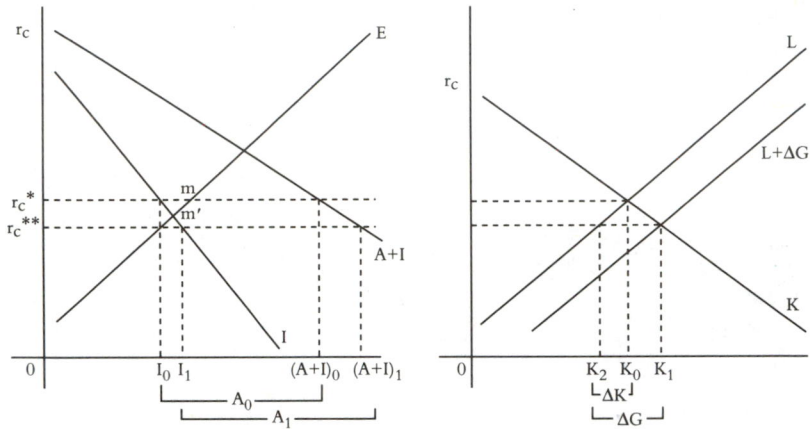

Figure 3. Impact of injection of government funds in the credit market (adapted from Sayad, 1977a, with permission)

making available an amount of resources ΔG. Assume also that the interest rate on that amount is small enough to make investors attracted to those resources (there is no harm to the analysis if the interest rate charged on those resources is assumed to be zero), and that they can be used only to finance purchases of A.

The impact of this increase in funds is illustrated in Figure 3. The government resources induce a shift of the L curve to the right and the new equilibrium in the loan market is established at a lower interest rate, r_c^{**}, and consequently, at a higher level of K, K_1. But in this model there is crowding-out of the loans from the private sector, which are reduced to K_2. Moreover, there is also a smaller amount of own capital used in the financing of the individual's portfolio. This reduction is represented by the difference between segments mr_c^* and $m'r_c^{**}$ in the diagram on the left.

Notice, also, that there is an increase in the total amount invested, represented by $(A+I)_1$ [greater than $(A+I)_0$] in the diagram on the left. Given that A/I is independent of r_c, in this new position both A and I should increase. But one can easily note that the total increase in A, ΔA, will be smaller than ΔG. This is so because there was a reduction in the volume or financing carried out by the individual's own capital. Symbolically, we have that:

$$\Delta A < \Delta A + \Delta I = \Delta E + \Delta K + \Delta G < \Delta G$$

since $\Delta I > 0$, $\Delta E < 0$ (as shown by the negative value of $m'r_c^{**} - mr_c^*$) and $\Delta K < 0$ (since $\Delta K = K_2 - K_0$).

This result has been obtained with the aid of the assumption that A/I is independent of r_c. This means that holders of A and I are aware of differences between changes in the relative return of A and I (r_a and r_i), real changes, and financial changes as reflected by changes in r_c. Their response to these latter types of changes is an increase in applications in both activities, keeping the aggregate proportions fixed, however. In short, there is no financial illusion in this model and this induces an increase in the aggregate applications of both A and I in response to the increased supply of government credit. The conclusions of the model would be different if this assumption were changed.

Based on the above results, following Sayad [1977a], a measure of efficiency of the program can be defined:

$$\frac{\Delta A}{\Delta G} = e$$

which is a number between zero and one. The extreme case of inefficiency, that is $e = 0$, would occur if both A and I were independent of r_c. In this situation, the reduction in own capital and borrowed capital from the private sector exactly offsets the increase in government resources.

One should note that each additional unit of government credit that is made available induces a reduction in the use of own capital and capital borrowed from the private sector. Let

$s = \dfrac{-(\Delta E + \Delta L)}{\Delta G}$ and note that it can be rewritten as:

$$s = \frac{\eta_{E+L}}{\eta_{A+I} + \eta_{E+L}}$$

where η_{A+I} is the absolute value of the elasticity of A + I with respect to r_c and η_{E+L} is the elasticity of E + L with respect to r_c. Moreover, since A/I is constant, say $a/(1+a)$, $a \leqslant 1$, we can write $e = a(1-s)$.

The closer a is to 1, the more efficient is the policy. That is, the greater A/I, the greater the efficiency of the policy.

Some special cases can now be considered to evaluate the efficiency of alternative policies.

(1) If $\eta_{A+I} = 0$, $s = 1$ (provided, of course, that $\eta_{E+L} \neq 0$). There is perfect substitution and the policy would not achieve its goal of increasing applications in A.

(2) As η_{A+I} increases, s decreases and, in the limit s = 0, provided again that $\eta_{E+L} \neq 0$. That is, the efficiency of the policy increases with the elasticity of the curve A + I. In the limiting case, e = a, which is the maximum possible value for the efficiency measure.

(3) If $\eta_{E+L} = 0$ (assume $\eta_{A+I} \neq 0$), then s = 0, and again, e = a. This is a very interesting case, since this assumption is equivalent to a shortage of capital loaned or borrowed (from private sources) or a lack of access to the capital market. In this situation the government will ease this constraint and no crowding out will take place. The government credit will be allocated to the acquisition of new assets and the applications in agriculture will be at their maximum possible value in this model, that is, a.

As noted above, one of the reasons for the emphasis on small farmers in agricultural credit programs has been their assumed lack of access to the capital or financial markets. This situation is typical of financial market conditions in Latin America. For present purposes, one can think of these farmers as having practically no own-capital and no—or very little—access to the capital markets. These farmers will clearly face a totally inelastic E + L curve at a very low level of E + L. Thus, government credit programs may well be justified for these groups. Perhaps a better way to put this would be to say that, given a commitment to credit policies, there would be a greater assurance resources allocated to this group would be devoted to the appropriate activity.

One should keep in mind, however, that credit programs are usually implemented through the banking system and consequently usual bank criteria are used to select the borrowers. Therefore, the prevailing institutional arrangements may be seen as ineffective in allocating the credit to the intended recipients. This is probably even more the case when large amounts of credit are being considered.

In summary, there has been a general failure to recognize conditions in financial markets in the implementation of many credit programs for small farmers. The fungibility of money and profit maximization are very real reasons for some major setbacks in these programs [Von Pischke and D. W. Adams, 1980].

(4) As η_{E+L} increases (provided $\eta_{A+I} > 0$), s goes to 1 and the efficiency of policy goes to zero. This means that as the possibilities of the individual in the capital market are greater, substitution of own and external high-cost resources for the "cheaper" government resources will take place.

To summarize, Sayad [1977a] and Brandão and Magalhães [1982], by specifying the important economic relationships which describe the credit or financial markets, make it possible

to understand the conditions under which an injection of
financial resources into the system will actually increase the
external resources moving into the sector. The framework
provides measures of efficiency and indicates the conditions
under which government policies will attain their objectives.

It should be noted that the model developed above assumes that the resources
applied in A will be applied in an efficient way. This should not come as surprise,
although it appears that some of the earlier justifications given for credit programs amounted precisely to a denial of such an assumption, especially in the case
of small farmers. It is, perhaps, a failure to consider how financial markets operate (which is well illustrated by the Sayad [1977a] and Brandão and Magalhães
[1982] models), and a failure to consider farmers' behavior (both kinds of agents
seeking maximum profits), that has led to some of the misconceptions with respect to credit and its effects on agricultural production. This is a crucial issue in
the evaluation of credit programs and we will return to it below.

There remains one theoretical issue to be considered, which is the effects of
concessionary interest rates on returns to factors of production. This has been neglected to some extent in the literature, although the Brazilian literature has some
interesting contributions [see Sayad, 1977a, b; Rezende, 1982a, b; Rabelo de
Castro, 1977; and Brandão and Magalhães, 1982].

The traditional analysis of *price* policies has shown that the gains occasioned
by such policies tend to be capitalized in the value of land, the factor in most inelastic supply [Floyd, 1965]. In the same way, landowners should be expected to
capture a substantial portion of the subsidy provided through credit. But given
the imperfect nature of rural financial markets in Latin America, there is also an
important distribution effect of such subsidized credit programs. Consider first
the effect on land values. The usual banking criterion for the concession of a loan
requires a collateral from the borrower, with land the collateral *par excellence*.
Therefore, in the presence of a subsidy and in a situation in which credit is ra-
tioned, the collateral becomes the most important criterion for the concession of
the loan and the consequent access to the subsidy. It follows that the demand for
land should increase since now there is an additional reason for a person to hold
land; that is, to have a share of the government subsidy. This will drive land
prices up until the rate of return on this asset is again in line with that on other
assets in the economy.

The effects of the application of this banking criterion, however, is to increase
further the income inequalities in the rural sector. Large land owners have access
to larger shares of the subsidy and therefore can finance the acquisition of still
more land, with the small owner finding it difficult to gain access to the subsidized credit. This further exacerbates the already unequal distribution of wealth
and income [Sayad, 1977a; Rezende, 1982a, b have discussed this issue].

An important issue in credit and rural finance is the role of credit in engendering the accumulation of capital at the farm level. Dale Adams and his colleagues in the Department of Agricultural Economics and Rural Sociology at the Ohio State University undertook an ambitious study of this issue in southern Brazil. A summary of this project can be found in D. W. Adams *et al.* [1975]. A list of the large number of individual reports can also be found in that summary.

INFORMAL CREDIT MARKETS

The literature surveys for Asia and Africa give much attention to informal credit markets. The Latin American literature, however, has given less attention to such markets. We review in this section two papers by Nisbet [1967, 1969] for Chile and some findings from the Ohio State's capital formation project in Brazil [D. W. Adams *et al.*, 1975].

Nisbet found an informal market in Chile that operated under rules quite distinct from the formal market. An important feature of this market was the reduced amount of "red tape" compared to the formal market. In addition, given the limited spatial dimension of these markets the suppliers of credit have good information on borrowers so they can avoid many operations that would increase the transaction costs in formal markets. This appears to be an especially important factor for small borrowers. Based on data from Brazil and Colombia, D. W. Adams and Nehman [1979] show that borrowing costs (nominal interest plus transaction costs plus changes in the purchasing power of money) are substantially higher for small compared to large borrowers.

Nisbet divided the informal market, from the supply side, into two segments: commercial lenders, and noncommercial lenders. The first group includes village stores, itinerant traders, and money lenders who usually charge high interest rates (see Table 55). The second group, however, is made up of friends, neighbors, relatives, *patrones*, and, in general, people who charge negative real rates of interest for their customers (see Table 55).

Another interesting finding by Nisbet was that in general there was very little competition among lenders in the informal market. Commercial lenders usually operated in relatively small geographic areas in which they knew (had information about) the borrowers and their financial situation. According to Nisbet this lack of competition was responsible, to a large extent, for the high interest rates on commercial loans.

In a later paper, Nisbet [1969] showed that the institutional (or formal) credit market was used mostly by the more well-to-do and the more educated borrowers. He also showed that the number of farmers seeking credit in the informal market exceeded those in the formal market, and that the volume of resources in the informal market was also larger.

Table 55. The structure of reported rates of interest with money interest rates deflated to real interest rates and rates on loans in kind expressed in real terms, Chile, 1964-65[a]

Type of lender[b] and type of loan	Interest rate distribution (in annual percentage rates)																			Total Cases
	-33	-22	-20	-13	-7	-3	0	18	27	30	33	40	46	60	75	90	128	165	360	
Friends																				
Cash	4	1			1															6
Kind							6	1												7
Neighbors																				
Cash																				0
Kind							4													4
Relatives																				
Cash	3	1																		4
Kind							7													7
Patrones																				
Cash	16																			16
Kind							3													3
Village stores																				
Cash	1	1	1	2	1	1							2					1		10
Kind							9								1	1				11
Itinerant traders																				
Cash																				0
Kind										1		1		1	1					4
Money lenders																				
Cash	1								1		1	1								4
Kind							1						5	4	1	1	1		1	14
Total Cases[c]	25	3	1	2	2	1	30	1	1	1	1	2	7	5	3	2	1	1	1	90

Source: Nisbet, 1967. With permission.

[a] All rates are on actual loans for the agricultural year May 1964 through May 1965, and money interest rates were deflated by using the consumer price index. The terms ranged from one month to over one year, so all rates were adjusted to annual figures.

[b] The same lender extends credit to different borrowers in some cases.

[c] In seven cases, borrowers did not know the interest rate charged, and there were eight cases of *recargos* (charges in addition to the original sum of value lent).

D. W. Adams *et al.* [1975] found a somewhat more limited role for informal markets in their study of southern and central Brazil. Even there, however, they observed that "in some situations, informal financial transactions are significant in terms of number of loans and number of people serviced. It will also be argued, however, that in terms of total value, informal credit markets make up a minor portion of total financial transactions in at least the highly commercialized agricultural areas of Brazil" [pp. 10-11].

The evidence reported by D. W. Adams *et al.* [1975], which covers areas in which agricultural production is mainly market-oriented, shows that: the informal credit market was used roughly by one-quarter to one-half of the farmers in the areas studied; informal credit made up one-third of the number of loans held by the farmers in the sample; in value terms, formal credit far exceeded informal credit; and informal credit was more important in the 1965 data set than in the 1969 data set. This was attributed to greater availability of formal credit in the later year, as well as to a higher number of small subsistence farmers in the 1965 samples.

It is not clear why the studies of informal credit markets are so few in number in the Latin American literature. It may be that we just missed them, although it is also the case that money lenders are not a major policy issue in Latin America. We suspect that part of the problem is due to the failure to study the interlinkages among factor markets. The *latifúndio-minifúndio* complex certainly involves borrowing and lending in an informal way [Teixeira, 1976]. Finally, the fact that Latin American economies have tended to be inflationary, with abundant liquidity and at large negative real rates of interest, may explain it. We suspect all these factors may be involved.

CONCLUDING COMMENTS

During the 1960s and 1970s subsidized rural credit was seen as a means to increase productivity and reduce poverty in the agricultural sectors of developing countries. Currently there is an emerging consensus among the development community that rural credit subsidies are not effective. Increases in agricultural output have not been obtained cost-effectively, and rural savings rates have deteriorated.

Attention has now turned to the financial markets, to the institutional arrangements that define these markets, and to the reform of these institutions. On these issues the reader is directed to an excellent recent paper by Braverman and Guasch [1984]. They noted that many of the financial institutions created to channel credit to rural areas have been shown to be inept and to lack accountability. They then discussed credit rationing in competitive markets and the interlinking of credit contracts with labor and land contracts. The policy implications of these theories were reviewed, with the authors finding them insufficient to account for the empirical evidence at hand. They concluded that a more systematic and rigorous analysis of institutions and institutional environments is essential for understanding and implementing effective policy reforms of rural credit markets, and presented some suggestions for undertaking such analyses.

It should also be noted that many of the rural credit programs in Latin America were based on the rural extension model of development. It was generally assumed that farmers did not know about available production technology, and that even if they did they would not have the wherewithal to acquire the modern inputs required. Hence, supervised credit programs were implemented not only to provide the credit but also, more importantly, to teach the new technology. The weakness in these programs has often been the failure of the supposedly available technology to be effective under available policy regimes and resource endowments. They may also have failed because of incomplete markets in other inputs which are the mechanisms through which landless and small farmers have access to credit.

An important puzzle is why policy makers have persisted in such heavy dependence on subsidized credit programs in Latin America when the evidence was abundant that such programs did not attain their avowed goals. The continued emphasis may lie in a perception that subsidized credit offsets the discriminatory policies against agriculture, in the fact that the benefits of such policies accrue to those with special political power, or that running the printing press is just a politically "easy" way to channel resources to farmers.

The "lost link" in reform of agricultural policies in Latin America is the failure to see the importance of investment in the capacity to produce new production technology, in expanding educational opportunities for the rural population, and in removing distortions in exchange rates and trade. Such investments would raise the productivity of resources in agriculture, and this in turn could induce a flow of credit and capital. The problem of institutional reform would remain, however.

An important problem, and one that has received insufficient attention in the literature on Latin America, is the failure to develop true financial intermediaries as part of the credit system. The financial system is viewed all too frequently by policy makers as a means of transferring resources to the private sector, and especially to targeted groups, and all too infrequently as a means of mobilizing savings and intermediating them to borrowers.

Finally, if it is desired to subsidize small producers, the provision of public services such as health, education, and production technology may be more effective than subsidized credit. Moreover, instead of implicit subsidies in the form of cheap credit, it may be far more effective to provide outright grants to small producers than to distort the credit system.

6. Labor Markets and Migration

Economic growth requires major adjustments in the use of resources. These resources have to be reallocated among sectors in response to changing conditions of demand and supply, and also geographically within the nation. Since labor is the most important factor of production, usually accounting for roughly two-thirds of the cost of producing the flow of final goods and services, it has traditionally been expected to bear the major burden of adjustment to changing conditions of demand and supply. We will note below that recent writers have challenged the view that labor inherently has to bear the full burden.

The shift of labor out of agriculture as development proceeds has been described by Johnston [1970] as the one "iron law" of economics. It is a combination of demand and supply forces that causes this to be the case. The income elasticity of demand for food is inherently lower than the elasticity of demand for nonfood goods and services. Population growth rates tend to be higher in rural areas than in urban areas because children tend to have a higher value in production in rural areas than they do in urban areas, and because the cost of "producing" children tends to be lower in rural areas.

This inherent need for labor to adjust out of agriculture in the course of economic development is often exacerbated by technical change which contributes to immiserizing development. Food commodities, especially those that are necessities, tend to have low price elasticities of demand. Technical change which raises labor productivity can under these conditions have a depressing effect on farm incomes since the increase in output it engenders may have a larger than proportional negative effect on the price of the commodity. These income effects provide a "push" effect to labor migration, and generate pressures for farm enlargement to create units of sufficient size to produce incomes comparable to those received by labor of similar quality in the nonfarm sector. Of course, expansion of the nonfarm sector may proceed at a rate to "pull" labor out of agriculture at a pace that avoids these income-depressing effects. In general, the intersectoral labor market is subject to both "push" and "pull" effects.

Migration, both occupational and geographic, is thus generally viewed as a desirable characteristic of economic development, and as a necessary condition for efficient allocation of resources and for a reasonably equitable distribution of income. The neoclassical migration model implies that migration will be equilibrating, since it is expected to raise the return to labor in the supplying region or sector and to reduce it in the receiving region or sector. The recent literature has also challenged this perspective, as we will note below.

The labor "surplus" in agriculture was for a significant period of time a major focus of the post-World War II literature on economic development. A vast literature evolved from Lewis's seminal article [1954], with sophisticated models formulated to explain how the excess labor in agriculture was transferred to the modern or industrial sector and to analyze the consequences of the transfer. The vitality of this literature was such that for a considerable period of time the potential importance of agriculture itself was forgotten, as was the importance of generating a commodity surplus [à la Nicholls, 1963b] and its mobilization for development purposes.

Hayami and Ruttan [1985] provide a critique of this literature in the context of their discussion of dual economy models [pp. 22-33], and especially that of dynamic dualism [pp. 24-30]. The key models in this literature are those by Lewis [1954], Fei and Ranis [1964, 1966], Ranis and Fei [1961], and Jorgenson [1961, 1966, 1969].

Lewis's model is described as classical since it assumes the marginal product of labor is zero and postulates that the agricultural wage is determined by institutional means. In this model labor can be removed from agriculture without having any effect on agricultural output or on the agricultural wage rate. The models by Jorgenson and by Ranis and Fei are described as neoclassical. Ranis and Fei retained Lewis's institutional wage for agriculture, with the wage approximating the average productivity of labor in the subsistence sector. However, they postulated that the marginal productivity of labor is positive, but lower than the wage rate. Labor can still be removed from agriculture at early states of development without reducing agricultural output.

In the Jorgenson model, the assumption of zero marginal productivity of labor and an institutionally determined wage rate in the subsistence sector are dropped. Wage rates are assumed to be determined in an intersectoral labor market even during the initial stages of development. As a result labor is never available to the industrial sector without sacrificing agricultural output, and the terms of trade move against the industrial sector continuously throughout the development process rather than after substantial development in the commercial-industrial sector.

The theory of labor markets has evolved significantly over the past twenty years, as has empirical research on such markets. Not many of these developments have emerged from the Latin American literature, however. This is part of the general lack of research on agricultural factor markets in that region, and the failure to capitalize on the recent theoretical developments on market linkages and the new tenure economics.

Some of the literature on labor markets was referred to in the discussion above on contractual arrangements and linkages, and will not be repeated here. Perhaps the best single source for the reader to survey the literature on labor markets in developing countries is in *Migration and the Labor Market in Developing Countries*, the results of a World Bank conference, edited by Sabot [1982a]. The titles of the chapters in this volume provide a good flavor of its content: "The Structure of

Labor Markets and Shadow Prices in LDCs," by J. E. Stiglitz; "Urban Unemployment in LDCs: Towards a More General Search Model," by J. R. Harris and R. H. Sabot; "Notes on the Estimation of Migration Decision Functions," by T. P. Schultz; "Measuring the Difference between Rural and Urban Incomes: Some Conceptual Issues," by P. Collier and R. H. Sabot; "Out-Migration, Rural Productivity, and the Distribution of Income," by G. E. Schuh; "Migration from Rural Areas of Poor Countries: The Impact on Rural Productivity and Income Distribution," by M. Lipton; and "Conclusion: Some Themes and Unresolved Issues," by R. H. Sabot.

Sabot's "Introduction" provides a good overview of the issues covered in the book. We choose to focus on the chapter by Schuh, however, because of the importance of its policy implications, and to contrast his perspective with that of Lipton in the same volume, who reaches somewhat similar conclusions via a different route.

The puzzle that Schuh tried to resolve is why, if migration is expected to be equilibrating, it takes so long for regional and sectoral income differentials to decline even though labor mobility may be relatively high. Agricultural incomes chronically lag behind those in the nonfarm sector, even at fairly late stages of development, and regional pockets of poverty persist for long periods of time. In the case of the U.S. South, it took approximately 100 years to reach some semblance of equilibrium, despite large and persistent migratory flows. The same chronic regional disparities have been features of the South of Italy, the Northeast of Brazil, certain regions of Mexico, and so on.

Schuh finds the answer to this question in the selectivity of migration, which has been documented in the migration literature [Greenwood, 1975]. The migrants tend to be the younger members of the population, the more highly educated, those with marketable skills, the healthy, and the more entrepreneurial and the risk takers. In effect, migration tends to extract the human capital from a region or sector and to give it to the receiving region as a "free" good. If the investment in education is financed by the local community in the "donating" region, the effect of this transfer is exacerbated even further. Reliance on such a policy hardly seems to be the way to narrow either sectoral or regional disparities.

The reader will recognize that this is precisely the issue raised in the brain-drain literature, and what Schuh does is to apply the models developed in that literature to the problem of domestic migration. In addition to a loss of the most productive workers and those with entrepreneurial skills, there is a loss of economic "surplus" previously generated by migrants whose contribution to output exceeded marginal product while their wage equalled it. There are migration-induced capital flows and changes in the terms of trade. And there are losses of economies of scale in the provision of public services due to a decline in population.

In effect, out-migration imposes negative externalities on the supplying region. Interestingly enough, the negative externalities imposed on the receiving (typically) urban center in the form of congestion and overcrowded infrastructure have long been recognized. Except for the brain-drain literature, however, the significance of negative externalities on the supplying region has been ignored.

Schuh argues that the solution to these problems is to decentralize the industrial development process. If this were done, sectoral mobility would be promoted at the same time that geographic mobility would be reduced. Improving the infrastructure in rural areas is one way of promoting such decentralization. But Schuh argues that if the negative externalities in the supplying region are as large as experience suggests they are, significant subsidies could be paid to induce industrial and other economic activities to locate close to or within regions of rural poverty. These subsidies could be in the form of training and schooling the local labor force. More generally, he notes that there are typically many implicit subsidies which encourage economic activities to locate in large urban agglomerations. A great deal could be accomplished towards decentralization by simply reducing or eliminating these subsidies. An important contribution of the Schuh paper is a first step in the construction of a taxonomy of the various economic changes in the rural sector induced by out-migration.

The contrast of Schuh's paper [1982] with that of Lipton [1982] is interesting. While Schuh assumes that labor markets work reasonably well, though far from perfectly, Lipton assumes the labor market is an extremely imperfect mechanism. In Lipton's analysis, these imperfections are important because they enhance the role financing plays in migration decisions, thus increasing the influence of the imperfections of capital markets on labor allocation in LDCs.

In this context, Lipton's principal concern is with the financing by rural residents of investment in education, not solely to the financing of the direct costs of migration. If capital market imperfections keep poor rural residents from financing their investment in education, there will be fewer potential students and existing students will have a higher supply price. The supply of educated "school learners" will shift to the left, and for a given demand schedule for educated workers, this implies a greater difference in earnings between educated and uneducated than if finance were not a constraint on education.

Turning to other important general literature on the intersectoral labor market, Harberger's [1971] paper on measuring the opportunity cost of labor is important, as are the papers by Todaro and Harris [Todaro, 1969; Harris and Todaro, 1970]. By taking account of being unemployed when the migrant moves to urban centers, Harris and Todaro provided a rationale for the apparent disequilibrium between urban and rural sectors. Access to the brain-drain literature can be had through R. A. Berry [1974]; R. A. Berry and Soligo [1969]; Bhagwati and Hamada [1974]; Bhagwati and Rodriguez [1975]; Grubel and Scott [1966]; Harry G. Johnson [1967a]; and Kenen [1970].

Some useful sources on the literature, more specifically on Latin America, include Elizago [1963, 1965] and Shaw [1974], Latin America; Sahota [1968], Brazil; T. P. Schultz [1969, 1971], Colombia; T. P. Schultz [1975], and Chen [1968], Venezuela; Cisneros [1959], Ecuador; Feindt and Browning [1972], Hancock [1959], and Weist [1970], Mexico; Wilkie [1973], Argentina; and Schmidt [1967], Guatemala. De Camargo [1960] is a classic study of the rural exodus in Brazil. Graham's [1970] study of migration in Brazil is also important.

Three studies from the literature on Brazil are of interest. The first is Alvés's [1972] test and development of subsistence and commercial farm models of the labor supply behavior of farmers. These models are in the subjective equilibrium tradition of Nakajima [1969]. The second is Ignez Lopes's [1977] Ph.D. dissertation, also published in Portuguese as I. G. V. Lopes and Schuh [1979], which estimates the parameters of a multiple job holding model for low-income families. This research was based on time-allocation data collected from one of the first such samples of data collected in the developing countries.

The third is Whitaker's thesis [1970], which focused on the off-farm market for labor. He estimated the parameters of models of the market for manufacturing labor as the basis for examining the role of economic policy in limiting the labor absorptive capacity of the nonfarm sector of the economy. Quite generally in Latin America the subsidies for development were put in place in such a way as to lower the price of capital, while large payroll taxes, minimum wage laws, and other distortions raised the private cost of labor. This helps explain why the labor absorptive capacity of so many Latin American economies has been so limited. Policy has had a strong antiemployment bias to it.

Huffman and Coltrane [1986] made a detailed and important study of international trade and immigration in the context of U.S. and Mexican agriculture. A series of empirical studies of fresh vegetable production provides estimates of parameters that are useful for understanding the relationships between immigrant labor and the fresh vegetable market. Their results show that if the labor services supplied to the United States by aliens should be reduced, U.S. wage rates for labor services of low-skill workers, including agricultural workers, would be higher than otherwise. Moreover, a higher U.S. farm wage rate would in general reduce the U.S.-produced supply of labor-intensive fresh vegetables, and increase their market prices.

In the absence of protectionist measures, the U.S. would become a more attractive market for Mexican exports of these commodities. A reduction in the inflow of aliens would also reduce U.S. exports of labor-intensive commodities. For example, a higher farm wage rate would reduce exports of fresh lettuce.

These relationships between U.S. immigration policy and its trade performance are in general not given sufficient attention in policy debates, either from the U.S. or Mexican side of the issue.

J. R. Ramos [1970] provides a comprehensive analysis of the role of the labor force in postwar Latin American development. To this end he attempted to identify the nature of shifts in the composition of the labor force, and the economic determinants of such changes. Coming as it did when there was much pessimism about the performance of Latin American economies and in particular their ability to absorb labor, some of his results are striking. For example, he concluded that the fall in participation rates observed in the postwar period in Latin America was the result not of population pressure (a sign of weak economic performance), but of economic development (a sign of strength) [p. 5].

Ramos found that the growth of labor quality in the postwar period was even faster than the strong growth of the population, and that the quality of labor in the secondary sector increased more rapidly than that of the tertiary sector. Moreover, skills were important to the development of Latin America even at this early stage of labor abundance, because of the availability and importation of a modern skill-using technology. Thus trends in Latin American employment, especially movements in the quality and sectoral distribution of the labor force, were similar to *recent* United States behavior. This divergence from historical patterns, for comparable stages of development, indicated in Ramos's view that the effects of a commonly available technology swamped the effects that arose from a similar demand structure at comparable stages of development [Ramos, 1970, p. 9]. Production technology for the manufacturing sector is highly transferable, but requires a particular configuration of labor skills. The adoption of this technology appears to have a greater effect on the structure of skills demanded than does the structure of demand for final goods and services.

Ramos's results were at variance with the ECLA interpretation at the time — that the postwar population surge in Latin America into the cities ran up against the employment-limiting, factor proportions rigidity of the secondary sector, and was thus forced into marginally productive tertiary employment. His model is based on the introduction of a more skill-using modern technology which raised the demand for skilled labor, resulting in an abrupt rise in the quality of the labor force, especially in manufacturing, and a reduction of disguised unemployment in manufacturing ranks. Thus he finds neither "overurbanization," nor "underexpansion" of secondary sector employment, nor increased urban disguised unemployment, once he takes into account the strong increase in labor quality attending the introduction of modern technology into postwar Latin America [Ramos, 1970, pp. 10-11].

Ramos's analysis of the data provides a sharp counterpart to de Janvry's thesis [1975] of severe marginalization and exploitation of labor in Latin America.

Moreover, his policy prescription is in sharp contrast to both de Janvry and ECLA. He argued that the policy emphasis should be shifted from providing employment to generating labor skills (i.e., training and education). He further argued that there was great need to raise the quality of the labor force at a rate faster than that in the historical experience of the West, if Latin America was to profit fully from having a modern technology already available [Ramos, 1970, p. 11]. In our view this argument prevails with equal if not greater force in the second half of the 1980s as it did in 1970. Ramos's study also has important implications for agricultural development policy. Increased investment in formal schooling and training for the rural population will not only help to accelerate out-migration from the sector, it will also make the migrant employable at an earlier date after migration and reduce the burden on social services in the cities.

Finally, many countries in Latin America have in recent years been using their frontiers as a safety valve for their pressing employment problems. This strategy meets a number of important national objectives: it attains the political objective of filling up the frontier and thus protecting it from potential invaders; it brings unused land into the national economy and combines it with unemployed or underemployed labor to produce a greater national product; and it alleviates social pressures both in traditional rural areas and in urban centers.

The policy makers' rationale for this strategy is very much along the lines of Frederick Jackson Turner's frontier thesis for the United States, a perspective long discredited among academicians. Nevertheless, the potential of the frontier in Latin America has generated a plethora of studies [see, for example, Van Es, Wilkening, and Pinto, 1968; Dozier, 1969; J. R. Taylor, 1969; Crist and Nissly, 1973; Thiesenhusen, 1974; N. J. H. Smith, 1982; Sandner, 1962, 1964; Margolis, 1973; Dambaugh, 1959; Moran, 1981; and Havens and Flinn, eds. 1970].

Sewastynowicz's [1986] perceptive analysis of the colonization process on the Costa Rican frontier provides a revisionist interpretation of the role of the frontier as a safety valve and as a means of providing income opportunities for the *migrant*. He noted that the typical conclusion of current research is that the poor migrant fails to improve his lot by migration to the frontier.

Sewastynowitz argued that this pessimistic conclusion is a result of the failure to recognize the two-step nature of the migratory process. It is by a *series* of migrations, usually limited to two steps in the lifetime of a given cohort, that a family is able to attain substantial upward mobility.

The author found that upward mobility on the frontier depends on two factors: time of arrival and initial capital. The earlier the arrival, the greater the opportunities to the colonist. Similarly, the more capital at his or her disposal, the greater his or her ability to take advantage of opportunities at any given time. The early arrival is able to obtain land at low prices and thus a modest amount of capital will enable him or her to get started. As in-migration continues, land values are bid up from population pressures, and in the meantime the colonist's family has increased the value of the land by its work activities. By selling out and moving to a new frontier, the family can have another leveraged effect.

Sewastynowicz believed the process works in the same way on the Brazilian and Colombian frontiers. His thesis was not that everybody becomes rich in this way, but that those who obtain the right piece of land and who have the proper entrepreneurial characteristics can benefit in a significant way. Thus he views it as a viable strategy for social improvement, subject to the availability of suitable soils and an adequate infrastructure.

CONCLUDING COMMENTS

Agricultural labor bears the brunt of the adjustments associated with economic development because in the normal course of events labor has to be transferred to nonfarm activities as per capita incomes rise. In Latin America, import-substituting industrialization policies have tended to have a strong anti-employment bias to them, with the result that the manufacturing sector has had a fairly limited potential to absorb the out-migrants. While there has been considerable rhetoric about making agriculture the employer of last resort, economic policies which have shifted the domestic terms of trade strongly against agriculture have provided strong incentives to push labor out of agriculture, often prematurely [see M. de R. Lopes, 1977; and M. de R. Lopes and Schuh, 1979].

As the above review suggests, research on the intersectoral labor market in Latin America is fairly limited. Moreover, as noted in the section on land and land tenure, there has been very little research on the interlinkages between land, labor, and credit markets, nor on the design of adjustment policies that could either help to make the labor market more efficient, or to facilitate the adjustment process. Similarly, policy makers have given this issue very little attention. There is much to be done on both the research and policy side. A major share of the labor adjustment problem is still ahead of most Latin American countries, Argentina being an important exception. Both the equity and the efficiency of the economic development process in the region will be largely determined by the efficiency of the intersectoral labor market.

7. Concluding Remarks on Product and Factor Market Research and Policy

The literature surveyed in this chapter elicits a number of general remarks. First, agricultural commodity policies as implemented in the developed countries are rare in Latin America. Policy makers have chosen to influence the domestic terms of trade largely by means of trade and exchange rate policies. Moreover, these policies have tended to shift the domestic terms of trade *against* agriculture rather than in favor of it as occurs in the developed countries. Exceptions to these general propositions include Brazilian wheat policy, in which case policy has attempted to keep domestic prices above border price levels at the prevailing (overvalued) exchange rate.

Second, the rapidly growing literature on contractual arrangements and interlinked markets, which has emerged so strongly in the literature on Asian agricultural development, has had little influence on research and policy in Latin America. Given the highly skewed distribution of income in the region, and especially the seriousness of the problem in agriculture, this literature would seem to have a great deal of relevance in addressing problems in the region.

Third, both policy makers and intellectuals in Latin America have tended to lack confidence in the ability of markets to allocate resources efficiently and distribute income equitably. Government interventions, by creating public monopolies and parastatals, have tended to exacerbate these problems rather than to alleviate them.

Fourth, given the extent of rural poverty in Latin America, a great deal more research is needed on agricultural factor markets, especially on the intersectoral labor market. Associated with such research, there is an important place for more creative institutional design work to improve the performance of institutions which serve agriculture.

Fifth and finally, the use of a broad macroeconomic perspective as the analytical framework for the research would be useful. Studies and institutional design questions cast in a narrow microeconomic perspective all too often end up dealing with the symptoms of more basic economic problems, with the result that these more basic economic problems are seldom addressed.

Chapter IX. Human Capital: A Policy Perspective

Human capital as an input to the development process has become increasingly recognized as a critical factor in promoting economic development. This kind of capital is ultimately pervasive in a society, and appears in a variety of forms. One of the most obvious, widespread, and yet controversial is the genetic endowment of a society. This endowment determines the physical ability, cognitive skills, disease resistance, life expectancy, and endurance of a society's population. Although significant sums are spent in modern society to breed improved plants and animals, to date ethical proscriptions have precluded doing the same thing with people. Equally important, racial and other sensitivities have prevented researchers from exploring the extent to which genetic endowments influence the physical, mental and economic characteristics of human agents. Growing applications of biotechnology and genetic engineering may change this situation, however.

The culture and values of a society are other forms of human capital which condition the potential of society for economic development. Language is still another form since it is the basis of communication, and languages differ in their capacity to communicate ideas and concepts.

An important feature of some forms of human capital is that they are both producible and reproducible. Perhaps one of the more basic forms of human capital in this respect is knowledge, important forms of which are produced by use of the scientific method (research) and other procedures. Closely related to knowledge is the technology society uses in transforming inputs into goods and services. Technology takes a variety of forms, sometimes being imbedded in physical capital, as is the case with much of product and process technology, but other times merely imbedded in the minds of human beings, such as when it refers to the ways of doing things. The institutions which govern how members of society relate to each other are an important form of social (and human) technology. The policies and laws of society are, in particular, important forms of human capital.

Education and training are activities which add to the stock of human capital since they are the means by which physical and cognitive skills are developed. The health of a population is another important form of human capital which, within limits, is reproducible. Nutrition is still another form, since it influences the productivity of members of the society. Like education, research and extension are the means by which new knowledge is generated and transmitted from one person to another.

Nobel Laureate T. W. Schultz perceptively saw the relevence of education and research for Latin America far earlier than most observers. See his *The Economic Test in Latin America* [1956]; "Education and Research in Rural Development in Latin America [1967a]; and "Economic Growth Theory and the Profitability of Farming in Latin America" [1967b].

In this chapter we focus on three dimensions of human capital important to agriculture and agricultural development. These include education, nutrition, and research and extension. Other forms of human capital, such as policies and institutions, have been discussed in earlier chapters. Part of the discussion of education overlaps somewhat with the discussion of labor markets in the previous chapter, but logically fits more closely with the other material on human capital and so is included in this chapter.

1. Education

Both Adam Smith and Marshall wrote quite a bit about education. The rediscovery of this important factor of production began in the second half of the 1950s when empirical investigations in the United States revealed that output was growing much faster than inputs as conventionally measured.[96] That part of the growth of output unaccounted for by conventional inputs came to be known as the "residual" or the "coefficient of our ignorance."

In the beginning, the residual was attributed to "technical change" or to "shifts of the production function," concepts that were not very helpful either analytically or from a policy perspective. Subsequently, efforts were made to explain the residual, or to reduce the size of the unexplained residual. These efforts led to the identification of knowledge and of changes in the quality of labor and other inputs as key factors accounting for the residual.

Today, these factors are generally carried under the rubric of human capital, and include the output of the knowledge industry *per se*, health, nutrition, formal and informal training and schooling, culture, genetic resources, and institutional arrangements. G. S. Becker [1964] consolidated the thinking on these issues in his book on human capital. Professor T. W. Schultz [1964] consolidated the state of the arts in terms of agriculture with his famous little book on *Transforming Traditional Agriculture*, published in the same year.

Griliches added importantly to our knowledge by noting that the quality of physical inputs was changed as new knowledge was imbedded in them; helping us to understand the economics of modern inputs such as fertilizer and machinery and equipment; and estimating the contribution of these various factors to the growth in U.S. agricultural output [see Griliches, 1957, 1963a, b, 1964, and 1970]. Of special significance was Griliches's research on fertilizer, in which case he showed that the rapid growth in fertilizer consumption in the United States was largely induced by a decline over time in its real price, brought about in large part by a technological breakthrough in the fertilizer industry [Griliches, 1958].

Hayami and Ruttan [1971, 1985] are essentially extensions of T. W. Schultz [1964] and Griliches [1963a, b, 1964], with the extensions being to bring in consideration of the institutional arrangements needed to focus agricultural research onto an efficient growth path, and to include a model of how the institutional arrangements respond to changing economic forces. Hayami and Ruttan's empirical results also point to the importance of new knowledge and education as factors accounting for agricultural growth.

This section addresses two main themes: education as a factor of growth, and education as a factor influencing the distribution of income. The economics of education is an important subtheme, as is the significant underinvestment in the education of rural people, in Latin America as well as in other countries.

THE RATIONALE FOR INVESTING IN EDUCATION

Before turning to the literature on Latin America, it is useful to consider the various rationales that have been provided for investing in education and some of the empirical research that supports them. Welch [1970] developed an important analytical model which explains how education has an effect on production. In broad terms he distinguished between a production effect *per se*, an allocatory effect, and a decoding effect for new knowledge. Empirical support for a direct production effect is found from estimates of the aggregate agricultural production function which include education as an "environmental" variable [Griliches, 1963a, b, 1964; and Hayami and Ruttan, 1971, 1985]. Griliches's early work with this production function, using U.S. data, found the production elasticity of the education variable to be the same as that for the labor input, implying that they were perfect substitutes in production.

The allocatory effect which Welch refers to is the improvement in allocation of resources (towards a more efficient configuration) which the cognitive skills developed by education make possible. This improvement in efficiency is reflected in an increase in output from a given bundle of resources. The "decoding" effect is the ability to make use of new knowledge made available from research programs and which leads to improvements in technical efficiency.

A second pathbreaking study of the role of education in agricultural development was Gisser's [1965] introduction of education into models describing the intersectoral labor market. Education was hypothesized to shift the demand curve for agricultural labor to the right, and the supply curve of labor offered to agriculture to the left. As long as general equilibrium effects are ignored, the expectation is that the net effects of education would be to transfer labor out of agriculture and to raise agricultural wage rates, other things being equal. Using cross-sectional data from U.S. agriculture to estimate the parameters for the reduced forms of this labor market model, Gisser found support for his hypotheses. This provides an important role for education in helping to deal with the so-called "farm problem," the traditional excess of labor in agriculture and the associated differentials in per capita incomes between agriculture and the rest of the economy.

An important point in this regard is the frequent (naive) argument that farmers do not need schooling to be good farmers. Wharton [1965] provided a perceptive discussion of the role of education in early-stage agriculture. The point of the Gisser [1965] study is that schooling and education are important in helping to facilitate agriculture's chronic labor adjustment problem by making it possible for the worker to adjust to new, alternative employment. But Welch's analysis [1970] provides a role for education in agriculture, *per se*, since he shows that when new technology is being introduced into agriculture the cognitive skills developed by means of formal schooling are needed to decode the new knowledge. The production and allocatory effects of education may also be important.

T. D. Wallace and Hoover [1966], again using data from the United States, provided additional evidence on why the labor market aspect of education is important. Their statistical results indicated that as new production technology is introduced into agriculture it has the effect of increasing the demand for labor so long as general equilibrium effects are ignored. When these effects are taken into account, however, new production technology will increase the demand for labor so long as the price elasticity of demand for the output is greater than one. But if the price elasticity for the output is less than one, it will reduce the demand (shift the demand curve back to the left), even though it has a productivity-enhancing effect. We thus see an intricate relationship between new production technology and education, and potentially a powerful role for education so long as new technology is being introduced into agriculture. Education, together with expansion of the nonfarm sector of the economy, keeps technical change from producing immiserizing growth for agriculture when the price elasticity of demand for agricultural output is less than one.

T. W. Schultz [1975c] provided still another perspective on the role of education. He saw education and entrepreneurship as having economic value in providing the ability to exploit disequilibrium situations. This perceptive and important paper provides a counterpoint to T. W. Schultz's concept of traditional agriculture [1964], which is defined as one in which equilibrium adjustments have been made to an agricultural world free of external shocks. Once these shocks are introduced from whatever source, there are significant economic gains to be had from exploiting the disequilibria which result. Education *and* entrepreneurship have an important role to play in this exploitation.

EVIDENCE ON THE RATES OF RETURN TO INVESTMENTS IN EDUCATION

For those interested in the returns to education as an investment, perhaps the best single reference is Psacharopoulos [1973]. Psacharopoulos reviewed fifty-three case studies of the returns to education representing the experience of thirty-two countries. He attempted to answer two major questions: How does the profitability of investment in education compare with the profitability of investment in physical capital? Can intercountry differences in the stock of human capital help to explain differences in the level of per capita income? In addition, he provided information on the following issues: the structure of the rates of return by level of education; the degree of public subsidization education receives in different countries; a new index of educational development based on costs; the earnings ratios of people with different levels of education within a given country; and the economic returns of higher education graduates who emigrate to work in a foreign country.

The social and private rates of return Psacharopoulos [1973] obtained for five Latin American countries are presented in Table 56, together with estimates for the United States for comparison. These estimates represent an effort to make the returns as comparable as possible across countries. They also include some of the author's own estimates of rates of return. The original studies from which the

Table 56. Social and private rates of return by educational level and country (in %)

Country	Year	Social			Private		
		Primary	Secondary	Higher	Primary	Secondary	Higher
United States	1959	17.8	14.0	9.7	155.1	19.5	13.6
Mexico	1963	25.0	17.0	23.0	32.0	23.0	29.0
Venezuela	1957	82.0	17.0	23.0	–	18.0	27.0
Colombia	1966	40.0	24.0	8.0	>50.0	32.0	15.5
Chile	1959	24.0	16.9	12.2	–	–	–
Brazil	1962	10.7	17.2	14.5	11.3	21.4	38.1

Source: Psacharopoulos, 1973, p. 62.

estimates of the rates of return are taken include Carnoy [1964] for Mexico; Shoup [1959] for Venezuela; Franco [1964], T. P. Schultz [1968], Selowsky [1967], and Dougherty [1971] for Colombia; Harberger and Selowsky [1966] and Selowsky [1967] for Chile; and Hewlett [1970], Castro [1970], Lerner [1970], and A. J. Rogers III [1969] for Brazil.

Both the social and private rates of return to investments in schooling in Latin America are almost uniformly high. Psacharopoulos [1973] estimated averages across all countries in his sample by level of schooling. The social rates of return were 25.1, 13.5, and 11.3 percent, respectively, for primary, secondary, and higher education. The corresponding private rates of return were 23.7, 16.3, and 17.5 [p. 65]. Except at the primary level, the private rates of return tend to be higher than the social rates of return, a reflection for the most part of subsidies to education.

The major findings from Psacharopoulos's study are ten in number. First, both in social and private terms education has a monetary payoff and this payoff is substantially higher in less developed than in advanced countries. Second, education is most profitable in most countries at the primary level, while higher education shows only a modest payoff, particularly in advanced countries. Third, returns to investment in human capital are well above the returns to physical capital in less developed countries. Fourth, per capita income differences can be better explained by differences in the endowments of human rather than physical capital. Fifth, investment in education contributes substantially to the rate of growth of output in most countries, particularly in the less developed group. Sixth, labor with secondary educational qualifications seems to contribute more to output than that labor is paid. Seventh, higher education is very expensive in relation to the other levels of education, particularly in less developed countries. Eighth, earnings inequality by educational level decreases as the level of development rises. Ninth, there is a high degree of substitution in production between different types of educated labor. This suggests that expansion of the educational system should be based on calculations of relative costs and benefits rather than on manpower needs. And tenth, there is a high return for those who graduate in the home country and subsequently emigrate to the United States. This is the incentive for the brain drain.

Psacharopoulos [1985] later updated his survey of rates of return. In this study he reported that average (private) returns to primary, secondary, and postsecondary schooling in Latin America were 0.32, 0.23, and 0.23, respectively, very respectable rates of return. Psacharopoulos's findings (especially [1973]) that the rate of return to investment in higher education is low on the average deserves some comment. It is easy to overgeneralize from this finding, and Psacharopoulos sometimes does so. This casts doubts on the value of investments in the establishment of graduate programs in the agricultural services in the region. We doubt whether this inference is merited, in large part because it is difficult to pick up the effects of such programs from aggregate data.

Unfortunately, there have been no studies to our knowledge which have made careful studies of the contribution of these graduate programs. The closest thing which comes to it is a study by Avila *et al.* [1983] which showed the rates of return to investment in its own human capital to be quite high. This investment included both graduate study abroad and in domestic graduate programs, however.

This issue deserves more attention, difficult as it may be to address empirically. Domestic graduate programs would appear to have a double payoff since they produce the research and teaching cadre for the future, but generate research in their own right. Presumably, investments in such programs would have a high social payoff.

Turning to other studies of the rates of return to investments in education, Stelcner, Arriagada, and P. R. Moock [1987] reported estimates for Peru that are significantly below the above averages (0.14, 0.09, and 0.11), and lower than for any other Latin American country cited. These authors cited estimates based on more recent data for wage earners which suggest that the returns to schooling in Peru are not out of line, however [p. 38].

Corbo and Stelcner [1983], using data for Chile, report a rate of return to a year of schooling of about 14 percent, while Heckman and Hotz [1986], using data from Panama, estimate a rate of 12 percent. Steir [1987] reports a rate of 10 percent for workers in Caracas, Venezuela, using 1984 data. A rate of return of 11 percent is estimated by Psacharopoulos, Arriagada and Velez [1987] using 1984 Colombian data. Mohan [1986], using 1978 data for Bogotá, Colombia, reports rates of 7 percent for primary schooling, 9 percent for secondary schooling and about 13 percent for postsecondary schooling.

Some additional general sources on education include Adelman and Morris [1973], Bowman and Anderson [1963], Easterlin, ed. [1981], Fields [1980b], Harry G. Johnson [1964], Krueger [1968], Psacharopoulos [1984], T. W. Schultz

[1963, 1972], Selowsky [1976], Weisbrod [1964], and Wheeler [1980]. Some additional references on Latin America include Behrman and Birdsall [1983a,b], Carnoy [1967a,b], Correa [1970], IADB [1978], Jallade [1974], and Psacharopoulos and Loxley [1985]. A. J. Rogers III [1969] and Roy, Waisanen, E. M. Rogers and UNESCO [1969] view education in the context of the diffusion model.

Psacharopoulos and Woodhall [1985] is a review of some twenty years of World Bank experience with educational programs. The treatment is comprehensive, and each chapter has a useful list of references. Bachmura, ed. [1968] is a useful collection of papers from about the time development economists were just beginning to use the concept of human resources, with special reference to Latin America. Waggoner and Waggoner [1971] provided a comprehensive description of educational institutions in Central America, related to the physical and cultural environment of the region, and with the analysis showing the complexity of the problems which confront the region.

Myers [1965] provided a perceptive analysis for Mexico of the challenge many developing countries face: should it allocate its scarce development resources to the elimination of ignorance and poverty in the backward regions, or should it allocate them to promoting rapid growth nationally. This is both a regional and a sectoral problem. Mexico in the beginning put the emphasis on agricultural education and in the poor regions. But later it shifted the emphasis to the nonagricultural sector and the more advanced regions. Myers argued that Mexico was correct in following this strategy, and that it is a policy other less developed countries may need to follow, since it is only by obtaining more rapid growth that it will have the resources to solve the problem of regional inequalities.

EDUCATION AND THE DISTRIBUTION OF INCOME

An important contribution to the education literature relative to Latin America is Carnoy's [1979] study with the title *Can Educational Policy Equalise Income Distribution in Latin America?* After bringing together earlier evidence on the subject, including some evidence on Chile and Cuba, the author compared the results obtained in empirical studies carried out by three Ph.D. students under his supervision, on Brazil [Velloso, 1975], Peru [Toledo M., 1976], and Mexico [Lobo, 1978].[97] These studies distinguished themselves by making comparisons between data for different points in time. Changes in earnings distributions between two census years were decomposed into those that were due to changes in the distribution of characteristics of the labor force, the major part of which was the distribution of educational qualifications, and into those due to changes in the payoffs to various characteristics of the labor force. Thus the study established policy instruments for a more general income distribution policy. It was this emphasis on policy that led to the discussion of this study in this section rather than in the earlier section on income distribution.

The authors distinguished between neoclassical models of the labor market, which emphasize characteristics of the supply side of labor markets in explaining how income distribution changes, and the segmented labor market argument, which emphasizes the demand side of the market in explaining changes in income distribution. From the supply side perspective, it was expected that the distribution of earnings would become more equal if the distribution of labor characteristics became more equal. Thus, the policy derived from that model stresses the equalization of personal attributes of workers, particularly their schooling and training. Equalizing the supply of these attributes is expected to equalize the productivity of workers and in turn their incomes.

The segmented labor market model, on the other hand, claims that the most important differences in earnings and changes in the distribution of earnings are the result of changes in the income paid to worker characteristics, not the distribution of the characteristics themselves. The segmented labor market perspective stresses that the changes in income paid for these different characteristics, including schooling and other worker attributes, are not necessarily connected to changes in productivity, even in the longer run, but are instead based on income policies based on the power of labor unions and other political factors.

When the distinction between the alternative models is cast in this sharp fashion, the reader can anticipate the results. The empirical results indicated that schooling and age (experience) have an important influence on earnings in Brazil, Peru, and Mexico. But the change in the distribution of schooling in the labor force had much less influence on the distribution of earnings than did changes in the value of different amounts of schooling over time (Brazil) and changes in the values (coefficients) of other independent variables (occupational segment in Peru and economic sector plus regions-worked-in for Mexico) [Carnoy et al., 1979].

Carnoy and his colleagues emphasized that these changes in values of worker characteristics, or sector or region of work, are the result of public policies which change the distribution of income directly. Moreover, they are essentially demand side changes. They also emphasized that in the three countries studied, the distribution of income changed in large part because higher income groups gained income at the expense of lower income groups, not because more people in the labor force acquired characteristics which enjoyed greater payoffs relative to characteristics which enjoyed less.

Carnoy and his colleagues concluded as follows:

> [T]here is an apparent paradox in income distribution policy;
> schooling apparently plays a very important role in
> determining individual earnings in Latin America, but the
> distribution of education in the labor force is not very
> important in influencing earnings distributions. Rather,

> government incomes policy affecting the reward to different
> levels of schooling, different work sectors, different types of
> occupations, and different regions of the country may be a
> much more important factor in understanding changes in
> income distribution. We suggest that educational policy can
> only contribute to the more equal distribution of earnings
> when it is carried out in concert with an incomes policy which
> attempts to equalize the earnings of workers with lower levels
> of schooling, in lower paying occupations, economic activities,
> and regions, and with less experience in the labor force to
> those in the higher paying categories of each of those variables.
> [Carnoy et al., 1979, p. 98]

We would underline the importance the authors give to the complementarity between labor market policies and investments in education, and would generalize it a bit further. It isn't just the complementarity between wage and income policies with education policy that matters, but the complementarity between the full panoply of economic policies. For example, economic policies which shift the domestic terms of trade against agriculture reduce the incentives to adopt new production technology. A more rational price policy would lead to a more extensive adoption of the new technology, and the adoption of the new technology would increase the rate of return to education since there would be an increased need to decode the information required for its adoption. Thus, one sees the complementarity between price policy (and implicitly trade and exchange rate policy), science and technology policy, and education policy.

Although not discussed until later, one could include health and nutrition policy as well, since healthy and well-fed workers and decision makers are more productive and better able to absorb and use cognitive skills. Thus, we believe there is a *policy* "package" that is far more important in obtaining agricultural development than the frequently stressed technological package that is recommended for agricultural modernization.

We believe there is a similar relationship that arises when the goal is to increase income or to redistribute incomes. For example, Carnoy and his colleagues [1979] noted that labor's share of national product in Latin American countries is closer to 50 percent than to the 60 to 70 percent share it makes up in industrialized countries. This at one and the same time shows the limited potential investments in education can have in altering the distribution of income in the short run, but the significant potential it has in the long run.

This low share of income attributed to labor also suggests why redistributing assets such as land may need to play an important role in redistributing income under Latin American conditions, especially since the size distribution of land tends to be so highly skewed. But a clear implication of what we have said above is that land redistribution alone is not likely to be very effective. Economic policies have to provide the proper incentives; a new, productive technology will add a great deal to productivity; and the recipients need to have schooling or receive training programs to handle their entrepreneurial responsibilities and challenges. Without the latter, the new enterprises will not be productive and it will be difficult to sustain the redistribution of physical assets since the land will drift back into the hands of those better able to make use of it. Moreover, we believe a freeholder system in which these skills are given to the individual will work better than group farming systems where there is group decision making.

A study of rural poverty in Brazil found support for this proposition [G. L. da S. Dias, ed., 1979]. Examination of alternative policies for raising family incomes found that land redistribution alone, new productive technology alone, or education alone, added only marginally to family incomes. But when all were done together, the impact was significant.

We conclude this discussion of education and its potential for influencing the distribution of income with two final comments. First, redistributing land under Latin American conditions tends to create political instability. One reason for this is that high and unstable rates of inflation and inadequately developed capital markets provide strong incentives to hold land as an asset, as a hedge against inflation, and not necessarily for productive purposes alone. Another reason is that buying the land at market prices for redistribution, plus providing the other inputs needed to make the reforms effective, including infrastructure, is very costly [see T. Barbosa, 1973]. Hence, it is almost inevitable that some degree of confiscation be involved in the process, either explicitly, or implicitly, if the redistribution is to be on any significant scale.

Second, this raises the issue of whether alternative means can be used to obtain a more equitable distribution of land holdings. Although it will take more time, we believe such means can be found. An important first step would be to create more stable monetary conditions and reduce the rate of inflation, thus reducing the incentive to hold land as a protection against inflation. A second step would be to create additional viable instruments for the capital markets and to make these markets more robust. A third would be to reduce the fiscal incentives to hold land, which Binswanger [1987] has so perceptively identified in Brazil.

Perhaps as important as anything, however, would be to give more attention to education for rural people. Increasing family incomes involves increasing the stock of resources it controls, raising the productivity of those resources, and/or raising the price (or rent) these resources earn in the market. Augmenting the stock of human capital the family has may ultimately be far more important in raising family income than augmenting its stock of physical capital, in part because it helps raise the productivity of the conventional resources, especially if new production technology is available. But the dynamic effects of education may be far more important. By opening nonfarm employment opportunities to the worker and his or her family, it can help to make labor scarce in agriculture and thus help to raise the wage rate. Ultimately, this will lead to a change in the size distribution of land holdings.

The skeptic will reply that the dominant land-owning class will resist the provision of expanded educational opportunities as well as the more commonly proposed redistribution of land. This may be the case. But it should also be easier to obtain general political support for expanding educational opportunities than it is for the confiscation and redistribution of land which, with its inherent challenge to private property, goes to the heart of the capitalist system. Moreover, landowners may see that education helps to raise the productivity of their workers.

Some additional references on education and the distribution of income include Marin and Psacharopoulos [1976] and Selowsky [1979].

EDUCATION AND PRODUCTIVITY

Few of the studies cited above have focused directly on the education of rural people and its impact on productivity or production efficiency. Jamison and Lau [1982] have provided a global survey of studies using micro data to test this hypothesis. The existing literature suggests that more educated farmers are more productive, particularly, as T. W. Schultz [1979] hypothesized, in an environment of agricultural modernization. Jamison and Lau [1982] is useful as a general source on this issue, for the authors provide a careful and detailed discussion of the methodological problems entailed in isolating the effect of education on production and productivity.

The research on income distribution has considered the role of education, with special attention to the role differences in education play in explaining the income disparities between agriculture and the nonfarm sector. Hayami and Ruttan [1971, 1985] give education a significant role in their empirical work, and show it to be a major factor in explaining intercountry differences in agricultural productivity.

To our knowledge there are only four studies that have used farm-level data from Latin America to investigate these issues. One of these refers to Colombia [Haller, 1972], and includes samples from four *municípios*. In no case was the coefficient for education statistically significant (at usually accepted levels) in this study. The other studies referred to Brazil, and included those by Pachico and Ashby [1976] for southern Brazil, Patrick and Kehrberg [1973] for eastern Brazil, and R. D. Singh *et al.* [1979] for the states of Ceará, Minas Gerais, and São Paulo.

Pachico and Ashby [1976] found that education was positively related to output among highly commercialized farms, but that in one region less than five years of schooling had no significant effect on output. Patrick and Kehrberg [1973] found that the returns to schooling were negative in the traditional agricultural regions, but became positive and increased as the regions became more modern among the five sample areas. The research reported in R. D. Singh *et al.* [1979] attempted to find a relationship between education and production on small farms, but there was so little variation in the education variable that a statistically significant relationship could not be identified. When education was included in an earnings function, however, a statistically significant relationship was obtained.

The Patrick and Kehrberg [1973] study is important from another perspective. One of the puzzles related to education is why school attendance is so low among rural children. Patrick and Kehrberg showed that this is due in large part to the high opportunity costs of school attendance. These costs swamped the direct cost of schooling, and are very high relative to family income among the poor.

Valdés [1971] examined the role of education in determining the structure of wages among agricultural workers in Chile. Using a 1965 sample of data from two Chilean provinces, he found strong statistical support for the proposition that schooling differences were important in understanding wage, income, and hence productivity differences among sample workers.

EDUCATION, DEMAND FOR CHILDREN, AND POPULATION
GROWTH RATES

Another important issue with education is the interrelationship between the economic value of children, household fertility decisions, and the demand for schooling. R. D. Singh and Schuh [1986] and R. D. Singh, Schuh, and Kehrberg [1978] have examined these interrelationships with data from the sample of poor households, analyzed and reported on more extensively in R. D. Singh *et al.* [1979]. These data were drawn from a survey of approximately 500 low income households from three rural regions of Brazil.

The new household economics, as pioneered by G. S. Becker [1965], G. S. Becker and H. G. Lewis [1973, 1975], de Tray [1973, 1975] and T. W. Schultz [1973b, 1975a, 1976] draws attention to the value of time in the household, and to the opportunity cost of that time, as factors affecting the behavior of members of the household. Considering the household as a production unit, the theory of the firm becomes relevant in understanding important activities in the household. [For a useful "state of the art" on this perspective, see T. W. Schultz, ed., 1973b, 1975a.] Importantly, it provides a framework for understanding the economics of the family, participation of family members in household production, farm production, and the labor market, and the decisions families make with respect to the quantity of children.

G. S. Becker and H. G. Lewis [1973, 1975] and de Tray [1973, 1975] hypothesized that families have an underlying demand for child services, based on the value of the children to the household as consumption goods and on their value in production or as a source of income. They further hypothesized that there is a negative relationship between quantity (number) and quality (as represented by years of schooling, for example). As economic incentives encourage more investment in children in the form of schooling, for example, they cause a reduction in the number of children demanded. Hence, there is a relationship between level of schooling and numbers of children demanded, with the costs and benefits of the children influencing both the level of investment in the children and their numbers.

Lack of household data has limited the use of this new perspective in developing countries. The only examples we have found for Latin America have been those referred to above, plus Jabara's [1977] study of the demand for education of children among small farmers in rural areas of Brazil. Each of these studies provide support for the underlying model, with Jabara [1977] and R. D. Singh, Schuh, and Kehrberg [1978] finding the demand for schooling influenced by economic variables in the expected way, and R. D. Singh and Schuh [1986] and R. D. Singh, Schuh and Kehrberg [1978] finding statistical support for the expected relationships in the case of fertility decisions. R. D. Singh, and Schuh [1986] in particular found a statistically strong (and negative) effect of the education of the father on child quantity. This indicates that parental education can be vital in lowering fertility rates and thereby slowing down population growth in low income regions.

T. W. Schultz [1974] and Nerlove [1974] used the new household economics and Becker's time allocation model to speculate about the long-term equilibrium of a country experiencing economic growth. Schultz argued that the ultimate constraint to economic development is the limit that a twenty-four-hour day puts on the development process. Schultz virtually stood the classical world on its head by postulating that the ultimate constraint to development comes from limitations on time for the consumption of household-produced goods and services. This is a constraint within the household, and does not arise from limits on physical resources or rising costs in production, as did the classical model. Moreover, the Schultz equilibrium is at a high income level, not the low level postulated by the classicists. The Schultz perspective is particularly rich, since it includes a population as well as an income equilibrium.

Nerlove [1974], although obviously in the Schultz tradition, provided a somewhat different perspective. He argued that productivity in the household, where both production and consumption take place, can be raised on a continual basis. Because human capital is one of the main outputs of the household, further investments in human capital actually increase the efficiency with which it can be produced. Hence, there is no reason for an equilibrium level of per capita incomes or population to exist. Nerlove's model does predict declining rates of population growth and declining rates of infant mortality, however, and these are the main features of the demographic transition.

EDUCATION AND COMPARATIVE ADVANTAGE

Finally, education and other forms of human capital, including health, nutrition, knowledge and institutional arrangements, are important in determining a country's comparative advantage in international markets. The important points on this issue are twofold. First, standard neoclassical trade theory has to be extended to take account of differences across countries in human capital endowments, as Kenen [1965, 1968, 1970], among others, has done. Second, one needs to recognize that economic development, with its associated rising value of human time, drives the economy to an increasingly human capital-intensive configuration. Hence, as an economy develops, its investments in education, in the production of new production technology, and in new institutional arrangements become increasingly important in determining the structure of a nation's imports and exports and its ability to compete internationally. These issues are discussed in Schuh [1981].

CONCLUDING COMMENTS

To conclude this section, education is seen to be a crucial form of investment not only for the development of agriculture, but also for the development of the economy as a whole. Unfortunately, Latin American countries have significantly underinvested in this important form of human capital, and in the case of its rural population they have seriously underinvested [Schuh and Angeli-Schuh, 1989]. The disparity in educational attainment between rural and urban populations is quite great. This problem is serious at all levels. It contributes to low resource productivity at the farm level, the failure to adopt new production technology as it becomes available, the piling up of unskilled labor in urban centers, increasingly skewed distributions of income, and relatively high population growth rates. High population growth rates contribute to a vicious circle, for they cause a large share of the population to be young and thus increase the burden of providing education to the next generation.

An issue not discussed above is the failure to develop undergraduate and graduate programs in the agricultural sciences that attain international standards. The failure to attain these levels will continue to be a constraint on the performance of all institutions that serve agriculture, of national agricultural research systems in the region (including the social sciences needed for policy analysis), and on the performance of higher level educational institutions in producing the supply of trained manpower for the future. Unfortunately, the international debt problem and the economic stagnation of the 1980s, and the crimp they have put in the availability of public resources, have caused many of the existing programs to deteriorate below levels they had once attained.

2. Nutrition

Nutrition is another important form of human capital. Malnutrition can reduce the level of productivity of the work force in a number of important and pervasive ways; Ward and Sanders [1980] cite four. For example, nutritional inadequacies can lead to irreparable damage to the normal physical and mental development of children, thus reducing or eliminating the returns to other types of investment in human capital, such as education or on-the-job training [Selowsky and Taylor, 1973]. Malnutrition among lactating or gestating women can result in the permanent mental and physical impairment of their children. Malnutrition in adults leads to increased susceptibility to infection and eventually to physical disabilities. And finally, malnutrition sets in motion a cyclical pattern of illnesses which reduces productivity by both reducing physical output and by increasing job absenteeism [Mata, Urrutia, and Garcia, 1975].

Important studies of nutrition and productivity include Leibenstein [1957] and Oshima [1967]. More general discussions of the nutrition-productivity relationship can be found in Belli [1971], Bliss and Stern [1978], with some important criticisms of previous nutrition-productivity studies in Franke and Barrett [1975]. Bruce Johnston's survey [1977] of the literature on food, health and population is also important. A. D. Berg [1973, 1981] is important on policy.

Measuring the extent of malnutrition is a difficult task, and often becomes controversial when attempts are made to infer its extent from secondary data. A widely cited source which makes a serious effort to make such estimates is Reutlinger and Selowsky [1976]. Dissenting views on how the data were used and what they mean can be found in Poleman [1983] and Srinivasan [1983]. In general, one of three methodologies are used to estimate the extent of malnutrition: the aggregate food balance method, the family food consumption recall method, and the direct individual observation method.

Ward and Sanders's [1980] study of nutritional determinants and migration in Brazil's Northeast is important on a number of counts. First, they cite a number of studies which show that despite recent rapid economic growth in Brazil the nutritional inadequacies in the Northeast at the time of their studies were still very substantial. They refer to the first major study of nutritional levels in Brazil, which was conducted by the Getulio Vargas Foundation [FGV, 1970] in the early sixties. Using the food consumption recall method, this study estimated that 44 percent of the national population and 75 percent of the population of the Northeast suffered from caloric deficits. Diets in the Northeast were also found to be deficient in protein and fat consumption, and urban nutrient intakes were found to be superior to those in rural areas. This study also found a relatively low income elasticity of demand for calories in Brazil (approximately 0.2-0.3), which implies that very large increases in income would be necessary to eliminate caloric deficits.

Ward and Sanders also provided information from other studies on Brazil. Their main concern was the serious nutritional deficiencies in the poverty-stricken Northeast, the apparent failure of these deficiencies to be reduced despite a period of rapid economic growth, and the obvious need for direct intervention programs. The authors also used a sample of household data to estimate the parameters of a model which related the consumption of nutrients to various socioeconomic variables.

One of the interesting findings in the Ward and Sanders paper is that rural-urban migrants in the Northeast experienced greater nutritional deficiencies in their new urban setting than did the rural poor, even though their per capita income levels were higher. This finding was consistent with an earlier study for Brazil as a whole [McCarthy, 1975], which showed that 68 percent of the urban poor had caloric deficits in 1975 while only 47 percent of the rural poor had such deficits [Ward and Sanders, 1980, p. 154].

These findings of Ward and Sanders point to the difficulties of doing something about the malnutrition problem. Berg [1987] described the situation very well:

> Malnutrition is a problem that defies pat solutions. It has many roots: inadequate food supply, limited purchasing power, poor health conditions, and incomplete knowledge about nutrition. These causes combine in different ways over time and place. In any combination, they are often aggravated by uncertain political commitment. This makes it difficult for government workers who must develop strategies and programs to combat malnutrition, and for foreign assistance agencies that try to help them.
>
> Adding to the complexity is the lack of an organizational locus for carrying out such programs, because nutrition is not a sector in the conventional sense. Rather, it is a condition, like unemployment. Solutions must cut across discipline and organization charts. Government officials have no ministry or department of nutrition to turn to. Nor does any United Nations organization have a primary mandate to help countries contend with malnutrition.
>
> As such, malnutrition is everybody's business but nobody's main responsibility. [p. 1]

The confusion over malnutrition can perhaps best be illustrated by the surge of literature on food security that emerged after the food crisis of the early and mid-1970s, when agricultural prices rose to unprecedented levels in international commodity markets and policy makers scurried to acquire supplies. This episode engendered widespread interest among academicians and international agencies in the food security issue. Unfortunately, the policy recommendations that came out of this concern gave unwarranted attention to the supply side of the problem: increasing production, obtaining food self-sufficiency at the national level, and augmenting stocks, with a great deal of emphasis on the latter. Other studies focused on compensatory finance [Goreux, 1980], general issues of international finance [Huddleston et al., 1984], and insurance schemes [Huddleston and Konandreas, 1981].

A. K. Sen [1981] eloquently noted some years ago that hunger and malnutrition have very little connection with the supply of food, even in times of famine. Instead, the problem is the lack of means among those affected by malnutrition and hunger to buy the food they need. At the national level, the problem is often the lack of foreign exchange to acquire the imports needed. Self-sufficiency in food production at the national level offers little or no assurance that malnutrition and hunger will be eliminated, or even reduced, as the experience of countries such as India so effectively illustrate.

There has also been insufficient recognition of the costs of carrying stocks, or of the difficulty of managing them to insure that supplies will be available when needed. Well-developed international commodity markets now offer a more secure and less costly source of supplies when shortfalls occur. Modest foreign exchange reserves can be carried to assure that resources are available to acquire imports when they are needed.

These issues are discussed by Reutlinger and van Holst Pellekaan [1986] in *Poverty and Hunger: Issues and Options for Food Security in Developing Countries*, a policy paper by the World Bank. This paper identifies policy options that can be used under alternative circumstances. While emphasizing that broad-based economic growth can ultimately solve the hunger and malnutrition problem, they argue that there is much governments can do in the short term to alleviate the problem. Short-term alleviation is needed to forestall the destruction of human capital that occurs from malnutrition, to produce a more productive labor force, and to improve the learning capacity of affected groups.

The adjustment processes many developing countries are currently going through as a consequence of the international debt crisis provide another important reason for addressing the food security problem. The adjustment policies typically entail significant realignments of exchange rates and a shift in the domestic terms of trade in favor of agriculture. These raise food prices in the domestic economy, often to the disadvantage of the poor. Targeted feeding programs are needed under these circumstances if political difficulties associated with food riots are to be avoided. Such programs not only make it possible to implement such policies, but to sustain them once they are in place. Targeted feeding programs also make it possible to phase out costly *general* food subsidies, which is often necessary as part of adjustment programs.

The World Bank undertook four experimental nutrition projects in the 1977-80 period, largely in an attempt to understand and deal with the administrative, economic, and political realities of nutrition interventions. These four projects were in Brazil, Colombia, India, and Indonesia. A. D. Berg [1987] reviews the experience with these projects and discusses what has been learned. In his opening chapter he identifies some of the unexpected findings that have emerged from Bank experience, many of which refute conventional notions about nutrition.

According to A. D. Berg, the Bank's experience supports the following propositions: although malnutrition is closely linked to a country's level of economic development, nutrition improvements need not await that development; large food programs—ranging from consumer food subsidies to child feeding through institutions—need not be prohibitive in cost; even though malnutrition is closely related to poverty, malnutrition need not be bound by family income; women's lack of schooling need not pose the insurmountable constraint to improved nutrition that it is widely believed to pose; vitamin and mineral deficiencies may be caused by a rapid shift from traditional, locally produced grains such as millet and sorghum to polished rice and refined wheat, both often imported (such processing causes loss of certain natural nutrients); improved nutrition appears to increase the capacity to work and learn; social and psychological dependence among recipients need not be an inevitable outcome of feeding programs; and increasing efficiencies in the food marketing system make it possible to reduce substantially the prices that low-income families must pay for food [A. D. Berg, 1987, pp. 6-7]. Berg concludes that the findings he reports suggest that measures that will enhance the access to food, particularly for women and small children, are at hand and that they are both efficacious and affordable.

Cuadernos de Economia, a journal of the Catholic University of Chile in Santiago, published a special edition in 1985 devoted to the economics of health and nutrition. Valdés [1985a, pp. 169-173] provides an overview of the papers in this collection. One of the papers he discussed reported an attempt to explain how the child mortality rate in Chile declined from 116 per 1,000 live births in 1955, one of the highest in the world, to twenty-one deaths per thousand live births in 1983. The study attributes this decline in large part to public interventions— potable water, improved sewage systems, public health services, and subsidized food. An important feature of the public health and subsidized food program was that access to the food was associated with access to the public health services.

This section would not be complete without references to the Institute of Nutrition of Central America and Panama (INCAP), which has now celebrated its 25th anniversary. The Institute is funded by the six Central American countries and is the most outstanding research center on nutrition in Latin America. In addition to research, the Institute provides training in nutrition and provides technical advisory services to the governments of the member countries.

The reader interested in policy-related sources can find a selection in A. D. Berg [1987]. Publications by Berg [1973, 1981] and by Reutlinger and Selowsky [1976] are of general interest. Pinstrup-Andersen has also contributed importantly, alone and with others, both on the general problem and on Latin America [Pinstrup-Andersen, 1977, 1984, 1985, ed. 1988; Pinstrup-Andersen, A. D. Berg, and Forman, eds. 1984; Pinstrup-Andersen and Calcedo, 1978; and Pinstrup-Andersen, de Londoño, and Hoover, 1976]. Pinstrup-Andersen's analysis [1983] of the impact of export crop production is of special interest in light of the growing tendency in Latin America to shift to more externally-oriented development policies. The use of food aid to deal with problems of malnutrition is another important issue. Singer and Maxwell [1983] provided a review of twenty years' experience with food aid. Knudsen and Scandizzo [1979] made an assessment of the problem of food supply and distribution in selected developing countries and evaluated policies to meet nutrition goals.

Schuh [1981] has argued for the use of targeted feeding programs as the means to encourage children from poor families to attend school. As noted earlier, the opportunity costs of attending school are quite high for such children since they have to work to help earn the subsistence of the family. Paying these children the equivalent of their opportunity cost of going to school in food, on condition they attend school, should increase the rate of schooling. If food aid should be used for this purpose, disincentive effects would be small since the food would be an income transfer. Food transferred in this way would add to human capital in three ways: improving nutrition; improving health; and increasing schooling.

We conclude this section by noting that there is a significant literature on nutrition problems in Latin America. There has also been a variety of attempts at public intervention to deal with the problem, including countries as disparate as Argentina, Brazil, Chile, and Mexico. Because of the extent of poverty in the region, however, hunger and malnutrition is still a serious problem. The wastage of human capital and the reduction in productivity that is a logical consequence of this malnutrition is a serious challenge to policy makers in the region. The time is ripe to match political commitment and resources with the growing knowledge on the nature of the problem and on how effective targeted programs can be implemented.

3. Research and Extension

Investments in agricultural research and extension date back some time in Latin America. For example, a very significant program in cotton research was initiated by the state of São Paulo in Brazil in 1924. During the 1930s this program, supported by the state government, was of roughly the same order of magnitude as the research program on hybrid corn in the United States. During the latter part of the 1930s, expenditures on the program were greater than for all hybrid corn research in the United States [Ayer and Schuh, 1972, pp. 557-558].

Another significant research program was that for corn and wheat in Mexico, supported by the Rockefeller Foundation, which began in the immediate post-World War II period [for background, see Myren, 1969]. The Rockefeller component of this program eventually evolved into the CIMMYT, one of the thirteen international agricultural research centers now in existence. There are two other such centers in Latin America, the International Center for Tropical Agriculture (CIAT), located in Palmira, Colombia, and the International Potato Center (CIP), located in Lima, Peru.

In a very real sense, however, governments in the region tended to give attention to extension services prior to building an effective research capacity, with a few exceptions such as the case of São Paulo, Brazil. This emphasis on extension services was motivated in part by the view prevailing in the immediate postwar period that agricultural technology was highly transferable. Added impetus was given by President Truman's Point IV program, which was also premised on the transferability assumption. Mosher's [1955] landmark study provides an evaluation of these early programs.

As Thiesenhusen [1987a] notes, however, up to 1960 little had been written on the application of technology to agriculture. It was in the 1960s that students of agriculture recognized that little land existed for agricultural expansion in the region save the vast Amazon region and underutilized land in already settled regions. Growing recognition of the need for increases in productivity as the basis for increases in per capita incomes and as the basis for making agriculture a more effective contributor to general economic growth were also important factors. In the period since, there has been a significant amount of research on research in the region.

A summary of the estimated rates of return from investments in agricultural research in the region is provided in Table 57. These rates of return are for the most part quite high—higher than those found for investments in schooling in Latin America — which suggests a substantial underinvestment in agricultural research. The puzzle is why governments do not allocate more resources to such

Table 57. Estimates of the rates of return from investment in agricultural research obtained by using index numbers and regression analysis

Investigators	Year	Country	Commodity	Time period	Annual internal rate of return (%)
Ardito–Barletta	1970	Mexico	Wheat	1943–63	90
Ardito–Barletta	1970	Mexico	Maize	1943–63	35
Ayer	1970	Brazil	Cotton	1924–67	>77
Ayer and Schuh	1972	Brazil	Cotton	1924–67	77–110
Scobie and Posada–Torres	1978	Colombia	Rice	1957–64	79–96
Hines	1972	Peru	Maize	1954–67	35–55
Hertford et al.	1977	Colombia	Rice	1957–72	60–82
Hertford et al.	1977	Colombia	Soybeans	1960–71	79–96
Hertford et al.	1977	Colombia	Wheat	1953–73	11–12
Hertford et al.	1977	Colombia	Cotton	1953–72	none
Wennergren and Whitaker	1977	Bolivia	Sheep	1966–75	44
Wennergren and Whitaker	1977	Bolivia	Wheat	1966–75	−48
Scobie and Posada–Torres	1978	Bolivia	Rice	1957–64	79–96
Ardito–Barletta	1970	Mexico	Crops	1943–63	45–93

Source: Hayami and Ruttan, 1985. With permission. The original sources can be found in the List of References.

high payoff investments. In this context it is important to note that the analysts have in each case chosen procedures that produce conservative estimates.

Oehmke [1986] has recently provided a behavioral theory which attempts to explain this persistent underinvestment. He notes that previous attempts to explain this phenomenon have rested on a fragmentation hypothesis. This hypothesis suggests that research resources are misallocated because the research bureaucracy is fragmented and allocation decisions are made by a number of subagencies in an uncoordinated fashion [Ruttan, 1982; J. Davis, 1981; Bonnen, 1983; and Glenn L. Johnson, 1985]. The fragmentation is complicated by the existence of discrepancies between the agency's goals and the incentives provided to administrative and other actors in the budget process [Wade, 1973a, b. References in this paragraph are from Oehmke, 1986].

Oehmke's [1986] theory explains the underfunding as the failure of funding agencies to respond adequately to secular changes in the value of research. A model of farmer and funding agency behavior is developed, and shown to imply that actual research funding will be consistently smaller than optimal funding. The assumptions and results of the model are then explained in terms of the institutional literature on public agricultural research agencies. This is the first attempt to draw on formal behavioral theory to explain this phenomenon.

There have been fewer attempts to assess the returns to or impact of agricultural extension programs in Latin America. Patrick and Kehrberg [1973], as part of their study of rural education in Brazil, introduced an extension variable (number of direct contacts of farmers with extension agents during the study year) into a production function for agriculture. This variable had a positive but generally not statistically significant effect on value added in farm production.

The ACAR (Associação de Crédito e Assistência Rural) extension program in Brazil has been formally assessed by a number of researchers. This program was based on a program of supervised credit similar to that developed by the U.S. Farm Security Administration. The program included four activities: supervised credit, general farm and home extension education, medical care and health education, and distribution of materials. In the early years, approximately 80 percent of the total activity involved supervised credit and general extension.

Mosher [1955] concluded that the ACAR program had exerted a substantial impact on the levels of living and the agricultural resources of the families with which it had worked. The program had not, however, appreciably increased agricultural production in Brazil, in the state of Minas Gerais in which it began, or even in the municípios in which the program had operated. Wharton [1960] calculated input/output ratios to assess the impact of the program on total factor productivity, and found similar disappointing results, as did Alves [1968], who estimated the parameters of a production function which made it possible to es-

Table 58. Agricultural research expenditures and manpower, by region,
1959, 1970, and 1980

Region/subregion	Expenditures (constant 1980 U.S.$ millions)			Manpower (000 SMY)[a]		
	1959	1970	1980	1959	1970	1980
Western Europe	248.0	918.6	1,490.0	6.3	12.5	19.5
Northern Europe	94.7	230.1	409.5	1.8	4.4	8.0
Central Europe	141.1	563.3	871.2	2.9	5.7	8.8
Southern Europe	39.2	125.2	208.8	1.5	2.4	2.7
Eastern Europe and the Soviet Union	568.3	1,282.2	1,492.8	17.7	43.7	51.6
Eastern Europe	195.9	436.1	553.4	5.7	16.0	20.2
Soviet Union	372.4	846.1	939.4	12.0	27.7	31.4
North America and Oceania	760.5	1,485.0	1,222.4	8.4	11.7	13.6
North America	688.9	1,221.0	1,335.6	6.7	8.6	10.3
Oceania	91.6	264.0	386.8	1.8	3.1	3.3
Latin America	79.6	216.0	462.6	1.4	4.9	8.5
Temperate South America	31.1	57.1	80.2	0.4	1.0	1.5
Tropical South America	34.8	129.0	269.4	0.6	2.7	4.8
Caribbean and Central America	15.7	30.0	112.9	0.5	1.2	2.2
Africa	119.1	251.6	424.8	1.9	3.8	8.1
North Africa	20.8	49.7	62.0	0.6	1.1	2.3
West Africa	44.3	91.9	205.7	0.4	1.0	2.5
East Africa	12.7	49.2	75.2	0.2	0.7	1.6
Southern Africa	41.3	60.8	81.8	0.7	1.1	1.7
Asia	261.1	1,205.1	1,797.9	11.4	31.8	46.7
West Asia	24.4	70.7	125.5	0.5	1.6	2.3
South Asia	32.0	72.6	190.9	1.4	2.6	5.7
Southeast Asia	9.0	37.4	103.2	0.4	1.7	4.1
East Asia	141.5	522.0	734.7	7.8	13.7	17.3
China	54.2	502.5	643.6	1.2	12.2	17.3
World	2,063.6	5,358.6	7,390.0	47.2	108.5	148.0

Source: Judd, Boyce, and Evenson, 1986, pp. 82–83. With permission.
[a] Scientist man-years.

timate the separate effects of technical and economic efficiency [see Ribeiro and
Wharton, 1969, for still another assessment]. Each of these authors pointed to the
thin technical base on which extension recommendations were given.

Data from Judd, Boyce, and Evenson [1986] on agricultural research and extension expenditures in Latin America are presented in Tables 58 and 59, with
comparisons to Africa and Asia. These data show significant increases over time
in the real expenditures on both research and extension, with extension having
somewhat the better of it (a 6.5-fold increase *versus* a 5.8-fold increase for research). This increase in expenditures explains why, as noted earlier, productivity

Table 59. Agricultural extension expenditures and manpower, by region,
1959, 1970, and 1980

Region/subregion	Expenditures (constant 1980 U.S.$ millions)			Manpower (000 SMY)[a]		
	1959	1970	1980	1959	1970	1980
Western Europe	234.0	457.7	514.3	16.0	24.4	27.9
Northern Europe	113.0	187.1	201.4	4.8	5.6	6.2
Central Europe	103.1	199.2	236.8	7.9	13.0	14.4
Southern Europe	17.9	71.3	76.1	3.3	5.7	7.2
Eastern Europe and the Soviet Union	367.3	562.9	750.3	29.0	43.0	55.0
Eastern Europe	126.6	191.5	278.1	9.3	15.7	21.5
Soviet Union	240.7	371.5	472.2	19.7	27.3	33.5
North America and Oceania	383.4	601.9	760.2	13.6	15.1	15.0
North America	332.9	511.8	634.2	11.5	12.5	12.2
Oceania	50.5	90.1	126.0	2.1	2.6	2.7
Latin America	61.5	206.0	396.9	3.4	10.8	22.8
Temperate South America	5.7	44.2	44.4	0.2	1.1	1.3
Tropical South America	47.3	136.9	294.7	2.4	7.6	16.0
Caribbean and Central America	8.4	24.8	57.9	0.8	2.1	5.5
Africa	237.9	481.1	514.7	28.7	58.7	79.9
North Africa	84.6	176.5	172.9	7.5	14.8	22.5
West Africa	53.6	181.3	205.0	9.0	22.0	29.5
East Africa	39.5	86.1	106.0	9.0	18.7	24.2
Southern Africa	60.2	37.2	30.8	3.2	3.2	3.7
Asia	143.9	412.9	507.1	86.9	142.5	148.8
West Asia	28.2	97.3	119.8	7.0	18.8	16.5
South Asia	56.4	87.7	82.2	57.0	74.0	81.0
Southeast Asia	19.7	55.4	64.0	9.5	30.5	34.0
East Asia	39.5	172.5	241.2	13.4	19.2	17.3
China	na	na	na	na	na	na
World	1,427.9	2,722.6	3,443.5	177.5	294.5	349.3

Source: Judd, Boyce, and Evenson, 1986, pp. 84–85. With permission.
[a] Scientist man-years.

growth accounts for a larger share of output growth in recent years. Nevertheless, the rates of return studies suggest a significant underinvestment in agricultural research. Perhaps equally as important, many countries in the region do not
have an adequate national research capacity to capitalize on the IARC's supported
by the international donor community.

There has been a number of studies which attempted to devise improved
methods for allocating resources in applied agricultural research in Latin America. Pinstrup-Andersen and Byrnes, eds. [1975] summarized the results of a conference on this topic sponsored by CIAT. Pinstrup-Andersen [1977] devised a
method of analyzing the distribution of consumer benefits from agricultural re-

search among income strata and discusses the implications of his empirical results for establishing research priorities. Ramalho de Castro and Schuh [1977] developed a formal model for establishing research priorities *ex ante*. Pinstrup-Andersen *et al.* [1976] attempted to take nutritional factors into account in establishing research priorities. Hertford *et al.* [1977] provided an example of how domestic price policy can affect the rate of return to investments in agricultural research.

Valdés, Scobie, and Dillon [1979] has a number of important papers on the economics and design of small-farmer technology that draw on Latin American experience. Scobie and Posada-Torres [1977, 1978] make a penetrating analysis of the impact of high-yielding varieties in Latin America, with special emphasis on rice in Colombia.

Another important study is Villa-Issa's [1976] analysis of Plan Pueblo in Mexico. This program was an intensive rural development project for small farmers in the State of Pueblo, Mexico (later duplicated in other parts of Mexico) [see Jiménez-Sanchez, 1970, and CIMMYT, 1974, for details on early experience with the program]. The central thrust of the program was the diffusion of a technological package consisting of increased use of fertilizer, a higher plant density, and a split application of fertilizer. The project also had an applied agronomic research program associated with it, and attempted to organize farmers so they could have greater access to credit and fertilizer. Since maize was the principal crop in the region, the technological package focused on this crop.

An interesting result from Villa-Issa's research [1976] is that he found there to be a labor constraint to the adoption of this technological package. The recommended technological package was labor intensive. Although the region in which it was introduced had overt symptoms of excess labor, families in the region engaged extensively in off-farm employment, in many cases commuting as far away as Mexico City. Villa-Issa found that those who did engage in off-farm employment in general did not adopt the technology, while those who concentrated their labor resources on their small holding did adopt it. Interestingly enough, family incomes were not greatly different between the two groups.

This research emphasizes the importance of taking local economic factors into account in designing new technology. It also shows the value of the new household economics and its emphasis on the value of time as a framework for understanding farmer behavior generally.

INSTITUTIONAL ISSUES

Martin E. Piñeiro and Eduardo J. Trigo provided leadership for a comprehensive research program on Latin American research institutes, entitled Cooperative Research Project on Agricultural Technology in Latin America. The results of this project were brought together in a volume edited by Piñeiro and Trigo [1983a,b], which contains references to many of the studies generated by the project. A synthesis of many of the results and a discussion of the policy implications is available in Piñeiro and Trigo [1986], which we draw on in the paragraphs which follow. A companion volume [Piñeiro et al., 1982] focuses on the Colombian sugar industry and has social articulation and technical change as its theme.

Piñeiro and Trigo [1986] focused on two main topics: the importance of the international and national efforts of the last few decades towards creation and development of national research institutes; and the recent history of technological change in Latin American agriculture. In retrospect, the authors found that relatively little effort had been made to understand the implications of the international nature of the innovative process, or the role of what Edquist and Edquist [1979] have called the social carriers of technology. They also noted that although publicly supported extension programs have received attention as the chief mechanisms of technology dissemination inside national borders, the role of the private sector and of other informal mechanisms of technological diffusion have been almost completely ignored.

Piñeiro and Trigo [1986] hypothesized that the presence and nature of social carriers of technology are an extremely important element in the process of agricultural modernization in the developing world. Social carriers are institutional arrangements, such as private sector institutions or public institutions which do research or extension. They believe this element will be even more important in the future.

The authors defined three main historical periods for the region, each characterized by a different institutional mechanism by means of which the productive structure had access to technical innovations and which for the most part had originated in the developed world. The first period extended through the Second World War. Technology was disseminated in this period through three main mechanisms: immigrant workers, the initiative and efforts of large production units, and the immigration of university scientists. The second period lasted until the mid-1970s, and was a period of strong participation by the public sector through national research institutes. Diffusion mechanisms were formalized in this period and autonomous research became the main policy objective. Finally, they defined a third period in which the private sector became increasingly predominant, and in which transnational firms that manufactured the inputs and capital goods implied by technological innovations became especially important.

Piñeiro and Trigo [1986] described and analyzed each of these three periods, identifying in each the principal social carriers of technology, the social and economic forces that brought these carriers into existence, and the consequences of their efforts. However, the analysis stressed the second phase and the national research institutes which were so important in that period.

Two different perspectives were used for the analysis. They first tried to put the national research institutes in an appropriate historical perspective, highlighting their relationships with the institutional setting that preceded them and with the institutional developments that were eventually set in motion, in part by their very existence. Second, they briefly traced the research contributions of the national research institutes and their real impact on agricultural development. In this part, they concluded that the real contribution of these institutes was primarily in diffusing international technology and in developing human capital, rather than in making direct contributions to new knowledge.

While recognizing that most of their analysis is highly tentative, the authors concluded that the whole process of technical change in the developing world is increasingly determined by developments that take place in the industrialized world and by the existence of institutional mechanisms that facilitate the emergence and operation of appropriate social carriers of international technology. They further suggested that in economies whose activities are organized with markets, social forces underlying development make the emergence of such carriers possible and, indirectly, will bring about drastic changes in the nature of public sector institutions.

Piñeiro and Trigo [1986] have perceptive comments on science and technology policy for agriculture. First, they note that such policy has been dominated in the region by T. W. Schultz's emphasis on the role and nature of technology in the process of agricultural modernization, and by the role assigned to the state in the technological process. These perspectives created a tendency for agriculture to be separated from the rest of the scientific and technological system and provided for extensive participation in research by the public sector. Because the private sector was not participating in research, resource allocations in the public sector organizations dictated research priorities and, indirectly, the supply of technology.

Second, they noted that the traditional approach to agricultural modernization has been to manipulate the supply of technology. Piñeiro and Trigo believe, however, that the important issue is not the type of technology that can be offered but, to the contrary, the ability to influence and guide the demand for new technological know-how. This gives a greater role to be played by price, credit and input policies in defining the technological paths to be followed.

Third, they believe the basic orientation of the institutional model adopted in Latin America for technology generation has been to improve the diffusion of technology by adapting innovations already available in the developed countries. The technological system in the region has thus taken shape within the boundaries of knowledge that is circumscribed by the priorities of the countries donating the technology. This knowledge is therefore adapted to the relative resource availability in the developed countries, and not to those in Latin America. This tendency is in their view further exacerbated by the lack of integration of the research systems with the productive sectors, often due to the public nature of the organizations. They believe the design of research institutions needs to be better adapted to the socioeconomic, political, and ecological characteristics of the region. In operational terms, key issues are to improve coordination with the productive sectors and to develop the capability for making fuller use of native productive potential.

Fourth, Piñeiro and Trigo are concerned about the gradual breakdown of the governments' ability to guide the technological process, and with the role that the governments *should* play under present circumstances. In their view the play of market forces has become the major force that governs how technological patterns determine production conditions, priorities of clients, and types of technology available. They believe there are no policy tools to guide these private activities, and as a result the technological variable has lost its influence as an active tool of agrarian policy.

To deal with this problem, Piñeiro and Trigo believe it is necessary to redefine how and to what degree scientific and technical policies for agriculture will be implemented. At the action level, one initiative in their view would be to introduce coordination between public agencies and the new institutions emerging in the private sector. This could take place at the general level of technological policy coordination by means of national science and technology councils, or at the sectoral level through councils or coordinating committees for agricultural science and technology. A second initiative would be to establish or adopt specific tools that would enable the government to coordinate and direct technological change. These tools include patent laws, technology imports, monitoring and auditing the financial mechanisms for research investments, and so forth.

Fifth, the authors note that developments in the world at large also have a strong influence on scientific and technological policy and on the role of the public sector in the technological process. The importance of modern inputs has grown steadily and international trade is responsible for providing these inputs. Therefore, in their view the mechanisms that control the flow of international trade must be able to give adequate consideration to the implicit technological advantages and functional limitations that both national and international research institutions present. Moreover, a reorientation of national research priorities and a change in organizational structures is needed so that national efforts can take greater advantage of new knowledge that flows from the international private and public systems.

Sixth and finally, Piñeiro and Trigo have some things to say about the concept of appropriate technology as a guide to research policy. They note that the thrust of this concept is the search for a technological pattern adapted to the relative availability of factors in the lesser developed countries, which are characterized by abundant labor, scarce capital, and small production units. They have two concerns about this perspective. The first is the need for the new technologies to be efficient for market economies. This means in their view that these technologies would not be capable of generating an average factor productivity equal to that of capital-intensive technologies, and needed in order that the production units can remain competitive in the domestic market. The second concern is that the technologies must be efficient for open economies so production units will also be competitive in international markets.

Piñeiro and Trigo caution that any restrictions on how much capital can be used or how it is to be used serve to impose restrictions on the range of possible scientific discoveries. As noted above, the authors are already concerned that the technology available to developing countries does not constitute the entire universe of theoretically possible technologies since that technology has been developed in the industrialized countries in accordance with relative factor prices in those countries. Consequently, the imposition of restrictions on the type of technology can only further reduce the utilization of scientific discovery.

Piñeiro and Trigo and their colleagues have significantly advanced our understanding of agricultural research in Latin America. Their perspective that technology is a part of broader social processes, and their sensitivity to the increasing interdependence of research at a global level and to the growing role of private research institutions in the region, are all well taken. As they note, all of this suggests that "it is necessary to take a hard look at the present organization and functional roles of public research institutions—both their role as producers of technology and their role as formulators of technological policy" [Piñeiro and Trigo, 1986, p. 250].

It is useful to contrast Piñeiro and Trigo's [1983a, b, c] perspective on factors influencing the direction of technical change with that of Hayami and Ruttan [1971, 1985].[98] In arguing that technology should be thought of as a social and not as a technical issue, they see technology as a source of conflict, which they believe is not recognized by Hayami and Ruttan. This perspective of Piñeiro and Trigo is very much in the Latin American structuralist tradition.

The case studies reported in Piñeiro and Trigo [1983a,b,c] indicate that Latin America has recently been following a rather capital-intensive agricultural development path. These cases also show that the reasons for this derive from the fact that the prices of factors of production do not reflect their scarcity value. If this is the case, the Hayami-Ruttan inducement mechanism cannot function, at least as a means to identify an efficient path for new technology. Moreover, they believe the whole issue is moot because Latin America for the most part imports its agricultural technology rather than producing it locally.

Thiesenhusen [1987a, p. 198] notes that if one believes, with Piñeiro and Trigo, that technology has something to do with power relationships, it is also true that power coalitions shift in Latin America and that a simple landlord-peasant paradigm is not satisfactory for the 1980s. He notes in particular that the legitimacy of the landlord as the mainstay of the rural community may be challenged as technology changes. In support of his argument Thiesenhusen draws on Barlett's [1982] study of Paso in Costa Rica. Barlett shows that with improvements in transportation, communications, markets, and farm-level technology over the last twenty years, the community has largely shifted from growing corn and beans for subsistence to growing beef and tobacco for export. This shift caused a major realignment of Paso's social structure away from the landlord-peasant paradigm and toward more commercial, market-oriented operations.

We conclude this discussion by making three points. First, like de Janvry when he is writing in the structuralist-dependency tradition, Piñeiro and Trigo, writing in the same tradition, suggest policy recommendations that are significantly less than revolutionary, and instead much in the neoclassical tradition. Second, to the extent that Latin America is made up of relatively open societies, it is natural that private decision makers should adopt technology available from abroad, especially when the relative factor prices *they face* give them the incentive to do so.

We agree that factor markets in the regions are imperfect, in large part because of deficiencies in the communication and transportation infrastructures, and that that limits the applicability of the Hayami-Ruttan model both in explaining the direction factor-saving technology takes and as a guide to technology policy. We also agree that economic policies have strongly distorted factor price ratios. They have done this by making the private price of labor in many cases higher than its true scarcity value by means of payroll taxes, minimum wage laws, and support for unions, by providing highly subsidized credit, and by grossly over-valuing national currencies. These distortions in factor price ratios have given an anti-employment bias to technical change in the region. As the World Bank's policy paper on mechanization [1986b] notes, general economic policy tends to have a far greater influence on the pace of mechanization than does more narrowly defined mechanization policies.

Finally, we are struck by the failure of Piñeiro and Trigo to note what may be the main reason for Latin America's dependence on externally supplied technology—its failure to develop its own national research institutions to international standards. Back of that failure is the failure to develop its graduate centers of teaching and research in the agricultural disciplines to international standards. Considerable progress was made in developing these centers during the 1960s and the first half of the 1970s. These efforts declined as institutional development became unfashionable in the international development community. The economic crisis of the 1980s has given further impetus to decline, and what were once relatively strong graduate and research centers are now in a state of decline for lack of investment in training abroad, inadequate salaries to retain quality staff, and lack of investment in laboratories and equipment.

We fully agree with Piñeiro and Trigo that Latin America's research capacity should focus on the local *realidad* and not just adopt technology that is produced abroad. We believe, however, that the solution to the problem goes beyond research and technology policy *per se*, and needs to focus equally as much on bringing graduate training centers in the region up to international standards.

OTHER STUDIES

We now turn to some other studies pertinent to research and extension policy. In the mid-1980s the CGIAR conducted a comprehensive "impact" study to assess the impact of its programs on agricultural research institutions and agricultural production in the Third World. A part of this study consisted of case studies of the impact of the CGIAR system in individual countries. Seven such studies have been published on Latin American countries: Costa Rica [Stewart, 1985a], Guatemala [Stewart, 1985b], Brazil [Homen de Melo, 1986], Ecuador [Posada-Torres, 1986], Peru [Paz-Silva, 1986], Cuba [Sanches and Scobie, 1986], and Chile [Venezian, 1987]. These studies are a good source of background on productivity in the agricultural sector of these countries, on domestic institutional arrangements for research, and on trends in productivity. CGIAR [1985] is a summary of the impact study. Scobie [1987] discusses the CGIAR in Latin America.

De Janvry and Dethier [1985], in one of the study papers produced by this large effort, examined the role of market and nonmarket forces in affecting the rate and bias of technical change in agriculture. Drawing mostly on Latin American experience, they argued that a theory of the rate and bias of technological innovation must go beyond the analysis of market forces because these forces explain only a fraction of changes in investment and productivity in agriculture. Such a theory must also, in their view, take into account institutional forces which, on the one hand, distort and supplant market mechanisms and, on the other hand, act independently of prices on the determination of investment and productivity. They further argued that the roles played by the various actors involved in agricultural research—the state and the national agricultural research institutes (NARIs), the IARCs, and the private sector—are being redefined as research moves into the "Post-Green Revolution" era. In particular, they argued that the private sector is being increasingly involved in research, and the work done at the IARCs modifies the research priorities of the NARIs.

R. E. Evenson [1987] examined the impact of the IARCs on spending for national agricultural research and extension. He found that they have had a positive impact in each of the crops for which the CGIAR crop research programs exist except cassava. Estimates for livestock and horticultural crop research programs show a significant positive CGIAR impact as well. There is also a positive impact on national extension spending, and the Center programs have had a significant impact on crop productivity. The final analytical report for this large project, which synthesizes the results from all the individual studies, is Anderson, Herdt, and Scobie [1988].

Some other useful references on research and technology policy, with special reference to Latin America, include ISNAR [1984] and CIMMYT [1984]. More general references include Pinstrup-Andersen [1982], the World Bank [1985b], and Evenson and Kislev [1975]. G. Feder *et al.* [1985] provide a valuable survey of the literature on the adoption of agricultural innovations in developing countries. For a national perspective, Alves e Outros [1985] is a collection of papers Eliseu Alves and his colleagues did on agricultural research in Brazil. Alves provided much of the thinking for the design of EMBRAPA, the new agricultural research system in Brazil, and served as its president for some five years. O'Mara [1983] and Cutié [1976] are useful studies of the diffusion of particular innovations.

De Janvry, Runsten, and Sadoulet [1987] provide a broad coverage of technological innovations in Latin American agriculture. R. F. dos Santos [1986] is an excellent analysis of the biases in technological change in Brazilian agriculture. Da Silva [1984] is an analysis of the contribution of research and extension to agricultural productivity in São Paulo, Brazil, with da Silva [1986] being a more recent analysis of the same issue. Alves [1988] is an analysis of the contributions extension and research can make to alleviating rural poverty in Brazil.

CONCLUDING COMMENTS

As in other cases noted above, the literature on research and extension has tended to follow developments in the economy of the region. Once technical change began to be an important source of output growth in the region, researchers began to address the issue of technical change and the institutions needed to bring it about. This literature is in both the neoclassical and structuralist traditions. Perhaps in the area of research, moreover, the two traditions tend to converge more than in any of the other subject matter areas reviewed above. In this case the convergence is especially productive. A disappointing feature of the literature on research and extension, however, is how little analysis and evaluation there has been of extension programs. Governments in the region invest rather large sums in these systems.

4. Conclusions

Latin American countries have significantly underinvested in human capital. This is true whether one considers the nutritional and health status of its population, the education and training of its labor force, the capacity for research—and thus the stock of knowledge and technology adapted to its local conditions—or the policies and other institutional arrangements needed for a modern economy.

Nutritional levels among the poor are low; health services are inadequate, especially in rural areas; illiteracy is pervasive, especially among the rural population; research institutions are weak and for the most part far below international standards; and economic policies are inadequate and often counterproductive.

Ironically, policy makers often claim the countries of the region cannot afford investments in human capital. This implies that these are consumption goods and services, instead of the investments they truly are. Moreover, it flies in the face of the demonstrably high social rates of return to such investments.

The literature dealing with human capital in the region is significant and growing. Unfortunately, much of it has been done by students trained abroad and has been rooted in international centers, not in indigenous Latin American institutions. Valuable as that literature is, much of it fails to address the important institutional reasons for the lack of investment when measured rates of return are high, and in the face of experiences in other countries that have moved more quickly down the development path and with a broader sharing of the benefits of that development.

Chapter X. Suggestions for Agricultural Development Policy and Concluding Comments

Agriculture and rural Latin America suffered serious neglect from policy makers in most countries of the region during the 1950s and the 1960s. The emphasis was on forced-draft, import-substituting industrialization, aided and abetted by the ideology of the Prebisch thesis and the work of the ECLA. Governments intervened extensively in their economies, subsidies were focused largely on physical capital and on urban workers and consumers, and there was a significant underinvestment in human capital for agriculture and rural development in terms of the capacity to produce new production technology, modern institutional arrangements, nutrition and health, and in the education and training of the rural population. With such serious discrimination against agriculture, offset only in part and ineffectively by subsidized credit and other measures, it is a wonder that agriculture has performed as well as it has.

As the 1970s and the 1980s have unfolded, however, many of the severe discriminatory policies against agriculture have been eased and economic policy has turned more outward. There have been limited but significant attempts to build the capacity for agricultural research in the region, and a greater dependence on market forces to allocate resources. The agricultural sector has responded accordingly, with productivity becoming increasingly important as a source of output growth.

Policy is still far from being on track, however. The tendency for governments to intervene, and in improper ways, is still pervasive in the region; national agricultural research systems are still far from being up to international standards; and the underinvestment in human resources for agriculture is still very large by almost any standards. The important point, however, is that both the theory and the accumulated empirical evidence point the direction that policies should take, and important strides are being taken in moving them in that direction. This bodes well for Latin America's future, for it has already demonstrated in numerous ways that it can compete internationally, that it has a lively and creative people, and that it can close the gap in per capita incomes with the more developed countries.

The trends in the international economy are towards a more equitable distribution of political and economic power. The days of two superpowers struggling with each other in the midst of a sea of economic and political dwarfs appear to be over. Among other things, this means that development assistance from abroad is likely to continue at present low levels, and that the countries of Latin America, as well as other developing countries, will have to pull themselves up by their own economic bootstraps. The experience of the newly industrializing countries (the NICs), which until the difficulties of the 1980s included Brazil and Mexico among their numbers, suggests that this can be done if economic policies foment economic efficiency and provide incentives and if the proper investments are made in human capital.

This chapter is devoted to a brief overview of policies that can help promote a more rapid rate of agricultural development in the region and that will help improve the economic lot of the rural disadvantaged. This overview reflects our own views, but is based on the survey of literature and analysis that has gone before, with the goal being to relate policy recommendations to conditions identified in the literature.

This chapter is divided into five parts. The first focuses on functional policy issues. The second focuses on important cross-cutting policy issues. The third addresses issues associated with international institutional arrangements, while the fourth discusses learning by doing. The fifth and final part presents some concluding comments on the literature on Latin American agricultural development itself.

1. Functional Policy Issues

Policies to promote agricultural development must be based on a clear concept of the process itself and of how agriculture relates to the larger economy of which it is a part. All too often those attempting to understand agricultural development, or to devise policies to promote it, treat the sector as if it were a self-contained unit, isolated from the rest of the economy. The factor-market linkages with the rest of the economy thus tend to be ignored, as are the interlinkages among the factor markets themselves. Similarly, agriculture is often viewed as if it were part of a closed economy, although the Latin American literature in general—with its tradition of international dependency theory—is not guilty of that charge. The economics of a closed economy is different than that of an open economy, however. All too much of the literature on agricultural development is predicated on closed economy assumptions.

Policy makers are often faced with two conflicting objectives in promoting agricultural development: promoting growth in agricultural output to feed their populations and to earn foreign exchange, and attempting to raise the incomes of rural people so as to reduce the wide disparity in per capita incomes between the rural and urban populations. If economic conditions are such that technical change leads to immiserizing growth, the conflict in these objectives can be real. However, if a proper mix of policies is used to promote agricultural development, the conflict may be more apparent than real. Policies that facilitate diversification in commodity mix, adjustment of labor out of the sector, and farm enlargement—all necessary to keep the technology from being immiserizing—not only help to raise per capita incomes within the sector, but help to realize the full benefits in an efficiency or output sense from the new production technology.

At present stages of economic and agricultural development in Latin America, policy makers can probably best promote the development of their agricultural sector and the development of their economies as a whole by focusing on productivity improvement as the main objective of their agricultural and other public policies. Many countries in the region will want to bring unoccupied frontier lands into production, both for political or national security reasons and to bring additional resources into their economy. That decision, to the extent that it is based on economic considerations, should be based on a careful consideration of the relative costs and benefits of investments required to open up new areas compared to the costs and benefits of alternative uses of the investment resources. The environmental consequences of opening the new lands should also be a part of the calculus.

Policies to promote agricultural development, especially those with the objectives of enhancing productivity and per capita incomes of rural people, can be classified under four headings: price or incentive policies; factor market policies; investment in human capital; and investment in physical infrastructure. The challenge in designing policies under each of these headings is to identify the proper role for governments and the proper role of markets. This issue all too often becomes entangled in ideological wrangles rather than being viewed as the empirical and analytical issue it truly is [for more detail, see Schuh, 1983]. Moreover, there is a general failure to recognize that what is proper policy at one stage of development may not be proper at a different stage of development.

PRICE AND INCENTIVE POLICY

The objective of price and incentive policy should be to have domestic prices, to the extent possible, reflect border price opportunity costs. This applies to modern (purchased) inputs as much as it does to commodities. This perspective is controversial in Latin America (and elsewhere in the world), but it is difficult to arrive at alternative pricing strategies so long as efficiency and equity are important policy goals.

Most Latin American governments have severely discriminated against their agricultural sectors by using trade and exchange rate policies to shift the domestic terms of trade against agriculture, although less so today than twenty years ago. The fundamental difficulties with such policies are that they significantly undervalue the resources in the agricultural sector and this is often the most important set of resources in the economy; they reduce the incomes of rural people, typically to the relative disadvantage of the poor; and they impede the adoption of new production technology by reducing the payoff to investments in research and reducing the incentives to adopt available technology.

Policies which protect industries for import-substituting purposes have equally damaging consequences, whether they refer to agricultural or manufacturing sectors. The infant industry argument has a great deal of intellectual appeal; the problem is that the infant seldom grows up if it is raised behind protectionist barriers. The protected sectors tend not to become more efficient, and the protection itself generates large economic rents which are captured by upper income groups and often used as bribes to perpetuate the protectionist measures.

There seems to be little alternative to using border prices as a means to induce efficiency and to promote adjustments to the ever changing conditions in the international economy. The exception is when dumping by other countries causes international prices to be lower than they would otherwise be. Under these circumstances some amount of protection of the domestic industry may be appropriate, although this depends importantly on whether the dumping is expected to continue for an extended period of time, and on a careful assessment of the costs and benefits to the domestic economy of the dumping by others. When such protection of the domestic economy is appropriate, the goal should be to set the domestic terms of trade equal to those prevailing in international markets.

Instability in commodity markets is a frequent issue in agricultural policy, and government interventions are often motivated by a desire to have more stability. Stability can probably most effectively and most efficiently be obtained by policies that shape the economic environment of the sector, rather than by direct intervention in commodity markets. Monetary stability should be the highest priority goal. Free trade, so the international economy can help bear the burden of adjustments, is also important. Lack of government intervention in commodity markets will create an environment in which the private sector will carry stocks at levels sufficient to smooth out price fluctuations. And an efficient credit system can provide the means by which private economic agents even out their income and consumption flow over time.

An important reason governments become so concerned with instability is the lack of risk markets by means of which private agents can insure against risks. Governments might be better advised to help develop their risk markets rather than to intervene extensively in markets. The establishment of futures markets is an important example.

FACTOR MARKET POLICIES

The second set of policies include those addressed to the factor markets, especially land, labor, and capital (or credit). Government interventions in these markets are also extensive in Latin America. The goal of policy makers in this case should be to have the prices of the factors of production reflect their shadow prices or scarcity values. A wide range of government policies often keeps this from happening. Minimum wage laws and payroll taxes are two important distortions in Latin America. If minimum wages are effectively set above market clearing levels, they can create unemployment, often for those most disadvantaged in the market. Payroll taxes are an especially pernicious form of government intervention since they typically are rationalized on grounds the capitalist is being taxed to provide some public service to the worker, when in fact the worker often bears the burden of the cost and adjustment in the form of reduced employment or unemployment.

Labor unions can also create distortions in labor markets, especially if they are backed by the government and push wage rates above what otherwise would be market clearing levels. Like minimum wage laws which set institutional wages above market clearing levels, unions that artificially raise wages have a (positive) supply effect as well as a (negative) demand effect. This contributes to queuing and makes observed unemployment greater than it otherwise would be.

Given the general need to transfer labor out of agriculture as development proceeds, and the prevalence of long-term swings in trade opportunities, governments should in general have positive adjustment policies for the agricultural labor force. Broad-based formal schooling should be part of such policies. But training programs, labor market information systems, and relocation subsidies are all appropriate to facilitate the adjustment process.

In the case of land markets, the goal in general should be to reduce or eliminate government interventions. These include size limits on land, regulation of tenure share arrangements, and rental ceilings. As noted in an earlier section, sharecropping fulfills a useful economic function. To fulfill that function, the arrangements need to be determined in such a way as to reflect prevailing economic conditions.

There may be cases in which redistribution of land and changes in the size distribution of land holdings are appropriate. At low levels of development, when human capital is relatively less important than at higher levels of development, land is relatively more important as a productive asset. In general, however, changing the size distribution of land holdings alone will not be sufficient, nor is it costless either in economic or political terms. Redistribution of land needs to be done in conjunction with proper price policy, the introduction of new production technology, training or retraining of the labor force, and the provision of adequate supplies of credit.

Credit policy should have as its goal the development of true financial intermediaries. But credit policy should be part, and only part, of a broader capital market policy. Capital market instruments which permit a wide variety of ways for economic agents to participate in capital markets are important. Similarly, instruments which generate savings at socially optimal rates are an imperative, both for agriculture and for the rest of the economy. Subsidized credit should be avoided. If it is desired to subsidize small producers, this should be done by transparent means such as grants rather than by implicit subsidies. The goal should be to have capital (and credit) markets which attract resources to agriculture (or retain them there) and which channel these resources to their highest uses, including in longer-term investments. The role of the government in this sector is to provide stable monetary conditions, an adequate information system, and the legal and institutional arrangements which are needed for the markets to function efficiently.

INVESTMENT IN HUMAN CAPITAL

The third set of policies includes those which promote investments in human capital. High on the priority list for most, if not all, countries in Latin America is the need to strengthen their capacity for agricultural research. New production technology is truly an engine of economic growth, and most countries significantly underinvest in it. There is really no justification for sacrificing such an important and powerful source of economic growth. Given the location specificity of agricultural technology, the goal of policy makers should be to have an effective agricultural research station in each of the important ecological zones in their country. In addition, a proper institutional environment should be established so that the private sector has incentives to invest in producing new technology. This requires patent rights and other means of capturing the return to their investments.

National governments should also insure that their research system can capture and adapt new knowledge generated abroad, and thus benefit from research that is done in other countries and in the international agricultural research centers. This requires an outward-looking research system, one that develops collaborative relationships with research centers in other parts of the world. It also requires that staff capabilities be kept on the frontier of knowledge so that domestic scientists can understand, participate in, and benefit from developments in other parts of the world.

In general there are three issues in having a research system that can contribute efficiently to agricultural growth. The first is to have institutional arrangements that effectively articulate applied, strategic, and basic research; that adequately reflect back to researchers the problems that emerge in society; and that reward researchers so that they are motivated to work in the system and to invest in their own training and development. The second issue is to fund the system at a level that will enable it to be productive and to produce new technology at a *socially optimal* rate. The third issue is the need to have a technical staff trained to international standards. That requires either the capacity to train scientists at a high level within the country, the willingness to send scientists abroad for training, or the willingness to hire *ex patriate* scientists. The third alternative may in the short term be the most efficient route in many countries, controversial as it is in most Latin American countries.

Effective extension systems are a very important complement to an effective domestic research capability. As an economy develops, the private sector can do a great deal of the diffusion of new knowledge in the agricultural sector. At early stages of development, however, most extension services will need to be provided by the public sector. At later stages of development the public component of the extension system can teach principles and more general issues of policy and public affairs. In general, it is important to design the extension system so that it serves both as a mechanism for diffusing new knowledge and as a mechanism for identifying problems in the rural sector and feeding that information back to researchers.

Most, if not all, Latin American countries also significantly underinvest in the formal schooling and training of their rural people. At present levels of development, the greatest deficiency is probably at the primary and secondary levels, and in technical or vocational training. Formal schooling will be increasingly important as investments in agricultural research are increased since cognitive skills will be needed to decode the new knowledge. But cognitive skills are also needed to take advantage of existing disequilibria, and both cognitive and vocational skills are needed to facilitate adjustment to nonfarm employment.

An important problem in promoting increased schooling is the high opportunity costs of going to school. This is especially serious with the first child, and to a lesser extent with the second child. Using internationally provided food aid in targeted feeding programs which offset these opportunity costs can have a triple payoff to society since it will help attain higher schooling rates, improve nutrition and health, and raise labor productivity generally. Even if international food aid is not available, targeted feeding programs directed to this end should have a high social payoff.

The proper level of higher level education is less clear. Certainly almost every country needs high level professionals to staff its research centers and other institutions which provide services to agriculture. Not all of these professionals have to be trained domestically, however, and not all countries will find it rational to sustain high level graduate programs. When the latter is the case, staff can be trained in the developed countries, use can be made of expatriate staff, or countries can collaborate in establishing regional centers of graduate training. Presently, most countries in the region are not giving enough attention to the need to have highly trained professionals in their teaching and research institutions.

The third component of human capital needing more attention in the region is nutrition and health. Adequate nutrition is the most basic form of human capital. Without it, other forms of human capital will make less than their full contribution to economic development, either in a physical input sense or in terms of cognitive and operational skills. Most countries in the region have all too many malnourished people. The nutrition of these people will not be improved by economic growth alone in the foreseeable future. Targeted feeding programs will be needed.

The disparity in health services between the rural and urban sectors of Latin American countries is also quite great. This gap needs to be closed if agriculture is to have a productive, efficient agricultural labor force.

Finally, there are institutional arrangements as a particular form of human capital. Some of these have been alluded to in the preceding paragraphs, although the emphasis tended to be on obtaining adequate levels of funding rather than on the institutions *per se*. The institutional design questions in the region are pervasive. Much more attention needs to be given to these issues, and to adapting institutions to changing economic conditions, especially the rising value of time as development proceeds [T. W. Schultz, 1968].

Related to the institutional design issues is the need to have an adequate research capacity in the social sciences (economics, political science, anthropology, and sociology). This capacity is needed to evaluate economic policies, to design new economic policies, to design and redesign new institutional arrangements, and to provide information to decision makers, both public and private. The "technology" output of the social sciences is, among other things, new institutional arrangements. In general, most countries in the region significantly underinvest in the design and implementation of new institutional arrangements.

INVESTMENT IN PHYSICAL INFRASTRUCTURE

Finally, there is the chronic tendency to underinvest in rural infrastructure. This covers the full gambit of farm to market roads, penetration roads and thoroughfares, railroads, schools, health clinics, rural electrification, rural telephone service, ports and port facilities, and so on. An adequate rural infrastructure is essential to having a modern agricultural sector. It is also critical to decentralizing the development process, which will make it possible to diversify out of agriculture in a more efficient way at later stages of modernization, while reducing the negative externalities on both the sending and receiving end of the migratory process.

2. Cross-Cutting Policy Issues

In addition to the above functional issues of policy, there are four issues that cut across functional lines. These include the mobilization of resources from agriculture; improving the distribution of income; dealing with declines in the external terms of trade; and population policy.

MOBILIZATION OF RESOURCES FROM AGRICULTURE

An important policy issue in most Latin American countries is how to mobilize resources from agriculture to promote the expansion of the rest of the economy. Many of the economic policies that have had such deleterious consequences for agriculture in the region were basically attempts to mobilize resources from agriculture and transfer them to the nonfarm sector to meet a variety of policy objectives. These policies have worked for the most part by shifting the domestic terms of trade against agriculture. This is probably the least efficient way of transferring resources out of the sector, since it has such strong and adverse output effects.

More desirable means of extracting resources from agriculture would be by means of income taxes and land taxes. In many respects, land taxes are ideal because, if set at proper levels, they have minimal resource allocation effects [M. Lopes, 1977; M. Lopes and Schuh, 1979]. In addition, they can have a great advantage in taxing away economic rents created by the subsidization of new production technology.

Establishing land taxes, however, or setting them at anything other than innocuous levels, is politically difficult to do. Most farmers and land owners view them as anathema, and as only slightly less threatening than land reform. Policy makers are thus left with the need either to reform their income tax systems so as to make them more effective for agriculture, or to implement major reforms which tax value added in an equitable way across sectors or consumption taxes which are not regressive against low income groups. Whichever policy is chosen, fiscal policies need a major reform in most countries if agriculture is to be taxed by more transparent means and if implicit taxation by distorting the domestic terms of trade is to be eliminated.

The credit system can also provide an important means of transferring resources from agriculture. To do this, the credit systems in the region need to be transformed into true financial intermediaries. The goal should not be to "tax" agriculture by means of such a system, but rather to have a system which transfers the resources in response to market forces.

IMPROVING THE DISTRIBUTION OF INCOME

Policies to improve the distribution of income is the second cross-cutting issue. It is difficult to imagine that the present highly skewed distributions of income in the region can be sustained much longer without major political disturbances. It is equally as important to recognize that the widely perceived trade-off between equity and efficiency is for the most part a myth. Countries can improve their distribution of income by investing in the human capital of the disadvantaged without sacrificing economic growth. In fact, such investments, if complemented with more favorable economic policy, can do much to *promote* economic growth while at the same time improving the distribution of income.

Import-substituting industrialization policies have done much to exacerbate income distribution problems in the region. Shifting away from these policies and turning to more outward oriented policies will do much to broaden employment and income-earning possibilities in most Latin American countries. If such a switch in policy were to be complemented with greater investments in the human capital of the poor, especially among the rural population, a more equal distribution of income could be obtained by something less than revolutionary means.

Investing in new production technology for agriculture, especially in the food crops, will also improve the distribution of income generally in most Latin American countries. Consumers are the ultimate beneficiaries of this new production technology, and they benefit in a progressive fashion since low income consumers spend a larger share of their budget on food. The decline in food prices which the production and diffusion of new production technology makes possible thus benefits these groups in a relative sense.

Finally, reform of fiscal policy is a key element in improving the distribution of income. Resources must be captured by the public sector, rather than be transformed implicitly by changes in the domestic terms of trade, if investments are to be made at socially optimal levels in human capital.

DECLINE IN THE EXTERNAL TERMS OF TRADE

Chronic and persistent declines in the external terms of trade are important features of the international economy in which most Latin American countries must now compete. Policy makers are challenged by such declines, when they occur, because they threaten their balance of payments and their ability to service their foreign debt. Such declines (and increases!) also can create serious adjustment problems in the domestic economy. At the same time, however, declines in the external terms of trade can be of considerable benefit to domestic consumers and to users of important raw materials.

To the extent that secular or long-term shifts in the external terms of trade reflect differential growth rates in sectoral productivity in the external economy, one way of dealing with them is to raise productivity domestically at the same pace it is growing in the international economy. If this is done, the loss of foreign exchange earnings from the price decline in the international economy will be reduced, although a number of parameters become relevant in determining the change in export earnings. Moreover, if productivity growth in the domestic export sector is faster than that in the international economy, an increase in foreign exchange earnings can result even when the external terms of trade are declining.

In any case, in analyzing policy alternatives under these circumstances it is important to distinguish between what happens to national welfare and what happens to foreign exchange earnings. As Brandão [1978] and Brandão and Schuh [1979] have shown, these are not necessarily the same thing. It is entirely possible for national welfare to improve while the external terms of trade are shifting against a country, and *vice versa*.

The classic policy response to an adverse shift in the external terms of trade is a devaluation of the nation's currency. This deals with the balance of payment problems such an adverse shift creates, and not with the problem of changes in national welfare. A flexible exchange rate policy has much to recommend it in this context. With a fixed exchange rate policy, balance of payment problems quickly burgeon out of control when the external terms of trade shift against a country. With flexible exchange rates, adjustment to the changed conditions in the international economy start immediately and the adjustment is spread widely in the domestic and international economy. Crises are thus avoided.

A frequent prescription for dealing with declines in the external terms of trade is by means of international commodity agreements, the solution advocated so widely by Prebisch and his followers and a major plank in UNCTAD's political platform. Three factors make one less than sanguine about this approach to dealing with the problem. First, the track record of past commodity agreements is anything but good. If such agreements are successful in raising prices above what they would otherwise be, they tend to contain the seeds of their own destruction.

Second, in a world of flexible, or bloc-floating exchange rates, commodity agreements are not feasible. As the value of national currencies change, the demand and supply conditions in individual countries change, and management of the system becomes very complex. This probably explains why even long-standing commodity agreements have broken down in recent years. Third, international commodity agreements tend to become highly politicized. Consequently, it becomes very difficult to adjust the main provisions of the agreement to changing economic conditions. And economic conditions in today's world change rapidly.

In conclusion, the only kind of commodity agreement that seems feasible in today's world is one that provides market information services to participants in the market. The current international wheat agreement is an example of such an arrangement. Such an organization can usefully collect data and disseminate information on prices, production, trade, and other pertinent information such as that pertaining to changes in exchange rates and international capital markets. Special studies on market and sectoral issues would also be of value. Disseminating the information collected, synthesized, and analyzed by a secretariat would be of special value to small participants in international markets.

POPULATION POLICY

Population policy is especially important to agriculture, since economic conditions are such that agriculture typically is a producer of population for the economy as a whole as well as a producer of agricultural commodities. This tends to aggravate the whole set of adjustment problems faced by the sector, and contributes to the chronic tendency of per capita incomes in agriculture to lag behind those in the nonfarm sector. In this regard it is interesting to note how seldom population control or family planning programs focus specifically on the agricultural sector.

We would make two main points about family planning and population programs. First, an important part of public policy should be directed to making family planning knowledge and technology accessible to all members of the society. There need be nothing coercive about this. The main goal should be to make available to low income families the information and technology that is available to upper income families.

Second, increases in per capita income — economic growth — play a major role in changing the economic alternatives that families face, and thus leading them to substitute quality of children for quantity of children. The problem with depending on economic growth alone to solve the problem is that it takes a long time to influence population growth rates in this manner. The proper role of government is to intervene on the side of human capital. This involves expanded support for education and training, for improved nutrition, and for improved health. The demand for children tends to decline as expected child mortality rates decline. Hence, nutrition and health programs are important as parts of a larger population and family planning program.

Education programs are critical to accelerating economic growth, especially if combined with an increase in the flow of new technology. Razin's [1977] cross-country analysis of the relationship between the growth of real per capita gross national product and the percent of population of school age children enrolled in the secondary level of education found a positive and highly significant association. Razin also has done pertinent work on other aspects of human capital and economic growth, especially on the issue of optimum investment in human capital [see Razin, 1969, 1972a, b].

In conclusion, our perspective is that rather than let the forces of economic growth work themselves out and eventually bring population growth rates down by inducing more private investment in human capital, government policy should be directed to accelerating the rate of social investment in human capital as a means of reducing population growth rates. This has the great advantage of investing in those forms of capital that are known to be cheap (high payoff) sources of economic growth.

3. International Institutional Issues

Another important issue in economic and development policy is the institutional arrangements for the international economy. National economic policy has to be the first priority of national governments, and no amount of "pushing problems off" on the international economy can substitute for sound domestic, trade, and exchange rate policies. But there should be little doubt that international institutional arrangements are in serious disarray, and not adequate for the way the international economy has evolved. The period encompassing the end of World War II and the beginning of the postwar era was a period of enormous creativity in terms of international arrangements. The UN system with its various institutions such as the Food and Agriculture Organization was created at that time, as were the Bretton Woods twins (the World Bank and the IMF), agreements on how to manage international monetary issues, and the GATT.

Unfortunately, many of these arrangements have broken down in response to changing economic conditions—the Bretton Woods conventions being an outstanding example. Other agencies and "rules" have become ossified or irrelevant to the changed circumstances. The IMF, for example, was created in part to provide balance of payments support for countries in a fixed exchange rate system. While one would have expected this function to decline with the shift to a bloc flexible exchange rate system, the World Bank has essentially added a balance of payment support program (adjustment lending) to its investment program in order to help deal with the international debt problem. That says a great deal about the true nature of that problem.

In any case, the important point is that the world's economic integration— driven by rapid technological developments in the transportation, communication, and computer sectors—has far outpaced its political and institutional development and integration. Ten years ago, for example, it was not technically possible to participate in international capital markets in the same way it is today.

A major effort is needed to update and modernize institutional arrangements for the emerging international economy, including the creation of new arrangements. This is not the place to review all that is needed. Schuh [1986] has outlined some of the reforms and innovations needed in some detail. Social scientists need to give a great deal more attention to these issues; political leaders need to develop the political will to address the issues.

It is on these issues that the developed countries can use their political and economic power to further the interests of the developing countries. They can provide the leadership to reform existing international arrangements and to design new arrangements more consistent with the changed international economy. High on the priority list for institutional innovations and reform are arrangements that lead to more monetary stability and to freer trade. Measures to these ends will lead to both a more equitable distribution of the world's income, while reducing the massive disequilibria in global agriculture which result in the sacrifices of so much production and income. Japan, the European Economic Community, and the United States can start by reducing the extensive and high protection of their own agricultural sectors.

Despite the emergence of aid-tiredness on the part of developed countries, there is also still much these countries could do to assist the developing countries directly. What limited foreign aid the developed countries are willing to provide should be channeled almost completely to helping build the human capital in the developing countries. Rapid population growth causes the investment demands for human capital to be great, and difficult to meet because of high dependency rates. In addition, the CGIAR system could be significantly strengthened and expanded to increase the flow of new production technology to the developing countries and to help them address their growing environmental problems. Helping to fill these gaps, both within the developing countries and in the international system at large, will redound to the developing countries in multiple and positive ways [Schuh, 1986].

4. Learning by Doing

Reference to Johnston and Clark's [1982] *Redesigning Rural Development* is a useful way to close this discussion of policy issues. Rather than the "grand design" which has been offered up (and pursued!) so many times as the panacea for solving the problem of rural poverty, Johnston and Clark make a plea for learning by doing. Their point is that mistakes in policy design and implementation are inevitable. Moreover, they note that the tragedy of mistakes is not that they happen, but that societies and policy makers learn so little from them, or from those of their neighbors, for their future policy designs. Fundamentally, this is a plea for policy analysts to learn from experience and not to rely on theoretical or ideological models.

Johnston and Clark do not plead for problem solving from below or social interaction, however. They believe that approach has failed also, largely because it cannot capture the benefits of outside technologies and the development experiences of others. Instead, they see the pressing need to be for a more effective integration of what they call "intellectual cogitation" and social interaction as complementary approaches to strategic problem solving.

Johnston and Clark stress the need for a long-run perspective in devising a strategy for rural development. They believe that resource constraints are not fixed in the long run, and in fact are very much a function of policy choices in the short run. In this context, they argue that one of the most important functions of policy analysts lies not in making policy recommendations *per se* but in forcing policy makers to revise priorities and objectives in the face of existing constraints. In fact, Johnston and Clark have one of the most perceptive descriptions of policy analysis we have seen:

> Doing this job well involves a delicate balance between realistic recognition of the political realities which constrain decision makers and a broader view of constraints and opportunities based on detailed and systematic analysis. There is an art and craft to this form of policy analysis, just as there is a requirement for a certain humility in professed objectives. Wildavsky is right in warning that if the analyst is seen to be maneuvering for a *de jure* powerful position, he is likely to disappear or arouse far stronger opposition then he can cope with. Successful practice means not flailing about for power one is incompetent to take or hold but instead seeking to understand how the powers in a particular situation work, and how support for desirable and feasible policies can be mobilized. [Johnston and Clark, 1982, p. 14]

5. Concluding Comments on the Literature of Latin American Agricultural Development

Turning to a final appraisal of the literature on agricultural development in Latin America, we would make three points in closing this survey. First, much of the general theoretical literature on agricultural and rural development comes from analyses of Asian data. Conditions in Latin America are greatly different than those in Asia and those theoretical models need to be adapted to Latin American conditions. Second, as we noted early in this paper, the Latin American literature tends to be eclectic, with individual writers drawing on various disciplines for their analysis. This often causes the analysis to appear to be less rigorous than more disciplinary-based studies. In compensation, the analysis tends to be much richer and more insightful.

Third, analysts of Latin American agriculture and rural development have not always drawn on theoretical literature from other parts of the world to the extent they could. Two bodies of theoretical work that have been neglected and which we believe have a great deal to offer in understanding problems of the region are the interlinking market literature and the new household economics. Both of these are important in understanding problems of poverty and in designing measures to alleviate it. We hope researchers concerned with the problems in the region will draw more extensively on this literature in the future.

One of our goals from the beginning was to relate the literature on agriculture and rural development to changes in the economies of the region. That complicated our task, and made this a far more ambitious undertaking than it would otherwise have been. We feel that we have been only partially successful in this effort. But we hope the reader, as a consequence of our efforts, has some appreciation of the problems of the region, its stage of development, and an appreciation of the literature itself. Importantly, the evidence is rather strong that the literature *has* responded to changing economic conditions.

Finally, we extend our apologies to the many capable people and the many excellent pieces of literature we have undoubtedly missed or neglected. One can view a paper such as this as either the proverbial half-filled or half-empty glass. We hope others will view the glass as half-filled—and accept the challenge of further synthesizing and integrating what is an interesting and rapidly growing literature.

Notes

1. Much of the material in this section is drawn from Schuh, 1975a.

2. It should be noted, however, that at the turn of the 20th century, Argentina had a per capita income that was greater than that of the United States. This is suggestive of the wide disparities among Latin American countries, a point to which we will return, and the extent of sluggish economic growth and stagnation in the economies of individual countries in the region during this century.

3. For an excellent analysis of the role of natural resources in the development of the region, see Grunwald and Musgrove, 1970.

4. For a perceptive analysis of this problem, see Schydlowsky, 1972.

5. For the evidence on the Brazil case, see the important paper by Fishlow, 1972a.

6. For a careful analysis of the data on rice, see Scobie and Posada-Torres, 1977.

7. There are important examples of modernization, such as in certain parts of Mexico, Colombia, and São Paulo in Brazil, and Argentina in the last decade. These tend to be exceptions to the general rule, however.

8. AID has provided major support to the University of Wisconsin LTC over the years, and also supported the documentation center at CIAT.

9. For a more detailed analysis, see Fienup *et al.*, 1978.

10. An important exception was William Nicholls at Vanderbilt University, who was a long-time student of Brazilian agriculture. The culmination of Nicholls's lifetime work on Brazil was in his impressive study, *Ninety-Nine Fazendas* [1963/1973], done in collaboration with his long-time Brazilian colleague, Ruy Miller Paiva. Regrettably, Nicholls's premature death precluded the completion of this important study. Other studies related to *Ninety-Nine Fazendas* include Nicholls, 1963a, Nicholls and Paiva, 1965, and Paiva and Nicholls, 1965.

11. There is a stream of thought which argues that large landowners have had an inordinate influence on economic policy in Latin America. It may be that they have influenced policy and the allocation of public resources *within* agriculture, but in the larger context one would have to explain why they imposed (or acquiesced to) policies which discriminated so severely against themselves by shifting the domestic terms of trade against agriculture and raising prices of producer and consumer goods they would want to import. In this regard, also see Leff, 1968.

12. The material in this and the following four paragraphs is taken from the excellent study by Grunwald and Musgrove, 1970, especially pp. 7-24. The reader is referred to this source for more detail and supporting evidence.

13. For evidence of the effects of these exchange rate realignments on Brazil and Mexico, see Schuh, 1984a, 1987.

14. Prebisch also had a significant influence on LAFTA, the CACM, the Alliance for Progress, and the development programs of several Latin American governments, such as the Kubitschek administration in Brazil (1956-61). For a perceptive and interesting analysis of how his ideas evolved over time, see Love, 1980.

15. In the context of the literature on the Prebisch Thesis the periphery countries broadly refer to the less-developed countries while the center refers to countries of the industrialized West, with special emphasis in most of the literature on the United States as the key economy of the center.

16. The other "classic" statement of the Prebisch Thesis was in a paper he presented at the annual meetings of the American Economic Association about ten years later [Prebisch, 1959a]. It turns out that Hans Singer articulated ideas very similar to those of Prebisch in a paper presented to the American Economic Association in 1949 [Singer, 1950]. Given the similarity of ideas and the almost simultaneous publication of the original papers, the thesis about the declining terms of trade for the low-income countries is often referred to as the Prebisch-Singer Thesis. We use Prebisch alone as the reference because the thesis is usually identified with him and because we try to trace the evolution of his thought.

17. The history of the post-World War II period has in fact been different, of course. The price of major agricultural commodities has declined relative to manufacturing products. This creates a presumption that technical progress has been faster in agriculture than in manufacturing. The usual caveat about the difficulty of correcting for changes in the quality of manufacturing products has to be kept in mind, however.

18. Prebisch appears to have been influenced by the lack of technological progress in Latin American agriculture at the time of his writings. Unfortunately, he appears to have generalized this experience to the world as a whole, given his assumption noted earlier that technical progress took place only in the manufacturing sector, or at least occurred at a faster rate in that sector.

19. See also Brandão, 1978.

20. At one time, for example, Brazil was a dominant producer of coffee.

21. See Spraos, 1980, for some critical but not totally convincing comments on this question of changes in quality.

22. This point was originally made by Kindleberger *et al.*, 1956. Also, see Spraos [1980], who argues: "Perhaps there exists other evidence which points the other way, but for the time being the conclusion must be that had the terms of trade specific to those primary products in which developing countries have a major interest been available, they would not have weakened the statistical foundation of the deterioration hypothesis" [p. 115].

23. F. H. Cardoso, 1977, provides a perceptive description and analysis of how dependency theory was absorbed by U.S. radicals.

24. Jackson *et al.*, 1979, provides an assessment of the empirical research on dependency.

25. For a model which gives explicit attention to agriculture, but which is more in the structuralist tradition, see Chichilnisky and Taylor, 1980. Also, see the comments on this model by Gardner, 1980, and Schuh, 1980.

26. We should note that in commenting on an earlier version of this chapter de Janvry denied that his paper was in the dependency tradition. We believe it *is* because of the emphasis it gives to volitional actions by social groups. The reader will have to decide for her or himself.

27. *Minifundistas* are small farmers tied by social relations to large farmers, referred to as *latifundistas*. It is inherent in the *latifúndio-minifundista* concept that this relationship is exploitive.

28. Alliances between elites in the periphery countries and groups in the center are an important feature of dependency models.

29. For a sampling of the literature which has a more optimistic view about what markets can accomplish, see *Chicago Essays in Economic Development* [Wall, ed., 1972].

30. For sources on the vast import-substituting literature and some evaluations, see Hirschman, 1968; Tavares, 1964; Fishlow, 1972b; and Baer and Samuelson, 1977.

31. For an attempt to assess the dependency of the center on the Latin American periphery, see Muñoz, 1981.

32. This study, as the authors note at the outset, is an empirical analysis of "monetarist" and "structuralist" factors. Therefore, it does not accept the label of monetarist. However, one of its basic conclusions is that public deficits were the major explanatory variable in the equations explaining inflation. As they put it, "Vimos que pelo modêlo estimado que a ponderaçao dos défices públicos na explicaçao dos aumentos de preços era sensivelmente maior do que a das demais causas" [p. 148].

33. *Latifundiários* are often assumed not to maximize profits, or to maximize on diversified portfolios. They may be pure land speculators, or hold land as a hedge against inflation rather than as a direct source of income.

34. This in essence is the perspective of Furtado, 1972. It is worth emphasizing, however, that the structuralists never demonstrated much concern with the issue of modernization of agriculture, as perhaps, the following passage makes clear:

> In the publications it has issued in the twenty years of its existence, ECLA has been gradually building up a fund of interpretations, criteria, and suggestions which cover many of the most important aspects of the theoretical and practical problems of Latin American development. The present document is intended to be a compendium of the ideas expressed on some of the questions which have been sources of concern to ECLA.
>
> In preparing this book it has been necessary, first, to focus attention on some basic subjects, and to disregard many others which, however

important, were not of such permanent concern or which lie mainly within the competence of other international agencies (FAO, IMF, UNESCO, the ILO, and others). This applies to questions in such fields as agriculture, education, and employment. [Furtado, 1976 p. xi]

For the ECLA economic planners, agriculture was to be left to other agencies.

35. These arguments are the same usually given in connection with the declining terms of trade thesis.

36. The following statement is illuminating and captures the structuralist view:

Ambos factores pressionam constantemente sobre el tipo de cambio provocando la devaluación crónica del peso. Las devaluaciones inducen a su vez al reajuste de los niveles de costos y ingresos en el país. Esta última reacción es particularmente sensible en Chile debido a los seguintes fatores: a) la producción industrial depende en gran medida de los insumos importados; b) hasta fecha reciente si importaban integralmente los combustibles y lubricantes consumidos en el país y; c) la exagerada ampliación de la importación de alimentos (alrededor de 60 millones de dollares anuales en 1955 y 1956. [Sünkel, 1958, p. 580]

37. Sünkel, 1958, p. 58, mentions an earthquake in Chile in 1939 that was responsible for the creation of two government institutions that were financed by the Central Bank in an amount that was close to 20 percent of the total money supply.

38. A redistribution of land is presumed to raise agricultural productivity by breaking up large farms that are unproductive and giving it to more productive smallholders. Those who oppose land reform, however, believe it will have a serious negative effect on production. They argue that many productive farms will be broken up, and that the effect of the reform will be to reduce the incentive to invest on other productive farms. In addition, they doubt whether the government will have the capacity to organize the *asentamientos*, and to finance and instruct the small farmers so they become technologically efficient.

39. This fact is undisputed in recent discussion of inflation in Latin America, but there was a time when it wasn't.

40. In addition to references cited in this section, other references on Brazilian agricultural development include Nicholls, 1969 a, b, c, and 1972; Nicholls and Paiva, 1963/73; Thomazinni-Ettori, 1964; Schuh, 1970b; and R. Dias *et al.*, 1972.

41. The classic studies of these cycles are Prado, 1969; R. C. Simonsen, 1977; and Furtado, 1968c, 1984.

42. See Ayer and Schuh, 1972. The other data referred to in this paragraph are taken from G. W. Smith, 1969; and Knight, 1971.

43. For tests of the urban-industrial impact model with data from Brazil, see Nicholls, 1969a; Rios, 1969; and Katzman, 1974, 1975 a, b.

44. There was a modest research capacity in the Northeastern state of Paranámbuco and in Rio Grande do Sul in the South. These capabilities did generate modest degrees of modernization in their respective states, but they were not on the same scale as the capability in São Paulo.

45. This article appears in a two-volume collection of the writings of Alves and his colleagues, 1985.

46. The social rate of return to investments in cotton research in Brazil has been estimated at over 80 percent, for example. See Ayer and Schuh, 1972. For additional evidence, see Thompson, 1974.

47. This paper generated an unusual volume of debate in professional journals in Brazil. For example, see Nicholls, 1973; Schuh, 1973; Paiva, 1973; Contador, 1974a; Pastore, Alves, and Rizzieri, 1976; Paiva, 1975; and Riff, 1976.

48. For a review and exposition of the staple theory see Watkins, 1963. For an interpretation of regional development in terms of the staple theory see North, 1955.

49. Gerhard, 1959, and Mikesell, 1960, provide syntheses of the comparative research on the Turner thesis in frontier societies.

50. Furtado has written extensively on the general problems of economic underdevelopment in Brazil and Latin America [1968a, b, 1970b, 1976/1985, 1981]. Furtado was also the first Superintendent of SUDENE, the ambitious regional agency created to promote the development of the Northeast [see Robock, 1963, and Hirschman, 1963a], and for a short time was Minister of Planning, when that agency was created as a superministry to coordinate economic development.

51. Furtado seems to have been strongly influenced by the agriculture of the Northeast of Brazil. As Nicholls, 1969c, noted, agriculture in the south of Brazil has been more open with considerable social and geographic mobility. The same applies today to the rapidly developing frontier areas of Brazil's Central-West. Nicholls would agree with Furtado that *historically* the Northeast dominated Brazil politically, socially, and economically, and that this left a strong class consciousness in Brazil. However, he would argue that the later opening of the frontier first in the South and later in the Central-West, together with rapid industrialization, changed that. Moreover, in explaining the pattern of development in the United States, Nicholls gives a greater role to the *frontier* in making for a more open, flexible society and economy, and does not rely only on the role of New England, as does Furtado.

52. Supply can, of course, be changed by means other than price policies. See below.

53. If both β and γ are assumed to be different from one, the model will not be identifiable unless further restrictions are imposed. The inclusion of Z_t in the model allows for identification, as Nerlove has pointed out. An account of the various restrictions that have been imposed can be found in Askari and Cummings, 1976; for an overall evaluation of this and other econometric problems of estimation, see specifically chapters three and four. See also F. de H. Barbosa, 1978.

54. See Askari and Cummings, 1976, pp. 43, 53, and 70, for example. We do not address the marketable surplus approach to agriculture supply since there are few such studies in the Latin American literature.

55. A summary of this work can be found in Askari and Cummings, 1976, pp. 237-245.

56. Arak does not present this estimate in her article. It was taken from Askari and Cummings, 1976, p. 24.

57. Estimates of expected prices were generated from four alternative specifications of a price expectation model. Each of these estimates was then used in an ordinary least squares estimate of the slaughter equation and a generalized least squares estimate.

58. This is necessary since it is required by law to keep females in the herd until they first calve. This usually occurs when they are about four years old.

59. See Fisher, 1982, for an interesting discussion of this issue.

60. The studies by Carvalho, 1972, and Mascolo, 1979 and 1980, are not subject to this criticism.

61. This is not what the author states, however. Langoni, 1973, p. 89 states that all indexes show a smaller increase for this group. In his words: "Além do mais, qualquer que seja o índice utilizado, o menor acréscimo é justamente nos indivíduos com educação superior."

62. To update the definition to 1970, it was necessary to choose a national price index for the period. Fox chose the São Paulo Cost of Living Index. The results are sensitive to the choice, but Fox claims her results to be an upper limit.

63. "Com efeito, região é a única variável que apresenta queda na sua contribuição marginal, em termos absolutos. Este resultado contrasta efetivamente com o aumento, ainda que pequeno, da interdesigualdade regional, estimada exclusivamente pelo índice de Theil. Podemos concluir que êste aumento da desigualdade está fortemente associado ao comportamento da estrutura educacional, etária, setorial e de sexo dentro das regiões e, quando o impacto dessas variáveis é mantido constante, verifica-se que houve efetivamente redução nos diferenciais 'puros' de renda entre regiões" [Langoni, 1973, p. 113].

64. The corrected distribution includes imputed values for nonmonetary income as calculated by Fishlow.

65. In most Latin American countries, and especially in those reviewed here, trade, exchange rate, and general macroeconomic policy have been the most important policies affecting agriculture, and with a few exceptions (such as land reform efforts in Chile and Mexico), far more important than what are generally viewed as agricultural policies.

66. The descriptive material in this and the next paragraph is taken from Valdés, 1984a.

67. Reca, 1980. This section draws heavily on this source.

68. Exchange rates are quoted throughout this paper in monetary units of each country's currency per U.S. dollar.

69. Diaz-Alejandro, 1970, attempts to understand these contrasting performances. In addition, see Ballesteros, 1958; and Reca, 1967.

70. The domestic resource coefficients are often interpreted as a measure of a country's comparative advantage. This is in general incorrect, and for two basic reasons. First, if a sector purchases its inputs from national, inefficient producers, it may appear low in the rank of DRC's. But this use of domestic inputs may be the result of regulations and distortions in trade policy, and thus not a reflection of the underlying competitive potential of the sector. This point is expanded upon in Mendonça de Barros, 1974. The other reason is that comparative advantage is an issue of demand and supply, and there is nothing in the calculation of DRCs that takes demand considerations into account.

71. If P_1^e is the international price of exports in units of the domestic currency, then the domestic price is $P_1 = P_1^e (1 - t_x)$ where t_x is the direct or explicit export tax. The direct effect is calculated as $t_x P_1^e x_1^e$. The indirect effect is calculated as $t_x P_1^e (x_1 - x_1^e)/P_1 x_1$ where x_1 is per capita production of agriculture and x_1^e is the per capita exports of agricultural goods. Both of these calculations ignore supply and demand responses to the export tax [Cavallo and Mundlak, 1982, p. 59].

72. If P_1 and P_2 are the domestic prices of the exported and imported goods, respectively, if P_1^* and P_2^* are their international counterparts, and if RE is the real exchange rate (as defined earlier), then $P_1 = P_1^* RE(1 - t_x)$ and $P_2 = P_2^* RE(1 + t_m)$ where t_x and t_m are the export tax and import tax, respectively.

73. Policy makers were also obviously influenced by Prebisch's thesis of the declining terms of trade and the precepts of import-substituting industrialization, both of which justified to adherents of these views the pursuit of autarchy and the isolation of domestic economies from the international economy. That policy makers opted for the configuration of policy they did suggests something about the relative political power of agricultural interests. De Janvry, 1973, of course, argued that *within* agriculture large farmers were able to influence the direction of technological change so as to benefit them at the expense of small producers and workers.

74. A comprehensive analysis of Brazilian agriculture trade policy was made by Veiga, 1974. See also Homen de Melo, 1979; Pastore, 1979; Alves and Pastore, 1978; and M. de R. Lopes, 1977.

75. Brazil had also launched a policy of import substitution for petroleum by developing a domestic alcohol industry based on sugar cane. See below.

76. Agricultural products "in natura" did not benefit from export subsidies, but processed agricultural exports such as frozen orange juice, poultry, soybean oil, and soybean meal did. It was these sectors which expanded rapidly in this period.

77. The distinction between export and domestic goods is based, according to the authors, on the mechanism of price determination. Those goods for which the price was mainly determined by considerations of domestic demand and supply were said to constitute the domestic goods subsector. On a regional basis, the sectors were classified as follows: *exportables*: cocoa, tobacco, castor beans, cotton, and sugar cane in the Northeast; soybeans, coffee, castor beans, cotton and peanuts in the Center-South; coffee, cotton, sugar cane, peanuts, soybeans, oranges, and castor beans in São Paulo; *domestic goods*: rice, sweet potatoes, onions, beans, cassava, corn, bananas, oranges, pineapple, coconut, and tomatoes in the Northeast; pineapple, rice, bananas, sweet potatoes, potatoes, onions, beans, cassava, corn, and tomatoes in São Paulo.

78. This observed response in the agricultural exports of Brazil has been strongly influenced by the soybean phenomenon of the early 1970s, but expansion of orange juice and poultry exports has also been significant, especially in more recent years.

79. The DRCs presented here are calculated as implicit exchange rates. To make them comparable with the previous DRCs, they should be divided by the exchange rate.

80. See J. G. Garcia, 1981. There were exceptions to this prohibition. In general, if a food item was in short supply, it would be allowed to enter the country in that particular period. The only food item consistently imported by Colombia, however, has been wheat.

81. Minor exports included both agricultural and nonagricultural goods. Typically, the agricultural commodities included were tobacco, sugar cane, cotton, bananas, and livestock. The reader interested in more details is referred to Diaz-Alejandro [1976].

82. V. Thomas et al., 1984, vol. I, pp. 67-83. See also vol. II, Annex 8, pp. 54-58.

83. The agriculture of Venezuela has long suffered from that country's strong petroleum sector which has kept the Venezuelan currency strong, imports of agricultural commodities cheap, and potential agricultural exports noncompetitive. Mexican agriculture experienced similar difficulties in the aftermath of the discovery of large petroleum reserves and growing exports in the late 1970s.

84. The remainder of this paragraph is based on Behrman, 1976, chapter 1.

85. The theoretical underpinning of the stabilization policy was rooted in the monetary theory of the balance of payments. Similar policies were implemented in Argentina and Uruguay. Some sources which contain discussions of these policies include Diaz-Alejandro, 1981; Dornbusch, 1982; and Machinea, 1983. An overview of these stabilization policies will be provided below.

86. See also Jeanneret, 1971. The results obtained by this author are in agreement with Behrman's results if we consider only those measures of protection that do not correct for the overvaluation of the currency. When the overvaluation is taken into account, the author obtains results more consistent with those obtained by Valdés, 1973. For agriculture, the result changed from a positive coefficient of protection of fifty, equal to that calculated by Behrman, 1976, to a negative coefficient that ranged from −6 to −11 depending on the assumption made. See especially pp. 165-167.

87. For a brief overview of the SAM system in English, see Luiselli and Cruz-Serrano, 1986.

88. CONASUPO stands for Compania Nacional de Subsistencias Populares (National Commission of Popular Subsistence). Lustig, 1988, has made an analysis of the fiscal costs and welfare effects of the maize subsidy implemented by this agency.

89. See, for example, Bale and Lutz [1981]; Lutz and Scandizzo, 1980; Peterson, 1979 and 1983; World Bank, 1986c; and T. W. Schultz, ed., 1978.

90. The material which follows is drawn from Corbo, de Melo, and Tybout, 1986. Other references include Fernandez, 1985; Corbo, 1985a, b; Hanson and J. de Melo, 1983 and 1985; Dornbusch, 1982; Rodriguez, 1982; Blejer and Mathieson, 1981; S. Edwards, 1985; Wogart and Marques, 1984; J. de Melo, Pascale, and Tybout, 1985; and Blejer, 1981.

91. For additional references on Peru, with particular emphasis on trade and trade policy, see Nogues, 1986; Cebrecos and Castro, 1979; Fitzgerald, 1976; and Schydlowsky, Hunt, and Mezzera, 1983. For studies which evaluate the effect of these policies on agriculture, see Valdés, 1985b, and Orden, Greene, Roe, and Schuh, 1982.

92. We want to express our thanks to William C. Thiesenhusen for helping us to identify key references in the Latin American land reform literature.

93. We express our thanks to Dale W. Adams for his assistance in helping us to identify key references in the credit literature.

94. The Brazilian experience with credit subsidies is interesting in this respect. See Brandão and Magalhães, 1982.

95. David and Meyer, 1980, provide a summary of existing evaluations of farm-level impacts. We return to this study later.

96. The classic studies were by Abramovitz, 1956; Solow, 1957; Fabricant, 1959; and T. W. Schultz, 1961a.

97. Lobo's dissertation proposal was approved in 1978 with the title "Change in Earnings Distribution in Mexico between 1960-1970: An Assessment of Human Capital, Labor Market Segmentation and Government Income Policies."

98. We draw on Thiesenhusen's perceptive essay, 1987a, for the material in this and the next two paragraphs.

References

Abramovitz, M. [1956]: *Resources and Output Trends in the United States since 1870*. New York, NBER, Occasional Paper No. 52.

Adams, D. W. [1971]: "Agricultural Credit in Latin America: A Critical Review of External Funding Policy." *Amer. J. Agr. Econ.* 53:163-172.

———— [1975]: "Rural Financial Markets, Farm Level Growth and Capital Formation in Brazil." In D. W. Adams, *et al.*, 1975, pp. 10-1 through 10-47.

———— [1978]: "Mobilizing Household Savings through Rural Financial Markets." *Econ. Dev. and Cultural Change* 26:547-560.

Adams, D. W., C. Y. Ahn, D. G. Francis, T. F. Glover, D. H. Graham, D. W. Larson, R. L. Meyer, N. Rusk, and I. J. Singh [1975]: *Farm Growth in Brazil*, Final Research Report Prepared for AID, Washington. Columbus, Ohio State Univ. Press.

Adams, D. W., and D. H. Graham [1981, 1984]: "A Critique of Traditional Agricultural Projects and Policies." *J. Dev. Econ.* 8:347-366. Also in Eicher and Staatz, eds., 1984, pp. 313-328.

Adams, D. W., and G. I. Nehman [1979]: "Borrowing Costs and the Demand for Rural Credit." *J. Dev. Studies* 15:165-176.

Adams, D. W., R. Peña A., and A. Giles [1966]: *El Crédito Supervisado en la Reforma Agraria Colombiana—Un Estudio Evaluativo* (Supervised Credit in Colombian Agrarian Reform—An Evaluative Study). Bogotá, Colombia, Centro Inter-americano de Reforma Agraria, Materiales de Ensenanza para Reforma Agraria, No. 9.

Adams, D. W., and J. L. Tommy [1974]: "Financing Small Farms: The Brazilian Experience, 1965-1969." *Agr. Finance Rev.* 35:36-41.

Adams, F. G., and J. R. Behrman, eds. [1978]: *Econometric Modeling of World Commodity Policy*. Lexington, MA, Heath.

Adams, R. I., and N. Rask [1979]: "Regional and Farm Level Adjustments to the Production of Energy from Agriculture—Brazil's Alcohol Plan." Paper presented at the IAAE Conference, Banff, Alberta.

Adelman, I. [1984]: "Beyond Export-Led Growth." *World Dev.* 12:937-949.

Adelman, I., and C. T. Morris [1973]: *Economic Growth and Social Equity in Developing Countries*. Stanford, CA, Stanford Univ. Press.

Adelman, I., and E. Thorbecke, eds. [1966]: *The Theory and Design of Economic Development*. Baltimore, Johns Hopkins Press.

Agency for International Development [AID, 1970]: *Spring Review of Land Reform: Background and Country Papers*. Washington.

——— [AID, 1973]: *Spring Review of Small Farmer Credit*. Washington, D.C.

Ahluwalia, M. S., J. H. Duloy, G. Pyatt, and T. N. Srinivasan [1980]: "Who Benefits from Economic Development? Comment." *Amer. Econ. Rev.* 70:242-245.

Ahmad, S. [1966]: "On the Theory of Induced Invention." *Econ. J.* 76:344-357.

Alberts, T. [1983]: *Agrarian Reform and Rural Poverty: A Case Study of Peru*. Boulder, CO, Westview Press.

Alderman, H. [1984]: *Impact of Income and Food Price Changes in Food Acquisition by Low-Income Households*. Washington, IFPRI, mimeo.

Alexander, M., ed. [1975]: *Agricultural Policy: A Limiting Factor in the Development Process*. Washington, IADB.

Allen, F. [1982]: "On Share Contracts and Screening." *Bell J. Econ.* 13:541-547.

Altimir, O. [1982]: *The Extent of Poverty in Latin America*. Washington, World Bank, Staff Working Paper No. 522.

Altimir, O., and J. Sourrouille [1980]: *Measuring Levels of Living in Latin America: An Overview of Main Problems*. Washington, World Bank LSMS, Working Paper No. 3.

Alves, E. R. de A. [1968]: "An Economic Evaluation of an Extension Program." Unpublished M.S. thesis, Purdue Univ., West Lafayette, IN.

——— [1972]: "An Econometric Study of the Agricultural Labor Market in Brazil: A Test of Subsistence and Commercial Family Farm Models." Unpublished Ph.D. dissertation, Purdue Univ., West Lafayette, IN.

——— [1988]: *Pobreza Rural no Brasil: Desafios da Extensão e da Pesquisa* (Rural Poverty in Brazil: Challenges for Extension and Research). Brasilia, Ministério da Irrigação, CODEVASP.

Alves, E. R. de A., and others [1985]: *Pesquisa Agropecuária: Novos Rumos* (Agricultural Research: New Paths). Brasília, EMBRAPA, Organização L. Yeganiantz.

Alves, E. R. de A., and A. C. Pastore [1978]: "Import Substitution and Implicit Taxation of Agriculture in Brazil." *Amer. J. Agr. Econ.* 60:865-871.

———— [1985]: "A Política Agrícola do Brasil e a Hipótese de Inovação Induzida" (Brazil's Agricultural Policy and the Hypothesis of Induced Innovation). In Alves, E. R. de A. and others, 1985, pp. 289-300.

Alves, M., and R. Fiorentino [1983]: "Modernización Agropecuária en el Sertão de Pernambuco" (Agricultural Modernization of Pernambuco's Sertão). In Piñeiro and Trigo, eds., 1983b, pp. 353-417.

Amin, S. [1974]: *Accumulation on a World Scale: A Critique of the Theory of Underdevelopment*, 2 vols, tr. B. Pearce. New York, Monthly Review Press.

Ampuero, E. [1981]: "Organization of Agricultural Research for the Benefit of Small Farmers in Latin America." Ithaca, NY, Cornell Univ., Program in International Agriculture.

Anderson, C. A., and M. J. Bowman, eds. [1965]: *Education and Economic Development*. Chicago, Aldine.

Anderson, J. R., R. W. Herdt, and G. M. Scobie [1988]: *Science and Food: The CGIAR and its Partners*. Washington, World Bank publication for the CGIAR.

Anderson, K., and R. Tyers [1986]: "Agricultural Policies of Industrial Countries and their Effects on Traditional Food Exporters." *Econ. Record* 62(179):385-399.

Anschel, K. R., R. H. Brannon, and E. D. Smith, eds. [1969]: *Agricultural Cooperatives and Markets in Developing Countries*. New York, Praeger.

Arak, M. V. [1967]: "The Supply of Brazilian Coffee." Unpublished Ph.D. thesis, Cambridge, MA, MIT.

———— [1968]: "The Price Responsiveness of São Paulo Coffee Growers." *Food Research Institute Studies* 8:211-223.

———— [1969]: "Estimation of Assymetric [sic] Long-Run Supply Functions: The Case of Coffee." *Canadian J. Agr. Econ.* 17(1):15-22.

Ardito-Barletta, N. [1970]: "Costs and Social Benefits of Agricultural Research in Mexico." Unpublished Ph.D. dissertation, Univ. of Chicago.

Arndt, T. M., D. G. Dalrymple, and V. W. Ruttan, eds. [1977]: *Resource Allocation and Productivity in National and International Agricultural Research*. Minneapolis, Univ. of Minnesota Press.

Askari, H., and J. T. Cummings [1976]: *Agricultural Supply Response: A Survey of the Econometric Evidence*. New York, Praeger.

Aspra, L. A. [1977]: "Import Substitution in Mexico: Past and Present." *World Dev.* 5:111-123.

Ávila, A. F. D., J. E. Borges-Andrade, L. J. M. Írias, and B. R. Quirino [1983]; Formação do Capital Humano e Retôrno dos Investimentos em Treinamentona EMBRAPA (Human Capital Formation and Investment Returns to EMBRAPA's Training Programs). Brasília, EMBRAPA, D. I. D., DDM Documento 4.

Axline, W. A. [1981]: "Latin American Regional Integration: Alternative Perspective on a Changing Reality." *Latin Amer. Research Rev.* 16(1):167-186.

Ayer, H. W. [1970]: "The Costs, Returns and Effects of Agricultural Research in a Developing Country: The Case of Cotton Seed Research in São Paulo, Brazil." Unpublished Ph.D. dissertation, Purdue Univ., West Lafayette, IN.

Ayer, H. W., and G. E. Schuh [1972]: "Social Rates of Return and Other Aspects of Agricultural Research: The Case of Cotton Research in São Paulo, Brazil." *Amer. J. Agr. Econ.* 54:557-569.

Ayres, C. H. S. [1976]: "Excedente Comercializavel de Produtos Selecionados no Estado do Espírito Santo" (Marketable Surplus of Selected Products in the State of Espírito

Santo)." Unpublished M.S. thesis, Universidade Federal de Viçosa, Minas Gerais, Brazil.

Bacha, E. L. [1968]: "An Econometric Model for the World Coffee Market: The Impact of Brazilian Price Policy." Unpublished Ph.D. dissertation, Yale Univ., New Haven, CT.

―――― [1977]: "Issues and Evidence on Recent Brazilian Economic Growth." *World Dev.* 5:47-67.

―――― [1979]: "Crescimento Econômico, Salários Urbanos e Rurais: O Caso do Brasil, (Economic Growth, Urban and Rural Wages: The Brazilian Case)." *Pesquisa e Planejamento Econômico* 9(3):585-627.

Bacha, E. L., and L. Taylor [1971]: "Foreign Exchange Shadow Prices: A Critical Review of Current Theories." *Quart. J. Econ.* 85:197-224.

―――― [1978]: "Brazilian Income Distribution in the 1960s: 'Facts,' Model Results and the Controversy." *J. Dev. Studies* 14:271-297.

Bachmura, F. T., ed. [1968]: *Human Resources in Latin America: An Interdisciplinary Focus.* Bloomington, Indiana Univ., Bureau of Business Research.

Baer, W. [1962]: "The Economics of Prebisch and ECLA." *Econ. Dev. and Cultural Change* 10(2, Part I): 169-182.

―――― [1964a]: "Regional Inequality and Economic Growth in Brazil." *Econ. Dev. and Cultural Change* 12:268-285.

―――― [1964b]: "The Economics of Prebisch and ECLA: Reply." *Econ. Dev. and Cultural Change* 12:315.

―――― [1967]: "The Inflation Controversy in Latin America: A Survey." *Latin Amer. Research Rev.* 2(2):3-25.

Baer, W., and I. Kerstenetzky, eds. [1964]: *Inflation and Growth in Latin America.* Homewood, IL, Irwin.

Baer, W., and L. Samuelson [1977]: "Editors' Introduction." *World Dev.* 5:1-6.

Balassa, B., [1983]: "Trade Policy in Mexico." *World Dev.* 11:795-811.

―――― [1985]: *Outward Orientation.* Washington, World Bank, Development Research Department, Report No. DRD148.

Balassa, B., and associates [1971]: *The Structure of Protection in Developing Countries.* Baltimore, Johns Hopkins Press for IBRD and IADB.

Balassa, B., and C. Michalopoulos [1985]: *Liberalizing World Trade.* Washington, World Bank, Office of the Vice President, Report No. VPERS4.

Bale, M. D., and E. Lutz [1981]: "Price Distortions in Agriculture and their Effects: An International Comparison." *Amer. J. Agr. Econ.* 63:8-22.

Ballesteros, M. A. [1958]: "Argentine Agriculture, 1908-54: A Study in Growth and Decline." Unpublished Ph.D. dissertation, Univ. of Chicago.

Baran, P. A. [1957]: *The Political Economy of Growth.* New York, Monthly Review Press.

Barbato de Silva, C. [1983]: "El Proceso de Generación, Difusion y Adopción de Tecnologia en la Ganaderia Vacúna Uruguaya: Un Estudio de Caso, 1950-1977" (The Process of Generation, Diffusion and Adoption of Technology in the Beef Sector in Uruguay: A Case Study, 1950-1977). In Piñeiro and Trigo, eds., 1983b, pp. 287-352.

Barbosa, F. de H. [1978]: "Expectativa Adaptada e Ajustamento Parcial: Identificação e Discriminação Entre os Dois Processos" (Adaptive Expectations and Partial Adjustment: Identification and Discrimination between the Two Processes). *Revista Brasileira de Economia* 32(3):399-418.

―――― [1983]: *A Inflação Brasileira no Pós-Guerra: Monetarismo vs. Estruturalismo* (Brazilian Post-War Inflation: Monetarism vs. Structuralism). Rio de Janeiro, IPEA/INPES, Série PNPE 8.

Barbosa, T. [1973]: "A Normative Analysis of Land Reform Measures in the Priority Area of Rio de Janeiro, Brazil." Unpublished Ph.D. Thesis, Purdue Univ., West Lafayette, IN.

Bardhan, P. K. [1977]: "Variations in Forms of Tenancy in a Peasant Economy." *J. Dev. Econ.* 4(2): 105-118.

_____ [1979]: "Agricultural Development and Land Tenancy in a Peasant Economy: A Theoretical and Empirical Analysis." *Amer. J. Agr. Econ.* 61:48-57.

_____ [1980]: "Interlocking Factor Markets and Agrarian Development: A Review of Issues." *Oxford Econ. Papers* NS 32:82-98.

Bardhan, P. K., and T. K. Srinivasan [1971]: "Cropsharing Tenancy in Agriculture: A Theoretical and Empirical Analysis." *Amer. Econ. Rev.* 61:48-64.

Barlett, P. F. [1982]: *Agricultural Choice and Change: Decision Making in a Costa Rican Community.* New Brunswick, NJ, Rutgers Univ. Press.

Barraclough, S. L. [1973]: *Agrarian Structure in Latin America.* Lexington, MA, Lexington Books.

_____ [1984]: "Review of de Janvry, 'The Agrarian Question and Reformism in Latin America.' " *Econ. Dev. and Cultural Change* 32:639-649.

Barraclough, S. L. and A. L. Domike [1966]: "Agrarian Structure in Seven Latin American Countries." *Land Econ.* 42:321-424.

Barros de Castro, A. [1969]: *Ensaios sôbre a Economia Brasileira* (Essays on the Brazilian Economy). Rio de Janeiro, Foruse.

Barsky, O., and G. Cosse [1983]: "Iniciativa Terrateniente, Cambio Técnico y Modelo Institucional: El Caso de la Producción Lechera en la Sierra Equatoriana" (Landholder Initiatives, Technical Change and Institutional Model: The Case of Milk Production in the Equatorian Sierra). In Piñeiro and Trigo, eds., 1983b, pp. 101-150.

Barzelay, M., and S. R. Pearson [1982]: "The Efficiency of Producing Alcohol for Energy in Brazil." *Econ. Dev. and Cultural Change* 31:131-144.

Bateman, M. J. [1969]: *Supply Responses in the Colombian Coffee Sector.* Santa Monica, CA, Rand Corp., Memorandum RM-5780-RC/AID.

Bath, C. R., and D. D. James [1976]: "Dependency Analysis of Latin America: Some Criticism, Some Suggestions." *Latin Amer. Research Rev.* II(3):3-54.

Becker, G. S. [1964]: *Human Capital.* Princeton, NJ, Princeton Univ. Press.

_____ [1965]: "A Theory of the Allocation of Time." *Econ. J.* 75:493-517.

Becker, G. S., and H. G. Lewis [1973, 1975]: "Interaction between Quantity and Quality of Children." In T. W. Schultz, ed., 1973b, 1975a, pp. 81-90.

Becker, W. S. [1970]: "Agricultural Credit and Colombia's Economic Development." Unpublished Ph.D. thesis, Louisiana State Univ., Baton Rouge.

Beckerman, P., and D. Coes [1980]: "Who Benefits from Economic Development? Comment." *Amer. Econ. Rev.* 70:246-249.

Behrman, J. R. [1968a]: *Supply Response in Underdeveloped Agriculture: A Case Study of Four Major Annual Crops in Thailand, 1937-63.* Amsterdam, North-Holland.

_____ [1968b]: "Monopolistic Cocoa Pricing." *Amer. J. Agr. Econ.* 50: 702-719.

_____ [1976]: *Foreign Trade Regimes and Economic Development: Chile.* New York, Columbia Univ. Press for the NBER.

Behrman, J. R., and N. Birdsall [1983a]: *The Implicit Equity-Productivity Tradeoff in the Distribution of Public School Resources in Brazil.* Washington, World Bank, Country Policy Discussion Paper No. 1983-1.

_____ [1983b]: "The Quality of Schooling: Quantity Alone is Misleading." *Amer. Econ. Rev.* 73:928-946.

———— [1988a]: "The Equity-Productivity Tradeoff: Public School Resources in Brazil." *European Econ. Rev.* 32:1585-1601.

———— [1988b]: " The Reward for Good Timing: Cohort Effects and Earning Functions for Brazilian Males." *Rev. Econ. and Stat.* 70:129-135.

Behrman, J. R., and L. R. Klein [1970]: "Econometric Growth Models for the Developing Economy." In Eltis, M. F. Scott, and J. N. Wolfe, eds., 1970, pp. 167-187.

Behrman, J. R., B. L. Wolfe, and D. M. Blau [1985]: "Human Capital and Earnings Distribution in a Developing Country: The Case of Prerevolutionary Nicaragua." *Econ. Dev. and Cultural Change* 34:1-30.

Bejarano, J. A. [1980]: "Crescimiento, Distribución, y Política Económica." (Growth, Distribution, and Political Economy). Paper presented at the *Congreso de Economistas de la Universidad Nacional*. Melgar, Colombia.

Bell, C., and P. Zusman [1979]: *New Approaches to the Theory of Rental Contracts in Agriculture*. Washington, World Bank, Development Research Center, mimeo.

Belli, P. [1970]: " 'Farmers' Response to Price in Underdeveloped Areas: The Nicaraguan Case." *Amer. Econ. Rev.* 60:385-392.

———— [1971]: The Economic Implications of Malnutrition: The Dismal Science Revisited." *Econ. Dev. and Cultural Change* 20:1-23.

Berg, A. D. [1973]: *The Nutrition Factor: Its Role in National Development*. Washington, Brookings Institution.

———— [1981]: *Malnourished People: A Policy View*. Washington, World Bank, Poverty and Basic Needs Series.

———— [1987]: *Malnutrition: What Can Be Done? Lessons from World Bank Experience*. Baltimore, Johns Hopkins Univ. Press.

Bergsman, J. [1980]: *Income Distribution and Poverty in Mexico*. Washington, World Bank, Staff Working Paper No. 395.

Berry, A., and F. Thoumi [1977]: "Import Substitution and Beyond: Colombia." *World Dev.* 5:89-109.

Berry, A., and M. Urrutia [1976]: *Income Distribution in Colombia*. New Haven, CT, Yale Univ. Press. Also in Spanish, *La Distribución del Ingreso en Colombia*. Medellin, Editorial Carreta, 1975.

Berry, R. A. [1974]: "Impact of Factor Emigration on the Losing Region." *Econ. Record* 50:405-422.

Berry, R. A., and W. R. Cline [1979]: *Agrarian Structure and Productivity in Developing Countries*. Baltimore, Johns Hopkins Univ. Press.

Berry, R. A., and R. Soligo [1969]: "Some Welfare Aspects of International Migration." *J. Polit. Economy* 77:778-794.

————, eds. [1980]: *Economic Policy and Income Distribution in Colombia*. Boulder, CO, Westview Press.

Bhaduri, A. [1973]: "Agricultural Backwardness under Semi-Feudalism." *Econ. J.* 83:120-137.

———— [1977]: "On the Formation of Usurious Interest Rates in Backward Agriculture." *Cambridge J. Econ.* 1:341-352.

Bhagwati, J., and K. Hamada [1974]: "The Brain Drain, International Integration of Markets for Professionals and Unemployment: A Theoretical Analysis." *J. Dev. Econ.* 1:19-42.

Bhagwati, J., R. W. Jones, R. A. Mundell, and J. Vanek, eds. [1971]: *Trade Balance of Payments and Growth. Papers in International Economics in Memory of C.P. Kindleberger*. Amsterdam, North-Holland.

Bhagwati, J., and C. Rodriguez [1975]: "Welfare-Theoretical Analyses of the Brain Drain." *J. Dev. Econ.* 2:195-222.

Bianchi, A. [1964a]: "Agriculture—The Revolutionary Background." In Seers, ed., 1964, pp. 65-99.

———— [1964b]: "Agriculture—The Post-Revolutionary Development." In Seers, ed., 1964, pp. 100-157.

Billington, J. A., ed. [1961]: *Frontier and Section: Selected Essays of Frederick Jackson Turner.* Englewood Cliffs, NJ, Prentice-Hall.

Binswanger, H. P. [1987]: *Fiscal and Legal Incentives with Environmental Effects on the Brazilian Amazon.* Washington, World Bank, Discussion Paper Report No. ARU69, mimeo.

Binswanger, H.P., and J. McIntire [1987]: "Behavioral and Material Determinants of Production Relations in Land-Abundant Tropical Agriculture." *Econ. Dev. and Cultural Change* 36:73-99.

Binswanger, H. P., Y. Mundlak, M. C. Yang, and A. Bowers [1985]: *Estimate of Aggregate Agriculture Supply Response.* Washington, World Bank, Agriculture and Rural Development Department, Report ARU No. 48, revised version.

Binswanger, H. P., and M. R. Rosenzweig, eds. [1984a]: *Contractual Arrangements, Employment, and Wages in Rural Labor Markets in Asia.* New Haven, CT, Yale Univ. Press.

———— [1984b]: "Contractual Arrangements, Employment, and Wages in Rural Labor Markets: A Critical Review." In Binswanger and Rosenzweig, eds., 1984a, pp. 1-40.

Binswanger, H. P., V. W. Ruttan and others, eds. [1978]: *Induced Innovation: Technology, Institutions, and Development.* Baltimore, Johns Hopkins Univ. Press.

Birdsall, N. [1985]: "Public Inputs and Child Schooling in Brazil." *J. Dev. Econ.* 18:67-86.

Birdsall, N., and J. R. Behrman [1984]: "Does Geographical Aggregation Cause Overestimates of the Returns to Schooling?" *Oxford Bull. Econ. and Stat.* 46:55-72.

Biswas, M., and P. Pinstrup-Andersen, eds. [1985]: *Nutrition and Development.* New York, Oxford Univ. Press.

Blasier, C. [1981]: "The Soviet Latin Americanists." *Latin Amer. Research Rev.* 16(1):107-123.

Blejer, M. I. [1981]: "The Dispersion of Relative Commodity Prices under Very Rapid Inflation." *J. Dev. Econ.* 9:347-356.

Blejer, M. I., and D. Matthieson [1981]: "The Preannouncement of Exchange Rate Changes as a Stabilization Instrument." *IMF Papers* 28:760-792.

Bliss, C. J., and N. H. Stern [1978]: "Productivity, Wages, and Nutrition." *J. Dev. Econ.* 5:331-398.

———— [1982]: *Palanpur: The Economy of an Indian Village.* Oxford, England, Clarendon Press.

Bonnen, J. T. [1983]: "Historical Sources of U.S. Agricultural Productivity: Implications for R&D Policy and Social Science." *Amer. J. Agr. Econ.* 65:958-966.

Boserup, E. [1965]: *The Conditions of Agricultural Growth: The Economics of Agrarian Change under Population Pressure.* Chicago, Aldine.

Bourguignon, F. [1980]: "The Role of Education in the Urban Labor Market during the Process of Development: The Case of Colombia." Mexico City, Paper presented at the Sixth World Congress of the International Economics Association.

Bowman, M. J., and C. A. Anderson [1963]: "Concerning the Role of Education in Development." In Geertz, ed., 1963, pp. 247-279.

Boyce, J. K., and R. E. Evenson [1975]: *National and International Agricultural Research and Extension Programs.* New York, ADC.

Brada, J. C., and J. A. Mendez [1983]: "Regional Economic Integration and the Volume of Intra-Regional Trade: A Comparison of Developed and Developing Country Experience." *Kyklos* 36:589-603.

Brandão, A. S. P. [1978]: "New Perspectives on the Terms of Trade and the Gains from Trade: A Case Study of Brazil." Unpublished Ph.D. thesis, Purdue Univ., West Lafayette, IN.

_____ [1981]: "Agricultural Development Models Commonly Advocated in Latin America." In Glenn L. Johnson and Maunder, eds., 1981, pp. 346-357.

_____ [1986]: *OPreço da Terra no Brasil: Verificação de Algumas Hipóteses* (Land Price in Brazil: Verification of Some Hypotheses). Rio de Janeiro, EPGE, Ensaios Econômicos No. 79.

Brandão, A. S. P., and U. Magalhães [1982]: *"Crédito Rural: Problemas Econômicos e Sugestões de Mudança"* (Rural Credit: Economic Problems and Suggestions for Change). EPGE Ensaios Econômicos, June.

Brandão, A. S. P., and G. E. Schuh [1979]: "Têrmos de Troca e Bem-Estar Econômico: Algumas Proposições Qualitativas" (Terms of Trade and Economic Welfare: Some Qualitative Propositions). *Revista Brasileira de Economia* 33(1):3-24.

_____ [1980]: "Têrmos de Troca para o Brasil: Uma Análise Empírica" (Terms of Trade for Brazil: An Empirical Analysis). *Revista de Economia Rural* 18(2):205-220.

Brandt, S. A., L. Hirata, F. C. Carvalho, and O. Cintra Filho [1968]: "Funções de Oferta Agrícola: Variações Estacionais e Regionais" (Agricultural Supply Functions: Stationary and Regional Variations). *Agricultura em São Paulo* 15 (1/2): 1-11.

Brannon, R. H. [1968]: *The Agricultural Development of Uruguay: Problems of Government Policy*. New York, Praeger.

Brasil, Ministério das Minas e Energia [1984]: *Auto-Suficiência Energética: Um Cenário de Extensão ao Modêlo Energético Brasileiro* (Energy Self-Sufficiency: A Projected Scenario to the Brazilian Energy Model). Brasilia.

Braverman, A., and J. L. Guasch [1984]: "Capital Requirements, Screening, and Interlinked Sharecropping and Credit Contracts." *J. Dev. Econ.* 14:359-374.

_____ [1986]: "Rural Credit Markets and Institutions in Developing Countries: Lessons for Policy Analysis from Practice and Modern Theory." *World Dev.* 14:1253-1267.

Braverman, A., and T. N. Srinivasan [1981]: "Credit and Sharecropping in Agrarian Societies." *J. Dev. Econ.* 9:289-312.

Braverman, A., and J. E. Stiglitz [1982]: "Sharecropping and the Interlinking of Agrarian Markets." *Amer. Econ. Rev.* 72: 695-715.

_____ [1986]: "Cost-Sharing Arrangements under Sharecropping: Moral Hazard, Incentive Flexibility, and Risk." *Amer. J. Agr. Econ.* 68:642-652.

Brown, C., G. Fane, and J. Medoff [1973]: "The Income Distribution as a Pure Public Good: Comment." *Quart. J. Econ.* 87:296-303.

Brundenius, C. [1984]: *Revolutionary Cuba: The Challenge of Economic Growth with Equity*. Boulder and London, Westview Press.

Brunner, H. [1977]: *Cuban Sugar Policy from 1963 to 1970*. Pittsburgh, PA, Univ. of Pittsburgh Press.

Bueno, G. M. [1971]: "The Structure of Protection in Mexico." In Balassa and associates, 1971, pp. 169-202.

_____ , ed. [1977]: *Opciones de la Política Económica en México Despues de la Devaluación* (Economic Policy Options in Mexico after the Devaluation). Mexico, Editorial Tecnos S.A.

Bunker, S. G. [1985]: *Underdeveloping the Amazon: Extraction, Unequal Exchange, and the Failure of the Modern State*. Urbana, Univ. of Illinois Press.

Byerlee, D. [1986]: "The Political Economy of Third World Food Imports: The Case of Wheat." *Econ. Dev. and Cultural Change* 35:307-328.

Byerlee, D., and G. Sain [1986]: "Food Pricing Policy in Developing Countries: Bias against Agriculture or for Urban Consumers?" *Amer. J. Agr. Econ.* 68:961-969.

Caballero, J. M., and H. Maletta [1983]: *Estilos de Desarrollo y Políticas Agroalimentarias — Tendencias y Dilemas en América Latina* (Experts' Consultations on Styles of Development and Agricultural Policies, Trends, and Dilemmas in Latin America). Santiago, Chile/Rome, CEPAL/FAO.

Calegar, G. M. [1984]: "Brazilian Wheat Policy and its Income Distribution and Trade Effects: A Case Study." Unpublished Ph.D. thesis, Univ. of Minnesota, St. Paul.

Calegar, G. M., and G. E. Schuh [1988]: *The Brazilian Wheat Policy: Its Costs, Benefits, and Effects on Food Consumption*. Washington, IFPRI Research Report. Also in Pinstrup-Andersen, ed. [1988], pp. 267-276.

Campos, R. de O. [1960]: "La Inflación y el Crecimiento Equilibrado" (Inflation and Balanced Growth). In Ellis and Wallich, eds., 1961, pp. 94-126.

———— [1961]: "Two Views on Inflation in Latin America." In Hirschman, ed., 1961, 1964, pp. 69-79.

Canavese, A. J. [1982]: "The Structuralist Explanation in the Theory of Inflation." *World Dev.* 10:523-529.

Cappi, C., L. B. Fletcher, R. Norton, C. Pomerada, and M. Wainer [1978]: "A Model of Agricultural Production and Trade in Central America." In Cline and Delgado, eds., 1978, Chapter 7, pp. 317-370.

Cardoso, E. A. [1980a]: "Brazilian Growth and Distribution in the 1960s: An Identity-Based Post-Mortem." In Taylor, Bacha, Cardoso and Lysy, 1980, pp. 77-101.

———— [1980b, 1981]: "Oferta de Alimento e Inflação." *Pesquisa e Planejamento Econômico* 10(1):45-69. Also in English as "Food Supply and Inflation." *J. Dev. Econ.* 8:269-284, 1981.

Cardoso, F. H. [1972]: "Dependency and Development in Latin America." *New Left Rev.* 74:83-95.

———— [1977]: "The Consumption of Dependency Theory in the United States." *Latin Amer. Research Rev.* 12(3): 7-29.

Cardoso, F. H., and E. Faletto [1969]: *Dependência y Desarrollo en América Latina* (Dependency and Development in Latin America). Santiago, ILPES.

Carnoy, M. [1964]: "The Cost and Returns to Schooling in Mexico: A Case Study." Unpublished Ph.D. dissertation, Univ. of Chicago.

———— [1967a]: "Earnings and Schooling in Mexico." *Econ. Dev. and Cultural Change* 15:408-419.

———— [1967b]: "Rates of Return to Schooling in Latin America." *J. Human Resources* 2:359-374.

Carnoy, M., in collaboration with J. Lobo, A. Toledo, and J. Velloso [1979]: *Can Educational Policy Equalise Income Distribution in Latin America?* Farnborough, United Kingdom, Saxon House.

Carnoy, M., and J. Werthein [1979, 1980]: *Cuba: Economic Change and Reform 1955-1978*. World Bank Staff Working Paper, 1979, N:317. Also in Spanish as: *Cuba: Cambio Económico y Reforma Educativa (1955-1978)*. Sacramento, Mexico, Editorial Nueva Imagem, 1980.

Carroll, T. F. [1961]: "The Land Reform Issue in Latin America." In Hirschman, ed., 1961, pp. 161-201.

Carvalho, J. L. [1972]: "Production, Investment and Expectations: A Study of the U.S. Cattle Industry." Unpublished Ph.D. dissertation, Univ. of Chicago.

Castenado, T. [1984]: *Determinantes de la Reducción de la Mortalidad Infantil en Chile, 1955-83* (Determinants of the Reduction in Infant Mortality in Chile, 1955-83). Santiago, Universidad de Chile, Department of Economics.

Castro, C. de M. [1970]: "Investment in Education in Brazil: A Study of Two Industrial Communities." Unpublished Ph.D. dissertation, Vanderbilt Univ., Nashville, TN.

Castro, C. de M., G. Frigotto, R. C. R. Martins, and R. de A. Córdova [1980]: *A Educação na América Latina: Um Estudo Comparativo de Custo e Eficiência* (Education in Latin America: A Comparative Study of Cost and Efficiency). Rio de Janeiro, Editôra da FGV.

Cavallo, D., and Y. Mundlak [1982]: *Agriculture and Economic Growth in an Open Economy: The Case of Argentina.* Washington, IFPRI Research Report No. 36.

Caves, R. E. [1968]: " 'Vent for Surplus' Models of Trade and Growth." In Theberge, ed., 1968, pp. 211-230.

Cebrecos, R., and J. V. Castro [1979]: *Los Efectos de una Nueva Política de Protección en el Perú* (The Effects of a New Protection Policy in Peru). Lima, Universidad Católica del Perú, CISEPA, Documento de Trabajo No. 40.

Cehelsky, M. [1979]: *Land Reform in Brazil: The Management of Social Change.* Boulder, CO, Westview Press.

Centro de Documentación Económica para América Latina [CEDEAL, 1976]: *Resumenes Analíticos en Economía Agrícola Latino-Americana* (Analytical Resumes in the Agricultural Economy of Latin America), Vol.1. Cali, Colombia.

Centro Internacional de Mejoramiento de Maíz y Trigo [CIMMYT, 1974]: *The Pueblo Project, Seven Years of Experience: 1967-73.* El Baton, Mexico, D.F.

_____ [CIMMYT, 1984]: *Proceedings of the Workshop on Strengthening Agricultural Research in Latin America and the Caribbean.* Mexico, D.F.

Centro Latinoamericano de Demografía [1963]. Santiago, Chile.

Chaigneau, S., and R. Szalachman [1977]: "Estimaciones Preliminares Elasticidad-Gasto y la Elasticidad-Ingreso." (Preliminary Estimates of Expenditure Elasticity and Income Elasticity). *Estudios de Economía* 10 (2nd Semestre), Santiago, Chile.

Chao, K. [1983]: "Tenure Systems in Traditional China." *Econ. Dev. and Cultural Change* 31:295-314.

Chavez, A., H. Bourges, and S. Busta, eds.[1975]: *Proceedings of the Ninth International Congress of Nutrition,* Mexico City, 1972, Vol. 2. Basel, Switzerland, Karger.

Chen, C. Y. [1968]: *Movimientos Migratorios en Venezuela* (Migratory Movements in Venezuela). Caracas, Instituto de Investigaciones Económicas de la Universidad Católica Andrea Bello.

Chenery, H. B., ed. [1971]: *Studies in Development Planning.* Cambridge, MA, Harvard Univ. Press.

Chenery, H. B., M. S. Ahluwalia, C. L. G. Bell, J. H. Duloy, and R. Jolly [1974]: *Redistribution with Growth: Policies to Improve Income Distribution in Developing Countries in the Context of Economic Growth.* London, Oxford Univ. Press for the World Bank.

Cheung, S. N. S. [1968]: "Private Property Rights and Sharecropping." *J. Polit. Economy* 76:1107-1122.

_____ [1969]: *The Theory of Share Tenancy, with Special Application to Asian Agriculture and the First Phase of Taiwan Land Reform.* Univ. of Chicago Press.

Chichilnisky, G., and L. Taylor [1980]: "Agriculture and the Rest of the Economy: Macro-connections and Policy Constraints." *Amer. J. Agr. Econ.* 62:303-309.

Chilcote, R. [1974]: "Dependency: A Critical Synthesis of the Literature." *Latin Amer. Perspectives* 1:4-29.

_____ [1978]: "A Question of Dependency." *Latin Amer. Research Rev.* 12(2):55-68.

Chilcote, R., and J. Edelstein [1974]: *Latin America: The Struggle with Dependency and Beyond.* New York, Wiley.

Chonchol, J. [1977]: "Social and Economic Organization of the Chilean Reformed Sector during the Popular Unity Government 1971-September 1973. In Dorner, ed., 1977, pp. 199-216.

Christ, C., ed. [1963]: *Measurement in Economics: Studies in Mathematical Economics and Econometrics in Memory of Yehuda Grunfeld.* Stanford, CA, Stanford Univ. Press.

Cisneros, C. C. [1959]: "Indian Migration from the Andean Zone of Ecuador." *America Indigena* 19:225-231.

Cline, W. R. [1970]: *Economic Consequences of a Land Reform in Brazil.* Amsterdam, North-Holland.

_____ [1978]: "Benefits and Costs of Economic Integration in Central America." In Cline and Delgado, eds., 1978, Chapter 3, pp. 59-121.

_____ [1980]: "Income Distribution and Economic Development: A Survey and Tests for Selected Latin American Cities." In Robert Ferber, ed., 1980, pp. 205-255.

Cline, W. R., and E. Delgado, eds. [1978]: *Economic Integration in Central America.* Washington, Brookings Institution.

Cline, W. R., and S. Weintraub, eds. [1981]: *Economic Stabilization in Developing Countries.* Washington, Brookings Institution.

Cochrane, W. W. [1958]: *Farm Prices: Myth and Reality.* Minneapolis, Univ. of Minnesota Press.

Cockcroft, J. D., A. G. Frank, and D. L. Johnson [1972]: *Dependence and Underdevelopment: Latin America's Political Economy.* Garden City, NY, Doubleday Anchor.

Coellho, C. N. de A. [1979]: *A Política de Preços Mínimos dentro de uma Perspectiva de Desenvolvimento Econômico* (The Policy of Minimum Prices in an Economic Development Perspective). Brasilia, Ministério da Agricultura, CFP, Coleção Análise e Pesquisa, Vol. 12.

Collier, P., and P. H. Sabot [1982]: "Measuring the Difference between Rural and Urban Incomes: Some Conceptual Issues." In Sabot, ed., 1982a, pp. 127-160.

Colyer, D., and G. Jiménez [1971]: "Supervised Credit as a Tool in Agricultural Development." *Amer. J. Agr. Econ.* 53:639-642.

Comisión Económica para América Latina [CEPAL, 1976]: *Reactivacio del Mercado Comum Centroamericano* (Reactivation of the Central American Common Market). Santiago, Chile.

Comité Interamericano de Desarollo Agrícola [CIDA, 1966]: *Land Tenure Conditions and Socio-Economic Development of the Agricultural Sector: Brazil.* Washington, Pan American Union.

Companhia de Financiamento da Produção [CFP, 1983]: *Análise das Distorções dos Preços Domésticos em Relação aos Preços de Fronteira: Um Estudo Preliminar* (Analysis of the Distortions of Domestic Prices in Relation to Frontier Prices: A Preliminary Study). Brasilia, Ministério da Agricultura, CFP, Coleção Análise e Pesquisa, Vol. 30.

Conjunctura Econômica [various issues]: Rio de Janeiro, Brazil, Fundação Getulio Vargas.

Consultative Group on International Agricultural Research [CGIAR, 1985]: *Summary of International Agricultural Research Centers: A Study of Achievements and Potential.* Washington, World Bank.

Contador, C. R. [1974a]: "Dualismo Technológico na Agricultura: Novos Comentários" (Technologic Dualism in Agriculture: New Comments)." *Pesquisa e Planejamento Econômico* 4(1):119-138.

———— [1974b]: "Trigo Nacional: O Custo Social da Auto-Suficiência" (National Wheat: The Social Cost of Self-Sufficiency). *Estudos Econômicos* 4(3):53-83.

————, ed. [1975]: *Tecnologia e Desenvolvimento Agrícola* (Technology and Agricultural Development). Rio de Janeiro, IPES, Instituto de Pesquisas, Monografia No. 17.

Corbo, V. [1985a]: "International Prices, Wages, and Inflation in an Open Economy: A Chilean Model." *Rev. Econ. and Stat.* 67:564-573.

———— [1985b]: "Reforms and Macroeconomic Adjustment in Chile during 1974-84." *World Dev.* 13:893-916.

———— [1986]: *Problems, Development Theory, and Strategies of Latin America.* Washington, World Bank, Development Research Department, Report No. DRD190.

Corbo, V., and J. de Melo [1987]: "Lessons from the Southern Cone Policy Reforms." *World Bank Research Observer* 2:111-142.

Corbo, V., J. de Melo, and J. Tybout [1986]: "What Went Wrong with the Recent Reforms in the Southern Cone?" *Econ. Dev. and Cultural Change* 34:606-640.

Corbo, V., and M. Stelcner [1983]: "Earnings Determination and Labour Markets: Gran Santiago, Chile—1978." *J. Dev. Econ.* 12:251-266.

Correa, H. [1970]: "Sources of Economic Growth in Latin America." *Southern Econ. J.* 37:17-31.

Costa Rego, A. J., and C. L. Wright [1981]: "Uma Análise da Distribuição do Crédito Rural no Brasil" (An Analysis of the Allocation of Rural Credit in Brazil). *Revista de Economia Rural* 19(2):217-38.

Cotiler, J., and R. R. Fagen, eds. [1974]: *Latin America and the United States: The Changing Political Realities.* Stanford, CA, Stanford Univ. Press.

Coutu, A. J., and R. A. King [1969]: *The Agricultural Development of Peru.* New York, Praeger.

Crist, R. E., and C. M. Nissly [1973]: *East from Andes.* Gainesville, Univ. of Florida Press, Social Sciences Monographs No. 1.

Crockett, J., and I. Friend [1980]: "Consumption and Savings in Economic Development." In Robert Ferber, ed., 1980, pp. 15-53.

Crowder, L. V., ed. [1967]: *Rural Development in Tropical Latin America.* Ithaca, Cornell Univ.

Cutié, T. J. [1976]: *Diffusion of Hybrid Corn Technology: The Case of El Salvador.* Mexico, CIMMYT, abridged by CIMMYT.

Daines, S. R. [1975]: *Guatemalan Farm Policy Analysis: The Impact of Small Farm Credit on Income, Employment and Production.* Washington, AID, Bureau for Latin America, Analytical Working Document No. 10.

Daines, S. R., A. Le Baron, and M. Whitaker [1971]: *Bibliography on the Economics of Agricultural Production and Irrigation in Latin America. Provisional Draft.* Logan, Utah State Univ.

Da Mata, M. [1982]: "Crédito Rural: Caracterização do Sistema e Estimativas dos Subsidios Implicitos (Rural Credit: Characterization of the System and Estimates of the Implicit Subsidies)." *Revista Brasileira de Economia* 36(3):215-245.

Dambaugh, L. N. [1959]: *The Coffee Frontier in Brazil.* Gainesville, Univ. of Florida Press, Latin American Monographs No. 7.

Da Silva, G. L. S. P. [1984]: "Contribuição da Pesquisa e Extensão Rural para a Productividade Agrícola: O Caso de São Paulo (The Contribution of Research and Extension to Agricultural Productivity: The Case of São Paulo)." *Estudos Econômicos* 14(2):315-353.

————— [1986]: *Pesquisa, Tecnologia, e Rendimento dos Principais Produtos da Agricultura Paulista* (Research, Technology and Yields of Principal Crops in São Paulo). São Paulo, SA, IEA.

David, C., and R. L. Meyer [1980]: "Measuring the Farm Level Impact of Agricultural Loans." In Howell, ed., 1980, pp. 201-234.

Davis, J. [1981]: "A Comparison of Procedures for Estimating Returns to Research Using Production Functions." *Australian J. Agr. Econ.* 25:60-72.

Davis, R. F. [1977]: "The Andean Pact: A Model of Economic Integration for Developing Countries." *World Dev.* 5:137-153.

Day, R. H. [1967]: "The Economics of Technological Change and the Demise of the Sharecropper." *Amer. Econ. Rev.* 57:427-449.

Dean, W. [1969]: *The Industrialization of São Paulo, 1880-1945.* Austin, Univ. of Texas Press for the Institute of Latin American Studies.

De Araujo, J. D. [1980]: *A Consistência da Política de Preços de Produtos Agrícolas: O Caso do Mercado de Rações* (The Consistency of Price Policy of Agricultural Products: The Case of Market for Rations). Brasilia, Ministério da Agricultura, CFP, Coleção Análise e Pesquisa, Vol. 19.

De Araujo, P. F. C. [1967]: "An Economic Study of Factors Affecting the Demand for Agricultural Credit at the Farm Level." Unpublished M.S. thesis, Ohio State Univ., Columbus.

De Araujo, P. F. C., and R. L. Meyer [1977]: "Agricultural Credit Policy in Brazil: Objectives and Results." *Amer. J. Agr. Econ.* 59:957-961. Also in *Savings and Development* 2:169-194, 1978.

De Camargo, J. F. [1960]: *Êxodo Rural no Brasil: Formas, Causas, e Consequências Econômicas Principais* (Rural Exodus in Brazil: Forms, Causes and Principal Economic Consequences). Rio de Janeiro, Brazil, Conquista.

Deere, C. D., and A. de Janvry [1979]: "A Conceptual Framework for the Empirical Analysis of Peasants." *Amer. J. Agr. Econ.* 4:601-611.

De Janvry, A. [1972]: "Optimum Levels of Fertilization under Risk: The Potential for Corn and Wheat Fertilization under Alternative Price Policies in Argentina." *Amer. J. Agr. Econ.* 54:1-10.

————— [1973]: "A Socioeconomic Model of Induced Innovations for Argentine Agricultural Development." *Quart. J. Econ.* 87:410-435.

————— [1975]: "The Political Economy of Rural Development in Latin America: An Interpretation." *Amer. J. Agr. Econ.* 57:490-499. Also in Eicher and Staatz, eds., 1984, pp. 82-95.

————— [1977]: "Inducement of Technological and Institutional Innovations: An Interpretative Framework." In Arndt, Dalrymple and Ruttan, eds., 1977, pp. 551-563.

————— [1978]: "Social Structure and Biased Technical Change in Argentine Agriculture." In Binswanger and Ruttan, eds., 1978, pp. 297-323.

————— [1981a]: *The Agrarian Question and Reformism in Latin America.* Baltimore, Johns Hopkins Univ. Press.

————— [1981b]: "The Role of Land Reform in Economic Development: Policies and Politics." *Amer. J. Agr. Econ.* 2:384-392.

————— [1987a]: "Latin American Agriculture from Import Substitution Industrialization to Debt Crisis." *International Political Economy Yearbook*, Vol. 3, 1987.

———— [1987b]: "Peasants, Capitalism, and the State in Latin American Agriculture." In Shanin, ed., 1987, pp. 391–404.

———— [1989]: "The Debt Crisis and Rural Development in Latin America." *Institute of International Studies*, Working Paper No. 3, Series 1.

De Janvry, A., J. Bieri, and A. Nunez [1972]: "Estimation of Demand Parameters under Consumer Budgeting: An Application to Argentina." *Amer. J. Agr. Econ.* 54:422-430.

De Janvry, A., and L. Crouch [1980a]: "Beyond Dependency Theory: New Directions In Latin American Political Economy." Paper presented at the Colloquium Series on Latin American Political Economy of the Berkeley and Stanford Latin American Centers, May 8, 1980.

———— [1980b]: "Technological Change and Peasants in Latin America." Proceedings of a Conference of the IICA, San José, Costa Rica.

De Janvry, A., and J. J. Dethier [1985]: *Technological Innovation in Agriculture: The Political Economy of its Rate and Bias.* Washington, World Bank, CGIAR Study Paper No. 1.

De Janvry, A., and C. Garramon [1977a]: "Accumulación de Capital y Miseria Rural en América Latina (Capital Accumulation and Rural Misery in Latin America)." *Problemas del Desarrollo* 8(29):65-94.

———— [1977b]: "The Dynamics of Rural Poverty in Latin America." *J. Peasant Studies* 4(3):206-216.

———— [1977c]: "Laws of Motion of Capital in the Center-Periphery Structure." *Rev. of Radical Political Econ.* 9(2):29-39.

De Janvry, A., and E. P. Le Veen [1983]: "Aspects of the Political Economy of Technical Change in Developed Economies." In Piñeiro and Trigo eds., 1983a, pp. 25-36.

De Janvry, A., E. P. Le Veen, and D. Runsten [1983]: "La Economía Politica del Cambio Técnico: La Mecanización de la Cosecha de Tomates en California" (The Political Economy of Technical Change: The Mechanization of the Tomato Harvest in California). In Piñeiro and Trigo, eds., 1983b, pp. 151-184.

De Janvry, A., D. Runsten, and E. Sadoulet [1987]: *Technological Innovations in Latin American Agriculture.* San José, Costa Rica, IICA, Program Papers Series, No. 4.

De Janvry, A., and E. Sadoulet [1986]: "The Conditions for Harmony between Third World Agricultural Development and U.S. Farm Exports." *Amer. J. Agr. Econ.* 5:1340-1346.

———— [1988]: "The Conditions for Compatibility between AID and Trade in Agriculture." *Econ. Dev. and Cultural Change* 37:1-30.

Delfim Netto, A. [1966]: "Agricultura e Desenvolvimento no Brasil" (Agriculture and Development in Brazil). *Estudos ANPES-5.* São Paulo.

Delfim Netto, A., A. C. Pastore, E. P. de Carvalho [1966]: "*Agricultura e Desenvolvimento no Brasil*" (Agriculture and Development in Brazil). Estudos ANPES-5. São Paulo.

Delfim Netto, A., A. C. Pastore, P. Cippolari, and E. P. de Carvalho [1965]: *Alguns Aspectos da Inflação Brasileira* (Some Aspects of Brazilian Inflation). Estudos ANPES-1. São Paulo.

Delgado, E. [1978]: "Institutional Evolution of the Central American Common Market and the Principle of Balanced Development." In Cline and Delgado, eds., 1978, Chapter 2, pp. 17-58.

De Melo, J., R. Pascale and J. Tybout [1985]: "Microeconomic Adjustments in Uruguay during 1973-81: The Interplay of Real and Financial Shocks." *World Dev.* 13:995-1015.

Denslow D., Jr., and W. G. Tyler [1983]: *Perspective on Poverty and Income Inequality in Brazil: An Analysis of the Changes during the 1970s.* Washington, World Bank, Staff Working Paper No. 601.

De Pablo, J. C. [1977]: "Beyond Import Substitution: The Case of Argentina." *World Dev.* 5:7-17.

De Rubinstein, E. M., and C. Budge C. [1988]: *North-South Grain Policies and Internal Trade: The Case of Chile*. DEA, Pontifícia Universidad Católica de Chile, Serie de Investigación, No. 56.

De Tray, D. N. [1973, 1975]: "Child Quality and the Demand for Children." In T. W. Schultz, ed., 1973b, 1975a, pp. 91-116.

Dias, G. L. da S. [1972]: "Avaliação da Política Econômica para a Pecuária de Corte no Brasil" (Evaluation of Economic Policy for the Beef Sector in Brazil). Unpublished Ph.D. dissertation, Univ. of São Paulo.

———— [1976]: *Mercado de Capital, Adoção de Tecnologia, e o Ciclo de Vida* (Capital Market, Adoption of Technology, and the Life Cycle). FIPE, São Paulo.

————, ed. [1979]: "Pobreza Rural no Brasil: Caracterização do Problema e Recomendações da Política" (Rural Poverty in Brazil: Characterization of the Problem and Policy Recommendations). Brasilia, CFP, Coleção Análise e Pesquisa—Vol. 16.

Dias, G. L. da S., and I. G. V. Lopes [1983]: *Avaliação do Comportamento dos Preços Domésticos em Relação Aos Preços de Importação e de Exportação de Algodão, Arroz, Milho, e Soja 1979/83* (Evaluation of the Behavior of Domestic Prices in Relation to Import and Export Prices of Cotton, Rice, Corn, and Soybeans, 1979/83). Brasilia, Ministério da Agricultura, CFP, Coleção Análise e Pesquisa. Vol. 27.

Dias, R., G. E. Schuh, and P. F. Warnken [1972]: *O Desenvolvimento da Agricultura Paulista* (The Development of Paulista Agriculture). São Paulo, SA, IEA.

Diaz-Alejandro, C. F. [1965]: *Exchange Rate Devaluation in a Semi-Industrialized Country: The Experience of Argentina, 1955-1961*. Cambridge, MA, MIT Press, chapter 4.

———— [1970]: *Essays on the Economic History of the Argentine Republic*. New Haven, CT, Yale Univ. Press, chapter 3.

———— [1973]: "The Andean Common Market: Gestation and Outlook." In Eckaus and Rosenstein-Rodan, eds., 1973, pp. 293-326.

———— [1976]: *Foreign Trade Regimes and Economic Development: Colombia*. New York, Columbia Univ. Press for the NBER.

———— [1981]: "Southern-Cone Stabilization Plans." In Cline and Weintraub, eds., 1981, pp. 119-147.

Di Marco, L. E., ed. [1972]: *International Economics and Development: Essays in Honor of Raúl Prebisch*. New York, Academic Press.

Dominguez, J. I. [1971]: "Sectoral Clashes in Cuban Politics and Development." *Latin Amer. Research Rev.* 6(3):61-87.

Donald, G. [1976]: *Credit for Small Farmers in Developing Countries*. Boulder, CO, Westview Press.

Dornbusch, R. [1982]: "Stabilization Policies in Developing Countries: What Have We Learned?" *World Dev.* 10:701-708.

Dorner, P., ed. [1971]: *Land Reform in Latin America: Issues and Cases*. Madison, Univ. of Wisconsin, LTC, Land Economics Monograph No. 3.

———— [1972]: *Land Reform and Economic Development*. Baltimore, Penguin Books.

————, ed. [1977]: *Cooperative and Commune*. Madison, Univ. of Wisconsin Press.

Dorner, P., and D. Kanel [1971]: "The Economic Case for Land Reform." In Dorner, ed., 1971, pp. 21-35.

Dorrance, G. S. [1964]: "The Effect of Inflation on Economic Development." In Baer and Kerstenetzky, eds., 1964, pp. 37-88.

Dougherty, C. R. S. [1971]: "Optimal Allocation of Investment in Education." In Chenery, ed., 1971, pp. 270-292.

Dovring, F. [1968]: "El Papel de la Agricultura dentro de las Poblaciones en Crescimiento, México: Un Caso de Desarrollo Económico Reciente" (The Role of Agriculture in Increasing Populations, Mexico: A Recent Case of Economic Development). *El Trimestre Económico* 35(1):25-50.

———— [1970]: "Land Reform and Productivity in Mexico." *Land Econ.* 46:264-274.

Dozier, C. C. [1969]: *Land Development and Colonization in Latin America: Case Studies of Peru, Bolivia, Mexico.* New York, Praeger.

Dudley, L., and R. J. Sandilands [1975]: "The Side Effects of Foreign Aid: The Case of Public Law 480 Wheat in Colombia." *Econ. Dev. and Cultural Change* 23:325-336.

Dumont, R. [1970, 1964]: *Cuba: Socialism and Development.* New York, Grove. Originally published in French as *Cuba: Socialisme et Développement.* Paris, France, E'ditions du Seuil, 1964.

Duncan, K., and J. Rutledge with the collaboration of C. Harding, eds. [1977]: *Land and Labour in Latin America: Essays on the Development of Agrarian Capitalism in the Nineteenth and Twentieth Centuries.* Cambridge, England, Cambridge Univ. Press.

Duvall, R. D. [1978]: "Dependence and Dependência Theory: Notes towards Precision of Concept and Argument." *International Organization* 32(1): 51-78.

Easterlin, R. A., ed. [1980]: *Population and Economic Change in Developing Countries.* Univ. of Chicago Press for NBER.

———— [1981]: "Why Isn't the Whole World Developed?" *J. Econ. Hist.* 41:1-19.

Echeverria, R. P. [1969]: *The Effect of Agricultural Price Policies on Intersectoral Income Transfers.* Ithaca, NY, Cornell Univ., Latin American Studies Program, Dissertation Series No. 13.

Echeverria, R. P., and J. Soto [1968]: *Respuesta de los Productores Agrícolas: Auto Cambio en los Precios* (Response of Agricultural Producers: Automatic Changes in Prices). ICIRA, Informe Técnico No. 1.

Eckaus, R. S., and P. N. Rosenstein-Rodan, eds. [1973]: *Analysis of Development Problems. Studies of the Chilean Economy.* Amsterdam, North-Holland.

Eckstein, S. [1969]: "El Macro Económico del Problema Agrario Mexicano" (The Principal Macroeconomics of the Mexican Agrarian Problem). Mexico, CIDA, Trabajo No. 11.

———— [1981]: "The Socialist Transformation of Cuban Agriculture: Domestic and International Constraints." *Social Problems* 29:178-196.

———— [1983]: "Domestic and International Constraints on Private and State Sector Agricultural Production." *Cuban Studies* 13(2):41-64.

Eckstein, S., G. Donald, D. Horton, and T. Carroll [1978]: *Land Reform in Latin America: Bolivia, Chile, Mexico, Peru, and Venezuela.* Washington, World Bank, Staff Working Paper No. 275.

Eckstein, Z. [1981]: "Rational Expectations Modeling of Agricultural Supply: The Egyptian Case." Unpublished Ph.D. thesis, Univ. of Minnesota, Minneapolis.

———— [1984]: "A Rational Expectations Model of Agricultural Supply." *J. Polit. Economy* 92:1-19.

Economic Commission for Latin America (ECLA) [1967]: "The Measurement of Latin American Real Income in U.S. Dollars." *Econ. Bul. for Latin America* 12:107-142.

———— [1970a]: *Economic Survey of Latin America, 1969.* New York.

———— [1970b]: "Income Distribution in Latin America." In ECLA, 1970a, pp. 364-417.

———— [1970c]: *Development Problems in Latin America.* Austin, Univ. of Texas Press.

_____ [1972]: *Economic Survey of Latin America, 1970*. New York.

Edquist, C. H., and O. Edquist [1979]: *Social Carriers of Techniques for Development*. Stockholm, Sweden, unpublished SAREC Report R3, 1979.

Edwards, E. O., ed. [1974]: *Employment in Developing Nations: Report on a Ford Foundation Study*. New York, Columbia Univ. Press.

Edwards G. G., and A. Ducci V. [1988]: *Alternativas de Estabilização de Preços Agropecuários* (Alternatives of Crop and Livestock Price Stabilization). Brasilia, Ministério da Agricultura, CFP, Coleção Análise e Pesquisa Vol. 36.

Edwards, S. [1985]: "Stabilization with Liberalization: An Evaluation of Ten Years of Chile's Experiment with Free Market Policies, 1973-1983." *Econ. Dev. and Cultural Change* 33:223-254.

Eicher, C. K., and J. M. Staatz, eds. [1984]: *Agricultural Development in the Third World*. Baltimore, Johns Hopkins Univ. Press.

Elias, V. [1981]: *Government Expenditures on Agriculture in Latin America*. Washington, IFPRI, Research Report No. 23.

Elizago, J. C. [1963]: *Migración Diferencial en Algunas Regiones y Ciudades de la América Latina, 1940-1950*. (Differential Migration in some Latin American Regions and Cities). Santiago, Chile, Centro Latinoamericano de Demografia.

_____ [1965]: "Assessment of Migration Data in Latin America." *Milbank Memorial Fund Quarterly* 43:76-106.

Ellis, H. S., ed. [1969]: *The Economy of Brazil*. Berkeley, Univ. of California Press.

Ellis, H. S., and H. C. Wallich, eds. [1960, 1961]: *El Desarrollo Económico y América Latina* (Economic Development and Latin America). México, BA, Fondo de Cultura Económica. English edition published 1961 by St. Martin's Press, New York.

Ellsworth, P. T. [1956]: "The Terms of Trade between Primary Producing and Industrial Countries." *Inter-Amer. Econ. Affairs* 10(1):47-65.

Eltis, W. A., M. F. Scott, and J. N. Wolfe, eds. [1970]: *Induction, Growth and Trade: Essays in Honour of Sir Roy Harrod*. Oxford, England, Clarendon Press.

Ely, R. T. [1914]: *Property and Contract in their Relations to the Distribution of Wealth*. New York, Macmillan.

Ely, R. T., and G. S. Wehrwein [1940]: *Land Economics*. New York, Macmillan.

Emmanuel, A. [1972]: *Unequal Exchange: A Study of the Imperialism of Trade*. New York, Monthly Review Press.

Epstein, E. C. [1975]: "Politicization and Income Distribution in Argentina: The Case of the Peronist Worker." *Econ. Dev. and Cultural Change* 23:615-631.

Erber, F. S. [1981]: "Science and Technology Policy in Brazil: A Review of the Literature." *Latin Amer. Research Rev.* 16(1):3-56.

Eswaran, M., and A. Kotwal [1985]: "A Theory of Contractual Structure in Agriculture." *Amer. Econ. Rev.* 75:352-367.

Evenson, D. D., and R. E. Evenson [1983]: "Legal Systems and Private Sector Incentives for the Invention of Agricultural Technology in Latin America." In Piñeiro and Trigo, eds., 1983a, pp. 189-216.

Evenson, R. E. [1987]: *The International Agricultural Research Centers: Their Impact on Spending for National Agricultural Research and Extension*. Washington, World Bank, Study Paper No. 22.

Evenson, R. E., and Y. Kislev [1975]: *Agricultural Research and Productivity*. New Haven, CT, Yale Univ. Press.

Evenson, R. E., P. E. Waggoner, and V. W. Ruttan [1979]: "Economic Benefits from Research: An Example from Agriculture." *Science* 205 (4411):1101-1107.

Fabricant, S. [1959]: *Basic Facts on Productivity*. New York, NBER, Occasional Paper No. 63.

Fagen, R. R. [1969]: *The Transformation of Political Culture in Cuba*. Stanford, CA, Stanford Univ. Press.

Falcon, W. P. [1970]: "The Green Revolution: Generations of Problems." *Amer. J. Agric. Econ.* 52:698-710.

Faminous, M. D., and J. S. Hillman [1986]: *Brazil's Response to the U.S. Soybean Embargo*. Washington, USDA, ERS/IED.

Feder, E. [1965]: "Land Reform under the Alliance for Progress." *J. Farm Econ.* 47:652-668.

Feder, G., R. E. Just, and D. Zilberman [1985]: "Adoption of Agricultural Innovations in Developing Countries: A Survey." *Econ. Dev. and Cultural Change* 33:255-298.

Fei, J. C. H., and G. Ranis [1964]: *Development of the Labor Surplus Economy: Theory and Policy*. Homewood, IL, Irwin.

———— [1966]: "Agrarianism, Dualism and Economic Development." In Adelman and Thorbecke, eds., 1966, pp. 3-43.

Feindt, W., and H. L. Browning [1972]: "Return Migration: Its Significance in an Industrial Metropolis and an Agricultural Town in Mexico." *International Migration Rev.* 6(2):158-165.

Felix, D. [1961]: "An Alternative View of the 'Monetarist Structuralist' Controversy." In Hirschman, ed., 1961, pp. 81-93.

Ferber, Robert, ed. [1980]: *Consumption and Income Distribution in Latin America*. Washington, OAS for ECIEL Program.

Fernandez, R. B. [1985]: "The Expectations Management Approach to Stabilization in Argentina during 1976-82." *World Dev.* 13:871-892.

Ferrer, A. [1967]: *The Argentina Economy*, tr. M. M. Urquidi. Berkeley, Univ. of California Press, pp. 100-101.

Ferris, E. G. [1979]: "National Political Support for Regional Integration: The Andean Pact." *International Organization* 33:83-104.

Ffrench-Davis, R. [1977]: "The Andean Pact: A Model of Economic Integration for Developing Countries." *World Dev.* 5:137-153.

Fields, G. S. [1977]: "Who Benefits from Economic Development? A Reexamination of Brazilian Growth in the 1960s." *Amer. Econ. Rev.* 67:570-582.

———— [1980a]: "Who Benefits from Economic Development? Reply." *Amer. Econ. Rev.* 70:257-262.

———— [1980b]: "Education and Income Distribution in Developing Countries: A Review of the Literature." In T. King, ed., 1980, pp. 231-315.

Fienup, D. F., A. L. Bandeira, D. O. Hansen, I. Martins do Carmo, I. A. Schneider, and W. L. Peterson [1978]: *Higher Education: Programs in Agricultural Economics and Rural Sociology in Brazil*. East Lansing, MSU/Brazil/MEC Project, Survey Team Report No. 85.

Fienup, D. F., R. H. Brannon, and F. A. Fender [1969]: *The Agricultural Development of Argentina: A Policy and Development Perspective*. New York, Praeger.

Figueroa, A., and R. Weisskoff [1980]: "Viewing Social Pyramids: Income Distribution in Latin America." In Robert Ferber, ed., 1980, pp. 257-294.

Fink, W. H. [1955]: "Trends in Latin America to Import and the Gains from Trade." *Inter-Amer. Econ. Affairs* 9(1):61-77.

Fisher, B. S. [1982]: "Rational Expectations in Agricultural Economics Research and Policy Analysis." *Amer. J. Agr. Econ.* 64:260-265.

Fishlow, A. [1972a]: "Brazilian Size Distribution of Income." *Amer. Econ. Rev.* 62:391-402. For a technical appendix to this paper, see Fishlow and Mesook, 1972.

_____ [1972b]: "Origins and Consequences of Import Substitution in Brazil." In di Marco, ed., 1972, pp. 311-365.

_____ [1973, 1977]: *Brazilian Income Size Distribution: Another Look*. Berkeley, Univ. of California, Department of Economics, mimeo. Also in Portuguese as "Distribuição da Renda no Brazil: Um Novo Exame." *Dados* 11:10-80, 1973.

_____ [1980]: "Who Benefits from Economic Development? Comment." *Amer. Econ. Rev.* 70:250-256.

Fishlow, A., and A. Mesook [1972]: *Technical Appendix — Brazilian Size Distribution of Income, 1960*. Berkeley, Univ. of California, Department of Economics, mimeo. This is the Technical Appendix to Fishlow, 1972a.

Fitchett, D. A. [1971]: "The Price Responsiveness of Farmers in Latin America: An Empirical Test for Cereal and Potato Producers in Chile." In Hunt, ed., 1971, pp. 432-440.

Fitzgerald, E. V. K. [1976]: *The State and Economic Development in Peru since 1968*. Cambridge, England, Cambridge Univ. Press.

Flanders, M. J. [1964a]: "Prebisch on Protectionism: An Evaluation." *Econ. J.* 74:305-326.

_____ [1964b]: "The Economics of Prebisch and ECLA: A Comment." *Econ. Dev. and Cultural Change* 12:312-314.

_____ [1969]: "Agriculture *versus* Industry in Development Policy: The Planner's Dilemma Re-Examined." *J. Dev. Studies* 5:171-189.

Flanders, M. J., and A. Razin, eds. [1981]: *Development in an Inflationary World*. New York, Academic Press.

Fletcher, L., and W. C. Merrill [1968]: *Latin American Agricultural Development and Policies*. Ames, ISU, International Studies in Economics, Monograph 8.

Floyd, J. E. [1965]: "The Effects of Farm Price Supports on the Returns to Land and Labor in Agriculture." *J. Polit. Economy* 73:148-158.

Food and Agriculture Organization (FAO) [various issues]: *Trade Yearbook*.

Forster, N. [1982]: "Cuban Agricultural Productivity: A Comparison of State and Private Farm Sectors." *Cuban Studies* 11(2), 12(1):105-125.

Forster, N., and H. Handelman [1984]: "Food Production and Distribution in Cuba: The Impact of Revolution." In Super and Wright, eds., 1985, pp. 174-198.

Foster, P., and J. R. Sheffield [1973]: *Education and Rural Development*. London, Evans Brothers Limited, in *The World Year Book of Education 1974*.

Foweracker, J. [1981]: *The Struggle for Land: A Political Economy of the Frontier in Brazil from 1930 to the Present*. Cambridge, England, Cambridge University Press.

Fox, M. L. [1983]: "Income Distribution in Post-1964 Brazil: New Results." *J. Econ. Hist.* 43:261-271.

Foxley, A., E. Aninat, and J. P. Arellano [1977]: "Chile: The Role of Asset Redistribution in Poverty Focused Development Strategies." *World Dev.* 5:69-88.

Franco, G. [1964]: "Rendimiento de la Inversión en Educación en Colombia" (Returns from Investment in Education in Colombia). Bogotá, CEPE, Universidad de los Andes, mimeo.

Frank, A. G. [1967]: *Capitalism and Underdevelopment in Latin America: Historical Studies of Chile and Brazil*. New York, Monthly Review Press.

_____ [1977]: "Dependence is Dead, Long Live Dependence, and the Class Struggle: An Answer to Critics." *World Dev.* 5(4):355-370.

Franke, R. H., and G. V. Barrett [1975]: "The Economic Implications of Malnutrition: Comment." *Econ. Dev. and Cultural Change* 23:341-350.

Friedman, M. [1968]: "The Role of Monetary Policy." *Amer. Econ. Rev.* 58:1-17.

Friere, R. [1966]: *Price Incentives in Argentine Agriculture.* Cambridge, MA, Harvard Univ., HIID, Center for International Affairs, Development Advisory Service Report.

Furtado, C. [1954]: *A Economia Brasileira: Contribuição a Análise do seu Desenvolvimento* (The Brazilian Economy: A Contribution to the Analysis of its Development). Rio de Janeiro, Editôra A. Norte.

———— [1968a]: *Dialética do Desenvolvimento* (The Dialectics of Development). Rio de Janeiro, Editôra Fundo de Cultura.

———— [1968b]: *Subdesenvolvimento e Estagnação da América Latina* (The Underdevelopment and Stagnation of Latin America). Rio de Janeiro, Civilização Brasileira.

———— [1968c, 1984]: *The Economic Growth of Brazil: A Survey from Colonial to Modern Times.* Westport, CT, Greenwood Press. Tr. by R. W. de Aguiar and E. C. Drysdale from *Formação Econômica do Brasil.* Berkeley, Univ. of California Press.

———— [1970a]: *Formação Econômica do Brasil* (The Economic Formation of Brazil). 10a edição, São Paulo, Editôra Nacional.

———— [1970b]: *Obstacles to Development in Latin America.* New York, Doubleday.

———— [1972]: *Análise do "Modelo" Brasileiro* (Analysis of the Brazilian "Model"). Rio de Janeiro, Civilização Brasileira.

———— [1976, 1985]: *Economic Development of Latin America: Historical Background and Contemporary Problems.* New York, Cambridge Univ. Press. 2nd ed. of translation by S. Macedo of *Formação Econômica da América Latina.*

———— [1981]: *O Brasil Pós-Milagre* (Post-Miracle Brazil). Rio de Janeiro, Editôra Paz e Terra S. A.

Garcia, J. C. [1975]: "Análise da Alocação de Recursos por Proprietários e Parceiros em Áreas de Agricultura de Subsistência" (Analysis of Resource Allocation by Landowners and Sharecroppers in Areas of Subsistence Agriculture). Unpublished M.S. thesis, Universidade Federal de Viçosa, Minas Gerais, Brazil.

Garcia, J. G. [1981]: *The Effects of Exchange Rates and Commercial Policy on Agricultural Incentives in Colombia, 1953-1978.* Washington, IFPRI, Research Report No. 24.

Garcia, J. G., and G. M. Llamas [1988]: *Coffee Boom, Government Expenditure, and Agricultural Prices: The Colombian Experience.* Washington, IFPRI, Research Report 68.

Gardner, B. [1980]: "Post-Keynesian Economics and Agriculture: Discussion." *Amer. J. Agr. Econ.* 62:325-327.

Geertz, C., ed. [1963]: *Old Societies and New States: The Quest for Modernity in Asia and Africa.* Glencoe, IL, Free Press of Glencoe.

Gemma, M. [1983]: "The World Soybean Diffusion and Price Responsiveness of Brazilian Soybean Supply." Unpublished M.S. thesis, Univ. of Minnesota, St. Paul.

Gerhard, D. [1959]: "The Frontier in Comparative View." *Comparative Studies in Society and History* 1:205-229.

Getulio Vargas Foundation [FGV, 1970]: *Food Consumption in Brazil: Family Budget Survey in the Early 1960s.* Rio de Janeiro, Instituto Brasileiro de Economia.

Gisser, M. [1965]: "Schooling and the Farm Problem." *Econometrica* 33:582-592.

Gonçalves, R., and A. C. Barros [1982]: "Tendência dos Têrmos de Troca: A Tese de Prebisch e a Economia Brasileira, 1850-1979" (Trends in the Terms of Trade: The Prebisch Thesis and the Brazilian Economy, 1850-1979). *Pesquisa e Planejamento Econômico* 12(1):109-132.

Gonzalez-Vega, C. [1977]: "Interest Rate Restrictions and Income Distribution." *Amer. J. Agr. Econ.* 59:973-976. Also in Eicher and Staatz, eds., 1984, pp.329-334.

Goodman, D. E. [1976]: "Estrutura Rural, Excedente Rural, e Modos de Produção no Nordeste Brasileiro" (Rural Structure, Rural Surplus, and Modes of Production in the Brazilian Northeast). *Pesquisa e Planejamento Econômico* 6(2):489-534.

Goodman, D. E., and M. Redclift [1982]: *From Peasant to Proletarian Capitalistic Development and Agrarian Transitions.* New York, St. Martin's Press, chapter 5.

Goreux, L. M. [1980]: *Compensatory Financing Facilities.* Washington, IMF, Pamphlet No. 34.

Government of the Netherlands [1983]: *Report of the World Food Program: Seminar on Food Aid.* Rome, Gov. of the Netherlands Publication.

Graber, K. L. [1975]: "Factors Explaining Farm Production and Family Earnings of Small Farmers in Brazil." Unpublished M.S. thesis, Purdue Univ., West Lafayette, IN.

Graham, D. H. [1970]: "Divergent and Convergent Regional Economic Growth and Internal Migration in Brazil, 1940-1960." *Econ. Dev. and Cultural Change* 18:362-382.

Graham, D. H., H. Gauthier, and J. R. Mendonça de Barros [1987]: "Thirty Years of Agricultural Growth in Brazil: Crop Performance, Regional Profile, and Recent Policy Review." *Econ. Dev. and Cultural Change* 36:1-34.

Gray, C. W. [1982]: *Food Consumption Parameters for Brazil and their Application to Food Policy.* Washington, IFPRI, Research Report 32.

Greenwood, M. J. [1975]: "Research on Internal Migration in the United States: A Survey." *J. Econ. Literature* 13:397-433.

Greenwood, M. J., and P. L. Stuart [1986]: "International Migration within the Pacific Basin: Characteristics, Causes, and Consequences." In Schuh and McCoy, eds., 1986, pp. 144-158.

Gregory, P. [1986]: *The Myth of Market Failure: Employment and the Labor Market in Mexico.* Baltimore, Johns Hopkins Univ. Press.

Griffin, K. [1974]: *The Political Economy of Agrarian Change: An Essay on the Green Revolution.* Cambridge, MA, Harvard Univ. Press.

_____ [1976]: *Land Concentration and Rural Poverty.* New York, Holmes and Meier.

Griliches, Z. [1957]: "Hybrid Corn: An Exploration in the Economics of Technological Change." *Econometrica* 25:231-252.

_____ [1958]: "Demand for Fertilizer: An Economic Interpretation of a Technical Change." *J. Farm Econ.* 40:591-606.

_____ [1963a]: "Estimates of Aggregate Agricultural Production Functions from Cross-Sectional Data." *J. Farm Econ.* 45:419-428.

_____ [1963b]: "The Sources of Measured Productivity Growth: United States Agriculture, 1940-60." *J. Polit. Economy* 71:331-346.

_____ [1964]: "Research Expenditures, Education, and the Aggregate Agricultural Production Functions." *Amer. Econ. Rev.* 54:961-974.

_____ [1970]: "Note on the Role of Education in Production Function and Growth Accounting." In W.L. Hansen, ed., 1970, pp. 71-127.

Grubel, H. B., and A. D. Scott [1966]: "The International Flow of Human Capital." *Amer. Econ. Rev., Papers and Proceedings* 56(2):268-274.

Grunwald, J. [1961]: "The 'Structuralist' School on Price Stabilization and Economic Development: The Chilean Case." In Hirschman, ed., 1961, pp. 95-123.

_____ [1964]: "Invisible Hands in Inflation and Growth." In Baer and Kerstenetzky, eds., 1964, pp. 290-318.

Grunwald, J., and P. Musgrove [1970]: *Natural Resources in Latin American Development.* Baltimore, Johns Hopkins Univ. Press.

Gudin, E. [1962]: "Inflation in Latin America." In Hague, ed., 1962, pp. 342-358.

Guimarães, A. P. [1963]: *Quatro Séculos de Latifúndio* (Four Centuries of Latifúndio). São Paulo, Editôra Fulgor.

Gurrieri, A., ed. [1982]: *La Obra de Prebisch en la CEPAL* (The Work of Prebisch in ECLA). México, D. F., Fondo de Cultura Econômica.

Gutierrez, A. N., and R. Hertford [1974]: *Una Evaluación de la Intervención del Gobierno en el Mercado de Arroz en Colombia* (An Evaluation of Government Intervention in the Rice Market in Colombia). Cali, Colombia, CIAT, Folleto Técnico No. 4.

Guzmán, G. [1976]: *El Desarrolo Latinoamericano y la CEPAL* (Latin American Develoment and CEPAL). Barcelona, España, Editorial Planeta.

Haberler, G. [1968]: "Terms of Trade and Economic Development." In Theberge, ed., 1968, pp.323-343. Also in Ellis and Wallich, eds., 1961, pp. 275-297.

Hagelberg, G. B. [1979, 1981]: "Cuba's Sugar Policy." In Weinstein, ed., 1979, pp. 31-50. Also in Horowitz, ed., Fourth Edition, 1981, pp. 141-162.

Hague, D. C., ed. [1962]: *Inflation: Proceedings of a Conference Held by the International Economic Association.* New York, St. Martin's Press.

Hall, L. L. [1980]: "Evaluating the Effects of P.L. 480 Wheat Imports on Brazil's Grain Sector." *Amer. J. Agr. Econ.* 62:19-28.

Hallagan, W. [1978]: "Self-Selection by Contractual Choice and the Theory of Sharecropping." *Bell J. Econ.* 9:344-354.

Haller, T. E. [1972]: "Education and Rural Development in Colombia." Unpublished Ph.D. dissertation, Purdue Univ., West Lafayette, IN.

Hancock, R. H. [1959]: *The Role of the Bracero in the Economic and Cultural Dynamics of Mexico.* Stanford, CA, Stanford Univ. Press.

Hansen, W. L., ed. [1970]: *Education, Income, and Human Capital.* New York, Columbia Univ. Press.

Hanson, J., and J. de Melo [1983]: "The Uruguayan Experience with Liberalization, 1974-1981." *J. Inter-Amer. Studies and World Affairs* 25:477-508.

―――― [1985]: "External Shocks, Financial Reforms and Stabilization Attempts in Uruguay during 1974-83." *World Dev.* 13:917-939.

Harberger, A. C. [1963]: "The Dynamics of Inflation in Chile." In Christ, ed., 1963, pp. 219-250.

―――― [1964]: "Some Notes on Inflation." In Baer and Kerstenetzky, eds., 1964, pp. 319-351.

―――― [1971]: "On Measuring the Social Opportunity Cost of Labor." *International Labor Rev.* (103):559-579.

―――― [1981]: "In Step and Out of Step with the World Inflation: A Summary History of Countries, 1952-1976." In Flanders and Razin, eds., 1981, pp. 35-46.

Harberger, A., and M. Selowsky [1966]: "Key Factors in the Economic Growth of Chile." Presented to the Conference at Cornell Univ. on *The Next Decade of Latin American Development,* mimeo.

Harris, J. R., and R. H. Sabot [1982]: "Urban Unemployment in LDCs: Towards a More General Search Model." In Sabot, ed., 1982a, pp. 65-89.

Harris, J. R., and M. P. Todaro [1970]: "Migration, Unemployment, and Development: A Two-Sector Analysis." *Amer. Econ. Rev.* 60:126-142.

Harrison, K., D. Henley, H. Riley, and J. Shaffer [1974]: *Improving Food Marketing Systems in Developing Countries: Experiences from Latin America.* East Lansing, MSU, LASC, Research Report No. 6.

Harvard Business School [1984]: *Colloquium on World Food Policy: 75th American Colloquium of the Harvard Business School.* Cambridge, MA, Harvard Univ.

Hathaway, D. E. [1964]: "Urban-Industrial Development and Income Differentials between Occupations." *J. Farm Econ.* 46:56-66.

Havens, A.E. [1965]: *Education in Rural Colombia: An Introduction on Human Resources.* Madison, Univ. of Wisconsin, LTC.

Havens, A. E., and W. L. Flinn, eds. [1970]: *Internal Colonialism and Structural Change in Colombia.* New York, Praeger.

Hayami, Y., and V.W. Ruttan [1971, 1985, 1989]: *Agricultural Development: An International Perspective.* Baltimore, Johns Hopkins Univ. Press. Also in Spanish as *Desarrollo Agrícola. Una Perspectiva Internacional.* Mexico, Fondo de Cultura Económica, 1989.

Hazell, P., C. Pomerada, and A. Valdés, eds. [1986]: *Crop Insurance for Agricultural Development.* Baltimore and London, Johns Hopkins.

Heady, E. O., and L. R. Whiting, eds. [1975]: *Externalities in the Transformation of Agriculture: Distribution of Benefits and Costs from Development.* Ames, ISU Press.

Heaton, J. E. [1969]: *The Agricultural Development of Venezuela.* New York, Praeger.

Heckman, J. J., and V. J. Hotz [1986]: "An Investigation of the Labor Market Earnings of Panamanian Males: Evaluating the Sources of Inequality." *Human Resources* 21:507-542.

Hemerly, F. X. [1975]: "Modêlo Econométrico dos Mercados Interno e de Exportação de Amendoim" (Econometric Model of Domestic and Foreign Markets of Peanuts). Unpublished M.S. thesis, Universidade Federal de Viçosa, Minas Gerais, Brazil.

Hertford, R. [1971]: *Sources of Change in Mexican Agricultural Production, 1940-65.* Washington, USDA, FAER No. 73.

Hertford, R., J. Ardila, A. Rocha, and C. Trujillo [1977]: "Productivity of Agricultural Research in Colombia." In Arndt, Dalrymple and Ruttan, eds., 1977, pp. 86-123.

Hewlett, S. A. [1970]: "Rate of Return Analysis: Its Role in Determining the Significance of Education in the Development of Brazil," mimeo.

Hibbard, B. H. [1924, 1965]: *A History of the Public Land Policies.* New York, Macmillan; Madison, Univ. of Wisconsin Press.

Hill, J. K. [1978]: "Imperfect Capital Markets and Life-Cycle Consumption." Unpublished Ph.D. dissertation, Rice Univ., Houston, TX.

Hines, J. [1972]: "The Utilization of Research for Development: Two Case Studies in Rural Modernization and Agriculture in Peru." Unpublished Ph.D. dissertation, Princeton Univ. Princeton, NJ.

Hirschman, A. O., ed. [1961, 1964]: *Latin American Issues: Essays and Comments.* New York, Twentieth Century Fund; Homewood, IL, Irwin.

———— [1963a]: *Journeys Toward Progress: Studies of Economic Policy-Making in Latin America.* New York. The Twentieth Century Fund.

———— [1963b]: "Brazil's Northeast." In Hirschman, 1963a, pp. 13-91.

———— [1968]: "The Political Economy of Import-Substituting Industrialization in Latin America." *Quart. J. Econ.* 82:1-31.

———— [1973]: "The Changing Tolerance for Income Inequality in the Course of Economic Development." *Quart. J. Econ.* 87:544-566.

———— [1977]: "A Generalized Linkage Approach to Development, with Special Reference to Staples." In Nash, ed., 1977, pp. 67-98. Also in Spanish in *El Trimestre Económico* 1977, 44:199-236.

Hoffman, R. [1984]: "A Pobreza no Brasil: Análise dos Dados dos Censos Demográficos de 1970 e 1980" (Poverty in Brazil: Analysis of Data from the Population Census of 1970 and 1980). *Proceedings, VI Encontro Brasileiro de Econometria,* São Paulo, 1984, pp. 175-213.

Hoffman, R., and A. A. Kageyama [1984]: "Distribuição da Renda no Brasil, entre Famílias e entre Pessoas, em 1970 e 1980" (Income Distribution in Brazil, among Families and among Individuals in 1970 and 1980). *Proceedings, XII Encontro Nacional de Economia*, Vol. III, São Paulo, 1984, pp. 799-834.

_____ [1985]: "Modernização da Agricultura e Distribuição de Renda no Brasil" (Modernization of Agriculture and Income Distribution in Brazil). *Pesquisa e Planejamento Econômico* 15:171-208.

Homen de Melo, F. B. [1979]: "A Política Econômica e o Setor Agrícola no Período Pós-Guerra" (Economic Policy and the Agricultural Sector in the Postwar Period). *Revista Brasileira de Economia* 33:25-63.

_____ [1980]: "Disponibilidade da Tecnologia entre Productos da Agricultura Brasileira" (Availability of Technology among Products of Brazilian Agriculture). *Revista de Economia Rural* 18(2):221-249.

_____ [1981]: "Política Comercial, Tecnologia, e Preços de Alimentos no Brasil" (Trade Policy, Technology, and the Price of Food in Brazil). *Estudos Econômicos* 11:123-142.

_____ [1982]: "A Contribuição da Agricultura: Alimentos Exportações e Energia." (The Contribution of Agriculture, Food, Exports, and Energy). *Revista de Economia Rural* 20(2):525-540.

_____ [1983a]: "Trade Policy Technology and Food Prices in Brazil." *Quart. Rev. Econ. and Business.* 23(1)58-78.

_____ [1983b]: "Export Agriculture and the Problem of Food Production." In *Brazilian Economic Studies* No. 7, pp. 1-19. Rio de Janeiro, IPEA/INPES.

_____ [1986]: *Brazil and the CGIAR Centers: A Study of their Collaboration in Agricultural Research.* Washington, World Bank, CGIAR Study Paper No. 9.

Homen de Melo, F. B., and E. R. Pelin [1984]: *As Soluções Energéticas e a Economia Brasileira* (Energy Solutions and the Brazilian Economy). São Paulo, HUCITEC.

Horowitz, I. L., ed. [1970]: *Masses in Latin America.* New York, Oxford Univ. Press.

_____ , ed. [1972]: *Cuban Communism*, second edition. New Brunswick, New Jersey, Transaction Books.

_____ , ed. [1981]: *Cuban Communism*, fourth edition. New Brunswick and London, Transaction Books.

Horton, D. [1977]: "Land Reform and Group Farming in Peru." In Dorner, ed. 1977, pp. 213-238.

Hough, J. F. [1981]: "The Evolving Soviet Debate on Latin America." *Latin Amer. Research Rev.* 16(1):124-143.

Howe, H. J. [1974]: "Estimation of the Linear and Quadratic Expenditure Systems: A Cross-Section Case for Colombia." Unpublished Ph.D. dissertation, Univ. of Pennsylvania, Philadelphia.

Howe, H. J., and P. Musgrove [1977]: "An Analysis of ECIEL Household Budget Data for Bogotá, Caracas, Guayaquil and Lima." In Lluch, Powell and Williams, eds., 1977, pp. 155-198.

Howell, J., ed. [1980]: *Borrowers and Lenders: Rural Financial Markets and Institutions in Developing Countries.* London, ODI.

Huberman, L., and P. M. Sweezy [1969]: *Socialism in Cuba.* New York and London. Monthly Review Press.

Huddleston, B., D. Gale Johnson, S. Reutlinger, and A. Valdés [1984]: *International Finance for Food Security.* Baltimore, Johns Hopkins Univ. Press.

Huddleston, B., and P. Konandreas [1981]: "Insurance Approach to Food Security: Simulation of Benefits for 1970/71-1975/76 and for 1978/82." In Valdés, ed., 1981, pp. 241-254.

Huffman, W. E. [1986]: "The U.S.-Mexican Labor Market." In Schuh and McCoy, eds., 1986, pp. 121-143.

Huffman, W. E., and R. Coltrane [1986]: *U.S.-Mexican Trade and Immigration.* Final Report of Research Agreement between ISU and USDA. Ames, ISU, mimeo.

Hunt, K. E., ed. [1971]: *Policies, Planning and Management for Agricultural Development. Papers and Reports, 14th ICAE, Minsk, August 23-September 2, 1970.* Oxford, England, Institute of Agrarian Affairs for IAAE.

Hyde, G. L. [1963]: "A Critique of the Prebisch Thesis." *Economia Internazionale* 16:463-487.

Iamaguchi, L. C. T. [1982]: "Matrizes de Elasticidade de Oferta: Uma Aplicação de Técnicas Modificadas de Regressão de Cume—Estado de São Paulo" (Supply Elasticity Matrix: An Application of Modified Ridge Regression). Unpublished M.S. thesis, Universidade Federal de Viçosa, Minas Gerais, Brazil.

Immink, M. D. C., and F. E. Viteri [1981]: "Energy Intake and Productivity of Guatemalan Sugarcane Cutters: An Empirical Test of the Efficiency Wage Hypothesis, Part I, Part II." *J. Dev. Econ.* 9:251-271 and 273-287.

Innis, H. A. [1927]: *The Fur Trade of Canada.* Toronto, Univ. of Toronto Library.

———— [1933]: *Problems of Staple Production in Canada.* Toronto, Ryerson Press.

———— [1940]: *The Cod Fisheries: The History of an International Economy.* New Haven, CT, Yale Univ. Press.

Instituto Interamericano de Estadística [1969]: *Boletin Estadístico de América Latina.* New York, United Nations.

Inter-American Development Bank (IADB) [1967]: *Agricultural Development in Latin America: The Next Decade.* Washington.

———— [IADB, 1978]: *The Financing of Education in Latin America.* Washington.

International Monetary Fund. [IMF, various issues]: *International Financial Statistics.*

International Political Economy Yearbook [1987]: *In Pursuing Food Security: Strategies and Obstacles in Africa, Asia, Latin America, and the Middle East.* Boulder, CO, Vol. 3.

International Service for National Agricultural Research [ISNAR, 1984]: *Selected Issues in Agricultural Research in Latin America.* The Hague, Netherlands.

Isaacmen, A., and T. Sunseri [no date]: *Third World Peasants and Agrarian Change in the Twentieth Century: A Select Bibliography.* Minneapolis, IIS, Univ. of MN, College of Liberal Arts.

Jabara, C. [1977]: "Demand for Education of Children among Small Farms in a Rural Region of Brazil." Unpublished M. Sc. thesis, Purdue Univ., West Lafayette, IN.

Jackson, S., B. Russett, D. Snidal, and D. Sylvan [1979]: "An Assessment of Empirical Research on Dependencia." *Latin Amer. Research Rev.* 14:(3):7-28.

Jaguaribe, H., A. Ferrer, M.S. Wionczek, and T. dos Santos [1970]: *La Dependencia Político-Económica de América Latina* (The Political-Economic Dependency of Latin America). México, Siglo Veintiuno Editores.

Jallade, J. P. [1973, 1974]: *Public Expenditure on Education and Income Distribution in Colombia.* Baltimore, Johns Hopkins Univ. Press. Also World Bank Staff Occasional Paper No. 18.

Jamison, D. T., and L. S. Lau [1982]: *Farmer Education and Farm Efficiency.* Baltimore, Johns Hopkins Univ. Press for the World Bank.

Jarvis, L. S. [1969]: "Supply Response in the Cattle Industry: The Argentine Case, 1937-38 and 1966-67." Unpublished Ph.D. dissertation, MIT, Cambridge, MA.

——— [1974]: "Cattle as Capital Goods and Ranchers as Portfolio Managers: An Application to the Argentine Cattle Sector." *J. Polit. Economy* 82:489-520.

——— [1985]: *Chilean Agriculture under Military Rule: From Reform to Reaction, 1973-1980.* Berkeley, Univ. of California, Institute of International Studies.

Jaynes, G. D. [1984]: "Economic Theory and Land Tenure." In Binswanger and Rosenzweig, eds., 1984, pp. 43-62.

Jeanneret, T. [1971]: "The Structure of Protection in Chile." In Balassa and associates, 1971, pp. 137-168, 349-362.

Jiménez, Sanchez L. [1970]: "The Puebla Project: A Regional Program for Rapidly Increasing Corn Yields among 50,000 Small Holders." In Myren, ed., 1970, pp. 11-18. Also as "El Plano Puebla, un Programa Regional para Aumentar los Rendimientos de Maíz entre Agricultores con Pequeñas Explotaciones." In Martinez, ed., 1970.

Johnson, A. W. [1971]: *Sharecroppers of the Sertão: Economics and Dependence on a Brazilian Plantation.* Stanford, CA, Stanford Univ. Press.

Johnson, D. Gale [1950]: "Resource Allocation under Share Contracts." *J. Polit. Economy* 58:111-123.

Johnson, D. Gale, and G. E. Schuh, eds. [1983]: *The Role of Markets in the World Food Economy.* Boulder, CO, Westview Press.

Johnson, Glenn L. [1985]: *Agricultural Economics—Dwindling Support and Expanding Opportunities: Theodore Brinkman Lecture.* Bonn, West Germany, Univ. of Bonn, mimeo.

Johnson, Glenn L., and A. Maunder, eds. [1981]: *Rural Change: The Challenge for Agricultural Economics. Proceedings, 17th ICAE.* Allanheld Osmun for the IAAE, Oxford, England, Institute of Agricultural Economics.

Johnson, Harry G. [1960]: "The Cost of Protection and the Scientific Tariff." *J. Polit. Economy* 68:327-345. Also in Harry G. Johnson, 1972, pp. 187-218.

——— [1964]: "Towards a Generalized Capital Accumulation Approach to Economic Development. Comments." In OECD, 1964, pp. 219-227.

——— [1967a]: "Some Economic Aspects of Braindrain." *Pakistan Dev. Rev.* 7:379-411.

——— [1967b]: *Economic Policies Toward Less Developed Countries.* Washington, Brookings Institution.

——— [1972]: *Aspects of the Theory of Tariffs.* Cambridge, MA, Harvard Univ. Press.

Johnston, B. F. [1970]: "Agriculture and Structural Transformation in Developing Countries: A Survey of Research." *J. Econ. Literature* 8:369-404.

——— [1977]: "Food, Health, and Population in Development." *J. Econ. Literature* 15:879-907.

Johnston, B. F., and W. C. Clark [1982]: *Redesigning Rural Development: A Strategic Perspective.* Baltimore, Johns Hopkins Univ. Press.

Johnston, B. F., C. Luiselli, C. C. Contreras, and R. D. Norton, eds. [1987]: *U.S.-Mexico Relations: Agriculture and Rural Development.* Stanford, CA, Stanford Univ. Press.

Jolly, R. [1964]: "Education." In Seers, ed., 1964, Part II, Chapters IV through VIII, pp. 161-280.

Jones, S., P. C. Joshi, and M. Murmis, eds. [1982]: *Rural Poverty and Agrarian Reform.* New Delhi, Allied Publishers.

Jorgenson, D. W. [1961]: "The Development of a Dual Economy." *Econ. J.* 71:309-334.

——— [1966]: "Testing Alternative Theories of the Development of a Dual Economy." In Adelman and Thorbecke, eds., 1966, pp. 45-66.

——— [1969]: "The Role of Agriculture in Economic Development: Classical *versus* Neo-classical Models of Growth." In Wharton, ed., 1969, pp. 320-342.

Judd, M. A., J. K. Boyce, and R. E. Evenson [1986]: "Investing in Agricultural Supply: The Determinants of Agricultural Research and Extension Investment." *Econ. Dev. and Cultural Change* 35:77-114.

Junta del Acuerdo de Cartagena [1975]: "Pacto Andino, Estrutura de Consumo y Distri-bución de Ingreso" (Andean Pact, Structure of Consumption and Income Distribution). Lima, Peru.

Kahil, R. [1963]: *Inflation and Economic Development in Brazil, 1946-1963*. Oxford, Oxford Univ. Press.

Katzman, M. T. [1974]: "The Von Thuenen Paradigm, the Industrial Urban Hypothesis, and the Spatial Structure of Agriculture." *Amer. J. Agr. Econ.* 56:683-696.

——— [1975a]: "The Brazilian Frontier in Comparative Perspective." *Comparative Studies in Society and History* 17:266-285.

——— [1975b]: "Regional Development Policy in Brazil: The Role of Growth Poles and Development Highways in Goiás." *Econ. Dev. and Cultural Change* 24:75-107.

——— [1977]: *Cities and Frontiers in Brazil: Regional Dimensions of Economic Development*. Cambridge, MA, Harvard Univ. Press.

——— [1987]: "Ecology, Natural Resources, and Economic Growth: Understanding the Amazon." *Econ. Dev. and Cultural Change* 35:425-436.

Kay, C. [1974]: "La Participación Campesina Bajo el Gobierno de la U.P." (Unidad Popular, Chile Peasant Participation under the government of the U.P.) *Revista Mexicana de So-ciología*, 36(2):279-295.

——— [1982]: "Achievements and Contradictions of the Peruvian Agrarian Reform." *J. Dev. Studies* 18:141-170.

Kendrick, J. W., ed. [1984]: *International Comparisons of Productivity and Causes of the Slow-down*. Cambridge, MA, Ballinger.

Kenen, P. B. [1965]: "Nature, Capital, and Trade." *J. Polit. Economy* 78:437-460.

——— [1968]: "Toward a More General Theory of Capital and Trade." In Kenen and Lawrence, eds., 1968, pp. 100-123.

——— [1970]: "Skills, Human Capital, and Comparative Advantage." In W.L. Hansen, ed., 1970, pp. 195-240.

——— [1971]: "Migration, the Terms of Trade, and Economic Welfare in the Source Country." In Bhagwati *et al.*, eds., 1971, pp. 238-260.

Kenen, P. B., and R. Lawrence, eds. [1968]: *The Open Economy: Essays on International Trade and Finance*. New York, Columbia Univ. Press.

Kindleberger, C. P., H. G. van der Tak, and U. Vanek [1956]: *The Terms of Trade: A Euro-pean Case Study*. New York, Technology Press of MIT and Wiley.

King, Roger [1977]: *Land Reform: A World Survey*. Boulder, CO, Westview Press.

King, T., ed. [1980]: *Education and Income*. Washington, World Bank, Staff Working Paper No. 402.

Knight, P. T. [1971]: *Brazilian Agricultural Technology and Trade: A Study of Five Commodities*. New York, Praeger.

Knudsen, O., and P. L. Scandizzo [1979]: *Nutrition and Food Needs in Developing Countries*. Washington, World Bank, Staff Working Paper No. 328.

Koester, U. [1982]: *Policy Options for the Grain Economy of the European Community: Implica-tions for Developing Countries*. Washington, IFPRI, Research Paper No. 35.

Krause, W., and F. J. Mathis [1968]: *The Latin American Common Market: Economic Disparity and Benefit Diffusion*. Atlanta, Georgia State College, School of Business Administration Studies in Business and Economics, Bulletin No. 15.

Krishna, R. [1982]: "Some Aspects of Agricultural Growth, Price Policy, and Equity in Developing Countries." *Food Research Institute Studies* 18:219-260.

Krueger, A. O. [1968]: "Factor Endowments and per Capita Income Differences among Countries." *Econ. J.* 78:641-659.

Krueger, A. O., C. Michalopoulos, V. W. Ruttan [1989]: *Aid and Development*. Baltimore and London, Johns Hopkins Univ. Press.

Krueger, A. O., and V. W. Ruttan [1983]: *The Development Impact of Economic Assistance to LDCs*, 2 vols. St. Paul, Univ. of Minnesota, Economic Development Center for AID. These two volumes were eventually published as Krueger, Michalopoulos, and Ruttan [1989].

Krueger, A. O., M. Schiff, and A. Valdés [1988]: "Agricultural Incentives in Developing Countries: Measuring the Effect of Sectoral and Economywide Policies." *The World Bank Econ. Rev.* 2(3):255-271.

Kugler, B., A. Reyes, and I. de Gomez [1979]: *Educación y Mercado de Trabajo Urbano en Colombia: Una Comparación entre Setores Moderno y no Moderno* (Education and the Urban Labor Market in Colombia: A Comparison between Modern and Traditional Sectors). Bogotá, Corporación Centro Regional de Población, Monograph No. 10.

Kutcher, G. P., and P. L. Scandizzo [1981]: *The Agricultural Economy of Northeast Brazil*. Baltimore, Johns Hopkins Univ. Press for the World Bank.

Kuznets, S. [1955]: "Economic Growth and Income Inequality." *Amer. Econ. Rev.* 45:1-28.

Labor Mobility and Population in Agriculture [1961]. Ames, ISU Press, CAEA.

Labys, W. C. [1973]: *Dynamic Commodity Models: Specification, Estimation and Simulation*. Lexington, MA, Heath.

———, ed. [1975]: *Quantitative Models of Commodity Markets*. Cambridge, MA, Ballinger.

Labys, W. C., I. Nadiri, and J. N. del Arco [1980]: *Commodity Markets and Latin American Development*. New York, NBER.

Labys, W. C., and P. K. Pollak [1984]: *Commodity Models for Forecasting and Policy Analysis*. New York, Nichols.

Ladman, J. R., and R. L. Tinnermeier [1981]: "The Political Economy of Agricultural Credit: The Case of Bolivia." *Amer. J. Agr. Econ.* 63:66-72.

Lambert, J. [1970]: *Os Dois Brasis* (The Two Brazils). São Paulo, Companhia Editôra Nacional.

Land Tenure Center [LTC, 1974]: *Agrarian Reform in Latin America: An Annotated Bibliography*. Madison, Univ. of Wisconsin.

Langoni, C. G. [1973]: *Distribuição de Renda e Desenvolvimento Econômico do Brasil* (Income Distribution and Economic Development of Brazil). Rio de Janeiro, Editôra Expressão e Cultura.

Lattimore, R. G. [1974]: "An Econometric Study of the Brazilian Beef Sector." Unpublished Ph.D. dissertation, Purdue Univ., West Lafayette, IN.

Lattimore, R., and G. E. Schuh [1976]: "Un Modelo de Política para la Industria Brasileña de Ganado Vacuno" (A Policy Model for the Brazilian Beef Sector). *Cuadernos de Economía* 13:51-75.

Le Baron, A., S. Daines, P. Aitken, R. Johnson, and A. Ely [1973a]: *Bibliografía Latino Americana de Producción de Desarrollo Agrícola. Vol. I. Insumos y Producción en Agricultura y Ganaderia* (Latin American Bibliography of Production Side of Agricultural Development. Vol. I. Inputs and Production in Agriculture and Livestock). Logan, Utah State Univ.

―――― [1973b]: *Bibliografía Latino Americana de Producción de Desarrollo Agrícola. Vol. II. Planificación Agrícola General, Estudios Estadísticas y de Medio* (Latin American Bibliography of the Production Side of Agricultural Development. Vol. II. General Agricultural Planning, Statistical Studies). Logan, Utah State Univ.

Le Baron, A., P. Aitken, R. Johnson, and A. Ely [1973c]: *Producción y Desarrollo Latino Americano. Catálogo Instituciones Nacionales de Investigación y Lista de Investigadores Americanos* (Latin American Production and Development. Catalogue, National Institutions of Research and List of American Researchers). Logan, Utah State Univ.

Leff, N. H. [1967]: "Export Stagnation and Autarkic Development." *Quart. J. Econ.* 81:286-301.

―――― [1968]: *Economic Policy-Making and Development in Brazil, 1947-1964.* New York, Wiley.

―――― [1969]: "The 'Exportable Surplus' Approach to Foreign Trade in Underdeveloped Countries." *Econ. Dev. and Cultural Change* 17:346-355.

Lehmann, D. [1972]: "Agriculture Chilena en el Periodo de Transición (Chilean Agriculture in a Period of Transition). *Sociedade y Desarrollo* 3:101-144.

―――― [1973]: "Agrarian Reform in Chile, 1965-72." Unpublished doctoral thesis, Oxford Univ., Oxford, England.

Leibenstein, H. [1957]: *Economic Backwardness and Economic Growth.* New York, Wiley.

Leite, C. A. M. [1975]: "Modêlo Econométrico dos Mercados Interno e de Exportação de Soja no Brasil" (Econometric Model of Domestic and Foreign Markets of Soybeans in Brazil). Unpublished M.S. thesis, Universidade Federal de Viçosa, Minas Gerais, Brasil.

Lekachman, R. [1961]: *National Policy for Economic Welfare at Home and Abroad.* New York, Russell of Russell Inc.

Lemgruber, A. C. [1974]: "Inflação: O Modêlo da Realimentação e o Modêlo da Aceleração" (Inflation: The Feedback Model and the Model of Acceleration). *Revista Brasileira de Economia* 28:(3)55-56.

Lerner, M. O. [1970]: "Determinants of Educational Attainment in Brazil, 1960." Unpublished Ph.D. dissertation, Univ. of California, Berkeley.

Levinson, J., and J. de Onís [1970]: *The Alliance That Lost Its Way: A Critical Report on the Alliance for Progress.* Chicago, Quadrangle Books.

Levy, S. [1970]: "An Economic Analysis of Investment in Education in the State of São Paulo." São Paulo, IPE, unpublished paper.

Lewis, W. A. [1954]: "Economic Development with Unlimited Supplies of Labour." *Manchester School Econ. and Social Studies* 22:139-191.

―――― [1965]: "Economic Development and World Trade." In Robinson, ed., 1965, pp. 483-497.

―――― [1984]: "Development Economics in the 1950s." In Meier and Seers, eds., 1984, pp. 121-137.

Lima, D. M. A., and J. H. Sanders [1975]: "Selecting and Evaluating New Technology for Small Farmers in the Central Sertão of Ceará." Paper presented at the *Conference on Growth, Productivity and Equity Issues in Brazilian Agriculture*, Columbus, Ohio State Univ.

Lipsey, R. E. [1963]: *Price and Quantity Trends in the Foreign Trade of the United States.* Princeton, NJ, Princeton Univ. Press for the NBER.

Lipton, M. [1982]: "Migration from Rural Areas of Poor Countries: The Impact on Rural Productivity and Income Distribution." In Sabot, ed., 1982a, pp. 191-228.

Lizano-Fait, E., ed. [1975a]: *La Integración Económica Centro-Americana* (The Economic Integration of Central America). México, Fondo de Cultura Económica, 2 vols.

———— [1975b]: "El Sector Agropecuário y la Integración Económica" (The Agricultural Sector and Economic Integration). In Lizano-Fait, ed., 1975a, Vol. I, pp. 253-271.

Lluch, C., A. A. Powell, and R. R. Williams, eds. [1977]: *Patterns in Household Demand and Savings.* New York, Oxford Univ. Press, for the World Bank.

Lobo, J. [1978]: "Change in Earnings Distribution in Mexico between 1960-1970: An Assessment of Human Capital Labor Market Segmentation and Government Income Policies." (Ph.D. Thesis that was not completed)

Lombardo, H. A. [1969]: *Research on Agricultural Development in Central America.* New York, ADC.

Lopes, I.G.V. [1977]: "Time Allocation of Low-Income Rural Brazilian Households: A Multiple Job Holding Model." Unpublished Ph.D. dissertation, Purdue Univ., West Lafayette, IN.

Lopes, I. G. V., and G. E. Schuh [1979]: *Alocação do Tempo de Famílias Rurais de Baixa Renda no Brasil: Um Modêlo de Engajamento em Empregos Múltiplos* (Time Allocation of Low-Income Rural Brazilian Households: A Multiple Job Holding Model). Brasilia, Ministério da Agricultura, CFP, Coleção Análise e Pesquisa, Vol. 17.

Lopes, M. de R. [1977]: "The Mobilization of Resources from Agriculture: A Policy Analysis for Brazil." Unpublished Ph.D. thesis, Purdue Univ. West Lafayette, IN.

———— [1983]: *Formação e Estabelização dos Preços Agrícolas: A Especulação nos Mercados Agrícolas e Formação da Renda do Produtor* (Formation and Stabilization of Agricultural Prices: Speculation in Agricultural Markets and the Formation of the Producer Income). Brasilia, Ministério da Agricultura, CFP, Coleção Análise e Pesquisa, Vol. 28.

———— [1986]: *A Intervenção do Governo nos Mercados Agrícolas no Brasil. O Sistema de Regras de Interferência no Mecanismo de Preços* (Government Intervention in the Agricultural Markets in Brazil. The system of Interference Rules in the Price Mechanisms.) Brasília Ministerio da Agricultura, CFP, Coleçao Análise e Presquisa Vol. 33.

Lopes, M. de R., and G. E. Schuh [1979]: *A Mobilização de Recursos da Agricultura: Uma Análise de Política para o Brasil* (The Mobilization of Resources from Agriculture: A Policy Analysis for Brazil). Brasília, Ministério da Agricultura, CFP, Coleção Análise e Pesquisa, Vol. 8.

Love, J. L. [1980]: "Raúl Prebisch and the Origins of the Doctrine of Unequal Exchange." *Latin Amer. Research Rev.* 15(3): 45-72.

———— [1990]: "The Origins of Dependency Analysis." *J. Latin Amer. Studies* 22:143-168.

Lubbert, J. [1980]: "Economic Integration and the Distribution of Income." In Robert Ferber, ed., 1980, pp. 387-416.

Lucas, R. E. B. [1979]: "Sharing, Monitoring and Incentives: Marshallian Misallocation Reassessed." *J. Polit. Economy* 87:501-521.

Luiselli, C., and A. Cruz-Serrano [1986]: "The SAM Approach to Food Security." In Schuh and McCoy, eds., 1986, pp. 56-70.

Lustig, N. [1988]: "Fiscal Cost and Welfare Effects of the Maize Subsidy in Mexico." In Pinstrup-Andersen, ed., 1988, pp. 277-288.

Lutz, E., and P. L. Scandizzo [1980]: "Price Distortions in Developing Countries: A Bias against Agriculture." *European Rev. Agr. Econ.* 7:5-27.

Lysy, F. J., and L. Taylor [1980]: "Income Distribution Simulations, 1959-71." In L. Taylor, Bacha, Cardoso, and Lysy, 1980, pp. 224-295.

MacEwan, A. [1981a]: *Revolution and Economic Development in Cuba*. New York, St. Martin's Press.

———— [1981b]: *Agricultural Development in Cuba*. New York, St. Martin's Press.

———— [1982]: "Revolution, Agrarian Reform, and Economic Transformation in Cuba." In Jones, Joshi, and Murmis, eds., 1982, pp. 162-182.

Machinea, J. L. [1983]: "Relative Price Behavior under Alternative Trade Liberalization Attempts." Unpublished Ph.D. dissertation, Univ. of Minnesota, Minneapolis.

Maffucci, E. A. [1969]: "Wheat and Corn Production Response to Price Incentives in the Pampean Region of Argentina, 1952-1956." Unpublished M.S. thesis, North Carolina State Univ., Raleigh.

Malan, P. S., and R. Bonelli [1977]: "The Brazilian Economy in the Seventies: Old and New Developments." *World Dev.* 5:19-45.

Malan, P.S. and J. Wells [1973]: "Distribuição da Renda e Desenvolvimento Econômico do Brasil" (Income Distribution and Economic Development of Brazil). *Pesquisa e Planejamento Econômico* 3:1103-1124. Also in Tolipan and Tinelli, eds., 1975, pp. 241-262.

Mandell, P. I. [1969]: "The Development of the Southern Goiás Brazilian Region: Development in a Land-Rich Economy." Unpublished Ph.D. dissertation, Columbia Univ., New York.

———— [1971, 1972]: "The Rise of Modern Rice Industry: Demand Expansion in a Dynamic Economy." *Food Research Institute Studies* 10:161-219. Also in Portuguese as "A Expansão da Moderna Rizicultura Brasileira: Crescimento da Oferta Numa Economia Dinâmica." *Revista Brasileira de Economia* 26(3):169-236, 1972.

Mantel, R. R., and A. M. Martirena-Mantel [1980]: "Economic Integration, Income Distribution and Consumption: A New Rationale for Economic Integration." In Robert Ferber, ed., 1980, pp. 349-386.

Margolis, M. L. [1973]: *The Moving Frontier: Social and Economic Change in a Southern Brazilian Community*. Gainesville, Univ. of Florida Press, Latin American Monographs, 2d series, No. 11.

Marin, A., and G. Psacharopoulos [1976]: "Schooling and Income Distribution." *Rev. Econ. and Stat.* 58:332-338.

Martin, L. R., ed. [1977]: *A Survey of Agricultural Economics Literature, Vol. I. Traditional Fields of Agricultural Economics, 1940s to 1970s*. Minneapolis, Univ. of Minnesota Press for the AAEA.

Martin, M. A. [1976]: "The Modernization of Brazilian Agriculture: An Analysis of Unbalanced Development." Unpublished Ph.D. thesis, Purdue Univ., West Lafayette, IN.

Martinez, G., ed. [1970]: *Estrategias para Aumentar la Productividad Agrícola en Zonas de Minifúndio* (Strategies for Increasing Agricultural Productivity on Small Holdings). El Baton, Mexico, CIMMYT.

Mascolo, J. L. [1979, 1980]: "Um Estudo Econométrico da Pecuária de Corte no Brasil." (An Econometric Study of Meat Animals in Brazil). Unpublished Ph.D. thesis, EPGE, FGV. Also in 1979 *Revista Brasileira de Economia* 33:65-105.

Mata, L. J., J. J. Urrutia, and B. Garcia [1975]: "Malnutrition and Infection in a Rural Village of Guatemala." In Chavez, Bourges and Busta, eds., 1975, Vol. 2, pp. 175-192.

Matsunaga, M. H. [1983]: "Currency Devaluation: Its Effect on the Corn Sector in Brazil." Unpublished M.S. thesis, Univ. of Minnesota, St. Paul.

Matus-Gardea, J. A. M. [1981]: "Trade Policy and Some Aspects of the Distribution of Income in Agriculture: Mexico," Unpublished Ph.D. thesis, Purdue Univ., West Lafayette, IN.

Maunder, A., and U. Renborg, eds. [1986]: *Agriculture in a Turbulent World Economy: Proceedings of the Nineteenth International Conference of Agricultural Economists in Málaga, Spain, 1985*, Gower.

Mazumdar, D. [1959]: "The Marginal Productivity Theory of Wages and Disguised Unemployment." *Rev. Econ. Studies* 26:190-197.

—— [1975]: "The Theory of Share-Cropping with Labour Market Dualism." *Economica* 42:261-271.

McCarthy, F. D. [1975]: *Planejamento Nutricional para o Brasil: Um Programa Multidisciplinário Orientado a Política* (Nutritional Planning for Brazil: A Multidisciplinary Program Oriented to Policy). Brasilia, Ministério de Agricultura, SUPLAN.

McGrath, M. G. [1973]: "Ariel or Caliban?" *Foreign Affairs* 52:75-95.

McGreevey, W. P., ed. [1980]: *Third World Poverty: New Strategies for Measuring Development Progress*. Lexington, MA, Heath.

McIntire, J. [1983]: "International Farm Prices and the Social Cost of Cheap Food Policies: Comment." *Amer. J. Agr. Econ.* 65:823-826.

McMullen, N. [1982]: *The Newly Industrializing Countries: Adjusting to Success*. Washington, British-North American Committee.

Meier, G. M., and D. Seers, eds. [1984]: *Pioneers in Development*. Oxford, England, Oxford Univ. Press.

Meiselman, D., ed. [1970]: *Varieties of Monetary Experience*. Univ. of Chicago Press.

Mellor, J. W. [1966]: *The Economics of Agricultural Development*. Ithaca, NY, Cornell Univ. Press.

—— [1976]: *The New Economics of Growth: A Strategy for India and the Developing World*. Ithaca, NY, Cornell Univ. Press.

—— [1978]: "Food Price Policy and Income Distribution in Low-Income Countries." *Econ. Dev. and Cultural Change* 27: 1-26.

Mellor, J. W., and B. F. Johnston [1967]: "Towards a Theory of Agricultural Development." In Southworth and Johnston, eds., 1967, pp. 21-60; plus "Comment," by T. W. Schultz, pp. 61-65.

Mendonça de Barros, J. R. [1974]: *Desenvolvimento da Agricultura e Exportações de Produtos Primários Não Tradicionais* (Agricultural Development and Exportation of Non-Traditional Primary Products). São Paulo, Séries IPE, Monograph Vol. 4.

Mendonça de Barros, J. R., L. R. Ferreira, C. T. Yamaguishi, L. Moricochi, and G. Toscano [1983]: "Agricultura e Produção de Energia: Avaliação do Custo de Matéria-Prima para a Produção de Álcool" (Agriculture and the Production of Energy: Evaluation of the Cost of Raw Materials for the Production of Alcohol). *Revista de Economia Rural* 21:439-469.

Mendonça de Barros, J. R., and D. H. Graham [1978]: "A Agricultura Brasileira e o Problema da Produção de Alimentos" (Brazilian Agriculture and the Problem of Food Production). *Pesquisa e Planejamento Econômico* 8:695-725.

Mendonça de Barros, J. R., H. D. Lobato, M. A. Travolo, and M. H. G. P. Zockun [1975]: "Sistema Fiscal e Incentivos as Exportações" (Fiscal System and Export Incentives). *Revista Brasileira de Economia* 29:3-23.

Merrill, W. C. [1967]: "Setting the Price of Peruvian Rice." *J. Farm Econ.* 49:389-402.

Mesa-Lago, C., ed. [1971]: *Revolutionary Change in Cuba*. Pittsburgh, Univ. of Pittsburgh Press.

—— [1972]: "The Labor Force Employment, Unemployment, and Underemployment in Cuba: 1899-1970," *Sage Professional Papers in International Studies*, Volume 1, Part 3. Beverly Hills, CA, Sage Publications.

———— [1978, 1979]: *Cuba in the 1970s: Pragmatism and Institutionalization*. Albuquerque, Univ. of New Mexico Press, (Revised edition of 1974 version). Also in Spanish as *Dialética de la Revolución Cubana: Del Idealismo Carismático al Pragmatismo Institucionalista*. Madrid, Spain, Editorial Playor, 1979.

———— [1981, 1983]: *The Economy of Socialist Cuba: A Two-Decade Appraisal*. Albuquerque, Univ. of New Mexico Press. Also in Spanish as *La Economía en Cuba Socialista: Una Evaluación de dos décadas*. Madrid, Editorial Playor, 1983.

Meyer, A. C. [1980]: "Patterns of Consumption in Latin America." In Robert Ferber, ed., 1980, pp. 171-200. Also in Spanish as "Comparaciones Internacionales de Patrones de Consumo." *Ensayos ECIEL-I*, 1974.

Meyer, R. L., D. W. Adams, N. Rask, and P. F. C. de Araujo [1973]: "Rural Capital Markets and Small Farmers in Brazil, 1960-1972." In AID, 1973, Vol. III, pp. 1-57.

Mikesell, N. [1960]: "Comparative Studies in Frontier History." *Annals Amer. Association of Geographers* 50:62-74.

Mintz, S. W. [1964]: *Sugar and Society in the Caribbean* (Foreword to Ramiro Guerria y Sánchez). New Haven, Yale Univ. Press.

———— [1966]: "The Industrialization of Sugar Production and its Relationship to Social and Economic Change." In R. F. Smith, ed., 1966, pp. 176-186.

Mirrlees, J. A. [1975]: "A Pure Theory of Underdeveloped Economies." In L. G. Reynolds, ed., 1975, pp. 84-106.

Mitchell, W. C. [1949]: *Lecture Notes on Types of Economic Theory*, 2 vols. New York, A. M. Kelley.

Mitra, P. K. [1983]: "A Theory of Interlinked Rural Transactions." *J. Public Econ.* 20:167-191.

Mohan, R. [1986]: *Work, Wages, and Welfare in a Developing Metropolis: Consequences of Growth in Bogotá, Colombia*. New York, Oxford Univ. Press for the World Bank.

Mohan, R., J. Garcia, and M. W. Wagner [1981]: *Measuring Urban Malnutrition and Poverty: A Case Study of Bogotá and Cali, Colombia*. Washington, World Bank, Staff Working Paper No. 447.

Montgomery, J. D., ed. [1984]: *International Dimensions of Land Reform*. Boulder, CO, Westview Press.

Moran, E. [1981]: *Developing the Amazon*. Bloomington, Indiana Univ. Press.

Morawetz, D. [1974]: *The Andean Group: A Case Study in Economic Integration among Developing Countries*. Cambridge, MA, MIT Press, p. 171.

Morley, S. A. [1971]: "Inflation and Stagnation in Brazil." *Econ. Dev. and Cultural Change* 19:184-203.

Morley S. A., and J. G. Williamson [1974]: "Demand, Distribution, and Employment: The Case of Brazil." *Econ. Dev. and Cultural Change* 23:33-60.

———— [1975]: *Growth, Wage Policy and Inequality: Brazil in the Sixties*. Madison, Univ. of Wisconsin, Social Systems Research Institute, Workshop Series No. 7519.

Mosher, A. T. [1955]: *Technical Cooperation in Latin America: Case Study of the Agricultural Program of ACAR in Brazil*. Washington, National Planning Association.

Moura, H. [1968]: *O Consumo Alimentar no Nordeste Urbano* (Food Consumption in the Urban Northeast). Fortaleza, Ceará, Banco do Nordeste do Brasil, Departmento de Estudos Econômicos do Nordeste (ETENE).

Mueller, C. C. [1980]: "Frontier Based Agricultural Expansion: The Case of Rondônia." In *Land, People, and Planning in Contemporary Amazônia*, ed. F. B. Scazzocchia. Cambridge, England, Cambridge Univ. Press.

Mundlak, Y. [1988]: *The Aggregate Agricultural Supply*. Washington, IFPRI, working paper no. 8511.

Muñoz, H. [1981]: "The Strategic Dependency of the Centers and the Economic Importance of the Latin American Periphery." *Latin Amer. Research Rev.* 16(3):3-30.

Musgrove, P. [1974]: "Determination and Distribution of Permanent Household Income in Urban South America." Unpublished Ph.D. dissertation, MIT, Cambridge, MA.

_____ [1975, 1977b, 1980b]: "Permanent Household Income and Consumption in Urban South America." *Amer. Econ. Rev.* 69:355-368, 1977b. Also as "The Distribution of Long Term Income in Urban South America," 1980b, in Robert Ferber, ed., 1980, pp. 117-148, and published with some changes in translation as "El Ingreso y el Consumo Permanente de las Familias Urbanas." *Ensayos ECIEL* No. 2, 1975.

_____ [1977a]: "The Structure of Household Spending in South American Cities: Indexes of Dissimilarity and Causes of Inter-City Differences." *Rev. Income and Wealth* 23:365-384.

_____ [1977c, 1980d]: "Household Size and Composition, Employment and Poverty in Urban Latin America." *Econ. Dev. and Cultural Change* 28:249-266, 1980d. Also published 1977c in Spanish as "Tamaño y Composición del Hogar, Ocupación y Pobreza en América Latina Urbana." *Estudios de Economía* 10(2).

_____ [1978a]: "Determinants of Urban Household Consumption in Latin America: A Summary of Evidence from the ECIEL Surveys." *Econ. Dev. and Cultural Change* 26:441-465.

_____ [1978b, 1980a]: *Consumer Behavior in Latin America*. Washington, Brookings Institution. Also published in Spanish, 1980a, as *Ingreso y Consumo Familiar Urbano*. Washington, OAS for ECIEL Program.

_____ [1980d]: "Food Needs and Absolute Poverty in Urban South America." Washington, ECIEL.

_____ [1982]: *The ECIEL Study of Household Income and Consumption in Urban Latin America: An Analytical History*. Washington, World Bank, LSMS Working Papers No. 12.

Muth, J. F. [1961]: "Rational Expectations and the Theory of Price Movements." *Econometrica* 29:315-335.

Myers, C. A. [1965]: *Education and National Development in Mexico*. Princeton, NJ, Princeton Univ., Industrial Relations Section.

Myint, H. [1958]: "The Classical Theory of International Trade and the Underdeveloped Countries." *Econ. J.* 68:317-337.

Myrdal, G. [1961]: "Toward a More Closely Integrated Free-World Economy." In Lekachman, 1961, pp. 235-263.

Myren, D. T. [1969]: "The Rockefeller Foundation Program in Corn and Wheat in Mexico." In Wharton, Jr., ed., 1969, pp. 438-452.

_____ , ed. [1970]: *Strategies for Increasing Agricultural Production on Small Holdings*. International Conference. Puebla, México, CIMMYT.

Nakajima, C. [1969]: "Subsistence and Commercial Family Farms: Some Theoretical Models of Subjective Equilibrium." In Wharton, ed., 1969, pp. 165-185.

Nash, M., ed. [1977]: *Essays on Economic Development and Cultural Change in Honor of Bert F. Hoselitz*. Univ. of Chicago Press.

Nelson, M. [1973]: *Tropical Lands: Policy Issues in Latin America*. Baltimore, Johns Hopkins Univ. Press.

Nelson, R. R., T. P. Schultz, and R. L. Slighton [1971]: *Structural Change in a Developing Economy: Colombia's Problems and Prospects*. Princeton, NJ, Princeton Univ. Press.

Nerlove, M. [1956]: "Estimates of the Elasticities of Supply of Selected Agricultural Commodities." *J. Farm Econ.* 38:496-509.

_____ [1958]: *The Dynamics of Supply: Estimation of Farmers' Response to Price.* Baltimore, Johns Hopkins Press.

_____ [1974]: "Household and Economy: Towards a New Theory of Population and Economic Growth." *J. Polit. Economy* 82(2, Part II):S200-S218.

_____ [1979]: "The Dynamics of Supply: Retrospect and Prospect." *Amer. J. Agr. Econ.* 61:874-888.

Nerlove, M., D. M. Grether, and J. L. Carvalho [1979]: *Analysis of Economic Time Series: A Synthesis.* New York, Academic Press.

Newbery, D. M. G. [1975]: "The Choice of Rental Contract in Peasant Agriculture." In L. G. Reynolds, ed., 1975, pp. 109-137.

Newbery, D. M. G., and J. E. Stiglitz [1979]: "Sharecropping, Risk Sharing, and the Importance of Imperfect Information." In Roumasset, Boussard and I. J. Singh, eds., 1979. pp. 311-339.

Nicholls, W. H. [1961]: "Industrialization, Factor Markets, and Agricultural Development." *J. Polit. Economy* 69:319-340.

_____ [1963a]: "Perspectiva Estatística da Estrutura Agrária do Brasil" (Statistical Perspective on the Agrarian Structure of Brazil). *Revista Brasileira de Economia* 17(2): 5-32.

_____ [1963b]: "An 'Agricultural Surplus' as a Factor in Economic Development." *J. Polit. Economy* 71:1-29.

_____ [1969a]: "The Transformation of Agriculture in a Semi-Industrialized Country: The Case of Brazil." In Thorbecke, ed., 1969, pp. 311-378.

_____ [1969b]: "The Changing Structure of Farm Product and Input Markets in Brazil." In Anschel, Brannon, and E. D. Smith, eds., 1969, pp. 63-78.

_____ [1969c]: "The Agricultural Frontier in Modern Brazilian History: The Case of Paraná, 1920-65." In Rippy, ed., 1969, pp. 36-64.

_____ [1972]: "The Brazilian Agricultural Economy: Recent Performance and Policy." In Roett, ed., 1972, pp. 147-184.

_____ [1973]: "Paiva e o Dualismo Tecnólogico na Agricultura: Um Comentário" (Paiva and Technologic Dualism in Agriculture: A Comment). *Pesquisa e Planejamento Econômico* 3(1):15-50.

Nicholls, W. H., and R. M. Paiva [1963/1973]: *Ninety-nine Fazendas: Structure and Productivity of Brazilian Agriculture.* Nashville, TN, Vanderbilt Univ. Center for Latin American Studies, 1965-1967, Vols. 2 (Caxias), 3 (Crato), 4 (Caruarú), 5 (The Triangle of Minas Gerais: Ituiutaba), 6 (The North of Paraná: Maringá), 7 (The Middle Paraíba Valley of São Paulo: Taubaté).

_____ [1965]: "The Structure and Productivity of Brazilian Agriculture." *J. Farm Econ.* 47: 347-361. Also in Portuguese in *Revista Brasileira de Economia* 19(2):5-27 and Annexes, 1965.

Nisbet, C. [1967]: "Interest Rates and Imperfect Competition in the Informal Credit Market of Rural Chile." *Econ. Dev. and Cultural Change* 16:73-90.

_____ [1969]: "The Relationship between Institutional and Informal Credit Markets in Rural Chile." *Land Econ.* 45:162-173.

Nogues, J. J. [1986]: *Peru's Trade Liberalization Policies of the 80s: A Historical Perspective.* Washington, World Bank, Report No. DRD 168.

Nores, G. A. [1972]: "Quarterly Structure of the Argentine Beef Cattle Economy—A Short-Run Model, 1960-70." Unpublished Ph.D. thesis, Purdue Univ., West Lafayette, IN.

North, D. C. [1955]: "Location Theory and Regional Economic Growth." *J. Polit. Economy* 63:243-258.

Norton, R. D., and L. Solís M., eds. [1983]: *The Book of CHAC: Programming Studies for Mexican Agriculture.* Baltimore, Johns Hopkins Univ. Press for the World Bank.

"Nutrición y Salud [1985]: Evaluaciones Socioeconómicos" (Nutrition and Health: Socioeconomic Evaluations). *Cuadernos de Economía* 66, Santiago, Special Number.

O'Connor, J. [1970]: *The Origins of Socialism in Cuba.* Ithaca and London, Cornell Univ. Press.

Oehmke, J. F. [1986]: "Persistent Underinvestment in Public Agricultural Research." *Agr. Econ.* 1:52-65.

Oliveira, A. J. [1977]: "Derived Demand for Agricultural Credit—A Multiperiod Investment Model." Unpublished Ph.D. dissertation, Purdue Univ., West Lafayette, IN.

Oliveira, J. do C. [1984]: "Incidência da Taxação Implícita Sôbre Produtos Agrícolas no Brasil: 1950-1974" (Incidence of Implicit Taxation on Agricultural Products in Brazil, 1950-1974). *Pesquisa e Planejamento Econômico* 14(2):399-452.

Olivera, J. H. G. [1964]: "On Structural Inflation and Latin American Structuralism." *Oxford Econ. Papers* NS 16:321-333.

――― [1970]: "On Passive Money." *J. Polit. Economy* 28:805-814.

O'Mara, G. T. [1983]: "The Microeconomics of Technique Adoption by Smallholding Mexican Farmers." In Norton and Solís M., eds., 1983, pp. 250-289.

Orden, D., D. Greene, T. Roe, and G. E. Schuh [1982]: *Policies Affecting the Food and Agricultural Sector in Peru, 1970-1982: An Evaluation and Recommendation.* Washington, AID.

Organization for Economic Co-operation and Development [OECD, 1964]: *The Residual Factor and Economic Growth.* Paris, Study Group in Economics Education.

Ortiz, F. [1947]: *Cuban Counterpoint: Tobacco and Sugar.* New York, Alfred A. Knopf, Inc.

――― [1966]: "Tobacco and Sugar: The Blending Which Produced the Culture of Cuba." In R. F. Smith, ed., 1966, pp. 168-175.

Oshima, H. T. [1967]: "Food Consumption, Nutrition, and Economic Development in Asian Countries." *Econ. Dev. and Cultural Change* 15:385-397.

Pachico, D. H., and J. A. Ashby [1976]: *Investments in Human Capital and Farm Productivity: Some Evidence from Brazil.* Ithaca, NY, Cornell Univ.

Paiva, R. M. [1971, 1975]: "Modernização e Dualismo Tecnológico na Agricultura" (Modernization and Technological Dualism in Agriculture). *Pesquisa e Planejamento Econômico* 1(2):171-234. Also in English in *Brazilian Economic Studies*, IPEA/INPES, Rio de Janeiro, 1975.

――― [1973]: "Modernização e Dualismo Tecnológico na Agricultura: Resposta aos Comentários dos Professores Nicholls e Schuh" (Modernization and Technological Dualism in Agriculture: Response to the Comments of Profs. Nicholls and Schuh). *Pesquisa e Planejamento Econômico* 3(1):95-116.

――― [1975]: "Modernização e Dualismo Tecnológico na Agricultura: Uma Reformulação" (Modernization and Technological Dualism in Agriculture: A Reformulation). *Pesquisa e Planejamento Econômico* 5(1):117-161.

――― [1979]: *A Agricultura no Desenvolvimento Econômico: Suas Limitações como Fator Dinâmico* (Agriculture in Economic Development: Its Limitations as a Dynamic Factor). Rio de Janeiro, Ministério da Agricultura, IPEA/INPES.

――― [1988]: Personnal Correspondence.

Paiva, R. M., and W. H. Nicholls [1965]: "Estágio do Desenvolvimento Técnico da Agricultural Brasileira" (The Stage of Technical Development of Brazilian Agriculture). *Revista Brasileira de Economia* 19(3):27-63 and annexes.

Paniago, E. [1969]: "An Evaluation of Agricultural Price Policies for Selected Food Products: Brazil." Unpublished Ph.D. dissertation, Purdue Univ., West Lafayette, IN.

Pant, C. [1981]: *Tenancy in Semi-Arid Tropical Villages of South India: Determinants and Effects on Cropping Patterns and Input Use.* Hyderabad, India, ICRISAT, Progress Report No. 20.

_____ [1983]: "Tenancy and Family Resources: A Model and Some Empirical Analysis." *J. Dev. Econ.* 12:27-40.

Pardo, J. R. S. [1984]: "An Economic Evaluation of Government Intervention in the Mexican Agriculture Sector: The Corn and Wheat Sector." Unpublished Ph.D. thesis, Univ. of Minnesota, St. Paul.

Parsons, K. H., R. J. Penn, and P. M. Raup, eds. [1956]: *Land Tenure: Proceedings of the International Conference on Land Tenure and Related Problems in World Agriculture.* Madison, Univ. of Wisconsin Press.

Pastore, A. C. [1973a]: "Observações Sôbre a Política Monetária no Programa Brasileiro de Estabelização" (Observations on Monetary Policy in the Brazilian Stabilization Program). Unpublished "Livre Docência" thesis, Univ. of São Paulo.

_____ [1973b]: *A Resposta da Produção Agrícola aos Preços no Brasil* (The Response of Agricultural Production to Prices in Brazil). São Paulo, APEC Editôra.

_____ [1974]: "Aspectos da Política Monetária Recente no Brasil" (Aspects of Recent Monetary Policy in Brazil). Univ. of São Paulo, IPE.

_____ [1979]: "Exportações Agrícolas e Desenvolvimento Econômico" (Agricultural Exports and Economic Development). In Veiga, ed., 1979, pp. 207-231.

Pastore, A. C., and R. D. Almonacid [1975]: "Gradualismo ou Tratamento de Choque: Considerações Sôbre os Custos da Estabilização" (Gradualism vs. Shock Treatment: Considerations on the Costs of Stabilization). *Pesquisa e Planejamento Econômico* 5(2):332-384.

Pastore, A. C., E. R. de A. Alves, and J. B. Rizzieri [1976]: "Inovação Induzida e os Limites a Modernização na Agricultura Brasileira" (Induced Innovation and the Limits of Modernization in Brazilian Agriculture). *Rev. de Economia Rural* 14(1):257-285.

Pastore, J., and E. R. de A. Alves [1977]: "Reforming the Brazilian Agricultural Research System." In Arndt, Dalrymple and Ruttan, eds., 1977, pp. 394-403.

Pastore, J., G. L. de S. Dias, M. C. de Castro [1976]: "Condicionantes da Productividade da Pesquisa Agrícola no Brasil" (Conditioners of the Productivity of Agricultural Research in Brazil). *Estudos Econômicos* 6(3):147-181.

Patrick, G. F. [1975]: "Fontes de Crescimento na Agricultura Brasileira: O Setor de Culturas" (Sources of Growth in Brazilian Agriculture: The Crops Sector). In Contador, ed., 1975, pp. 89-110.

Patrick, G. F., L. J. Brainard, and F. W. Obermiller, eds. [1975]: *Small Farm Agriculture: Studies in Developing Nations.* West Lafayette, IN, Purdue Univ., Station Bulletin No. 101.

Patrick, G. F., and K. L. Graber [1977]: "Income Generation among Small Farmer Households in Brazil." *J. Developing Areas* 11:465-478.

Patrick, G. F., and E. W. Kehrberg [1973]: "Costs and Returns of Education in Five Agricultural Areas of Eastern Brazil." *American J. Agr. Econ.* 55:145-153.

Patrick, H. T. [1966]: "Financial Development and Economic Growth in Underdeveloped Countries." *Econ. Dev. and Cultural Change* 14:174-189.

Paz-Silva, L. J. [1986]: *Peru and the CGIAR Centers: A Study of their Collaboration in Agricultural Research.* Washington, World Bank, CGIAR Study Paper No. 12.

Pearse, A. [1975]: *The Latin American Peasant.* London, Ed Frank Cass.

Peek, P., and G. Standing, eds. [1982]: *State Policies and Migration: Studies in Latin America and the Caribbean*. London, Croom Helm for ILO.

Peres, F. C. [1976]: "Derived Demand for Credit under Conditions of Risk." Unpublished Ph.D. thesis, Ohio State Univ., Columbus.

Pérez-López, J. F. [1987]: *Measuring Cuban Economic Performance*. Austin, Univ. of Texas Press.

Perloff, H. S. [1969]: *Alliance for Progress: A Social Invention in the Making*. Baltimore and London, Johns Hopkins Press.

Personnel of the Integrated Program of Agricultural Marketing [PIMA, 1973]: *Fomenting Improvements in Food Marketing in Costa Rica*. East Lansing, MSU, LASC.

Peterson, W. L. [1979]: "International Farm Prices and the Social Cost of Cheap Food Policies." *Amer. J. Agr. Econ.* 61:12-21.

——— [1983]: "International Farm Prices and the Social Cost of Cheap Food Policies: Reply." *Amer. J. Agr. Econ.* 65:827-828.

Petras, J., and M. Zeithin, eds. [1968]: *Latin America: Reform or Revolution? A Reader*. Greenwich, CT, Fawcett.

Pfefferman, G., and R. Webb [1983]: "Poverty and Income Distribution in Brazil." *Rev. Income and Wealth* 29(2):101-124. Also in World Bank Reprint Series No. 259, and published in Portuguese as "Pobreza e Distribuição de Renda no Brasil: 1960-1980." *Revista Brasileira de Economia* 37:147-175, 1983.

Piñeiro, M. E., R. Fiorentino, E. J. Trigo, A. Balcázar, and A. Martínez [1982]: *Articulación Social y Cambio Técnico: La Produción de Azúcar en Colombia*" (Social Articulation and Technical Change: The Production of Sugar in Colombia). San José, Costa Rica, IICA.

——— [1983d]: "Social Relations of Production, Conflict, and Technical Change: The Case of Sugar Production in Colombia." In Piñeiro and Trigo, eds., 1983a, pp. 47-69.

Piñeiro, M. E., and E. J. Trigo, eds. [1983a]: *Technical Change and Social Conflict in Agriculture: Latin American Perspectives*. Boulder, CO, Westview Press.

———, eds. [1983b]: *Procesos Sociales e Innovación Tecnología en la Agricultura de América Latina* (Social Processes and Technological Innovation in Latin American Agriculture). San José, Costa Rica, IICA.

——— [1983c]: *Cambio Técnico en el Agro Latino Americano: Situación y Perspectivas en la Década de 1980* (Technical Exchange in Latin American Agriculture: Situation and Perspectives of the 1980s). San José, Costa Rica, IICA.

——— [1985]: "Latin American Agricultural Research—The Public Sector: Problems and Perspectives." The Hague, Netherlands, ISNAR, Working Paper No. 1.

——— [1983]: "The Changing Institutional Nature of Technology Diffusion in Latin America: Policy Implications." In Schuh and McCoy, eds., 1986, pp. 223-252.

Piñeiro, M. E., E. J. Trigo, and R. Fiorentino [1983]: "Technical Change in Latin American Agriculture: A Conceptual Framework for its Interpretation." In Piñeiro and Trigo, eds., 1983a, pp. 37-44.

Pinstrup-Andersen, P. [1977]: "Decision Making on Food and Agricultural Research Policy: The Distribution of Benefits from New Agricultural Technology among Consumer Income Strata." *Agricultural Administration* 4(1): 13-28.

——— [1982]: *Agricultural Research and Technology in Economic Development*. London, Longman, 1982.

——— [1983]: *Export Crop Production and Malnutrition*. Chapel Hill, Univ. of North Carolina, Institute of Nutrition, Occasional Paper, Vol. II, No. 10.

_____ [1984]: "Food Subsidies: The Concern to Provide Consumer Welfare While Assuring Producer Incentives." Paper prepared for IFPRI Workshop on Food and Agricultural Policy, Elkridge, Maryland, April 29-May 2.

_____ [1985]: "Food Prices and the Poor in Developing Countries." *European Rev. Agr. Econ.* 12 (1-2): 69-81.

_____, ed. [1988]: *Food Subsidies in Developing Countries: Costs, Benefits and Policy Options.* Baltimore, Johns Hopkins Univ. Press.

Pinstrup-Andersen, P., A. D. Berg, and M. Forman, eds. [1984]: *International Agricultural Research and Human Nutrition.* Washington, IFPRI.

Pinstrup-Andersen, P., and F. C. Byrnes, eds. [1975]: *Methods for Allocating Resources in Applied Agricultural Research in Latin America.* Cali, Colombia, CIAT.

Pinstrup-Andersen, P., and E. Calcedo [1978]: "The Potential Impact of Changes in Income Distribution of Food Demand and Human Nutrition." *Amer. J. Agr. Econ.* 60: 402-415.

Pinstrup-Andersen, P., N. R. de Londoño, and E. Hoover [1976]: 'The Impact of Increasing Food Supply on Human Nutrition: Implications for Commodity Priorities in Agricultural Research and Policy." *Amer. J. Agr. Econ.* 58:131-142.

Pinto, A. [1974]: "Economic Relations between Latin America and the United States: Some Implications and Perspectives." In Cotiler and Fagen, eds., 1974, pp. 100-116.

Pinto, A., and J. Knakal [1973]: *América Latina y el Cambio en la Economía Mundial* (Latin America and the Change in the World Economy). Lima, Instituto de Estudios Peruanos.

Poleman, T. T. [1983]: "World Hunger: Extent, Causes, and Cures." In D. Gale Johnson and Schuh, eds., 1983, pp. 41-75.

Posada-Torres, R. [1986]: *Ecuador and the CGIAR Centers: A Study of their Collaboration in Agricultural Research.* Washington, World Bank, CGIAR Study Paper No. 11.

Prado, C., Jr. [1969]: *Formação do Brasil Contemporâneo: Colônia* (The Colonial Background of Modern Brazil), tr. from Portuguese by S. Macedo. Berkeley, Univ. of California Press.

_____ [1972]: *A Revolução Brasileira* (The Brazilian Revolution). São Paulo, Editôra Brasiliense.

Prebisch, R. [1950, 1962]: *The Economic Development of Latin America and its Principal Problems.* Lake Success, NY, UN ECLA. Also, see *Economic Bulletin for Latin America* 1962, 7(1):1-22.

_____ [1959a]: "Commercial Policy in the Underdeveloped Countries." *Amer. Econ. Rev., Papers and Proceedings* 49:251-273.

_____ [1959b]: *El Mercado Comun Latinoamericano* (The Latin American Common Market). New York, United Nations.

_____ [1961a]: "Economic Development or Monetary Stability: The False Dilemma." *Econ. Bull. for Latin America* 6(1):1-25.

_____ [1961b]: "Discussion of Myrdal." In Lekachman, 1961, pp. 277-280.

_____ [1963]: *Towards a Dynamic Development Policy for Latin America.* New York, United Nations, E/CN 12/680, Rev.1.

_____ [1971]: *Change and Development—Latin America's Great Task.* New York, Praeger.

Psacharopoulos, G. [1973]: *Returns to Education: An International Comparison.* New York, Elsevier.

_____ [1984]: "The Contribution of Education to Economic Growth: International Comparisons." In Kendrick, ed., 1984, pp. 335-355.

_____ [1985]: "Returns to Education: A Further International Update and Implications." *J. Human Resources* 20:583-604.

Psacharopoulous, G., A. M. Arriagada, and E. Velez [1987]: *Earnings and Education among the Self-Employed in Colombia*. Washington, World Bank, Education and Training Department.

Psacharopoulos, G., and W. Loxley [1985]: *Diversified Secondary Education and Development: Evidence from Colombia and Tanzania*. Baltimore, Johns Hopkins Univ. Press.

Psacharopoulos, G., and M. Woodhall [1985]: *Education for Development: An Analysis of Investment Choices*. New York, Oxford Univ. Press.

Pyatt, G [1977]: "Economic Strategies for Growth with Equity." *Econ. Dev. and Cultural Change* 25:581-587.

Quezada, N. [1981]: "Indigenous Agricultural Price and Trade Policy in the Dominican Republic." Unpublished Ph.D. thesis, Purdue Univ., West Lafayette, IN.

Quibria, M. G., and S. Rashid [1984]: "The Puzzle of Sharecropping: A Survey of Theories." *World Dev.* 12:103-114.

⸺ [1986]: "Sharecropping in Dual Agrarian Economies: A Synthesis." *Oxford Econ. Papers* NS 38:94-111.

Quiros-Guardia, R. [1973a]: *Agricultural Development in Central America: Its Origins and Nature*. Madison, Univ. of Wisconsin, LTC, Research Paper No. 49.

⸺ [1973b]: *Agricultural Development and the Central American Common Market*. Madison, Univ. of Wisconsin, LTC, Research Paper No. 50.

Rabelo de Castro, P. [1977]: *Agro-Analysis: Edição Especial* (Agricultural Analysis: Special Edition). Rio de Janeiro, FGV, Instituto Brasileiro de Economia.

Ramalho de Castro, J. P. [1974]: "An Economic Model for Establishing Priorities for Agricultural Research and a Test for the Brazilian Economy." Unpublished Ph.D. Dissertation, Purdue Univ., West Lafayette, IN.

Ramalho de Castro, J. P., and G. E. Schuh [1977]: "An Empirical Test of an Economic Model for Establishing Research Priorities: A Brazil Case Study." In Arndt, Dalrymple, and Ruttan, eds., 1977, pp. 498-525.

Ramos, J.R. [1970]: *Labor and Development in Latin America*. New York, Columbia Univ. Press.

Rangel, I. [1979]: *A Questão Agrária Brasileira* (The Brazilian Agrarian Question). Rio de Janeiro, Conselho de Desenvolvimento.

Ranis, G. [1980]: "Distribución del Ingreso y Crecimiento en Colombia" (Distribution of Income and Growth in Colombia). *Desarrollo y Sociedad* Jan., 1980, pp. 67-96.

Ranis, G., and J. C. H. Fei [1961]: "A Theory of Economic Development." *Amer. Econ. Rev.* 51: 533-565.

Rao, B. P. [1973]: *The Economics of Agricultural Credit—Use in Southern Brazil*. India, Andhra Pradesh Univ. Press.

Rao, C. H. H. [1971]: "Uncertainty, Entrepreneurship and Sharecropping in India." *J. Polit. Economy* 79: 578-595.

⸺ [1977]: *Technological Change and Distribution of Gains in Indian Agriculture*. Delhi, Macmillan.

Rao, J. M. [1989]: "Agricultural Supply Response: A Survey." *Agr. Econ.* Vol. 3, No. 1, pp. 1-22.

Raup, P. M. [1967]: "Land Reform and Agricultural Development." In Southworth and Johnston eds., 1967, pp. 267-314.

Rausser, G. C., ed. [1982]: *New Directions in Econometric Modeling and Forecasting in U.S. Agriculture*. Amsterdam, North-Holland.

Rausser, G. C., E. Lichtenberg, and R. Lattimore [1982]: "Developments in Theory and Empirical Applications of Endogenous Governmental Behavior." In Rausser, ed., 1982, pp. 547-614.

Razin, A. [1969]: "Investment in Human Capital and Economic Growth: A Theoretical Study." Unpublished Ph.D. dissertation, Univ. of Chicago.

_____ [1972a]: "Optimum Investment in Human Capital." Rev. Econ. Studies 39:455-460.

_____ [1972b]: "Investment in Human Capital and Economic Growth." Metroeconomica 24:101-116.

_____ [1977]: "Economic Growth and Education: New Evidence." Econ. Dev. and Cultural Change 25:317-324.

Rebello, A. da P. P. [1973]: "Estruturas de Excedente Comercializável, Oferta, e Demanda de Arroz em Áreas Selecionadas do Estado do Pará" (Structure of Marketable Surplus, Supply, and Demand for Rice in Selected Areas of the State of Pará). Unpublished M.S. thesis, Universidade Federal de Viçosa, Minas Gerais, Brazil.

Reca, L. G. [1967]: "The Price and Production Duality within Argentine Agriculture, 1923-1965." Unpublished Ph.D. dissertation, Univ. of Chicago.

_____ [1980]: Argentina: Country Case Study of Agricultural Prices, Taxes, and Subsidies. Washington, World Bank, Staff Working Paper No. 386.

Redclift, M. [1977]: Agrarian Reform and Peasant Organization on the Ecuadorian Coast. London, Athlone Press.

Reid, J. D., Jr. [1976]: "Sharecropping and Agricultural Uncertainty." Econ. Dev. and Cultural Change 24:549-577.

_____ [1977]: "The Theory of Share Tenancy Revisited—Again." J. Polit. Economy 85: 403-407.

_____ [1979]: "Sharecropping in American History." In Roumasset, Boussard, and I. J. Singh, eds., 1979, pp. 283-309.

Reutlinger, S., and M. Selowsky [1976]: Malnutrition and Poverty: Magnitude and Policy Options. Baltimore, Johns Hopkins Univ. Press.

Reutlinger, S., and J. van H. Pellekaan [1986]: Poverty and Hunger: Issues and Options for Food Security in Developing Countries. Washington, World Bank.

Reynolds, C. W. [1970]: The Mexican Economy: Twentieth Century Structure and Growth. New Haven, CT, Yale Univ. Press.

Reynolds, L. G., ed. [1975]: Agriculture Development Theory. New Haven, CT, Yale Univ. Press.

Rezende, G. C. [1975]: "Estrutura e Nível Técnico da Agricultura Brasileira Segundo Furtado" (Structure and the Technical Level of Brazilian Agriculture According to Furtado). Pesquisa e Planejamento Econômico 5(1):219-230.

_____ [1982a]: "Crédito Rural Subsidiado e Preço da Terra no Brasil" (Subsidized Rural Credit and Land Prices in Brazil). Estudos Econômicos 12(2):117-137.

_____ [1982b]: "Política Agrícola, Preço da Terra e Estrutura Agrária" (Agricultural Policy, Land Prices, and Agrarian Structure). Revista de Economia Rural 20(1):73-100, Special Number.

Ribeiro, F. B. [1975]: "Estrutura da Oferta na Agricultura Tradicional—O Caso do Estado do Piauí" (Structure of Supply in Traditional Agriculture—The Case of the State of Piauí). Unpublished M.S. thesis, Universidade Federal de Viçosa, Minas Gerais, Brazil.

Ribeiro, J. P., and C. R. Wharton, Jr. [1969]: "The ACAR Program in Minas Gerais, Brazil." In Wharton, ed., 1969, pp. 424-438.

Riff, T. B. [1976]: "A Difusão da Inovação Tecnológica na Agricultura: 'Mecanismos de Autocontrôle' versus 'Modernização Induzida' " (The Diffusion of Technological Innovation in Agriculture: "Mechanisms of Endogenous Control" *versus* "Induced Modernization"). *Revista Brasileira de Economia* 30(3):295-327.

Riley, H. M., K. Harrison, N. Suarez, J. Shaffer, D. Henley, D. Larson, C. Guthrie, and D. Lloyd-Clare [1970b]: *Market Coordination in the Development of the Cauca Valley Region—Colombia.* East Lansing, MSU, LASC, Research Report No. 5.

Riley, H. M., C. C. Slater, K. Harrison, J. Wish, J. Griggs, V. Parad, J. Santiago, I. Rodriguez [1970a]: *Food Marketing in Economic Development of Puerto Rico.* East Lansing, MSU, LASC.

Rios, P. L. [1969]: "Urban-Rural Developmental Interrelationships in Minas Gerais, Brazil, 1940-1960." Unpublished Ph.D. thesis, ISU, Ames.

Rippy, M, ed. [1969]: *Cultural Change in Brazil: Papers from the Midwest Association for Latin American Studies.* IN, Ball State Univ.

Ritter, A. R. M. [1974]: *The Economic Development of Revolutionary Cuba: Strategy and Performance.* New York, Praeger.

Robinson, E. A. G., ed. [1965]: *Problems in Economic Development.* London, Macmillan.

Robock, S. H. [1963]: *Brazil's Developing Northeast.* Washington, Brookings Institution.

Roca, S. [1976]: *Cuban Economic Policy and Ideology: The Ten Million Ton Sugar Harvest.* Beverly Hills, CA, Sage Professional Paper, International Studies Series, Vol. 4, Sage Publications.

Rodriguez, C. A. [1982]: "The Argentine Stabilization Plan of December 20th." *World Dev.* 10:801-811.

Roett, R., ed. [1972]: *Brazil in the Sixties.* Nashville, TN, Vanderbilt Univ. Press.

Rogers, A. J. III [1969]: "Professional Incomes and Rates of Return to Higher Education in Brazil." Unpublished Ph.D. dissertation, MSU, East Lansing.

Rogers, E. M. [1962]: *Diffusion of Innovations.* New York, The Free Press.

Rogers, E. M., in association with L. Svenning [1969]: *Modernization among Peasants: The Impact of Communication.* New York, Holt, Rinehart and Winston.

Rogge, B. A. [1956]: "Economic Development in Latin America: The Prebisch Thesis." *Inter-Amer. Econ. Affairs* 9(4):24-49.

Rollemberg-Mollo, M. de. L. [1983]: *Política de Garantia de Preços Mínimos: Uma Avaliação* (Guarantee Policy of Minimum Prices: An Evaluation). Brasília, Ministério da Agricultura, CFP, Coleção Análise e Pesquisa, Vol. 29.

Roumasset, J. A., J. M. Boussard, and I. J. Singh, eds. [1979]: *Risk, Uncertainty and Agricultural Development.* New York, ADC.

Roxborough, I. [1974]: "Agrarian Policy in the Popular Unity Government." *Occasional Papers* No. 14, Institute of Latin American Studies, Univ. of Glasgow.

Roy, P., F. B. Waisanen, E. M. Rogers, and UNESCO [1969]: *The Impact of Communication on Rural Development: An Investigation in Costa Rica and India.* Paris, UNESCO.

Ruttan, V. W. [1955]: "The Impact of Urban-Industrial Development on Agriculture in the Tennessee Valley and the Southeast." *J. Farm Econ.* 37:38-56.

_____ [1982]: *Agricultural Research Policy.* Minneapolis, Univ. of Minnesota Press.

Sabato, J. F. [1983]: "Agriculture in the Argentine Pampas: Technology Adoption in Corn Cultivation from 1950 to 1978." In Piñeiro and Trigo, eds., 1983a, pp. 71-124. Also in Spanish in Piñeiro and Trigo, eds., 1983b, pp. 185-286.

Sabot, R. H., ed. [1982a]: *Migration and the Labor Market in Developing Countries.* Boulder, CO, Westview Press.

_____ [1982b]: "Conclusion: Some Themes and Unresolved Issues." In Sabot, ed., 1982a, pp. 229-241.

Sacks, R. E. G. [1979]: "Educational Needs of Farmers: A Cross-National Study." *Sociologia Ruralis* 19:29-42.

Sahota, G. S. [1968]: "An Economic Analysis of Internal Migration in Brazil." *J. Polit. Economy.* 76:218-245.

Sahota, G. S., and C. A. Rocca [1981]: "Process of Production and Distribution in Brazilian Agriculture." *Econ. Dev. and Cultural Change* 29:683-721.

———— [1985]: *Income Distribution: Theory, Modeling, and Case Study of Brazil.* Ames, ISU Press.

Saint, W. S. [1981]: "The Wages of Modernization: A Review of the Literature on Temporary Labor Arrangements in Brazilian Agricultures." *Latin Amer. Research Rev.* 16(3):91-110.

Sanches, P. A., and G. M. Scobie [1986]: *Cuba and the CGIAR Centers: A Study of their Collaboration in Agricultural Research.* Washington, World Bank, CGIAR Study Paper No. 14.

Sanders, J. H. [1973]: "Mechanization and Employment in Brazilian Agriculture, 1950-1971." Unpublished Ph.D. thesis, Univ. of Minnesota, St. Paul.

Sanders, J. H., and J. K. Lynam [1982]: "Evaluation of New Technology on Farms: Methodology and Some Results from Two Crop Programmes at CIAT." *Agricultural Systems* 9(2):92-112.

Sanders, J. H., and V. W. Ruttan [1978]: "Biased Choice of Technology in Brazilian Agriculture." In Binswanger, Ruttan, *et al.*, eds., 1978, pp. 276-296.

Sanderson, S. [1986]: *Transformation of Mexican Agriculture.* Princeton, NJ, Princeton Univ. Press.

Sandner, G. [1962, 1964]: *La Colonización Agrícola de Costa Rica* (The Agricultural Colonization of Costa Rica) 2 vols. San José, Costa Rica, IICA.

Sanint, L. R., and USDA, ERS [1983]: *Demand for Carbohydrate Foods in Colombia and Venezuela.* USDA, FAER No. 187, pp. 19-23.

Santana, C. A. M. [1984]: "The Impact of Economic Policies on the Soybean Sector of Brazil: An Effective Protection Analysis." Unpublished Ph.D. thesis, Univ. of Minnesota, St. Paul.

Santos, D. B. dos [1974]: "Incentivos de Preços na Agricultura do Estado de Pernambuco" (Price Incentives in the Agriculture of the State of Pernambuco). Unpublished M.S. thesis, Universidad Federal de Viçosa, Minas Gerais, Brazil.

Santos, L. F. dos [1971]: "Estimativa da Oferta de Arroz, Milho e Feijão em Minas Gerais, 1947/1969" (Estimates of the Supply of Rice, Corn and Beans for Minas Gerais, 1947/1969). Unpublished M.S. thesis, Universidade Federal de Viçosa, Minas Gerais, Brazil.

Santos, R. F. dos [1986]: *Presença de Viéses de Mudança Técnica na Agricultura Brasileira* (The Presence of Biases on the Technical Change in Brazilian Agriculture). São Paulo, IPE/USP, Ensaios Econômicos, 63.

Santos, T. dos [1970a]: "The Structure of Dependence." *Amer. Econ. Rev.* 60:231-236.

Santos, T. dos [1970b]: "La Crisis de la Teoria del Desarrollo y las Relaciones de Dependencia en América Latina" (The Crisis of Development Theory and Dependency Relationships in Latin America). In Jaguaribe *et al.*, 1970, pp. 147-187.

São Paulo, Secretaria da Agricultura [1972]: *Desenvolvimento da Agricultura Paulista* (The Development of Paulista Agriculture). SA, Instituto de Economia Rural.

Sayad, J. [1977a]: "Planejamento, Crédito, e Distribuição de Renda" (Planning, Credit, and Income Distribution). *Estudos Econômicos* 7(1):9-34.

———— [1977b]: "Preço da Terra e Mercados Financeiros" (Land Prices and Financial Markets). *Pesquisa e Planejamento Econômico* 3(7):623–662.

———— [1981]: *Rural Credit and Rural Rates of Interest.* Univ. of São Paulo/IPE, mimeo.

———— [1983]: "The Impact of Rural Credit on Production and Income Distribution in Brazil." In von Pischke, D. W. Adams and Donald, eds., 1983, pp. 379–386.

———— [1984]: *Crédito Rural no Brasil. Avaliação das Críticas e das Propostas de Reforma* (Rural Credit in Brazil. An Evaluation of Critiques and Reform Proposals). São Paulo, FIPE/Pioneira.

Saylor, R. G. [1974]: "Alternative Measures of Supply Elasticities: The Case of São Paulo Coffee." *Amer. J. Agr. Econ.* 56:98–106.

Scandizzo, P. L., and C. Bruce [1980]: *Methodologies for Measuring Agricultural Price Intervention Effects.* Washington, World Bank, Staff Working Paper No. 394.

Schmidt, L. J. [1967]: "The Role of Migratory Labor in the Economic Development of Guatemala." Unpublished Ph.D. dissertation, Univ. of Wisconsin, Madison.

Schuh, G. E. [1968]: "Effects of Some General Economic Development Policies on Agricultural Development." *Amer. J. Agr. Econ.* 50:1283–1293.

———— [1969]: "Comments on Nicholls." In Thorbecke, ed., 1969, pp. 379–385.

———— [1970a]: *Research on Agricultural Development in Brazil.* New York, ADC.

———— [1970b, 1971]: *The Agricultural Development of Brazil.* New York, Praeger. Also in Portuguese, *O Desenvolvimento da Agricultura no Brasil*, Rio de Janeiro, APEC Editôra S.A., 1971.

———— [1973]: "Modernização e Dualismo Tecnológico na Agricultural: Alguns Comentários" (Modernization and Technological Dualism in Agriculture: Some Comments). *Pesquisa e Planeiamento Econômico.* 3:51–94.

———— [1974]: "The Exchange Rate and U.S. Agriculture." *Amer. J. Agr. Econ.* 56:1–13.

———— [1975a]: "Patterns of Equity under Agricultural Development in Latin America." In Heady and Whiting, eds., 1975, pp. 234–273.

———— [1975b]: "The Exchange Rate and U.S. Agriculture: Reply." *Amer. J. Agr. Econ.* 57:696–700.

———— [1975c]: "A Modernização da Agricultura Brasileira: Uma Interpretação" (The Modernization of Brazilian Agriculture: An Interpretation). In Contador, ed., 1975, pp. 7–45.

———— [1975d]: "General Economic Policy as a Constraint to the Development of Agriculture." In M. Alexander, ed., 1975.

———— [1976a]: "The New Macroeconomics of Agriculture." *Amer. J. Agr. Econ.* 58:802–811.

———— [1976b]: "Theoretical Considerations for Cost of Production Studies." *Proceedings, International Seminar on the Cost of Production.* IEA, S.A. of São Paulo, Brazil.

———— [1978]: "Approaches to Basic Needs and 'Equity' that Distort Incentives to Agriculture." In T. W. Schultz, ed., 1978, pp. 307–327.

———— [1980]: "Post-Keynesian Economics and Agriculture: Discussion." *Amer. J. Agr. Econ.* 62: 328–330.

———— [1981a]: "Economics and International Relations: A Conceptual Framework." *Amer. J. Agr. Econ.* 63:767–778.

———— [1981b]: "Food Aid and Human Capital Formation." In *Food Aid and Development.* New York: Agricultural Development Council, pp. 49–61.

———— [1982]: "Out-Migration, Rural Productivity, and the Distribution of Income." In Sabot, ed., 1982a, pp. 161–190.

_____ [1983]: "The Role of Markets and Governments in the World Food Economy." In D. Gale Johnson and Schuh, eds., 1983, pp. 277-301.

_____ [1984a]: *Third Country Monetary Disturbances in a Changed International Economy: The Case of Brazil and Mexico.* Minnesota Dept. of Agricultural and Applied Economics, 1984, Staff Papers Series.

_____ [1984b]: "The Political Economy of Rural Development in Latin America: Comment." In Eicher and Staatz, eds., 1984, pp. 96-109.

_____ [1986]: *The United States and the Developing Countries: An Economic Perspective.* Washington, D.C., The National Planning Association.

_____ [1987]: "Monetary Disturbances in a Changed International Economy: The Case of Mexico's Agriculture and Mexican-U.S. Trade." In Johnston, Luiselli, Contreras, and Norton, eds., 1987, pp. 145-158.

Schuh, G. E., and M. I. Angeli-Schuh [1989]: "Human Capital for Agricultural Development in Latin America." San José, Costa Rica, IICA, Série de Documentos de Programas No. 11, PLANALC.

Schuh, G. E., and J. L. McCoy, eds. [1986]: *Food, Agriculture, and Development in the Pacific Basin.* Boulder, CO, Westview Press.

Schuh, G. E., and R. L. Thompson [1980]: "Assessing Agricultural Progress and the Commitment to Agriculture." In McGreevey, ed., 1980 pp. 121-156.

Schultz, T. P. [1968]: *Returns to Education in Bogotá, Colombia.* Santa Monica, CA, Rand Memorandum RM-5645.

_____ [1969]: *Population Growth and Internal Migration in Colombia.* Santa Monica, CA, Rand Corporation RM-5765.

_____ [1971]: "Rural-Urban Migration in Colombia." *Rev. Econ. and Stat.* 53:157-163.

_____ [1975]: "Determinants of Internal Migration in Venezuela." Paper presented at *Econometric Society, Third World Congress,* Toronto.

_____ [1982]: "Notes on the Estimation of Migration Decision Functions." In Sabot, ed., 1982a, pp. 91-126.

Schultz, T. W. [1951]: "Framework for Land Economics—The Long View." *J. Farm Econ.* 33:204-215.

_____ [1953]: *The Economic Organization of Agriculture.* New York, McGraw-Hill.

_____ [1956]: *The Economic Test in Latin America.* Ithaca, N.Y. New York State School of Industrial and Labor Relations, Cornell Univ., Bulletin 35.

_____ [1961a]: "Education and Economic Growth." In *Social Forces Influencing American Education.* Chicago, National Society for the Study of Education.

_____ [1961b]: "A Policy to Redistribute Losses from Economic Progress." *J. Farm Econ.* 43:554-565. Also in *Labor Mobility and Population in Agriculture,* 1961, pp. 158-168.

_____ [1963]: *The Economic Value of Education.* New York, Columbia Univ. Press.

_____ [1964]: *Transforming Traditional Agriculture.* New Haven, CT, Yale Univ. Press.

_____ [1967a]: "Education and Research in Rural Development in Latin America." In Crowder, ed., 1967, pp. 391-402.

_____ [1967b]: "Economic Growth Theory and The Profitability of Farming in Latin America." In IADB, 1967, AB-80-6, pp. 169-188.

_____ [1968]: "Institutions and the Rising Economic Value of Man." *Amer. J. Agr. Econ.* 50:1113-1122.

_____ , ed., [1972]: *Investment in Education: The Equity—Efficiency Quandary. J. Polit. Economy* 80(3, Part II):S1-S292, separate paperback.

_____ [1973a]: "The Education of Farm People: An Economic Perspective." In Foster and Sheffield, 1973, in *World Year Book of Education,* 1974, pp. 50-68.

———, ed. [1973b, 1975a]: *Economics of the Family: Marriage, Children and Human Capital.* Published in 1973 in *J. Polit. Economy* 81(2, Part II):S1-S299. Also published in book form in 1975 by the Univ. of Chicago Press.

——— [1973c, 1975b]: "Fertility and Economic Values." In T. W. Schultz, ed., 1973b, 1975a, pp. 3-22.

——— [1974]: "The High Value of Human Time: Population Equilibrium." *J. Polit. Economy* 82(2, Part II):S2-S10.

——— [1975c]: "The Value of the Ability to Deal with Disequilibrium." *J. Econ. Literature* 13:827-846.

——— [1976, 1977]: *The Economic Value of Human Time over Time.* Univ. of Chicago. Human Capital Paper No. 76:2. Also published in USDA, ERS publication ERS-35, 1977.

———, ed. [1978]: *Distortions of Agricultural Incentives.* Bloomington, IN, Indiana Univ. Press.

——— [1979, 1980]: *The Economics of Being Poor: Nobel Lecture in Economics.* Stockholm, Nobel Foundation. Also in *Bulletin of the Atomic Scientists* 36(9):32-37, 1980, and in *J. Polit. Economy* 88:639-651, 1980.

Schydlowsky, D. M. [1972]: "Latin American Trade Policies in the 1970s: A Prospective Appraisal." *Quart. J. Econ.* 86:263-289.

Schydlowsky, D., S. J. Hunt, and J. Mezzera [1983]: *La Promoción de Exportaciones no Tradicionales en el Perú* (The Promotion of Nontraditional Exports in Peru). Lima, Perú, Asociación de Exportadores del Perú.

Scobie G. M. [1987]: *Partners in Research: The CGIAR in Latin America.* Washington, World Bank, Study Paper No. 24.

Scobie, G. M., and R. Posada-Torres [1977]: *The Impact of High-Yielding Rice Varieties in Latin America: With Special Emphasis on Colombia.* Cali, Colombia, CIAT, CEDEAL Series JE:01.

——— [1978, 1984]: "The Impact of Technical Change on Income Distribution: The Case of Rice in Colombia." *Amer. J. Agr. Econ.* 60:85-92. Also in Eicher and Staatz, ed., 1984, pp. 378-388.

Scott, G. J. [1985a]: *Mercados, Mitos, e Intermediarios. Un Estudio de la Comercialización de la Papa en la Zona Central de Perú* (Markets, Myths, and Middlemen. A Case Study of Potato Marketing in Central Peru.) Lima Univ. del Pacífico.

——— [1985b]: *Markets, Myths, and Middlemen: A Case Study of Potato Marketing in Central Peru.* Lima, International Potato Center.

Scott, G. J., and M. G. Costello, eds. [1985]: *Comercialización Interna de los Alimentos en América Latina: Problemas, Productos y Políticas* (Domestic Marketing of Foods in Latin America: Problems, Products and Policies). Ottawa, Canada, IDRC.

Secretaria da Agricultura [SA, 1972]: *Desenvolvimento da Agricultura Paulista* (The Development of Paulista Agriculture). São Paulo, Gov. do Estado de S. Paulo, Instituto de Economia Rural.

Seers, D. [1962]: "A Theory of Inflation and Growth in Underdeveloped Economies Based on the Experience of Latin America." *Oxford Econ. Papers* NS14:173-195.

———, ed. [1964]: *Cuba: The Economic and Social Revolution.* Chapel Hill, Univ. of North Carolina Press.

Selowsky, M. [1967]: "Education and Economic Growth: Some International Comparisons." Unpublished Ph.D. dissertation, Univ. of Chicago.

——— [1968]: *The Effect of Unemployment and Growth on the Rate of Return to Education: The Case of Colombia.* Univ. Center for International Affairs, Report No. 116.

_____ [1976]: "A Note on Pre-School Age Investment in Human Capital in Developing Countries." *Econ. Dev. and Cultural Change* 24:707-720.

_____ [1979]: *Who Benefits from Government Expenditure? A Case Study of Colombia.* New York, Oxford Univ. Press.

Selowsky, M., and L. Taylor [1973]: "The Economics of Malnourished Children: An Example of Disinvestment in Human Capital." *Econ. Dev. and Cultural Change* 22:17-30.

Sen, A. K. [1981]: *Poverty and Famine: An Essay on Entitlement and Deprivation.* Oxford, Clarendon Press.

Serôa da Motta, R. [1985]: "Alcohol as Fuel: A Cost-Benefit Study of the Brazilian National Alcohol Program." Unpublished Ph.D. Dissertation, University of London, London.

_____ [1986]: "A Social-Cost Benefit Study of Ethanol Production in Brazil." London, University College London, Department of Economics, Discussion Paper No. 86.02.

Serôa da Motta, R., and L. da R. Ferreira [1988]: "The Brazilian National Alcohol Programme: An Economic Reappraisal and Adjustments." *Energy Economics* (July):229-234.

Sewastynowicz, J. [1986]: " 'Two-Step' Migration and Upward Mobility on the Frontier: The Safety Valve Effect in Pejibaye, Costa Rica." *Econ. Dev. and Cultural Change* 34:731-753.

Shanin, T., ed. [1987]: *Peasants and Peasant Societies.* New York, Viking Penguin.

Shaw, R. P. [1974]: "Land Tenure and the Rural Exodus in Latin America." *Econ Dev. and Cultural Change* 23:123-132.

Shoup, C. [1959]: *The Fiscal System of Venezuela.* Baltimore, Johns Hopkins Press.

Silva, G. S. P. [1979]: "Pesquisa e Produção Agrícola no Brasil' (Research and Agricultural Production in Brazil). *Agricultura em São Paulo* 26(2): 175-256.

_____ [1980]: "Investimento na Geração e Difusão da tecnologia Agrícola no Brasil" (Investment in the Generation and Diffusion of Agricultural Technology in Brazil). *Revista de Economia Rural* 18(2):328-335.

Simonsen, M. H. [1970]: *Inflação: Gradualismo versus Tratamento de Choque* (Inflation: Gradualism *versus* Shock Treatment). Rio de Janeiro, APEC.

Simonsen, R. C. [1977]: *História Econômica do Brasil: 1500-1820.* (Economic History of Brazil: 1500-1820). 7th edition. São Paulo, Companhia Editôra Nacional. Also in Brasilia, Instituto Nacional de Livros, Coleção Brasiliana, Vol. 10.

Singer, H. W. [1950]: "The Distribution of Gains between Investing and Borrowing Countries." *Amer. Econ. Rev.* 40(2):473-485.

Singer, H. W. with the collaboration of S. J. Maxwell [1983]: "Development through Food Aid: Twenty Years' Experience." *Report of the World Food Programme.* The Hague, Government of the Netherlands, 1983, pp. 31-46.

Singh, G. [1974]: "Farm Level Determinants of Credit Allocation and use in Southern Brazil, 1965-1969." Unpublished Ph.D. thesis, Ohio State Univ., Columbus.

Singh, I. J., and C. Y. Ahn [1978]: "A Dynamic Multi-Commodity Model of the Agricultural Sector: A Regional Application in Brazil." *European Econ. Rev.* 11(2):155-179.

Singh, R. D., E. W. Kehrberg, G. F. Patrick, and G. E. Schuh [1979]: *Poor Rural Households, Technical Change, and Income Distribution in LDCs—Brazil.* West Lafayette, IN., Purdue Univ., DAE.

Singh, R. D., and G. E. Schuh [1986]: "The Economic Contribution of Farm Children and the Household Fertility Decisions: Evidence from a Developing Country, Brazil." *Indian J. Agr. Econ.* 41(1):29-41.

Singh, R. D., G. E. Schuh, and E. W. Kehrberg [1978]: *Economic Analysis of Fertility Behavior and the Demand for Schooling among Poor Households in Rural Brazil*. West Lafayette, IN, Purdue Univ., AES Bulletin No. 214.

Sisler, D. G. [1959]: "Regional Differences in the Impact of Urban-Industrial Development on Farm and Nonfarm Income." *J. Farm Econ.* 41:1100-1112.

Slater, C., D. Henley, J. Wish, V. Farace, L. Jacobs, D. Lindley, A. Mercado, and M. Moran [1969a]: *Market Processes in La Paz, Bolivia*. East Lansing, MSU, LASC, Research Report No. 3.

Slater, C., H. Riley, V. Farace, K. Harrison, F. Neves, A. Bogatay, N. Doctoroff, D. Larson, R. Nason, and T. Webb [1969b]: *Market Processes in the Recife Area of Northeast Brazil*. East Lansing, MSU, LASC, Research Report No. 2.

Smith, G. W. [1965]: "Agricultural Marketing and Economic Development: A Brazilian Case Study." Unpublished Ph.D. dissertation, Harvard Univ., Cambridge, MA.

―――― [1969]: "Brazilian Agricultural Policy: 1950-1967." In Ellis, ed., 1969, pp. 213-265.

Smith, N. J. H. [1982]: *Rainforest Corridors: The Tranzamazon Colonization Scheme*. Berkeley, Univ. of California Press.

Smith, R. F., ed. [1966]: *Background to Revolution: The Development of Modern Cuba*. New York, Alfred A. Knopf.

Soares, A. C. [1977]: "Resource Allocation and Choice of Enterprise under Risk on Cotton Farms in Northeast Brazil." Unpublished Ph.D. thesis, Ohio State Univ., Columbus.

Solimano, G., and L. Taylor [1980]: *Food Price Policies and Nutrition in Latin America*. Tokyo, United Nations Univ.

Solow, R. M. [1957]: "Technical Change and the Aggregate Production Function." *Rev. Econ. and Stat.* 39:312-320.

Southworth, H. M., and B. F. Johnston, eds. [1967]: *Agricultural Development and Economic Growth*. Ithaca, NY, Cornell Univ. Press.

Souza, G. da S. [1984]: "Sôbre a Validade da Tese de Prebisch para a Série de Relações de Troca da Economia Brasileira" (On the Validity of the Prebisch Thesis for a Series of Terms of Trade Relations in Brazil). *Pesquisa e Planejamento Econômico* 14(2):561-568.

Spraos, J. [1980]: "The Statistical Debate on the Net Barter Terms of Trade between Primary Commodities and Manufactures." *Econ. J.* 90:107-128.

Srinivasan, T. N. [1979]: "Agricultural Backwardness under Semi-Feudalism—Comment." *Econ. J.* 89:416-419.

―――― [1983]: "Hunger: Defining it, Estimating its Global Incidence, and Alleviating it." In D. Gale Johnson and Schuh, eds., 1983, pp. 77-108.

Stanfield, D. [1976]: *The Chilean Agrarian Reform, 1975*. Madison, Univ. of Wisconsin, LTC.

Stavenhagen, R. [1968]: "Seven Fallacies about Latin America." In Petras and Zeithin, eds., 1968, pp. 13-31.

―――― , ed. [1970]: *Agrarian Problems and Peasant Movements in Latin America*. Garden City, NY, Doubleday Anchor.

―――― [1974]: "The Future of Latin America: Between Underdevelopment and Revolution." *Latin Amer. Perspectives* 1(1): 124-148.

―――― [1975]: *Las Clases Sociales en las Sociedades Agrarias* (Social Classes in Agrarian Societies), tr. J. A. Helman. New York, Doubleday Anchor.

Steir, F. [1987]: "Schooling, Experience, and Earnings: Issues in Venezuelan Development, 1975-84." Unpublished Ph.D. dissertation, Columbia Univ., NY.

Stelcner, M., A. M. Arriagada, and P. R. Moock [1987]: *Wage Determinants and School Attainment among Men in Peru*. Washington, World Bank, LSMS, Working Paper No. 38.

Stern, S. J. [1988]: "Feudalism, Capitalism, and the World-System in the Perspective of Latin America and the Caribbean." *American Historical Rev.* 93(4):836-872.

Stewart, R. [1985a]: *Costa Rica and the CGIAR Centers: A Study of their Collaboration in Agricultural Research.* Washington, World Bank, CGIAR Study Paper No. 4.

―――― [1985b]: *Guatemala and the CGIAR Centers: A Study of their Collaboration in Agricultural Research.* Washington, World Bank, CGIAR Study Paper No. 5.

Stiglitz, J. E. [1974]: "Incentives and Risk Sharing in Sharecropping." *Rev. Econ. Studies* 41:219-255.

―――― [1976]: "The Efficiency Wage Hypothesis, Surplus Labour, and the Distribution of Income in LDCs." *Oxford Econ. Papers* NS 28:185-207.

―――― [1982]: "The Structure of Labor Markets and Shadow Prices in LDCs." In Sabot, ed., 1982a, pp. 13-63.

Sünkel, O. [1958]: "La Inflación Chilena: Un Enfoque Heterodoxo." (Chilean Inflation: An Unorthodox Perspective). *Trimestre Económico* 25(4):570-599.

―――― [1966]: "The Structural Background of Development Problems in Latin America." *Weltwirtschaftliches Archiv* 97(II):22-63.

―――― [1973]: "Transnational Capitalism and National Disintegration in Latin America." *Social and Economic Studies* 22(1):132-176.

Super, J. C., and T. C. Wright, eds. [1985]: *Food, Politics, and Society in Latin America.* Lincoln, Univ. of Nebraska Press.

Swift, J. [1969]: "Economic Study of the Chilean Agrarian Reform." Unpublished Ph.D. thesis, MIT, Cambridge, MA.

Tallone-Rosso, W. I. [1965]: "Estimativas Estruturais das Relações de Oferta de Milho no Estado de Minas Gerais 1944/1962" (Structural Estimates for Supply Relations of Corn in the State of Minas Gerais, 1944/1962). Minas Gerais, Brazil, Universidade Federal de Viçosa.

Tang, A. M. [1958]: *Economic Development in the Southern Piedmont, 1860-1950: Its Impact on Agriculture.* Chapel Hill, Univ. of North Carolina Press.

Tannenbaum, F. [1929]: *The Mexican Agrarian Revolution.* New York, Macmillan.

Tavares, M. C. [1964]: "The Growth and Decline of Import Substitution in Brazil." *Econ. Bull. for Latin America* 9(1):1-59.

Taylor, J. R. [1969]: *Agricultural Settlement and Development in Eastern Nicaragua.* Madison, Univ. of Wisconsin, LTC, Research Paper No. 33

Taylor, L., E. L. Bacha, E. A. Cardoso, and F. J. Lysy [1980]: *Models of Growth and Distribution for Brazil.* New York, Oxford Univ. Press for the World Bank.

Taylor, W. B. [1985]: "Between Global Process and Local Knowledge: An Inquiry Into Early Latin American Social History 1500-1900." In Zunz, ed., 1985, pp. 115-190.

Teixeira, T. [1976]: "Resource Efficiency and the Market for Family Labor: Small Farms in the Sertão of Northeast Brazil." Unpublished Ph.D. dissertation, Purdue Univ., West Lafayette, IN.

Tendler, J. [1962, 1970]: *Agricultural Credit in Brazil, Parts I and II.* Washington, AID.

Theberge, J. D., ed. [1968]: *Economics of Trade and Development.* New York, Wiley.

Thiesenhusen, W. C. [1969]: "Population Growth and Agricultural Employment in Latin America, with Some U.S. Comparisons." *Amer. J. Agric. Econ.* 51:735-753.

―――― [1971]: "Latin America's Employment Problem." *Science* 171(3974):868-874.

―――― [1972]: "A Suggested Policy for Industrial Reinvigoration in Latin America." *J. Latin Amer. Studies* 4(1):85-104.

―――― [1974]: "Chile's Experiments in Agrarian Reform: Four Colonization Projects Revisited." *Amer. J. Agr. Econ.* 56:323-530.

_____ [1982]: "Land Reform in Latin America: Some Current Literature." *Latin Amer. Research Rev.* 17(2):199-211.

_____ [1987a]: "Rural Development Questions in Latin America." *Latin Amer. Research Rev.* 22(1):171-203.

_____ [1987b]: "Incomes of Some Agrarian Reform Asentamientos in Panama." *Econ. Dev. and Cultural Change* 35:809-832.

Thirsk, W. R. [1972]: "The Economics of Farm Mechanization in Colombia." Unpublished Ph.D. dissertation, Yale Univ., New Haven, CT.

_____ [1980]: "A General Equilibrium Analysis of the Effects of Subsidized Farm Mechanization on Output and Income Generation in Colombia." In R. A. Berry and Soligo, eds., 1980, pp. 177-201.

_____ [1985]: *The Growth and Import of Farm Mechanization in Latin America.* Washington, World Bank Report.

Thomas, R. N., ed. [1973]: *Population Dynamics of Latin America: A Review and Bibliography.* East Lansing, MI, Conference of Latin American Geographers, April 1971.

Thomas, V. [1987]: "Differences in Income and Poverty within Brazil." *World Dev.* 15:263-273.

Thomas, V., S. Edwards, J. Nash, M. Thobani, G. Rioseco, and A. C. Wee [1984]: *Colombia: External Sector and Agriculture Policies for Adjustment and Growth*, 2 vols. Washington, World Bank.

Thomazinni-Ettori, O. J. [1964]: "Produtividade Física da Agricultura em São Paulo: Causas das Variações, Tendências, Medidas para sua Elevação" (Physical Productivity of Agriculture in São Paulo: Causes of the Variations, Trends, Measures for its Increase). *Agricultura em São Paulo* 11(7):3-48.

Thome, J. R., and D. Kaimowitz [1985]: "Agrarian Reform." In T. W. Walker, ed., 1985, pp. 299-315.

Thompson, R. L. [1974]: "The Metaproduction Function of Brazilian Agriculture: An Analysis of Productivity and Other Aspects of Growth." Unpublished Ph.D. dissertation, Purdue Univ., West Lafayette, IN.

_____ [1975]: "Agricultural Price Policy as a Factor in Economic Development." In Alexander, ed., 1975, pp. 71-85.

Thompson, R. L., and G. E. Schuh [1978]: "Política Comercial e Exportação: O Caso do Milho no Brasil" (Trade and Export Policy: The Case of Corn in Brazil). *Pesquisa e Planejamento Econômico* 8(3):663-693.

Thompson, R. L., and G. W. Williams [1984]: *The Brazilian Soybean Industry: Economic Structure and Policy Interventions.* Washington, USDA, FAER No. 200.

Thorbecke, E., ed. [1969]: *The Role of Agriculture in Economic Development.* New York, Columbia Univ. Press for the NBER.

Thorp, R. [1977]: "The Post-Import Substitution Era: The Case of Peru." *World Dev.* 5(1/2):125-136.

Thurow, L. C. [1971]: "The Income Distribution as a Pure Public Good." *Quart. J. Econ.* 85:327-336.

_____ [1973]: "The Income Distribution as a Pure Public Good: A Response." *Quart J. Econ.* 87:316-319.

Todaro, M. P. [1969]: "A Model of Labor Migration and Urban Unemployment in Less Developed Countries." *Amer. Econ. Rev.* 59:138-148.

Toledo M., A. C. [1976]: "Schooling and the Distribution of Labor Income in Peru, 1961-1972." Unpublished Ph.D. dissertation, Stanford, CA, Stanford Univ.

Tolipan, R., and A. C. Tinelli, eds. [1975]: *A Controversia sôbre a Distribuição de Renda e Desenvolvimento* (The Controversy over Income Distribution and Development). Rio de Janeiro, Zahar Editôres.

Tollini, H. [1977]: *Small Farmers in Brazil and the Adoption of Agricultural Technology*. West Lafayette, IN, Purdue Univ., DAE.

Torrealba, P. [1971]: *Agricultural Marketing in Economic Development — An Annotated Bibliography*. East Lansing, MSU, LASC.

Tourinho, O. A. F., L. R. Ferreira, and R. F. Pimentel [1985]: *Agricultura e Produção de Energia: Um Modêlo de Programação Linear para Avaliação Econômica do PROALCOOL* (Agriculture and Production of Energy: An LP Model for the Economic Evaluation of the PROALCOHOL). Rio de Janeiro, Instituto de Pesquisas do IPEA.

Toyama, N. K., and R. M. C. Pescarin [1970]: "Projeções da Oferta Agrícola do Estado de São Paulo" (Agricultural Supply Projections for the State of São Paulo). *Agricultura em São Paulo* 17(9/10):3-97.

Trigo, E., and M. Piñeiro [1983]: "Foundations of a Science and Technology Policy for Latin American Agriculture." In Pineiro and Trigo, eds., 1983a, pp. 165-173.

Trigo, E., M. Piñeiro, and J. F. Sabato [1983]: "Technology as a Social Issue: Agricultural Research Organization in Latin America." In Piñeiro and Trigo, eds., 1983a, pp. 125-137.

Turner, F. J. [1920]: *The Significance of the Frontier in American History*. New York, Henry Holt.

Turner, L., and N. McMullen [1982]: *The Newly Industrialising Countries: Trade and Adjustment*. London, Allen and Unwin for the Royal Institute of International Affairs.

Tyers, R., and K. Anderson [1985]: *Economic Growth and Agricultural Protection in East and Southeast Asia: Implications for International Grain and Meat Trade*. Kuala Lumpur and Canberra, ASEAN-Australia Joint Research Project.

United Nations [1951]: *Land Reform: Defects in Agrarian Structure as Obstacles to Economic Development*. New York, Department of Economic Affairs.

United Nations Conference on Trade and Development [UNCTAD, 1964]: *Towards a New Trade Policy for Development. Report by the Secretary General of the UNCTAD*. New York, UN Document E/Conf. 46/3.

Urrutia, M. [1981]: *Winners and Losers in Colombia's Recent Growth Experience*. Bogotá, Colombia, presented to the World Bank, mimeo.

—————— [1985]: *Winners and Losers in Colombia's Economic Growth of the 1970s*. Washington, Oxford Univ. Press for the World Bank.

Vaitsos, C. V. [1978]: "Crisis in Regional Economic Cooperation among Developing Countries: A Survey." *World Dev.* 6:719-769.

Valderama, M. [1979]: "Effecto de las Exportaciones Norteamericanas de Trigo en Bolivia, Perú y Colombia" (Effect of North American Exports of Wheat in Bolivia, Peru and Colombia). *Estudios Rurales Latinoamericanos* 2(2):173-198.

Valdés, A. [1971]: "Wages and Schooling of Agricultural Workers in Chile." *Econ. Dev. and Cultural Change* 19:313-329.

—————— [1973]: "Trade Policy and its Effect on the External Agricultural Trade of Chile: 1945-1965." *Amer. J. Agr. Econ.* 55:154-164.

—————— [1974]: "The Transition to Socialism: Observations on the Chilean Agrarian Reform." In E. O. Edwards, ed., 1974, pp. 405-418.

—————— , ed. [1981]: *Food Security for Developing Countries*. Boulder, CO, Westview Press.

_____ [1983]: "La Protección Agrícola en los Países Industrializados: Su Costo para la América Latina" (Agricultural Protection in the Industrialized Countries: Its Costs to Latin America). *El Trimestre Económico* 50(3):1963-1720.

_____ [1984a]: "Trade in Agricultural Products between Developing Countries: Latin American Exports during 1962-1979." *Materie Prime* 3(2):96-107.

_____ [1984b]: "Comercio de Productos Agrícolas entre Países en Desarrollo: América Latina durante 1962-1979" (The Commerce of Agricultural Products among Developing Countries: Latin America during 1962-1979). *Cuadernos de Economía* 21(63):169-206.

_____ [1985a]: "Subsidios Alimentarios en Países en Desarrollo: Estimaciones de Sus Costos y Effectos Distributivos" (Food Subsidies in Developing Countries: Estimates of its Costs and Distributive Effects). *Cuadernos de Economía* 22(66):329-336.

_____ [1985b, 1986]: *Exchange Rates and Trade Policy: Help or Hindrance to Agricultural Growth?* Washington, IFPRI, mimeo. Also in Maunder, A. and U. Renborg, eds., 1986, pp. 624-637.

_____ [1985c]: "Introducción." *Cuadernos de Economía* 22(66)169-173.

_____ [1987]: "Agriculture in the Uruguay Round: Interests of Developing Countries." *World Bank Econ. Rev.* 1(4):571-593. Also in IFPRI, Washington, Reprint No. 111, 1987.

Valdés, A., and J. León A. [1987]: "Politica Comercial Industrialización y Su Sesgo Anti-exportador: Peru 1940-1983" (Trade Policy, Industrialization, and Anti-Export Bias). *Cuadernos de Economia* 71:3-28.

Valdés, A., and E. Muchnik de R. [1984]: "Structure and Tendencies in Production, Consumption and Export Trade of Agricultural Products in Latin America." In *Proceedings of the Workshop Strengthening Agricultural Research in Latin America and the Caribbean.* Mexico, CIMMYT, pp. 6-33.

Valdés, A., G. M. Scobie, and J. L. Dillon [1979]: *Economics and the Design of Small-Farmer Technology.* Ames, ISU Press.

Valdés, A., and J. Zietz [1980]: *Agricultural Protection in OECD Countries: Its Costs to Less Developed Countries.* Washington, IFPRI, Research Report No. 21.

Van Es, J. C., E. A. Wilkening, and J. B. G. Pinto [1968]: *Rural Migrants in Central Brazil: A Study of Itumbiara, Goiás.* Madison, Univ. of Wisconsin, LTC Research Paper No. 29.

Veiga, A. [1974]: "The Impact of Trade Policy on Brazilian Agriculture, 1947-1967." Unpublished Ph.D. thesis, Purdue Univ., West Lafayette, IN.

_____ [1975]: "Efeitos da Política Comercial Brasileira no Setor Agrícola" (The Effects of Brazilian Trade Policy on Brazilian Agriculture). In Contador, ed., 1975, pp. 285-308.

_____ , ed. [1979]: *Ensaios Sôbre Política Agrícola Brasileira* (Essays on Brazilian Agricultural Policy). São Paulo, SA.

Veiga, A., and G. E. Schuh [1975]: "Politica Cambial e Exportações Agrícola no Brazil" (Exchange Rate Policy and Agricultural Exports in Brazil). *Agricultura em São Paulo* 22(1,2):1-47.

Vellianitis-Fidas, A. [1975]: "The Exchange Rate and U.S. Agriculture: Comment." *Amer. J. Agr. Econ.* 57:692-695.

Velloso, J. [1975]: "Human Capital and Market Segmentation: An Analysis of the Distribution of Earnings in Brazil, 1970." Unpublished Ph.D. dissertation, Stanford Univ., Stanford, CA.

Venezian, E. [1987]: *Chile and the CGIAR Centers: A Study of their Collaboration in Agricultural Research.* Washington, World Bank, CGIAR Study Paper No. 20.

Venezian, E. L., and W. K. Gamble [1969]: *The Agricultural Development of Mexico: Its Structure and Growth since 1950.* New York, Praeger.

Vilas, A. T. [1972]: "Estimativas de Funções de Oferta de Arroz para o Estudo de Goiás e suas Implicações Econômicas, Período 1948-1968" (Estimates of Rice Supply Functions for the State of Goiás and their Economic Implications, the 1948-68 Period). Unpublished M.S. thesis, Universidade Federal de Viçosa, Minas Gerais, Brazil.

———— [1975]: "A Spatial Equilibrium Analysis of the Rice Economy in Brazil." Unpublished Ph.D. dissertation, Purdue Univ., West Lafayette, IN.

Villa-Issa, M. R. de [1976]: "The Effect of the Labor Market on the Adoption of New Production Technology in a Rural Development Project: The Case of Plan Puebla, Mexico." Unpublished Ph.D. thesis, Purdue Univ., West Lafayette, IN.

Virmani, A. [1985]: *Credit Markets and Credit Policy in Developing Countries: Myths and Reality*. Washington, World Bank, Discussion Paper, Report No. DRD175.

Vogel, R. C. [1974]: "The Dynamics of Inflation in Latin America, 1950-1969." *Amer. Econ. Rev.* 64:102-114.

Von Gersdorff, R. [1960]: "Agricultural Credit Problems in Brazil." *Indian J. Economics* 41(161):151-171.

Von Pischke, J. D., and D. W. Adams [1980]: "Fungibility and the Design and Evaluation of Agricultural Credit Projects." *Amer. J. Agr. Econ.* 62:719-726.

Von Pischke, J. D., D. W. Adams, and G. Donald, eds. [1983]: *Rural Financial Markets in Developing Countries: Their Use and Abuse*. Baltimore, John Hopkins Univ. Press.

Von Thuenen, J. H. [1842, 1966]: *Der Isolierte Staat in Beziehung auf Landwirtschaft und National Oekonomic* (The Isolated State with Regard to Agriculture and National Economics). Hamburg, Perthes, first edition, 1842.

Wachter, S. M. [1976]: *Latin American Inflation*. Lexington, MA, Lexington Books.

Wade, N. [1973a]: "Agriculture: Social Sciences Oppressed and Poverty Stricken." *Science* 180(4087):719-722.

———— [1973b]: "Agriculture: Research Planning Paralyzed by Pork Barrel Politics." *Science* 180(4089):932-937.

Waggoner, G. R., and B. A. Waggoner [1971]: *Education in Central America*. Manhattan, Univ. Press of Kansas.

Walinsky, L. J., ed. [1977]: *Agrarian Reform as Unfinished Business: The Selected Papers of Wolf Ladejinsky*. New York, Oxford Univ. Press.

Walker, T. W., ed. [1985]: *Nicaragua: The First Five Years*. New York, Praeger.

Wall, D., ed. [1972]: *Chicago Essays in Economic Development*. Univ. of Chicago Press.

Wallace, R., and A. Ten Kate with the collaboration of A. Waarts and M. D. R. de Wallace [1979]: *La Política de Protección en el Desarollo Económico de México* (Protection Policy in the Economic Development of Mexico). México, Fondo de Cultura Económico.

Wallace, T. D., and D. M. Hoover [1966]: "Income Effects of Innovation: The Case of Labor in Agriculture." *J. Farm Econ.* 48:325-335.

Wallerstein, I. [1974]: *The Modern World-System: Capitalist Agriculture and the Origins of the European World-Economy in the Sixteenth Century*. New York, Academic Press.

Ward, J. O., and J. H. Sanders [1980]: "Nutritional Determinants and Migration in the Brazilian Northeast: A Case Study of Rural and Urban Ceará." *Econ. Dev. and Cultural Change* 29:141-163.

Watkins, M. H. [1963]: "A Staple Theory of Economic Growth." *Canadian J. Econ. and Polit. Science* 29(2): 141-158.

Weekes-Vagliani, W. [1985]: *Actors and Institutions in the Food Chain: The Case of Peru*. Paris, OECD.

Weinstein, M., ed. [1979]: *Revolutionary Cuba in the World Arena*. Philadelphia, PA, Institute for the Study of Human Issues.

Weisbrod, B. A. [1964]: *External Benefits of Public Education: An Economic Analysis*. Princeton, NJ, Princeton Univ., Industrial Relations Section.

Weiskoff, R. [1980]: "The Growth and Decline of Import Substitution in Brazil—Revisited." *World Dev.* 8:647-675.

Weist, R. [1970]: "Wage Labor, Migration and Household Maintenance in a Central Mexican Town." Unpublished Ph.D. thesis, University of Oregon, Eugene.

Welch, F. [1970]: "Education in Production." *J. Polit. Economy* 78:35-59.

Wennergren, E. B., and M. D. Whitaker [1977]: "Social Returns to U.S. Technical Assistance in Bolivian Agriculture: The Case of Sheep and Wheat." *Amer. J. Agr. Econ.* 59:565-569.

Wharton, C. R., Jr. [1960]: "The Economic Impact of Technical Assistance: A Brazilian Case Study." *J. Farm Econ.* 42:252-267.

―――― [1965]: "Education and Agricultural Growth: The Role of Education in Early-Stage Agriculture." In Anderson and Bowman, eds., 1965, pp. 202-228.

―――― , ed. [1969]: *Subsistence Agriculture and Economic Development*. Chicago, Aldine.

Wheeler, D. [1980]: *Human Resources Development and Economic Growth in Developing Countries: A Simultaneous Model*. Washington, World Bank, Staff Working Paper No. 407.

Whitaker, M. D. [1970]: "Labor Absorption in Brazil: An Analysis of the Industrial Sector." Unpublished Ph.D. dissertation, Purdue Univ., West Lafayette, IN.

Whitaker, M. D., and G. E. Schuh [1974]: "Problemas Relacionados com a Absorção da Mão de Obra no Brasil: Uma Análise do Setor Industrial" (Problems Related with Labor Absorption in Brazil: An Analysis of the Industrial Sector). *Revista de Economia Rural* 3(3):31-56.

―――― [1977]: "O Mercado de Trabalho Industrial no Brasil e suas Implicações para a Absorção de Mão-de-Obra" (The Market for Industrial Labor in Brazil and its Implications for Absorption). *Pesquisa e Planejamento Econômico* 7(2):333-366.

White, T. K. [1975]: "Credit and Agricultural Development—Some Observations on a Brazilian Case." In Patrick, Brainard, and Obermiller, eds., 1975, pp. 66-91.

Whyte, W. F., and D. Boynton, eds. [1983]: *Higher Yielding Human Systems for Agriculture*. Ithaca, NY, Cornell Univ. Press.

Wilford, W. T., and G. Christon [1973]: "A Sectoral Analysis of Disaggregated Trade Flows in the Central American Common Market, 1962-1970." *J. Common Market Studies* 12(2):159-175.

Wilkie, R. W. [1973]: "Toward a Behavioral Model of Peasant Migration: An Argentine Case of Spatial Behavior by Social Class Level." In R. N. Thomas, ed., 1973, pp. 83-114.

Willems, E. [1972]: "The Rise of a Rural Middle Class in a Frontier Society." In Roett, ed., 1972, pp. 325-344.

Williamson, J. G. [1977]: "Strategic Wage Goods, Prices, and Inequality." *Amer. Econ. Rev.* 67:29-41.

Willmore, L. N. [1976]: "Trade Creation, Trade Division, and Effective Protection in the Central American Common Market." *J. Dev. Studies* 12:396-414.

Wionczek, M. S. [1969-70]: "United States Investment and the Development of Middle America." *Studies in Comparative International Dev.* 5(2):3-17.

―――― [1971a]: "The Pacific Market for Capital, Technology, and Information and its Possible Opening for Latin America." *J. Common Market Studies* 10(1):78-95.

―――― [1971b]: *Inversión y Tecnología Extranjera en América Latina* (Foreign Investment and Technology in Latin America). México, Joaquin Mortiz.

_____ [1972]: "Un Punto de Vista Latino-Americano Sobre los Problemas de Ciencia y Tecnología" (A Latin American Point of View about the Problems of Science and Technology). *Commercio Exterior* 22(4):346-349.

_____ , ed. [1974a]: *La Sociedad Mexicana: Presente y Futuro* (The Mexican Society: Present and Future). México, Fondo de Cultura Economia, second edition.

_____ [1974b]: "El Subdesarrollo Cientifico y Tecnológico: Sus Consecuencias" (Scientific and Technological Underdevelopment: Its Consequences). In Wionczek, ed. 1974a, pp. 359-384.

_____ [1981]: "On the Viability of a Policy for Science and Technology in Mexico." *Latin Amer. Research Rev.* 16(1):57-78.

Witherell, W. H. [1969]: "A Comparison of the Determinants of Wool Production in the Six Leading Producing Countries: 1949-1965." *Amer. J. Agr. Econ.* 51:138-158.

Wogart, J. P., and J. S. Marques [1984]: "Trade Liberalization, Tariff Redundancy, and Inflation: A Methodological Exploration Applied to Argentina." *Weltwirtschaftliches Archiv* 120:18-39.

World Bank [1979]: *World Development Report, 1979.* Washington.

_____ [1981]: *Accelerated Development in Sub-Saharan Africa; An Agenda for Action.* Washington.

_____ [1985a]: *World Development Report, 1985.* Washington, Oxford Univ. Press.

_____ [1985b]: *Agricultural Research and Extension: An Evaluation of the World Bank's Experience.* Washington.

_____ [1986a]: *Poverty in Latin America: The Impact of Depression.* Washington.

_____ [1986b]: *Agricultural Mechanization: Issues and Policies.* Washington, Agriculture and Rural Development Report.

_____ [1986c]: *World Development Report, 1986.* Washington, Oxford Univ. Press.

Yates, P. L. [1981]: *Mexico's Agricultural Dilemma.* Tucson, Univ. of Arizona Press.

Youmans, R., and G. E. Schuh [1968]: "An Empirical Study of the Agricultural Labor Market in a Developing Country, Brazil." *Amer. J. Agr. Econ.* 50:943-961.

Yver, R. [1971]: "The Investment Behavior and the Supply Response of the Cattle Industry in Argentina." Unpublished Ph.D. dissertation, Univ. of Chicago.

Zietz, J., and A. Valdés [1986]: *The Costs of Protectionism to Developing Countries: An Analysis for Selected Agricultural Products.* Washington, World Bank, Staff Working Paper No. 769.

_____ [1988]: *Agriculture in the GATT: An Analysis of Alternative Approaches to Reform.* Washington, IFPRI, Research Report 70.

Zimbalist, A., and C. Brundenius [1989]: *The Cuban Economy: Measurement and Analysis of Socialist Performance.* Baltimore, Johns Hopkin Univ. Press.

Zunz, O., ed. [1985]: *Reliving the Past: The Worlds of Social History.* Chapel Hill, NC

PART FOUR

Philosophic Foundations of Agricultural Economic Thought
from World War II to the Mid-1970s

Philosophic Foundations of Agricultural Economic Thought from World War II to the Mid-1970s

Glenn L. Johnson*

Chapter I. Introduction

Philosophic considerations helped to shape the history and literature of agricultural economics since World War II. The philosophic orientation of agricultural economists determined the kind of literature they produced, and in turn that work and literature determined their philosophic orientations. Agricultural economics literature is better understood when one is sensitive to its philosophic orientations.

Scholars writing on philosophy and methodology sometimes apologize for discussing a boring, semantic, unproductive subject [Harrod, 1938]. No such apology is made here. I do not find it boring or unproductive to understand the philosophic foundations that guide our work and literature, and I do find it important to understand how those foundations are, themselves, changed by our work.

1. Structure of This Review

This review of philosophy in agricultural economics since World War II will be carried out in a farm management tradition, incongruous as that may seem.

---

*The author would like readers to know that this chapter was written in the late 1970s and therefore does not include current research nor does it reflect his most current points of view.

First we will define terms and basic concepts. Then we will take a *beginning inventory* as of the end of World War II. Following that, what happened between World War II and 1976 will be examined. The beginning inventory, plus *the analysis of what went on in this highly productive, controversial, and dynamic period*, will provide the basis for an *ending inventory* that will indicate the present condition and changes taking place in the way agricultural economists work with philosophy.

2. Coverage

The geographic area of analysis will be primarily the area served by the journal of the American Agricultural Economics Association. The *American Journal of Agricultural Economics* (AJAE) has worldwide significance and serves areas also served by the English, Canadian, and Australian journals. World War II seriously disrupted agricultural economics in Germany, Japan, France, and Italy. Professor Nou [1967] has written an excellent summary of European agricultural economics that covers the immediate prewar period in contrast to the U.S. history by the Taylors [1952] that ends its coverage in 1932. The 50th anniversary issue of the English *Journal of Agricultural Economics* contains useful review articles for our subject [Hunt, 1976; Coats, 1976; Giles, 1976]. Emphasis on literature reported or considered in the *AJAE* with attention to complementary literature permits one to cover many of the important developments in the world's agricultural economics literature from the end of World War II to the early 1970s. Such a procedure will not work in the future as other journals are now functioning well. Fledgling journals in many countries are becoming increasingly important, and the new *European Review of Agricultural Economics* will undoubtedly become highly important in world agricultural economics circles.

It should also be noted that some important writings on methodology and philosophy have not been printed; instead, they have appeared as "phantom" literature in mimeographed form, and unbound, unprinted proceedings of seminars and conferences where intense interactions influenced the philosophies and methodologies of leaders. Examples include the crucial Land o' Lakes and Black Duck farm management workshop proceedings,[1] as well as papers presented at the various regional land-tenure and marketing research committees and at the University of Chicago conference on efficiency.

Work, philosophy, and methodology are interrelated. Our work is guided by our philosophies and their accompanying methodology. And, in turn, successes and failures in our work influence both our selection among philosophies and our own philosophic development. Thus, the other review articles of the AAEA survey of agricultural economic literature are most valuable summaries of the work which has both conditioned and been guided by our philosophic thought.

3. Meanings and Basic Concepts

In reply to the frequent charge that philosophical and methodological discussions "are merely semantics," one must note that communications depend on our understanding of the words we use. If we are going to communicate with each other about the interrelationships between our philosophical thought and our work, we must have a vocabulary with which to discuss such matters. This section provides that vocabulary.

Agricultural economists are concerned with the generation, distribution, and use of knowledge. Central to this concern is epistemology—study of the theory of science and the grounds of knowledge especially with respect to its limits and validity [Runes, ed., 1960]. Concern for truth is important in generating, distributing, and utilizing knowledge [Knight, 1940].

THE POSITIVE, NONNORMATIVE, AND PRESCRIPTIVE

Agricultural economics literature contains at least three kinds of knowledge: nonnormative or positive, normative, and prescriptive knowledge. Machlup's [1969] excellent discussion of the meaning of positivism and normativism in economics literature indicates that we should define these three terms. By *positive (or nonnormative)* knowledge, we mean knowledge about conditions, situations, and things—knowledge that does not have to do with their goodness and badness. The use of the word nonnormative, instead of positive, is preferable in contexts where the word positive could imply an unintended acceptance of positivistic philosophy. The word positivistic is used to denote tendencies toward positivism and the positive. The word normative, on the other hand, is used here to denote the goodness and badness, *per se*, of conditions, situations, and things. The word normativistic is used to denote tendencies toward normativism and the normative. The normative is to be distinguished from the prescriptive in that the prescriptive indicates which action or goal "ought to" be sought or attained. The prescriptive depends on both the positive and the normative in a functional way. The function relating the positive and normative to the prescriptive is a decision-making rule. As contrasted to the normative, the *prescriptive* deals with "rightness and wrongness," not with goodness and badness. Following Lewis [1955] we note that it is not always *right* to do that which is good because it may be possible to do something which is better (or more good). And, the *wrong* is not necessarily associated with the bad because it may be right to do (or try to do) that which is bad if it is the least bad which can be done. When there is no alternative course of action which is good, the decision rule becomes a loss minimization rule that makes it *right* to do *bad* (the least possible bad). Conversely, when maximizing good it is *wrong* to do *good* (less good than one could do).

ANALYTIC AND SYNTHETIC

From an epistemological standpoint, it is also worthwhile to distinguish between *analytic* and *synthetic* [Carnap, 1953; Kemeny, 1959, ch. 2]. Purely analytical statements are "formal statements." They have a logical form but contain no empirical knowledge. Logical statements are a consequence of axioms and grammatical rules which permit one to derive other statements from initial axioms. By contrast, synthetic statements and some applied logical statements [Carnap, 1953] purport to describe the real world. Their empirical content comes from primitive terms which are substituted for formal terms in analytical statements. *Primitive terms* are not defined and are not matters of logic; instead they emerge out of the experience of people [Popper, 1959, p. 83; Rudner, 1966, p. 19]. We know what a primitive term means from experience. Eventually, we are able to use such terms to communicate about similar or mutually shared experiences. Once primitive terms are introduced into analytical systems, the systems are said to become synthetic or to be descriptive of the real world—to have empirical content. In agricultural economics literature, this distinction was discussed well by Halter and Jack [1961] and by Schmitt and Timmermann [1969] in commenting on Castle's [1968] observations on objectivity.

Despite the efforts of econometricians, Mini [1974] is concerned that economists overemphasize the analytical (theoretical) at the expense of the synthetic (empirical). Leontief [1971] gives agricultural economists higher marks in this regard than he gives general economists.

The descriptive or empirical content of a synthetic sentence or concept is no better than the primitive terms on which it is based. Similarly, the logic is no better than the axioms which form the basis for the analytical sentences which are transformed into synthetic sentences [Popper, 1959, pp. 81ff.]. Axioms are taken as given and are "unproven." Primitive terms are experienced but never proven. Thus, synthetic or descriptively empirical sentences are always subject to error [Carnap, 1953, p. 125] and dependent on *subjective interpretations* of sense impressions as those interpretations are conditioned by the current stage of development of science and society.

DISCIPLINARY, SUBJECT-MATTER, AND PROBLEM-SOLVING EFFORTS

In reviewing agricultural economics research literature especially, but also agricultural literature about teaching and extension, it is worthwhile distinguishing among three broad types of efforts. The *first* is purely *disciplinary* having to do with improving and teaching the theories, data, and quantitative techniques of the discipline of economics. Disciplinary knowledge can be classified into two categories: those of known and unknown relevance. Those of known relevance are necessarily related to problem-solving research or to subject-matter research which is, in turn, related to problem-solving research. The *second* kind is *subject-*

matter oriented. Examples of subject-matter research include land tenure, food and nutrition, unemployment, energy, and farm management. Subject-matter research is concerned with a kind of information that is broader than the information organized systematically in an existing discipline. The practical reason for dealing with a subject is its relevance to a set of problems. For example, work is often done on the national accounts of a country or of a subsector. Such knowledge goes beyond economics to deal with political subdivisions, technical aspects of society, and human interests and is useful in solving a wide range of problems. However, such knowledge is typically inadequate for solving any specific problem in the set of problems involved. Typically subjects are multidisciplinary because solving most practical problems requires different types of knowledge. Besides, as pointed out above, if not multidisciplinary, a subject-matter area would be disciplinary. There is some circularity involved in defining a subject and the set of problems for which it is relevant. The kind of information considered in defining a subject is part, at least, of the information required in solving a specific set of problems. On the other hand, the set of problems is that set that requires the specific kind of information considered for the subject definition. Important subjects correspond with important sets of problems. *Problem-solving* research deals with a particular practical problem—one problem—before a specific decision maker or, perhaps, set of decision makers. A practical problem is time, space, and decision maker specific. Practical problems are typically multidisciplinary with the mix of disciplines specific to the problem at hand. The solutions to problems are prescriptions as to what ought to be done to solve them. Positive and normative knowledge contribute to the solution of practical problems but are individually inadequate to solve them.

COHERENCE, CORRESPONDENCE, CLARITY, AND WORKABILITY

In science and in human activity, in general, there are four tests which, if failed, establish the falsity or unacceptability of knowledge [Northrop, 1947; Feigl, 1953]. A concept that has been subjected to and has passed these tests can be referred to as objective. An investigator willing to subject his concepts to such tests and to abide by the results can be said to be objective. The less ambiguous and, hence, more falsifiable our knowledge statements are, the greater "faith" we have in those that survive testing [Popper, 1959, ch. 4]. In any event we do not prove empirical knowledge—we only fail to disprove it, which means that all accepted empirical knowledge is accepted on the basis of some "faith" or risk of being disproven later. (See subsection entitled "The Meaning of Truth in Economics.")

The *first* test is that of *coherence or logical consistency* [Runes, ed. 1960, p. 58]. Analytical knowledge as defined above is expected to be logically coherent. So are synthetic systems which include primitive terms as well as axioms. Logical co-

herence among concepts includes comprehensiveness; i.e., the wider the range of logical consistency of a tested concept with other accepted concepts, the greater its acceptability. Thus, the coherence test is somewhat more than a test for analytical truth. Leibniz and Descartes stressed the coherence test [Mitroff and Turoff, 1973]. Mini [1974] has argued in a recent book that economists overemphasize coherence and that they should place more emphasis on *correspondence*, which is the *second* test of truth [Runes, ed. 1960, p. 68]. Such scientists and philosophers as Bacon and Locke placed primary emphasis on correspondence with experience as a test of truth or falsity [Mitroff and Turoff, 1973]. As pointed out above, empirical knowledge is never proven. It must be stressed that it is not given to mankind to know empirical truth beyond question. About all one, either as citizen or scientist, can do is establish empirical concepts "well enough" for the purposes at hand. Descriptive or empirical concepts are not compared directly with reality to find out if they are true or false. Instead, the tested concept is compared with a new concept (not reality) based on new experiences not used in formulating the tested concept to see if the two correspond [Rudner, 1966]. If there are sufficient "degrees of freedom" in the new experiences (or even in the observations used to construct the original concept) and if the two concepts agree, the tested concept passes the correspondence test. The test is one of correspondence between the tested concept and a new concept and of degrees of freedom, not of correspondence between a concept and reality—our minds can only compare concepts. Some analysts and students of research methods regard a conception that has passed the coherence test as *validated* and one which has passed the correspondence test as *verified*. The *third* test is that of *clarity*. Linguistic analysts and others have pointed out that it is difficult to know if ambiguous statements are false because we do not know precisely what they mean. It is not until statements are clear and unambiguous that one can apply the tests of either correspondence or coherence. Ambiguous statements are less clear and less falsifiable [Popper, 1959] than unambiguous ones because they have more than one possible meaning. A *fourth* criterion used by Dewey [1938, p. 160] in establishing the falsity of statements is the test of workability [Mitroff and Turnoff, 1973, pp. 69ff.; Northrop, 1947, ch. 1]. A precondition for applying such a test is objective, descriptive, normative, or prescriptive knowledge. When either a normative or positive concept is used to solve a problem, the workability test consists of checking to see whether the prescribed solution "works" in the sense of solving the problem. The test of workability is one of the cornerstones of pragmatism and has been stressed by Dewey [1938] and John R. Commons [1934] in institutional economics, and recently by Georgescu-Roegen [1971].

In application, the above four tests for truth and objectivity are social; hence, the pursuit of science is a social activity and the knowledge it produces is a social product. If knowledge must be free of social considerations in order to be objec-

tive, then all knowledge (scientific or otherwise) must be unobjective [Lenzen, 1955; Georgescu-Roegen, 1971]. And as disciplines develop through time as integral parts of society, they and the knowledge they generate are "time" and "society" dependent. For a somewhat different point of view on objectivity, see Breimyer [1967] and Grove [1968].

Chapter II. The Beginning Inventory

Our beginning inventory for the postwar era covers activity *circa* World War II with respect to methods based on three or four broad philosophic approaches. These approaches include Wisconsin pragmatic institutionalism, positivism, and utilitarianism [Salter, 1948]. Positivism, in turn, can be divided in agricultural economics into Cornell empiricism and the more theoretical statistical developments which found pre- and post-World War II expression in the works of Henry Schultz on price analysis at the University of Chicago, Mordecai Ezekiel in the U.S. Department of Agriculture, and George Snedecor at Iowa State College, not to mention members of the Cowles Commission then at the University of Chicago. See Judge *et al.*, eds. [1977] for a review of the statistical literature of this period. Recognizing the desirability of distinguishing between Cornell empiricism and the more theoretical work of statisticians and econometricians, this inventory is divided into four parts: Wisconsin institutionalism, Cornell empiricism, utilitarianism, and theoretical statistics and econometrics.

The order in which these four subjects are discussed is not related to their importance or dominance. Utilitarianism is really a special form of normativism. Neoclassical utilitarianism is the special form of utilitarianism which evolved out of the Marshall-Clark synthesis of supply and demand determining forces to explain "value and exchange." Normativism, a broader classification, also includes nonutilitarian philosophies *vis-à-vis* goodness and badness. A special form of positivism somewhat paradoxically dubbed "conditional normativism" by the author [Glenn L. Johnson, 1960b] and picked up by Castle [1968] will be discussed in connection with the influence of Gunnar Myrdal [1944]. The work of the International Association of Agricultural Economists (IAAE) was brought to a standstill by World War II and will not be discussed as an aspect of the beginning inventory except under the above four headings, but it will be discussed as important for the post-World War II period.

1. Wisconsin Institutionalism

At the end of World War II, Wisconsin institutionalism was intact, strong, and exerting a major influence. Its then grand old men, its work horses, and its students helped make major institutional changes to improve American agriculture in the depression and World War II era. Richard T. Ely, George S. Wehrwein,

Asher Hobson, O. C. Stine, and many others had carried John Dewey's pragmatic philosophy, which had been built into Wisconsin institutionalism by John R. Commons [1934], into the worlds of politics, administration, and institutional change. Wisconsin institutionalism is, of course, related to the German historical school with its emphasis on "society" and "the state" instead of on the activities of individual consumers and producers. Because of his interest in the state, the influence of Hegel on the German historical school was much greater than his influence on Wisconsin institutionalism. John Dewey's impact on Wisconsin institutionalism made it (relative to the German historical school) more democratic and more participatory for affected persons than German historicism. It is hard to conceive of Wisconsin institutionalists supporting the authoritarianism of either a Hitler or a communist party elite composed of a small proportion of a country's population. Neither are we likely to find it supporting control of large blocks of a country's working population by either a corporation or a union bureaucracy. The decades between the two world wars was a period of great institutional change in the United States and the Wisconsin scholars contributed much. Two of the first post-World War II meetings of the American Farm Economic Association were held in Wisconsin at Green Lake. Then younger institutionalists such as Kenneth Parsons, Ray Penn, Leonard Salter, and Erven Long came to the fore at Wisconsin while John Timmons, Rainer Schickele, and others held forth at other institutions. Charles Stewart, Marshall Harris, and Joseph Ackerman, while not trained at Wisconsin, were influenced by Wisconsin viewpoints.

Historically, Wisconsin institutionalism is pragmatic, the pragmatism coming from John Dewey *via* John R. Commons. Dewey in turn was influenced by Peirce [Dewey, 1938, pp. 9n, 12, 14, 156, 468, 470; Mitroff and Turoff, 1973; Northrop, 1947, ch. 1; Churchman, 1961, ch. 9]. Pragmatism is characterized, first, by the use of the test of workability as a criterion of truth and, second, by the belief that knowledge about good and bad (referred to as *normative* in this review) is dependent for its truth upon the truth of positivistic knowledge (having nothing to do with good and bad), the converse also being true. Pragmatists are interested in right actions to solve specific problems. In their discussion of problems, they do not distinguish between practical problems of decision makers and disciplinary questions (problems); indeed, they rule out by implication the positivistic questions of the biological and physical sciences and the normative questions of the humanities by positing, metaphysically, that answers to positive and normative questions are mutually interdependent. Pragmatists are practical and action-oriented and it was this orientation that Dewey contributed both to institutionalism and American education. The pragmatic interest in knowing which act is the "right" solution to a problem and the belief that purely normative and positivistic sets of information are mutually interdependent make it nonsensical, within the pragmatic philosophy, to pursue answers of questions concerning goodness and badness independently of the pursuit of answers to positivistic questions [Dewey, 1939; Mitroff and Turoff, 1973, p. 69], and vice versa. Thus, both positivism (which holds metaphysically that there is nothing objective to know about good and bad) and normativism (which holds metaphysically that knowledge of good and bad can be held independently of positive knowledge) are related to pragmatism. Pragmatists address themselves to problems and their solutions—to them knowledge is relevant and true or false in the context of practical problems.

Agricultural economics lost an articulate, competent expositor of pragmatic methods and institutional economics when Leonard Salter [1948] perished in the LaSalle Hotel fire early in the period covered by this review.

2. Cornell Empiricism

At the onset, it must be recognized that Cornell was not monolithic—it had its theorists, normativists, and prescribers. With all that, however, it is also true that Cornell farm management was a center of positivism in agricultural economics before and after World War II. Cornell empiricists were in a strong national position, particularly *vis-à-vis* farm management. Generally speaking, U.S. farm management prior to World War II was dominated by Cornell's empiricism, which can be more nearly characterized as "pure" than as "logical" positivism although such terminology was not used by such critics as John D. Black, T. W. Schultz, and others. Instead of saying that the Cornell empiricists were pure rather than logical positivists, these critics argued that the meaning of empirical observations is greater if theoretically interpreted. The theory Black, Schultz, and others advocated for use in interpreting data was neoclassical utilitarian economics [Schultz, 1939b; Black, 1939].

Cornell empiricism tended to be based upon the writings of Karl Pearson as expressed in his 1900 *Grammar of Science* [Salter, 1948]. Positivism is based on the presupposition that empirical knowledge is derivable only from experience. If restricted only to experience without the use of logic, positivism can be regarded as "pure." If experience is interpreted with logic, "logical positivism" results. A restrictive characteristic of positivism is its acceptance of the metaphysical presupposition that there are no normative experiences of goodness or badness from which to develop primitive terms to use in making empirical (synthetic) normative concepts. The Cornell economists tended to be pure as contrasted to logical positivists. Cornell empiricists did well both administratively and practically. At the end of World War II many of them were heads of departments of agricultural economics and/or farm management, deans, and extension directors throughout the land-grant colleges. Overseas, the Cornell empiricists had important interactions with English, German, French, Scandinavian, and Italian agricultural economists. They were farm- and farmer-oriented both by background and as a result of data collection activities that kept them in close touch with farmers and their problems. They played important practical roles despite their positivism, which should have precluded their use of normative knowledge both to define and to solve practical problems. Although Schultz [1939a] did criticize them for failing to change the kinds of data they collected as the agriculture economy went through the changing problems of the great depression, the Cornell empiricists were relevant.

3. Neoclassical Utilitarianism

Neoclassical economics found its first full expression, of course, in the works of Marshall in England and Clark in the United States. It explains value in exchange in terms of supply functions based on production costs and demand functions based on utility. As production costs are based on the value of resources in producing alternative products that possess utility, its neoclassical analysis is so strongly influenced by utilitarianism as to make it reasonable to call it "neoclassical utilitarianism." The adjective "neoclassical" distinguishes neoclassical utilitarianism from the utilitarianism of Bentham and others of importance in law, political science, and philosophy which did not explain "value in exchange."

At the end of World War II, neoclassical utilitarianism was not nearly as well established in agricultural economics as it is now. John D. Black was firmly ensconced in a chair at Harvard University as a neoclassical, utilitarian, agricultural, production economist. Henry C. Taylor was primarily a classical or neoclassical utilitarian with emphasis on consumption as well as production economics [Nou, 1967]. Earlier at Cornell, Davenport taught classical and neoclassical economics to agricultural economists including those nurtured in Cornell farm management empiricism. T. N. Carver was in the neoclassical tradition. At Iowa State College, the Department of Economics and Sociology came forward with emphasis on neoclassical, utilitarian economics under the leadership of T. W. Schultz before the oleomargarine blowup scattered its personnel.

It is interesting to note that both Schultz and Black were Wisconsin trained. Neither produced much significant post-World War II literature on research methods although both greatly influenced agricultural economics research methods by their pre-World War II publications and their lifelong teaching. Their teaching was mainly in the neoclassical, market-adjustment tradition. As consultants, however, both were pragmatic. Neither became keenly sensitive to Pareto optimality. Black's methodological contributions were mainly *via* the Social Science Research Council (SSRC) prior to World War II. Schultz's article [1939b] on scope and method was also pre-World War II as was Black's reply [1939]. After World War II, Black wrote on value judgments [1953a]. In doing so he did not consider Pareto optimality. Perhaps because of his contact with pragmatism at Wisconsin, Black did not distinguish in his writings between good and bad on the one hand, and right and wrong on the other [1953a, p. 293]. At another point [p. 294], he implied that economic value judgments deal with values in exchange while intrinsic value judgments are noneconomic. He distinguished three roles for economists in dealing with values—as investigators, generalists, and administrators.

At the end of World War II, William Murray, Kenneth Boulding, Earl Heady, Geoffrey Shepherd, and Gerhard Tintner were more or less holding the neoclassical, utilitarian fort together at Iowa State pending postwar recovery from the oleomargarine affair. Within agricultural economics, personnel interested in marketing had a sustained interest in neoclassical utilitarian economics. F. L. Thomsen, Warren Waite, Ray Bressler, Geoffrey Shepherd, and marketing economists and price analysts in the USDA operated more or less within the neoclassical utilitarian tradition. At national policy levels, both the Wisconsin pragmatic institutionalists and the utilitarians were important. John D. Black, T. W. Schultz, Murray Benedict, Henry C. Taylor, Howard Tolley, Frederick Waugh, O. C. Stine, and others employed neoclassical reasoning in reaching policy conclusions. The neoclassical utilitarianism used was mainly of the Pigouvian [1932] welfare economics variety [Robinson, 1968]. As the questions about the interpersonal validity of welfare measurements raised by Pareto and worked into economic theory by John R. Hicks [1939] were not yet of concern, the neoclassical utilitarian apparatus was used to reach welfare conclusions without questioning the interpersonal validity of welfare comparisons. In the pre-Hicks decades, progressive income taxation was introduced into the United States and Western democracies. Further, U. S. agricultural economists had just emerged from more than a decade during which they contributed much to decisions about institutional changes designed to redistribute the ownership of many rights and privileges as well as monetary incomes within agriculture and between the farm and nonfarm sectors. While John R. Hicks's book was being read at such places as Iowa State College, it was not generally used and the introduction of "modern welfare economics" was something for the future of agricultural economics [Brownlee, 1948; Tweeten, 1970]. Nonetheless, there were very uneasy feelings among the neoclassical utilitarians; enough positivism had gotten through to agricultural economists *via* Cornell empiricism and contact with physical scientists that agricultural economists were wary about the "scientific objectivity" of dealing with questions of goodness or badness. One way out was to assume answers to questions of goodness and badness and then use the neoclassical utilitarian apparatus to determine how to maximize the difference between the good (sometimes money and sometimes utility) and the bad (again sometimes money and sometimes utility). The tendency to do this was greatly strengthened by Gunnar Myrdal [1944], who recognized that values (knowledge of "good and bad") had to be taken into account in solving practical problems. Though he yielded, perhaps naively, to the positivistic position that values cannot be investigated objectively, he advocated that all value premises be explicitly stated beforehand to develop a method of analysis conditioned by normative assumptions. Analyses so based have been dubbed "conditionally normative" by the author of this paper. Appendix II of Myrdal's *Dilemma* [1944] was assigned by professors of agricultural policy and research methodology in the immediate post-World War II period.

4. Theoretical Statistics and Econometrics

In addition to the Cornell empiricists, groups of statisticians and economists particularly interested in agricultural data collection, parameter estimation, and analysis of problems were developing at a number of locations. At Iowa State College, this development received assistance from Snedecor. At the University of Chicago, Henry Schultz took an early lead with his pre-World War II groundbreaking analyses of agricultural prices. During the war, the Cowles Commission, then at the University of Chicago, extended the domain of econometrics. In the USDA, Elmer Working made substantial progress in demand analysis followed by the work of Mordecai Ezekiel, Richard Foote, F. L. Thomsen, Russell Ives, Meyer Girschick, and others. Mordecai Ezekiel [1930] brought out his widely used book on correlation analysis prior to World War II [Judge *et al.*, eds., 1977]. Earlier at Minnesota and after the war in California, Holbrook Working made substantial progress [Judge *et al.*, eds., 1977]. The statisticians and budding agricultural econometricians responsible for these developments tended toward positivism in their underlying philosophies. They sought to produce positivistic quantitative estimates and, in doing so, avoided the normative as something essentially unknowable or in any event not amenable to quantification and application of probability calculus and mathematics. For the most part econometricians and price analysts seemed to regard prices as positive (probably because they were quantifiable) rather than normative even though anyone with common sense knows that prices are "*values* in exchange." In the immediate post-World War period, this group proved itself so competent in improving estimates and manipulating data that its success and tendency toward positivism concentrated the attention of agricultural economists on positivism.

Chapter III. Developments from World War II through 1976

The postwar period started with a rush. Older members of the profession were anxious to re-establish teaching and research programs while younger men who had not been in the military were anxious to get such work under way and World War II veterans, financed by the G.I. Bill, were anxious to enter graduate study to catch up for the years they had lost professionally to the military. Prior to and during the war period, many experiences pointed to the inadequacies of old ways of doing things. Leonard Salter [1948] wrote effectively on needs to improve research methods in land economics and in doing so provided an excellent review of the role of philosophy in agricultural economics prior to World War II. There was a widespread desire to improve agricultural economics. These desires and needs were documented in the SSRC's prewar reports on agricultural economics [T. W. Schultz, 1939b]. They were shared by some administrators and were particularly strong among working staff members and students. Graduate work in agricultural economics was re-established at Harvard. T. W. Schultz, W. H. Nicholls, and D. Gale Johnson at Chicago picked up many students who would have gone to Ames before the war and work got under way again at St. Paul, Madison, Berkeley, Urbana, Bozeman, Storrs, Lexington, Raleigh, Ames, etc. At Harvard, John D. Black continued and expanded his group of graduate students. The Farm Foundation, taking a lesson from a successful prewar regional study of livestock marketing, decided to work regionally to improve research methods in the rural social sciences. In this connection, there was close collaboration between Frank Peck, managing director of the Farm Foundation, Noble Clark from the University of Wisconsin, and Joseph Ackerman, who became the new managing director of the Farm Foundation. Under Peck's and then Ackerman's leadership, a number of agricultural economics committees were established. The work of these committees involved philosophic points of view that substantially influenced research methods and philosophy of agricultural economics work [Glenn L. Johnson, 1955; Jensen, 1977]. The land tenure committee tended to be a rallying point or stronghold for pragmatic Wisconsin institutionalists while the farm management research committee became the battle ground *first* between farm management empiricists and Wisconsin pragmatists, and *later* (and much more fundamentally) between empiricists and neoclassical utilitarians [Jensen, 1977]. The farm management extension committee was more dominated by farm management empiricism although pragmatism had its impact on the work of that committee as a result of Dewey's impact on American education and hence, agricultural extension. NCR-20, the North Central Regional Marketing Committee, started out with a heavy emphasis on firm adjustment and efficiency following a market-adjustment, neoclassical, utilitarian approach, modified with an engineering slant originating in part with the engineering training of Ray Bressler [French, 1977].

1. A Short Summary of Agricultural Economics Work, 1946-76

As a prelude to examining the interdependencies between work (research and teaching) and philosophy, a short summary is needed of the work of agricultural economists from 1946 to 1976.

Early in the period there was a heavy emphasis on national policy and commodity programs. William H. Nicholls, D. Gale Johnson, and Frederick Waugh won the first three of eighteen awards in the *Journal of Farm Economics* price policy essay contest [Nicholls and D. Gale Johnson, 1946]. The essays dealt with price and income measures, and complementary measures in the postwar transition period. T. W. Schultz's *Agriculture in an Unstable Economy* [1945] and D. Gale Johnson's book on forward prices [1947] started the postwar emphasis on policy. This interest in policy problems was continued through the 1950s and early 1960s at the national level by Cochrane, Brandow, Hathaway, and others. John D. Black along with Bonnen and others studied the productive capacity of American agriculture. They placed heavy emphasis on the subject of food, both internationally and domestically. In the same period USDA research focused on subjects relevant to the national agricultural problems and issues of the times as well as on the problems and issues themselves.

Post-World War II farm management research was exciting and dynamic. Neoclassical utilitarian economics was brought to bear on whole farm, enterprise, and enterprise combination problems. Production functions were fitted to cross-sectional, time series, and experimental data [Jensen, 1977; Woodworth, 1977; Day and Sparling, 1977]. Managerial processes were studied and the results transferred to textbooks, experiment station bulletins, and extension reports [Glenn L. Johnson *et al.*, eds., 1961]. Farm management workers also became interested in what their research could contribute to macro policy studies. This interest led to a series of adjustment studies by the different regional committees and to the establishment of the Center for Agricultural Adjustment (CAA) at Iowa State University (ISU) [Jensen, 1977].

The prewar interest in statistics expanded to econometrics, mathematical economics, linear programming, input/output analysis, simulation, etc. By the late fifties, a number of people including T. W. Schultz [1959], the author [Glenn L. Johnson, 1957], Brinegar, Bachman and Southworth [1959], and others sensed that many farm management researchers were losing touch with the problems of farmers and that their work was, first, increasingly disciplinary in nature and technique-oriented, and second, not particularly relevant at either micro or macro levels.

Agricultural marketing received impetus in the postwar period from the Research and Marketing Act. There was emphasis on problem solving through greater efficiency at the individual firm level, i.e., on market adjustment to external changes in institutions, technology, and demand. Bressler's influence led to an early postwar expansion of the economic/engineering approach [French, 1977]. By 1959, an unease appeared about marketing research. Such persons as Robert Clodius, Ray Bressler, and Paul Farris became interested in the impacts of market adjustments on the distribution of property ownership and other forms of market power. An "industrial organization" interest in "structure, conduct, and performance"—in structuralism—developed [Farris, ed., 1964]. Though this interest was pragmatic in some senses, it was somewhat more academic or disciplinary than the earlier firm efficiency work in marketing.

George Brandow headed the Food Commission Study of food marketing in which subject matter relevant to national marketing problems was addressed [1966].

In the 1960s, domestic agricultural economics tended to lose its practical problem-solving orientation. The farm bloc no longer kept practical pressure on the agricultural economists of the USDA. Indeed, O. V. Wells disbanded the old Bureau of Agricultural Economics (BAE), in part, to deconcentrate the agricultural economics budget into a less vulnerable target for congressional cost cutters [Wells, 1954; Hardin, 1946]. Later, disciplinary and technical interests of farm management, econometrics, marketing, and policy researchers tended to eclipse interests in practical problems, either public or private, while the number of farmers declined to the point that people feared farmers might be outnumbered by the agrarian bureaucracy. Later, Roger Gray [1970] wrote a delightful allegory about black-footed ferrets (agricultural economists), an endangered species which fed on prairie dogs (farmers). Gray noted that there was a problem because a federal agency (the USDA) was responsible for protecting the ferrets while it eradicated the prairie dogs. Some feared agricultural economics was dead — the *Journal of Farm Economics* was renamed the *American Journal of Agricultural Economics* in 1968 — some wanted to go further and call it a journal of applied social science. The loss of perspective, purpose, and sense of relevance in agricultural economics was probably at its worst when the racial, social, antiwar, and anti-establishment unrest broke out in the late 1960s. This was accompanied and/or followed by environmental concerns, consumerism, and anti-agrarian establishment views. Then came Watergate in 1972, the Egyptian invasion of the Sinai, energy shortages called to our attention by the consequent oil embargo, the elimination of market surpluses of food in the United States, worldwide food shortages, and the grain sales to Russia followed by U. S. export embargoes on grain.

In the remainder of this paper, the changes in agricultural economics that were induced by the upheavals of the late sixties will be referred to as "responses to the crisis of the late 1960s." Seldom does a study area or discipline encounter such an abrupt change. Agricultural economics seemed to go overnight from a shortage to a surfeit of practical problems. Moreover, many of the problems were not amenable to solution by "making the market work better"; instead, their solutions required nonmarket changes in institutions, technology, and people — changes that were not Pareto optimal. It was not that no agricultural economists had worked on problems requiring non-Pareto optimal solutions or that no one was working on them. There was the development work overseas, the poverty commission's work, and the work on the food sector, to mention a few examples. In the late 1960s and earlier, however, agricultural economists had to respond to: legislatively mandated directives to examine the roles of women; political and social demands to help do something about rural poverty; pesticide and fertilizer pollution; malnutrition among the domestic poor; environmental quality; criticism from labor unions, blacks, women, the aged, and nutritionists; small farmers; land settlement; human capital formation; the extended use of fossil fuels relative to renewable resources; worker alienation; and recycling. In some senses the problems and subjects were not new as many had been considered before and not always in a peripheral way. There was, however, a sharp change from the preceding ten to twenty years. A high proportion of agricultural economists in the late 1960s had spent two decades (or had been trained by people whose main experience was in those decades) seeking Pareto optimal solutions within the market; these analysts now faced new problems requiring nonmarket solutions that were not Pareto optimal. Whether the analysts of the late 1960s and 1970s realized it or not, they were faced with the task of helping decision makers attain enough interpersonal validity in welfare measures to make the essential non-Pareto optimal decisions required to solve the problems facing society.

There was a change from the pre-1967 stream of disciplinary and subject-matter articles in the literature to a concern with the practical. Although this new line of literature is seldom problem-solving in nature, it treats *subjects* important in dealing with sets of practical problems involving the subjects under consideration. This new stream represents a return to relevance with consequent impact on the philosophic interests and/or practices of agricultural economists. This impact of the crisis of the late 1960s is discussed in still more detail in later subsections dealing with "Unease and Loss of Purpose" and "Research Work on Private and Public Decision Making" (see pp. 997-998 and 1000-1008).

2. The Philosophical and Methodological Significance of Selected Developments in Agricultural Economics, 1946-76

In order to examine the crucial relationship between work and philosophy this section examines developments in agricultural economics which have philosophic significance using the above chronological summary of the 1946-76 period as background.

THE POST-WORLD WAR II RISE OF NEOCLASSICAL UTILITARIANISM IN AGRICULTURAL ECONOMICS

The prewar emphasis on neoclassical utilitarianism (defined earlier) of Henry C. Taylor, T. N. Carver, Richard T. Ely, and a small number of other people continued after World War II under the leadership of John D. Black, T. W. Schultz, and others.

The 1946-54 period was a time of administrative disillusionment with farm management, and most departments of farm management were either converted into agricultural economics departments or merged into subsections of agricultural departments [Jensen, 1977]. For example, at Michigan State College, the old farm management *department* became part of a new Department of Agricultural Economics formed from an agricultural economics *section* that had existed in the Department of Economics. In the new department, the stress was on the use of neoclassical utilitarian economics data. In Kentucky separate departments of farm management and of markets and prices were merged in the early fifties into one department.

At the end of World War II, utilitarianism was well established in marketing and policy work prior to World War II and continued to gain ground in these two areas in the postwar years. The upsurgence of neoclassical utilitarianism was even more pronounced academically than administratively. Almost without exception the postwar departments of agricultural economics found it essential to have staffs and graduate students well trained in neoclassical, utilitarian theory and younger staff members were quick to oblige. Both individuals and departments, too, exhibited considerable pride in attaining such competence. Some empiricists and some others in administrative posts resisted or dragged their feet as agricultural economics adjusted to the increased strength of neoclassical utilitarian economics.

Neoclassical utilitarian economics is, of course, normative as well as positive. Its calculus is devoted to defining optima based upon normative information, both monetary (prices) and nonmonetary (utility), and positive information. The theory prescribes as the "right action" the one that equates marginal costs and marginal returns measured in terms of prices or utility, both of which are normative. The theory maximizes the difference between good and bad, i.e., profits or net utility. The theory is also used to predict or project micro, semiaggregative, and aggregative behavior on the assumption that households and firms behave in a maximizing manner. The neoclassical calculus is normative as well as positive whether used to: define problems and *derive prescriptions* for individual firms or households [Heady, 1949]; *predict* aggregative and semiaggregative behavior at sector, subsector, national or international levels [Swanson, 1971]; or *evaluate* present and past solutions [Brownlee, 1948]. These three uses can be subsumed under *evaluation* (prescription, evaluation, and problem definition) and *prediction* [Spitze, 1965].

Evaluative uses of neoclassical utilitarian economics were drastically constrained by the questions raised by Pareto and by the reformulation of neoclassical utilitarian theory carried out by John R. Hicks in his *Value and Capital* [1939]. Because of Pareto's questions concerning the interpersonal validity of utility or welfare measurements, it became clear that it was empirically dangerous to prescribe neoclassical utilitarian equilibria as maximizing welfare when reaching such equilibria imposes damages on some in order to confer benefits on others. It was also seen as equally dangerous to use such equilibria to evaluate past and existing situations and to define problems. This narrowed the area within which problems could be defined and solved to those instances in which solutions do not involve imposition of losses on anyone in order to confer benefits on others [Brownlee, 1948, 1950]; such solutions are in theory attainable by the market mechanism. Logically, this kept analysts from using the theory to define and solve some of the most important problems facing society, a situation that became abundantly clear after the social crises of the late 1960s [Buchanan, 1959, 1962]. The compensation principle, if applied, negates attempts to carry out non-Pareto better redistributions. If not applied, the compensation principle leaves uncertainty about the interpersonal validity of a hypothetical substitution of money for an interpersonally valid measure of utility [Dorfman and Dorfman, eds., 1972].

Restrictions on the evaluative use of utilitarian economics resulting from Pareto's argument were not heavily emphasized by practicing agricultural economists in the postwar years when the stress was on agricultural market *adjustment* problems. However, ways were sought to make the overall agricultural economy, farms, and marketing firms and agencies operate more efficiently without redistributing the ownership of resources, rights, and privileges. In instances of damage, resource owners were compensated in various market relocation, irrigation, flood control, and other projects; redistribution of capital ownership did not occur unless there was over- or undercompensation.

The empiricists in farm management had been attacked for not working on the current problems of farmers. Neoclassical utilitarian economics provided a way of defining adjustment problems and an agricultural adjustment center was established at Ames [Heady, 1949]. The deliberations of the North Central Farm Management Research Committee and, at the policy level, the University of Chicago conference on efficiency did not face up to evaluation of nonmarket adjustments to changes in technology, institutions, or human beings in ways that impose damages on some in order to confer benefits on others. Neither did the North Central Marketing Committee (NCR-20) nor did the North Central Soil Conservation Committee (NC-12). Indeed, there seemed to be little general awareness of the difficulties involved in obtaining interpersonally valid welfare measures and of the more general need to obtain objective normative information.

Heady [1962] in a chapter entitled "Criteria for Policy" presented a clear statement on Pareto optimality with some discussion of its limitation and the need to attain interpersonally valid utility measures in dealing with poverty problems. Many accepted Friedman's assertion [1953] that the problems and issues of the day were resolvable mainly with factual information about the positive and that answers to normative questions (presumably other than those concerning market-determined prices) were unimportant; this reduced attention to policy questions that involved restructuring the ownership of property (in its wide sense, the ownership of rights and privileges). Hathaway [1953] and others dealt with income and freedom as if they had an interpersonally valid welfare measure [Kutish, 1954; Long, 1953, 1954].

Even in the 1950s and early 1960s, there was concern about the valuational particularly with respect to public policy research. In 1950, the conference on efficiency held at the University of Chicago with support from the SSRC produced papers on the subject that materialize (when one cleans out old file drawers) as part of agricultural economics' "phantom literature." O. H. Brownlee's papers [1948, 1950] on the meaning of efficiency made essentially the points later covered by the theory of second best as well as those necessary to reject the Coase theorem. In doing this, Brownlee drew on the literature about Pareto optimality by Reder [1947]. At the same conference John Brewster [1950] presented a paper on efficiency, justice, and freedom in which he went, pragmatically, beyond Pareto optimality. Brewster's and John Baker's [1950] somewhat dissonant papers did not divert the Chicago group from concentration on market adjustment, a productive (but limited) approach that led eventually to theories and empirical work on induced innovations [Hayami and Ruttan, 1970, 1971], induced institutional change, and the formation of human capital [T. W. Schultz, 1971]. It did not, however, lead to evaluation of nonmarket changes in the ownership of property (including all rights and privileges). In a sense, Friedman's conclusion [1953, p. 5] was accepted. He wrote: "Differences about economic policy among disinterested citizens derive predominantly from different predictions about the economic consequences of taking action—differences that in principle can be eliminated by the progress of positive economics—rather than from fundamental differences in basic values, differences about which men can ultimately only fight." Even T. W. Schultz [1959, p. 189] basically agreed:

> Let me simply pay my respects to something called "the objectives of farm policy." I know that there is so much that could be said, but I resist saying it because it has been said over and over again. Then, too the working staff of agricultural economists is, I assume, fairly sophisticated on these issues. True, the beginning graduate student is well advised to take stock of the concepts and thought that have been propounded, for example, on Valuations and Beliefs, the Means-End Schema, Change and Order, Learning Theory, Game Theory, Decision-Making, Policy Formation, and, by no means least for economists, Welfare Economics. Altogether, it is a big cup, good for an early breakfast. It is also quickly drained. It

assuredly will open one's eyes, get one going, and, as a rule, better oriented than would otherwise be the case. But the hard core of the particular analytical work that needs to be done on farm policy and agricultural adjustments cannot be undertaken with these concepts.

In general economics, the Pareto question was only slightly blunted by the development of the so-called Coase theorem [1960] that asserts the allocative (resource) neutrality of non-Pareto optimal redistributions of rights and privileges (property) ownership. The nonneutrality *vis-à-vis* the distribution of income by redistributing property ownership is obvious. It should have been equally obvious that only under special restrictive assumptions about income demand elasticities could income be redistributed without affecting relative prices and, hence, the allocation of resources. Such allocative nonneutrality cannot be evaluated in the absence of an interpersonally valid measure of welfare (a normative common denominator). This was recognized in the development of the theory of second best [Lipsey and Lancaster, 1956; Harry G. Johnson, 1960; Fishlow and David, 1961], which notes that the Pareto optimum after a change that is not Pareto better is not demonstrably either inferior or superior to the Pareto optimum before the change. This is, of course, consistent with Arrow's analysis of individual preferences and social choices [Arrow, 1951].

Some attention was paid to Pareto, Reder, and Arrow in teaching courses in agricultural economics. For instance, in his teaching at both the University of Kentucky and Michigan State University (MSU), this reviewer expanded Knight's pre-Hicksian assumptions to confine static economics and its evaluative and prescriptive power to Pareto optima. Agricultural policy texts, however, were slow to recognize the strictures of Pareto optimality. The questions raised by Pareto still permit the use of static economics in predicting (as contrasted to evaluating) consequences of non-Pareto better changes in people, institutions, and technology. Though some analysts have been accused of using the Coase theorem [1960] to support the *status quo*, such support would be hard to justify logically as Pareto optimal economics cannot judge the *status quo* to be either inferior or superior to alternatives that are not Pareto optimal.

Though agricultural economists were slow to adopt Pareto optimality, they became increasingly aware of it. This uneasiness slowly reduced the willingness of agricultural economists to define and prescribe solutions that were not Pareto optimal.

In the mid-1960s the deleterious impact of positivism on problem-oriented work continued but was masked by positivism itself. When one believes, metaphysically, that goodness and badness are not experienced and that objective empirical (descriptive) knowledge of them is impossible, a choice between the following is predetermined: try to do objective descriptive and empirical work with concepts of goodness and badness, or accept arbitrarily asserted or assumed con-

cepts of goodness and badness. When forced, a positivist must deal with goodness and badness by arbitrary assumption or assertion. When he does this, he becomes a conditional normativist à la Myrdal [1944] or a mere advocate. The social
upheavals of the late 1960s and early 1970s found many agricultural economists
choosing between the above alternatives. In effect the powerful arguments of
positivism underscored Pareto's questions by asserting that all normative knowledge is nonobjective—not just that which compares the welfare of two individuals. The student activists and concerned noneconomists of the late 1960s saw
that serious normative questions about environmental quality, poverty, racism,
war, inflation, etc., were going unattended. They acted as if they felt that much
more interpersonally valid welfare knowledge was available than was being used.
Often without understanding the niceties of economic theory and its associated
quantitative methods, they struck directly at the key difficulty.

From the end of World War II to 1967, positivism had expanded its influence
in agricultural economics despite the statement in the *International Encyclopedia of
the Social Sciences* that "the beginning of World War II marked the beginning of
the end of logical positivism as a movement" [Kaplan, 1968, p. 394]. Positivistic
tendencies were reflected partially in increased conditionally normative analysis
and partially in avoidance of normative investigations. The strictures of Pareto
optimally reduced the evaluative power of neoclassical economics [Robinson,
1968]. As will be noted later, the importance of pragmatism, with its emphasis on
workability and the interdependence of the positive and normative, decreased in
this period.

It is also important to note that relaxation of Knight's assumptions of perfect
knowledge and foresight leads to the conclusion that imperfectly informed market decisions which appear Pareto better, *ex ante*, have *ex post* consequences that
are not Pareto optimal, in the presence of transfer costs, acquisition costs in excess of selling prices, or salvage value [Glenn L. Johnson, 1960a, 1960c]. This
leads to the important conclusion that evaluation of the consequences of market
phenomena under uncertainty requires interpersonally valid (normative) welfare
measures [Glenn L. Johnson and Quance, eds., 1972].

The need for interpersonally valid normative data took on even greater practical importance with the social and economic unrest of the late 1960s. Market
adjustments were obviously inadequate in the face of demands to redistribute
ownership of rights and privileges from whites to blacks and other minorities,
from rich to poor, from developed to OPEC nations, from developed to Third
World nations, from the USDA/land-grant establishment to the biological and
physical scientists outside that establishment, from males to females, and from
traditional users of the environment to newly emerging claimants. Increasingly,
the choice became one of:

(1) trying to work objectively in establishing interpersonally valid concepts of goodness and badness; or

(2) accepting arbitrarily asserted or assumed concepts of the goodness or badness of certain redistributions, without recourse to experience, or logic.

This difficulty is discussed in more detail in later sections of this review.

THE POST-WORLD WAR II UPSURGE IN STATISTICS AND ECONOMETRICS — A STRENGTHENING OF POSITIVISTIC EMPIRICISM

Throughout U. S. agricultural economics, there was a substantial postwar strengthening of statistics and a great expansion in econometric teaching and analysis [Leontief, 1971]. This was evidenced by increased acceptance of journal articles reflecting competence in statistics and econometric analysis. While the initial center of interest in econometrics was in the Cowles Commission, then at the University of Chicago, competence soon developed at Iowa State College, at the University of California, and elsewhere in the land-grant system. There had been a substantial pre-World War II competence in statistics and econometrics in the USDA [Working, 1927; Girschick, 1946] that continued to develop in the immediate postwar years.

This upsurge in statistics and econometrics changed the relationships between agricultural economics and philosophy in part by more than offsetting the reduced attention to positivism that resulted from reduced emphasis on Cornell positivistic farm management. Perhaps this change is best seen by discussing statistics and econometrics separately.

The increased interest in and emphasis on theoretical statistics grew out of the long-standing quantitative work of agricultural economists. Statistics as a discipline had much to offer to improve the quantitative techniques of agricultural economics in the post-World War II period. At the same time, agricultural economists were being influenced increasingly by positivists and positivistic arguments holding that there are no normative experiences and that, hence, normative information cannot be descriptive or quantified for statistical treatment. From this point, it is but a short *non sequitur* to the unthinking conclusion that statistics is positivistic and the unconscious acceptance of positivism as an apparent accompaniment of statistics.

Econometricians have — as an avowed purpose — the joint use of mathematics, statistics, and economic theory as appropriate in studying economic phenomena [Marschak, 1953]. The economic theory employed by econometricians was largely of the neoclassical utilitarian and Keynesian varieties. When econometricians use this theory to *develop* aggregative or semiaggregative models of (and to *predict*) supply and demand responses, they assume that the maximization behavior of neoclassical, utilitarian economics is characteristic of firms, households, and resource owners. These theories are normative in that utility functions and prices are normative. Maximization and minimization processes deal with the attainment of utility (or some other good or subgood) and the avoidance of disutility (or some other bad or sub-bad). Thus, the work of econometricans is not as positivistic as some of the earlier joint work of statisticians and farm management researchers.

However, it is interesting to note that the supply response studies of econometricians were often labeled normative, with the word being used as an epithet by positivistically inclined supply and demand analysts. These analysts wanted what they termed "predictive" or "behavioral" supply and demand estimates rather than "normative" or "prescriptive" ones [Glenn L. Johnson, 1960a; Swanson, 1971]. As econometric supply and demand functions predicted behavior on the basis of the maximizing theory of neoclassical economics, the critics probably wanted different kinds of predictions and different behavioral assumptions. They seemed to believe that predictive or behavioral estimates could be obtained but that estimates of the consequences of human behavior that take motivation into account can be neither behavioral nor predictive! Their positivism seemed to preclude the possibility that anything normative could be empirical, descriptive, or predictive of anything "real" or objective.

In agricultural economics, developing quantitative expertise augmented the tendency toward positivism. Together they seemed only another important part of attaining academic excellence in the late 1940s, the 1950s, and early 1960s. Policy, farm management, and marketing analysts wanted reliable estimates of the important parameters of neoclassical utilitarian theory to use in their adjustment studies. Prior to the late 1960s many agricultural economists became increasingly positivistic without noting the adverse impact of their positivism on the ability to define and solve problems [Buchanan, 1962] whose solutions required nonmarket adjustments.

PRAGMATISM AND INSTITUTIONALISM ON THE DEFENSE

Despite its strong prewar record and the substantial contribution of pragmatic Wisconsin institutionalists during the World War II period, pragmatism and institutionalism were put on the defensive by the postwar strength (as described above) of the neoclassical utilitarians and positivistic statisticians, and econometricians.

The neoclassical utilitarians took an early post-World War II offensive in the North Central Farm Management Research Committee. This offensive overcame the empiricism of the farm management workers and tended to stifle pragmatic interests. By the end of the second North Central Farm Management Research Workshop at Black Duck in 1949,[2] the stage was set for a heavy emphasis on neoclassical utilitarianism in farm management research. In turn, the progress and/or changes made in farm management by the neoclassical utilitarians created aggressive competition with academic pressure on the pragmatists in the land tenure committee and on the practicing Wisconsin institutionalists and land tenure researchers throughout the land-grant system. A rather natural affinity between neoclassical utilitarians, on one hand, and statisticians and econometricians, on the other, strengthened neoclassical utilitarianism while the lack of such affinity weakened the pragmatic position. The pragmatists had trouble combining their somewhat clumsy though comprehensive methods with the relatively simple, straightforward methods of statistics and econometrics. Their trouble stemmed in part from their use of the workability criterion and in part from their somewhat haphazard stress on the interdependence between normative and positive truth. The methods of the pragmatists were well suited for working on ill-structured problems but unduly complicated for relatively well-structured problems of interest to the utilitarians [Mitroff and Blankenship, 1973].

Conditional normativism was a natural sort of adjustment to positivism which was superficially easier than pragmatism for persons concerned with simple, well-structured problems involving stable, well-known, and noncontroversial values. K. H. Parsons later [1958] attacked conditional normativism as "opening the way for a reversion to the medieval view that the world of thought and action should be organized around social values presented to mankind as dogma." As a pragmatist, K. H. Parsons objected to normative assumptions—he believed that both normative and positive truth are revealed interdependently in the *process* of defining, studying, and solving problems [K. H. Parsons, 1949]. Parsons was uncomfortable with the positivism that causes conditional normativists to avoid normative investigations by making normative assumptions instead of letting normative knowledge emerge, interactively and interdependently (dialectically), with positive knowledge out of problem-solving processes.

The literature of the late 1940s, 1950s, and early 1960s contains a substantial number of statements expressing pragmatic, institutional points of view at some variance with those of the upsurging utilitarians and econometricians. Included here are works by Bushrod Allin [1948, 1949], Erven Long [1952, 1953], Maurice Kelso [1949, 1965a, 1965b], John Timmons [1959], and John Brewster and H. L. Parsons [1946], to mention a few examples.

DISCIPLINARY EXCELLENCE VS. PROBLEM-SOLVING AND SUBJECT-MATTER EFFORTS

By the mid-1950s, much greater disciplinary excellence as economists had been attained by agricultural economists than in the prewar years. Knowledge of economic theory was greater and more extended. Quantitative work including estimation of the parameters commonly encountered in economic theory was greatly improved; however, as disciplinary excellence increased, problem-solving and subject-matter research probably lost ground absolutely as well as relatively.

Farm management, marketing, land tenure, and price research, etc., lost academic status unless "enduring contributions" could be made to economic theory and quantitative techniques. Problem-solving research was increasingly denigrated as short-term "brush fire" effort. These trends developed in the late 1950s and 1960s when the profession was moving toward neoclassical utilitarianism and was defining problems increasingly in terms of market disequilibria and trying to solve them by re-establishing equilibria. This emphasis led to the neglect of a whole class of problems by agricultural economists, i.e., those not solvable by the market. The emphasis on market adjustment was compatible with excellence in economics and associated quantitative techniques. This stress on market adjustment combined with the additional tranquilizing effect of a lack of interest on the part of political farm organization leaders in nonmarket adjustments to mask the consequences of reduced attention to problems beyond the market place.

In 1956, an SSRC meeting was held under the chairmanship of Brooks James. It was based on reports from earlier seminars and conferences in eight agricultural economics departments or sections in academia and in the USDA. The main emphasis was on problems solvable through market adjustments. However, reports from Minnesota, Wisconsin, and Michigan State did call attention to distributional and equity considerations (other than agriculture *vs.* the remainder of society) and valuations and values. The summary, which is part of the phantom literature, noted that "the philosophic literacy of agricultural economists with respect to value problems might well be examined." It was also noted that "some of the younger men . . . had been giving increasing attention to valuation problems and valuation conflicts as they bear on policy and policy making" [SSRC, 1956, pp. 54–56].

In the late 1950s, George Brinegar chaired a special subcommittee of the SSRC on agricultural economics. That subcommittee worked cooperatively with the executive committee of the American Farm Economics Association. An article entitled "Reorientation in Research in Agricultural Economics" [Brinegar, Bachman, and Southworth, 1959] was published. That article diagnosed the ills of agricultural economics as being the consequences of its low productivity. It was argued that "we are failing to measure up to the present challenge, and will continue to do so unless and until we can direct our thinking to new and broader formulations of problems as they now press upon us" [p. 602]. Reasons for the failure were ascribed to fragmentation into disciplinary and subdisciplinary interests—fragmentation which doesn't develop "systems of thought" *vis-à-vis* problems; inadequate research methods; and lack of attempts to develop "subject-matter compilations." F. S. C. Northrop's [1947] book was highly recommended by the Brinegar subcommittee. The report did not deal with the relationship between problem definitions and solutions (prescriptions), on the one hand, and concepts of goodness and badness, *per se*, on the other. Perhaps because this relationship was ignored, little stress was placed on methods for working with the normative in defining and solving problems. Northrop's failure [1947, pp. 328ff.] to distinguish between prescriptive knowledge and our past and present normative experiences was not seen as a limitation when his book was recommended to agricultural economists by Brinegar, Bachman, and Southworth [1959].

Shepherd [1956] wrote meaningfully about what researchers can say about values. His somewhat eclectic statement reflected pragmatic as well as positivistic and conditionally normative methods. When Ciriacy-Wantrup [1956] wrote about policy considerations in farm management research with attention to Pigou and Pareto, he did not distinguish between prescriptive and normative as was done earlier in this review. As a result his article was somewhat ambiguous. He labeled the maximization principle a "scientific fiction" and noted shortcomings of Pareto optimality. His article had the distinct merit of recognizing the inadequacies of national income and Pareto optimality as criteria for policy choices *vis-à-vis* nonmarket adjustments. Brandow [1955] recognized the need to work with values but was of little operational help.

Subject-matter research on such subjects as land tenure, farm records, time and motion, econometrics, farm management, and marketing (as subjects), etc., lost ground to disciplinary interests. Even the "revolutions" in these subject areas were disciplinary; witness the "structure, conduct and performance" transformation in marketing. In econometrics, as practiced by agricultural economists, the interests were in more advanced parameter estimation techniques, lagged adjustment coefficients, distributed lags, etc. Farm management researchers were more interested in integer programming, decision making, fitting production functions, recursive linear programming, and asset fixity than in the problems of farmers solvable outside of the market place. Simple, well-structured systems [Mitroff and Blankenship, 1973] were researched with well-established techniques. Complex, ill-structured but important problems were avoided or neglected.

When U.S. agricultural economists changed their work, they also changed the name of their professional journal from the *Journal of Farm Economics* to the *American Journal of Agricultural Economics*. The *AJAE*, continuing a situation that had developed before its name change, concentrated increasingly on the disciplinary—the theoretical and the quantitative—and became smaller than the JFE of the more immediate postwar years. Articles dealing with solutions of specific problems of farmers, agribusinessmen, and public officials occupy a smaller proportion of space than in the "Sears Roebuck catalog" proceedings and regular issues of the old *JFE*. Disciplinary quality has clearly increased. Significant changes in the quality of subject-matter and problem-solving research are hard to detect in *AJAE* articles, perhaps because the sample size is so small! Since 1967, as will be seen later, the quantity and quality of problem-solving and subject-matter research published outside the *AJAE* have increased.

In the late 1950s and early 1960s, concentration on the disciplinary increased the emphasis on the positive and, if values were considered at all, on monetary as opposed to nonmonetary values. One of the exceptions to the emphasis on monetary values was the interstate managerial study [Glenn L. Johnson *et al.*, 1961] within which Halter's disciplinary work measured the utility of wealth and income cardinally with techniques developed by von Neumann and Morgenstern [1947] and Friedman and Savage [1948]. These measures were not interpersonally valid insofar as the unit of measurement was concerned; however, interpersonally valid comparability of inflection points and other characteristics of the utility function was attained. Halter spread his interest in utility measurement to California and Australia during sabbatical leaves [Halter and Dean, 1971]. Since then Dillon and Anderson [1971], Anderson, Dillon, and Hardaker [1977], and others have published extensively on utility measurement and the expected utility hypothesis as a decision rule.

It is interesting to note that utility measurement is sometimes regarded as positivistic despite the fact that a nonmonetary value, utility, is being measured. Part of the confusion arises from failing to distinguish the normative (good and bad) from the prescriptive (right and wrong) or what "ought" or "ought not" to be done [Machlup, 1969; Lewis, 1955; Moore, 1903]. Another part of the confusion arises from the positivistic presupposition that nothing normative can be experienced and, hence, measured. This leads to the strange conclusion that utility must be positive if it is measurable. The same argument seems to apply to the conclusion that prices are positivistic.

UNEASE AND LOSS OF PURPOSE

Though the literature contains earlier statements of unease with the trend toward the disciplinary and the positive [Allin, 1948, 1949; Mitchell, 1949; Conklin, 1947], the criticism of the 1960s became more telling. As previously pointed out, the pragmatists were always uneasy with the trends. K. H. Parsons [1958] expressed his uneasiness at a North Central Farm Management Research Committee meeting. Maurice Kelso's critical appraisal of agricultural economics [1965a] elicited methodological and philosophical comments from Brown [1965], Grove [1965], Mighell [1965], Reinsel [1965], Schmid [1965], Spitze [1965], and, in turn, further comments from Kelso [1965b]. After noting remarkable disciplinary advances, Kelso indicated that we had not advanced our predictive and problem-solving capacity correspondingly. He felt that we had become "more exact, more positive, more quantitative, more complex in our rationalistic analysis of hypothetical, simplified, imaginary systems from which man—as a partly irrational, unpredictable, emotional animal—is banished to be replaced by the lightening [*sic*] calculator in human form."

The "adjustment" orientation of the North Central Farm Management Committee [Heady, Diesslin, Jensen, and Glenn L. Johnson, eds., 1958] left problems involving nonmarket solutions unattended. Cases in point are problems involving conservation, taxation, land-use regulation, publicly supported technological research, price supports, income taxes, social security for farmers, unemployment compensation, foreign commodity and technical aid, regulation of pesticide use, etc. The regional farm management extension committees did better than the research committees on taxation, social security, and labor use regulations. These and other considerations left the author of this review uneasy and led to his investigation of philosophic questions concerning how to research the normative objectively [Glenn L. Johnson, 1960b, 1961b, c, 1963a]. This investigation indicated that essentially the same kind of objectivity was attainable for normative as for positive knowledge.

By the early 1970s, questions about the worth of the discipline-like work of the regional research committees led to their partial abandonment. As will be seen later, this was part of the response of agricultural economists to the crises of the late 1960s. Recent attempts to re-establish the regional research committees in the midwestern states, like their earlier abandonment, have done little to improve the problem-solving and subject-matter work of regional committees. Perhaps this failure is due to perceiving the difficulty as one of inappropriate personnel on committees rather than of administrators unable or uninterested enough to shift the emphasis to problem-solving work.

THE RESURGENCE OF PRAGMATISM AND INSTITUTIONALISM

Kenneth Boulding is reputed to have once characterized institutional economics as a combination of poor sociology and bad economics. However, by the time he delivered his presidential address before the American Economic Association in 1968 he had changed his mind and was essentially an institutionalist [Boulding, 1969]. What happened to Boulding also happened to some agricultural economists. Solving problems with market adjustments within a neoclassical Pareto better context left serious problems of conservation, poverty, racial inequality, environmental quality, minority rights, etc., unattended by the utilitarians and market adjusters. This permitted pragmatic institutionalists to score points again with their ability, cumbersome as it was and is, to define and at least participate in the processes of solving problems whose solutions inevitably seem not to be Pareto better. As problems involving environmental quality, poverty, and ownership of rights and privileges by minority and majority groups, etc., came to greater prominence in the latter part of the 1960s, the limitations of conditionally normative and/or Pareto better, market-adjustment studies became more and more apparent.

In this period, John Brewster in the USDA placed heavy emphasis on pragmatism. Earlier, Brewster had participated with others in bringing G. H. Mead's pragmatic papers to posthumous publication under the title *Philosophy of the Act* [Mead, 1938]. In turn, in 1970, J. Patrick Madden and David E. Brewster brought John Brewster's writings together in a book entitled *A Philosopher among Economists* [J. M. Brewster, 1970] under the sponsorship of a committee including such USDA agricultural economic leaders as Frederick Waugh, Kenneth L. Bachman, Willard W. Cochrane, and Harry Trelogan and with the support of the Farm Foundation. While the preface denies, for reasons not made clear, that Brewster was a pragmatist, his essays and bibliographies reflect much of the pragmatic thought of G. H. Mead, Charles S. Pierce, William James, and John Dewey, particularly the emphasis on "social awareness and commitment" and on beliefs and values as being as interdependent as "the two sides of a coin" [J. M. Brewster, 1970, p. 11]. Though at times conditionally normative, Brewster was pragmatic in relating values to particular problem solutions, crises, and periods of time. He also showed how values affect one's positivistic views of reality. His 1964 seminar, "Philosophy: Principles of Reasoning Especially Applicable to Science," contained little pragmatism and, indeed, ignored the normative while concentrating on creative discovery of conflicts between "prevailing generalizations and exceptional observations" [J. M. Brewster, 1970]. In his chapter on philosophy, in his life, and in other writings, however, Brewster concentrated on conflicts with respect to the problems and issues of society in a manner consistent with pragmatism. In Brewster's lifework, pragmatism and/or normativism were used in working with the value dimensions of the social problems and issues with which he was concerned. In practice, Brewster went beyond conditional normativism to pragmatism. He did not extensively consider the possibility of objective nonmonetary, normative knowledge independent of positive knowledge, following Moore [1903] and Lewis [1955]; hence, he was more of a pragmatist than a normativist. Brewster's vagueness *vis-à-vis* working with the normative in considering societal problems and issues was continued in a USDA publication entitled *Beliefs and Values in American Farming* [Gulley, 1974]. That report contrasts factual with normative beliefs thereby implying, positivistically, that normative beliefs (such as those about price levels or about such nonmonetary values as the goodness of life or the badness of racial inequality) are not "facts." Further, "goodness" is confused with what "ought not to be done" (the prescriptive) without seeing that we sometimes "ought not to" bring into existence that which is good if something better can be brought about at the same cost. Both Brewster's and Gulley's works reflect the confusion in our discipline, so well summarized by Machlup [1969], of values (the normative) with the prescriptive. Each, however, has the virtue of concern with societal issues and problems supported with enough pragmatism to go beyond positivism and conditional normativism in helping public (and private) decision makers solve their problems.

Boulding's recognition of institutional economics and pragmatism also appears in his book *The Image* [1956]. He states that the processes whereby one forms positive and normative images are not essentially different [p. 173] and are pragmatic [Boulding, 1969]. Work by this reviewer also represented a considerable acceptance of pragmatism [Glenn L. Johnson, 1970]. The resurgence of pragmatism in the late 1960s was substantially constrained by positivism (including conditional normativism).

RESEARCH WORK ON PRIVATE AND PUBLIC DECISION-MAKING PROCESSES

In the postwar period there was considerable interest in decision-making processes at both private and public levels. Much of the postwar interest in private decision-making processes grew out of Frank Knight's *Risk, Uncertainty and Profit* [1921]. Before the war, T. W. Schultz [1939b] had urged farm management advocates and production economists such as John D. Black to pay more attention to risk, uncertainty, and management processes. It was not until he published his *Introduction to Economics for Agriculture* [Black, 1953b, pp. 72ff.] that Black had a section on risk and uncertainty. His earlier farm management effort [Black, Clawson, Sayre, and Wilcox, 1947] was devoid of risk and uncertainty theory. Nor did it deal with how expectations affect decisions or how farm managers learn (either positively or normatively).

D. Gale Johnson [1947], however, responded to the Schultz [1939b] admonition by using the Knight analysis of risk, uncertainty, and profit to develop the idea of forward prices for American agriculture as a means of reducing price risk. Also, empirical work at Iowa State on risk and uncertainty was contained in Heady's production economics text [1952].

The neoclassical production economists, who replaced the Wisconsin pragmatists in the Land o' Lakes/Black Duck debate with farm management empiricists, also developed interests in risk, uncertainty, expectations and managerial processes [Glenn L. Johnson, 1950]. Their work went forward largely under the auspices of the North Central Farm Management Research Committee, which inaugurated [1] the Inter-State Managerial Study (IMS). North Central farm management researchers interested in managerial processes were largely conditionally normative. At the beginning at least [Glenn L. Johnson, Halter, Jensen, and Thomas, eds., 1961], their interest in managerial processes was not a pragmatic, dialectical one. As a result of a conditionally normative orientation, IMS investigations of the managerial process dealt largely with how managers accumulate positivistic kinds of information and use it in making decisions, the main exception being price information. In subsequent stages of the IMS, it became clear that information about nonmonetary values as well as positivistic information is accumulated by managers in reaching prescriptions as to how to solve their problems. It also became clear that the managerial processes were sometimes pragmatic in nature with normative and positivistic truths being interdependent in the context of the problem being defined and solved. The IMS did much to deepen interest in normative and positive epistemologies and in the processes for reaching prescriptive decisions.

At the University of Missouri, the "balanced farming" approach to extension took hold. This approach to agricultural extension recognized the interrelationships between production and consumption in farming. When generalized to other states in the 1950s, the approach was known as "farm and home development." The emphasis on the home or consumption side of farming stressed the normative and, within the normative, nonmonetary as well as monetary values. The necessity of dealing with consumption as well as production—with expenditures on living as well as expenditures on production and investments—in farm planning and budgeting clearly involved the normative. By the 1960s the farm and home development approach to farm management extension had more or less withered away administratively, especially in the principal Corn Belt states where agriculture is more profitable. In these states, production and consumption are more separable and, whether separable or not, production is important enough to be studied independently and in its own right. This permitted considerable positivistic and conditionally normative farm management work to proceed with some success, uncomplicated by pragmatic and normativistic philosophies. Such work produced little formal understanding of decision making. However, extension farm managers less hampered by philosophic constraints did much to develop in *practice* the processes of working interactively and iteratively with farmers in answering both the positive and normative questions that determine prescriptions as to what "ought to be done."

Though small farms were studied more or less continuously from 1946 to the present [Heady, 1952, ch. 13; Heady, Back, and Peterson, 1953; Wilt, 1957], the "small farm" has recently been "rediscovered" [Thompson and Hepp, 1976]. In a research and extension movement reminiscent of the O. E. Baker, Borsodi, and Wilson depression book, *Agriculture in Modern Life* [1939], part-time farms, rural residences, and hobby farms were being studied. Such studies required the capacity to work objectively with the normative, particularly nonmonetary, values, if one was to understand the economics of such farms, which involve consumption as well as production.

At the public level, John D. Black at Harvard had become deeply interested in agricultural extension administration and conducted a program for training agricultural administrators that stressed administrative processes. Being concerned with administration, the Harvard program was less positivistic (and therefore less scientific in some views) and less disciplinary than the mainstream of agricultural economics of that time. As such, it involved an attention to values and prescriptions that were probably at some variance with the research methods Black taught if not with his practices as consultant and adviser. There was a connection with Wisconsin institutionalism *via* Gaus (in the Department of Government at Harvard), who, like Black, was trained at Wisconsin [Gaus and Wolcott, 1940].

A somewhat different situation prevailed in the Department of Agricultural Economics at Purdue University under the leadership of Lowell Hardin (and his predecessors and successors). Purdue's eclectic, multidisciplinary approach to farm management made that department a leading trainer of personnel for farms and agribusiness firms. This development seems to have been an outgrowth of close contact with farmers and agribusinessmen and their *processes* for solving problems rather than of commitment to pragmatism or more normativistic philosophies. At Michigan State, the new Department of Agricultural Economics under Thomas Cowden and the extension leadership of John Doneth preserved much that was of value out of the Cornell farm management approach.

In the late 1940s, 1950s, and early 1960s, such then newcomers to the fields of agricultural policy as D. Gale Johnson, Cochrane, Halcrow, Brandow, Hathaway, and Bonnen joined such stalwarts as Black, Schultz, Wilcox, Stine, Benedict, Clawson, and others. Also, policy work was done both domestically and abroad by persons such as Heady, Back, Sorenson, Gray, Hillman, this reviewer, and others.

In the 1950s and 1960s, most books on agricultural policy did not consider Pareto optimality and the need for interpersonally valid welfare measures. This included policy books by Hathaway [1963]; Clawson [1968]; Heady, Haroldsen, Mayer, and Tweeten [1965]; Halcrow [1953]; and Wilcox, Cochrane, and Herdt [1974]. Schickele [1954] paid more attention to values. Hathaway's book distinguished between values and prescriptions, as defined in this review, but then translated values into a list of goals (prescriptions still to be executed) virtually the same as the values from which they were supposedly derived in view of the constraining nature of positive reality. Walter Wilcox [1956] also wrote on ethics. In summary, the policy literature of the period provides few clues as to how to work objectively with values in solving policy problems.

A conference on goals and values in agricultural policy, sponsored by Iowa State University's Center for Agricultural and Economic Development [ISU, CAED, 1963], gave only slight attention to Pareto optimality and still less to important methodological issues of how normative questions can be objectively researched. Brownlee [1961], in discussing Markham's paper [1961], did bring out a little about the limitations of Pareto optimality. This reviewer, in discussing a paper by Bishop and Bachman [1961], expressed concern that the conference was not well organized to address the normative [Glenn L. Johnson, 1961b]. Kaldor and Hines [1961] noted that they were unable to coauthor an assigned paper with Ward Bauder on goal conflicts in agriculture because integration of "sociological and economic approaches" was not possible in "the time at 'their' disposal." Kaldor and Hines did not deal with Pareto optimality, lack of interpersonally valid welfare measurements, or, for that matter, methods for working objectively with values—theirs was mainly a market-adjustment approach. A subsequent CAED conference [ISU, CAED, 1963] attempted to rectify the shortcomings of the first conference. That conference placed heavy emphasis on religion and religious leaders but did not investigate the question of how to derive normative knowledge from experience and the use of logic in an objective manner.

Tweeten [1970] produced an agricultural policy book that fully incorporated the Pareto concepts of Hicks [1939] as expounded by Reder [1947], Scitovsky [1951], and Arrow [1951]. In addition it had substantial empirical normative content [Tweeten, 1970, ch. 1-4]. This descriptive normative content drew heavily on sociological [Burchinal, 1961], political [Talbot and Wiggins, 1967], and historical studies [Benedict, 1953]. Unfortunately, and unlike Hathaway's earlier book [1963], Tweeten's effort did not distinguish between the normative and prescriptive; i.e., he used the word "goals" ambiguously to mean both or either. However, his goals were similar to those delineated by Hathaway. Because Hathaway tended to lose the distinction by the time he translated values (the normative) into goals (prescriptions not yet executed), the similarity is not surprising. The theoretical presentation [Tweeten, 1970, ch. 16] concentrated largely on Pareto optimality and, hence, market adjustments to attain "economic efficiency." The limitations of Pareto optimality and economic efficiency as stated by Brownlee [1950] at the Chicago efficiency conference were not stressed by Tweeten—neither was the re-expression of those limitations as the theory of second best [Lipsey and Lancaster, 1956]. Further, normative philosophies and epistemologies were not seriously considered [Moore, 1956; Lewis, 1955; Dewey, 1938]. Again, the reader and student were left without assistance in working objectively with the normative. In a bow to positivism, Tweeten [pp. 502-503] implied that the valuable normative contents of his chapters 1-4 are of questionable objectivity. In effect he backed off from Pareto optimality and its modifications but not to an objective normativism or pragmatism for handling problems "solvable" only with nonmarket changes in institutions, humans, and technology, changes that violate the Pareto optimality criteria.

In addition to the literature on national agricultural policy considered above, there is a literature on state, local, and international decisions and policies. This literature deals with market adjustments and with nonmarket changes in institutions, technology, and people. Such literature ranges from problem solving through subject matter to the disciplinary with, perhaps, its greatest concentration falling on subject matter.

By the first half of the 1970s, four different streams were detectable. *One* stream involved systems analysis in its broad, general sense [Halter and Miller, 1966; Manetsch *et al.*, 1971]. This stream's intellectual ancestry includes cybernetics and the work of systems scientists and is related to optimal control theory. In the general systems-simulation approach, emphasis is placed on "state, policy, behavioral and criteria" variables, the latter being normative and of use in evaluating systems performance. This stream becomes pragmatic and dialectic but is still scientific in the broad sense.

A *second* stream was generated by the market structuralists [Farris, ed., 1964]. It was concerned with "structure, conduct and performance," first with respect to markets and later with respect to government as well as markets (note the correspondences between state and structure, behavior and conduct, criterion and performance). A notable application in this stream was a study of the northern California water industry by Bain, Caves, and Margolis [1966], published by Resources for the Future (RFF). In effect the word "market" was magnified to include all of economics in changing institutional, technological, and behavioral settings.

The *third* stream was developed by persons responding to the crises of the late 1960s [Castle, 1972; Libby, 1971; Randall, 1972; Dorner, 1971; Bawden, 1972; Bieri, de Janvry, and Schmitz, 1972; Kelso, 1968]. As noted above, pragmatically oriented agricultural and general economists were more prepared philosophically, and better equipped methodologically, to address themselves to these issues than their more disciplinary counterparts who emphasized market adjustment, Pareto optimality, positivistic quantitative techniques, and conditional normativism. Their preparation was based on earlier works such as those of Margolis [1957] and Buchanan [1962] as well, of course, as the much earlier work of the Wisconsin institutionalists. Some pointed out that agricultural economics had in fact been working on small farms, technological change, poverty, energy, environmental quality long before 1967—and they are correct. However, the emphasis changed in the late 1960s—after 1967 these topics become proportionally more important and a part of the mainstream of agricultural economics thought.

The *fourth* stream of work involved agricultural information systems as one aspect of public and private capacity to analyze and solve problems. After 1969, a committee of the AAEA chaired by James Bonnen addressed itself to the adequacy of the U.S. agricultural economics information systems [AAEA Committee on Economic Statistics, 1972; Bonnen, 1975]. Results of this work were presented in congressional testimony [U.S. Congress, JEC, 1974; U.S. Congress, OTA, 1976a] and were considered in the Office of Technology Assessment's review of agricultural information systems [U.S. Congress, OTA, 1976b]. The information in the U. S. agricultural information system is both positive and normative, the latter dealing mainly with prices, income, and expenditures. Karl Fox

[1974] attempted to measure nonmonetary values (in terms of dollars) to permit him to expand GNP into a concept of gross social product (GSP). Fox's work on the normative helped to repair neglect of that aspect of information systems by general and agricultural economists. Other neglected aspects included the role of markets and political systems as mechanisms for transmitting normative information. Also, little of the literature on information systems has considered iterative interactions between investigators, on the one hand, and decision makers, executives, and those responsible for decisions and action, on the other. At the private level, practical farm management advisers know the importance of iterative interaction as a source of both positive and normative knowledge. At the public level, practical consultants and advisers have a similar awareness. Perhaps the connection between information systems and cybernetics (with its positivistic background) is too close to expect much consideration of the normative in the iterative, interactive processes of reaching prescriptions to solve problems [Dunn, 1971].[3]

The first three of the above four streams were concerned with structure (state), conduct (behavioral), and performance (criteria) variables. All four were normative as well as positivistic. As the general systems-simulation approach derives from cybernetics and engineering (with its close association with positivistic physics and chemistry), it is not surprising that the approach often takes several criterion valuables as given in a sort of "multiple conditional normativism." Similarly, in view of the upsurge of positivism in agricultural economics after World War II, it is not surprising that market structuralism tends to take performance variables as "givens." The similar tendency of the more pragmatic investigators, who responded to the crisis of the late 1960s and 1970s with structuralist or industrial organization approaches, seems to reflect the defensive posture of pragmatism and the resurgence of positivism in the 1960s. All three streams produced research results on such subjects as poverty, environmental quality, transportation, and development in general. The conditional normativism involved in taking a preselected list of performance or criterion variables as given is less of a constraint on subject-matter research than on problem-solving research where participatory *interaction* between investigators and decision makers (including affected people) *modifies* the list of criterion or performance variables *in the dialectical process of* solving the problem.

The U. S. developed its national agricultural accounts and ability to make associated macro agricultural projections in the 1920s and especially in the 1930s. This was in response to urgent agrarian problems of direct concern to the then dominant "farm bloc." As the bloc controlled USDA appropriations, interactions among bloc leaders, members of the executive branch, and researchers were close and iterative. The national accounts, associated indexes, and other measures were, in effect, a "U. S. agricultural sector model." This model related state, behavior, and policy variables to *projections* of criteria or performance variables. The performance variables were numerous and were used to indicate consequences of non-Pareto better changes in price supports, production controls, food stamp programs, credit subsidies, and new credit institutions. In general, this model and its associated data systems have remained the principal way the USDA has provided congressional and executive branch decision makers with projections. This general model has been supplemented with numerous more specific econometric models within the USDA and in studies done largely in association with ISU and to a lesser extent elsewhere. Parts of the USDA's general model were formalized on computers, an example being the National Interregional Agricultural Projection (NIRAP) model developed under the leadership of Quance [Boutwell et al., 1976].

Econometric and linear programming (LP) models are more highly specialized on economics and maximization than are the U. S. agricultural accounts and associated indexes and measures. The econometric and LP models assume maximizing behavior in making projections or producing prescriptions, both of which involve the normativism of neoclassical economics. Further, economic models incorporate fewer biological and institutional variables. They also pay less attention to the consequences of nonmaximizing behavior than do the more eclectic "models" composed of the national agricultural accounts and associated indexes and measures. Perhaps it was these limitations that kept modern, more specialized models from more fully replacing the projections based on agricultural accounts and associated statistics.

Somewhat similar to the national accounts "model" was the work of John D. Black and James Bonnen [1956] in projecting the productive capacity of American agriculture. Their eclectic effort drew heavily on technological, institutional, and behavioral information. Like most 1950-70 textbooks on agricultural policy, it tended to be in the neoclassical tradition yet it did not subject itself to the strictures of Pareto optimality or conditional normativism and did not attempt to reduce all values to a common denominator. The joint RFF/MSU 1917-65 study of the U. S. agricultural economy [Glenn L. Johnson and Quance, eds., 1972] was somewhat similar to the Black/Bonnen effort in that it too avoided heavy emphasis on maximization, the strictures of Pareto optimality, and conditional normativism. At a less macro level, the Mighell/Black study of dairy adjustment was also philosophically and methodologically eclectic [1951]. The author is under the impression that the national accounts model, the Black/Bonnen effort, and the Mighell/Black effort attained greater credibility with decision makers than more specialized, less eclectic efforts such as the Lakes State Dairy Adjustment Study, the North Central Feedgrain-Livestock Study [Sharples, Miller, and L. M. Day, 1968], and similar studies in other regions.

The overseas research of agricultural economists followed the same two patterns — one eclectic and general, the other specialized both philosophically and on the discipline of economics. Studies specialized on utilitarian economics included: a Guatemala study [Fletcher, Graber, Merrill, and Thorbecke, 1970], Day's and Singh's work in India [R. H. Day and Singh, 1977], and Heady's work in Thailand [Nicol, Striplung, and Heady, eds., 1982]. Other studies were more eclectic and less specialized on maximization either for purposes of prediction or prescription; these include studies in Nigeria [Glenn L. Johnson, Scoville, Dike, and Eicher, 1969], in Korea [Rossmiller et al., 1972], and in Latin America [Harrison et al., 1974]. In some of these (and other) less disciplinary, more eclectic studies, substantial participatory interaction took place between decision makers and analysts so that the results reflected dialectic interaction in the process of developing and using the studies.

Domestically, there were studies dealing with problems and issues involving environmental quality [Castle, 1972; Schmid, 1972]; poverty [Bawden, 1972]; and discrimination, rural development, and community services [R. J. Hildreth and Schaller, 1972]. These studies tended to be eclectic and relatively unspecialized philosophically and with respect to disciplines in part because agricultural economists involved were often faced with the necessity of making a place for themselves on problem-solving or subject-matter teams including biophysical scientists, sociologists, engineers, political scientists, and others. Many of these studies were of a subject-matter rather than problem-solving nature. Relatively little participatory interaction with decision makers and affected people was involved. Lack of problem-solving interaction made this research less pragmatic than might be expected; in fact much of it can be characterized as multiconditionally normative as several values were often taken as given. In this connection the reader may want to refer to the section on efficiency considerations in the AAEA review article on rural development [Jansma et al., 1981].

In international work, the late 1950s and early 1960s were characterized by little attention to agriculture, and where agriculture was attended to at all, the concentration was mainly on large-scale, more commercial farming possibilities in the less developed world. By the mid-1960s it finally became clear to general economists and central planners that agriculture was of fundamental importance and there was a rush to rediscover agriculture. This movement was followed a short time later by a rush on the part of agricultural economists who had been engaged in developmental work to rediscover farm management. Currently, general economists and agricultural economists who were not previously interested in farm management are working with theories of the firm which take into account firm-household relationships and the normative as well as the positive. One familiar with the balanced farming program of the University of Missouri and with the other domestic farm and home development studies and programs

of the late 1940s and the 1950s feels that he is seeing a summer rerun on television. In any event, there is a deep interest in decision-making theory at the farm level that encompasses the firm and the household and, hence, the normative, positive, and prescriptive.

Also in international work, many agricultural economists came in contact with public decision making as they served as advisers, consultants, and problem-solving researchers in the U. S. Agency for International Development (AID) and its predecessors, the International Bank for Reconstruction and Development (IBRD, the World Bank), UN Food and Agriculture Organization (FAO), Ford Foundation, Rockefeller Foundation, and other agencies, missions, and projects. These practical contacts served to stimulate interest in public decision making as part of a problem-solving process. Interest shifted from that of a detached study of decision making to a participatory interest. Some agricultural economists became interested in how to participate in making decisions rather than in merely how decisions are made. Such participation involved problem definition and solution, and hence an interest in prescriptive knowledge as it is related to positive and normative knowledge.

While some agricultural economists were participating in problem-solving activities abroad, others were having similar experiences domestically as a result of the problems and issues that upset U. S. tranquillity in the late 1960s and early 1970s. By then, it was clear that the agricultural economics profession was not dead as Gray [1970] feared; instead, many of its members were grappling with practical problems requiring philosophic underpinnings to help them work objectively with both the positive and normative in finding non–Pareto optimal prescriptions as solutions to problems. With the exhilarating experiences of participating in obviously useful problem-solving efforts came an increased interest in disciplinary and subject-matter research relevant to problem solving. Agricultural economics became, indeed, alive and well.

3. The Need to Study the Normative

As the constituencies of agricultural economists became concerned with real world problems and began to insist on practical relevance for the discipline [Glenn L. Johnson, 1971], the need to study the normative received increasing attention through expressions of dissatisfaction with price-weighted measures of gross national product, national income, and national indexes of productivity. Karl Fox [1974], among agricultural and general economists, became concerned with social indicators and produced a book in which the GNP concept was expanded to include monetary values for the nonmonetary incomes generated in the full twenty-four hours available to each person in the society. In addition to agricultural economists, many other persons became concerned with the measurement of value. Sometimes, the measures were much more specialized as

when nutritionists measure calorie consumption, protein consumption, and when engineers and others become involved in energy accounting.

Neoclassical utilitarian economics underwent a transformation during and after World War II. It was transformed to neoclassical utilitarian "market economics" as a result of the questions raised by Pareto. As Hicks [1939] and later Reder [1947] restricted neoclassical utilitarian economics to Pareto optimality, its evaluative and predictive capacity was restricted to the adjustments which take place in the market. Neoclassical utilitarian economics was changed from the *"old" Pigouvian welfare economics to the "new" Pareto optimal welfare economics.* Formal economics lost the logical structure it had used to evaluate the consequences of institutional, technological, and human changes which damage some individuals in order to benefit others. Without such a structure there was no way for economics to evaluate attempts to alleviate poverty, to restrict the activities of polluters, to redistribute rights and privileges among minority groups, or to do much to solve problems not solvable in the market place.

Positivism did not provide the answers [Pirsig, 1974]. Conditional normativism left answers to normative questions in an arbitrary state. For both positivism and conditional normativism, one set of values was as appropriate, as objective, and as true as any other set. This arbitrariness did not satisfy such pragmatists as Kenneth Parsons [1958]. Nor did it satisfy more normative persons inclined to believe that justice, equality, environmental quality, etc., really do have basic values which cannot be ignored in solving problems. Alternatives to characteristics of the real world, can be either directly or indirectly perceived. Moore was one of the major influences in John Maynard Keynes's student life [Moore, 1956, dust jacket].

C. I. Lewis dealt with the problem of converting normative and positivistic information into prescriptive statements about "how things ought to be done" — with what is right as contrasted to wrong — *vis-à-vis* both proposed and actual actions. The normativism of Moore and Lewis provides a philosophical alternative to positivism, pragmatism, and conditional normativism. Such an alternative would help solve the problem of interpersonally valid welfare measures and make Pareto optimality less necessary.

Agricultural economists, as a group, have not taken Pareto optimality seriously; they have not done so, perhaps, because they, in their common sense, know that the values of conditions, situations, and things are experienced and that terms describing these experiences can be used in logical discussions to arrive at non-Pareto optimal solutions of problems. Probably, it was a crude but inadequate recognition of this commonsense position regarding the normative that underlay the use of conditional normativism by agricultural economists during the period. Pareto optimality did not find full expression in agricultural policy textbooks until Tweeten's book was published in 1970. Clearly, agricultural economists have been more normativistic in practice than in their methodological and philosophic pronouncements. In their methodological pronouncements they tend to be positivistic, conditionally normative and pragmatic; in practice they are more normative—along the above interpretation of the Moore-Lewis line of thought. The normativism they practice seems based more on necessity than on knowledge of normative philosophy. This reviewer, some of his students, and a few others began to take outright normativism seriously, particularly after 1965 [Glenn L. Johnson and Zerby, 1973; Glenn L. Johnson, 1976] but also before the late 1960s [Glenn L. Johnson and Zerby, 1961; Glenn L. Johnson, 1960b, 1961a, 1963b].

PRAGMATISM

This philosophy is an alternative to positivism, conditional normativism, and Pareto optimality, and has long been advocated and practiced by a substantial number of agricultural economists. As repeatedly noted before, the Wisconsin institutionalists have long held a strong interest in practical problems solvable mainly with institutional adjustments outside the market place. This interest on the part of the pragmatists and their willingness to address practical problems has made them more successful in working on *institutional* aspects of the problems of the late 1960s than the positivists, the conditional normativists, and Pareto optimalists. Unfortunately, they have not demonstrated similar competencies *vis-à-vis* nonmarket *changes in technology and people.*

The interest of the pragmatists focuses heavily on prescriptive knowledge. Hypotheses tested by pragmatists tend to be prescriptive whereas those tested by positivists tend to be positivistic. As prescriptions are functions of both the positive and normative, the pragmatic metaphysical presupposition that the positive and normative are interdependent does little direct damage to the pragmatists' problem-solving activities. Pragmatism does, however, come into conflict with the positivistic philosophies of the physical and biological scientists who are working on technological change [Glenn L. Johnson, 1977b]. It also comes into conflict with outright normativists who hold the possibility at least that there may be knowledge of the values of such things as freedom, equality, justice, etc., that are independent of the positive. Pragmatism as an alternative to positivism, conditional normativism, and Pareto optimality is somewhat limited by the metaphysical presupposition of interdependence between the normative and positive on the part of the pragmatists. The question of interpersonal validity of welfare measurements hardly arises for the pragmatists. They regard the normative and positive as interdependent in the context of the problem at hand. Prescriptive knowledge—derived from both positive and normative knowledge—is always conditioned by the problematic situation at hand.

A constructive post-World War II effort noted earlier was the development of a training program in public administration for agriculturalists at Harvard University under the leadership of John D. Black and with the assistance of John M. Gaus. This deepened the interest in *process* in agricultural economics, which meant that Black's emphasis on neoclassical utilitarianism was modified to complement and supplement the interest in process that had long characterized the Wisconsin institutionalists. This in not surprising inasmuch as both Black and Gaus were trained at Wisconsin [Gaus and Wolcott, 1940].

A QUANTIFICATION OF PRAGMATISM

A quantification of pragmatism has been taking place that has hardly been recognized. Historically pragmatists tended to be suspicious of the quantitative techniques of statisticians and econometricians probably because those techniques often treat positive knowledge as independent of normative knowledge. As the lives of O. C. Stine, Wesley Mitchell, and others attest, pragmatic institutionalists could also be leaders in developing data and in making projections as to the consequences of alternative courses of action, particularly projections emerging iteratively out of interactions between investigators and decision makers or affected persons.

In recent years, the interactive, iterative process of making projections has been viewed in somewhat more formal terms by some agricultural economists. By contrast, systems scientists (such as Jay Forrester at MIT) drawing on cyberneticists were positivistic and conceived of a system as closed if the positive feedback loops were complete. When the systems-science simulation approach has been used by agricultural economists in practical, problem-solving contexts, there have been substantial interactions and iterations between researchers and decision makers and/or affected persons [Rossmiller *et al.*, 1972]. These iterations and interactions have, in effect, closed *normative* as well as *positive* loops and are so viewed by some agricultural economists working with systems-simulation models [Glenn L. Johnson, 1977a].

Probably the most significant philosophic tendency of the mid-1970s in agricultural economics is this quantification of pragmatism, which originated in cybernetic work that, ironically, was more positivistic than pragmatic. From cybernetics and systems science came an eclectic approach developed for modeling the domains of problems. Work in cybernetics also spawned the work of the information theorists, which has had important impacts on agricultural economics *via* the joint committee of the AAEA and the American Statistical Association, chaired by James Bonnen [1975]. In practical contexts, the system scientists began to make contributions to agricultural sector analyses as well as to the modeling of subnational and private systems [Halter and Dean, 1971; Manetsch *et al.*, 1971; Rossmiller *et al.*, 1972], not to mention the global modeling efforts of the Club of

Rome reported in the book entitled *Limits to Growth* [D. Meadows *et al.*, 1972]. These models dealt with multiple criterion variables and avoided premature use of the maximization techniques of neoclassical utilitarian economics. Because such models were mainly problem-solving in nature or relevant subject-matter efforts, agricultural economic analysts using such models began to interact with decision makers and affected persons. As the approach was already iterative, the interaction converted the whole approach to a pragmatic one in which information about criterion (normative) variables and positivistic variables emerged interactively and iteratively out of the problem-solving process.

The development described above is of substantial philosophic significance. While Churchman has not been involved directly in work with agricultural economists, his writings [Churchman and Ackoff, 1950; Churchman, 1968; Mitroff and Turoff, 1973] display an understanding of the fundamental importance of what is being done by agricultural economists using the system-science simulation approach. Mitroff and Blankenship [1973] argue for a pragmatic, dialectic approach to the conceptualization of large-scale, social experiments. They consider the difference between conceptualizing well- and ill-structured systems. A well-structured system may be easily conceptualized in terms of, say, a given discipline such as physics using a positivistic philosophy or alternatively in terms of a given discipline such as ethics using a normative philosophy. An ill-structured system may involve more than one and sometimes unknown disciplines and hence may have to be conceptualized in terms of different philosophies. Conceptualizing such a system involves "defining the state of nature" of the problematic system. For conceptualizing such systems, Mitroff and Blankenship [1973, pp. 345ff.] offer the following guidelines:

Guideline 1. AT LEAST TWO "radically distinct" disciplines of knowledge must be brought to bear on the conceptualization of any potential holistic experiment.

Guideline 2. AT LEAST TWO "radically distinct" kinds of conceptualizers (personality types) must be brought to bear on the conceptualization of any potential holistic experiment.

Guideline 3. AT LEAST TWO "radically distinct" philosophical inquiry models (conceptualizations) must be brought to bear.

Guideline 4. The subjects (general populace) of any potential holistic experiment must be included within the class of experimenters; the professional experimenters must become part of the system on which they are experimenting—in effect the experimenters must become the subjects of their own experiments.

Guideline 4'. The reactions of the subjects to the experiment and to the experimenters (and vice versa) are part of the experiment and as such must be swept into its design (i.e., conceptualization).

Guideline 5. The epistemic design rule for resolving the disparity between conflicting conceptualizations is CONFLICT—NOT "agreement" or "consensus."

Guideline 6. The methodological (i. e., philosophical inquiry) system for handling conflict is that of a Dialectical Inquirer.

Guideline 7. An appropriate design tool for modeling any conceptualization is simulation but—IF AND ONLY IF—it includes LIVE HUMAN PLAYERS chosen in accordance with all the previous guidelines.

Note the dialectic, pragmatic, eclectic, interactive, and iterative nature of these guidelines. Mitroff's pragmatic connection is *via* Singer rather than Dewey who was the source for Commons's institutionalism.

When agricultural economists followed such guidelines in developing a computerized model of the Korean agricultural sector, they in effect quantified pragmatism [Rossmiller *et al.*, 1972]. The iterations and the interactions with decision makers and affected persons provided an important source of normative and positive information which may be interdependent. This, and the postponement of maximization until the preconditions for carrying out maximization are met, make the practicing agricultural economists and systems scientists such as Mitroff and Blankenship essentially pragmatic in their approach and in the philosophy.

4. Individualism, Liberalism, Existentialism, Reactionaryism, and Processes

So far, this review has implied that agricultural economists have been only vaguely aware of the role such conflicting philosophies as pragmatism, positivism, outright normativism, and conditional normativism have played in guiding their thinking and activities since World War II. Awareness levels with respect to the topics of this subsection—individualism, liberalism, authoritarianism, existentialism, and reactionaryism—were even lower; hence, it is even harder to discern consistent patterns in the literature, thinking, and activities of agricultural economists from the end of World War II through the mid-1970s.

The neoclassical utilitarian tradition placed heavy emphasis on individualism. This emphasis was reinforced to a certain extent by the new welfare economics that tended to leave individuals and the *status quo* in a somewhat more dominant position by refusing to grant interpersonal validity to welfare measurements. As part of this rather illogical tendency, however, we find modern welfare economics also used illogically to defend the *status quo* instead of recognizing that it cannot be used either to defend or attack the *status quo*. It has become fashionable for younger agricultural economists to argue that older agricultural economists spent the 1930s, 1940s, 1950s, and 1960s defending the *status quo*. Such arguments reflect an ignorance of history and perhaps failure to recognize that many who are well off today were the disadvantaged of former years who were helped at that time with subsidized credit, public education, new technology, social security transfers, disaster relief, resettlement schemes, public irrigation, and drainage projects.

The 1946-76 period started out with a liberal rejection of the authoritarianism of the right and with military victories over the totalitarian rightist powers of Germany, Italy, and Japan. Paradoxically, the period later included a so-called liberalism that viewed sympathetically the authoritarianism of the left, both abroad and domestically! In fact, it has been difficult, at times, to distinguish a *liberalism* that would impose drastic damages on some individuals in order to confer rights and privileges on others from *authoritarianism*. During the period, the word liberal became almost meaningless—sometimes it stood for preserving the rights of individuals while at other times it stood for efforts to redistribute rights and privileges from some individuals to other individuals in such ways that the individuality of the losers (and even the gainers) would be greatly impaired, the central city welfare programs being important cases in point. In general, central city welfare recipients have been granted little control over programs designed to help them. As a consequence, their individualities have been diminished as they have had to deal with paid welfare workers. By contrast the subsidized production credit, land bank, and soil conservation associations, the 4H clubs, and the county extension service programs were placed under local control in ways which preserved local pride and individual identities. Many of the radical "New Left," who are not so new anymore, reject both the traditional and the current liberalism as protecting the rights and privileges of the "haves" [Zerby, 1971].

During the postwar period, a few agricultural economists became interested in existentialism. Existentialism is based on the conviction that knowledge of one's existence is the most empirically based knowledge one has. It also assigns high value to establishment and maintenance of one's existence or "identity"—individualism is important!

The pragmatic educational philosophy encountered by extension workers and vocational agricultural workers in their training is reinforced by Wisconsin institutionalism in agricultural economics. In the 1950-76 period, agricultural economists working abroad learned again and again the existentialist importance of the individual and of working interactively with individuals so as to develop, not constrain, their personal identities.

Many of the above "isms" are concerned with processes both in historical and current contexts. The concern is often normative and prescriptive, i.e., with whether "things are working out well." There is an extended concern with history, with processes, and with control over the processes that partially determine destinies and, hence, future history.

5. Marxism

In the 1960s and early 1970s, one could observe in agricultural economics circles a renewed interest in the teaching and philosophy of Marx although that interest has not become as widespread or popular in the United States as in France, for example. This interest was strengthened by the meeting of the International Association of Agricultural Economists in Minsk in 1970 and by the activities of leftist groups both outside of economics and within the American Economic Association. Kuhn's book—perhaps because of its title, *The Structure of Scientific Revolutions* [1970]—attracted some attention among agricultural economists. When a major change takes place in the questions to which a scientific discipline addresses itself, it becomes ill-structured and its laws, theories, models, and data have to be changed to re-establish its structure. Kuhn regards such major changes as a revolutionary change in "paradigm." It is now a fad to use the word paradigm (both with and without prior consultation of a dictionary) to upgrade even minor changes in thought to the status of *revolutionary paradigmatic shifts*, which Kuhn would probably call "improvements for puzzle solving" rather than (revolutionary) paradigmatic shifts! Some of the so-called New Left have asserted that the questions raised for economics by the social unrest of the late 1960s require abandonment of what they call the neoclassical paradigm and use of a Marxist paradigm based on the classical labor theory of value, social ownership of the means of production, the perfectibility of man, and so on.

The range of problems and subjects researched by agricultural economists in both foreign and domestic locations seems to have become too complex to be handled by a single, large paradigmatic change going (retrogressing?) from neoclassical and post-neoclassical utilitarian economics to the pre-neoclassical labor theory of value in Marxism. The labor theory of value seems too simplistic to deal with: the values of individualism; equity in the distribution of market, political, military and police power; imperfections in human beings. Instead, numerous, smaller, more adaptive, hardly revolutionary changes are required when proceeding from problem to problem and subject to subject relative to changing issues. To this reviewer, the assertions of the antique labor theory of value seem to cry out for objective normative research (both analytic and synthetic) rather than dogmatic adherence. One is reminded of Parsons's fears of reversion to medieval dogma and the assured ends of conduct [K. H. Parsons, 1958]. Unfortunately, positivists and conditional normativists are in weak positions to complain about arbitrary unobjective endorsement of values (right, left, center, or otherwise), for they insist that *all* normative concepts are arbitrary and unobjective. In fact, and as Parsons feared, it is their refusal to grant the possibility of objective normative research that opens the way to capricious use of arbitrary values (right, left, and otherwise) including adherence to such "religious-like" values as those of conservationism, agricultural fundamentalism, central city fundamentalism, environmentalism, energy fundamentalism, and consumerism.

The New Left has also advocated dialectics as opposed to analytics. Methodologically, Carnap [1953] and Popper [1962, 1972] have stressed: the relationship between analytic and synthetic knowledge; the tentative, questionable (dialectical) nature of all synthetic knowledge based on primitive terms (as subjective interpretations of sense impression); and the dependence of purely analytic or logical truth on axioms bearing no known relationship to reality. Thus, as empirical scientific truth is always subject to question and challenge, modern science is dialectic. To put one's knowledge above the tests of logic and experience—i.e., above question—is to sin in the eyes of the scientific community. Georgescu-Roegen [1971] argues that even physics is normative (at least in the pragmatic sense) [Glenn L. Johnson, 1973, pp. 492ff.] and dialectic. Modern science has no important disagreement with Marxism on dialectics though some Marxists attack modern science as nondialectic.

6. The International Association of Agricultural Economists

The proceedings of the triennial meetings of the International Association of Agricultural Economists reflect the work of agricultural economists, albeit belatedly, and in turn affect their work. The first postwar IAAE conference, held in Dartington Hall in 1947, served to re-establish the Association after the World War II interruption. Subsequent meetings at Stressa in 1949, East Lansing in 1952 and Helsinki in 1955 helped to reunite the agricultural economists of the world. An important accomplishment of the first four meetings was reunification and re-establishment of dialogue between the agricultural economists of the Allied and Axis powers. Important also was the participation of agricultural economists from Eastern bloc (including persons from the People's Republic of China) and less developed countries. By the time of the Helsinki meetings, communications among the agricultural economists in the different countries was being extended from older, well-established persons to include the younger group of post-World War II agricultural economists.

From the Helsinki meetings in 1955 to the 1970 meetings in Minsk, there was a steady, slow growth in the influence of IAAE but not a great deal of change in the discussions at meetings of *philosophies, research approaches, or techniques*. The Association had been influenced early by Cornell empiricism as had English agricultural economics. That influence persisted in the International Association after World War II until younger economists began to play a more important role. Despite this emphasis, the Association meetings were eclectic, moderately multidisciplinary, and, above all, concerned with the lives and welfare of rural people. Leonard Elmhirst, the founding father, was as much a rural sociologist as an agricultural economist and was keenly aware of technology and the human element. Although the Association was oriented to rural people, not farmers alone, the problems of farmers were always high priority.

Particularly at the 1967 meeting in Sydney and again at the 1970 meeting in Minsk, but also earlier at Helsinki, Cuernavaca, and Dijon, younger members of the Association voiced demands for greater attention to economic theory and modern quantitative techniques—in short, for more disciplinary excellence. There was also a demand for greater participation on the part of younger persons. The demand was forcefully expressed at the Minsk meeting by John Dillon, who expressed disappointment that he did not have an opportunity to hear more from competent, young Soviet theorists and statisticians. This demand for greater disciplinary excellence and for wider participation was reflected in the contributed paper sections at the 1973 São Paulo meetings and particularly at the Nairobi meeting in 1976.

Although the Nairobi conference remained eclectic and multidisciplinary, it did give considerable attention to disciplinary excellence with respect to dynamics—particularly that part of dynamics dealing with decision making (both public and private). The decision-making theories discussed dealt with the normative and prescriptive as well as the positive information-gathering and -processing activities of decision makers. Prior to the Nairobi meeting there was a conference on risk and uncertainty sponsored by the ADC at CIMMYT. In contrast to the Nairobi meeting, the CIMMYT meeting dealt largely with risk aversion and preference and the expected utility hypothesis, particularly as handled by Arrow [1971], Hull et al. [1973], and Dillon and Anderson [1971]. The Nairobi approach was broader and dealt with the learning, execution, and responsibility phases of management as well as with decision making and the mathematical niceties of the expected utility hypothesis. Thus, the Nairobi meeting explicitly involved normative philosophies and pragmatism to a much greater extent than the CIMMYT meeting.

Chapter IV. The Ending Inventory

By the mid-1970s, the philosophic orientations of agricultural economists had changed substantially from those held at the beginning of the post-Word War II period. Their mid-1970s positions will be discussed under the following headings: positivism, neoclassical utilitarian economics, the residual impacts of the late 1960s, normativism, normativistic and/or positivistic subject-matter research, prescription, the meaning of truth in economics, and the emerging quantification of pragmatism.

1. Positivism

By the mid-1970s positivism was probably stronger in agricultural economics than in the late 1940s; however, the positivism of the mid-1970s was much different from that of the early 1940s. In the 1940s, for instance, much of the positivism in agricultural economics was found in farm management; of that in farm management, much was nearer to pure positivism or pure empiricism than logical positivism. Since the late 1940s, agricultural economists and, particularly, econometrically inclined agricultural economists have paralleled the development in the physical sciences summarized by Carnap [1953]. The econometricians, like the physical scientists, combined the work of logicians and theorists with that of economic empiricists and clarified the relationships among the logical (or analytic), primitive terms, and the empirical (synthetic). In the hard sciences, important logicians were Leibniz and Descartes and important empiricists were Bacon and Locke [Mitroff and Turoff, 1973]. Modern science put the logical together with primitive terms to form the synthetic or empirical in the manner described by Carnap. Similarly, the econometricians put the theories of economics, statistics, and mathematics together with the empirical work of economists to form econometrics, also in the manner detailed by Carnap. The result is a form of positivism known as *logical* positivism rather than the "*pure*" or "*straight*" positivism of the traditional farm managers of the 1930s and early 1940s. Mini [1974] in his book on philosophy and economics seems unaware of this development in either general or agricultural economics and pleads for its occurrence.

One current anomaly is that the positivistic work of econometricians and of such positivistic economists as Friedman deals with price, income, expenditures, gross national product, and other variables that are normative in the sense that they measure monetary *values*. Perhaps because these normative variables are so quantifiable, they are not regarded as normative!

Positivistic econometric techniques have been greatly improved and were at a much higher state of development by the mid-1970s than in the late 1940s. Techniques important here involved programming with all of its modifications and variations, the estimation of parameters of simultaneous equations, input/output analysis, benefit/cost ratios, and so on.

Two strong pieces of evidence of the continued strength of positivism among agricultural economists are the use of conditional normativism and Pareto optimality. Conditional normativism à la Myrdal was commonly practiced in the late 1940s and early 1950s and is probably no less commonly practiced now than then. Pareto optimality, which is intellectually related to positivism, had very little impact on agricultural economics for much of the post-World War II period but is now more widely but not universally used among agricultural economists. Our review of what went on in the period between World War II and the mid-1970s indicates that policy analysts among agricultural economists were particularly slow in moving to Pareto optimality. Fortunately (in the mind of this reviewer) that slowness delayed the impact of positivism on agricultural policy analysis.

Another avenue whereby positivism may have strengthened its grip on agricultural economics is *via* cybernetics with its concern about information systems and data processing. Further evidence of the impact of positivism on agricultural economics is the current tendency to regard production functions as positive even though many of the older members of the agricultural economics profession learned, early in their careers, that production was "the creation of time, form, and place *utility*."

Positivism, despite the sharp criticism it has received at the hands of philosophers and many others, is not dead in agricultural economics by any means. In fact it may be peaking some thirty years after the date Kaplan [1968] gave as the beginning of the end for positivism—the end of World War II.

Agricultural economists in their work as disciplinarians have made substantial progress under positivistic influences. Ironically, some of this progress has been based on the methods of positivism in working with such normative variables as production, prices, incomes, expenditures, indexes of output, and indexes of input. Also positivistic methods have been productive when employed by economists doing subject-matter research. It will be recalled that subject-matter research was defined earlier as the accumulation of a set of multidisciplinary information (multidisciplinary because it would be disciplinary if only one discipline were involved) useful in solving a *defined set* of problems but not adequate to solve any *one* of the problems completely. When the defined set of information in a subject-matter research effort is positivistic or largely positivistic, positivism with its highly effective associated methodologies has much to contribute. It is in the realm of problem-solving research that the weaknesses of positivism are revealed—for instance, its inability to work objectively with the normative. This inability means that positivistically inclined workers have difficulty both in defining problems and in determining "what ought to be done" as a prescription to solve a particular practical problem.

2. Neoclassical Utilitarian Economics

Utilitarian economics is, of course, normative and prescriptive. However, utilitarianism is so specialized to economics that it is commonly considered separately from other more general forms of normativism (to be discussed in a subsequent subsection).

Neoclassical utilitarian economics has strengthened its philosophic grip on agricultural economists and is now in a stronger position than at the end of World War II. However, the utilitarianism of agricultural economists, like their positivism, changed significantly between the beginning and ending inventory. The change can be described as a change from "neoclassical economics" to "neoclassical *market* economics."

The change resulted from Pareto's questions introduced into economics by Hicks in his *Value and Capital* [1939]. As neoclassical utilitarian economics came under the influence of the Hicks questions, the evaluative and prescriptive power of neoclassical utilitarian economics was increasingly restricted to Pareto optimal adjustments that, of course, are attainable in a market; by contrast, nonmarket adjustments that are executed by the government and that are not Pareto optimal can no longer be evaluated [Reder, 1947]. Thus, much of the neoclassical utilitarian economics that existed and was used by agricultural economists in the mid-1970s was weaker than the neoclassical utilitarian economics being practiced by agricultural economists at the end of World War II. Fortunately neoclassical utilitarian economics, even though weakened by Pareto optimality, can be (and still is) used to predict the consequences of nonmarket adjustments. However, it withholds evaluative judgments and refuses to make welfare statements concerning nonmarket adjustments that damage some in order to benefit others. This, of course, had been recognized in the theory of second best [Lipsey and Lancaster, 1956]. Econometricians, operations researchers, and others employing Pareto optimal, neoclassical, utilitarian economics can thus predict the consequences of nonmarket adjustments but cannot evaluate them.

Another consequence of Pareto optimality is inability to compare production aggregates involving more than one person when they result from nonmarket (coerced) changes in technology, people, and institutions. This weakening of neoclassical utilitarian economics by Pareto optimality is, of course, a valid explanation of the inability of post-Pareto neoclassical economics to respond constructively and effectively to the crises of the late 1960s. It is not a bias of Pareto optimality in favor of the *status quo* that does the damage; instead, it is impotence for handling redistributive problems. Student activists seemed to have sensed this more acutely than did disciplinary economists!

3. Residual Impacts of the Late 1960s

The problems and issues of the late 1960s have left their impact on the philosophic orientation of agricultural economics. There has been an increased demand for relevance on the part of students, research clientele, and the public in general. This has shown up in demands for greater accountability on the part of research organizations.

The upshot of these demands has been an increased awareness of the inadequacy of Pareto optimality as a basis for welfare decisions concerning nonmarket changes in technology, institutions, and people. Early in the crisis years of the late 1960s, there was a tendency to turn to rather arbitrary, superficially defined values. It was not long, however, before the results of this procedure ran into conflicts with both logic and experience. This led to a deeper search in logic and experience for normative knowledge.

As part of the uneasiness with neoclassical market or Pareto optimal economics, some of the younger agricultural economists developed a substantial interest in Marxism and in the labor theory of value. There is some evidence that these younger economists now see the "antique" nature of the labor theory of value in the history of economic thought. They have also encountered logical conflicts among the labor theory of value, the Marxist assumption that man is perfectible, and the severe force by elitist Marxist regimes to bring "imperfect men" from the masses into line with the objectives of small party elites. Marxism has, perhaps, been more prevalent among French agricultural economists than among any other national group outside the Communist bloc nations. This may be why penetrating thinking in France concerning Marxism seems to focus increasingly on the conflicts noted above. In terms of G. E. Moore [1903], the Marxist adherence to the labor theory of value amounts to a naturalistic fallacy, i. e., the fallacy of stating that that which possesses the characteristic of goodness, in this case laboring, is goodness. One might hypothesize that economists and agricultural economists reflecting logically on their own normative experiences might see increasingly that the labor theory of value is an inadequate source of knowledge about the value of racial justice, individual freedom, environmental quality, minority rights, civil rights, and so on.

The residual impact of having to respond to the crises of the late 1960s is a demand for more objective, less arbitrary research on values. This demand can be met with an outright normativism that can view good and bad as knowable independently of positivistic knowledge, and a pragmatism that views the normative and positive as interdependent in the context of a problem and expressible mainly in the form of prescriptions to solve that particular problem.

4. Normativism

As noted above, there has been a strengthening interest in outright normativism since the crises of the late 1960s. Before concluding, however, that normativism is in a stronger position than in the immediate postwar period, it must be remembered that neoclassical utilitarian economics has been substantially weakened by the Pareto optimality restriction. Then, too, conditional normativism is at least as strong as it was in the immediate post-World War II period. Probably the position of that particular form of normativism labeled "neoclassical utilitarian economics" is weakest among the most disciplinary of the agricultural economists as they tend to be positivistic. Pareto optimality has been followed less consistently by agricultural than by general economists. And among agricultural economists, Pareto optimality has been ignored more consistently by agricultural economists doing problem-solving and subject-matter research than by those with disciplinary interests and concern for peer group approval, tenure, and "refereed" publications; ironically, those individuals with these latter concerns include some of the young people who stressed the normative aspects of the problems and issues in the late 1960s.

Operationally, a substantial amount of normativism exists among agricultural economists who make their living measuring and analyzing income, prices, expenditures, the contribution of agriculture to GNP, and other factors. Such work is fundamentally normative even though it is confined to monetary as opposed to nonmonetary values. Many agricultural economists also specialize in estimating exchange values not provided satisfactorily by the market as when they compute opportunity costs and shadow prices. We note, of course, that the interest of economists in prices is an interest in values in exchange. Much less attention is paid to total or intrinsic value as opposed to exchange values by agricultural economists [Black, 1953a]. Agricultural economists also make extensive use of the normative concepts of consumer and producer surpluses. Although some would initially deny it, agricultural economists also deal with the normative when they deal with production—the creation of time, form, and place *utility*.

In employing benefit/cost techniques, agricultural economists often become involved in estimating nonmonetary values in order to obtain a common denominator in terms of which to express benefits and costs as a ratio. Such work is more than conditionally normative because it requires the economist to establish the value of one condition, situation, or thing in terms of other conditions, situations, or things. When used to evaluate nonmarket adjustments, benefit/cost ratios also imply interpersonal comparability.

Then, too, agricultural economists have become involved in the social indicators movement, a prominent example being the work of Karl Fox [1974]. Again, outright normative work is being done when attempts are made either to estimate nonmonetary values or to find a common denominator (*numeraire*) other than money among various values. Agricultural economists have also been involved in such social indicators as those used in energy accounting, measurement of nutritional status, levels of living, and environmental quality. Fortunately, the role of agricultural economists in the latter connection has often been that of critics.

Outright normative work tends to be either disciplinary or subject matter in nature. The fact that outright normative work is ordinarily disciplinary or subject matter in nature does not indicate that such work was irrelevant in responding to the crises of the late 1960s. We must recognize, however, that problems are not solved with normative information alone, any more than with positive information alone. We reach prescriptive knowledge—knowledge about what ought to be done—on the basis of both positive and normative knowledge.

5. Pragmatism

This philosophy is in a stronger position in agricultural economics in the mid-

1970s than in the late 1950s and 1960s and it continues to be a major alternative to positivism. Its greater recent strength grows in part out of recognition that an alternative to positivism is essential in designing non-market-induced changes in institutions, technologies, and people to handle issues and solve problems and, in part, out of the increased generality of quantitative methods that permit quantification of pragmatism, so that we begin now to have computational capacity to match and handle the complexity of pragmatic methodologies.

6. Normativism and Positivism in Subject-Matter Research

When agricultural economists conduct research in such subjects as energy, employment generation, and food and nutrition, they often deal with positive and/or normative information and, in the case of the latter, with both monetary and nonmonetary values. Positivistic or conditionally normative subject-matter researchers tend to confine themselves to the positive with and without normative assumptions. Many such positivistic and conditionally normative workers regard monetary values as positivistic probably because such values are readily quantifiable.

7. Prescriptive Research

Economics and agricultural economics in particular are decision-making disciplines. They are concerned with prescribing "what ought or ought not to be done." They also evaluate "what ought or ought not to be done" in terms of whether what was done was a justifiable prescription. Also, economists, along with engineers, architects, physicians, and others, often go into the "design mode" to conceive of new institutional arrangements, technological advances, and changes in human behavior to solve a problem. Further, agricultural economists often assume that producers, consumers, and resource owners are maximizers in the neoclassical utilitarian sense and use the results of computations based on that assumption in predicting the behavior of consumers, producers, and resource owners.

We have already seen that the use of neoclassical utilitarian economics for evaluative purposes was weakened by Pareto optimality. In this paper, I have labeled this weaker form of neoclassical utilitarian economics neoclassical "market" utilitarian economics. The use of conditionally normative techniques has also reduced the evaluative power of agricultural economics. Nonetheless, agricultural economists have gone beyond such market economics and Pareto optimality to evaluate and prescribe changes made outside the market in institutions, technology, and people [Libby, 1971; Randall, 1972; Dorner, 1971].

Some of these prescriptions are arrived at with the methods implied by pragmatism. Others are arrived at under the presupposition that both normative and positive information are independently attainable but jointly processible *via* a decision rule into prescriptions as to what ought to be done or evaluations as to what ought or ought not to have been done.

In general, there has been a strengthening of pragmatism and it is in a stronger position in the agricultural economics of the mid-1970s than at most any other time since World War II [Boulding, 1969].

8. The Meaning of Truth in Economics

With apologies to Frank Knight, who used a similar title in writing an important review of pronouncements by Hutchinson *vis-à-vis* positivism [Knight, 1940], we proceed at this point to review the meaning of truth for agricultural economists as of the mid-1970s.

As indicated in the early pages of this review, there are at least four tests for truth employed by researchers and the public in general. Not all agricultural economists apply all of these tests [Glenn L. Johnson and Zerby, 1973]. *One* test is that of logical consistency (referred to as *coherence* by philosophers and internal consistency by some agricultural economists, including this reviewer earlier). A *second* is the test of *correspondence* (referred to by some agricultural economists, including this reviewer earlier, as external consistency). A *third* is the test of *clarity* or lack of ambiguity, which must be met before either the correspondence or coherence tests can be applied and is closely related to Popper's concept of falsifiability. The *fourth* is the test of *workability*. Some analysts treat models and equations that pass the coherence test as *validated* and those that pass the correspondence test as *verified*.

Use of these four tests makes truth social, as Knight [1940] pointed out. *This social nature of truth* has also been stressed by Georgescu-Roegen in his work on entrophy [1971] and by Popper [1959; 1962, ch. 10, sec. 3; 1972, ch. 2, sec. 8-11] in his related concept of "verisimilitude." That which we accept as true is that which has not yet flunked one or more of the above tests. As our empirical knowledge changes, the new concepts may be inconsistent with other previously accepted concepts. Accepted knowledge (thesis) is always to be confronted with its denial (antithesis); hence, truth is the result of a dialectic process and attempts to dichotomize science and dialectics are probably false. Similarly, when we change reality as a result of previous problem-solving efforts and decisions, concepts that previously passed the test of correspondence may no longer pass it. Also, as our ability to conceptualize and measure increases, previously acceptable descriptive information may flunk the test of correspondence. Additionally, what is clear at one point in a society's development may not be clear at a later point. Science, itself, is a social phenomenon, and hence scientific truth is a function of the state of science as a social activity. Two papers presented at the 1977 AAEA meetings in San Diego dealt significantly with the normative [Hartman, 1977; Moles, 1977]. While these papers may become part of the phantom literature of the 1970s, they should be noted here. Both stress that knowledge, including scientific knowledge, is social and cultural. Both papers identify science with positivism and scientific knowledge with positivistic knowledge. Modern economics is then identified as scientific and, hence, positivistic; unfortunately this then causes both Hartman and Moles to conclude that neither economics nor science can deal with "quality"—i. e., with value—although Moles is less definite on this conclusion. If positivism were identified not as *the* philosophy of science but rather as *a* philosophy of science, *science* would come off better. As institutional economics, welfare economics, and indeed neoclassical economics are normative as well as positive, the Hartman and Moles papers do seem to involve some *non sequiturs* even though they are basically correct about the need for agricultural economics to go beyond the positive (scientific in their view) to the normative. Also Hartman's paper has the distinct merit of recognizing empirical normative knowledge based on experience. They do not discuss dialectics and ill-structured systems [Mitroff and Blankenship, 1973; Runes, ed., 1960, pp. 77-78].

The social nature of truth means that truth is arrived at through a process and that this process involves successive iterations. Truth is arrived at iteratively and interactively as societies of investigators, decision makers, and affected people interact. Iteration and interaction are important with respect to the normative positive as well as the prescriptive. The process is also dialectic as truth is always opposed or being subjected to one of these tests. The oft-discussed dichotomy between science and dialectics seems to be unsustainable. The dichotomy between science and dialectics fails on two counts: (1) positivism, which in its earlier form was less dialectic than now, has never been *the* only philosophy of science; and (2) dialectics is quite characteristic of most modern thought in the sciences and in thought about the philosophic foundations of science. Popper [1959] has stressed the importance of falsifiability in science—the more falsifiable a concept, the more testable it is and the more reliable it is if it survives testing. It is desirable that concepts be confrontable dialectically with their opposites. Similarly, pragmatism and John Dewey's scientific pragmatism have always been dialectic. We should note, in passing, that some feel that the dichotomy between science and dialectics is one between analytics and dialectics; this cannot be, as modern science is both analytic and dialectic [Carnap, 1953]. Much more fundamentally, however, physics, chemistry, biochemistry, and biology are changing so rapidly that their dialectic nature would force science to be dialectic even if it were not already so. The second law of thermodynamics, according to Georgescu-Roegen [1971], requires that physics be dialectic—low-level entropy is valuable and differences in the value of different kinds and levels of entropy determine the variables we consider as well as the classes into which we divide these variables in physics. In turn, penumbra develop between variables and classes. Tensions then develop as alternative variables and classification schemes confront each other dialectically in the development of a discipline such as physics. When the conceptual structure of a discipline becomes obsolete for its purposes, new conceptual structures emerge as paradigmatic changes [Kuhn, 1970] to confront existing structures in a revolutionary, dialectic manner. Science is part of culture—as such it changes—and tensions build as alternative formulations develop to confront their predecessors. If science were not dialectic, it would have to become so. Agricultural economists are well advised not to defend either their science or their analyses against dialectics.

In terms employed by Mitroff and Blankenship [1973], Kuhn's paradigmatic revolutions involve ill-structured systems that must be studied dialectically. Well-structured systems within disciplines are less dialectic—for these nondialectic methods may do well—for the time being. It is the routine well-structured systems which the "hewers of wood" handle well in a discipline. In agricultural economics, for instance, such workers do their linear programming, Cobb-Douglas, input/output, operations research, and program evaluation and review technique (PERT) analyses until, as T. W. Schultz commented, "the *AJAE* runneth over."

Chapter V. In Conclusion

Karl Brandt [1955, p. 806] concluded his 1955 presidential address with the following words:

> Agricultural economics will gain in stature and influence if, as one of the disciplines in the realm of humanities, it sets its sights high and keeps aware of the fact that its subject is concerned with cause and effect relations in human and social actions, and that this involves far more than material needs. Let us suppose that we have a generation without a major war ahead of us, and that the imagination, energy, and drive of the nations can to a large extent be allocated to and absorbed by efforts toward accelerated economic development. The changes brought about in the economic and social spheres will be breathtaking, and call for bold perception of the macroeconomic problems. The actual pace of economic progress may overtake the economic profession just as the stalling of investment, exchange, and employment caught it unprepared in the great depression. In such a period of economic growth as may lie ahead, problems of maladjustment may become even more severe, but their nature will be dynamic—such as disparity in place of development—and their susceptibility to remedial action will be greater. All this argues for more alertness in our profession to the strategy to be

employed in allocating our human research resources, a firmed
understanding of the economic order as a whole, and
strengthening of the will to create theory, or, as I prefer to say:
to complement analysis by synthesis.

Brandt's projection of breathtaking changes in economic and social spheres and
the overtaking of the economics profession by economic progress has certainly
been fulfilled. The period of growth and relative peace that followed his address
did indeed create problems and maladjustments—and they were severe. There
have been unfulfilled demands for a firmer "understanding of the economic order
as a whole," for "theory . . . and to complement analysis by synthesis."

In a preceding paragraph, however, Professor Brandt [1955, p. 806] noted, "If
it should be a sound endeavor to orient economics more towards theory, I doubt
that concentration on methodology will help. While it has been much stressed in
the past that any science worthy of its name must be concerned with methodology, it also seems possible that the preoccupation with methodology may be a
sign of science's decay." Our review of the years since World War II has indicated
that responding to the crises of the late 1960s challenged not only our methods
but also the underlying philosophies that structure them. Even Brandt probably
failed to envision the severity of the problems that were to appear in the late
1960s. Had he realized their seriousness, I believe he would have seen that they
would challenge our underlying philosophies—yes, even his positivistic philosophy of science—to such an extent that we would have to consider methods beyond the normal tool kits of agricultural economists in the mid-1950s. Another
sentence from Karl Brandt [1955, p. 806] indicates that he may have glimpsed this
need: "More exchange of thought with general economists and discourse on
problems of economic development with researchers in other disciplines such as
philosophy, logic, philosophy of law, jurisprudence, political science, economics,
history, and anthropology will not only widen horizons but give by analogy or
transposition a firmer grasp of what economics is and what it cannot be." Then
skipping a sentence we find: "To establish this contact with other disciplines a
deliberate effort towards orientation may be made and our Association could be
the catalyst." The intervening sentence read as follows: "It [economics] cannot be
the arbiter of values for society and it cannot decide what ought to be done." If
that sentence means that we cannot research values and contribute the results, it
does not have a basis in either logic or experience—instead, it is a metaphysical
presupposition of positivism which is not accepted by pragmatists and normativists. Clearly, the experiences from World War II to 1967 do not confirm this
last quoted sentence from Brandt. Brandt's own positivism was itself a casualty
of the late 1960s. This is not surprising, for Kaplan noted [1968] that the end of
World War II was the beginning of the end (for the dominance) of positivism.
Pragmatists and normativists do have much needed philosophic and, hence,

methodological contributions to make to agricultural economics in the years ahead.

We can look forward to continued philosophic reorientation and, hence, continual restructuring of our research methods. This reorientation should broaden our capacity to do disciplinary research by strengthening our ability to work with the normative while not diminishing the capacity we have acquired from the positivists to work with the positive. Similarly it should add a normative dimension to our subject-matter research without loss of its positive dimensions. *Vis-à-vis* problem-solving research, it can be anticipated that the presently perceived dichotomy between science (largely positivistic) and dialectics will be seen to be false. The same will probably be seen with respect to the supposed dichotomy between science and the humanities. Along with this realization is likely to come an eclecticism. When the forces of change cause us to perceive serious structural flaws [Mitroff and Blankenship, 1973] in our discipline, dialectics and eclecticism are likely to lead to appropriate changes in our discipline's paradigms [Kuhn, 1970]. With respect to problem solving, this eclecticism and dialectism will help us convert ill-structured definitions of problems into better-structured ones. Such better-structured problem definitions will help us to recognize more clearly the unique multidisciplinary dimensions of each problem and the unique appropriate mix of philosophic orientations and research methods to use in solving each specific problem.

Notes

1. These references are in the hands of only a few individuals. One set is available in the offices of the Farm Foundation, Oak Brook, Ill.

2. Part of the "phantom literature."

3. For additional discussion of agricultural information systems, see the following AAEA literature reviews in volume 2 [Judge *et al.*, eds., 1977]: Bonnen [1977]; Bryant [1977]; Trelogan *et al.* [1977]; and Upchurch [1977].

References

Allin, B. W. [1948]: "The Objectives and Methods of Agricultural Economics." *J. Farm Econ.* 30: 545-552.

————— [1949]: "Theory: Definition and Purpose." *J. Farm Econ.* 31: 409-417.

AAEA Committee on Economic Statistics [1972]: "Our Obsolete Data Systems: New Directions and Opportunities." *Amer. J. Agr. Econ.* 54: 867-875.

Anderson, J. R., J. L. Dillon, and J. B. Hardaker [1977]: *Agricultural Decision Analysis.* Ames, ISU Press.

Arrow, K. J. [1951]: *Social Choice and Individual Values.* New York, Wiley (2nd ed., 1963).

————— [1971]: *Essays in the Theory of Risk Taking.* Chicago, Markham.

Bain, J. S., R. E. Caves, and J. Margolis [1966]: *Northern California's Water Industry: The Comparative Efficiency of Public Enterprise in Developing a Scarce Natural Resource.* Baltimore, Johns Hopkins Press for RFF.

Baker, J. A. [1950]: *The Place and Functions of Economic Research in a Democracy*. SSRC Project in Agricultural Economics, Economic Efficiency Series, Univ. of Chicago, Paper 2.

Baker, O. E., R. Borsodi, and M. L. Wilson [1939]: *Agriculture in Modern Life*. New York, Harper & Bros.

Bawden, D. L. [1972]: "Welfare Analysis of Poverty Programs." *Amer. J. Agr. Econ.* 54: 809-814.

Benedict, M. R. [1953]: *Farm Policies of the U. S., 1790-1950: A Study of their Origins and Development*. New York, Twentieth Century Fund.

Bieri, J., A. de Janvry, and A. Schmitz [1972]: "Agricultural Technology and the Distribution of Welfare Gains." *Amer. J. Agr. Econ.* 54: 801-808.

Bishop, C. E., and K. L. Bachman [1961]: "Structure of Agriculture." In ISU, CAED, 1961, pp. 237-250.

Black, J. D. [1939]: "A Reply." [to T. W. Schultz, 1939b] *J. Polit. Economy* 47: 717-721.

———— [1953a]: "Should Economists Make Value Judgments?" *Quart. J. Econ.* 67: 286-297.

———— [1953b]: *Introduction to Economics for Agriculture*. New York, Macmillan.

Black, J. D., and J. T. Bonnen [1956]: *A Balanced United States Agriculture in 1965*. Washington, National Planning Association, Special Report 42.

Black, J. D., M. Clawson, C. R. Sayre, and W. W. Wilcox [1947]: *Farm Management*. New York, Macmillan.

Bonnen, J. T. [1975]: "Improving Information on Agriculture and Rural Life." *Amer. J. Agr. Econ.* 57: 753-763.

———— [1977]: "Assessment of the Current Agricultural Data Base: An Information System Approach." In Judge, *et al.*, 1977, pp. 368-407.

Boulding, K. E. [1956]: *The Image*. Ann Arbor, Univ. of Michigan Press.

———— [1969]: "Economics as a Moral Science." *Amer. Econ. Rev.* 59: 1-12.

Boutwell, W., C. Edwards, R. Haidacher, H. Hogg, W. E. Kost, J. B. Penn, J. M. Ropp, and L. Quance [1976]: "Comprehensive Forecasting and Projection Models in the Economic Research Service." *Agr. Econ. Res.* 28: 41-51.

Brandow, G. E. [1955]: "Methodological Problems in Agricultural Policy Research." *J. Farm Econ.* 37: 1316-1324.

———— [1966]: "The Food Commission: Its Product and Its Role." *J. Farm Econ.* 48: 1319-1327.

Brandt, K. [1955]: "The Orientation of Agricultural Economics." *J. Farm Econ.* 37: 793-806.

Breimyer, H. F. [1967]: "The Stern Test of Objectivity for the Useful Science of Agricultural Economics." *J. Farm Econ.* 49: 339-350.

Brewster, J. M. [1950]: *Efficiency, Justice and Freedom*. SSRC Project in Agricultural Economics, Economic Efficiency Series, Univ. of Chicago, Paper 4.

———— [1970]: *A Philosopher among Economists. Selected Works*, ed. J. P. Madden and D. E. Brewster. Philadelphia, J. T. Murphy.

Brewster, J. M., and H. L. Parsons [1946]: "Can Prices Allocate Resources in American Agriculture?" *J. Farm Econ.* 28: 938-960.

Brinegar, G. K., K. L. Bachman, and H. M. Southworth [1959]: "Reorientations in Research in Agricultural Economics." *J. Farm Econ.* 41: 600-619.

Brown, W. G. [1965]: "An Appraisal of Critical Appraisal." *J. Farm Econ.* 47: 843-844.

Brownlee, O. H. [1948]: "Marketing Research and Welfare Economics." *J. Farm Econ.* 30: 55-68.

_____ [1950]: *The Meaning of Economic Efficiency in Terms of Possibilities and Choices.* SSRC Project in Agricultural Economics, Economic Efficiency Series, Univ. of Chicago, Paper 1.

_____ [1961]: "Discussion." [of 'Goals for Economic Organization,' by J. W. Markham] In ISU, CAEA, 1961, pp. 110-113.

Bryant, W. K. [1977]: "Rural Economic and Social Statistics." In Judge, *et al.*, eds., 1977, pp. 408-420.

Buchanan, J. M. [1959]: "Positive Economics, Welfare Economics, and Political Economy." *J. Law and Econ.* 2: 124-138.

_____ [1962]: "The Relevance of Pareto Optimality." *J. Conflict Resolution* 6: 341-354.

Buchanan, J. M., and G. Tullock [1962]: *The Calculus of Consent: Logical Foundations of Constitutional Democracy.* Ann Arbor, Univ. of Michigan Press.

Burchinal, L. G. [1961]: "Differences in Educational and Occupational Aspirations of Farm, Small-Town, and City Boys." *Rural Sociology* 26: 107-121.

Carnap, R. [1953]: "Formal and Factual Science." In Feigl and Brodbeck, eds., pp. 123-128.

Castle, E. N. [1968]: "On Scientific Objectivity." *Amer. J. Agr. Econ.* 50: 809-814.

_____ [1972]: "Economics and the Quality of Life." *Amer. J. Agr. Econ.* 54: 723-735.

Churchman, C. W. [1961]: *Prediction and Optimal Decision: Philosophical Issues of a Science of Values.* Englewood Cliffs, NJ, Prentice-Hall.

_____ [1968]: *The Systems Approach.* New York, Dell.

Churchman, C. W., and R. L. Ackoff [1950]: "Purposive Behavior and Cybernetics." *Social Forces* 29: 32-39.

Ciriacy-Wantrup, S. V. [1956]: "Policy Considerations in Farm Management Research in the Decade Ahead." *J. Farm Econ.* 38: 1301-1311.

Clawson, M. [1968]: *Policy Directions for U. S. Agriculture: Long-Range Choices in Farming and Rural Living.* Baltimore, Johns Hopkins Univ. Press for RFF.

Coase, R. H. [1960]: "The Problem of Social Cost." *J. Law and Econ.* 3: 1-44. Also in Staaf and Tannian, eds., 1972, pp. 119-161.

Coats, A. W. [1976]: "The Development of the Agricultural Economics Profession in England." *J. Agr. Econ.* 27: 381-392.

Commons, J. R. [1934]: *Institutional Economics.* New York, Macmillan. Reprinted in two paperback volumes, Univ. of Wisconsin Press, 1959.

Conklin, H. E. [1947]: "A Neglected Point in the Training of Agricultural Economists." *J. Farm Econ.* 29: 925-937.

Day, R. H., and I. J. Singh [1977]: *Economic Development as an Adaptive Process: A Green Revolution Case Study.* London, Cambridge Univ. Press.

Day, R. H., and E. Sparling [1977]: "Optimization Models in Agricultural and Resource Economics." In Judge, *et al.*, eds., 1977, pp. 93-127.

Dewey, J. [1938]: *Logic: The Theory of Inquiry.* New York, Henry Holt.

_____ [1939]: "The Continuum of Ends-Means," Chapter V in "Theory of Valuation." In *International Encyclopedia of Unified Science*, Vol. II, No. 4. Univ. of Chicago Press, pp. 40-50. Also in *Ethical Theories: A Book of Readings*, ed. A. I. Melden, New York, Prentice-Hall, 1950, pp. 360-366.

Dillon, J. L., and J. R. Anderson [1971]: "Allocative Efficiency, Traditional Agriculture, and Risk." *Amer. J. Agr. Econ.* 53: 26-32.

Dorfman, R., and N. S. Dorfman, eds. [1972]: *Economics of the Environment: Selected Readings.* New York, Norton.

Dorner, P. [1971]: "Needed Directions in Economic Analysis for Agricultural Development Policy." *Amer. J. Agr. Econ.* 53: 8-16.

Dunn, E. S., Jr. [1971]: *Economic and Social Development: A Process of Social Learning.* Baltimore, Johns Hopkins Press for RFF.

Ezekiel, M. [1930]: *Methods of Correlation Analysis.* New York, Wiley.

Farris, P. L., ed. [1964]: *Market Structure Research: Theory and Practice in Agricultural Economics.* Ames, ISU Press.

Feigl, H. [1953]: "The Scientific Outlook: Naturalism and Humanism." In Feigl and Brodbeck, eds., 1953, pp. 8-18.

Feigl, H., and M. Brodbeck, eds. [1953]: *Readings in the Philosophy of Science.* New York, Appleton-Century-Crofts.

Fishlow, A., and P. A. David [1961]: "Optimal Resource Allocation in an Imperfect Market Setting." *J. Polit. Economy* 69: 529-546.

Fletcher, L. B., E. Graber, W. C. Merrill, and E. Thorbecke [1970]: *Guatemala's Economic Development: The Role of Agriculture.* Ames, ISU Press.

Fox, K. A. [1974]: *Social Indicators and Social Theory: Elements of an Operational System.* New York, Wiley.

French, B. C. [1977]: "The Analysis of Productive Efficiency in Agricultural Marketing: Models, Methods, and Progress." In Martin, ed., 1977, pp. 91-206.

Friedman, M. [1953]: *Essays on Positive Economics.* Univ. of Chicago Press.

Friedman, M., and J. L. Savage [1948]: "The Utility Analysis of Choices Involving Risk." *J. Polit. Economy* 56: 279-304.

Gaus, J. M., and L. O. Wolcott [1940]: *Public Administration and the U. S. Department of Agriculture.* Chicago, Public Administration Service for the SSRC.

Georgescu-Roegen, N. [1971]: *The Entropy Law and the Economic Process.* Cambridge, MA, Harvard Univ. Press.

Giles, A. K. [1976]: "The A.E.S.: A Commentary on Its Past, Present and Future." *J. Agr. Econ.* 27: 393-413.

Girschick, M. A. [1946]: "Contributions to the Theory of Sequential Analysis: I, II and III." *Annals of Math. Stat.* 17: 123-143, 282-298.

Gray, R. W. [1970]: "Agricultural Economics: An Orientation for the 70s." *Proceedings, Western Agricultural Economics Association, 1970,* pp. 22-26.

Grove, E. W. [1965]: "Basic Research in a Quasi Science." *J. Farm Econ.* 47: 846-848.

_____ [1968]: "The Stern Test of Objectivity for the Useful Science of Agricultural Economics: Comment." *Amer. J. Agr. Econ.* 50: 153-155.

Gulley, J. L. [1974]: *Beliefs and Values in American Farming.* USDA, ERS-558.

Halcrow, H. G. [1953]: *Agricultural Policy of the United States.* New York, Prentice-Hall.

Halter, A. N., and G. W. Dean [1971]: *Decisions under Uncertainty with Research Applications.* Cincinnati, South-Western.

Halter, A. N., and H. H. Jack [1961]: "Toward a Philosophy of Science for Agricultural Economic Research." *J. Farm Econ.* 43: 83-95.

Halter, A. N., and S. F. Miller [1966]: *River Basin Planning: A Simulation Approach.* Oregon State Univ., AES, Special Report No. 224.

Hardin, C. M. [1946]: "The Bureau of Agricultural Economics under Fire: A Study in Valuation Conflicts." *J. Farm Econ.* 28: 635-668.

Harrison, K., D. Henley, H. Riley, and J. Shaffer [1974]: *Improving Food Marketing Systems in Developing Countries: Experiences from Latin America.* East Lansing, MSU, LASC, Research Report No. 6.

Harrod, R. F. [1938]: "Scope and Method of Economics." *Econ. J.* 48: 383-412.

Hartman, L. M. [1977]: "Economics as Science and as Culture." *Amer. J. Agr. Econ.* 59: 925-930.

Hathaway, D. E. [1953]: "Agricultural Policy and Farmers' Freedom: A Suggested Framework." *J. Farm Econ.* 35: 496-510.

———— [1963]: *Government and Agriculture.* New York, Macmillan.

Hayami, Y., and V. W. Ruttan [1970]: "Factor Prices and Technical Change in Agricultural Development: The United States and Japan, 1880-1960." *J. Polit. Economy* 78: 1115-1141.

———— [1971]: *Agricultural Development: An International Perspective.* Baltimore, Johns Hopkins Press.

Heady, E. O. [1949]: "Implications of Particular Economics in Agricultural Economics Methodology." *J. Farm Econ.* 31: 837-850.

———— [1952]: *Economics of Agricultural Production and Resource Use.* Englewood Cliffs, NJ, Prentice-Hall.

———— [1962]: *Agricultural Policy under Economic Development.* Ames, ISU Press.

Heady, E. O., W. B. Back, and G. A. Peterson [1953]: *Interdependence between the Farm Business and the Farm Household with Implications on Economic Efficiency.* ISU, AES, Research Bulletin 398.

Heady, E. O., H. G. Diesslin, H. R. Jensen, and Glenn L. Johnson, eds. [1958]: *Agricultural Adjustment Problems in a Growing Economy.* Ames, ISU Press.

Heady, E. O., E. O. Haroldsen, L. V. Mayer, and L. G. Tweeten [1965]: *Roots of the Farm Problem: Changing Technology, Changing Capital Use, Changing Labor Needs.* Ames, ISU Press.

Hicks, J. R. [1939]: *Value and Capital: An Inquiry into Some Fundamental Principles of Economic Theory.* Oxford, England, Oxford Univ. Press.

Hildreth, R. J., and W. N. Schaller [1972]: "Community Development in the 1970s." *Amer. J. Agr. Econ.* 54: 764-773.

Hull, J., P. G. Moore, and H. Thomas [1973]: "Utility and Its Measurement." *J. Royal Statistical Society, Series A* 136: 226-247.

Hunt, K. E. [1976]: "The Concern of Agricultural Economists in Great Britain since the 1920s." *J. Agr. Econ.* 27: 285-296.

Iowa State Univ., Center for Agricultural Adjustment [1954]: *Problems and Policies of American Agriculture.* Ames.

Iowa State Univ., Center for Agricultural and Economic Development [1961]: *Goals and Values in Agricultural Policy.* Ames, ISU Press.

———— [1963]: *Farm Goals in Conflict: Family Farm, Income, Freedom, Security.* Ames, ISU Press.

Jansma, J. D., H. B. Gamble, J. P. Madden, and R. H. Warland [1981]: "Rural Development: A Review of Conceptual and Empirical Studies." In Martin, ed., 1981, pp. 285-390.

Jensen, H. R. [1977]: "Farm Management and Production Economics, 1946-70." In Martin, ed., 1977, pp. 1-89.

Johnson, D. Gale [1947]: *Forward Prices for Agriculture.* Univ. of Chicago Press.

Johnson, Glenn L. [1950]: "Needed Developments in Economic Theory as Applied to Farm Management." *J. Farm Econ.* 32: 1140-1158.

———— [1955]: "Results from Production Economics Analysis." *J. Farm Econ.* 37: 206-222.

———— [1957]: "Agricultural Economics, Production Economics and the Field of Farm Management." *J. Farm Econ.* 39: 441-450.

———— [1960a]: "The State of Agricultural Supply Analysis." *J. Farm Econ.* 42: 435-452.

———— [1960b]: "Value Problems in Farm Management." *J. Agr. Econ.* 14: 13-31.

_____ [1960c]: "The Labour Utilisation Problem in European and American Agriculture." *J. Agr. Econ.* 14: 73-87.

_____ [1961a]: "Budgeting and Engineering Analysis of Normative Supply Functions." In *Agricultural Supply Functions*, ed. E. O. Heady, C. B. Baker, H. G. Diesslin, E. Kehrberg, and S. Staniforth. Ames, ISU Press, pp. 170-176.

_____ [1961b]: "Discussion." [of 'Structure of Agriculture,' by C. E. Bishop and K. L. Bachman] In ISU, CAED, 1961, pp. 254-259.

_____ [1961c]: *Some Philosophic Thoughts about NCR(4)'s Work*, presented at the Fall Meeting of NCR(4). Chicago, Farm Foundation.

_____ [1963a]: "Methodology for the Managerial Input." In *The Management Input in Agriculture*. North Carolina State Univ., Agricultural Policy Institute, Southern Farm Management Research Committee and Farm Foundation.

_____ [1963b]: "Stress on Production Economics." *Australian J. Agr. Econ.* 7: 12-26.

_____ [1970]: *The Role of the University and Its Economists in Economic Development*. J. S. McLean Visiting Professor Lecture, Univ. of Guelph, DAE.

_____ [1971]: "The Quest for Relevance in Agricultural Economics." *Amer. J. Agr. Econ.* 53: 728-739.

_____ [1973]: "Review of 'The Entropy Law and the Economic Process' by N. Georgescu-Roegen." *J. Econ. Issues* 7: 492-499.

_____ [1976]: *Economics, Ethics, Food and Energy*. The Second James C. Snyder Memorial Lecture in Agricultural Economics, Purdue Univ., West Lafayette, IN.

_____ [1977a]: "General Systems Simulation Analyses (GSSA) of the Nigerian and Korean Agricultural Sectors and Related Efforts." *European Rev. Agr. Econ.* 3: 391-410.

_____ [1977b]: "Recent U. S. Research Priority Assessments for Food and Nutrition: The Neglect of the Social Sciences." *Canadian J. Agr. Econ., Proceedings Issue* 25: 76-89.

Johnson, Glenn L., A. N. Halter, H. R. Jensen, and D. W. Thomas, eds. [1961]: *A Study of Managerial Processes of Midwestern Farmers*. Ames, ISU Press.

Johnson, Glenn L., and C. L. Quance, eds. [1972]: *The Overproduction Trap in U. S. Agriculture: A Study of Resource Allocation from World War I to the Late 1960s*. Baltimore, Johns Hopkins Univ. Press for RFF.

Johnson, Glenn L., O. J. Scoville, G. K. Dike, and C. K. Eicher [1969]: *Strategies and Recommendations for Nigerian Rural Development. 1969/1985*. East Lansing, MSU, Consortium for the Study of Nigerian Rural Development, CSNRD-33.

Johnson, Glenn L., and L. K. Zerby [1961]: "Values in the Solution of Credit Problems." In *Capital and Credit Needs in a Changing Agriculture*, ed. E. L. Baum, H. G. Diesslin, and E. O. Heady. Ames, ISU Press, pp. 271-290.

Johnson, Glenn L., and L. K. Zerby [1973]: *What Economists Do About Values: Case Studies of Their Answers to Questions They Don't Dare Ask*. East Lansing, MSU, Center for Rural Manpower and Public Affairs and DAE.

Johnson, Harry G. [1960]: "The Cost of Protection and the Scientific Tariff." *J. Polit. Economy* 68: 327-345.

Judge, G. G. [1977]: "Estimation and Statistical Inference in Economics." In Judge, *et al.*, eds., 1977, pp. 3-53.

Judge, G. G., R. H. Day, S. R. Johnson, G. C. Rausser, and L. R. Martin, eds. [1977]: *A Survey of Agricultural Economics Literature, Vol. 2, Quantitative Methods in Agricultural Economics, 1940s to 1970s*. Minneapolis, Univ. of Minnesota Press for the AAEA.

Kaldor, D. R., and H. H. Hines [1961]: "Goal Conflicts in Agriculture." In ISU, CAED, 1961, pp. 184-202.

Kaplan, A. [1968]: "Positivism." In *International Encyclopedia of the Social Sciences*, Vol. 12, ed. D. L. Sills. New York, Macmillan and Free Press, pp. 389-395.

Kelso, M. M. [1949]: "New Directions for Land Economics Research: West." *J. Farm Econ.* 31: 1035-1042.

―――― [1965a]: "A Critical Appraisal of Agricultural Economics in the Mid-Sixties." *J. Farm Econ.* 47: 1-16.

―――― [1965b]: "The Author's Last Word—For Now!" *J. Farm Econ.* 47: 854-857.

―――― [1968]: "Public Land Policy in the Context of Planning-Programming-Budgeting Systems." *Amer. J. Agr. Econ.* 50: 1671-1685.

Kemeny, J. G. [1959]: *A Philosopher Looks at Science*. Princeton, NJ, Van Nostrand, pp. 14-35 (chapter on 'mathematics').

Knight, F. H. [1921]: *Risk, Uncertainty and Profit*. Boston, Houghton Mifflin.

―――― [1940]: " 'What is Truth' in Economics?" [Review of T. W. Hutchinson, *The Significance and Basic Postulates of Economic Theory*, London, Macmillan, 1938] *J. Polit. Economy* 48: 1-32.

Kuhn, T. S. [1970]: *The Structure of Scientific Revolutions*, 2nd ed., enlarged. Univ. of Chicago Press.

Kutish, L. J. [1954]: "The Value Question." *J. Farm Econ.* 36: 666-671.

Lenzen, V. F. [1955]: "Procedures of Empirical Science." In *International Encyclopedia of Unified Science*, Vol. I. Part I., ed. O. Neurath, R. Carnap and C. Morris. Univ. of Chicago Press, pp. 279-339.

Leontief, W. [1971]: "Theoretical Assumptions and Nonobserved Facts." *Amer. Econ. Rev.* 61: 1-7.

Lewis, C. I. [1955]: *The Ground and Nature of the Right*. New York, Columbia Univ. Press.

Libby, L. W. [1971]: "Needed Redirections in Economics: Comment." *Amer. J. Agr. Econ.* 53: 658-660.

Lipsey, R. G., and K. Lancaster [1956]: "The General Theory of Second Best." *Rev. Econ. Studies* 24: 11-32.

Long, E. J. [1952]: "Some Theoretical Issues in Economic Development." *J. Farm Econ.* 34: 723-733.

―――― [1953]: "Freedom and Security as Policy Objectives." *J. Farm Econ.* 35: 317-322.

―――― [1954]: "A Reply to Kutish's 'The Value Question'." *J. Farm Econ.* 36: 671-674.

Machlup, F. [1969]: "Positive and Normative Economics: An Analysis of the Ideas." In *Economic Means and Social Ends: Essays in Political Economics*, ed. R. L. Heilbroner. Englewood Cliffs, NJ, Prentice-Hall, pp. 99-129.

Manetsch, T. J., M. L. Hayenga, A. N. Halter, T. W. Carroll, M. H. Abkin, D. R. Byerlee, K.-Y. Chong, G. Page, E. Kellogg, and Glenn L. Johnson [1971]: *A Generalized Simulation Approach to Agricultural Sector Analysis with Special Reference to Nigeria*. East Lansing, MSU.

Margolis, J. [1957]: "Secondary Benefits, External Economies, and the Justification of Public Investments." *Rev. Econ. and Stat.* 39: 284-91.

Markham, J. W. [1961]: "Goals for Economic Organization: A Theoretical Analysis." In ISU, CAED, 1961, pp. 88-110.

Marschak, J. [1953]: "Economic Measurements for Policy and Prediction." In *Studies in Econometric Method*, ed. W. C. Hood and T. C. Koopmans. New York, Wiley, pp. 1-26.

Martin, L. R., ed. [1977]: *A Survey of Agricultural Economics Literature, Vol. 1. Traditional Fields of Agricultural Economics, 1940s to 1970s*. Minneapolis, Univ. of Minnesota Press for the AAEA.

_____ ed. [1981]: *A Survey of Agricultural Economics Literature, Vol. 3. Economics of Welfare, Rural Development, and Natural Resources in Agriculture, 1940s to 1970s.* Minneapolis, Univ. of Minnesota Press for the AAEA.

Mead, G. H. [1938]: *The Philosophy of the Act,* ed. C. W. Morris *et al.*, Univ. of Chicago Press.

Meadows, D. H., D. L. Meadows, J. Randers, and W. W. Behrens III [1972]: *The Limits to Growth: A Report for the Club of Rome's Project on the Predicament of Mankind.* New York, New American Library.

Mighell, R. L. [1965]: "Is Agricultural Economics a Science?" *J. Farm Econ.* 47: 848-849.

Mighell, R. L., and J. D. Black [1951]: *Interregional Competition in Agriculture: With Special Reference to Dairy Farming in the Lake States and New England.* Cambridge, MA, Harvard Univ. Press.

Mini, P. V. [1974]: *Philosophy and Economics.* Gainesville, Univ. of Florida Press.

Mitchell, C. C. [1949]: "A Comment on 'Planning and Control'." *J. Farm Econ.* 31: 708-711.

Mitroff, I. I., and L. V. Blankenship [1973]: "On the Methodology of the Holistic Experiment: An Approach to the Conceptualization of Large-Scale Social Experiments." *Technological Forecasting and Social Change* 4: 339-353.

Mitroff, I. I., and M. Turoff [1973]: "Technology Forecasting and Assessment: The Whys Behind the Hows." *IEEE Spectrum* 10(3): 62-71.

Moles, J. A. [1977]: "The Creation of 'Truth' by Social Scientists and Planners: Assumptions, Decisions, and the Negotiation of Reality." *Amer. J. Agr. Econ.* 59: 918-924.

Moore, G. E. [1903]: *Principia Ethica.* Cambridge, England, Cambridge Univ. Press. Reprinted in 1956.

Myrdal, G. [1944]: *An American Dilemma.* New York, Harper & Row. McGraw-Hill paperback, 1964.

Nicholls, W. H., and D. Gale Johnson [1946]: "The Farm Policy Awards, 1945: A Topical Digest of the Winning Essays." *J. Farm Econ.* 28: 267-283.

Nicol, K. J., S. Striplung, and E. O. Heady, eds. [1982]: *Agricultural Development Planning in Thailand.* Ames, ISU Press.

Northrup, F. S. C. [1947]: *The Logic of the Sciences and the Humanities.* New York, Macmillan.

Nou, J. [1967]: *Studies in the Development of Agricultural Economics in Europe.* Uppsala, Sweden, Almquist and Wiksells.

Parsons, K. H. [1949]: "The Logical Foundations of Economic Research." *J. Farm Econ.* 31: 656-686.

_____ [1958]: "The Value Problem in Agricultural Policy." In Heady, Diesslin, Jensen, and Glenn L. Johnson, eds., 1958, pp. 285-299.

Pearson, K. [1900]: *The Grammar of Science,* 2nd edition. London, Adam and Charles Black.

Pigou, A. C. [1932]: *Economics of Welfare,* 4th edition. London, Macmillan.

Pirsig, R. M. [1974]: *Zen and the Art of Motorcycle Maintenance.* New York, Morrow.

Popper, K. R. [1959]: *The Logic of Scientific Discovery.* New York, Harper & Row. First published in German, 1934.

_____ [1962]: *Conjectives and Refutations: The Growth of Scientific Knowledge.* New York, Basic Books.

_____ [1972]: *Objective Knowledge: An Evolutionary Approach.* London, Oxford Univ. Press.

Randall, A. [1972]: "Market Solutions to Externality Problems: Theory and Practice." *Amer J. Agr. Econ.* 54: 175-183.

Reder, M. W. [1947]: *Studies in the Theory of Welfare Economics.* New York, Columbia Univ. Press.

Reinsel, E. I. [1965]: "Agricultural Economics Is an Inexact Science." *J. Farm Econ.* 47: 849-851.

Robinson, A. [1968]: "Pigou, Arthur Cecil." In *International Encyclopedia of the Social Sciences,* Vol. 12, ed. D. L. Sills. New York, Macmillan and Free Press, pp. 90-97.

Rossmiller, G. E., T. W. Carroll, S. G. Kim, Y. S. Kim, T. J. Manetsch, H. H. Suh, D. H. Kim, and Glenn L. Johnson [1972]: *Korean Agricultural Sector Analysis and Recommended Development Strategies, 1971-1985.* MSU, DAE and Republic of Korea, Ministry of Agriculture and Forestry, Agricultural Economics Research Institute.

Rudner, R. S. [1966]: *Philosophy of Social Science.* Englewood Cliffs, NJ, Prentice-Hall.

Runes, D. D., ed. [1960]: *Dictionary of Philosophy,* 15th edition, rev. New York, Philosophical Library.

Salter, L. A., Jr. [1948]: *A Critical Review of Research in Land Economics.* Minneapolis, Univ. of Minnesota Press.

Schickele, R. [1954]: *Agricultural Policy: Farm Programs and National Welfare.* New York, McGraw-Hill.

Schmid, A. A. [1965]: "Science, Art, and Agricultural Economics." *J. Farm Econ.* 47: 851-852.

———— [1972]: "Analytical Institutional Economics: Challenging Problems in the Economics of Resources for a New Environment." *Amer. J. Farm Econ.* 54: 893-901.

Schmitt, G., and W. Timmermann [1969]: "On Scientific Objectivity: A Comment." *Amer. J. Agr. Econ.* 51: 921-924.

Schultz, T. W. [1939a]: "Theory of the Firm and Farm Management Research." *J. Farm Econ.* 21: 570-586.

———— [1939b]: "Scope and Method in Agricultural Economics Research." *J. Polit. Economy* 47: 705-717.

———— [1945]: *Agriculture in an Unstable Economy.* New York, McGraw-Hill.

———— [1959]: "Omission of Variables, Weak Aggregates, and Fragmentation in Policy Adjustment Studies." In ISU, CAA, 1959, pp. 189-203.

———— [1971]: *Investment in Human Capital: The Role of Education and of Research.* New York, Free Press.

Scitovsky, T. [1951]: "The State of Welfare Economics." *Amer. Econ. Rev.* 41: 303-315.

Sharples, J. A., T. A. Miller, and L. M. Day [1968]: *Evaluation of a Firm Model in Estimating Aggregate Supply Response.* North Central Regional Research Publication No. 179, ISU, AES, Research Bulletin 558.

Shepherd, G. [1956]: "What Can a Research Man Say about Values?" *J. Farm Econ.* 38: 8-16.

Social Science Research Council [1956]: *American Agriculture and Agricultural Economics, 1955-75: A Report of the Committee on Agricultural Economics to the Social Science Research Council.* Raleigh, NC, mimeograph.

Spitze, R. G. F. [1965]: "Agricultural Economics: Predictive or Prescriptive." *J. Farm Econ.* 47: 852-854.

Staaf, R. J., and F. X. Tannian, eds. [1972]: *Externalities: Theoretical Dimensions of Political Economy.* New York, Dunellen.

Swanson, E. R. [1971]: "Programmed Normative Agricultural Supply Response: Establishing Farm-Regional Links." In *Economics Models and Quantitative Models for Decision and Planning in Agriculture: Proceedings of an East-West Seminar,* ed. E. O. Heady. Ames, ISU Press, pp. 229-242.

Talbot, R., and C. Wiggins [1967]: *Political Forces in American Agriculture*. Ames, ISU, Department of History, Government and Philosophy, mimeograph.

Taylor, H. C., and A. D. Taylor [1952]: *The Story of Agricultural Economics in the United States, 1840-1932: Men-Services-Ideas*. Ames, ISU Press.

Thompson, R. L., and R. E. Hepp [1976]: *Description and Analysis of Michigan Small Farms*. MSU, AES, Research Report No. 296.

Timmons, J. F. [1959]: "Land Institutions Impeding and Facilitating Agricultural Adjustment." In ISU, CAA, 1959, pp. 166-188.

Trelogan, H. C., C. E. Caudill, H. F. Huddleston, W. E. Kibler, and E. Brooks [1977]: "Technical Developments in Agricultural Estimates Methodology." In Judge, *et al.*, eds., 1977, pp. 373-385.

Tweeten, L. [1970]: *Foundations of Farm Policy*. Lincoln, Univ. of Nebraska Press.

U. S. Congress, Joint Economic Committee [1974]: *Government Price Statistics: Hearings before the Subcommittee on Economic Statistics*. Washington, 87th Congress, 1st Session, Part 1.

U. S. Congress, Office of Technology Assessment [1976a]: *Food Information Systems: Hearings before the Technology Assessment Board of OTA*. Washington.

_____ [1976b]: *Food Information Systems — Summary and Analysis*. Washington, OTA-F-35.

Upchurch, M. L. [1977]: "Developments in Agricultural Economic Data." In Judge, *et al.*, eds., 1977, pp. 305-372.

Von Neumann, J., and O. Morgenstern [1947]: *Theory of Games and Economic Behavior*. Princeton, NJ, Princeton Univ. Press.

Wells, O. V. [1954]: "Agricultural Economics under the USDA Reorganization of November 2, 1953." *J. Farm Econ*. 36: 1-5.

Wilcox, W. W. [1956]: *Social Responsibility in Farm Leadership: An Analysis of Farm Problems and Farm Leadership in Action*. New York, Harper.

Wilcox, W. W., W. W. Cochrane, and R. W. Herdt [1974]: *Economics of American Agriculture*, 3rd edition. Englewood Cliffs, NJ, Prentice-Hall.

Wilt, H. S. [1957]: *Managing the Small, Part-time Farm*. East Lansing, MSU, Cooperative Extension Bulletin No. 341.

Woodworth, R. C. [1977]: "Agricultural Production Function Studies." In Judge, *et al.*, eds. 1977, pp. 128-154.

Working, E. J. [1927]: "What Do Statistical 'Demand Curves' Show?" *Quart. J. Econ*. 41: 212-235.

Zerby, L. K. [1971]: "What Is New about the New Left?" *The Torch (J. of International Torch Clubs)* 44: 11-115.